More Things in Heaven and Earth

A Celebration of Physics at the Millennium

Springer-Verlag Berlin Heidelberg GmbH

BENJAMIN BEDERSON
EDITOR

More Things in Heaven and Earth

A Celebration of Physics at the Millennium

 Springer   **APS**
The American Physical Society

Benjamin Bederson
Department of Physics
New York University
New York, NY 10003
USA

Library of Congress Cataloging-in-Publication Data
More things in heaven and earth : a celebration of physics
at the millennium / edited by Ben Bederson.
　　p.　cm.
　　Includes bibliographical references.
　　ISBN 978-1-4612-7174-1　　ISBN 978-1-4612-1512-7 (eBook)
　　DOI 10.1007/978-1-4612-1512-7
　　1. Physics—History—20th century.　2. Astrophysics—History—20th century.
　　I. Bederson, Benjamin.　II. Title: A celebration of physics at the millennium.
　QC7.M568　1999　　　　　　　　　　09-53116
　530'.09'04—dc21　　　　　　　　　　CIP

Printed on acid-free paper.

©1999 Springer-Verlag Berlin Heidelberg
Originally published by Springer-Verlag Berlin Heidelberg New York in 1999

All rights reserved. This work may not be translated or copied in whole or in part without the written permission of The American Physical Society, One Physics Ellipse, College Park, MD 20740-3844, USA, except for brief excerpts in connection with reviews or scholarly analysis. Use in connection with any form of information storage and retrieval, electronic adaptation, computer software, or by similar or dissimilar methodology now known or hereafter developed is forbidden.
The use of general descriptive names, trade names, trademarks, etc., in this publication, even if the former are not especially identified, is not to be taken as a sign that such names, as understood by the Trade Marks and Merchandise Marks Act, may accordingly be used freely by anyone.

Production managed by MaryAnn Cottone; manufacturing supervised by Nancy Wu.
Photocomposed pages prepared by American Institute of Physics, Woodbury, NY.

9 8 7 6 5 4 3 2 1

ISBN 978-1-4612-7174-1　　　　　　　　　　　　　　　SPIN 10697621

Contents

- ix   Contributors
- xv   Foreword
  Martin Blume
- xvii   Preface
  Benjamin Bederson
- xxi   Introduction
  Hans A. Bethe

Historic Perspectives—Personal Essays on Historic Developments

- 3   Quantum Theory
  Hans A. Bethe
- 12   Nuclear Physics
  Hans A. Bethe
- 29   Theoretical Particle Physics
  A. Pais
- 43   Elementary Particle Physics: The Origins
  Val L. Fitch
- 56   Astrophysics
  George Field
- 69   A Century of Relativity
  Irwin I. Shapiro
- 89   From Radar to Nuclear Magnetic Resonance
  Robert V. Pound
- 96   An Essay on Condensed Matter Physics in the Twentieth Century
  W. Kohn
- 129   A Short History of Atomic Physics in the Twentieth Century
  Daniel Kleppner

Particle Physics and Related Topics

- 143   Quantum Field Theory
  Frank Wilczek
- 161   The Standard Model of Particle Physics
  Mary K. Gaillard, Paul D. Grannis, and Frank J. Sciulli

188 String Theory, Supersymmetry, Unification, and All That
John H. Schwarz and Nathan Seiberg

203 Accelerators and Detectors
W. K. H. Panofsky and M. Breidenbach

223 Anomalous g Values of the Electron and Muon
V. W. Hughes and T. Kinoshita

234 Neutrino Physics
L. Wolfenstein

Astrophysics

245 Cosmology at The Millennium
Michael S. Turner and J. Anthony Tyson

278 Cosmic Rays: The Most Energetic Particles in the Universe
James W. Cronin

291 Cosmic Microwave Background Radiation
Lyman Page and David Wilkinson

302 Black Holes
Gary T. Horowitz and Saul A. Teukolsky

313 Gravitational Radiation
Rainer Weiss

329 Deciphering the Nature of Dark Matter
Bernard Sadoulet

Nuclear Physics

345 Nuclear Physics at the End of the Century
E. M. Henley and J. P. Schiffer

370 Stellar Nucleosynthesis
Edwin E. Salpeter

Atomic, Molecular, and Optical Physics

377 Atomic Physics
Sheldon Datz, G. W. F. Drake, T. F. Gallagher, H. Kleinpoppen, and G. zu Putlitz

408 Laser Spectroscopy and Quantum Optics
T. W. Hänsch and H. Walther

426 Atom Cooling, Trapping, and Quantum Manipulation
Carl E. Wieman, David E. Pritchard, and David J. Wineland

442 Laser Physics: Quantum Controversy in Action
W. E. Lamb, W. P. Schleich, M. O. Scully, and C. H. Townes

460 Quantum Effects in One-Photon and Two-Photon Interference
    L. Mandel

474 From Nanosecond to Femtosecond Science
    N. Bloembergen

482 Experiment and Foundations of Quantum Physics
    Anton Zeilinger

Condensed Matter Physics

501 The Fractional Quantum Hall Effect
    Horst L. Stormer, Daniel C. Tsui, and Arthur C. Gossard

515 Conductance Viewed as Transmission
    Yoseph Imry and Rolf Landauer

526 Superconductivity
    J. R. Schrieffer and M. Tinkham

534 Superfluidity
    A. J. Leggett

543 In Touch with Atoms
    G. Binnig and H. Rohrer

555 Materials Physics
    P. Chaudhari and M. S. Dresselhaus

563 The Invention of the Transistor
    Michael Riordan, Lillian Hoddeson, and Conyers Herring

Statistical Physics and Fluids

581 Statistical Mechanics: A Selective Review of Two Central Issues
    Joel L. Lebowitz

601 Scaling, Universality, and Renormalization: Three Pillars of Modern Critical Phenomena
    H. Eugene Stanley

617 Insights from Soft Condensed Matter
    Thomas A. Witten

629 Granular Matter: A Tentative View
    P. G. de Gennes

644 Fluid Turbulence
    Katepalli R. Sreenivasan

665 Pattern Formation in Nonequilibrium Physics
    J. P. Gollub and J. S. Langer

Plasma Physics

679 The Collisionless Nature of High-Temperature Plasmas
T. M. O'Neil and F. V. Coroniti

Chemical Physics and Biological Physics

693 Chemical Physics: Molecular Clouds, Clusters, and Corrals
Dudley Herschbach

706 Biological Physics
Hans Frauenfelder, Peter G. Wolynes, and Robert H. Austin

726 Brain, Neural Networks, and Computation
J. J. Hopfield

Computational Physics

741 Microscopic Simulations in Physics
D. M. Ceperley

Applications of Physics to Other Areas

755 Physics and Applications of Medical Imaging
William R. Hendee

767 Nuclear Fission Reactors
Charles E. Till

775 Nuclear Power—Fusion
T. Kenneth Fowler

781 Physics and U.S. National Security
Sidney D. Drell

798 Laser Technology
R. E. Slusher

812 Physics and the Communications Industry
W. F. Brinkman and D. V. Lang

827 Index

Contributors

Robert H. Austin
Department of Physics
Princeton University
Princeton, New Jersey 08544

Hans A. Bethe
Floyd R. Newman Laboratory
 of Nuclear Studies
Cornell University
Ithaca, New York 14853

Gerd K. Binnig
IBM Research Division
Zurich Research Laboratory
Säumerstrasse 4
8803 Rüschlikon, Switzerland

Nicolaas Bloembergen
Department of Physics
Harvard University
Cambridge, Massachusetts 02138

Martin Breidenbach
Stanford Linear Accelerator Center
Stanford University
Stanford, California 94305

William F. Brinkman
Physical Sciences Research
AT&T Bell Laboratories
Lucent Technologies
Murray Hill, New Jersey 07974

David M. Ceperley
National Center for Superconducting
 Applications
University of Illinois
Urbana, Illinois 61801

Praveen Chaudhari
IBM Thomas J. Watson Research Center
Yorktown Heights, New York 10598

Ferdinand V. Coroniti
Department of Physics and Astronomy
University of California
Los Angeles, California 90095

James W. Cronin
Enrico Fermi Instiute
University of Chicago
Chicago, Illinois 60637

Sheldon Datz
Atomic Physics Division
Oak Ridge National Laboratory
Oak Ridge, Tennessee 37831

Pierre Giles de Gennes
College de France
Physique de la Matiere Condensee
75231 Paris Cedex 05
France

Gordon W. F. Drake
Department of Physics
University of Windsor
Windsor, Ontario
Canada

Sidney D. Drell
Stanford Linear Accelerator Center
Stanford University
Stanford, California 94309

Mildred S. Dresselhaus
Massachusetts Institute of Technology
Cambridge, Massachusetts 02139

George B. Field
Harvard-Smithsonian Center for Astrophysics
Cambridge, Massachusetts 02138

Val L. Fitch
Department of Physics
Princeton University
Princeton, New Jersey 08544

T. Kenneth Fowler
Nuclear Engineering
University of California
4155 Etcheverry Hall
Berkeley, California 94720

Hans Frauenfelder
Center for Nonlinear Studies
Los Alamos National Laboratory
Los Alamos, New Mexico 87545

Mary K. Gaillard
Physics Department
Lawrence Berkeley Laboratory
University of California
Berkeley, California 94720

Thomas F. Gallagher
Department of Physics
University of Virginia
Charlottesville, Virginia 22901

Jerry P. Gollub
Physics Department
Haverford College
Haverford, Pennsylvania 19041

Arthur C. Gossard
Materials Department
University of California
Santa Barbara, California 93106

Paul D. Grannis
Department of Physics
State University of New York
Stony Brook, New York 11794

Theodore W. Hänsch
Max-Planck-Insitut fur Quantenoptik
D-85740 Garching
Germany

William R. Hendee
Medical College of Wisconsin
8701 Watertown Plank Road
Milwaukee, Wisconsin 53226

Ernest M. Henley
Department of Physics
University of Washington
Seattle, Washington 98195

Conyers Herring
Department of Applied Physics
Stanford University
Stanford, California 94305

Dudley R. Herschbach
Chemistry and Biochemistry Departments
Harvard University
Cambridge, Massachusetts 02138

Lillian Hoddeson
Department of Physics
University of Illinois
Urbana, Illinois 61801

John J. Hopfield
Department of Molecular Biology
Princeton University
Princeton, New Jersey 08544

Gary T. Horowitz
Department of Physics
University of California
Santa Barbara, California 93106

Vernon W. Hughes
Department of Physics
Yale University
New Haven, Connecticut 06520

Yoseph Imry
Department of Physics
Weizmann Institute of Science
Rehovot 76 100
Israel

Thomas Kinoshita
Department of Physics
Cornell University
Ithaca, New York 14853

Hans Kleinpoppen
Department of Physics
University of Stirling
Stirling FK9 4LA
Scotland

Daniel Kleppner
Department of Physics
Massachusetts Institute of Technology
Cambridge, Massachusetts 02139

Walter Kohn
Department of Physics
University of California
Santa Barbara, California 93106

Willis E. Lamb
Optical Sciences Center
University of Arizona
Tucson, Arizona 85721

Rolf W. Landauer
IBM Thomas J. Watson Research Center
Yorktown Heights, New York 10598

David V. Lang
AT&T Bell Laboratories
Lucent Technologies
Murray Hill, New Jersey 07974

James S. Langer
Department of Physics
University of California
Santa Barbara, California 93106

Joel L. Lebowitz
Department of Mathematics and Physics
Rutgers University
Piscataway, New Jersey 08854

Anthony J. Leggett
Department of Physics
University of Illinois
Urbana, Illinois 61801

Leonard Mandel
Department of Physics and Astronomy
University of Rochester
Rochester, New York 14627

Thomas M. O'Neil
Department of Physics
University of California at San Diego
La Jolla, California 92093

Lyman A. Page
Department of Physics
Princeton University
Princeton, New Jersey 08544

Abram Pais
Department of Physics
Rockefeller University
New York, New York 10021

Wolfgang K. H. (Pief) Panofsky
Stanford Linear Accelerator Center
Stanford University
Stanford, California 94309

Robert V. Pound
Department of Physics
Harvard Univerity
Cambridge, Massachusetts 02138

David E. Pritchard
Department of Physics
Massachusetts Institute of Technology
Cambridge, Massachusetts 02140

Michael Riordan
4532 Cherryvale
Soquel, California 95073

Heinrich Rohrer
IBM Research Division
Zurich Research Laboratory
8805 Richterswil
Zurich, Switzerland

Bernard Sadoulet
Center for Particle Astrophysics
Lawrence Berkeley National Laboratory and
   Physics Department
University of California
Berkeley, California 94720

Edwin E. Salpeter
Center for Radiophysics and Space Research
Cornell University
Ithaca, New York 14853

John P. Schiffer
Physics Division
Argonne National Laboratory
Argonne, Illinois 60439

Wolfgang P. Schleich
Abteilung für Quantenphysik
Universität Ulm
D-89069 Ulm
Germany

J. Robert Schrieffer
National High Magnetic Field Laboratory
Florida State University
Tallahassee, Florida 32310

John H. Schwarz
Department of Theoretical Physics
California Institute of Technology
Pasadena, California 91125

Frank J. Sciulli
Professor of Physics
Columbia University
New York, New York 10027

Marlan O. Scully
Department of Physics
Texas A&M University
College Station, Texas 77843 and
Max-Planck-Institut fur Quantenoptik
Hans-Kopfermann Strasse 1
D-85748 Garching
Germany

Nathan Seiberg
Institute for Advanced Study
Princeton, New Jersey 08854

Irwin I. Shapiro
Harvard-Smithsonian Center for Astrophysics
Harvard University
Cambridge, Massachusetts 02138

Richard E. Slusher
AT&T Bell Laboratories
Lucent Technologies
Murray Hill, New Jersey 07974

Katepalli R. Sreenivasan
Department of Mechanical Engineering
Yale University
New Haven, Connecticut 06520

H. Eugene Stanley
Department of Physics
Boston University
Boston, Massachusetts 02215

Horst L. Stormer
Departments of Physics and of Applied
  Physics
Columbia University
New York, New York 10027 and
AT&T Bell Laboratories
Lucent Technologies
Murray, Hill New Jersey 07974

Saul A. Teukolsky
Floyd R. Newman Laboratory of Nuclear
  Studies
Cornell University
Ithaca, New York 14853

Charles E. Till
Argonne National Laboratory
Argonne, Illinois 60439

Michael Tinkham
Department of Physics
Harvard University
Cambridge, Massachusetts 02138

Charles H. Townes
Department of Physics
University of California
Berkeley, California 94720

Daniel C. Tsui
Department of Electrical Engineering
Princeton University
Princeton, New Jersey 08544

Michael S. Turner
Departments of Physics and of Astronomy &
  Astrophysics
Enrico Fermi Institute
University of Chicago
Chicago, Illinois 60637

J. Anthony Tyson
AT&T Bell Laboratories
Lucent Technologies
Murray Hill, New Jersey 07974

Herbert Walther
Max-Planck-Institut fur Quantenoptik
D-85740 Garching
Germany

Rainer Weiss
Department of Physics
Massachusetts Institute of Technology
Cambridge, Massachusetts 02139

Carl E. Wieman
Joint Institute for Laboratory Astrophysics
 and Department of Physics
University of Colorado
Boulder, Colorado 80309

Frank A. Wilczek
Institute for Advanced Study
School of Natural Sciences
Princeton, New Jersey 08540

David Todd Wilkinson
Department of Physics
Princeton University
Princeton, New Jersey 08544

David Wineland
National Institute of Standards and
 Technology
Time and Frequency Division
Boulder, Colorado 80309

Thomas A. Witten
James Franck Institute
University of Chicago
Chicago, Illinois 60637

Lincoln Wolfenstein
Department of Physics
Carnegie-Mellon University
Pittsburgh, Pennsylvania 15213

Peter G. Wolynes
Department of Chemistry
University of Illinois
Urbana, Illinois 61801

Anton Zeilinger
Institut für Experimentalphysik
University of Vienna
Boltzmanngasse 5
A-1000 Vienna
Austria

Gisbert zu Putlitz
Physikalisches Institut der Universität
 Heidelberg
D-6900 Heidelberg
Germany

Foreword

It is most appropriate that the Centennial of the American Physical Society be marked by a special issue of *Reviews of Modern Physics*, here published in a hardcover edition. RMP was begun when the American Physical Society was thirty years old, as the Society began its rise to importance on the international scene of physics. Many of the splendid review articles that appeared in RMP have chronicled the development of physics since that time, and several of the distinguished contributors to this volume have authored some of those articles. The forward march of physics continues unabated. There has been some hesitation, however, on looking ahead for the next hundred years—the example of predictions made at the end of the last century should be sufficient to give pause to all but the most determined prognosticators. I will only say that startling new ideas and phenomena will continue to appear, and I hope and expect that the most novel of these will continue to be revealed in *Physical Review* and summarized in *Reviews of Modern Physics*. We are all deeply indebted to Ben Bederson, my predecessor as Editor-in-Chief, for taking on the job of special editor for this issue. He deserves our thanks and gratitude for this splendid volume.

Martin Blume
Editor-in-Chief
The American Physical Society

Preface

Inspired by the 100th anniversary of the founding of the American Physical Society, we undertook this project in order to display the vast canvas that encompasses a century of magnificent accomplishment in physics as we enter the new millennium. Our intent was to present a contemporary portrait of physics, based on its past achievements and its current vitality. With regard to its future, maybe the best statement we can offer is contained in the last sentence of Hans Bethe's Introduction: *Looking at the predictions of 100 years ago, it would be foolish to make predictions for the next 100 years*. But, even so, scattered throughout the volume will be found somewhat more modest extrapolations, on a more limited time scale.

The scientific content of this volume is the same as the special Centenary issue of *Reviews of Modern Physics*. In a sense it can be considered as a companion to the collection edited by H. Henry Stroke, *The Physical Review: The First Hundred Years*, a book and CD-ROM which contains about 1000 articles from the *Physical Review* and *Physical Review Letters*, published in 1995 by the American Institute of Physics (with an upgraded CD to be issued by Springer-Verlag in 1999). That volume commemorated the slightly earlier centenary of the *Physical Review*, whose birth preceded that of the APS by three years.

We were all too painfully aware that a single volume, however ambitious its scope, could not hope to present a full portrait of all of physics—an understatement of the limitations that this project would surely be up against. As a result, we set ourselves a somewhat more limited goal. While we invited articles in the major subfields of physics, we asked authors to deal with these in a relatively informal, even personal way. We encouraged them not to try to include everything, and where appropriate to discuss subjects that were of particular interest to themselves. In some cases we chose to deal with specific aspects of a subject. These should be considered as "case studies." As a result readers will find many favorite topics omitted. Emphatically, it does not pretend to be comprehensive and readers should not be surprised or chagrined to discover significant omissions. This is the price we had to pay to accomplish our limited goals.

We asked authors to write their articles at the level of a good departmental colloquium. How successful our authors have been in abiding by this guideline is for the reader to determine.

The first order of business was to form an advisory group, a special "Editorial Board." The group, all of whom readily agreed to serve, consisted of six very special individuals: Kurt Gottfried (Cornell University), representing theoretical particle, nuclear physics, and general theoretical physics; Walter Kohn (University of California Santa Barbara), for theoretical condensed matter physics; Eugene Merzbacher (University of North Carolina), a past APS president, for atomic, molecular, and optical physics; Myriam Sarachik (City College of New York), for experimental condensed matter physics; David Schramm (University of Chicago), representing astrophysics and cosmology; and Andrew Sessler (University of California Berkeley), for experimental high energy physics and a variety of other subfields and, incidentally, President of APS in 1998. Sadly David Schramm perished in an airplane accident about halfway through the project, although he played a major role during its initial phase in developing the outline and in identifying appropriate authors. We were very fortunate to have George Field, Harvard Smithsonian Center for Astrophysics, then agree to serve in his stead. He has played an equally invaluable role in overseeing the astrophysics and cosmology aspects of the volume.

Inevitably, in reviewing past achievements, the history of physics, as a subject in itself, had to play a major role. In order to present this history in the same spirit as for the subject matter of the volume we decided to have a special history section, prepared by what one might characterize as "eyewitnesses"— though the contributors to this section are by no means as old as the APS. Peter Galison, Professor of the History of Science at Harvard University, has served as coordinator of this section and has participated in the reviewing of these articles. Of course the history of physics cannot be totally separated from physics itself, so history appears interwoven with scientific presentations throughout the volume, although with a different emphasis in the main body of the volume.

Physics is a living subject—reshaping, developing, evolving even as you look at it. Just within the brief period during which we have been receiving manuscripts major developments have occurred in many fields which warrant mention, even without our expressed goal of not being all-inclusive. Our authors have continued to revise their articles up to the very last possible minute.

To simplify the difficult task facing our authors we have encouraged them to keep citations to original research articles to a minimum, referring instead where possible to review articles and books. This necessarily results in omissions of worthy citations. Please bear this in mind, especially if you happen to be one of those omitted. However, authors did not uniformly abide by this rule, and we did not attempt to enforce it rigorously; this explains the inconsistencies with regard to references in some of the articles. It should not surprise anyone that there is an enormous variety of styles, range of material covered, and historical material included. This reflects, we believe, the enormous range of styles and personal taste of physicists themselves. On the other hand all articles were reviewed by at least one expert, either an editorial board member or an outside referee, although primarily for scientific content rather than uniformity of presentation. We are also aware that the volume possesses a daunting size, much larger than that which was originally intended. The editor must bear the brunt of possible criticism in this regard—in the last analysis he was unwilling to insist on Draconian measures to cut articles down to previously allotted lengths.

The time constraint—eighteen months from inception to completion—has doubtlessly resulted in the appearance of some scientific and possibly historic errors. We have of course worked diligently to keep these to a minimum; still we cannot claim that they do not exist.

Acknowledgments

It is literally true that there would have been no volume at all were it not for the continuing and generous advice, criticism, and encouragement of the extraordinary *ad hoc* Editorial Board, whose members are listed above. They deserve very special thanks.

Obviously such a project could not have come to fruition without the full cooperation, assistance, and encouragement of a large number of people, beyond the members of the

Editorial Board and, of course, the authors. Among these are the many other physicists whose advice we sought and received and anonymous reviewers of submitted articles.

I wish to acknowledge the following individuals and groups for their specific roles in producing this volume:

First, the authors. On very short notice, and with very severe constraints in time and space, our authors responded magnificently to this challenge. They good-naturedly acquiesced to the editor's haggling and imposed deadlines; they generally agreed to what we recognize as outrageous requests to cut out mention of important work, reduce the number of references to a painful few, generally making their articles possibly less authoritative and comprehensive than they would have been with more reasonable space allotments. We like to believe that their extraordinary efforts have been worth their while.

The APS Publications Oversight Committee, and the APS Executive Board, for their support and encouragement, especially Marty Blume, APS Editor-in-Chief, and Tom McIlrath, APS Treasurer, for their enthusiastic support, and George Bertsch, the Editor of *Reviews of Modern Physics*, for his willingness to relinquish his oversight authority for what appears within his journal for this single supplementary issue.

Karie Friedman, Assistant Editor of RMP, for playing the dominant role in editing final copy, performing for us what she has done with distinction for the regular issues of the journal over many years, continuing to earn the thanks and appreciation of our authors.

Maria Taylor, Executive Editor of Springer-Verlag, New York, for her invaluable assistance in arranging for and in producing this companion hardcover version of the volume.

We received enormous help from both the American Institute of Physics and Springer-Verlag, New York, in implementing remarkably tight production schedules—a crucial element in satisfying our requirement of having the supplement available for mail distribution to subscribers before the Centennial meeting and the hardcover volume available for distribution at the meeting itself.

Finally, I must state that as essential as was the advice and assistance received from all the groups and individuals noted above, the final responsibility for what appears in this volume rests with the Editor.

Benjamin Bederson
Professor of Physics Emeritus, New York University
Editor-in-Chief Emeritus, The American Physical Society

Introduction

A hundred years ago, some of the great physicists in England and Continental Europe predicted that physics was at an end. We know what actually happened, and we are proud of the contribution that The American Physical Society and *Reviews of Modern Physics*, where the material appearing in this volume first appeared, have made to it.

This volume tries to summarize some of the important areas of twentieth century physics. We have not covered all areas, and undoubtedly the essays on individual areas are not complete. But we hope some of them may be useful.

Looking at the predictions of 100 years ago, it would be foolish to make predictions for the next 100 years.

Hans A. Bethe
Cornell University

Introduction

Historic Perspectives—Personal Essays on Historic Developments

This section presents articles describing historic developments in a number of major areas of physics, prepared by authors who played important roles in these developments. The section was organized and coordinated with the help of Peter Galison, professor of the History of Science at Harvard University.

- 3   Quantum Theory
  Hans A. Bethe
- 12 Nuclear Physics
  Hans A. Bethe
- 29 Theoretical Particle Physics
  A. Pais
- 43 Elementary Particle Physics: The Origins
  Val L. Fitch
- 56 Astrophysics
  George Field
- 69 A Century of Relativity
  Irwin I. Shapiro
- 89 From Radar to Nuclear Magnetic Resonance
  Robert V. Pound
- 96 An Essay on Condensed Matter Physics in the Twentieth Century
  W. Kohn
- 129 A Short History of Atomic Physics in the Twentieth Century
  Daniel Kleppner

Quantum Theory
Hans A. Bethe

Twentieth-century physics began with Planck's postulate, in 1900, that electromagnetic radiation is not continuously absorbed or emitted, but comes in quanta of energy $h\nu$, where ν is the frequency and h Planck's constant. Planck got to this postulate in a complicated way, starting from statistical mechanics. He derived from it his famous law of the spectral distribution of blackbody radiation,

$$n(\nu) = [e^{h\nu/kT} - 1]^{-1}, \qquad (1)$$

which has been confirmed by many experiments. It is also accurately fulfilled by the cosmic background radiation, which is a relic of the big bang and has a temperature $T = 2.7$ K.

Einstein, in 1905, got to the quantum concept more directly, from the photoelectric effect: electrons can be extracted from a metal only by light of frequency above a certain minimum, where

$$h\nu_{\min} = w, \qquad (2)$$

with w the "work function" of the metal, i.e., the binding energy of the (most loosely bound) electron. This law was later confirmed for x rays releasing electrons from inner shells.

Niels Bohr, in 1913, applied quantum theory to the motion of the electron in the hydrogen atom. He found that the electron could be bound in energy levels of energy

$$E_n = -\frac{\text{Ry}}{n^2}, \qquad (3)$$

where n can be any integer. The Rydberg constant is

$$\text{Ry} = \frac{me^4}{2\hbar^2}. \qquad (4)$$

Light can be emitted or absorbed only at frequencies given by

$$h\nu = E_m - E_n, \qquad (5)$$

where m and n are integers. This daring hypothesis explained the observed spectrum of the hydrogen atom. The existence of energy levels was later confirmed by the experiment of J. Franck and G. Hertz. Ernest Rutherford, who had earlier proposed the nuclear atom, declared that now, after Bohr's theory, he could finally believe that his proposal was right.

In 1917, Einstein combined his photon theory with statistical mechanics and found that, in addition to absorption and spontaneous emission of photons, there had to be stimulated emission. This result, which at the time seemed purely theoretical, gave rise in the 1960s to the invention of the laser, an eminently practical and useful device.

A. H. Compton, in 1923, got direct evidence for light quanta: when x rays are scattered by electrons, their frequency is diminished, as if the quantum of energy $h\nu$ and momentum $h\nu/c$ had a collision with the electron in which momentum and energy were conserved. This Compton effect finally convinced most physicists of the reality of light quanta.

Physicists were still confronted with the wave/particle duality of light quanta on the one hand and the phenomena of interference, which indicated a continuum theory, on the other. This paradox was not resolved until Dirac quantized the electromagnetic field in 1927.

Niels Bohr, ever after 1916, was deeply concerned with the puzzles and paradoxes of quantum theory, and these formed the subject of discussion among the many excellent physicists who gathered at his Institute, such as Kramers, Slater, W. Pauli, and W. Heisenberg. The correspondence principle was formulated, namely, that in the limit of high quantum numbers classical mechanics must be valid. The concept of oscillator strength f_{mn} for the transition from level m to n in an atom was developed, and dispersion theory was formulated in terms of oscillator strength.

Pauli formulated the exclusion principle, stating that only one electron can occupy a given quantum state, thereby giving a theoretical foundation to the periodic system of the elements, which Bohr had explained phenomologically in terms of the occupation by electrons of various quantum orbits.

A great breakthrough was made in 1925 by Heisenberg, whose book, *Physics and Beyond* (Heisenberg, 1971), describes how the idea came to him while he was on vacation in Heligoland. When he returned home to Göttingen and explained his ideas to Max Born the latter told him, "Heisenberg, what you have found here are matrices." Heisenberg had never heard of matrices.

Born had already worked in a similar direction with P. Jordan, and the three of them, Born, Heisenberg, and Jordan, then jointly wrote a definitive paper on "matrix mechanics." They found that the matrices representing the coordinate of a particle q and its momentum p do not commute, but satisfy the relation

$$qp - pq = i\hbar 1, \qquad (6)$$

where 1 is a diagonal matrix with the number 1 in each diagonal element. This is a valid formulation of quantum mechanics, but it was very difficult to find the matrix elements for any but the simplest problems, such as the harmonic oscillator. The problem of the hydrogen atom was soon solved by the wizardry of W. Pauli in 1926. The problem of angular momentum is still best treated by matrix mechanics, in which the three components of the angular momentum are represented by noncommuting matrices.

Erwin Schrödinger in 1926 found a different formulation of quantum mechanics, which turned out to be most useful for solving concrete problems: A system of n particles is represented by a wave function in $3n$ dimensions, which satisfies a partial differential equation, the "Schrödinger equation." Schrödinger was stimulated by the work of L. V. de Broglie, who had conceived of particles as being represented by waves. This concept was beautifully confirmed in 1926 by the experiment of Davisson and Germer on electron diffraction by a crystal of nickel.

Schrödinger showed that his wave mechanics was equivalent to Heisenberg's matrix mechanics. The elements of Heisenberg's matrix could be calculated from Schrödinger's wave function. The eigenvalues of Schrödinger's wave equation gave the energy levels of the system.

It was relatively easy to solve the Schrödinger equation for specific physical systems: Schrödinger solved it for the hydrogen atom, as well as for the Zeeman and the Stark effects. For the latter problem, he developed perturbation theory, useful for an enormous number of problems.

A third formulation of quantum mechanics was found by P. A. M. Dirac (1926), while he was still a graduate student at Cambridge. It is more general than either of the former ones and has been used widely in the further development of the field.

In 1926 Born presented his interpretation of Schrödinger's wave function: $|\psi(x_1,x_2,...,x_n)|^2$ gives the probability of finding one particle at x_1, one at x_2, etc.

When a single particle is represented by a wave function, this can be constructed so as to give maximum probability of finding the particle at a given position x and a given momentum p, but neither of them can be exactly specified. This point was emphasized by Heisenberg in his uncertainty principle: classical concepts of motion can be applied to a particle only to a limited extent. You cannot describe the orbit of an electron in the ground state of an atom. The uncertainty principle has been exploited widely, especially by Niels Bohr.

Pauli, in 1927, amplified the Schrödinger equation by including the electron spin, which had been discovered by G. Uhlenbeck and S. Goudsmit in 1925. Pauli's wave function has two components, spin up and spin down, and the spin is represented by a 2×2 matrix. The matrices representing the components of the spin, σ_x, σ_y, and σ_z, do not commute. In addition to their practical usefulness, they are the simplest operators for demonstrating the essential difference between classical and quantum theory.

Dirac, in 1928, showed that spin follows naturally if the wave equation is extended to satisfy the requirements of special relativity, and if at the same time one requires that the differential equation be first order in time. Dirac's wave function for an electron has four components, more accurately 2×2. One factor 2 refers to spin, the other to the sign of the energy, which in relativity is given by

$$E = \pm c(p^2 + m^2c^2)^{1/2}. \tag{7}$$

States of negative energy make no physical sense, so Dirac postulated that nearly all such states are normally occupied. The few that are empty appear as particles of positive electric charge.

Dirac first believed that these particles represented protons. But H. Weyl and J. R. Oppenheimer, independently, showed that the positive particles must have the same mass as electrons. Pauli, in a famous article in the *Handbuch der Physik* (Pauli, 1933), considered this prediction of positively charged electrons a fundamental flaw of the theory. But within a year, in 1933, Carl Anderson and S. Neddermeyer discovered positrons in cosmic radiation.

Dirac's theory not only provided a natural explanation of spin, but also predicted that the interaction of the spin magnetic moment with the electric field in an atom is twice the strength that might be naively expected, in agreement with the observed fine structure of atomic spectra.

Empirically, particles of zero (or integral) spin obey Bose-Einstein statistics, and particles of spin $\frac{1}{2}$ (or half-integral), including electron, proton, and neutron, obey Fermi-Dirac statistics, i.e., they obey the Pauli exclusion principle. Pauli showed that spin and statistics should indeed be related in this way.

Applications

1926, the year when I started graduate work, was a wonderful time for theoretical physicists. Whatever problem you tackled with the new tools of quantum mechanics could be successfully solved, and hundreds of problems, from the experimental work of many decades, were around, asking to be tackled.

Atomic physics

The fine structure of the hydrogen spectrum was derived by Dirac. Energy levels depend on the principal quantum number n and the total angular momentum j, orbital momentum plus spin. Two states of orbital momentum $\ell = j + \frac{1}{2}$ and $j - \frac{1}{2}$ are degenerate.

The He atom had been an insoluble problem for the old (1913–1924) quantum theory. Using the Schrödinger equation, Heisenberg solved it in 1927. He found that the wave function, depending on the position of the two electrons $\Psi(\mathbf{r}_1, \mathbf{r}_2)$, could be symmetric or antisymmetric in $\mathbf{r}_1$ and $\mathbf{r}_2$. He postulated that the complete wave function should be antisymmetric, so a Ψ symmetric in $\mathbf{r}_1$ and $\mathbf{r}_2$ should be multiplied by a spin wave function antisymmetric in σ_1 and σ_2, hence belonging to a singlet state (parahelium). An antisymmetric spatial wave function describes a state with total spin $S = 1$, hence a triplet state (orthohelium). Heisenberg thus obtained a correct qualitative description of the He spectrum. The ground state is singlet, but for the excited states, the triplet has lower energy than the singlet. There is no degeneracy in orbital angular momentum L.

Heisenberg used a well-designed perturbation theory and thus got only qualitative results for the energy levels. To get accurate numbers, Hylleraas (in 1928 and later) used a variational method. The ground-state wave function is a function of r_1, r_2, and r_{12}, the distance of the two electrons from each other. He assumed a "trial function" depending on these variables and on some parameters, and then minimized the total energy as a function of these parameters. The resulting energy was very accurate. Others improved the accuracy further.

I also was intrigued by Hylleraas's success and applied his method to the negative hydrogen ion H^-. I showed that this ion was stable. It is important for the outer layers of the sun and in the crystal LiH, which is ionic: Li^+ and H^-.

For more complicated atoms, the first task was to obtain the structure of the spectrum. J. von Neumann and E. Wigner applied group theory to this problem, and could reproduce many features of the spectrum, e.g., the feature that, for a given electron configuration, the state of highest total spin S and highest total orbital momentum L has the lowest energy.

In the late 1920's J. Slater showed that these (and other) results could be obtained without group theory, by writing the wave function of the atom as a determinant of the wave functions of the individual electrons. The determinant form ensured antisymmetry.

To obtain the electron orbitals, D. R. Hartree in 1928 considered each electron as moving in the potential produced by the nucleus and the charge distribution of all the other electrons. Fock extended this method to include the effect of the antisymmetry of the atomic wave function. Hartree calculated numerically the orbitals in several atoms, first using his and later Fock's formulation.

Group theory is important in the structure of crystals, as had been shown long before quantum mechanics. I applied group theory in 1929 to the quantum states of an atom inside a crystal. This theory has also been much used in the physical chemistry of atoms in solution.

With modern computers, the solution of the Hartree-Fock system of differential equations has become straightforward. Once the electron orbitals are known, the energy levels of the atom can be calculated. Relativity can be included. The electron density near the nucleus can be calculated, and hence the hyperfine structure, isotope effect, and similar effects of the nucleus.

Molecules

A good approximation to molecular structure is to consider the nuclei fixed and calculate the electron wave function in the field of these fixed nuclei (Born and Oppenheimer, 1927). The eigenvalue of the electron energy, as a function of the position of nuclei, can then be considered as a potential in which the nuclei move.

Heitler and F. London, in 1927, considered the simplest molecule, H_2. They started from the wave function of two H atoms in the ground state and calculated the energy perturbation when the nuclei are at a distance R. If the wave function of the electrons is symmetric with respect to the position of the nuclei, the energy is lower than that of two separate H atoms, and they could calculate the binding energy of H_2 and the equilibrium distance R_0 of the two nuclei. Both agreed reasonably well with observation. At distances $R < R_0$, there is repulsion.

If the wave function is antisymmetric in the positions of the two electrons, there is repulsion at all distances. For a symmetric wave function, more accurate results can be obtained by the variational method.

Linus Pauling was able to explain molecular binding generally, in terms of quantum mechanics, and thereby helped create theoretical chemistry—see Herschbach (1999).

An alternative to the Heitler-London theory is the picture of molecular orbitals: Given the distance R between two nuclei, one may describe each electron by a wave function in the field of the nuclei. Since this field has only cylindrical symmetry, electronic states are described by two quantum numbers, the total angular momentum and its projection along the molecular axis; for example, $p\sigma$ means a state of total angular momentum 1 and component 0 in the direction of the axis.

Solid state

In a metal, the electrons are (reasonably) free to move between atoms. In 1927 Arnold Sommerfeld showed that the concept of free electron obeying the Pauli principle could explain many properties of metals, such as the relation between electric and thermal conductivity.

One phenomenon in solid-state physics, superconductivity, defied theorists for a long time. Many wrong theories were published. Finally, the problem was solved by John Bardeen, Leon Cooper, and Robert Schrieffer. Pairs of electrons are traveling together, at a considerable distance from each other, and are interacting strongly with lattice vibrations [see Schrieffer and Tinkham (1999)].

Collisions

The old (pre-1925) quantum theory could not treat collisions. In quantum mechanics the problem was solved by Born. If a particle of momentum $\mathbf{p}_1$ collides with a system Ψ_1, excites that system to a state Ψ_2, and thereby gets scattered to a momentum $\mathbf{p}_2$, then in first approximation the probability of this process is proportional to the absolute square of the matrix element,

$$\mathcal{M} = \int \exp[i(\mathbf{p}_1 - \mathbf{p}_2) \cdot \mathbf{r}/\hbar] \Psi_1 \Psi_2^* V d\tau, \tag{8}$$

where V is the interaction potential between particle and system, and the integration goes over the coordinates of the particle and all the components of the system. More accurate prescriptions were also given by Born.

There is an extensive literature on the subject. Nearly all physics beyond spectroscopy depends on the analysis of collisions see Datz *et al.* (1999).

Radiation and electrodynamics

The paradox of radiation's being both quanta and waves is elucidated by second quantization. Expanding the electromagnetic field in a Fourier series,

$$F(\mathbf{r},t) = \sum a_k \exp i(\mathbf{k} \cdot \mathbf{r} - \omega t), \tag{9}$$

one can consider the amplitudes a_k as dynamic variables, with a conjugate variable $a_k^\dagger$. They are quantized, using the commutation relation

$$a_k a_k^\dagger - a_k^\dagger a_k = 1. \tag{10}$$

The energy of each normal mode is $\hbar\omega(n + \tfrac{1}{2})$.

Emission and absorption of light is straightforward. The width of the spectral line corresponding to the transition of an atomic system from state m to state n was shown by E. Wigner and V. Weisskopf to be

$$\Delta\omega = \frac{1}{2}(\gamma_m + \gamma_n), \tag{11}$$

where γ_m is the rate of decay of state m (reciprocal of its lifetime) due to spontaneous emission of radiation.

Heisenberg and Pauli (1929, 1930) set out to construct a theory of quantum electrodynamics, quantizing the electric field at a given position $\mathbf{r}_m$. Their theory is self-consistent, but it had the unfortunate feature that the electron's self-energy, i.e., its interaction with its own electromagnetic field, turned out to be infinite.

E. Fermi (1932) greatly simplified the theory by considering the Fourier components of the field, rather than the field at a given point. But the self-energy remained infinite. This problem was only solved after World War II. The key was the recognition, primarily due to Kramers, that the self-energy is necessarily included in the mass of the electron and cannot be separately measured. The only observable quantity is then a possible *change* of that self-energy when the electron is subject to external forces, as in an atom.

J. Schwinger (1948) and R. Feynman (1948), in different ways, then constructed relativistically covariant, and finite, theories of quantum electrodynamics. Schwinger deepened the existing theory while Feynman invented a completely novel technique which at the same time simplified the technique of doing actual calculations. S. Tomonaga had earlier (1943) found a formulation similar to Schwinger's. F. J. Dyson (1949) showed the equivalence of Schwinger and Feynman's approaches and then showed that the results of the theory are finite in any order of $\alpha = e^2/\hbar c$. Nevertheless the perturbation series diverges, and infinities will appear in order $\exp(-\hbar c/e^2)$. An excellent account of the development of quantum electrodynamics has been given by Schweber (1994).

It was very fortunate that, just before Schwinger and Feynman, experiments were performed that showed the intricate effects of the self-interaction of the electron. One was the discovery, by P. Kusch and H. M. Foley (1948) that the magnetic moment of the electron is slightly (by about 1 part in 1000) greater than predicted by Dirac's theory. The other was the observation by W. Lamb and R. Retherford (1947) that the $2s$ and the $2p_{1/2}$ states of the H atom do not coincide, $2s$ having an energy higher by the very small amount of about 1000 megaHertz (the total binding energy being of the order of 10^9 megaHertz).

All these matters were discussed at the famous Shelter Island Conference in 1947 (Schweber, 1994). Lamb, Kusch, and I. I. Rabi presented experimental results, Kramers his interpretation of the self-energy, and Feynman and Schwinger were greatly stimulated by the conference. So was I, and I was able within a week to calculate an approximate value of the Lamb shift.

After extensive calculations, the Lamb shift could be reproduced within the accuracy of theory and experiment. The Lamb shift was also observed in He$^+$, and calculated for the $1s$ electron in Pb. In the latter atom, its contribution is several Rydberg units.

The "anomalous" magnetic moment of the electron was measured in ingenious experiments by H. Dehmelt and collaborators. They achieved fabulous accuracy, viz., for the ratio of the anomalous to the Dirac moments

$$a = 1\ 159\ 652\ 188\ (4) \times 10^{-12}, \tag{12}$$

where the 4 in parenthesis gives the probable error of the last quoted figure. T. Kinoshita and his students have evaluated the quantum electrodynamic (QED) theory with equal accuracy, and deduced from Eq. (12) the fine-structure constant

$$\alpha^{-1} = \hbar c/e^2 = 137.036\ 000. \tag{13}$$

At least three other, independent methods confirm this value of the fine-structure constant, albeit with less precision. See also Hughes and Kinoshita (1999).

Interpretation

Schrödinger believed at first that his wave function gives directly the continuous distribution of the electron charge at a given time. Bohr opposed this idea vigorously.

Guided by his thinking about quantum-mechanical collision theory (see earlier section on Collisions) Born proposed that the absolute square of the wave function gives the probability of finding the electron, or other particle or particles, at a given position. This interpretation has been generally accepted.

For a free particle, a wave function (wave packet) may be constructed that puts the main probability near a position x_0 and near a momentum p_0. But there is the uncertainty principle: position and momentum cannot be simultaneously determined accurately, their uncertainties are related by

$$\Delta x \Delta p \geqslant \frac{1}{2}\hbar. \tag{14}$$

The uncertainty principle says only this: that the concepts of classical mechanics cannot be directly applied in the atomic realm. This should not be surprising because the classical concepts were derived by studying the motion of objects weighing grams or kilograms, moving over distances of meters. There is no reason why they should still be valid for objects weighing 10^{-24} g or less, moving over distances of 10^{-8} cm or less.

The uncertainty principle has profoundly misled the lay public: they believe that everything in quantum theory is fuzzy and uncertain. Exactly the reverse is true. Only quantum theory can explain why atoms exist at all. In a classical description, the electrons hopelessly fall into the nucleus, emitting radiation in the process. With quantum theory, and only with quantum theory, we can understand and explain why chemistry exists—and, due to chemistry, biology.

(A small detail: in the old quantum theory, we had to speak of the electron "jumping" from one quantum state to another when the atom emits light. In quantum *mechanics*, the orbit is sufficiently fuzzy that no jump is needed: the electron can move continuously in space; at worst it may change its velocity.)

Perhaps more radical than the uncertainty principle is the fact that you cannot predict the result of a collision but merely the probability of various possible results. From a practical point of view, this is not very different from statistical mechanics, where we also only consider probabilities. But of course, in quantum mechanics the result is unpredictable *in principle*.

Several prominent physicists found it difficult to accept the uncertainty principle and related probability predictions, among them de Broglie, Einstein, and Schrödinger. De Broglie tried to argue that there should be a deterministic theory behind quantum mechanics. Einstein forever thought up new examples that might contradict the uncertainty principle and confronted Bohr with them; Bohr often had to think for hours before he could prove Einstein wrong.

Consider a composite object that disintegrates into $A+B$. The total momentum $P_A + P_B$ and its coordinate separation $x_A - x_B$ can be measured and specified simultaneously. For simplicity let us assume that $P_A + P_B$ is zero, and that $x_A - x_B$ is a large distance. If in this state the momentum of A is measured and found to be P_A, we know that the momentum of B is definitely $-P_A$. We may then measure x_B and it seems that we know both P_B and x_B, in apparent conflict with the uncertainty principle. The resolution is this: the measurement of x_B imparts a momentum to B (as in a γ-ray microscope) and thus destroys the previous knowledge of P_B, so the two measurements have no predictive value.

Nowadays these peculiar quantum correlations are often discussed in terms of an "entangled" spin-zero state of a composite object AB, composed of two spin-one-half particles, or two oppositely polarized photons (Bohm and Aharonov). Bell showed that the quantum-mechanical correlations between two such separable systems, A and B, cannot be explained by any mechanism involving hidden variables. Quantum correlations between separated parts A and B of a composite system have been demonstrated by some beautiful experiments (e.g., Aspect *et al.*). The current status of these issues is further discussed by Mandel (1999) and Zeilinger (1999), in this volume.

References

Born, M., and J. R. Oppenheimer, 1927, Ann. Phys. (Leipzig) **84**, 457.
Datz, S., G. W. F. Drake, T. F. Gallagher, H. Kleinpoppen, and G. zu Putlitz, 1999, Rev. Mod. Phys. **71**, 223; pp. 377–407 in this book.
Dirac, P. A. M., 1926, Ph.D. Thesis (Cambridge University).
Dyson, F. J., 1949, Phys. Rev. **75**, 486.
Fermi, E., 1932, Rev. Mod. Phys. **4**, 87.
Feynman, R. P., 1948, Rev. Mod. Phys. **76**, 769.
Heisenberg, W., 1971, *Physics and Beyond* (New York, Harper and Row).
Heisenberg, W., and W. Pauli, 1929, Z. Phys. **56**, 1.
Heisenberg, W., and W. Pauli, 1930, Z. Phys. **59**, 168.
Herschbach, D., 1999, Rev. Mod. Phys. **71**, 411; pp. 693–705 in this book.
Hughes, V. W., and T. Kinoshita, 1999, Rev. Mod. Phys. **71**, 133; pp. 223–233 in this book.
Kusch, P., and H. M. Foley, 1948, Phys. Rev. **73**, 412; **74**, 250.
Lamb, W. E., and R. C. Retherford, 1947, Phys. Rev. **72**, 241.
Mandel, L., 1999, Rev. Mod. Phys. **71**, 274; pp. 460–473 in this book.
Pauli, W., 1933, *Handbuch der Physik*, 2nd Ed. (Berlin, Springer).
Schrieffer, J. R., and M. Tinkham, 1999, Rev. Mod. Phys. **71**, 313; pp. 526–553 in this book.
Schweber, S. S., 1994, *QED and the Men who Made It* (Princeton University Press, Princeton, NJ), pp. 157–193.
Schwinger, J., 1948, Phys. Rev. **73**, 416.
Tomonaga, S., 1943, Bull. IPCR (Rikenko) **22**, 545 [Eng. Translation 1946].
Zeilinger, A., 1999, Rev. Mod. Phys. **71**, 288; pp. 482–498 in this book.

Nuclear Physics
Hans A. Bethe

Nuclear physics started in 1894 with the discovery of the radioactivity of uranium by A. H. Becquerel. Marie and Pierre Curie investigated this phenomenon in detail: to their astonishment they found that raw uranium ore was far more radioactive than the refined uranium from the chemist's store. By chemical methods, they could separate (and name) several new elements from the ore which were intensely radioactive: radium ($Z=88$), polonium ($Z=84$), a gas they called emanation ($Z=86$) (radon), and even a form of lead ($Z=82$).

Ernest Rutherford, at McGill University in Montreal, studied the radiation from these substances. He found a strongly ionizing component which he called α rays, and a weakly ionizing one, β rays, which were more penetrating than the α rays. In a magnetic field, the α rays showed positive charge, and a charge-to-mass ratio corresponding to ^{4}He. The β rays had negative charge and were apparently electrons. Later, a still more penetrating, uncharged component was found, γ rays.

Rutherford and F. Soddy, in 1903, found that after emission of an α ray, an element of atomic number Z was transformed into another element, of atomic number $Z-2$. (They did not yet have the concept of atomic number, but they knew from chemistry the place of an element in the periodic system.) After β-ray emission, Z was transformed into $Z+1$, so the dream of alchemists had become true.

It was known that thorium ($Z=90$, $A=232$) also was radioactive, also decayed into radium, radon, polonium and lead, but obviously had different radioactive behavior from the decay products of uranium ($Z=92$, $A=238$). Thus there existed two or more forms of the same chemical element having different atomic weights and different radioactive properties (lifetimes) but the same chemical properties. Soddy called these isotopes.

Rutherford continued his research at Manchester, and many mature collaborators came to him. H. Geiger and J. M. Nuttall, in 1911, found that the energy of the emitted α particles, measured by their range, was correlated with the lifetime of the parent substance: the lifetime decreased very rapidly (exponentially) with increasing α-particle energy.

By an ingenious arrangement of two boxes inside each other, Rutherford proved that the α particles really were He atoms: they gave the He spectrum in an electric discharge.

Rutherford in 1906 and Geiger in 1908 put thin solid foils in the path of a beam of α particles. On the far side of the foil, the beam was spread out in angle—not surprising because the electric charges in the atoms of the foil would deflect the α particles by small angles and multiple deflections were expected. But to their surprise, a few α particles came back on the front side of the foil, and their number increased with increasing atomic weight of the material in the foil. Definitive experiments with a gold foil were made by Geiger and Marsden in 1909.

Rutherford in 1911 concluded that this backward scattering could not come about by multiple small-angle scatterings. Instead, there must also occasionally be single deflections by a large angle. These could only be produced by a big charge concentrated somewhere in the atom. Thus he conceived the nuclear atom: each atom has a nucleus with a positive charge equal to the sum of the charges of all the electrons in the atom. The nuclear charge Ze increases with the atomic weight.

Rutherford had good experimental arguments for his concept. But when Niels Bohr in 1913 found the theory of the hydrogen spectrum, Rutherford declared, "Now I finally believe my nuclear atom."

The scattering of fast α particles by He indicated also a stronger force than the electrostatic repulsion of the two He nuclei, the first indication of the strong nuclear force. Rutherford and his collaborators decided that this must be the force that holds α particles inside the nucleus and thus was attractive. From many scattering experiments done over a decade they concluded that this attractive force was confined to a radius

$$R = 1.2 \times 10^{-13} A^{1/3} \text{ cm}, \tag{1}$$

which may be considered to be the nuclear radius. This result is remarkably close to the modern value. The volume of the nucleons, according to Eq. (1), is proportional to the number of particles in it.

When α particles were sent through material of low atomic weight, particles were emitted of range greater than the original α particle. These were interpreted by Rutherford and James Chadwick as protons. They had observed the disintegration of light nuclei, from boron up to potassium.

Quantum mechanics gave the first theoretical explanation of natural radioactivity. In 1928 George Gamow, and simultaneously K. W. Gurney and E. U. Condon, discovered that the potential barrier between a nucleus and an α particle could be penetrated by the α particle coming from the inside, and that the rate of penetration depended exponentially on the height and width of the barrier. This explained the Geiger-Nuttall law that the lifetime of α-radioactive nuclei decreases enormously as the energy of the α particle increases.

On the basis of this theory, Gamow predicted that protons of relatively low energy, less than one million electron volts, should be able to penetrate into light nuclei, such as Li, Be, and B, and disintegrate them. When Gamow visited Cambridge, he encouraged the experimenters at the Cavendish Laboratory to build accelerators of relatively modest voltage, less than one million volts. Such accelerators were built by M. L. E. Oliphant on the one hand, and J. D. Cockcroft and E. T. S. Walton on the other.

By 1930, when I spent a semester at the Cavendish, the Rutherford group understood α particles very well. The penetrating γ rays, uncharged, were interpreted as high-frequency electromagnetic radiation, emitted by a nucleus after an α ray: the α particle had left the nucleus in an excited state, and the transition to the ground state was accomplished by emission of the γ ray.

The problem was with β rays. Chadwick showed in 1914 that they had a continuous spectrum, and this was repeatedly confirmed. Rutherford, Chadwick, and C. D. Ellis, in their book on radioactivity in 1930, were baffled. Bohr was willing to give up conservation of energy in this instance. Pauli violently objected to Bohr's idea, and suggested in 1931 and again in 1933 that together with the electron (β-particle) a neutral particle was emitted, of such high penetrating power that it had never been observed. This particle was named the neutrino by Fermi, "the small neutral one."

The Neutron and the Deuteron

In 1930, when I first went to Cambridge, England, nuclear physics was in a peculiar situation: a lot of experimental evidence had been accumulated, but there was essentially no theoretical understanding. The nucleus was supposed to be composed of protons and electrons, and its radius was supposed to be $<10^{-12}$ cm. The corresponding momentum, according to quantum mechanics, was

$$P > P_{min} = \frac{\hbar}{R} = \frac{10^{-27}}{10^{-12}} = 10^{-15} \text{ erg}/c, \qquad (2)$$

while from the mass m_e of the electron

$$m_e c = 3 \times 10^{-17} \text{ erg}/c. \qquad (3)$$

Thus the electrons had to be highly relativistic. How could such an electron be retained in the nucleus, indeed, how could an electron wave function be fitted into the nucleus?

Further troubles arose with spin and statistics: a nucleus was supposed to contain A protons to make the correct atomic weight, and $A-Z$ electrons to give the net charge Z. The total number of particles was $2A-Z$, an odd number if Z was odd. Each proton and electron was known to obey Fermi statistics, hence a nucleus of odd Z should also obey Fermi statistics. But band spectra of nitrogen, N_2, showed that the N nucleus, of $Z=7$, obeyed Bose statistics. Similarly, proton and electron had spin $\frac{1}{2}$, so the nitrogen nucleus should have half-integral spin, but experimentally its spin was 1.

These paradoxes were resolved in 1932 when Chadwick discovered the neutron. Now one could assume that the nucleus consisted of Z protons and $A-Z$ neutrons. Thus a nucleus of mass A would have Bose (Fermi) statistics if A was even (odd) which cleared up the ^{14}N paradox, provided that the neutron obeyed Fermi statistics and had spin $\frac{1}{2}$, as it was later shown to have.

Chadwick already showed experimentally that the mass of the neutron was close to that of the proton, so the minimum momentum of 10^{-15} erg/c has to be compared with

$$M_n c = 1.7 \times 10^{-24} \times 3 \times 10^{10} = 5 \times 10^{-14} \text{ erg}/c, \qquad (4)$$

where M_n is the mass of the nucleon. $P_{min} = 10^{-15}$ is small compared to this, so the wave function of neutron and proton fits comfortably into the nucleus.

The discovery of the neutron had been very dramatic. Walther Bothe and H. Becker found that Be, bombarded by α particles, emitted very penetrating rays that they interpreted as γ rays. Curie and Joliot exposed paraffin to these rays, and showed that protons of high energy were ejected from the paraffin. If the rays were actually γ rays, they needed to have extremely high energies, of order 30 MeV. Chadwick had dreamed about neutrons for a decade, and got the idea that here at last was his beloved neutron.

Chadwick systematically exposed various materials to the penetrating radiation, and measured the energy of the recoil atoms. Within the one month of February 1932 he found the answer: indeed the radiation consisted of particles of the mass of a proton, they were neutral, hence neutrons. A beautiful example of systematic experimentation.

Chadwick wondered for over a year: was the neutron an elementary particle, like the proton, or was it an excessively strongly bound combination of proton and electron? In the latter case, he argued, its mass should be less than that of the hydrogen atom, because of the binding energy. The answer was only obtained when Chadwick and Goldhaber disintegrated the deuteron by γ rays (see below): the mass of the neutron was 0.8 MeV greater than that of the H atom. So, Chadwick decided, the neutron must be an elementary particle of its own.

If the neutron was an elementary particle of spin $\frac{1}{2}$, obeying Fermi statistics, the problem of spin and statistics of ^{14}N was solved. And one no longer needed to squeeze electrons into the too-small space of a nucleus. Accordingly, Werner Heisenberg and Iwanenko independently in 1933 proposed that a nucleus consists of neutrons and protons. These are two possible states of a more general particle, the nucleon. To emphasize this, Heisenberg introduced the concept of the isotopic spin τ_z, the proton having $\tau_z = +\frac{1}{2}$ and the neutron $\tau_z = -\frac{1}{2}$. This concept has proved most useful.

Before the discovery of the neutron, in 1931 Harold Urey discovered heavy hydrogen, of atomic weight 2. Its nucleus, the deuteron, obviously consists of one proton and one neutron, and is the simplest composite nucleus. In 1933, Chadwick and Goldhaber succeeded in disintegrating the deuteron by γ rays of energy 2.62 MeV, and measuring the energy of the proton resulting from the disintegration. In this way, the binding energy of the deuteron was determined to be 2.22 MeV.

This binding energy is very small compared with that of ^{4}He, 28.5 MeV, which was interpreted as meaning that the attraction between two nucleons has very short range and great depth. The wave function of the deuteron outside the potential well is then determined simply by the binding energy ε. It is

$$\psi = \exp(-\alpha r)/r, \tag{5}$$

$$\alpha = (M\varepsilon)^{1/2}/\hbar, \tag{6}$$

with M the mass of a nucleon.

The scattering of neutrons by protons at moderate energy can be similarly determined, but one has to take into account that the spins of the two nucleons may be either parallel (total $S=1$) or antiparallel ($S=0$). The spin of the deuteron is 1. The $S=0$ state is not bound. The scattering, up to at least 10 MeV, can be described by two parameters for each value of S, the scattering length and the effective range r_0. The phase shift for $L=0$ is given by

$$k \cot \delta = -\frac{1}{a} + \frac{1}{2}k^2 r_0, \tag{7}$$

where k is the wave number in the center-of-mass system, δ the phase shift, a the scattering length, and r_0 the effective range. Experiments on neutron-proton scattering result in

$$a_t = 5.39 \text{ fm}, \quad r_{ot} = 1.72 \text{ fm},$$

$$a_s = -23.7 \text{ fm}, \quad r_{os} = 2.73 \text{ fm}, \tag{8}$$

where t and s designate the triplet and singlet $L=0$ states, ^3S and ^1S. The experiments at low energy, up to about 10 MeV, cannot give any information on the shape of the potential. The contribution of $L>0$ is very small for $E<10$ MeV, because of the short range of nuclear forces.

Very accurate experiments were done in the 1930s on the scattering of protons by protons, especially by Tuve and collaborators at the Carnegie Institution of Washington, D.C., and by R. G. Herb et al. at the University of Wisconsin. The theoretical interpretation was mostly done by Breit and collaborators. The system of two protons, at orbital momentum $L=0$, can exist only in the state of total spin $S=0$. The phase shift is the shift relative to a pure Coulomb field. The scattering length resulting from the analysis is close to that of the ^1S state of the proton-neutron system. This is the most direct evidence for charge independence of nuclear forces. There is, however, a slight difference: the proton-neutron force is slightly more attractive than the proton-proton force.

Before World War II, the maximum particle energy available was less than about 20 MeV. Therefore only the S-state interaction between two nucleons could be investigated.

The Liquid Drop Model

Energy

The most conspicuous feature of nuclei is that their binding energy is nearly proportional to A, the number of nucleons in the nucleus. Thus the binding per particle is nearly constant, as it is for condensed matter. This is in contrast to electrons in an atom: the binding of a $1S$ electron increases as Z^2.

The volume of a nucleus, according to Eq. (1), is also proportional to A. This and the binding energy are the basis of the liquid drop model of the nucleus, used especially by Niels Bohr: the nucleus is conceived as filling a compact volume, spherical or other shape, and its energy is the sum of an attractive term proportional to the volume, a repulsive term proportional to the surface, and another term due to the mutual electric repulsion of the positively charged protons. In the volume energy, there is also a positive term proportional to $(N-Z)^2 = (A-2Z)^2$ because the attraction between proton and neutron is stronger than between two like particles. Finally, there is a pairing energy: two like particles tend to go into the same quantum state, thus decreasing the energy of the nucleus. A combination of these terms leads to the Weizsäcker semi-empirical formula

$$E = -a_1 A + a_2 A^{2/3} + a_3 Z^2 A^{-1/3} + a_4 (A-2Z)^2 A^{-1} + \lambda a_5 A^{-3/4}. \tag{9}$$

Over the years, the parameters $a_1, \ldots, a_5$ have been determined. Green (1954) gives these values (in MeV):

$$a_1 = 15.75, \quad a_2 = 17.8, a_3 = 0.710, \quad a_4 = 23.7, \quad a_5 = 34. \tag{10}$$

The factor λ is $+1$ if Z and $N = A - Z$ are both odd, $\lambda = -1$ if they are both even, and $\lambda = 0$ if A is odd. Many more accurate expressions have been given.

For small mass number A, the symmetry term $(N-Z)^2$ puts the most stable nucleus at $N=Z$. For larger A, the Coulomb term shifts the energy minimum to $Z<A/2$.

Among very light nuclei, the energy is lowest for those which may be considered multiples of the α particle, such as ^{12}C, ^{16}O, ^{20}Ne, ^{24}Mg, ^{28}Si, ^{32}S, ^{40}Ca. For $A=56$, ^{56}Ni ($Z=28$) still has strong binding but ^{56}Fe ($Z=26$) is more strongly bound. Beyond A

=56, the preference for multiples of the α particle ceases.

For nearly all nuclei, there is preference for even Z and even N. This is because a pair of neutrons (or protons) can go into the same orbital and can then have maximum attraction.

Many nuclei are spherical; this giving the lowest surface area for a given volume. But when there are many nucleons in the same shell (see later section on the Shell Model), ellipsoids, or even more complicated shapes (Nielsen model), are often preferred.

Density distribution

Electron scattering is a powerful way to measure the charge distribution in a nucleus. Roughly, the angular distribution of elastic scattering gives the Fourier transform of the radial charge distribution. But since $Ze^2/\hbar c$ is quite large, explicit calculation with relativistic electron wave functions is required. Experimentally, Hofstadter at Stanford started the basic work.

In heavy nuclei, the charge is fairly uniformly distributed over the nuclear radius. At the surface, the density falls off approximately like a Fermi distribution,

$$\rho/\rho_0 \approx [1+\exp(r-R)/a]^{-1}, \tag{11}$$

with $a \approx 0.5$ fm; the surface thickness, from 90% to 10% of the central density, is about 2.4 fm.

In more detailed studies, by the Saclay and Mainz groups, indications of individual proton shells can be discerned. Often, there is evidence for nonspherical shapes. The neutron distribution is more difficult to determine experimentally; sometimes the scattering of π mesons is useful. Inelastic electron scattering often shows a maximum at the energy where scattering of the electron by a single free proton would lie.

α radioactivity

Equation (9) represents the energy of a nucleus relative to that of free nucleons, $-E$ is the binding energy. The mass excess of Z protons and $(A-Z)$ neutrons is

$$\Delta M = 7.3Z + 8.1(A-Z) \text{ MeV}, \tag{12}$$

which complies with the requirement that the mass of ^{12}C is 12 amu. The mass excess of the nucleus is

$$E+\Delta M = E+7.3Z+8.1(A-Z) \text{ MeV}. \tag{13}$$

The mass excess of an α particle is 2.4 MeV, or 0.6 MeV per nucleon. So the excess of the mass of nucleus (Z,A) over that of $Z/2$ α particles plus $A-2Z$ neutrons is

$$E' = E+\Delta M - (Z/2)0.6 \text{ MeV} = E+7.0Z+8.1(A-Z). \tag{14}$$

The (smoothed) energy available for the emission of an α particle is then

$$E''(Z,A) = E'(Z,A) - E'(Z-2,A-4). \tag{15}$$

This quantity is negative for small A, positive from about the middle of the periodic table on. When it becomes greater than about 5 MeV, emission of α particles becomes observable. This happens when $A \geq 208$. It helps that $Z=82$, $A=208$ is a doubly magic nucleus.

Fission

In the mid 1930s, Fermi's group in Rome bombarded samples of most elements with neutrons, both slow and fast. In nearly all elements, radioactivity was produced. Uranium yielded several distinct activities. Lise Meitner, physicist, and Otto Hahn, chemist, continued this research in Berlin and found some sequences of radioactivities following each other. When Austria was annexed to Germany in Spring 1938, Meitner, an Austrian Jew, lost her job and had to leave Germany; she found refuge in Stockholm.

Otto Hahn and F. Strassmann continued the research and identified chemically one of the radioactive products from uranium ($Z=92$). To their surprise they found the radioactive substance was barium, ($Z=56$). Hahn, in a letter to Meitner, asked for help. Meitner discussed it with her nephew, Otto Frisch, who was visiting her. After some discussion, they concluded that Hahn's findings were quite natural, from the standpoint of the liquid drop model: the drop of uranium split in two. They called the process "fission."

Once this general idea was clear, comparison of the atomic weight of uranium with the sum of the weights of the fission products showed that a very large amount of energy would be set free in fission. Frisch immediately proved this, and his experiment was confirmed by many laboratories. Further, the fraction of neutrons in the nucleus, $N/A = (A-Z)/A$, was much larger in uranium than in the fission products hence neutrons would be set free in fission. This was proved experimentally by Joliot and Curie. Later experiments showed that the average number of neutrons per fission was $\nu=2.5$. This opened the prospect of a chain reaction.

A general theory of fission was formulated by Niels Bohr and John Wheeler in 1939. They predicted that only the rare isotope of uranium, U-235, would be fissionable by slow neutrons. The reason was that U-235 had an odd number of neutrons. After adding the neutron from outside, both fission products could have an even number of neutrons, and hence extra binding energy due to the formation of a neutron pair. Conversely, in U-238 one starts from an even number of neutrons, so one of the fission products must have an odd number. Nier then showed experimentally that indeed U-235 can be fissioned by slow neutrons while U-238 requires neutrons of about 1 MeV.

The chain reaction

Fission was discovered shortly before the outbreak of World War II. There was immediate interest in the chain reaction in many countries.

To produce a chain reaction, on average at least one of the 2.5 neutrons from a U-235 fission must again be captured by a U-235 and cause fission. The first chain reaction was established by Fermi and collaborators on 2 December 1942 at the University of Chicago. They used a "pile" of graphite bricks with a lattice of uranium metal inside.

The graphite atoms served to slow the fission neutrons, originally emitted at about 1 MeV energy, down to thermal energies, less than 1 eV. At those low energies, capture by the rare isotope U-235 competes favorably with U-238. The carbon nucleus absorbs very few neutrons, but the graphite has to be very pure C. Heavy water works even better.

The chain reaction can either be controlled or explosive. The Chicago pile was controlled by rods of boron absorber whose position could be controlled by the operator. For production of power, the graphite is cooled by flowing water whose heat is then used to make steam. In 1997, about 400 nuclear power plants were in operation (see Till, 1999).

In some experimental "reactors," the production of heat is incidental. The reactor serves to produce neutrons which in turn can be used to produce isotopes for use as tracers or in medicine. Or the neutrons themselves may be used for experiments such as determining the structure of solids.

Explosive chain reactions are used in nuclear weapons. In this case, the U-235 must be separated from the abundant U-238. The weapon must be assembled only immediately before its use. Plutonium-239 may be used instead of U-235 (see Drell, 1999).

The Two-Nucleon Interaction

Experimental

A reasonable goal of nuclear physics is the determination of the interaction of two nucleons as a function of their separation. Because of the uncertainty principle, this requires the study of nuclear collisions at high energy. Before the second World War, the energy of accelerators was limited. After the war, cyclotrons could be built with energies upward of 100 MeV. This became possible by modulating the frequency, specifically, decreasing it on a prescribed schedule as any given batch of particles, e.g., protons, is accelerated. The frequency of the accelerating electric field must be

$$\omega \sim B/m_{\text{eff}},$$

in order to keep that field in synchronism with the orbital motion of the particles. Here B is the local magnetic field which should decrease (slowly) with the distance r from the center of the cyclotron in order to keep the protons focused; $m_{\text{eff}} = E/c^2$ is the relativistic mass of the protons which increases as the protons accelerate and r increases. Thus the frequency of the electric field between the dees of the cyclotron must decrease as the protons accelerate.

Such frequency modulation (FM) had been developed in the radar projects during World War II. At the end of that war, E. McMillan in the U.S. and Veksler in the Soviet Union independently suggested the use of FM in the cyclotron. It was introduced first at Berkeley and was immediately successful. These FM cyclotrons were built at many universities, including Chicago, Pittsburgh, Rochester, and Birmingham (England).

The differential cross section for the scattering of protons by protons at energies of 100 to 300 MeV was soon measured. But since the proton has spin, this is not enough: the scattering of polarized protons must be measured for two different directions of polarization, and as a function of scattering angle. Finally, the change of polarization in scattering must be measured. A complete set of required measurements is given (Walecka, 1995). The initial polarization, it turns out, is best achieved by scattering the protons from a target with nuclei of zero spin, such as carbon.

Proton-proton scattering is relatively straightforward, but in the analysis the effect of the Coulomb repulsion must, of course, be taken into account. It is relatively small except near the forward direction. The nuclear force is apt to be attractive, so there is usually an interference minimum near the forward direction.

The scattering of neutrons by protons is more difficult to measure, because there is no source of neutrons of definite energy. Fortunately, when fast protons are scattered by deuterons, the deuteron often splits up, and a neutron is projected in the forward direction with almost the full energy of the initial proton.

Phase shift analysis

The measurements can be represented by phase shifts of the partial waves of various angular momenta. In proton-proton scattering, even orbital momenta occur only together with zero total spin (singlet states), odd orbital momenta with total spin one (triplet states). Phase shift analysis appeared quite early, e.g., by Stapp, Ypsilantis, and Metropolis in 1957. But as long as only experiments at one energy were used, there were several sets of phase shifts that fitted the data equally well. It was necessary to use experiments at many energies, derive the phase shifts and demand that they depend smoothly on energy.

A very careful phase shift analysis was carried out by a group in Nijmegen, Netherlands, analyzing first the pp and the np (neutron-proton) scattering up to 350 MeV (Bergervoet et al., 1990). They use np data from well over 100 experiments from different laboratories and energies. Positive phase shifts means attraction.

As is well known, S waves are strongly attractive at low energies, e.g., at 50 MeV, the ^3S phase shift is 60°, ^1S is 40°. ^3S is more attractive than ^1S, just as, at $E=0$, there is a bound ^3S state but not of ^1S. At high energy, above about 300 MeV, the S phase shifts become repulsive, indicating a repulsive core in the potential.

The P and D phase shifts at 300 MeV are shown in Table I (Bergervoet et al., 1990). The singlet states are attractive or repulsive, according to whether L is even or odd. This is in accord with the idea prevalent in early nuclear theory (1930s) that there should be exchange forces, and it helps nuclear forces to saturate. The triplet states of $J=L$ have nearly the same phase shifts as the corresponding singlet states. The triplet states show a tendency toward a spin-orbit force, the higher J being more attractive than the lower J.

Potential

In the 1970s, potentials were constructed by the Bonn and the Paris groups. Very accurate potentials, using the Nijmegen data base were constructed by the Nijmegen and Argonne groups.

We summarize some of the latter results, which include the contributions of vacuum polarization, the magnetic moment interaction, and finite size of the neutron and proton. The longer range nuclear interaction is one-pion exchange (OPE). The shorter-range potential is a sum of central, L^2, tensor, spin-orbit and quadratic spin-orbit terms. A short range core of $r_0 = 0.5$ fm is included in each. The potential fits the experimental data very well: excluding the energy interval 290–350 MeV, and counting both pp and np data, their $\chi^2 = 3519$ for 3359 data.

No attempt is made to compare the potential to any meson theory. A small charge dependent term is found. The central potential is repulsive for $r < 0.8$ fm; its minimum is -55 MeV. The maximum tensor potential is about 50 MeV, the spin-orbit potential at 0.7 fm is about 130 MeV.

Table 1. P and D phase shifts at 300 MeV, in degrees.

^1P	-28	$^1D_2$	$+25$
$^3P_0$	-10	$^3D_1$	-24
$^3P_1$	-28	$^3D_2$	$+25$
$^3P_2$	$+17$	$^3D_3$	$+4$

Inclusion of pion production

Nucleon-nucleon scattering ceases to be elastic once pions can be produced. Then all phase shifts become complex. The average of the masses of π^+, π^0, and π^- is 138 MeV. Suppose a pion is made in the collision of two nucleons, one at rest (mass M) and one having energy $E > M$ in the laboratory. Then the square of the invariant mass is initially

$$(E+M)^2 - P^2 = 2M^2 + 2EM. \tag{16}$$

Suppose in the final state the two nucleons are at rest relative to each other, and in their rest system a pion is produced with energy ε, momentum π, and mass μ. Then the invariant mass is

$$(2M+\varepsilon)^2 - \pi^2 = 4M^2 + 4M\varepsilon + \mu^2. \tag{17}$$

Setting the two invariant masses equal,

$$E - M = 2\varepsilon + \mu^2/2M, \tag{18}$$

a remarkably simple formula for the initial kinetic energy in the laboratory. The absolute minimum for meson production is 286 MeV. The analysts have very reasonably chosen $E - M = 350$ MeV for the maximum energy at which nucleon-nucleon collision may be regarded as essentially elastic.

Three-Body Interaction

The observed binding energy of the triton, ^{3}H, is 8.48 MeV. Calculation with the best two-body potential gives 7.8 MeV. The difference is attributed to an interaction between all three nucleons. Meson theory yields such an interaction based on the transfer of a meson from nucleon i to j, and a second meson from j to k. The main term in this interaction is

$$V_{ijk} = A Y(mr_{ij}) Y(mr_{jk}) \sigma_i \cdot \sigma_j \sigma_j \cdot \sigma_k \tau_i \cdot \tau_j \tau_j \cdot \tau_k, \tag{19}$$

where Y is the Yukawa function,

$$Y(mr) = \frac{\exp(-mcr/\hbar)}{mcr/\hbar}. \tag{20}$$

The cyclic interchanges have to be added to V_{123}. There is also a tensor force which has to be suitably cut off at small distances. It is useful to also add a repulsive central force at small r.

The mass m is the average of the three π mesons, $m = \frac{1}{3}m_{\pi^0} + \frac{2}{3}m_{\pi^\pm}$. The coefficient A is adjusted to give the correct ^{3}H binding energy and the correct density of nuclear matter. When this is done, the binding energy of ^{4}He automatically comes out correctly, a very gratifying result. So no four-body forces are needed.

The theoretical group at Argonne then proceed to calculate nuclei of atomic weight 6 to 8. They used a Green's function Monte Carlo method to obtain a suitable wave function and obtained the binding energy of the ground state to within about 2 MeV. For very unusual nuclei like ^{7}He or ^{8}Li, the error may be 3–4 MeV. Excited states have similar accuracy, and are arranged in the correct order.

Nuclear Matter

"Nuclear matter" is a model for large nuclei. It assumes an assembly of very many nucleons, protons, and neutrons, but disregards the Coulomb force. The aim is to calculate the density and binding energy per nucleon. In first approximation, each nucleon moves independently, and because we have assumed a very large size, its wave function is an exponential, $\exp(i\mathbf{k}\cdot\mathbf{r})$. Nucleons interact, however, with their usual two-body forces; therefore, the wave functions are modified wherever two nucleons are close together. Due to its interactions, each nucleon has a potential energy, so a nucleon of wave vector $\mathbf{k}$ has an energy $E(k) \neq (\hbar^2/2m)k^2$.

Consider two particles of momenta $\mathbf{k}_1$ and $\mathbf{k}_2$; their unperturbed energy is

$$W = E(k_1) + E(k_2), \tag{21}$$

and their unperturbed wave function is

$$\phi = \exp[i\mathbf{P}\cdot(\mathbf{r}_1 + \mathbf{r}_2)] \times \exp[i\mathbf{k}_0\cdot(\mathbf{r}_1 - \mathbf{r}_2)], \tag{22}$$

where $\mathbf{P} = (\mathbf{k}_1 + \mathbf{k}_2)/2$ and $\mathbf{k}_0 = 1/2(\mathbf{k}_1 - \mathbf{k}_2)$. We disregard the center-of-mass motion and consider

$$\phi = e^{i\mathbf{k}_0\cdot\mathbf{r}}, \tag{23}$$

as the unperturbed wave function. Under the influence of the potential v this is modified to

$$\psi = \phi - (Q/e)v\psi. \tag{24}$$

Here $v\psi$ is considered to be expanded in plane wave states k_1', k_2', and

$$e = E(k_1') + E(k_2') - W, \tag{25}$$

$Q = 1$ if states k_1' and k_2' are both unoccupied,

$$Q = 0 \text{ otherwise.} \tag{26}$$

Equation (26) states the Pauli principle and ensures that $e > 0$ always. It is assumed that the occupied states fill a Fermi sphere of radius k_F.

We set

$$v\psi = G\phi, \tag{27}$$

and thus define the reaction matrix G, which satisfies the equation

$$\langle \mathbf{k}|G|\mathbf{k}_0;\mathbf{P},W\rangle = \langle \mathbf{k}|v|\mathbf{k}_0\rangle - (2\pi)^{-3}\int d^3k'$$
$$\langle \mathbf{k}|v|\mathbf{k}'\rangle \frac{Q(\mathbf{P},k')}{E(\mathbf{P}+\mathbf{k}') + E(\mathbf{P}-\mathbf{k}') - W}\langle \mathbf{k}'|G|\mathbf{k};\mathbf{P},W\rangle. \tag{28}$$

This is an integral equation for the matrix $\langle k|G|k_0\rangle$. P and W are merely parameters in this equation.

The diagonal elements $\langle k|G|k_0,P\rangle$ can be transcribed into the k_1, k_2 of the interacting nucleons. The one-particle energies are then

$$W(k_1) = \sum_{k_2} \langle k_1 k_2 | G | k_1 k_2 \rangle + (\hbar^2/2M)k_1^2. \tag{29}$$

With modern computers, the matrix Eq. (28) can be solved for any given potential v. In the 1960s, approximations were used. First it was noted that for states outside the Fermi sphere, G was small; then $E(\mathbf{P} \pm \mathbf{k}')$ in the denominator of Eq. (28) was replaced by the kinetic energy. Second, for the occupied states, the potential energy was approximated by a quadratic function,

$$W(k) = (\hbar^2/2M^*)k^2, \tag{30}$$

M^* being an effective mass.

It was then possible to obtain the energy of nuclear matter as a function of its density. But the result was not satisfactory. The minimum energy was found at too high a density, about 0.21 fm^{-3} instead of the observed 0.16 fm^{-3}. The binding energy was only 11 MeV instead of the observed 16 MeV.

Modern theory has an additional freedom, the three-body interaction. Its strength can be adjusted to give the correct density. But the binding energy, according to the Argonne-Urbana group, is still only 12 MeV. They believe they can improve this by using a more sophisticated wave function.

In spite of its quantitative deficiencies nuclear matter theory gives a good general approach to the interaction of nucleons in a nucleus. This has been used especially by Brown and Kuo (1966) in their theory of interaction of nucleons in a shell.

Shell Model

Closed shells

The strong binding of the α particle is easily understood; a pair of neutrons and protons of opposite spin, with deep and attractive potential wells, are the qualitative explanation. The next proton or neutron must be in a relative p state, so it cannot come close, and, in addition, by the exchange character of the forces (see earlier section on Potential), the interaction with the α particle is mainly repulsive: thus there is no bound nucleus of $A = 5$, neither ^{5}He nor ^{5}Li. The α particle is a closed unit, and the most stable light nuclei are those which may be considered to be multiples of the α particles, ^{12}C, ^{16}O, ^{20}Ne, ^{24}Mg, etc.

But even among these α-particle nuclei, ^{16}O is special: the binding energy of α to ^{12}C, to form ^{16}O, is considerably larger than the binding of α to ^{16}O. Likewise, ^{40}Ca is special: it is the last nucleus "consisting" of α particles only which is stable against β decay.

The binding energies can be understood by considering nuclei built up of individual nucleons. The nucleons may be considered moving in a square well potential with rounded edges, or more conveniently, an oscillator potential of frequency ω. The lowest state for a particle in that potential is a $1s$ state of energy ε_0. There are two places in the $1s$ shell, spin up and down; when they are filled with both neutrons and protons, we have the α particle.

The next higher one-particle state is $1p$, with energy $\varepsilon_0 + \hbar\omega$. The successive eigenstates are

$$(1s), \quad (1p), \quad (1d2s), \quad (1f2p), \quad (1g2d3s)$$

with energies

$$(\varepsilon_0), \quad (\varepsilon_0 + \hbar\omega), \quad (\varepsilon_0 + 2\hbar\omega), \quad (\varepsilon_0 + 3\hbar\omega).$$

The principal quantum number is chosen to be equal to the number of radial nodes plus one. The number of independent eigenfunctions in each shell are

$$(2), \quad (6), \quad (12), \quad (20), \quad (30),$$

so the total number up to any given shell are

$$(2), \quad (8), \quad (20), \quad (40), \quad (70), \quad \ldots .$$

The first three of these numbers predict closed shells at ^{4}He, ^{10}O, and ^{40}Ca, all correct. But $Z = 40$ or $N = 40$ are not particularly strongly bound nuclei.

The solution to this problem was found independently by Maria Goeppert-Mayer and H. Jensen: nucleons are subject to a strong spin-orbit force which gives added attraction to states with $j = \ell + 1/2$, repulsion to $j = \ell - 1/2$. This becomes stronger with increasing j. The strongly bound nucleons beyond the $1d2s$ shell, are

$$(1f_{7/2}), \quad (2p1f_{5/2}1g_{9/2}), \quad (2d3s1g_{7/2}1h_{11/2}), \quad (2f3p1h_{9/2}1i_{13/2}).$$

The number of independent eigenfunctions in these shells are, respectively,

$$(8), \quad (22), \quad (32), \quad (44).$$

So the number of eigenstates up to $1f_{7/2}$ is 28, up to $1g_{9/2}$ is 50, up to $1h_{11/2}$ is 82, and up to $1i_{13/2}$ is 126. Indeed, nuclei around $Z = 28$ or $N = 28$ are particularly strongly bound. For example, the last α particle in ^{56}Ni ($Z = N = 28$) is bound with 8.0 MeV, while the next α particle, in ^{60}Zn ($Z = N = 30$) has a binding energy of only 2.7 MeV. Similarly, ^{90}Zr ($N = 50$) is very strongly bound and Sn, with $Z = 50$, has the largest number of stable isotopes. ^{208}Pb ($Z = 82, N = 126$) has closed shells for protons as well as neutrons, and nuclei beyond Pb are unstable with respect to α decay. The disintegration ^{212}Po $\rightarrow$ ^{208}Pb $+ \alpha$ yields α particles of 8.95 MeV while ^{208}Pb $\rightarrow$ ^{204}Hg $+ \alpha$ would release only 0.52 MeV, and an α particle of such low energy could not penetrate the potential barrier in 10^{10} years. So there is good evidence for closed nucleon shells.

Nuclei with one nucleon beyond a closed shell, or one nucleon missing, generally have spins as predicted by the shell model.

Open shells

The energy levels of nuclei with partly filled shells are usually quite complicated. Consider a nucleus with the 44-shell about half filled: there will be of the order of $2^{44} \approx 10^{13}$ different configurations possible. It is obviously a monumental task to find the energy eigenvalues.

Some help is the idea of combining a pair of orbitals of the same j and m values of opposite sign. Such pairs have generally low energy, and the pair acts as a boson. Iachello and others have built up states of the nucleus from such bosons.

Collective Motions

Nuclei with incomplete shells are usually not spherical. Therefore their orientation in space is a significant observable. We may consider the rotation of the nucleus as a whole. The moment of inertia θ is usually quite large; therefore, the rotational energy levels which are proportional to $1/\theta$ are closely spaced. The lowest excitations of a nucleus are rotations.

Aage Bohr and Ben Mottleson have worked extensively on rotational states and their combination with intrinsic excitation of individual nucleons. There are also vibrations of the nucleus, e.g., the famous vibration of all neutrons against all protons, the giant dipole state at an excitation energy of 10–20 MeV, depending on the mass number A.

Many nuclei, in their ground state, are prolate spheroids. Their rotations then are about an axis perpendicular to their symmetry axis, and an important characteristic is their quadrupole moment. Many other nuclei have more complicated shapes such as a pear; they have an octopole moment, and their rotational states are complicated.

Weak Interactions

Fermi, in 1934, formulated the the first theory of the weak interaction on the basis of Pauli's neutrino hypothesis. An operator of the form

$$\bar{\phi}_e \phi_\nu \bar{\psi}_p \psi_n \tag{31}$$

creates an electron ϕ_e and an antineutrino $\bar{\phi}_\nu$, and converts a neutron ψ_n into a proton ψ_p. The electron and the neutrino are not in the nucleus, but are created in the β process. All operators are taken at the same point in space-time.

Fermi assumed a vector interaction in his first β-decay paper.

The Fermi theory proved to be essentially correct, but Gamov and Teller later introduced other covariant combinations allowed by Dirac theory. Gamov and Teller said there could be a product of two 4-vectors, or tensors, or axial vectors, or pseudoscalars. Experiment showed later on that the actual interaction is

$$\text{Vector minus Axial vector,} \tag{32}$$

and this could also be justified theoretically.

The β-process, Eq. (31), can only happen if there is a vacancy in the proton state ψ_p. If there is in the nucleus a neutron of the same orbital momentum, we have an allowed transition, as in $^{13}\text{N} \rightarrow {}^{13}\text{C}$. If neutron and proton differ by units in angular momentum, so must the leptons. The wave number of the leptons is small, then the product $(kR)^L$ is very small if L is large: such β transitions are highly forbidden. An example is ^{40}K which has angular momentum $L=4$ while the daughter ^{40}Ca has $L=0$. The radioactive ^{40}K has a half-life of 1.3×10^9 years.

This theory was satisfactory to explain observed β decay, but it was theoretically unsatisfactory to have a process involving four field operators at the same space-time point. Such a theory cannot be renormalized. So it was postulated that a new charged particle W was involved which interacted both with leptons and with baryons, by interactions such as

$$\bar{\phi}_e W \bar{\phi}_\nu, \quad \bar{\psi}_p W \psi_n.$$

This W particle was discovered at CERN and has a mass of 80 GeV. These interactions, involving three rather than four operators, are renormalizable. The high mass of W ensures that in β-decay all the operators ψ_n, ψ_p, ϕ_ν, ϕ_e have to be taken essentially at the same point, within about 10^{-16} cm, and the Fermi theory results.

A neutral counterpart to W, the Z particle, was also found at CERN; it can decay into a pair of electrons, a pair of neutrinos, or a pair of baryons. Its mass has been determined with great accuracy,

$$m(Z) = 91 \text{ GeV}. \tag{33}$$

The difference in masses of Z and W is of great theoretical importance. The mass of Z has a certain width from which the number of species of neutrinos can be determined, namely three: ν_e, ν_μ, and ν_τ.

Nucleosynthesis

It is an old idea that matter consisted "originally" of protons and electrons, and that complex nuclei were gradually formed from these (see Salpeter, 1999). (Modern theories of the big bang put "more elementary" particles, like quarks, even earlier, but this is of no concern here.) At a certain epoch, some neutrons would be formed by

$$H + e^- \to N + \nu. \tag{34}$$

These neutrons would immediately be captured by protons,

$$N + H \to D + \gamma, \tag{35}$$

and the deuterons would further capture protons, giving ^{3}He and ^{4}He. This sequence of reactions, remarkably, leads to a rather definite fraction of matter in ^{4}He nuclei, namely

$$^4\text{He} \approx 23\%, \tag{36}$$

nearly all the rest remaining H. Traces of D, ^{3}He, and ^{7}Li remain.

Again remarkably, there exist very old stars (in globular clusters) in which the fraction of ^{4}He can be measured, and it turns out to be just 23%. This fraction depends primarily on the number of neutrino species which, as mentioned at the end of the section on Weak Interactions, is three.

In stars like the sun and smaller, nuclear reactions take place in which H is converted into He at a temperature of the order of 10–20 million degrees, and the released energy is sent out as radiation. If, at later stages in the evolution, some of the material of such a star is lost into the galaxy, the fraction of ^{4}He in the galaxy increases, but very slowly.

In a star of three times the mass of the sun or more, other nuclear processes occur. Early in its life (on the main sequence), the star produces energy by converting H into He in its core. But after a long time, say a billion years, it has used up the H in its core. Then the core contracts and gets to much higher temperatures, of the order of 100 million degrees or more. Then α particles can combine,

$$3 \, ^4\text{He} \to ^{12}\text{C} + \gamma. \tag{37}$$

Two ^{4}He cannot merge, since ^{8}Be is slightly heavier than two ^{4}He, but at high temperature and density, ^{8}Be can exist for a short time, long enough to capture another ^{4}He.

Equation (37) was discovered in 1952 by E. E. Salpeter; it is the crucial step.

Once ^{12}C has formed, further ^{4}He can be captured and heavier nuclei built up. This happens especially in the inner part of stars of 10 or more times the mass of the sun. The buildup leads to ^{16}O, ^{20}Ne, ^{24}Mg, ^{28}Si, and on to ^{56}Ni. The latter is the last nucleus in which the α particle is strongly bound (see the earlier section on the Shell Model). But it is unstable against β decay; by two emissions of positrons it transforms into ^{56}Fe. This makes ^{56}Fe one of the most abundant isotopes beyond ^{16}O. After forming all these elements, the interior of the star becomes unstable and collapses by gravitation. The energy set free by gravitation then expels all the outer parts of the star (all except the innermost $1.5 M_\odot$) in a supernova explosion and thus makes the elements formed by nucleosynthesis available to the galaxy at large.

Many supernovae explosions have taken place in the galaxy, and so galactic matter contains a fair fraction Z of elements beyond C, called "metals" by astrophysicists, viz., $Z \approx 2\%$. This is true in the solar system, formed about 4.5 billion years ago. New stars should have a somewhat higher Z, old stars are known to have smaller Z.

Stars of $M \geqslant 3 M_\odot$ are formed from galactic matter that already contains appreciable amounts of heavy nuclei up to ^{56}Fe. Inside the stars, the carbon cycle of nuclear reactions takes place, in which ^{14}N is the most abundant nucleus. If the temperature then rises to about 100 million degrees, neutrons will be produced by the reactions

$$^{14}\text{N} + ^{4}\text{He} \rightarrow ^{17}\text{F} + n,$$

$$^{17}\text{O} + ^{4}\text{He} \rightarrow ^{20}\text{Ne} + n. \tag{38}$$

The neutrons will be preferentially captured by the heavy nuclei already present and will gradually build up heavier nuclei by the s-process described in the famous article by E. M. and G. R. Burbidge, Fowler, and Hoyle in *Reviews of Modern Physics* (1957).

Some nuclei, especially the natural radioactive ones, U and Th, cannot be built up in this way, but require the r-process, in which many neutrons are added to a nucleus in seconds so there is no time for β decay. The conditions for the r-process have been well studied; they include a temperature of more than 10^9 K. This condition is well fulfilled in the interior of a supernova a few seconds after the main explosion, but there are additional conditions so that it is still uncertain whether this is the location of the r-process.

Special Relativity

For the scattering of nucleons above about 300 MeV, and for the equation of state of nuclear matter of high density, special relativity should be taken into account. A useful approximation is mean field theory which has been especially developed by J. D. Walecka.

Imagine a large nucleus. At each point, we can define the conserved baryon current $i\bar{\psi}\gamma_\mu\psi$ where ψ is the baryon field, consisting of protons and neutrons. We also have a scalar baryon density $\bar{\psi}\psi$. They couple, respectively, to a vector field V_μ and a scalar field ϕ with coupling constants g_w and g_s. The vector field is identified with the ω meson, giving a repulsion, and the scalar field with the σ meson, giving an attraction. Coupling constants can be adjusted so as to give a minimum energy of -16 MeV per nucleon and equilibrium density of 0.16 fm^{-3}.

The theory can be generalized to neutron matter and thus to the matter of neutron stars. It can give the charge distribution of doubly magic nuclei, like ^{208}Pb, ^{40}Ca, and ^{16}O, and these agree very well with the distributions observed in electron scattering.

The most spectacular application is to the scattering of 500 MeV protons by ^{40}Ca, using the Dirac relativistic impulse approximation for the proton. Not only are cross section minima at the correct scattering angles, but polarization of the scattered protons is almost complete, in agreement with experiment, and the differential cross section at the second, third, and fourth maximum also agree with experiment.

References

Bergervoet, J. R., P. C. van Campen, R. A. M. Klomp, J. L. de Kok, V. G. J. Stoks, and J. J. de Swart, 1990, Phys. Rev. C **41**, 1435.

Brown, G. E., and T. T. S. Kuo, 1966, Nucl. Phys. **85**, 140.

Burbidge, E. M., G. R. Burbidge, W. A. Fowler, and F. Hoyle, 1957, Rev. Mod. Phys. **29**, 547.

Drell, S. D., 1999, Rev. Mod. Phys. **71**, 460; pp. 781–797 in this book.

Green, E. S., 1954, Phys. Rev. **95**, 1006.

Pudliner, B. S., V. R. Pandharipande, J. Carlson, S. C. Pieper, and R. B. Wiringa, 1997, Phys. Rev. E **56**, 1720.

Rutherford, E., J. Chadwick, and C. D. Ellis, 1930, *Radiations from Radioactive Substances* (Cambridge, England, Cambridge University).

Salpeter, E. E., 1999, Rev. Mod. Phys. **71**, 220; pp. 370–374 in this book.

Siemens, P. J., 1970, Nucl. Phys. A **141**, 225.

Stoks, V. G. J., R. A. M. Klomp, M. C. M. Rentmeester, and J. J. de Swart, 1993, Phys. Rev. C **48**, 792.

Till, C., 1999, Rev. Mod. Phys. **71**, 451; pp. 767–774 in this book.

Walecka, J. D., 1995, *Theoretical Nuclear and Subnuclear Physics* (Oxford, Oxford University).

Wiringa, R. B., V. G. J. Stoks, and R. Schiavilla, 1995, Phys. Rev. E **51**, 38.

Theoretical Particle Physics
A. Pais

"Gentlemen and Fellow Physicists of America: We meet today on an occasion which marks an epoch in the history of physics in America; may the future show that it also marks an epoch in the history of the science which this Society is organized to cultivate!" (Rowland, 1899).[1] These are the opening words of the address by Henry Rowland, the first president of the American Physical Society, at the Society's first meeting, held in New York on October 28, 1899. I do not believe that Rowland would have been disappointed by what the next few generations of physicists have cultivated so far.

It is the purpose of these brief preludes to give a few glimpses of developments in the years just before and just after the founding of our Society.

First, events just before: Invention of the typewriter in 1873, of the telephone in 1876, of the internal combustion engine and the phonograph in 1877, of the zipper in 1891, of the radio in 1895. The *Physical Review* began publication in 1893. The twilight of the 19th century was driven by oil and steel technologies.

Next, a few comments on "high-energy" physics in the first years of the twentieth century:

Pierre Curie in his 1903 Nobel lecture: "It can even be thought that radium could become very dangerous in criminal hands, and here the question can be raised whether mankind benefits from the secrets of Nature."[1]

From a preview of the 1904 International Electrical Congress in St. Louis, found in the *St. Louis Post Dispatch* of October 4, 1903: "Priceless mysterious radium will be exhibited in St. Louis. A grain of this most wonderful and mysterious metal will be shown." At that Exposition a transformer was shown which generated about half a million volts (Pais, 1986).

In March 1905, Ernest Rutherford began the first of his Silliman lectures, given at Yale, as follows:

The last decade has been a very fruitful period in physical science, and discoveries of the most striking interest and importance have followed one another in rapid succession . . . . The march of discovery has been so rapid that it has been difficult even for those directly engaged in the investigations to grasp at once the full significance of the facts that have been brought to light . . . . The rapidity of this advance has seldom, if ever, been equalled in the history of science (Rutherford, 1905, quoted in Pais, 1986).

---

[1]Quoted in Pais, 1986. Individual references not given in what follows are given in this book, along with many more details.

The text of Rutherford's lectures makes clear which main facts he had in mind: X rays, cathode rays, the Zeeman effect, α, β, and γ radioactivity, the reality as well as the destructibility of atoms, in particular the radioactive families ordered by his and Soddy's transformation theory, and results on the variation of the mass of β particles with their velocity. There is no mention, however, of the puzzle posed by Rutherford's own introduction of a characteristic lifetime for each radioactive substance. Nor did he touch upon Planck's discovery of the quantum theory in 1900. He could not, of course, refer to Einstein's article on the light-quantum hypothesis, because that paper was completed on the seventeenth of the very month he was lecturing in New Haven. Nor could he include Einstein's special theory of relativity among the advances of the decade he was reviewing, since that work was completed another three months later. It seems to me that Rutherford's remark about the rarely equaled rapidity of significant advances driving the decade 1895–1905 remains true to this day, especially since one must include the beginnings of quantum and relativity theory.

Why did so much experimental progress occur when it did? Largely because of important advances in instrumentation during the second half of the nineteenth century. This was the period of ever improving vacuum techniques (by 1880, vacua of 10^{-6} torr had been reached), of better induction coils, of an early type of transformer, which, before 1900, was capable of producing energies of 100 000 eV, and of new tools such as the parallel-plate ionization chamber and the cloud chamber.

All of the above still remain at the roots of high-energy physics. Bear in mind that what was high energy then (~ 1 MeV) is low energy now. What was high energy later became medium energy, 400 MeV in the late 1940s. What we now call high-energy physics did not begin until after the Second World War. At this writing, we have reached the regime of 1 TeV$=10^{12}$ eV$=1.6$ erg.

To do justice to our ancestors, however, I should first give a sketch of the field as it developed in the first half of this century.

The Years 1900–1945

The early mysteries of radioactivity

High-energy physics is the physics of small distances, the size of nuclei and atomic particles. As the curtain rises, the electron, the first elementary particle, has been discovered, but the reality of atoms is still the subject of some debate, the structure of atoms is still a matter of conjecture, the atomic nucleus has not yet been discovered, and practical applications of atomic energy, for good or evil, are not even visible on the far horizon.

On the scale of lengths, high-energy physics has moved from the domain of atoms to that of nuclei to that of particles (the adjective "elementary" is long gone). The historical progression has not always followed that path, as can be seen particularly clearly when following the development of our knowledge of radioactive processes, which may be considered as the earliest high-energy phenomena.

Radioactivity was discovered in 1896, the atomic nucleus in 1911. Thus even the simplest qualitative statement—radioactivity is a nuclear phenomenon—could not be made until fifteen years after radioactivity was first observed. The connection between nuclear binding energy and nuclear stability was not made until 1920. Thus some twenty-five years would pass before one could understand why some, and only some, elements are radioactive. The concept of decay probability was not properly formulated until 1927.

Until that time, it remained a mystery why radioactive substances have a characteristic lifetime. Clearly, then, radioactive phenomena had to be a cause of considerable bafflement during the early decades following their first detection. Here are some of the questions that were the concerns of the fairly modest-sized but elite club of experimental radioactivists: What is the source of energy that continues to be released by radioactive materials? Does the energy reside inside the atom or outside? What is the significance of the characteristic half-life for such transformations? (The first determination of a lifetime for radioactive decay was made in 1900.) If, in a given radioactive transformation, all parent atoms are identical, and if the same is true for all daughter products, then why does one radioactive parent atom live longer than another, and what determines when a specific parent atom disintegrates? Is it really true that some atomic species are radioactive, others not? Or are perhaps all atoms radioactive, but many of them with extremely long lifetimes?

One final item concerning the earliest acquaintance with radioactivity: In 1903 Pierre Curie and Albert Laborde measured the amount of energy released by a known quantity of radium. They found that 1 g of radium could heat approximately 1.3 g of water from the melting point to the boiling point in 1 hour. This result was largely responsible for the worldwide arousal of interest in radium.

It is my charge to give an account of the developments of high-energy theory, but so far I have mainly discussed experiments. I did this to make clear that theorists did not play any role of consequence in the earliest stages, both because they were not particularly needed for its descriptive aspects and because the deeper questions were too difficult for their time.

As is well known, both relativity theory and quantum theory are indispensable tools for understanding high-energy phenomena. The first glimpses of them could be seen in the earliest years of our century.

Re relativity: In the second of his 1905 papers on relativity Einstein stated that

if a body gives off the energy L in the form of radiation, its mass diminishes by L/c^2 .... The mass of a body is a measure of its energy .... It is not impossible that with bodies whose energy content is variable to a high degree (e.g., with radium salts) the theory may be successfully put to the test (Einstein 1905, reprinted in Pais, 1986).

The enormous importance of the relation $E = mc^2$ was not recognized until the 1930s. See what Pauli wrote in 1921: "Perhaps the law of the inertia of energy will be tested *at some future time* on the stability of nuclei" (Pauli, 1921, italics added).

Re quantum theory: In May 1911, Rutherford announced his discovery of the atomic nucleus and at once concluded that α decay is due to nuclear instability, but that β decay is due to instability of the peripheral electron distribution.

It is not well known that it was Niels Bohr who set that last matter straight. In his seminal papers of 1913, Bohr laid the quantum dynamical foundation for understanding atomic structure. The second of these papers contains a section on "Radioactive phenomena," in which he states: "On the present theory it seems also necessary that the nucleus is the seat of the expulsion of the high-speed β-particles" (Bohr, 1913). His main argument was that he knew enough by then about orders of magnitude of peripheral electron energies to see that the energy release in β decay simply could not fit with a peripheral origin of that process.

In teaching a nuclear physics course, it may be edifying to tell students that it took 17 years of creative confusion, involving the best of the past masters, between the discovery of radioactive processes and the realization that these processes are all of nuclear origin—time spans not rare in the history of high-energy physics, as we shall see in what follows.

One last discovery, the most important of the lot, completes the list of basic theoretical advances in the pre-World-War-I period. In 1905 Einstein proposed that, under certain circumstances, light behaves like a stream of particles, or light quanta. This idea initially met with very strong resistance, arriving as it did when the wave picture of light was universally accepted. The resistance continued until 1923, when Arthur Compton's experiment on the scattering of light by electrons showed that, in that case, light does behave like particles—which must be why their current name, photons, was not introduced until 1926 (Lewis, 1926).

Thus by 1911 three fundamental particles had been recognized: the electron, the photon, and the proton [so named only in 1920 (Author unnamed, 1920)], the nucleus of the hydrogen atom.

Weak and strong interactions: Beginnings

In the early decades following the discovery of radioactivity it was not yet known that quantum mechanics would be required to understand it nor that distinct forces are dominantly responsible for each of the three radioactive decay types:

Process	Dominant interaction
α decay	strong
β decay	weak
γ decay	electromagnetic

The story of α and γ decay will not be pursued further here, since they are not primary sources for our understanding of interactions. By sharpest contrast, until 1947—the year μ-meson decay was discovered—β decay was the *only* manifestation, rather than one among many, of a specific type of force. Because of this unique position, conjectures about the nature of this process led to a series of pitfalls. Analogies with better-known phenomena were doomed to failure. Indeed, β decay provides a splendid example of how good physics is arrived at after much trial and many errors—which explains why it took twenty years to establish that the *primary* β process yields a continuous β spectrum. I list some of the false steps—no disrespect intended, but good to tell your students.

(1) It had been known since 1904 that α rays from a pure α emitter are monochromatic. It is conjectured (1906) that the same is true for β emitters.

(2) It is conjectured (1907) that the absorption of monoenergetic electrons by metal forces satisfies a simple exponential law as a function of foil thickness.

(3) Using this as a diagnostic, absorption experiments are believed to show that β emitters produce homogeneous energy electrons.

(4) In 1911 it is found that the absorption law is incorrect.

(5) Photographic experiments seem to claim that a multiline discrete β spectrum is present (1912–1913).

(6) Finally, in 1914, James Chadwick performs one of the earliest experiments with counters, which shows that β rays from RaB (Pb^{214}) and RaC (Bi^{214}) consist of a continuous spectrum, and that there is an additional line spectrum. In 1921 it is under-

stood that the latter is due to an internal conversion process. In 1922 the first nuclear energy-level diagram is sketched.

Nothing memorable relevant to our subject happened between 1914 and 1921. There was a war going on. There were physicists who served behind the lines and those who did battle. In his obituary to Henry Moseley, the brilliant physicist who at age 28 had been killed by a bullet in the head at Suvla Bay, Rutherford (1915) remarked: "His services would have been far more useful to his country in one of the numerous fields of scientific inquiry rendered necessary by the war than by the exposure to the chances of a Turkish bullet," an issue that will be debated as long as the folly of resolving conflict by war endures.

Continuous β spectra had been detected in 1914, as said. The next question, much discussed, was: are these primary or due to secondary effects? This issue was settled in 1927 by Ellis and Wooster's difficult experiment, which showed that the continuous β spectrum of RaE (Bi^{210}) was primary in origin. "We may safely generalize this result for radium E to all β-ray bodies and the long controversy about the origin of the continuous spectrum appears to be settled" (Ellis and Wooster, 1927).

Another three years passed before Pauli, in December 1930, gave the correct explanation of this effect: β decay is a three-body process in which the liberated energy is shared by the electron and a hypothetical neutral particle of very small mass, soon to be named the neutrino. Three years after that, Fermi put this qualitative idea into theoretical shape. His theory of β decay, the first in which quantized spin-$\frac{1}{2}$ fields appear in particle physics, is the first quantitative theory of weak interactions.

As for the first glimpses of strong-interaction theory, we can see them some years earlier.

In 1911 Rutherford had theoretically deduced the existence of the nucleus on the assumption that α-particle scattering off atoms is due to the $1/r^2$ Coulomb force between a pointlike α and a pointlike nucleus. It was his incredible luck to have used α particles of moderate energy and nuclei with a charge high enough so that his α's could not come very close to the target nuclei. In 1919 his experiments on α-hydrogen scattering revealed large deviations from his earlier predictions. Further experiments by Chadwick and Etienne Bieler (1921) led them to conclude,

The present experiments do not seem to throw any light on the nature of the law of variation of the forces at the seat of an electric charge, but merely show that the forces are of very great intensity . . . . It is our task to find some field of force which will reproduce these effects" (Chadwick and Bieler, 1921).

I consider this statement, made in 1921, as marking the birth of strong-interaction physics.

The early years of quantum field theory

Apart from the work on β decay, all the work we have discussed up to this point was carried out before late 1926, in a time when relativity and quantum mechanics had not yet begun to have an impact upon the theory of particles and fields. That impact began with the arrival of quantum field theory, when particle physics acquired, one might say, its own unique language. From then on particle theory became much more focused. A new central theme emerged: how good are the predictions of quantum field theory? Confusion and insight continued to alternate unabated, but these ups and downs mainly occurred within a tight theoretical framework, the quantum theory of fields. Is this theory the ultimate framework for understanding the structure of matter and the description of elementary processes? Perhaps, perhaps not.

Quantum electrodynamics (QED), the earliest quantum field theory, originated on the heels of the discoveries of matrix mechanics (1925) and wave mechanics (1926). At that time, electromagnetism appeared to be the only field relevant to the treatment of matter in the small. (The gravitational field was also known by then but was not considered pertinent until decades later.) Until QED came along, matter was treated like a game of marbles, of tiny spheres that collide, link, or disconnect. Quantum field theory abandoned this description; the new language also explained how particles are made and how they disappear.

It may fairly be said that the theoretical basis of high-energy theory began its age of maturity with Dirac's two 1927 papers on QED. By present standards the new theoretical framework, as it was developed in the late twenties, looks somewhat primitive. Nevertheless, the principal foundations had been laid by then for much that has happened since in particle theory. From that time on, the theory becomes much more technical. As Heisenberg (1963) said: "Somehow when you touched [quantum mechanics] ... at the end you said 'Well, was it that simple?' Here in electrodynamics, it didn't become simple. You could do the theory, but still it never became that simple" (Heisenberg, 1963). So it is now in all of quantum field theory, and it will never be otherwise. Given limitations of space, the present account must become even more simple-minded than it has been hitherto.

In 1928 Dirac produced his relativistic wave equation of the electron, one of the highest achievements of twentieth-century science. Learning the beauty and power of that little equation was a thrill I shall never forget. Spin, discovered in 1925, now became integrated into a real theory, including its ramifications. Entirely novel was its consequence: a new kind of particle, as yet unknown experimentally, having the same mass and opposite charge as the electron. This "antiparticle," now named a positron, was discovered in 1931.

At about that time new concepts entered quantum physics, especially quantum field theory: groups, symmetries, invariances—many-splendored themes that have dominated high-energy theory ever since. Some of these have no place in classical physics, such as permutation symmetries, which hold the key to the exclusion principle and to quantum statistics; a quantum number, parity, associated with space reflections; charge conjugation; and, to some extent, time-reversal invariance. In spite of some initial resistance, the novel group-theoretical methods rapidly took hold.

A final remark on physics in the late 1920s: "In the winter of 1926," K. T. Compton (1937) has recalled, "I found more than twenty Americans in Goettingen at this fount of quantum wisdom." Many of these young men contributed vitally to the rise of American physics. "By 1930 or so, the relative standings of *The Physical Review* and *Philosophical Magazine* were interchanged" (Van Vleck, 1964). Bethe (1968) has written: "J. Robert Oppenheimer was, more than any other man, responsible for raising American theoretical physics from a provincial adjunct of Europe to world leadership . . . . It was in Berkeley that he created his great School of Theoretical Physics." It was Oppenheimer who brought quantum field theory to America.

The 1930s

Two main themes dominate high-energy theory in the 1930s: struggles with QED and advances in nuclear physics.

QED

All we know about QED, from its beginnings to the present, is based on perturbation theory, expansions in powers of the small number $\alpha = e^2/\hbar c$. The nature of the struggle was this: To lowest order in α, QED's predictions were invariably successful; to higher order, they were invariably disastrous, always producing infinite answers. The tools were those still in use: quantum field theory and Dirac's positron theory.

Infinities had marred the theory since its classical days: The self-energy of the point electron was infinite even then. QED showed (1933) that its charge is also infinite—the vacuum polarization effect. The same is true for higher-order contributions to scattering or annihilation processes or what have you.

Today we are still battling the infinities, but the nature of the attack has changed. All efforts at improvement in the 1930s—mathematical tricks such as nonlinear modifications of the Maxwell equation—have led nowhere. As we shall see, the standard theory is very much better than was thought in the 1930s. That decade came to an end with a sense of real crisis in QED.

Meanwhile, however, quantum field theory had scored an enormous success when Fermi's theory of β decay made clear that electrons are not constituents of nuclei—as was believed earlier—but are *created* in the decay process. This effect, so characteristic of quantum field theory, brings us to the second theme of the thirties.

Nuclear physics

It was only after quantum mechanics had arrived that theorists could play an important role in nuclear physics, beginning in 1928, when α decay was understood to be a quantum-mechanical tunneling effect. Even more important was the theoretical insight that the standard model of that time (1926–1931), a tightly bound system of protons and electrons, led to serious paradoxes. Nuclear magnetic moments, spins, statistics—all came out wrong, leading grown men to despair.

By contrast, experimental advances in these years were numerous and fundamental: The first evidence of cosmic-ray showers (1929) and of billion-eV energies of individual cosmic-ray particles (1932–1933), the discoveries of the deuteron and the positron (both in 1931) and, most trail-blazing, of the neutron (1932), which ended the aggravations of the proton-electron nuclear model, replacing it with the proton-neutron model of the nucleus. Which meant that quite new forces, only glimpsed before, were needed to understand what holds the nucleus together—the strong interactions.

The approximate equality of the number of p and n in nuclei implied that short-range nn and pp forces could not be very different. In 1936 it became clear from scattering experiments that pp and pn forces in $1s$ states are equal within the experimental errors, suggesting that they, as well as nn forces, are also equal in other states. From this, the concept of charge independence was born. From that year dates the introduction of isospin for nucleons (p and n), p being isospin "up," neutron "down," the realization that charge independence implies that nuclear forces are invariant under isospin rotations, which form the symmetry group SU(2).

With this symmetry a new lasting element enters physics, that of a *broken symmetry*: SU(2) holds for strong interactions only, not for electromagnetic and weak interactions.

Meanwhile, in late 1934, Hideki Yukawa had made the first attack on describing nuclear forces by a quantum field theory, a one-component complex field with charged massive quanta: mesons, with mass estimated to be approximately $200m$ (where

m=electron mass). When, in 1937, a particle with that order of mass was discovered in cosmic rays, it seemed clear that this was Yukawa's particle, an idea both plausible and incorrect. In 1938 a neutral partner to the meson was introduced, in order to save charge independence. It was the first particle proposed on theoretical grounds, and it was discovered in 1950.

To conclude this quick glance at the 1930s, I note that this was also the decade of the birth of accelerators. In 1932 the first nuclear process produced by these new machines was reported: $p + \text{Li}^7 \rightarrow 2\alpha$, first by Cockroft and Walton at the Cavendish, with their voltage multiplier device, a few months later by Lawrence and coworkers with their first, four-inch cyclotron. By 1939 the 60-inch version was completed, producing 6-MeV protons. As the 1930s drew to a close, theoretical high-energy physics scored another major success: the insight that the energy emitted by stars is generated by nuclear processes.

Then came the Second World War.

Modern Times

As we all know, the last major prewar discovery in high-energy physics—fission—caused physicists to play a prominent role in the war effort. After the war this brought them access to major funding and prepared them for large-scale cooperative ventures. Higher-energy regimes opened up, beginning in November 1946, when the first synchrocyclotron started producing 380-MeV α particles.

QED triumphant

High-energy theory took a grand turn at the Shelter Island Conference (June 2–4, 1947), which many attendees (including this writer) consider the most important meeting of their career. There we first heard reports on the Lamb shift and on precision measurements of hyperfine structure in hydrogen, both showing small but most significant deviations from the Dirac theory. It was at once accepted that these new effects demanded interpretation in terms of radiative corrections to the leading-order predictions in QED. So was that theory's great leap forward set in motion. The first "clean" result was the evaluation of the electron's anomalous magnetic moment (1947).

The much more complicated calculation of the Lamb shift was not successfully completed until 1948. Here one meets for the first time a new bookkeeping in which all higher-order infinities are shown to be due to contributions to mass and charge (and the norm of wave functions). Whereupon mass and charge are *renormalized*, one absorbs these infinities into these quantities, which become *phenomenological parameters*, not theoretically predictable to this day—after which corrections to all physical processes are finite.

By the 1980s calculations of corrections had been pushed to order α^4, yielding, for example, agreement with experiment for the electron's magnetic moment to ten significant figures, the highest accuracy attained anywhere in physics. QED, maligned in the 1930s, has become theory's jewel.

Leptons

In late 1946 it was found that the absorption of negative cosmic-ray mesons was ten to twelve orders of magnitude weaker than that of Yukawa's meson. At Shelter Island a way out was proposed: the Yukawa meson, soon to be called a pion (π), decays into another weakly absorbable meson, the muon (μ). It was not known at that time that a Japanese group had made that same proposal before, nor was it known that evidence for the two-meson idea had already been reported a month earlier (Lattes et al., 1947).

The μ is much like an electron, only ~200 times heavier. It decays into $e + 2\nu$. In 1975 a still heavier brother of the electron was discovered and christened τ (mass ~1800 MeV). Each of these three, e, μ, τ, has a distinct, probably massless neutrino partner, ν_e, ν_μ, ν_τ. The lot of them form a *particle family*, the leptons (name introduced by Møller and Pais, 1947), subject to weak and electromagnetic but not to strong interactions. In the period 1947–1949 it was found that β decay, μ decay, and μ absorption had essentially equal coupling strength. Thus was born the *universal* Fermi interaction, followed in 1953 by the law of lepton conservation.

So far we have seen how refreshing and new high-energy physics became after the war. And still greater surprises were in store.

Baryons, more mesons, quarks

In December 1947, a Manchester group reported two strange cloud-chamber events, one showing a fork, another a kink. Not much happened until 1950, when a CalTech group found thirty more such events. These were the early observations of new mesons, now known as K^0 and $K^\pm$. Also in 1950 the first hyperon (Λ) was discovered, decaying into $p + \pi^-$. In 1954 the name "baryon" was proposed to denote nucleons (p and n) and hyperons collectively (Pais, 1955).

Thus began baryon spectroscopy, to which, in 1952, a new dimension was added with the discovery of the "33-resonance," the first of many nucleon excited states. In 1960 the first hyperon resonance was found. In 1961 meson spectroscopy started, when the ρ, ω, η, and $K*$ were discovered.

Thus a new, deeper level of submicroscopic physics was born, which had not been anticipated by anyone. It demanded the introduction of new theoretical ideas. The key to these was the fact that hyperons and K's were very long-lived, typically $\sim 10^{-10}$ sec, ten orders of magnitude larger than the guess from known theory. An understanding of this paradox began with the concept of associated production (1952, first observed in 1953), which says, roughly, that the production of a hyperon is always associated with that of a K, thereby decoupling strong production from weak decay. In 1953 we find the first reference to a hierarchy of interactions in which strength and symmetry are correlated and to the need for enlarging isospin symmetry to a bigger group. The first step in that direction was the introduction (1953) of a phenomenological new quantum number, strangeness (s), conserved in strong and electromagnetic, but not in weak, interactions.

The search for the bigger group could only succeed after more hyperons had been discovered. After the Λ, a singlet came, Σ, a triplet, and Ξ, a doublet. In 1961 it was noted that these six, plus the nucleon, fitted into the octet representation of SU(3), the ϱ, ω, and $K*$ into another 8. The lowest baryon resonances, the quartet "33" plus the first excited Σ's and Ξ's, nine states in all, would fit into a decuplet representation of SU(3) if only one had one more hyperon to include. Since one also had a mass formula for these *badly broken* multiplets, one could predict the mass of the "tenth hyperon," the Ω^-, which was found where expected in 1964. SU(3) worked.

Nature appears to keep things simple, but had bypassed the fundamental 3-representation of SU(3). Or had it? In 1964 it was remarked that one could imagine baryons to be made up of three particles, named quarks (Gell-Mann, 1964), and mesons to be made up of one quark (q) and one antiquark ($\bar{q}$). This required the q's to have fractional charges (in units of e) of 2/3 (u), $-1/3$ (d), and $-1/3$ (s), respectively. The idea of a new deeper level of fundamental particles with fractional charge initially seemed a bit rich, but today it is an accepted ingredient for the description of matter, including an explanation of why these quarks have never been seen. More about that shortly.

K mesons, a laboratory of their own

In 1928 it was observed that in quantum mechanics there exists a two-valued quantum number, parity (P), associated with spatial reflections. It was noted in 1932 that no quantum number was associated with time-reversal (T) invariance. In 1937, a third discrete symmetry, two-valued again, was introduced, charge conjugation (C), which interchanges particles and antiparticles.

K particles have opened quite new vistas regarding these symmetries.

Particle mixing

In strong production reactions one can create $K^0(S=1)$ or $\bar{K}^0(S=-1)$. Both decay into the same state $\pi^+ + \pi^-$. How can charge conjugation transform the final but not the initial state into itself? It cannot do so as long as S is conserved (strong interactions) but it can, and does, when S is not conserved (weak interactions). Introduce $K^1 = (K^0 + \bar{K}^0)/\sqrt{2}$ and $K^2 = (K^0 - \bar{K}^0)/\sqrt{2}$. We find that K^1 can and K^2 cannot decay into $\pi^+ + \pi^-$. These states have different lifetimes: K^2 should live much longer (unstable only via non-2π modes). Since a particle is an object with a unique lifetime, K^1 and K^2 are particles and K^0 and $\bar{K}^0$ are *particle mixtures*, a situation never seen before (and, so far, not since) in physics. This gives rise to bizarre effects such as regeneration: One can create a pure K^0 beam, follow it downstream until it consists of K^2 only, interpose an absorber that by strong interactions absorbs the $\bar{K}^0$ but not the K^0 component of K^2, and thereby regenerate K^1: 2π decays reappear.

Violations of P and C

A K^+ can decay into π^+ and π^0, the "θ mode," or into $2\pi^+ + \pi^-$, the "τ mode." Given the spin (zero) and parity (odd) of pions, a τ (spin zero) must have odd parity but a θ even parity! How can that be? Either θ and τ are distinct particles rather than alternative decay modes of the same particle, or there is only one K but parity is not conserved in these weak decays. This was known as the θ-τ puzzle.

In 1956, a brilliant analysis of all other weak processes (β decay, μ decay) showed that P conservation had never been established in any of them (Lee and Yang, 1956). In 1957 it was experimentally shown that in these processes both parity and charge conjugation were violated! (Wu et al., 1957; Friedman and Telegdi, 1957). Up until then these invariances had been thought to be universal. They were not, a discovery that deeply startled the pros.

This discovery caused an explosion in the literature. Between 1950 and 1972, 1000 experimental and 3500 theoretical articles (in round numbers) appeared on weak interactions. New theoretical concepts appeared: two-component neutrino theory; the V-A (vector minus axial-vector) theory of weak interactions, the remarkable link between its A-part and strong interactions, which in turn led to the concept of a partially conserved axial current; the insight that, while C and P were violated, their product CP still held—which sufficed to save the concept of particle mixture.

Violations of CP and T

In 1964, a delicate experiment showed that, after all, K^2 *does* decay into π^+ and π^-, at a rate of ~0.2 percent of all decay modes, a rate weaker than weak. CP invariance had fallen by the wayside; its incredibly weak violation made the news even harder to digest. (Particle mixing remained substantially intact.) The following thirty years of hard experimental labor have failed so far to find any other CP-violating effect—but has shown that T is also violated!

That, in a way, is a blessing. In the years 1950–1957 the "CPT theorem" was developed, which says that, under very general conditions, any relativistic quantum field theory is necessarily invariant under the product operation CPT—which means that, if CP is gone, T separately must also be gone.

Downs and ups in mid-century

The postwar years as described so far were a period of great progress. It was not all a bed of roses, however.

Troubles with mesons

It seemed reasonable to apply the methods so successful in QED to the meson field theory of nuclear forces, but that led to nothing but trouble. Some meson theories (vector, axial-vector) turned out to be unrenormalizable. For those that were not (scalar, pseudoscalar), the analog of the small number $e^2/\hbar c$ was a number larger than 10—so that perturbation expansions made no sense.

S-matrix methods

Attention now focused on the general properties of the scattering matrix, the S matrix, beginning with the successful derivation of dispersion relations for π-nucleon scatterings (1955). This marked the beginning of studies of analytic properties of the S matrix, combined with causality, unitarity, and crossing, and culminating in the bootstrap vision which says that these properties (later supplemented by Regge poles) should suffice to give a self-consistent theory of the strong interactions. This road has led to interesting mathematics but not to much physics.

Current algebra

More fertile was another alternative to quantum field theory but closer to it: current algebra, starting in the mid-sixties, stimulated by the insights that weak interactions have a current structure and that quarks are basic to strong interactions. Out of this grew the proposal that electromagnetic and weak vector currents were members of an SU(3) octet, axial currents of another one, both taken as quark currents. Current algebra, the commutator algebra of these currents, has led to quite important sum rules.

New lepton physics

In the early sixties design began of high-energy neutrino beams. In the late sixties, experiments at SLAC revealed that high-energy "deep"-inelastic electron-nucleon scattering satisfied scaling laws, implying that in this régime nucleons behaved like boxes filled with hard nuggets. This led to an incredibly simple-minded but successful model for inelastic electron scattering as well as neutrino scattering, as the incoherent sum of elastic lepton scatterings off the nuggets, which were called partons.

Quantum field theory redux

Quantum chromodynamics (QCD)

In 1954 two short brilliant papers appeared marking the start of non-Abelian gauge theory (Yang and Mills, 1954a, 1954b). They dealt with a brand new version of strong interactions, mediated by vector mesons of zero mass. The work was received with considerable interest, but what to do with these recondite ideas was another matter. At that time there were no vector mesons, much less vector mesons with zero mass. There the matter rested until the 1970s.

To understand what happened then, we must first go back to 1964, when a new symmetry, static SU(6), entered the theory of strong interactions. Under this symmetry SU(3) and spin were linked, a generalization of Russell-Saunders coupling in atoms, where spin is conserved in the absence of spin-orbit coupling. The baryon octet and decouplet together formed one SU(6) representation, the "56," which was totally symmetric in all three-quark variables. This, however, violated the exclusion principle. To save that, the u, d, and s quarks were assigned a new additional three-valued degree of freedom, called *color*, with respect to which the 56 states were totally antisymmetric. The corresponding new group was denoted $SU(3)_c$, and the "old" SU(3) became flavor SU(3), $SU(3)_f$.

Out of gauges and colors grew quantum chromodynamics (QCD), a quantum field theory with gauge group $SU(3)_c$, with respect to which the massless gauge fields, gluons, form an octet. In 1973 the marvelous discovery was made that QCD is *asymptotically free*: strong interactions diminish in strength with increasing energy—which explains the parton model for scaling. All the earlier difficulties with the strong interactions residing in the low-energy region ($\lesssim$ few GeV) were resolved.

A series of speculations followed: $SU(3)_c$ is an unbroken symmetry, i.e., the gluons are strictly massless. The attractive potential between quarks grows with increasing distance, so that quarks can never get away from each other, but are *confined*, as are single gluons. Confinement is a very plausible idea but to date its rigorous proof remains outstanding.

Electroweak unification

In mid-century the coupling between four spin-1/2 fields, the Fermi theory, had been very successful in organizing β-decay data, yet it had its difficulties: the theory was unrenormalizable, and it broke down at high energies ($\lesssim$300 GeV). In the late 1950s the first suggestions appeared that the Fermi theory was an approximation to a mediation of weak interactions by heavy charged vector mesons, called $W^{\pm}$. That would save the high-energy behavior, but not renormalizability.

There came a time (1967) when it was proposed to unify weak and electromagnetic interactions in terms of a SU(2)×U(1) gauge theory (Weinberg, 1967), with an added device, the Higgs phenomenon (1964), which generates masses for three of the four gauge fields—and which introduces one (perhaps more) new spinless boson(s), the Higgs particle(s). One vector field remains massless: the photon field; the massive fields are $W^{\pm}$, as conjectured earlier, plus a new neutral field for the "Z," coupled to a hypothesized neutral current.

During the next few years scant attention was paid to this scheme—until 1971, when it was shown that this theory is renormalizable, and with a small expansion parameter!

There now followed a decade in particle physics of a kind not witnessed earlier in the postwar era, characterized not only by a rapid sequence of spectacular experimental discoveries but also by intense and immediate interplay between experiment and fundamental theory. I give a telegraph-style account of the main events.

1972: A fourth quark, charm (c), is proposed to fill a loophole in the renormalizability of SU(2)×U(1).

1973: First sighting of the neutral current at CERN.

1974: Discovery of a new meson at SLAC and at Brookhaven, which is a bound $\bar{c}\,c$ state.

1975: Discovery at SLAC that hadrons produced in high-energy e^+e^- annihilations emerge more or less as back-to-back jets.

1977: Discovery at Fermilab of a fifth quark, bottom, to be followed, in the 1990s, by discovery of a sixth quark, top.

1983: Discovery at CERN of the W and the Z at mass values that had meanwhile been predicted from other weak-interaction data.

Thus was established the validity of unification, a piece of reality of Maxwellian stature.

Prospects

The theory as it stands leaves us with several desiderata.

$SU(3)_c$ and SU(2)×U(1) contain at least eighteen adjustable parameters, whence the very strong presumption that the present formalism contains too much arbitrariness. Yet to date SU(2)×U(1) works very well, including its radiative corrections.

Other queries. Why do P and C violation occur only in weak interactions? What is the small CP violation trying to tell us? Are neutrino masses strictly zero or not? What can ultrahigh-energy physics learn from astrophysics?

The search is on for the grand unified theory which will marry QCD with electroweak theory. We do not know which is the grand unified theory group, though there are favored candidates.

New options are being explored: global supersymmetry, in which fermions and bosons are joined within supermultiplets and known particles acquire "superpartners." In its local version gravitons appear with superpartners of their own. The most recent phase of this development is superstring theory, which brings us to the Planck length ($\sim 10^{-33}$ cm), the inwardmost scale of length yet contemplated in high-energy theory. All this has led to profound new mathematics but not as yet to any new physics.

High-energy physics, a creation of our century, has wrought revolutionary changes in science itself as well as in its impact on society. As we reach the twilight of 20th-century physics, now driven by silicon and software technologies, it is fitting to conclude with the final words of Rowlands's 1899 address with which I began this essay:

Let us go forward, then, with confidence in the dignity of our pursuit. Let us hold our heads high with a pure conscience while we seek the truth, and may the American Physical Society do its share now and in generations yet to come in trying to unravel the great problem of the constitution and laws of the Universe (Rowland, 1899).

References

Author unnamed (editorial contribution), 1920, Nature **106**, 357.
Bethe, H. A., 1968, Biogr. Mem. Fellows R. Soc. **14**, 391.
Bohr, N., 1913, Philos. Mag. **26**, 476.
Chadwick, J., and E. S. Bieler, 1921, Philos. Mag. **42**, 923.
Compton, K. T., 1937, Nature (London) **139**, 238.
Ellis, C. D., and W. A. Wooster, 1927, Proc. R. Soc. London, Ser. A **117**, 109.
Friedman, J., and V. Telegdi, 1957, Phys. Rev. **105**, 1681; **106**, 1290.
Gell-Mann, M., 1964, Phys. Lett. **8**, 214.
Heisenberg, W., 1963, interview with T. Kuhn, February 28, Niels Bohr Archive, Blegdamsvej 17, DK-2100, Copenhagen.
Lattes, C., C. H. Muirhead, G. Occhialini, and C. F. Powell, 1947, Nature (London) **159**, 694.
Lee, T. D., and C. N. Yang, 1956, Phys. Rev. **104**, 1413.
Lewis, G. N., 1926, Nature (London) **118**, 874.
Møller, C., and A. Pais, 1947, in *Proceedings of the International Conference on Fundamental Particles* (Taylor and Francis, London), Vol. 1, p. 184.
Pais, A., 1955, in *Proceedings of the International Physics Conference*, Kyoto (Science Council of Japan, Tokyo), p. 157.
Pais, A., 1986, *Inward Bound* (Oxford University Press, New York).
Pauli, W., 1921, in *Encykl. der Math. Wissenschaften* (Teubner, Leipzig), Vol. 5, Part 2, p. 539.
Rowland, H., 1899, Science **10**, 825.
Rutherford, E., 1915, Nature (London) **96**, 331.
Van Vleck, J. H., 1964, Phys. Today, June, p. 21.
Weinberg, S., 1967, Phys. Rev. Lett. **19**, 1264.
Wu, C. S., E. Ambler, R. Hayward, D. Hoppes, and R. Hudson, 1957, Phys. Rev. **105**, 1413.
Yang, C. N., and R. Mills, 1954a, Phys. Rev. **95**, 631.
Yang, C. N., and R. Mills, 1954b, Phys. Rev. **96**, 191.

Elementary Particle Physics: The Origins
Val L. Fitch

With the standard model summarizing everything that has been learned about elementary particles in the past 50 to 60 years, it is perhaps difficult to remember that physics remains a subject that has its foundations in experiment. Not only is it because particle physics can be conveniently encapsulated in a theoretical model that we fail to remember, but it is also true that most physics textbooks devoted to the subject and popular accounts are written by theorists and are colored with their particular point of view. From a plethora of texts and memoirs I can point to relatively few written by experimental physicists. Immediately coming to mind is Perkins' *Introduction to High Energy Physics* and the fascinating memoir of Otto Frisch, "What Little I Remember." Bruno Rossi contributed both texts and a lively memoir. We can also point to Alvarez, Segrè, and Lederman. But still this genre by experimentalists is relatively rare. One can speculate why this is the case—that theorists are naturally more contemplative, that experimentalists are people of action (they have to be—the vacuum system always has a leak, there is always an excess of noise and cross-talk in the electronics, there is always something to be fixed).

In the late 1940s when it became clear that the muon was nothing more than a heavy brother of the electron with no obvious role in the scheme of things, Rabi made his oft-quoted remark, "Who ordered that?" In time, he also could have questioned who ordered strange particles, the tau-theta puzzle, CP violation, the avalanche of hadron and meson resonances and the tau lepton. Initially, these discoveries appeared on the scene unwanted, unloved, and with deep suspicion. Now they are all incorporated in the standard model.

It is probably with this in mind that the editor of this volume has asked me to write about the history of particle physics from the point of view of an experimentalist. In the limited space available I have decided to restrict myself to the early days when a large fraction of the new particles were discovered in cosmic rays, starting with Anderson's positron. Those who became interested in cosmic rays tended to be rugged individualists, to be iconoclastic, and to march to the drummer in their own heads rather than some distant one. After all, this was the period when nuclear physics was coming into its own, it was the fashionable subject, it was the subject that had the attention of the theorists, it was the subject for which great accelerators were being built. The cosmic-ray explorers eschewed all that and found their satisfactions in what might be called the backwater of the time.

The Mists of Scotland

Just as modern biology was launched with the invention of the microscope, in physics, too, areas for investigation have been opened with the development of new observational tools. The Wilson cloud chamber is one of these. What would inspire anyone to want to study the behavior of water vapor under unusual conditions? In his Nobel lecture Wilson (1927) answers the question. His curiosity about the condensation of water droplets in moist air was piqued through having watched and wondered about the "wonderful optical phenomena shown when the sun shone on the clouds" that engulfed his Scottish hilltops.

"Cosmic Rays Go Downward, Dr. Anderson"

The discovery of tracks in a cloud chamber associated with cosmic rays was made by Skobelzyn (1929) in the Soviet Union. Almost immediately Auger in France and Anderson and Milliken in the U.S. took up the technique (see Auger and Skobelzyn, 1929). Using electroscopes and ion chambers, Milliken and his students had already resolved a number of important questions about cosmic rays, e.g., that their origin was in the heavens and not terrestrial. Milliken was a forceful person, a skillful popularizer, and an excellent lecturer. He had a knack for memorable phrases. It was Milliken who had coined the name "cosmic rays." Referring to his pet theory on their origin, he called them the "birth cries" of the atoms. Carl Anderson had been a graduate student of Milliken's, and Milliken insisted that he remain at Caltech to build a cloud chamber for studying this new corpuscular radiation from space. As President of Caltech, Milliken was in an excellent position to supply Anderson with the resources required to design and construct a chamber to be operated in a high magnetic field, 17 000 gauss. The chamber was brought into operation in 1932, and in a short time Anderson had many photographs showing positive and negative particles. Blind to the fact that the positives had, in general, an ionization density similar to the negative (electron) tracks, Milliken insisted that the positive particles must be protons. Anderson was troubled by the thought that the positives might be electrons moving upwards but Milliken was adamant. "Cosmic rays come down!" he said, "they are protons." Anderson placed a 0.6-cm lead plate across the middle of the chamber. Almost at once he observed a particle moving upward and certainly losing energy as it passed through the plate; its momentum before entering the plate was 63 MeV/c and 23 MeV/c on exiting. It had to be a positive electron. And irony of ironies, with the history of Milliken's insistence that "cosmic rays go downwards," this first example of a positron was moving upwards.[1]

On Making a Particle Take a Photograph of Itself

Shortly afterward, in England, a stunning improvement in the use of cloud chambers led to a whole array of new discoveries. This was the development of the counter-controlled cloud chamber.

Bruno Rossi, working in Florence, had considerably refined the coincidence counter technique initiated by Bothe in Berlin, and he had launched an experimental program

---

[1] Anderson's paper in *The Physical Review* is entitled "The Positive Electron." In the abstract, written by the editors of the journal, it is said, "these particles will be called positrons."

studying cosmic rays. In Italy, no one had yet operated a cloud chamber and Rossi was anxious to introduce the technique. Accordingly, he arranged for a young assistant, Giuseppe Occhialini, to go to England to work with Patrick Blackett. Blackett had already become widely known for his cloud-chamber work studying nuclear reactions (Lovell, 1975).

As they say, the collaboration of Blackett and Occhialini was a marriage made in heaven. Both men were consummate experimentalists. Both took enormous pleasure in working with their hands, as well as their heads. They both derived much satisfaction in creating experimental gear from scratch and making it work as planned. In Solley (Lord) Zuckerman's collection (1992) of biographical profiles, *Six Men Out of the Ordinary*,[2] Blackett is described as "having a remarkable facility of thinking most deeply when working with his hands." Occhialini has been described as a man with a vivid imagination and a tempestuous enthusiasm: a renaissance man with a great interest in mountaineering, art, and literature as well as physics.

Occhialini arrived in England expecting to stay three months. He remained three years. It was he who knew about the Rossi coincidence circuits and the (then) black art needed to make successful Geiger counters. It was Blackett who must have known that the ion trails left behind by particles traversing a cloud chamber would remain in place the 10 to 100 milliseconds it took to expand the chamber after receipt of a pulse from the coincidence circuit.

In Blackett's own words (1948), "Occhialini and I set about, therefore, to devise a method of making cosmic rays take their own photographs, using the recently developed Geiger-Muller counter as detectors of the rays. Bothe and Rossi had shown that two Geiger counters placed near each other gave a considerable number of simultaneous discharges, called coincidences, which indicated, in general, the passage of a single cosmic ray through both counters. Rossi developed a neat valve circuit by which such coincidences could easily be recorded."

"Occhialini and I decided to place Geiger counters above and below a vertical cloud chamber, so that any ray passing through the two counters would also pass through the chamber. By a relay mechanism the electric impulse from the coincident discharge of the counters was made to activate the expansion of the cloud chamber, which was made so rapid that the ions produced by the ray had no time to diffuse much before the expansion was complete."

After an appropriate delay to allow for droplet formation, the flash lamps were fired and the chamber was photographed. Today, this sounds relatively trivial until it is realized that not a single component was available as a commercial item. Each had to be fashioned from scratch. Previously, the chambers had been expanded at random with the obvious result, when trying to study cosmic rays, that only 1 in about 50 pictures (Anderson's experience) would show a track suitable for measurement. Occhialini (1975), known as Beppo to all his friends, described the excitement of their first success. Blackett emerged from the darkroom with four dripping photographic plates in his hands exclaiming for all the lab to hear, "one on each, Beppo, one on each!" He was, of course, exalting over having the track of at least one cosmic-ray particle in each picture instead of the one in fifty when the chamber was expanded at random. This work (Blackett and Occhialini, 1932) was first reported in *Nature* in a letter dated Aug. 21, 1932 with the title, "Photography of Penetrating Corpuscular Radiation."

---

[2] Of the "six men out of the ordinary," two are physicists, I. I. Rabi and P. M. S. Blackett.

Shortly after this initial success they started observing multiple particles: positive and negative electrons, which originated in the material immediately above the chamber. This was just a few months after Anderson (1932) had reported the existence of a positive particle with a mass much less than the proton. Here they were seeing pair production for the first time. Furthermore, they occasionally observed the production of particles showering from a metal (lead or copper) plate which spanned the middle of their chamber. These were clearly associated with particles contained in showers that had developed in the material above their chamber. The paper in which they first discuss these results is a classic and should be required reading by every budding experimental physicist (Blackett and Occhialini, 1933). In this paper they describe in detail their innovative technique. They also analyze the new and surprising results from over 500 photographs. Their analysis is an amazing display of perspicacity. It must be remembered that this was nearly two years before the Bethe-Heitler formula (1934) and five years before Bhabha and Heitler (1937) and Carlson and Oppenheimer (1937) had extended the Bethe-Heitler formula to describe the cascade process in electromagnetic showers.

Blackett, Occhialini, and Chadwick (1933), as well as Anderson and Neddermeyer (1933), studied the energetics of the pairs emitted from metals when irradiated with the 2.62-MeV γ rays from thorium-C. They found, as expected, that no pair had an energy greater than 1.61 MeV. This measurement also permitted the mass of the positron to be determined to be the same as the electron, to about 15%. The ultimate demonstration that the positive particle was, indeed, the antiparticle of the electron came with the detection of 2 γ's by Klemperer (1934), the annihilation radiation from positrons coming to rest in material.

Blackett and Occhialini[3] must have been disappointed to have been scooped in the discovery of the positron, but they graciously conclude that to explain their results it was "necessary to come to the same remarkable conclusion" as Anderson.

The Slow Discovery of the Mesotron

In contrast to the sudden recognition of the existence of the positron from one remarkable photograph, the mesotron had a much longer gestation, almost five years. It was a period marked by an extreme reluctance to accept the idea that the roster of particles could extend beyond the electron-positron pair, the proton and neutron, and the neutrino and photon. It was a period of uncertainty concerning the validity of the newly minted quantum theory of radiation, the validity of the Bethe-Heitler formula. The second edition of Heitler's book, *The Quantum Theory of Radiation* (1944) serves, still, as a *vade mecum* on the subject. The first edition (1935), however, carries a statement revealing the discomfort many theorists felt at the time, to wit, the "theory of radiative energy loss breaks down at high energies." The justification for this reservation came from measurements of Anderson and Neddermeyer and, independently, Blackett and Wilson, who showed that cosmic-ray particles had a much greater penetrating power than predicted by the theory which pertained to electrons, positrons, and their radiation. The threshold energy at which a deviation from theoretical expectations appeared was around 70 MeV, highly suggestive that things were breaking down at the mass of the electron divided by the fine-structure

---

[3]There is an unusual symmetry associated with these men. The Englishman, Blackett, had an Italian wife; the Italian, Occhialini, had an English wife.

constant, 1/137. However, the theoretical predictions hardened in 1934 when C. F. von Weizsacker and, independently, E. J. Williams showed that in a selected coordinate system both bremsstrahlung and pair production involved energies of only a few mc^2, independent of the original energy. Finally, the ionization and range measurements, primarily by Anderson and Neddermeyer (1937) and Street and Stevenson (1937), forced the situation to the following conclusion: that the mass of the penetrating particles had to be greater than that of the electron and significantly less than that of the proton. In this regard, it is noted that Street and Stevenson were first to employ a double cloud-chamber arrangement that later was to become widely used, i.e., one chamber above the other with the top chamber in a magnetic field for momentum measurements and the lower chamber containing multiple metal plates for range measurements.[4]

About a month after the announcement of the new particle with a mass between that of the electron and the proton, Oppenheimer and Serber (1937) made the suggestion "that the particles discovered by Anderson and Neddermeyer and Street and Stevenson are those postulated by Hideki Yukawa (1935) to explain nuclear forces." [5] Yukawa's paper had been published in 1935 in a Japanese journal, but there had been no reference to it in western physics journals until Oppenheimer and Serber called attention to it. Here at last was the possibility of some theoretical guidance. If the new particle discovered in cosmic rays was that postulated by Yukawa to explain nuclear forces, it would have a mass of the order of 200 electrons, it should be strongly interacting, it should have a spin of 0 or 1, and it should undergo β decay, most likely to an electron and a neutrino.[6]

Blackett, who with Wilson had made some of the earliest and best measurements on the penetrating particles, was curiously reluctant to embrace the new particle. He found it easier to believe that the theory was faulty than that a brand new particle existed.

The first evidence of mesotron decay came from the cloud-chamber pictures of Williams and Roberts (1940). These stimulated Franco Rasetti (1941) to make the first direct electronic measurements of the mean life. He obtained 1.5 ± 0.3 microseconds.

Earlier Rossi, now in America (another one of those marvelous gifts of the Fascist regimes in Europe to the United States), had measured the mean decay length of the mesotrons in the atmosphere by comparing the attenuation in carbon with an equivalent thickness of atmosphere. With measurements performed from sea level to the top of Mt. Evans in Colorado (14 000 ft) he determined the mean decay length to be 9.5 km. Blackett had measured the sea-level momentum spectrum. From that Rossi could obtain an average momentum and, assuming a mass, obtain a proper lifetime. Using his own best estimate of the mass of the mesotron, 80 MeV, he obtained a mean life of 2 microseconds. A bit later Rossi and Neresson (1942) considerably refined the direct method of Rasetti and obtained a lifetime value of 2.15 ± 0.07 microseconds, remarkably close to today's value. And talk about experimental ingenuity, how does one measure a time of the order of microseconds with a *mechanical* oscillograph? They first produced a pulse the amplitude of which was proportional to the time interval between the arrival of a stopping mesotron, as determined by one set of counters, and the appearance of the decay product from a

---

[4]Originally Anderson and Neddermeyer had suggested mesoton for the name of this new particle. Milliken, still a feisty laboratory director, objected and at his insistence the name became mesotron. With usage and time the name evolved into meson.

[5]Serber (1983) has commented, "Anderson and Neddermeyer were wiser: they suggested 'higher mass states of ordinary electrons'."

[6]A highly illuminating and interesting account of post-mesotron theoretical developments has been provided by Robert Serber (1983).

separate set. Considerably stretched in time, these pulses could be displayed on the oscillograph. The distribution in pulse heights then gave the distribution in time, a beautiful exponential.

At about this time research in cosmic rays was essentially stopped because of W.W.II. One summary of the state of knowledge about the subject at that time was provided by Heisenberg. In 1943 he edited a volume of papers devoted to cosmic rays. In this volume the best value for the mass of the mesotron came from the mean decay length in the atmosphere determined by Rossi as well as his direct lifetime measure. The mass was quoted as 100 MeV, which "can be incorrect by 30%, at most." Furthermore, the authors in this volume still accepted, without question, the mesotron to be the Yukawa particle with spin 0 or 1 decaying to electron and neutrino.[7]

The Mesotron Is Not the Yukon

In naming the new particle, serious consideration was given to honoring Yukawa with the obvious appelation, the Yukon. However, this was considered too frivolous and mesotron was adopted. Now out of ravaged war-torn Italy came an astonishing new result: the mesotron was *not* the particle postulated by Yukawa. There had been disquieting indications of this. Despite numerous photographs of their passing through plates in chambers, never had mesotrons shown an indication that they had interacted. Furthermore, the best theoretical estimate of their lifetime was around 10^{-8} seconds, whereas the measured lifetime was 100 times longer. These discrepancies were largely ignored.

As far back as 1940 Araki and Tomonaga (later of QED fame) had published a paper in which they observed that a positively charged Yukawa particle, on coming to rest in matter, would be repelled by the Coulomb field of the nucleus and simply decay as though it were in free space. The negative particles, on the other hand, would interact with the nucleus long before they had a chance to decay. Fortunately, the paper was published in the *Physical Review* (Tomonaga and Araki, 1940), rather than in a Japanese journal, so the conclusions were disseminated widely and quickly.

Three Italians working in Rome, Conversi, Pancini, and Piccioni, set out to test the Araki-Tomonaga result. This was during the time the Germans, under the pressure of the allied armies, were withdrawing from central Italy. At one time or another, while setting up the experiment, Pancini was in northern Italy with the partisans; Piccioni, an officer in the Italian army, was arrested by the retreating Germans (but shortly released), while Conversi, immune to military service because of poor eyesight, was involved in the political underground. Despite the arduous circumstances and many interruptions, they managed to perform an elegant experiment. Data taking started in the spring of 1945 near the end of the war. Using a magnetic spectrometer of novel design, they selected first positive then negative stopping mesotrons and found that essentially no negative particles were observed to decay when stopped in iron, but, contrary to Araki and Tomonaga, those that stopped in carbon did decay and at the same rate as the positives (Conversi *et al.*, 1947). Fermi, Teller, and Weisskopf (1947) quickly showed that this implied the time for

---

[7]The book was originally published to mark the 75th birthday of Heisenberg's teacher, Arnold Sommerfield. On the very day which the book was intended to commemorate, bombs fell on Berlin, destroying the plates and all the books that had not been distributed, nearly the entire stock. The English version, *Cosmic Radiation*, Dover Publications, New York (1946) is a translation by T. H. Johnson from a copy of the German edition loaned by Samuel Goudsmit.

capture was of the order of 10^{12} longer than expected for a strongly interacting particle. It was the experiment that marked the end of the identification of the mesotron with the Yukawa particle.

"Even a Theoretician Might Be Able to Do It"

In Bristol in 1937 Walter Heitler showed Cecil Powell a paper by Blau and Wambacher (1937), which exhibited tracks produced by the interaction of cosmic-ray particles with emulsion nuclei. He made the remark that the method appeared so simple that "even a theoretician might be able to do it." Powell and Heitler set about preparing a stack of photographic plates (ordinary lantern slide material) interspersed with sheets of lead. Heitler placed this assembly on the Jungfraujoch in the Alps for exposure in the summer of 1938. The plates were retrieved almost a year later and their scanned results led to a paper on "Heavy cosmic-ray particles at Jungfraujoch and sea level."

The photographic technique had had a long and spotty history which had led most people to the conclusion that it was not suitable for quantitative work. The emulsions swelled on development and shrank on drying. The latent images faded with time, so particles arriving earlier were more faint than those, with the same velocity, that arrived later. The technique was plagued by nonuniform development. Contrary to the unanimous advice of others, Powell became interested; he saw that what was needed was precise microscopy, highly controlled development of the emulsions, and emul- sions, which up till then had been designed for other purposes, tailored to the special needs of nuclear research, richer in silver content and thicker. Powell attended to these things and convinced the film manufacturers to produce the special emulsions (Frank *et al.*, 1971). Initially, the new emulsions were not successful. Then W.W.II intervened. During the war, Powell was occupied with measuring neutron spectra by examining the proton recoils in the emulsions then available.

The Reservoir of Ideas Runneth Over

Except for the highly unusual cases like that just described, most physicists had their usual research activities pushed aside by more pressing activities during W.W.II.[8] Some, disgusted with the political situation at home, found refuge in other countries. However, ideas were still being born to remain latent and await development at the end of the war.

Immediately after the war the maker of photographic materials, Ilford, was successful in producing emulsions rich in silver halide, 50 microns thick, and sensitive to particles that ionized a minimum of six times. These were used by Perkins (1947), who flew them in aircraft at 30 000 ft. He observed "stars" when mesons came to the end of their range. It was assumed that these were negative mesotrons, which would interact instead of decay.

Occhialini[9] took these new plates to the Pic-du-Midi in the Pyrenees for a one-month exposure. On development and scanning back in Bristol, in addition to the "stars" that

---

[8]For example, in the U.K. Blackett was to become "the father of operations research" and was to be a bitter (and losing) foe of the policy of strategic bombing. In the U.S. Bruno Rossi was recruited by Oppenheimer to bring his expertise in electronics to Los Alamos.

[9]Occhialini had gone to the University of Sao Paulo in Brazil in 1938 but returned to England in 1945 to work with Powell at Bristol.

Perkins had observed, the Powell group discovered two events of a new type. A meson came to rest but then a second appeared with a range of the order of 600 microns.[10] This was the first evidence (Lattes et al., 1947a) suggesting two types of mesons. The authors also conclude in this first paper that if there is a difference in mass between primary and secondary particles it is unlikely to be greater than 100 m_e.[11] More plates were exposed, this time by Lattes at 18 600 ft in the Andes in Bolivia and, on development back in Bristol, 30 events were found of the type seen earlier. Here it was possible to ascertain the mass ratio of the two particles, and they state that it is unlikely to be greater than 1.45. We now know it to be 1.32. The first, the π meson, was associated with the Yukawa particle and the second with the mesotron of cosmic rays, the μ meson.[12]

The work on emulsions continued, and by 1948 Kodak had produced an emulsion sensitive to minimum ionizing particles. The Powell group took them immediately to the Jungfraujoch for exposure under 10 cm of Pb for periods ranging from eight to sixteen days. They were immediately rewarded with images of the complete π-μ-e decay sequence. More exciting was the observation of the first tau-meson decay to three π mesons (Brown et al., 1949) and like the Rochester and Butler particles, discussed below, its mass turned out to be around 1000 m_e. The emulsion technique continued to evolve. Emulsions 0.6-mm thick were produced. Dilworth, Occhialini, and Payne (1948) found a way to ensure uniform development of these thick pieces of gelatin richly embedded with silver halides, and problems associated with shrinkage were solved. Stripped from their glass plates, stacks of such material were exposed, fiducial marks inscribed, and the emulsions returned to the glass plates for development. Tracks could then be followed from one plate to another with relative ease. Emulsions became genuine three-dimensional detectors.

"There Is No Excellent Beauty That Hath Not Some Strangeness in the Proportion"[13]

Concurrent with the development of the emulsion technique by Occhialini and Powell, Rochester and Butler were taking pictures using the Blackett magnet chamber, refurbished, and with a new triggering arrangement to make it much more selective in favor of penetrating showers: Very soon, in October 1946 and May 1947, they had observed two

---

[10]One of the worries of the Powell group was that, on stopping, the first meson had somehow gained energy in a nuclear interaction and then continued on. This question was considered in depth by C. F. Frank (1947) who concluded that this would only happen if the mesotron fused a deuteron and a proton which would release 5.6 MeV. Frank concluded that it was "highly improbable that the small amount of deuterium in an emulsion could account for the observed phenomena." It was to be another ten years before "cold fusion" was discovered in a hydrogen bubble chamber by the Alvarez group in Berkeley. They were unaware of the previous work by Frank.

[11]The two-meson hypothesis was actively discussed by Bethe and Marshak at the famous Shelter Island conference, June 2–4, 1947 with no knowledge of the experimental evidence already obtained by Lattes, Muirhead, Occhialini, and Powell in Nature (1947a). This issue was on its way across the Atlantic, by ship in those days, at the time of the conference. The mesons are named m_1 and m_2 in the first paper and π and μ in the second and third papers (1947b)

[12]There is a story, perhaps apocryphal, that they were called the π and μ mesons because these were the only two Greek letters on Powell's typewriter. I am willing to believe it because I had such a typewriter myself (the author).

[13]Francis Bacon, 1597, "Of Beauty."

unique events, forked tracks appearing in the chamber which could not have been due to interactions in the gas. It became clear that they were observing the decay of particles with a mass of the order of half the proton mass, about 1000 m_e, (Rochester and Butler, 1947). These were the first of a new class of particles, the so-called strange particles. They created a sensation in Blackett's laboratory. However, no more such events were seen in more than a year of running. It was then decided to move the chamber to the high mountains for a higher event rate. But where? Two sites were possible, the Aiguille-du-Midi near Chamonix or the Pic-du-Midi in the Pyrenees. The Blackett magnet was much too massive to be transported to the Aiguille; this could be solved by building a new magnet that could be broken down into small pieces for transport on the téléférique up the mountain. The Pic-du-Midi was at a much lower altitude. It was accessible in winter only on skis, and supplies had to be carried in. However, the heavy Blackett magnet could be installed and adequate power for it was promised. They chose the site in the Pyrenees and were in operation in the summer of 1950. Almost immediately they began observing forked tracks similar to those observed in Manchester.[14]

Somewhat before, the Anderson group at Caltech had also observed events like those originally seen by Rochester and Butler. It was at this time that Anderson and Blackett got together and decided that these new types of particles should be called V particles.

And So Was Born the Tau-Theta Puzzle

It was Thompson at Indiana University (he had earlier been a student of Rossi's at MIT) who singlehandedly brought the cloud-chamber technique to its ultimate precision. His contribution to the field has been tellingly described by Steinberger (1989):

Because many new particles were being observed, the early experimental situation was most confused. I would like to recall here an incident at the 1952 Rochester conference, in which the puzzle of the neutral V's was instantly clarified. It was the session on neutral V particles. Anderson was in the chair, but J. Robert Oppenheimer was dominant. He called on his old friends, Leighton from Caltech and W. B. Fretter from Berkeley, to present their results, but no one was much the wiser after that. Some in the audience, clearly better informed than I was, asked to hear from Robert W. Thompson from Indiana, but Oppenheimer did not know Thompson, and the call went unheeded. Finally there was an undeniable insistence by the audience, and reluctantly the lanky young midwesterner was called on. He started slowly and deliberately to describe his cloud chamber, which in fact was especially designed to have less convection than previous chambers, an improvement crucial to the quality of the measurements and the importance of the results. Oppenheimer was impatient with these details, and sallied forth from his corner to tell this unknown that we were not interested in details, that he should get on to the results. But Thompson was magnificently imperturbable: 'Do you want to hear what I have to say, or not?' The audience wanted to hear, and he continued as if the great master had never been there. A few minutes later, Oppenheimer could again no longer restrain himself, and tried again, with the same effect. The young man went on, exhibited a dozen well-measured V^0's, and, with a beautiful and original analysis, showed that there were two different particles, the $\Lambda^0 \to p + \pi^-$ and $\theta^0 \to \pi^+ + \pi^-$. The θ^0 (θ for Thompson) is the present K^0.

When the events of the Rochester conference of 1952 were unfolding, additional examples of tau-meson decay had been observed in photographic emulsions. In the next

---

[14]Not without a price. One young researcher suddenly died when skiing up the mountain to the laboratory.

three years several hundred fully reconstructed decays were observed worldwide, largely in emulsions. Almost immediately, a fundamental problem presented itself. A τ^+ decays to $\pi^+ + \pi^+ + \pi^-$. A few instances were seen where the π^- had very little energy, i.e., was carrying away no angular momentum. In that the $\pi^+ + \pi^+$ system must be in an even state of angular momentum (Bose statistics) and that the π has an odd intrinsic parity, there was no way the τ and the θ could have the same parity. These rather primitive observations were borne out by detailed analyses prescribed by Dalitz (1954). So was born the tau-theta puzzle.

What appeared to be a clear difference in the tau and theta mesons made it imperative to know just how many different mesons existed with a mass of about $1000\ m_e$. To answer this question an enormous stack of emulsion was prepared, large enough to stop any of the charged secondaries from the decay. The experiment was the culmination of the development of the photographic technique. The so-called "G stack" collaboration, Davies *et al.* (1955), involved the Universities of Bristol, Milan, and Padua. In this 1954 experiment 250 sheets of emulsion, each 37×27 cm and 0.6 mm thick were packed together separated only by thin paper. The package was 15 cm thick and weighed 63 kg. It was flown over northern Italy supported by a balloon at 27 km for six hours. Because of a parachute failure on descent about 10% of the emulsion stack was damaged but the remainder was little affected. This endeavor marked the start of large collaborative efforts. In all, there were 36 authors from 10 institutions.

Cloud-chamber groups in Europe and the United States were discovering new particles. There were, in addition to Thompson working at sea level at Indiana, the Manchester group at the Pic-du-Midi and the French group under Louis Leprince-Ringuet from the Ecole Polytechnique working at the Aiguille-du-Midi and the Pic-du-Midi. Rossi's group from MIT and a Princeton group under Reynolds were on Mt. Evans in Colorado; the group of Brode was at Berkeley, and Anderson's at Caltech was on Mt. Wilson. The camaraderie of this international group was remarkable, perhaps unique. Sharing data and ideas, this collection of researchers strove mightily to untangle the web being woven by the appearance of many new strange particles, literally and figuratively.

The role of cosmic rays in particle physics reached its apex at the time of the conference in the summer of 1953 at Bagnères-de-Bigorre in the French Pyrenees, not far from the Pic-du-Midi. It was a conference characterized by great food and wines and greater physics, a truly memorable occasion. All of the distinguished pioneers were there: Anderson, Blackett, LePrince-Ringuet, Occhialini, Powell, and Rossi. It was a conference at which much order was achieved out of a rather chaotic situation through nomenclature alone. For example, it was decided that all particles with a mass around $1000\ m_e$ were to be called K mesons. There was a strong admonition from Rossi (1953) that they were to be the same particle until proven otherwise. All particles with a mass greater than the neutron and less than the deuteron were to be called hyperons. And finally, at the end, Powell announced, "Gentlemen, we have been invaded... the accelerators are here."

The Crepuscular Years for Cloud Chambers

The study of cosmic rays with cloud chambers and emulsions remained the only source of information about strange particles through most of 1953. That information was enough for Gell-Mann (1953) and Nakano *et al.* (1953) to see a pattern based on isotopic spin that was to be the forerunner of SU(3) and the quark model. Then data from the new accelerators started to take over, beginning with the observation of associated production by Shutt and collaborators at Brookhaven (Fowler et al., 1953). It was an experiment that still used the cloud chamber as the detector, in this case a diffusion chamber. The continuously sensitive diffusion chamber had been developed by Alex Langsdorf (1936) before W.W.II but had never found use studying cosmic rays because the sensitive volume was a relatively thin horizontal layer of vapor whereas, as Milliken said, "cosmic rays come down." However, with the high-energy horizontal π^- beams at the Brookhaven cosmotron, the diffusion chamber had a natural application.

In these last years of the cloud chamber one more magnificent experiment was performed. In one of the transcendent theoretical papers of the decade, M. Gell-Mann and A. Pais (1955) proposed a resolution of a puzzle posed by Fermi two years before, i.e., if one observes a $\pi^+ + \pi^-$ pair in a detector, how can one tell if the source is a θ^0 or its antiparticle, the $\bar{\theta}^0$? The conclusion of the Gell-Mann and Pais analysis was that the particles which decay are two linear combinations of θ^0 and $\bar{\theta}^0$ states, one short lived and decaying to the familiar $\pi^+ + \pi^-$ and the other, long lived. It was a proposal so daring in its presumption that many leading theorists were reluctant to give it credence. However, Lederman and his group accepted the challenge of searching for the long-lived neutral counterpart. And they were successful in discovering the θ_2 which lives 600 times longer than the θ_1, the object that decays to $\pi^+ + \pi^-$.

This was the last great experiment performed using the Wilson cloud chamber, which had had its origins in the curiosity of a man ruminating about the mists over his beloved Scottish hillsides. Glaser's bubble chamber, the inspiration for which came from a glass of beer in a pub, was ideally suited for use with accelerators and soon took over as the visual detector of choice. By 1955 K mesons were being detected by purely counter techniques at the Brookhaven Cosmotron and the Berkeley Bevatron, and the antiproton was discovered at the Bevatron. Data from large emulsion stacks exposed in the beams from the accelerators quickly surpassed the cosmic-ray results in quality and quantity.

The big questions, which were tantalizingly posed by the cosmic-ray results, defined the directions for research using accelerators. The tau-theta puzzle was sharpened to a major conundrum. Following the edict of Hippocrates that serious diseases justify extreme treatments, Lee and Yang were to propose two different remedies: the first, that particles exist as parity doublets; and the second, much more revolutionary than the first, that a cherished conservation principle, that of parity, was violated in the weak interactions. They suggested a number of explicit experimental tests which, when carried out, revealed a new symmetry, that of *CP*. This, too, was later shown to be only approximate.[15] Indeed, within the framework of our current understanding, the preponderance of matter over antimatter in our universe is due to a lack of *CP* symmetry. Furthermore, as we have already noted, a large fraction of the discoveries that were key to the theoretical developments in the 1950s and early 1960s, discoveries which led to the quark model, also were made in cosmic-ray studies. Most were unpredicted, unsolicited, and in many cases, unwanted at their birth. Nonetheless, these formed the foundations of the standard model.

Today, discoveries in cosmic rays continue to amaze and confound. The recent evidence (Fukuda et al., 1998) that neutrinos have mass has been the result of studying the nature of the neutrinos originating from the π-μ-e decay sequence in the atmosphere. This is a story that remains to be completed.

---

[15] See chapter by Henley and Schiffer on pp. 345–369 in this book.

General Reading

Brown, L. M., and L. Hoddeson, 1983, Eds., *The Birth of Particle Physics* (Cambridge University, Cambridge, England).
Colston Research Society, 1949, *Cosmic Radiation*, Symposium Proceedings (Butterworths, London). American edition (Interscience, New York).
Marshak, R., 1952, *Meson Theory* (McGraw-Hill, New York).
Occhialini, G. P. S., and C. F. Powell, 1947, *Nuclear Physics in Photographs* (Oxford University, New York).
Pais, A., 1986, *Inward Bound* (Oxford University, New York).
Peyrou, C., 1982, International Colloquium on the History of Particle Physics, J. Phys. (Paris) Colloq. **43**, C-8, Suppl. to No. 12, 7.
Powell, C. F., P. H. Fowler, and D. H. Perkins, 1959, *The Study of Elementary Particles by the Photographic Method* (Pergamon, London).
Rochester, G. D., and J. G. Wilson, 1952, *Cloud Chamber Photographs of the Cosmic Radiation* (Pergamon, London).

References

Anderson, C. D., 1932, Science **76**, 238.
Anderson, C. D., 1933, Phys. Rev. **43**, 491.
Anderson, C. D., and S. H. Neddermeyer, 1933, Phys. Rev. **43**, 1034.
Anderson, C. D., and S. H. Neddermeyer, 1937, Phys. Rev. **51**, 884.
Auger, P., and D. Skobelzyn, 1929, C. R. Acad. Sci. **189**, 55.
Bhabha, H., and W. Heitler, 1937, Proc. R. Soc. London, Ser. A **159**, 432.
Blackett, P. M. S., 1948, Nobel Address.
Blackett, P. M. S., and G. P. S. Occhialini, 1933, Proc. R. Soc. London, Ser. A **139**, 699.
Blackett, P. M. S., G. P. S. Occhialini, and J. Chadwick, 1933, Nature (London) **131**, 473.
Blau, M., and Wambacher, 1937, Nature (London) **140**, 585.
Brown, R., U. Camerini, P. H. Fowler, H. Muirhead, C. F. Powell, and D. M. Ritson, 1949, Nature (London) **163**, 82.
Carlson, J. F., and J. R. Oppenheimer, 1937, Phys. Rev. **51**, 220.
Conversi, M., E. Pancini, and D. Piccioni, 1947, Phys. Rev. **71**, 209.
Dalitz, R. H., 1954, Phys. Rev. **94**, 1046.
Davis, *et al.*, 1955, Nuovo Cimento, Series X, Vol. **2**, 1063.
Dillworth, C. C., G. P. S. Occhialini, and R. M. Payne, 1948, Nature **162**, 102.
Fermi, E., E. Teller, and V. Weisskopf, 1947, Phys. Rev. **71**, 314.
Fowler, W. B., *et al.*, 1953, Phys. Rev. **98**, 12 1.
Frank, F. C., 1947, Nature (London) **160**, 525.
Frank, F. C., D. H. Perkins, and A. M. Tyndall, 1971, Biogr. Mem. Fellows R. Soc. **17**, 541.
Fukuda, Y., *et al.*, 1998, Phys. Rev. Lett. **81**, 1562.
Gell-Mann, M., 1953, Phys. Rev. **92**, 833.
Gell-Mann, M., and A. Pais, 1955, Phys. Rev. **97**, 1387.
Heisenberg, W., 1946, Ed., *Cosmic Radiation* (Dover, New York), translated by T. H. Johnson from the 1943 German edition.
Heitler, W., 1935, *The Quantum Theory of Radiation* (Oxford, University Press, London). Second Ed. 1944.
Klemperer, O., 1934, Proc. Cambridge Philos. Soc. **30**, 347.
Langsdorf, A., 1936, Phys. Rev. **49**, 422.
Lattes, C. M. G., H. Muirhead, G. P. S. Occhialini, and C. F. Powell, 1947a, Nature (London) **159**, 694.

Lattes, C. M. G., H. Muirhead, G. P. S. Occhialini, and C. F. Powell, 1947b, Nature (London) **160**, 453, 486.
Lovell, B., 1975, Biogr. Mem. Fellows R. Soc. **21**, 1.
Nakano, T. and K. Nishijima, 1953, Prog. Theor. Phys. **10**, 581.
Occhialini, G. P. S., 1975, Notes Rec. R. Soc. **29**, 144.
Oppenheimer, J. R., and R. Serber, 1937, Phys. Rev. **51**, 113.
Perkins, D. H., 1947, Nature (London) **159**, 126.
Rasetti, F., 1941, Phys. Rev. **60**, 198.
Rochester, G. D., and C. C. Butler, 1947, Nature (London) **160**, 855.
Rossi, B., and N. Neresson, 1942, Phys. Rev. **62**, 417.
Serber, Robert, 1983, *The Birth of Particle Physics*, edited by Laurie M. Brown and Lillian Hoddeson (Cambridge University, Cambridge, England), p. 206.
Skobelzyn, D., 1929, Z. Phys. **54**, 686.
Steinberger, J., 1989, in *Pions to Quarks*, edited by L. M. Brown, M. Dresden, and L. Hoddeson (Cambridge University, Cambridge, England), p. 317.
Street, J. C., and E. C. Stevenson, 1937, Phys. Rev. **51**, 1005.
Tomonaga, S., and G. Araki, 1940, Phys. Rev. **58**, 90.
Williams, E. J., and G. E. Roberts, 1940, Nature (London) **145**, 102.
Yukawa, H., 1935, Proc. Phys. Math. Soc. Jpn. **17**, 48.
Zuckerman, J., 1992, *Six Men Out of the Ordinary* (Peter Owen, London).

Astrophysics
George Field

Astrophysics interprets astronomical observations of stars and galaxies in terms of physical models. During this century new classes of objects were discovered by astronomers as novel instruments became available, challenging theoretical interpretation.

Until the 1940s, astronomical data came entirely from optical ground-based telescopes. Photographic images enabled one to study the morphology of nebulae and galaxies, filters permitted the colors of stars and hence their surface temperatures to be estimated, and spectrographs recorded atomic spectral lines. After World War II, physicists developed radio astronomy, discovering relativistic particles from objects like neutron stars and black holes. NASA enabled astronomers to put instruments into earth orbit, gathering information from the ultraviolet, x-ray, infrared, and gamma-ray regions of the spectrum.

As the century opened, astrophysicists were applying classical physics to the orbits and internal structure of stars. The development of atomic physics enabled them to interpret stellar spectra in terms of their chemical composition, temperature, and pressure. Bethe (1939) demonstrated that the energy source of the sun and stars is fusion of hydrogen into helium. This discovery led astrophysicists to study how stars evolve when their nuclear fuel is exhausted and hence contributed to an understanding of supernova explosions and their role in creating the heavy elements. Study of the interstellar medium is allowing us to understand how stars form in our Galaxy, one of the billions in the expanding universe. Today the chemical elements created in supernova explosions are recycled into new generations of stars. A question for the future is how the galaxies formed in the first place.

Stellar Energy and Evolution

A key development in astrophysics was Bethe's proposal that the carbon cycle of nuclear reactions powers the stars. H fuses with ^{12}C to produce ^{13}N, then ^{14}N, ^{15}O, and ^{15}N. The latter reacts with H to form ^{12}C again, plus ^{4}He. Thus each kilogram of H fuses to form slightly less than a kilogram of He, with the release of 6×10^{14} joules. Bethe was trying to find an energy source that would satisfy three conditions: (a) Eddington's finding (1926) that the central temperature of main-sequence stars is of the order of 10^7 K, (b) that the earth is Gigayears (Gy) old, and (c) that the sun and stars are mostly hydrogen. Bethe's cycle works on hydrogen at about 10^7 K, and the luminosity of the sun can be balanced for 10 Gy by burning only 10% of it.

The stage had been set by Hertzsprung (1911) and Russell (1914), who had found that, in a diagram in which the luminosity of a star is plotted against its surface temperature, most stars are found along a "main sequence" in which the hotter stars are brighter and the cooler are fainter. A sprinkling of stars are giants, which greatly outshine their main-sequence counterparts, or white dwarfs, which though hot, are faint. Eddington (1924) had found that the masses of main-sequence stars correlate well with their luminosities, as he had predicted theoretically, provided the central temperatures were all about the same. Bethe's proposal fitted that requirement, because the fact that only the Maxwell-Boltzmann tail of the nuclear reactants penetrates the Coulomb barrier makes the reaction rate extremely sensitive to temperature. But Bethe's discovery did not explain the giants or the white dwarfs.

Clues to this problem came with the application of photoelectric photometry to the study of clusters of stars like the Pleiades, which were apparently all formed at the same time. In such clusters there are no luminous—hence massive—main-sequence stars, while giants are common. In 1952 Sandage and Schwarzschild showed that main-sequence stars increase in luminosity as helium accumulates in the core, while hydrogen burns in a shell. The core gradually contracts, heating as it does so; in response, the envelope expands by large factors, explaining giant stars. Although more massive stars have more fuel, it is consumed far faster because luminosity increases steeply with mass, thus explaining how massive stars can become giants, while less massive ones are still on the main sequence.

The age of a cluster can be computed from the point at which stars leave the main sequence. Sandage found that ages of clusters range from a few million to a few billion years. In particular, globular star clusters—groups of 10^5 stars distributed in a compact region—all have the same age, about 10 Gy, suggesting that this is the age of the Galaxy. The article by Turner and Tyson in this volume explains why the age of globular clusters is a key datum in cosmology.

As more helium accumulates, the core of a star contracts and its temperature increases. When it reaches 10^8 K, ^{4}He burns to ^{12}C via the triple-α process discovered by Salpeter (1952); the core shrinks until the density is so high that every cell in phase space is occupied by two electrons. Further compression forces the electron momenta to increase according to the Pauli principle, and, from then on, the gas pressure is dominated by such momenta rather than by thermal motions, a condition called electron degeneracy. In response, the envelope expands to produce a giant. Then a "helium flash" removes the degeneracy of the core, decreasing the stellar luminosity, and the star falls onto the "horizontal branch" in the Hertzsprung-Russell diagram, composed of giant stars of various radii. Formation of a carbon core surrounded by a helium-burning shell is accompanied by an excursion to even higher luminosity, producing a supergiant star like Betelgeuse.

If the star has a mass less than eight solar masses, the central temperature remains below the 6×10^8 K necessary for carbon burning. The carbon is supported by degeneracy pressure, and instabilities of helium shell burning result in the ejection of the stellar envelope, explaining the origin of well-known objects called planetary nebulae. The remaining core, being very dense ($\sim 10^9$ kg m^{-3}), is supported by the pressure of its degenerate electrons. Such a star cools off at constant radius as it loses energy, explaining white dwarfs.

Chandrashekhar (1957) found that the support of massive white dwarfs requires such high pressure that electron momenta must become relativistic, a condition known as relativistic degeneracy. "Chandra," as he was called, found that for stars whose mass is nearly 1.5 times the mass of the sun (for a helium composition), the equation of state of relativistic degenerate gas requires that the equilibrium radius go to zero, with no solutions for larger mass. Though it was not realized at the time, existence of this limiting mass was pointing to black holes (see later section on Compact Objects).

Stars of mass greater than eight solar masses follow a different evolutionary path. Their cores do reach temperatures of 6×10^8 K at which carbon burns without becoming degenerate, so that contraction of the core to even higher temperatures can provide the thermal pressure required as nuclear fuel is exhausted. Shell burning then proceeds in an onion-skin fashion. As one proceeds inward from the surface, H, He, C, O, Ne, Mg, and Si are burning at successively higher temperatures, with a core of Fe forming when the temperature reaches about 2×10^9 K. When the mass of Fe in the core reaches a certain value, there is a crisis, because it is the most stable nucleus and therefore cannot release energy to balance the luminosity of the core. The core therefore turns to its store of gravitational energy and begins to contract. Slow contraction turns to dynamical collapse, and temperatures reach 10^{10} K. Heavier elements are progressively disintegrated into lighter ones, until only free nucleons remain, sucking energy from the pressure field in the process and accelerating the collapse. As the density approaches nuclear values (10^{18} kg m^{-3}) inverse β decay ($p + e \rightarrow n + \mu$) neutronizes the material and releases about 10^{46} J of neutrinos, which are captured in the dense layers above, heating them to $\sim 10^9$ K and reversing their inward collapse to outward expansion. Most of the star is ejected at 20 000 km s^{-1}, causing a flash known to astronomers as a supernova of Type II. This scenario was confirmed in 1987 when Supernova 1987A exploded in the Large Magellanic Cloud, allowing 19 neutrinos to be detected by underground detectors in the U.S. and Japan. If the core is not too massive, neutrons have a degeneracy pressure sufficient to halt the collapse, and a neutron star is formed. Analogous to a white dwarf but far denser, about 10^{18} kg m^{-3}, it has a radius of about 10 km. The "bounce" of infalling material as it hits the neutron star may be a major factor in the ensuing explosion.

Ordinary stars are composed mostly of hydrogen and helium, but about 2% by mass is C, N, O, Mg, Si, and Fe, with smaller amounts of the other elements. The latter elements were formed in earlier generations of stars and ejected in supernova explosions. As the supernova shock wave propagates outward, it disintegrates the nuclei ahead of it, and as the material expands and cools again, nuclear reactions proceed, with the final products being determined by how long each parcel of material spends at what density and temperature. Numerical models agree well with observed abundances.

However, there is a serious problem with the above description of stellar evolution. In 1964 John Bahcall proposed that it be tested quantitatively by measuring on earth the neutrinos produced by hydrogen burning in the core of the sun, and that available models of the sun's interior be used to predict the neutrino flux. Raymond Davis took up the challenge and concluded (Bahcall et al., 1994) that he had detected solar neutrinos, qualitatively confirming the theory, but at only 40% of the predicted flux, quantitatively contradicting it. Since then several other groups have confirmed his result. A new technique, helioseismology, in which small disturbances observed at the surface of the sun are interpreted as pressure waves propagating through its interior, allows one to determine the run of density and temperature in the interior of the sun. Increasingly accurate measurements indicate that Bahcall's current models and hence theoretical neutrino fluxes are accurate to about 1%, so the neutrino discrepancy remains.

The best solution to the solar neutrino problem may be that the properties of electron neutrinos differ from their values in the standard model of particle physics. Specifically, they may oscillate with tau neutrinos, and thus would have to have a rest mass. An upper limit of 20 eV on the neutrino mass deduced from the near-simultaneous arrival of the 19 neutrinos from Supernova 1987A is consistent with this hypothesis. Experiments are now under way to measure the energy spectrum of solar neutrinos and thereby check whether new physics beyond the standard model is needed.

Compact Objects

Three types of compact stellar objects are recognized: white dwarfs, neutron stars, and black holes. White dwarfs are very common, and their theory is well understood. Models of neutron stars were presented by Oppenheimer and Volkoff in 1939. The gravitational binding energy in a neutron star is of the order of $0.1c^2$ per unit mass, so general relativity, rather than Newtonian physics, is required. As in the case of white dwarfs, neutron stars cannot exist for masses greater than a critical limiting value which depends upon the equation of state of bulk nuclear matter, currently estimated to be three solar masses.

If the evolution of a massive star produces a core greater than three solar masses, there is no way to prevent its collapse, presumably to the singular solution of general relativity found by Karl Schwarzschild in 1916, or that found for rotating stars by Kerr, in which mass is concentrated at a point. Events occurring inside spheres whose circumference is less than 2π times the "Schwarzschild radius," defined as $R_S = 2GM/c^2$ (=3 km for 1 solar mass), where G is Newton's constant, are forever impossible to view from outside R_S. In 1939 Oppenheimer and Snyder found a dynamical solution in which a collapsing object asymptotically approaches the Schwarschild solution. Such "black holes" are the inevitable consequence of stellar evolution and general relativity.

While optical astronomers despaired of observing an object as small as a neutron star, in 1968 radio astronomers Anthony Hewish, Jocelyn Bell, and their collaborators discovered a neutron star by accident, when they noticed a repetitive pulsed radio signal at the output of their 81-MHz array in Cambridge, England (Hewish et al., 1968). The pulses arrive from pulsar PSR 1919+21 with great regularity once every 1.337 sec. Hundreds of pulsars are now known.

Conservation of angular momentum can explain the regularity of the pulses if they are due to beams from a rotating object. The only type of star that can rotate once per second without breaking up is a neutron star. In 1975 Hulse and Taylor showed that PSR 1913+16 is in a binary system with two neutron stars of nearly the same mass, 1.4 solar masses. The slow decrease in orbital period they observed is exactly that predicted by the loss of orbital energy to gravitational radiation, providing the most stringent test yet of strong-field general relativity.

Giacconi et al. (1962) launched a rocket capable of detecting cosmic x rays above the atmosphere. They detected a diffuse background that has since been shown to be the superposition of thousands of discrete cosmic x-ray sources at cosmological distances. They also observed an individual source in the plane of the Milky Way, subsequently denoted Scorpius X-1. Later study by the *Uhuru* satellite revealed many galactic sources that emit rapid pulses of x rays, and the frequency of these pulses varies as expected for Doppler shifts in a binary system. X-ray binaries are systems in which a neutron star or black hole is accreting matter from a normal star and releasing gravitational energy up to 10^5 times the luminosity of the sun as x rays. Regular pulses are due to magnetized neutron stars in which accretion is concentrated at the magnetic poles. Even a tiny amount of angular momentum in the accreting gas prevents direct accretion, so the incoming material must form a Keplerian disk orbiting the compact object, supported by rotation in the plane of the disk and by much smaller thermal pressure normal to it. Solutions for thin disks give the rate at which angular momentum flows outward via turbulent viscosity, allowing material to accrete, and predict surface temperatures in the keV range, in agreement with observation.

In the 1960s, military satellites detected bursts of 100-keV gamma rays. Declassified in 1973 (Klebesadel, Strong, and Olson, 1973), gamma-ray bursts proved to be one of the most intriguing puzzles in astronomy, with theories proliferating. It is difficult to test them, because bursts last only seconds to minutes, usually do not repeat, and are hard to locate on the sky because of the lack of directionality in high-energy detectors. In 1997, the x-ray observatory *Beppo Sax* observed a flash of x rays coinciding in time with a gamma-ray burst from the source GRB 970228. The x-ray position was determined to within a minute of arc (IAU, 1997), allowing optical telescopes to detect a faint glow at that position. An absorption line originating in a galaxy in the same direction shows that the source is behind it, and hence at a cosmological distance (see later section on The Interstellar Medium and Star Formation). Other x-ray afterglows have now confirmed that gamma-ray bursts are at cosmological distances, showing that the typical energy in a burst is 10^{45} joules. As this energy is 10% of the binding energy of a neutron star, a possible explanation is the collision of two neutron stars, inevitable when the neutron stars in a binary of the type discovered by Hulse and Taylor spiral together as a result of the loss of energy to gravitational radiation. Estimates of the frequency with which this happens agree with the frequency of gamma-ray bursts.

Galaxies

Our Galaxy, the Milky Way, is a thin disk of stars, gas, and dust, believed to be embedded in a much larger ball of dark matter. The nearby stars are arranged in a thin layer. Interstellar dust extinguishes the light of distant stars, and, until this was realized and allowed for, it appeared that the disk was centered on the sun and not much wider than it was thick.

In 1918 and 1919 Harlow Shapley used stars of known luminosities to estimate the distances to individual globular star clusters and found that they form an approximately spherical system whose center is 50 000 light years away in the constellation of Sagittarius (newer data yield a value closer to 30 000 light years). We now realize that the Milky Way is a disk about 30 000 light years in radius and 3000 light years thick, together with a thicker bulge of stars surrounding the center, which tapers off into a roughly spherical halo of stars. Many of the halo stars are located in globular star clusters (described in the earlier section on Stellar Energy and Evolution) of which there are several hundred. The sun revolves around the center once in 250 million years, and Kepler's third law applied to its orbit implies that mass inside it is about 10^{11} suns. We are prevented from seeing the galactic center by the enormous extinction of visible light by interstellar dust, but infrared radiation, which penetrates dust more easily, reveals that there is a very dense cluster of stars located right at the center of the Galaxy. To explain the very high observed velocities of the stars, there must be a compact object there with a mass of over 10^6 suns. This object is probably a massive black hole, a cousin of even more massive ones in active galactic nuclei (see below).

The morphology of the Galaxy reflects its origin and evolutionary history. Halo stars, in particular those in globular star clusters, are deficient in heavy elements. Since such elements are created in supernova explosions, this can be understood if halo stars formed early in the history of the Galaxy, when the number of supernovae was still small. Since globular star clusters all have about the same age, 10 Gy, it would fit the observations if the Galaxy formed by collapse of primordial material that long ago. Walter Baade gave the name Population II to halo stars, which are deficient in heavy elements and are old, and the name Population I to stars in the disk, which are younger and have normal abundances. The fact that Population I stars are confined to the disk of the Galaxy suggests that the interstellar gas from which they are still forming up to the present time was prevented from rapid star formation by its high angular momentum compared to that of the gas which formed the halo stars. Recent study suggests that the Galaxy is a barred spiral, in which the stellar orbits in the interior have coalesced into a bar like those seen in other galaxies. The spiral arms in the outer parts of the Galaxy are driven by the gravitational field of the bar, as explained theoretically in 1967 by C. C. Lin and Frank Shu.

In 1925 Edwin Hubble showed from the apparent brightness of stars of known luminosity that the distance of the nearby spiral galaxy M31 is about 1 million light years (now revised to 2 million), far larger than the 60 000-light-year diameter of our Galaxy, indicating that M31 is far outside it. Surveying galaxies to the limits of the 100-inch telescope on Mt. Wilson, Hubble concluded that the system of galaxies is at least 500 million light years in radius. Galaxies are distributed in groups of a few up to a few dozen galaxies, like the Local Group of which the Galaxy and M31 are a part, in clusters of 1000 galaxies or more, and in superclusters containing a dozen or more clusters. The modern view is that galaxies are clustered on all scales up to 300 million light years, while the distribution on even larger scales is nearly uniform.

Hubble discovered something else of monumental importance. Measuring Doppler shifts of galaxies at various distances (which he inferred from indicators like bright stars and exploding stars known as novae), he announced in 1929 that the radial velocities v of galaxies inferred from their Doppler shifts are always positive, indicating recession, and, further, that they are proportional to their distances r. This discovery caused a sensation. Hubble's law, $v = Hr$ (where H is called the Hubble constant), suggests a time in the past, $r/v = H^{-1}$, when the expansion apparently started. The expansion of the universe as a whole had been predicted on the basis of general relativity by Alexander Friedmann in 1922, but Hubble made no reference to that prediction in his 1929 paper. The implications of Friedmann's models for cosmology are described in the article by Turner and Tyson in this volume. Here we note that Hubble's law enables one to estimate the distance of any galaxy for which the Doppler shift is known, once the Hubble constant H is determined from galaxies of known distance.

Back to the origin of the elements. George Gamow proposed in 1946 that all of the elements were created in the first few minutes of the cosmological expansion, when, he calculated, the energy density was dominated by radiation rather than by matter. Under these conditions, Friedmann's equations show that the temperature T is related to the age t of the universe by $T \simeq 1$ MeV$/\sqrt{t}$ if t is in seconds. With Alpher and Bethe, Gamow showed in 1948 that the protons and free neutrons at such temperatures would react to form helium, but because of the lack of a stable nucleus of mass 5, nuclear reactions would stop there, contradicting his hypothesis that all of the elements are created in the first few minutes. As we have seen above, conditions in supernovae produce the elements heavier than helium, and in 1957 Burbidge, Burbidge, Fowler, and Hoyle were able to assign the origin of all the heavier elements to one of several processes occurring in evolved stars, such as the triple α reactions described earlier, the slow addition of free neutrons, and the rapid addition of neutrons that occurs in supernova explosions. Fred Hoyle recognized that the helium must be made in Gamow's "Big Bang," as Hoyle referred to it, and with Roger Tayler, he made an early calculation of the expected helium abundance based on modern nuclear data. It is now believed that H and He were formed in the big bang, and that all the other elements are formed in supernovae.

Penzias and Wilson (1965) discovered an isotropic microwave background radiation at 7.3-cm wavelength, having a brightness temperature 3.5 ± 1 K, since then referred to as the CMB, for cosmic microwave background (see the article by Wilkinson in this volume) and interpreted this as radiation from the big bang. The discovery of the CMB solved a puzzle I had known about for years. Optical absorption by interstellar CN molecules in the upper rotational level is observed in the spectra of distant stars, so that level is somehow excited. In an unpublished note I had shown that the excitation must be due to radiation at a wavelength of 2.6 mm having a brightness temperature of about 3 K. I was thus able, in 1966 and with John Hitchcock, to use his recent CN measurements to show that the spectrum of the CMB is that of a blackbody over the wavelength interval from 7.3 cm to 2.6 mm, a factor of 30.

One discovery in extragalactic astronomy was completely unexpected—supermassive black holes. The story goes back to Seyfert (1943), who noticed that about 1% of spiral galaxies are anomalously bright. He found that the emission is highly concentrated at the nucleus of the galaxy, rather than spread over its surface, and that the optical spectrum consists of emission lines of hot gas. Moreover, these lines are so wide that if they are due to the Doppler effect, atoms must be moving at an unprecedented several thousand km s^{-1}.

Later, radio astronomers began to localize radio sources using Michelson interferometry, enabling a number of sources to be identified optically. It was a great surprise when Cygnus A, the second brightest radio source, was identified by Rudolph Minkowski with a galaxy 500 million light years from the earth. Although 10^{11} times more distant than the sun, it appears brighter at radio wavelengths. Cygnus A opened an era of extragalactic radio astronomy in which each new survey revealed larger numbers of ever-fainter radio galaxies distributed isotropically and therefore most likely at cosmological distances. Their spectra and, later, their polarization properties revealed that the emission is synchrotron radiation, that is, high harmonics of cyclotron radiation by relativistic electrons trapped in magnetic fields. Geoffrey Burbidge showed that to explain the observations there must be a minimum energy in particles and fields. For Cygnus A and other powerful extragalactic sources, this minimum energy is 10^{53} joules, equivalent to the rest mass of 10^6 suns. Nuclear energy cannot be released fast enough to account for the powerful radio galaxies.

Improved interferometers revealed that a radio galaxy typically has two synchrotron clouds disposed on either side, with a point source at the nucleus of the galaxy and jets leading to the lobes, demonstrating that the galactic nucleus is the energy source. Optical astronomers discovered that the spectra of the nuclei of radio galaxies are similar to those of Seyfert galaxies, while radio astronomers studying Seyfert galaxies discovered that they also emit radio waves.

Schmidt (1963) obtained an optical spectrum of a compact radio source, 3C273, and found the Balmer spectrum of hydrogen in emission, shifted to longer wavelengths by 15%, corresponding to a recession velocity of 45 000 km s^{-1}. Many other such objects were found and given the name "quasistellar radio source," or quasar. Their high redshifts mean that quasars are at cosmological distances, and they appear to be compact because of their great distances. The nearer ones have now been resolved by the Hubble Space Telescope, showing that this is correct.

Today we recognize that Seyfert galaxies, radio galaxies, quasars, and QSO's (luminous pointlike extragalactic objects that have little radio emission) belong to a single class of object called active galactic nuclei, which are energized by a powerful engine in the nucleus. The luminosity, radio-to-optical ratio, and presence or absence of a jet are determined by local circumstances. X-ray and even γ-ray emissions have now been observed from many such objects, and their characteristics can be understood in the context of such a model.

The key question is the nature of the central engine. The fact that nuclear energy does not suffice leads to the suggestion of gravitational energy, released as material accretes onto a compact object, as in x-ray binary stars. Because the total energy requires masses of millions of suns, which vastly exceeds the mass of a neutron star, only a supermassive black hole will do.

Rarely seen in Seyfert galaxies, but common in quasars, jets are remarkably well collimated, maintaining their structure for hundreds of thousands of light years. As the polarization is aligned with the jet, coherent magnetic structures must be involved. In the model of Blandford and Znajec, a spinning black hole briefly interacts with the magnetic field of accreting material, setting up a vortex that leads to jet formation. Only such a mechanism seems capable of accelerating those jets for which the Lorentz factor must be 10 to 100.

How are supermassive black holes formed? The "best-buy model," favored by Martin Rees, is that collisions among stars in the dense nucleus of a galaxy produce gas that spirals in by turbulent viscosity to feed a black hole through an accretion disk.

The Interstellar Medium and Star Formation

In the 1940s, Lyman Spitzer, Jr., recognized that the youth of Population I star clusters demands that stars must be forming "now," and that the only available mass reservoir is the interstellar gas recognized at that time from optical interstellar absorption lines in the spectra of distant stars. Strömgren (1939) had shown that the emission nebulae observed near hot giant stars can be explained by the photoionization of nearby clouds of hydrogen. Photons above the 13.6-eV ionization limit of hydrogen ionize hydrogen from HI to HII out to a distance such that the number of recombinations balances the number of ionizing photons. Photoionization heating balances emission-line cooling at a kinetic temperature of about 10^4 K, as confirmed by the strength of emission lines in HII regions.

The absence of ionizing photons outside of HII regions leaves hydrogen in the atomic state (HI), and Spitzer and Savedoff (1950) calculated that the kinetic temperature of HI regions should be of the order of 100 K. However, measurements of temperature and density were not possible at that time, because the absorption lines of HI in its ground state are in the far ultraviolet. I showed (Field, 1965) that the thermal equilibrium of high-density HI is stable, but that low-density gas is unstable to the formation of dense clouds surrounded by hot gas. Such multiphase gases are observed in a variety of astrophysical contexts.

In 1945 van de Hulst calculated that the 21-cm line in the ground state of interstellar HI, due to the $F = 3/2 \rightarrow 1/2$ hyperfine transition, would be observable. In 1951 Ewen and Purcell detected 21-cm emission from the Milky Way. Soon confirmed by groups in Holland and Australia, the 21-cm line provided an important tool for measuring physical conditions in interstellar space as well as the dynamics of the Galaxy. Spitzer's prediction that $T \sim 100$ K was verified. The emission at various angular distances from the galactic center reveals a roughly circular distribution of HI, forming spiral arms that correlate with the distribution of Pop I stars, consistent with the formation of such stars from interstellar gas. The total mass of HI, 2×10^9 solar masses, is about 2% of the gravitational mass inside the sun's orbit.

The 21-cm line has been used to observe HI in other spiral galaxies. Typically, for distances comparable to the sun's orbit, the results are not unlike the Galaxy. But whereas our position in the Galaxy makes it difficult to estimate the distance of HI outside the Sun's orbit, this is not true of external galaxies. As stellar emission is observed to decrease sharply away from the center of a spiral galaxy, most of the stellar mass is concentrated within 30 000 light years, so that at large distances R there should be Keplerian rotation, with $V \propto R^{-1/2}$. However, it is observed that $V(R) \sim$ const out to the largest distances measured. To balance the larger implied centrifugal force, gravitation must be stronger, implying the presence of large amounts of dark matter, either baryonic (faint stars, black holes, etc.) or perhaps nonbaryonic matter created in the big bang (see the article by Sadoulet in this volume). The baryon-to-photon ratio implied by the deuterium abundance is so low that the former is not very credible, so most cosmologists lean toward the latter, with the accompanying implications for the standard model of particle physics described in the article by Turner and Tyson in this volume.

In 1937 Zwicky found that the masses of individual galaxies, deduced by application of the virial theorem to clusters of galaxies, are much greater than those inferred from studies of the inner, stellar, parts of galaxies. He proposed that galaxies had unseen matter in their outer parts, and as stated above, this has now been verified for spiral galaxies from their rotation curves. More recent studies of the distribution of hot x-ray-emitting gas in elliptical galaxies confirms the need for dark matter there too.

In Spitzer's day astronomers had noted the presence of dust clouds in the Milky Way that extinguish the light of stars behind them. Although densities must be thousands of times greater than the interstellar average, 21-cm observations detected no HI, and Salpeter proposed that in such clouds hydrogen is converted to H_2 molecules by reactions on the surfaces of interstellar dust particles. H_2 was found in a rocket ultraviolet absorption spectrum by George Carruthers in 1970, and NASA's *Copernicus* ultraviolet satellite observed H_2 in all regions where the extinction is high enough to shield it from photodissociating photons. The amount of H_2 in the Galaxy is about equal to that of HI.

As the youngest stars are found in or near dark clouds, these clouds, which must be molecular, are the places where stars are born. To assess conditions within them, a type of emission is needed that can penetrate the dust. Emission by the OH radical at 1665 MHz, to which dark clouds are transparent, was found in 1965 by Weinreb, Barrett, Weeks, and Henry. Since then, over 100 different molecular species have been found in the microwave and mm-wave regions, the latest with 11 carbon atoms. Carbon monoxide is widespread and has been detected in distant galaxies. It has been found that individual "cores" of molecular clouds, having masses of the order of one solar mass, are contracting at several 10^2 m s^{-1}, presumably to form new stars like those observed in their vicinity, confirming that stars are forming today in molecular clouds.

Throughout the interstellar medium there are enough free electrons that the medium can support electric currents and hence magnetic fields. That such fields are present was inferred in 1949 by Hiltner and by Hall from the alignment of interstellar dust particles discovered to polarize the radiation of distant stars. The necessity of galactic magnetic fields was recognized by physicists, who demonstrated that there was a high degree of isotropy in the arrival directions of cosmic rays. This isotropy can be explained by gyration around a field. In 1977–1978 it was shown independently by several groups that shock waves in the magnetized interstellar medium can accelerate charged particles to cosmic-ray energies. Such shocks are observed around old supernovae, and the energetics work out.

Hannes Alfvén demonstrated that the motion of a gas with high electrical conductivity like the interstellar medium drags the magnetic field along with it, conserving magnetic flux as it does so. This effect is important for the contraction of molecular cloud cores because it increases the magnetic stress, resisting further contraction. Star formation is delayed while the neutral gas molecules slowly drift through the ions stuck to the field. When enough molecules have accumulated, self-gravitation takes over, and the core collapses to form one or more stars.

It is now appreciated that the Galaxy is a dynamic place, with supernova explosions heating and accelerating surrounding gas and providing fresh heavy elements to it. Subsequently some of this gas collapses to form new stars, some of which explode as new supernovae. In this way, a spiral galaxy like our own, born in the collapse of dark matter from the expanding universe, slowly converts its gas to stars, a process that will cease only when all the gas has collapsed to form stars whose masses are too small for them to explode.

A Personal Perspective

My career in astrophysics started when I was about 12 years old at the William H. Hall Free Library in Edgewood, Rhode Island. It housed, among other things, a small group of books by Sir James Jeans and Sir Arthur Eddington, theoretical astrophysicists in Cambridge, England.

Eddington, who was the first to show from first principles why the stars shine with the power they do, wrote lucidly for the public about stars, galaxies, and the universe. I was hooked; when asked about my future career for the high school yearbook, I unhesitatingly replied, "theoretical astrophysicist." Why theory, not observation? Jeans and Eddington certainly influenced me this way; but there was also the fact that I never seemed to get a big enough explosion in my basement chemistry lab, while mathematics is precise, powerful, and predictable. At MIT I majored in physics and mathematics. For my bachelor's thesis I did an experiment under Hans Mueller on the alignment of small particles in a fluid by an electric field to simulate the then recently discovered polarization of starlight by interstellar dust. As I recall, the data were nearly useless, but the theoretical analysis was not bad.

I had read George Gamow's *Creation of the Universe*, so when I found myself working in a Department of Defense lab in Washington during the Korean War, in 1951–1952, I attended night courses Gamow gave at George Washington University. I admired

Gamow's quick mind and fun-loving personality. He speculated in class about the structure of DNA, and explained how his theory of nucleosynthesis in the big bang accounted for the chemical elements.

To decide on a graduate school, I read articles in the Astrophysical Journal at the Library of Congress, and found those by Lyman Spitzer, Jr., at Princeton the most interesting. I applied and got in, in part, I learned later, because Gamow had sent a postcard saying "Field is OK."

At that time the Princeton Observatory was housed in a Victorian building attached to the Director's home. Departmental teas around a big table in the room which joined the house to the Observatory were memorable for the sparkling conversations between Lyman Spitzer and Martin Schwarzschild, two of the world's leading astrophysicists. I recall their discussing the recent discovery by Nancy Roman that the abundance of heavy elements in stars, as judged by the intensity of absorption lines in their spectra, is correlated with their space velocities relative to the sun, with the low-abundance stars moving faster—an intriguing puzzle. Spitzer's and Schwarzschild's discussion that day led to a paper in which they showed that recently discovered massive clouds of interstellar gas perturb the orbits of stars in the Galaxy over time, so that the oldest stars would have the largest spread in velocities, as observed. I had witnessed the solution of a problem in theoretical astrophysics, in which theoretical dynamics joined nuclear physics and stellar spectroscopy to explain an observed fact. Great stuff!

Radio astronomy was then a relatively new field, and I was excited when Ewen and Purcell discovered 21-cm-line emission from interstellar atomic hydrogen in the Galaxy. I suggested searching for 21-cm absorption against a suitable background radio source and began to think about applications to cosmology. I had taken a course in relativity with John Wheeler and had listened to lectures by Fred Hoyle, a visitor to the Observatory, expounding his steady-state theory, which made a firm prediction for the mean density of the universe. As the stars in galaxies contribute far less than the predicted amount, I decided to search for hydrogen residing between the galaxies in intergalactic space.

So as a postdoc at Harvard I searched for an absorption "trough" between 21 cm and $21(1+z)$ cm, the wavelength of the line shifted to the velocity $v = cz$ of the source, where intergalactic hydrogen should absorb according to Robertson-Walker cosmology. I measured the brightness of the Cygnus A radio source, a radio galaxy with a redshift $z = 0.057$ and observed no effect, and so could only put an upper limit on intergalactic hydrogen. Much later, when quasars were discovered with redshifts about 2, the Lyman-α resonance line at 122 nm of intergalactic HI near the quasar was redshifted to 365 nm, where it can be observed from the ground. Soon a whole "forest" of intergalactic Ly-α lines were discovered, with many implications for cosmology. Intergalactic HI is there after all, but it is clumped into clouds to which my experiment was insensitive. In any event, the amount of HI is not sufficient to contribute significantly to the mass of the universe.

One of the inputs I needed to calculate the excitation of the 21-cm line was the background radiation temperature T_R at 21-cm wavelength. Doc Ewen told me that an absolute measurement of T_R was not feasible at that time. I discussed the same issue with Arno Penzias much later, but neither of us, as I recall it, made any connection with Alpher and Herman's (1948) prediction that there should be a background radiation from the big bang with a temperature of about 5 K. Penzias and Robert Wilson later discovered the 3-degree microwave background, which had such an impact on cosmology.

I then returned to teach at Princeton and worked on problems of the interstellar medium, cosmology, and planetary science. Over the years I have written papers on the Moon, Mercury, Jupiter, and comets, the latter stimulated by the collision of the Shoemaker-Levy cometary fragments with Jupiter in 1993. I found that the literature on hypersonic reentry contains conceptual errors and was able to interpret modern numerical simulations in terms of the growth and saturation of Kelvin-Helmholtz and Rayleigh-Taylor instabilities in the melted fragment material. My paper with Andrea Ferrara on the subject, with its predictions of post impact phenomena, was accepted on 15 July, 1994, the day before the impacts began. The planets present a fascinating variety of solvable physical problems, which provide a welcome respite from the relatively intractable problems presented by the interstellar medium.

The Future

World astronomy is now building new telescopes at an unprecedented rate. Following the success of the two 10-m Keck telescopes on Mauna Kea, a dozen large (6-to-10-m) telescopes are under construction. Planning is under way for a Next Generation Space Telescope, to be launched in 2007, and the Hubble Space Telescope will continue operations until 2010. In 1999, NASA will launch the Advanced X-Ray Astrophysics Facility, capable of sub-arcsecond imaging in the 1-to-10-keV band. Two satellites will measure the cosmic microwave background with sufficient precision to determine most cosmological parameters. LIGO, a ground-based laser inferometer, will search for gravitational waves emitted by collisions of black holes and neutron stars. Neutrino experiments will test whether neutrino oscillations are responsible for the famous deficit of solar neutrinos. Additional observing facilities at millimeter, submillimeter, and infrared wavelengths will come on-line.

It is astonishing that citizens of the world are willing to pay for these instruments, which have little prospect of improving their material lives. Something else is at work here, probably widespread curiosity about the natural world in which we find ourselves. While supernova explosions are far away, and galaxies forming in the early universe remoter still, taxi drivers and bartenders pause when they hear that until one of those supernova explosions occurred, most of the material on earth did not exist. Astrophysicists are trying to figure out how it all happened, and to do so with the best physics they can command.

References

Alpher, R. A., H. Bethe, and G. Gamow, 1948, Phys. Rev. **73**, 803.
Alpher, R. A. and R. C. Herman, 1948, Nature (London) **162**, 774.
Bahcall, J. N., 1964, Phys. Rev. Lett. **12**, 300.
Bahcall, J. N., R. Davis, P. Parker, A. Smirnow, and R. Ulrich, 1994, *Solar Neutrinos: The First Thirty Years* (Addison-Wesley, Reading, MA).
Bethe, H. A., 1939, Phys. Rev. **55**, 434.
Burbidge, E. M., G. R. Burbidge, W. A. Fowler, and F. Hoyle, 1957, Rev. Mod. Phys. **29**, 547.
Chandrasekhar, S., 1957, *An Introduction to the Study of Stellar Structure* (Dover, New York).
Eddington, A. S., 1924, Mon. Not. R. Astron. Soc. **84**, 308.
Eddington, A. S., 1926, *The Internal Constitution of the Stars* (Cambridge University, Cambridge).
Ewen, H. I., and E. M. Purcell, 1951, Nature (London) **168**, 356.
Field, G. B., 1965, Astrophys. J. **142**, 531.

Giacconi, R., H. Gursky, F. R. Paolini, and B. B. Rossi, 1962, Phys. Rev. Lett. **9**, 442.
Hertzsprung, E., 1911, Publ. Astrophys. Obs. Potsdam **22**, No. 63.
Hewish, A., S. J. Bell, J. D. H. Pilkington, P. F. Scott, and R. A. Collins, 1968, Nature (London) **217**, 709.
Hubble, E. P., 1929, Proc. Natl. Acad. Sci. USA **15**, 168.
Hulse, R. A., and J. H. Taylor, 1975, Astrophys. J. **195**, L51.
IAU, 1997, International Astronomical Union Circular No. 6576, March 6, 1997.
Klebesadel, R., I. B. Strong, and R. A. Olson, 1973, Astrophys. J., Lett. Ed. **182**, L85.
Oppenheimer, J. R., and G. M. Volkoff, 1939, Phys. Rev. **55**, 374.
Oppenheimer, J. R., and H. Snyder, 1939, Phys. Rev. **56**, 455.
Penzias, A. A., and R. W. Wilson, 1965, Astrophys. J. **142**, 419.
Russell, H. N., 1914, Popular Astron. **22**, 275.
Salpeter, E. E., 1952, Astrophys. J. **115**, 326.
Sandage, A. R., and M. Schwarzschild, 1952, Astrophys. J. **116**, 463.
Schmidt, M., 1963, Nature (London) **197**, 1040.
Schwarzschild, K., 1916, Sitzungsber. K. Preuss. Akad. Wiss. Berlin **1**, 189.
Seyfert, C. K., 1943, Astrophys. J. **97**, 28.
Shapley, H., 1918, Astrophys. J. **48**, 154; 1919, **49**, 311.
Spitzer, Jr., L., and M. P. Savedoff, 1950, Astrophys. J. **111**, 593.
Strömgren, B., 1939, Astrophys. J. **89**, 526.
van de Hulst, H. C., 1945, Ned. Tijdschr. Natuurkd. **11**, 210.
Zwicky, F., 1937, Astrophys. J. **86**, 217.

A Century of Relativity
Irwin I. Shapiro

Except for quantum mechanics—a more than modest exception—relativity has been the most profound conceptual advance in 20th century physics. Both in developing special and general relativity, Albert Einstein's hallmark was to anchor his theory on a few simple but profound principles. The results have provided endless fascination and puzzlement to the general public, and have had an enormous impact on our conceptual framework for understanding nature.

In this brief review, I note the rise and spread of special and general relativity throughout physics and astrophysics. This account is quasihistorical, first treating special and then general relativity. In each case, I consider theory, experiment, and applications separately, although in many respects this separation is definitely not "clean." Responding to the request of the editors of this volume, I have included my personal research in matters relativistic. As a result, the recent is emphasized over the remote, with the coverage of the recent being rather slanted towards my involvement.

Special Relativity

The roots of special relativity were formed in the 19th century; we pick up the story near the beginning of this century.

Theory

Hendrik Lorentz regarded his 1904 set of transformations among space and time variables—the (homogeneous) Lorentz transformation—as a mathematical device; he believed in the ether and also in the inability to observe any effects from the motion of light with respect to the ether, which he attributed to dynamical effects caused by motion through the ether.

Henri Poincaré adopted the notion that no motion with respect to the ether was detectable. He also suggested in 1902 that the ether is a hypothesis that might someday be discarded as pointless; in fact, he gave a physical interpretation of "frame time," in terms of the synchronization with light signals of clocks at rest in that frame, as distinct from ether-frame time. Poincaré did not, however, develop a comprehensive theory that postulated a new interpretation of space and time, and he did not discard the concept of an ether. Those tasks were left to Einstein.

The state of Einstein's knowledge of these issues, both theoretical and experimental, and the thinking that undergirded his development of his special theory of relativity in 1905, remain elusive; even historians have failed to reach a consensus. It is nonetheless clear that he was thinking seriously about these issues as early as 1899. He based his new kinematics of moving bodies on two now well-known principles: (1) the laws of nature are the same in all frames moving with constant velocity with respect to one another; and (2) the speed of light in vacuum is a constant, independent of the motion of the light source. He used these postulates to define simultaneity for these (nonaccelerating) frames in a consistent way and to derive transformation equations identical to Lorentz's, but following from quite different underlying reasoning. Einstein also derived a new composition law for the "addition" of relative velocities and from it new formulas for the Doppler effect and for aberration.

Poincaré in 1905 showed that the transformation equations formed a group and named it the Lorentz group.[1] The extension of this group to the inhomogeneous group, which included spatial and temporal translations as well, is now known as the Poincaré group.

Also in 1905, Einstein concluded that the inertial mass is proportional to the energy content for all bodies and deduced perhaps the most famous equation in all of science: $E=mc^2$. Although this type of relation had been proposed somewhat earlier for a specific case, Einstein was apparently the first to assert its universality.

Experiment

Special relativity has among its roots the famous Michelson-Morley experiment.[2] This experiment, based on clever use of optical interferometry, found no evidence, at the few percent level, for the effect expected were the Earth moving through a ("stationary") ether. The round-trip average—both group and phase—speed of light in vacuum has been demonstrated in many experiments this past century to be independent of direction and of the motion of the source. In addition, just recently, analysis of the radio signals from the Global Positioning System (GPS) satellites—all of whose clocks were, in effect, governed by a single atomic standard—yielded a verification of the independence of direction of the one-way speed of light, at the level of about 3 parts in 10^9.

The first experimental tests of special relativity verified the velocity-momentum relation for electrons produced in beta decay. During the 1909–1919 decade a sequence of experiments resulted in verification reaching the 1% level.[3]

The time dilation effect for moving clocks is a major prediction of special relativity. Its experimental verification had to await the discovery of unstable elementary particles, e.g., mesons, whose measured lifetimes when in motion could be compared to the corresponding measurements with the particles at rest (or nearly so). First, in the late 1930s this predicted effect of special relativity was used by Bruno Rossi and his colleagues to infer

---

[1] He did not mention Einstein's paper and may not yet have been aware of it; in any case, he seems never to have referred in print to Einstein's work on special relativity.

[2] Although the extent to which this experiment influenced Einstein's development of special relativity is not clear, it is clear that he knew of its existence: A paper by Wien, mentioned by Einstein in an early letter to Mileva Maric, referred to this experiment, allowing one to conclude with high reliability that Einstein was aware of it. In any event, it was definitely a major factor early on in the acceptance of special relativity by the physics community (John Stachel, private communication).

[3] A comprehensive review of these experiments is given in Walter Gerlach's 1933 Handbuch article (volume 20/1).

the at-rest lifetime of mesons from cosmic-ray observations, following a 1938 suggestion by Homi Bhabha.

Another effect—the so-called "twin paradox"—gave rise to a huge literature over a period of over two decades, before the "opponents," like old generals, just faded away: If twin member B leaves twin member A, who is in an inertial frame, and moves along another world line and returns to rest at the location of A, B will have aged less than A in the interim. Such an effect has been demonstrated experimentally to modest accuracy: the predicted difference in clock readings of a clock flown around the world from one remaining at "home," matched the observed difference to within the approximately 1% standard error of the comparison.

Another of the many verifications, and one of the most important, was of the equivalence of mass and energy. A quantitative check was first made in 1932 via a nuclear reaction by John Cockcroft and Ernest Walton.

Applications

After the invention of quantum mechanics, the need to make it consistent with special relativity led Paul Dirac to create the relativistic wave equation for the electron in 1928. This equation eventually led Dirac to propose that its negative-energy solutions describe a particle with the same mass as the electron, but with opposite charge. The discovery of the positron shortly thereafter in 1932 ranks as one of the major discoveries in 20th century physics. Dirac's equation was soon incorporated into the developing formulation of quantum field theory.

Before and after Dirac's work on the relativistic wave equation, relativistic treatments and their refinements were developed for a wide variety of domains such as classical electrodynamics, thermodynamics, and statistical mechanics, while Newtonian gravitation was replaced by an entirely new relativistic theory: general relativity.

On the experimental side, special relativity has also left indelible marks, as witnessed by its important application in the design of high-energy particle accelerators. The equivalence of mass and energy, coupled with developments in nuclear physics, formed the basis for the solution of the previously perplexing problem of the generation of energy by stars. This work reached an apex with Hans Bethe's development and detailed analysis of the carbon-nitrogen cycle of nuclear burning.

A striking contribution of special relativity to the flowering of astrophysics in the 1970s was discovered serendipitously: "superluminal" expansion. My group and I used very-long-baseline (radio) interferometry (VLBI) in October 1970 to observe two powerful extragalactic radio sources, 3C279 ($z \approx 0.5$)[4] and 3C273 ($z \approx 0.2$), to measure the deflection of light by solar gravity (see below). To our surprise, we noticed that the time variation of the 3C279 fringe pattern with the diurnally changing resolution of our two-element, cross-continental interferometer, matched very well that for a model of two equally bright point sources.

Comparison observations taken four months later, in February 1971, showed an even more dramatic result: these two bright pointlike sources had moved apart at an apparent speed of about 10 c. I developed a simple model of this behavior that showed that if a radio-bright "jet" were ejected from a radio-visible "core" within a few degrees of our

---

[4]The redshift z is the fractional increase in the observed wavelength of an electromagnetic signal emitted from an object moving away from the observer.

line of sight at nearly the speed of light, the speed of separation of the two sources on the plane of the sky could match that observed (the derivation is simplicity itself and depends, in essence, only on the speed of light being a constant, independent of the motion of the source). We later became aware of a related analysis having been published in 1969 in the then Soviet Union by Leonid Ozernoy and Vladimir Sasanov, and of Martin Rees' even earlier (1966) corresponding analysis for a uniformly radiating, relativistically expanding spherical shell. After this discovery of superluminal motion,[5] of which there had been earlier hints, many other radio sources were discovered that exhibited similar behavior, albeit with core and jet components having brightnesses different from one another and exhibiting discernible fine structure.

General Relativity

Theory

The action-at-a-distance implicit in Newton's theory of gravitation is inconsistent with special relativity. Einstein therefore set out to develop a successor theory of gravitation that would not suffer from this defect. He began this development no later than 1907. As a heuristic guide he used one main principle, the principle of equivalence, which states that the direct effect of mass ("gravitation") was indistinguishable from uniform acceleration, except for tidal effects: inside an "Einstein elevator" the behavior of nature is the same, whether the (small) elevator is at rest in a gravitational field or is uniformly accelerating in a field-free region. Another guide was the principle of general covariance: The form of the field equations for a new theory would be invariant under general (space-time) coordinate transformations. However, this principle waxed and waned as an influence on Einstein's development but ended up consistent with his final 1915 form of the theory.[6]

There have been impressive advances in developing solutions to the field equations of general relativity. The first, still the staple, was the 1916 Schwarzschild—exterior and interior—solution for a spherically symmetric mass distribution, followed soon by several others such as the Reissner-Nordstrom solution for a spherically symmetric charge distribution. For the next several decades mostly approximate, perturbative solutions were developed. For example, Lorentz and Johannes Droste in 1917 and Einstein, Leopold Infeld, and Banesh Hoffmann in 1938 developed expansions to solve the dynamical equations of motion for a collection of mass points.[7] In the solar system, perturbations accurate to the post-Newtonian level are still quite adequate (see below) for comparison with the most exquisitely accurate interplanetary measurements present technology allows, e.g., fractional standard errors of a part in 10^{10} and occasionally smaller for

---

[5]During this discovery period, Roger Blandford dubbed this phenomenon "superluminal" motion; the appellation immediately took hold within the astronomical community.

[6]Until 1997, it had been generally accepted that David Hilbert had submitted a paper containing a form of the field equations, essentially equivalent to Einstein's, several days before Einstein had submitted his in final form. However, the proofs of Hilbert's paper survive and show in his handwriting, that Hilbert made essential changes to his originally submitted paper that, with other information, substantiate Einstein's primacy in the development of general relativity.

[7]These equations flow directly from the field equations due primarily to the inherent conservation identities; in Isaac Newton's theory of gravity, by contrast, the equations of motion and those for the gravitational potential follow from separate assumptions.

measurements of echo time delays and angular positions, the former by radar and radio transponders and the latter by VLBI. However, technology is poised to allow much more accurate measurements in the next decade so that, at least in the solar system, some post-post-Newtonian effects should be detectable.

Progress in obtaining approximate solutions to the field equations has been dramatic in the last decade due to the development of useful asymptotic expansions and clever numerical techniques coupled with the availability of ever more powerful computers, including, especially, parallel processors. Spurred by the possibility of detecting gravitational waves, physicists have been applying these tools to the complicated analyses of collisions between black holes, and similar catastrophic events, with prime attention being given to accompanying bursts of gravitational radiation (see below). The reliability of these results will remain open to some question, at least until checked wholly independently.

The general relativistic effects of the rotation of massive bodies were first studied in 1918 by Lense and Thirring who noted that the rotation of a central mass would cause the orbit of a test particle to precess about the spin vector of that central mass, an effect dubbed "frame dragging." This rotation would cause the spin vector of a test gyroscope to precess similarly. A major advance in exact solutions encompassed this central-body rotation, but was not discovered until the early 1960s, by Roy Kerr: the "Kerr metric." It pertains to a rotating axially symmetric mass distribution.

Also in the 1960s and continuing in the 1970s, Roger Penrose, Stephen Hawking, George Ellis, and others developed new mathematical techniques to study global properties of space-time, based on the field equations of general relativity. Singularity theorems were developed that described the conditions for "naked" singularities, i.e., those not shielded by a horizon. Although such singularities exist mathematically, such as for the Schwarzschild solution with negative mass, many physicists, especially Penrose, believe that in nature singularities would always be shielded. Speculations by John Wheeler, Kip Thorne, and others roamed widely and included discussions of "worm holes" which might connect our Universe to others and, perhaps, allow time travel.

The early treatments of gravitational radiation, including the original one by Einstein, were based on the linearized field equations. It was not until the 1960s that Hermann Bondi, Ray Sachs, and others carried out a rigorous treatment far from the source, establishing that gravitational waves follow from the full, nonlinear, theory of general relativity.

The vexing problem of "unifying" the classical theory of general relativity, which stands apart from the rest of fundamental physics, with quantum mechanics remains unsolved, despite enormous effort by extraordinarily talented theorists. This unification remains a holy grail of theoretical physics. The infinities that plagued quantum electrodynamics are not removable by the renormalization techniques that worked so well for the spin-1 photon; they are not applicable to the spin-2 graviton. However, the development of string theory has led many to believe that its unavoidable incorporation of both quantum mechanics and general relativity is a synthesis that will solve the problem of unifying gravitation and the other three known types of interactions (Schwartz and Seiberg, this volume). Unfortunately, tests of predictions unique to string theory or to the newer "M theory" are far beyond the grasp of present experimental virtuosity.

A forced marriage of general relativity with quantum mechanics was begun in midcentury. Rather than a unification of the two, quantum-mechanical reasoning was applied on the four-dimensional space-time (Riemannian) background of general relativity, somewhat akin to grafting the former theory onto the latter—a semiclassical approach. The first dramatic result of this development was Hawking's argument in the context of this "grafted" model, that vacuum fluctuations would lead to black-body radiation just outside the horizon of a black hole and thence to its evaporation. Jacob Bekenstein's pioneering work, and the later work of others, yielded the corresponding theory of the thermodynamics of black holes, with the temperature of a black hole being inversely proportional to its mass. Thus the evaporation rate would be greater the smaller the mass, and the lifetime correspondingly shorter. For the last stages of evaporation, Hawking predicted a flash of high-energy gamma rays. As yet, no gamma-ray burst has been observed to have the properties predicted for the end stage of black-hole evaporation. Other thermodynamic properties of black holes were also adduced, for example, the entropy of a black hole being proportional to its (proper) surface area. None of these beautiful theoretical results is yet near being testable.

Experiment

Principle of equivalence

The principle of equivalence in its weak form—the indistinguishability of gravitational from inertial mass—is a profound statement of nature, of interest at least since the 5th century and demonstrated by Newton in the 17th century to hold to a fractional accuracy of about 1 part in 10^2, via observations of the moons of Jupiter and measurements of pendulums made from different materials. At the beginning of this century, using a torsion balance, Baron von Eötvös in Hungary balanced the effect of the rotational acceleration of the Earth and its gravitational effect (and the Sun's), and established the principle of equivalence for a variety of materials to a fractional accuracy of about 1 part in 10^8; this great achievement—given the technology of that time—was published in exquisite detail in 1922, some years after Eötvös' death. Robert Dicke and his group, in the late 1950s and early 1960s used essentially the same approach as Eötvös, but based on a half century more of technology development. Their results, also in agreement with the (weak) principle of equivalence, had an estimated standard error in the fractional difference between the predicted and observed values for aluminum versus gold of "a few parts in 10^{11}." In 1972 Vladimir Brazinsky and Vladimir Panov stated about a tenfold better result from a similar torsion-balance experiment, with the materials being aluminum and platinum.

With these laboratory tests of the principle of equivalence, including the recent and more accurate ones of Eric Adelberger and his colleagues, the equivalence of gravitational and inertial mass has been established for comparisons of a large number of materials. We infer from these null results that the various forms of binding energy, specifically those due to electrical and strong nuclear interactions, contribute equally to gravitational and inertial mass. However, a comparable test of the binding energy associated with the weak nuclear interaction and, especially, with the gravitational interaction is beyond the grasp of these experiments. The latter is more than 10 orders of magnitude too small for a useful such test to be made with a laboratory-sized body. Planetary-sized bodies are needed, since the effect scales approximately with the square of the linear dimension of the objects whose binding energies are to be compared. But a two-body system is ineffective, unless there is an independent means to determine the bodies' masses; otherwise a violation of the principle of equivalence could not be distinguished from a rescaling of the relative masses of the two bodies. A three-body system can yield an unambiguous result and a detailed proposal for such an experiment was made by Kenneth Nordtvedt in 1968. The placement, starting in 1969, of corner reflectors on the Moon by the Apollo astronauts provided the targets for a suitable three-body system: the Sun-Earth-Moon system. Lunar laser ranging (LLR) from the Earth to these corner reflectors initially yielded echo delays of the laser signals with about 10 nsec standard errors (i.e., about 4 parts in 10^9 of the round-trip signal delays). For proper interpretation, such accuracies required the development of elaborate models of the translational and rotational motions of the Moon, far more critical here than for the interpretation of radar data (see below). By the mid-1970s, sufficient and sufficiently accurate data had been accumulated to make a useful test.[8] With the further accumulation of LLR data, more stringent results were obtained; the latest shows the principle of equivalence to be satisfied to about 1 part in 10^3. Continued decreases in this standard error will require additional modeling, such as representing the reflecting properties of the Moon as a function of aspect to properly account for solar radiation pressure, which is now only about an order of magnitude away from relevancy in this context.

Redshift of spectral lines

It is often claimed that the predicted redshift of spectral lines, which are generated in a region of higher magnitude of gravitational potential than is present at the detector, is more a test of the principle of equivalence than of general relativity. But it is perforce also a prediction of general relativity. A test of this prediction was first proposed by Einstein in 1907, on his road to developing general relativity, in the context of measuring on Earth the frequencies of spectral lines formed in the Sun's photosphere. The difficulty here is primarily to discriminate between the sought-after gravitational effects and the contributions from the Sun's rotation ("ordinary" Doppler effect) and, especially, from motion-related fluctuations. The most accurate determination, in the late 1950s, was in agreement with prediction to within an estimated five-percent standard deviation.

In the early 1960s, soon after the discovery of the Mössbauer effect, Robert Pound realized that he could utilize this effect to measure the shift of the gamma-ray line from Fe^{57} in the Earth's gravitational field. In a carefully designed and brilliantly executed experiment, Pound and Glen Rebka (and later, in 1965, Pound and Joseph Snyder), used a facility somewhat over 20 m high between the basement and roof of Harvard's Jefferson Physical Laboratory, periodically interchanging the location of source and detector to eliminate certain systematic errors. The Mössbauer effect produces an extremely narrow gamma-ray line, allowing Pound and Snyder to achieve a measurement accuracy of 1% of the predicted effect, redshift and blueshift, despite the minute fractional change in gravitational potential over a vertical distance of about 20 m.

---

[8]The LURE (lunar ranging experiment) team, sponsored by NASA, at first obtained a result at variance with the predictions of the principle of equivalence, finding the trajectory of the Moon "off" by a meter or so, far larger than measurement uncertainties would allow. Independently, my colleagues, Charles Counselman and Robert King, and I had analyzed the same LLR data, which were freely available, with our Planetary Ephemeris Program (see below) and found no violation of the principle of equivalence. We agreed to withhold our results from publication until the LURE team completed a review of its analysis. It turned out that an approximation made in their analysis software was responsible for their nonnull result; once fixed, the LURE team's result was consistent with ours and by agreement both groups submitted papers simultaneously to Physical Review Letters, which published them back-to-back.

In 1976, using hydrogen-maser frequency standards, first developed in Norman Ramsey's laboratory at Harvard, Robert Vessot and his colleagues conducted a suborbital test of this prediction. One hydrogen maser was launched in a rocket and continually compared with two virtually identical masers on the ground; the rocket's apogee was 10 000 km above the Earth's surface. The results of this flight agreed fractionally with predictions to within the 1.4 parts in 10^4 estimated standard error. This accuracy will remain the gold standard at least through the end of this century for this type of experiment.

Deflection of light by solar gravity

The "classical" test of the predicted deflection of light by solar gravity was first carried out successfully in 1919 in expeditions led by Arthur Eddington and Andrew Crommelin. In these observations, the relative positions of stars in a field visible around the Sun during a total eclipse were measured on photographic glass plates and compared with similar measurements made from plates exposed several months later when the Sun was far from the star field. This approach is fraught with systematic errors, especially from the need to accurately determine the plate scale over the relevant area for each plate. Although repeated a number of times during subsequent solar eclipses, no application of this technique, through 1976, the last such attempt, succeeded in lowering the "trademark" standard error below 0.1 of the predicted magnitude of the effect. In 1967, I suggested that this deflection might be measured more accurately using radar interferometry or, more generally, radio interferometry, the former via observations of planets near superior conjunction, and the latter via observations of compact, extragalactic radio sources with the technique of VLBI. The second suggestion bore fruit, with the ground-based standard errors having just recently been reduced to about 1 part in 10^4 by Marshall Eubanks and his colleagues. This result has been achieved through a progression of estimates of almost monotonically decreasing standard errors, from the late 1960s to the present. The next major advance may come from space interferometers, operating at visible wavelengths, and/or from laser signals propagating near the Sun.

Time-delay by gravitational potential

Before describing the time-delay experiment, I present some of the background, primarily the development of radar astronomy and my involvement in considering its potential for testing general relativity.

Dicke resurrected experimental relativity in the 1950s from near total neglect, starting as noted above with his work on the refinement of the Eötvös experiment. I became interested near the end of the 1950s, through the advent of radar astronomy, which was being actively pursued at the MIT Lincoln Laboratory, mostly through the foresight of Jack Harrington, then a Division Head there. Powerful radar systems were being developed to track Soviet intercontinental missiles; the systems might also be capable, he thought, of detecting radar echoes from planets. Because of the inverse fourth-power dependence of the radar echo on the distance to the target, Venus at its closest approach to the Earth provides echoes about 10^7 times weaker in intensity than those from the Moon, despite the approximately twelvefold larger (geometric) cross section of Venus.[9] The detection of Venus by radar at its furthest point from Earth, near superior conjunc-

---

[9]The Moon was first detected by radar from Earth in 1946.

tion, provides echoes weaker by another factor of about 10^3. Mercury at its greatest distance from the Earth provides echoes threefold weaker again, due to its smaller cross section more than offsetting its distance advantage over Venus at superior conjunction. Despite these depressing numbers, it appeared that over the coming years, the diameter of radar antennas and the power of transmitters could be increased substantially while the noise of receivers might be decreased dramatically. Hence before the first radar echoes were reliably obtained from Venus at its inferior conjunction, I began to think about testing general relativity with this new technique. My first thought, in 1959, was to check on the perihelion advance of Mercury (see next subsection); standard errors of order 10 μsec in measurements of round-trip travel-time between the Earth and the inner planets seemed feasible; the fractional errors affecting such data would then be at the parts in 10^8 level, far more accurate than the corresponding optical data—errors of about five parts in 10^6—and of a different type. Despite the long temporal base of the latter, important for accurate measurement of a secular effect such as the orbital perihelion advance, the increased accuracy of individual echo time ("time-delay") measurements would allow about tenfold higher accuracy to be achieved in estimating the perihelion advance after a few decades of radar monitoring. Of course, correspondingly detailed modeling was required to interpret properly the results of these measurements. I therefore decided to abandon the time-honored tradition of using analytic theories of planetary motion, carried to the needed higher level of accuracy (I was influenced by my remembrance of being told in the only astronomy course I had taken as an undergraduate about the more than 500 terms in Brown's analytic theory of the Moon's motion). I decided that a wholly numerical approach would be the way to go. With the group I built for the purpose, we—especially Michael Ash and Menasha Tausner—created a model accurate through post-Newtonian order of the motions of the Moon, planets, and Sun, as well as of many asteroids. Detailed models were also required for the rotational and orbital motion of the Moon that involved the few lowest orders of the spherical-harmonic expansion of its gravitational field as well as the second zonal harmonic of the Earth's field. In addition, (elaborate) modeling of the surfaces of the target inner planets was needed—the then major source of systematic error. In principle, the topography of each inner planet's surface can be substantially reduced as a source of error by making repeated radar observations of the same (subradar) point on the planet, each from a different relative orbital position of the Earth and planet. Such opportunities are, however, relatively rare, and scheduling and other realities prevent a bountiful supply of such observations. Again, in principle, high-resolution topographic and reflectivity mapping of an inner-planet surface is feasible via use of a radar system on a spacecraft orbiting that planet; only Venus has so far been mapped at relevant accuracy and resolution, but the practicalities of applying these results to the ground-based radar problem are formidable. The observables—the round-trip signal propagation times—also need to be modeled accurately; they involve the precession, nutation, rotation, and polar motion of the Earth; the geographic location of the effective point of signal reception; and the propagation medium, primarily the interplanetary plasma and the Earth's ionosphere and troposphere. The needed software codes under the rubric Planetary Ephemeris Program (PEP), rapidly reached over 100 000 lines.

The first successful planetary radar observations, of Venus, determined the astronomical unit—in effect the mean distance of the Earth from the Sun—in terms of the terrestrial distance unit, with about three orders of magnitude higher accuracy than previously known, disclosing in the process that the previous best value deduced solely from optical observations was tenfold less accurate than had been accepted.[10]

Before any improvement in determining perihelia advances could be made, indeed before even the first detection of Mercury by radar, I attended an afternoon of presentations c. 1961–1962 by MIT staff on their progress on various research projects, conducted under joint services (DOD) sponsorship. One was on speed-of-light measurements by George Stroke who mentioned something about the speed depending on the gravitational potential. This remark surprised me and I pursued it via "brushing up" on my knowledge of general relativity and realized the obvious: whereas the speed of light measured locally in an inertial frame will have the same value everywhere, save for measurement errors, the propagation time of light along some path will depend on the gravitational potential along that path. Thus it seemed to me that one might be able to detect this effect by timing radar signals that nearly graze the limb of the Sun on their way to and from an inner planet near superior conjunction. At the time, however, this idea seemed far out; the possibility of detecting radar echoes from Mercury, the nearest planet at superior conjunction, or even Venus, seemed far off.

In 1964 the Arecibo Observatory, with its 305 m-diameter antenna, was then under development and began radar observations of Venus. Unfortunately, the possibility of testing this prediction of general relativity on echo time delay was not feasible to do at Arecibo because the radar transmitted at a frequency of 430 MHz, sufficiently low that the effects on the echo delays of the plasma fluctuations in the solar corona would swamp any general relativistic signal.

That October the new Haystack Observatory at MIT's Lincoln Laboratory was dedicated. At a party, the day after the birth of Steven, my first child, I was telling Stanley Deser about this new radar facility, when I realized it was going to operate at a frequency of 7.8 GHz, high enough so that the coronal effect, which scales approximately as the inverse square of the frequency, would not obscure a general relativistic signal. I then got quite excited and decided to both submit a paper describing this test and "push" for Lincoln Laboratory to undertake the experiment. Given the new—for me—responsibilities of fatherhood, plus the Lincoln review process, the paper was not received by Physical Review Letters until two weeks after the precipitating party. Colleagues at the Laboratory, most notably John Evans and Bob Price, from a detailed analysis of the system parameters, concluded that to do the experiment well we needed about fourfold more transmitter power—a nontrivial need. I went to Bill Radford, then the director of the Laboratory, to plead the case for the more powerful transmitter, pointing out, too, its obvious advantages for the other planned uses of the Haystack radar system. Radford, not knowing how to evaluate my proposed general-relativity experiment, called on Ed Purcell for advice. Purcell said he knew little about general relativity but opined that "Shapiro has a knack for being right." (He was referring, I suspected when the quotation was repeated to me, to my then recent work on the "artificial ionosphere" created by the Project West Ford dipoles, whose orbits, greatly influenced by solar radiation pressure, followed my colleagues' and my predictions extraordinarily well.) In any event, Radford called an Air Force general at the Rome Air Development Center and succeeded in getting a $500 000 budget increase for building this new transmitter and its

---

[10]This relation, in fact known only to about 1 part in 10^3 from optical data at that early 1960s time, was needed more accurately to ease the problem of navigating interplanetary spacecraft. Over the succeeding two decades the radar value increased in accuracy a further three orders of magnitude.

associated microwave plumbing, protective circuits, and other technical intricacies. A nice holiday present for December 1964. A year and a half later, the team of Lincoln Laboratory engineers assigned to this project and led by Mel Stone, completed the new transmitter system; the first radar observations of Mercury under the guidance of Gordon Pettengill and others were made soon thereafter. By early 1968, we had published the first result, a 10% confirmation of the time-delay predictions of general relativity. Controversy both before and after the test centered partly on the observability of the effect (was it simply a coordinate-system mirage?) and partly on the accuracy of my calculations (this latter part lasted for about 30 years).

The experiment, which I labeled the fourth test of general relativity, was refined over the following years, with the standard error reduced to 1 part in 10^3. This accuracy was achieved with essential contributions from Robert Reasenberg and Arthur Zygielbaum in the years 1976–1978 with the four Viking spacecraft that were deployed in orbit around Mars and on its surface.

Until very recently, this accuracy exceeded that from the closely related experiments involving VLBI measurements of the deflection of radio waves; but now the accuracy pendulum has swung decisively toward these latter measurements.

"Anomalous" perihelion advance

The first inkling that Newton's "laws" of motion and gravitation might not be unbreakable came in the mid-nineteenth century with the carefully documented case by Urbain LeVerrier of an anomalous advance in the perihelion of Mercury's orbit, reinforced and refined near the end of that century by Simon Newcomb. Never explained satisfactorily by alternative proposals—is there a planet (Vulcan) or cloud of planetesimals inside the orbit of Mercury; an unexpectedly large solar gravitational oblateness; and/or a slight change in the exponent of Newton's inverse square law?—this advance, as Einstein first showed, followed beautifully and directly from his theory of general relativity. The agreement between observation and theory was remarkably good, to better than one percent, and within the estimated standard error of the observational determination of the anomalous part of this advance at that time. The analysis of the new radar data, alone, now yield an estimate for this advance about tenfold more accurate than that from the several centuries of optical observations.

A main problem has been in the interpretation of the radar—and optical—measurements, in the following sense: How much of the advance could be contributed by the solar gravitational oblateness? Although this secular Newtonian effect falls off more rapidly, by one power of the distance, than does the post-Newtonian general relativistic effect, neither is detectable with sufficient accuracy in the orbit of any planet more distant from the Sun than is Mercury. There are short-term orbital effects that offer a less demanding, but by no means easy, means of discrimination. In any event, the correlation between the estimates of the magnitudes of the relativistic and solar-oblateness contributions to the advance will likely remain high until interplanetary measurement errors are substantially lower. An independent measurement of the oblateness through direct study of the Sun's mass distribution is thus highly desirable. Dicke, who built on earlier ideas of Pascual Jordan, developed a scalar-tensor theory alternative to general relativity, with his (Dicke's) student Carl Brans. This theory had an adjustable parameter, representing, in effect, the relative scalar and tensor admixtures, and could account for a smaller advance. Thus Dicke set out to measure the solar visual oblateness, which could then be used via straightforward classical theory to deduce the gravitational oblateness, i.e., the coefficient, J_2, of the second zonal harmonic of the Sun's gravitational field. In the late 1960s Dicke and Mark Goldenberg using a very clever, but simple, instrument to estimate the Sun's shape, deduced a value for the visual oblateness that would account for 10% of Mercury's anomalous perihelion advance, thus implying that general relativity was not in accord with observation and that the previous precise accord was a coincidence. The Brans-Dicke theory's adjustable parameter could accommodate this result. These solar-oblateness measurements, in the fashion of the field, were scrutinized by experts from various disciplines resulting in many questions about the accuracy of the oblateness determination: extraordinary claims must be buttressed by extraordinary evidence. The net result was a rejection by a large majority of the scientific community of the accuracy claimed by Dicke for Goldenberg's and his solar-oblateness measurements, leaving their claim of a higher-than-expected value for J_2 unsubstantiated.

More recently, a new field—helioseismology—has been developed to probe the mass distribution of the Sun: optical detection of the oscillations of the Sun's photosphere allows the solar interior to be deeply probed. The net result leaves the agreement between observation and general relativity in excellent accord.[11]

Possible variation of the gravitational constant

In 1937, Dirac noticed a curious coincidence, which he dubbed the law of large numbers. It was based on the fact that the ratio of the strengths of the electrical and the gravitational interactions of, say, an electron and a proton—about 10^{39}—was, within an order of magnitude or two, equal to the age of the universe measured in atomic units of time (e.g., light crossing time for an electron). Dirac noted that this near identity could be a mere coincidence of the present age of the universe or could have a deeper meaning. If the latter were true, Dirac reasoned, the relation between gravitational and atomic units should be a function of time to preserve this (near) equality. The most reasonable proposal, he concluded, was to assume that the gravitational constant G decreased with (atomic) time. This proposal lay dormant for several decades. In the 1950s calculations were made of the brightness history of the Sun and its effects on the Earth that might be discernible in the geologic record. These deductions were quite controversial because, for example, of their reliance on (uncertain) aspects of stellar evolution and the difficulty in separating atomic from gravitational effects in determining such a brightness history. More modern calculations of the same type are similarly afflicted. In 1964, in thinking of other possibilities for radar tests of general relativity, I considered the obvious check on any change in G through monitoring the evolution of planetary orbits with atomic time t. Expanding $G(t)$ about the present epoch t_0, I sought evidence for $\dot{G}_0 \neq 0$. Were $\dot{G}_0 < 0$ as would follow from Dirac's hypothesis, then the orbits of the planets would appear to spiral out, an effect most noticeable in the (relative) longitudes of the planets. The main limitation of this test at present is the systematic error due to incomplete modeling of the effects of asteroids.

---

[11]The analysis of the Sun's pressure modes, both from the ground-based network of observatories and the space-based, Solar Heliospheric Observatory (SOHO), allows a rather robust estimation: $J_2 = (2.3 \pm 0.1) \times 10^{-7}$, of more than adequate accuracy for the interpretation of the solar-system data. Because of the high correlation between the orbital effects of the solar gravitational quadrupole moment and of the postNewtonian terms in the equations of motion, such an accurate independent determination of J_2 allows "full" use of the solar-system data for checking on the relativistic contributions to the orbital motion.

Nonetheless, the radar data are able to constrain any fractional change in G (i.e., constrain $\dot{G}_0/G_0$) to be under a few parts in 10^{12} per year. A similar level of accuracy has been achieved with the LLR data; here the main source of systematic error is probably the modeling of the tidal interaction between the Earth and the Moon as it affects the spiraling out of the orbit of the latter. There have been publications of similar accuracies based on the analysis of the pulse timing data from a binary neutron-star system (see below). This bound, however, is of the self-consistency type in that this effect is very highly correlated with the main orbital effect of gravitational radiation: the two are inseparable; the results for one must be assumed to be correct to test for the other, save for self-consistency. The solar-system tests are free from such a fundamental correlation, but are limited in accuracy for the near future by other correlations, at about the level of the bound already achieved.

Frame dragging

One quantitative test of the Lens-Thirring effect has just been published: an apparent verification of the prediction that the orbital plane of a satellite will be "dragged" (precess) around the spinning central body in the direction of rotation of that body. Specifically, in 1998, Ignazio Ciufolini and his colleagues analyzed laser-ranging data for two nearly spherical Earth satellites, Lageos I and II, each with a very low area-to-mass ratio. These authors concluded that the precession agreed with the predictions to 10%, well within their estimated standard error of 20%. There are, however, an awesome number of potentially obscuring effects, such as from ocean tides, that are not yet well enough known to be reasonably certain of the significance of this test. With continued future gathering and analysis of satellite-tracking data from the increasing number of satellites that are designed, at least in part, to improve knowledge of both the static and time-varying contributions to the gravitational potential of the Earth, this Lens-Thirring test will doubtless improve.

A definitive quantitative verification of the effect of frame dragging on orbiting gyroscopes is promised by the Stanford-NASA experiment. This experiment was developed, based on Leonard Schiff's original (1959) suggestion, by William Fairbank, Francis Everitt, and others at Stanford, starting in the early 1960s. The experiment will involve a "drag-free" satellite containing four extraordinarily spherical quartz "golf ball"-sized gyroscopes, coated with niobium, cryogenically cooled, and spun up with their direction of spin monitored by "reading" the gyroscope's London moment with superconducting quantum interference devices. The directions of the London moments will be compared to that of a guide star whose proper motion with respect to a (quasi) inertial frame is being determined to sufficient accuracy for this purpose by my group via VLBI, following a suggestion I made to the Stanford team in the mid-1970s. For the orbit and the guide star chosen, the predicted gyroscope precession is about 40"/yr, with the anticipated standard error being 0."2/yr. This Stanford-NASA experiment will also measure the so-called geodetic precession with at least two orders of magnitude smaller standard error than the 2% value we obtained a decade ago from analysis of the LLR data. This truly magnificent physics experiment is now scheduled for launch in the year 2000.

Gravitational radiation

For about the last third of this century, physicists have addressed with great experimental and theoretical virtuosity the problem of detecting gravitational radiation. The pioneer experimenter, Joseph Weber, developed the first cylindrical bar detectors; his claim in the early 1970s to have detected gravitational waves from the center of our Galaxy, despite being wrong, awakened great interest, resulting in a relentless pursuit of this holy grail of experimental gravitational physics. Very significant human and financial resources have been expended in this hunt, which doubtless will eventually be successful and will also provide profound insights into astrophysical processes. But not this century.

The now-classic neutron-star binary system discovered in 1974 by Russell Hulse and Joseph Taylor has exhibited the orbital decay expected from gravitational quadrupole radiation for these objects, which are in a (noncircular) orbit with a period of just eight hours. This decay is a striking confirmation of the general relativistic prediction of gravitational radiation, with the observed changes in orbital phase due to this decrease in period matching predictions to within about one percent.[12] The sensitivity of these measurements to this decay increases approximately with the five-halves power of the time base over which such measurements extend, since the effect of the radiation on orbital phase grows quadratically with that time base, and the effect of the random noise drops as the square root of that base, given that the measurements are spaced approximately uniformly. However, systematic errors now limit the achievable accuracy to about the present level, the chief villain being the uncertainty in the Galactic acceleration of the binary system, which mimics in part the effect on pulse arrival times of the orbital decay.[13]

The larger universe is the hoped-for source of gravitational waves which will be sought by the laser interferometer gravitational-wave observatory (LIGO) and its counterparts in Germany, Italy, and Japan—all currently in various stages of planning and construction. The two LIGO sites, one each in the states of Louisiana and Washington, will each have two 4-km-long evacuated tubes, perpendicular to each other, and forming an "L"; a test mass is at each far end and at the intersection of the two arms. LIGO will be sensitive to gravitational waves with frequencies $f \geq 30$ Hz. The first generation of laser detectors should be sensitive to the difference in strains in the two arms at the fractional level of about one part in 10^{21}. No one expects gravitational waves that would cause strains at this sensitivity level or greater to pass our way while detectors of some orders of magnitude greater sensitivity are being developed for deployment on LIGO and on the other instruments. However, Nature often fools us, especially in the variety and characteristics of the macroscopic objects in the universe. So I personally would not be totally shocked were this first generation of laser interferometer detectors to pick up bona fide signals of gravitational waves. As a counterpoise, note that some of the best gravitation theorists have worked for several decades to conjure and analyze scenarios that might lead to

---

[12] It is often argued that this detection of gravitational waves is "indirect" because we detect only the consequences of the radiation in the orbital behavior. However, one could argue as well that a similar criticism applies to any detection since the presence of the waves must be inferred from observations of something else (e.g., the vibrations of a massive bar or the oscillatory changes in distance between suspended masses as measured by laser interferometers). The key difference is whether we infer the properties of the radiation from its effects on the sources or on the detectors; in the latter cases, of course, the experimenter-observers have much greater control.

[13] The arrival-time data are also rich enough to measure with reasonable accuracy other predicted relativistic effects and to determine the masses of the neutron-star components of the binary; the mass of each is about 1.41 solar masses, in splendid accord with the Chandrasekhar limit (see below).

Applications

General relativity was at first of interest only in a small subfield of physics—aside from the profound impression it made on the psyche of the general public. Still irrelevant for applications in everyday terrestrial life and science,[14] general relativity now provides a key tool in the armamentarium of theoretical—and observational—astrophysicists. It is employed to tackle problems from the largest to the smallest macroscopic scales encountered in our studies of the universe.

Cosmology

The first and perhaps still the most important application of general relativity is to cosmology. Einstein, thinking that the universe was static, found a corresponding cosmological solution. Since the universe is nonempty, Einstein had to first tamper with his field equations by introducing on the "geometry side," a term with a constant coefficient—the so-called cosmological constant—whose value was not specified by the theory, but which would provide the large-scale repulsion needed to keep a nonempty universe static. Later, after Alexandre Friedmann and Georges Lemaitre exhibited expanding-universe solutions and Hubble presented the first evidence of expansion, Einstein reputedly called his introduction of the cosmological-constant term "the greatest scientific blunder of my life."[15] This term did not, however, fade away forever, but was resurrected recently when exploration of the implications of vacuum fluctuation energy on a cosmological scale uncovered the so-called cosmological-constant paradox: for some modern theories the (nonzero) value is about 120 orders of magnitude larger than the upper bound from the observational evidence.

The 1920s provided the main thrust of the program in cosmology for the rest of the century: Under the assumption of a homogeneous isotropic universe, cosmologists attempt to measure the Hubble constant H_0 and the deceleration parameter q_0. Values of these two parameters would provide, respectively, a measure of the size scale (and, hence the age) of the universe and a determination of whether the universe is open, closed, or on the border (and, hence, whether the average mass density of the universe is below, above, or at the "closure" density).

In the 1930s, the estimate of H_0 was quite high, about 500 km s^{-1} Mpc^{-1}, implying an age for the universe of only a few billion years. Even before radioactive dating techniques were able to disclose that the age of the Earth was about 4.5 billion years old, astronomers discovered a serious problem with their method of inferring H_0 from the distances of "standard candles,"[16] leading, after further revisions, to the conclusion that H_0 was severalfold smaller and the universe correspondingly older. Over the following decades,

---

[14] Except insofar as the Newtonian limit serves us admirably.
[15] There is, however, no known written evidence supporting this (apocryphal?) quotation (John Stachel, private communication, 1998).
[16] The main candles used were Cepheid variable stars. There were two principal problems: recognition only later that there were two classes of such stars with different period-luminosity relations and mistaken identification in the most distant indicators of, e.g., unresolved star clusters for a single ("most luminous") star.

there was no appreciable improvement in accuracy; however, for the past several decades, there has been a schism among the practitioners: those such as Allan Sandage claiming H_0 to be about 50 km s^{-1} Mpc^{-1} and those such as Gerard DeVaucouleurs proclaiming a value of 100 km s^{-1} Mpc^{-1}, each with estimated uncertainty of the order of 10%. The methods they used depend on the accurate calibration of many steps in the so-called-cosmic distance ladder, making it difficult to obtain reliable estimates of the overall errors. With some exceptions, more modern values have tended to cluster between 65 and 75 km s^{-1} Mpc^{-1}, still a distressingly large spread. Also, new methods have joined the fray, one depending directly on general relativity: gravitational lensing, discussed below.

The pursuit of q_0 has until recently led to no result of useful accuracy. Now a wide variety of techniques has indicated that the universe does not have sufficient mass to stop its expansion, and hence is "open." Most recently, two large independent groups have obtained the tantalizing result from observations of distant ($z \approx 1$) type 1a supernovae that the universe is not only open but its expansion is accelerating. This result is now at the "two sigma" level; if confirmed by further data and analysis, it will have a profound effect on theory: Is the cosmological constant nonzero after all and if so, how does one reconcile that result with current quantum-field-theory models? Or, for example, are vacuum fluctuations causing some strange locally weak, but globally strong, repulsion that produces this acceleration? These problems, doubtless, will not be fully resolved until the next millenium.

Black holes

Beyond the structure and evolution of the universe on large scales, probably the most profound effect of general relativity on astrophysics in the past century has been through the prediction of black holes. The name was coined by John Wheeler in the early 1960s, but the concept was, in effect, conceived over two centuries earlier by the Reverend John Michell who reasoned, based on Newton's corpuscular theory of light, that light could not escape an object that had the density of the Sun but a diameter 500 times larger. Early in the 1930s, based on a quantum-mechanical analysis, Lev Landau predicted that so-called neutron stars could exist and Subramanian Chandrasekhar showed that the mass of such a collapsed stellar object could not exceed about 1.4 solar masses—the now famous Chandrasekhar limit whose existence was vehemently, and unreasonably, opposed by Eddington.

In 1938 Einstein analyzed his "thought" analog of a collapsing stellar object and concluded that a black hole would not form. However, he did not carry out a dynamical calculation, but treated the object as a collection of particles moving in ever-smaller circular orbits; he deduced that the speed of these particles would reach the velocity of light barrier before reaching the Schwarzschild radius, and thereby drew an incorrect conclusion.

Soon thereafter, in 1939, J. Robert Oppenheimer and Hartland Snyder made a major advance in understanding gravitational collapse. They showed that in principle, according to general relativity, black holes could be produced from a sufficiently massive stellar object that collapsed after consuming its full complement of nuclear energy. Basing their analysis on the Schwarzschild metric, and thus neglecting rotation and any other departure from spherical symmetry, they deduced correctly that with the mass of the star remaining sufficiently large—greater than about one solar mass (their value)—this collapse would continue indefinitely: the radius of the star would approach its gravitational radius asymptotically, as seen by a distant observer. This discussion was apparently the first (correct) description of an event horizon. Oppenheimer and Snyder specifically contrasted the possibly very short collapse time that would be seen by a comoving observer, with the corresponding infinite time for the collapse that would be measured by a distant observer. They also described correctly, within the context of general relativity, the confinement of electromagnetic radiation from the star to narrower and narrower cones about the surface normal, as the collapse proceeds.

In many ways establishing the theoretical existence of black holes, within the framework of general relativity, was easier and less controversial than establishing their existence in the universe. However, after well over a decade of controversy and weakly supported claims, there is now a widespread consensus that astronomers have indeed developed persuasive evidence for the existence of black holes. As in most astronomic taxonomy there are two classes: the stellar-mass black holes and the $10^6 - 10^9$ times larger mass black holes. Evidence for the former consists of estimates for binary star systems of the mass, or of a lower bound on the mass, of a presumably collapsed member of each such system. In these systems Doppler measurements allow the determination of the so-called mass function, which sets a lower bound on the mass of this invisible, and likely collapsed, member of the binary. (A point estimate of this mass cannot be determined directly because of the unknown inclination of the orbit of the binary with respect to the line of sight from Earth.) These observations show in the "best" case that the black-hole candidate has a mass greater than eight solar masses, far in excess of the Chandrasekhar limit and more than twice the largest conceivable nonblack-hole collapsed object that quantum mechanics and a maximally "stiff" equation of state seem to allow.

The evidence for large ("supermassive") black holes became almost overwhelming just a few years ago from the partly serendipitous study, via combined radio spectroscopy and VLBI, of the center of the galaxy NGC 4258, i.e., the 4 258th entry of the New General (optical sky) Catalog, which stems from the early part of this century. This 1995 study yielded strong kinematic evidence for material in Keplerian orbits. In turn, these orbits implied average mass densities interior to these orbits, of at least 10^9 solar masses per cubic parsec, a density so high that no configuration of mass consistent with current understanding could be responsible other than a supermassive black hole. There is also growing evidence, albeit not yet as convincing, for the presence at the center of our own galaxy of a black hole of mass of the order of 3×10^6 solar masses.

Another relevant and impressive result from astrophysics relating to predictions from general relativity concerns "evidence for"—the phrase of choice when astrophysical results are under scrutiny—an event horizon. As shown in 1997 by Michael Garcia, Jeff McClintock, and Ramesh Narayan, the luminosity of x-ray emissions from a sample of neutron stars and candidate black holes shows a tendency to separate into two clusters, with the luminosity of the neutron stars larger than those for black holes, as would be expected as radiating material "blinks off" as it approaches the event horizon.

Gravitational lenses

The idea that mass could, like glass lenses, produce images was apparently first articulated in print in 1919 by Oliver Lodge, but not pursued in any systematic way, either theoretically or experimentally, for nearly two decades. Then in 1936, at the urging of a

Czech engineer, Einstein analyzed such lensing,[17] demonstrating that in the case of collinearity of a source, a (point-mass) lens, and an observer, the image seen by the observer would appear as a circle—now known as the Einstein ring.[18] Its radius depends directly on the mass of the lens and on a function of the relevant lengths. For an asymmetric geometry the ring breaks into two images, one formed inside and one outside the corresponding Einstein ring. Einstein dismissed the possibility of observing this phenomenon—a ring or double image in the case of noncollinear geometry—based on two arguments, neither supported in the paper by calculations: (1) the probability of a chance alignment was negligible; and (2) the light from the lens star would "drown out" the light from the distant lensed star, despite the magnification of this latter light. Fritz Zwicky, an astronomer-physicist who often had insights 50 years ahead of his time, was quick to point out, in a paper submitted barely six weeks later, that whereas Einstein's conclusions about observability might well be correct for stars in the Milky Way, the (more distant) nebulae—now called "galaxies"—offer far greater prospects for observability; two months later still, he noted that the existence of lensing with multiple images was virtually a certainty and pointed out the basic importance of such imaging to cosmology.

Despite Zwicky's upbeat conclusions, the field then lay fallow for nearly thirty more years, until, independently, in 1964, Steven Liebes in the United States and Sjur Refsdal in Norway, published analyses of gravitational lensing, the former focusing more on image shapes and characteristics, the latter more on cosmological uses, most importantly the prospects for determining a value for the Hubble constant. This determination, being based on very distant sources, would be independent of the rungs of the conventional cosmic distance ladder. After these articles by Leibes and Refsdal appeared, the theoretical astrophysics literature on gravitational lenses started to mushroom with a number of papers pointing out the consequences of lensing on various statistical questions, such as the effect of magnifications in distorting "unbiased" samples of galaxy luminosities. Not until March 1979, however, within about two weeks of the 100th anniversary of Einstein's birth, did Dennis Walsh in England serendipitously discover the first gravitational lens. He had been trying to find optical identifications for sources discovered in a low-angular-resolution radio survey; two optical objects about six arcseconds apart on the sky were candidates for one such radio source. Following up with telescopes in the southwest United States, Walsh and his colleagues measured the spectra from the two objects and noticed a striking similarity between them, aside from the then-puzzling difference in mean slope in the infrared, later identified as due to the main lensing galaxy, which was not separately visible from those observations. Walsh and his colleagues thus took the courageous step of claiming, correctly, that their two objects were in fact images of the same quasar, since each had the same redshift ($z \approx 1.41$, indicating a very distant object) and nearly the same optical spectrum. Further lens discoveries of multiple images of a single object were somewhat slow in coming, but the pace has quickened. A number of rings, sets of multiple images, and arcs were discovered in radio, optical, and infrared images of the sky, and astrophysical applications have been tumbling out at an awe-

---

[17]Recently, however, Jürgen Renn, Tilman Sauer, and Stachel examined Einstein's notebooks from 1912, prior to the completion of general relativity; these showed that he had developed all of the resultant formulas at that earlier time, although the values for the deflections were half those of his completed theory.

[18]The formula for the ring was apparently first published in 1924 by Otto Chowlson.

inspiring rate. Statistical analyses of results from observations of (faint) arcs from very distant objects have even allowed Christopher Kochanek to place a (model-dependent) bound on the cosmological constant.

The search for the value of the Hubble constant also took a new tack, along with the old ones, based on Refsdal's noting that multiple images of a distant light source produced by a point-mass lens could be used to infer the distance to the source. The idea is elegantly simple: if the light source were to vary in its intensity, then these variations would be seen by an Earth observer to arrive at different times in the different images. Such a difference in the time of arrival of a feature in the light curve in two images is proportional to the light travel time from source to observer and, hence, to H_0^{-1}. This difference is also proportional to the mass of the lens. And therein lies the rub: independent estimates of this mass or, more accurately, of the mass distribution, are difficult to come by, either from other properties of the images, such as their optical shapes and the locations of their centroids, or from other types of astronomical measurements.[19] Thus this general method of estimating H_0 has been notoriously difficult to apply, both because of the difficulty with the time-delay measurement and, especially, because of the difficulty in determining the lens' mass distribution with useful accuracy. As we near the millenium only about a few dozen gravitational lenses have been confirmed—a subtle process in itself—and only three have yielded reasonably reliable time delays. The first gravitational lens system discovered has led to the estimate $H_0 = 65 \pm 10 \text{ km s}^{-1} \text{ Mpc}^{-1}$; however, this standard error does not account fully for possibly large model errors.

Perhaps the most spectacular results obtained so far followed from a suggestion in the late 1980s by Bohdan Paczyinski to make wide-field observations with modest-sized optical telescopes of about a million stars simultaneously and repeatedly. The purpose was to detect, with charge-coupled devices and modern computers, color-independent brightening and subsequent dimming among members of this star collection, such variations being the hallmark of an intervening (dark) lens passing by on the sky. (One can forgive Einstein for not envisioning this multipronged development of technology.) The durations of such "events" can vary from minutes to months. The duration and brightening factors depend on the mass of the (invisible) lens and the geometry of the lens system, unknowns that most often preclude a useful point estimate being made of this mass. For long duration events, the parallax afforded by the Earth's orbital motion allows an estimate to be made of the lens' distance; for all events, long or short, such determinations could be obtained were a telescope in orbit far from the Earth observing the same parts of the sky simultaneously (were two or more such spacecraft employed, these observations would be freed from ground-based weather, but at some less-than-modest cost). Attempts to observe such "microlensing" effects were proposed initially to detect invisible mass ("dark matter") in our Galaxy that could be in the form of compact objects, so-called MACHOs: massive compact halo objects. This monitoring project, started in the early 1990s, is being carried out by three independent collaborations and has been remarkably successful: over 100 events have so far been detected. The results show, for example, that the "dark matter" problem cannot be solved by MACHOs alone.

---

[19] These latter are needed in any event because of a fundamental degeneracy noted in 1985 by Marc Gorenstein, Emilio Falco, and myself. From measurements of the images alone, one cannot distinguish between the actual mass distribution and a different (scaled) one in which a uniform surface density "sheet" is also present, with the light source being correspondingly smaller, yielding the same image sizes. A separate type of measurement is needed to break this degeneracy.

Future

Predicting the future is easy; predicting it accurately for the next century is a tad more difficult. One can at the least anticipate that advances in experimental and theoretical gravitation will yield profound and unexpected insights into astrophysical phenomena, especially those that are invisible in electromagnetic-wave observations. One could even conceive of black holes being "tamed" to extract energy from infalling matter. More likely, the issues and the problems offering the greatest challenges and rewards over the next century are not now conceivable or at least not yet conceived.

Acknowledgments

I thank George Field, Kurt Gottfried, and John Stachel for their extensive and very helpful comments, and Stephen Brush, Thibault Damour, Stanley Deser, Gerald Holton, Martin Rees, Stuart Shapiro, Saul Teukolsky, and Robert Vessot, for their useful suggestions.

From Radar to Nuclear Magnetic Resonance
Robert V. Pound

As war engulfed Europe from September 1939, and the Axis powers overran most of the western European countries in 1940, the United States undertook to build up quickly its military capabilities. By late 1940, a full year before the U.S. was drawn into war as a combatant, there began a massive move of scientists, especially physicists, temporarily into new organizations set up to develop technologically advanced weapons. One of the first of these, destined to become one of the largest, was the "Radiation Laboratory," so named with an intent to obscure its purpose, at the Massachusetts Institute of Technology, established in November of 1940 by the National Defense Research Committee (NDRC). Its mission was to develop radio detection and ranging (to become known as Radar), inspired by the startling performance of the pulsed cavity magnetron revealed to U.S. military officers and members of the National Defense Research Committee by the British "Tizard Mission." Although magnetrons as generators of electromagnetic energy of short wavelength had been developed in several places many years earlier, the new magnetron was a breakthrough in that it could produce microwave pulses many orders of magnitude more intense than could anything else then in existence. It was an ideal device for the development of radar. The visit from the beleaguered British scientists and high military officers, headed by Sir Henry Tizard, had entered into this exchange of secret new military technology to try to obtain technical help and manufacturing support from US industry, relatively insulated from air attack. The cavity magnetron has been described, in view of its developed use in the war, as possibly the most important cargo ever to cross the Atlantic, although many other secret developments were included in the exchange. For example, the British progress toward releasing nuclear energy was also revealed.

Physicists, and other scientists, were recruited in very great numbers into the new Radiation Laboratory and other emergency organizations, resulting in a nearly complete shutdown of basic research. The scale is apparent from the evidence that the *Physical Review* published a single issue of only 54 pages for the whole month of January, 1946, its publishing nadir. That issue contained only two regular articles, both of which were reports of researches completed before the shutdown, but which saw print only after the war had ended. A large part of that issue was the "Letters" section that reported some early postwar activity, including our first report (Purcell, Torrey, and Pound, 1946) of the successful detection of nuclear magnetic resonance through its absorption of radio-frequency energy. An early step in the postwar rebirth of research in basic physics was thereby announced. For scale it is worth noting that in the corresponding month of

January 1996, the *Physical Review* published nearly 8000 pages in its several parts.

Although the Radiation Laboratory was eminently successful in its technical goals and is widely recognized as having contributed importantly to the Allied victory, there were several less generally recognized effects on the postwar world of physics. Many physicists not previously greatly concerned with electronic instrumentation developed a much deeper understanding especially of the fundamental limits to instrumental sensitivity set by various sources of interfering "noise." Beyond atmospheric noise, often called "static," wider recognition of shot effect noise, from the electronic granularity of electrical currents, and thermal noise, explained by the equipartition theorem of statistical mechanics, were more widely understood. It was demonstrated quite early that the most sensitive device for the initial signal detection in a microwave (wavelength 10 cm and less) receiver was a developed version of the archaic "crystal diode" detector. This consisted of a semiconductor and a metal "cat's whisker" and had been superceded by thermionic vacuum tubes for serious radio uses more than twenty years before. As a result, an extensive program of research and development of semiconductors, principally silicon and germanium, was underwritten by the NDRC. The succeeding development of solid-state electronics in the postwar era is deeply rooted in the importance attached to finding and making reliable the best detectors for microwave radar in the war years.

Perhaps as important as any direct technical progress was a long-lasting effect of the personal interactions produced by this temporary relocation to one large organization of a large fraction of the active research physicists who had already achieved important results in widely diverse areas at many institutions. In the two decades between the world wars there had been much progress in such fields as nuclear physics, where accelerators, including Van de Graaf's belt electrostatic generator and cyclotrons, as devised by E. O. Lawrence, were coming into use in several institutions. The study of cosmic rays was the focus of several groups and, of course, the discovery of neutron-induced fission of ^{235}U, just as W.W. II was beginning, is well known. The elegant rf resonance techniques in atomic and molecular beams developed at Columbia University in the group headed by I. I. Rabi for the study of nuclear spins and moments (Rabi *et al.*, 1938; Kellog *et al.*, 1939) had not spread significantly to groups at other institutions. An analogous technique had been developed by Felix Bloch and Luis Alvarez in California to determine the magnetic moment of the neutron (Alvarez and Bloch, 1940). Rabi was an early participant in the Radiation Laboratory and became an influential Associate Director and head of the Research Division for most of the five years of life of the Laboratory. Many of the members of his former group at Columbia also joined the MIT Laboratory and others at both Columbia and the Bell Laboratories were in close communication as they pursued projects in aid of those at MIT. A consequence was a much wider recognition of the achievements of the Columbia group among the main body of physicists, especially highlighted by the award of the Nobel Prize for Physics for 1944 to I. I. Rabi at a special ceremony in New York.

When the war finally ended on August 14, 1945, there began an exodus of scientists to return to their old institutions or to take up new civilian positions. However, many of the active members were asked to remain to contribute to technical books describing the advances made in secret during those five years. There was created the Office of Publications, headed by physicist Louis Ridenour, which kept many of us at MIT until June 30, 1946, to contribute to the resulting Radiation Laboratory Series of 28 volumes. Understandably our concentration on the writing projects was often diluted with thoughts about new possibilities for research that would grow from our special experiences. Henry C.

Torrey, one early veteran of the Rabi team at Columbia, shared an office with me during this period. One day in September we invited Edward M. Purcell, who had headed the Fundamental Developments Group, a part of the Division headed by Rabi, to join us for lunch. As we walked from MIT to Central Square, Ed asked Henry what he would think about the possibility of detecting resonant absorption of radio-frequency energy by nuclei in condensed materials by their magnetic dipole moments, if their states of spin orientation were split by a strong applied magnetic field. Purcell has indicated that his thoughts were led in this direction by his close association with Rabi and other alumni of the Columbia group. In addition, he was writing up the discovery and the explanation of the absorption of 1.25-cm microwaves by atmospheric water vapor, a discovery that had decreased the enthusiasm for intensive development of radar systems operating on that wavelength by his group. The H_2O molecule was found to possess two energy levels, among a very large number, with an energy difference just matching the quantum energy, $h\nu$, of 1.25-cm microwaves.

Torrey's initial response at our luncheon was actually pessimistic but, after giving the question some more quantitative attention at home that evening, he convinced himself that it should be possible. When he so informed Ed the next morning, Purcell immediately proposed that we three undertake experimentally to detect such an absorption. Thus we began a spare-time project that helped lighten the dullness of the writing assignments which remained as our committed full-time employment.

Our five-year immersion in microwave technology led us to design and construct a cavity resonator to contain the test sample in a strong magnetic field. We could plan on a field only large enough to bring the resonant frequency of protons to 30 MHz, or ten meters' wavelength, hardly to be described as a microwave. I include as Fig. 1 a photograph of the coaxial cavity, cut open to reveal its innards, and now held by the Smithsonian Museum. As an open-ended coaxial resonator it would be a quarter wavelength long, 2.5 meters, but a disk insulated from the lid by a thin sheet of mica as a loading capacitance shortened it to about 10 cm. The space below the disk that should contain circumferential rf magnetic flux at the cavity resonance was filled with about two pounds of paraffin wax, chosen because of its high concentration of hydrogen and its negligible dielectric loss. Purcell's search for a suitable magnet at MIT was unrewarded, but Ed was offered by our colleague J. Curry Street the use of a magnet at Harvard he had built in the mid 1930s for bending the tracks of cosmic-ray particles in his cloud chamber. It was with this magnet that Street had measured the mass of the cosmic-ray muon (Street and Stevenson, 1937). He had joined the Radiation Laboratory at its beginning, and the magnet had been collecting cobwebs for five years. Thus our project was moved to Harvard and was carried out mostly in evenings and on weekends. The basic concept was a balanced bridge, with the cavity resonator in one arm, excited with a 30-MHz signal generator. Its transmitted signal was nearly balanced out by adjustment of the phase and amplitude sent to a common junction through a parallel arm. In this way we were able to look for very small changes in the signal transmitted through the cavity when the magnetic field was adjusted through the magnitude that should result in magnetic resonance of the protons in the cavity arm. The net signal from the bridge was amplified in a low-noise 30-MHz preamplifier, borrowed from the Radiation Laboratory, where it had been developed for the intermediate frequency amplifiers of radar receivers. The amplified signal fed into a 30-MHz communication receiver, also borrowed from my laboratory at MIT, for further amplification and was then detected and observed on a micro-ammeter as an output meter. The magnetic-field strength required for resonance was calculated from the measurements of the Rabi group of the proton magnetic moment, as about 7 kilogauss. We had added new pole caps to Street's magnet and used a flip coil and ballistic galvanometer to calibrate the field vs current, as adjusted by a rheostat in our laboratory that controlled the field current of the remote *dc* generator.

Attempts to see the absorption as a deflection on the micro-ammeter failed during a frustrating lengthy Thursday night effort and again for most of the following Saturday afternoon. However, a final run, intended to be a preliminary part of a discouraged shutdown for further thought, was made taking the magnet to the highest current available from the generator. This amounted to some 40% more than had been called for in our calibration. As we lowered the current slowly, suddenly, at about 15% above our calibration value, there occurred just such a meter deflection as we had anticipated seeing at the proton resonant field. The "signal" was clearly caused by the dip in cavity transmission we sought. We almost failed to achieve our goal because we had assumed our calibration was not in error by more than 10%. In fact, it proved to be in error by only 2%, but we had failed to appreciate that our calibration data showed that our scan over plus and minus 10% in current covered less than a 2% excursion in field strength. That resulted from the iron core of the magnet severely saturating at these currents. So, happily, the project succeeded on Saturday afternoon, December 15, 1945, as a pencilled note in the hand of Henry Torrey in my small, now rather tattered, notebook testifies (Fig. 2).

In the course of our preparation for the trial we had become concerned about whether the two-level proton spin system would come to thermal equilibrium in a reasonable time, as required to obtain the magnetic polarization, or reach an energy state population difference, needed to have absorption. Without such a difference, induced emission and

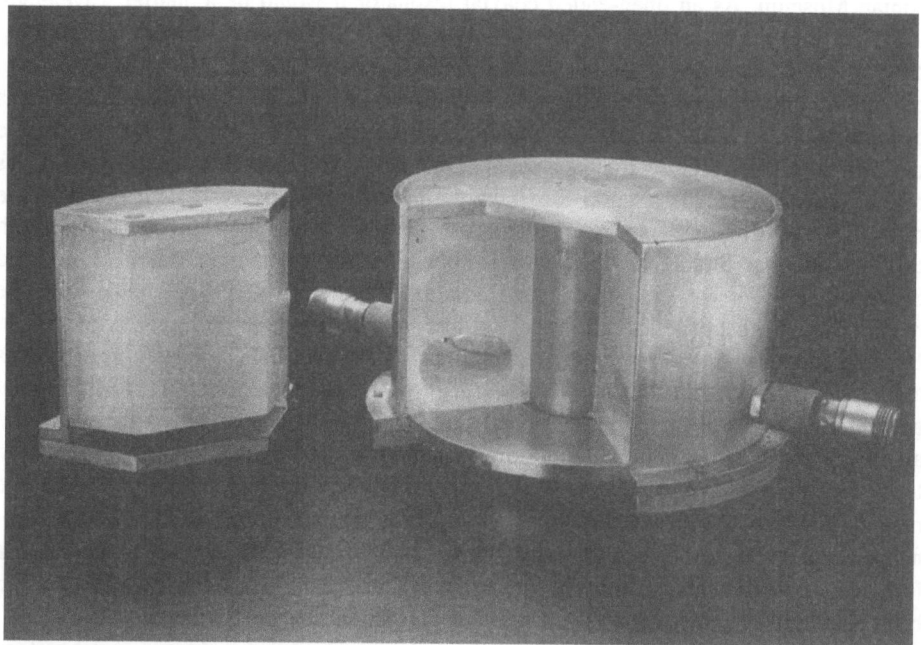

Figure 1. The 30-MHz resonant cavity filled with paraffin as a proton sample. It is held at the Smithsonian Museum and has been cut open to reveal its inner structure. (See Color Plate 1.)

absorption would exactly balance and no signal would be produced. Torrey had estimated a relaxation time of some hours by adapting to apply to nuclei a theory of I. Waller (1932) developed for electronic spins in crystalline materials. We had estimated that our large sample and weak rf field would allow some hours of observation of the resonant absorption without seriously overriding the thermal population differential, once that equilibrium had been established. We feared that our initial failures might be a sign that the relaxation time was too long, but were greatly relieved when we found that those failures were so easily explained. Some quick experiments showed that the thermal relaxation time of the protons in paraffin wax was shorter than the shortest time, perhaps fifteen seconds, required to bring the magnet current up from zero. This was explained in the course of the ensuing researches in this new field, as conducted at Harvard, when Nicolaas Bloembergen, a graduate student newly arrived from the Netherlands, joined Purcell and me in the enterprise. We carried out the work that formed Bloembergen's thesis at Leiden, which became widely known as BPP (Bloembergen, Purcell, and Pound, 1948), and which explains the spin-lattice relaxation and averaging out of line-broadening interactions by internal motions and fluctuations especially dominant in liquids. Paraffin wax

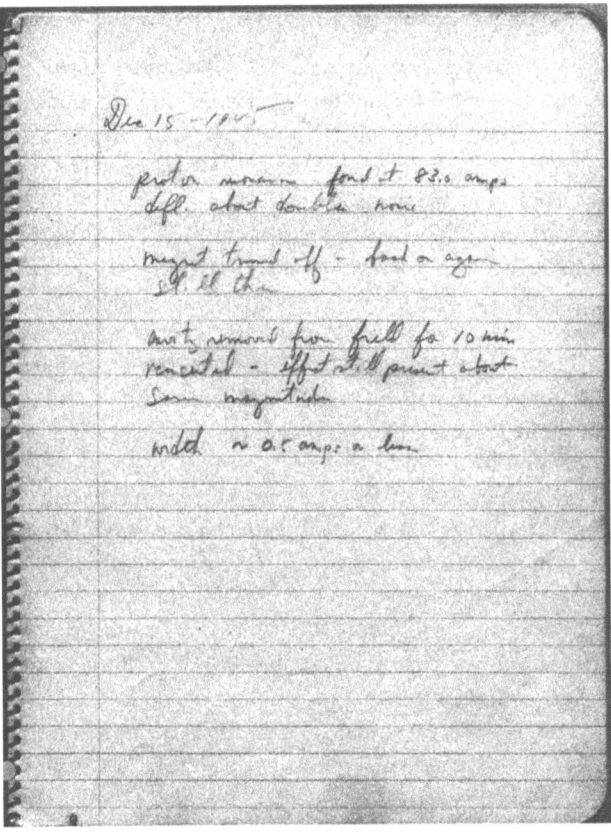

Figure 2. The note, in the hand of H. C. Torrey, that records our successful detection of proton NMR absorption. It appears on a yellowing page of the small spiral bound notebook that served to record a miscellany of notes during the years 1942 to 1948.

clearly possesses considerable internal molecular motions and has a spin-lattice relaxation time on the order of milliseconds.

In the course of our initial experiment we learned that C. J. Gorter in Holland had reported failed efforts to detect NMR, first calorimetrically in 1936 (Gorter, 1936), and then by electronic frequency pulling resulting from the dispersion that should accompany absorption (Gorter and Broer, 1942). Had we been aware of these reports at the beginning we might have been discouraged from undertaking the project, but access to the wartime Dutch publications in 1945 was limited because of the only recently ended German occupation. We also heard, near the culmination of our project, that Felix Bloch and W. W. Hansen were making a similar effort at Stanford University. Hansen had served as a consultant on microwaves to the MIT Radiation Laboratory from its beginning, giving weekly lectures, traveling up from his wartime base at the Sperry Gyroscope Company on Long Island, New York. Bloch spent a couple of years at the radar countermeasures laboratory at Harvard. Both had, however, returned to Stanford and had begun their nuclear resonance project early in 1945. They reported observing the resonance of protons by a technique they named "nuclear induction" in a short letter (Bloch, Hansen, and Packard, 1946) submitted to the editor about six weeks after ours. Their approach sensed the precession of the macroscopic nuclear polarization rather than absorption. The relationship between these different ways of observing the same basic resonance was only properly understood sometime later. Not all two-level systems that can be observed in absorption actually involve macroscopic precession, however. For example, absorption at energy state differences created by nuclear electric quadrupole interactions with crystalline electric fields does not involve precession. Studies of "nuclear electric quadrupole resonance" in solids have much in common with NMR for nuclei of spin greater than 1/2, which includes a majority of the nuclear isotopes.

In the beginning the main motivation for pursuing NMR had been as a way to study spins and moments of nuclear isotopes. Although that aspect occupied the attention of many researchers, properties of the resonance provide evidence of the effects of the environment of the nuclear spins in materials. Because of the motional narrowing, as explained by BPP, results in resonance lines for hydrogen often fractions of a Hertz in width, very small chemical shifts, and complex resonance line structures have turned out to provide a tremendous source of new information on molecular and crystalline structure. NMR has become a powerful tool in chemical analysis, materials science, and, recently, as the basis of MRI (magnetic resonance imaging) in medicine. Through linewidth studies, relaxation-time observations, and small frequency shifts of chemical and condensed-matter origin, NMR has come to provide a window into the workings of many materials, including the human body.

An extension of ideas originating in NMR led to a realization that the effects that give rise to line structures and to spin-lattice equilibrium play a role in a seemingly unconnected field, the study of the correlation in direction of the emission of successive radiations from excited nuclei (Abragam and Pound, 1950). In addition to magnetic interactions, NMR allowed observation of the effects of electric fields on nuclei through the nuclear electric quadrupole moments of nuclei of spin one or more (Pound, 1950). Thus, if intermediate states in nuclear decay are not extremely short lived, those interactions cause reorientations before a succeeding decay, resulting in a departure from the expected directional correlation. Study of these disturbances, especially as a function of time after the initial decay, has yielded a way to determine properties of the nuclear moments and of the environment of short-lived isotopes. That work goes under the acronym TDPAC for time-delayed perturbed angular correlations.

Another phenomenon relating to NMR came from the discovery of recoil-free resonance of gamma rays by nuclei bound in crystals, as reported by Rudolf Mössbauer (1958a, 1958b) for ^{191}Ir in iridium metal. In 1959 the technique was extended to a 14-keV gamma ray from an isotope of iron, ^{57}Fe, that follows the decay of 270-day ^{57}Co to iron (Pound and Rebka, 1959; Schiffer and Marshall, 1959) in a stable ground state. In this case the resonance phenomenon was truly "nuclear," in that gamma rays were emitted and absorbed by nuclei. The binding of the nuclei to crystalline sites in a lattice turned out to reduce to negligible importance the broadening from the Doppler effect from thermal vibration, because the component of the atomic velocity along the direction of emission or absorption, averaged over the life of the excited nuclear state involved, vanishes for the atom bound to a lattice site. In this scheme gamma rays emitted from a crystalline source are sent through a crystalline absorber containing bound nuclei in the ground state to which the radiation leads. The transmission is observed as a function of applied relative motion, which Doppler shifts the gamma-ray frequency. These resonances turn out to be fractionally so narrow that hyperfine structures are resolved and evaluated. Even the minute effect of gravity on the relative frequency of gamma rays of nuclei held at elevations differing by only a few meters, an energy shift of only a part in 10^{15}, has been demonstrated with a precision of one percent (Pound and Rebka, 1960; Pound and Snyder, 1965). With the introduction especially of ^{57}Fe, gamma-ray resonance spread into many areas of physics, chemistry, and even biology, providing yet another window on the detailed inner secrets of materials.

References

Abragam, A., and R. V. Pound, 1953, Phys. Rev. **92**, 943.
Alvarez, L., and F. Bloch, 1940, Phys. Rev. **57**, 111.
Bloch, F., W. W. Hansen, and Martin Packard, 1946, Phys. Rev. **69**, 127.
Bloembergen, N., E. M. Purcell, and R. V. Pound, 1948, Phys. Rev. **73**, 679.
Gorter, C., 1936, Physica (Utrecht) **3**, 995.
Gorter, C. J., and L. J. F. Broer, 1942, Physica (Utrecht) **9**, 591.
Kellog, J. M. B., I. I. Rabi, N. F. Ramsey, Jr., and J. R. Zacharias, 1939, Phys. Rev. **56**, 728.
Mössbauer, R. L., 1958a, Naturwissenschaften **45**, 538.
Mössbauer, R. L., 1958b, Z. Phys. **151**, 124.
Mössbauer, R. L., 1959, Z. Naturforsch. A **14A**, 211.
Pound, R. V., 1950, Phys. Rev. **79**, 685.
Pound, R. V., and G. A. Rebka, Jr., 1959, Phys. Rev. Lett. **3**, 554.
Pound, R. V., and G. A. Rebka, Jr., 1960, Phys. Rev. Lett. **4**, 337.
Pound, R. V., and J. L. Snider, 1965, Phys. Rev. **140**, 788B.
Purcell, E. M., H. C. Torrey, and R. V. Pound, 1946, Phys. Rev. **69**, 37.
Rabi, I. I., J. R. Zacharias, S. Millman, and P. Kusch, 1938, Phys. Rev. **53**, 318.
Schiffer, J. P., and W. Marshall, 1959, Phys. Rev. Lett. **3**, 556.
Street, J. C., and E. C. Stevenson, 1937, Phys. Rev. **52**, 1003.
Waller, I., 1932, Z. Phys. **79**, 370.

An Essay on Condensed Matter Physics in the Twentieth Century
W. Kohn

When the 20th century opened, the fields of crystallography, metallurgy, elasticity, magnetism, etc., dealing with diverse aspects of solid matter, were largely autonomous areas of science. Only in the 1940s were these and other fields consolidated into the newly named discipline of "solid state physics" which, two decades later, was enlarged to include the study of the physical properties of liquids and given the name "condensed matter physics" (CMP). At Harvard, for example, J. H. Van Vleck had several times taught a graduate course on magnetism in the 1930s and 1940s. However, the first time a course called "solid state physics" was taught there was in 1949 (by the writer, at Van Vleck's suggestion); it was based largely on the influential, comprehensive monograph *Modern Theory of Solids* by F. Seitz, which had appeared in 1940. In the early 1950s only a handful of universities offered general courses on solids and only a small number of industrial and government laboratories, most notably Bell Telephone Laboratories, conducted broad programs of research in CMP.

Today condensed matter physics is by far the largest subfield of physics. The writer estimates that at least a third of all American physicists identify themselves with CMP and with the closely related field of materials science. A look at the 1998 Bulletin of the March Meeting of the American Physical Society shows about 4500 papers in these fields.

Over the course of this century condensed matter physics has had a spectacular evolution, often by revolutionary steps, in three intertwined respects: new experimental discoveries and techniques of measurement; control of the compositions and atomic configurations of materials; and new theoretical concepts and techniques. To give a brief and readable account of this evolution is immensely difficult due to CMP's extraordinary diversity and many interconnections. Nevertheless, in the following pages the reader will find one theorist's broad-brush—and necessarily very incomplete—attempt at this task. The writer (not a historian of science) had to make many difficult, often rather arbitrary, choices: how to organize this very short essay on a very broad subject and—most painful—what to include and what important material to omit. He begs the reader's indulgence.

The Last Years of the Classical Era

The very first Nobel Prize was awarded in 1901 to W. C. Roentgen for the discovery of penetrating, so-called x rays. A few years later in 1912, M. von Laue and collaborators demonstrated that these rays were electromagnetic waves of very short wavelengths, which could be diffracted by the atoms of crystals. This discovery dramatically proved beyond a doubt the reality of atomic lattices underlying crystalline solids and at the same time yielded quantitative geometric information about the relative positions of the atoms in solids. It constituted the beginning of CMP on a microscopic scale of 10^{-8} cm.

Building on von Laue's work, M. Born and co-workers (in the 1910s) developed a simple, classical, partially predictive theory of the cohesive energy of alkali halide crystals. Their chemical composition was known to be of the form A^+B^- (where A = alkali and B = halogen), and the geometric arrangements were known from x-ray experiments. The theory postulated the existence of pairwise interactions consisting of the known long-range Coulomb interactions between the charged ions and short-range repulsions between nearest neighbors, phenomenologically characterized by two parameters: strength and range.

When the two parameters were fitted to the known lattice parameters and elastic bulk moduli, the calculated cohesive energies were in quantitative agreement with experiment on the ~3% level.

Another success based on von Laue's demonstration of the existence of atomic crystal lattices was Born's theory of classical lattice vibrations (1910s). It was based on a Hamiltonian of the form

$$H = \sum_l \frac{M_l}{2} \dot{u}_l^2 + \Phi(u_1, u_2, \ldots, u_N), \tag{2.1}$$

where M_l and u_l are the masses and displacements of the atoms, labeled by l, the first sum is the kinetic energy, and Φ is the potential energy, which was expanded up to second order in u, making use of periodicity and other symmetries. Together with convenient periodic boundary conditions, this led to propagating normal modes of vibration with wave vectors q and frequencies $\omega_j(q)$, where j is an additional label. (For monatomic crystals $j = 1,2,3$, corresponding to the three so-called acoustic modes; for polyatomic crystals there are also so-called optical modes.) This theory successfully unified the theory of the static elasticity and of long-wavelength sound waves in crystals, and further yielded, in terms of the expansion coefficients of Φ, the normal-mode frequencies $\omega_j(q)$ for arbitrarily large wave vectors q, which were not directly observed until half a century later by inelastic neutron scattering.

Attempts by P. Drude and H. A. Lorentz in the first decade of the century to understand the salient properties of metals in classical terms could not avoid major inconsistencies and had only very limited success. A crucial feature was the (correct) postulate that in a metal some atomic electrons are not attached to specific atoms but roam throughout the entire system. Their scattering by the atomic nuclei was regarded as the cause of electrical resistance. However, this theory could not explain why the resistance of metals generally dropped towards zero linearly as a function of the temperature, or why the expected substantial classical heat capacity, $\frac{3}{2}k$ for each free electron, was never observed.

On the question of ferromagnetism, the observed dependence of the magnetization density M on the applied field H and temperature T could be explained in terms of classical statistical mechanics by assuming a phenomenological effective field acting on an atomic dipole, given by $H_{\text{eff}} = H + \alpha M$, where $\alpha \approx 0 \ (10^3 - 10^4)$. The form of the so-called Weiss field αM (1907), could be roughly understood as due to classical dipole interactions, but the required high magnitude of α was some three orders of magnitude larger than the classical value, of order unity.

The experimental achievement of lower and lower temperatures, culminating in the liquefaction of He at 4.2 K by K. Onnes in 1908, dramatically brought to light the insufficiency of classical concepts. Thus while the classical law of Dulong and Petit, which assigned a heat capacity of $3k$ to each atom in a solid, was generally in rather good agreement with experiment at sufficiently high temperatures (typically room temperature or above), all measured heat capacities were found to approach zero as the temperature was reduced towards zero. This was recognized as a critical failure of the concepts of classical statistical mechanics.

The dramatic discovery by Onnes in 1911 of superconductivity, a strictly vanishing resistivity below a critical temperature of a few degrees K, remained a major puzzle for more than four decades.

Thermionic emission from hot metal surfaces or filaments, an important subject during the infancy of electric light bulbs, was partially understood. The velocity distribution of the emerging electrons was, as expected, the classical Maxwell-Boltzmann distribution, $A \exp(-mv^2/2kT)$, where m is the electron mass and k is Boltzmann's constant, but the magnitude of A was not understood.

The photoelectric effect, the emission of electrons from solid surfaces in response to incident light, was also very puzzling. Light of low frequency ω caused *no* emission of electrons, no matter how high the intensity; however, when ω exceeded a threshold frequency ω_0, electrons were emitted in proportion to the intensity of the incident light.

Thus we see that while the classical theory of CMP at the beginning of this century had some impressive successes, it also had two major, general deficiencies:

(1) When classical theory was successful in providing a satisfactory phenomenological description, it usually had no tools to calculate, even in principle, the system-specific parameters from first principles.
(2) Some phenomena, such as the vanishing of heat capacities at low temperatures and the behavior of the photoemission current as a function of the frequency and intensity of the incident light, could not be understood at all.

Both deficiencies were to be addressed by quantum theory with dramatic success.

Early Impacts of the Quantum of Action on Condensed Matter Physics

As is widely known, Max Planck (1900) ushered in the new century with the introduction of the quantum of action, h, into the theory of blackbody radiation. Much less known is the fact that Einstein received the Nobel Prize not specifically for his work on relativity theory but for "his services to theoretical physics and especially for his discovery of the law of the photoelectric effect." In fact, it was in his considerations of the photoelectric effect in 1905 that Einstein developed the concept of the photon, the quantum of excitation of the radiation field with energy $\hbar\omega$, where $\hbar$ is Planck's constant divided by 2π and ω is the circular frequency. This concept was one of the most important ideas in the early history of quantum theory. It also led naturally to the resolution of the photoelectric effect conundrum: for any particular emitting metal surface the photon must have a surface-specific minimum energy, the so-called work function W, to lift an electron out of the metal into the vacuum. Thus a minimum light frequency is required.

Shortly after this great insight Einstein (1907), not surprisingly, also understood the reason why the lattice heat capacity of a solid approached zero at low temperatures. He modeled each atom as a three-dimensional (3D) harmonic oscillator of frequency $\bar{\omega}$. Again he quantized the excitation energies of each vibrational mode in units of $\hbar\bar{\omega}$, which directly yielded $3\hbar\bar{\omega}/(e^{\hbar\bar{\omega}/kT}-1)$ for the mean energy per mode at temperature T. At high temperatures, $kT \gg \hbar\bar{\omega}$, this yielded the classical result $3k$ for the heat capacity per atom, in agreement with the empirical high-temperature law of Dulong and Petit. But at low temperatures the Einstein heat capacity correctly approached zero.

By choosing an appropriate mean frequency $\bar{\omega}$ for a given solid, one could fit experimental results very well, except at the lowest temperatures, where the experimental lattice heat capacity behaved as T^3, while Einstein's theory gave an exponential behavior. This deficiency was repaired by P. Debye (1912), who quantized Born's lattice modes and realized that at low temperatures T only long-wavelength modes with frequencies $\hbar\omega \lesssim kT$ would be appreciably excited. The number of these modes behaves as T^3 and their typical excitation energy is of the order kT. This immediately gave the empirical T^3 law at low temperatures for the lattice heat capacity. The excitation quanta of the normal modes, characterized by a wave vector q, a frequency ω, and an energy $\hbar\omega$, were called phonons and became an indispensable component of CMP.

In these developments we observe (1) the decisive role played by the quantum of action $\hbar$; (2) the importance (in Debye's work) of long-wavelength/low-energy collective modes; and (3) the mutually fruitful interplay between CMP and other fields of science. (For example, in Einstein's work on the photoelectric effect, with quantum electrodynamics.) These features have marked much of CMP for the rest of the century.

The Quantum-Mechanical Revolution

The advent of quantum mechanics, particularly in the form of the Schrödinger equation (1926), coupled with the discovery of the electron spin and the Pauli exclusion principle (1925), totally transformed CMP, as it did all of chemistry. While the Bohr theory of the hydrogen atom had brilliantly and accurately described this one-electron system, it proved to be quantitatively powerless even in the face of the two-electron systems He and H_2 let alone condensed matter systems consisting of $\sim 10^{23}$ interacting nuclei and electron. The Schrödinger equation changed all this. The ground-state energy of He was soon calculated by E. Hylleraas (1929) with a fractional accuracy of 10^{-4}, the binding energy and internuclear separation of H_2 was calculated first by W. Heitler and F. London (1927), and then by others, with accuracies of about 10^{-2} to 10^{-3}. This left no reasonable doubt that the Schrödinger equation, applied to both electrons and nuclei, *in principle* was the correct theory for CMP systems.

A very useful organizing principle, the Born-Oppenheimer approximation (1927), was soon articulated: because of the small mass ratio of electrons and nuclei, usually $m/M \sim 10^{-5}$, typical electronic time scales in molecules and presumably also in solids were much shorter than those of nuclei, in proportion to $(m/M)^{1/2}$. This led to the conclusion that the dynamics of electrons and nuclei could, to a good approximation, be decoupled.

In the first stage the nuclei are considered fixed in positions $R_1, R_2, \ldots$ and the ground-state electronic energy $E_{el}(R_1, R_2, \ldots)$ is determined. In the second stage the electrons no longer appear explicitly and the dynamics of the nuclei are determined by the sum of their kinetic energy and an effective potential energy given by $E_{el}(R_1, R_2, \ldots) + E_{nuc}(R_1, R_2, \ldots)$, where the last term describes the internuclear Coulomb repulsion.

Since all of condensed matter consists of nuclei and electrons, the field henceforth could, for most purposes, be divided into two parts: one dealing with electron dynamics for fixed nuclear positions (e.g., total energies, magnetism, optical properties, etc.), the other dealing with nuclear dynamics (e.g., lattice vibrations, atomic diffusion, etc.). Important exceptions were phenomena that critically involved the electron-phonon interaction, such as the temperature-dependent part of electrical resistance (F. Bloch, 1930) and, as discovered much later, the phonon-dependent so-called Bardeen-Cooper-Schrieffer (BCS) superconductivity (1957).

Another consequence of the small value of m/M was that, whereas typical electronic energies in solids were of the order of 1–10 eV,[1] those related to the nuclear dynamics were of the order of $10^{-2} - 10^{-1}$ eV. Thus room temperature with $kT \approx 0.025$ eV was generally very cold for electrons but quite warm for nuclear dynamics.

Several of the major failures of classical theory when applied to metals, as described in Sec. II, were soon remedied by the combination of the new quantum mechanics with the Pauli exclusion principle. Of course, a straightforward solution of the Schrödinger equation for $\sim 10^{23}$ strongly interacting electrons was out of the question. But by boldly proposing that, at least roughly, the forces on a given electron due to the other electrons canceled those due to the nuclei, W. Pauli (1927) and, very extensively, A. Sommerfeld (1928) were led to the quantum-mechanical free-electron model of metals: Each electron was described by a plane wave $\varphi_q(r) \equiv \exp(iqr)$ and an up- or down-spin function $\chi_\sigma (\sigma = \pm 1)$. Coupled with the Pauli exclusion principle and the resulting Fermi-Dirac statistics, this model naturally explained the following experimental facts: that in many simple metals, e.g., the alkalis, the magnetic susceptibility due to electronic spins was weak and nearly temperature independent (instead of, classically, large and proportional to T^{-1}); and that the electronic specific heat at low temperatures was small and proportional to T (instead of, classically, 3/2 k per electron and independent of T).

The Pauli-Sommerfeld theory represented major, fundamental progress for metals, but at the same time it left a host of observed phenomena still unexplained: For example, the fact that metallic resistance decreases linearly with temperature and that in some materials the Hall coefficient has the counterintuitive, "wrong" sign. Many of these puzzles were soon greatly clarified by replacing the *uniform* effective potential of the Sommerfeld model by a *periodic* potential reflecting the periodic arrangements of the ions, as will be discussed in the next section. A deeper understanding of the effects of the electron-electron interaction evolved much more slowly.

Another early, spectacular success of the new quantum mechanics was the unexpected explanation by W. Heisenberg (1928) of the "enormous" magnitude of the Weiss effective magnetic field mentioned in the Introduction. Heisenberg realized that the Pauli principle, which prevents two electrons of the same spin from occupying the same state, generates an effective interaction between the spin magnetic moments, quite unrelated to the classical magnetic dipole interaction and typically several orders of magnitude larger.

---

[1] However, for metals, electronic *excitation* energies begin at zero.

The Band-Structure Paradigm

Two very significant physical effects were omitted from the Sommerfeld model of metals: the effects of the periodicity and other symmetries of the lattice, and the effects of the electron-electron interaction beyond the Hartree approximation. This section deals with the remarkable consequences of lattice periodicity.

In 1928 F. Bloch posited that electrons could be treated as independent particles moving in some effective potential $v(r)$, which of course had to reflect the periodicity and other symmetries of the lattice. This led to the important concepts of Bloch waves and energy bands, the eigenfunctions and eigenvalues of the single-particle Schrödinger equation,

$$\left(-\frac{\hbar^2}{2m}\nabla^2 + v_{\text{per}}(r)\right)\psi_{n,k}(r) = \epsilon_{n,k}\psi_{n,k}(r), \tag{5.1}$$

where $v_{\text{per}}(r)$ satisfies $v_{\text{per}}(r) = v_{\text{per}}(r+\tau)$ (τ = lattice translation vector); $\psi_{n,k}$ is a quasi-periodic Bloch wave of the form $\psi_{n,k}(r) = u_{n,k}(r)e^{ik\cdot r}$, with $u_{n,k}$ periodic; k is the wave vector, a continuous quantum number describing the phase change from one unit cell to another, and n is an additional discrete quantum number, the so-called band index; the eigenvalues $\epsilon_{n,k}$ as a function of k reflect the periodicity and other symmetries of the lattice. They are periodic functions of k. In terms of these so-called energy bands the essence of most metallic and, as a bonus, insulating and semiconducting behavior (A. H. Wilson, 1930s) could be understood. This is illustrated for a one-dimensional crystal of periodicity a in Fig. 1. One observes that metallic electrons have excitation energies starting from zero; those of insulators and semiconductors have finite gaps. This simple categorization provided a powerful orientation for most simple solids.

The Bloch theory also gave a beautiful explanation of why metallic resistance approached zero at low temperatures, in spite of the presence of *individually* strongly scattering ion cores: it was quantum-mechanical coherence that caused the eigenfunctions in a periodic array of scatterers to remain unscattered. At higher temperatures, ionic positions deviated more and more from perfect periodicity, giving rise to increasing resistance.

This picture also allowed an elegant explanation by Peierls (1929) of the paradoxical Hall effects with the "wrong" sign. For example, if some traps capture electrons from the

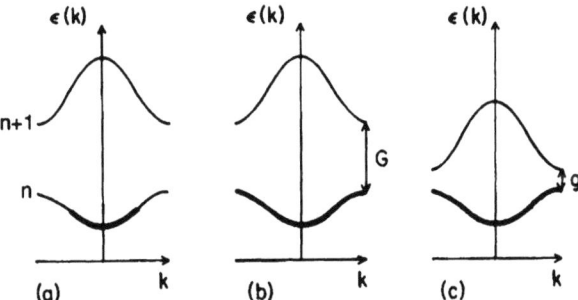

Figure 1. Schematic energy bands: (a) metal; (b) insulator; (c) semiconductor. Heavy lines denote occupation and light lines nonoccupation by electrons at $T=0$ K. G is the insulating energy gap [$\sim O(5\ eV)$]; g is the semiconducting gap ($\lesssim 1\ eV$).

valence band of a semiconductor [Fig. 1(c)] into localized bound states, this introduces holes into the top of the previously filled valence band, which behave precisely like particles with positive charge $+e$.

Even in the absence of quantitative knowledge of $v_{\text{per}}(r)$ a whole host of phenomena could now be studied and, with the help of some experimental input, understood qualitatively or better. A major tool was the quantum-mechanical transport equation (modeled after the classical Boltzmann equation):

$$\frac{\partial f(r,v,t)}{\partial t} = \frac{\partial f(r,v,t)}{\partial t}\bigg|_{\text{drift}} + b - a. \tag{5.2}$$

Here $f(r,v)d\tau$ is the number of electrons in the phase-space element $d\tau = dr dv$, $(\partial f/\partial t)_{\text{drift}} d\tau$ is their net drift into a fixed $d\tau$ due to their velocity v and to their acceleration $\dot{v}$, produced by external fields; $b-a$ describes changes in f due to collisions with lattice vibrations or defects which take electrons into and out of $d\tau$.

This equation gave considerable microscopic insight into electrical and thermal conductivities, σ and K. For the venerable universal Wiedemann-Franz constant, first experimentally discovered in 1853, it led to the result

$$\frac{K}{\sigma T} = \frac{\pi^2}{3}\left(\frac{k}{e}\right)^2, \tag{5.3}$$

in good agreement with classical theory and experiment. While σ and K individually depend strongly on specifics, including the collision processes, which are roughly describable by a mean free path l between collisions, the ratio depends only on classical fundamental constants. Electrothermal effects, named after Thomson and Seebeck, could also be successfully described.

Finally, optical properties of solids, including the origin of color, could be understood as due to transitions of electrons between occupied and unoccupied states of the same k but different band quantum number n (see Fig. 1).

Total energies

While, as we have seen, the Bloch picture was extremely useful for many purposes, it did not seriously address the extremely important issue of total electronic energies $\mathcal{E}(R_1, R_2, \ldots)$ as a function of the nuclear configuration. For insulators and semiconductors there existed some alternative strategies, e.g., the approach by Born for ionic crystals like Na^+C^- (see Sec. II) and, since the 1930s, L. Pauling's concept of the chemical bond for covalent crystals like Si. This left as a major challenge the work of understanding the total energies of metals.

Here a great advance was achieved by E. Wigner and F. Seitz in their work on the alkali metals beginning in 1933. By a bold, physical argument they proposed that the periodic potential for the valence electrons in any crystal cell l be taken as $v(r - R_l)$, equal to the effective potential for the valence electron in an isolated atom located at R_l. The latter had been accurately determined by comparison with the observed energy spectra of isolated atoms. This was a major step beyond the formal theory of Bloch electrons: The abstract $v_{\text{per}}(r)$ was replaced by a specific, independently determined potential.

In order to obtain the total energy as a function of the lattice parameter, they first calculated the sum of the noninteracting Bloch energies in the periodic potential $\Sigma_l v(r$

$-R_l$) and argued that a Hartree-like intracell Coulomb interaction energy E_H was approximately canceled by the so-called exchange and correlation energy E_{xc}. (By definition E_{xc} is the difference between the exact physical energy and the energy calculated in the Hartree approximation.) Their results were generally in semiquantitative agreement with experiment for cohesive energies, lattice parameters, and compressibility.

Subsequently they actually estimated the neglected energies E_H and E_{xc} for a uniform electron gas of the appropriate density, confirmed the near-cancellation, and obtained similar results. This involved the first serious many-body study of an infinite system, a uniform interacting electron gas, by E. Wigner in 1938. He arrived at the estimate of $\epsilon_c = -0.288/(r_s + 5.1 a_0)$ atomic units for the correlation energy per particle, which needs to be added to the Hartree-Fock energy. [r_s is the so-called Wigner-Seitz radius given by $(4\pi/3)r_s^3 = $(density)$^{-1}$, and a_0 is the Bohr radius.] This result has withstood the test of time extremely well. At this time the best available results have been obtained by numerical, so-called Monte Carlo methods (D. Ceperley and others, 1980s) with an accuracy of $\sim 1 \times 10^{-2}$.

The Wigner-Seitz approach was, of course, very soon tried out on other metals, e.g., the noble metals and Be, which were not so similar to uniform electron gases, but generally with much less success. Not until the advent of density-functional theory 30 years later, in the form of the Kohn-Sham theory were the Bloch and Wigner-Seitz approaches unified in great generality (see later section on Moderate and Radical Effects of the Electron-Electron Interaction).

From the time of Bloch's original paper in 1928 up until the 1950s the band-structure paradigm provided an invaluable conceptual framework for understanding the electronic structure of solids, but very little was known quantitatively about the band structures of specific materials, except the very simplest, like the alkali metals. This now changed dramatically.

Fermi surfaces of metals

In a beautiful short note L. Onsager (1952) considered the dynamics of a crystal electron in a (sufficiently weak) magnetic field $B = (0,0,B_z)$. In momentum space it is governed by the semiclassical equations

$$\hbar \dot{k} = \frac{e}{c}[v(k) \times B], \qquad (5.4)$$

where $v(k)$ is the velocity,

$$v(k) = \hbar^{-1} \nabla_k \epsilon_k. \qquad (5.5)$$

Combining this with purely geometric considerations, Onsager showed that an electron starting at a point k_0 with energy $\epsilon(k_0)$ will return to k_0 cyclically with a so-called cyclotron period

$$T_c = \left(\frac{c}{eB_2}\right)\hbar^2 \frac{dS}{d\epsilon}, \qquad (5.6)$$

where S is the area in k space enclosed by the curve C, which is generated by the intersection of the plane $k_z = k_{0z}$ and the surface $\epsilon(k) = \epsilon$. (See Fig. 2.) Using this result he showed further that for *any* band structure, no matter how complex, the magnetization is an oscillatory function of B_z^{-1} (the so-called de Haas-van Alphen-Shubnikoff oscillation) with a period (and this was new) given by

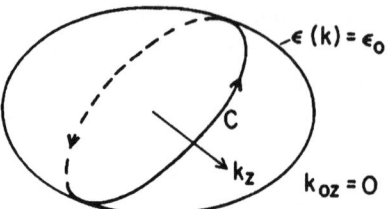

Figure 2. The Onsager orbit C on the plane $k_x = k_{0z}$ and on the constant-energy surface $\epsilon(k) = \epsilon_0$.

$$\Delta\left(\frac{1}{B_z}\right) = \frac{2\pi e}{c}\frac{1}{S}; \tag{5.7}$$

here S is a maximum, minimum, or other stationary cross-sectional area, perpendicular to B, of the so-called *Fermi surface*, in k space, which, by definition, encloses all occupied k vectors. Thus, by tilting the direction of the magnetic field, one could measure geometrically cross-sectional areas with different normals! This was impressively accomplished by D. Schoenberg and his group in Cambridge in the 1950s.

These cross-sectional areas, combined with known symmetries, some rough guidance from approximate band calculations, and the general Luttinger theorem (see later section on Moderate and Radical Effects of the Electron-Electron Interaction), which fixed the volume enclosed by the Fermi surface, generally permitted unique and accurate determination of the entire shape of the Fermi surfaces (see Fig. 3) and ushered in a geometric/topological era of CMP. Since, because of the Pauli exclusion principle,

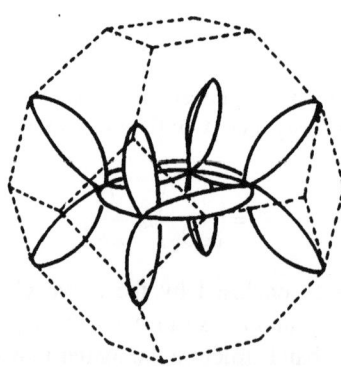

Figure 3. "Weird" topologies of Fermi surfaces: a portion of the Fermi surface of aluminum.

low-energy/low-temperature[2] electronic excitations of metals involve electrons and holes near the Fermi surface, its empirical determination represented a major advance in metal physics.

Angle-resolved photoemission and inverse photoemission

Another experimental technique that has shed great light on band structures, both of metals and of nonmetals, is angle-resolved photoemission, which began with the work of W. E. Spicer (1958). This was followed, in 1983, by inverse photoemission. The former explores occupied states, the latter unoccupied states. Photoemission and inverse photoemission have been used to study bulk bands and surface bands. (See also the next section.)

In bulk photoemission a photon of energy $\hbar\omega$ is absorbed by an electron in an occupied state (n,k), which makes a transition to an unoccupied state (n',k'). Here $k'=k$ because the photon momentum can be neglected. Energy conservation requires that

$$\epsilon_{n',k} = \epsilon_{nk} + \hbar\omega. \tag{5.8}$$

$\hbar\omega$ is chosen large enough so that $\epsilon_{n'/k}$ can be taken as free-electron-like and (apart from an additive constant) as known. The external momenta of those final electrons that reach the surface and surmount the dipole barrier give direct information about the initial momenta k and energies $\epsilon_{n,k}$. In this way occupied energy bands of many materials have been directly determined by photoemission.

In inverse photoemission external electrons of known k_e and $\epsilon_e = \hbar^2 k_e^2 / 2m$ may penetrate the surface, occupy unoccupied Bloch states (n',k), and then emit a photon whose frequency ω is again given by Eq. (5.8). From such measurements direct information about the unoccupied $\epsilon_{n',k}$ can be obtained.

A most helpful theoretical tool, at least for so-called simple metals whose valence electrons have at least some resemblance to free electrons, was the concept of the weak, effective pseudopotential $v_{ps}(r)$, due to H. Hellmann (1936) and especially to J. C. Phillips (1958), and widely used by the group of V. Heine (1960s). For the valence electrons, the weak v_{ps} had an effect equivalent to the actual Bloch potential, which is very strong near the nuclei. v_{ps} could be characterized by two or three independent Fourier coefficients and, as in the case of aluminum (see Fig. 3), even a very complicated Fermi surface could be accurately fitted everywhere.

The band-structure paradigm has remained the most important basis for understanding the electronic structure of solids. Even when there are significant, but not radical, effects of interaction and/or nonperiodicity it is usually an indispensable starting point. Theoretical materials science since the 1930s, and especially since about 1970, has increasingly made quantitative use of it. The great silicon revolution of the second half of this century might not have been possible without it.

Surfaces and Interfaces

The 20th century has seen a transformation of surface science and, more generally, of two-dimensional (2D) science, made possible by major advances in vacuum technology combined with various techniques like atomic and molecular deposition, beam writing, and etching.

---

[2] As explained in Sec. IV, "low" means typically $\ll 1$ eV or 10^4 K, in fact rather "high."

In parallel with the dramatic advances in surface science, the physics of interfaces between two bulk phases has also made major progress. Perhaps its most important practical applications are to highly controlled artificial layer structures in semiconductors and magnetic materials.

At the beginning of the century surfaces were already of great practical interest for the mitigation of corrosion, heterogeneous catalysis, thermionic emission in light bulbs and vacuum tubes, electrochemistry at solid-liquid interfaces, friction, etc. However, the best available vacua were only $\sim 10^{-4}$ torr, and most surfaces at room temperature and below were covered by unknown layers of adsorbed atoms and/or molecules. Atomically clean surfaces were generally achievable only at the highest temperatures when, in favorable circumstances, adsorbates would evaporate.

In 1927 C. J. Davisson, and L. H. Germer demonstrated diffraction patterns in the scattering of electrons by crystal surfaces, an experiment of double significance: it was a direct demonstration of the reality of the wave nature of electrons and the beginning of surface diagnostics on an Angstrom scale—analogous to von Laue scattering of x rays by bulk crystals.

Today, at the end of the century, vacua of 10^{-10} torr can be routinely generated. This has made possible the preparation of atomically clean surfaces or of surfaces covered in a controlled way with an accuracy of ~ 0.01 monolayers. Since the 1980s we have acquired the ability to check the structural and chemical condition of surfaces point by point with an accuracy of order 10^{-1} Å(!), using electron tunneling and force microscopies (G. Binnig and H. Rohrer, 1980s). A host of other, less local, but powerful diagnostic techniques such as Auger electron spectroscopy, x-ray and neutron reflectometry, low- and high-energy electron diffraction, x-ray and ultraviolet photoelectron spectroscopy (XPS, UPS), have also been highly developed; so have a wide variety of surface treatments.

A major factor driving surface science since the middle of this century has been the dramatic rush to greater and greater miniaturization. While in a structural steel beam surface atoms constitute a fraction $\sim 10^{-9}$ of all the atoms, in a miniaturized semiconducting structure used in contemporary devices the ratio of interface atoms to bulk atoms is on the order of $\sim 10^{-2}$, and surface properties often dominate device performance. For example, in the late 1940s, surface and interface physics was a major aspect of the invention of the first point-contact transistors as well as of later versions, and it has remained a critical element of the subsequent quantum electro-optical revolution, which continues in full swing.

The band-structure paradigm for bulk solids, when appropriately modified, also became a valuable guide for understanding the physics of surfaces and interfaces. A crystal terminated in a plane parallel to $z=0$ remains periodic in the x-y plane so that the electronic states have the 2D Bloch property

$$\psi_{n,k}(r+\tau_j) = e^{ik\cdot\tau_j}\psi_{n,k}(r), \quad j=1,2, \tag{6.1}$$

where the τ_j are 2D lattice vectors in the x-y plane.

Two important qualitatively new features were observed. First, the atoms near the surface may reconstruct so that the symmetry is lowered not only in the z direction (no surprise) but also in the x-y plane. Dimerization is the simplest example, but important cases with much larger supercells have been observed, for example, on the surfaces of silicon. Such reconstruction can radically affect surface electronic structure and interactions with adsorbed atoms or molecules.

Secondly, some electronic wave functions near a surface penetrate into the bulk and some are localized near the surface. The localized surface states, first theoretically proposed by I. Tamm in 1932, are occupied or empty depending on whether they are below or above the electronic chemical potential μ. They have three quantum numbers, k_x, k_y, and m, which describe Bloch-type propagation in the x-y plane and localization along the z direction. If, for example, only states with $m=0$ have energies below μ, we can have a 2D metal. (The motion in the z direction is "frozen out.")

There are two important, microscopically averaged surface properties, the surface energy per unit area ($\frac{1}{2}$ of the cleavage energy) and the electric dipole barrier. They play important roles for thermodynamic and electronic considerations. An early, rather rough, theory for these quantities constitutes the thesis of John Bardeen (1936). Much later, beginning with work by N. Lang and W. Kohn (1970), good quantitative results were obtained by the use of density-functional theory (see the next section).

Another major area of surface science is the joint domain of chemistry and physics: the study of atoms and molecules in various interactive relationships with surfaces—collisions, adsorption, desorption, and, for adsorbed atoms and molecules, diffusion, physical and chemical interactions, and chemical reactions. Again, controlled ultrahigh vacua and deposition methods, combined with the battery of mostly post-1950 diagnostic techniques, have led to spectacular advances. Improved catalytic conversion of noxious automobile emissions and cracking of crude oil are examples of important applications.

Systems involving both surfaces and molecules present a difficult challenge to theory because of the absence of symmetry simplifications and the significant involvement of many (10–10^3) atoms. The first detailed quantitative results using density-functional theory for H_2 on solid surfaces have just begun to appear (A. Gross, M. Scheffler, and others, 1995). These computations would not be possible without the use of the highest state-of-the-art computing power.

The foregoing paragraphs have treated surfaces essentially as two-dimensional versions of bulk crystals. However, we shall see shortly (in the section on Mesoscopics and Nanoscience) that, in the presence of disorder and/or electron-electron interaction, entirely new phenomena, such as the quantum Hall effect, exist in two dimensions. In fact, the recognition of the highly nontrivial role of *dimensionality* is one of the hallmarks of CMP in this century. Probably the first example of this recognition is the surprising observation by R. Peierls (1935) that while a harmonic 3D crystal has positional correlation over an infinite range at any finite temperature, this is true of a (free-floating) 2D crystal *only* at $T=0$.

A final remark about surface science and surface technology. The former, both in the laboratory and in the theorist's thinking, deals mostly with as perfect systems as possible. On the other hand, in technological applications surfaces are generally very imperfect, both structurally and chemically. Nevertheless, concepts developed by idealized surface science have been very important guides for practical applications.

Moderate and Radical Effects of the Electron-Electron Interaction

The many great successes of the band-structure paradigm in accounting, at least qualitatively, for electronic properties and phenomena in solids strongly implied that interactions often only alter the Bloch picture quantitatively without changing its qualitative features. This turned out to be partly true and partly false. We begin with an account of some moderate interaction effects.

Landau Fermi-liquid theory

In 1956, a few years after the first reliable experiments on He^3, the rare *fermionic* isotope of helium, L. Landau published his famous, largely heuristic, theory on the low-energy properties of a uniform gas of mutually repelling neutral fermions. He concluded that there are low-lying excited states that can be described as arising from the ground state by the addition of quasiparticles and quasiholes with momenta $k(-k)$ and energies $\epsilon_k = (\hbar^2/2m^*)|k^2 - k_0^2|$, where k_0 is the Fermi momentum of the noninteracting gas, $|k| \gtreqless k_0$ for electrons and holes, respectively, and $|k - k_0| \ll k_0$; m^* is an effective mass, which for He^3 turned out to be 3.0 times the mass of the bare He^3 atom. Although the quantitative renormalization of the mass was large, these excitations were in 1-to-1 correspondence with those of noninteracting fermions.

Much more interestingly, Landau also introduced an effective spin-dependent interaction $f(\theta)$ between low-energy quasiparticles with momenta k and k', where θ is the angle between k and k'. $f(\theta)$ is usually parametrized in terms of a few spin-dependent angular expansion coefficients $F_0, F_1, \ldots$. While these coefficients are not needed for the low-temperature specific heat, they do enter significantly into the spin susceptibility and compressibility, which, for He^3, are also strongly renormalized by factors of 9.1 and 3.7, respectively. But the most interesting result of Landau's theory was that these interactions lead to a new dynamical collective mode of coherent, interacting quasiparticle-quasihole pairs, the so-called zeroth sound mode, with a linear dispersion relation, $\omega = sq$. The velocity s was also expressible in terms of the F_l. The experimental confirmation of this mode by J. Wheatley and co-workers (1966), and the consistency of the experimentally overdetermined parameters F_0 and F_1 (F_2, etc. are very small) was a great triumph for this theory.

Implications for an interacting electron gas were immediately recognized, the important differences being the presence of the periodic potential due to the nuclei and the long range of the Coulomb interactions. The effect of the latter on a degenerate uniform electron gas had, in fact, been previously shown by D. Pines and D. Bohm (1952) to lead to the collective plasma mode, with dispersion approximately given by $\omega^2 = \omega_p^2 + \frac{3}{5} v_F^2 q^2$; here ω_p is the classical plasma frequency and v_F is the Fermi velocity. Landau's theory provided a unification of the theories of neutral and charged uniform Fermi systems.

Electrons under the influence of both a periodic potential and the Coulomb interaction were soon studied perturbatively using the newly developed machinery of many-body theory. A key result was obtained by J. M. Luttinger (1960), who showed formally that, to all orders in perturbation theory, a sharply defined Fermi surface $k(\theta, \varphi)$ persisted in k space and, though its shape was altered by the interactions, the k-space volume enclosed by it remained unchanged, determined entirely by the mean density $\bar{n}$ of the electrons. This so-called "Luttinger theorem" has been very helpful in studies of metals with complex Fermi surfaces.

Strong magnetism

We have already mentioned Heisenberg's qualitative realization that the Pauli exclusion principle combined with the electron-electron interaction can bring about a strong effective interaction between spins. By a dimensional argument it is of the form $\gamma e^2/a$, where a is an effective interelectronic distance and γ is a dimensionless constant. Heisenberg's approach was well suited for insulators describable by localized orbitals, but the traditional ferromagnetic materials, Fe, Ni, and Co are metals.

For these F. Bloch contributed an early insight (1929). He compared the energies of two possible Sommerfeld ground states: (1) A paramagnetic state with each plane wave $k \leq k_0$ (k_0 = Fermi wave number), occupied by both a spin-up and a spin-down electron. (2) A ferromagnetic state, in which all spins are pointing in the same direction, say z, and hence each plane wave is occupied by at most one electron. Thus the maximum occupied k is now increased by a factor of $2^{1/3}$, and the kinetic energy by a factor of $2^{2/3}$. However, the exchange energy due to the Coulomb repulsion of the electrons favors the ferromagnetic state, since the Pauli exclusion principle keeps *all* electrons apart from each other, whereas in the paramagnetic state electrons with opposite spin are not kept apart. At sufficiently low density this effect prevails and the ferromagnetic state has the lower energy. (Bloch ignored correlation effects, which, in fact, change the conclusion for the uniform electron gas.) This was the beginning of the concept of itinerant magnetism in metals, soon considerably developed by E. C. Stoner, J. C. Slater, and others.

Bloch soon returned to another aspect of magnetism (1934), this time Heisenberg's localized type. Starting from N atoms, in a perfectly spin-aligned ground state Ψ_0 with a total z spin equal to $\frac{1}{2}N$, he observed that there were gapless, propagating excited states, spin waves of the form $\Phi_k = \Sigma_l \exp(ikR_l) S_l^- \Phi_0$, where S_l^- is the spin operator, which turns over the spin at the site R_l. Their energy spectrum had the form $\epsilon_k \propto k^2$ and their contribution to the low-temperature heat capacity was $\propto T^{3/2}$. The most compelling confirmation of Bloch's spin waves came in the 1960s by means of inelastic neutron scattering, which directly measured the dispersion relation ϵ_k and found remnants of spin waves even above the critical temperature where the average magnetization vanishes.

Since its quantum-mechanical beginning in the late 1920s, the field of strong magnetism has had an explosive growth. One of the most interesting events was the prediction by L. Néel, and the subsequent experimental confirmation (1930s), of a new kind of magnetism, later called antiferromagnetism: the lattice consists of two equivalent sublattices A and B, with all A atoms carrying a magnetic moment m_A and all B atoms carrying the moment $m_B = -m_A$. Thus the total magnetization, in contrast to ferromagnetism, vanishes. However, both the magnetic susceptibility and low-temperature specific heat reveal the "hidden" strong sublattice magnetizations. Again, the incontrovertible proof was provided by direct observation of two magnetic sublattices in elastic neutron scattering (C. G. Shull and J. S. Smart, 1949). Following Néel's work many complex magnetic structures were discovered, especially among heavy metals and metal compounds.

In the first half of the century the most important practical applications of magnetism were electromagnets: generators, motors, electromagnetic relays, etc. Today these are joined by magnetic memory devices, read-in and read-out devices, magnetic layer structures with "giant" magnetoresistance, etc. The field has entered a new, very active phase.

Density-functional theory

By the 1960s quantum-chemical methods had been very successful in calculating properties of N-electron systems, with N up to $O(10)$. However, in condensed matter physics $N = O(10^{23})$, and even in the smallest representative clusters $N = O(10^2 - 10^3)$. Density-functional theory (DFT), introduced by P. Hohenberg, W. Kohn, and L. J. Sham in 1964–1965, provided a practical, new approach to electronic structure, applicable also to large-N systems. Density-functional theory is couched in terms of the electron density $n(r)$ [or, for magnetized systems, spin densities $n_\sigma(r)$, $\sigma = \pm 1$] instead of the many-electron wave function Ψ. It leads to the Kohn-Sham self-consistent equations, similar to the Hartree equations, in which, however, exchange and correlation effects are included (in principle, exactly) by the addition of the exchange-correlation potential of $v_{xc}(r)$.

The theory allows parameter-free calculations of densities and spin densities, ground-state energies, as well as related quantities such as lattice structures and constants; elastic coefficients; work functions, surface energies, and atom-surface interaction energies; phonon dispersion relations; magnetic moments, etc. Accuracies typically range from 1–20 % depending on the context; geometries emerge very accurately, typically ±1%. Density-functional theory is, in principle, exact, but in practice requires an approximation for the exchange-correlation energy E_{xc}. The simplest, the "local-density approximation" (LDA), rests on accurate Monte Carlo calculations of a uniform, interacting electron gas. The scaling of the computation time with N is a relatively very favorable N^α, where $1 \leq \alpha \leq 3$, so that calculations for finite systems with $N \approx 100$–1000 have been quite feasible.

Density-functional theory has become the method of choice for calculating electron densities and energies of most condensed matter systems. It also leads to nominal energy bands, which are usually a very useful approximation to the physical bands. In the 1980s and 1990s the LDA was greatly improved by density-gradient corrections (A. Becke, J. P. Perdew, and others). Since about 1990, density-functional theory has also been widely used by theoretical chemists, particularly for large, complex molecules and clusters.

Collective excitations

Collective excitations are an important hallmark of many-body systems. They depend for their very existence on particle-particle interactions and are delocalized excitations of the entire system. Familiar examples are the vibrations of molecules or of crystal lattices, whose nature has been well understood since about 1910.

We have already mentioned a few other condensed matter examples: zeroth sound in He^3 and plasmons in a uniform electron gas, as well as spin waves in magnetic systems. The latter represent a separation of electronic spin and charge, already well understood by F. Bloch, who about 1930 is said to have remarked: "If electrons can hop from one atom to another why not spins?" (the condensed matter version of Lewis Carroll's Cheshire cat and his grin).

A collective excitation in insulators was proposed in 1931 by J. Frenkel, now called the Frenkel exciton. It is most easily visualized for a lattice of distinct neutral atoms, say Ar, and is formally analogous to spin waves. Let Ψ_0 be the ground state of the system, with all atoms in their ground state, and let Ψ_l be the state in which the atom at site R_l is in the first excited state. The states $\Psi_1, \Psi_2, \ldots$ are degenerate and, because of the proximity of the atoms, they interact. The correct linear combinations reflecting the lattice periodicity are the excitation waves, or *excitons*, $\Psi_k \equiv A e^{ik \cdot R_l} \Psi_l$ with wave vector k, whose energies ϵ_k are k dependent. Excitons are the lowest excited states of insulators. Being neutral, they carry no electric current. They were first clearly identified in optical spectra in the 1940s.

A different view of excitons is due to G. Wannier (1937). If one ignores interactions between electrons, the lowest-lying excited states, Ψ_{k_c,k_v}, have an electron (k_c) near the bottom of the conduction band and a hole (k_v), with a positive charge, near the top of the valence band. Because of their Coulomb attraction the electron and hole can form lower-lying, traveling bound states with total wave number k, a kind of condensed matter positronium. These are again the excitons. The Frenkel picture is appropriate for small-excitons, where the electron and hole are tightly bound to each other (e.g., in Na^+Cl^-); the Wannier exciton is more appropriate in the opposite limit (e.g., in Si).

Excitons, consisting of two fermions, are bosons and in principle should exhibit Bose condensation (L. V. Keldysh, 1960s); however, so far this has escaped clear identification.

Radical effects

The foregoing paragraphs dealt with what I call moderate effects of the electron-electron (e-e) interaction, when a model of noninteracting effective electrons and/or holes is a good starting point. There are, however, many condensed matter systems in which this is not the case, whose history will now be briefly addressed. E. Wigner (1938), considering the ground state of a dilute gas of electrons moving in a neutralizing positive charge background, observed that the free-electron kinetic energy per electron behaved as r_s^{-2} while the e-e repulsive energy behaved as r_s^{-1}, and thus in the dilute limit the latter would prevail [r_s, previously defined, is proportional to (density)$^{-1/3}$]. He concluded that the electrons would form an ordered lattice and perform small zero-point vibrations around their equilibrium positions. This so-called *Wigner lattice* was an early indication that there may be condensed matter systems or regimes for which the band paradigm is overwhelmed by the effects of the e-e interaction. (Much later a 2D Wigner crystal was observed for electrons trapped on the surface of liquid He^4.)

In 1949 N. Mott noted that the compound NiO_2 was an insulator, although, based on the number of electrons per unit cell and the Bloch band paradigm, it should be a metal with a half-filled band. This led him to consider a model consisting of H atoms forming a simple cubic lattice with adjustable lattice parameter a. He adopted a tight-binding point of view, in which the many-body wave function is entirely described in terms of atomic $1s$ orbitals ω_l, centered on the nuclei R_l. He then estimated the effects on the total energy due to electrons' hopping onto neighboring sites. By giving electrons more room, one would cause their kinetic energy to be reduced, while the Coulomb repulsion energy of the electrons would increase by an energy U for each double occupancy. He concluded that, when a exceeded a critical value a_c, this system, which in band language has a half-full band, would nevertheless become an insulator, now called a *Mott insulator*. The internal structure of this insulator is quite different from that of the filled-band Bloch insulator. (There is an obvious relationship between Wigner's and Mott's considerations.)

These ideas were further developed by J. Hubbard (1963) in the so-called Hubbard model of interacting electrons with the Hamiltonian

$$H = \sum_{l\sigma} \epsilon_{l\sigma} n_{l\sigma} + t \sum_{\substack{l,l' \\ (nn)}} c^*_{l\sigma} c_{l'\sigma} + U \sum_i n_{l\uparrow} n_{l\downarrow} . \tag{7.1}$$

Here l and σ denote sites and spins; the sum over l and l' is over nearest neighbors; $\epsilon_{l\sigma}$ and t are the site-diagonal and hopping energies; U describes the additional energy due to double occupation; and $n_{l\uparrow}, n_{l\downarrow}$ denote the numbers of spin-up (-down) electrons on site

t. This Hamiltonian interpolates between isolated atoms ($t=0$) and noninteracting, itinerant electrons ($U=0$). Approximate solutions for U/t finite do indeed yield the Mott metal-insulator transition for a critical value of U/t. But the model has allowed many extensions to more complex systems (e.g., high-T_c superconductors), excitations, defects, effective spin Hamiltonians, magnetic phenomena, longer-range interactions, etc. It has been a valuable guide for understanding systems such as oxides, sulfides, and many other compounds which, under the band paradigm, would be described as narrow-band materials.

In a similar spirit P. W. Anderson (1961) had earlier proposed that an isolated impurity atom, immersed in and hybridized with a sea of conduction electrons, could, due to an intra-atomic e-e repulsion U, develop a finite magnetization.

In 1964 T. Kondo considered the effect of such a localized impurity spin on the scattering of conduction electrons and surprisingly found (in low-order perturbation theory) very unusual behavior (paralleling earlier experimental findings) below what is now called the Kondo temperature T_K. In fact, for $T \ll T_K$ the impurity spin forms a singlet state with the conduction electrons, and its magnetic susceptibility vanishes.

Since the 1980s so-called "heavy-fermion" materials have attracted much attention. They are associated with incompletely filled $4f$ and $5f$ shells such as in the Ce and U compounds $CeAl_3$ and UPt_3. At very low temperatures [$T = O(1\text{ K})$] this class of materials displays a linear specific heat, γT, with γ values corresponding to an enormously enhanced effective mass, $m^* = (10^2 - 10^3)m$! They exhibit a great variety of electronic behavior, including paramagnetism (with a huge magnetic susceptibility), various forms of cooperative magnetism with very small magnetic moments, insulating behavior, and superconductivity. It is generally believed that their properties reflect the opposing tendencies of the Kondo mechanism, which tends to suppress localized f moments, and an indirect, so-called RKKY (Ruderman-Kittel-Kasuya-Yosida), interaction between f moments on different sites, via the conduction electrons. Attempts to understand their behavior usually employ a generalization of the Anderson/Hubbard Hamiltonians, including a repulsive energy U for f electrons on the same site and an f-electron/conduction-band hybridization term. In addition to more standard techniques, μSR (positive muon spin rotation and relaxation) has provided a wealth of information about local magnetic fields in these compounds.

At century's end, heavy-fermion systems, together with high-T_c superconductors, represent major challenges to condensed matter theorists.

Moderate and Radical Breakdowns of Lattice Periodicity

Condensed matter consists, by its nature, of very many significantly interacting atoms. The *periodicity* of crystal lattices was the simplifying feature which, beginning in the decades 1910–1930, gave physicists and chemists the courage to undertake experiments and construct theories that ultimately led to an impressive understanding of crystalline solids on an atomic scale. The periodicity paradigm has remained invaluable ever since. At the same time perfect periodicity for an extensive system is a thermodynamic fiction. For, as is easily shown, at any finite temperature T, the introduction of a small concentration of nonperiodic defects, while raising the internal energy U, also increases the entropy S in such a way that the free energy, $F \equiv U - TS$, is reduced. Thus it is not surprising that, over the course of the century, nonperiodic systems have also been intensively studied, starting with dilute point defects in periodic lattices (1920s) and later including systems for which the periodicity paradigm has little if any relevance, for example, liquids, amorphous solids, and fractals.

A crucial opposite development also took place. Poorly controlled, high levels of structural and/or chemical disorder can prevent meaningful scientific studies or dependable applications. The semiconductors Si and Ge are a case in point. Their electrical and low-frequency optical properties are largely due to very small concentrations of chemical impurities, which could not be adequately controlled until the middle of the century. As a result their applications, e.g., to crystal radios, were highly unreliable. With the advent of zone refining (1940s) structurally excellent crystals of Si and Ge were grown in which the concentration of critical impurities, like B and P, could be controlled at the unprecedented fractional level of $\sim 10^{-8}$. This dramatic accomplishment was indispensable for the semiconductor revolution of this century. Also, the writer recalls seeing (in about 1955) a Si whisker with a *single* structural defect, a so-called screw dislocation, around which it had grown, whose resistance to fracture was many orders of magnitude higher than that of "normally" grown Si.

Point defects

Ionic crystals are among the easiest to grow with high structural and chemical perfection and therefore were early subjects of study. The most common point defects are so-called vacancies (missing atoms or ions), followed by interstitials (additional atoms or ions, located in interstices of the periodic lattices). Local electronic neutrality is energetically very strongly favored. Thus in Na^+Cl^- a Cl^- vacancy is typically either paired with a nearby neutralizing Na^+ vacancy or it traps a neutralizing electron. The latter defect was identified as the previously empirically discovered F-center (F = Farbe, i.e., color), which lent a distinctive color to Na^+Cl^- crystals containing them. This and other similar centers became the subject of detailed optical studies and concomitant theoretical work—perhaps the first quantitative application of quantum mechanics to a complex condensed matter system (experiments by R. W. Pohl and co-workers, 1920s; theory by J. H. De Boer and others, 1930s). In roughest approximation the F-center may be regarded as a condensed matter version of a hydrogen atom, with the net positive charge $+e$, due to the removal of Cl^-, playing the role of the proton in hydrogen. Of course the positive charge is effectively spread out over the volume of the vacancy, and beyond it the electron moves not through a vacuum but through the dynamical ions of the Na^+Cl^- lattice. The primary absorption at 2.7 eV, in the visible spectrum is the greatly modified analog of the 10.2 eV $1s \to 2p$ line in H. The temperature dependence of the linewidth could be quite well explained as due to the vibrations of the nearby ions. Detailed optical studies of this and other so-called color centers, associated with a single structural point defect or with complexes of several structural defects, were later very effectively complemented by the invention of nuclear and electron spin-resonance techniques and remained an important, highly quantitative field of CMP into the 1960s.

Beginning in the 1950s, analogous but even more precise studies were undertaken of the so-called donor and acceptor defects in covalent semiconductors. These studies were greatly stimulated by the invention of the transistor and decisively aided by the independent measurement of the effective masses of low-energy electrons and holes using so-called cyclotron resonance in external magnetic fields (B. Lax and others, 1950s). Donor and acceptor defects in Si are created by replacement of a four-valent Si atom by, say, a five-valent P atom or a three-valent B atom. Four electrons from the P atom become part of the bonding structure (or filled "band") of the Si matrix. The extra charge $+e$ on the phosphorous nucleus can weakly trap the extra electron of the P atom in one of several hydrogenlike donor states, or "donate" it to the continuum of conduction-band states. An analogous situation obtains for a B acceptor and positive holes. At room temperature the donors and acceptors provide conducting electrons and holes, while in their absence most semiconductors are effectively insulating. The solid-state spectroscopy of trapped donor electrons and acceptor holes has now an astonishing accuracy of $\sim 10^{-3}$. Excited states can be calculated with similarly high accuracy by parameter-free so-called "effective mass theory," in good agreement with experiment. The theory for the more tightly bound donor or acceptor ground states is not so precise.

Substitutional alloys

Another class of moderately nonperiodic systems is made up of the substitutional alloys $A_x B_{1-x}$, where A and B are elements with the same valency, similar atomic radii, and, in their pure forms, the same crystal structure. Alloys of Cu and Au are an example. Except for special values of x, such as $x = 1/4$, when ordered superlattices can form, such alloys display some degree of disorder. Early in this century the theory of such alloys was based on a simple but successful phenomenological mean-field model for the energy (W. L. Bragg and E. J. Williams, 1934): $E = N_{AA} v_{AA} + N_{BB} v_{BB} + N_{AB} v_{AB}$, where N_{AA} are the number of AA-"bonds" and v_{AA} is the corresponding bond energy, etc. Today the energies of many alloy systems have been successfully calculated by parameter-free density-functional theory (Sec. VII) with accuracies of a few percent.

The thermodynamics of the Bragg-Williams model, which is mathematically isomorphic with that of the so-called Ising model for magnetism (1925), has been the subject of intensive theoretical study ever since the 1920s. A major theoretical breakthrough was the exact analytical solution of this model in 2D by L. Onsager in 1944, showing a logarithmic singularity in the specific heat at a critical temperature T_c where long-range magnetic order disappears. For the 3D Ising model, although very precise numerical results are available today, the intensive quest for exact analytical results has so far not succeeded.

Dislocations and grain boundaries

In all the examples above, while there are local distortions of the lattice structure, the topology of the underlying periodic lattice remains intact. The major new concept of *dislocation*, a topological defect, was put forward in 1934, independently by G. I. Taylor, E. Orowan, and M. Polanyi, to explain the fact that permanent deformation occurs in metals (e.g., in a Cu wire) under stresses about three orders of magnitude smaller than estimated for a perfect crystal. Figure 4 shows a so-called edge dislocation, which can be thought of as arising from the insertion of an extra half plane of atoms into an otherwise perfect periodic lattice. The *local* properties of the lattice are significantly disturbed only near the terminating edge of the half plane, the so-called dislocation line. However, even far away from this line the topology of the lattice is altered. Thus any circuit enclosing this line, consisting of N steps to the right, N upwards, N to the left, and N down, fails to close by 1 step, no matter how large the number N.

Edge dislocations, just like vacancies and interstitials, are naturally present at finite temperatures. They can be made to slip sideways under much smaller stresses than are

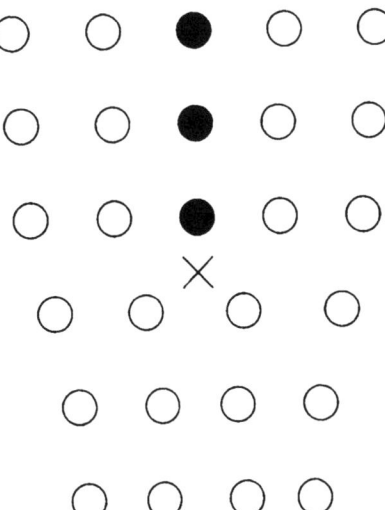

Figure 4. Edge dislocation line (perpendicular to paper). The solid circles are part of the additional half plane.

needed for the simultaneous slippage of the entire upper half-crystal $z>0$ over the lower half-crystal $z<0$. Slippages of many dislocation lines under stress result in a so-called plastic deformation, which, unlike an elastic deformation, remains when the stress is removed. The concept of dislocations completely revolutionized our understanding of the strength of materials. Single-crystal materials of macroscopic dimensions result only under conditions of extremely slow growth. Otherwise even structurally "good" materials are usually polycrystalline, consisting of small microcrystals (or grains) with different orientations. Grain boundaries can be conceptualized as accumulations of dislocation lines. Their properties are of critical importance in metallurgy.

The year 1984 brought a big surprise in the field of crystallography. Mathematical crystallography had been regarded as a closed subject since the work of Schoenflies in the 19th century. All possible point groups consistent with periodicity had been listed. In particular the icosahedral point group was not allowed. Yet D. Schechtman and co-workers reported a beautiful x-ray pattern with unequivocal icosahedral symmetry for rapidly quenched AlMn compounds. The appropriate theory was independently developed by D. Levine and P. Steinhardt, who coined the words *quasicrystal* and *quasiperiodic*. Even more curious was the fact that R. Penrose (1984) had anticipated these concepts in purely geometric, so-called Penrose tilings (Fig. 5).

Quasiperiodicity has raised new questions about vibrational lattice modes and electronic structure and growth mechanisms, which continue to attract interest.

Totally nonperiodic systems

In all of the preceding examples, the perfect periodic lattice has been the appropriate background against which to understand the nature of the defects. There are other condensed matter systems in which periodicity is totally or largely absent.

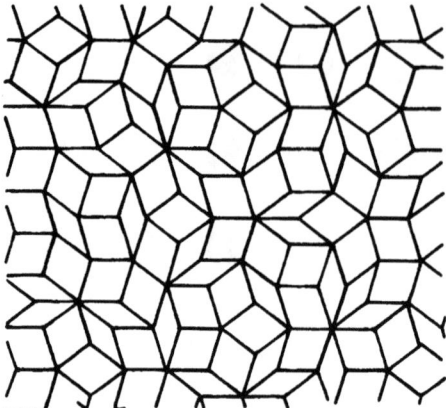

Figure 5. Quasiperiodic two-dimensional Penrose tiling.

Classical liquids, of course, are nonperiodic. Their macroscopic properties have been and are the subject of mature, specialized fields such as hydrodynamics. On a microscopic scale they are described by classical mechanics and thermodynamics, using a potential-energy function $V(R_1, R_2, \ldots, R_N)$, where the R_l are the positions of the nuclei. For some simple liquids, e.g., Ar, V is simply the sum of molecular pair potentials, accurately known from chemistry. But for metallic liquids pair potentials are generally inadequate. In the 1960s quantum theory began to provide good potential functions also for metals. We mention here further that molecular dynamics, i.e., computer simulation of the classical motions of the constituent atoms, has played an important clarifying role for finite-temperature properties and phenomena since the 1960s (B. Alder, A. Rahman, and others).

Quantum liquids consist of light elements, especially the isotopes of H and He. For these liquids the nuclear dynamics at low temperatures must be described by quantum mechanics and exhibit measurable, striking quantum effects: Rotational phase transitions for liquid H_2; Fermi and Bose statistics for He^3 and He^4; diffusion by quantum tunneling even as $T \to 0$; superfluidity of ultracold He^3.

Critical phenomena refer to the behavior of a thermodynamic system near its critical point. They constitute one of the most important parts of the discipline of statistical mechanics, with major applications to CMP. Since this essay cannot attempt to deal adequately with the general principles of statistical mechanics we must limit ourselves to very cursory remarks.

The 1910 Nobel Prize in physics was awarded to J. D. van der Waals "for his work on the equation of state for liquids," $P = F(\rho, T)$, where P, ρ, and T denote pressure, density, and temperature. The form of the function F, when expressed in system-dependent, dimensionless variables, was universal for all liquid-gas systems. In particular the theory accounted for the universal existence of the so-called critical point (P_c, ρ_c, T_c), where the line in the (T,P) plane, which separates the liquid and gas phases, terminates.

A magisterial generalization of van der Waals' concepts to a general theory of phase transitions was put forward by L. Landau in the 1930s, couched in terms of the key concept of the order parameter S (spatially uniform or slowly varying), such as the difference between liquid and gas densities at given pressure, or the magnetization per unit volume M of a ferromagnet for a given external magnetic field B. Near a critical point—where the order parameter vanishes—Landau expanded the free energy in powers of S and $(T-T_c)$. The theory predicted a simple analytic behavior [e.g., $(T_c-T) \propto (\rho - \rho_c)^{1/2}$] for the coexistence of liquid and gas. However, beginning in the 1940s, and especially in the 1960s, the exponent was found to be not the rational number $\frac{1}{2}$ but, universally, 0.34—very near the critical point. Similarly, other so-called nonclassical universal exponents were found for other classes of thermodynamic systems. These developments eventually led to the major, radically new concept of the *renormalization group* (K. Wilson, 1971) in which spatial correlation functions of the order parameter at or near the critical point $G(r,r') = \langle S(r)S(r') \rangle$ play a central role. Renormalization-group theory, with its new concepts of scaling, universality, stable and unstable fixed points, and basins of attraction, has led to entirely new thinking about the physics of phase transitions, especially near the critical point, and has had far-reaching impacts not only elsewhere in physics, but beyond.

Glasses, known since antiquity and of enormous practical importance, are still not fully understood. A key advance was made by W. H. Zachariason (1932), who proposed that glassy silica, SiO_2, has the same local bonding structure as crystalline quartz (each Si atom being bonded to four oxygen atoms and each oxygen being bonded to two Si atoms), but that the overall topology of the 3D bonding network has random elements and no periodicity. The nature of the so-called glass transition, in which the viscosity changes by many orders of magnitude over a narrow temperature interval, remains subject to study and controversy.

Spin glasses have received a great deal of attention since the 1960s. They are dilute alloys in which spin-carrying ions, e.g., Mn^{++}, are randomly distributed in a nonmagnetic metallic matrix and interact with each other by long-range oscillatory forces. Spin glasses exhibit novel kinds of dynamic magnetic susceptibilities. They have been explored as possible models of neural networks.

The physics of radically nonperiodic systems is infinitely diversified and the foregoing remarks of necessity could touch on only a very small fraction of the important developments during this century. In particular, the interesting electronic transport process in such systems have not been addressed.

Bose-Einstein Condensation: Superfluidity and Superconductivity

The history of superfluidity and superconductivity in this century is, in the view of many, the most remarkable chapter in the 20th-century history of condensed matter physics. These systems not only behave in radically nonclassical, counterintuitive ways, but also for a long time could not be understood in terms of those quantum-mechanical concepts that had been highly successful in explaining the properties of atoms, molecules, and "normal," i.e., nonsuperconducting/fluid matter. This history had its beginning in 1908, when Kammerlingh Onnes had reached a sufficiently low temperature to liquify He^4 at 4.2 K, thereby creating the important new field of "low-temperature physics" including superconductivity/fluidity. Today, 90 years later, we believe that we understand most of the essential characteristics of these systems, except that the mechanism underlying so-called high-temperature (or high-T_c) superconductivity, which was discovered in 1986, still remains a mystery.

The story of superconductivity/fluidity is a wonderful example of the complexity of scientific progress, involving a mixture of serendipity and planned research, of phenomenology and *a priori* microscopic calculations, of well-designed experiments (in line with Onnes' motto "through measurement to knowledge") and of simple models. This in spite of the fact that in the Schrödinger equation we have, but only in principle, the "theory of everything" in CMP. The history of superfluidity is a good example of the need to be alert to the significance of only rough agreement between theory and experiment, as in the case of the experimental transition temperature of superfluid He^4 (2.2 K) and the theoretical condensation temperature (3.3 K) of a model of noninteracting bosons with the mass of He^4 atoms, as well as the need to be alert to the possible significance of tiny unexpected "blips," as in the experimental discovery of the superfluidity of He^3.

The close relationship between superconductivity of metals, first discovered in 1911, and the superfluidity of He^4, fully established in 1938, was first clearly grasped by F. London (partly in collaboration with his brother Heinz), who saw in both phenomena an underlying long-range order in momentum space. His two books *Superfluids I—Macroscopic Theory of Superconductivity* (1950) and *Superfluids II—Macroscopic Theory of Superfluid Helium* (1954) are marvels of what can be achieved in physics by the application of general fundamental physical principles—in this case, thermodynamics, classical electrodynamics, and general quantum concepts—in a thoughtful analysis of often quite "bizarre" experimental results. I have just re-read London's introduction to the first of these volumes (which actually deals in a coherent way with both superconductivity and He^4 superfluidity), written just a few years before the microscopic theories of Bardeen, Cooper, and Schrieffer (1957) of phonon-mediated superconductivity and the Bogoliubov theory of an interacting Bose gas for He^4. I found it exhilarating to see how very much of the most essential physics London had grasped without any microscopic knowledge of the underlying mechanism for superconductivity or of the effects of interactions on Bose-Einstein condensation.

Both phenomena are now understood to reflect the occurrence of a quasi-Bose-Einstein (BE) condensation, of the bosonic atoms in superfluid He^4, and of the bosonic electron pairs in superconducting metals. (He^3 superfluidity, though much more complex, is fundamentally analogous to superconductivity.) I write "quasi" because, although in superconductors/superfluids the original BE condensation for noninteracting bosons is very strongly modified, the most essential aspects of the BE condensation survive.

The interwoven histories of these systems, both experimental and theoretical, extending over almost a century and still evolving, is extraordinarily complex. Disrupted by two World Wars, it captivated the minds and hands of many of the world's best physicists. The 1957 Bardeen-Cooper-Schrieffer (BCS) microscopic theory of superconductivity, taking advantage of a half century of often inspired experimental and theoretical advances, in one blow provided a coherent quantitative explanation of the key properties and phenomena of simple superconductors. It may be regarded as the dramatic high point of this history, which was, however, followed by many additional major and unanticipated advances.

In this essay I must be content to give a very cursory account of the most crucial milestones before about 1950 and then only to list the subsequent high points. I refer the reader to the article in this volume by J. R. Schrieffer and M. Tinkham and that of A. Leggett for more extended accounts.

The beginning of it all was the totally unexpected discovery in 1911 by K. Onnes and G. Hulst of the sudden vanishing of the electrical resistance, or so-called superconductivity, of Hg when the temperature was reduced to below 4.2 K. Soon some other metals were found to be superconducting at a few degrees K, while others remained normal.

Liquid He^4, the cooling liquid in these experiments, was itself found to have the most unusual properties. In 1930 W. Keesom and J. N. van der Emde discovered accidentally that at very low temperatures liquid He passed through extremely small cracks which, at higher temperatures, were quite impervious to liquid or gaseous He. This suggested a vanishing or extremely small viscosity. At 2.2 K, He^4 exhibited a mysterious liquid/liquid phase transition, with no latent heat but a singularity in the specific heat, most clearly established by W. Keesom and A. P. Keesom in 1932. In 1936 they observed an extraordinarily high apparent heat conductivity. In 1938 P. Kapitza, who conducted closely related experiments, coined the term "superfluid."

On the superconducting front W. Meissner showed in 1933 that superconductors, in addition to the remarkable transport property of vanishing resistance, also had the remarkable thermodynamic property of perfect diamagnetism, i.e., the complete expulsion of a weak magnetic field from the interior.

Beginning in 1935 Fritz London, partly with his brother Heinz, using Meissner's discovery of perfect diamagnetism and general physical principles, proposed the new "London equation"

$$\lambda_L^2 \operatorname{curl} j + H = 0,$$

as the appropriate constitutive equation for the current density j in superconductors. Here λ_L is the so-called London penetration depth. He also brilliantly put forward the notion (later fully confirmed) that the electrons in a superconductor display a long-range order in momentum space.

Returning to superfluid He^4, F. London's insight (1938) that its extraordinary properties are the reflection of a BE condensation proved to be most fruitful. It became the basis of a highly successful two-fluid phenomenological model put forward by L. Tisza (1938) to describe the available experiments: One fluid was the superfluid, the other a normal fluid, each fluid having its own thermodynamic and dynamical variables like density ρ_s, ρ_n, velocity v_s, v_n, and specific entropy $s_s = 0, s_n$. (The vanishing of s_s is a consequence of the occupation of a single quantum state by the macroscopic condensate.) The two fluids were viewed as completely interpenetrating each other. Landau, who at first appears not to have accepted the concept of BE condensation, developed his own two-fluid model (1941 and later) in which the normal fluid consisted of the gas of elementary excitations, phonons, and—a new concept—rotons with vorticity. The two-fluid model led to a new collective mode, called second sound, which eventually was experimentally confirmed.

In 1946 Andronikashvili conducted an experiment which beautifully supported the two-fluid model. The moment of inertia of a slowly rotating stack of discs immersed in superfluid He^4 agreed with the picture that only the normal fluid is dragged along.

The first low-temperature experiments (1949) on the then exceedingly rare isotope He^3, which obeys Fermi statistics, showed no sign of superfluidity down to 0.5 K. This strongly supported F. London's contention that the superfluidity of He^4 depended critically on its Bose statistics.

Following is a mere listing of some of the most important developments since about 1950.

Around 1950, A. B. Pippard, in his microwave experiments on the London penetration depth λ_L, was led to the notion of a second length parameter, the coherence length ξ

entering a nonlocal generalized London equation in which the current $j(r)$ is proportional to an average of $A(r')$ over a range $|r'-r| \lesssim \xi$. ξ was found to have a strong dependence on the mean free path l. At about the same time L. D. Landau and V. L. Ginzburg put forward a phenomenological theory of superconductivity, which also included the coherence length, in term of a complex wave function $\psi(r)$, playing the role of a space-dependent order parameter. This theory grew out of Landau's general theory of phase transitions, coupled to general principles of electrodynamics. A crucial parameter was the ratio $\kappa \equiv \lambda_L / \xi$; values of $\kappa < (2)^{-1/2}$ resulted in the "usual" kind of superconductivity with a complete Meissner effect. In 1957 A. A. Abrikosov showed that when $\kappa > (2)^{-1/2}$, in so-called type-II superconductors, magnetic-field tubes can penetrate the superconductor, forming a vortex lattice.

In 1950, H. Fröhlich put forward a (nonpairing) theory of superconductivity, depending on electron-phonon interactions, and several experimental groups discovered independently an isotopic mass dependence of $T_c \propto M^{-1/2}$, consistent with Fröhlich's electron/phonon coupling concept. Independent of the specifics of Fröhlich's theory, the empirical isotope effect showed persuasively that lattice vibrations played an essential role in superconductivity.

The discovery of the isotope effect greatly fired up John Bardeen's old interest in superconductors and in the middle 1950s, he embarked on an intensive research program with two young collaborators, L. Cooper and J. R. Schrieffer. In 1956, Cooper showed that a normal electron gas with attractive electron-electron interaction is unstable with respect to the formation of electronic bound pairs (bosonic, so-called Cooper pairs). In 1957 this led to the microscopic BCS theory of phonon-mediated superconductivity, which gave a coherent and wonderfully successful description of a wide variety of properties and phenomena and has become the main paradigm for superconductivity.

The year 1961 saw the prediction and confirmation of what is now called the Josephson effect: A dc voltage V across a superconducting tunnel junction (superconductor/normal metal/superconductor) gives rise to an ac (!) current of frequency $2eV/\hbar$. (The charge $2e$ reflects the electron pairing; the frequency is independent of material properties.)

During more than three decades of painstaking materials research by B. Matthias and many others the highest known superconducting transition temperature rose by about 8 K to ~25 K. Suddenly in 1986 A. Mueller and G. Bednorz, studying a new class of materials containing stacks of hole-doped CuO_2 planes, discovered superconductivity at 30–40 K. Further studies of related compounds, now called high-T_c materials, have taken T_c up to about 160 K! While there is no doubt that the carriers are again electron pairs, there is a wide consensus, consistent with generally small isotope effects, that one or more mechanisms beyond electron-phonon coupling are at work; but there is no consensus about their nature. At century's end, this is a major challenge, as is the experimentalists' dream of reaching room-temperature T_c's.

Returning to the He isotopes: the Bose-Einstein condensation of He^4 was confirmed by painstaking analyses of neutron-scattering experiments, with a small macroscopic occupation of ~6% in the zero-momentum state at $T=0$, according to the best recent estimates. However, in 1995 BE condensates of over 90% of bosonic atoms of certain *dilute gases* with very weak interactions were produced by laser cooling and selective evaporation down to below microdegrees Kelvin and exquisitely studied.

The long-searched-for superfluidity of He^3, analogous to the superconductivity of electrons, but much more complex, was found below 3 millidegrees K by D. M. Lee, D. Osheroff, and R. C. Richardson in 1971.

Finally we mention that the concepts of pairing and of superfluidity/conductivity have had important applications in the theory of nuclear structure and of neutron stars.

All in all a heroic chapter, still unfinished, in the history of science.

Mesoscopics and Nanoscience

Sometime in the 1960s Richard Feynman is said to have given a talk entitled "There is always room at the bottom," in which he articulated a then new frontier of science, miniaturization of man-made structures down to dimensions of a few atoms, i.e., 1 nanometer = 10 Å. Today, we have in some respects reached this frontier, in others we are close; even single atoms have been successfully observed and manipulated. Far from being simply a matter of setting new records, this journey has led to some of the most exciting physics and most important technological advances of the last several decades. There can be no question that the journey will continue well into the next century.

The conception, successful fabrication (1948), and dramatic applications of the transistor (see the article by Riordan *et al.* in this issue) highlighted the possibility of controlling the dynamics of electrons in very thin surface and interface layers. This no doubt was a major impulse for what today is called mesoscopics and nanoscience.

A few words about the terminology: mesoscopic systems are "in the middle" between microscopic and macroscopic systems, i.e., they contain between about 10^3 and 10^6 atoms. These limits are very rough and depend on the context. Nanoscience, often overlapping with mesoscopics, emphasizes small dimensions, typically 1–100 nanometers, in one, two, or three dimensions.

Strict control of the chemical, structural, and geometric perfection of the samples has been of the essence. Among the numerous techniques used we mention, in particular, molecular-beam epitaxy (MBE) going back to the 1970s, which has allowed the fabrication of layer structures that are atomically flat and compositionally controllable to an accuracy of about 1%. Combined with lithographic techniques it has permitted fabrication of complex semiconductor/metal structures on a nanoscale.

Mesoscopic and nanosystems highlight the qualitative difference between 1D, 2D, and 3D systems. Of course, literally speaking, all physical systems are three dimensional. However, if, for example, electrons are trapped in a layer of sufficiently small thickness, the motion normal to the layer is quantum-mechanically "frozen out" and the system behaves like a 2D gas. Similarly in a "quantum wire" the electrons are confined to a small cross-sectional area in the x-y plane and constitute a 1D electron gas.

We list here some of the interesting results associated with lower dimensions.

(1) Long-range order: In 1935 Peierls had noticed, by direct exact calculations for harmonic lattices, that while in 3D there is long-range (really infinite) order, even at finite T, in 2D it exists only at $T=0$, and in 1D not even then. Later this observation was generalized by N. D. Mermin and H. Wagner (1966) to include other order parameters like magnetization and the superconducting gap function.

(2) Localization: Calling l the nominal elastic mean free path, an arbitrarily weak static disorder localizes electrons over a distance l in 1D (N. F. Mott and W. T. Twose, 1960) and over generally longer distances in 2D (D. Thouless, 1980s). In 3D, for strong disorder all states are localized, for weak disorder only those near the band edges, so-called Anderson localization (P. W. Anderson, 1958).

(3) Luttinger liquid: As Landau first realized, the effect of electron-electron interaction in

3D is rather mild. In particular the Fermi surface in k space remains sharp at $T=0$. However, in 1D, as shown by J. M. Luttinger (1966), interaction effects smooth out the momentum distribution at the Fermi energy. They also have dramatic effects on low-temperature transport processes. (It has been suggested that a 2D electron gas may also be a Luttinger liquid.)

(4) The 2D quantum Hall effect: The ordinary 3D Hall effect, discovered at the end of the last century, occurs when a dc electronic current, flowing in the x direction, is subjected to a magnetic field B_z in the z direction. The resulting Lorentz force in the y direction is balanced by an electric field E_y (due to induced surface charges). It is given by $E_y = \rho_{yx} J_x$, where J_x is the x-current density and ρ_{yx}, the so-called Hall resistivity, is given by $\rho_{yx} = -B_z/nec$, where n is the electron density. This result is robust under the action of moderate periodic potentials, electron-electron interactions, and electron scattering by impurities or phonons. The 3D Hall effect is very useful for determining the electron density n which, when combined with the conductivity, also yields the electron mobility—a critical figure of merit.

The Hall effect has also been observed for 2D electron gases moving freely in a confining surface layer of a semiconductor but being "frozen out" in the perpendicular, z direction. The expected result was $V_H = -B_z I_x/n_2 ec$, where n_2 is the electron number/unit area and I_x the electron current in the x direction. By means of a perpendicular gate voltage V_G the chemical potential μ and hence n_2 could be changed. In 1986, Von Klitzing and co-workers, working with a 2D electron gas of very high perfection, made the startling discovery that the Hall voltage V_h, as a function of V_G, had a series of steps for discrete values of n_2 given by $n_2 = \nu(eB_z/h)$ where ν was an integer. This is called the integral quantum Hall effect. Furthermore, associated with the Hall steps, the voltage V_x across the sample *parallel* to the current drops to 0, i.e., the diagonal conductivity in the x direction becomes infinite! Shortly afterwards Hall steps were also found for $\nu = 1/3$ and other fractional values $\nu = p/q$: the fractional quantum Hall effect. What is the physical meaning of these numbers? It was first shown by L. Landau in 1933 that a magnetic field B_z bunches the eigenvalues of 2D free electrons into discrete, highly degenerate levels given by $\epsilon_n = (n + \frac{1}{2})\hbar\omega$, where $\omega = (eB_z/mc)$. A value of $\nu = 2$ indicates that the electrons exactly fill the two lowest Landau levels, and $\nu = 1/3$ corresponds to 1/3 filling of the lowest level.

A beautiful explanation of these experiments was provided by R. B. Laughlin (1981, 1983) in which gauge invariance, localization of electrons by disorder, and so-called fractional statistics (a generalization of the "integral" Fermi and Bose statistics) play critical roles.

(5) Universal conductance fluctuations: The conductance of a macroscopic disordered metal will, of course, vary very slightly from sample to sample because of differences in the precise configuration of the atoms. Provided that the inelastic scattering length is much greater than the length of the sample, these conductance fluctuations are of the order of $(e^2/\pi\hbar)$ irrespective of the sample resistance, which can be easily measured (R. A. Webb and others, 1985). Diffusion of a single impurity, from one equilibrium position to another, produces such a fluctuation. Underlying this and many other mesoscopic phenomena is the fact that, while inelastic electron scattering destroys phase coherence, elastic scattering does not.

(6) The Aharonov-Bohm effect and related effects: The so-called Aharonov-Bohm effect (1959) has found interesting applications in nanoscience. The basic geometry is

shown in Fig. 6. Even when the magnetic flux Φ is entirely inside the ring and there is thus no magnetic field acting on the electrons, the conductance dI/dV depends periodically on Φ/Φ_0, where $\Phi_0 (\equiv hc/e)$ is the flux quantum. This is due to the fact that the flux introduces *differential phase shifts*, between the parts of an electron wave function propagating in the upper and lower halves of the ring, which affect their interference at the exit point. When Φ/Φ_0 is an integer, the differential phase shift is a multiple of 2π, equivalent to 0. For the observation of this effect, inelastic scattering, which destroys phase coherence, must be negligible. Thus ultralow temperatures and highly miniaturized systems are required ($T \approx 1$ K, dimensions≈ 1 μ). There is another novel periodicity of $\Phi_0/2$ associated with electron paths and their time-reversed partners. For these path pairs the condition for interference is strictly independent of the particular positions of the impurities [B. Altshuler and others (theory), D. and Yu. Sharvin (experiment), 1981].

(7) Quantum dots: By a variety of experimental techniques it has become possible (1980s) to fabricate so-called quantum dots, singly or in arrays, of lateral dimensions of $\sim 10-1000$ nanometers. The number of mobile electrons in such a dot is typically $O(1-10^4)$. They are sometimes called "artificial atoms" because, due to the small dimensions of the dot, level spacings are correspondingly large. The conductance of such a dot shows enormous fluctuations as a function of gate voltage. These reflect successive resonances of electronic energy levels with the Fermi energy of the attached leads. The positions of these levels are strongly affected by the Coulomb repulsions between electrons, which is the origin of the so-called Coulomb blockade. As miniaturization progresses, it leads to greater spacings Δ between the electronic energy levels. Thus the necessary temperatures, $kT \approx \Delta$, for strong mesoscopic effects is rising. It seems not unreasonable to expect that nanoscience, in the next few decades, will have practical applications at liquid-nitrogen or even room temperature.

Soft Matter

The traditional conceptual paradigm of condensed matter physics, going back to the early part of this century, has been the picture of a dense periodic lattice of ions with valence

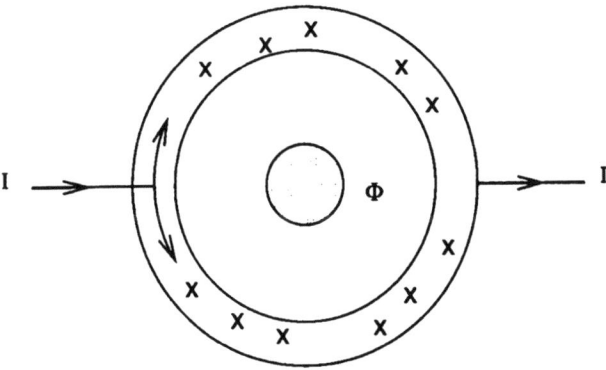

Figure 6. The Aharonov-Bohm geometry (schematic). Φ is a magnetic flux threading a metallic ring with impurities.

electrons described by a band structure. This paradigm, with its elaborations and modifications, has been spectacularly successful and continues to underlie much of the ongoing work in CMP, especially for systems with conducting electrons. Let us call it the Bloch paradigm. For insulators a paradigm viewing condensed matter as a collection of interacting atoms or ions (which has been shown to be equivalent to the Bloch paradigm) is often more natural. Both standard metals and insulators have in common periodicity and dense packing. They are very stable and resistant to small perturbations. Let us call them "hard."

Simple liquids of course do not have an underlying periodic lattice. Yet local atomic configurations, densities, cohesive energies, and static compressibilities are very similar above and below the melting temperature. But, unlike crystalline solids, liquids offer no resistance to a static shear stress. In the present context we call them hard/supersoft.

There is, however, another major class of materials, in recent years called "soft matter," whose properties and behavior are covered by different paradigms. Their study has been shared between chemists and physicists. They are characterized by the fact that, unlike gases and ordinary liquids, they do have some shape (or other) stability, but unlike "hard" materials, they respond very strongly to small external disturbances—mechanical, electrical, etc. For example, most edibles, like Jell-O, or fibers, like wood or wool, fall into this category. The bonds between the relevant constituents are usually weak (van der Waals, hydrogen) or easily swiveled or both.

Polymers

Polymers are the best known and most extensively studied subclass of soft matter. The simplest ones consist of a chain or "necklace" of identical units, called monomers, strongly and rigidly bonded internally, but flexibly bonded to each other. Monomers on different chains interact by strong short-range repulsions and weak long-range attractions.

Polyethylene,

$$CH_2-CH_2-CH_2-\cdots \text{ or } [CH_2]_N \ (N \approx 10^{3\pm 1}), \tag{11.1}$$

is one of the simplest and most thoroughly analyzed. While examples of polymers, e.g., rubber, were known in the 19th century, their nature was clarified only in this century. In 1920 H. Staudinger proved chemically the linear character of polymers. Their flexibility was first demonstrated in 1934 by W. Kuhn for independent polymer chains in dilute

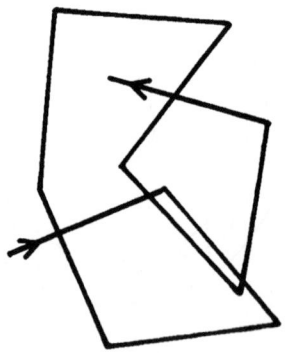

Figure 7. A polymer or a random walk (in 2D).

solution. He recognized that at finite temperatures a correspondence could be made between the instantaneous shape of a polymer and the trace of a random walk, with the constituent monomers corresponding to the successive steps. This pointed to the very important role of entropy, much higher for a configuration with random orientations of the constituent monomers than for a completely aligned configuration (Fig. 7). From this emerged the picture of a long polymer chain like a tangled ball of yarn after a long period of being "cat-tangled." The elementary theory of random walks led Kuhn to the famous scaling law that the end-to-end distance of a long polymer in solution, and also the effective radius of the tangled 3D ball, obey the scaling law, $R = aN^{1/2}$. Here a is of the order of the length of a monomer, whose precise value depends on the flexibility of the bonds between monomers, etc., but the exponent 1/2 is universal. (A far cry from the paradigms of crystalline solids!)

Kuhn's random-walk analysis was substantially deepened by P. J. Flory (1949), who included the effect of the strong intermonomer repulsions, leading to the analogy with a self-avoiding random walk. A simplified analysis led him to the modified result $R \propto N^{3/5}$, refined by subsequent numerical work to $N^{0.588}$. (On account of these fractional exponents, polymers are examples of so-called fractals, which attracted much attention in the 1980s.)

Because of their enormous and growing scientific and practical importance (as fibers, structural and biological materials, packaging, adhesives, etc.), research and development of polymers has become a rapidly progressing subfield of CMP. The theories of single linear polymer chains have been extended in many directions: polymers consisting of more than one monomer (co-polymers); nonlinear, branched chains; mutually entangled chains; polymeric crystals, melts, and glasses; gels; diffusion of polymer chains through the tangle of other chains in a melt (reptation, which means snaking); polymers at interfaces and adhesion.

Polymer physics has enormously benefited from the development of elastic and especially inelastic neutron-scattering techniques led by the work of C. Shull and B. Brockhouse in the middle decades of this century, and these techniques have been applied to polymers since the 1970s by G. Jannink and others. This is because polymers consist largely of light elements (H,C,O,...), which interact weakly with x rays but strongly with neutrons. Furthermore, typical frequencies of collective modes of polymers are of the order of $10-10^2$ K, comparable to the energies of cold or thermal neutrons. Finally isotopic replacement of H by D, with very different masses and neutron-scattering properties, has provided a useful research "handle." Thus we owe much of our knowledge of both the structure and the dynamics of polymers to neutron experiments.

Membranes

Since the 1970s membranes, another important category of soft matter, have received increasing attention from physicists. Their building blocks are molecules with a hydrophilic (water-loving) head and a hydrophobic (water-fearing) tail.

When dissolved in water these molecules tend to aggregate in flexible two-dimensional membrane structures that "protect" the hydrophobic tails from water (Fig. 8).

For a given flexed shape the elastic free energy of a single free membrane is given by

$$F = \frac{K}{2} \int dS \left(\frac{1}{R_1} + \frac{1}{R_2} \right)^2, \qquad (11.2)$$

Figure 8. A free-floating flexible membrane.

where K is the bending stiffness and the R_j are the radii of curvature. Because of thermal fluctuations, memory of the direction of the normal is lost at a characteristic decorrelation distance L_p proportional to $\exp(2\pi K/kT)$, beyond which the membrane will be crumpled.

Membranes are, of course, very important in biology. In fact, the entire broad field of soft matter has become an important bridge between physics and biology.

Liquid crystals

The term "liquid crystals" seems to be self-contradictory, but in fact these fascinating materials in some ways strongly resemble conventional, ordered crystals and in other ways conventional liquids. Liquid crystals consist of highly anisotropic weakly coupled molecules. They were first discovered and their essence understood by the French chemist G. Friedel at the end of the 19th century. After a long period of relative neglect their study was actively resumed in the 1960s. There are two main classes, nematics, and smectics. Figure 9 shows a liquid crystal of the "smectic A" type. It clearly exhibits substantial orientational order, in the orientation both of the molecules (along the z direction) and of the layers (in the x-y plane). Furthermore, the spacing of the layers in the z direction exhibits translational order in the z direction extending over many planes. All these characteristics are reminiscent of conventional crystals. However, the positions and motions of the molecules within any one layer in the x-y plane are highly disordered and resemble those of a two-dimensional liquid.

In some liquid crystals the constituent anisotropic molecules can be realigned by very weak electric fields, which in turn strongly affects optical properties. Such liquid crystals have found extensive use in the displays of electronic watches and other devices (see the chapter by Witten on pp. 617–628 of this book).

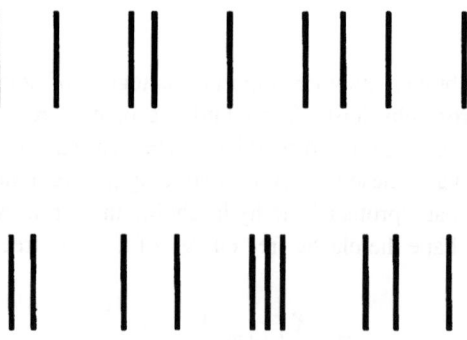

Figure 9. A "smectic A" liquid crystal.

General Comments

It is perhaps interesting to look at the history of condensed matter physics from the viewpoint of T. S. Kuhn, as expressed in *The Structure of Scientific Revolutions* (Chicago, 1962). He sees scientific history as a succession of (1) periods of "normal" science, governed by serviceable scientific paradigms, followed by (2) transitional, troubled periods in which existing paradigms are found to be seriously wanting, which in turn lead to (3) "scientific revolutions," i.e., the establishment of new paradigms, which may or may not be accompanied by the rejection of the old ones.

Such a linear view seems applicable to the whole field of CMP for some of the broadest revolutions, which directly or indirectly affected a large fraction of the field: x-ray diagnostics yielding crystal structures (1910s); achievement of low temperatures allowing the observation of calmed condensed matter (1900s); quantum mechanics, (1920s); the bandstructure paradigm (1920s, 1930s); nuclear and electron spin magnetic-resonance diagnostics (1940s and 50s); neutron elastic and inelastic diagnostics (1950s); many-body electron theories (beginning in the 1930s, with major revolutionary steps in the 1950s and 60s); electronic computer-assisted theory and experiments (1960s–); soft matter (1960s–); and nanoscience (1980s–).

The indicated decades are for rough orientation only. I have included not only conceptual revolutions in the sense of Kuhn but also experimental and technical ones that transformed existing areas of inquiry or opened up important new ones.

Within subfields of CMP there are many additional revolutions. For example: Heisenberg's theory of strong magnetic interactions for magnetism (1920s); scaling and renormalization group for critical phenomena (1960s–); Bose-Einstein condensation and the BCS pairing theory for He^4 and superconductors (1930s and 1950s); masers/lasers for high-intensity, coherent radiation studies (1960s); dislocations for the strength of materials (1930s); high vacua for studies of clean surfaces (1950s).

Others would no doubt choose differently, but few would disagree that the 20th century has been revolutionary for CMP. The fertilizing influence of CMP, conceptual and technical, for other subfields of physics and other sciences has been repeatedly mentioned. Further, condensed matter physics has been a major factor in reshaping technology so that the human experience today is, for most of mankind, very different from what it was 100 years ago.

Looking back over the last century, we see major shifts to the use of more and more sophisticated, man-designed and -fabricated materials, more and more miniaturization, and radically more sensitive diagnostic techniques. Prognostications are, fortunately, beyond the scope of this essay. But it is obvious that the future holds many promises.

Acknowledgments

I am grateful to Drs. D. Clougherty, A. Gossard, C. Herring, L. Hoddeson, J. Langer, Y. Meir, and P. Pincus for helpful suggestions and comments.

References

Brown, Laurie, M. Abraham Pais, and Sir Brian Pippard, 1995, Eds., *Twentieth Century Physics* (Institute of Physics, Philadelphia).

Hoddeson, Lillian, Ernst Braun, Jurgen Teichmann, and Spencer Weart, Eds., 1992, *The History of Solid State Physics* (Oxford University Press, New York).

London, F., 1950, *Superfluids I—Macroscopic Theory of Superconductivity* (Wiley, New York); revised 2nd Ed. (Dover, New York, 1961).

London, F., 1954, *Superfluids II—Macroscopic Theory of Superfluid Helium* (Wiiley, New York); revised 2nd Ed. (Dover, New York, 1964).

Riordan, Michael, Lillian Hoddeson, and Conyers Herring, 1999, Rev. Mod. Phys. **71**, 336; pp. 563–578 in this book.

Warnow-Blewett, Joan, and Jürgen Teichmann, 1992, *Guide to Sources for History of Solid State Physics* (AIP, New York).

A Short History of Atomic Physics in the Twentieth Century
Daniel Kleppner

This brief account describes some of the developments of atomic physics during the twentieth century that particularly appeal to the author's sensibility, that of an experimental physicist. It makes no pretense to being comprehensive. On the contrary, major theoretical and experimental areas have been omitted because of the space limitations. Several excellent historical studies that can help to fill out the picture are listed in the Bibliography.

In 1943 the American Physical Society established the first of the many divisions by which physics is now split into subfields. This was the Division of Electron Physics, later to become the Division of Electron and Atomic Physics, and later yet to become the Division of Atomic, Molecular, and Optical Physics. At the beginning of the century, however, such distinctions were unnecessary. What we now call atomic physics was then the very core of physical science.

Here are some of the concepts that were on hand at the turn of the century. There was overwhelming but indirect evidence for the existence of atoms, including the success of kinetic theory, Mendeleev's periodic table, the association of spectral lines with elements, the existence of electrons and ions, and an understanding of the electromagnetic origin of radiation by matter. (If the Nobel Prize can be regarded as the ultimate sanction of scientific credibility, however, then official recognition of the existence of atoms came surprisingly late, in 1926, when the prize was given to Jean Baptiste Perrin for research on the "discontinuous structure of matter.") In the background were the great edifices of Newtonian mechanics and electromagnetic theory, and, on a somewhat less firm pedestal, thermodynamics and statistical mechanics.

Towards the close of the 19th century the accomplishments of physics were so astonishing that Oliver Lodge exclaimed that "the whole subject of radiation is working out splendidly," and at the opening of the Ryerson Laboratory of the University of Chicago in 1894 A. A. Michelson stated that "The more important fundamental laws and facts of physical science have all been discovered...." Nevertheless, there were vexing problems, for instance the failure of simple gases to have the predicted heat capacities, the lack of any real understanding of atoms, and the failure to discover a key for interpreting the thousands of pages of accurate spectral data that had been accumulated over the decades. Then as the century drew to a close a revolution was precipitated by the discoveries of radioactivity, x rays, electrons, and the electrical nature of matter, and particularly the recognition of the complete failure of statistical mechanics to describe a thermal radiation field.

The First Thirty-Five Years

The scientific revolution that led to the creation of modern physics was largely accomplished in the first three decades of this century. Its major achievements were Einstein's theories of relativity and gravitation, and the creation of quantum mechanics. It is quantum mechanics that plays the most important role in this history, for before its creation atomic theory was crude and fundamentally empirical—which is to say there really was no theory—while afterwards there existed a comprehensive theory that provided a new language for describing nature and could account for atomic and molecular structure and dynamical processes in exquisite detail. The major figures in the development of quantum mechanics are well known: Planck, Einstein, Bohr, and later de Broglie, Heisenberg, Schrödinger, Dirac, Pauli, and Born. Such a confluence of theoretical genius represents one of those remarkable episodes in history when great minds profoundly change our world view, but their achievements were inspired and guided by the discoveries of experimenters who were also scientists of genius.

The first intimations that the foundations of physics might be fundamentally flawed surfaced in 1900 when Planck first introduced the concept of quantization. Planck's proposal was directly inspired by an experiment. An accurate spectrum in the near infrared of energy radiated by a hot body had been obtained in 1897 by E. Paschen and G. Wien, who discovered that the data could be accurately described by an expression that decreased exponentially with frequency. In October, 1900, while attempting to find a physical justification for Wien's exponential rule, Planck learned of surprising results from two groups, O. Lummer and E. Pringsheim, and H. Rubens and F. Kurlbaum. Using new techniques for infrared detection they were able to extend the radiation measurements farther into the infrared regime. To his confusion, Planck found that the new data seriously departed from the exponential behavior predicted by Wien. Before the end of the year, however, Planck found a new empirical expression that fitted the thermal spectrum throughout the infrared and visible range. He pointed out that one could "derive" his expression from statistical mechanics by simply quantizing the energies of the fictitious oscillators with which he modeled matter. This quantum hypothesis was so outlandish, however, that Planck regarded it as little more than a mathematical trick.

Planck's hypothesis had essentially no impact until 1905 when Einstein treated it seriously, pointing out that it implied that light itself must have quantum properties. Planck's hypothesis was motivated by experiment, but it had little direct consequence. Characteristically, Einstein's theory seemed to be motivated by no experimental evidence, but it had lots of consequences. One was that the energy of a photoelectron should depend only on the frequency of light, not its intensity. The photoelectric effect had been discovered by H. Hertz in 1887. In 1899 J. J. Thomson showed that the effect resulted from the ejection of electrons. In 1909 R. A. Millikan carried out the first of a series of studies of photoelectron energy and the results were consistent with Einstein's hypothesis. Nevertheless, the hypothesis itself remained controversial. However, in a 1916 paper, Einstein showed that light quanta also carry momentum and this was experimentally confirmed by A. Compton in 1923. Compton measured the energy loss in x-ray scattering due to the electron recoil. His experiment left little doubt as to the physical reality of light quanta.

The driving force for the quantum theory of radiation was the problem of thermal radiation, but the driving force for the actual creation of quantum mechanics was the need to understand atoms. The crucial event was Bohr's 1913 paper on the hydrogen atom, in which he introduced the concept of stationary energy states and quantum jumps accompanied by the emission of monochromatic radiation. The paper is remarkable for its daring introduction of radical ideas and its cavalier disregard of classical electromagnetic theory.

Bohr's starting point was the discovery of the atomic nucleus. In 1911 E. Rutherford, building on his studies of radioactive transformations, carried out the classic experiment on alpha-particle scattering from gold, which resulted in his discovery of the nucleus and led him to propose that atoms have planetary-like properties. By combining the classical description of an electron moving in the field of a proton with rules that were absurd by contemporary standards, Bohr accounted for the existence of atomic spectral lines, the exact form of the hydrogen spectrum, and the precise numerical value for the single constant in Balmer's empirical formula—the arguably misnamed Rydberg constant. This was one of those rare syntheses in physics in which apparently unrelated data are combined to describe phenomena that previously seemed unrelated, such as Newton's derivation of the acceleration of gravity on earth from the period of the moon, or Maxwell's deduction of the speed of light from the electric and magnetic force constants. All of these achievements led to a flowering of activity that confirmed the theory, but while Newton's gravitational theory and Maxwell's electromagnetic theory were essentially complete (at least, for their own epochs), Bohr's model of the atom was fundamentally incomplete. It was intended to serve as a guide and an imperative for a revolutionary new mechanics. The esteem in which Bohr was held by those who knew him can be traced to the vision with which he saw what was to come and to his role in guiding the revolution.

At the heart of Bohr's model was his concept of stationary energy states, an idea totally incompatible with traditional physics. Nevertheless, within a year its physical reality was demonstrated by J. Franck and G. Hertz in a study of the energy loss of electrons in a gas. Spatial quantization of angular momentum, a concept proposed by A. Sommerfeld that was equally at odds with traditional theory, was demonstrated in 1921 by O. Stern and W. Gerlach in an experiment on the deflection of atoms in an inhomogeneous magnetic field.

Early attempts by Bohr, Sommerfeld, and others to describe these phenomena, the "old" quantum theory, ultimately failed. The correct theory came in what seems, in retrospect, like a series of thunderbolts. In 1923 de Broglie pointed out that energy quantization could be achieved by associating a wavelength with the electron, in 1924 Heisenberg published his theory of matrix mechanics, and within a half year Schrödinger published his theory of wave mechanics. There was deep confusion about the interpretation of these theories until Born, in 1926, showed how to interpret them in terms of probability theory. In 1928, when Dirac presented his relativistic theory for the electron, quantum mechanics came of age.

Throughout the period of these developments, the major features of the nucleus were identified: the relation between nuclear charge and nuclear mass, isotopes, nuclear spin and statistics, and nuclear magnetic moments. The final nuclear constituent, the neutron, was discovered by Chadwick in 1932. With this understanding of the nucleus, and the creation of quantum mechanics, the foundations of atomic physics were complete.

Nearly all the major players in this history received the Nobel Prize in physics: Thomson (1906), Wien (1911), Planck (1919), Einstein (1921), Bohr (1922), Millikan (1923), Franck and Hertz (1926), Perrin (1926), Compton (1927), de Broglie (1929), Heisenberg (1932), Schrödinger (1933), Dirac (1933), and Born (1955). Rutherford received the Nobel Prize in chemistry in 1908, and in 1935 Chadwick received the prize in physics.

The Next Thirty Years

We break this chronological narrative for a moment and skip ahead to June, 1947, when a group of physicists met to discuss fundamental problems in physics at Shelter Island, New York. High on the agenda were questions about the validity of the Dirac theory of the electron, particularly the problem of the electron's self-energy and the possibility that there might be observable effects of the vacuum, issues that had been raised by H. Bethe, H. Kramers, and others. The problems had been in the air since the early 1930s, and late in the decade the possibility of an experimental test had been raised. According to the Dirac theory, the principal optical spectral line of hydrogen has two components, separated by the small fine-structure interval. There was some suggestion that a third component might exist, but the evidence—a possible substructure with a splitting much smaller than the width of the spectral line—was hardly definitive.

The Shelter Island meeting was devoted to theory, but three new experimental results were reported whose impact was profound. I. I. Rabi described the first atomic-beam resonance measurements of the hyperfine structure of hydrogen. The hyperfine interval was found to be larger than predicted by the Dirac theory by a little over one part in a thousand. In spite of this small size the discrepancy was big, for the experiment was accurate to about one part in a hundred thousand. G. Breit suggested that the discrepancy could signify a discrepancy in the size of the magnetic moment of the electron from the value predicted by the Dirac theory—in other words, a breakdown of the Dirac theory. Rabi also described a series of atomic-beam resonance experiments by P. Kusch, which confirmed that the electron's magnetic moment was indeed anomalous.

W. Lamb reported results of an experiment that left no doubt that the Dirac theory was in error. Lamb showed that there was indeed a third component in the fine-structure spectrum of hydrogen, using a radio-frequency resonance technique with a resolution hundreds of times superior to the best that could be achieved optically. The extra component was due to an energy splitting between two states which, according to the Dirac theory, should have had identical energy.

All three of these experiments gave precise values for effects that one decade earlier would have been unobservably small. Their impact was immediate: they triggered the creation of the modern relativistic theory of quantum electrodynamics (QED) by J. Schwinger, and R. P. Feynman, both of whom were at the Shelter Island meeting, and S.-I. Tomanaga. For these advances the Nobel Prize was awarded to Lamb and Kusch in 1955, to Feynman, Schwinger, and Tomanaga (who was not at Shelter Island) in 1965. Rabi received the prize in 1944 for the invention of molecular-beam magnetic resonance. Rabi's prize was for the experimental advance that made these experiments possible and that catapulted discoveries and new technologies for decades to come.

Molecular-beam magnetic resonance had its origin in the magnetic deflection technique developed by O. Stern to demonstrate spatial quantization. In 1933 Rabi set up a laboratory at Columbia University to apply the deflection method to improve on Stern's measurement of the proton's magnetic moment. In attempting to understand some problems in this experiment, Rabi realized that by applying a magnetic field that oscillates at the frequency with which the proton precesses in an applied magnetic field, one could reorient the proton's spin. The reorientation would be detectable because it would alter the trajectory of the molecule in a subsequent field gradient. In short, Rabi made it possible to determine a magnetic moment by measuring a frequency. Furthermore, the moment could be measured to a precision incomparably higher than had been obtainable by any previous method.

Molecular-beam magnetic resonance could achieve breathtaking precision by making it possible to observe atomic and molecular systems free from collisions or other perturbations, essentially in total isolation. The method revealed internal interactions in atoms and molecules and provided a wealth of information not only on atomic and molecular structure, but also on nuclear properties. Among the very first discoveries by Rabi's group was that the deuteron possesses a quadrupole moment. This was the very first evidence that the force between nucleons is noncentral. War abruptly brought the research to a halt, but at the war's conclusion, the research rushed forward. One new stream of studies was devoted to determining the spins, magnetic moments, and higher-order moments in nuclei by atomic-beam magnetic resonance; another to determining magnetic and electronic interactions in molecules using magnetic and electric molecular-beam resonance.

The inherent resolution of molecular-beam magnetic resonance is determined by the uncertainty principle. The resolution increases directly with the time during which the atom interacts with the oscillating field. In principle one can increase this time simply by making the apparatus longer, but this strategy soon ran into technical difficulties. However, these were largely overcome by N. F. Ramsey's invention of the separated oscillatory field method in 1950. Ramsey's method opened the way to a wealth of studies on the internal interactions in molecular hydrogen and other molecules, and on hyperfine structure in atoms. When this method was used to measure the hyperfine interval of cesium, the transition frequency could be determined with such high accuracy that it could be employed as a frequency standard, providing the basis of an atomic clock. The first cesium atomic clock was operated in a standards laboratory by L. Essen and J. V. L. Parry in 1955, and J. R. Zacharias pioneered the construction of a practical, portable, cesium atomic-beam clock. Cesium clocks were soon constructed in the world's major standards laboratories. These clocks have been steadily refined over the decades and now provide the timing basis and the satellite-borne clocks that made possible the Global Positioning System. Ramsey was awarded the Nobel Prize for the separated oscillatory field method in 1989.

Increasingly sensitive tests of QED have been carried out up to the present day. The free electron and the hydrogen atom continue to provide principal testing grounds, as will be described, but high-precision studies have also been carried out on the spectrum of helium and high hydrogenlike and heliumlike heavy ions. Quantum electrodynamic tests using hydrogen are eventually limited by uncertainties in the structure of the proton, and to overcome this problem V. W. Hughes created muonium (the muon-electron atom) in 1960. Studies of the hyperfine structure of muonium by Hughes and V. L. Telegdi, and later studies of the optical spectrum, are among the critical tests of QED. Positronium (the electron-positron atom), first created by M. Deutsch in 1952, has been similarly employed, with early measurement of its hyperfine structure followed by later studies of its optical spectrum.

A related development during this period was the creation of nuclear magnetic resonance by F. Bloch and, independently, by E. M. Purcell, in 1946. Nuclear magnetic resonance (NMR) provided a new method for measuring the spins and magnetic moments of nuclei. The principle applications have been to the study of molecular structure, the structure and dynamics of solids, liquids and gases, and to biological and medical applications, including the technique of magnetic-resonance imaging. However, because these applications are somewhat distinct from atomic physics, we mention these developments only in passing.

The wartime advances in electronics, particularly in radar, played a profound role in advancing the magnetic-resonance experiments of the late 1940s. Concerns about the absorption of microwave signals by the atmosphere, particularly by water vapor, had stimulated microwave absorption measurements in gases. At the war's end these methods were employed to measure molecular rotational and vibrational structure. Working at Columbia, C. H. Townes realized that if he selected molecules occupying the upper of two energy levels and applied a field oscillating at the transition frequency, the radiated energy would stimulate the molecules to radiate, adding to the energy in the applied field and thereby amplifying it. If the fields were large enough, the device would oscillate. For such a device the name "maser" was coined, an acronym for microwave amplification by stimulated emission of radiation.

Stimulated emission—the physical process at the heart of maser operation—was first recognized by Einstein in 1916. Under normal conditions of thermal equilibrium, however, in populations of atoms or molecules, the number of particles in a lower state always exceeds that in a higher energy state, and radiation is absorbed rather than amplified. The first maser was demonstrated by Townes in 1954, using a microwave transition in ammonia which had been prepared in an excited state by molecular-beam methods. The principle of maser operation was recognized independently in the Soviet Union by N. G. Basov and A. M. Prokhorov, who shortly afterward also achieved maser operation with ammonia. The statistical properties of the maser's radiation were of immediate interest: these studies inaugurated the field of quantum optics. The maser was also investigated as a frequency standard and as an amplifier. In 1956 N. Bloembergen proposed a solid-state three-level maser in which microwave pumping created a population inversion, operating on paramagnetic ions in a host lattice. Such solid-state masers were soon developed by a number of groups and found immediate application as low-noise amplifiers by radio astronomers. Among the discoveries made with these masers was the existence of the 3-degree cosmic background radiation by A. Penzias and R. W. Wilson in 1965.

The possibilities for achieving maser action for frequencies extending into the optical region were discussed by A. L. Schawlow and Townes in 1958. In 1960 the first optical maser, soon to be dubbed a laser, was demonstrated by T. H. Maiman, using a ruby system that was optically pumped with a flashlamp. Shortly afterwards continuous laser action was achieved by A. Javan, using a gaseous discharge of helium and neon to generate infrared radiation in an inverted population of neon ions. Soon thereafter the helium-neon was operated in the optical region on a red transition, forming the familiar red "HeNe" laser that has been a workhorse ever since.

Two other advances in this period deserve mention, for they established a major theme for atomic physics in the decades to come: control of the motions and the internal states of atoms and ions. A. Kastler devised a method for polarizing atoms by absorption of circularly polarized optical resonance light. Atoms in a thermal distribution of internal magnetic states could be transformed into a single state. The state could be changed by applying a radio-frequency field and detected by monitoring the transmitted light. This effect, called optical double resonance, was first observed by J. Brossel and Kastler in 1953. The method opened the way to new measurements of atomic interactions and to the creation of magnetometers and optically pumped atomic clocks.

In attempting to develop new ways to guide and focus ions, W. Paul discovered that by combining an oscillating and a static quadrupole electric field he could achieve regions of stability in which ions of a certain charge-to-mass ratio are efficiently channeled. The method provided the basis of an extremely simple and sensitive mass spectrometer, which is now widely used in research and industrial applications. He went on to operate the device in three dimensions, creating a trap for ions that would hold the particles almost indefinitely. This work established a theme that has continued ever since—increasingly precise control of the motions of ions and atoms.

Townes, Basov, and Prokhorov received the Nobel Prize in 1964 for the maser and the laser. Kastler received the Prize in 1966 for his method of radio-frequency spectroscopy with optically pumped atoms, and Paul received the Nobel Prize in 1989 for development of the ion trap technique.

Atomic Physics since 1965

During the final third of this century lasers became ubiquitous in daily life. They revolutionized communications and found applications from heavy manufacturing to eye surgery. Lasers also became ubiquitous throughout the sciences, with applications ranging from aligning great telescopes and gargantuan accelerators to measuring sizes and shapes of macromolecules. In atomic physics the advent of tunable lasers caused a fundamental change in the concept of spectroscopy. Initially lasers "merely" increased spectroscopic resolution by several powers of ten, but then they opened the way to the creation of new atomic species, the extension of spectroscopy from the frequency to the time domain, the development of nonlinear optics, and the creation of powerful ways to manipulate and control atoms. In addition, the generation of laser light precipitated new studies in the statistical properties of light, the nature of light-matter interactions, and nonlinear optics. It created the field that grew into quantum optics. In this brief history one can only pick among some of the highlights.

Following the creation of the first ruby laser and the gaseous helium-neon laser, an arsenal of other types of lasers was developed and rapidly employed in atomic physics: gaseous lasers operating on rare-gas ions and various molecular species, solid-state lasers operating in the infrared and visible regimes, ultraviolet excimer lasers, and semiconductor diode lasers. All of these emit radiation at one of a series of discrete frequencies. Laser spectroscopy, however, requires continuously tunable radiation. This became a reality in 1965 when P. P. Sorokin invented the dye laser.

In traditional spectroscopy the resolution is limited by the thermal motion of the atoms—the first-order Doppler effect. The high spectral purity of a laser does not by itself overcome this problem. However, as pointed out by Lamb, the Doppler effect can be eliminated by using one laser beam to excite atoms that happen to be at rest, and a second to probe them. This technique, known as saturation spectroscopy, was applied by T. W. Hänsch in 1974 to study spectra in alkali atoms, the workhorses of atomic physics. Hänsch employed a relatively simple tunable dye-laser design that was quickly taken up by other laboratories, essentially opening a floodgate of new research.

Once Doppler broadening is eliminated, spectral resolution is often limited by the time available for the particle to interact with the radiation field. In the 1970s J. Hall and V. Chebotayev constructed a spectrometer designed to lengthen this time for a molecular gas by employing a wide-diameter radiation field with carefully controlled optical properties. With such spectrometers a series of spectral "atlases" were created that provided ultraprecise frequency markers across wide spectral regions. Laser stabilization techniques have been steadily refined by Hall, Hänsch, and others, and stability of greater than one part in 10^{14} over a period of many seconds has been achieved.

Schawlow played a major role not only in the creation of the laser but in many of the innovations of laser spectroscopy. For these contributions he received the Nobel Prize in 1981.

The intense fields of laser light make it possible to observe high-order radiation processes such as multiphoton transitions that are essentially unobservable with conventional light sources. Hydrogen, which continues to serve as a touchstone for spectroscopy, has yielded the most precise test of QED in an atom though study of such a transition—the two-photon transition from the ground state to the metastable 2S state. V. Chebotayev pointed out that by exciting the hydrogen in counterpropagating laser beams, one could excite every atom in the gas with no broadening due to the first-order Doppler effect. Hänsch observed the Doppler-free transition in hydrogen in 1975, and in a continuing series of advances in the control of atoms, the stabilization of lasers, and optical frequency metrology he eventually measured the transition to an absolute accuracy of four parts in 10^{13}. Combining this result with other ultraprecise measurements of hydrogen yields a value for the Lamb shift in which the comparison with QED is limited only by uncertainty in the charge distribution in the proton.

The most stringent of all low-energy tests of QED is a comparison of the experimental and theoretical values of the magnetic moment of the free electron. In the initial measurements of Rabi and Kusch, the magnetic moment anomaly—the discrepancy with the Dirac value—was precise to one percent. H. Dehmelt achieved a precision of three parts in 10^9 by observing a single electron confined in a trap consisting of a static quadrupole electric field and a magnetic field (the Penning trap). The electron in such a trap executes both cyclotron and spin precessional motions at frequencies which should be identical according to Dirac. Transitions between the two motions are induced by a weak oscillating field, and the state of the electron is monitored by measuring its vibrational amplitude through the current it induces in the electrodes. The difference between the experimental value for the anomaly and the prediction of QED, as calculated by T. Kinoshita, is $51 \pm 30 \times 10^{-9}$. Whether the small discrepancy is real or due to a possible error in the fine-structure constant, which sets the scale for all the QED effects, remains to be determined. Within this uncertainty, this result represents the most precise low-energy test of QED and indeed the most precise test of any theory in physics. For this achievement, Dehmelt was awarded the Nobel Prize in 1989.

The tradition of extracting nuclear interactions from atomic measurements dates back to early studies of hyperfine structure in the 1920s, but a new line of research was created in 1974 when C. and M.-A. Bouchiat pointed out the possibility of measuring effects from parity-nonconserving electron-nucleon interactions predicted by the electroweak theory. Experimental searches were carried out by several groups, as were major theoretical efforts to calculate the effect of the electroweak interactions on atomic structure. In 1996 C. E. Wieman succeeded in measuring the ratio of two electron-quark parity-violating interactions in cesium. The ratio adds a further constraint to the standard model, taking its place with the large body of data from high-energy physics on which the standard model is built.

In a work published the year after his 1916 paper that introduced stimulated emission, Einstein pointed out the intimate connection between momentum exchange and energy exchange in establishing the motional equilibrium of atoms and radiation. Fifty years were required for stimulated emission to be exploited in the creation of the maser and the laser; another twenty years were needed for atom-radiation momentum interchange to be exploited to manipulate and control atomic motion, and then to cool atoms to the microdegrees kelvin regime. A number of streams of research converged to achieve these advances.

Studies carried out by A. Ashkin in the early 1970s on the force of light on dielectric particles helped to stimulate research on the force of light by atoms. The research also produced the technology of "optical tweezers" for manipulating small particles, making it possible, for instance, to manipulate not only cells, but also the material within them. In the mid 1970s V. S. Letokhov and V. G. Minogin demonstrated effects of the alteration of atomic velocities with laser light.

The possibility of cooling atoms with radiation was proposed in 1975 by D. J. Wineland and Dehmelt, as well as by Hänsch and Schawlow. The key idea is to provide a portion of the energy needed for an atom to absorb radiation from the atom's kinetic energy by tuning the laser slightly to the red of the transition wavelength. (Alternatively, one can think of exploiting the Doppler effect to shift the radiation into resonance.) If the atom returns to its initial state by spontaneous emission, then as the process is repeated the atom cools. The process was demonstrated on a cloud of trapped ions by Wineland and also by Dehmelt in 1978. Applied to a gas, the method is known as Doppler cooling. Cooling ceases when the Doppler shift due to thermal motion becomes comparable with the natural linewidth for the transition, a situation called the Doppler limit, typically at a temperature of a few hundred microdegrees kelvin.

In 1982 W. D. Phillips and H. Metcalf slowed an atomic beam of sodium and cooled its longitudinal motion, using a counterpropagating beam of laser light and a spatially varying magnetic field to maintain the resonance condition. S. Chu demonstrated three-dimensional Doppler cooling with orthogonal laser beams in 1985. The motion of atoms at the intersection of the beams is so heavily damped that the gas behaves like a viscous fluid, dubbed "optical molasses." Phillips measured the temperature of optical molasses and found it to be far below the predicted Doppler limit. The full theoretical explanation was provided by C. Cohen-Tannoudji, who showed that sub-Doppler cooling arises from an interplay between an atom's internal and translational states, involving energy shifts induced by the radiation field and optical pumping effects. When this "polarization gradient" cooling occurs, the temperature approaches the so-called recoil limit, typically one microdegree kelvin, set by the momentum kick due to the emission of a single photon. For these advances in laser cooling and trapping, the 1997 Nobel Prize was awarded to Chu, Phillips, and Cohen-Tannoudji.

For experimental studies cold atoms generally need to be confined, and of the various optical and magnetic traps that have been used for this purpose, the magneto-optical trap emerged as a workhorse because of its great strength and the ease of loading. The trap, created in 1987 by D. E. Pritchard and colleagues, employs a combination of magnetic-field gradients and circularly polarized standing waves to provide a relatively simple and open geometry.

With these tools for cooling and trapping atoms it was possible to study processes such as ultracold collisions, molecular photo-association, and the tunneling of atoms in optical lattices. In dense atomic clouds, scattering and absorption prevents laser cooling. The atoms can nevertheless be cooled efficiently by evaporation, as demonstrated by H. Hess, T. J. Greytak, and D. Kleppner in 1987. In 1995 Bose-Einstein condensation of an atomic gas was achieved by E. Cornell and Wieman, and W. Ketterle, using a strategy of laser cooling and trapping followed by evaporative cooling. Shortly thereafter R. Hulet demonstrated Bose-Einstein condensation in an atom with attractive interactions, previously believed not capable of condensing. The creation of Bose-Einstein condensates opened a new field of quantum fluids, attracting wide theoretical and experimental interest, and enabling studies of collective motions, atomic coherence, sound propagation, condensation dynamics, interactions between multicomponent condensates, and the demonstration of an atom laser.

In parallel with these developments, the field of atom optics was created, in which matter waves are manipulated coherently with the tools of geometrical and wave optics. A seminal experiment in this advance was the diffraction of a matter wave from a grating composed of light by Pritchard in 1983. Pritchard and several other groups demonstrated atom interferometers in 1991. Atom interferometers have been employed to measure the refractive index of atoms, to study decoherence in quantum systems, to monitor geophysical effects revealed by variations in the acceleration of gravity, and to create a matter-wave gyroscope. Familiar components of optics that have now been replicated for matter waves include lenses, mirrors, gratings (composed of light waves and also fabricated structures), and waveguides.

With the creation of the laser, the field of quantum optics came into being. The seminal experiment in this field actually predated the laser. This was the R. Hanbury Brown and P. Q. Twiss experiment of 1954 in which the diameter of a radio source was measured by observing intensity fluctuations. Brown and Twiss demonstrated that the amount of coherence between two points in a radiation field could be inferred from the intensity correlations from two radio antennas. The spectral properties of radiation from a maser were analyzed in 1955 by J. P. Gordon, H. J. Zeiger, and C. H. Townes, and those from a laser in 1958 by Schawlow and Townes. A seminal work on the quantum theory of optical coherence was presented by R. J. Glauber in 1962, and in 1965 F. T. Arecchi experimentally characterized the counting statistics from a laser source and a pseudo-Gaussian source. In 1966 Lamb and M. O. Scully presented a quantum theory of the laser. During that same period the foundations of nonlinear optics were developed in a series of papers by Bloembergen, for which he received the Nobel Prize in 1981. The light from a laser operating far above the threshold for oscillation has the statistical properties of a classical radiation source, but nonclassical light rapidly moved to center stage in quantum optics. In a series of experiments L. Mandel generated light with nonclassical statistics and demonstrated purely quantum entanglement phenomena using correlated photons. The so-called "squeezed states" of light, in which quantum fluctuations in two conjugate variables are divided nonsymmetrically, were demonstrated in an atomic system by H. J. Kimble.

The Lamb shift and other QED effects can be pictured as arising from the interactions of an electron and the vacuum. Vacuum effects are unimportant in laser fields where the photon occupation number is very high. However, dynamical effects of the vacuum can be important for atoms in cavities, where only one or a small number of vacuum modes are important. The study of atom-radiation systems in cavities in low-lying quantum states has become known as cavity quantum electrodynamics. Starting in the early 1980s cavity QED effects were observed with Rydberg atoms in cavities, including suppressed spontaneous emission (Kleppner), the micromaser—a maser in which the number of radiating atoms is less than one—(H. Walther), and experiments on atom-cavity interactions, including the entanglement of single atoms with the fields of cavities (S. Haroche). Such experiments were later extended to the optical regime by H. J. Kimble and M. S. Feld.

Looking back over the century, each of the three stages of this short history advanced with its own particular element of drama. In the first third of the century, quantum mechanics itself came into being, providing a new language and an arsenal of theoretical tools. In the second, a series of powerful experimental methods were developed on the basis of elementary quantum ideas, including molecular-beam magnetic resonance, the maser, and the laser. The new techniques were applied to basic problems in quantum electrodynamics, to studies of atomic and nuclear properties, and to devices such as atomic clocks. In the final third of the century, an explosion of new studies—far too many to summarize here—occurred, many of them made possible by lasers. Prominent among these were basic studies of the radiation field and the manipulation of atoms, culminating in the achievement of Bose-Einstein condensation of an atomic gas.

This brief and biased history has omitted major areas of theoretical and experimental development: advances in relativistic many-body theory, electron correlations, transient states and collision dynamics, multiply charged ions, atoms in intense radiation fields, and more. Also neglected are the applications of atomic physics—save brief mention of the role of atomic clocks in the Global Positioning System. Applications for the concepts and techniques of atomic physics are to be found in chemistry, astronomy, atmospheric science, space science, surface science, nuclear physics, and plasma physics, to name some areas. It has numerous applications in defense scenarios and environmental science. Practically every aspect of energy production involves some component of atomic physics. Metrological techniques from atomic physics are of broad importance in science, industry, and the military.

Perhaps one concrete example provides a more useful summary than a list. Kastler's method of optical pumping led to a flowering of activity in the 1960s that largely subsided when laser spectroscopy was introduced. However, W. Happer continued to use optical pumping to study the mechanism of polarization transfer between alkali-metal atoms and rare-gas atoms. From these studies he developed methods for polarizing rare-gas nuclei at high density that found applications in nuclear physics. The techniques also found an application in medicine: a new type of magnetic-resonance imaging based on the production of polarized rare gases at high density. By providing detailed images of the lung, rare-gas magnetic-resonance imaging provides a powerful diagnostic tool for pulmonary problems.

It is tempting to predict the future direction of atomic physics. However, recognizing that in each of these periods the progress far exceeded the most optimistic vision at its commencement, the author will forbear.

References

Bloembergen, N., 1996, *Nonlinear Optics* (World Scientific, Singapore).
Bertolotti, M., 1983, *Masers and Lasers, An Historical Approach* (Adam Hilger, Bristol, England).
Brown, L. M., Abraham Pais, and Sir Brian Pippard, 1995, Eds., *Twentieth Century Physics* (Institute of Physics, Bristol, England and AIP, New York).
Mattson, J., and M. Simon, 1996, *The Story of MRI* (Dean Books, Jericho, New York).
Pais, A., 1982, *Subtle is the Lord* (Oxford University, New York).
Pais, A., 1986, *Inward Bound* (Oxford University, New York).
Ramsey, N.F., 1998, *Spectroscopy with Coherent Radiation* (World Scientific, Singapore).
Rigden, John S., 1987, *Rabi, Scientist and Citizen* (Basic Books, Inc., New York).

Townes, C.H., 1999, *How the Laser Happened—Adventures of a Scientist* (Oxford University Press, London).

van Linden van den Heuvel, H. B., J. T. M. Walraven, and M. W. Reynolds, 1997, Eds., *Atomic Physics 15: Proceedings of the 15th International Conference on Atomic Physics, Zeeman-Effect Centenary, Amsterdam, 1996* (World Scientific, Singapore). This series of conferences reports on all the major advances in atomic physics since the mid 1960s.

Particle Physics and Related Topics

Quantum Field Theory
Frank Wilczek

Quantum field theory is the framework in which the regnant theories of the electroweak and strong interactions, which together form the standard model, are formulated. Quantum electrodynamics (QED), besides providing a complete foundation for atomic physics and chemistry, has supported calculations of physical quantities with unparalleled precision. The experimentally measured value of the magnetic dipole moment of the muon,

$$(g_\mu - 2)_{\text{exp}} = 233\ 184\ 600\ (1680) \times 10^{-11}, \tag{1}$$

for example, should be compared with the theoretical prediction

$$(g_\mu - 2)_{\text{theor}} = 233\ 183\ 478\ (308) \times 10^{-11} \tag{2}$$

(see the chapter by Hughes and Kinoshita on pp. 223–233 in this book).

In quantum chromodynamics (QCD) we cannot, for the forseeable future, aspire to comparable accuracy. Yet QCD provides different, and at least equally impressive, evidence for the validity of the basic principles of quantum field theory. Indeed, because in QCD the interactions are stronger, QCD manifests a wider variety of phenomena characteristic of quantum field theory. These include especially running of the effective coupling with distance or energy scale and the phenomenon of confinement. QCD has supported, and rewarded with experimental confirmation, both heroic calculations of multiloop diagrams and massive numerical simulations of (a discretized version of) the complete theory.

Quantum field theory also provides powerful tools for condensed-matter physics, especially in connection with the quantum many-body problem as it arises in the theory of metals, superconductivity, the low-temperature behavior of the quantum liquids He^3 and He^4, and the quantum Hall effect, among others. Although for reasons of space and focus I shall not attempt to do justice to this aspect here, the continuing interchange of ideas between condensed-matter and high-energy theory, through the medium of quantum field theory, is a remarkable phenomenon in itself. A partial list of historically important examples includes global and local spontaneous symmetry breaking, the renormalization group, effective field theory, solitons, instantons, and fractional charge and statistics.

It is clear, from all these examples, that quantum field theory occupies a central position in our description of Nature. It provides both our best working description of fundamental physical laws and a fruitful tool for investigating the behavior of complex systems. But the enumeration of examples, however triumphal, serves more to pose than to answer more basic questions: What are the essential features of quantum field theory? What does quantum field theory add to our understanding of the world, that was not already present in quantum mechanics and classical field theory separately?

The first question has no sharp answer. Theoretical physicists are very flexible in adapting their tools, and no axiomization can keep up with them. However, I think it is fair to say that the characteristic, core ideas of quantum field theory are twofold. First, that the basic dynamical degrees of freedom are operator functions of space and time—quantum fields, obeying appropriate commutation relations. Second, that the interactions of these fields are local. Thus the equations of motion and commutation relations governing the evolution of a given quantum field at a given point in space-time should depend only on the behavior of fields and their derivatives at that point. One might find it convenient to use other variables, whose equations are not local, but in the spirit of quantum field theory there must always be some underlying fundamental, local variables. These ideas, combined with postulates of symmetry (e.g., in the context of the standard model, Lorentz and gauge invariance) turn out to be amazingly powerful, as will emerge from our further discussion below.

The field concept came to dominate physics starting with the work of Faraday in the mid-nineteenth century. Its conceptual advantage over the earlier Newtonian program of physics, to formulate the fundamental laws in terms of forces among atomic particles, emerges when we take into account the circumstance, unknown to Newton (or, for that matter, Faraday) but fundamental in special relativity, that influences travel no faster than a finite limiting speed. For then the force on a given particle at a given time cannot be deduced from the positions of other particles at that time, but must be deduced in a complicated way from their previous positions. Faraday's intuition that the fundamental laws of electromagnetism could be expressed most simply in terms of fields filling space and time was, of course, brilliantly vindicated by Maxwell's mathematical theory.

The concept of locality, in the crude form that one can predict the behavior of nearby objects without reference to distant ones, is basic to scientific practice. Practical experimenters—if not astrologers—confidently expect, on the basis of much successful experience, that after reasonable (generally quite modest) precautions to isolate their experiments, they will obtain reproducible results. Direct quantitative tests of locality, or rather of its close cousin causality, are afforded by dispersion relations.

The deep and ancient historic roots of the field and locality concepts provide no guarantee that these concepts remain relevant or valid when extrapolated far beyond their origins in experience, into the subatomic and quantum domain. This extrapolation must be judged by its fruits. That brings us, naturally, to our second question.

Undoubtedly the single most profound fact about Nature that quantum field theory uniquely explains is *the existence of different, yet indistinguishable, copies of elementary particles*. Two electrons anywhere in the universe, whatever their origin or history, are observed to have exactly the same properties. We understand this as a consequence of the fact that both are excitations of the same underlying ur-stuff, the electron field. The electron field is thus the primary reality. The same logic, of course, applies to photons or quarks, or even to composite objects such as atomic nuclei, atoms, or molecules. The indistinguishability of particles is so familiar, and so fundamental to all of modern physical science, that we could easily take it for granted. Yet it is by no means obvious. For example, it directly contradicts one of the pillars of Leibniz' metaphysics, his "principle of the identity of indiscernibles," according to which two objects cannot differ solely in number. And Maxwell thought the similarity of different molecules so remarkable that he devoted the last part of his *Encyclopedia Brittanica* entry on atoms—well over a thousand words—to discussing it. He concluded that "the formation of a molecule is therefore an event not belonging to that order of nature in which we live ... it must be referred to the epoch, not of the formation of the earth or the solar system ... but of the establishment of the existing order of Nature."

The existence of classes of indistinguishable particles is the necessary logical prerequisite to a second profound insight from quantum field theory: *the assignment of unique quantum statistics to each class*. Given the indistinguishability of a class of elementary particles, and complete invariance of their interactions under interchange, the general principles of quantum mechanics teach us that solutions forming any representation of the permutation symmetry group retain that property in time, but do not constrain which representations are realized. Quantum field theory not only explains the existence of indistinguishable particles and the invariance of their interactions under interchange, but also constrains the symmetry of the solutions. For bosons only the identity representation is physical (symmetric wave functions), for fermions only the one-dimensional odd representation is physical (antisymmetric wave functions). One also has the spin-statistics theorem, according to which objects with integer spin are bosons, whereas objects with half odd-integer spin are fermions. Of course, these general predictions have been verified in many experiments. The fermion character of electrons, in particular, underlies the stability of matter and the structure of the periodic table.

A third profound general insight from quantum field theory is *the existence of antiparticles*. This was first inferred by Dirac on the basis of a brilliant but obsolete interpretation of his equation for the electron field, whose elucidation was a crucial step in the formulation of quantum field theory. In quantum field theory, we reinterpret the Dirac wave function as a position- (and time-) dependent operator. It can be expanded in terms of the solutions of the Dirac equation, with operator coefficients. The coefficients of positive-energy solutions are operators that destroy electrons, and the coefficients of the negative-energy solutions are operators that create positrons (with positive energy). With this interpretation, an improved version of Dirac's hole theory emerges in a straightforward way. (Unlike the original hole theory, it has a sensible generalization to bosons and to processes in which the number of electrons minus positrons changes.) A very general consequence of quantum field theory, valid in the presence of arbitrarily complicated interactions, is the *CPT* theorem. It states that the product of charge conjugation, parity, and time reversal is always a symmetry of the world, although each may be—and is—violated separately. Antiparticles are strictly defined as the *CPT* conjugates of their corresponding particles.

The three outstanding facts we have discussed so far, the existence of indistinguishable particles, the phenomenon of quantum statistics, and the existence of antiparticles, are all essentially consequences of *free* quantum field theory. When one incorporates interactions into quantum field theory, two additional general features of the world immediately become brightly illuminated.

The first of these is *the ubiquity of particle creation and destruction processes*. Local interactions involve products of field operators at a point. When the fields are expanded into creation and annihilation operators multiplying modes, we see that these interactions correspond to processes wherein particles can be created, annihilated, or changed into different kinds of particles. This possibility arises, of course, in the primeval quantum field theory, quantum electrodynamics, where the primary interaction arises from a product of the electron field, its Hermitian conjugate, and the photon field. Processes of radiation and absorption of photons by electrons (or positrons), as well as electron-positron pair creation, are encoded in this product. Just because the emission and absorption of light is such a common experience, and electrodynamics such a special and familiar classical field theory, this correspondence between formalism and reality did not initially make a big impression. The first conscious exploitation of the potential for quantum field theory to describe processes of transformation was Fermi's theory of beta decay. He turned the procedure around, inferring from the observed processes of particle transformation the nature of the underlying local interaction of fields. Fermi's theory involved creation and annihilation not of photons, but of atomic nuclei and electrons (as well as neutrinos)—the ingredients of "matter." It began the process whereby classic atomism, involving stable individual objects, was replaced by a more sophisticated and accurate picture. In this picture it is only the fields, and not the individual objects they create and destroy, that are permanent.

The second feature that appears from incorporating interaction into quantum field theory is *the association of forces and interactions with particle exchange*. When Maxwell completed the equations of electrodynamics, he found that they supported source-free electromagnetic waves. The classical electric and magnetic fields thus took on a life of their own. Electric and magnetic forces between charged particles are explained as due to one particle's acting as a source for electric and magnetic fields, which then influence other particles. With the correspondence of fields and particles, as it arises in quantum field theory, Maxwell's discovery corresponds to the existence of photons, and the generation of forces by intermediary fields corresponds to the exchange of virtual photons. The association of forces (or, more generally, interactions) with exchange of particles is a general feature of quantum field theory. It was used by Yukawa to infer the existence and mass of pions from the range of nuclear forces, more recently in electroweak theory to infer the existence, mass, and properties of W and Z bosons prior to their observation, and in QCD to infer the existence and properties of gluon jets prior to their observation.

The two additional outstanding facts we just discussed, the possibility of particle creation and destruction and the association of particles with forces, are essentially consequences of classical field theory, supplemented by the connection between particles and fields that we learn from free field theory. Indeed, classical waves with nonlinear interactions will change form, scatter, and radiate, and these processes exactly mirror the transformation, interaction, and creation of particles. In quantum field theory, they are properties one sees already in tree graphs.

The foregoing major consequences of free quantum field theory, and of its formal extension to include nonlinear interactions, were all well appreciated by the late 1930s. The deeper properties of quantum field theory, which will form the subject of the remainder of this paper, arise from the need to introduce infinitely many degrees of freedom, and the possibility that all these degrees of freedom are excited as quantum-mechanical fluctuations. From a mathematical point of view, these deeper properties arise when we consider loop graphs.

From a physical point of view, the potential pitfalls associated with the existence of an infinite number of degrees of freedom first showed up in connection with the problem that led to the birth of quantum theory, that is, the ultraviolet catastrophe of blackbody radiation theory. Somewhat ironically, in view of later history, the crucial role of the quantum theory here was to remove the disastrous consequences of the infinite number of degrees of freedom possessed by classical electrodynamics. The classical electrodynamic field can be decomposed into independent oscillators with arbitrarily high values of the wave vector. According to the equipartition theorem of classical statistical mechanics, in thermal equilibrium at temperature T each of these oscillators should have average energy kT. Quantum mechanics alters this situation by insisting that the oscillators of frequency ω have energy quantized in units of $\hbar\omega$. Then the high-frequency modes are exponentially suppressed by the Boltzmann factor, and instead of kT receive

$$\frac{\hbar\omega e^{-(\hbar\omega/kT)}}{1-e^{(-\hbar\omega/kT)}}.$$

The role of the quantum, then, is to prevent accumulation of energy in the form of very-small-amplitude excitations of arbitrarily high frequency modes. It is very effective in suppressing the *thermal* excitation of high-frequency modes.

But while removing arbitrarily small-amplitude excitations, quantum theory introduces the idea that the modes are always intrinsically excited to a small extent, proportional to $\hbar$. This so-called zero-point motion is a consequence of the uncertainty principle. For a harmonic oscillator of frequency ω, the ground-state energy is not zero, but $\frac{1}{2}\hbar\omega$. In the case of the electromagnetic field this leads, upon summing over its high-frequency modes, to a highly divergent total ground-state energy. For most physical purposes the absolute normalization of energy is unimportant, and so this particular divergence does not necessarily render the theory useless.[1] It does, however, illustrate the dangerous character of the high-frequency modes, and its treatment gives a first indication of the leading theme of renormalization theory: we can only require—and generally will only obtain—sensible, finite answers when we ask questions that have direct, operational physical meaning.

The existence of an infinite number of degrees of freedom was first encountered in the theory of the electromagnetic field, but it is a general phenomenon, deeply connected with the requirement of locality in the interactions of fields. For in order to construct the local field $\psi(x)$ at a space-time point x, one must take a superposition

$$\psi(x)=\int \frac{d^4k}{(2\pi)^4}e^{ikx}\tilde{\psi}(k) \tag{3}$$

that includes field components $\tilde{\psi}(k)$ extending to arbitrarily large momenta. Moreover, in a generic interaction

$$\int \mathcal{L}=\int \psi(x)^3=\int \frac{d^4k_1}{(2\pi)^4}\frac{d^4k_2}{(2\pi)^4}\frac{d^4k_3}{(2\pi)^4}\tilde{\psi}(k_1)\tilde{\psi}(k_2)\tilde{\psi}(k_3)(2\pi)^4\delta^4(k_1+k_2+k_3) \tag{4}$$

we see that a low-momentum mode $k_1\approx 0$ will couple without any suppression factor to high-momentum modes k_2 and $k_3\approx -k_2$. Local couplings are "hard" in this sense. Because locality requires the existence of infinitely many degrees of freedom at large

---

[1] One would think that gravity should care about the absolute normalization of energy. The zero-point energy of the electromagnetic field, in that context, generates an infinite cosmological constant. This might be cancelled by similar negative contributions from fermion fields, as occurs in supersymmetric theories, or it might indicate the need for some other profound modification of physical theory.

momenta, with hard interactions, ultraviolet divergences similar to the ones cured by Planck, but driven by quantum rather than thermal fluctuations, are never far off-stage. As mentioned previously, the deeper physical consequences of quantum field theory arise from this circumstance.

First of all, it is much more difficult to construct nontrivial examples of interacting relativistic quantum field theories than purely formal considerations would suggest. One finds that *the consistent quantum field theories form a quite limited class, whose extent depends sensitively on the dimension of space-time and the spins of the particles involved.* Their construction is quite delicate, requiring limiting procedures whose logical implementation leads directly to renormalization theory, the running of couplings, and asymptotic freedom.

Secondly, *even those quantum theories that can be constructed display less symmetry than their formal properties would suggest.* Violations of naive scaling relations—that is, ordinary dimensional analysis—in QCD, and of baryon number conservation in the standard electroweak model are examples of this general phenomenon. The original example, unfortunately too complicated to explain fully here, involved the decay process $\pi^0 \to \gamma\gamma$, for which chiral symmetry (treated classically) predicts much too small a rate. When the correction introduced by quantum field theory (the so-called "anomaly") is retained, excellent agreement with experiment results.

These deeper consequences of quantum field theory, which might superficially appear rather technical, largely dictate the structure and behavior of the standard model—and therefore of the physical world. My goal in this preliminary survey has been to emphasize their profound origin. In the rest of the article I hope to convey their main implications, in as simple and direct a fashion as possible.

Formulation

The physical constants $\hbar$ and c are so deeply embedded in the formulation of relativistic quantum field theory that it is standard practice to declare them to be the units of action and velocity, respectively. In these units, of course, $\hbar = c = 1$. With this convention, all physical quantities of interest have units which are powers of mass. Thus the dimension of momentum is $(\text{mass})^1$ or simply 1, since $\text{mass} \times c$ is a momentum, and the dimension of length is $(\text{mass})^{-1}$ or simply -1, since $\hbar c/\text{mass}$ is a length. The usual way to construct quantum field theories is by applying the rules of quantization to a continuum field theory, following the canonical procedure of replacing Poisson brackets by commutators (or, for fermionic fields, anticommutators). The field theories that describe free spin-0 or free spin-1/2 fields of mass m, μ, respectively, are based on the Lagrangian densities

$$\mathcal{L}_0(x) = \frac{1}{2} \partial_\alpha \phi(x) \partial^\alpha \phi(x) - \frac{m^2}{2} \phi(x)^2, \tag{5}$$

$$\mathcal{L}_{1/2}(x) = \bar{\psi}(x)(i\gamma^\alpha \partial_\alpha - \mu)\psi(x). \tag{6}$$

Since the action $\int d^4x \, \mathcal{L}$ has mass dimension 0, the mass dimension of a scalar field like ϕ is 1 and of a spinor field like ψ is $\frac{3}{2}$. For free spin-1 fields the Lagrangian density is that of Maxwell,

$$\mathcal{L}_1(x) = -\frac{1}{4}(\partial_\alpha A_\beta(x) - \partial_\beta A_\alpha(x))(\partial^\alpha A^\beta(x) - \partial^\beta A^\alpha(x)), \tag{7}$$

so that the mass dimension of the vector field A is 1. The same result is true for non-Abelian vector fields (Yang-Mills fields).

Thus far all our Lagrangian densities have been quadratic in the fields. Local interaction terms are obtained from Lagrangian densities involving products of fields and their derivatives at a point. The coefficient of such a term is a coupling constant and must have the appropriate mass dimension, so that the Lagrangian density has mass dimension 4. Thus the mass dimension of a Yukawa coupling y, which multiplies the product of two spinor fields and a scalar field, is zero. Gauge couplings g arising in the minimal coupling procedure $\partial_\alpha \to \partial_\alpha + igA_\alpha$ are also clearly of mass dimension zero.

The possibilities for couplings with non-negative mass dimension are very restricted. This fact is quite important, for the following reason. Consider the effect of treating a given interaction term as a perturbation. If the coupling κ associated with this interaction has negative mass dimension $-p$, then successive powers of it will occur in the form of powers of $\kappa \Lambda^p$, where Λ is some parameter with dimensions of mass. Because, as we have seen, the interactions in a local field theory are hard, we can anticipate that Λ will characterize the largest mass scale we allow to occur (the cutoff) and will diverge to infinity as the limit on this mass scale is removed. So we expect that it will be difficult to make sense of fundamental interactions having negative mass dimensions, at least in perturbation theory. Such interactions are said to be nonrenormalizable.

The standard model is formulated entirely using renormalizable interactions. It has been said that this is not in itself a fundamental fact about nature. For if nonrenormalizable interactions occurred in the effective description of physical behavior below a certain mass scale, it would simply mean that the theory must change its nature—presumably by displaying new degrees of freedom—at some larger mass scale. If we adopt this point of view, the significance of the fact that the standard model contains only renormalizable operators is that it does not require modification up to arbitrarily high scales (at least on the grounds of divergences in perturbation theory). Whether or not we call this a fundamental fact, it is certainly a profound one.

Moreover, all the renormalizable interactions consistent with the gauge symmetry and multiplet structure of the standard model do seem to occur—"what is not forbidden is mandatory." There is a beautiful agreement between the symmetries of the standard model, allowing arbitrary renormalizable interactions, and the symmetries of the world. One understands why strangeness is violated, but baryon number is not. (The only discordant element is the so-called θ term of QCD, which is allowed by the symmetries of the standard model but is measured to be quite accurately zero. A plausible solution to this problem exists. It involves a characteristic very light *axion* field.)

The power counting rules for estimating divergences assume that there are no special symmetries cancelling off the contribution of high-energy modes. They do not apply, without further consideration, to supersymmetric theories, in which the contributions of boson and fermionic modes cancels, nor to theories derived from supersymmetric theories by soft supersymmetry breaking. In the latter case the scale of supersymmetry breaking plays the role of the cutoff Λ.

The power counting rules, as discussed so far, are too crude to detect divergences of the form $\ln \Lambda^2$. Yet divergences of this form are pervasive and extremely significant, as we shall now discuss.

Running Couplings

The problem of calculating the energy associated with a constant magnetic field, in the more general context of an arbitrary non-Abelian gauge theory coupled to spin-0 and spin-1/2 charged particles, provides an excellent concrete illustration of how the infinities of quantum field theory arise and of how they are dealt with. It introduces the concept of running couplings in a natural way and leads directly to qualitative and quantitative results of great significance for physics. The interactions of concern to us appear in the Lagrangian density

$$\mathcal{L} = -\frac{1}{4g^2} G^I_{\alpha\beta} G^{I\alpha\beta} + \bar{\psi}(i\gamma^\nu D_\nu - \mu)\psi + \phi^\dagger(-D_\nu D^\nu - m^2)\phi, \qquad (8)$$

where $G^I_{\alpha\beta} \equiv \partial_\alpha A^I_\beta - \partial_\beta A^I_\alpha - f^{IJK} A^J_\alpha A^K_\beta$ are the standard field strengths and $D_\nu \equiv \partial_\nu + iA^I_\nu T^I$ the covariant derivative. Here the f^{IJK} are the structure constants of the gauge group, and the T^I are the representation matrices appropriate to the field on which the covariant derivative acts. This Lagrangian differs from the usual one by a rescaling $gA \rightarrow A$, which serves to emphasize that the gauge coupling g occurs only as a prefactor in the first term. It parametrizes the energetic cost of nontrivial gauge curvature or, in other words, the stiffness of the gauge fields. Small g corresponds to gauge fields that are difficult to excite.

From this Lagrangian itself, of course, it would appear that the energy required to set up a magnetic field B^I is just $1/2g^2(B^I)^2$. This is the classical energy, but in the quantum theory it is not the whole story. A more accurate calculation must take into account the effect of the imposed magnetic field on the zero-point energy of the charged fields. Earlier, we met and briefly discussed a formally infinite contribution to the energy of the ground state of a quantum field theory (specifically, the electromagnetic field) due to the irreducible quantum fluctuations of its modes, which mapped to an infinite number of independent harmonic oscillators. Insofar as only differences in energy are physically significant, we could ignore this infinity. But the change in the zero-point energy as one imposes a magnetic field cannot be ignored. It represents a genuine contribution to the physical energy of the quantum state induced by the imposed magnetic field. As we shall soon see, the field-dependent part of the energy also diverges.

Postponing momentarily the derivation, let me anticipate the form of the answer and discuss its interpretation. Without loss of generality, I will suppose that the magnetic field is aligned along a normalized, diagonal generator of the gauge group. This allows us to drop the index and to use terminology and intuition from electrodynamics freely. If we restrict the sum to modes whose energy is less than a cutoff Λ, we find for the energy

$$\mathcal{E}(B) = \mathcal{E} + \delta\mathcal{E} = \frac{1}{2g^2(\Lambda^2)} B^2 - \frac{1}{2}\eta B^2 (\ln(\Lambda^2/B) + \text{finite}), \qquad (9)$$

where

$$\eta = \frac{1}{96\pi^2}[-(T(R_0) - 2T(R_{1/2}) + 2T(R_1))] + \frac{1}{96\pi^2}[3(-2T(R_{1/2}) + 8T(R_1))], \qquad (10)$$

and the terms not displayed are finite as $\Lambda \rightarrow \infty$. The notation $g^2(\Lambda^2)$ has been introduced for later convenience. The factor $T(R_s)$ is the trace of the representation for spin s, and basically represents the sum of the squares of the charges for the particles of that spin. The denominator in the logarithm is fixed by dimensional analysis, assuming $B \gg \mu^2, m^2$.

The most striking, and at first sight disturbing, aspect of this calculation is that a cutoff is necessary in order to obtain a finite result. If we are not to introduce a new fundamental scale, and thereby (in view of our previous discussion) endanger locality, we must remove reference to the arbitrary cutoff Λ in our description of physically meaningful quantities. This is the sort of problem addressed by the renormalization program. Its guiding idea is the thought that if we are working with experimental probes characterized by energy and momentum scales well below Λ, we should expect that our capacity to affect, or be sensitive to, the modes of much higher energy will be quite restricted. Thus we expect that the cutoff Λ, which was introduced as a calculational device to remove such modes, can be removed (taken to infinity). In our magnetic energy example, for instance, we see immediately that the difference in susceptibilities

$$\mathcal{E}(B_1)/B_1^2 - \mathcal{E}(B_0)/B_0^2 = \text{finite} \qquad (11)$$

is well behaved—that is, independent of Λ as $\Lambda \to \infty$. Thus once we measure the susceptibility, or equivalently the coupling constant, at one reference value of B, the calculation gives sensible, unambiguous predictions for all other values of B.

This simple example illustrates a much more general result, the central result of the classic renormalization program. It goes as follows. A small number of quantities, corresponding to the couplings and masses in the original Lagrangian, that if calculated formally would diverge or depend on the cutoff, are chosen to fit experiment. They define the physical, as opposed to the original, or bare, couplings. Thus, in our example, we can define the susceptibility to be $1/2g^2(B_0)$ at some reference field B_0. Then we have the physical or renormalized coupling

$$\frac{1}{g^2(B_0)} = \frac{1}{g^2(\Lambda^2)} - \eta \ln(\Lambda^2/B_0). \qquad (12)$$

(In this equation I have ignored, for simplicity in exposition, the finite terms. These are relatively negligible for large B_0. Also, there are corrections of higher order in g^2.) This of course determines the "bare" coupling to be

$$\frac{1}{g^2(\Lambda^2)} = \frac{1}{g^2(B_0)} + \eta \ln(\Lambda^2/B_0). \qquad (13)$$

In these terms, the central result of diagrammatic renormalization theory is that after bare couplings and masses are reexpressed in terms of their physical, renormalized counterparts, the coefficients in the perturbation expansion of any physical quantity approach finite limits, independent of the cutoff, as the cutoff is taken to infinity. (To be perfectly accurate, one must also perform wave-function renormalization. This is no different in principle; it amounts to expressing the bare coefficients of the kinetic terms in the Lagrangian in terms of renormalized values.)

The question of whether this perturbation theory converges, or is some sort of asymptotic expansion of a soundly defined theory, is left open by the diagrammatic analysis. This loophole is no mere technicality, as we shall soon see.

Picking a scale B_0 at which the coupling is defined is analogous to choosing the origin of a coordinate system in geometry. One can describe the same physics using different choices of normalization scale, so long as one adjusts the coupling appropriately. We capture this idea by introducing the concept of a running coupling defined, in accordance with Eq. (12), to satisfy

$$\frac{d}{d \ln B} \frac{1}{g^2(B)} = \eta. \tag{14}$$

With this definition, the choice of a particular scale at which to define the coupling will not affect the final result.

It is profoundly important, however, that the running coupling does make a real distinction between the behavior at different mass scales, even if the original underlying theory was formally scale invariant (as is QCD with massless quarks), and even at mass scales much larger than the mass of any particle in the theory. Quantum zero-point motion of the high-energy modes introduces a hard source of scale symmetry violation.

The distinction among scales, in a formally scale-invariant theory, embodies the phenomenon of *dimensional transmutation*. Rather than a range of theories parametrized by a dimensionless coupling, we have a range of theories differing only in the value of a dimensional parameter, say (for example), the value of B at which $1/g^2(B) = 1$.

Clearly, the qualitative behavior of solutions of Eq. (14) depends on the sign of η. If $\eta > 0$, the coupling $g^2(B)$ will get smaller as B grows, or in other words as we treat more and more modes as dynamical, and approach closer to the "bare" charge. These modes were enhancing, or antiscreening, the bare charge. This is the case of *asymptotic freedom*. In the opposite case of $\eta < 0$ the coupling formally grows and even diverges as B increases. $1/g^2(B)$ goes through zero and changes sign. On the face of it, this would seem to indicate an instability of the theory, toward formation of a ferromagnetic vacuum at large field strength. This conclusion must be taken with a big grain of salt, because when g^2 is large the higher-order corrections to Eqs. (13) and (14), on which the analysis was based, cannot be neglected.

In asymptotically free theories, we can complete the renormalization program in a convincing fashion. There is no barrier to including the effect of very large energy modes and removing the cutoff. We can confidently expect, then, that the theory is well defined, independent of perturbation theory. In particular, suppose the theory has been discretized on a space-time lattice. This amounts to excluding the modes of high energy and momentum. In an asymptotically free theory one can compensate for these modes by adjusting the coupling in a well-defined, controlled way as one shrinks the discretization scale. Very impressive nonperturbative calculations in QCD, involving massive computer simulations, have exploited this strategy. They demonstrate the complete consistency of the theory and its ability to account quantitatively for the masses of hadrons.

In a non-asymptotically free theory the coupling does not become small, there is no simple foolproof way to compensate for the missing modes, and the existence of an underlying limiting theory becomes doubtful.

Now let us discuss how η can be calculated. The two terms in Eq. (10) correspond to two distinct physical effects. The first is the convective, diamagnetic (screening) term. The overall constant is a little tricky to calculate, and I do not have space to do it here. Its general form, however, is transparent. The effect is independent of spin, and so it simply counts the number of components (one for scalar particles, two for spin-1/2 or massless spin-1 particles, both with two helicities). It is screening for bosons, while for fermions there is a sign flip, because the zero-point energy is negative for fermionic oscillators.

The second is the paramagnetic spin susceptibility. For a massless particle with spin s and gyromagnetic ratio g_m the energies shift, giving rise to the altered zero-point energy

$$\Delta \mathcal{E} = \int_0^{E=\Lambda} \frac{d^3k}{(2\pi)^3} \frac{1}{2} (\sqrt{k^2 + g_m s B} + \sqrt{k^2 - g_m s B} - 2\sqrt{k^2}). \tag{15}$$

This is readily calculated as

$$\Delta \mathcal{E} = -B^2 (g_m s)^2 \frac{1}{32\pi^2} \ln\left(\frac{\Lambda^2}{B}\right). \tag{16}$$

With $g_m = 2$, $s = 1$ (and $T = 1$) this is the spin-1 contribution, and with $g_m = 2$, $s = \frac{1}{2}$, after a sign flip, it is the spin-1/2 contribution. The preferred moment $g_m = 2$ is a direct consequence of the Yang-Mills and Dirac equations, respectively.

This elementary calculation gives us a nice heuristic understanding of the unusual antiscreening behavior of non-Abelian gauge theories. It is due to the large paramagnetic response of charged vector fields. Because we are interested in very-high-energy modes, the usual intuition that charge will be screened, which is based on the electric response of heavy particles, does not apply. Magnetic interactions, which can be attractive for like charges (paramagnetism), are, for highly relativistic particles, in no way suppressed. Indeed, they are numerically dominant.

Though I have presented it in the very specific context of vacuum magnetic susceptibility, the concept of running coupling is much more widely applicable. The basic heuristic idea is that, in analyzing processes whose characteristic energy-momentum scale (squared) is Q^2, it is appropriate to use the running coupling at Q^2, i.e., in our earlier notation $g^2(B=Q^2)$. For in this way we capture the dynamical effect of the virtual oscillators, which can be appreciably excited, while avoiding the formal divergence encountered if we tried to include all of them (up to infinite mass scale). At a more formal level, use of the appropriate effective coupling allows us to avoid large logarithms in the calculation of Feynman graphs, by normalizing the vertices close to where they need to be evaluated. There is a highly developed, elaborate chapter of quantum field theory which justifies and refines this rough idea into a form in which it makes detailed, quantitative predictions for concrete experiments. I am able to do proper justice to the difficult, often heroic, labor that has been invested, on both the theoretical and the experimental sides, to yield Fig. 1; but it is appropriate to remark that quantum field theory gets a real workout, as calculations of two- and even three-loop graphs with complicated interactions among the virtual particles are needed to do justice to the attainable experimental accuracy.

An interesting feature visible in Fig. 1 is that the theoretical prediction for the coupling focuses at large Q^2, in the sense that a wide range of values at small Q^2 converge to a much narrower range at larger Q^2. Thus even crude estimates of what are the appropriate scales [e.g., one expects $g^2(Q^2)/4\pi \sim 1$ where the strong interaction is strong, say for $100 \text{ MeV} \lesssim \sqrt{Q^2} \lesssim 1 \text{ GeV}$] allow one to predict the value of $g^2(M_Z^2)$ with $\sim 10\%$ accuracy. The original idea of Pauli and others that calculating the fine-structure constant was the next great item on the agenda of theoretical physics now seems misguided. We see this constant as just another running coupling, neither more nor less fundamental than many other parameters, and not likely to be the most accessible theoretically. But our essentially parameter-free approximate determination of the observable strong-interaction analog of the fine-structure constant realizes a form of their dream.

The electroweak interactions start with much smaller couplings at low mass scales, so the effects of their running are less dramatic (though they have been observed). Far more spectacular than the modest quantitative effects we can test directly, however, is the conceptual breakthrough that results from application of these ideas to unified models of the strong, electromagnetic, and weak interactions.

The different components of the standard model have a similar mathematical structure, all being gauge theories. Their common structure encourages the speculation that they are different facets of a more encompassing gauge symmetry, in which the different strong and weak color charges, as well as electromagnetic charge, would all appear on the same footing. The multiplet structure of the quarks and leptons in the standard model fits beautifully into small representations of unification groups such as $SU(5)$ or $SO(10)$. There is the apparent difficulty, however, that the coupling strengths of the different standard model interactions are widely different, whereas the symmetry required for unification requires that they share a common value. The running of couplings suggests an escape from this impasse. Since the strong, weak, and electromagnetic couplings run at different rates, their inequality at currently accessible scales need not reflect the ultimate state of affairs. We can imagine that spontaneous symmetry breaking—a soft effect—has hidden the full symmetry of the unified interaction. What is really required is that the fundamental, bare couplings be equal, or in more prosaic terms, that the running couplings of the different interactions should become equal beyond some large scale.

Using simple generalizations of the formulas derived and tested in QCD, we can calculate the running of couplings, to see whether this requirement is satisfied in reality. In doing so one must make some hypothesis about the spectrum of virtual particles. If there are additional massive particles (or, better, fields) that have not yet been observed,

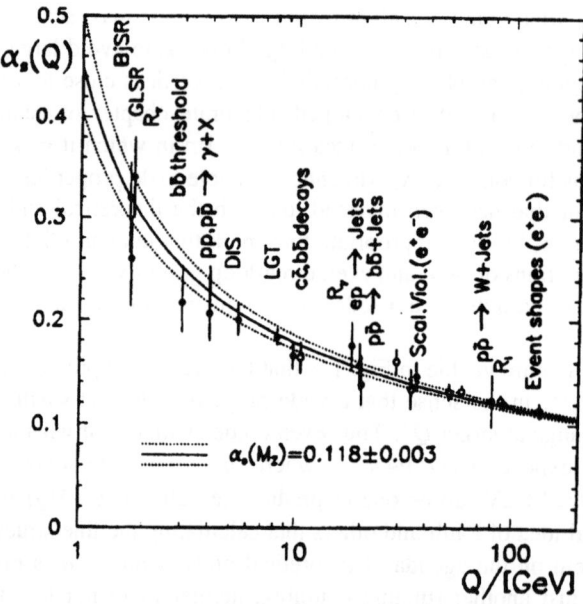

Figure 1. Comparison of theory and experiment in QCD, illustrating the running of couplings. Several of the points on this curve represent hundreds of independent measurements, any one of which might have falsified the theory. From Schmelling (1997).

they will contribute significantly to the running of couplings once the scale exceeds their mass. Let us first consider the default assumption, that there are no new fields beyond those that occur in the standard model. The results of this calculation are displayed in Fig. 2.

Considering the enormity of the extrapolation, this calculation works remarkably well, but the accurate experimental data indicate unequivocally that something is wrong. There is one particularly attractive way to extend the standard model, by including supersymmetry. Supersymmetry cannot be exact, but if it is only mildly broken (so that the superpartners have masses $\lesssim 1$ TeV), it can help explain why radiative corrections to the Higgs mass parameter, and thus to the scale of weak symmetry breaking, are not enormously large. In the absence of supersymmetry, power counting would indicate a hard, quadratic dependence of this parameter on the cutoff. Supersymmetry removes the most divergent contribution, by cancelling boson against fermion loops. If the masses of the superpartners are not too heavy, the residual finite contributions due to supersymmetry breaking will not be too large.

The minimal supersymmetric extension of the standard model, then, makes semiquantitative predictions for the spectrum of virtual particles starting at 1 TeV or so. Since the running of couplings is logarithmic, it is not extremely sensitive to the unknown details of the supersymmetric mass spectrum, and we can assess the impact of supersymmetry on the unification hypothesis quantitatively. The results, as shown in Fig. 3, are quite encouraging.

With all its attractions, there is one general feature of supersymmetry that is especially challenging, and it deserves mention here. We remarked earlier how the standard model, without supersymmetry, features a near-perfect match between the generic symmetries of its renormalizable interactions and the observed symmetries of the world. With supersymmetry, this feature is spoiled. The scalar superpartners of fermions are represented by fields of mass dimension one. This means that there are many more possibilities for low-dimension (including renormalizable) interactions that violate flavor symmetries

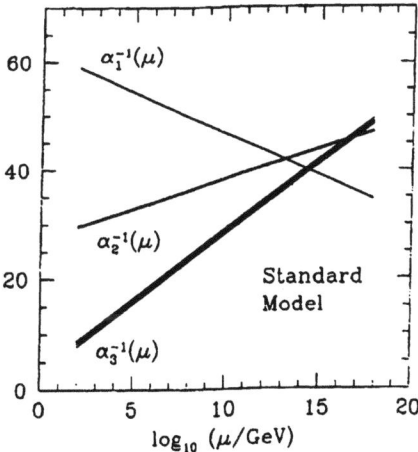

Figure 2. Running of the couplings extrapolated toward very high scales, using just the fields of the standard model. The couplings do not quite meet. Experimental uncertainties in the extrapolation are indicated by the width of the lines. Figure courtesy of K. Dienes.

including lepton and baryon number. It seems that some additional principles, or special discrete symmetries, are required in order to suppress these interactions sufficiently.

A notable result of the unification of couplings calculation, especially in its supersymmetric form, is that the unification occurs at an energy scale that is enormously large by the standards of traditional particle physics, perhaps approaching 10^{16-17} GeV. From a phenomenological viewpoint, this is fortunate. The most compelling unification schemes merge quarks, antiquarks, leptons, and antileptons into common multiplets and have gauge bosons mediating transitions among all these particle types. Baryon-number-violating processes almost inevitably result, whose rate is inversely proportional to the fourth power of the gauge boson masses, and thus to the fourth power of the unification scale. Only for such large values of the scale is one safe from experimental limits on nucleon instability. From a theoretical point of view the large scale is fascinating because it brings us from the internal logic of the experimentally grounded domain of particle physics to the threshold of quantum gravity, as we shall now discuss.

Limitations?

So much for the successes, achieved and anticipated, of quantum field theory. The fundamental limitations of quantum field theory, if any, are less clear. Its application to gravity has certainly, to date, been much less fruitful than its triumphant application to describe the other fundamental interactions.

All existing experimental results on gravitation are adequately described by a very beautiful, conceptually simple classical field theory—Einstein's general relativity. It is easy to incorporate this theory into our description of the world based on quantum field theory, by allowing a minimal coupling to the fields of the standard model—that is, by changing ordinary into covariant derivatives, multiplying with appropriate factors of $\sqrt{g}$, and adding an Einstein-Hilbert curvature term. The resulting theory—with the convention

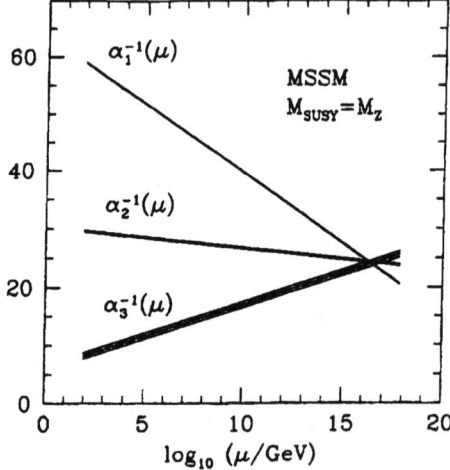

Figure 3. Running of the couplings extrapolated to high scales, including the effects of supersymmetric particles starting at 1 TeV. Within experimental and theoretical uncertainties, the couplings do meet. Figure courtesy of K. Dienes.

that we simply ignore quantum corrections involving virtual gravitons—is the foundation of our working description of the physical world. As a practical matter, it works very well indeed.

Philosophically, however, it might be disappointing if it were too straightforward to construct a quantum theory of gravity. One of the great visions of natural philosophy, going back to Pythagoras, is that the properties of the world are determined uniquely by mathematical principles. A modern version of this vision was formulated by Planck, shortly after he introduced his quantum of action. By appropriately combining the physical constants $c, \hbar$ as units of velocity and action, respectively, and the Planck mass

$$M_{\text{Planck}} = \sqrt{\frac{\hbar c}{G}}$$

as the unit of mass, one can construct any unit of measurement used in physics. Thus the unit of energy is $M_{\text{Planck}} c^2$, the unit of electric charge is $\sqrt{\hbar c}$, and so forth. On the other hand, one cannot form a pure number from these three physical constants. Thus one might hope that in a physical theory where $\hbar$, c, and G were all profoundly incorporated, all physical quantities could be expressed in natural units as pure numbers.

Within its domain, QCD achieves something very close to this vision—actually, in a more ambitious form. Indeed, let us idealize the world of the strong interaction slightly, by imagining that there were just two quark species with vanishing masses. Then from the two integers 3 (colors) and 2 (flavors), $\hbar$, and c—with no explicit mass parameter—a spectrum of hadrons, with mass ratios and other properties close to those observed in reality, emerges by calculation. The overall unit of mass is indeterminate, but this ambiguity has no significance within the theory itself.

The ideal Pythagorean/Planckian theory would not contain any pure numbers as parameters. (Pythagoras might have excused a few small integers.) Thus, for example, the value $m_e/M_{\text{Planck}} \sim 10^{-22}$ of the electron mass in Planck units would emerge from a dynamical calculation. This ideal might be overly ambitious, yet it seems reasonable to hope that significant constraints among physical observables will emerge from the inner requirements of a quantum theory that consistently incorporates gravity. Indeed, as we have already seen, one does find significant constraints among the parameters of the standard model by requiring that the strong, weak, and electromagnetic interactions emerge from a unified gauge symmetry; so there is precedent for results of this kind.

The unification of couplings calculation provides not only an inspiring model, but also direct encouragement for the Planck program, in two important respects. First, it points to a symmetry-breaking scale remarkably close to the Planck scale (though apparently smaller by 10^{-2}–10^{-3}), so there are pure numbers with much more "reasonable" values than 10^{-22} to shoot for. Second, it shows quite concretely how very-large-scale factors can be controlled by modest ratios of coupling strength, due to the logarithmic nature of the running of couplings—so that 10^{-22} may not be so unreasonable after all.

Perhaps it is fortunate, then, that the straightforward, minimal implementation of general relativity as a quantum field theory—which lacks the desired constraints—runs into problems. The problems are of two quite distinct kinds. First, the renormalization program fails, at the level of power counting. The Einstein-Hilbert term in the action comes with a large prefactor $1/G$, reflecting the difficulty of curving space-time. If we expand the Einstein-Hilbert action around flat space in the form

$$g_{\alpha\beta} = \eta_{\alpha\beta} + \sqrt{G} h_{\alpha\beta}, \tag{17}$$

we find that the quadratic terms give a properly normalized spin-2 graviton field $h_{\alpha\beta}$ of mass dimension 1, as the powers of G cancel. But the higher-order terms, which represent interactions, will be accompanied by positive powers of G. Since G itself has mass dimension -2, these are nonrenormalizable interactions. Similarly for the couplings of gravitons to matter. Thus we can expect that ever-increasing powers of $\Lambda/M_{\text{Planck}}$ will appear in multiple virtual graviton exchange, and it will be impossible to remove the cutoff.

Second, one of the main qualitative features of gravity—the weightlessness of empty space, or the vanishing of the cosmological constant—is left unexplained. Earlier we mentioned the divergent zero-point energy characteristic of generic quantum field theories. For purposes of nongravitational physics only energy differences are meaningful, and we can sweep this problem under the rug. But gravity ought to see this energy. Our perplexity intensifies when we recall that according to the standard model, and even more so in its unified extensions, what we commonly regard as empty space is full of condensates, which again one would expect to weigh far more than observation allows. The failure, so far, of quantum field theory to meet these challenges might reflect a basic failure of principle or merely the fact that the appropriate symmetry principles and degrees of freedom, in terms of which the theory should be formulated, have not yet been identified.

Promising insights toward construction of a quantum theory including gravity are coming from investigations in string/M theory, as discussed elsewhere in this volume. Whether these investigations will converge toward an accurate description of Nature, and if so whether this description will take the form of a local field theory (perhaps formulated in many dimensions and including many fields beyond those of the standard model), are questions not yet decided. It is interesting, in this regard, to consider briefly the rocky intellectual history of quantum field theory.

After the initial successes of the 1930s, already mentioned above, came a long period of disillusionment. Initial attempts to deal with the infinities that arose in calculations of loop graphs in electrodynamics, or in radiative corrections to beta decay, led only to confusion and failure. Similar infinities plagued Yukawa's pion theory, and it had the additional difficulty that the coupling required to fit experiment is large, so that tree graphs provide a manifestly poor approximation. Many of the founders of quantum theory, including Bohr, Heisenberg, Pauli, and (for different reasons) Einstein and Schrodinger, felt that further progress required a radically new innovation. This innovation would be a revolution of the order of quantum mechanics itself and would introduce a new fundamental length.

Quantum electrodynamics was resurrected in the late 1940s, largely stimulated by developments in experimental technique. These experimental developments made it possible to study atomic processes with such great precision that the approximation afforded by keeping tree graphs alone could not do them justice. Methods to extract sensible finite answers to physical questions from the jumbled divergences were developed, and spectacular agreement with experiment was found—all without changing electrodynamics itself or departing from the principles of relativistic quantum field theory.

After this wave of success came another long period of disillusionment. The renormalization methods developed for electrodynamics did not seem to work for weak-interaction theory. They did suffice to define a perturbative expansion of Yukawa's pion theory, but the strong coupling made that limited success academic (and it came to seem utterly implausible that Yukawa's schematic theory could do justice to the wealth of newly discovered phenomena). In any case, as a practical matter, throughout the 1950s and 1960s a flood of experimental discoveries, including new classes of weak processes and a rich spectrum of hadronic resonances with complicated interactions, had to be absorbed and correlated. During this process of pattern recognition, the elementary parts of quantum field theory were used extensively, as a framework, but deeper questions were put off. Many theorists came to feel that quantum field theory, in its deeper aspects, was simply wrong and would need to be replaced by some S-matrix or bootstrap theory; perhaps most thought it was irrelevant, or that its use was premature, especially for the strong interaction.

As it became clear, through phenomenological work, that the weak interaction is governed by current×current interactions with universal strength, the possibility of ascribing it to exchange of vector gauge bosons became quite attractive. Models incorporating the idea of spontaneous symmetry breaking to give mass to the weak gauge bosons were constructed. It was conjectured, and later proved, that the high degree of symmetry in these theories allows one to isolate and control the infinities of perturbation theory. One can carry out a renormalization program similar in spirit, though considerably more complex in detail, to that of QED. It is crucial, here, that spontaneous symmetry breaking is a very soft operation. It does not significantly affect the symmetry of the theory at large momenta, where the potential divergences must be cancelled.

Phenomenological work on the strong interaction made it increasingly plausible that the observed strongly interacting particles—mesons and baryons—are composites of more basic objects. The evidence was of two disparate kinds: on the one hand, it was possible in this way to make crude but effective models for the observed spectrum with mesons as quark-antiquark, and baryons as quark-quark-quark, bound states; and on the other hand, experiments provided evidence for hard interactions of photons with hadrons, as would be expected if the components of hadrons were described by local fields. The search for a quantum field theory with appropriate properties led to a unique candidate, which contained both objects that could be identified with quarks and an essentially new ingredient, color gluons.

These quantum field theories of the weak and strong interactions were dramatically confirmed by subsequent experiments, and have survived exceedingly rigorous testing over the past two decades. They make up the standard model. During this period the limitations, as well as the very considerable virtues, of the standard model have become evident. Whether the next big step will require a sharp break from the principles of quantum field theory or, like the previous ones, a better appreciation of its potentialities, remains to be seen.

For further information about quantum field theory, the reader may wish to consult Cheng and Li (1984), Peskin and Schroeder (1995), and Weinberg (1995, 1996).

Acknowledgments

I wish to thank S. Treiman for extremely helpful guidance and M. Alford, K. Babu, C. Kolda, and J. March-Russell for reviewing the manuscript. F.W. was supported in part by DOE grant DE-FG02-90ER40542.

References

Cheng, T. P., and L. F. Li, 1984, *Gauge Theory of Elementary Particle Physics* (Oxford University Press, London).

Peskin, M., and D. Schroeder, 1995, *Introduction to Quantum Field Theory* (Addison-Wesley, Redwood City, CA).

Schmelling, M. 1997, preprint hep-ex/9701002.

Weinberg, S., 1995, *The Quantum Theory of Fields, I* (Cambridge University, Cambridge, England).

Weinberg, S., 1996, *The Quantum Theory of Fields, II* (Cambridge University, Cambridge, England).

The Standard Model of Particle Physics
Mary K. Gaillard, Paul D. Grannis, and Frank J. Sciulli

Over the past three decades a compelling case has emerged for the now widely accepted standard model of elementary particles and forces. A "Standard Model" is a theoretical framework built from observation that predicts and correlates new data. The Mendeleev table of elements was an early example in chemistry; from the periodic table one could predict the properties of many hitherto unstudied elements and compounds. Nonrelativistic quantum theory is another standard model that has correlated the results of countless experiments. Like its precursors in other fields, the standard model of particle physics has been enormously successful in predicting a wide range of phenomena. And, just as ordinary quantum mechanics fails in the relativistic limit, we do not expect the standard model to be valid at arbitrarily short distances. However, its remarkable success strongly suggests that the standard model will remain an excellent approximation to nature at distance scales as small as 10^{-18} m.

In the early 1960s particle physicists described nature in terms of four distinct forces, characterized by widely different ranges and strengths as measured at a typical energy scale of 1 GeV. The strong nuclear force has a range of about a fermi or 10^{-15} m. The weak force responsible for radioactive decay, with a range of 10^{-17} m, is about 10^{-5} times weaker at low energy. The electromagnetic force that governs much of macroscopic physics has infinite range and strength determined by the fine-structure constant, $\alpha \approx 10^{-2}$. The fourth force, gravity, also has infinite range and a low-energy coupling (about 10^{-38}) too weak to be observable in laboratory experiments. The achievement of the standard model was the elaboration of a unified description of the strong, weak, and electromagnetic forces in the language of quantum gauge-field theories. Moreover, the standard model combines the weak and electromagnetic forces in a single electroweak gauge theory, reminiscent of Maxwell's unification of the seemingly distinct forces of electricity and magnetism.

By midcentury, the electromagnetic force was well understood as a renormalizable quantum field theory (QFT) known as quantum electrodynamics or QED, described in the preceeding article. "Renormalizable" means that once a few parameters are determined by a limited set of measurements, the quantitative features of interactions among charged particles and photons can be calculated to arbitrary accuracy as a perturbative expansion in the fine-structure constant. QED has been tested over an energy range from 10^{-16} eV to tens of GeV, i.e., distances ranging from 10^8 km to 10^{-2} fm. In contrast, the nuclear force was characterized by a coupling strength that precluded a perturbative expansion. Moreover, couplings involving higher spin states (resonances), which appeared to be on the same footing as nucleons and pions, could not be described by a renormalizable theory, nor could the weak interactions that were attributed to the direct coupling of four fermions to one another. In the ensuing years the search for renormalizable theories of strong and weak interactions, coupled with experimental discoveries and attempts to interpret available data, led to the formulation of the standard model, which has been experimentally verified to a high degree of accuracy over a broad range of energies and processes.

The standard model is characterized in part by the spectrum of elementary fields shown in Table I. The matter fields are fermions and their antiparticles, with half a unit of intrinsic angular momentum, or spin. There are three families of fermion fields that are identical in every attribute except their masses. The first family includes the up (u) and down (d) quarks that are the constituents of nucleons as well as pions and other mesons responsible for nuclear binding. It also contains the electron and the neutrino emitted with a positron in nuclear β decay. The quarks of the other families are constituents of heavier short-lived particles; they and their companion charged leptons rapidly decay via the weak force to the quarks and leptons of the first family.

The spin-1 gauge bosons mediate interactions among fermions. In QED, interactions among electrically charged particles are due to the exchange of quanta of the electromagnetic field called photons (γ). The fact that the γ is massless accounts for the long range of the electromagnetic force. The strong force, quantum chromodynamics or QCD, is mediated by the exchange of massless gluons (g) between quarks that carry a quantum number called color. In contrast to the electrically neutral photon, gluons (the quanta of the "chromo-magnetic" field) possess color charge and hence couple to one another. As a consequence, the color force between two colored particles increases in strength with increasing distance. Thus quarks and gluons cannot appear as free particles, but exist only inside composite particles, called hadrons, with no net color charge. Nucleons are composed of three quarks of different colors, resulting in "white" color-neutral states. Mesons contain quark and antiquark pairs whose color charges cancel. Since a gluon inside a nucleon cannot escape its boundaries, the nuclear force is mediated by color-neutral bound states, accounting for its short range, characterized by the Compton wavelength of the lightest of these: the π meson.

The even shorter range of the weak force is associated with the Compton wavelengths of the charged W and neutral Z bosons that mediate it. Their couplings to the "weak charges" of quarks and leptons are comparable in strength to the electromagnetic coupling. When the weak interaction is measured over distances much larger than its range, its effects are averaged over the measurement area and hence suppressed in amplitude by a factor $(E/M_{W,Z})^2 \approx (E/100 \text{ GeV})^2$, where E is the characteristic energy transfer in the measurement. Because the W particles carry electric charge they must couple to the γ, implying a gauge theory that unites the weak and electromagnetic interactions, similar to QCD in that the gauge particles are self-coupled. In distinction to γ's and gluons, W's couple only to left-handed fermions (with spin oriented opposite to the direction of motion).

The standard model is further characterized by a high degree of symmetry. For example, one cannot perform an experiment that would distinguish the color of the quarks involved. If the symmetries of the standard-model couplings were fully respected in nature, we would not distinguish an electron from a neutrino or a proton from a neutron; their detectable differences are attributed to "spontaneous" breaking of the symmetry. Just as the spherical symmetry of the earth is broken to a cylindrical symmetry by the earth's magnetic field, a field permeating all space, called the Higgs field, is invoked to

Table 1. Elementary particles of the standard model: $S\hbar$ is spin, Qe is electric charge, and $m(\text{GeV}/c^2)$ is mass. Numerical subscripts indicate the distinct color states of quarks and gluons.

Quarks: $S=\frac{1}{2}$				Leptons: $S=\frac{1}{2}$				Gauge bosons: $S=1$	
$Q=\frac{2}{3}$	m	$Q=-\frac{1}{3}$	m	$Q=-1$	m	$Q=0$	m	Quanta	m
$u_1 u_2 u_3$	$(2-8)10^{-3}$	$d_1 d_2 d_3$	$(5-15)10^{-3}$	e	5.11×10^{-4}	ν_e	$<1.5\times10^{-8}$	$g_1\cdots g_8$	$<$ a few $\times10^{-3}$
$c_1 c_2 c_3$	$1.0-1.6$	$s_1 s_2 s_3$	$0.1-0.3$	μ	0.10566	ν_μ	$<1.7\times10^{-4}$	γ	$<6\times10^{-25}$
$t_1 t_2 t_3$	173.8 ± 5.0	$b_1 b_2 b_3$	$4.1-4.5$	τ	1.7770	ν_τ	$<1.8\times10^{-2}$	$W^\pm, Z^0$	$80.39\pm0.06, 91.187\pm0.002$

explain the observation that the symmetries of the electroweak theory are broken to the residual gauge symmetry of QED. Particles that interact with the Higgs field cannot propagate at the speed of light, and acquire masses, in analogy to the index of refraction that slows a photon traversing matter. Particles that do not interact with the Higgs field—the photon, gluons, and possibly neutrinos—remain massless. Fermion couplings to the Higgs field not only determine their masses; they induce a misalignment of quark mass eigenstates with respect to the eigenstates of the weak charges, thereby allowing all fermions of heavy families to decay to lighter ones. These couplings provide the only mechanism within the standard model that can account for the observed violation of CP, that is, invariance of the laws of nature under mirror reflection (parity P) and the interchange of particles with their antiparticles (charge conjugation C).

The origin of the Higgs field has not yet been determined. However, our very understanding of the standard model implies that the physics associated with electroweak symmetry breaking must become manifest at energies of present-day colliders or at the LHC under construction. There is strong reason, stemming from the quantum instability of scalar masses, to believe that this physics will point to modifications of the theory. One shortcoming of the standard model is its failure to accommodate gravity, for which there is no renormalizable quantum field theory because the quantum of the gravitational field has two units of spin. Recent theoretical progress suggests that quantum gravity can be formulated only in terms of extended objects like strings and membranes, with dimensions of order of the Planck length 10^{-35} m. Experiments probing higher energies and shorter distances may reveal clues connecting the standard-model physics to gravity and may shed light on other questions that it leaves unanswered. In the following we trace the steps that led to the formulation of the standard model, describe the experiments that have confirmed it, and discuss some outstanding unresolved issues that suggest a more fundamental theory underlies the standard model.

The Path to QCD

The invention of the bubble chamber permitted the observation of a rich spectroscopy of hadron states. Attempts at their classification using group theory, analogous to the introduction of isotopic spin as a classification scheme for nuclear states, culminated in the "Eightfold Way" based on the group SU(3), in which particles are ordered by their "flavor" quantum numbers: isotopic spin and strangeness. This scheme was spectacularly confirmed by the discovery at Brookhaven National Laboratory (BNL) of the Ω^- particle, with three units of strangeness, at the predicted mass. It was subsequently realized that the spectrum of the Eightfold Way could be understood if hadrons were composed of three types of quarks: u, d, and the strange quark s. However, the quark model presented a dilemma: each quark was attributed one-half unit of spin, but Fermi statistics precluded the existence of a state like the Ω^- composed of three strange quarks with total spin $\frac{3}{2}$. Three identical fermions with their spins aligned cannot exist in an an s-wave ground state. This paradox led to the hypothesis that quarks possess an additional quantum number called color, a conjecture supported by the observed rates for π^0 decay into $\gamma\gamma$ and e^+e^- annihilation into hadrons, both of which require three different quark types for each quark flavor.

A combination of experimental observations and theoretical analyses in the 1960s led to another important conclusion: pions behave like the Goldstone bosons of a spontaneously broken symmetry, called chiral symmetry. Massless fermions have a conserved quantum number called chirality, equal to their helicity: $+1$ for right-handed fermions and -1 for left-handed fermions. The analysis of pion scattering lengths and weak decays into pions strongly suggested that chiral symmetry is explicitly broken only by quark masses, which in turn implied that the underlying theory describing strong interactions among quarks must conserve quark helicity—just as QED conserves electron helicity. This further implied that interactions among quarks must be mediated by the exchange of spin-1 particles.

In the early 1970s, experimenters at the Stanford Linear Accelerator Center (SLAC) analyzed the distributions in energy and angle of electrons scattered from nuclear targets in inelastic collisions with momentum transfer $Q^2 \approx 1$ GeV/c from the electron to the struck nucleon. The distributions they observed suggested that electrons interact via photon exchange with pointlike objects called partons—electrically charged particles much smaller than nucleons. If the electrons were scattered by an extended object, e.g., a strongly interacting nucleon with its electric charge spread out by a cloud of pions, the cross section would drop rapidly for values of momentum transfer greater than the inverse radius of the charge distribution. Instead, the data showed a "scale-invariant" distribution: a cross section equal to the QED cross section up to a dimensionless function of kinematic variables, independent of the energy of the incident electron. Neutrino-scattering experiments at CERN and Fermilab (FNAL) yielded similar results. Comparison of electron and neutrino data allowed a determination of the average squared electric charge of the partons in the nucleon, and the result was consistent with the interpretation that they are fractionally charged quarks. Subsequent experiments at SLAC showed that, at center-of-mass energies above about two GeV, the final states in e^+e^- annihilation into hadrons have a two-jet configuration. The angular distribution of the jets with respect to the beam, which depends on the spin of the final-state particles, is similar to that of the muons in an $\mu^+\mu^-$ final state, providing direct evidence for spin-$\frac{1}{2}$ partonlike objects.

The Path to the Electroweak Theory

A major breakthrough in deciphering the structure of weak interactions was the suggestion that they may not conserve parity, prompted by the observation of K decay into both 2π and 3π final states with opposite parity. An intensive search for parity violation in other decays culminated in the establishment of the "universal $V-A$ interaction." Weak processes such as nuclear β decay and muon decay arise from quartic couplings of fermions with negative chirality; thus only left-handed electrons and right-handed positrons are weakly coupled. Inverse β decay was observed in interactions induced by electron antineutrinos from reactor fluxes, and several years later the muon neutrino was demonstrated to be distinct from the electron neutrino at the BNL alternating-gradient synchrotron.

With the advent of the quark model, the predictions of the universal $V-A$ interaction could be summarized by introducing a weak-interaction Hamiltonian density of the form

$$H_w = \frac{G_F}{\sqrt{2}} J^\mu J^\dagger_\mu,$$

$$J_\mu = \bar{d}\gamma_\mu(1-\gamma_5)u + \bar{e}\gamma_\mu(1-\gamma_5)\nu_e + \bar{\mu}\gamma_\mu(1-\gamma_5)\nu_\mu, \tag{1}$$

where G_F is the Fermi coupling constant, γ_μ is a Dirac matrix, and $\frac{1}{2}(1-\gamma_5)$ is the negative chirality projection operator. However, Eq. (1) does not take into account the observed β decays of strange particles. Moreover, increasingly precise measurements, together with an improved understanding of QED corrections, showed that the Fermi constant governing neutron β decay is a few percent less than the μ-decay constant. Both problems were resolved by the introduction of the Cabibbo angle θ_c and the replacement $d \to d_c = d\cos\theta_c + s\sin\theta_c$ in Eq. (1). Precision measurements made possible by high-energy beams of hyperons (the strange counterparts of nucleons) at CERN and FNAL have confirmed in detail the predictions of this theory with $\sin\theta_c \approx 0.2$.

While the weak interactions maximally violate P and C, CP is an exact symmetry of the Hamiltonian (1). The discovery at BNL in 1964 that CP is violated in neutral-kaon decay to two pions at a level of 0.1% in amplitude could not be incorporated into the theory in any obvious way. Another difficulty arose from quantum effects induced by the Hamiltonian (1) that allow the annihilation of the antistrange quark and the down quark in a neutral kaon. This annihilation can produce a $\mu^+\mu^-$ pair, resulting in the decay $K^0 \to \mu^+\mu^-$, or a $\bar{d}s$ pair, inducing K^0-$\bar{K}^0$ mixing. To suppress processes like these to a level consistent with experimental observation, a fourth quark flavor called charm (c) was proposed, with the current density in Eq. (1) modified to read

$$J_\mu = \bar{d}_c \gamma_\mu (1-\gamma_5) u + \bar{s}_c \gamma_\mu (1-\gamma_5) c + \bar{e}\gamma_\mu(1-\gamma_5)\nu_e + \bar{\mu}\gamma_\mu(1-\gamma_5)\nu_\mu,$$

$$s_c = s\cos\theta_c - d\sin\theta_c. \qquad (2)$$

With this modification, contributions from virtual $c\bar{c}$ pairs cancel those from virtual $u\bar{u}$ pairs, up to effects dependent on the difference between the u and c masses. Comparison with experiment suggested that the charmed-quark mass should be no larger than a few GeV. The narrow resonance J/ψ with mass of about 3 GeV, found in 1974 at BNL and SLAC, was ultimately identified as a $c\bar{c}$ bound state.

The Search for Renormalizable Theories

In the 1960s the only known renormalizable theories were QED and the Yukawa theory—the interaction of spin-$\frac{1}{2}$ fermions via the exchange of spinless particles. Both the chiral symmetry of the strong interactions and the $V-A$ nature of the weak interactions suggested that all forces except gravity are mediated by spin-1 particles, like the photon. QED is renormalizable because gauge invariance, which gives conservation of electric charge, also ensures the cancellation of quantum corrections that would otherwise result in infinitely large amplitudes. Gauge invariance implies a massless gauge particle and hence a long-range force. Moreover, the mediator of weak interactions must carry electric charge and thus couple to the photon, requiring its description within a Yang-Mills theory that is characterized by self-coupled gauge bosons.

The important theoretical breakthrough of the early 1970s was the proof that Yang-Mills theories are renormalizable, and that renormalizability remains intact if gauge symmetry is spontaneously broken, that is, if the Lagrangian is gauge invariant, but the vacuum state and spectrum of particles are not. An example is a ferromagnet for which the lowest-energy configuration has electron spins aligned; the direction of alignment spontaneously breaks the rotational invariance of the laws of physics. In quantum field theory, the simplest way to induce spontaneous symmetry breaking is the Higgs mechanism. A set of elementary scalars ϕ is introduced with a potential-energy density function $V(\phi)$ that is minimized at a value $\langle\phi\rangle \neq 0$ and the vacuum energy is degenerate. For example, the gauge-invariant potential for an electrically charged scalar field $\phi = |\phi|e^{i\theta}$,

$$V(|\phi|^2) = -\mu^2|\phi|^2 + \lambda|\phi|^4, \quad (3)$$

has its minimum at $\sqrt{2}\langle|\phi|\rangle = \mu/\sqrt{\lambda} = v$, but is independent of the phase θ. Nature's choice for θ spontaneously breaks the gauge symmetry. Quantum excitations of $|\phi|$ about its vacuum value are massive Higgs scalars: $m_H^2 = 2\mu^2 = 2\lambda v^2$. Quantum excitations around the vacuum value of θ cost no energy and are massless, spinless particles called Goldstone bosons. They appear in the physical spectrum as the longitudinally polarized spin states of gauge bosons that acquire masses through their couplings to the Higgs field. A gauge-boson mass m is determined by its coupling g to the Higgs field and the vacuum value v. Since gauge couplings are universal this also determines the Fermi constant G for this toy model: $m = gv/2, G/\sqrt{2} = g^2/8m^2 = v^2/2$.

The gauge theory of electroweak interactions entails four gauge bosons: $W^{\pm 0}$ of SU(2) or weak isospin $\vec{I}_w$, with coupling constant $g = e \sin\theta_w$, and B^0 of U(1) or weak hypercharge $Y_w = Q - I_w^3$, with coupling $g' = e \cos\theta_w$. Symmetry breaking can be achieved by the introduction of an isodoublet of complex scalar fields $\phi = (\phi^+ \phi^0)$, with a potential identical to Eq. (3) where $|\phi|^2 = |\phi^+|^2 + |\phi^0|^2$. Minimization of the vacuum energy fixes $v = \sqrt{2}|\phi| = 2^{1/4} G_F^{1/2} = 246$ GeV, leaving three Goldstone bosons that are eaten by three massive vector bosons: $W^\pm$ and $Z = \cos\theta_w W^0 - \sin\theta_w B^0$, while the photon $\gamma = \cos\theta_w B^0 + \sin\theta_w W^0$ remains massless. This theory predicted neutrino-induced neutral-current interactions of the type $\nu +$ atom $\rightarrow \nu +$ anything, mediated by Z exchange. The weak mixing angle θ_w governs the dependence of the neutral-current couplings on fermion helicity and electric charge, and their interaction rates are determined by the Fermi constant G_F^Z. The ratio $\rho = G_F^Z/G_F = m_W^2/m_Z^2 \cos^2\theta_w$, predicted to be 1, is the only measured parameter of the standard model that probes the symmetry-breaking mechanism. Once the value of θ_w was determined in neutrino experiments, the W and Z masses could be predicted: $m_W^2 = m_Z^2 \cos^2\theta_w = \sin^2\theta_w \pi\alpha/\sqrt{2}G_F$.

This model is not renormalizable with three quark flavors and four lepton flavors because gauge invariance is broken at the quantum level unless the sum of electric charges of all fermions vanishes. This is true for each family of fermions in Table I, and could be achieved by invoking the existence of the charmed quark, introduced in Eq. (2). However, the discovery of charmed mesons ($c\bar{u}$ and $c\bar{d}$ bound states) in 1976 was quickly followed by the discovery of the τ lepton, requiring a third full fermion family. A third family had in fact been anticipated by efforts to accommodate CP violation, which can arise from the misalignment between fermion gauge couplings and Higgs couplings provided there are more than two fermion families.

Meanwhile, to understand the observed scaling behavior in deep-inelastic scattering of leptons from nucleons, theorists were searching for an asymptotically free theory—a theory in which couplings become weak at short distance. The charge distribution of a strongly interacting particle is spread out by quantum effects, while scaling showed that at large momentum transfer quarks behaved like noninteracting particles. This could be understood if the strong coupling becomes weak at short distances, in contrast to electric charge or Yukawa couplings that decrease with distance due to the screening effect of vacuum polarization. QCD, with gauged SU(3) color charge, became the prime candidate for the strong force when it was discovered that Yang-Mills theories are asymptotically free: the vacuum polarization from charged gauge bosons has the opposite sign from the fermion contribution and is dominant if there are sufficiently few fermion flavors. This qualitatively explains quark and gluon confinement: the force between color-charged particles grows with the distance between them, so they cannot be separated by a distance much larger than the size of a hadron. QCD interactions at short distance are characterized by weak coupling and can be calculated using perturbation theory as in QED; their effects contribute measurable deviations from scale invariance that depend logarithmically on the momentum transfer.

The standard model gauge group, $SU(3) \times SU(2) \times U(1)$, is characterized by three coupling constants $g_3 = g_S$, $g_2 = g$, $g_1 = \sqrt{5/3} g'$, where g_1 is fixed by requiring the same normalization for all fermion currents. Their measured values at low energy satisfy $g_3 > g_2 > g_1$. Like g_3, the coupling g_2 decreases with increasing energy, but more slowly because there are fewer gauge bosons contributing. As in QED, the U(1) coupling increases with energy. Vacuum polarization effects calculated using the particle content of the standard model show that the three coupling constants are very nearly equal at an energy scale around 10^{16} GeV, providing a tantalizing hint of a more highly symmetric theory, embedding the standard-model interactions into a single force. Particle masses also depend on energy; the b and τ masses become equal at a similar scale, suggesting the possibility of quark and lepton unification as different charge states of a single field.

Brief Summary of the Standard-Model Elements

The standard model contains the set of elementary particles shown in Table I. The forces operative in the particle domain are the strong (QCD) interaction, responsive to particles carrying color, and the two pieces of the electroweak interaction, responsive to particles carrying weak isospin and hypercharge. The quarks come in three experimentally indistinguishable colors and there are eight colored gluons. All quarks and leptons, as well as the γ, W, and Z bosons, carry weak isospin. In the strict view of the standard model, there are no right-handed neutrinos or left-handed antineutrinos. As a consequence the simple Higgs mechanism described in Sec. IV cannot generate neutrino masses, which are posited to be zero.

In addition, the standard model provides the quark mixing matrix which gives the transformation from the basis of the strong-interaction charge $-\frac{1}{3}$ left-handed quark flavors to the mixtures which couple to the electroweak current. The elements of this matrix are fundamental parameters of the standard model. A similar mixing may occur for the neutrino flavors, and if accompanied by nonzero neutrino mass, would induce weak-interaction flavor-changing phenomena that are outside the standard-model framework.

Finding the constituents of the standard model spanned the first century of the American Physical Society, starting with the discovery by Thomson of the electron in 1897. Pauli in 1930 postulated the existence of the neutrino as the agent of missing energy and angular momentum in β decay; only in 1953 was the neutrino found in experiments at reactors. The muon was unexpectedly added from cosmic-ray searches for the Yukawa particle in 1936; in 1962 its companion neutrino was found in the decays of the pion.

The Eightfold Way classification of the hadrons in 1961 suggested the possible existence of the three lightest quarks (u, d, and s), though their physical reality was then regarded as doubtful. The observation of substructure of the proton, the 1974 observation of the J/ψ meson interpreted as a $c\bar{c}$ bound state, and the observation of mesons with a single charm quark in 1976 cemented the reality of the first two generations of quarks. This state of affairs, with two symmetric generations of leptons and quarks, was theoretically tenable and the particle story very briefly seemed finished.

In 1976, the τ lepton was found in a SLAC experiment, breaking new ground into the third generation of fermions. The discovery of the Y at FNAL in 1979 was interpreted as the bound state of a new bottom (b) quark. The neutrino associated with the τ has not been directly observed, but indirect measurements certify its existence beyond reasonable doubt. The final step was the discovery of the top (t) quark at FNAL in 1995. Despite the completed particle roster, there are fundamental questions remaining; chief among these is the tremendous disparity of the matter particle masses, ranging from the nearly massless neutrinos, the 0.5-MeV electron, and few-MeV u and d quarks, to the top quark whose mass is nearly 200 GeV. Even the taxonomy of particles hints at unresolved fundamental questions!

The gauge particle discoveries are also complete. The photon was inferred from the arguments of Planck, Einstein, and Compton early in this century. The carriers of the weak interaction, the W and Z bosons, were postulated to correct the lack of renormalizability of the four-Fermion interaction and given relatively precise predictions in the unified electroweak theory. The discovery of these in the CERN $p\bar{p}$ collider in 1983 was a dramatic confirmation of this theory. The gluon which mediates the color-force QCD was first demonstrated in the e^+e^- collider at DESY in Hamburg.

The minimal version of the standard model, with no right-handed neutrinos and the simplest possible electroweak symmetry-breaking mechanism, has 19 arbitrary parameters: nine fermion masses; three angles and one phase that specify the quark mixing matrix; three gauge coupling constants; two parameters to specify the Higgs potential; and an additional phase θ that characterizes the QCD vacuum state. The number of parameters is larger if the electroweak symmetry-breaking mechanism is more complicated or if there are right-handed neutrinos. Aside from constraints imposed by renormalizability, the spectrum of elementary particles is also arbitrary. As will be discussed in Sec. VII, this high degree of arbitrariness suggests that a more fundamental theory underlies the standard model.

Experimental Establishment of the Standard Model

The current picture of particles and interactions has been shaped and tested by three decades of experimental studies at laboratories around the world. We briefly summarize here some typical and landmark results.

Establishing QCD

Deep-inelastic scattering

Pioneering experiments at SLAC in the late 1960s directed high-energy electrons on proton and nuclear targets. The deep-inelastic scattering process results in a deflected electron and a hadronic recoil system from the initial baryon. The scattering occurs through the exchange of a photon coupled to the electric charges of the participants. Deep-inelastic scattering experiments were the spiritual descendents of Rutherford's scattering of α particles by gold atoms and, as with the earlier experiment, showed the existence of the target's substructure. Lorentz and gauge invariance restrict the matrix element representing the hadronic part of the interaction to two terms, each multiplied by phenomenological form factors or structure functions. These in principle depend on the two independent kinematic variables; the momentum transfer carried by the photon (Q^2) and energy loss by the electron (ν). The experiments showed that the structure functions were, to good approximations, independent of Q^2 for fixed values of $x = Q^2/2M\nu$. This "scaling" result was interpreted as evidence that the proton contains subelements, originally called partons. The deep-inelastic scattering occurs when a beam electron scatters with one of the partons. The original and subsequent experiments established that the struck partons carry the fractional electric charges and half-integer spins dictated by the quark model. Furthermore, the experiments demonstrated that three such partons (valence quarks) provide the nucleon with its quantum numbers. The variable x represents the fraction of the target nucleon's momentum carried by the struck parton, viewed in a Lorentz frame where the proton is relativistic. The deep-inelastic scattering experiments further showed that the charged partons (quarks) carry only about half of the proton momentum, giving indirect evidence for an electrically neutral partonic gluon.

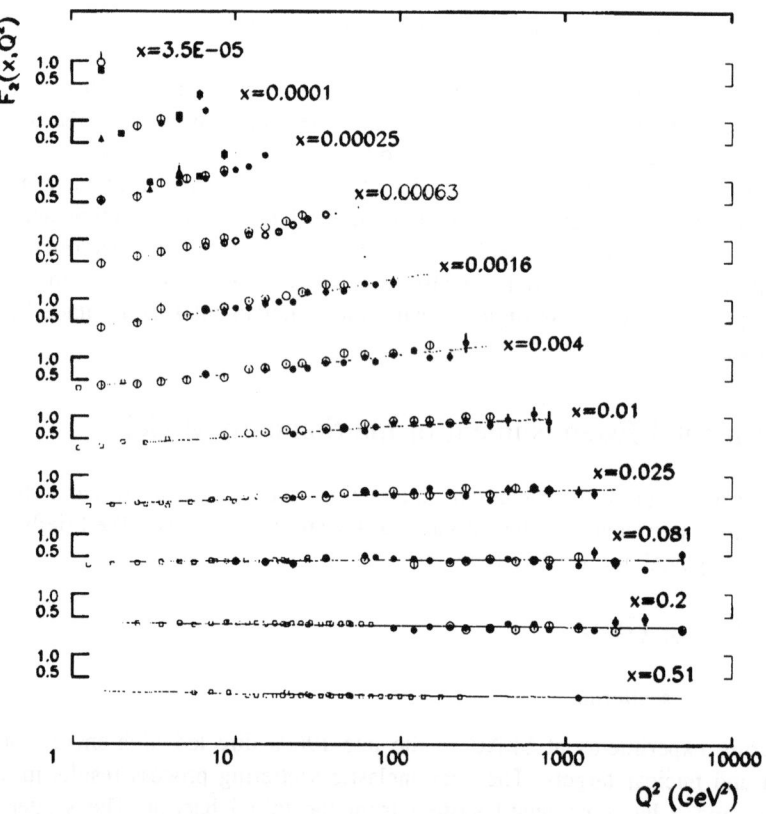

Figure 1. The proton structure function F_2 vs Q^2 at fixed x, measured with incident electrons or muons, showing scale invariance at larger x and substantial dependence on Q^2 as x becomes small. The data are taken from the HERA ep collider experiments H1 and ZEUS, as well as the muon-scattering experiments BCDMS and NMC at CERN and E665 at FNAL.

Further deep-inelastic scattering investigations using electrons, muons, and neutrinos and a variety of targets refined this picture and demonstrated small but systematic nonscaling behavior. The structure functions were shown to vary more rapidly with Q^2 as x decreases, in accord with the nascent QCD prediction that the fundamental strong-coupling constant α_S varies with Q^2 and that, at short distance scales (high Q^2), the number of observable partons increases due to increasingly resolved quantum fluctuations. Figure 1 shows sample modern results for the Q^2 dependence of the dominant structure-function, in excellent accord with QCD predictions. The structure-function *values* at all x depend on the quark content; the *increases* at larger Q^2 depend on both quark and gluon content. The data permit the mapping of the proton's quark and gluon content exemplified in Fig. 2.

Quark and gluon jets

The gluon was firmly predicted as the carrier of the color force. Though its presence had been inferred because only about half the proton momentum was found in charged constituents, direct observation of the gluon was essential. This came from experiments at the DESY e^+e^- collider (PETRA) in 1979. The collision forms an intermediate virtual photon state, which may subsequently decay into a pair of leptons or pair of quarks. The colored quarks cannot emerge intact from the collision region; instead they create many quark-antiquark pairs from the vacuum that arrange themselves into a set of colorless hadrons moving approximately in the directions of the original quarks. These sprays of roughly collinear particles, called jets, reflect the directions of the progenitor quarks. However, the quarks may radiate quanta of QCD (gluons) prior to formation of the jets, just as electrons radiate photons. If at sufficiently large angle to be distinguished, the gluon radiation evolves into a separate jet. Evidence for the "three-pronged" jet topologies expected for events containing a gluon was found in the event energy-flow patterns. Experiments at higher-energy e^+e^- colliders illustrate this gluon radiation even better, as

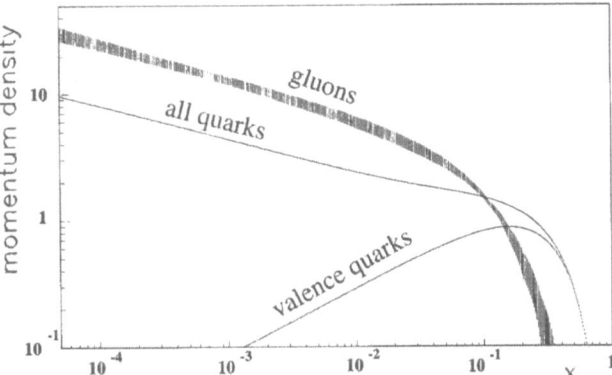

Figure 2. The quark and gluon momentum densities in the proton vs x for $Q^2=20$ GeV^2. The integrated values of each component density give the fraction of the proton momentum carried by that component. The valence u and d quarks carry the quantum numbers of the proton. The large number of quarks at small x arises from a "sea" of quark-antiquark pairs. The quark densities are from a phenomenological fit (CTEQ collaboration) to data from many sources; the gluon density bands are the one-standard-deviation bounds to QCD fits to ZEUS data (low x) and muon-scattering data (higher x).

shown in Fig. 3. Studies in e^+e^- and hadron collisions have verified the expected QCD structure of the quark-gluon couplings and their interference patterns.

Strong-coupling constant

The fundamental characteristic of QCD is asymptotic freedom, dictating that the coupling constant for color interactions decreases logarithmically as Q^2 increases. The coupling α_S can be measured in a variety of strong-interaction processes at different Q^2 scales. At low Q^2, deep-inelastic scattering, tau decays to hadrons, and the annihilation rate for e^+e^- into multihadron final states give accurate determinations of α_S. The decays of the Υ into three jets primarily involve gluons, and the rate for this decay gives $\alpha_S(M_\Upsilon^2)$. At higher Q^2, studies of the W and Z bosons (for example, the decay width of the Z, or the fraction of W bosons associated with jets) measure α_S at the 100-GeV scale. These and many other determinations have now solidified the experimental evidence that α_S does indeed "run" with Q^2 as expected in QCD. Predictions for $\alpha_S(Q^2)$, relative to its value at some reference scale, can be made within perturbative QCD. The current information from many sources is compared with calculated values in Fig. 4.

Strong-interaction scattering of partons

At sufficiently large Q^2 where α_S is small, the QCD perturbation series converges sufficiently rapidly to permit accurate predictions. An important process probing the highest accessible Q^2 scales is the scattering of two constituent partons (quarks or gluons) within colliding protons and antiprotons. Figure 5 shows the impressive data for the inclusive

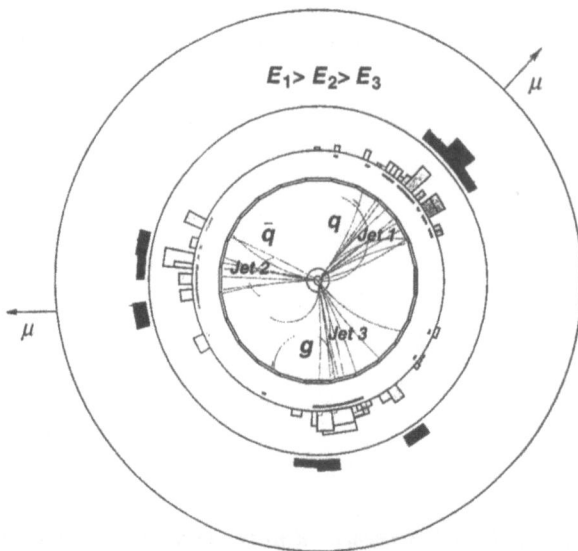

Figure 3. A three-jet event from the OPAL experiment at LEP. The curving tracks from the three jets may be associated with the energy deposits in the surrounding calorimeter, shown here as histograms on the middle two circles, whose bin heights are proportional to energy. Jets 1 and 2 contain muons as indicated, suggesting that these are both quark jets (likely from b quarks). The lowest-energy jet 3 is attributed to a radiated gluon.

production of jets due to scattered partons in $p\bar{p}$ collisions at 1800 GeV. The QCD NLO predictions give agreement with the data over nine orders of magnitude in the cross section.

The angular distribution of the two highest-transverse-momentum jets from $p\bar{p}$ collisions reveals the structure of the scattering matrix element. These amplitudes are dominated by the exchange of the spin-1 gluon. If this scattering were identical to Rutherford scattering, the angular variable $\chi = (1 + |\cos\theta_{cm}|)/(1 - |\cos\theta_{cm}|)$ would provide $d\sigma/d\chi$ = constant. The data shown in Fig. 6 for dijet production show that the spin-1 exchange process is dominant, with clearly visible differences required by QCD, including the varying α_S. These data also demonstrate the *absence* of further substructure (of the partons) to distance scales approaching 10^{-19} m.

Many other measurements test the correctness of QCD in the perturbative regime. Production of photons and W and Z bosons occurring in hadron collisions are well described by QCD. Production of heavy quark pairs, such as $t\bar{t}$, is not only sensitive to perturbative processes, but also reflects additional effects due to multiple-gluon radiation from the scattering quarks. Within the limited statistics of current data samples, the top quark production cross section is also in good agreement with QCD.

Nonperturbative QCD

Many physicists believe that QCD is a theory "solved in principle." The basic validity of QCD at large Q^2, where the coupling is small, has been verified in many experimental studies, but the large coupling at low-Q^2 makes calculation exceedingly difficult. This low-Q^2 region of QCD is relevant to the wealth of experimental data on the static properties of nucleons, most hadronic interactions, hadronic weak decays, nucleon and nucleus structure, proton and neutron spin structure, and systems of hadronic matter with

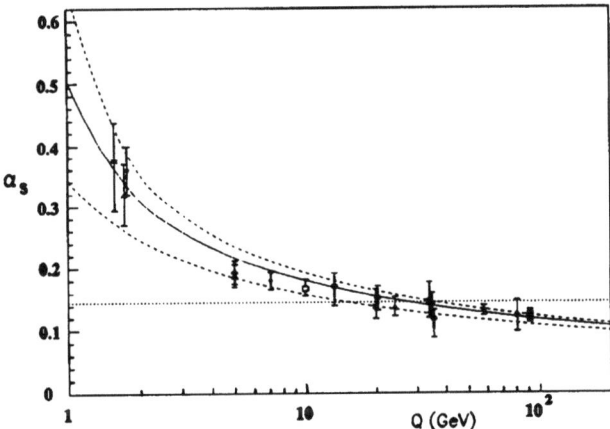

Figure 4. The dependence of the strong-coupling constant α_S vs Q, using data from deep-inelastic-scattering structure functions from e, μ, and ν beam experiments as well as ep collider experiments, production rates of jets, heavy-quark flavors, photons, and weak vector bosons in ep, e^+e^-, and $p\bar{p}$ experiments. The data are in clear disagreement with a strong coupling independent of Q (horizontal line). All data agree with the dependence expected in QCD. The curves correspond to next-to-leading-order calculations of $\alpha_S(Q)$ evaluated using values for $\alpha_S(M_Z)$ of 0.1048, 0.1175, and 0.1240.

very high temperature and energy densities. The ability of theory to predict such phenomena has yet to match the experimental progress.

Several techniques for dealing with nonperturbative QCD have been developed. The most successful address processes in which some energy or mass in the problem is large. An example is the confrontation of data on the rates of mesons containing heavy quarks (c or b) decaying into lighter hadrons, where the heavy quark can be treated nonrelativistically and its contribution to the matrix element is taken from experiment. With this phenomenological input, the ratios of calculated partial decay rates agree well with experiment. Calculations based on evaluation at discrete space-time points on a lattice and extrapolated to zero spacing have also had some success. With computing advances and new calculational algorithms, the lattice calculations are now advanced to the stage of calculating hadronic masses, the strong-coupling constant, and decay widths to within roughly (10–20)% of the experimental values.

The quark and gluon content of protons are consequences of QCD, much as the wave functions of electrons in atoms are consequences of electromagnetism. Such calculations require nonperturbative techniques. Measurements of the small-x proton structure functions at the HERA ep collider show much larger increases in parton density with decreasing x than were extrapolated from larger x measurements. It was also found that a large fraction ($\sim 10\%$) of such events contained a final-state proton essentially intact after collision. These were called "rapidity gap" events because they were characterized by a large interval of polar angle (or rapidity) in which no hadrons were created between the emerging nucleon and the jet. More typical events contain hadrons in this gap due to the exchange of the color charge between the struck quark and the remnant quarks of the proton. Similar phenomena have also been seen in hadron-hadron and photon-hadron scattering processes. Calculations that analytically resum whole categories of higher-order subprocesses have been performed. In such schemes, the agent for the elastic or quasi-elastic scattering processes is termed the "Pomeron," a concept from the Regge theory of a previous era, now viewed as a colorless conglomerate of colored gluons. These ideas

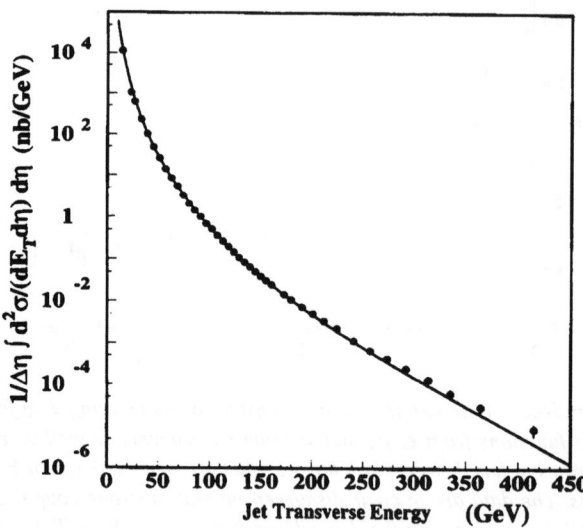

Figure 5. Inclusive jet cross section vs jet transverse momentum. The data points are from the CDF experiment. The curve gives the prediction of next-to-leading-order QCD.

have provided semiquantitative agreement with data coming from the *ep* collider at DESY and the Tevatron.

Establishing the electroweak interaction

Neutral currents in neutrino scattering

Though the electroweak theory had been proposed by 1968, it received little experimental attention until early in the next decade, when it was shown that all such gauge theories are renormalizable. The electroweak theory specifically proposed a new neutral-current weak interaction.

For virtually any scattering or decay process in which a photon might be exchanged, the neutral-current interaction required added Feynman diagrams with Z exchange and predicted modifications to known processes at very small levels. However, Z exchange is the only mechanism by which an electrically neutral neutrino can scatter elastically from a quark or from an electron, leaving a neutrino in the final state. The theory predicted a substantial rate for this previously unanticipated ν-induced neutral-current process. The only competitive interactions were the well-known charged-current processes with exchange of a W and a charged final-state lepton.

The neutral-current interactions were first seen at CERN in 1973 with scattering from nuclei at rates about 30% of the charged-current scattering (as well as hints of a purely

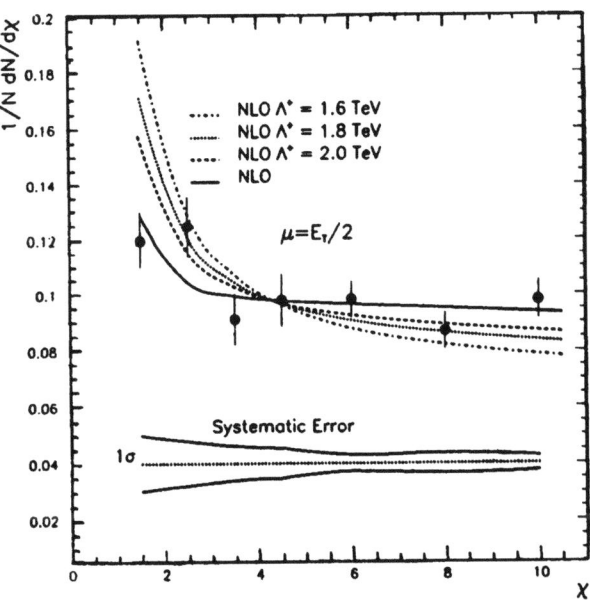

Figure 6. The dijet angular distribution from the DØ experiment plotted as a function of χ (see text) for which Rutherford scattering would give $d\sigma/d\chi = $ constant. The predictions of next-to-leading-order QCD (at scale $\mu = E_T/2$) are shown by the curves. Λ is the compositeness scale for quark/gluon substructure, with $\Lambda = \infty$ for no compositness (solid curve); the data rule out values of $\Lambda < 2$ TeV.

leptonic neutrino interaction with electrons). The results were initially treated with skepticism, since similar experiments had determined limits close to and even below the observed signal, and other contemporary experiments at higher energy obtained results that were initially ambiguous. By 1974, positive and unambiguous results at FNAL had corroborated the existence of the neutral-current reaction using high-energy ν's. In subsequent FNAL and CERN measurements using $\bar{\nu}$'s as well as ν's, the value of ρ was determined to be near unity, and the value of the weak angle, $\sin^2\theta_w$, was established. With time, the values of these parameters have been measured more and more accurately, at low and high energies, in ν reactions with electrons as well as with quarks. All are consistent with the electroweak theory and with a single value of $\sin^2\theta_w$. Figure 7 shows the characteristics of these charged-current and neutral-current events.

Photon and Z interference

The neutral current was found at about the anticipated level in several different neutrino reactions, but further verification of its properties were sought. Though reactions of charged leptons are dominated by photon exchange at accessible fixed-target energies, the parity-violating nature of the small Z-exchange contribution permits very sensitive experimental tests. The vector part of the neutral-current amplitude interferes constructively or destructively with the dominant electromagnetic amplitude. In 1978, the first successful

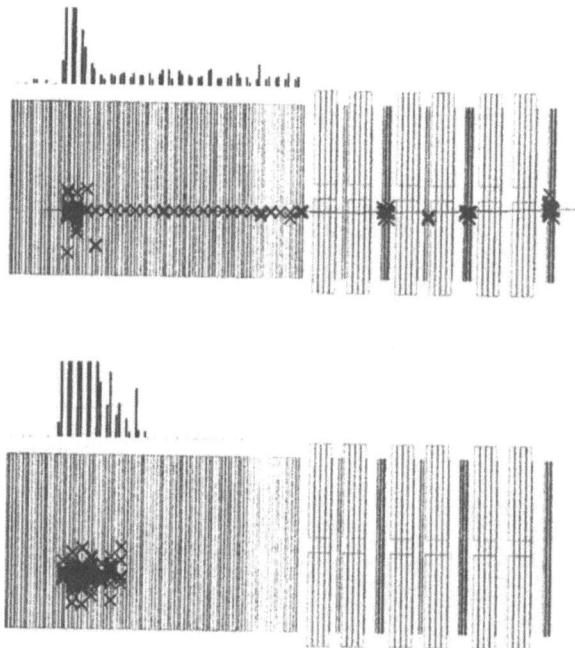

Figure 7. Displays of events created by ν_μ's in the CCFR experiment at Fermilab. The upper picture is a charged-current interaction, the lower a neutral-current interaction. In each case, the ν enters from the left and interacts after traversing about 1 m of steel. The charged-current event contains a visible energetic μ, which penetrates more than 10 m of steel; the neutral-current event contains an energetic final state ν, which passes through the remainder of the apparatus without trace. Each (×) records a hit in the sampling planes, and the histogram above the display shows the energy deposition in the scintillator planes interspersed in the steel. The energy near the interaction vertex results from produced hadrons.

such effort was reported, using the polarized electron beam at SLAC to measure the scattering asymmetry between right-handed and left-handed beam electrons. Asymmetries of about 10^{-4} were observed, using several different energies, implying a single value of $\sin^2\theta_w$, in agreement with neutrino measurements.

High-energy e^+e^- collisions provided another important opportunity to observe $\gamma - Z$ interference. By 1983 several experiments at DESY had observed the electromagnetic-weak interference in processes where the e^- and e^+ annihilate to produce a final-state μ pair or τ pair. The asymmetry grows rapidly above a center-of-mass (c.m.) energy of 30 GeV, then changes sign as the energy crosses the Z resonance. The weak electromagnetic interference is beautifully confirmed in the LEP data, as shown in Fig. 8.

Discovery of W and Z

With the corroborations of the electroweak theory with $\rho \sim 1$ and several consistent measurements of the one undetermined parameter, $\sin^2\theta_w$, reliable predictions existed by 1980 for the masses of the vector bosons W and Z. The predicted masses, about 80 and 90 GeV, respectively, were not accessible to e^+e^- colliders or fixed-target experiments, but adequate c.m. energy was possible with existing proton accelerators, so long as the collisions were between two such beams. Unfortunately, none had the two rings required to collide protons with protons.

A concerted effort was mounted at CERN to find the predicted bosons. To save the cost and time of building a second accelerating ring, systems were constructed to produce and accumulate large numbers of antiprotons, gather these and "cool" them into a beam, and then accelerate them in the existing accelerator to collide with a similar beam of protons. In 1983, the W and Z decays were observed with the anticipated masses. Present-day

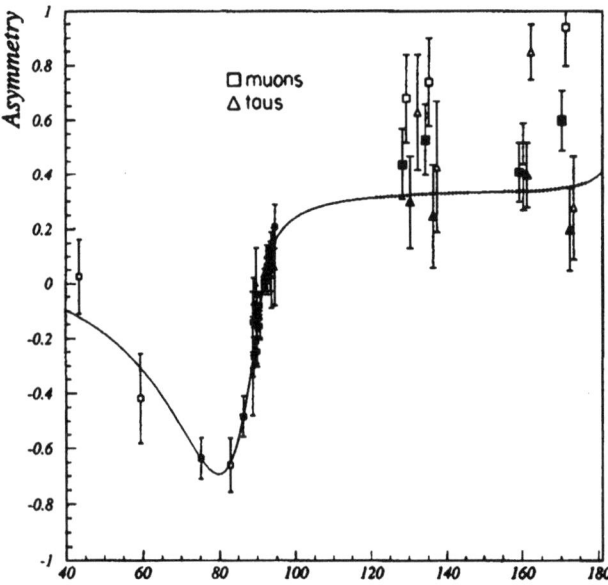

Figure 8. Forward-backward asymmetry in $e^+e^- \to \mu^+\mu^-$ and $e^+e^- \to \tau^+\tau^-$ as a function of energy from the DELPHI experiment at LEP. The interference of γ and Z contributions gives the asymmetry variation with energy, as indicated by the standard-model curve.

measurements from LEP (Fig. 9) give a fractional Z mass precision of about 10^{-5} and studies at the FNAL $p\bar{p}$ collider give a fractional W mass precision of about 10^{-3} (Fig. 10).

Z properties and precision tests of the electroweak standard model

The LEP and SLAC linear collider experiments have made many precise measurements of the properties of the Z, refining and testing the electroweak model. The asymmetries due to weak electromagnetic interference discussed above were extended to include all lepton species, c- and b-quark pairs, and light-quark pairs, as well as polarization asymmetries involving τ pairs and initial-state left- or right-handed electrons. From these data, the underlying vector and axial couplings to fermions have been extracted and found to be in excellent agreement with the standard model and with lepton universality. The fundamental weak mixing parameter, $\sin^2 \theta_w$, has been determined from these and other inputs to be 0.23152 ± 0.00023.

The total width of the Z is determined to be 2.4948 ± 0.0025 GeV; the invisible decay contributions to this total width allow the number of light ($m_\nu < m_Z/2$) neutrino generations to be measured: $N_\nu = 2.993 \pm 0.011$, confirming another aspect of the standard model. The partial widths for the Z were measured, again testing the standard model to the few-percent level and restricting possible additional non-standard-model particle contributions to the quantum loop corrections. The electroweak and QCD higher-order corrections modify the expectations for all observables. Figure 11 shows the allowed values in the $\sin^2 \theta_w$ vs Γ_{lepton} plane under the assumption that the standard model is valid. Even accounting for uncertainties in the Higgs boson mass, it is clear that the higher-order electroweak corrections are required.

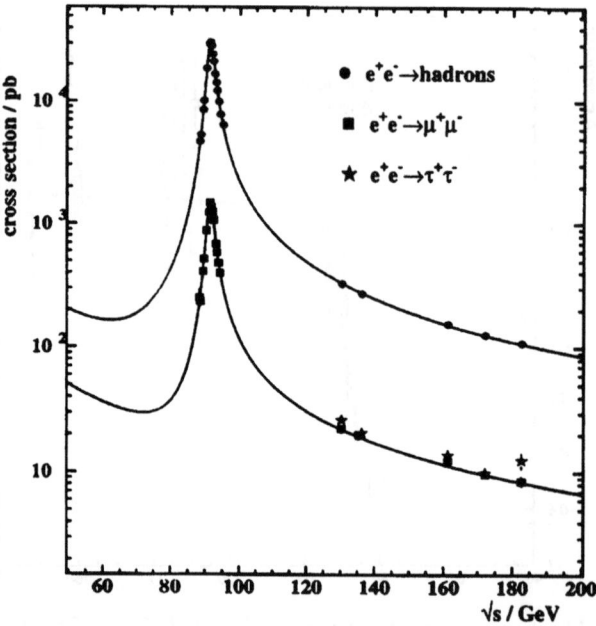

Figure 9. Dielectron invariant-mass distribution for $ee \rightarrow$ hadrons and $ee \rightarrow \mu\mu$ from the LEP collider experiments. The prominent Z resonance is clearly apparent.

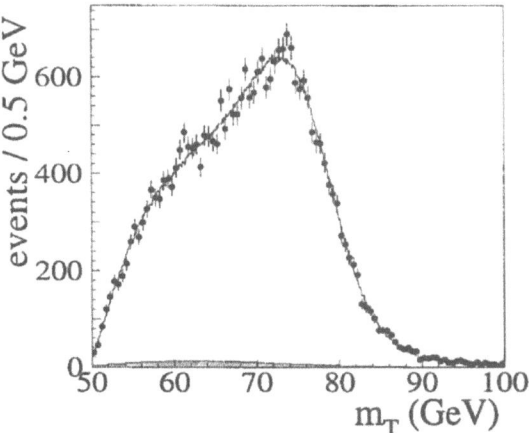

Figure 10. Transverse mass distribution for $W \to e\nu$ from the DØ experiment. The transverse mass is defined as $M_T = [2E_T^e E_T^\nu (1 - \cos \phi^{e\nu})]^{1/2}$ with E_T^e and E_T^ν the transverse energies of electron and neutrino and $\phi^{e\nu}$ the azimuthal angle between them. M_T has its Jacobian edge at the mass of the W boson.

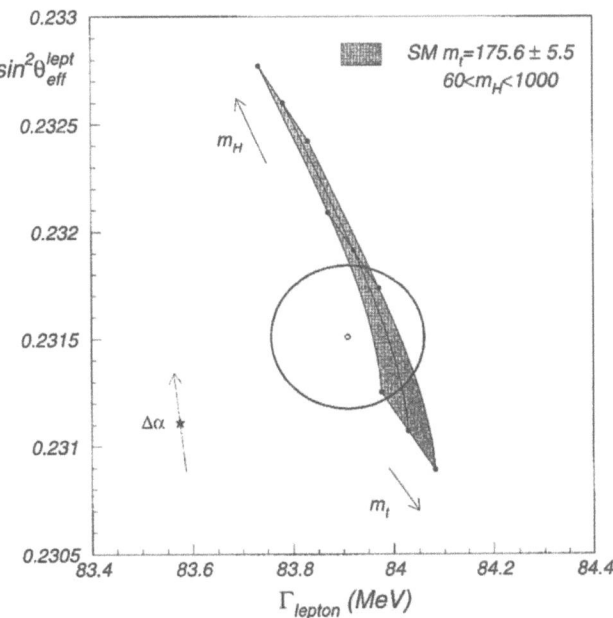

Figure 11. The allowed region for $\sin^2 \theta_w$ vs Γ_{lepton} in the context of the standard model, showing the need for the higher-order electroweak corrections. The region within the ellipse is allowed (at 1 standard deviation) by the many precision measurements at the LEP and SLC ee colliders and the FNAL $p\bar{p}$ collider; the shaded region comes from the measurements of the top mass at FNAL, for a range of possible Higgs masses. The star, well outside the allowed region, gives the expected value in the standard model without the higher-order electroweak corrections.

Taken together, the body of electroweak observables tests the overall consistency of the standard model. Extensions of the standard model would result in modification of observables at quantum loop level; dominant non-standard-model effects should modify the vacuum polarization terms and may be parametrized in terms of weak-isospin-conserving (S) and weak-isospin-breaking (T) couplings. S and T may be chosen to be zero for specific top quark and Higgs mass values in the minimal standard model; Fig. 12 shows the constraints afforded by several precision measurements and indicates the level to which extensions to the standard model are ruled out.

The top quark

The top quark was expected even before measurements in e^+e^- scattering unambiguously determined the b quark to be the $I_3 = -\frac{1}{2}$ member of an isospin doublet. In 1995, the two FNAL $p\bar{p}$ collider experiments reported the first observations of the top. Though expected as the last fermion in the standard model, its mass of about 175 GeV is startlingly large compared to its companion b, at about 4.5 GeV, and to all other fermion masses. The t decays nearly always into a W and a b, with final states governed by the subsequent decay of the W. The large top quark mass gives it the largest fermionic coupling to the Higgs sector. Since its mass is of order the Higgs vacuum expectation value $\langle |\phi| \rangle$, it is possible that the top plays a unique role in electroweak symmetry breaking. The top quark mass is now measured with a precision of about 3%. Together with other precision electroweak determinations, the mass gives useful standard-model constraints on the unknown Higgs boson mass, as shown in Fig. 13. At present, measure-

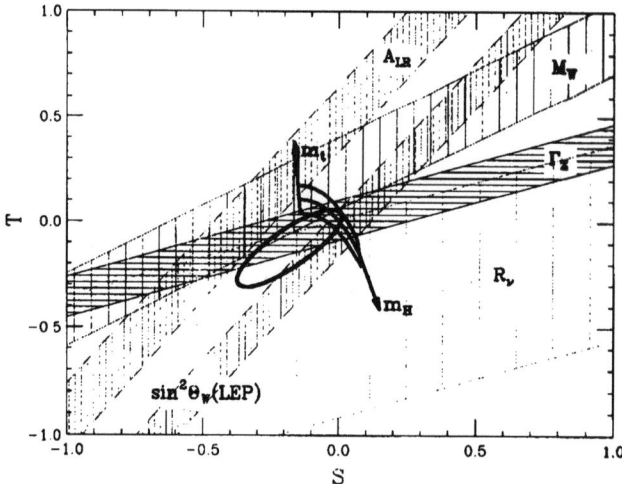

Figure 12. Several precise electroweak measurements in terms of the S and T variables which characterize the consistency of observables with the standard model. The bands shown from the experimental measurements of A_{LR} (SLC), Γ_Z (LEP), $\sin^2\theta_w$ (LEP), M_W (FNAL and CERN), and R_ν (ν deep-inelastic scattering experiments at CERN and FNAL) indicate the allowed regions in S, T space. The half-chevron region centered on $S=T=0$ gives the prediction for top mass = 175.5±5.5 GeV and Higgs mass between 70 and 1000 GeV, providing the standard model is correct. A fit to all electroweak data yields the 68% confidence region bounded by the ellipse and shows the consistency of the data and the agreement with the minimal standard-model theory.

ments require a standard-model Higgs boson mass less than 420 GeV at 95% confidence level. Such constraints place the Higgs boson, if it exists, within the range of anticipated experiments.

Trilinear gauge couplings

The gauge symmetry of the electroweak standard model exactly specifies the couplings of the W, Z, and γ bosons to each other. These gauge couplings may be probed through the production of boson pairs: WW, $W\gamma$, WZ, $Z\gamma$, and ZZ. The standard model specifies precisely the interference terms for all these processes. The diboson production reactions have been observed in FNAL collider experiments and the WW production has been seen at LEP. Limits have been placed on possible anomalous couplings beyond the standard model. For $WW\gamma$, the experiments have shown that the full electroweak gauge structure of the standard model is necessary, as shown in Fig. 14, and constrain the anomalous magnetic dipole and electric quadrupole moments of the W.

Quark mixing matrix

The generalization of the rotation of the down-strange weak-interaction eigenstates from the strong-interaction basis indicated in Eq. (2) to the case of three generations gives a 3×3 unitary transformation matrix $\mathbf{V}$, whose elements are the mixing amplitudes among the d, s, and b quarks. Four parameters—three real numbers (e.g., Euler angles) and one phase—are needed to specify this matrix. The real elements of this Cabibbo-Kobayashi-Maskawa (CKM) matrix are determined from various experimental studies of weak flavor-changing interactions and decays. The decay rates of c and b quarks depend on the CKM elements connecting the second and third generation. These have been extensively explored in e^+e^- and hadronic collisions which copiously produce B and charmed

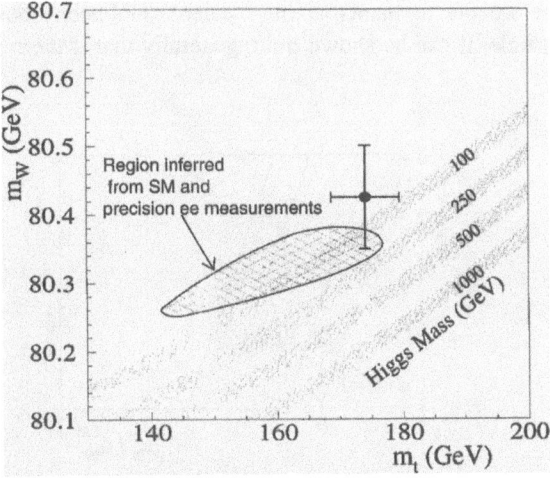

Figure 13. W boson mass vs top quark mass. The data point is the average of FNAL data for the top quark mass and FNAL and CERN data for the W boson mass. The shaded bands give the expected values for specific conventional Higgs boson mass values in the context of the minimal standard model. The cross-hatched region shows the predictions for m_W and m_{top}, at 68% confidence level, from precision electroweak measurements of Z boson properties.

mesons at Cornell, DESY, and FNAL. The pattern that emerges shows a hierarchy in which the mixing between first and second generation is of order the Cabibbo angle, $\lambda = \sin\theta_c$, those between the second and third generation are of order λ^2, and those between first and third generation are of order λ^3.

A nonzero CKM phase would provide CP-violating effects such as the decay $K_L^0 \to \pi\pi$, as well as different decay rates for B^0 and $\bar{B}^0$ into CP-eigenstate final states. CP violation has only been observed to date in the neutral-K decays, where it is consistent with (though not requiring) the description embodied in the CKM matrix. Well-defined predictions of the CKM phase for a variety of B-decay asymmetries will be tested in experiments at SLAC, KEK in Japan, Cornell, DESY, and FNAL in the coming few years. The unitarity relations $V_{ij}^{\dagger}V_{jk} = \delta_{ik}$ impose constraints on the observables that must be satisfied if CP violation is indeed embedded in the CKM matrix and if there are but three quark generations. Figure 15 shows the current status of the constraints on the real and imaginary parts (ρ, η) of the complex factor necessary if the origins of CP violation are inherent to the CKM matrix.

Unresolved Issues: Beyond the Standard Model

While the standard model has proven highly successful in correlating vast amounts of data, a major aspect of it is as yet untested, namely, the origin of electroweak symmetry breaking. The Higgs mechanism described in Sec. IV is just the simplest ansatz that is compatible with observation. It predicts the existence of a scalar particle, but not its mass; current LEP data provide a lower limit: $m_H > 80$ GeV. The Higgs mass is determined by its coupling constant λ [cf. Eq. (3)] and its vacuum value v: $m_H \approx \lambda \times 348$ GeV. A Higgs mass of a TeV or more would imply strong coupling of longitudinally polarized W and Z bosons that are the remnants of the "eaten" Goldstone boson partners of the physical Higgs particle. It can be shown quite generally that if there is no Higgs particle

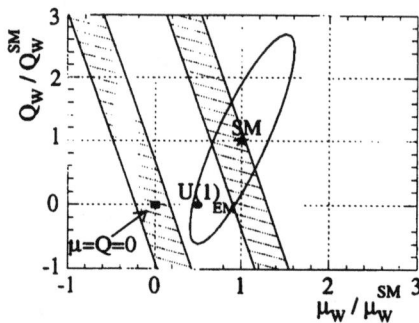

Figure 14. The W boson electric quadrupole moment vs magnetic dipole moment from $W\gamma$ production relative to their standard-model values. The ellipse shows the 95% confidence level limit from the DØ experiment with both Q and μ allowed to vary. Limits from $b \to s\gamma$ from CLEO at Cornell and ALEPH at LEP are shown as the hatched bands. The star shows the moments if the standard-model couplings are correct; the filled circle labeled $U(1)_{EM}$ corresponds to a standard-model $SU(2)$ coupling of zero.

with a mass less than about a TeV, strong W,Z scattering will occur at TeV c.m. energies; the observation of this scattering requires multi-TeV proton-proton c.m. energies, as will be achieved at the LHC.

However, the introduction of an elementary scalar field in quantum field theory is highly problematic. Its mass is subject to large quantum corrections that make it difficult to understand how it can be as small as a TeV or less in the presence of large scales in nature like the Planck scale of 10^{19} GeV or possibly a scale of coupling-constant unification at 10^{16} GeV. Moreover, a strongly interacting scalar field theory is not self-consistent as a fundamental theory: the coupling constant grows with energy and therefore any finite coupling at high energy implies a weakly coupled theory at low energy. There is therefore strong reason to believe that the simple Higgs mechanism described in an earlier section is incorrect or incomplete and that electroweak symmetry breaking must be associated with fundamentally new physics. Several possibilities for addressing these problems have been suggested; their common thread is the implication that the standard model is an excellent low-energy approximation to a more fundamental theory and that clues to this theory should appear at LHC energies or below.

For example, if quarks and leptons are composites of yet more fundamental entities, the standard model is a good approximation to nature only at energies small compared with the inverse radius of compositeness Λ. The observed scale of electroweak symmetry breaking, $v \sim \frac{1}{4}$ TeV, might emerge naturally in connection with the compositeness scale. A signature of compositeness would be deviations from standard-model predictions for high-energy scattering of quarks and leptons. Observed consistency (e.g., Fig. 6) with the standard model provides limits on Λ that are considerably higher than the scale v of electroweak symmetry breaking.

Another approach seeks only to eliminate the troublesome scalars as fundamental fields. Indeed, the spontaneous breaking of chiral symmetry by a quark-antiquark condensate in QCD also contributes to electroweak symmetry breaking. If this were its only source, the W,Z masses would be determined by the 100-MeV scale at which QCD is strongly coupled: $m_W = \cos\theta_w m_Z \approx 30$ MeV. To explain the much larger observed masses, one postulates a new gauge interaction, called technicolor, that is strongly

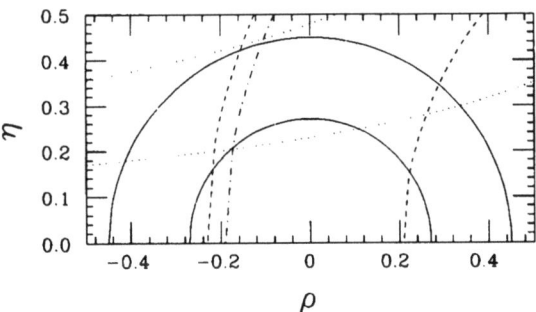

Figure 15. Experimentally allowed regions in the $\rho\eta$ plane from experiments. The region between the solid semicircles is from the ratio of b quark decays into u or c quarks. The CP-violating amplitudes from K_L^0 decays give the band between the dotted hyperbolae. The region between the dashed semicircles is allowed by measurements of B^0-$\bar{B}^0$ mixing. The constraint imposed from current limits on B_s^0-$\bar{B}_s^0$ mixing is to the right of the dot-dashed semicircle. Current experiments thus are consistent, and favor nonzero values of the CP-violating parameter η.

coupled at the scale $v \sim \frac{1}{4}$ TeV. At this scale fermions with technicolor charge condense, spontaneously breaking both a chiral symmetry and the electroweak gauge symmetry. The longitudinally polarized components of the massive W and Z are composite pseudoscalars that are Goldstone bosons of the broken chiral symmetry, analogous to the pions of QCD. This is a concrete realization of a scenario with no light scalar particle, but with strong W,Z couplings in the TeV regime, predicting a wealth of new composite particles with TeV masses. However, it has proven difficult to construct explicit models that are consistent with all data, especially the increasingly precise measurements that probe electroweak quantum corrections to W and Z self-energies; these data (Figs. 12 and 13) appear to favor an elementary scalar less massive than a few hundred GeV.

The quantum instability of elementary scalar masses can be overcome by extending the symmetry of the theory to one that relates bosons to fermions, known as supersymmetry. Since quantum corrections from fermions and bosons have opposite signs, many of them cancel in a supersymmetric theory, and scalar masses are no more unstable than fermion masses, whose smallness can be understood in terms of approximate chiral symmetries. This requires doubling the number of spin degrees of freedom for matter and gauge particles: for every fermion f there is a complex scalar partner $\tilde{f}$ with the same internal quantum numbers, and for every gauge boson v there is a spin-$\frac{1}{2}$ partner $\tilde{v}$. In addition, the cancellation of quantum gauge anomalies and the generation of masses for all charged fermions requires at least two distinct Higgs doublets with their fermion superpartners. Mass limits on matter and gauge superpartners ($m_{\tilde{\gamma},\tilde{W}} > 50$ GeV, $m_{\tilde{q},\tilde{g}} > 200$ GeV) imply that supersymmetry is broken in nature. However, if fermion-boson superpartner mass splittings are less than about a TeV, quantum corrections to the Higgs mass will be suppressed to the same level. For this scenario to provide a viable explanation of the electroweak symmetry-breaking scale, at least some superpartners must be light enough to be observed at the LHC.

Another untested aspect of the standard model is the origin of CP violation, conventionally introduced through complex Yukawa couplings of fermions to Higgs particles, resulting in complex parameters in the CKM matrix. This ansatz is sufficient to explain the observed CP violation in K decay, is consistent with limits on CP violation in other processes, and predicts observable CP-violating effects in B decay. Planned experiments at new and upgraded facilities capable of producing tens of millions of B mesons will determine whether this model correctly describes CP violation, at least at relatively low energy. A hint that some other source of CP violation may be needed, perhaps manifest only at higher energies, comes from the observed predominance of matter over antimatter in the universe.

While in the minimal formulation of the standard model neutrinos are massless and exist only in left-handed states, there have been persistent indirect indications for both neutrino masses and mixing of neutrino flavors. Nonzero neutrino mass and lepton flavor violation would produce spontaneous oscillation of neutrinos from one flavor to another in a manner similar to the strangeness oscillations of neutral-K mesons. Solar neutrinos of energies between 0.1 and 10 MeV have been observed to arrive at the earth at a rate significantly below predictions from solar models. A possible interpretation is the oscillation of ν_e's from the solar nuclear reactions to some other species, not observable as charged-current interactions in detectors due to energy conservation. Model calculations indicate that both solar-matter-enhanced neutrino mixing and vacuum oscillations over the sun-earth transit distance are viable solutions. A deficit of ν_μ relative to ν_e from the decay products of mesons produced by cosmic-ray interactions in the atmosphere has been seen in several experiments. Recent data from the Japan-U.S. SuperKamiokande experiment, a large water Cerenkov detector located in Japan, corroborate this anomaly. Furthermore, their observed ν_μ and ν_e neutrino interaction rates plotted against the relativistic distance of neutrino transit (Fig. 16) provide strong evidence for oscillation of ν_μ into ν_τ—or into an unseen "sterile" neutrino. An experimental anomaly observed at Los Alamos involves an observation of ν_e interactions from a beam of ν_μ. These indications of neutrino oscillations are spurring efforts worldwide to resolve the patterns of flavor oscillations of massive neutrinos.

The origins of electroweak symmetry breaking and of CP violation, as well as the issue of the neutrino mass, are unfinished aspects of the standard model. However, the very structure of the standard model raises many further questions, strongly indicating that this model provides an incomplete description of the elementary structure of nature.

The standard model is characterized by a large number of parameters. As noted above, three of these—the gauge coupling constants—approximately unify at a scale of about 10^{16} GeV. In fact, when the coupling evolution is calculated using only the content of the standard model, unification is not precisely achieved at a single point: an exact realization of coupling unification requires new particles beyond those in the standard model spectrum. It is tantalizing that exact unification can be achieved with the particle content of the minimal supersymmetric extension of the standard model if superpartner masses lie in a range between 100 GeV and 10 TeV (Fig. 17).

Coupling unification, if true, provides compelling evidence that, above the scale of unification, physics is described by a more fundamental theory incorporating the standard-

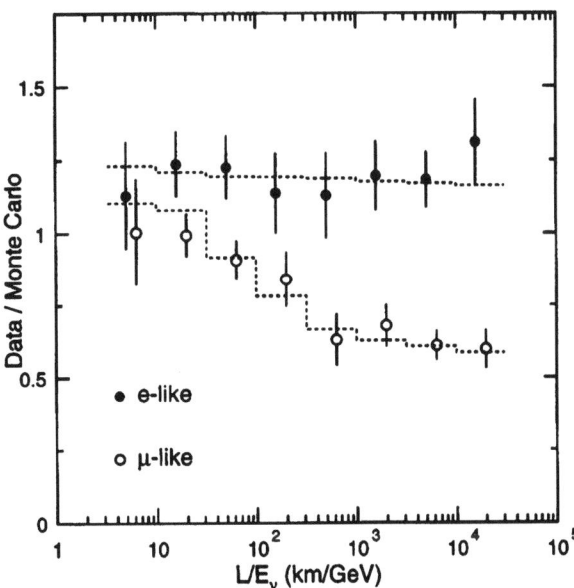

Figure 16. The ratio of the number of ν_e and ν_μ interactions in the SuperKamiokande detector to the Monte Carlo expectations for each, as a function of L/E_ν, where L is the distance of travel from neutrino production in the earth's atmosphere and E_ν is the neutrino energy. Neutrinos produced on the far side of the earth and going upwards in the detector contribute at the largest L/E_ν. The Monte Carlo curves are computed for the best-fit difference in mass squared between oscillating neutrinos of 2.2×10^{-3} eV^2 and maximal mixing.

model interactions in a fully unified way. One possibility, grand unified theory, invokes a larger gauge group, characterized by a single coupling constant, which is broken to the standard-model gauge group by a Higgs vacuum value, $v \sim 10^{16}$ GeV. Couplings differ at low energies because some particles acquire large masses from this Higgs field; symmetry is restored at energy scales above 10^{16} GeV, where these masses are unimportant. Another possibility is that a completely different theory emerges above the scale of unification, such as a superstring theory in ten-dimensional spacetime—perhaps itself an approximation to a yet more fundamental theory in eleven dimensions (see the following article). In string-derived models, coupling unification near the string scale is due to the fact that all gauge coupling constants are determined by the vacuum value of a single scalar field.

Most of the remaining parameters of the standard model, namely, the fermion masses and the elements of the CKM matrix (including a CP-violating phase), are governed by Yukawa couplings of fermions to the Higgs fields. The observed hierarchies among quark fermion masses and mixing parameters are strongly suggestive that new physics must be at play here as well. If there are no right-handed neutrinos, the standard model, with its minimal Higgs content, naturally explains the absence, or very strong suppression, of neutrino masses. However, many extensions of the standard model, including Grand Unified Theory and string-derived models, require right-handed neutrinos, in which case additional new physics is needed to account for the extreme smallness of neutrino masses.

Many models have been proposed in attempts to understand the observed patterns of fermion masses and mixing. These include extended gauge or global symmetries, some in the context of Grand Unified Theory or string theory, as well as the possibility of quark and lepton compositeness. Unlike the issues of electroweak symmetry breaking and CP violation, there is no well-defined energy scale or set of experiments that is certain to provide positive clues, but these questions can be attacked on a variety of fronts, including precision measurements of the CKM matrix elements, searches for flavor-changing transitions that are forbidden in the standard model, and high-energy searches for new particles such as new gauge bosons or excited states of quarks and leptons.

The standard model has another parameter, θ, that governs the strength of CP violation induced by nonperturbative effects in QCD. The experimental limit on the neutron electric dipole moment imposes the constraint $\theta < 10^{-9}$, again suggestive of an additional symmetry that is not manifest in the standard model. Many other questions remain unre-

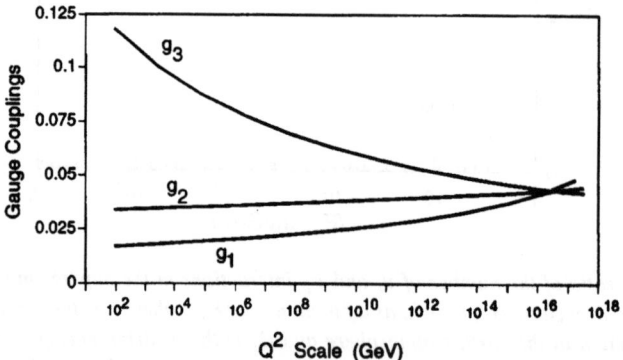

Figure 17. Gauge couplings g_1, g_2, g_3 as a function of Q^2 in the context of the minimal supersymmetric model, showing unification around 10^{16} GeV.

solved; some have profound implications for cosmology, discussed earlier. Is the left/right asymmetry of the electroweak interaction a fundamental property of nature, or is mirror symmetry restored at high energy? Is the proton stable? Grand Unified Theory extensions of the standard model generally predict proton decay at some level, mediated by bosons that carry both quark and lepton numbers. Why are there three families of matter? Some suggested answers invoke extended symmetries; others conjecture fermion compositeness; in string theory the particle spectrum of the low-energy theory is determined by the topology of the compact manifold of additional spatial dimensions. Why is the cosmological constant so tiny, when, in the context of quantum field theory, one would expect its scale to be governed by other scales in the theory, such as the electroweak symmetry-breaking scale of a TeV, or the Planck scale of 10^{19} GeV? The standard model is incomplete in that it does not incorporate gravity. Superstrings or membranes, the only candidates at present for a quantum theory of gravity, embed the standard model in a larger theory whose full content cannot be predicted at present, but which is expected to include a rich spectrum of new particles at higher energies.

Future experiments can severely constrain possible extensions of the standard model, and the discovery of unanticipated new phenomena may provide a useful window into a more fundamental description of nature.

Thousands of original papers have contributed to the evolution of the standard model. We apologize for omitting references to these, and for the necessarily incomplete coverage of many incisive results. We list some recent reviews (Quigg, 1983; Weinberg, 1993; Darriulat, 1995; Veneziano, 1997; Dawson, 1998), which give an entry into this illuminating and impressive literature.

References

Darriulat, P., in *Proceedings of the XXVII International Conference on High Energy Physics*, Glasgow, Scotland, 1995, edited by P. J. Bussey and I. G. Knowles (Institute of Physics, Bristol), p. 367.

Dawson, S., 1998, in *Proceedings of the 1996 Annual Divisional Meeting of the Division of Particles and Fields*, Minneapolis, Minnesota, 1996, in press.

Quigg, C., 1983, *Gauge Theories of the Strong, Weak, and Electromagnetic Interactions* (Benjamin/Cummings, New York).

Veneziano, G., 1996, in *Proceedings of the XXVII International Conference on High Energy Physics*, Warsaw, Poland, 1996, edited by Z. Ajduk and A. K. Wroblewski (World Scientific, Singapore), p. 449.

Weinberg, S., 1993, in *Proceeding of the XXVI International Conference on High Energy Physics*, Dallas, Texas 1992, AIP Conf. Proc. No. 272, edited by J. R. Sanford (AIP, New York), p. 346.

String Theory, Supersymmetry, Unification, and All That
John H. Schwarz and Nathan Seiberg

The standard model of particle physics (see the article by Gaillard, Grannis, and Sciulli in this volume) is a beautiful theory that accounts for all known phenomena up to energies of order 100 GeV. Its consistency relies on the intricacies of quantum field theory (see Wilczek's article), and its agreement with experiment is spectacular. However, there are many open problems with the standard model. In particular, we would like to know what lies beyond the standard model. What is the physics at energies above 100 GeV?

One suggestion for physics at nearby energies of order 1 TeV ($= 1000$ GeV), which we shall review below, is supersymmetry. At higher energies the various interactions of the standard model can be unified into a grand unified theory. Finally, at energies of the order of the Planck energy, $M_P c^2 = (c\hbar/G)^{1/2} c^2 \sim 10^{19}$ GeV, the theory must be modified. This energy scale is determined on dimensional grounds using Newton's constant G, the speed of light c, and Planck's constant $\hbar$. It determines the characteristic energy scale of any theory that incorporates gravitation in a relativistic and quantum-mechanical setting. At this energy scale the gravitational interactions become strong and cannot be neglected. How to combine the elaborate structure of quantum field theory and the standard model with Einstein's theory of gravity—general relativity—is one of the biggest challenges in theoretical physics today. String theory is the only viable attempt to achieve this!

There are various problems that arise when one attempts to combine general relativity and quantum field theory. The field theorist would point to the breakdown of renormalizability—the fact that short-distance singularities become so severe that the usual methods for dealing with them no longer work. By replacing pointlike particles with one-dimensional extended strings, as the fundamental objects, *superstring theory* certainly overcomes the problem of perturbative nonrenormalizability. A relativist might point to a different set of problems including the issue of how to understand the causal structure of space-time when the metric has quantum-mechanical fluctuations. There are also a host of problems associated with black holes, such as the fundamental origin of their thermodynamic properties and an apparent loss of quantum coherence. The latter, if true, would imply a breakdown in the basic structure of quantum mechanics. The relativist's set of issues cannot be addressed properly in a perturbative setup, but recent discoveries are leading to nonperturbative understandings that should help in addressing them. Most string theorists expect that the theory will provide satisfying resolutions of these problems without any revision in the basic structure of quantum mechanics. Indeed, there are indications that someday quantum mechanics will be viewed as an implication of (or at least a necessary ingredient of) superstring theory.

String theory arose in the late 1960s in an attempt to describe strong nuclear forces. In 1971 it was discovered that the inclusion of fermions requires world-sheet supersymmetry. This led to the development of space-time supersymmetry, which was eventually recognized to be a generic feature of consistent string theories—hence the name *superstrings*. String theory was a quite active subject for about five years, but it encountered serious theoretical difficulties in describing the strong nuclear forces, and QCD came along as a convincing theory of the strong interaction. As a result the subject went into decline and was abandoned by all but a few diehards for over a decade. In 1974 two of the diehards (Joël Scherk and John Schwarz) proposed that the problems of string theory could be turned into virtues if it were used as a framework for realizing Einstein's old dream of unification, rather than as a theory of hadrons and strong nuclear forces. In particular, the massless spin-two particle in the string spectrum, which had no sensible hadronic interpretation, was identified as the graviton and shown to interact at low energies precisely as required by general relativity. One implication of this change in viewpoint was that the characteristic size of a string became the Planck length, $L_P = \hbar/cM_P = (\hbar G/c^3)^{1/2} \sim 10^{-33}$ cm, some 20 orders of magnitude smaller than previously envisaged. More refined analyses lead to a string scale L_S that is a couple of orders of magnitude larger than the Planck length. In any case, experiments at existing accelerators cannot resolve distances shorter than about 10^{-16} cm, which explains why the point-particle approximation of ordinary quantum field theories is so successful.

Supersymmetry

Supersymmetry is a symmetry relating bosons and fermions, according to which every fermion has a bosonic superpartner and vice versa. For example, fermionic quarks are partners of bosonic *squarks*. By this we mean that quarks and squarks belong to the same irreducible representation of the supersymmetry. Similarly, bosonic gluons (the gauge fields of QCD) are partners of fermionic *gluinos*. If supersymmetry were an unbroken symmetry, particles and their superpartners would have exactly the same mass. Since this is certainly not the case, supersymmetry must be a broken symmetry (if it is relevant at all). In supersymmetric theories containing gravity, such as supergravity and superstring theories, supersymmetry is a gauge symmetry. Specifically, the superpartner of the graviton, called the *gravitino*, is the gauge particle for local supersymmetry.

Fermionic dimensions of space-time

Another presentation of supersymmetry is based on the notion of *superspace*. We do not change the structure of space-time but we add structure to it. We start with the usual four coordinates, $X^\mu = t, x, y, z$, and add four odd dimensions, θ_α ($\alpha = 1, \ldots, 4$). These odd dimensions are fermionic and anticommute:

$$\theta_\alpha \theta_\beta = -\theta_\beta \theta_\alpha.$$

They are quantum dimensions that have no classical analog, which makes it difficult to visualize or to understand them intuitively. However, they can be treated formally.

The fact that the odd directions are anticommuting has important consequences. Consider a function of superspace,

$$\Phi(X,\theta) = \phi(X) + \theta_\alpha \psi_\alpha(X) + \cdots + \theta^4 F(X).$$

Since the square of any θ is zero and there are only four different θ's, the expansion in powers of θ terminates at the fourth order. Therefore a function of superspace includes only a finite number of functions of X (16 in this case). Hence we can replace any function of superspace $\Phi(X,\theta)$ with the component functions $\phi(X), \psi(X), \ldots$. These include bosons $\phi(X), \ldots$ and fermions $\psi(X), \ldots$. This is one way of understanding the pairing between bosons and fermions.

A supersymmetric theory looks like an ordinary theory with degrees of freedom and interactions that satisfy certain symmetry requirements. Indeed, a supersymmetric quantum field theory is a special case of a more generic quantum field theory rather than being a totally different kind of theory. In this sense, supersymmetry by itself is not a very radical proposal. However, the fact that bosons and fermions come in pairs in supersymmetric theories has important consequences. In some loop diagrams, like those in Fig. 1, the bosons and the fermions cancel each other. This boson-fermion cancellation is at the heart of most of the applications of supersymmetry. If superpartners are present in the TeV range, this cancellation solves the gauge hierarchy problem (see below). This cancellation is also one of the underlying reasons for our ability to analyze supersymmetric theories exactly.

Supersymmetry in the TeV range

There are several indications (discussed below) that supersymmetry is realized in the TeV range, so that the superpartners of the particles of the standard model have masses of the order of a few TeV or less. This is an important prediction, because the next generation of experiments at Fermilab and CERN will explore the energy range where at least some of the superpartners are expected to be found. Therefore, within a decade or two, we should know whether supersymmetry exists at this energy scale. If supersymmetry is indeed discovered in the TeV range, this will amount to the discovery of the new odd dimensions and will be a major change in our view of space and time. It would be a remarkable success for theoretical physics—predicting such a deep notion without any experimental input!

The gauge hierarchy problem

The *gauge hierarchy problem* is essentially a problem of dimensional analysis. Why is the characteristic energy of the standard model, which is given by the mass of the W boson $M_W \sim 100$ GeV, so much smaller than the characteristic scale of gravity, the Planck mass $M_P \sim 10^{19}$ GeV? It should be stressed that in quantum field theory this problem is not merely an aesthetic problem, but also a serious technical problem. Even if such a hier-

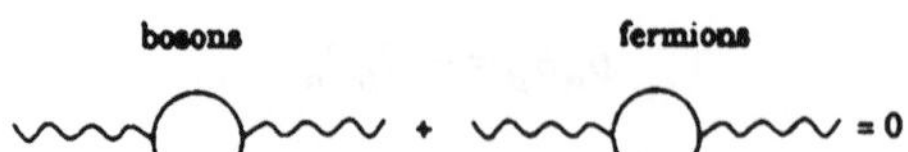

Figure 1. Boson-fermion cancellation in some loop diagrams.

archy is present in some approximation, radiative corrections tend to destroy it. More explicitly, divergent loop diagrams restore dimensional analysis and move $M_W \to M_P$.

The main theoretical motivation for supersymmetry at the TeV scale is the hierarchy problem. As we mentioned, in supersymmetric theories some loop diagrams vanish—or become less divergent—due to cancellations between bosons and fermions. In particular the loop diagram restoring dimensional analysis is canceled as in Fig. 1. Therefore, in its simplest form, supersymmetry solves the technical aspects of the hierarchy problem. More sophisticated ideas, known as dynamical supersymmetry breaking, also solve the aesthetic problem.

The supersymmetric standard model

The minimal supersymmetric extension of the standard model (the MSSM) contains superpartners for all the particles of the standard model, as we have already indicated. Some of their coupling constants are determined by supersymmetry and the known coupling constants of the standard model. Most of the remaining coupling constants and the masses of the superpartners depend on the details of supersymmetry breaking. These parameters are known as *soft breaking terms*. Various phenomenological considerations already put strong constraints on these unknown parameters but there is still a lot of freedom in them. If supersymmetry is discovered, the new parameters will be measured. These numbers will be extremely interesting as they will give us a window into physics at higher energies.

The MSSM must contain two electroweak doublets of Higgs fields. Whereas a single doublet can give mass to all quarks and charged leptons in the standard model, the MSSM requires one doublet to give mass to the charge-2/3 quarks and another to give mass to the charge-1/3 quarks and charged leptons. Correspondingly, electroweak symmetry breaking by the Higgs mechanism involves two Higgs fields' obtaining vacuum expectation values. The ratio, called $\tan \beta$, is an important phenomenological parameter. In the standard model the Higgs mass is determined by the Higgs vacuum expectation value and the strength of Higgs self-coupling (coefficient of the ϕ^4 term in the potential). In supersymmetry the latter is related to the strength of the gauge interactions. This leads to a prediction for the mass of the lightest Higgs boson h in the MSSM. In the leading semiclassical approximation one can show that $M_h \leq M_Z |\cos 2\beta|$, where $M_Z \sim 91$ GeV is the mass of the Z boson. Due to the large mass of the top quark, radiative corrections to this bound can be quite important. A reasonably safe estimate is that $M_h \leq 130$ GeV, which should be compared to current experimental lower bounds of about 80 GeV. The discovery of a relatively light Higgs boson, which might precede the discovery of any superparticles, would be encouraging for supersymmetry. However, it should be pointed out that there are rather mild extensions of the MSSM in which the upper bound is significantly higher.

It is useful to assign positive R parity to the known particles (including the Higgs) of the standard model and negative R parity to their superpartners. For reasonable values of the new parameters (including the soft breaking terms) R parity is a good symmetry. In this case the lightest supersymmetric particle (called the LSP) is absolutely stable. It could be an important constituent of the dark matter of the universe.

Supersymmetric grand unification

The second motivation for supersymmetry in the TeV range comes from the idea of gauge unification. Recent experiments have yielded precise determinations of the strengths of the SU(3)×SU(2)×U(1) gauge interactions—the analogs of the fine-structure constant for these interactions. They are usually denoted by α_3, α_2, and α_1 for the three factors in SU(3)×SU(2)×U(1). In quantum field theory these values depend on the energy at which they are measured in a way that depends on the particle content of the theory. Using the measured values of the coupling constants and the particle content of the standard model, one can extrapolate to higher energies and determine the coupling constants there. The result is that the three coupling constants do not meet at the same point. However, when one repeats this extrapolation with the particles belonging to the minimal supersymmetric extension of the standard model, the three gauge-coupling constants meet at a point, M_{GUT}, as sketched in Fig. 2. At that point the strengths of the various gauge interactions become equal and the interactions can be unified into a *grand unified theory*. Possible grand unified theories embed the known SU(3)×SU(2)×U(1) gauge group into SU(5) or SO(10).

How much significance should we assign to this result? Two lines must meet at a point. Therefore, there are only two surprises here. The first is that the third line intersects the same point. The second more qualitative one is that the unification scale, M_{GUT}, is at a reasonable value. Its value is consistent with the experimental bound from proton decay, and it is a couple of orders of magnitude below the Planck scale, where gravity would need to be taken into account. One could imagine that that there are other modifications of the standard model that achieve the same thing, so this is far from a proof of supersymmetry, but it is certainly encouraging circumstantial evidence. It is an independent indication that superpartner masses should be around a TeV.

Supersymmetric quantum field theories

Quantum field theory is notoriously complicated. It is a nonlinear system of an infinite number of coupled degrees of freedom. Therefore, until recently when the power of supersymmetry began to be exploited, there were few exact results for quantum field theories (except in two dimensions). However, it has been realized recently that a large

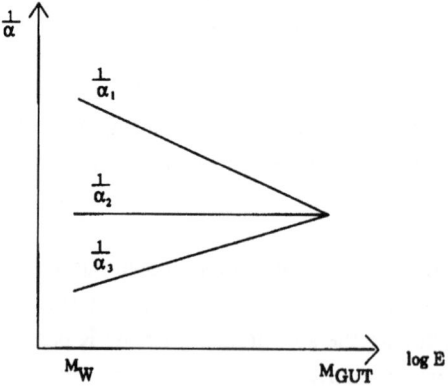

Figure 2. Coupling-constant unification in supersymmetric theories.

class of physical quantities in many supersymmetric quantum field theories can be computed exactly by analytic methods!

The main point is that these theories are very constrained. The dependence of some observables on the parameters of the problem is so constrained that there is only one solution that satisfies all of the consistency conditions. More technically, because of supersymmetry some observables vary holomorphically (complex analytically) with the coupling constants, which are complex numbers in these theories. Due to Cauchy's theorem, such analytic functions are determined in terms of very few data: the singularities and the asymptotic behavior. Therefore, if supersymmetry requires an observable to depend holomorphically on the parameters, and we know the singularities and the asymptotic behavior, we can determine the exact answer. The boson-fermion cancellation, which we mentioned above in the context of the hierarchy problem, can also be understood as a consequence of a constraint following from holomorphy.

Families of vacua

Another property of many supersymmetric theories that makes them tractable is that they have a family of inequivalent vacua. To understand this fact we should contrast it with the situation in a ferromagnet, which has a continuum of vacua, labeled by the common orientation of the spins. These vacua are all equivalent; i.e., the physical observables in one of these vacua are exactly the same as in any other. The reason is that these vacua are related by a symmetry. The system must choose one of them, which leads to spontaneous symmetry breaking.

We now study a situation with inequivalent vacua in contrast to the ferromagnet. Consider the case in which degrees of freedom, called x and y, have the potential $V(x,y)$ shown in Fig. 3. The vacua of the system correspond to different points along the valley of the potential, $y=0$ with arbitrary x. However, as we tried to make clear in the figure, these points are inequivalent—there is no symmetry that relates them. More explicitly, the potential is shallow around the origin but becomes steep for large x. Such *accidental degeneracy* is usually lifted by quantum effects. For example, if the system corresponding to the potential in the figure has no fermions, the zero-point fluctuations around the different vacua would be different. They would lead to a potential along the valley, pushing the minimum to the origin. However, in a supersymmetric theory the zero-point energy of the fermions exactly cancels that of the bosons, and the degeneracy is not lifted. The valleys persist in the full quantum theory. Again, we see the power of the boson-fermion cancellation. We see that a supersymmetric system typically has a continuous family of vacua. This family, or manifold, is referred to as a *moduli space of vacua*, and the modes of the system corresponding to motion along the valleys are called *moduli*.

The analysis of supersymmetric theories is usually simplified by the presence of these manifolds of vacua. Asymptotically, far along the flat directions of the potential, the analysis of the system is simple and various approximation techniques are applicable. Then, by using the asymptotic behavior along several such flat directions, as well as the constraints from holomorphy, one obtains a unique solution. This is a rather unusual situation in physics. We perform approximate calculations, which are valid only in some regime, and this gives us the exact answer. This is a theorist's heaven—exact results with approximate methods!

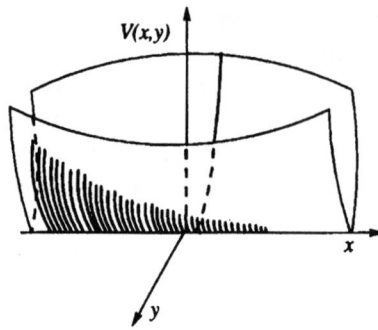

Figure 3. Typical potential in supersymmetric theories exhibiting "accidental vacuum degeneracy."

Electric-magnetic duality

Once we know how to solve such theories, we can analyze many examples. The main lesson that has been learned is the fundamental role played by *electric-magnetic duality*. It turns out to be the underlying principle controlling the dynamics of these systems.

When faced with a complicated system with many coupled degrees of freedom it is common in physics to look for weakly coupled variables that capture most of the phenomena. For example, in condensed-matter physics we formulate the problem at short distance in terms of interacting electrons and nuclei. The desired solution is the macroscopic behavior of the matter and its possible phases. It is described by weakly coupled effective degrees of freedom. Usually they are related in a complicated, and in most cases unknown, way to the microscopic variables. Another example is hydrodynamics, where the microscopic degrees of freedom are molecules and the long-distance variables are properties of a fluid that are described by partial differential equations.

In one class of supersymmetric field theories, the long-distance behavior is described by a set of weakly coupled effective degrees of freedom. These are composites of the elementary degrees of freedom. As the characteristic length scale becomes longer, the interactions between these effective degrees of freedom become weaker, and the description in terms of them becomes more accurate. In other words, the long-distance theory is a "trivial" theory in terms of the composite effective degrees of freedom.

In another class of examples there are no variables in terms of which the long-distance theory is simple—the theory remains interacting. Because it is scale invariant, it is at a nontrivial fixed point of the renormalization group. In these situations there are two (or more) dual descriptions of the physics leading to identical results for the long-distance interacting behavior.

In both classes of examples an explicit relation between the two sets of variables is not known. However, there are several reasons to consider these pairs of descriptions as being electric-magnetic duals of one another. The original variables at short distance are referred to as the electric degrees of freedom and the other set of long-distance variables as the magnetic ones. These two dual descriptions of the same theory give us a way to address strong-coupling problems. When the electric variables are strongly coupled, they fluctuate rapidly and their dynamics are complicated (see Table I). However, then the magnetic degrees of freedom are weakly coupled. They do not fluctuate rapidly and their dynamics is simple. In the first class of examples the magnetic degrees of freedom are the macroscopic ones, which are free at long distance. They are massless bound states of the elementary particles. In the second class of examples there are two valid descriptions of the long-distance theory: electric and magnetic. Since both of them are interacting, neither of them gives a "trivial" description of the physics. However, as one of them becomes more strongly coupled, the other becomes more weakly coupled (see Table I).

Finally, using this electric-magnetic duality we can find a simple description of complicated phenomena associated with the phase diagram of the theories. For example, as the electric degrees of freedom become strongly coupled, they can lead to confinement. In the magnetic variables, this is simply the Higgs phenomenon (superconductivity), which is easily understood in weak coupling. The electric-magnetic relations are summarized in Table I.

Apart from the "practical" application to solving quantum field theories, the fact that a theory can be described in terms of either electric or magnetic variables has deep consequences:

(i) For theories belonging to the first class of examples it is natural to describe the magnetic degrees of freedom as composites of the elementary electric ones. The magnetic particles typically include massless gauge particles reflecting a new magnetic gauge symmetry. These massless composite gauge particles are associated with a new gauge symmetry which is not present in the fundamental electric theory. Since this gauge symmetry is not a symmetry of the original, short-distance theory, it is generated by the dynamics rather than being "put in by hand." We see that, in this sense, *gauge invariance cannot be fundamental.*

(ii) For theories of the second class the notion of elementary particle breaks down. *There is no invariant way of choosing which degrees of freedom are elementary and which are composite.* The magnetic degrees of freedom can be regarded as composites of the electric ones and vice versa.

Superstring Theory

Perturbative string theory

All superstring theories contain a massless scalar field, called the *dilaton* ϕ, that belongs to the same supersymmetry multiplet as the graviton. In the semiclassical approximation, this field defines a flat direction in the moduli space of vacua, so that it can take any value ϕ_0. Remarkably, this determines the string coupling constant $g_S = e^{\phi_0}$, which is a dimensionless parameter on which one can base a perturbation expansion. The perturbation expansions are power-series expansions in powers of the string coupling constant like those customarily used to carry out computations in quantum field theory.

Table 1. Dual electric and magnetic descriptions.

	Electric	Magnetic
Coupling	strong	weak
Fluctuations	large	small
Phase	confinement	Higgs

Structure of the string world sheet and the perturbation expansion

A string's space-time history is described by functions $X^\mu(\sigma,\tau)$ that map the string's two-dimensional *world sheet* (σ,τ) into space-time X^μ. There are also other world-sheet fields that describe other degrees of freedom, such as those associated with supersymmetry and gauge symmetries. Surprisingly, *classical* string theory dynamics is described by a conformally invariant 2D *quantum* field theory. What distinguishes one-dimensional strings from higher-dimensional analogs (discussed later) is the fact that this 2D theory is renormalizable. Perturbative quantum string theory can be formulated by the Feynman sum-over-histories method. This amounts to associating a genus h Riemann surface (a closed and orientable two-dimensional surface with h handles) to a Feynman diagram with h loops. It contains a factor of g_S^{2h}. For example, the string world sheet in Fig. 4 has one handle.

The attractive features of this approach are that there is just one diagram at each order h of the perturbation expansion and that each diagram represents an elegant (though complicated) finite-dimensional integral that is ultraviolet finite. In other words, they do not give rise to the severe short-distance singularities that plague other attempts to incorporate general relativity in a quantum field theory. The main drawback of this approach is that it gives no insight into how to go beyond perturbation theory.

Five superstring theories

In 1984–1985 a series of discoveries convinced many theorists that superstring theory is a very promising approach to unification. This period is now sometimes referred to as *the first superstring revolution*. Almost overnight, the subject was transformed from an intellectual backwater to one of the most active areas of theoretical physics, which it has remained ever since. By the time the dust settled, it was clear that there are five different superstring theories, each requiring ten dimensions (nine space and one time), and that each has a consistent perturbation expansion. The five theories are denoted type I, type IIA, type IIB, $E_8 \times E_8$ heterotic (HE, for short), and SO(32) heterotic (HO, for short). The type-II theories have two supersymmetries in the ten-dimensional sense, while the other three have just one. The type-I theory is special in that it is based on unoriented open and closed strings, whereas the other four are based on oriented closed strings. Type-I strings can break, whereas the other four are unbreakable. The type-IIA theory is nonchiral (i.e., it is parity conserving), and the other four are chiral (parity violating).

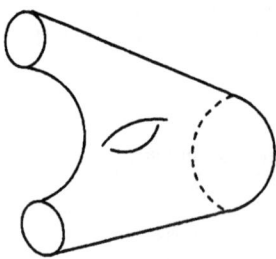

Figure 4. An example of a string world sheet with two initial strings, one final string, and a handle.

Compactification of extra dimensions

To have a chance of being realistic, the six extra space dimensions must somehow curl up into a tiny geometrical space as in Kaluza-Klein theory. The linear size of this space is presumably comparable to the string scale L_S. Since space-time geometry is determined dynamically (as in general relativity), only geometries that satisfy the dynamical equations are allowed. Among such solutions, one class stands out: The $E_8 \times E_8$ heterotic (HE) string theory, compactified on a particular kind of six-dimensional space, called a Calabi-Yau manifold, has many qualitative features at low energies that resemble the supersymmetric extension of the standard model of elementary particles. In particular, the low-mass fermions occur in suitable representations of a plausible unifying gauge group. Moreover, they occur in families whose number is controlled by the topology of the Calabi-Yau manifold. These successes have been achieved in a perturbative framework and are necessarily qualitative at best, since nonperturbative phenomena are essential to an understanding of supersymmetry breaking and other important details.

T duality and stringy geometry

The basic idea of T duality can be illustrated by considering a compact spatial dimension consisting of a circle of radius R. In this case there are two kinds of excitations to consider. The first, which is not unique to string theory, is due to the quantization of the momentum along the circle. These Kaluza-Klein excitations contribute $(n/R)^2$ to the energy squared, where n is an integer. The second kind are winding-mode excitations, which arise due to a closed string's being wound m times around the circular dimension. They are unique to string theory, though there are higher-dimensional analogs. When one lets $T = (2\pi L_S^2)^{-1}$ denote the fundamental string tension (energy per unit length), the contribution of a winding mode to the energy squared is $(2\pi R m T)^2$. T duality exchanges these two kinds of excitations by mapping $m \leftrightarrow n$ and $R \leftrightarrow L_S^2/R$. This is part of an exact map between a T-dual pair of theories A and B.

We see that the underlying geometry is ambiguous—there is no way to tell the difference between a compactification on a circle of radius R and a compactification on a circle of radius L_S^2/R. This ambiguity is clearly related to the fact that the objects used to probe the circle are extended objects—strings—which can wind around the circle.

One implication of this ambiguity is that usual geometric concepts break down at short distances, and classical geometry is replaced by *stringy geometry*, which is described mathematically by 2D conformal field theory. It also suggests a generalization of the Heisenberg uncertainty principle according to which the best possible spatial resolution Δx is bounded below not only by the reciprocal of the momentum spread, Δp, but also by the string size, which grows with energy. This is the best one can do using fundamental strings as probes. However, by probing with certain nonperturbative objects called D-branes, which we shall discuss later, it is sometimes possible (but not in the case of the circle discussed above) to do better.

A closely related phenomenon is that of mirror symmetry. In the example of the circle above the topology was not changed by T duality. Only the size was transformed. In more complicated compactifications, such as those on Calabi-Yau manifolds, there is even an ambiguity in the underlying topology—there is no way to tell on which of two mirror pairs of Calabi-Yau manifolds the theory is compactified. This ambiguity can be useful because it is sometimes easier to perform some calculations with one Calabi-Yau manifold than with its mirror manifold. Then, using mirror symmetry, we can infer what the answers are for different compactifications.

Two pairs of ten-dimensional superstring theories are T dual when compactified on a circle: the type-IIA and IIB theories and the HE and HO theories. The two edges of Fig. 5 labeled T connect vacua related by T duality. For example, if the IIA theory is compactified on a circle of radius R_A, leaving nine noncompact dimensions, this is equivalent to compactifying the IIB theory on a circle of radius $R_B = L_S^2/R_A$. The T duality relating the two heterotic theories, HE and HO, is essentially the same, though there are additional technical details in this case.

Another relation between theories is the following. A compactification of the type-I theory on a circle of radius R_I turns out to be related to a certain compactification of the type-IIA theory on a line interval I with size proportional to L_S^2/R_I. The line interval can be thought of as a circle with some identification of points $I = S^1/\Omega$. Therefore we can say that the type-I theory on a circle of radius R_I is obtained from the type IIA on a circle of radius L_S^2/R_I by acting with Ω. Since by T duality the IIA theory on a circle of radius L_S^2/R_I is the same as the IIB theory on a circle of radius R_I, we conclude that upon compactification on a circle type I is obtained from IIB by the action of Ω. By taking R_I to infinity we ensure that this relation is also true in ten dimensions. This is the reason for the edge denoted by Ω in Fig. 5.

These dualities reduce the number of (apparently) distinct superstring theories from five to three, or if we also use Ω to two. The point is that the two members of each pair are continuously connected by varying the compactification radius from zero to infinity. Like the string coupling constant, the compactification radius arises as the value of a scalar field. Therefore varying this radius is a motion in the moduli space of quantum vacua rather than a change in the parameters of the theory.

Nonperturbative string theory

The *second superstring revolution* (1994–??) has brought nonperturbative string physics within reach. The key discoveries were various dualities, which show that what was viewed previously as five distinct superstring theories is in fact five different perturbative expansions of a single underlying theory about five different points in the moduli space of

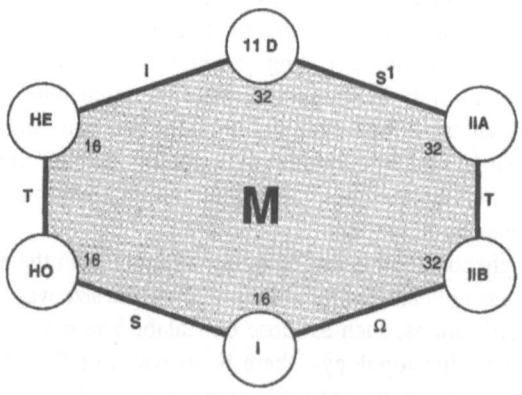

Figure 5. The M theory moduli space.

consistent vacua! It is now clear that there is a unique theory, though it allows many different vacua. A sixth special vacuum involves an 11-dimensional Minkowski space-time. Another lesson we have learned is that, nonperturbatively, objects of more than one dimension (membranes and higher p-branes) play a central role. In most respects they appear to be on an equal footing with strings, but there is one big exception: a perturbation expansion cannot be based on p-branes with $p > 1$.

A schematic representation of the relationship between the five superstring vacua in 10D and the 11D vacuum is given in Fig. 5. The idea is that there is some large moduli space of consistent vacua of a single underlying theory—here denoted by M. The six limiting points, represented as circles, are special in the sense that they are the ones with (super) Poincaré invariance in ten or eleven dimensions. The letters on the edges refer to the type of duality relating a pair of limiting points. The numbers 16 or 32 refer to the number of unbroken supersymmetries. In 10D the minimal spinor has 16 real components, so the conserved supersymmetry charges (or *supercharges*) correspond to just one spinor in three cases (type I, HE, and HO). Type-II superstrings have two such spinorial supercharges. In 11D the minimal spinor has 32 real components.

S duality

Suppose now that a pair of theories (A and B) are S dual. This means that if $f_A(g_S)$ denotes any physical observable of theory A, where g_S is the coupling constant, then there is a corresponding physical observable $f_B(g_S)$ in theory B such that $f_A(g_S) = f_B(1/g_S)$. This duality relates one theory at weak coupling to the other at strong coupling. It generalizes the electric-magnetic duality of certain field theories, discussed earlier in the section on Electric-Magnetic Duality. S duality relates the type-I theory to the HO theory and the IIB theory to itself. This determines the strong-coupling behavior of these three theories in terms of weakly coupled theories. Varying the strength of the string coupling also corresponds to a motion in the moduli space of vacua.

The edge connecting the HO vacuum and the type-I vacuum is labeled by S in Fig. 5, since these two vacua are related by S duality. It had been known for a long time that the two theories had the same gauge symmetry [SO(32)] and the same kind of supersymmetry, but it was unclear how they could be equivalent, because type-I strings and heterotic strings are very different. It is now understood that SO(32) heterotic strings appear as nonperturbative excitations in the type-I description.

M theory and the eleventh dimension

The understanding of how the remaining two superstring theories, type IIA and HE, behave at strong coupling came as quite a surprise. In each case there is an 11th dimension whose size R becomes large at strong string coupling g_S. In the IIA case the 11th dimension is a circle, whereas in the HE case it is a line interval. The strong-coupling limit of either of these theories gives an 11-dimensional Minkowski space-time. The 11-dimensional description of the underlying theory is called *M theory*.[1]

The 11D vacuum, including 11D supergravity, is characterized by a single scale—the 11D Planck scale L_P. It is proportional to $G^{1/9}$, where G is the 11D Newton constant. The connection to type-IIA theory is obtained by taking one of the ten spatial dimensions to

---

[1] The letter M could stand for a variety of things such as magic, mystery, meta, mother, or membrane.

be a circle (S^1 in the diagram) of radius R. As we pointed out earlier, the type-IIA string theory in 10D has a dimensionless coupling constant g_S, given by the value of the dilaton field, and a length scale L_S. The relationship between the parameters of the 11D and IIA descriptions is given by

$$L_P^3 = RL_S^2, \qquad (1)$$

$$R = L_S g_S. \qquad (2)$$

Numerical factors (such as 2π) are not important for present purposes and have been dropped. The significance of these equations will emerge later. However, one point can be made immediately. The conventional perturbative analysis of the IIA theory is an expansion in powers of g_S with L_S fixed. The second relation implies that this is an expansion about $R=0$, which accounts for the fact that the 11D interpretation was not evident in studies of perturbative string theory. The radius R is a modulus—the value of a massless scalar field with a flat potential. One gets from the IIA point to the 11D point by continuing this value from zero to infinity. This is the meaning of the edge of Fig. 5 labeled S^1.

The relationship between the HE vacuum and 11D is very similar. The difference is that the compact spatial dimension is a line interval (denoted I in Fig. 5) instead of a circle. The same relations in Eqs. (1) and (2) apply in this case. This compactification leads to an 11D spacetime that is a slab with two parallel 10D faces. One set of E_8 gauge fields is confined to each face, whereas the gravitational fields reside in the bulk. One of the important discoveries in the first superstring revolution was a mechanism that cancels quantum-mechanical anomalies in the Yang-Mills and Lorentz gauge symmetries. This mechanism works only for SO(32) and $E_8 \times E_8$ gauge groups. There is a nice generalization of this 10D anomaly cancellation mechanism to the setting of 11 dimensions with a 10D boundary. It works only for E_8 gauge groups!

p-branes and D-branes

In addition to the strings the theory turns out to contain other objects, called *p-branes*. A p-brane is an extended object in space with p spatial dimensions. (The term p-brane originates from the word membrane, which describes a 2-brane.) For example, the 11D M theory turns out to contain two basic kinds of p-branes with $p=2$ and $p=5$, called the M2-brane and the M5-brane. A simpler example of a brane is readily understood in the type-IIA theory when it is viewed as a compactification of the 11D theory on a circle. Eleven-dimensional particles with momentum around the circle appear as massive particles in 10D, whose masses are proportional to $1/R$. Since they are point particles, they are referred to as 0-branes. Using Eq. (2), we find $1/R = 1/L_S g_S$, and we see that in the perturbative string region, where $g_S \ll 1$, these 0-branes are much heavier than the ordinary string states whose masses are of order $1/L_S$. The type-IIA string in ten dimensions can be identified as the M2-brane wrapping the compact circle.

These p-branes are crucial in the various dualities discussed above—since they are states in the theory, they should be mapped correctly under T and S dualities. This is particularly interesting for S duality, which maps the fundamental string of one theory to a heavy 1-brane of the other. For example, the heterotic string is such a heavy 1-brane in the weakly coupled type-I theory. We therefore see that the notion of an elementary (or fundamental) string is ill defined. The string that appears fundamental at one boundary of Fig. 5 is a heavy brane at another boundary and vice versa. We have already encountered a similar phenomenon in our discussion of electric-magnetic duality in field theory, where there was an ambiguity in the notion of elementary objects.

A special class of p-branes is called *Dirichlet p-branes* (or *D-branes* for short). The name derives from the boundary conditions assigned to the ends of open strings. The usual open strings of the type-I theory have Neumann boundary conditions at their ends. More generally, in type-II theories, one can consider an open string with boundary conditions at the end given by $\sigma = 0$:

$$\frac{\partial X^\mu}{\partial \sigma} = 0, \quad \mu = 0, 1, \ldots, p,$$

$$X^\mu = X_0^\mu, \quad \mu = p+1, \ldots, 9,$$

and similar boundary conditions at the other end. The interpretation of these equations is that strings end on a p-dimensional object in space—a D-brane. The description of D-branes as a place where open strings can end leads to a simple picture of their dynamics. For weak string coupling this enables the use of perturbation theory to study nonperturbative phenomena!

D-branes have found many interesting applications. One of the most remarkable of these concerns the study of black holes. Specifically, D-brane techniques can be used to count the quantum microstates associated with classical black-hole configurations and to show that in suitable limits the entropy (defined by $S = \log N$, where N is the number of quantum states the system can be in) agrees with the Bekenstein–Hawking prediction: 1/4 the area of the event horizon. For further details, see the article by Horowitz and Teukolsky in this volume.

D-branes also led to new insights and new results in quantum field theory, arising from the realization that the open strings which end on D-branes are described at low energies by a local quantum field theory "living" on the brane. The dynamics of quantum field theories on different branes must be compatible with the various dualities. One can use this observation to test the dualities. Alternatively, assuming the various string dualities and the consistency of the theory, one can easily derive known results in quantum field theory from a new perspective as well as many new results.

Conclusion

During the last 30 years the structure of string theory has been explored both in perturbation theory and nonperturbatively with enormous success. A beautiful and consistent picture has emerged. The theory has also motivated many other developments, such as supersymmetry, which are interesting in their own right. Many of the techniques that have been used to obtain exact solutions of field theories were motivated by string theory. Similarly, many applications to mathematics have been discovered, mostly in the areas of topology and geometry. The rich structure and the many applications are viewed by many people as indications that we are on the right track. However, the main reason to be interested in string theory is that it is the only known candidate for a consistent quantum theory of gravity.

There are two main open problems in string theory. The first is that the underlying conceptual principles of the theory—the analog of curved space-time and general covariance for gravity—are not yet understood. Unlike other fields, string theory is not yet a mature field with a stable framework. Instead, the properties of the theory are being discovered with the hope that eventually they will lead to an understanding of the principles and the framework. The various revolutions that the field has undergone in recent years have completely changed our perspective on the theory. It is likely that there will be a few other revolutions and our perspective will change again. Indeed, fascinating connections to large-N gauge theories are currently being explored, which appear to be very promising. In any case, the field is developing very rapidly and it is clear that an article about string theory for the next centenary volume will look quite different from this one.

The second problem, which is no less important, is that we should like to make contact with experiment. We need to find unambiguous experimental confirmation of the theory. Supersymmetry would be a good start.

Acknowledgments

We have benefited from many discussions with many colleagues throughout the years. Their insights and explanations were extremely helpful in shaping our point of view on the subjects discussed here. The work of J.H.S. was supported in part by DOE Grant No. DE-FG03-92-ER40701 and that of N.S. by DOE Grant No. DE-FG02-90ER40542.

References

(i) Selected books and reviews on supersymmetry:

Alvarez, L., and F. Zamora, 1997, "Duality in Quantum Field Theory and String Theory," CERN-TH-97-257.

Gates, S. J., Jr., M. T. Grisaru, M. Rocek, and W. Siegel, 1983, *Superspace: or one thousand and one lessons in supersymmetry*, Frontiers in Physics Lecture Note Series No. 58 (Benjamin/Cummings, New York).

Haber, H. E., and G. L. Kane, 1985, "The Search for Supersymmetry: Probing Physics Beyond the Standard Model," Phys. Rep. **117**, 75.

Intriligator, K., and N. Seiberg, 1996, "Lectures on Supersymmetric Gauge Theories and Electric-Magnetic Duality," Nucl. Phys. B, Proc. Suppl. **45BC**, 1.

Nilles, H. P., 1984, "Supersymmetry, Supergravity and Particle Physics," Phys. Rep. **110**, 1.

Wess, J., and J. Bagger, 1983, *Supersymmetry and Supergravity* (Princeton University, Princeton, NJ).

West, P., 1986, *Introduction to Supersymmetry and Supergravity* (World Scientific, Singapore).

(ii) Selected books on string theory:

Efthimiou, C., and B. Greene, 1997, *Fields, Strings, and Duality (TASI 96)* (World Scientific, Singapore).

Green, M. B., J. H. Schwarz, and E. Witten, 1987, *Superstring Theory* (Cambridge University Press, Cambridge).

Polchinski, J., 1998, *String Theory* (Cambridge University Press, Cambridge), in 2 Vols.

Accelerators and Detectors
W. K. H. Panofsky and M. Breidenbach

The developing understanding of particle physics, especially in the past 60 or so years, has been largely paced by the evolution of high-energy accelerators and detectors. We restrict ourselves here to describing crucial developments in accelerators related to high-energy particle collisions. Similarly, discussion of detectors will be restricted to those associated with the accelerators and colliders covered.

There exist extensive reviews on our subject (Particle Data Group, 1996; see pp. 128 ff. for colliders and pp. 142 ff for detectors). While there are extensive reviews, detailed technical descriptions of accelerators and detectors in the peer reviewed literature are very incomplete; original source material is largely contained in laboratory reports and conference proceedings and most major accelerator installations have never been comprehensively documented.

Growth Patterns of Accelerators and Colliders

Accelerators and colliders can be parametrized by a number of characteristics. The energy of a particle as accelerated in the laboratory is not what is relevant in determining the threshold for initiating a particular elementary-particle process. The center-of-mass energy $E_{c.m.}$ of two colliding particles of rest masses m_1, and m_2 and total energies E_1 and E_2, respectively, is given by $E_{c.m.}^2 = p_i p^i$ where p_i is the total four-momentum of the particles. For instance, if a proton of energy $E_1 = \gamma m_1 c^2$ strikes a proton at rest, then $E_{c.m.} = [2(\gamma+1)]^{1/2} m_1 c^2$. In the nonrelativistic limit only one-half of the incident kinetic energy is available, while in the relativistic limit $\gamma \gg 1$ the center-of-mass energy grows with the square root of the energy of the incident protons. If two relativistic particles collide head on then $E_{c.m.} = 2(E_1 E_2)^{1/2}$ or $2E$ if the particles have identical energy. These relations demonstrate the energy advantage of colliding beams. But as investigations extend to smaller dimensions, the concept of what constitutes an elementary particle changes. At distances with the analyzing power of current colliders (10^{-18}m) quarks and leptons are "elementary." Thus the relevant energy of a high-energy accelerator or collider defining its "reach" in initiating elementary-particle processes is neither the laboratory beam energy nor the collision energy in the center-of-mass frame of *composite* colliding particles, but is the collision energy in the frame of the center of mass of colliding "elementary" *constituent* particles.

The luminosity, defined as the data rate per unit cross section of the process under investigation, is another critical parameter. For high-momentum-transfer events the cross section is expected to vary inversely as the square of the momentum transferred. Therefore the luminosity of colliders should increase quadratically with energy in order to yield a constant data rate for "interesting" or novel events. In addition other parameters are of relevance to the experimenter, such as the background conditions, that is particle fluxes other than those originating from the collision under investigation. Then there is the "duty cycle" of the machine, which is the time structure over which collisions occur. Few modern accelerators or colliders produce random collisions uniformly distributed in time. Some accelerators are pulsed and most colliders employ bunches of particles rather than continuous beams. The resulting "duty cycle" limits the ability to interpret the time relationship among products of interaction. Experiments with accelerators use either the primary collisions or secondary beams produced in these collisions; the quality and quantity of secondary beams differ among types of accelerators and colliders.

Last, but unfortunately not least, is the matter of cost. The scaling laws that relate costs to growth of each technology define the historical growth patterns of accelerators and colliders. Figure 1 describes the growth over time of laboratory energy for various particle

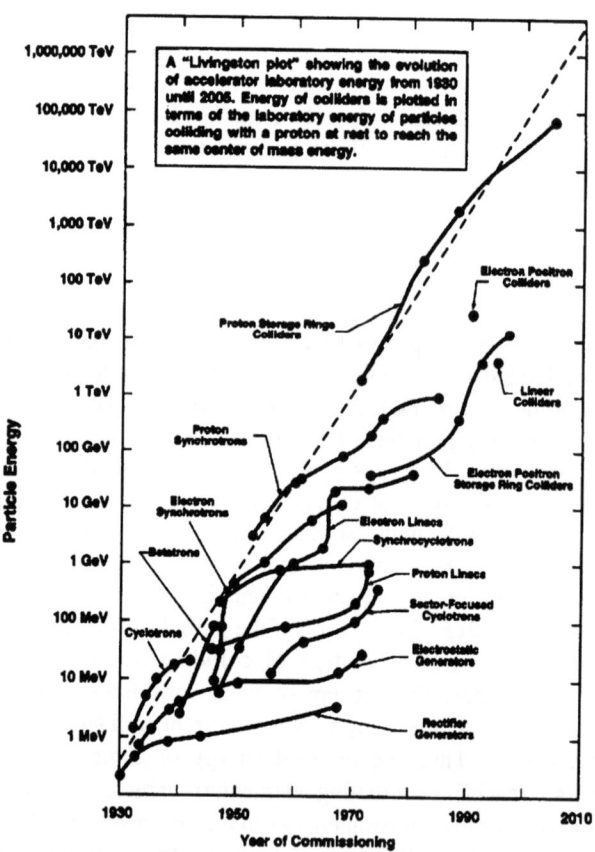

Figure 1. A "Livingston plot" showing the evolution of accelerator laboratory energy from 1930 until 2005. Energy of colliders is plotted in terms of the laboratory energy of particles colliding with a proton at rest to reach the same center-of-mass energy.

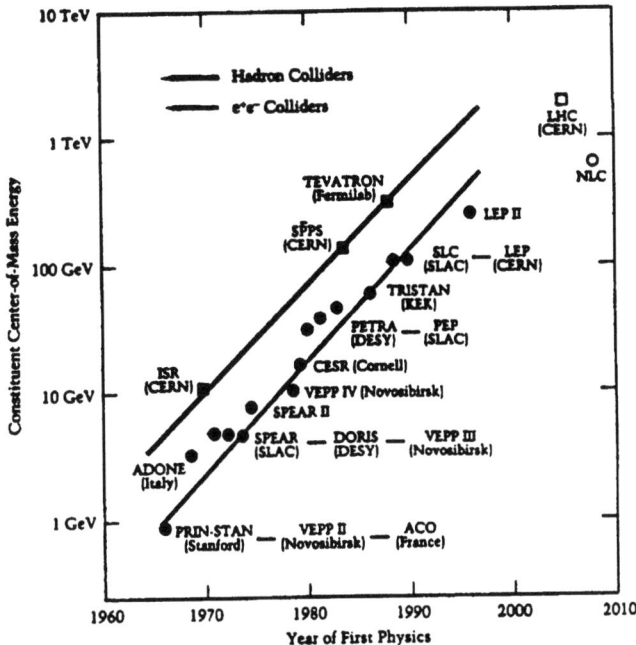

Figure 2. The energy in the constituent frame of electron-positron and hadron colliders: filled circles and squares, constructed; open circle and square, planned. The energy of hadron colliders has here been derated by factors of 6–10 in accordance with the fact that the incident-proton energy is shared among its quark and gluon constituents.

accelerators.[1] This pattern, first published by Livingston (1954), exhibits important features. An almost exponential growth of laboratory energy with time is fed by a succession of technologies; each technology saturates and is superseded by new technologies. In parallel with this pattern such new technologies have led to a decrease in cost per unit of laboratory energy by about four orders of magnitudes over the time period covered by Fig. 1.

The more relevant quantity describing the "reach" of accelerators into the unknown is the center-of-mass energy in the constituent frame, shown in Fig. 2. In this figure hadron colliders (proton-proton and proton-antiproton) and lepton colliders (electron-electron or positron) are plotted separately. The constituent center-of-mass energy of hadrons has been derated by a factor of about 6 relative to that of lepton colliders, to account for the hadron substructure of quarks and gluons. Needless to say, such a derating of colliders using "nonelementary" particles can only be an approximation. The internal dynamics of the substructure of composite particles can permit a lowering of reaction thresholds, albeit accompanied by a decrease in luminosity. Again, an exponential growth has apparently been sustained over the limited period of time over which colliding-beam devices have been successfully constructed, and that growth is comparable for hadron and lepton colliders.

---

[1] For a listing of the relevant machines, see the 1996 Review of Particle Physics (Particle Data Group, 1996). We shall cite here only machines at the frontiers of performance.

Particle beams striking stationary targets of condensed matter produce effective luminosities many orders of magnitude larger than those attainable by colliding beams. The luminosity growth of colliders is shown in Fig. 3. Thus far this growth has not matched the quadratic growth with energy required to maintain constant data rates.

Principles, Categorization, and Evolution of Accelerators and Colliders

Fundamentally accelerators are either electrostatic machines, in which particles are accelerated by traversing a difference in electrical potential once, or they are transformers, which repeatedly use high-current low-voltage circuit elements to supply energy to a high-voltage low-current accelerating path.

Electrostatic devices

Early accelerators were discharge tubes fed by conventional high-voltage sources. Limitations arose in the ability of the discharge tubes to sustain high voltages and in the availability of high-voltage sources. The Tesla-coil accelerators were resonant step-up transformers with both primary and secondary resonating at the same frequency. Cascade accelerators, pioneered by Cockroft and Walton (1932), were able to attain voltages in the several hundreds of kilovolts by charging capacitors in parallel and reconnecting them in series. In the Van-de Graaff generator (Van de Graaff, 1931) charges were sprayed onto a moving belt and then removed inside a high-voltage electrode; this device reached energies near ten million electron volts. The design of discharge columns evolved to permit better voltage distribution and focusing, and vacuum practices improved. Electrostatic generators continue to be produced for research in nuclear physics and for medical uses.

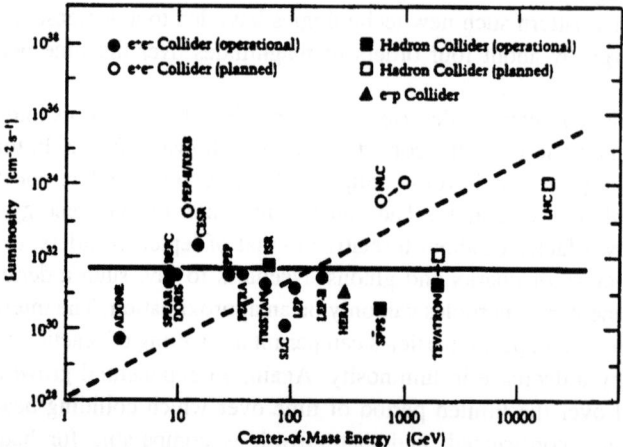

Figure 3. Peak luminosities achieved at existing colliders and values projected for planned or upgraded machines: dashed line, luminosity increasing as the square of the center-of-mass energy. Note that the rated machine energy has been used in calculating the abscissa. Data updated courtesy of Greg Loew, SLAC.

Transformers

Energies above about 10 MeV are not attainable electrostatically. The most important early development to exceed that limit was the cyclotron proposed by Lawrence and Edlefsen (1930) and put into practice by Lawrence and Livingston (1932) using the well-known principle that the orbital period of nonrelativistic charged particles circulating in a uniform magnetic field is independent of energy. Thus if a radio-frequency voltage matching the revolution frequency is applied across a gap placed in such a field, then the particle will gain energy and will spiral out in the magnet.

Cyclotrons developed rapidly in the period before World War II but the decrease in orbital frequency as the particles become relativistic limits the attainable energy. Focusing was first addressed by empirical "shimming" of the magnetic field. A more analytical approach initially by Steinbeck (1935) showed that focusing both horizontally and vertically could be obtained by a small radial decrease of the magnetic field, thus generating a further decrease in orbital frequency. These decreases in orbital frequency can only be overcome by extremely high radio-frequency voltages so that the desired energy can be attained in relatively few orbital turns. The 184-inch cyclotron in Berkeley was designed accordingly to attain deuteron energies above 100 MeV but the machine was diverted to military purposes as an isotope separator. In the meantime discovery of the phase stability principle discussed below made this brute force approach unnecessary.

The cyclotron principle fails for electrons, whose motion becomes relativistic at moderate energies. The betatron, invented by Wideroe[2] and first put into use by Kerst (1940) was a transformer in which the energy of electrons in circular orbits was increased by the induced electric field from an increasing flux in a central iron core driven by appropriate windings. The required average magnetic field in the drive core had to be twice that of the radically decreasing magnetic field at the orbit of the betatron. Betatrons reached an energy up to about 300 MeV, limited by the radiation loss per turn, which cannot be compensated in a betatron.

In a *linear induction accelerator* individual iron cores are stacked axially and excited through separate driving circuits. This principle permits acceleration of very-high-intensity electrons, in the kiloamp region (Christofilos *et al.*, 1964). Such devices are still used as x-ray-sources for diagnosis of rapid dynamic systems. A linear radio-frequency accelerator was developed by Sloan and Lawrence (1931), in which an alternating rf voltage was applied across a succession of gaps traversed by a beam, limited to low-velocity heavy ions by the low frequency of available rf sources.

A dramatic extension of accelerators and colliders to high energies was made possible by conceptual and technical developments.

Phase stability

Phase Stability was invented independently by McMillan (1945) and Veksler (1944).[3] In the pre-World War II accelerators synchronization between the rf fields and the particle bunches was achieved by "dead reckoning" McMillan and Veksler recognized that the phase of the accelerating rf voltage could be stably "locked" into synchronization with

---

[2] For a full discussion of the complex of inventions and demonstrations leading to the betatron see; Waloschek (1994).
[3] Described within the USSR in 1944; published in English in 1945.

the transit time of particle bunches under appropriate conditions. In a circular accelerator such stability is achieved by the particle bunch's crossing an accelerating gap during either a descending or ascending part of the radio-frequency voltage, depending on the relation between orbital path length and orbital momentum. This relation depends on the focusing mechanism and the relativistic mass. In a linear accelerator such stability is achieved by accelerating the bunch during the *ascending* part of the radio-frequency voltage. Such stability permits "synchrotron oscillation" about a stable phase.

The principle of phase stability led to diverse applications. In a synchrocyclotron particles are injected into a static magnetic field and are accelerated by a radio-frequency source whose frequency decreases to match the revolution frequency as the energy, and therefore the relativistic mass, increases and as the magnetic field weakens as the particle spirals out. Under this condition the particles remain phase locked to the electric field.

In a synchrotron particles are injected into a rising magnetic field and traverse a radio-frequency cavity excited at a near-constant frequency.[4] A magnet of only small radial aperture is needed. The particles remain locked in stable phase while their energy, but not their radius, increases with the magnetic field. This is the principle of all of today's high-energy circular electron and proton accelerators, including LEP (at CERN), the world's highest-energy electron collider (nearly 100 GeV per beam) and the Tevatron at Fermilab, the world's highest-energy proton collider. The latter is to be followed by the Large Hadron Collider at CERN, designed to attain a proton energy per beam of 7 TeV.

All modern *proton linear accelerators* use phase stability and continue to be the devices of choice as injectors into today's proton synchrotrons. They operate as accelerators in their own right up to about 1 GeV where high intensity is required.

Strong focusing

Strong focusing was adapted to synchrotrons independently by Christofilos (1950), and by Courant, Livingston, and Synder (1952). Focusing in earlier circular machines was attained through radial falloff in the magnetic field; electrostatic focusing or magnetic solenoids provided focusing in linear accelerators. "Strong focusing" originated from the realization that if a diverging and a converging lens of equal and opposite focal strength are separated by a finite distance, then the net focusing effect is positive. A magnetic quadrupole produces focusing in one plane and defocusing in the plane at right angles. Thus two quadrupoles separated by a finite distance and rotated by 90 degrees relative to one another focus in both planes. This focusing strength varies quadratically with the magnetic-field gradient in the quadrupoles and can be much stronger than that of solenoids or that of radial magnetic-field gradients.

Strong focusing drastically decreases the needed aperture of proton and electron synchrotrons and of linear accelerators. Thus strong focusing greatly extends the range of particle energies that can be economically attained.

Strong focusing results in particle oscillations about a central orbit whose wavelengths are generally shorter than the circumference of the circular accelerator. This creates the possibility of resonances between such focusing oscillations and harmonics of the basic orbital frequency. Moreover, the region in accelerating phase for which phase stability exists can change sign, leading to a transition energy at which phase stability vanishes.

---

[4]If injection into a synchrotron is at a particle velocity not fully relativistic, the frequency can be slightly modulated, for instance by loading of the accelerating cavity with ferrite.

Such problems can be avoided by appropriate design of the "lattice" of the focusing elements and by rapid passage through transition.

Proton and electron synchrotrons have been configured into colliders, leading to obvious center-of-mass advantages. Circular colliders are composed of storage rings, which are synchrotrons storing beams after the magnetic field has reached its final value. Stored electrons require compensation of the radiation loss by cavity reacceleration. Synchrotron radiation loss of protons is still negligible even at the highest energies attained today, but will become of future importance.

All modern circular colliders use both phase stability and strong focusing. Circular electron-positron colliders incur a radiation loss per turn which scales as the fourth power of the energy divided by the orbit radius. If the costs growing linearly with radius are matched with those scaling with the energy loss per turn, then the total cost of an electron-positron collider will grow with the square of the energy. The 27-km circumference electron-positron collider at CERN will probably be the highest-energy electron circular collider ever built. Collisions between linear accelerator beams and beams stored in a storage ring have been considered but thus far studies do not project competitive luminosity.

High-impedance microwave devices

W. W. Hansen (Ginzton, Hansen, and Kennedy, 1948) invented the electromagnetic cavity in 1937 with the goal of generating high voltage at moderate input power. The invention led to amplifiers, oscillators, cavities to compensate energy loss in circular accelerators, and linear accelerators, among them the disk-loaded waveguide. When such a waveguide is operating as an electron accelerator, the phase velocity of a propagating wave in this structure is matched to the particle velocity. The group velocity is tailored to provide a filling time compatible with the pulse length of the radio-frequency source, so as to provide an appropriate profile of accelerating voltage versus length. The highest-energy electron linear accelerator is the SLAC machine operating up to 50 GeV. Beyond that, linear colliders in which an electron beam from one linear accelerator collides with a beam accelerated by a separate machine are the most promising developments to exceed electron-positron energies attainable by circular storage rings. They require high average beam powers and exceedingly small beam cross sections in order to attain the required luminosity. The SLAC linear collider produces collisions between 50 GeV electrons and positrons.

Superconducting technology

The availability of superconducting materials made another gain in particle energy possible. For electromagnets the material of choice has been niobium-titanium, which can be fabricated into multistrand cables designed to minimize losses during magnetic field changes. Niobium-tin can sustain higher magnetic fields, but its mechanical brittleness has thus far prevented extensive use. The new high-temperature superconductors have only limited application in high-energy physics, restricted to connections and leadins. After extensive development, solid niobium or niobium coatings inside radio-frequency cavities have become practical and reliable and serve as accelerating cavities, both in linear accelerators and as accelerating elements in proton and electron synchrotrons.

Accelerator and Collider Limitations

Continued growth of accelerators and colliders is bounded by technical and economic factors. Technical limitations are in the following categories:

(1) *Material limits*. Vacuum breakdown and field emission are controlled by practical factors such as surface irregularities, dielectric inclusions, whisker growth, and so forth. These limit gradients in linear accelerators and in accelerating devices in synchrotrons.

Magnetic fields in "warm" magnets are limited by the saturation of iron, while those in superconducting magnets are limited by quenching of the superconducting materials in magnetic fields and by the problems inherent in restraining the large forces on conductors in such magnets. Frontiers in this respect have been advanced by the use of supercooled helium and by metallurgical advances in the production of superconductors and cables.

(2) *Nonlinear dynamics and collective effects*. The previous sketchy discussion has focused on the behavior of "free-space" single particles in "ideal" externally generated electric and magnetic fields. Such motions will be modified by the electromagnetic fields generated by induced currents in metallic envelopes, by deviations of fields from the ideal, generally linear, form, and by collective effects of particle groups on the motion of a single particle.

Induced fields and the collective fields of a bunch generate a "wake field" that affects individual particle orbits, both longitudinal along the motion of the particles and transverse to that motion. Wake fields not only affect the shape of a bunch of particles, in that the fields produced by the head of a bunch affect the motion of its tail, but they also can result in electromagnetic coupling between successive bunches in an accelerator. In the transverse direction such effects can produce decreased luminosity and outright instability. Luminosity decreases when wake fields dilute the phase-space density of the particles in a bunch. Instability can result if transverse displacement of preceding particles induces wake fields which successively deflect succeeding particles further. Such phenomena are complex.

Longitudinal wake fields result in the lengthening of the particle bunch in an accelerator. This can counteract efforts to maximize luminosity in a collider using very-short-focal-length magnets near the interaction region, since shortening of the focal length will be ineffective if the bunch length is too large. Transverse instabilities are particularly serious if the transverse displacement of the particle induces fields in either engineered or inadvertent resonant structures. The transverse displacement can induce so-called "higher-order modes" in such structures; any discontinuity in a vacuum envelope can enhance transverse wake fields.

The effect of transverse wake fields can be counteracted by a number of measures. The discontinuities in vacuum envelopes can be minimized; focusing strength can be enhanced, thus limiting transverse excursions; the frequency of transverse focusing oscillations can be dispersed among successive sections in a radio-frequency linear accelerator, thus damping a resonant buildup of transverse motion.

The coupling among particle bunches can become coherent as the wavelengths of the Fourier component of the electromagnetic field become comparable to the dimensions of the particle bunch. In this event fields will act coherently and the forces correspondingly increase. The principal countermeasure against coherent instabilities is external feedback: The electromagnetic field of the particle bunch is sensed by appropriate electrodes and is fed back to deflecting electrodes with a phase to damp the motion. Additionally structures can be designed which damp the relevant higher-order modes.

(3) Beam-beam interaction. Collisions between intense bunches of particles produce electromagnetic forces of one bunch of particles in the other. These forces shift the frequency of the focusing oscillations. If this "tune shift" becomes too large, then the frequency of radial focusing oscillations can shift into regions of instability, as discussed above. Actually the limiting tune shift is set, generally empirically, by nonlinear effects in the beam-beam interaction. Thus the permissible tune shift is subject to practical limits, which can be minimized by optimized design of the focusing lattice and shaping the beam profile during collision. In addition to the tune shift, the beam-beam interaction in electron-positron linear colliders also produces electromagnetic radiation as each particle experiences the collective electromagnetic field of the opposing bunch. Radiative effects broaden the energy spectrum of the colliding beams, thus making them less useful in elementary-particle physics experiments, and they also produce electromagnetic background.

(4) Beam-"vacuum" interactions. Interaction of the beams with residual gases or charge clouds in the vacuum can produce background. In addition, if electromagnetic radiation from synchrotron radiation or beam-beam interactions impact the vacuum wall, photoejected electron clouds affecting particle motions can be formed. Recent analyses (Raubenheimer and Zimmerman, 1995) show that this can lead to serious instabilities, in particular for the highest-energy proton-proton colliders.

(5) Injection. The design of ion sources in the case of hadron colliders and design of either thermionic or photocathodes in the case of electron machines can affect luminosity. In particular, space-charge effects at injection are limiting.

According to Liouville's theorem, the invariant emittance, that is, the phase-space density times the relativistic factor γ, cannot decrease during acceleration, storage, or final interaction in a nondissipative system. Liouville's theorem can be violated if damping takes place in the motion subsequent to injection. Such damping can be produced by emission of synchrotron radiation. This fact is utilized in damping rings inserted at an appropriate step in an accelerating cycle. Damping can also be accomplished by beam-to-beam cooling, in which an external beam of small phase volume is permitted to interact with the beam of the accelerator and exchange momentum. Finally damping can be accomplished by feedback in a circular machine by picking up signals from radial excursions and feeding those back onto the orbit at subsequent turns.

Thus the final luminosity may or may not be limited by the phase space at injection, depending on the presence of damping mechanisms.

Future Collider Possibilities

The previous discussion outlined the principles underlying past and present accelerator and collider systems and identified installations at the current frontier. Existing technology permits limited extension but major advance depends on new technology.

Along conventional lines further extension in energy attainable by large circular hadron-hadron colliders beyond the Large Hadron Collider and larger linear colliders fed by traditional electron and positron linear accelerators appears feasible. Such machines can also become the basis of electron-electron and photon-photon colliders at high energies.

Hadron colliders beyond the Large Hadron Collider face economic limitations and must take synchrotron radiation into account. Therefore such machines require large radio-frequency power and have to face the potential of charge cloud and other instabilities discussed above. At the same time synchrotron radiation will provide damping, which may be beneficial in reducing instabilities.

An international effort is addressing construction of a large linear collider, possibly approaching the TeV per beam range. Leading candidates to feed such a device are conventional microwave linear accelerators operating at higher frequency than now in use. In addition superconducting linear accelerators are being explored, aiming at improvements in economy and gradient beyond current experience. Finally, there exists the possibility of feeding a linear collider by variants of a two-beam-accelerator principle. Here a high-current, low-voltage linear accelerator fed by induction or low radio-frequency sources drives a high-energy, high-gradient machine. Energy from the driver is coupled through appropriate transfer structures into the high-energy accelerator.

Substituting muons for electrons in circular machines reduces radiation by a large factor, while strong and weak interactions of muons appear identical to those of electrons. While the idea is old (Tinlot, 1962; Budker *et al.*, 1969) optimism has grown that muon colliders of adequate luminosity and background conditions can be designed. Luminosity depends both on initial muon yields and on cooling of the muons resulting from pion decay in a practical manner. The background problem is serious due to the large electron fluxes originating from decay of muons in orbit, and even decay neutrinos pose a substantial hazard.

In addition to devices based on extrapolations of established practice, new technologies are being analyzed. All of these, to be useful for high-energy physics, would have to be configured into linear colliders and therefore would have to generate both high energy and high average beam powers. Current research focuses on acceleration by very large intrinsic voltage gradients. Among these are devices using the high fields in intense laser beams (Channell, 1982). The electromagnetic field in a laser wave in free space cannot accelerate charged particles and therefore research addresses special geometries which generate longitudinal electric-field components. Possibilities are the electric field when optical laser beams are diffracted from gratings, when coherent laser beams are crossed to generate a longitudinal-field component and similar geometrical arrangements. Other methods utilize the high electric fields contained in plasmas (Schoessow, 1994), the high gradients in the wake fields produced by intense particle bunches, and finally the extremely high electric fields that could be generated if plasma waves were excited in crystals (Chen and Noble, 1986); these could be used to accelerate particles channeled in such crystals.

Physical Processes in Particle Detection

Charged and neutral particles interact with detector material via limited processes. Charged particles ionize any medium and can radiate Čerenkov, synchrotron, or transition photons. The ionization density, and consequently the rate of energy loss (dE/dx) of charged particles in matter, is in essence a measure of particle velocity (Bethe, 1932; Bloch, 1933). Therefore measurements of ionization density in combination with deflection in a magnetic field (which determines the ratio of particle momentum to charge) can result in determination of rest mass. The ionization as a function of particle velocity βc has three regions: (a) a low-velocity region where the ionization decreases roughly as β^{-2} and then levels off to a region of (b) minimum ionization and (c) a region of logarithmic growth (relativistic rise) which reaches a plateau with $\beta\gamma$ defined by the dielectric properties of the material, affecting the relativistically contracted electromagnetic field in its ability to ionize remote atoms.

Neutral hadrons may interact strongly to produce charged particles. Photons may interact electromagnetically via Compton scattering, photoelectric effect, or pair production. Neutrinos can generate charged particles with very small cross sections via the weak interaction. At high energies strongly interacting particles produce cascades, and electrons and photons produce electromagnetic showers. Ultimately any detector either senses ionization caused by primary or secondary charged particles or detects secondary photons by photoelectric mechanisms.

Detector Components

Pictorial detectors

Pictorial Detectors utilize the particle track left by ionization and process the image of that track into a photographic or digital record.

The earliest track detector was the *cloud chamber*, which can either produce super saturation following an expansion of water vapor or it can be a continuously sensitive diffusion chamber in which a thermal gradient in water vapor leaves a region where condensation forms around an ionizing track. Subsequently photographic *emulsions* were specifically tailored through enhanced silver content to reveal, after development, particle tracks which are microscopically scanned. A *streamer chamber* produces conditions in which ionization in a gas generates enough light through ion recombination to permit photographic recording. In a *spark chamber* local breakdown occurs between high-voltage electrodes; the sparks can be photographed in a sequence of gaps between electrodes leading to a track in a photograph. In a *bubble chamber* a liquid is expanded leading to a superheated condition; gaseous bubbles will be formed along an ionizing track, which can be photographed.

All these devices greatly contributed to elementary-particle physics. Cloud chambers have been major tools in cosmic-ray research, including the discovery of the positron. Bubble chambers recorded associated strange-particle production and established the foundation of hadron spectroscopy. A limitation of bubble chambers and cloud chambers is that they cannot be "triggered"; they record *all* ionizing events irrespective of whether the events are novel or are signatures of well-known processes. However, photography can be triggered to select only events of current interest to limit labor in data analysis. Spark chambers and streamer chambers can be triggered but have inferior location accuracy. All pictorial devices other than the spark chamber permit measurements of ionization density.

Pictorial devices require substantial effort in data analysis; images have to be scanned either manually, semiautomatically, or totally automatically; tracks have to be reconstructed and hypotheses as to the event that may have occurred have to be fitted to the track pattern.

Pictorial devices have largely disappeared from use in elementary-particle physics. They cannot handle events produced with small cross sections in the presence of large uninteresting background. They tend to be expensive considering the data-analysis effort. Resolving time is generally long. Yet the slowest of these detectors—photographic emulsions—are still in use for elementary-particle physics because of their unexceeded track resolution of near one micrometer. Large emulsion stacks continue to be used in connection with neutrino experiments. Auxiliary electronic detectors can limit the emulsion area to be searched for precise vertex measurements.

Electronic detectors

Scintillation counting

Suitably doped plastics have long been utilized for position measurement, time-of-flight measurement, dE/dx, and calorimetry. The scintillation photons from inorganic crystals and wavelength-shifted photons from plastic scintillators can either be detected by photomultipliers as high-gain, low-noise amplifiers of the electrons emitted by the photocathode or by suitable solid-state photodetectors. Issues of the number of separate measurements required, operation in magnetic fields, quantum efficiency, size, and cost determine the choice of photodetector. Scintillators and photomultipliers still excel at precision timing in applications with modest spatial resolution requirements. For example, in the proposed Minos neutrino detector (Wojcicki, 1998), solid scintillator bars couple their scintillation light to optical fibers using wavelength-shifting dopants in the fibers. A variant is the Fiber Tracker, in which optical scintillator fibers form large arrays, with each fiber having an independent photodetector (DO Upgrade, 1996).

Wire drift chambers

Wire drift chambers (Charpak, 1976) amplify the few electrons produced by ionization by an avalanche near the anode wires. Electron multiplication near the anode wire produces an easily processed signal which can be timed to produce a variety of precision spatial measurement systems. Electrodes are designed to provide electric fields in which the drift velocity can be well understood and provide small regions of high field that generate the electron avalanche from a primary ionization electron. Chambers range from small detectors to planar or cylindrical arrays of many square meters; 100-micrometer spatial resolution is routinely attained, as is multitrack resolution of better than 1 mm. Wire drift chambers can operate at the extremely high rates necessary in many fixed-target experiments and have been radiation hardened to operate in the harsh environment of high-luminosity proton colliders (CDF II Detector Technical Design Report, 1996). The coordinate measured by the drift time is normal to the wire. Low-precision measurements along the wire can be made utilizing resistive charge division and measuring the signal on both ends of the wires. Higher precision is achieved by small-angle stereo, necessitating the association of wire hits with tracks, which can be difficult in a busy environment. These systems can be used in a magnetic field for momentum measurement.

Proportional wire systems

Proportional wire systems operate in a mode where the signal is proportional to the primary ionization, thus measuring the energy-loss rate of the primary particle in the gas. *Avalanche systems* amplify the primary ionization to saturation, yielding large, very noise-immune signals. Such systems using single anode wires in moderate-resistivity tubes can give position signals by induction to strips with any geometry on the tube surface (Iarocci, 1983). They are widely used for muon detection with active areas of order 1000 square meters.

Time-projection chamber

In a time-projection chamber (Nygren, 1974) a sensitive volume is filled with a gas mixture with very low electron-attachment cross section. An applied uniform electric field drifts ionization electrons towards a two-dimensional array of detectors at one end. Tracks of charged particles are reconstructed from this detector array, with time of arrival at the array providing the third coordinate. The time-projection chamber avoids most of the association difficulties of wire chambers and provides digitized pictorial images of events. The time-projection chamber has limitations in high-rate environments, as accumulated slow-moving positive ions distort the drift field.

Semiconductor detectors

The development of high-resistivity silicon led to pn diodes that can directly detect the ionization caused by the passage of a charged particle. A minimum ionizing particle yields about 80 electron-hole pairs per micrometer of depleted silicon, or of order 10^3 electrons in a typical detector. The geometry of electrodes is nearly arbitrary, and detectors range from large-area diodes to "microstrips," arrays several cm long divided into diode units of width 25 to 100 micrometers. The overall scale is set by the size of the silicon wafer. Elaborate arrays of microstrips, with sophisticated low-mass space frame structures supporting the silicon stable to a few micrometers in space are used as vertex detectors. Since it is not yet possible to process complex transistor arrays on the detector wafers and maintain high resistivity, many connections must be made to nearby readout electronics. A vertex detector may have 10^4 to 10^5 channels, so power and thermal management of the electronics, as well as of the wire-bonded connections, are challenging. Two-dimensional information may be gained by connections to different sets of strips on either side of the silicon, or by using several one-sided arrays.

True pixel arrays are desirable because they give unambiguous space points, even in a dense particle jet. One approach uses charge-coupled devices (CCD's) fabricated from high-resistivity silicon. The simultaneous advantage and disadvantage of CCD's is that they are read out serially from a small number of readout nodes, thus requiring relatively little electronics but requiring tens of milliseconds for the readout process. For low-interaction-rate environments (such as e^+e^- linear colliders) this situation is ideal, and CCD's can provide 20-micrometer-sided pixels. Arrays of pixel diodes that are bump bonded to readout electronics are now being developed for Tevatron and Large Hadron Collider experiments. Such devices incorporate local smart readout to compress data, and can operate at high rates.

Čerenkov radiation detectors

The simplest Čerenkov counters are velocity threshold devices using a medium whose index of refraction has been adjusted so that particles above some velocity generate radiation. The angle of radiation emission measures particle velocity. Focusing devices can send the radiation through a circular slit, allowing differential cuts on velocity. Such devices have relatively small acceptance and are used primarily in fixed-target experiments. A large step was taken with devices that actually image the cone of Čerenkov radiation on a sensitive focal plane to measure the Čerenkov angle. In composite large detectors on the scale of square meters, the focal planes follow the momentum measurement and must detect single photons. In high-rate environments, the focal plane might be a pixellated array. The DELPHI experiment (Aarnio et al., 1991) at LEP and the SLC Large Detector (SLD) at SLAC developed devices that contained low-electronegativity, high-photoabsorbtion organic molecules in large rectangular quartz-walled boxes. The Čerenkov photon converts to an electron in the organic vapor, and then drifts in an electric field of a few hundred V/cm to a wire chamber at the end of the box. The drift length is read out as time, and the conversion coordinates normal to the drift are read out by the wire number and by charge division on the wire.

Transition radiation detectors

A charged particle traversing a boundary between materials differing in dielectric constant will emit photons. Thus detector components of sensitivity sharply increasing with γ can be constructed of sandwiched layers of gas and foils, with a photon detector, usually a heavy-gas wire drift chamber, facing the exit surface. In practice such detectors require $\gamma > 10^4$ for adequate signals.

Detector Systems

Detector systems generally accomplish particle tracking, momentum measurement of charged particles, particle identification, and total-energy measurement of single particles or groups (jets) of particles. Additionally on-line and off-line data analysis is provided. Since optimal use of the accelerator has become important, detector systems have grown in geometric acceptance and measurement resolution to maximize information from each event and optimally use accelertor luminosity. Increases in the number of channels and data rates, and in measurement precision, have augmented costs and sizes.

Fixed-target experiments generally include beam definition, target, drift or decay region, and detectors. The size of such experiments ranges from emulsions to the long baseline of neutrino oscillations. While primary accelerated particles are protons and electrons (or their antiparticles), fixed-target experiments can utilize secondary beams of long-lifetime particles. The experimentalist controls, albeit within limits, beam momentum, momentum spread, spill time, intensity, backgrounds, and experimental geometry. For *collider experiments* almost all parameters save beam energy are fixed by the collider design; in some cases some control of beam polarization may be possible. As a generality, more luminosity, if consistent with background requirements, is always wanted.

The basic scale of collider detectors is set by the highest particle momentum to be analyzed. This is defined by the dimensions of the required magnetic analyzer and the range of the most penetrating particles (generally muons).

In general, at the energy frontier, e^+e^- detectors are smaller and simpler than the p-p (or $p\bar{p}$) detectors; e^+e^- machines operate at significantly lower energies, and total cross sections are much smaller, and there is usually less demand for forward acceptance with e^+e^- detectors. Radiation hardness and rate requirements are less challenging for e^+e^- detectors; such colliders produce far fewer than one interaction per crossing. In contrast, the Large Hadron collider proton collider is expected to have more than 10 events per crossing with a crossing rate of 25 MHz.

Momentum measurement: Magnet configurations for collider detectors

The detectors at the CERN Intersecting Storage Rings (Giccomelli and Jacob, 1981), the first of the large-scale pp colliders, used varied magnetic configurations, mostly of modest acceptance. The first large-scale cylindrically symmetric detector using a magnetic solenoid was the MKI at SPEAR at SLAC. All subsequent collider detectors except for the UA1 at the SPPS of CERN were cylindrically symmetric, mostly with solenoidal magnetic fields, although toroidal fields for muon momentum analysis have been used. This magnetic configuration leads to coaxial "barrels" of vertex detection, momentum measurement, particle identification, calorimetry, and muon measurement. Geometric variations include endcaps closing the barrels and additional downstream detectors to improve the forward acceptance, which is compromised in the solenoidal geometry.

Magnetic fields of 1.5 to 4 T produced by superconducting solenoids are now used or proposed, as are position resolutions of somewhat better than 100 micrometers, leading to tracker radii in the range of 1 to 3 meters.

Particle identification

Particle identification generally relies on measurements of velocity (or the relativistic factor γ) or rests on observation of interactions (or their lack). Velocity (or γ) measurements use time of flight (TOF) or the outputs from Čerenkov or transition radiation detectors and observation of ionization density (dE/dx). For sufficiently slow particles, measurement of time of flight (TOF) is straightforward. A plastic scintillator with reasonably good geometry coupled to a photomultiplier can give time resolution below 100 ps (Benlloch et al., 1990). The technique is limited by lengths of the flight path and by background.

Calorimetry (total-energy measurement)

Calorimeters are used to measure the energy and position of hadrons (charged or neutral), electrons, and jets, and to help identify leptons in hadronic jets. Electromagnetic calorimeters must be thick enough to develop, contain, and measure cascade showers induced by electrons and photons. Hadronic calorimeters must be substantially thicker to contain nuclear cascades. Angular resolution of the calorimeter can be critical, and since position resolution is limited by transverse cascade shower dimensions, the calorimeter may become rather large. Calorimeter design requires optimization among performance parameters of energy, angle, and time resolution with radiation hardness, size, and cost. Calorimeters can be sampling or nonsampling, i.e., homogeneous. The sampling calorimeters alternate high-atomic-number metals for shower development with layers of a sensitive medium, e.g., scintillator, to sample the shower development. Homogeneous devices, practical only for electromagnetic calorimeters, utilize a uniform ionization-sensitive medium both to develop and to measure the energy of a shower, such as crystals of NaI. Crystals of lead tungstanate have been developed for the Compact Muon Solenoid (CMS) electromagnetic calorimeter.

Statistical fluctuations in shower development and the corresponding fluctuations of the ionization in the sensitive medium limit the energy resolution of a sampling calorimeter. Thus the energy resolution will vary as $E^{1/2}$ and as $t^{1/2}$ where E is the incident-particle energy and t is the thickness of the radiator between samples. In homogeneous calorimeters, stochastic processes lead to a fractional energy resolution which varies as $1/(E)^{1/4}$, but leakage of the shower from the calorimeter, electronics noise, nonuniformity of light collection, and calibration errors add a constant term (which must be combined in quadrature). Sampling electromagnetic calorimeters achieve fractional energy resolutions in the range $10-15\ \%/E^{1/2}$, while crystal calorimeters achieve $1-5\ \%/E^{1/4} \oplus 1-3\ \%$, where E is measured in GeV.

An electromagnetic shower can usually be contained in about 25 radiation lengths, corresponding to 15 cm of lead. Hadronic calorimeters require roughly 10 interaction lengths for containment of hadronic jets, corresponding to 112 cm of uranium or 171 cm of iron. Economics usually dictate a sampling calorimeter with liquid argon, scintillator, or wire chambers as the active medium. In addition to sampling statistics, the resolution of hadronic calorimeters is affected by their relative response to electromagnetic and hadronic showers, usually resulting in fractional energy resolutions of $50-75\ \%/E^{1/2}$.

Many interesting variants on the basic designs have been developed. For example, liquid argon is extremely radiation hard, but the traditional electrodes of alternating layers of metal have relatively slow response because of their inductance. Folded electrodes in an accordion shape better approximate a transmission line and have been proposed for ATLAS and GEM (ATLAS Liquid Argon Calorimeter Technical Design Report, 1996). To improve resolution, scintillating fibers can be effectively cast into a lead matrix.

On-line analysis, data acquisition, and trigger systems

Increases in speed, density, and functionality in data-acquisition electronics, even *after* the widespread utilization of transistor circuits, are quite impressive. While most early and some modern experiments utilize standardized modular electronics, many larger experiments have improved performance and economics by using custom electronics integrated with the detectors proper. While channel densities are hard to compare, channel counts are shown versus proposal date for several detectors in Fig. 4. This growth with

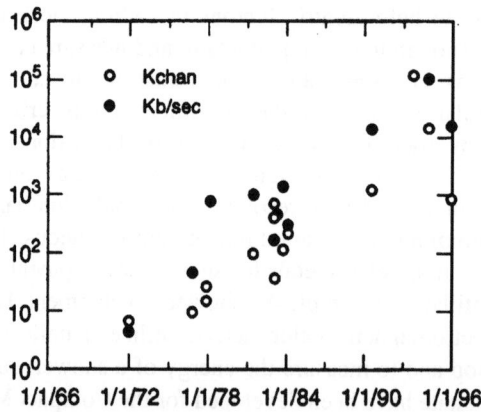

Figure 4. Evolution of the number of detector signal channels with time, indicating growth of collider detector instrumentation capability over the last 24 years: open circles, number of electronic instrumentation channels in thousands; closed circles, design data rate in kilobytes per second to permanent storage. The date is that of the detector proposal or Technical Design Report. e^+e^-, pp, and $\bar{p}p$ detectors are included.

only moderate cost increase rests largely on continuing developments in circuit integration and computing technologies.

Most experiments produce raw data rates from the first stages of their electronics far too great to be recorded and subsequently analyzed. Trigger systems select a subset of events for recording, and the data-acquisition system compresses and corrects the data and associates data with different detector subsystems for each event. Most trigger systems have a three stage architecture: Level 1 is a fast, relatively simple hardware process that operates at the basic interaction rate of the machine and buffers a subset of events for Level 2, which uses more complex, slower algorithms to further reduce the rate for Level 3. Level 3 is usually a set of processors executing much of the nominal event-reconstruction code, thus making the full set of analysis cuts available. Level 1 implies fast, synchronous buffering of the event data, perhaps with only a subset of the data available to the Level-1 trigger processor. Level 2 requires slower, asynchronous buffering of the event data, and may have much of the data available to the processor. Level 3 has complete access to all of the data. The output of Level 3 is stored for off-line analysis, with data rates of roughly a megabyte per second.

The computation demands of many collider detectors continue to require leading-edge computation technology. Reconstruction of an e^+e^- event may require of order 10^9 instructions, and hadron collider events may need an additional order of magnitude. Most analyses require calculations of acceptances and efficiencies, implying generation and reconstruction of Monte Carlo data sets several times the size of the real data set. Finally, event samples may exceed 10^9 events, and data storage facilities of petabytes (10^{15}) are proposed. Fortunately, event computation can easily run on arrays of computers on an event by event basis, and thus parallel "farms" are widely used.

The SLC Large Detector as an example of a collider detector system

The Stanford Linear Collider (SLC) collides bunches of about 4×10^{10} e^- and e^+ at 120 Hz, with a luminosity approaching 2×10^{30} cm^{-2} sec^{-1}. A linear collider implies a very low true event rate and beam crossing rate; a very small luminous region in all three dimensions; and almost negligible radiation damage load on detector components. The SLC final focus system produces beam spots of about 1 micrometer horizontally and 0.5 micrometer vertically. Synchrotron radiation backgrounds are minimized by a masking system requiring multiple reflections for a photon to enter the detector. These features are exploited in the SLD design, shown in Fig. 5, to permit a CCD vertex detector with about 3×10^8 pixels, a purely computational trigger, and a time-multiplexed data-acquisition system.

The vertex detector, consisting of 96 18×80-mm CCD's, is arrayed around a 25-mm-radius Be beampipe. The position resolution of the vertex detector is dominated by multiple scattering at lower momenta. The solenoidal magnet has an inner diameter of 3 m and produces a magnetic field of 0.6 T. Charged-particle momenta are measured by an 80-layer cylindrical drift chamber extending radially from 20 cm to 1 m, and with a total length of 2 m, arranged in 10 superlayers of alternating stereo angle. The longitudinal coordinate is first estimated by charge division on the anode wires and then fitted using the stereo information. Momentum resolution of $\Delta P/P = 0.01 \oplus 0.0026 P_\perp$ (GeV/c) is achieved. SLD utilizes a Čerenkov ring imaging detector for particle identification. Three-standard-deviation separation between P's and K's is achieved from momenta of 1.5 to 5 and 9.5 to 45 GeV/c, and between K's and π's between 0.35 and 25 GeV/c. Next comes a sampling calorimeter of lead plates in liquid argon, arranged as towers that point projectively towards the vertex. An electromagnetic section is 22 radiation lengths deep, followed by an hadronic section approximately 3 interaction lengths deep. The total calorimeter is not thick enough to contain hadronic showers, but the tails are measured in an iron calorimeter that follows the aluminum solenoid. The iron calorimeter consists of 5-cm sheets of steel interleaved with limited streamer-mode chambers (Iarocci tubes). The chamber cathode surfaces are read out, on one side as a continuation of the liquid-argon calorimeter towers, and on the other as strips for a muon tracking system. The barrels are closed by endcaps of similar instrumentation.

The luminosity is monitored by small-angle Bhabha scattering measured by a pair of highly segmented tungsten-silicon diode calorimeters arranged as cylinders capturing the

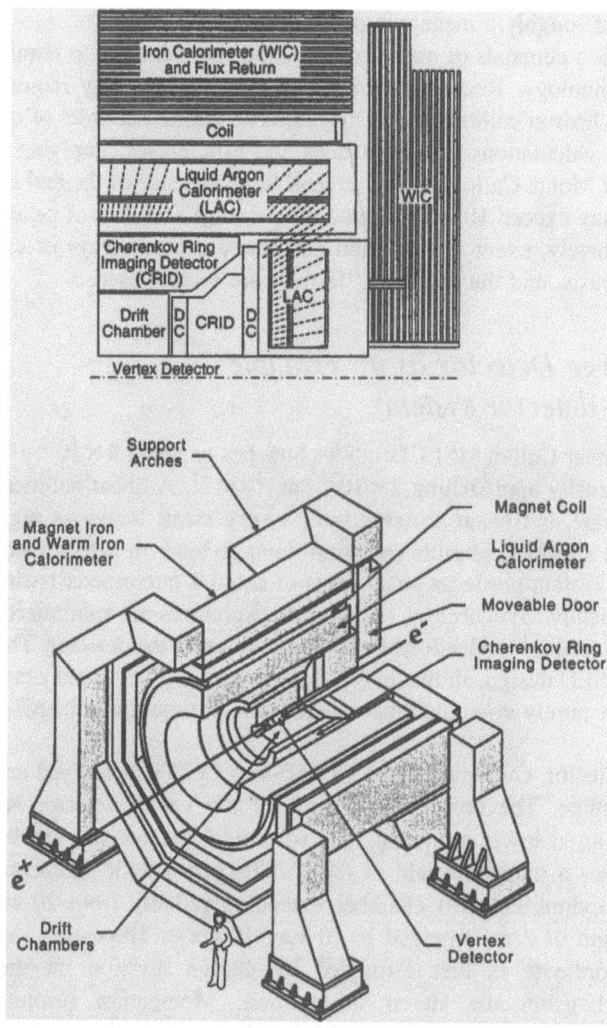

Figure 5. Quadrant and cut-away views of the SLD detector.

beampipe about 1.5 m from the interaction point. Electron-beam polarization is measured by scattering a circularly polarized laser beam from the electron beam exiting the detector and measuring the asymmetry of the Compton-scattered electrons as the longitudinal polarization of the electrons is changed.

Essentially all electronics were customized for SLD. The basic architecture consists of preamplifiers feeding application-specific integrated circuits of switched-capacitor arrays to record each signal wave form. Subsets of this data are fed to a network of microprocessors to compute a trigger. In the trigger architecture described previously, Level 1 is the intrinsic SLC crossing rate, Level 2 is the microprocessor network, and Level 3 was not implemented since an acceptable rate for permanent storage is achieved by Level 2. After a trigger, data still held in the capacitor arrays are multiplexed to digitizers and transmitted via optical fibers to a network of about 600 microprocessors for data correction and compression.

The Future

Extension of existing accelerator and collider principles to higher performance requires advances in magnet technologies, superconducting technologies, etc. Detector systems must be able to operate in even more severe backgrounds. Today, work on new collider technologies is focused primarily on high gradients, and further issues concerning efficient conversion of power from the primary source to the beam must be addressed. Detector and data analysis methods are likely to match this evolution. During the next century of the American Physical Society these proposals should lead to practical designs for collider-based physics.

Acknowledgments

We should like to thank J. Jaros, A. Odian, A. Sessler, R. Siemann, and D. Whittum, for critical reading and helpful insights. Work was supported by Department of Energy contract DE-AC03-76SF00515.

References

Aarnio, P., *et al.* (DELPHI Collaboration) 1991, Nucl. Instrum. Methods Phys. Res. A **303**, 233.
ATLAS Liquid Argon Calorimeter Technical Design Report, 1996, CERN/LHCC 96, 41.
Benlloch, J. M., M. V. Castillo, A. Ferrer, J. Fuster, E. Higon, A. Llopis, J. Salt, E. Sanchez, E. Sanchis, E. Silvestre, and J. Cuevas, 1990, Nucl. Instrum. Methods Phys. Res. A **290**, 327.
Bethe, H. A., 1932, Z. Phys. **76**, 293.
Bloch, F., 1933, Ann. Phys. **16**, 285.
Budker, G. I., 1969, Proceedings of the 7th International Accelerator Conference Program, Yerevan (1969).
CDF II Detector Technical Design Report, 1996, Fermilab Publ. 96/390-E.
Channell, P. J., 1982, Ed., *Laser Acceleration of Particles*, AIP Conf. Proc. No. 91 (AIP, New York).
Charpak, G., 1976, Nucl. Part. Phys. **6**, 157.
Chen, P., and R. J. Noble, 1986, in *Advanced Accelerator Concepts*, AIP Conf. Proc. No. 156, edited by F. E. Mills (AIP, New York), p. 222.

Christofilos, N., 1950, U.S. Patent #2,736,766.
Christofilos, N. C., R. E. Hester, W. A. S. Lamb, D. D. Reagan, W. A. Sherwood, and R. E. Wright, 1964, Rev. Sci. Instrum. **35**, 886.
Cockroft, J. D., and E. T. S. Walton, 1932, Proc. R. Soc. London, Ser. A **136**, 619.
Courant, E., M. S. Livingston, and H. Snyder, 1952, Phys. Rev. **88**, 1190.
DO Upgrade, 1996, Fermilab Pub. 96/357-E.
Giccomelli, G., and M. Jacob, 1981, in *CERN: 25 Years of Physics*, edited by M. Jacob (North-Holland, Amsterdam), p. 217.
Ginzton, E. L., W. W. Hansen, and W. R. Kennedy, 1948, Rev. Sci. Instrum. **19**, 89.
Iarocci, E., 1983, Nucl. Instrum. Methods Phys. Res. **217**, 30.
Kerst, D. W., 1940, Phys. Rev. **58**, 841.
Lawrence, E. O., and N. E. Edlefsen, 1930, Science **72**, 378.
Lawrence, E. O., and M. S. Livingston, 1932, Phys. Rev. **40**, 19.
Livingston, M. Stanley, 1954, *High-Energy Accelerators* (Interscience, New York), p. 151.
McMillan, E. M., 1945, Phys. Rev. **68**, 143.
Nygren, D. R., 1974, in Proceedings of the PEP Summer Study, Lawrence Berkeley Laboratory, "The Time Projection Chamber: A new 4 Pi Detector for Charged particles," p. 58.
Particle Data Group, 1996, Phys. Rev. D **54**, 128,142.
Raubenheimer, T. O., and F. Zimmermann, 1995, Phys. Rev. E **52**, 5487.
Schoessow, P., 1994, Ed., *Advanced Accelerator Concepts*, AIP Conf. Proc. No. 335 (AIP, New York).
Sloan, D. H., and E. O. Lawrence, 1931, Phys. Rev. **38**, 2021.
Steinbeck, M., 1935, U.S. Patent #2,103,303.
Tinlot, J., 1962, A storage ring for 10 BeV Mesons, SLAC-5, p. 165.
Van de Graff, R. J., 1931, Phys. Rev. **38**, 1919.
Waloschek, P., 1994, *The Infancy of Particle Accelerators: Life and Work of Rolf Wideroe* (Vieweg, Braunschweig).
Wojcicki, S., 1998, NuMI-L-337 TDR, FNAL.
Veksler, V. I., 1945, J. Phys (USSR) **9**, 153.

Anomalous g Values of the Electron and Muon

V. W. Hughes and T. Kinoshita

Historically, the spin magnetic moment of the electron μ_e or its g value g_e has played a central role in modern physics, dating from its discovery in atomic optical spectroscopy and its subsequent incorporation in the Dirac theory of the electron, which predicted the value $g_e = 2$. The experimental discovery in atomic microwave spectroscopy that g_e was larger than 2 by a multiplicative factor of about 1 part in 10^3, $g_e = 2.00238(10)$, together with the discovery of the Lamb shift in hydrogen ($S = 2^2S_{1/2} - 2^2P_{1/2}$), led to the development of modern quantum electrodynamics with its renormalization procedure. The theory enables us to calculate these effects precisely as finite radiative corrections. By now the experimental value of $g_e - 2$ has been measured to about 4 ppb, and the theoretical value, which is expressed as a power series in the fine-structure constant α, has been evaluated to better than 1 ppb, assuming the value of α is known.

For the muon, as well, g_μ is greater than 2 by a multiplicative factor of about 1 part in 10^3. This was found experimentally shortly after the discovery of parity nonconservation in the weak interaction, which provided the basic tools for the measurement of g_μ. This result provided one of the crucial pieces of evidence that the muon behaves like a heavy electron, i.e., there is $\mu - e$ universality. By now the value of $g_\mu - 2$ has been measured to 7 ppm. Treating the muon as a heavy electron, theorists have evaluated $g_\mu - 2$ to within better than 1 ppm. The main difference between g_μ and g_e is that the lepton vacuum-polarization contributions are very different for the muon and the electron. Furthermore, because the muon has a heavier mass than that of the electron, higher-mass particles—some perhaps not yet discovered—contribute much more to g_μ than to g_e by a factor of $\sim (m_\mu/m_e)^2 \approx 4 \times 10^4$.

The motivation for a continued study of electron and muon anomalous g values, $a \equiv (g-2)/2$, is twofold:

(1) Theoretically the anomalous g value is the simplest quantity calculable to an arbitrary precision. Note that quantities such as particle mass and the coupling constant $\propto$ are external parameters of the current standard theory and cannot be calculated from the theory itself. Precision measurements of a_e and a_μ therefore provide a crucial test of predictions of (renormalizable) quantum field theory. The firm theoretical basis for computing a_μ and a_e, taken together with more precise measurements of a_μ, will

not only test the standard model further but may open up a window into the study of entirely new physics.

(2) The measurement and theory of a_e have become so precise that a_e gives the most stringent test of QED if α is known precisely. Unfortunately, no available α is known with sufficient precision to enable such a test. This means, however, that the theory and measurement of a_e together will lead to the most precise value of the fine-structure constant α currently available. Comparison of α derived from a_e with other high-precision measurements of α based on condensed-matter physics, atomic physics, and other means offers an intriguing opportunity to introduce a quantitative measure of the success of quantum theory, which is at the root of all physics developed in the twentieth century. This topic will be discussed in greater detail in the section on Some Implications for Fundamental Physics.

Electron $g-2$ Experiments

The latest and most precise measurement of the electron $g-2$ value involves observation of microwave-induced transitions between Landau-Rabi levels of an electron in a magnetic field (Fig. 1) by Dehmelt and his collaborators (Van Dyck, Schwinburg, and Dehmelt, 1987; Van Dyck, 1990).

A single electron (or positron) moves in a Penning trap in a strong magnetic field of 5 T at a low temperature of 4 K, forming a "geonium" atom. Axial, cyclotron, and magnetron motions occur. The cyclotron frequency ω_c and the difference frequency ω_a (anomaly frequency) between the spin precession frequency ω_s and ω_c are measured. Their ratio determines a_e. The transitions are detected by changes in the axial frequency of the electron, observed through an induced voltage in an external circuit. This experiment has led to very precise values for electron and positron:

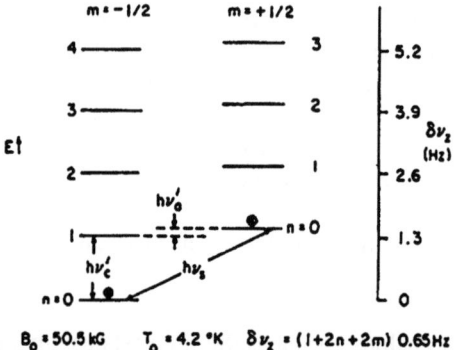

Figure 1. Lowest Landau-Rabi levels for a geonium atom. The axial frequency (shown in the right-hand scale) corresponds to the coupling via the axial magnetic bottle field. The quantities ν'_c and ν'_a are perturbed values of ν_c and ν_a. The lowest state ($n=0$) which is occupied by the electron or positron 80–90% of the time differs by 1.3 Hz depending on the exact spin state. This is the signature used to indicate that a spin has flipped. From Van Dyck (1990).

$$a_{e^-}(\text{expt}) = 1\,159\,652\,188.4\,(4.3) \times 10^{-12} \quad (4 \text{ ppb}),$$

$$a_{e^+}(\text{expt}) = 1\,159\,652\,187.9\,(4.3) \times 10^{-12} \quad (4 \text{ ppb}). \quad (1)$$

The values for a_{e^-} and a_{e^+} agree to within 1 ppb.

The statistical error in Eq. (1) is 0.62×10^{-12}, a systematic error of 1.3×10^{-12} is due to the uncertainty in a residual microwave power shift, and the largest uncertainty of 4×10^{-12} is assigned to a potential cavity-mode shift. This last error arises from a shift in the cyclotron frequency of the electron associated with image charges induced in the metallic Penning trap, an effect which depends on the cavity frequency modes and on the electron cyclotron frequency (Brown et al., 1985a, 1985b).

Studies to improve the experimental precision for a_e focus on the understanding and control of this cavity influence on the cyclotron frequency. For this purpose Mittleman et al. (1995) have produced and studied a many-electron (kiloelectron) cluster in the trap, which magnifies the shift of the cyclotron frequency. Gabrielse and Tan (1994) are studying the use of a cylindrical cavity where the cyclotron frequency shift can be better understood and controlled. Eventual reduction of experimental uncertainty by about an order of magnitude is the goal.

Muon $g-2$ Experiments

The muon $g-2$ value has been determined in a series of experiments at CERN (Bailey et al., 1979; Farley and Picasso, 1990). In the latest experiment, polarized muons from pion decays are captured in a storage ring with a uniform magnetic field and a weak-focusing electric quadrupole field. For a muon momentum of 3.09 GeV/c and $\gamma = 29.3$ the muon spin motion is unaffected by the electric quadrupole field and the difference frequency ω_a is given by

$$\omega_a = \omega_s - \omega_c = \frac{eB}{mc} a_\mu, \quad (2)$$

in which ω_s is the spin precession frequency and ω_c the orbital cyclotron frequency. Measurements of ω_a and B thus determine a_μ.

The stored μ^+ in the ring decay to e^+ via the parity-violating weak decay $\mu^+ \to e^+ + \nu_e + \bar{\nu}_\mu$, and the high-energy e^+ are emitted preferentially in the direction of the muon spin. Decay e^+ are detected with lead/scintillator detectors as a function of time after π injection. Of course μ^- can be treated in the same way. The time spectrum for the e^+ counts is given by

$$N_e = N_0 e^{-t/\gamma \tau_0} [1 + A \cos(\omega_a t + \phi)], \quad (3)$$

in which τ_0 is the muon lifetime at rest, γ is the relativistic time dilation factor, and A and ϕ are fitting parameters. The exponential muon decay is modulated at the frequency ω_a, which is determined from the fit of Eq. (3) to the data. The storage ring field B is measured by NMR.

The CERN results were

$$a_{\mu^-}(\text{expt}) = 1\,165\,936\,(12) \times 10^{-9} \quad (10 \text{ ppm}),$$

$$a_{\mu^+}(\text{expt}) = 1\,165\,910\,(11) \times 10^{-9} \quad (10 \text{ ppm}), \quad (4)$$

and for μ^+ and μ^- combined

$$a_\mu(\text{expt}) = 1\,165\,923\,(8.5) \times 10^{-9} \quad (7 \text{ ppm}), \tag{5}$$

in which the dominant error is statistical (Bailey *et al.*, 1979; Farley and Picasso, 1990). The largest systematic error of 1.5 ppm was due to uncertainty in the value of the magnetic field B.

At present a new experiment is in progress at Brookhaven National Laboratory with the goal of measuring a_μ to a precision of 0.35 ppm, which would represent an improvement by a factor of 20 over our present knowledge. The method of the BNL experiment is basically the same as that of the last CERN measurement of a_μ.

The important advances for the BNL experiment are

(1) An increase in primary proton-beam intensity by a factor of 200 with the present alternating-gradient synchroton as compared to the CERN PS used in the CERN experiment.
(2) A superferric magnet storage ring that provides a magnetic field of excellent stability and homogeneity, and an NMR system capable of field measurement to 0.1 ppm.
(3) A modern Pb/scintillating fiber detector system, incorporating a Loran frequency standard, capable of measuring time intervals with a precision of 20 ps.
(4) Muon as well as pion injection into the storage ring. Muon injection increases the number of stored muons and reduces background in the ring.

A photograph of the storage ring is shown in Fig. 2.

Figure 2. The superferric C-magnet storage ring for the muon $g-2$ experiment at Brookhaven National Laboratory. The ring diameter is 14 m and the central field is 1.45 T. Twenty-four detectors are placed around the inside of the ring.

During 1997 a run for experimental checkout and initial data taking with pion injection was made. Figure 3 shows a time spectrum of decay positrons where the expected decay of the muons and the $g-2$ precession frequency are apparent. A total of 11.8 M e^+ with energy greater than 1.8 GeV were detected.

The value obtained for a_{μ^+} is

$$a_{\mu^+}(\text{expt}) = 1\,165\,925\,(15) \times 10^{-9} \quad (13\ \text{ppm}), \tag{6}$$

in which the dominant error is statistical (Carey et al., 1998). This value agrees with the CERN value of Eq. (4).

Theory of the Electron $g-2$

The current status of the theoretical calculation of a_e may be summarized as (Kinoshita, 1996)

$$a_e(\text{th}) = 0.5\left(\frac{\alpha}{\pi}\right) - 0.328\,478\,965\ldots \left(\frac{\alpha}{\pi}\right)^2 + 1.181\,241\,456\ldots \left(\frac{\alpha}{\pi}\right)^3$$

$$- 1.509\,8\,(384)\left(\frac{\alpha}{\pi}\right)^4 + 4.393\,(27) \times 10^{-12}. \tag{7}$$

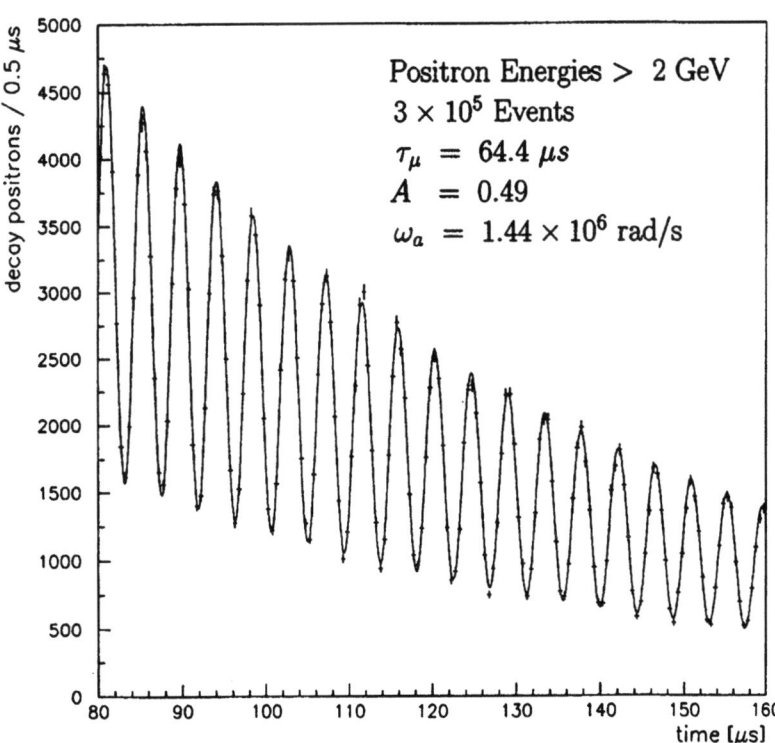

Figure 3. A positron time spectrum fit by Eq. (3). Statistical errors are indicated.

The analytic values of the α term and α^2 term have been known for a long time. The analytic value of the α^3 term has been obtained only recently (Laporta and Remeddi, 1996). It is in excellent agreement with the most recent numerical result, 1.181 259 (40), which was obtained shortly before the analytic result became available (see references in the review article of Kinoshita, 1996).

The α^4 term requires evaluation of 891 four-loop Feynman diagrams. This problem is so huge that analytic evaluation is prohibitively difficult even with the help of the fastest computers. Crude numerical evaluation of these integrals began around 1981 (for literature prior to 1990, see Kinoshita, 1990). It is only in the last few years that the calculation of this term began to move from a "qualitative" to a "quantitative" stage, thanks to the development of massively parallel computers. The coefficient of the α^4 term in Eq. (7) is the latest of the constantly improving values. Although it has a substantially higher precision than the best previous value, the old error estimate is used here pending completion of a more precise error analysis.

The last term of Eq. (7) consists of contributions from vacuum-polarization loops involving muons and taus and from hadronic and weak interactions. Evaluation of these quantities within the standard model gives

$$a_e(\mu \, \tau \text{v.p.}) = 2.721 \times 10^{-12},$$

$$a_e(\text{hadronic v.p.}) = 1.642(27) \times 10^{-12},$$

$$a_e(\text{weak}) = 0.030 \times 10^{-12}. \tag{8}$$

Although the non-QED effect on the electron anomaly a_e is very small, it must be included in the theory of the electron $g-2$ in view of the forthcoming experiments. These contributions are estimated assuming the validity of the standard model and indeed require that the theory be renormalizable and incorporates $\mu - e$ universality (Kinoshita, 1996).

To compare the theory of a_e with experiment, it is necessary to know the value of α. Currently the best measurements of α, with a relative uncertainty of less than 1×10^{-7}, are those based on the quantum Hall effect, the ac Josephson effect, the muonium hyperfine structure, and the de Broglie wavelength of a neutron beam (Kruger et al., 1995; Kinoshita, 1996; Jeffery et al., 1997; Liu et al., 1998):

$$\alpha^{-1}(\text{q. Hall}) = 137.036\,003\,7(33) \ [2.4 \times 10^{-8}],$$

$$\alpha^{-1}(\text{ac J}) = 137.035\,977\,0(77) \ [5.6 \times 10^{-8}],$$

$$\alpha^{-1}(\text{M}) = 137.035\,996\,3(80) \ [5.8 \times 10^{-8}],$$

$$\alpha^{-1}(h/m_n) = 137.036\,010\,62\,(503) \ [3.7 \times 10^{-8}], \tag{9}$$

where numbers within the brackets represent fractional precisions. Substituting these values in Eq. (7), one finds

$$a_e(\text{q. Hall}) = 1\,159\,652\,153.5\,(1.2)\,(28.0) \times 10^{-12},$$

$$a_e(\text{ac J}) = 1\,159\,652\,379.1\,(1.2)\,(65.3) \times 10^{-12},$$

$$a_e(\text{M}) = 1\,159\,652\,216.0\,(1.2)\,(67.8) \times 10^{-12},$$

$$a_e(h/m_n) = 1\,159\,652\,095.0\,(1.2)\,(42.7) \times 10^{-12}, \tag{10}$$

where the numbers enclosed in parentheses on each line are the uncertainty in the numerical integration result and in that of α used in the evaluation, respectively. The values in Eq. (10) are about $-1.3, +2.9, +0.14$, and -2.2 standard deviations away from the measured value in Eq. (1).

Theory of the muon $g-2$

The standard model prediction of a_μ consists of three parts (Kinoshita and Marciano, 1990):

(i) Pure QED contribution. If one uses $\alpha(a_e)$ from Eq. (17) one finds

$$a_\mu(\text{QED}) = 116\,584\,705.7\ (1.8) \times 10^{-11}. \tag{11}$$

Note that this does not agree with Eq. (4). This shows clearly that at least the effect of hadronic vacuum polarization must be taken into account. Furthermore, the goal of the new BNL muon $g-2$ experiment is to have the sensitivity to measure the weak-interaction effect. Hence, for comparison with experiment, a theory of the muon $g-2$ must deal with the strong and weak interactions as well as the electromagnetic interaction. The standard model satisfies this requirement.

(ii) Hadronic contribution, which itself consists of three parts:

(a) Hadronic vacuum-polarization contribution. This is obtained mainly from the measured hadron production cross section R in e^+e^- collisions. We quote here only the latest value that includes additional information obtained from the analysis of hadronic tau decay data (CLEO Collaboration, 1997; Davier and Höcker, 1998):

$$a_\mu(\text{had}_a) = 6\,951\ (75) \times 10^{-11}. \tag{12}$$

However, the CVC predictions for the τ-lepton branching ratios based on e^+e^- data are systematically lower than observed in τ decays (Eidelman and Ivanchenko, 1998). If the e^+e^- data alone are used to evaluate $a_\mu(\text{had}_a)$, the value of $a_\mu(\text{had}_a)$ decreases by about 60×10^{-11} and its error increases by about 50% (Alemany et al., 1998).

(b) Higher-order hadronic vacuum-polarization effect (Krause, 1997):

$$a_\mu(\text{had}_b) = -101\ (6) \times 10^{-11}. \tag{13}$$

(c) Hadronic light-by-light scattering contribution (Hayakawa and Kinoshita, 1998):

$$a_\mu(\text{had}_c) = -79.2\ (15.4) \times 10^{-11}. \tag{14}$$

(iii) Electroweak contribution of up to two-loop order (Kukhto et al., 1992; Czarnecki et al., 1995, 1996; Peris et al., 1995; Degrassi and Giudice, 1997):

$$a_\mu(\text{weak}) = 151\ (4) \times 10^{-11}. \tag{15}$$

Degrassi and Giudice (1997) employ an effective Lagrangian approach to derive the leading-logarithm two-loop electroweak contributions, which confirms the earlier explicit calculation of Kukhto et al. (1992), Czarnecki et al. (1995, 1996) and Peris et al. (1995). It estimates further the leading-logarithm three-loop electroweak contribution, which they find to be small. It also provides a useful parametrization for a certain class of new physics contribution to a_μ and estimates that the QED correction reduces such a new physics contribution by about 6%.

The sum of all these contributions, namely, the prediction of the standard model,

$$a_\mu(\text{th}) = 116\ 591\ 628\ (77) \times 10^{-11} \quad (0.66\ \text{ppm}), \tag{16}$$

is in good agreement with the measurements in Eqs. (4) and (6).

The uncertainty in Eq. (16) comes mainly from the hadronic vacuum-polarization contribution from Eq. (12). It must be improved by at least a factor of 2 before we can extract the full useful information from the new high-precision measurement of a_μ. Fortunately, this contribution is calculable from the measured value of R in e^+e^- collisions. Future measurements of R at VEPP-2M, VEPP-4M, DAΦNE, and BEPS, as well as analysis of the hadronic tau decay data, will reduce the uncertainty of this contribution to a satisfactory level (CLEO Collaboration, 1997; Davier and Höcker, 1998).

The contribution of the hadronic light-by-light scattering effect in Eq. (14) is smaller but is potentially a source of a serious problem because it is difficult to express it in terms of experimentally accessible observables. Evaluation of this term in QCD has not yet been attempted. The best approach available is to estimate it within the framework of chiral perturbation theory and the $1/N_c$ expansion (Bijnens et al., 1995, 1996; Hayakawa et al., 1995, 1996). Recently, however, an important part of this term was improved significantly (Hayakawa and Kinoshita, 1998) using the information obtained from new measurements of the $P\gamma\gamma^*$ form factors (Gronberg et al., 1998) where P stands for π^0, η, and η' mesons. The result of this work is included in Eq. (14).

Some Implications for Fundamental Physics

Because of the unusually high sensitivity of a precise experimental value of a_μ to physics beyond the standard model, theoretical predictions of the contributions to a_μ of speculative theories are of great interest. In general any new particles or interactions which couple to the muon or to the photon contribute to a_μ, whose value then provides a sum rule for physics. In comparison with experimental data from the higher-energy colliders (LEP II, Tevatron, LHC), an a_μ value with a precision of 0.35 ppm, as projected for the current BNL experiment, provides a comparable or greater sensitivity to a composite structure of the muon or W boson and also to the new particles in supersymmetric (SUSY) theories. For the muon a composite mass scale $\Lambda = 4$ TeV and for the W boson an anomalous magnetic moment $\kappa = 0.04$ would be observable. In supersymmetry theory a sparticle mass scale of about 130 GeV would be detected. Of course, any observation of physics beyond the standard model from a_μ would be indirect and would not by itself determine the process involved.

In the rest of this paper let us focus on a_e as a tool to test the validity of quantum mechanics. We note that the *intrinsic* uncertainty of theoretical values of a_e listed in Eq. (10) is already quite small, the overall uncertainty being dominated by those of α listed in Eq. (9). This means that we can obtain the most precise value of α from the theory and measurement of a_e. From the average of a_{e^-} and a_{e^+} in Eq. (1) and the theory one finds

$$\alpha^{-1}(a_e) = 137.035\ 999\ 58\ (14)\ (50) = 137.035\ 999\ 58\ (52)\ [3.8 \times 10^{-9}], \quad (17)$$

where the uncertainties on the first line are from the α^4 term and the measurement uncertainty of a_e given in Eq. (1), respectively.

Continuing theoretical work on a_e will reduce the theoretical uncertainty by a factor of 2 to 3 in the near future. If the experimental precision is improved by an order of magnitude, the precision of $\alpha(a_e)$ will exceed 1 part in 10^9 (Gabrielse and Tan, 1994; Mittleman et al., 1995). Besides these determinations of α, a powerful new approach using atom-beam interferometry of C_S is being developed (Weiss et al., 1993). Another new approach is based on single-electron tunneling which has achieved a precision of 15 ppb in counting the number of electrons (Kinoshiita, 1996). Spectroscopic measurements of the He atom fine structure in the 2^3P state is also a promising source of a very precise α value. The best values of α available at present are shown in Fig. 4.

It is fortunate that many independent ways are available for measuring α with high precision. This offers an opportunity to examine the theoretical bases of all these measurements on an equal footing. The precision of these measurements requires that the underlying theories be valid to the same extent. The theories are based on quantum mechanics extended to include relativistic effects, radiative corrections, and renormalization with respect to the electroweak and strong interactions.

Currently, such a theoretical basis is fully satisfied only by $\alpha(a_e)$ and by α determined by the muonium hyperfine structure and other atomic measurements. Although the principle of neutron de Broglie wavelength measurement looks very simple, it requires determination of the *free* neutron mass from nuclear physics, which can be fully justified only within the context of renormalizable quantum field theory. The α determined in

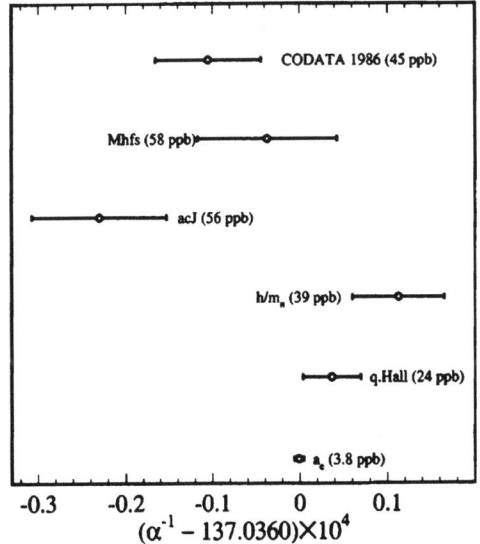

Figure 4. Values of the fine-structure constant determined by various means. The CODATA 1986 value of α (Kinoshita, 1996) is included for comparison.

condensed-matter physics has another unsettled problem. It is argued that, although the theories of the ac Josephson effect and the quantum Hall effect start from the condensed-matter physics Hamiltonian with its usual simplifying approximations, their predictions may in fact be valid to a higher degree than that of $\alpha(a_e)$ because they are derived from the gauge invariance and one-valuedness of the wave function and are not dependent on specific approximations adopted in condensed-matter physics. It is important to note, however, that this assertion has not yet been proven rigorously. In particular, the theory of condensed-matter physics in the present form is not renormalizable. The NRQED method of Caswell and Lepage (1986) may provide an approach for establishing a sounder basis for condensed-matter physics. (Note that NRQED is not a nonrelativistic approximation to QED. Rather, it is a systematic expansion of QED in the electron velocity and is fully equivalent to QED on resummation.)

Currently, the standard model is the simplest theory to represent extended quantum mechanics, and within its context all measurements of α that can be reduced to those of the charge form factor or the magnetic form factor at zero-momentum transfer must give the same answer.

An expectation that the α's obtained from the charge form factor may be affected by short-range interactions by $\sim (\alpha/\pi)^2 (m_e/m_\rho)^2 \simeq 2.4 \times 10^{-12}$, where m_ρ is the ρ meson mass, is not realized. This effect cannot be detected since it is absorbed by charge renormalization, which applies universally to all measurements of the charge form factor at threshold. The magnetic form factor, on the other hand, will be affected by the known short-range forces by $\sim 1.7 \times 10^{-12}$, which contributes about 1.5 ppb to $\alpha(a_e)$. But this effect is already taken into account in defining $\alpha(a_e)$. Thus $\alpha(a_e)$ determined from the magnetic form factor must have the same value as α's derived from the charge form factor. This equality is not affected by short-distance effects. This remark applies as well to α derived from the muonium hyperfine structure.

Effects beyond the standard model on $\alpha(a_e)$ can also be estimated using the measured a_μ insofar as the new interaction satisfies μ-e universality. Relative to known weak interactions, this effect will scale as $(m_W/m_X)^2$, where m_X is the mass scale of the new interaction. Such an effect will be too small to be significant at the present level of precision of $\alpha(a_e)$. Another useful constraint on a new interaction may come from a new measurement of the muon electric dipole moment.

The data shown in Fig. 4 cast some doubt on the likelihood that the α values determined by the different methods are the same. Improved precision in determining the α values, both experimental and theoretical, will provide a more sensitive test of the validity of (extended) quantum mechanics.

Acknowledgments

The work of V.W.H. was supported in part by the U.S. Department of Energy and that of T.K. by the U.S. National Science Foundation.

References

Alemahy, R., M. Davier, and A. Höcker, 1998, Eur. Phys. J. C **2**, 123.
Bailey, J., *et al.*, 1979, Nucl. Phys. B **150**, 1.
Bijnens, J., E. Pallante, and J. Prades, 1995, Phys. Rev. Lett. **75**, 1447; **75**, 3781 (E).

Bijnens, J., E. Pallante, and J. Prades, 1996, Nucl. Phys. B **474**, 379.
Brown, L.S., et al., 1985a, Phys. Rev. A **32**, 3204.
Brown, L.S., et al., 1985b, Phys. Rev. Lett. **55**, 44.
Carey, R.M., et al., 1998 (preprint).
Caswell, W.E., and G.P. Lepage, 1986, Phys. Lett. B **167** 437.
CLEO Collaboration, Report No. CLEO CONF97-31, EPS97, 368 (unpublished).
Czarnecki, A., B. Krause, and W.J. Marciano, 1995, Phys. Rev. D **52**, 2619.
Czarnecki, A., B. Krause, and W.J. Marciano, 1996, Phys. Rev. Lett. **76** 3267.
Davier, M., and A. Höcker, 1998, Phys. Lett. B **419**, 419.
Degrassi, G, and G.F. Giudice, 1997, CERN-TH/97-86, hep-ph/9803384.
Eidelman, S. I., and V. N. Ivanchenko, 1998 (preprint).
Farley, F.J.M., and E. Picasso, 1990, in *Quantum Electrodynamics*, edited by T. Kinoshita (World Scientific, Singapore), p. 479.
Gabrielse, G., and J. Tan, 1994, Adv. At. Mol., Opt. Phys. Suppl. **2**, 267.
Gronberg, J., et al., 1998, Phys. Rev. D **57**, 33.
Hayakawa, M., T. Kinoshita, and A.I. Sanda, 1995, Phys. Rev. Lett. **75**, 790.
Hayakawa, M., T. Kinoshita, and A.I. Sanda, 1996, Phys. Rev. D **54**, 3137.
Hayakawa, M., and T. Kinoshita, 1998, Phys. Rev. D **57**, 465.
Jeffery, A.-M., et al., 1997, IEEE Trans Instrum. Meas. **46**, 264.
Kinoshita, T., 1990, in *Quantum Electrodynamics*, edited by T. Kinoshita (World Scientific, Singapore), p. 218.
Kinoshita, T., 1996, Rep. Prog. Phys. **59**, 1459.
Kinoshita, T., and W.J. Marciano, 1990, in *Quantum Electrodynamics*, edited by T. Kinoshita (World Scientific, Singapore), p. 419.
Krause, B., 1997, Phys. Lett. B **390**, 392.
Krüger, E., W. Nistler, and W. Weirauch, 1995, Metrologia **32**, 117.
Kukhto, T.V., et al., 1992, Nucl. Phys. B **371**, 567.
Lapporta, S., and E. Remiddi, 1996, Phys. Lett. B **379**, 283.
Liu, W., et al., 1998, Phys. Rev. Lett. (in press).
Mittleman, R., et al., 1995, Phys. Rev. Lett. **75**, 2839.
Peris, S., M. Perrottet, and E. de Rafael, 1995, Phys. Lett. B **355**, 523.
Van Dyck, R.S., Jr., 1990, in *Quantum Electrodynamics*, edited by T. Kinoshita (World Scientific, Singapore), p. 322.
Van Dyck, R.S., Jr., P.B. Schwinberg, and H.G. Dehmelt, 1987, Phys. Rev. Lett. **59**, 26.
Weiss, D.S., B.C. Young, and S. Chu, 1993, Phys. Rev. Lett. **70**, 2706.

Neutrino Physics
L. Wolfenstein

The neutrino was invented by Wolfgang Pauli in 1930 to explain a problem concerning nuclear beta decay. The emitted electrons had a continuous energy spectrum, whereas they were expected to have one or possibly a few discrete energies corresponding to the energy of the nuclear transition. Pauli proposed that a neutral particle was emitted together with the electron so that the sum of the energies of this particle and the electron was essentially constant.

After the neutron was discovered, Enrico Fermi named this particle the neutrino (little neutron) and formulated in 1933 a theory of beta decay. This theory involved a new interaction in which the neutron changed into a proton, and an electron and an antineutrino. In the language of quantum field theory this was described by an interaction Hamiltonian,

$$H = G_F \bar{\psi}_p \gamma_\mu \psi_n \bar{\psi}_e \gamma_\mu \psi_\nu + \text{Hermitian conjugate,} \qquad (1.1)$$

where ψ_α is the field operator for particle α. In lowest order, this leads to the transitions

$$n \rightarrow p + e^- + \bar{\nu}, \qquad (1.2a)$$

$$p \rightarrow n + e^+ + \nu. \qquad (1.2b)$$

Fermi used to say that once he understood quantum electrodynamics (he gave a famous series of lectures on QED in Michigan published in 1932) he knew how to write the beta-decay interaction. Note that the e^- and $\bar{\nu}$ are created in the process of beta decay, just as an e^+e^- pair is created in QED.

This work of Pauli and Fermi truly marks the beginning of elementary-particle physics, which attempts to discover the fundamental particles in nature and their interactions. The neutrino was the first new particle proposed that was not a constituent of normal matter, and Fermi's weak interaction was the first proposed interaction that had no classical analog.

Much of the subsequent work on neutrinos follows directly from the work of Fermi. Using Eq. (1.1) Bethe and Peierls in 1934 calculated the interaction cross sections for neutrinos:

$$\nu + n \rightarrow e^- + p, \qquad (1.3a)$$

$$\bar{\nu} + p \rightarrow e^+ + n. \qquad (1.3b)$$

For a typical neutrino energy of 1 MeV the cross section was less than 10^{-43} cm^2; if a beam of 10^{10} neutrinos headed toward the earth all but one would emerge on the other side. Detection of neutrinos seemed impossible. Nevertheless, after World War II, Fred Reines and Clyde Cowan at Los Alamos took up the challenge. At first they thought to use the neutrinos from a nuclear bomb explosion, but then they decided to detect the large $\bar{\nu}$ flux from the fission products of a nuclear reactor. Their definitive experiments were done at the new Savannah River reactor in South Carolina and announced in 1956. In their experiment the neutron from reaction (1.3b) was detected via a capture gamma ray and the positron from the two annihilation gamma rays. In 1995, long after Cowan had died, Reines finally received the Nobel Prize.

Fermi's original paper discusses the effect of the neutrino mass on the shape of the electron spectrum at the high-energy end point. From this he deduced that the neutrino mass was much less than the electron mass. Many experiments over the past 50 years have studied the spectrum, particularly that of ^{3}H, which has an end-point energy of only 17 keV. These have produced an upper mass limit of about 10 eV on the neutrino from beta decay.

In 1947, Cecil Powell observed in emulsion the decay chain $\pi^+ \to \mu^+ \to e^+$. The positively charged π^+ (then called the pi meson and now the pion) stopped, and a single charged particle, the μ^+ (then called the mu meson and now the muon), emerged from the stopping point with a unique energy. In order to conserve momentum, it had to be assumed that an unseen neutral particle was emitted opposite to the μ^+. It was natural to assume this was a neutrino. But was it the same neutrino as the one in nuclear beta decay? With the development of high-energy accelerators, it was pointed out independently by Bruno Pontecorvo and Mel Schwartz that it would be possible to answer this question using beams of neutrinos from the decay of a focused beam of pions. The experiment was carried out at the Brookhaven Alternating Gradient Synchrotron (AGS) in 1962, and the neutrinos from pion decay were observed to interact, yielding muons but not electrons. These were a new type of neutrinos called ν_μ with interactions

$$\nu_\mu + n \to \mu^- + p, \tag{1.4a}$$

$$\bar{\nu}_\mu + p \to \mu^+ + n, \tag{1.4b}$$

while the old neutrinos interacting via reactions (1.3a) and (1.3b) must be labeled ν_e and $\bar{\nu}_e$. For the discovery of this second neutrino, Jack Steinberger, Leon Lederman, and Schwartz won the Nobel Prize in 1988. In the decay of the muon, the electron emerges with a continuous distribution of energy; it has been concluded that the decay is

$$\mu^- \to e^- + \bar{\nu}_e + \nu_\mu. \tag{1.5}$$

After the discovery of the τ lepton, a heavy partner to the electron and muon, by Martin Perl in 1975, it was concluded there is still a third type of neutrino, ν_τ, emitted in the decays of the τ. At the time of this writing, there has not been a direct detection of ν_τ. Direct limits on the masses of ν_μ and ν_τ are much weaker than that on ν_e; an upper limit on $m(\nu_\mu)$ is about 200 keV and on $m(\nu_\tau)$ about 20 MeV.

Fermi's original interaction, Eq. (1.1), involves two vector currents in analogy with the interaction between two electron currents in QED. In beta decay, this leads to a selection rule $\Delta \vec{J} = 0$ (when $\vec{J}$ is the total angular momentum) for the most allowed transitions. When it appeared the $\Delta \vec{J} = 1$ transitions could be equally probable, Gamow and Teller proposed adding an interaction between two axial-vector currents. The correct form of the interaction remained unclear until the discovery of parity violation in 1957. It was then proposed by Marshak and Sudarshan and by Feynman and Gell-Mann that H involved the coupling of V-A currents

$$H=(G_F/8^{1/2})\bar{\psi}_p\gamma_\mu(g_V-g_A\gamma_5)\psi_n\bar{\psi}_e\gamma_\mu(1-\gamma_5)\psi_{\nu_e}+\text{H.c.} \quad (1.6)$$

A variety of subsequent experiments verified this.

The factor $(1-\gamma_5)$ is in fact a helicity projection operator for the neutrino if it is massless. This means that ν_e is emitted only as a left-handed particle and $\bar{\nu}_e$ only as a right-handed particle. This leads to the possibility that only these two states exist and the neutrino is represented by a two-component spinor, a possibility originally suggested by Weyl, in contrast to the four-component Dirac spinor for the electron. An elegant experiment directly measuring the ν_e helicity was carried out by Goldhaber, Grodzins, and Sunyar in 1958. The electron capture on the europium nucleus

$$e^- + {}^{152}\text{Eu}(J=0) \rightarrow {}^{152}\text{Sm}^*(J=1) + \nu_e \quad (1.7)$$

was studied. Measurement of the J_z value of the final excited state of samarium from the circular polarization of the emitted gamma ray made it possible to deduce the helicity of ν_e.

A closer analogy to QED is to couple each of the V-A currents to a charged intermediate vector boson $W^\pm$ with the interaction then involving the exchange of such a boson. Fermi did not do this because the exchange of a massless particle like the photon leads to a long-range interaction, whereas Fermi's analysis of the beta spectrum required a short-range interaction which he chose as a delta function. With the demonstration by Yukawa that massive-particle exchange leads to a short-range interaction, it was natural to consider the possibility of a massive vector boson $W^\pm$ mediating weak interactions like beta decays. However, it was not until 1970 that a successful theory of this type was developed based on the idea of a spontaneously broken gauge theory, which proved to be renormalizable. A crucial feature of this theory was that in addition to the charged-current interactions mediated by $W^\pm$, there were neutral-current interactions mediated by a neutral vector boson Z.

The neutral-current reactions predicted were

$$\nu_x + n(\text{or } p) \rightarrow \nu_x + n(\text{or } p) \quad (1.8)$$

with the same cross section for ν_e, ν_μ, or ν_τ. In addition ν_μ and ν_τ were predicted to scatter elastically from electrons due to Z exchange, although with a smaller cross section than ν_e, which could scatter via both Z and W exchange. Although many experiments with ν_μ beams had been done over ten years, it was not until the theory was developed that neutral-current reactions were observed. The first observations were made at CERN in 1973 using the very large heavy-liquid bubble chamber called Gargamelle. Since the only signature of reaction (1.8) occurring on a nucleus was the hadronic recoil, it was necessary to demonstrate carefully that the observations were not due to a neutron background. It was these neutrino experiments that provided the first compelling evidence for what is now called the standard electroweak theory.

In the early 1980s the Z and $W^\pm$ bosons were produced in high-energy proton-antiproton collisions at CERN. This was followed by the design of electron-positron colliders at CERN and SLAC, tuned to the center-of-mass energy equal to the Z mass (93 GeV) so as to form the Z as a resonance. This allowed for many precision tests of the standard electroweak theory. By comparing the total width of the Z with the decay width observed into visible final states, it was possible to determine the "invisible decay width" into neutrinos. This agreed perfectly with the standard model, provided there were three—and only three—types of light neutrinos.

Neutrino Mass and Neutrino Oscillations

The elementary fermions in the standard model consist of six leptons and six quarks appearing in three generations or families:

Quarks

$$\begin{pmatrix} u \\ d \end{pmatrix} \begin{pmatrix} c \\ s \end{pmatrix} \begin{pmatrix} t \\ b \end{pmatrix}$$

Leptons

$$\begin{pmatrix} \nu_e \\ e \end{pmatrix} \begin{pmatrix} \nu_\mu \\ \mu \end{pmatrix} \begin{pmatrix} \nu_\tau \\ \tau \end{pmatrix}.$$

In the standard model these quarks and leptons are first introduced as left-handed Weyl particles arranged in doublets with respect to the gauge group SU(2) that interact with the $W^\pm$ in exactly the same way. There are also right-handed Weyl particles, except no right-handed neutrino is introduced. When the gauge symmetry is broken, there arises a mass term that couples the left-handed particles to the right-handed ones, yielding four-component massive Dirac particles. The neutrinos remain massless only because the quark-lepton symmetry is arbitrarily broken by leaving out the right-handed neutrinos.

If the right-handed neutrino were introduced, then it would be natural for neutrinos also to acquire mass, but there would be no explanation for their very small masses. A very interesting theoretical idea is that the quark-lepton symmetry is exact at some very high energy scale. One consequence of the breaking would be to give the right-handed neutrinos, which are neutral with respect to the standard model gauge symmetry, a very large mass M. As a result the normal mass term M_D that mixes left and right would produce light neutrinos with masses of the order M_D^2/M. This so-called "see-saw mechanism" is naturally implemented in the grand unified gauge theory (GUT) called SO(10), as originally suggested by Gell-Mann, Ramond, and Slansky.

In this see-saw picture, the light neutrinos are two-component Majorana particles rather than four-component Dirac. This means that the neutrino is its own antiparticle. Because the masses are so small, the motion is extremely relativistic, and ν appears as overwhelmingly left-handed while $\bar{\nu}$ is right-handed due to the $(1-\gamma_5)$ in the interaction that produced them. If one could detect the right-handed component of ν, it would be identical to the right-handed $\bar{\nu}$. However, this is so difficult that there seem to be no practical experiments to distinguish Dirac neutrinos from Majorana neutrinos.

An interesting possible consequence of neutrino mass is the phenomenon called neutrino oscillations, which may allow the detection of masses much smaller than could be detected directly kinematically. This was first proposed by Pontecorvo as an oscillation of ν into $\bar{\nu}$ in vacuum, in analogy with the $K^0-\bar{K}^0$ oscillation. However since ν and $\bar{\nu}$ have opposite helicities such an oscillation would violate angular momentum conservation. The interesting possibility suggested by Maki, Nakagawa, and Sakata in 1962 is the oscillation of one type of neutrino into another type. Such oscillations are expected if ν_e, ν_μ, and ν_τ are each coherent mixtures of the three neutrino mass eigenstates ν_1, ν_2, ν_3; such mixing is in fact expected in every theory of neutrino mass. This is closely analogous to the Cabibbo mixing of the quarks.

The basic idea can be explained considering only two neutrino types, say ν_e and ν_μ,

$$\nu_e = \cos\theta \nu_1 + \sin\theta \nu_2,$$

$$\nu_\mu = -\sin\theta \nu_1 + \cos\theta \nu_2, \qquad (2.1)$$

where ν_1, ν_2 are the mass eigenstates and θ is the mixing angle. As a function of time or distance, the relative phase of ν_1 and ν_2 changes because of the mass difference so that a neutrino originating as ν_e has a nonzero probability later of being detected as ν_μ. If the neutrino originating as ν_e is labeled $\nu_e(t)$, the oscillation probability is easily calculated to be

$$|\langle \nu_\mu | \nu_e(t) \rangle|^2 = \sin^2(2\theta) \sin^2\left(\frac{\pi x}{l_v}\right), \qquad (2.2a)$$

$$l_v = 4\pi p_\nu / (m_2^2 - m_1^2) = \frac{4\pi p_\nu}{\Delta m^2}, \qquad (2.2b)$$

where p_ν is the momentum of the neutrino. The formula is the same as that of a spin precessing in a magnetic field, and we may think of neutrino oscillation as a precession in generation space.

Many experiments have been carried out with accelerators and reactors searching for, but not finding, neutrino oscillations. With ν_μ beams at accelerators, one looks for the appearance of ν_e, although this is eventually limited in sensitivity by the presence of ν_e in the ν_μ beam. These experiments have ruled out values of $\sin^2 2\theta$ above about 10^{-2} but only for $\Delta m^2 \gtrsim 1$ eV^2. Reactor neutrinos can be used to explore lower values of Δm^2 because of their lower energy, but the only signature of oscillation is the disappearance of the ν_e, so that very small mixing angles cannot be explored. An interesting source of neutrinos for oscillation experiments is stopping pions. While the π^- are captured by nuclei before decaying, the π^+ undergo the decay chain

$$\pi^+ \to \mu^+ + \nu_\mu; \mu^+ \to e^+ + \nu_e + \bar{\nu}_\mu.$$

Thus a low-energy neutrino beam is produced in which there should be no $\bar{\nu}_e$. Using such a beam, researchers at the Los Alamos meson factory LAMPF have found events that look like $\bar{\nu}_e$, which they attribute to $\bar{\nu}_\mu \to \bar{\nu}_e$ oscillations with $\Delta m^2 \approx 1$ eV^2 but very small mixing. It remains to be seen whether this oscillation will be confirmed by future experiments.

For small values of Δm^2, it follows from Eq. (2.2b) that the oscillation length is large. Large oscillation lengths can be explored with neutrinos that arise from the decays of pions and muons resulting from the interactions of cosmic rays with the atmosphere. Since the neutrinos can penetrate the entire earth, upward-going neutrinos have traveled a distance of 13 000 km, in contrast to downward-going ones that have gone 10 to 30 km. Several experiments have given indications of oscillations. The most compelling evidence today for neutrino mass is a factor-of-2 suppression of upward-going ν_μ relative to downward-going ones observed recently in the Superkamiokande water Cerenkov detector. This result is interpreted as an oscillation of ν_μ to ν_τ, with a value of Δm^2 of order 2.10^{-3} eV^2 and a large mixing angle. In this detector, ν_μ and ν_e are identified from the Cerenkov cone produced by the muons and electrons resulting from neutrino interactions [Eqs. (1.3) and (1.4)] in 50 kilotons of water, but ν_τ cannot be seen. There are proposals at Fermilab and CERN to explore such small values of Δm^2 and identify ν_τ by sending ν_μ beams from accelerators to underground detectors 750 km away.

The big bang theory predicts that the universe is filled with very-low-energy neutrinos. Like the photons of the microwave background radiation, these are left over from an early time in the history of the universe when they were in thermal equilibrium. Since the calculated density, about 100/cm^3 for each type of neutrino, is about 10^9 times that of nucleons, these neutrinos would dominate the energy density of the universe if any type of neutrino had a mass greater than a few electron volts. There is strong evidence that most of the matter of the universe is nonluminous, so-called dark matter. Massive neutrinos are an important possibility for at least some of this dark matter.

No one has conceived of a practical way to detect the background neutrinos because of their extremely low energy. However, the dark matter problem has raised interest in the possibility that ν_τ, which might be the heaviest of the neutrinos, has a mass of a few eV. To search for this, experiments are being carried out at CERN (called CHORUS and NOMAD) on $\nu_\mu \rightarrow \nu_\tau$ oscillations. For values of Δm^2 greater than a few eV^2, there is a sensitivity to very small oscillation probabilities. The key is to look for the appearance of ν_τ, since there are practically no ν_τ in the beam and the τ decay has a unique signature. As of this time, no oscillations have been seen. This intersection between particle physics experiments and cosmology is one of the exciting developments of recent years.

Astrophysical Neutrinos

In the 1930s the work of von Weisszacker, Bethe, and Critchfield and others detailed the nuclear reactions that could provide the energy inside stars. This energy is slowly transported, mainly by radiative transfer, to the stellar surface. However, a small portion of the energy, about 3% in the case of the sun, is calculated to be emitted in the form of neutrinos that can come directly from the stellar interior to the earth.

The possibility of detecting these neutrinos from the sun was taken seriously by Raymond Davis, a radiochemist at Brookhaven National Laboratory, inspired by detailed calculations by John Bahcall. In 1967 Davis installed 600 tons of C$_2$Cl$_4$ in the Homestake gold mine in South Dakota. The goal was to collect and detect the argon atoms from a reaction first proposed by Pontecorvo in 1946:

$$\nu_e + {}^{35}\text{Cl} \rightarrow e^- + {}^{35}\text{A}. \qquad (3.1)$$

Only about one atom a day was expected to be produced from the calculated ν_e flux. But Raymond Davis is a very patient man. His successful detection of solar neutrinos confirmed our general picture of the nuclear reactions that power the stars. It is one of the great success stories of the last 50 years.

The next detection of solar neutrinos was made in the years 1988 through 1995 in Japan using a large water tank, surrounded by phototubes (Kamiokande). This detector was originally built to search for the proton decay predicted by grand unified theories. Recoil electrons from neutrino-electron scattering were detected via their Cerenkov light. A crucial feature of this experiment was that the recoil electron direction was approximately the direction of the neutrino so that there was direct evidence that the neutrinos observed were coming from the sun.

When stars much larger than the sun burn up their nuclear fuel, the central core collapses, leading to the spectacular event known as a type-II supernova. In fact, less than one percent of the energy of collapse emerges in the form of photons. The collapse leads to a region of extremely high density and temperature where electroweak interactions produce high-energy gamma rays, electron-positron pairs, and neutrino-antineutrino pairs roughly in thermal equilibrium. The density is so high that even the neutrinos cannot directly emerge. However, neutrinos, since they interact very weakly, have a much better chance of getting out than anything else. So it was calculated that nearly all the energy of collapse should be emitted in the form of neutrinos, 10^{53} ergs within a period of about 10 seconds.

In February of 1987, the first supernova visible to the naked eye in over 300 years was seen in the Southern Hemisphere. After traveling for more than 150 000 years, neutrinos from this supernova arrived just a couple of years after Kamiokande in Japan and another water Cerenkov detector, IMB in the United States, came on line, within a ten-second interval shortly before the supernova became visible. Eleven neutrinos were observed in Kamiokande and 8 in IMB. Of some 10^{58} neutrinos emitted 19 had been detected. It was reasonable to assume that the neutrinos observed were all $\bar{\nu}_e$ since the detectors were primarily sensitive to reaction (1.3b). However, it is expected that neutrinos of all types should carry away the energy. It is hoped that larger detectors and ones capable of detecting all types of neutrinos will be active when the next supernova erupts nearby.

The study of solar neutrinos led to an intriguing problem: the deficiency of the observed flux compared to theoretical calculations. The ratio of observation to theory was found by Davis to be about one-third and then by the Kamiokande group to be about one-half. There are three main sources for solar neutrinos. The starting point of the process of the solar energy cycles in the sun is the weak reaction

$$p+p \rightarrow d+e^{+}+\nu_e, \qquad (3.2a)$$

yielding a continuous spectrum of neutrinos with an end point of 420 keV. The deuterons quickly combine with a proton to form ^{3}He. This ^{3}He can combine with another one to form ^{4}He (plus two leftover protons), thus completing the conversion of hydrogen into helium. It is calculated that about 10% of the time the ^{3}He interacts with ^{4}He, which is left over from the big bang and is, after hydrogen, the main constituent of the primordial sun, to form ^{7}Be. In the hot plasma of the solar interior, ^{7}Be captures an electron,

$$e^{-}+{}^7\text{Be} \rightarrow \nu_e + {}^7\text{Li}, \qquad (3.2b)$$

yielding a line spectrum of ν_e, primarily at 790 keV. With a small probability (calculated as about one time in a thousand) the ^{7}Be interacts with a proton to produce ^{8}Be, which decays as

$$^8\text{B} \rightarrow {}^8\text{Be}+e^{+}+\nu_e, \qquad (3.2c)$$

yielding a continuous ν_e spectrum with an end point of about 14 MeV.

The Kamiokande experiment is sensitive only to the rare ^{8}B neutrinos. The calculated rate of these has the largest theoretical error, both because of the uncertainty in the nuclear cross section leading to ^{8}B and because of extreme sensitivity to the temperature. The Davis experiment is also sensitive in addition to ^{7}Be neutrinos; comparing the two experiments independently of any calculated ^{8}B neutrino flux suggests a large deficiency of ^{7}Be neutrinos.

Over the last decade two radiochemical experiments based on the reaction

$$\nu_e + {}^{71}\text{Ga} \to e^- + {}^{71}\text{Ge}$$

have been operating aimed at detecting the most abundant neutrinos, the low-energy pp neutrinos from reaction (3.2a). These experiments, Gallex in the Gran Sasso mine in Italy and SAGE in the Baksan in Russia, give similar results corresponding to a measured rate less than 60% of that calculated. What seems particularly significant is that the detected rate corresponds to that expected from the pp neutrinos alone, leaving out a sizable flux expected from ^{7}Be neutrinos. An exciting possibility is that a significant fraction of ν_e have oscillated into other forms of neutrinos that could not be detected.

The possibility of neutrino oscillations for solar neutrinos provides a sensitivity to values of Δm^2 much smaller than in terrestrial experiments because of the large distances involved [see Eq. (2.2b)]. An interesting possibility is that the oscillation takes place as the neutrino passes from the center of the sun to the surface. In this case the calculation of the oscillation probability must be modified to include the effect of the material. The point, first made by Wolfenstein in 1978, is that the index of refraction of ν_e is different from that of ν_μ due to scattering from electrons and that the phase associated with refraction must be included in the quantum-mechanical oscillation equations. Applying this idea to the case of varying density in the sun, Mikhayev and Smirnov showed that a large suppression of the ν_e signal was possible for very small values of θ. They called this the resonant amplification of neutrino oscillations, and it is now referred to as the MSW effect. All the present solar neutrino data are well explained by this effect for a value of Δm^2 around 10^{-5} eV^2 and $\sin^2 2\theta \approx 10^{-2}$. There are also explanations with large mixing angles.

New types of solar neutrino detectors are now coming on line or being planned. The Sudbury Neutrino Observatory (SNO), located in a deep nickel mine in Canada, is a water Cerenkov detector like Kamiokande but using heavy water so as to detect the reaction

$$\nu_e + d \to e^- + p + p,$$

which has a much larger cross section than neutrino-electron scattering. The goal is to detect as well, the neutral-current reaction

$$\nu_x + d \to \nu_x + n + p,$$

which is equally sensitive to all three types of neutrinos. If the flux of all types of neutrinos is greater than that of ν_e, then oscillations must have taken place.

The Borexino detector planned for the Gran Sasso is designed to detect ^{7}Be neutrinos from neutrino-electron scattering in a scintillator. Extreme purity of the materials (as low as one part in 10^{16} of uranium or thorium) is needed to cut the background when looking for neutrinos with such low energy. Still farther in the future are novel detectors being designed to detect the pp neutrinos in real time.

Solar and supernova neutrinos are the only astrophysical neutrinos detected so far. High-energy gamma rays have been observed from a variety of sources, particularly active galactic nuclei (AGNs). It is believed that some of these arise from the decay of neutral pions produced in high-energy collisions. In this case there should also be high-energy neutrinos from charged-pion decays. The neutrinos will emerge from much deeper inside the sources than the gamma rays and could reveal new information. To detect such neutrinos, extremely large volumes of water instrumented with photodetectors will be needed. Projects now in the prototype stage involve instrumenting the ice at the South Pole (AMANDA) or the Mediterranean Sea off Greece (NESTOR). Full scale neutrino astronomy remains as a challenge for the next century.

For further references see Mohapatra and Pal (1991), Sutton (1992), Raffaelt (1996), and Los Alamos Science No. 25 (1997). Winter (1991) includes reprints of many of the original papers mentioned.

Acknowledgment

This work was supported by the U.S. Department of Energy under grant number DE-FG02-91-ER-40682.

References

Los Alamos Science No. 25, 1997.
Mohapatra, R. N., and P. B. Pal, 1991, *Massive Neutrinos in Physics and Astrophysics* (World Scientific, Singapore).
Raffaelt, G. G., 1996, *Stars as Laboratories for Fundamental Physics* (University of Chicago, Chicago).
Sutton, C., 1992, *Spaceship Neutrino* (Cambridge University, Cambridge, England). This is a nontechnical book for the general public.
Winter, K., 1991, *Neutrino Physics* (Cambridge University, Cambridge, England).

Astrophysics

Cosmology at the Millennium
Michael S. Turner and J. Anthony Tyson

One hundred years ago, we did not know how stars shine and we had only a rudimentary understanding of one galaxy, our own Milky Way. Our knowledge of the Universe—in both space and time—was scant: Most of it was as invisible as the world of the elementary particles.

Today, we know that we live in an evolving universe filled with billions of galaxies within our sphere of observation, and we have recently identified the epoch when galaxies first appeared. Cosmic structures from galaxies increasing in size to the Universe itself are held together by invisible matter whose presence is only known through its gravitational effects (the so-called dark matter).

The optical light we receive from the most distant galaxies takes us back to within a few billion years of the beginning. The microwave echo of the big bang discovered by Penzias and Wilson in 1964 is a snapshot of the Universe at 300 000 years, long before galaxies formed. Finally, the light elements D, ^{3}He, ^{4}He, and ^{7}Li were created by nuclear reactions even earlier and are relics of the first seconds. (The rest of the elements in the periodic table were created in stars and stellar explosions billions of years later.)

Crucial to the development of our understanding of the cosmos were advances in physics—atomic, quantum, nuclear, gravitational, and elementary-particle physics. The hot big-bang model, based upon Einstein's theory of General Relativity and supplemented by the aforementioned microphysics, provides our quantitative understanding of the evolution of the Universe from a fraction a second after the beginning to the present, some 13 billion years later. It is so successful that for more than a decade it has been called the standard cosmology (see, e.g., Weinberg, 1972).

Beyond our current understanding, we are striving to answer fundamental questions and test bold ideas based on the connections between the inner space of the elementary particles and the deep outer space of cosmology. Is the ubiquitous dark matter that holds the Universe together composed of slowly moving elementary particles (called cold dark matter) left over from the earliest fiery moments? Does all the structure seen in the Universe today — from galaxies to superclusters and great walls—originate from quantum mechanical fluctuations occurring during a very early burst of expansion driven by vacuum energy (called "inflation")? Is the Universe spatially flat as predicted by inflation? Does the absence of antimatter and the tiny ratio of matter to radiation (around one part in 10^{10}) involve forces operating in the early Universe that violate baryon-number conservation and matter-antimatter symmetry? Is inflation the dynamite of the big bang, and if not, what is? Is the expansion of the Universe today accelerating rather than slowing, due to the presence of vacuum energy or something even more mysterious?

Our ability to study the Universe has improved equally dramatically. One hundred years ago our window on the cosmos consisted of visible images taken on photographic plates using telescopes of aperture one meter or smaller. Today, arrays of charge-coupled devices have replaced photographic plates, improving photon collection efficiency one-hundred fold, and telescope apertures have grown tenfold. Together, they have increased photon collection by a factor of 10^4. Wavelength coverage has widened by a larger factor. We now view the Universe with eyes that are sensitive from radio waves of length 100 cm to gamma rays of energy up to 10^{12} eV, from neutrinos to cosmic-ray particles, and perhaps someday via dark-matter particles and gravitational radiation.

At all wavelengths advances in materials and device physics have spawned a new generation of low-noise, high-sensitivity detectors. Our new eyes have opened new windows, allowing us to see the Universe 300 000 years after the beginning, to detect the presence of black holes, neutron stars, and extra-solar planets, and to watch the birth of stars and galaxies. One hundred years ago the field of spectroscopy was in its infancy; today, spectra of stars and galaxies far too faint even to be seen then, are revealing the chemical composition and underlying physics of these objects. The advent of computers and their dramatic evolution in power (quadrupling every three years since the 1970s) has made it possible to handle the data flow from our new instruments as well as to analyze and to simulate the Universe.

This multitude of observations over the past decades has permitted cross checks of our basic model of the Universe past as a denser, hotter environment in which structure forms via gravitational instability driven by dark matter. We stand on the firm foundation of the standard big-bang model, with compelling ideas motivated by observations and fundamental physics, as a flood of new observations looms. This is a very exciting time to be a cosmologist. Our late colleague David N. Schramm more than once proclaimed the beginning of a golden age, and we are inclined to agree with him.

Foundations

There is now a substantial body of observations that support directly and indirectly the relativistic hot big-bang model for the expanding Universe. Equally important, there are no data that are inconsistent. This is no mean feat: The observations are sufficiently constraining that there is no alternative to the hot big bang consistent with all the data at hand. Reports in the popular press of the death of the big bang usually confuse detailed aspects of the theory that are still in a state of flux, such as models of dark matter or scenarios for large-scale structure formation, with the basic framework itself. There are indeed many open problems in cosmology, such as the age, size, and curvature of the Universe; the nature of the dark matter, and details of how large-scale structures form and how galaxies evolve—these issues are being addressed by a number of current observations. But the evidence that our Universe expanded from a dense hot phase roughly 13 billion years ago is now incontrovertible (see, e.g., Peebles et al., 1991).

When studied with modern optical telescopes the sky is dominated by distant faint blue galaxies. To 30th magnitude per square arcsecond surface brightness (4×10^{-18} erg sec^{-1} cm^{-2} arcsec^{-2} in 100 nm bandwidth at 450 nm wavelength, or about five photons per minute per galaxy collected with a 4-meter mirror) there are about 50 billion galaxies over the sky. On scales less than around 100 Mpc galaxies are not distributed uniformly, but rather cluster in a hierarchical fashion. The correlation length for bright galaxies is

$8h^{-1}$ Mpc (at this distance from a galaxy the probability of finding another galaxy is twice the average). (1 Mpc = 3.09×10^{24} cm ≃ 3 million light years, and $h = H_0/100$ km s^{-1} Mpc^{-1} is the dimensionless Hubble constant.)

About 10 percent of galaxies are found in clusters of galaxies, the largest of which contain thousands of galaxies. Like galaxies, clusters are gravitationally bound and no longer expanding. Fritz Zwicky was among the first to study clusters, and George Abell created the first systematic catalogue of clusters of galaxies in 1958; since then, some four thousand clusters have been identified (most discovered by optical images, but a significant number by the x rays emitted by the hot intracluster gas). Larger entities called superclusters are just now ceasing to expand and consist of several clusters. Our own supercluster was first identified in 1937 by Holmberg, and characterized by de Vaucouleurs in 1953. Other features in the distribution of galaxies in three-dimensional space have also been identified: regions devoid of bright galaxies of size roughly $30h^{-1}$ Mpc (simply called voids) and great walls of galaxies that stretch across a substantial fraction of the sky and appear to be separated by about $100h^{-1}$ Mpc. Figure 1 is a three panel summary of our knowledge of the large-scale structure of the Universe.

In the late 1920s Hubble established that the spectra of galaxies at greater distances were systematically shifted to longer wavelengths. The change in wavelength of a spectral line is expressed as the "redshift" of the observed feature,

$$1 + z \equiv \lambda_{observed}/\lambda_{emitted}. \tag{1}$$

Interpreting the redshift as a Doppler velocity, Hubble's relationship can be written

$$z \simeq H_0 d/c \quad (\text{for } z \ll 1). \tag{2}$$

The factor H_0, now called the Hubble constant, is the expansion rate at the present epoch. Hubble's measurements of H_0 began at 550 km sec^{-1} Mpc^{-1}; a number of systematic errors were identified, and by the 1960s H_0 had dropped to 100 km s^{-1} Mpc^{-1}. Over the last two decades controversy surrounded H_0, with measurements clustered around 50 km s^{-1} Mpc^{-1} and 90 km s^{-1} Mpc^{-1}. In the past two years or so, much progress has been made because of the calibration of standard candles by the Hubble Space Telescope (see, e.g., Filippenko and Riess et al., 1998; Madore et al., 1998), and there is now a general consensus that $H_0 = (67 \pm 10)$ km s^{-1} Mpc^{-1} (where ± 10 km s^{-1} Mpc^{-1} includes both statistical and systematic error; see Fig. 2). The inverse of the Hubble constant—the Hubble time—sets a timescale for the age of the Universe: $H_0^{-1} = (15 \pm 2)$ Gyr.

By now, through observations of a variety of phenomena from optical galaxies to radio galaxies, the cosmological interpretation of redshift is very well established. Two recent interesting observations provide further evidence: numerous examples of high-redshift objects being gravitationally lensed by low redshift objects near the line of sight; and the fading of supernovae of type Ia, whose light curves are powered by the radioactive decay of Ni^{56}, and at high redshift exhibit time dilation by the predicted factor of $1 + z$ (Leibundgut et al., 1996).

An important consistency test of the standard cosmology is the congruence of the Hubble time with other independent determinations of the age of the Universe. (The product of the Hubble constant and the time back to the big bang, $H_0 t_0$, is expected to be between 2/3 and 1, depending upon the density of matter in the Universe; see Fig. 3.) Since the discovery of the expansion, there have been occasions when the product $H_0 t_0$ far exceeded unity, indicating an inconsistency. Both H_0 and t_0 measurements have been plagued by systematic errors. Slowly, the situation has improved, and at present there is consistency within the uncertainties. Chaboyer *et al.* (1998) date the oldest globular stars at 11.5 ± 1.3 Gyr; to obtain an estimate of the age of the Universe, another $1-2$ Gyr must be added to account for the time to the formation of the oldest globular clusters. Age estimates based upon abundance ratios of radioactive isotopes produced in stellar explosions, while dependent upon the time history of heavy-element nucleosynthesis in the Galaxy, provide a lower limit to the age of the Galaxy of 10 Gyr (Cowan *et al.*, 1991). Likewise, the age of the Galactic disk based upon the cooling of white dwarfs, >9.5 Gyr, is also consistent with the globular cluster age (Oswalt *et al.*, 1996). Recent type Ia

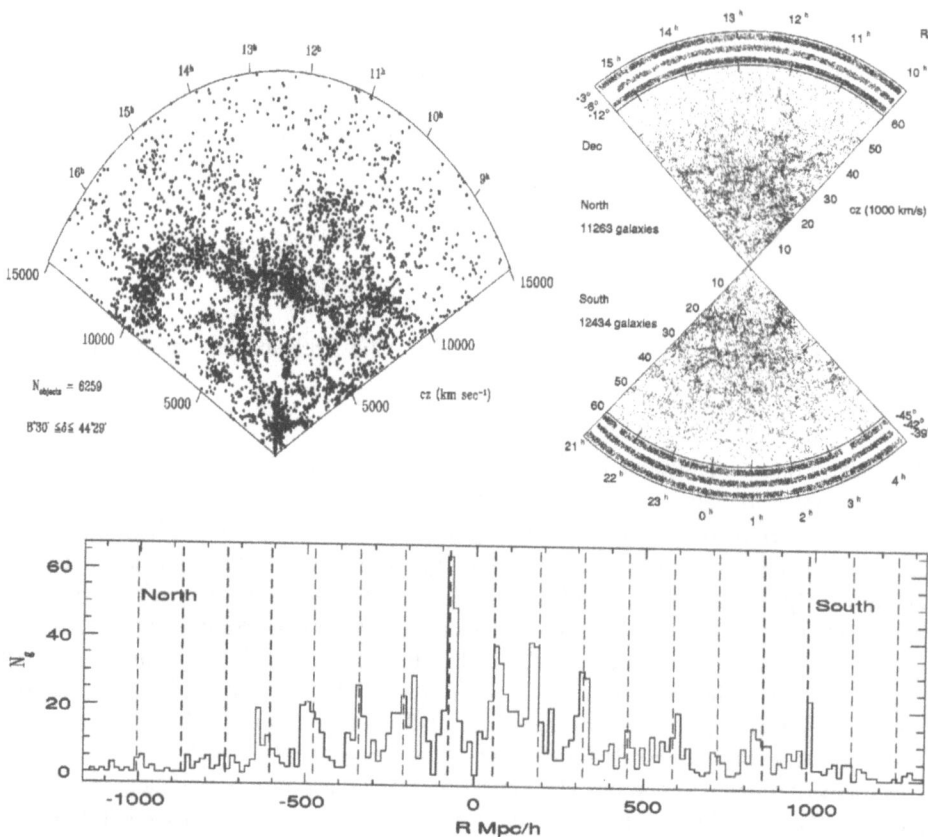

Figure 1. Large-scale structure in the Universe as traced by bright galaxies: (upper left) The Great Wall, identified by Geller and Huchra (1989) (courtesy of E. Falco). This coherent object stretches across most of the sky; walls of galaxies are the largest known structures (see Oort, 1983). We are at the apex of the wedge, galaxies are placed at their "Hubble distances," $d = H_0^{-1} zc$; note too, the regions devoid of galaxies ("voids"). (Upper right) Pie-diagram from the Las Campanas Redshift Survey (Shectman et al., 1996). Note the structure on smaller length scales including voids and walls, which on larger scales dissolves into homogeneity. (Lower) Redshift-histogram from deep, pencil-beam surveys (Willmer, et al., 1994; see also Broadhurst et al., 1990; courtesy of T. Broadhurst). Each pencil beam covers only a square degree on the sky. The narrow width of the beam "distorts" the view of the Universe, making it appear more inhomogeneous. The large spikes spaced by around $100h^{-1}$ Mpc are believed to be great walls.

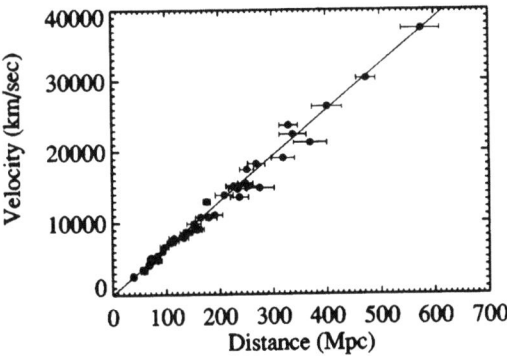

Figure 2. Hubble diagram based upon distances to supernovae of type Ia (SNe1a). Note the linearity; the slope, or Hubble constant, $H_0 = 64$ km s^{-1} Npc^{-1} (courtesy of A. Riess; see Filippenko and Riess, 1998).

supernova data yield an expansion age for the Universe of 14.0 ± 1.5 Gyr, including an estimate of systematic errors (Riess *et al.*, 1998).

Within the uncertainties, it is still possible that $H_0 t_0$ is slightly greater than one. This could either indicate a fundamental inconsistency or the presence of a cosmological constant (or something similar). A cosmological constant can lead to accelerated expansion and $H_0 t_0 > 1$. Recent measurements of the deceleration of the Universe, based upon the distances of high-redshift supernovae of type Ia (SNe1a), in fact show evidence for accelerated expansion; we will return to these interesting measurements later.

Another observational pillar of the big bang is the 2.73 K cosmic microwave background radiation (CMB) (see Wilkinson, 1999). The far infrared absolute spectrophotom-

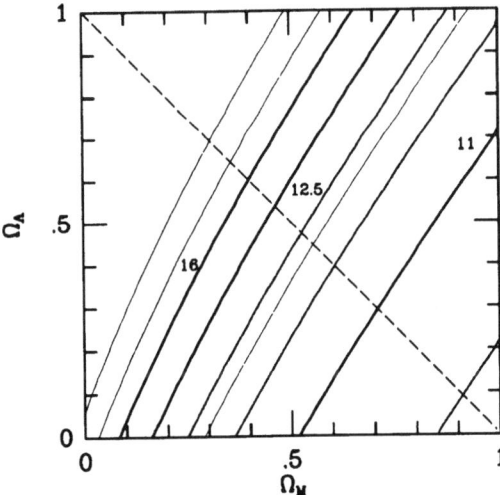

Figure 3. Contours of constant time back to the big bang in the $\Omega_M - \Omega_\Lambda$ plane. The three bold solid lines are for $h = 0.65$; the light solid lines are for $h = 0.7$; and the dotted lines are for $h = 0.6$. The diagonal line corresponds to a flat universe. Note, for $h \sim 0.65$ and $t_0 \sim 13$ Gyr a flat universe is possible only if $\Omega_\Lambda \sim 0.6$; $\Omega_M = 1$ is only possible if $t_0 \sim 10$ Gyr and $h \sim 0.6$.

eter (FIRAS) on the Cosmic Background Explorer (COBE) satellite has probed the CMB to extraordinary precision (Mather *et al.*, 1990). The observed CMB spectrum is exquisitely Planckian: any deviations are smaller than 300 parts per million (Fixsen *et al.*, 1996), and the temperature is 2.7277 ± 0.002 K (see Fig. 4). The only viable explanation for such perfect black-body radiation is the hot, dense conditions that are predicted to exist at early times in the hot big-bang model. The CMB photons last scattered (with free electrons) when the Universe had cooled to a temperature of around 3000 K (around 300 000 years after the big bang), and ions and electrons combined to form neutral atoms. Since then the temperature decreased as $1 + z$, with the expansion preserving the black body spectrum. The cosmological redshifting of the CMB temperature was confirmed by a measurement of a temperature of 7.4 ± 0.8 K at redshift 1.776 (Songaila *et al.*, 1994) and of 7.9 ± 1 K at redshift 1.973 (Ge *et al.*, 1997), based upon the population of hyperfine states in neutral carbon atoms bathed by the CMB.

The CMB is a snapshot of the Universe at 300 000 yrs. From the time of its discovery, its uniformity across the sky (isotropy) was scrutinized. The first anisotropy discovered was dipolar with an amplitude of about 3 mK, whose simplest interpretation is a velocity with respect to the cosmic rest frame. The FIRAS instrument on COBE has refined this measurement to high precision: the barycenter of the solar system moves at a velocity of 370 ± 0.5 km s^{-1}. Taking into account our motion around the center of the Galaxy, this translates to a motion of 620 ± 20 km s^{-1} for our local group of galaxies. After almost thirty years of searching, firm evidence for primary anisotropy in the CMB, at the level of 30 μK (or $\delta T/T \simeq 10^{-5}$) on angular scales of 10° was found by the differential microwave radiometer (DMR) on COBE (see Fig. 5). The importance of this discovery was twofold. First, this is direct evidence that the Universe at early times was extremely smooth since density variations manifest themselves as temperature variations of the same magnitude. Second, the implied variations in the density were of the correct size to account for the structure that exists in the Universe today: According to the standard

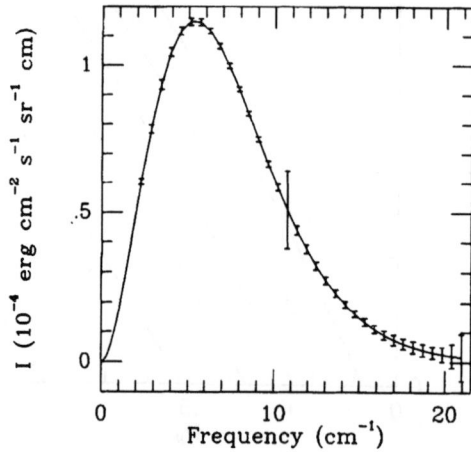

Figure 4. Spectrum of the Cosmic Microwave Background Radiation as measured by the FIRAS instrument on COBE and a black-body curve for $T=2.7277$ K. Note, the error flags have been enlarged by a factor of 400. Any distortions from the Planck curve are less than 0.005% (see Fixsen et al., 1996).

cosmology the structure seen today grew from small density inhomogeneities ($\delta\rho/\rho \sim 10^{-5}$) amplified by the attractive action of gravity over the past 13 Gyr.

The final current observational pillar of the standard cosmology is big-bang nucleosynthesis (BBN). When the Universe was seconds old and the temperature was around 1 MeV a sequence of nuclear reactions led to the production of the light elements D, ^{3}He, ^{4}He, and ^{7}Li. In the 1940s and early 1950s, Gamow and his collaborators suggested that nuclear reactions in the early Universe could account for the entire periodic table; as it turns out Coulomb barriers and the lack of stable nuclei with mass 5 and 8 prevent further nucleosynthesis. In any case, BBN is a powerful and very early test of the standard cosmology: the abundance pattern of the light elements predicted by BBN (see Fig. 6) is consistent with that seen in the most primitive samples of the cosmos. The abundance of deuterium is very sensitive to the density of baryons, and recent measurements of the deuterium abundance in clouds of hydrogen at high redshift (Burles and Tytler, 1998a,1998b) have pinned down the baryon density to a precision of 10%.

As Schramm emphasized, BBN is also a powerful probe of fundamental physics. In 1977 he and his colleagues used BBN to place a limit to the number of neutrino species (Steigman, Schramm, and Gunn, 1977), $N_\nu < 7$, which, at the time, was very poorly constrained by laboratory experiments, $N_\nu \lesssim$ a few thousand. The limit is based upon the fact that the big-bang ^{4}He yield increases with N_ν; see Fig. 7. In 1989, experiments done at $e^\pm$ colliders at CERN and SLAC determined that N_ν was equal to three, confirming the cosmological bound, which then stood at $N_\nu < 4$. Schramm used the BBN limit on N_ν to pique the interest of many particle physicists in cosmology, both as a heavenly laboratory

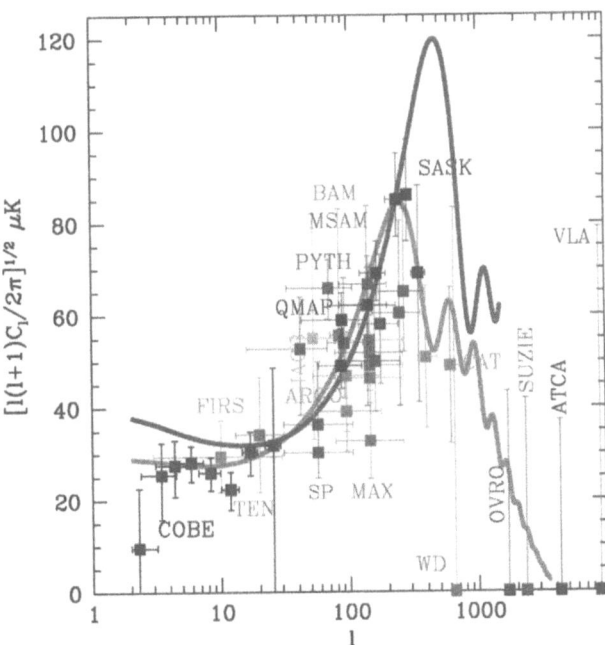

Figure 5. Summary of current measurements of the power spectrum of CMB temperature variations across the sky against spherical harmonic number l for several experiments. The first acoustic peak is evident. The light curve, which is preferred by the data, is a flat universe ($\Omega_0=1$, $\Omega_M=0.35$), and the dark curve is for an open universe ($\Omega_0=0.3$) (courtesy of M. Tegmark).

The Standard Cosmology

Most of our present understanding of the Universe is concisely and beautifully summarized in the hot big-bang cosmological model (see, e.g., Weinberg, 1972; Peebles, 1993). This mathematical description is based upon the isotropic and homogeneous Friedmann-Lemaitre-Robertson-Walker (FLRW) solution of Einstein's general relativity. The evolution of the Universe is embodied in the cosmic scale factor $R(t)$, which describes the scaling up of all physical distances in the Universe (separation of galaxies and wavelengths of photons). The conformal stretching of the wavelengths of photons accounts for the redshift of light from distant galaxies: the wavelength of the radiation we see today is

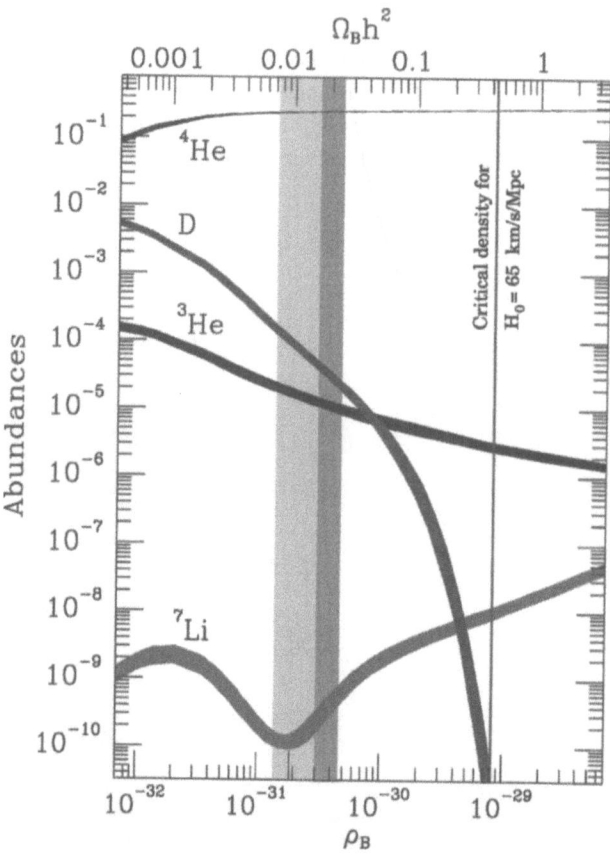

Figure 6. Predicted abundances of ^{4}He, (mass fraction), D, ^{3}He, and ^{7}Li (relative to hydrogen) as a function of the baryon density. The broader band denotes the concordance interval based upon all four light elements. The narrower, darker band highlights the determination of the baryon density based upon a measurement of the primordial abundance of the most sensitive of these—deuterium (Burles and Tytler, 1998a, 1998b), which implies $\Omega_B h^2 = 0.02 \pm 0.002$.

larger by the factor $R(now)/R(then)$. Astronomers denote this factor by $1+z$, which means that an object at "redshift z" emitted the light seen today when the Universe was a factor $1+z$ smaller. Normalizing the scale factor to unity today, $R_{emission} = 1/(1+z)$.

It is interesting to note that the assumption of isotropy and homogeneity was introduced by Einstein and others to simplify the mathematics; as it turns out, it is a remarkably accurate description at early times and today averaged over sufficiently large distances (greater than 100 Mpc or so).

The evolution of the scale factor is governed by the Friedmann equation for the expansion rate:

$$H^2 \equiv (\dot{R}/R)^2 = \frac{8\pi G\rho}{3} \pm \frac{1}{R_{curv}^2}, \qquad (3)$$

where $\rho = \Sigma_i \rho_i$ is the total energy density from all components of mass energy, and R_{curv} is the spatial curvature radius, which grows as the scale factor, $R_{curv} \propto R(t)$. (Hereafter we shall set $c=1$.) As indicated by the $\pm$ sign in Eq. (3) there are actually three FLRW models; they differ in their spatial curvature: The plus sign applies to the negatively curved model, and the minus sign to the positively curved model. For the spatially flat model the curvature term is absent.

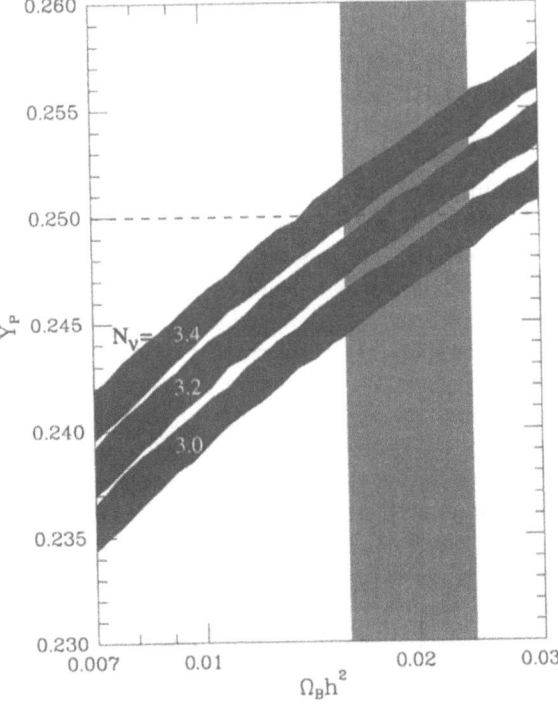

Figure 7. The dependence of primordial ^{4}He production, relative to hydrogen, Y_P, on the number of light neutrino species. The vertical band denotes the baryon density inferred from the Burles-Tytler measurement of the primordial deuterium abundance (Burles and Tytler, 1998a, 1998b); using $Y_P < 0.25$, based upon current ^{4}He measurements, the BBN limit stands at $N_\nu < 3.4$ (from Schramm and Turner, 1998).

The energy density of a given component evolves according to

$$d\rho_i R^3 = -p_i dR^3, \qquad (4)$$

where p_i is the pressure (e.g., $p_i \ll \rho_i$ for nonrelativistic matter or $p_i = \rho_i/3$ for ultrarelativistic particles and radiation). The energy density of matter decreases as R^{-3}, due to volume dilution. The energy density of radiation decreases more rapidly, as R^{-4}, the additional factor arising because the energy of a relativistic particle "redshifts" with the expansion, $E \propto 1/R(t)$. (This of course is equivalent to the wavelength of a photon growing as the scale factor.) This redshifting of the energy density of radiation by R^{-4} also implies that for black-body radiation, the temperature decreases as $T \propto R^{-1}$.

It is convenient to scale energy densities to the critical density, $\rho_{\rm crit} \equiv 3H_0^2/8\pi G = 1.88 h^2 \times 10^{-29}\,{\rm g\,cm}^{-3} \approx 8.4 \times 10^{-30}\,{\rm g\,cm}^{-3}$ or approximately 5 protons per cubic meter,

$$\Omega_i \equiv \rho_i/\rho_{\rm crit}, \qquad (5)$$

$$\Omega_0 \equiv \sum_i \Omega_i, \qquad (6)$$

$$R_{\rm curv} = H_0^{-1}/|\Omega_0 - 1|^{1/2}. \qquad (7)$$

Note that the critical-density universe ($\Omega_0 = 1$) is flat; the subcritical-density universe ($\Omega_0 < 1$) is negatively curved; and the supercritical-density universe ($\Omega_0 > 1$) is positively curved.

There are at least two components to the energy density: the photons in the 2.728 K cosmic microwave background radiation (number density $n_\gamma = 412\,{\rm cm}^{-3}$); and ordinary matter in the formation of neutrons, protons, and associated electrons (referred to collectively as baryons). The theory of big-bang nucleosynthesis and the measured primordial abundance of deuterium imply that the mass density contributed by baryons is $\Omega_B = (0.02 \pm 0.002) h^{-2} \approx 0.05$. In addition, the weak interactions of neutrinos with electrons, positrons, and nucleons should have brought all three species of neutrinos into thermal equilibrium when the Universe was less than a second old, so that today there should be three cosmic seas of relic neutrinos of comparable abundance to the microwave photons, $n_\nu = \frac{3}{11} n_\gamma \approx 113\,{\rm cm}^{-3}$ (per species). (BBN provides a nice check of this, because the yields depend sensitively upon the abundance of neutrinos.) Together, photons and neutrinos (assuming all three species are massless, or very light, $\ll 10^{-3}$ eV) contribute a very small energy density $\Omega_{\nu\gamma} = 4.17 h^{-2} \times 10^{-5} \approx 10^{-4}$.

There is strong evidence for the existence of matter beyond the baryons, as dynamical measurements of the matter density indicate that it is at least 20% of the critical density ($\Omega_M > 0.2$), which is far more than ordinary matter can account for. The leading explanation for the additional matter is long-lived or stable elementary particles left over from the earliest moments (see later section on Dark Matter and Structure Formation).

Finally, although it is now known that the mass density of the Universe in the form of dark matter exceeds 0.2 of the closure density, there are even more exotic possibilities for additional components to the mass-energy density, the simplest of which is Einstein's cosmological constant. Seeking static solutions, Einstein introduced his infamous cosmological constant; after the discovery of the expansion by Hubble he discarded it. In the quantum world it is no longer optional: the cosmological constant represents the energy density of the quantum vacuum (Weinberg, 1989; Carroll et al., 1992). Lorentz invariance implies the pressure associated with vacuum energy is $p_{\rm VAC} = -\rho_{\rm VAC}$, and this ensures that ρ_{VAC} remains constant as the Universe expands. Einstein's cosmological constant appears as an additional term $\Lambda/3$ on the right-hand side of the Friedmann equation [Eq. (3)]; it is equivalent to a vacuum energy $\rho_{VAC} = \Lambda/8\pi G$.

All attempts to calculate the cosmological constant have been unsuccessful to say the very least: due to the zero-point energies the vacuum energy formally diverges ("the ultraviolet catastrophe"). Imposing a short wavelength cutoff corresponding to the weak scale ($\sim 10^{-17}$ cm) is of little help: $\Omega_{VAC} \sim 10^{55}$. The mystery of the cosmological constant is a fundamental one that is being attacked from both ends: Cosmologists are trying to measure it, and particle physicists are trying to understand why it is so small.

Because the different contributions to the energy density scale differently with the cosmic scale factor, the expansion of the Universe goes through qualitatively different phases. While today radiation and relativistic particles are not significant, at early times they dominated the energy, since their energy density depends most strongly on the scale factor (R^{-4} vs R^{-3} for matter). Only at late times does the curvature term ($\propto R^{-2}$) become important; for a negatively curved universe it becomes dominant. For a positively curved universe, the expansion halts when it cancels the matter density term and a contraction phase begins.

The presence of a cosmological constant, which is independent of scale factor, changes this a little. A flat or negatively curved universe ultimately enters an exponential expansion phase driven by the cosmological constant. This also occurs for a positively curved universe, provided the cosmological constant is large enough,

$$\Omega_\Lambda > 4\Omega_M \left\{ \cos\left[\frac{1}{3}\cos^{-1}(\Omega_M^{-1}-1) + \frac{4\pi}{3}\right] \right\}^3. \tag{8}$$

If it is smaller than this, recollapse occurs. Einstein's static universe obtains for $\rho_M = 2\rho_{VAC}$ and $R_{curv} = 1/\sqrt{8\pi G \rho_{VAC}}$.

The evolution of the Universe according to the standard hot big-bang model is summarized as follows:

1. *Radiation-dominated phase.* At times earlier than about 10 000 yrs, when the temperature exceeded $k_B T \gtrsim 3$ eV, the energy density in radiation and relativistic particles exceeded that in matter. The scale factor grew as $t^{1/2}$ and the temperature decreased as $k_B T \sim 1$ MeV$(t/\text{sec})^{-1/2}$. At the earliest times, the energy in the Universe consists of radiation and seas of relativistic particle-antiparticle pairs. (When $k_B T \gg mc^2$ pair creation makes particle-antiparticle pairs as abundant as photons.) The standard model of particle physics, the $SU(3) \otimes SU(2) \otimes U(1)$ gauge theory of the strong, weak, and electromagnetic interactions, provides the microphysics input needed to go back to 10^{-11} sec when $k_B T \sim 300$ GeV. At this time the sea of relativistic particles includes six species of quarks and antiquarks (up, down, charm, strange, top, and bottom), six types of leptons and antileptons (electron, muon, and tauon, and their corresponding neutrinos), and twelve gauge bosons (photon, $W^\pm$, Z^0, and eight gluons). When the temperature drops below the mass of a particle species, those particles and their antiparticles annihilate and disappear (e.g., $W^\pm$ and Z^0 disappear when $k_B T \sim mc^2 \sim 90$ GeV). As the temperature fell below $k_B T \sim 200$ MeV, a phase transition occurred from a quark-gluon plasma to neutrons, protons, and pions, along with the leptons, antileptons, and photons. At a temperature of $k_B T \sim 100$ MeV, the muons and antimuons disappeared. When the temperature was around 1 MeV a sequence of events and nuclear reactions began that ultimately resulted in the synthesis of D, ^{3}He, ^{4}He,

and ^{7}Li. During BBN, the last of the particle-antiparticle pairs, the electrons and positrons, annihilated.

2. *Matter-dominated phase.* When the temperature reached around $k_BT \sim 3$ eV, at a time of around 10 000 yrs the energy density in matter began to exceed that in radiation. At this time the Universe was about 10^{-4} of its present size and the cosmic-scale factor began to grow as $R(t) \propto t^{2/3}$. Once the Universe became matter dominated, primeval inhomogeneities in the density of matter (mostly dark matter), shown to be of size around $\delta\rho/\rho \sim 10^{-5}$ by COBE and other anisotropy experiments, began to grow under the attractive influence of gravity ($\delta\rho/\rho \propto R$). After 13 billion or so years of gravitational amplification, these tiny primeval density inhomogeneities developed into all the structure that we see in the Universe today: galaxies, clusters of the galaxies, superclusters, great walls, and voids. Shortly after matter domination begins, at a redshift $1+z \simeq 1100$, photons in the Universe undergo their last-scattering off free electrons; last scattering is precipitated by the recombination of electrons and ions (mainly free protons), which occurs at a temperature of $k_BT \sim 0.3$ eV because neutral atoms are energetically favored. Before last-scattering, matter and radiation are tightly coupled; after last-scattering, matter and radiation are essentially decoupled.

3. *Curvature-dominated or cosmological constant dominated phase.* If the Universe is negatively curved and there is no cosmological constant, then when the size of the Universe is $\Omega_M/(1-\Omega_M) \sim \Omega_M$ times its present size the epoch of curvature domination begins (i.e., R_{curv}^{-2} becomes the dominant term on the right-hand side of the Friedmann equation). From this point forward the expansion no longer slows and $R(t) \propto t$ (free expansion). In the case of a cosmological constant and a flat universe, the cosmological constant becomes dominant when the size of the Universe is $[\Omega_M/(1-\Omega_M)]^{1/3}$. Thereafter, the scale factor grows exponentially. In either case, further growth of density inhomogeneities that are still linear ($\delta\rho/\rho < 1$) ceases. The structure that exists in the Universe is frozen in.

Finally, a comment on the expansion rate and the size of the "observable Universe." The inverse of the expansion rate has units of time. The Hubble time, H^{-1}, corresponds to the time it takes for the scale factor to roughly double. For a matter-, radiation-, or curvature-dominated universe, the age of the Universe (time back to zero scale factor) is: $\frac{2}{3}H^{-1}$, $\frac{1}{2}H^{-1}$, and H^{-1}, respectively. The Hubble time also sets the size of the observable (or causally connected) Universe: the distance to the "horizon," which is equal to the distance that light could have traveled since time zero, is $2t = H^{-1}$ for a radiation-dominated universe and $3t = 2H^{-1}$ for a matter-dominated universe. Paradoxically, although the size of the Universe goes to zero as one goes back to time zero, the expansion rate is larger, and so points separated by t are moving apart faster than light can catch up with them.

Inner Space and Outer Space

The "hot" in the hot big-bang cosmology makes fundamental physics an inseparable part of the standard cosmology. The time-temperature relation, $k_BT \sim 1$ MeV$(t/\text{sec})^{-1/2}$, implies that the physics of higher energies and shorter times is required to understand the Universe at earlier times: atomic physics at $t \sim 10^{13}$ sec, nuclear physics at $t \sim 1$ sec, and elementary-particle physics at $t < 10^{-5}$ sec. The standard cosmology model itself is based upon Einstein's general relativity, which embodies our deepest and most accurate understanding of gravity.

The standard model of particle physics, which is a mathematical description of the strong, weak, and electromagnetic interactions based upon the $SU(3) \otimes SU(2) \otimes U(1)$ gauge theory, accounts for all known physics up to energies of about 300 GeV (Gaillard, Grannis, and Sciulli, 1999). It provides the input microphysics for the standard cosmology necessary to discuss events as early as 10^{-11} sec. It also provides a firm foundation for speculations about the Universe at even earlier times.

A key feature of the standard model of particle physics is asymptotic freedom: at high energies and short distances, the interactions between the fundamental constituents of matter—quarks and leptons—are perturbatively weak. This justifies approximating the early Universe as hot gas of noninteracting particles (dilute gas approximation) and opens the door to sensibly speculating about times as early as 10^{-43} sec, when the framework of general relativity becomes suspect, since quantum corrections to this classical description are expected to become important.

The importance of asymptotic freedom for early-Universe cosmology cannot be overstated. A little more than 25 years ago, before the advent of quarks and leptons and asymptotic freedom, cosmology hit a brick wall at 10^{-5} sec because extrapolation to early times was nonsensical. The problem was twofold: the finite size of nucleons and related particles and the exponential rise in the number of "elementary particles" with mass. At around 10^{-5} sec, nucleons would be overlapping, and with no understanding of the strong forces between them, together with the exponentially rising spectrum of particles, thermodynamics became ill defined at higher temperatures.

The standard model of particle physics has provided particle physicists with a reasonable foundation for speculating about physics at even shorter distances and higher energies. Their speculations have significant cosmological implications, and—conversely— cosmology holds the promise to test some of their speculations. The most promising particle physics ideas (see, e.g., Schwarz and Seiberg, 1999) and their cosmological implications are:

1. *Spontaneous symmetry breaking (SSB)*. A key idea, which is not fully tested, is that most of the underlying symmetry in a theory can be hidden because the vacuum state does not respect the full symmetry; this is known as spontaneous symmetry breaking and accounts for the carriers of the weak force, the $W^{\pm}$ and Z^0 bosons, being very massive. (Spontaneous symmetry breaking is seen in many systems, e.g., a ferromagnet at low temperatures: it is energetically favorable for the spins to align thereby breaking rotational symmetry.) In analogy to symmetry breaking in a ferromagnet, spontaneously broken symmetries are restored at high temperatures. Thus, it is likely that the Universe underwent a phase transition at around 10^{-11} sec when the symmetry of the electroweak theory was broken, $SU(2) \otimes U(1) \to U(1)$.

2. *Grand unification*. It is possible to unify the strong, weak, and electromagnetic interactions by a larger gauge group, e.g., $SU(5)$, $SO(10)$, or $E8$. The advantages are twofold: the three forces are described as different aspects of a more fundamental force with a single coupling constant, and the quarks and leptons are unified as they are placed in the same particle multiplets. If true, this would imply another stage of spontaneous symmetry breaking, $G \to SU(3) \otimes SU(2) \otimes U(1)$. In addition, grand unified theories (or GUTs), predict that baryon and lepton number are violated—so

that the proton is unstable and neutrinos have mass—and that stable topological defects associated with SSB may exist, e.g., pointlike defects called magnetic monopoles, one-dimensional defects referred to as "cosmic" strings, and two-dimensional defects called domain walls. The cosmological implications of GUTs are manifold: neutrinos as a dark matter component, baryon and lepton number violation explaining the matter-antimatter asymmetry of the Universe, and SSB phase transitions producing topological defects that seed structure formation or a burst of tremendous expansion called inflation.

3. *Supersymmetry.* In an attempt to put bosons and fermions on the same footing, as well as to better understand the "hierarchy problem," namely, the large gap between the weak scale (300 GeV) and the Planck scale (10^{19} GeV), particle theorists have postulated supersymmetry, the symmetry between fermions and bosons. (Supersymmetry also appears to have a role to play in understanding gravity.) Since the fundamental particles of the standard model of particle physics cannot be classified as fermion-boson pairs, if correct, supersymmetry implies the existence of a superpartner for every known particle, with a typical mass of order 300 GeV. The lightest of these superpartners is usually stable and called "the neutralino." The neutralino is an ideal dark matter candidate.

4. *Superstrings, supergravity, and M-theory.* The unification of gravity with the other forces of Nature has long been the holy grail of theorists. Over the past two decades there have been some significant advances: supergravity, an 11-dimensional version of general relativity with supersymmetry, which unifies gravity with the other forces; superstrings, a ten-dimensional theory of relativistic strings, which unifies gravity with the other forces in a self-consistent, finite theory; and M-theory, an ill-understood, "larger" theory that encompasses both superstring theory and supergravity theory. An obvious cosmological implication is the existence of additional spatial dimensions, which today must be "curled up" to escape notice, as well as the possibility of sensibly describing cosmology at times earlier than the Planck time.

Advances in fundamental physics have been crucial to advancing cosmology: e.g., general relativity led to the first self-consistent cosmological models; from nuclear physics came big-bang nucleosynthesis; and so on. The connection between fundamental physics and cosmology seems even stronger today and makes realistic the hope that much more of the evolution of the Universe will be explained by fundamental theory, rather than the *ad hoc* theory that dominated cosmology before the 1980s. Indeed, the most promising paradigm for extending the standard cosmology, inflation+cold dark matter, is deeply rooted in elementary particle physics.

Dark Matter and Structure Formation

As successful as the standard cosmology is, it leaves important questions about the origin and evolution of the Universe unanswered. To an optimist, these questions suggest that there is a grander cosmological theory, which encompasses the hot big-bang model and resolves these questions. It can easily be argued that the most pressing issues in cosmology are the quantity and composition of energy and matter in the Universe, and the origin and nature of the density perturbations that seeded all the structure in the Universe. Cosmology is poised for major progress on these two questions. Answering these questions will provide a window to see beyond the standard cosmology.

Dark matter and dark energy

Our knowledge of the mass and energy content of the Universe is still poor, but is improving rapidly (see Sadoulet, 1999). We can confidently say that most of the matter in the Universe is of unknown form and dark (see, e.g., Dekel, Burstein, and White, 1997; Bahcall *et al.*, 1993): Stars (and closely related material) contribute a tiny fraction of the critical density, $\Omega_{lum} = (0.003 \pm 0.001)h^{-1} \simeq 0.004$, while the amount of matter known to be present from its gravitational effects contributes around ten times this amount, $\Omega_M = 0.35 \pm 0.07$ (this error flag is ours; it is meant to indicate 95% certainty that Ω_M is between 0.2 and 0.5). The gravity of dark matter is needed to hold together just about everything in the Universe—galaxies, clusters of galaxies, superclusters, and the Universe itself. A variety of methods for determining the amount of matter all seem to converge on $\Omega_M \sim 1/3$; they include measurements of the masses of clusters of galaxies and the peculiar motions of galaxies. Finally, the theory of big-bang nucleosynthesis and the recently measured primeval abundance of deuterium pin down the baryon density very precisely: $\Omega_B = (0.02 \pm 0.002)h^{-2} \simeq 0.05$. The discrepancy between this number and dynamical measurements of the matter density is evidence for nonbaryonic dark matter.

Particle physics suggests three dark-matter candidates (Sadoulet, 1999): a 10^{-5} eV axion (Rosenberg, 1998); a 10 GeV – 500 GeV neutralino (Jungman, Kamionkowski, and Griest, 1996); and a 30 eV neutrino. These three possibilities are highly motivated in two important senses: first, the axion and neutralino are predictions of fundamental theories that attempt to go beyond the standard model of particle physics, as are neutrino masses; and second, the relic abundances of the axion and neutralino turn out to be within a factor of ten of the critical density, and similarly for the neutrino—GUTs predict masses in the eV range, which is what is required to make neutrinos a significant contributor to the mass density.

Because measuring the masses of galaxy clusters has been key to defining the dark matter problems it is perhaps worth further discussion. Cluster masses can be estimated by three different techniques, which give consistent results. The first, which dates back to Fritz Zwicky (1935), uses the measured velocities of cluster galaxies and the virial theorem to determine the total mass (i.e., $KE_{gal} \simeq |PE_{gal}|/2$). The second method uses the temperature of the hot x-ray emitting intracluster gas and the virial theorem to arrive at the total mass. The third and most direct method is using the gravitational lensing effects of the cluster on much more distant galaxies. Close to the cluster center, lensing is strong enough to produce multiple images; farther out, lensing distorts the shape of distant galaxies. The lensing method allows the cluster (surface) mass density to be mapped directly. An example of mapping the mass distribution of a cluster of galaxies is shown in Fig. 8.

Using clusters to estimate the mean mass density of the Universe requires a further assumption: that their mass-to-light ratio provides a good estimate for the mean mass-to-light ratio. This is because the mean mass density is determined by multiplying the mean luminosity density (which is reasonably well measured) by the inferred cluster mass-to-light ratio. Using this technique, Carlberg *et al.* (1996) find $\Omega_M = 0.19 \times 0.06$. If clusters have more luminosity per mass than average, this technique would underestimate Ω_M.

There is another way to estimate Ω_M, using clusters, based upon a different, more physically motivated assumption. X-ray measurements more easily determine the amount of hot, intracluster gas; and as it turns out, most of the baryonic mass in a cluster resides here rather than in the mass of individual galaxies (this fact is also confirmed by lensing measurements). Together with the total cluster mass, the ratio of baryonic mass to total mass can be determined; a compilation of the existing data give $M_B/M_{tot} = (0.07 \pm 0.007)h^{-3/2} \simeq 0.15$ (Evrard, 1997, and references therein). Assuming that clusters provide a fair sample of matter in the Universe so that $\Omega_B/\Omega_M = M_B/M_{tot}$, the accurate BBN determination of Ω_B can be used to infer $\Omega_M = (0.3 \pm 0.05)h^{-1/2} \simeq 0.4$. [A similar result for the cluster gas to total mass ratio is derived from cluster gas measurements based upon the distortion of the CMB spectrum due to CMB photons scattering off the hot cluster gas (Sunyaev-Zel'dovich effect); see Carlstrom, 1999.]

Two other measurements bear on the quantity and composition of energy and matter in the Universe. First, the pattern of anisotropy in the CMB depends upon the total energy density in the Universe (i.e., Ω_0) (see, e.g., Jungman, Kamionkowski, Kosowsky, and Spergel, 1996). The peak in the multipole power spectrum is $l_{peak} \simeq 200/\sqrt{\Omega_0}$. The current data, shown in Fig. 5, are consistent with $\Omega_0 \simeq 1$, though $\Omega_0 \sim 0.3$ cannot be excluded. This together with the evidence that $\Omega_M \simeq 0.3$ leaves room for a component of energy that does not clump, such as a cosmological constant.

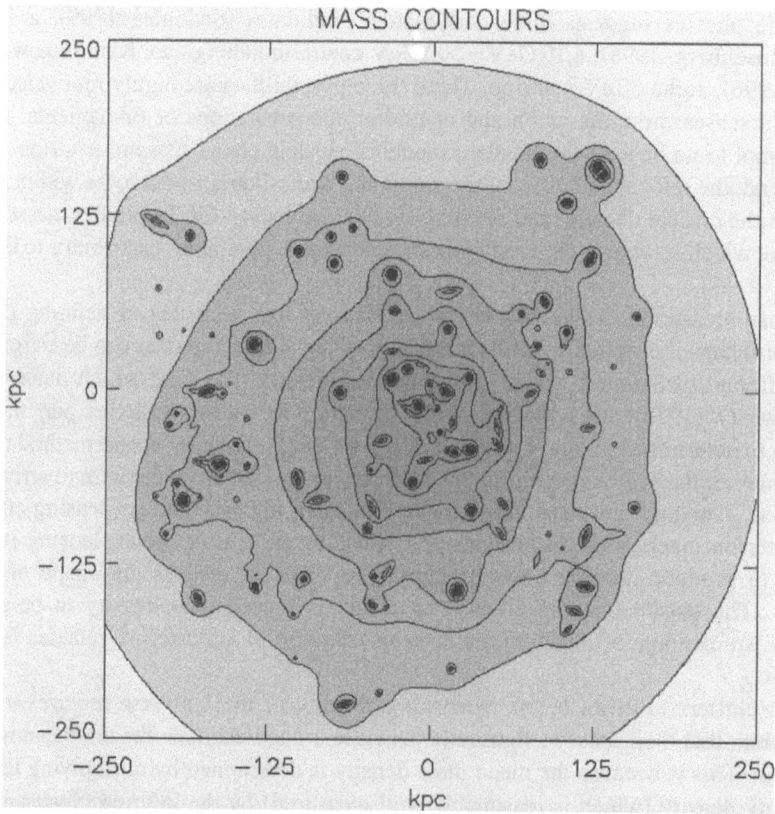

Figure 8. The reconstructed total mass density in the cluster of galaxies 0024+1654 at redshift $z=0.39$, based on parametric inversion of the associated gravitational lens. Projected mass contours are spaced by $430 M_\odot$ pc^{-2}, with the outer contour at $1460 M_\odot$ pc^{-2}. Excluding dark mass concentrations centered on visible galaxies, more than 98% of the remaining mass is represented by a smooth concentration of dark matter centered near the brightest cluster galaxies, with a 50 kpc soft core (Tyson et al., 1998).

The oldest approach to determining the total mass-energy density is through the deceleration parameter (Baum, 1957; Sandage, 1961), which quantifies the present slowing of the expansion due to gravity,

$$q_0 \equiv -\frac{(\ddot{R}/R)_0}{H_0^2} = \frac{\Omega_0}{2}[1+3p_0/\rho_0], \quad (9)$$

where subscript zero refers to quantities measured at the current epoch. Note, in a universe where the bulk of the matter is nonrelativistic ($p \ll \rho$), q_0 and Ω_0 differ only by a factor of two. The luminosity distance to an object at redshift $z \ll 1$ is related to q_0,

$$d_L H_0 = z + z^2(1-q_0)/2 + \cdots, \quad (10)$$

and thus accurate distance measurements can be used to determine q_0. (The luminosity distance to an object is defined as that inferred from the inverse square law: $d_L \equiv \sqrt{\mathcal{L}/4\pi\mathcal{F}}$.)

Recently, two groups (the Supernova Cosmology Project and the High-z Supernova Team) using type Ia supernovae (SNe1a) as standard candles (objects of known $\mathcal{L}$) and assuming that their flux measurements (i.e., $\mathcal{F}$) were not contaminated by sample selection, evolution, or dust systematics, both conclude that the expansion of the Universe is accelerating rather than decelerating (i.e., $q_0 < 0$) (Schmidt et al., 1998; Perlmutter et al., 1998). If correct, this implies that much of the energy in the Universe is in an unknown component, with negative pressure, $p_X \lesssim -\rho_X/3$ (Garnavich et al., 1998). The simplest explanation is a cosmological constant with $\Omega_\Lambda \sim 2/3$. [In fact, Eq. (10), which is deeply rooted in the history of cosmology, is not sufficiently accurate at the redshifts of the SNe1a being used, and the two groups compute $d_L \equiv (1+z)r(z)$ as a function of Ω_M and Ω_Λ and fit to the observations.]

Pulling this together, cosmologists for the first time have a plausible accounting of matter and energy in the Universe: stars contribute around 0.4% of the critical density, baryons contribute 5%, nonrelativistic particles of unknown type contribute 30%, and vacuum energy contributes 64%, for a total equaling the critical density (see Figs. 9 and 10). We should emphasize that plausible does not mean correct.

In addition to the fact that most of the matter and energy in the Universe is dark, most of the ordinary matter is dark (i.e., not in bright stars). The possibilities for the dark baryons include "dark stars" and diffuse hot or warm gas (recall, in clusters, most of the baryons are in hot, intracluster gas). Dark stars could take the form of faint, low-mass stars, failed stars (i.e., objects below the mass required for hydrogen burning, $M \lesssim 0.08 M_\odot$), white dwarfs, neutron stars, or black holes.

Most of the mass of our own Milky Way galaxy is dark, existing in an extended halo (an approximately spherical distribution of matter with density falling as $1/r^2$). Unsuccessful searches for faint stars in our galaxy have eliminated them as a viable candidate, and theoretical arguments disfavor white dwarfs, black holes, and neutron stars—all should lead to the production of more heavy elements than are observed. Further, the measured rate of star formation indicates that only a fraction of the baryons have formed into bright, massive stars.

Experimental searches for dark stars in our own galaxy have been carried out using the gravitational microlensing technique: dark stars along the light-of-sight to nearby galaxies (e.g., the Large and Small Magellanic Clouds and Andromeda) can gravitationally lens the distant bright stars, causing a well-defined, temporary brightening (Paczynski, 1986).

The results however are perplexing (see, e.g., Sadoulet, 1999). More than a dozen such brightenings of Large Magellanic Cloud (LMC) stars have been seen, suggesting that a significant fraction of our galaxy's halo exists in the form of half-solar mass white dwarfs. However, such a population of white dwarfs should be visible, and they have not been seen. Because of our imperfect knowledge of our own galaxy and the LMC it is possible that the lenses are not associated with the halo of our galaxy, but rather are low-mass stars in the LMC, in an intervening dwarf galaxy between us, or are actually in the disk of our galaxy, if the disk is warped enough to pass in front of the line to the LMC.

Structure formation and primeval inhomogeneity

The COBE detection of CMB anisotropy on angular scales of 10° was a major milestone (Smoot et al., 1992), providing the first evidence for the fluctuations that seeded all the structure in the Universe and strong evidence for the gravitational instability picture for structure formation, as the size of the inhomogeneity was sufficient to explain the structure observed today. It also ushered in a powerful new probe of structure formation and dark matter. An early implication of COBE was galvanizing: nonbaryonic dark matter is required to explain the structure seen today. Because baryons are tightly coupled to photons in the Universe and thereby supported against gravitational collapse until after decoupling, larger amplitude density perturbations are required, which in turn lead to larger CMB temperature fluctuations than are observed.

Two key issues are the character and origin of the inhomogeneity and the quantity and composition of matter, discussed above. It is expected that there is a spectrum of fluc-

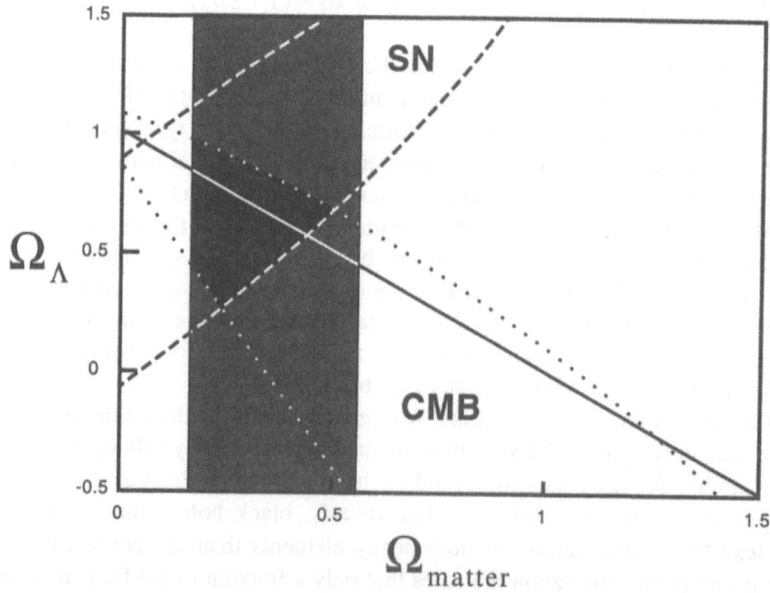

Figure 9. Constraints in the Ω_Λ vs Ω_M plane. Three different types of observations are shown: SNe Ia measures of expansion acceleration (SN); the CMB observations of the location of the first acoustic peak (CMB); and determinations of the matter density, $\Omega_M = 0.35 \pm 0.07$ (dark vertical band). Diagonal line indicates a flat universe, $\Omega_M + \Omega_\Lambda = 1$; regions denote "$3\sigma$" confidence. Darkest region denotes concordance region: $\Omega_\Lambda \sim 2/3$ and $\Omega_M \sim 1/3$.

tuations, described by its Fourier decomposition into plane waves. In addition, there are two generic types of inhomogeneity: curvature perturbations, fluctuations in the local curvature of the Universe, which by the equivalence principle, affect all components of the energy density alike; and isocurvature perturbations, which as their name indicates are not ingrained in the curvature but arise as pressure perturbations caused by local changes in the equation of state of matter and energy in the Universe.

The two most promising ideas for the fundamental origin of the primeval inhomogeneity are quantum fluctuations that become curvature fluctuations during inflation (Hawking, 1982; Guth and Pi, 1982; Starobinskii, 1982; Bardeen, Steinhardt, and Turner, 1983) and topological defects (such as cosmic strings) that are produced during a cosmological phase transition (see, e.g., Vilenkin and Shellard, 1994). The inflation scenario will be discussed in detail later on. Topological defects produced in a cosmological symmetry-breaking phase transition around 10^{-36} sec generate isocurvature fluctuations: the conversion of energy from radiation to defects leads to a pressure perturbation that propagates

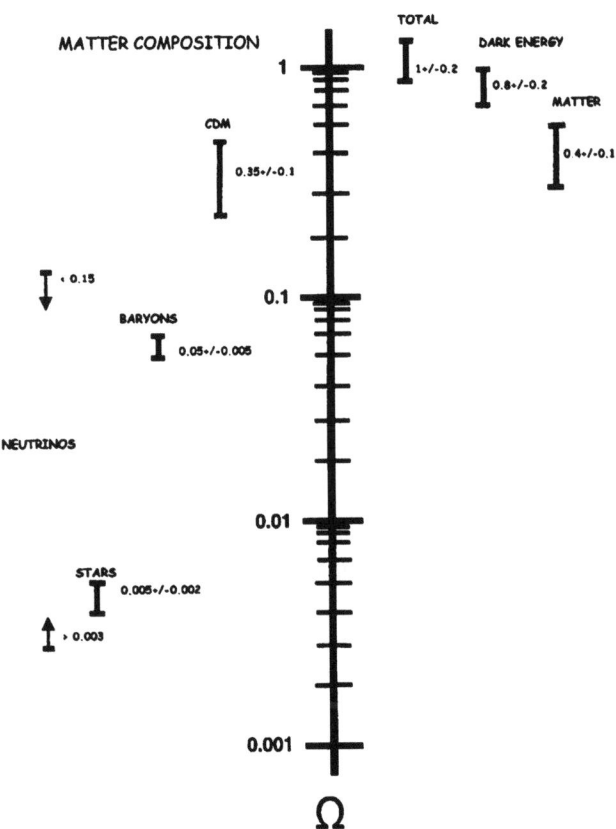

Figure 10. Summary of matter/energy in the Universe. The right side refers to an overall accounting of matter and energy; the left refers to the composition of the matter component. The upper limit to mass density contributed by neutrinos is based upon the failure of the hot dark matter model and the lower limit follows from the Super-K evidence for neutrino oscillations.

outward and ultimately leads to a density inhomogeneity. The defect scenario is currently disfavored by measurements of CMB anisotropy (Allen et al., 1997; Pen et al., 1997).

One graphic indicator of the progress being made on the large-scale structure problem is the number of viable models: the flood of data has trimmed the field to one or possibly two models. A few years ago the defect model was a leading contender; and another, more phenomenological model put forth by Peebles was also in the running (Peebles, 1987). Peebles' model dispensed with nonbaryonic dark matter, assumed $\Omega_B = \Omega_0 \sim 0.2$, and posited local variations in the distribution of baryons (isocurvature perturbations) of unknown origin. Its demise was CMB anisotropy: it predicted too much anisotropy on small angular scales. The one clearly viable model is cold dark matter + inflation, which is discussed below. The challenge to theorists is to make sure that at least one model remains viable as the quantity and quality of data improve.

More Fundamental Questions

Beyond the questions involving dark matter and structure formation, there is a set of more fundamental questions, ranging from the matter-antimatter asymmetry in the Universe to the origin of the expansion itself. For these questions there are attractive ideas, mainly rooted in the physics of the early Universe, which remain to be developed, suggesting that a more fundamental understanding of our Universe is possible.

Baryon-lepton asymmetry. While the laws of physics are very nearly matter-antimatter symmetric, the Universe is not. On scales as large as clusters of galaxies there is no evidence for antimatter. In the context of the hot big bang, a symmetric universe would be even more puzzling: at early times ($t \ll 10^{-5}$ sec) matter-antimatter pairs would be as abundant as photons, but as the Universe cooled matter and antimatter would annihilate until nucleons and antinucleons were too rare to find one another. This would result in only trace amounts of matter and antimatter, a few nucleons and antinucleons per 10^{18} photons, compared to the observed nucleon to photon ratio: $\eta \equiv n_N/n_\gamma = (5 \pm 0.5) \times 10^{-10}$.

In order to avoid the annihilation catastrophe the early Universe must possess a slight excess of matter over antimatter, i.e., a small net baryon number: $n_B/n_\gamma \equiv n_b/n_\gamma - n_{\bar{b}}/n_\gamma = \eta = 5 \times 10^{-10}$. Such an initial condition for the Universe seems as odd as having to assume the ^{4}He mass fraction is 25%. (Charge neutrality requires a similar excess of electrons over positrons; because lepton number can be hidden in the three neutrino species, it is not possible to say that the total lepton asymmetry is comparable to the baryon asymmetry.)

A framework for understanding the origin of the baryon asymmetry of the Universe was put forth in a prescient paper by Sakharov in 1967: baryon-number violating and matter-antimatter symmetry violating interactions occurring in a state of nonequilibrium allow a small, net baryon number to develop. If the idea of baryogenesis is correct, the explanation of the baryon asymmetry is not unlike that of the primeval ^{4}He abundance (produced by nonequilibrium nuclear reactions). The key elements of baryogenesis are all in place: baryon number is violated in the standard model of particle physics (by subtle quantum mechanical effects) and in GUTs; matter-antimatter symmetry is known to be violated by a small amount in the neutral Kaon system (CP violation at the level of 10^{-3}); and maintaining thermal equilibrium in the expanding and cooling Universe depends upon whether or not particle interactions proceed rapidly compared to the expansion rate. The details of baryogenesis have not been worked out, and may involve grand unification physics, but the basic idea is very compelling (see, e.g., Kolb and Turner, 1990).

The heat of the big bang. The entropy associated with the CMB and three neutrino seas is enormous: within the observable Universe, 10^{88} in units of k_B (the number of nucleons is 10 orders of magnitude smaller). Where did all the heat come from? As we discuss in the next section, inflation may provide the answer.

Origin of the smoothness and flatness. On large scales today and at very early times the Universe is very smooth. (The appearance of inhomogeneity today does belie a smooth beginning as gravity drives the growth of fluctuations.) Since the particle horizon at last-scattering (when matter and radiation decoupled) corresponds to an angle of only 1° on the sky, the smoothness could not have arisen via causal physics. (Within the isotropic and homogeneous FLRW model no explanation is required of course.)

In a sense emphasized first by Dicke and Peebles (1979) and later by Guth (1982), the Universe is very flat. Since Ω_0 is not drastically different from unity, the curvature radius of the Universe is comparable to the Hubble radius. During a matter or radiation dominated phase the curvature radius decreases relative to the Hubble radius. This implies that at earlier times it was even larger than the Hubble radius, and that Ω was even closer to one: $|\Omega - 1| < 10^{-16}$ at 1 sec. To arrive at the Universe we see today, the Universe must have begun very flat (and thus, must have been expanding very close to the critical expansion rate).

The flatness and smoothness problems are not indicative of any inconsistency of the standard model, but they do require special initial conditions. As stated by Collins and Hawking (1973), the set of initial conditions that evolve to a universe qualitatively similar to ours is of measure zero. While not *required* by observational data, the inflation model addresses both the smoothness and flatness problems.

Origin of the big bang, expansion, and all that. In naming the big-bang theory Hoyle tried to call attention to the colossal big-bang event, which, in the context of general relativity corresponds to the creation of matter, space, and time from a space-time singularity. In its success, the big-bang theory is a theory of the events *following* the big-bang singularity. In the context of general relativity the big-bang event requires no further explanation (it is consistent with "St. Augustine's principle," since time is created along with space, there is no *before* the big bang). However, many if not most physicists believe that general relativity, which is a classical theory, is not applicable any earlier than 10^{-43} sec because quantum corrections should become very significant, and further, that a quantum theory of gravity will eliminate the big-bang singularity allowing the "before the big-bang question" to be addressed. As we will discuss, inflation addresses the big-bang question too.

Beyond the Standard Model: Inflation+Cold Dark Matter

The 1980s were ripe with interesting ideas about the early Universe inspired by speculations about the unification of the forces and particles of Nature (see, e.g., Kolb and Turner, 1990): relic elementary particles as the dark matter; topological defects as the seeds for structure formation; baryon number violation and C, CP violation as the origin of the baryon asymmetry of the Universe (baryogenesis); and inflation. From all this, a compelling paradigm for extending the standard cosmology has evolved: inflation+cold dark matter. It is bold and expansive and is being tested by a flood of observations. It may even be correct!

The story begins with a brief period of tremendous expansion—a factor of greater than 10^{27} growth in the scale factor in 10^{-32} sec. The precise details of this "inflationary phase" are not understood, but in most models the exponential expansion is driven by the (potential) energy of a scalar field initially displaced from the minimum of its potential energy curve. Inflation blows up a small, subhorizon-sized portion of the Universe to a size much greater than that of the observable Universe today. Because this subhorizon-sized region was causally connected before inflation, it can be expected to be smooth— including the very small portion of it that is our observable part of the Universe. Likewise, because our Hubble volume is but a small part of the region that inflated, it looks flat, regardless of the initial curvature of the region that inflated, $R_{\mathrm{curv}} \gg H_0^{-1}$, which via the Friedmann equation implies that $\Omega_0 = 1$.

It is while this scalar field responsible for inflation rolls slowly down its potential that the exponential expansion takes place (Albrecht and Steinhardt, 1982; Linde, 1982). As the field reaches the minimum of the potential energy curve, it overshoots and oscillates about it: the potential energy of the scalar field has been converted to coherent scalar field oscillations (equivalently, a condensate of zero momentum scalar-field particles). Eventually, these particles decay into lighter particles which thermalize, thereby explaining the tremendous heat content of the Universe and ultimately the photons in the CMB (Albrecht et al., 1982).

Quantum mechanical fluctuations arise in such a scalar field that drives inflation; they are on truly microscopic scales ($\lesssim 10^{-23}$ cm). However, they are stretched in size by the tremendous expansion during inflation to astrophysical scales. Because the energy density associated with the scalar field depends upon its value (through the scalar field potential energy), these fluctuations also correspond to energy density perturbations, and they are imprinted upon the Universe as perturbations in the local curvature. Quantum mechanical fluctuations in the space-time metric give rise to a stochastic, low-frequency background of gravitational waves.

The equivalence principle holds that local acceleration cannot be distinguished from gravity; from this it follows that curvature perturbations ultimately become density perturbations in all species—photons, neutrinos, baryons, and particle dark matter. The shape of the spectrum of perturbations is nearly scale invariant. [Such a form for the spectrum was first discussed by Harrison (1970). Zel'dovich, who appreciated the merits of such a spectrum early on, emphasized its importance for structure formation.] Scale-invariant refers to the fact that the perturbations in the gravitational potential have the same amplitude on all length scales (which is not the same as the density perturbations having the same amplitude). When the wavelength of a given mode crosses inside the horizon ($\lambda = H^{-1}$), the amplitude of the density perturbation on that scale is equal to the perturbation in the gravitational potential.

The overall amplitude (or normalization) depends very much upon the specific model of inflation (of which there are many). Once the overall normalization is set, the shape fixes the level of inhomogeneity on all scales. The detection of anisotropy on the scale of 10° by COBE in 1992 and the subsequent refinement of that measurement with the full four-year data set permitted the accurate (10%) normalization of the inflationary spectrum of density perturbations; soon, the term COBE-normalized became a part of the cosmological vernacular.

On to the cold dark matter (CDM) part: inflation predicts a flat Universe (total energy density equal to the critical density). Since ordinary matter (baryons) contributes only about 5% of the critical density there must be something else. The leading candidate is elementary particles remaining from the earliest moments of particle democracy. Generically, they fall into two classes—fast moving, or hot dark matter; and slowly moving, or cold dark matter (see Sadoulet, 1999). Neutrinos of mass 30 eV or so are the prime example of hot dark matter—they move quickly because they were once in thermal equilibrium and are very light. Axions and neutralinos are examples of cold dark matter. Neutralinos move slowly because they too were once in thermal equilibrium and they are very heavy. Axions are extremely light but were never in thermal equilibrium (having been produced very, very cold).

If most of the matter is hot, then structure in the Universe forms from the top down: large things, like superclusters, form first, and fragment into smaller objects such as galaxies. This is because fast moving neutrinos smooth out density perturbations on small scales by moving from regions of high density into regions of low density (Landau damping or collisionless phase mixing). Observations very clearly indicate that galaxies formed at redshifts $z \sim 2-4$ (see Fig. 11), before superclusters which are just forming today. So hot dark matter is out, at least as a major component of the dark matter (White, Frenk, and Davis, 1983). This leaves cold dark matter.

Cold dark matter particles cannot move far enough to damp perturbations on small scales, and structure then forms from the bottom up: galaxies, followed by clusters of galaxies, and so on (see, e.g., Blumenthal *et al.*, 1984). For COBE-normalized cold dark matter we can be even more specific. The bulk of galaxies should form around redshifts $z \sim 2-4$, just as the observations now indicate.

At present, the cold dark matter+inflation scenario looks very promising—it is consistent with a large body of observations: measurements of the anisotropy of the CMB, redshift surveys of the distribution of matter today, deep probes of the Universe (such as the Hubble Deep Field), and more (see Liddle and Lyth, 1993, and Fig. 11). While the evidence is by no means definitive, and has hardly begun to discriminate between different inflationary models and versions of CDM, we can say that the data favor a flat Universe, almost scale-invariant density perturbations, and cold dark matter with a small admixture of baryons.

Precision Cosmology

The COBE DMR measurement of CMB anisotropy on the 10° angular scale and determination of the primeval deuterium abundance served to mark the beginning of a new era of precision cosmology. Overnight, COBE changed the study of large-scale structure: for theories like inflation and defects that specify the shape of the spectrum of density perturbations, the COBE measurement fixed the level of inhomogeneity on all scales to an accuracy of around 10%. Likewise, the measurement of the primeval deuterium abundance led to a 10% determination of the baryon density.

Within the next few years, an avalanche of data, driven by advances in technology, promises definitive independent observations of the geometry, mass distribution and composition, and detailed structure of the Universe. In a radical departure from its history, cosmology is becoming an exact science. These new observations span the wavelength range from microwave to gamma rays and beyond, and utilize techniques as varied as

CMB microwave interferometry, faint supernova photometry and spectroscopy, gravitational lensing, and massive photometric and spectroscopic surveys of millions of galaxies.

The COBE measurement of CMB anisotropy on angular scales from around 10° to 100° yielded a precise determination of the amplitude of mass fluctuations on very large scales, 10^3 Mpc – 10^4 Mpc. A host of experiments will view the CMB with much higher angular resolution and more precision than COBE, culminating in the two satellite experiments, NASA's MAP and ESA's Planck Surveyor, which will map the full sky to an angular resolution of 0.1°. In so doing, the mass distribution in the Universe at a simpler time, before nonlinear structures had formed, will be determined on scales from 10^4 Mpc down to 10 Mpc. (Temperature fluctuations on angular scale θ arise from density fluctuations on length scales $L \sim 100 h^{-1}$ Mpc[θ/deg]; fluctuations on scales ~ 1 Mpc give rise to galaxies, on scales ~ 10 Mpc give rise to clusters, and on scales ~ 100 Mpc give rise to great walls.)

The multipole power spectrum of CMB temperature fluctuations has a rich structure and encodes a wealth of information about the Universe. The peaks in the power spectrum are caused by baryon-photon oscillations, which are driven by the gravitational force of the dark matter. Since decoupling is essentially instantaneous, different Fourier modes are caught at different phases, which is reflected in multiple spectrum of anisotropy (see Fig. 5). The existence of the first peak is evident. If the satellite missions are as successful as cosmologists hope, and if foregrounds (e.g., diffuse emission from our own galaxy and extragalactic point sources) are not a serious problem, it should be possible to use the measured multipole power spectrum to determine Ω_0 and many other cosmological

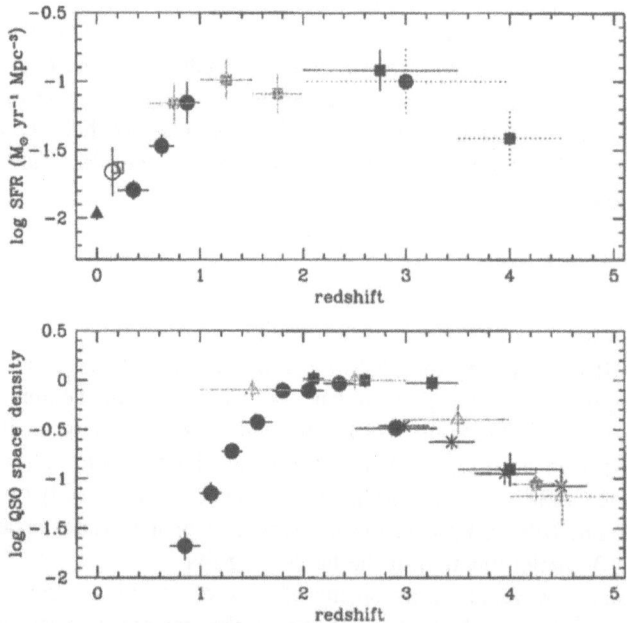

Figure 11. Top: *Star formation rate in galaxies is plotted vs redshift.* Bottom: *The number density of quasistellar objects (galaxies with accreting black holes) vs redshift. The points have been corrected for dust and the relative space density of QSOs (from Madau, 1999).*

parameters (e.g., $\Omega_B h^2$, h, n, level of gravitational waves, Ω_Λ, and Ω_ν) to precision of few percent or better in some cases (see Wilkinson, 1999).

Another impressive map is in the works. The present three-dimensional structure of the local Universe will also be mapped to unprecedented precision in the next few years by the Sloan Digital Sky Survey (SDSS) (see Gunn et al., 1998), which will obtain the redshifts of a million galaxies over 25% of the northern sky out to redshift $z \sim 0.1$, and the Two-degree Field Survey (2dF), which will collect 250 000 redshifts in many 2° patches of the southern sky (Colless, 1998). These surveys will cover around 0.1% of the observable Universe, and more importantly, will map structure out to scales of about $500 h^{-1}$ Mpc, well beyond the size of the largest structures known. This should be large enough to provide a typical sample of the Universe. The two maps—CMB snapshot of the Universe at 300 000 yrs and the SDSS map of the Universe today—when used together have enormous leverage to test cosmological models and determine cosmological parameters.

Several projects are underway to map smaller, more distant parts of the Universe to study the "recent" evolution of galaxies and structure. Using a new large spectrograph, the 10-meter Keck telescope will begin to map galaxies in smaller fields on the sky out to redshifts of 4 or so. Ultimately, the Next Generation Space Telescope, which is likely to have an 8-meter mirror and capability in the infrared (most of the light of high-redshift galaxies has been shifted into the infrared) will probe the first generation of stars and galaxies.

Much of our current understanding of the Universe is based on the assumption that light traces mass, because telescopes detect light and not mass. There is some evidence that light is not a terribly "biased" tracer of mass, at least on the scales of galaxies. However, it would be a convenient accident if the mass-to-light ratio were universal. It is possible that there is a lot of undiscovered matter, perhaps even enough to bring Ω_M to unity, associated with dim galaxies or other mass concentrations that are not correlated with bright galaxies.

Gravitational lensing is a powerful means of measuring cosmic mass overdensities in the linear regime directly (see, e.g., Blandford and Narayan, 1992; Tyson, 1993): dark matter overdensities at moderate redshift ($z \sim 0.2 - 0.5$) systematically distort background galaxy images (referred to as weak gravitational lensing). A typical random one-square-degree patch of the sky contains a million faint high-redshift galaxies. Using these galaxies, weak gravitational lensing may be used to map the intervening dark matter overdensities directly. This technique has been used to map known or suspected mass concentrations in clusters of galaxies over redshifts $0.1 < z < 0.8$ (see Fig. 8 and Clowe et al., 1998) and only recently has been applied to random fields. Large mosaics of CCDs make this kind of direct mass survey possible, and results from these surveys are expected in the coming years.

Crucial to taking advantage of the advances in our understanding of the distribution of matter in the Universe and the formation of galaxies are the numerical simulations that link theory with observation. Simulations now involve billions of particles, allowing a dynamical range of a factor of one thousand (see Fig. 12). Many simulations now involve not only gravity, but the hydrodynamics of the baryons. Advances in computing have been crucial.

Impressive progress has been made toward measuring the cosmological parameters H_0, q_0, and t_0, and more progress is on the horizon. A 5% or better measurement of the Hubble constant on scales that are a substantial fraction of the distance across the

Universe may be within our grasp. Techniques that do not rely upon phenomenological standard candles are beginning to play an important role. The time delay of a flare event seen in the multiple images of a lensed quasar is related only to the redshifts of the lens and quasar, the lens magnification, the angular separation of the quasar images, and the Hubble constant. Thanks to a recent flare, an accurate time delay between the two images of the gravitationally lensed quasar Q0957+561 has been reliably determined, but the lens itself must be mapped before H_0 is precisely determined (Kundic *et al.*, 1997). This technique is being applied to other lensed quasar systems as well. The pattern of CMB anisotropy has great potential to accurately determine H_0. Another technique (Sunyaev-Zel'dovich, or SZ), which uses the small distortion of the CMB when viewed through a cluster containing hot gas (due to Compton up-scattering of CMB photons), has begun to produce reliable numbers (Birkinshaw, 1998).

Currently, the largest gap in our knowledge of the mass content of the Universe is identifying the bulk of the matter density, $\Omega_? = \Omega_M - \Omega_B \sim 0.3$. The most compelling idea is that this component consists of relic elementary particles, such as neutralinos, axions, or neutrinos. If such particles comprise most of the dark matter, then they should account for most of the dark matter in the halo of our own galaxy and have a local mass density of around 10^{-24} g cm^{-3}. Several laboratory experiments are currently running with sufficient sensitivity to search directly for neutralinos of mass 10 GeV – 500 GeV and cross section that is motivated by the minimal supersymmetric standard model. While the supersymmetric parameter space spans more than three orders of magnitude in cross section, even greater sensitivities are expected in the near future. These experiments

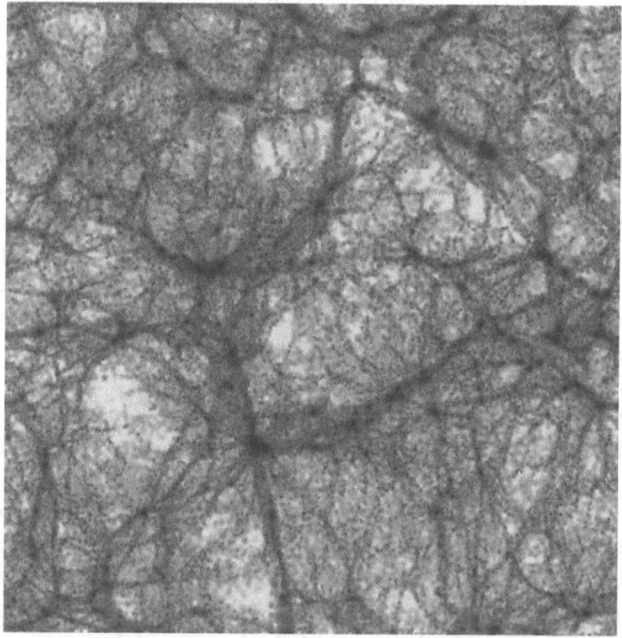

Figure 12. One of the largest simulations of the development of structure in the Universe (from Virgo Collaboration, 1998). Shown here is projected mass in a ΛCDM simulation with 256^3 particles and $\Omega_M=0.3$, 240 Mpc h^{-1} on a side. The map shown in Fig. 8 would correspond to a window 0.5 Mpc across, centered on one of the minor mass concentrations.

involve high-sensitivity, low-background detectors designed to detect the small (order keV) recoil energy when a neutralino elastically scatters off a nucleus in the detector; the small rates (less than one scattering per day per kg of detector) add to the challenge (Sadoulet, 1999).

An axion detector has achieved sufficient sensitivity to detect halo axions, and is searching the mass range 10^{-6} eV $- 10^{-5}$ eV where axions would contribute significantly to the mass density. This detector, based upon the conversion of axions to photons in strong magnetic field, consists of a hi-Q cavity immersed in a 7 Tesla magnetic field and is operating with a sensitivity of 10^{-23} W in the GHz frequency range. Within five years it is hoped that the entire theoretically favored mass range will be explored (Rosenberg, 1998).

While light neutrinos are no longer favored by cosmologists for the dark matter, as they would lead to structure in the Universe that is not consistent with what we see today, because of their large numbers, 113 cm^{-3}, they could be an important component of mass density even if only one species has a tiny mass:

$$\Omega_\nu = \sum_i (m_{\nu_i}/90h^2 \text{ eV}),$$

$$\Omega_\nu/\Omega_{\text{lum}} \simeq \sum_i (m_{\nu_i}/0.2 \text{ eV}),$$

$$\Omega_\nu/\Omega_B \simeq \sum_i (m_{\nu_i}/2 \text{ eV}). \qquad (11)$$

Even with a mass as small as one eV neutrinos would make an imprint on the structure of the Universe that is potentially detectable.

Particle theorists strongly favor the idea that neutrinos have small, but nonzero mass, and the see-saw mechanism can explain why their masses are so much smaller than the other quarks and leptons: $m_\nu \sim m_{q,l}^2/M$ where $M \sim 10^{10}$ GeV $- 10^{15}$ GeV is the very large mass of the right-handed partner(s) of the usual left-handed neutrinos (see, e.g., Schwarz and Seiberg, 1999). Because neutrino masses are a fundamental prediction of unified field theories, much effort is directed at probing neutrino masses. The majority of experiments now involve looking for the oscillation of one neutrino species into another, which is only possible if neutrinos have mass. These experiments are carried out at accelerators, at nuclear reactors, and in large-underground detectors such as Super-Kamiokande (Super-K) and the SNO facility.

Super-K detects neutrinos from the sun and those produced in the Earth's atmosphere by cosmic-ray interactions. For several years now the solar-neutrino data has shown evidence for neutrino oscillations, corresponding to a neutrino mass-difference squared of around 10^{-5} eV^2 or 10^{-10} eV^2, too small to be of cosmological interest (unless two neutrino species are nearly degenerate in mass). The Super-K Collaboration recently announced evidence for neutrino oscillations based upon the atmospheric neutrino data. Their results, which indicate a mass difference squared of around $10^{-3} - 10^{-2}$ eV^2 (Fukuda et al., 1998) and imply at least one neutrino has a mass of order 0.1 eV or larger, are much more interesting cosmologically. Over the next decade particle physicists will pursue neutrino mass with a host of new experiments, characterized by very long baselines (neutrino source and detector separated by hundreds of kilometers) and should clarify the situation.

Testing inflation+CDM in the precision era

As we look forward to the abundance (avalanche) of high-quality observations that will test inflation+CDM, we have to make sure the predictions of the theory match the precision of the data. In so doing, CDM+ inflation becomes a theory with ten or more parameters. For cosmologists, this is a bit daunting, as it may seem that a ten-parameter theory can be made to fit any set of observations. This will not be the case when one has the quality and quantity of data that are coming. The standard model of particle physics offers an excellent example: it is a 19-parameter theory, and because of the high quality of data from experiments at high-energy accelerators and other facilities it has been rigorously tested, with parameters measured to a precision of better than 1% in some cases.

In fact, the ten parameters of CDM+inflation are an opportunity rather than a curse: Because the parameters depend upon the underlying inflationary model and fundamental aspects of the Universe, we have the very real possibility of learning much about the Universe, inflation, and perhaps fundamental physics. The ten parameters can be split into two groups: cosmological and dark-matter.

Cosmological parameters

1. h, the Hubble constant in units of $100 \text{ km s}^{-1} \text{ Mpc}^{-1}$.
2. $\Omega_B h^2$, the baryon density.
3. n, the power-law index of the scalar density perturbations. CMB measurements indicate $n = 1.1 \pm 0.2$; $n = 1$ corresponds to scale-invariant density perturbations. Several popular inflationary models predict $n \simeq 0.95$; range of predictions runs from 0.7 to 1.2.
4. $dn/d \ln k$, "running" of the scalar index with comoving scale ($k =$ wave number). Inflationary models predict a value of $\mathcal{O}(\pm 10^{-3})$ or smaller.
5. S, the overall amplitude squared of density perturbations, quantified by their contribution to the variance of the quadrupole CMB anisotropy.
6. T, the overall amplitude squared of gravitational waves, quantified by their contribution to the variance of the quadrupole CMB anisotropy. Note, the COBE normalization determines $T+S$ (see below).
7. n_T, the power-law index of the gravitational wave spectrum. Scale invariance corresponds to $n_T = 0$; for inflation, n_T is given by $-\frac{1}{7}(T/S)$.

Dark-matter parameters

1. Ω_ν, the fraction of critical density in neutrinos ($= \Sigma_i m_{\nu_i}/90h^2$). While the hot dark matter theory of structure formation is not viable, it is possible that a small fraction of the matter density exists in the form of neutrinos.
2. Ω_X, the fraction of critical density in a smooth component of unknown composition and negative pressure ($w_X \lesssim -0.3$); the simplest example is a cosmological constant ($w_X = -1$).
3. g_*, the quantity that counts the number of ultrarelativistic degrees of freedom (at late times). The standard cosmology/standard model of particle physics predicts $g_* = 3.3626$ [photons in the CMB+ three massless neutrino species with temperature $(4/11)^{1/3}$ times that of the photons]. The amount of radiation controls when the Universe became matter dominated and thus affects the present spectrum of density fluctuations.

The parameters involving density and gravitational-wave perturbations depend directly upon the inflationary potential. In particular, they can be expressed in terms of the potential and its first two derivatives (see, e.g., Lidsey et al., 1997):

$$S \equiv \frac{5\langle |a_{2m}|^2\rangle}{4\pi} \simeq 2.2 \frac{V_*/m_{Pl}^4}{(m_{Pl}V'_*/V_*)^2}, \qquad (12)$$

$$n - 1 = -\frac{1}{8\pi}\left(\frac{m_{Pl}V'_*}{V_*}\right)^2 + \frac{m_{Pl}}{4\pi}\left(\frac{m_{Pl}V'_*}{V_*}\right)', \qquad (13)$$

$$T \equiv \frac{5\langle |a_{2m}|^2\rangle}{4\pi} = 0.61(V_*/m_{Pl}^4), \qquad (14)$$

where $V(\phi)$ is the inflationary potential, prime denotes $d/d\phi$, and V_* is the value of the scalar potential when the present horizon scale crossed outside the horizon during inflation.

As particle physicists can testify, testing a ten (or more) parameter theory is a long, but potentially rewarding process. To begin, one has to test the basic tenets and consistency of the underlying theory. Only then can one proceed to take full advantage of the data to precisely measure parameters of the theory. The importance of establishing a theoretical framework is illustrated by measurements of the number of light neutrino species derived from the decay width of the Z^0 boson: $N_\nu = 3.07 \pm 0.12$ (not assuming the correctness of the standard model); $N_\nu = 2.994 \pm 0.012$ (assuming the correctness of the standard model).

In the present case, the putative theoretical framework is inflation+CDM, and its basic tenets are a flat, critical density Universe, a nearly scale-invariant spectrum of Gaussian density perturbations, and stochastic background of gravitational waves. The first two predictions are much more amenable to testing, by a combination of CMB anisotropy and large-scale structure measurements. For example, a flat universe with Gaussian curvature perturbations implies a multipole power spectrum of well-defined acoustic peaks, beginning at $l \simeq 200$ (see Fig. 5). In addition, there are consistency tests: comparison of the precise BBN determination of the baryon density with that derived from CMB anisotropy; an accounting of the dark matter and dark energy by gravitational lensing; SNe1a measurements of acceleration, and comparison of the different determinations of the Hubble constant. Once the correctness and consistency of inflation+CDM has been verified—assuming it is—one can zero in on the remaining parameters (subset of the list above) and hope to determine them with precision.

Present status of inflation+CDM

A useful way to organize the different CDM models is by their dark-matter content; within each CDM family, the cosmological parameters can still vary: sCDM (for simple), only CDM and baryons; τCDM: in addition to CDM and baryons additional radiation (e.g., produced by the decay of an unstable massive tau neutrino); νCDM: CDM, baryons, and a dash of hot dark matter (e.g., $\Omega_\nu = 0.2$); and ΛCDM: CDM, baryons, and a cosmological constant (e.g., $\Omega_\Lambda = 0.6$). In all these models, the total energy density sums to the critical energy density; in all but ΛCDM, $\Omega_M = 1$.

Figure 13 summarizes the viability of these different CDM models, based upon CBR measurements and current determinations of the present power spectrum of fluctuations (derived from redshift surveys; see Fig. 14). sCDM is only viable for low values of the Hubble constant (less than 55 km s^{-1} Mpc^{-1}) and/or significant tilt (deviation from scale invariance); the region of viability for τCDM is similar to sCDM, but shifted to larger values of the Hubble constant (as large as 65 km s^{-1} Mpc^{-1}). νCDM has an island of viability around $H_0 \approx 60$ km s^{-1} Mpc^{-1} and $n \approx 0.95$. ΛCDM can tolerate the largest values of the Hubble constant.

Considering other relevant data too—e.g., age of the Universe, determinations of Ω_M, measurements of the Hubble constant, and limits to Ω_Λ—ΛCDM emerges as the "best-fit CDM model" (see, e.g., Krauss and Turner, 1995; Ostriker and Steinhardt, 1995). Moreover, its "key signature," $q_0 \sim -0.5$, may have been confirmed. Given the possible systematic uncertainties in the SNe1a data and other measurements, it is premature to conclude that ΛCDM is anything but the model at which to take aim.

The Next Hundred Years

The progress in cosmology over the last hundred years has been stunning. With the hot big-bang cosmology we can trace the history of the Universe to within a fraction of a second of the beginning. Beyond the standard cosmology, we have promising ideas, rooted in fundamental theory, about how to extend our understanding to even earlier times addressing more profound questions, e.g., inflation+cold dark matter. While it remains to

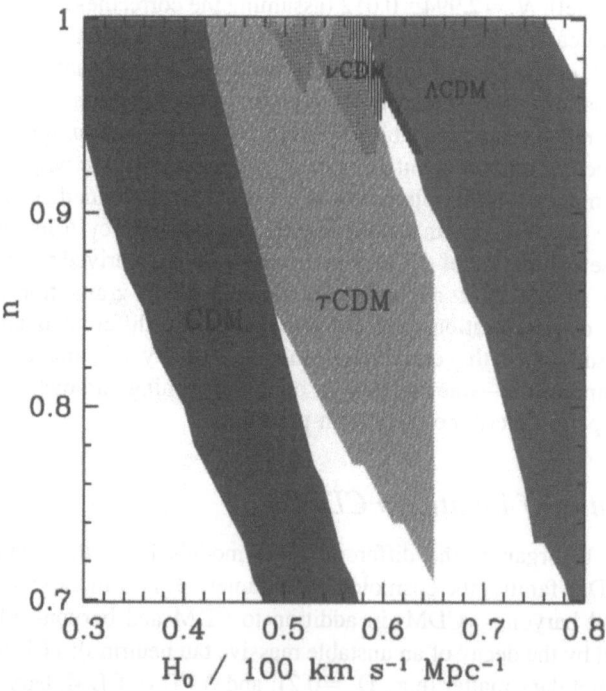

Figure 13. Acceptable cosmological parameters for different CDM models, as are characterized by their invisible matter content: simple CDM (CDM), CDM plus cosmological constant (ΛCDM), CDM plus some hot dark matter (νCDM), and CDM plus added relativistic particles (τCDM) (from Dodelson et al., 1996).

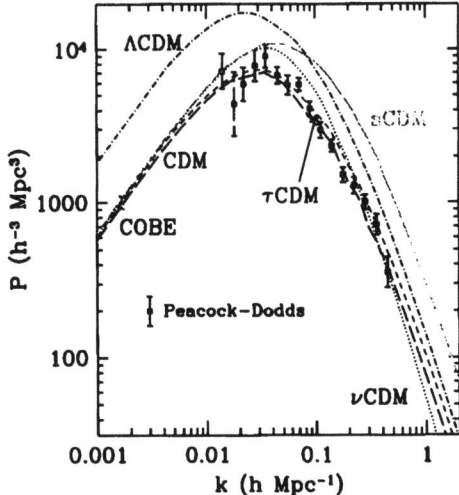

Figure 14. The power spectrum of fluctuations today, as traced by bright galaxies (light), as derived from redshift surveys assuming light traces mass (Peacock and Dodds, 1994). The curves correspond to the predictions of various cold dark matter models. The relationship between the power spectrum and CMB anisotropy in a ΛCDM model is different, and in fact, the ΛCDM model shown is COBE normalized.

be seen whether or not the expansion is accelerating, it is a fact that our knowledge of the Universe is accelerating, driven by new observational results. Cosmology seems to be in the midst of a golden age; within ten years we may have a cosmological theory that explains almost all the fundamental features of the Universe—the smoothness and flatness, the heat of the CMB, the baryon asymmetry, and the origin of structure.

There are still larger questions to be answered and to be asked. What is the global topology of the Universe? Did the Universe begin with more than four dimensions? Is inflation the dynamite of the big bang, and were there other such big bangs? Are there cosmological signatures of the quantum gravity epoch?

It is difficult—and dangerous—to speculate where cosmology will go in the next twenty years, let alone the next hundred. One can never predict the serendipitous discovery that radically transforms our understanding. In an age of expensive, complex and highly focused experiments, we must be especially vigilant and keep an open mind. And it can be argued that the two most important discoveries in cosmology—the expansion and the CMB—were unexpected. In astrophysics, it is usually a safe bet that things are more complicated than expected. But then again, Einstein's ansatz of large-scale homogeneity and isotropy—made to make the equations of general relativity tractable—turned out to be a remarkably good description of the Universe.

References

Albrecht, A., and P.J. Steinhardt, 1982, Phys. Rev. Lett. **48**, 1220.
Albrecht, A., *et al.*, 1982, Phys. Rev. Lett. **48**, 1437.
Allen, B., *et al.*, 1997, Phys. Rev. Lett. **79**, 2624.
Bahcall, N., *et al.*, 1993, Astrophys. J. Lett. **415**, L17.

Bardeen, J., P.J. Steinhardt, and M.S. Turner, 1983, Phys. Rev. D **28**, 679.
Baum, W. A., 1957, Astrophys. J. **62**, 6.
Birkinshaw, M., 1998, Phys. Rep. in press (astro-ph/9808050).
Blandford, R.D., and R. Narayan, 1992, Annu. Rev. Astron. Astrophys. **30**, 311.
Blumenthal, G., S. Faber, J. Primack, and M. Rees, 1984, Nature (London) **311**, 517.
Broadhurst, T.J., R. Ellis, D.C. Koo, and A.S. Szalay, 1990, Nature (London) **343**, 726.
Burles, S., and D. Tytler, 1998a, Astrophys. J. **499**, 699.
Burles, S., and D. Tytler, 1998b, Astrophys. J. **507**, 732.
Carlberg, R.G., H.K.C. Lee, E. Ellingson, R. Abraham, P. Gravel, S. Morris, and C.J. Pritchet, 1996, Astrophys. J. **462**, 32.
Carlstrom, J. 1999, Phys. Scr., in press.
Carroll, S.M., W.H. Press, and E.L. Turner, 1992, Annu. Rev. Astron. Astrophys. **30**, 499.
Chaboyer, B., P. Demarque, P. Kernan, and L. Krauss, 1998, Astrophys. J. **494**, 96.
Clowe, D., G.A. Luppino, N. Kaiser, J.P. Henry, and I.M. Gioia, 1998, Astrophys. J. **497**, L61.
Colless, M., 1998, Philos. Trans. R. Soc. London, Ser. A, in press (astro-ph/9804079).
Collins, C.B., and S.W. Hawking, 1973, Astrophys. J. **180**, 317.
Cowan, J., F. Thieleman, and J. Truran, 1991, Annu. Rev. Astron. Astrophys. **29**, 447.
Dekel, A., D. Burstein, and S.D.M. White, 1997, in *Critical Dialogues in Cosmology*, edited by N. Turok (World Scientific, Singapore).
Dicke, R.H., and P.J.E. Peebles, 1979, in *General Relativity: An Einstein Centenary Survey*, edited by S. Hawking and W. Israel (Cambridge University Press, Cambridge), p. 504.
Dodelson, S., E.I. Gates, and M.S. Turner, 1996, Science **274**, 69.
Evrard, A.E., 1997, Mon. Not. R. Astron. Soc. **292**, 289.
Filippenko A.V., and A.G. Riess, 1998, astro-ph/9807008.
Fixsen, D.J., *et al.*, 1996, Astrophys. J. **473**, 576.
Fukuda, Y., *et al.* (SuperKamiokande Collaboration), 1998, Phys. Rev. Lett. **81**, 1562.
Gaillard, M.K., P. Grannis, and F. Sciulli, 1999, Rev. Mod. Phys. **71**, 96; pp. 161–187 in this book.
Garnavich, P., *et al.*, 1998, Astrophys. J., in press (astro-ph/9806396).
Ge, J., J. Bechtold, and J.H. Black, 1997, Astrophys. J. **474**, 67.
Geller, M., and J. Huchra, 1989, Science **246**, 897.
Gunn, J.E., M. Carr, C. Rockosi, and M. Sekiguchi, 1998, Astron. J., in press.
Guth, A., 1982, Phys. Rev. D **23**, 347.
Guth, A., and S.-Y. Pi, 1982, Phys. Rev. Lett. **49**, 1110.
Harrison, E.R., 1970, Phys. Rev. D **1**, 2726.
Hawking, S.W., 1982, Phys. Lett. B **115**, 295.
Jungman, G., M. Kamionkowski, and K. Griest, 1996, Phys. Rep. **267**, 195.
Jungman, G., M. Kamionkowski, A. Kosowsky, and D.N. Spergel, 1996, Phys. Rev. Lett. **76**, 1007.
Kolb, E. W., and M. S. Turner, 1990, *The Early Universe* (Addison-Wesley, Redwood City).
Kundic, T., *et al.*, 1997, Astrophys. J. **482**, 75.
Krauss, L., and M.S. Turner, 1995, Gen. Relativ. Gravit. **27**, 1137.
Leibundgut, B., *et al.*, 1996, Astrophys. J. Lett. **466**, L21.
Liddle, A., and D. Lyth, 1993, Phys. Rep. **231**, 1.
Lidsey, J., *et al.*, 1997, Rev. Mod. Phys. **69**, 373.
Linde, A., 1982, Phys. Lett. B **108**, 389.
Madau, P., 1999, Phys. Scr. (to be published).
Madore, B., *et al.*, 1998, Nature (London) **395**, 47.
Mather, J.C., *et al.*, 1990, Astrophys. J. Lett. **354**, L37.
Oort, J.H., 1983, Annu. Rev. Astron. Astrophys. **21**, 373.
Ostriker, J. P., and P. J. Steinhardt, 1995, Nature (London) **377**, 600.
Oswalt, T.D., J.A. Smith, M.A. Wood, and P. Hintzen, 1996, Nature (London) **382**, 692.
Paczynski, B., 1986, Astrophys. J. **304**, 1.
Peacock, J., and S. Dodds, 1994, Mon. Not. R. Astron. Soc. **267**, 1020.

Peebles, P.J.E., 1987, Nature (London) **327,** 210.
Peebles, P.J.E., 1993, *Principles of Physical Cosmology* (Princeton University Press, Princeton).
Peebles, P.J.E., D.N. Schramm, E.L. Turner, and R.G. Kron, 1991, Nature (London) **352,** 769.
Pen, U.-L., *et al.*, 1997, Phys. Rev. Lett. **79,** 1611.
Perlmutter, S., *et al.*, 1998, Astrophys. J., in press (astro-ph/9812133).
Riess, A.G., W.H. Press, and R.P. Kirshner, 1996, Astrophys. J. **473,** 88.
Rosenberg, L.J., 1998, Proc. Natl. Acad. Sci. USA **95,** 59.
Sadoulet, B., 1999, Rev. Mod. Phys. **71,** 197; pp. 329–342 in this book.
Sandage, A., 1961, Astrophys. J. **133,** 355.
Schmidt, B., *et al.*, 1998, Astrophys. J. **507,** 46.
Schramm, D., and M. Turner, 1998, Rev. Mod. Phys. **70,** 303.
Schwarz, J., and N. Seiberg, 1999, Rev. Mod. Phys. **71,** 112; pp. 188–202 in this book.
Shectman, S., *et al.*, 1996, Astrophys. J. **470,** 172.
Sloan Digital Sky Survey (SDSS), see http://www.sdss.org
Smoot, G., *et al.*, 1992, Astrophys. J. Lett. **396,** L1.
Songaila, A., L.L. Cowie, M. Keane, A.M. Wolfe, E.M. Hu, A.L. Oren, D. Tytler, and K.M. Lanzetta, 1994, Nature (London) **371,** 43.
Starobinskii, A.A., 1982, Phys. Lett. B **117,** 175.
Steigman, G., D.N. Schramm, and J. Gunn, 1977, Phys. Lett. B **66,** 202.
Two-degree Field (2dF), see http://msoww.anu.edu.au/~colless/2dF/
Tyson, J.A., 1993, Phys. Today **5,** 24.
Tyson, J.A., G. Kochanski, and I.P. Dell'Antonio, 1998, Astrophys. J. Lett. **498,** L107.
Vilenkin, A., and E.P.S. Shellard, 1994, *Cosmic Strings and Other Topological Defects* (Cambridge University Press, Cambridge).
Virgo Collaboration, see http://www.mpa-garching.mpg.de/~jgc/sim_virgo.html
Weinberg, S., 1972, *Gravitation and Cosmology: Principals and Applications of the General Theory of Relativity* (Wiley, New York).
Weinberg, S., 1989, Rev. Mod. Phys. **61,** 1.
White, S.D.M., C. Frenk, and M. Davis, 1983, Astrophys. J. Lett. **274,** L1.
Wilkinson, D.T., 1999, Rev. Mod. Phys., in press.
Willmer, C.N.A., D.C. Koo, A.S. Szalay, and M.J. Kurtz, 1994, Astrophys. J. **437,** 560.

Cosmic Rays: The Most Energetic Particles in the Universe
James W. Cronin

Cosmic rays are a source of ionizing radiation incident on the whole earth. The intensity of this ionizing radiation varies with magnetic latitude, with altitude, and with solar activity. The attribution "cosmic rays" is misleading in that the radiation consists principally of fully ionized atomic nuclei incident on the earth from outer space.

The field of elementary-particle physics owes its origin to discoveries made in course of cosmic-ray research, and the study of cosmic rays has contributed to the understanding of geophysical, solar, and planetary phenomena. The existence of cosmic rays also has its practical side. An example is radio-carbon dating, first suggested by Libby (1965). Radioactive C^{14} is produced by the collisions of the cosmic rays with the N^{14} in the atmosphere. This produces an activity of 15 disintegrations per minute per gram of natural carbon in all living matter. On death, the C^{14} decays with a half-life of 5600 years. Thus the specific activity of C^{14} provides an accurate archeological clock for the dating of objects in history and prehistory.

This article presents a very personal view of the most important questions for future research. I restrict it to energies well above 1 TeV (10^{12} eV) where most of the observations are ground based due to low fluxes. As in many fields, new technologies permit unique investigations that could only be dreamed of in the past. If we take a broad definition of cosmic rays they consist not only of electrons and nuclei, but of other particles as well, particularly gamma rays and neutrinos, which, being neutral, point back to their source.

At present there are many programs under development around the world that seek to measure high-energy neutrinos in the primary cosmic radiation (Gaisser *et al.*, 1995). These are neutrinos that come directly from astrophysical sources, as distinct from being produced by ordinary cosmic rays in the atmosphere. The detectors consist of large volumes of antarctic ice or seawater instrumented with photomultipliers. At present there are major experimental efforts under way or proposed. It is expected that high-energy neutrino detectors will make discoveries in astronomy, cosmology, and fundamental-particle physics.

In recent years astronomy has been extended to sources emitting γ rays with energies more than 100 GeV. Numerous galactic and extragalactic sources have been observed with ground-based instruments which detect the Čerenkov radiation emitted by the showering of the high-energy γ rays. At these energies the satellite detectors, the Compton

Gamma-Ray Observatory, and even the new detector Gamma-ray Large Area Space Telescope (to be launched about 2005) do not have the sensitivity necessary to observe sources at energies above 100 GeV. This rapidly expanding area of astronomy has been the subject of a number of recent reviews (Weekes *et al.*, 1998; Ong, 1998).

In the remainder of this paper I shall concentrate on the cosmic rays above 10^{14} eV where most observations have been made with ground-based instruments.

A Brief History

The history of research in cosmic rays is a fascinating one, filled with serendipity, personal conflict, and experiments on a global scale. The discovery of cosmic rays, attributed to Victor Hess (1912), had its origin in the obsession of some scientists to understand why a heavily shielded ion chamber still recorded radiation. It was assumed that this was some residual radiation from the earth's surface and by placing the ion chamber at some distance above the earth's surface the detected radiation would be reduced. When Victor Hess took an ion chamber several thousand meters above the earth in a balloon, it was found that the radiation level actually rose, leading to the conclusion that the radiation was arriving from outer space.

It took more than 30 years to discover the true nature of the cosmic radiation, principally positively charged atomic nuclei arriving at the top of the atmosphere (Sekido and Elliot, 1985; Simpson, 1995). Many hypotheses were offered for the nature of these cosmic rays. One of the most interesting ideas was that of Robert A. Millikan (Millikan and Cameron, 1928). Millikan noted Aston's discovery of nuclear binding energies. He suggested that the cosmic rays were the result of the formation of complex nuclei from primary protons and electrons. In the 1920s electrons and ionized hydrogen were the only known elementary particles to serve as building blocks for atomic nuclei. The formation of atomic nuclei was assumed to be taking place throughout the universe, with the release of the binding energy in the form of gamma radiation, which was the "cosmic radiation." A consequence of this hypothesis was that the cosmic radiation was neutral and would not be influenced by the earth's magnetic field. A worldwide survey led by Arthur Compton demonstrated conclusively that the intensity of the cosmic radiation depended on the magnetic latitude (Compton 1933). The cosmic radiation was predominately charged particles. This result was the subject of an acrimonious debate between Compton and Millikan at an AAAS meeting that made the front page of the *New York Times* on December 31, 1932.

In 1938, Pierre Auger and Roland Maze, in their Paris laboratory, showed that cosmic-ray particles separated by distances as large as 20 meters arrived in time coincidence (Auger and Maze, 1938), indicating that the observed particles were secondary particles from a common source. Subsequent experiments in the Alps showed that the coincidences continued to be observed even at a distance of 200 meters. This led Pierre Auger, in his 1939 article in *Reviews of Modern Physics*, to conclude

One of the consequences of the extension of the energy spectrum of cosmic rays up to 10^{15} eV is that it is actually impossible to imagine a single process able to give to a particle such an energy. It seems much more likely that the charged particles which constitute the primary cosmic radiation acquire their energy along electric fields of a very great extension. (Auger *et al.*, 1939).

Auger and his colleagues discovered that there existed in nature particles with an energy of 10^{15} eV at a time when the largest energies from natural radioactivity or artificial acceleration were just a few MeV. Auger's amazement at Nature's ability to produce particles of enormous energies remains with us today, as there is no clear understanding of the mechanism of production, nor is there sufficient data available at present to hope to draw any conclusions.

In 1962 John Linsley observed a cosmic ray whose energy was 10^{20} eV (Linsley, 1962). This event was observed by an array of scintillation counters spread over 8 km² in the desert near Albuquerque, New Mexico. The energetic primary was detected by sampling some of the 5×10^{10} particles produced by its cascade in the atmosphere. Linsley's ground array was the first of a number of large cosmic-ray detectors that have measured the cosmic-ray spectrum at the highest energies.

Cosmic-Ray Spectrum

After 85 years of research, a great deal has been learned about the nature and sources of cosmic radiation (Zatsepin *et al.*, 1966; Berezinskii *et al.*, 1990; Watson, 1991; Cronin, 1992; Sokolsky *et al.*, 1992; Swordy, 1994; Nagano, 1996; Yoshida *et al.*, 1998). In Fig. 1 the spectrum of cosmic rays is plotted for energies above 10^8 eV. The cosmic rays are

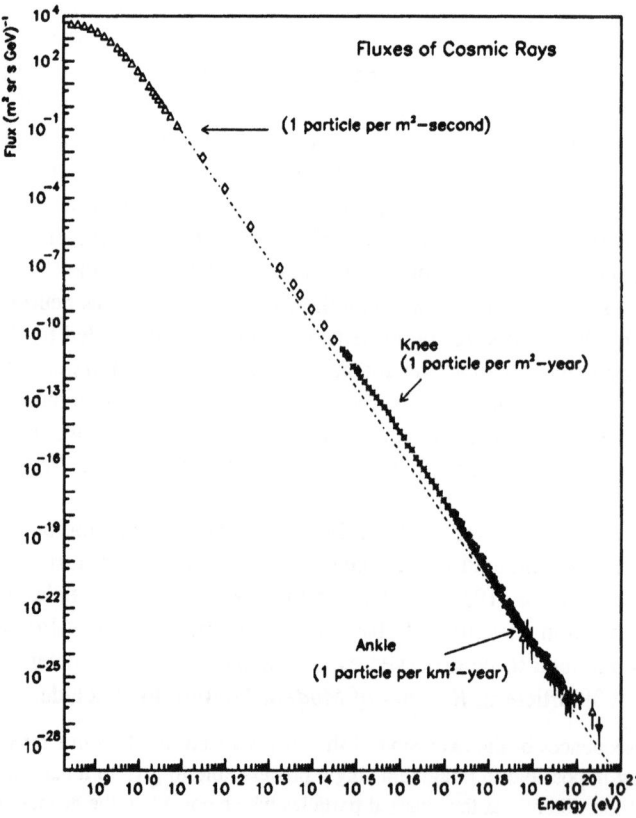

Figure 1. Spectrum of cosmic rays greater than 100 MeV. This figure was produced by S. Swordy, University of Chicago.

predominately atomic nuclei ranging in species from protons to iron nuclei, with traces of heavier elements. When ionization potential is taken into account, as well as spallation in the residual gas of space, the relative abundances are similar to the abundances of elements found in the sun. The energies range from less than 1 MeV to more than 10^{20} eV. The differential flux is described by a power law:

$$dN/dE \sim E^{-\alpha}, \qquad (3.1)$$

where the spectral index α is roughly 3, implying that the intensity of cosmic rays above a given energy decreases by a factor of 100 for each decade in energy. The flux of cosmic rays is about 1/cm^2/sec at 100 MeV and only of order 1/km^2/century at 10^{20} eV.

The bulk of the cosmic rays are believed to have a galactic origin. The acceleration mechanism for these cosmic rays is thought to be shock waves from supernova explosions. This basic idea was first proposed by Enrico Fermi (1949), who discussed the acceleration of cosmic rays as a process of the scattering of the charged cosmic-ray particles off moving magnetic clouds. Subsequent work has shown that multiple "bounces" off the turbulent magnetic fields associated with supernova shock waves is a more efficient acceleration process (Drury, 1983). At present there is no direct proof of this hypothesis. The argument for it is based on the fact that a fraction of the energy released by supernova explosions is sufficient to account for the energy being pumped into cosmic rays. A second point in favor of the hypothesis is that the index of the spectrum, 2.7 below 5×10^{15} eV, is consistent with shock acceleration when combined with the fact that the lifetime of the cosmic rays in our galaxy is about 10^7 years due to leakage of the rays out of the "bottle" provided by the magnetic field of our galaxy. Shock acceleration would provide an index of 2.0. The leakage out of the galaxy accounts for the steeper spectrum given by 2.7.

The spectrum steepens (the knee) to an index of 3.0 at about 5×10^{15} eV. In the most recent experiments, this bend in the spectrum is gradual. The conventional explanation of the knee is that the leakage of the cosmic rays from the galaxy depends on the magnetic rigidity E/Z. The knee results from the fact that, successively, the lighter components of cosmic rays are no longer contained in the galaxy as the energy increases. This hypothesis requires that the mean atomic number of the cosmic rays becomes progressively heavier as the energy rises. At the present time this prediction has not been convincingly demonstrated.

Techniques of Measurement

At energies below 10^{14} eV the flux of primary cosmic rays is sufficient to be measured directly with instruments on balloons and satellites. Above 10^{14} eV the flux is about 10/m^2/day. At this energy very-large-area detectors are required to measure the cosmic rays directly. But fortunately at this energy the cascades in the atmosphere produce a sufficient number of particles on the earth's surface so that the primary cosmic ray can be observed indirectly by sampling the cascade particles on the ground. This technique is just an application of Auger's experiment with modern technology. Observations made with a surface array of particle detectors can adequately measure the total energy and the direction of the primary cosmic ray. It should be noted that the atmosphere is an essential part of a surface detector. The technique has been extended to instruments that cover as much as 100 km^2 with individual detector spacings of 1 km. Much larger arrays will eventually be built. At energies above 10^{18} eV the density of particles at a fixed distance (500–1000 m) from the shower axis is proportional to the primary energy. The constant of proportionality is calculated by shower simulation.

A second technique has been used to measure the spectrum above 10^{17} eV. Optical photons in the range 300 nm to 400 nm are produced by the passage of the charged particles through the nitrogen of the atmosphere (Baltrusaitus et al., 1985; Kakimoto, 1996). About four fluorescence photons are produced per meter for each charged shower particle. With an array of photomultipliers, each focused on a part of the sky, the longitudinal development of a shower can be directly measured and the energy inferred from the total amount of fluorescence light. The limitation of this technique is that it can only function on dark moonless nights, which amounts to only 10% of the time. The positive aspect of the technique is that it rather directly measures the energy of the shower dissipated in the atmosphere, which in most cases is a large fraction of the primary energy. Absolute knowledge of the fluorescence efficiency of the nitrogen, the absorption of the atmosphere, and the quantum efficiency and gain of the photomultipliers is required.

Neither technique is particularly effective in identifying the nature of the primary (nucleon, nucleus, or photon). The mean fraction of energy contained in the muonic component of the shower particles increases as the primary becomes heavier. The mean depth in the atmosphere where the cascade is at its maximum moves higher as the primary becomes heavier. Because of fluctuations in these quantities, neither technique offers hope of identifying the nature of the primary on an event by event basis.

Properties of Cosmic Rays above 10^{17} eV

Above 10^{17} eV the cosmic-ray spectrum shows additional structure. This structure is displayed in Fig. 2, where the differential spectrum has been multiplied by E^3 to better expose the observed structures. These data are the combined results of four experiments that have operated over the past 20 years. They are from the Haverah Park surface array in England (Lawrence et al., 1991), the Yakutsk surface array in Siberia (Afanasiev et al., 1995), the Fly's Eye fluorescence detector in Utah (Bird et al., 1994), and the AGASA surface array in Japan (Yoshida et al., 1995). Before plotting, the energy scale of each experiment was adjusted by amounts ≤20% to show most clearly the common features. The method of energy determination in each of these experiments is quite different, and the fact that they agree within 20% is remarkable.

Above 5×10^{17} eV the spectrum softens from an index of 3.0 to an index of 3.3. Above 5×10^{18} eV the spectrum hardens, changing to an index of 2.7. Beyond 5×10^{19} eV the data are too sparse to be certain of the spectral index. There is no clear explanation of this structure. Above 10^{18} eV, the galactic magnetic fields are not strong enough to act as a magnetic "bottle" even for iron nuclei. If the cosmic rays continue to be produced in the galaxy, they should show an anisotropy that correlates with the galactic plane. No such anisotropy has been observed. The hardening of the spectrum to an index of 2.7 above 5×10^{18} eV may then be a sign of an extragalactic component emerging as the galactic component dies away.

The Difficulty of Acceleration

Above 10^{19} eV the precision of the spectrum measurement suffers from lack of statistics. There have been about 60 events recorded with energy greater than 5×10^{19} eV. Yet it is above this energy that the scientific mystery is the greatest. There is little understanding of how known astrophysical objects could produce particles of such energy. At the most primitive level, a necessary condition for the acceleration of a proton to an energy E in units of 10^{20} eV is that the product of the magnetic field B and the size of the region R be much larger than 3×10^{17} G-cm. This value is appropriate for a perfect accelerator such as might be scaled up from the Tevatron at Fermilab. The Tevatron has a product $BR = 3 \times 10^9$ G-cm and accelerates protons to 10^{12} eV. Analogous acceleration of cosmic rays to energies above 10^{19} eV seems difficult, and the literature is filled with speculations. Two reviews that discuss the basic requirements are those of Greisen (1965) and Hillas (1984). While these were written some time ago, they are excellent in outlining the basic problem of cosmic-ray acceleration. Biermann (1997) has recently reviewed all the ideas offered for achieving these high energies. Hillas in his outstanding review of 1984 presented a plot that graphically shows the difficulty of cosmic-ray acceleration to 10^{20} eV. Figure 3 is an adaptation of his figure. Plotted are the size and strength of possible acceleration sites. The upper limit on the energy is given by

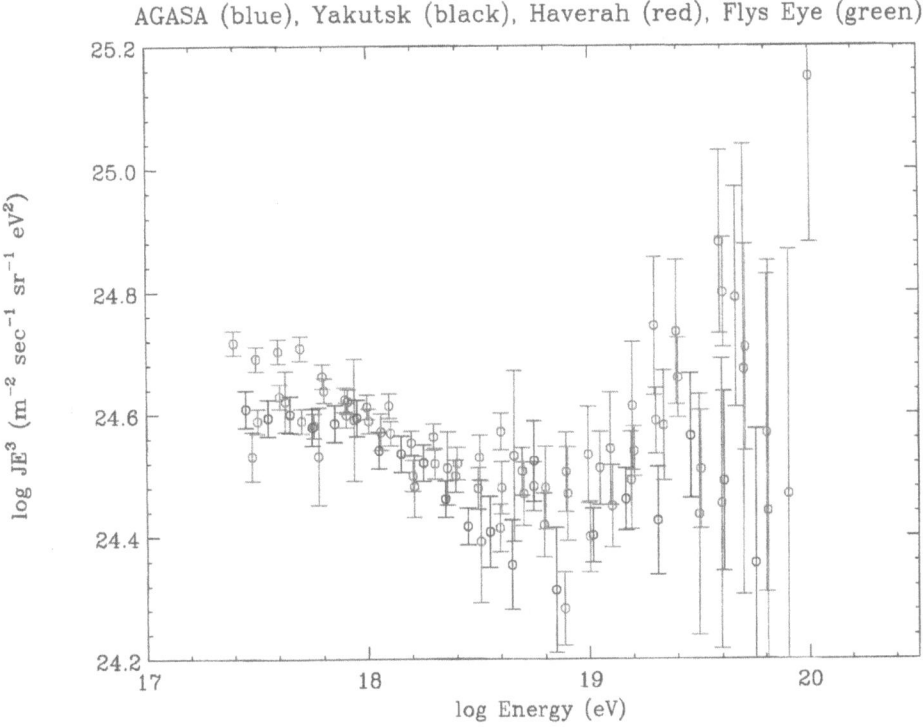

Figure 2. Upper end of the cosmic-ray spectrum. Haverah Park points (red; Lawrence et al., 1991) serve as a reference. Yakutsk points (black; Afanasiev et al., 1995) have been reduced in energy by 20%. Fly's Eye points (green; Bird et al., 1995) have been raised in energy by 10%. AGASA points (Yoshida et al., 1995) have been reduced by 10%. (See Color Plate 2.)

$$E_{18} \leq 0.5\beta ZB_{\mu G}L_{kpc}. \tag{6.1}$$

Here the E_{18} is the maximum energy measured in units of 10^{18} eV. L_{kpc} is the size of the accelerating region in units of kiloparsecs, and $B_{\mu G}$ is the magnetic field in μG. The factor β was introduced by Greisen to account for the fact that the effective magnetic field in the accelerator analogy is much less than the ambient field. The factor β in Hillas's discussion is the velocity of the shock wave (relative to c), which provides the acceleration. The plotted lines correspond to a 10^{20} eV proton with $\beta=1$ and 1/300. A line is also plotted for iron nuclei ($\beta=1$). With $Z=26$, iron is in principle easier to accelerate. Realistic accelerators should lie well above the dashed line. The figure is also relevant for "one-shot" acceleration, as it represents the electromotive force (emf) induced in a conductor of length L moving with a velocity β through a uniform magnetic field B.

Synchrotron energy loss is also important. For protons the synchrotron loss rate at 10^{20} eV requires that the magnetic field be less than 0.1 G for slow acceleration (the accelerator analogy; Greisen 1965). From Fig. 3 it can be seen that the acceleration of cosmic rays to 10^{20} eV is not a simple matter. Because of this, some authors have seriously postulated that cosmic rays are not accelerated but are directly produced by "top down" processes. For example, defects in the fabric of spacetime could have huge energy content and could release this energy in the form of high-energy cosmic rays (Bhattacharjee, Hill, and Schramm, 1992).

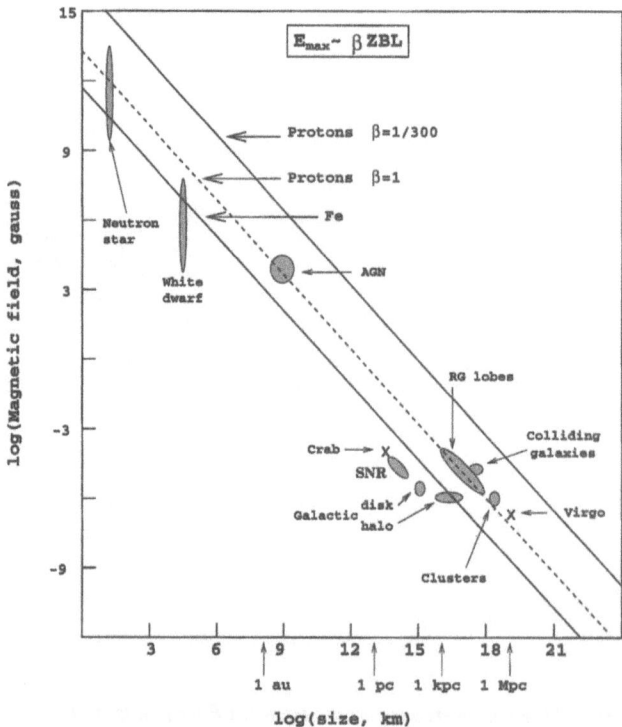

Figure 3. Modified Hillas plot (Hillas, 1984). Size and magnetic field of possible sites of acceleration. Objects below the dashed line cannot accelerate protons to 10^{20} eV.

Nature's Diagnostic Tools

There are some natural diagnostic tools that make the analysis of the cosmic rays above 5×10^{19} eV easier than at lower energies. The first of these is the 2.7-K cosmic background radiation (CBR). Greisen (1966) and Zatsepin and Kuz'min (1966) pointed out that protons, photons, and nuclei all interact strongly with this radiation, a phenomenon that has become known as the GZK effect. As an example, a collision of a proton of 10^{20} eV with a CBR photon of 10^{-3} eV produces several hundred MeV in the center-of-mass system. The cross section for pion production is quite large so that collisions are quite likely, resulting in a loss of energy for the primary proton. In Fig. 4 we plot the results of the propagation of protons through the CBR. Regardless of the initial energy of the proton, it will be found with less than 10^{20} eV after propagating through a distance of 100 Mpc (3×10^8 light years). Thus the observation of a cosmic-ray proton with energy greater than 10^{20} eV implies that its distance of travel is less than 100 Mpc. This distance corresponds to a redshift of 0.025 and is small compared to the size of the universe. Similar arguments can be made for nuclei or photons in the energy range considered. There are a limited number of possible sources that fit the Hillas criteria (Fig. 3) within a volume of radius 100 Mpc about the earth.

The fact that the cosmic rays, if protons, will be little deflected by galactic and extragalactic magnetic fields serves as the second diagnostic tool. The deflection of protons of energy 5×10^{19} eV by the galactic magnetic field (~ 2 μG) and the intergalactic magnetic fields ($\leq 10^{-9}$ G) is only a few degrees (Kronberg, 1994a, 1994b), so that above 5×10^{19} eV it is possible that the cosmic rays will point to their sources. We approach an astronomy, even for charged cosmic rays, in which the distance to the possible sources is limited.

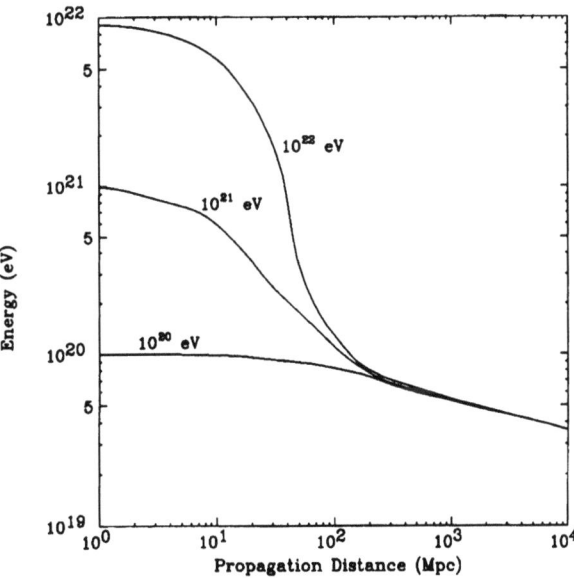

Figure 4. Proton energy as a function of propagation distance through the 2.7-K cosmic background radiation for the indicated initial energies.

Cosmic-Ray Astronomy

The energy 5×10^{19} eV represents a lower limit for which the notion of an astronomy of charged particles from "local" sources can be applied. The GZK effect enhances the number of events from sources within a distance of 100 Mpc. Of these events, two particularly stand out with energies reported to be 2×10^{20} eV by the AGASA experiment (Hayashida et al., 1994) and 3×10^{20} eV by the Fly's Eye experiment (Bird et al., 1995; Elbert and Sommers, 1995). More recently a total of six events with energy $\geq 10^{20}$ eV have been reported by the AGASA experiment (Takeda et al., 1998). For all these events the probable distance to the source is less than 50 Mpc.

The events above 5×10^{19} eV are too few to derive a spectral index. It is not clear that a single spectrum is even the proper way to characterize these events. Since they must come from "nearby," the actual number of sources may not form an effective continuum in space, so the spectrum observed may vary with direction. The matter within 100 Mpc is not uniformly distributed over the sky. It is probably more fruitful to take an astronomical approach and plot the arrival directions of these events on the sky in galactic coordinates.

Arrival-direction data are available for the Haverah Park experiment (Watson, 1997), the AGASA experiment (Hayashida et al., 1996), and for the most energetic event recorded by the Fly's Eye experiment (Bird et al., 1995; Elbert and Sommers, 1995). In Fig. 5 we plot the arrival directions of 20 AGASA events and 16 Haverah Park events. The size of the symbols corresponds to the angular resolution. In addition, the error box for the most energetic event recorded by the Fly's Eye experiment is plotted. What is remarkable in this figure is the number of coincidences of cosmic rays coming from the same direction in the sky. Of the 20 events reported by AGASA, there are two pairs. The probability of a chance coincidence for this is about 2%. The addition of the Haverah Park events shows a coincidence with one of the AGASA pairs. However, the Fly's Eye event coincides with one of the AGASA events. It is not possible to estimate properly the probability of chance overlaps, but the possibility that these overlaps may be real should not be ignored. The triple coincidence contains the AGASA event of 2×10^{20} eV, the Haverah Park event of about 1×10^{20} eV, and the AGASA event of 5×10^{19} eV. The Fly's Eye event of 3×10^{20} eV is in coincidence with the AGASA event of 6×10^{19} eV. The third pair contains AGASA events of 6×10^{19} eV and 8×10^{19} eV, respectively.

The triple coincidence is particularly interesting if it is not the result of pure chance. It contains cosmic rays separated by a factor of 4 in energy that have not been separated in space by more than a few degrees. This is an encouraging prospect for future experiments in which, with many more events, one may observe point sources, clusters, and larger-scale anisotropies in the sky. The crucial questions will be: Does the distribution of cosmic rays in the sky follow the distribution of matter within our galaxy or the distribution of "nearby" extragalactic matter, or is there no relation to the distribution of matter? Are there point sources or very tight clusters? What is the energy distribution of events from these clusters? Are these clusters associated with specific astrophysical objects? If there is no spatial modulation or no correlation with observed matter, what is the spectrum? This situation would imply an entirely different class of sources, which are visible only in the "light" of cosmic rays with energy $\geq 5 \times 10^{19}$ eV. Of course there may be a combination of these possibilities. If even crude data on primary composition are available, they can be divided into categories of light and heavy components, which may have different distributions. Crucial to these considerations is uniform exposure over the whole sky. And a final and fundamental question is: Is there an end to the cosmic-ray spectrum?

New Experiments

The flux of cosmic rays with energy $\geq 5 \times 10^{19}$ eV is about 0.03/km^2/sr/yr. It required five years for the AGASA array, with an acceptance of 125 km^2-sr, to collect 20 events above this energy. In 1999 an improved version of the Fly's Eye experiment (HiRes) will begin operation (Abu-Zayyad, 1997). It will have an acceptance of about 7000 km^2-sr above 5×10^{19} eV. With a 10% duty cycle it should collect about 20 events per year. The experiment will be located in northern Utah. Only half of the sky will be observed.

Experiments with far greater statistical power are required to make real progress. It is very likely that a combination of types of sources and phenomena are responsible for the highest-energy cosmic rays. Thus the experiment must be constructed so as not to have a bias towards a particular or single explanation for the cosmic rays. An ideal experiment should have uniform coverage of the entire sky. It should also be fully efficient at energies beginning at 10^{19} eV, as the present data available above that energy are very sparse. It

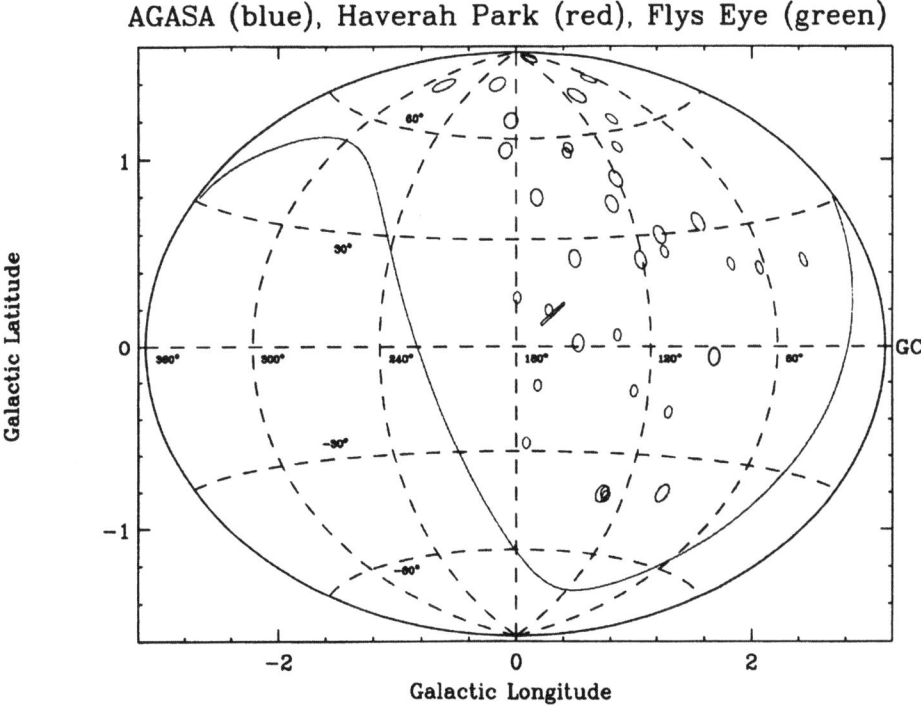

Figure 5. Plot of arrival directions of cosmic rays with energy $\geq 5 \times 10^{19}$: red points, Haverah Park (Lawrence et al., 1991); blue points, AGASA (Yoshida et al., 1995); green point, Fly's Eye event with energy 3×10^{20} eV. The size of the symbols represents the resolution of each experiment. The empty region marked by the blue line is the part of the sky not seen by the northern hemisphere location of the observations. (See Color Plate 3.)

should have the best possible means to identify the primary particle, although no experiment can make a unique identification on an event by event basis.

A number of experiments have been proposed or will be proposed in the next few years. These are all described in the Proceedings of the 25th International Cosmic-Ray Conference held in Durban in 1997. One of these seeks to satisfy all the general requirements outlined above. The experiment of the Pierre Auger Observatories (Boratav, 1997) consists of two detectors with acceptance 7000 km^2-sr. They will be located at midlatitude in the southern and northern hemispheres, which will provide nearly uniform sky coverage. An important feature of the Auger experiment is its use of a hybrid detector that combines both a surface array and a fluorescence detector. Such an experiment will collect ~450 events $\geq 5 \times 10^{19}$ eV each year. Some 20% of the events may originate from point sources or tight clusters if the AGASA results (Hayashida *et al.*, 1996) are used as a guide.

Also being proposed is an all-fluorescence detector called the Telescope Array (Telescope Array Collaboration, 1997) to be located in the northern hemisphere. It would have an aperture of 70 000 km^2-sr (7000 km^2-sr with the 10% duty cycle). It would also co-locate two of its fluorescence units with the northern Auger detector.

A visionary idea has been offered in which the fluorescence light produced by a cosmic ray in the atmosphere would be viewed from a satellite (Linsley, 1997; Krizmanic, Ormes, and Streitmatter, 1998). There are many technical difficulties in such a project. It would, however, represent a next step in the investigations if the projects above are realized and no end to the cosmic-ray spectrum is observed. The estimated sensitivity of such a satellite detector for cosmic rays with energy $\geq 10^{20}$ eV would be 10–100 times that of the Pierre Auger Observatories.

Conclusion

It is now widely recognized that the investigation of the upper end of the cosmic-ray spectrum will produce new discoveries in astrophysics or fundamental physics. There are a number of complementary proposals for new experiments that will provide the needed observations within the next ten years.

Acknowledgments

Over the years I have learned much from discussions with many individuals concerning the highest-energy cosmic rays. Prominent among these are V. Berezinskii, P. Biermann, M. Boratav, T. Gaisser, A. M. Hillas, P. P. Kronberg, M. Nagano, R. Ong, and A. A. Watson. Support from the U.S. National Science Foundation is gratefully acknowledged.

References

Abu-Zayyad, T., 1997, in *Proceedings of the 25th International Cosmic Ray Conference*, Durban, edited by M. S. Potgieter, B. C. Raubenheimer, and D. J. van der Walt (World Scientific, Singapore), Vol. 5, p. 321; this paper and the eleven that immediately follow describe various aspects of the HiRes detector.

Afanasiev, B. N., *et al.*, 1995, in *Proceedings of the 24th International Cosmic Ray Conference*, Rome, edited by N. Lucci and E. Lamanna (University of Rome, Rome), Vol. 2, p. 756.

Auger, P., and R. Maze, 1938, C. R. Acad. Sci. Ser. B **207**, 228.
Auger, P., P. Ehrenfest, R. Maze, J. Daudin, Robley, and A. Fréon, 1939, Rev. Mod. Phys. **11**, 288.
Baltrusaitis, R. M., *et al.*, 1985, Nucl. Instrum. Methods Phys. Res. A **240**, 410.
Berezinskii, V. S., S. V. Bulanov, V. A. Dogiel, V. L. Ginzburg, and V. S. Ptuskin, 1990, in *Astrophysics of Cosmic Rays*, edited by V. L. Ginzburg (Elsevier Science, New York/Amsterdam).
Bhattacharjee, P., C. T. Hill, and D. N. Schramm, 1992, Phys. Rev. Lett. **69**, 567.
Biermann, P., 1997, J. Phys. G **23**, 1.
Bird, D. J., *et al.*, 1994, Astrophys. J. **424**, 491.
Bird, D. J., *et al.*, 1995, Astrophys. J. **441**, 144.
Boratav, M., 1997, in *Proceedings of the 25th International Cosmic Ray Conference*, Durban, edited by M. S. Potgieter, B. C. Raubenheimer, and D. J. van der Walt (World Scientific, Singapore), Vol. 5, p. 205; this paper and the five that immediately follow describe various aspects of the Pierre Auger Observatories.
Compton, A. H., 1933, Phys. Rev. **43**, 387.
Cronin, J. W., 1992, Nucl. Phys. B (Proc. Suppl.) **28**, 213.
Drury, L. O'C., 1983, Rep. Prog. Phys. **46**, 973.
Elbert, J. W., and P. Sommers, 1995, Astrophys. J. **441**, 151.
Fermi, E., 1949, Phys. Rev. **75**, 1169.
Gaisser, T. K., F. Halzen, and T. Stanev, 1995, Phys. Rep. **258**, 173.
Greisen, K., 1965, in *Proceedings of the 9th International Cosmic Ray Conference*, London (The Institute of Physics and The Physical Society, London), Vol. 2, p. 609.
Greisen, K., 1966, Phys. Rev. Lett. **16**, 748.
Hayashida, N., *et al.*, 1994, Phys. Rev. Lett. **73**, 3491.
Hayashida, N., *et al.*, 1996, Phys. Rev. Lett. **77**, 1000.
Hillas, A. M., 1984, Astron. Astrophys. **22**, 425.
Hess, V. F., 1912, Z. Phys. **13**, 1084.
Kakimoto, F., E. C. Loh, M. Nagano, H. Okuno, M. Teshima, and S. Ueno, 1996, Nucl. Instrum. Methods Phys. Res. A **372**, 527.
Krizmanic, J. F., J. F. Ormes, and R. E. Streitmatter 1998, Eds., *Workshop on Observing Giant Cosmic Ray Air Showers from geq 10^{20} eV Particles from Space*, AIP Conf. Proc. No. 433 (AIP, New York). This volume is dedicated to the design of a space-based cosmic-ray observatory.
Kronberg, P. P., 1994a, Rep. Prog. Phys. **57**, 325.
Kronberg, P. P., 1994b, Nature (London) **370**, 179.
Lawrence, M. A., *et al.*, 1991, J. Phys. G **17**, 773.
Libby, W. F., 1965, *Radio Carbon Dating*, 2nd ed. (University of Chicago, Chicago).
Linsley, J., 1962, Phys. Rev. Lett. **10**, 146.
Linsley, J., 1997, in *Proceedings of the 25th International Cosmic Ray Conference*, Durban, edited by M. S. Potgieter, B. C. Raubenheimer, and D. J. van der Walt (World Scientific, Singapore), Vol. 5, p. 381.
Millikan, R. A., and G. H. Cameron, 1928, Phys. Rev. **32**, 533.
Nagano, M., 1996, Ed., *Proceedings of the International Symposium on Extremely High Energy Cosmic Rays: Astrophysics and Future Observations* (Institute for Cosmic Ray Research, University of Tokyo, Japan).
Ong, R. A., 1998, Phys. Rep. **305**, 93.
Sekido, Y., and H. Elliott, 1985, Eds., *Early History of Cosmic Ray Studies* (Reidel, Dordrecht).
Simpson, J., 1995, in *The Physical Review—The First Hundred years*, edited by H. Stroke (AIP, New York), p. 573.
Sokolsky, P., P. Sommers, and B. R. Dawson, 1992, Phys. Rep. **217**, 225.
Swordy, S., 1994, rapporteur talk, in *Proceedings of the 23rd International Cosmic Ray Conference*, Calgary, edited by R. B. Hicks, D. A. Leahy, and D. Venkatesan (World Scientific, Singapore), p. 243.

Takeda, M., *et al.*, 1998, Phys. Rev. Lett. **81**, 1163.
Telescope Array Collaboration, 1997, in *Proceedings of the 25th International Cosmic Ray Conference*, Durban, edited by M. S. Potgieter, B. C. Raubenheimer, and D. J. van der Walt (World Scientific, Singapore), Vol. 5, p. 369.
Watson, A. A., 1991, Nucl. Phys. B (Proc. Suppl.) **22**, 116.
Watson, A. A., 1997, University of Leeds, private communication.
Weekes, T. C., F. Aharnian, D. J. Fegan, and T. Kifune, 1997, in *Proceedings of the Fourth Compton Gamma-Ray Observatory Symposium*, AIP Conf. Proc. No. 410, edited by C. D. Dermer, M. Strikman, and J. D. Kurfess (AIP, New York), p. 361.
Yoshida, S., *et al.*, 1995, Astropart. Phys. **3**, 105.
Yoshida, S., and H. Dai, 1998, J. Phys. G **24**, 905.
Zatsepin, G. T., and V. A. Kuz'min, 1966, JETP Lett. **4**, 78.

Cosmic Microwave Background Radiation

Lyman Page and David Wilkinson

Most astronomers and physicists now believe that we live in an expanding universe that evolved from an early state of extremely high density and temperature. Measurements of the spectrum and anisotropy of the cosmic microwave background radiation (CMBR) provide strong evidence supporting this picture. Today, the spectrum of the CMBR matches that of a 2.728 K blackbody to within 0.01%, and the radiation is highly isotropic on the sky. Both of these properties are expected of the thermal radiation remnant of a hot, dense early universe (Peebles, 1993).

The cartoon in Fig. 1 shows how the CMBR evolves as the universe expands and cools. Local thermal equilibrium is established in the radiation epoch by radiative scattering

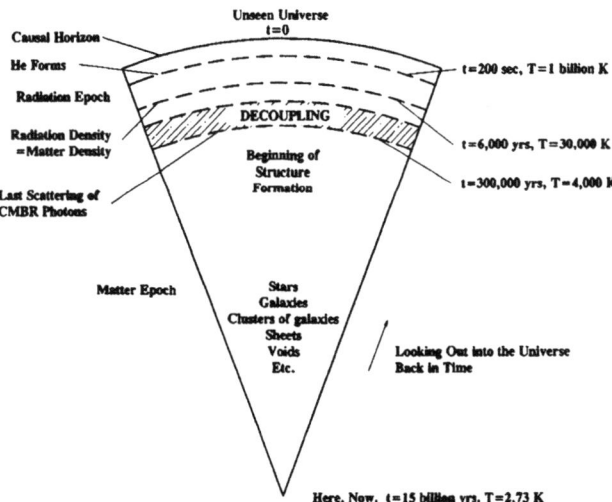

Figure 1. Significant events for the CMBR in the standard cosmological model. We are at the vertex looking out at past epochs. The big bang happened here, and everywhere else, about 15 billion years ago. We see it now on our causal light horizon at a radius of about 15 billion light years. After matter and radiation decouple, gravity aggregates the now neutral matter, and the universe is transparent to CMBR photons. Their anisotropy carries information about physical conditions in the universe at decoupling and may provide a new way to measure cosmological parameters.

processes, bremsstrahlung and radiative Compton scattering. Thus the CMBR thermal spectrum was established when the temperature of the universe was $T_{CMBR} > 10^7$ K, at an age $t < 1$ a day after the beginning. The accurately measured thermal spectrum indicates that standard physics was at work during this radiation epoch. As the density and temperature continue to fall, the dominant source of energy density in the universe changes from radiation to matter. (In an expanding universe with scale factor a, $T_{CMBR} \propto a^{-1}$, $\rho_{rad} \propto a^{-4}$, and $\rho_{mat} \propto a^{-3}$.) At $T_{CMBR} \approx 4000$ K, neutral hydrogen forms for the first time. The scattering cross section for photons off matter drops dramatically, and they decouple. This decoupling epoch (age $t \approx 3 \times 10^5$ years) is very important in the life of a typical CMBR photon. It scatters here for the last time (unless the universe reionizes soon after decoupling) and travels directly to us. However, slight perturbations in the matter density or motion impose a small temperature anisotropy on the scattered radiation. By measuring this tiny anisotropy (< 100 μK) across the sky, we can learn about the density perturbations that seeded the formation of cosmic structure such as galaxies and clusters of galaxies. Accurate measurements of the angular power spectrum of the anisotropy at scales of a degree and smaller might tell us about the detailed physics of decoupling and determine long-sought parameters of the cosmological model (see Hu, Sugiyama, and Silk, 1997 for a conceptual view).

Measurements of the CMBR

The spectrum

The temperature of the CMBR can be estimated by noting that a substantial fraction ($\approx 24\%$) of the matter in the universe is helium and assuming that it was made by nuclear reactions in the early universe. If so, at an age of $t \approx 100$ s, while neutrons were still around, a temperature of 10^9 K and a baryon density of 10^{27} m^{-3} would generate the observed He abundance. The mean baryon density today of about 1 m^{-3} gives an expansion factor of 10^9 and a temperature of 1 K. This rough estimate means that the CMBR is in the microwave band.

The essential ingredients of an experiment to measure the CMBR temperature at a given wavelength are (1) an antenna (usually a horn) with very low side-lobe response, (2) a cold emitter of known temperature (called the "cold load"), and (3) accurate knowledge of all sources of radio noise other than the CMBR. It is important to reject radiation from the ground, the galactic plane, moon, etc., so usually beam sizes of a few degrees are used with careful attention to shielding. Ideally, the cold load can be connected to the horn antenna aperture without disturbing the radiometer. It establishes an output reading at a known temperature, a zero-point calibration. Most experimenters use a good microwave absorber connected to a bath of boiling liquid helium, a temperature conveniently close to T_{CMBR}. The gain (output units per Kelvin), is usually measured by making a known change in the cold-load temperature. Gain fluctuations are removed by periodically switching between the antenna signal and a known, stable, source (Dicke, 1946). The more troublesome noise sources are generated skyward of the Dicke switch. These include atmospheric emission, Galactic radiation, or emission from within the instrument, for example, radiation from the inner walls of the horn antenna. As in any experiment measuring an absolute number, all these must be accurately measured and subtracted from the total measured radiation temperature. In early experiments these extraneous sources were as large as several K. In a well-designed modern experiment they are the order of a few tens of mK.

Many measurements of the CMBR temperature have been made from the ground, balloons, high-flying aircraft, and a satellite. (For comprehensive reviews see Weiss, 1980, and Partridge, 1995.) From the ground one needs to measure and subtract emission from atmospheric oxygen and water vapor. This can be hundreds of Kelvin near emission lines or a few Kelvin in the atmosphere's microwave windows. Long-wavelength measurements ($\lambda > 20$ cm) are still made from the ground because appropriate antennas are large and cumbersome. At centimeter wavelengths, balloons offer a good way of reducing atmospheric emission to an effective temperature of a few mK. Balloon-based instruments have achieved $\leq 1\%$ accuracies at wavelengths between 1 cm and 3 cm.

However, the field was changed forever by a CMBR spectrum measurement from the Cosmic Background Explorer (COBE) satellite. The Far-Infrared Absolute Spectrophotometer (FIRAS) compared the spectrum of the sky with that of a very black cold load (emissivity > 0.99997) whose temperature was accurately measured (Mather *et al.*, 1990). The instrument was a scanning Fourier-transform spectrometer with one input coupled to the sky through a tapered horn antenna with low side-lobe response. To minimize the effects of internal emission, the entire instrument was cooled to 1.5 K in a 500 liter dewar of liquid He. The entire satellite was oriented to always point away from the earth and sun. The crucial cold load could be moved into position over the FIRAS antenna, thus giving a blackbody reference source of accurately known temperature. Its temperature could be adjusted to closely match the signal from the sky, which gave a nearly zero output from the balanced spectrometer. The high accuracy of the measured CMBR spectrum is traceable to the differential nature of the instrument, and to the blackness and broad frequency range of the cold load.

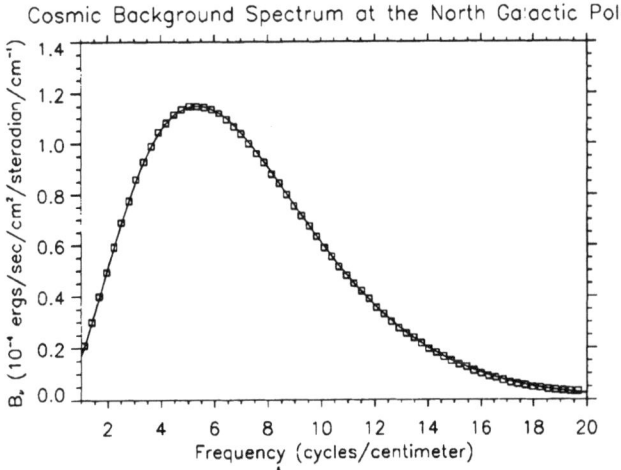

Figure 2. The initial sky spectrum from the FIRAS instrument aboard the COBE satellite. The solid line is a 2.735 K blackbody and the squares are the FIRAS results. There can be no question that the universe is filled with thermal radiation as required by the hot-big-bang model. This preliminary spectrum was based on nine minutes of data. The horizontal axis is wave number (cm^{-1}); the vertical axis is sky brightness. (Figure courtesy Edward S. Cheng, NASA/GSFC.)

One version of the COBE/FIRAS spectrum is shown in Fig. 2. This is the first spectrum seen by the COBE team only a month after COBE's launch in November, 1989. When John Mather (COBE Project Scientist and FIRAS Principle Investigator) showed it to a packed session of the American Astronomical Society's January 1990 meeting, the audience stood and applauded. Subsequent careful analysis of the FIRAS data has greatly improved the accuracy of the results. The FIRAS team has shown that the sky spectrum fits a 2.728 ± 0.004 K blackbody spectrum to an accuracy of 0.01% at wavelengths between 5 mm and 0.5 mm (Fixsen et al., 1996). A temperature measurement of this accuracy is not particularly useful in cosmology, but the spectral fit is very important. Only the hot-big-bang model predicts, or accounts for, such an accurate fit to a Planck spectrum.

Within two weeks of the announcement by the COBE/FIRAS team, a very similar result was reported by Gush, Halpern, and Wishnow (1990), who used a rocket-borne spectrometer. Their result would have predated the COBE/FIRAS result except that an erroneous setting in a vibration test damaged the instrument, postponing the successful rocket flight by about 5 months.

The anisotropy

Searches for anisotropy in the CMBR started soon after its discovery (Penzias and Wilson, 1965; Dicke et al., 1965). Anisotropy experiments compare the CMBR temperatures at two or more points in the sky. Such relative measurements are intrinsically more accurate than the absolute temperature measurements. However, the expected CMBR anisotropy signals are small, < 100 μK, so extraordinary detector sensitivity and stability are needed. Also, care must be taken to avoid changes in the signal from antenna sidelobes when moving the antenna beams, and the effects of atmospheric emission and galactic radiation must be minimized. Early experimenters spent many years developing receivers and techniques for observations from the ground, balloons, and aircraft. They successfully detected the dipole effect, a Doppler shift due to the sun's velocity with respect to the CMBR frame. The dipole amplitude is 3.3 mK, $\approx 10^{-3} T_{CMBR}$. However, the early experiments failed to separate the important intrinsic CMBR anisotropy from spurious effects.

By the late 1970s, the lessons learned from the early experiments led to the design of the COBE satellite's differential microwave radiometers (DMR) which conclusively detected the CMBR anisotropy after a year of orbital data had been analyzed (Smoot et al., 1992). The COBE/DMR experiment used differential radiometers at 31, 53, and 90 GHz to map the sky with beams of 7° full width. The three frequencies were needed to separate the CMBR anisotropy from Galactic radio emission. These sources have different, but known, frequency dependence. The rms amplitude of the CMBR on angular scales $>7°$ was found to be 30 ± 5 μK. One sees why terrestrially based experiments had so much trouble separating the CMBR anisotropy from the effects of the 300 K background of the Earth's environment. After four years of data the signal-to-noise ratio in regions of the COBE/DMR map away from the galactic plane was about 2, so the CMBR anisotropy was clearly detected (Bennett et al., 1996). Figure 3 shows a four-year COBE/DMR map in which the galactic signal has been subtracted using data from all three COBE/DMR frequencies. The subtraction is not complete near the galactic plane.

The COBE/DMR result was quickly confirmed using data from an earlier MIT balloon experiment, the Far Infrared Survey (FIRS). The measurement was made using bolom-

COBE-DMR 4-year Sky Map

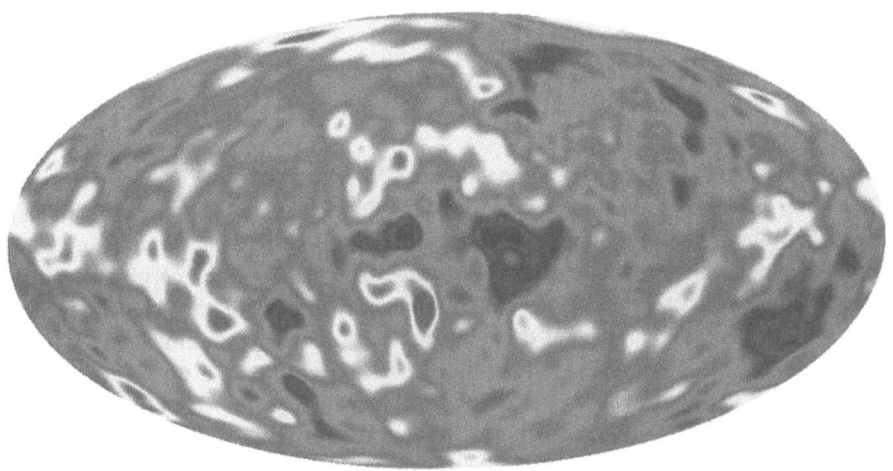

Figure 3. The final four-year sky map (in galactic coordinates) produced by the DMR experiment aboard COBE. Some of the stronger features seen off the galactic plane are due to real CMBR anisotropy. The 2.73 K CMBR level, the 3.3 mK dipole, and most of the galactic emission have been subtracted to obtain this map. The temperature range is ± 150 μK. (Figure courtesy COBE Science Working Group, NASA/GSFC.) (See Color Plate 4.)

etric detectors at 170 GHz and higher. On their own, the FIRS data were too noisy and contaminated by spurious effects to permit an unambiguous isolation of the CMBR anisotropy signal. However, cross correlation of the FIRS map with the COBE/DMR map showed a strong common signal. The much higher frequency of the FIRS data provided a measurement of the spectrum of the common signal; it matched that expected from a 2.73 K blackbody (Ganga et al., 1993).

After COBE, interest in the CMBR anisotropy turned to measurements at smaller angular scales. Again, experimenters have used ground- and balloon-based instruments with coherent receivers, bolometric detectors, and interferometers (see Page, 1997, for a review). Since the COBE/DMR measured anisotropy only at angles $>7°$, it was natural for the experiments to move to smaller angular scales. Also, theorists had found that, during the decoupling process of the standard model, certain angular scales are enhanced, embedding measurable features in the angular power spectrum of the CMBR anisotropy (e.g., Bond and Efstathiou, 1987). These were predicted to appear at scales from about 1° to 0.1°. Experimental progress at these intermediate angular scales has been rapid, with all three detection techniques contributing to the current picture. Figure 4 shows a plot of the average of many measurements of CMBR anisotropy amplitude vs angular scale. (Angles are represented by the spherical-harmonic index ℓ, angle $\approx 180°/\ell$.) The solid line shows the prediction of a representative theoretical model. One is impressed that the measurements indicate increased fluctuations at about the angular scales predicted by the models. However, even after averaging over all measurements available at this time, the accuracy is not yet sufficient to confirm a specific model or to reliably fit for parameters.

At large ℓ, radio telescopes and close-packed radio interferometers are needed to get the small beam sizes and sensitivity needed for CMBR anisotropy measurements. The measured upper limits at $\ell \approx 1700$ in Fig. 4 seem to show a falling spectrum. The effect is due to photon diffusion out of small regions and to multiple scattering of photons during the finite decoupling epoch, $\Delta z \approx 75$ at $z = 1400$.

What We've Learned from the CMBR Measurements

Before discussing the near-term prospects for CMBR anisotropy research, we review briefly what the CMBR measurements have taught us to date. The accurate fit of the CMBR spectrum to that of a blackbody shows that the universe went through an epoch of local thermal equilibrium when radiative processes thermalized the energies in radiation and matter. It also places limits on the amount of energy that could have been injected into the early universe, for example by decaying particles. Future experiments at $\lambda > 5$ mm will continue to search for distortions in the CMBR spectrum.

The isotropy of the CMBR created a causality problem for the original big-bang cosmological model. How did the different parts of the currently visible universe manage to come to the same temperature when they were not in causal contact at earlier times? The idea of inflation, an enormous expansion in the very early universe (Turner and Tyson, 1999, this volume), currently provides the best solution to this puzzle. One pictures our visible universe as a causally connected sphere (radius ≈ 15 billion light years) inside a much larger, inflated, region. The observed isotropy of the CMBR temperature supports the inflation hypothesis, since no other explanation exists.

Causality has an interesting implication for the CMBR anisotropy measurements. The angular scales of the anisotropy measured by COBE/DMR ($>7°$) are larger than a

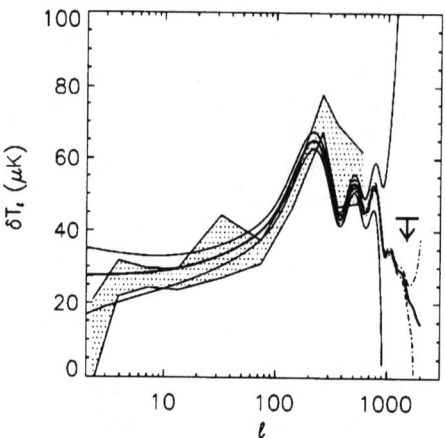

Figure 4. Current results from CMBR anisotropy experiments averaged into ten bins in ℓ, the spherical-harmonic index. The point near $\ell = 1700$ (highlighted by an arrow) is an upper limit. δT_ℓ is the rms temperature fluctuation per logarithmic interval in ℓ. The COBE/DMR results appear at $\ell < 20$. Recent experiments are trying to test the prediction of some cosmological models that a series of peaks should appear at intermediate angular scales, $20 < \ell < 1000$. The thick curve is the prediction for "standard CDM" (cold dark matter) with $\Omega_{baryon} = 0.05$, $\Omega_{CDM} = 0.95$, and $H_0 = 50$ km/sec/Mpc. The solid lines on either side of the model are an estimate of the noise per ℓ for the MAP satellite. At low ℓ, cosmic variance dominates; at high ℓ one is limited by MAP's angular resolution. The dot-dash line is a similar estimate for the Planck Surveyor.

causally connected patch on the decoupling surface. That is, at the time of decoupling a photon has a causal horizon, and any physical process that generates anisotropy at that time must lie within that horizon. Today, the angular size of those causal patches on the decoupling surface is about 1°, much smaller than angular size of anisotropy observed by the COBE/DMR. One explanation is that the large scale anisotropy was produced by quantum fluctuations in the preinflation epoch. Inflation can then produce anisotropy patches larger than the causal size at the decoupling epoch. The quantum fluctuation idea also predicts the slope of the angular spectrum for the anisotropy measured at large scales. The COBE/DMR measurements are in agreement with this prediction, lending further support to the inflation hypothesis.

From measurements of the anisotropy of the CMBR we also learn that the gravitational scenario for cosmic structure formation is plausible. The amplitude of the CMBR anisotropy corresponds to density fluctuations at decoupling, which are about the amplitude needed to seed gravitational growth of the cosmic structures seen today. It was by no means obvious that this would be the case. Indeed, the great notoriety that accompanied the announcement of the COBE/DMR detection of CMBR anisotropy was, in part, an expression of relief by cosmologists that the standard model was still alive.

Interpretation of More Accurate CMBR Measurements

The origin of the peaks in the theoretical angular spectrum of CMBR anisotropy (Fig. 4) offers some physical insight into how more accurate measurements might be used. The basic physics of peak generation is straightforward (Hu, Sugiyama, and Silk, 1997). During the epoch of decoupling, the baryons and photons are coupled by the baryon-electron Coulomb force and Compton scattering. To illustrate the peak-generation process, we use the cold dark matter (CDM) model, even though it is inconsistent with some modern observations. In this model, density fluctuations in the dominant dark matter provide gravitational potential wells into which baryons fall, compressing and raising the temperature of the photons. However, the collapsing baryon-photon fluid will bounce due to the increasing pressure of the photons. These acoustic oscillations continue until the matter is neutral and decoupling is complete. The sound-horizon size in the photon-baryon fluid at that time establishes a maximum physical scale size for a causally connected region, which corresponds to the angular scale of the first peak in the anisotropy spectrum. The peaks at smaller angular scales correspond to higher-frequency modes of the acoustic oscillations. Of course, the details of the spectrum generation at decoupling are more complicated. For example, the density and temperature of the compressed or rarefied fluid at last scattering is only one effect on the photon temperature. Two others are the gravitational redshift as the photons climb out of the potential wells and the Doppler shift due to the fluid motion at last scattering.

This simplified picture of peak generation gives some insight into how careful measurements of the anisotropy spectrum give values for cosmological constants. Space curvature is measured because one knows a physical scale size at an early epoch in the universe and measures the apparent angular size now. In a positively curved universe (closed), the angle will be larger than in a flat universe; the angle will be smaller in an open universe. So the angle corresponding to the first peak in the anisotropy spectrum should indicate the curvature of space over a distance nearly equal to our horizon size. Likewise, the amplitude of the peaks is sensitive to the baryon density in the universe. In the CDM model, the additional mass of the baryons in the potential wells increases the compression of the oscillating fluids and the amplitude of the peaks.

Polarization of the CMBR has not yet been observed. However, theoretical studies predict linear polarization at about 1 part in 10^6, a few μK. Only recently have detector technology and experimental technique reached levels where detection of such a small polarization might be possible. Polarization of the CMBR is generated when last scattering of photons occurs from a region bathed in a local quadrupole temperature anisotropy (Rees, 1968). Models predict an angular spectrum with many peaks for the CMBR polarization across the sky, and most models predict correlations between the patterns of polarization and temperature anisotropy. Polarization sky maps contain information complementary to that derived from temperature anisotropy maps, so much can be learned from accurate sky maps of the CMBR polarization.

A particularly attractive feature of probing cosmology with CMBR anisotropy and polarization is that the physics is relatively simple. Linear processes dominate and the time dependence is straightforward. So for a given set of model parameters, accurate calculations are possible.

Future Experiments

Currently, the primary needs in CMBR research are: (1) more accurate anisotropy spectra at scales smaller than 10°, and (2) the detection and mapping of CMBR polarization. Theoretical studies of physical processes during and after decoupling are producing detailed predictions of the spectral and statistical properties of the CMBR anisotropy and polarization. The predictions are highly dependent on choices of the cosmological model, the composition of dark matter, and the nature of the fluctuations. Thus accurate measurements of the anisotropy may determine cosmological parameters and allow us to study the detailed physics of decoupling and postdecoupling processes (Jungman et al., 1995). It remains to be seen whether the measured anisotropy spectrum will show complex features, and how many models can be made to fit them accurately.

Modern experiments aim to measure the angular spectrum of the CMBR to a percent, map the polarization of the radiation and correlate it with the temperature fluctuations, measure the frequency spectrum of the anisotropy, determine the statistics of the fluctuations, and measure the distortion of the anisotropy due to gravitational lensing of the photons on their way to our detectors. The future of experimentation on the CMBR lies primarily in the control and identification of instrumental systematic effects and in an increased understanding of contamination by the astrophysical foreground emission and by the atmosphere. While an increase in detector sensitivity is always beneficial for identifying the tiny signals, the trend in current experimental design is to sacrifice sensitivity for control over potential systematic error.

We look forward to the launch of two satellite missions in the first decade of the new millennium. The primary goal of both NASA's MAP[1] and ESA's Planck Surveyor,[2] is to make multifrequency, high-fidelity, polarization-sensitive maps of the CMBR anisotropy over the entire sky. These maps will be sensitive to angular scales ranging from the dipole (180°) to the instrument resolution ($\approx 0.2°$ for MAP, smaller for Planck Surveyor) and

---

[1] MAP, http://map.gsfc.nasa.gov, scheduled launch is late 2000.
[2] PLANCK, http://astro.estec.esa.nl/SA.general/Projects/Planck, expected launch is 2007.

will be calibrated to better than 1% accuracy. Both missions will observe from L2, the earth-sun Lagrange point. In addition, there will be a number of multielement ground-based and balloon experiments, to corroborate and extend the satellite results.

The statistical nature of the anisotropy is of fundamental importance and will not be well constrained without unbiased full-sky maps. In the basic inflationary model, the production of the anisotropy by fluctuations in a primordial quantum field results in a normal distribution of temperatures. Other models make different predictions. Should the measured temperatures follow a normal distribution, then the anisotropy is completely specified by the angular power spectrum, as shown in Fig. 4. In this case, we may determine the variance of the source of the fluctuations only to order $\sqrt{2/N_d}$, where N_d is the number of degrees of freedom in the spherical harmonic, 5 for the quadrupole, 10^3 for $\ell = 500$. This is called the cosmic variance limit. MAP will be "cosmic variance" limited up to $\ell \approx 700$ and Planck to $\ell \approx 1700$.

The other natural limit is set by the contamination of the anisotropy maps by emission from our galaxy and emission from distant galaxies, as shown in Fig. 5. Our knowledge of these sources is incomplete. Fortunately, over much of the angular spectrum, these contributions are small, ~ 3 μK, and add in quadrature to the primary anisotropy. In addition, the contaminants can be identified by their frequency spectra. MAP is expected to be able to extract the CMBR anisotropy to the 1% level with its frequency coverage of 20 to 100 GHz; measuring the CMBR anisotropy to greater accuracy will await the 30 to 850 GHz coverage of the Planck Surveyor.

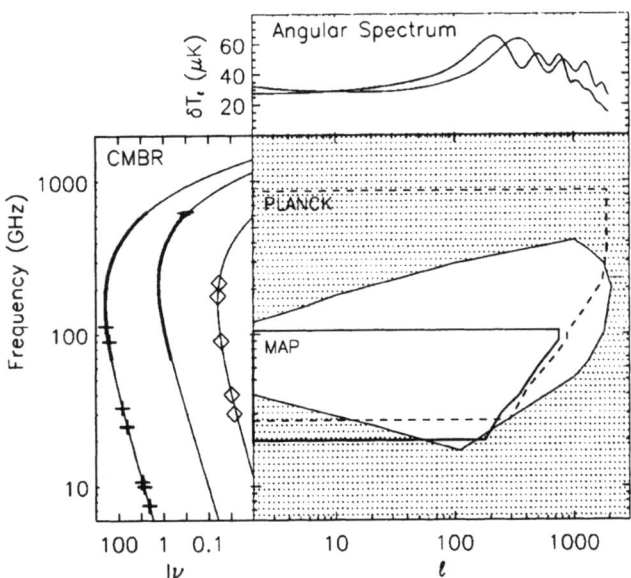

Figure 5. The frequency and ℓ-space coverage of the upcoming satellite missions. In the shaded region, foreground emission from our galaxy and distant galaxies is expected to dominate the anisotropy in the CMBR. The top panel shows two models of the anisotropy, the one peaking at smaller ℓ is "standard CDM," as in Fig. 4; the other is an open model with $\Omega_0 = 0.4$. Notice how the peak moves to smaller angular scales. The left panel shows the surface brightness spectrum of the CMBR in units of 10^{-26} W/m^2/Hz/sr vs frequency. Representative measurements are shown. The thickened lines are from the COBE satellite. The dipole is shown at $10^{-3} T_{CMBR}$ and the anisotropy is shown at $10^{-5} T_{CMBR}$.

It has become clear from years of experimentation that it is not possible to make full-sky maps suitable for detailed anisotropy analysis from balloons or the ground. This is primarily due to thermal variations in the radiometers, the anisotropy in the observing environment, and varying thermal gradients in the atmosphere. However, patches of order 10° can be mapped with precision. While various radiometric and beam-switching techniques continue to be productive, interferometers offer a proven method for minimizing atmospheric effects. Vigorous experimental efforts are underway to build arrays of bolometers at frequencies between 90 and 300 GHz and interferometers between 30 and 200 GHz.

If the pattern in the observed anisotropy should match one of the theoretical models, then the predicted polarization level and correlation between the polarization and temperature provide a built-in cross check. A number of groups are building radiometers and small arrays designed specifically to measure the linear polarization of the CMBR as a function of angular position. Because one measures the difference between orthogonal polarizations in a single beam, the receiver systems are intrinsically more immune to systematic error than traditional beam-switching experiments.

In addition to its intrinsic properties, the CMBR may be thought of as a distant light source that illuminates younger objects. Hot electrons in distant clusters of galaxies scatter low-frequency photons, for example at 30 GHz, to higher frequencies, for example at 240 GHz. This is known as the Sunyaev-Zel'dovich effect (Sunyaev and Zel'dovich, 1972). When viewed at 30 GHz and high resolution, the CMBR appears cold at the positions of galaxy clusters containing large amounts of ionized gas. This effect has been used to study distant clusters of galaxies. In the future, it may reveal their motions and, with a large number of clusters, it offers another way to ascertain the cosmological parameters.

In the beginning of the century, we learned of the existence of other galaxies. We are now probing the anisotropy of the oldest photons in the universe to a part in a million to discover how those galaxies came to be. The interlocking web of theory and constraints from observations of the distributions and velocities of galaxies, the abundances of the light elements, and the CMBR may single out a cosmological model. Should the evidence admit, it will perhaps be as convincing as the "standard model" of particle physics.

Acknowledgments

We would like to acknowledge the help of David Spergel, Gary Hinshaw, Mark Halpern, and Max Tegmark. This work was supported by grants from the NSF and NASA.

References

Bennett, C. L., *et al.*, 1996, Astrophys. J. Lett. **464**, L1.
Bond, J. R., and G. Efstathiou, 1987, Mon. Not. R. Astron. Soc. **226**, 655.
Dicke, R. H., 1946, Rev. Sci. Instrum. **17**, 268.
Dicke, R. H., P. J. E. Peebles, P. G. Roll, and D. T. Wilkinson, 1965, Astrophys. J. **142**, 413.
Fixsen, D. J., E. S. Cheng, J. M. Gales, J. C. Mather, R. A. Shafer, and E. L. Wright, 1996, Astrophys. J. **473**, 576.
Ganga, K. M., E. S. Cheng, S. Meyer, and L. A. Page, 1993, Astrophys. J. Lett. **410**, L57.
Gush, H. P., M. Halpern, and E. H. Wishnow, 1990, Phys. Rev. Lett. **65**, 537.

Hu, W., N. Sugiyama, and J. Silk, 1997, Nature (London) **386**, 37.
Jungman, G., M. Kamionkowski, A. Kosowsky, and D. Spergel, 1995, Phys. Rev. D **54**, 1332.
Mather, J. C., *et al.*, 1990, Astrophys. J. Lett. **354**, L37.
Page, L. A., 1997, in *Generation of Large-Scale Structure*, edited by D. N. Schramm and P. Galeotti (Kluwer, Netherlands), p. 75.
Partridge, R. B., 1995, *3K: The Cosmic Microwave Background Radiation* (Cambridge University, Cambridge).
Peebles, P. J. E., 1993, *Principles of Physical Cosmology* (Princeton University, Princeton, NJ)
Penzias, A. A., and R. W. Wilson, 1965, Astrophys. J. **142**, 419.
Rees, M. J., 1968, Astrophys. J. Lett. **153**, L1.
Smoot, G. F., *et al.*, 1992, Astrophys. J. Lett. **396**, L1.
Sunyaev, R. A., and Y. A. Zel'dovich, 1972, Comments Astrophys. Space Phys. **4**, 173.
Turner, M. S., and J. A. Tyson, 1999, Rev. Mod. Phys. **71**, 145; pp. 245–277 in this book.
Weiss, R., 1980, Annu. Rev. Astron. Astrophys. **18**, 489.

Black Holes
Gary T. Horowitz and Saul A. Teukolsky

Black holes are predicted by general relativity to be formed whenever sufficient mass is compressed into a small enough volume. In Newtonian language, the escape velocity from the surface becomes greater than the speed of light, so that nothing can escape. In general relativity, a black hole is defined as a region of spacetime that cannot communicate with the external universe. The boundary of this region is called the surface of the black hole, or the event horizon.

It appears impossible to compress matter on earth sufficiently to form a black hole. But in nature, gravity itself can compress matter if there is not enough pressure to resist the inward attractive force. When a massive star reaches the endpoint of its thermonuclear burning phase, nuclear reactions no longer supply thermal pressure, and gravitational collapse will proceed all the way to a black hole. By contrast, the collapse of a less massive star halts at high density when the core is transformed entirely into nuclear matter. The envelope of the star is blown off in a gigantic supernova explosion, leaving the core behind as a nascent neutron star.

The "modern" history of the black hole begins with the classic paper of Oppenheimer and Snyder (1939). They calculated the collapse of a homogeneous sphere of pressureless gas in general relativity. They found that the sphere eventually becomes cut off from all communication with the rest of the universe. Ultimately, the matter is crushed to infinite density at the center. Most previous discussions of the exterior gravitational field of a spherical mass had not taken into account the fact that the apparent singularity in the solution at the Schwarzschild radius was merely a coordinate artifact. Einstein himself claimed that one need not worry about the "Schwarzschild singularity" since no material body could ever be compressed to such a radius (Einstein, 1939). His error was that he considered only bodies in equilibrium. Even the usually sober Landau had been bothered by the prospect of continued gravitational collapse implied by the existence of a maximum stable mass for neutron stars and white dwarfs. To circumvent this, he believed at one time that " . . . all stars heavier than $1.5M_\odot$ certainly possess regions in which the laws of quantum mechanics . . . are violated" (Landau, 1932).

Despite the work of Oppenheimer and Snyder, black holes were generally ignored until the late 1950s, when Wheeler and his collaborators began a serious investigation of the problem of gravitational collapse (Harrison et al., 1965). It was Wheeler (1968) who coined the term "black hole." The discovery of quasars, pulsars, and compact x-ray sources in the 1960s finally gave observational impetus to the subject and ushered in the "golden age" of black-hole research.

Black holes are now believed to exist with a variety of masses. A current estimate for the dividing line between progenitor stars that produce neutron stars and those that produce black holes is around $25 M_\odot$. The resulting black holes are expected to have masses in the range $3-60\ M_\odot$. As discussed below, there is also good astrophysical evidence for supermassive black holes, with masses of order $10^6-10^9\ M_\odot$. There are a number of scenarios that could produce such large black holes: the gravitational collapse of individual supermassive gas clouds, the growth of a seed black hole capturing stars and gas from a dense star cluster at the center of a galaxy, or the merger of smaller black holes produced by collapse. There has also been speculation that black holes with a very wide range of masses might have been produced from density fluctuations in the early universe, but so far there is no convincing evidence for the existence of such primordial black holes.

This article provides just an overview of the astrophysical evidence for black holes, and discusses some recent theoretical developments in black-hole research. For a more complete discussion of the basic properties of black holes, see the books by Misner, Thorne, and Wheeler (1973), Shapiro and Teukolsky (1983), or Wald (1984).

Observational Evidence for Black Holes

The maximum mass of neutron stars

Neutron stars of small enough mass can exist happily in equilibrium, but beyond a certain critical mass, the inward pull of gravity overwhelms the balancing pressure force—the star is unstable and will collapse to a black hole. This provides one of the key observational signatures of a black hole astronomically: look for a system containing a dark, compact object. If you can determine that the mass of the object is greater than the maximum allowed mass of a neutron star, then it must be a black hole.

The value of the maximum neutron star mass is uncertain theoretically because we do not understand nuclear physics well enough to calculate it reliably (see, e.g., Baym, 1995). Current conventional nuclear equations of state predict a maximum mass around $2 M_\odot$ (see, e.g., the discussion and references in Cook, Shapiro, and Teukolsky, 1994, or Baym, 1995). (For some ''unconventional'' possibilities, see Brown and Bethe, 1994; Bahcall, Lynn, and Selipsky, 1990; Miller, Shahbaz, and Nolan, 1998.)

Because of these uncertainties, astrophysicists generally rely on a calculation that assumes we understand nuclear physics up to some density ρ_0 and then varies the pressure-density relation over all possibilities beyond this point to maximize the resulting mass (Rhoades and Ruffini, 1974). This procedure yields an upper limit to the maximum mass of

$$M_{max} \simeq 3.2 M_\odot \left(\frac{4.6 \times 10^{14}\ \mathrm{g\ cm^{-3}}}{\rho_0} \right)^{1/2}. \tag{1}$$

Kalogera and Baym (1996) have redone the Rhoades-Ruffini calculation with more up-to-date physics and obtained essentially the same numbers: a coefficient of $2.9 M_\odot$ for a preferred matching density of $5.4 \times 10^{14}\ \mathrm{g\ cm^{-3}}$. Rotation increases the amount of matter that can be supported against collapse, but even for stars rotating near breakup speed, the effect is only about 25% (see, e.g., Cook, Shapiro, and Teukolsky, 1994). The Rhoades-Ruffini calculation assumes the causality condition that the speed of sound is less than the speed of light: $dP/d\rho \leq c^2$. Abandoning this assumption increases the coefficient in Eq. (1) from 3.2 to 5.2 (Hartle and Sabbadini, 1977, and references therein). But it is not clear that this can be done without the material of the star becoming spontaneously unstable (Bludman and Ruderman, 1970; but see also Hartle, 1978). In summary, circumventing these mass limits would require us to accept some unconventional physics—much more unconventional than black holes!

Observational signatures of black holes

A black hole is the most compact configuration of matter possible for a given mass. The size of a black hole of mass M is given by the Schwarzschild radius, the radius of the event horizon:

$$R_S = \frac{2GM}{c^2} = 3 \text{ km}\left(\frac{M}{M_\odot}\right). \tag{2}$$

One way of verifying the compactness of a candidate black hole is to measure the speed of matter in orbit around it, which is expected to approach c near the horizon. This test is feasible since accretion flows of orbiting gas are common around gravitating objects in astrophysics. In a few objects, direct evidence for high orbital speeds is obtained by measuring the Doppler broadening of spectral lines from the accreting gas. More often, black-hole candidates exhibit gas outflows, or jets, with relativistic speeds. Another indication of compactness comes from observations of strong x-ray emission from the accreting gas, which imply high temperatures $>10^9$ K. Such temperatures are easily achieved by accretion onto a black hole or a neutron star, both of which have sufficiently deep potential wells.

When the radiation (typically x rays) from a compact object varies on a characteristic time scale t, without contrived conditions the size of the object must be less than ct. If this size limit is comparable to R_S (determined from an independent mass estimate) then the object is potentially a black hole. For solar-mass black holes, this implies looking for variability on the scale of less than a millisecond.

The demonstration of compactness alone, however, is not sufficient to identify a black hole; a neutron star, with a radius of about $3R_S$, is only slightly larger than a black hole of the same mass. Clear evidence that $M > M_{\max}$ is needed in addition to compactness.

Any gravitating object has a maximum luminosity, the Eddington limit, given by

$$L_{\text{Edd}} \simeq 10^{38} \text{ erg s}^{-1}\left(\frac{M}{M_\odot}\right) \tag{3}$$

(see, e.g., Shapiro and Teukolsky, 1983). Above this luminosity, the outward force due to escaping radiation on the accreting gas overwhelms the attractive force due to gravity, and accretion is no longer possible. Thus the observed luminosity sets a lower limit on the mass of the accreting object, which can often suggest the presence of a black hole.

Supermassive black holes in galactic nuclei

Quasars emit immense amounts of radiation, up to $\sim 10^{46}$ erg s^{-1}, from very small volumes. They are members of a wider class of objects, active galactic nuclei, all of which generally radiate intensely.

Nearly all active galactic nuclei emit substantial fractions of their radiation in x rays, and some emit the bulk of their radiation in even more energetic γ rays. Rapid variability of the flux has been observed in some active galactic nuclei. Many also have relativistic jets. These are all signatures of a compact relativistic object. If the observed radiation is powered by accretion, as is generally assumed, then the Eddington limit [Eq. (3)] implies masses in the range $10^6-10^{10} M_\odot$. This is well above the maximum mass of a neutron star, and so active galactic nuclei are considered secure black-hole candidates. Menou, Quataert, and Narayan (1998) give a summary of the current best supermassive black-hole candidates at the centers of nearby galaxies.

Direct evidence for the existence of a central relativistic potential well has come from the recent detection of broad iron fluorescence lines in x rays in a few active galactic nuclei. The line broadening can be interpreted as a combination of Doppler broadening and gravitational redshift. A spectacular example is the galaxy MCG-6-30-15, where a very broad emission line has been observed. The data can be interpreted as suggesting that the central mass is a rapidly rotating black hole, but this is still tentative. (See Menou, Quataert, and Narayan, 1998, for a discussion and references for this source and many others. See Rees, 1998, for a general discussion of astrophysical evidence for black holes.)

Black holes in x-ray binaries

In an x-ray binary, one of the stars is compact and accretes gas from the outer layers of its companion. Because of angular momentum conservation in the rotating system, gas cannot flow directly onto the compact object. Instead, it spirals towards the compact object and heats up because of viscous dissipation, producing x rays. In many cases, the compact star is known to be a neutron star, but there are also a number of excellent black-hole candidates.

The mass of the x-ray-emitting star M_X can be constrained by observations of the spectral lines of the secondary star. The Doppler shifts of these lines give an estimate of the radial velocity v_r of the secondary as it orbits the x-ray star. Combining v_r with the orbital period P of the binary and using Kepler's third law yields the "mass function" of the compact object,

$$f(M_X) \equiv \frac{M_X \sin^3 i}{(1+q)^2} = \frac{P v_r^3}{2\pi G} \tag{4}$$

(see, e.g., Shapiro and Teukolsky, 1983). The mass function does not give M_X directly because of its dependence on the unknown inclination i of the binary orbit and the ratio q of the two masses. However, it is a firm lower limit on M_X. Therefore, mass functions above $3 M_\odot$ suggest the presence of black holes. Additional observational data—absence or presence of eclipses, for instance, or information on the nature of the secondary star—can help to constrain i or q, so that a likely value of M_X can often be determined. The best stellar-mass black-hole candidates currently known are summarized in Menou, Quataert, and Narayan (1998).

The first black-hole candidate discovered in this way was Cyg X-1. Although its mass function is not very large, there are good observations that set limits on i and q and suggest that M_X is definitely greater than 3–4 $M_\odot$, with the likely value being 7–20 $M_\odot$. Even stronger evidence is provided by other x-ray binaries for which $f(M_X) > 3 M_\odot$. Without any further astrophysical assumptions, one can be pretty sure that these objects are not neutron stars. Currently, the most compelling black-hole candidate is V404 Cyg, with a mass function of $6M_\odot$.

Many of these sources show the key observational signatures of black holes described in Sec. II.B. Some display rapid variability in their x-ray emission. Many occasionally reach high luminosities, implying masses greater than that of a neutron star via the Eddington limit Eq. (3). A few exhibit relativistic jets.

Conclusive evidence for black holes

All the methods for finding black holes described above are indirect. They essentially say that there is a lot of mass in a small volume. Direct proof that a candidate object is a black hole requires a demonstration that the object has the spacetime geometry predicted by Einstein's theory. For example, we would like to have evidence for an event horizon, the one feature that is unique to a black hole.

One possible approach is via accretion theory (see Menou, Quataert, and Narayan, 1998, for a review). Two kinds of accretion are important for flow onto compact objects. The first is accretion from a thin disk. The accreting gas quickly radiates whatever energy is released through viscous dissipation. The gas stays relatively cool and so the disk remains thin, each gas element orbiting the central mass at the Keplerian velocity. Unlike the Newtonian case, the gravitational field of a compact mass in general relativity has a final stable circular orbit. The inner edge of the disk extends up to this radius. Observations such as those of the iron fluorescence lines described above provide information on the radius of the inner edge of the accretion disk. Since the radius of the last stable circular orbit depends on the spin of central mass, we may be able to measure the spin of black holes in this way.

Thin disks have oscillatory modes whose details depend on general relativity. Quasi-periodic oscillations have been detected in several x-ray binaries, and can be used to probe the spacetime geometry ("diskoseismology"; see Rees, 1998, for a review and references). In addition, if the disk is tilted with respect to the spin axis of the central mass, it will precess because of frame dragging (Lense-Thirring effect). This produces a periodic modulation of the x-ray luminosity, which may already have been seen in a few cases.

The second important kind of accretion is advection-dominated accretion flow. Here, the accreting gas advects most of the energy released by viscosity to the center. The gas becomes relatively hot and quasispherical. The spectrum is quite different from that of a thin disk. Advection-dominated accretion flows appear to be present in both galactic nuclei and in x-ray binaries when the accretion rate is relatively low. In an advection-dominated accretion flow, what happens to the energy advected to the center depends on the nature of the central object. If it is a black hole, the energy simply disappears behind the event horizon. If it is a neutron star or any object with a surface, the energy is reradiated from the surface and will dominate the spectrum. For those black-hole candidates that seem to be accreting in advection-dominated accretion flows, the evidence is that they lack surfaces. While not yet conclusive because of modeling uncertainties, this is the most direct evidence yet that black holes with event horizons are present in nature.

Is there any hope of a clean observation of black-hole geometry without the complications of dirty astrophysics? The best hope is from the observation of gravitational waves from black-hole collisions (see the article by Weiss in this volume). Laser interferometers now under construction, such as LIGO, VIRGO, and GEO (see, e.g., Abramovici *et al.*, 1992; Thorne, 1994) will be sensitive to black hole–black hole and black hole–neutron star collisions with black-hole masses up to a few tens of solar masses. The predicted event rate for such collisions is highly uncertain: estimates range from about one per year for the initial LIGO detector and thousands per year for the upgraded LIGO (Siggurdson and Hernquist, 1993; Lipunov, Postnov, and Prokhorov, 1997; Bethe and Brown, 1998), to essentially zero (Zwart and Yungelson, 1998). If nature is kind and we do detect such events, the wave form encodes a great deal of information about the spacetime geometry. The part of the wave form from the highly nonlinear merger phase is currently being calculated with large-scale supercomputer simulations (see, e.g., Finn, 1997), and it is expected that comparison of such calculations with observations should yield not only the masses and spins of the colliding objects, but also a check that the wave form is consistent with general relativity. The final part of the wave form is a "ring down," like a damped harmonic oscillator. This has been calculated by perturbation theory, and should provide another strong test.

There is also good reason to believe that, when two galaxies each containing supermassive black holes merge, the black holes will spiral together and coalesce. The frequency of the gravitational waves emitted is too low to be detectable on earth, where the waves would be swamped by seismic noise. However, such events should be readily detectable by a laser interferometer in space, such as the proposed LISA detector (see, e.g., Bender et al., 1996).

Black-Hole Uniqueness

The solution of Einstein's equations that describes a spherical black hole was discovered by Karl Schwarzschild only a few months after Einstein published the final form of general relativity:

$$ds^2 = -\left(1 - \frac{2M}{r}\right)dt^2 + \left(1 - \frac{2M}{r}\right)^{-1} dr^2 + r^2(d\theta^2 + \sin^2\theta \, d\phi^2). \tag{5}$$

(Here, and for the remainder of our discussion, we use units with $c = G = 1$.) This metric turns out to be the only spherically symmetric solution in the absence of matter. In general relativity, as in Newtonian gravity, the vacuum gravitational field outside any spherically symmetric object is the same as that of a point mass. The event horizon occurs at $r = 2M$ [cf. Eq. (2)]. Although the metric components are singular there, they can be made regular by a simple change of coordinates. In contrast, the singularity at $r = 0$ is real. An observer falling into a Schwarzschild black hole will be ripped apart by infinite tidal forces at $r = 0$.

One might expect that solutions of Einstein's equations describing realistic black holes that form in nature and settle down to equilibrium would be very complicated. After all, a black hole can be formed from collapse of all kinds of matter configurations, with arbitrary multipole distributions, magnetic fields, distributions of angular momentum, and so on. For most situations, after the black hole has settled down, it can be described by a solution of Einstein's vacuum field equations. Remarkably, one can show that the only stationary solution of this equation that is asymptotically flat and has a regular event horizon is a generalization of Eq. (5) known as the Kerr metric. This solution has only two parameters: the mass M and angular momentum J. All other information about the precursor state of the system is radiated away during the collapse. Astrophysical black holes are not expected to have a large electric charge since free charges are rapidly neutralized by plasma in an astrophysical environment. Nevertheless, there is an analog of this uniqueness theorem for charged black holes: all stationary solutions of the Einstein-Maxwell equations that are asymptotically flat and have a regular event horizon are known, and depend only on M, J, and the charge Q.

The simplicity of the final black-hole state is summarized by Wheeler's aphorism, "A black hole has no hair." This is supported not only by the above uniqueness theorems, but also by results showing that if one couples general relativity to simple matter fields, e.g., free scalar fields, there are no new stationary black-hole solutions. However, it has recently been shown that if more complicated matter is considered, new black-hole solutions can be found. Examples include Einstein-Yang-Mills black holes, black holes inside magnetic monopoles, and charged black holes coupled to a scalar "dilaton." Even these new black holes are characterized by only a few parameters, so the spirit of Wheeler's aphorism is maintained. (For a recent review and references, see Bekenstein, 1997.)

Cosmic Censorship

In the late 1960s, a series of powerful results were established in general relativity showing that, under generic conditions, gravitational collapse produces infinite gravitational fields, i.e., infinite spacetime curvature (see, e.g., Hawking and Ellis, 1973). However, these "singularity theorems" do not guarantee the existence of an event horizon. It is known that uniform-density, spherically symmetric gravitational collapse produces a black hole (the Oppenheimer-Snyder solution), and small perturbations do not change this. It is conceivable, however, that highly nonspherical collapse or, e.g., the collision of two black holes could produce singularities that are not hidden behind event horizons. These regions of infinite curvature would be visible to distant observers and hence are called "naked" singularities. Penrose (1969) proposed that naked singularities could not form in realistic situations, a hypothesis that has become known as cosmic censorship. If this is violated, general relativity could break down outside black holes, and would not be sufficient to predict the future evolution. On the positive side, this would open up the possibility of direct observations of quantum gravitational effects. Establishing whether cosmic censorship holds is perhaps the most important open question in classical general relativity today.

Despite almost 30 years of effort, we are still far from a general proof of cosmic censorship. (For a recent review and references, see Wald, 1997.) This seems to require analysis of the late time evolution of Einstein's equation in the strong-field regime. The much simpler problem of determining the global evolution of relatively weak (but still nonlinear) gravitational waves was achieved only in the late 1980s, and was hailed as a technical tour-de-force. In light of this, progress has been made by studying simpler systems, trying to find counterexamples, and by numerical simulations. The simpler systems are usually general relativity with one or two symmetries imposed. For example, cosmic censorship has been established for a class of solutions with two commuting symmetries. One class of potential counterexamples consists of time-symmetric initial data containing a minimal surface S. Assuming cosmic censorship, one can show that the area of this minimal surface must be related to the total mass M by $A(S) \leq 16\pi M^2$. Unsuccessful attempts were made to find initial data that violate this inequality. Recently, a general proof of this inequality has been found, showing that no counterexamples of this type exist. Numerical simulations of nonspherical collapse have found some indication that cosmic censorship may be violated in certain situations (Shapiro and Teukolsky, 1991), and suggest that any theorem might need careful specification of what is meant by "generic" initial data.

Perhaps the most effort and the most interesting results have come from studying spherically symmetric collapse. It was shown in the early 1970s that naked singularities could form in inhomogeneous dust collapse, but it was quickly realized that these "shell-crossing" or "shell-focusing" singularities also occurred in the absence of gravity and just reflected an unrealistic model of matter. It was believed at the time that any description of matter that did not produce singularities in flat spacetime would not produce naked singularities when coupled to gravity. This has recently been shown to be false. Consider spherically symmetric scalar fields coupled to gravity. If the initial amplitude is small, the waves will scatter and disperse to infinity. If the initial amplitude is large, the waves will collapse to form a black hole. As one continuously varies the amplitude, there is a critical value that divides these two outcomes. It has been shown that, at this critical value, the evolution produces a naked singularity. This is not believed to be a serious counterexample to cosmic censorship since it is not generic. But it again indicates that a true formulation of cosmic censorship is rather subtle.

Studies of spherical scalar-field collapse near the critical amplitude $\mathcal{A}_0$ have yielded a surprising result. The mass of the resulting black hole, for $\mathcal{A} > \mathcal{A}_0$, is

$$M_{BH} \sim |\mathcal{A} - \mathcal{A}_0|^\gamma, \tag{6}$$

where γ is a universal exponent that is independent of the initial wave profile. Gravitational collapse of other matter fields, or axisymmetric gravitational waves, exhibit similar behavior (with a different exponent). Furthermore, the solution with $\mathcal{A} = \mathcal{A}_0$, exhibits a type of scale invariance. These properties are similar to critical phenomena in condensed matter systems. They are not yet fully understood, but may turn out to be related to thermodynamic properties of black holes, which we discuss next. For recent reviews of critical phenomena in gravitational collapse, see Gundlach (1998) and Choptuik (1998).

Quantum Black Holes

For an equilibrium black hole, one can define a quantity called the surface gravity κ which can be thought of as the force that must be exerted on a rope at infinity to hold a unit mass stationary near the horizon of a black hole. During the early 1970s, it was shown that black holes have the following properties:

(0) The surface gravity is constant over the horizon, even for rotating black holes that are not spherically symmetric.

(1) If one throws a small amount of mass into a stationary black hole characterized by M, Q, and J, it will settle down to a new stationary black hole. The change in these three quantities satisfies

$$\delta M = \frac{\kappa \delta A}{8\pi} + \Omega \delta J, \tag{7}$$

where A is the area of the event horizon and Ω is the angular velocity of the horizon.

(2) The area of a black hole cannot decrease during physical processes.

It was immediately noticed that there was a close similarity between these "laws of black-hole mechanics" and the usual laws of thermodynamics, with κ proportional to the temperature and A proportional to the entropy. However, it was originally thought that this could only be an analogy, since if a black hole really had a nonzero temperature, it would have to radiate and everyone knew that nothing could escape from a black hole. This view changed completely when Hawking (1975) showed that if matter is treated quantum mechanically, black holes do radiate. This showed that black holes are indeed thermodynamic objects with a temperature and entropy given by

$$T_{bh} = \frac{\hbar \kappa}{2\pi}, \quad S_{bh} = \frac{A}{4\hbar}. \tag{8}$$

This turns out to be an enormous entropy, much larger than the entropy of a corresponding amount of ordinary matter. For a review of black-hole thermodynamics, see Wald (1998).

In all other contexts, we know that thermodynamics is the result of averaging over a large number of different microscopic configurations with the same macroscopic properties. So it is natural to ask, What are the microstates of a black hole that are responsible for its thermodynamic properties? This question has recently been answered in both of the dominant approaches to quantum gravity today: string theory and canonical quantization of general relativity. We will focus on the situation in string theory, since this is further developed. (String theory is discussed in more detail in the article by Schwarz and Seiberg in this volume.) Briefly, string theory is based on the idea that elementary particles are not pointlike, but are actually different excitations of a one-dimensional extended object—the string. Strings interact by a simple splitting and joining interaction that turns out to reproduce the standard interactions of elementary particles. The strength of the interactions is governed by a string coupling constant g. A crucial ingredient in string theory is that it is supersymmetric. In any supersymmetric theory, the mass and charge satisfy an inequality of the form $M \geq cQ$ for some constant c. States that saturate this bound are called BPS (Bogolmonyi-Prasad-Sommerfield) states and have the special property that their mass does not receive any quantum corrections.

Now consider all BPS states in string theory with a given large charge Q. At weak string coupling g, these states are easy to describe and count. Now imagine increasing the string coupling. This increases the force of gravity, and causes these states to become black holes. Charged black holes also satisfy the inequality $M \geq cQ$ and, when equality holds, the black holes are called extremal. So the BPS states all become extremal black holes. But there is only one black hole for a given mass and charge, so the BPS states all become identical black holes. This is the origin of the thermodynamic properties of black holes. When one compares the number of BPS states N to the area of the event horizon, one finds that, in the limit of large charge,

$$N = e^{S_{bh}}, \tag{9}$$

in precise agreement with black-hole thermodynamics. This agreement has been shown to hold for near-extremal black holes as well, where the mass is slightly larger than cQ.

Extremal black holes have zero Hawking temperature and hence do not radiate. But near-extremal black holes do radiate approximately thermal radiation at low temperature. Similarly, the interactions between near-BPS states in string theory produce radiation. Remarkably, it turns out that the radiation predicted in string theory agrees precisely with that coming from black holes. This includes deviations from the black-body spectrum, which arise from two very different sources in the two cases. In the black-hole case, the deviations occur because the radiation has to propagate through the curved spacetime around the black hole. This gives rise to an effective potential that results in a frequency-dependent "grey-body factor" in the radiation spectrum. The string calculation at weak coupling is done in flat spacetime, so there are no curvature corrections. Nevertheless, there are deviations from a purely thermal spectrum because there are separate left- and right-moving degrees of freedom along the string. Remarkably, the resulting spectra agree. Progress has also been made in understanding the entropy of black holes far from extremality. In both string theory and a canonical quantization of general relativity, there are calculations of the entropy of neutral black holes up to an undetermined numerical coefficient. For reviews of these developments in string theory, see Horowitz (1998) or Maldacena (1996). For the canonical quantization results, see Ashtekar et al. (1998).

Conclusions

Black holes connect to a wide variety of fields of physics. They are invoked to explain high-energy phenomena in astrophysics, they are the subject of analytic and numerical inquiry in classical general relativity, and they may provide key insights into quantum gravity. We also seem to be on the verge of verifying that these objects actually exist in nature with the spacetime properties given by Einstein's theory. Finding absolutely incontrovertible evidence for a black hole would be the capstone of one of the most remarkable discoveries in the history of science.

Acknowledgments

We thank Ramesh Narayan for helpful discussions. This work was supported in part by NSF Grants PHY95-07065 at Santa Barbara and PHY 94-08378 at Cornell University.

References

Abramovici, A., W. E. Althouse, R. W. P. Drever, Y. Gursel, S. Kawamura, F. J. Raab, D. Shoemaker, L. Sievers, R. E. Spero, K. S. Thorne, R. E. Vogt, R. Weiss, S. E. Whitcomb, and M. E. Zucker, 1992, Science **256**, 325.
Ashtekar, A., J. Baez, A. Corichi, and K. Krasnov, 1998, Phys. Rev. Lett. **80**, 904.
Baym, G., 1995, Nucl. Phys. A **590**, 233c.
Bahcall, S., B. W. Lynn, and S. B. Selipsky, 1990, Astrophys. J. **362**, 251.
Bekenstein, J. D., 1997, in *Second International A. D. Sakharov Conference on Physics*, edited by I. M. Dremin and A. Semikhatov (World Scientific, Singapore); expanded version at gr-gc/9605059.
Bender, P., *et al.*, 1996, LISA Pre-Phase A Report, Max-Planck-Institut fur Quantenoptik, Report MPQ 208, Garching, Germany. Available at http://www.mpq.mpg.de/mpq-reports.html.
Bethe, H. A., and G. E. Brown, 1998, preprint astro-ph/9802084.
Bludman, S. A., and M. A. Ruderman, 1970, Phys. Rev. D **1**, 3243.
Brown, G. E., and H. A. Bethe, 1994, Astrophys. J. **423**, 659.
Choptuik, M., 1998, preprint gr-qc/9803075.
Cook, G. B., S. L. Shapiro, and S. A. Teukolsky, 1994, Astrophys. J. **424**, 823.

Einstein, A., 1939, Ann. Math. **40**, 922.
Finn, L. S., 1997, in *Proceedings of the 14th International Conference on General Relativity and Gravitation*, edited by M. Francaviglia, G. Longhi, L. Lusanna, and E. Sorace (World Scientific, Singapore), p. 147. Also at gr-qc/9603004.
Gundlach, C., 1998, Adv. Theor. Math. Phys. **2**, 1; also at gr-qc/9712084.
Harrison, B. K., K. S. Thorne, M. Wakano, and J. A. Wheeler, 1965, *Gravitation Theory and Gravitational Collapse* (University of Chicago, Chicago).
Hartle, J. B., 1978, Phys. Rep. **46**, 201.
Hartle, J. B., and A. G. Sabbadini, 1977, Astrophys. J. **213**, 831.
Hawking, S., 1975, Commun. Math. **43**, 199.
Hawking, S. W., and G. F. R. Ellis, 1973, *The Large Scale Structure of Space-time* (Cambridge University, Cambridge).
Horowitz, G. T., 1998, in *Black Holes and Relativistic Stars*, edited by R. M. Wald (University of Chicago, Chicago), p. 241.
Kalogera, V., and G. Baym, 1996, Astrophys. J. Lett. **470**, L61.
Landau, L. D., 1932, Phys. Z. Sowjetunion **1**, 285.
Lipunov, V. M., K. A. Postnov, and M. E. Prokhorov, 1997, New Astron. **2**, 43.
Maldacena, J. M., 1996, preprint hep-th/9607235.
Menou, K., E. Quataert, and R. Narayan, 1998, *Proceedings of the Eighth Marcel Grossman Meeting on General Relativity* (World Scientific, Singapore, in press); also at astro-ph/9712015.
Miller, J. C., T. Shahbaz, and L. A. Nolan, 1998, Mon. Not. R. Astron. Soc. **294**, L25; also at astro-ph/9708065.
Misner, C. W., K. S. Thorne, and J. A. Wheeler, 1973, *Gravitation* (Freeman, San Francisco).
Oppenheimer, J. R., and H. Snyder, 1939, Phys. Rev. **56**, 455.
Penrose, R., 1969, Riv. Nuovo Cimento **1**, 252.
Rees, M. J., 1998, in *Black Holes and Relativistic Stars*, edited by R. M. Wald (University of Chicago, Chicago), p. 79.
Rhoades, C. E., and R. Ruffini, 1974, Phys. Rev. Lett. **32**, 324.
Schwarz, J. H., and N. Seiberg, 1999, Rev. Mod. Phys. **71**, 112; pp. 188–202 in this book.
Shapiro, S. L., and S. A. Teukolsky, 1983, *Black Holes, White Dwarfs, and Neutron Stars: The Physics of Compact Objects* (Wiley, New York).
Shapiro, S. L., and S. A. Teukolsky, 1991, Phys. Rev. Lett. **66**, 994.
Siggurdson, S., and L. Hernquist, 1993, Nature (London) **364**, 423.
Thorne, K. S., 1994, in *Relativistic Cosmology, Proceedings of the 8th Nishinomiya-Yukawa Memorial Symposium*, edited by M. Sasaki (Universal Academy, Tokyo), p. 67.
Wald, R. M., 1984, *General Relativity* (University of Chicago, Chicago).
Wald, R. M., 1997, preprint gr-qc/9710068.
Wald, R. M., 1998, in *Black Holes and Relativistic Stars*, edited by R. M. Wald (University of Chicago, Chicago), p. 155.
Weiss, R., 1999, Rev. Mod. Phys. **71**, 187; pp. 313–328 in this book.
Wheeler, J. A., 1968, Am. Sci. **56**, 1.
Zwart, S. F. P., and L. R. Yungelson, 1998, Astron. Astrophys. **332**, 173; also at astro-ph/9710347.

Gravitational Radiation
Rainer Weiss

There is an excellent prospect that early in the next century gravitational radiation emitted by astrophysical sources will be detected and that gravitational-wave astrophysics will become another method of observing the universe. The expectation is that it will uncover new phenomena as well as add new insights into phenomena now observed in electromagnetic astrophysics. Gravitational radiation will come from the accelerated motions of mass in the interior of objects, those regions obscured in electromagnetic, and possibly, even neutrino astronomy. It arises from the motion of large bodies and represents the coherent effects of masses moving together rather than individual motions of smaller constituents such as atoms or charged particles that create the electromagnetic astrophysical emissions. Over the past 20 years relativistic gravitation has been tested with high precision in the weak field, characterized by the dimensionless gravitational potential $Gm/rc^2 = \varphi_{\text{Newton}}/c^2 \ll 1$ in solar system and Earth orbital tests, and in the past decade, most spectacularly, in the Hulse-Taylor binary neutron star system (PSR 1913+16). Gravitational radiation will provide an opportunity to observe the dynamics in the regions of the strong field and thereby test the general relativity theory where Newtonian gravitation is a poor approximation—in the domain of black holes, the surfaces of neutron stars, and possibly, in the highly dense epochs of the primeval universe.

The basis for the optimism is the development and construction of sensitive gravitational-wave detectors on the ground, and eventually in space, with sufficient sensitivity and bandwidth at astrophysically interesting frequencies to intersect reasonable estimates for sources.

This short article shall provide the nonspecialist an entry to the new science including a cursory description of the technology (as well as its limits), a brief overview of the sources, and some understanding of the techniques used to establish confidence in the observations.

Brief History

(See the article by I. Shapiro in this issue for a comprehensive review of the history of relativity.) Newtonian gravitation does not have the provision for gravitational radiation although Newton did ponder in a letter to Lord Berkeley how the "palpable effects of gravity manage to maintain their influence." When the theory of special relativity was put forth in 1905 it was clearly necessary to determine the news function for gravitation and several Lorentz covariant gravitational theories were developed (scalar, vector, tensor theories), all with gravitational radiation.

The basis of most current thinking is the Einstein theory of general relativity, which was proposed in the teens of our century and whose subtlety and depth has been the subject of gravitation theories ever since. Gravitational radiation, the spreading out of gravitational influence, was first discussed in the theory by Einstein in 1916 in a paper given in the Proceedings of the Royal Prussian Academy of Knowledge (Sitzungsberichte der Königlich Preussichen Akademie der Wissenschaften). This paper deals with small field approximations to the general theory and is nestled between a paper describing the perception of light by plants and another that analyzes the authenticity of some writings attributed to Epiphanius as well as a commentary on the use of the first person in Turkish grammar. Einstein was still new at the development of the theory and made an algebra mistake, which resulted in the prediction of gravitational radiation from accelerating spherical mass distributions. In a later paper in 1918, in the proceedings of the same academy, this time preceded by a paper on the Icelandic *Eddas* and followed by one on the middle-age history of a cloister in Sinai, he corrected his mistake and showed that the first-order term was quadrupolar. He was troubled by the fact that he could only make a sensible formulation of the energy carried by the waves in a particular coordinate system (the theory is supposed to be covariant, able to be represented in a coordinate-independent manner) and that he had to be satisfied with a pseudotensor to describe the energy and momentum flow in the waves. He found solutions to the field equations for gravitational waves that carry energy but also ones that seemed not to, the so-called coordinate waves. This problem of deciding what is real (i.e., measurable) and what is an artifact of the coordinates has been an endless source of difficulty for many (especially the experimenters) ever since.

It was recognized early on that the emission of gravitational radiation is so weak and its interaction with matter so small, that there was no hope for a laboratory confirmation with a source and a neighboring receiver in the radiation zone, as was the case for electricity and magnetism in the famous Hertz experiment. If there was any chance to observe the effects of the radiation it would require the acceleration of astrophysical size masses at relativistic speeds, and even then, the detection would require the measurement of infinitesimal motions. This was going to be a field that would require the development of new technology and methods to observe the universe.

The weakness of the radiation, however, also leads to some profound benefits for astrophysics. The waves will not scatter, so they emanate undisturbed from the densest regions of the universe, from the inner cores of imploding stars, the earliest instants of the primeval universe, and from the formation of black holes, and even singularities of the classical theory (unmodified by quantum theory). They will provide information on the dynamics in these regions, which are impossible to reach by any other means. Furthermore, the gravitational waves, to be detectable, would have to come from regions of strong gravity and large velocity, just those regions where Newtonian gravitation is a poor approximation and relativistic gravitation has not been tested.

Strong Indirect Evidence for Gravitational Radiation

A radio survey for pulsars in our galaxy made by R. Hulse and J. Taylor (1974, 1975) uncovered the unusual system PSR 1913+16. During the past two decades the information inscribed in the small variations of the arrival times of the pulses from this system have revealed it to be a binary neutron star system, one star being a pulsar with a regular pulse period in its rest frame. The stars are hard, dense nuggets about the mass of the sun, but only 10 km in size. It takes so much energy to excite the internal motions of the stars that in their orbital motion around each other they can be considered rigid pointlike objects. Luckily, the system is also isolated from other objects. The separation of the neutron stars is small enough that the dimensionless gravitational potential of one star on the other is 10^{-6}, compared with 10^{-1} on the stellar surfaces. The system is made to order as a relativity laboratory; the proverbial moving proper clock in a system with "point" test masses [Taylor and Weisberg (1982, 1989)].

One has to marvel at how much is learned from so sparse a signal. The small changes in the arrival time of the pulses encode most of the dynamics of the two-body system. By modeling the orbital dynamics and expressing it in terms of the arrival time of the pulses, it is possible to separate and solve for terms that are dependent on the different physical phenomena involved in the motion. The motion of the pulsar radio waves in the field of the companion star experiences both the relativistic retardation (the Shapiro effect) and angular modulation (bending of light). The aphelion advance of the orbit, the analog to the perihelion advance of Mercury around the Sun, is 4 degrees per year rather than the paltry 40 seconds of arc per century. Finally, the unrelenting acceleration of the orbit as the two stars approach each other is due to the loss of energy to gravitational waves, explained by the Einstein quadrupole formula

$$\langle P \rangle = \frac{G}{45c^5}\left(\frac{d^3Q}{dt^3}\right)^2, \tag{1}$$

to a precision of a few parts in 1000; where $\langle P \rangle$ is the average power radiated, and Q is the gravitational quadrupole moment of the system characterized by the product of the mass and the square of the orbit size. As an additional bonus the various relativistic effects permit the solution for the masses of the individual stars (a pure Newtonian description could only provide the sum of the masses) and shows, remarkably, that the two stars are each at the anticipated value for a neutron star of 1.4 solar masses. One of the most elegant graphs in recent astrophysics shows the locus of points for the various relativistic effects as a function of m_1 and m_2 (Taylor and Weisberg, 1989).

The measurements of the binary neutron star system have laid to rest uncertainties about the existence of gravitational waves. Furthermore, the possibility of directly detecting the gravitational waves from the coalescence of such systems throughout the universe has had the effect of setting design criteria for some of the instruments coming into operation in the next few years.

Wave Kinematics and Description of the Interaction

Gravitational waves can be thought of as a tidal force field transverse to the wave propagation in a flat space (flat-space representation with a complex force field) or as a distortion of the spatial geometry transverse to the propagation direction (curved space with no forces). The former approach works best for bar detectors where one needs to consider other phenomena than gravitational forces. This is the approach taken by J. Weber (1961), who was the first to attempt the direct detection of gravitational radiation. The interferometric detectors both on the ground and in space are more easily understood in the latter approach. It is a matter of taste which representation is used, the only proviso being not to mix them, as that leads to utter confusion. Here, a heuristic application of the curved-space approach is taken.

Far from the sources the waves will be a small perturbation h on the Minkowski metric η of inertial space

$$g_{ij} = \eta_{ij} + h_{ij}.$$

The gravitational-wave perturbation is transverse to the propagation direction and comes in two polarizations, h_+ and $h_\times$. For a wave propagating in the x_1 direction, the metric perturbations have components in the x_2 and x_3 directions. The + polarization is distinguished by $h_{22} = -h_{33}$; a stretch in one direction and a compression in the other. The $\times$ polarization is rotated around x_1 by 45 degrees.

The gravitational wave can be most easily understood from a "gendanken" experiment to measure the travel time of a pulse of light through the gravitational wave. Suppose we lay out the usual special-relativistic assembly of synchronized clocks at all coordinate points and use the + polarization. The first event is the emission of the pulse, $E_1(x_2, t)$, and the second, $E_2(x_2 + \Delta x_2, t + \Delta t)$, is the receipt. Since the events are connected by the propagation of light, the interval between the events is zero. So writing the interval in terms of the coordinates of the events one gets

$$\Delta s^2 = 0 = g_{ij} dx_i dx_j = [1 + h_{22}(t)](\Delta x_2)^2 - c^2(\Delta t)^2.$$

The coordinate time and the proper time kept by the synchronized clocks are the same and not affected by the gravitational wave. The "real" distance between the end points of the events is determined by the travel time of the light as inferred by the clocks. There are two pieces to the inferred spatial distance between the events given by

$$c\Delta t = \left(1 + \frac{h_{22}(t)}{2}\right)\Delta x_2, \quad \text{where} \quad h_{22} \ll 1.$$

The larger part is simply the spatial separation of the events Δx_2, while the smaller is the spatial distortion due to the gravitational wave $[h_{22}(t)/2] = (\delta x_2 / \Delta x_2)$, the gravitational-wave strain. (A more formal calculation would take the integral over time; the result here is valid if h changes little during the transit time of the pulse.) The strain h is the wave amplitude analogous to the electric field in an electromagnetic wave and varies as the reciprocal of the distance from the source. The intensity in the wave is related to the time derivative of the strain by

$$I = \frac{c^3}{16\pi G}\left(\frac{dh_{22}}{dt}\right)^2. \tag{2}$$

The enormous coefficient in this equation is another way of understanding why gravitational waves are difficult to detect (space is very stiff, it takes a large amount of energy to create a small distortion). For example, a gravitational wave exerting a strain of 10^{-21} with a 10 millisec duration, typical parameters for the detectors operating in the next few years, carries 80 μ watts/meter^2 (about 10^{20} Jansky) past the detector.

A relation that is useful for estimating the gravitational-wave strain h from astrophysical sources, consistent with a combination of Eqs. (1) and (2), is

$$h \approx \frac{GM}{Rc^2}\left(\frac{v^2}{c^2}\right) = \frac{\varphi_{\text{Newton}}}{c^2}\beta^2. \tag{3}$$

v^2 is a measure of the nonspherical kinetic energy; for example, the tangential kinetic energy in a simple orbiting source. Now, to finally set the scale, the very best one could expect is a highly relativistic motion $\beta \sim 1$ of a solar mass placed at the center of our galaxy. Even with these extreme values, $h \sim 10^{-17}$. With this as "opener" it is easy to understand why this line of research is going to be a tough business. The initial goal for the new generation of detectors is $h \sim 10^{-21}$ for averaging times of 10 msec.

Techniques for Detection and their Limits

All the current detection techniques, as well as those planned, measure the distortions in the strain field directly, the "electric" interactions. One can conceive of dynamic detectors (high-energy particle beams) that interact with the "magnetic" terms in the gravitational wave, although there seems to be no compelling argument, at the moment, to develop them.

Most of the initial gravitational-wave searches have been carried out with acoustic detectors of the type initially developed by J. Weber (1961) and subsequently improved by six orders of magnitude in strain sensitivity in the hands of a dedicated international community of scientists. The detection concept, depicted in Fig. 1(a), is to monitor the amplitude of the longitudinal normal-mode oscillations of a cylinder excited by the passage of a gravitational wave. The detector is maximally sensitive to waves propagating in the plane perpendicular to the longitudinal axis. The frequency response of the detector is concentrated in a narrow band around the normal-mode resonance, although it is possible, by designing the motion transducer together with the resonator as a coupled system, to increase the bandwidth. Typical resonance frequencies have ranged from 800 to 1000 Hz with detection bandwidths ranging from 1 to 10 Hz.

The resonator is isolated from perturbations in the environment by being suspended in a vacuum, and if this is done successfully, the measurement will be limited only by fundamental noise terms. The fundamental limits come from thermal noise (thermal phonon excitations in the resonator and transducer), which can be reduced significantly by operating at cryogenic temperatures, and from the amplifier noise, which has both a broadband component that helps to mask the displacement measurement as well as a component that randomly drives the resonator through the transducer (back-action force). This combination of sensing noise and back-action-force noise is characteristic of all linear systems and ultimately results in the "naive" quantum limit of the measurement. ("Naive" since ideas have been proposed to circumvent the limit though these do not seem trivial to execute.) Current performance limits for the acoustic detectors is an rms strain sensitivity of approximately 5×10^{-19} near 1 kHz (see Fig. 4). The "naive" quantum limit is still a factor of 50 to 100 times smaller, so there is room for improvement in these systems.

Acoustic detectors have operated at lower frequencies ranging from 50–300 Hz in other resonator configurations such as tuning forks and disks but with a reduced overall sensitivity due to their limited size. Upper limits for a gravitational-wave background have been set in various narrow low-frequency bands by measuring the excitation of the normal modes of the Sun and the Earth. The Earth's prolate-to-oblate spheroidal mode at a period of 53 minutes was used to set an upper limit on a gravitational-wave background at about 10^{-14} strain. Laboratory spherical detectors with a higher sensitivity are currently being considered.

The high-sensitivity detectors being constructed now and planned for space are based on electromagnetic coupling. The underlying reason for their sensitivity comes from the fact that most of the perturbative noise forces affecting the relative displacement measurement used to determine the strain are independent of the detector baseline, while the gravitational-wave displacement grows with the baseline.

Figure 1(b) shows a schematic diagram of a laser gravitational-wave interferometer in a Michelson interferometer configuration with the Fabry-Perot cavity optical storage elements in the arms (Saulson, 1994). The test masses are mirrors, suspended to isolate them from external perturbative forces. Light from the laser is divided equally between the two arms by the symmetric port of the beam splitter; transmission is to the right arm and reflection is toward the left arm. The light entering the cavities in a storage arm can be thought of as bouncing back and forth b times before returning to the beam splitter. The storage time in the arms is $\tau_{st} = b(L/c)$. Light directed to the photodetector is a

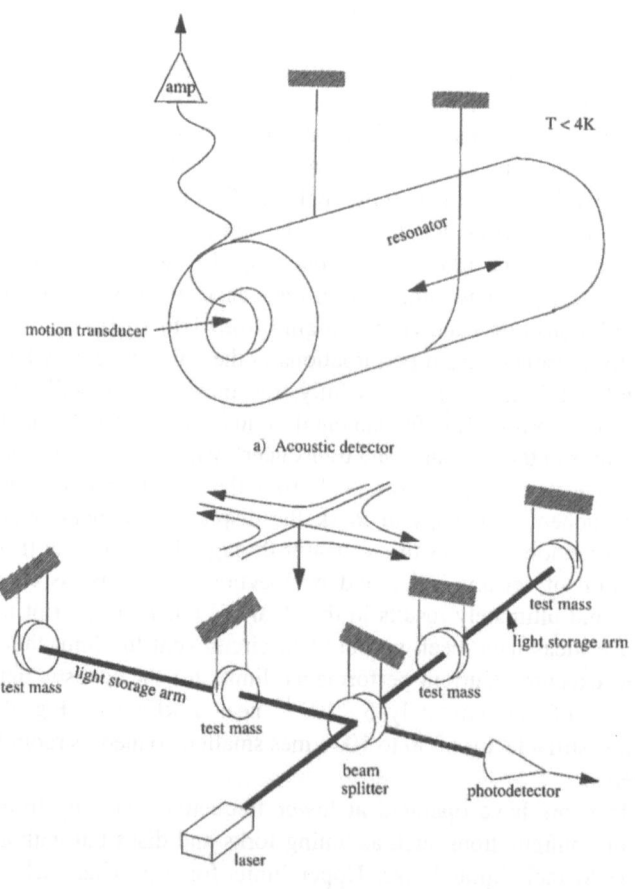

Figure 1. Schematic diagrams: (a) an acoustic bar detector, (b) a laser interferometer detector.

combination of the light from the right arm reflected by the splitter and transmitted light from the left arm. By choosing the path lengths properly and taking note of the sign change of the optical electric field on reflection from the detector side of the splitter (the antisymmetric port), it is possible to make the field vanish at the photodetector (destructive interference). At this setting, a stretch in one arm and a compression in the other, the motion induced by the polarization of the gravitational wave, will change the optical field at the photodetector in proportion to the product of the field at the symmetric port times h. The optical phase associated with this field becomes the gravitational-wave detector output. The interferometer is uniformly sensitive to gravitational-wave frequencies $<(1/4\tau_{st})$ and loses sensitivity in proportion to the frequency at higher values.

With little intensity at the photodetector, almost all the light entering the interferometer is reflected back to the laser. One can increase the light power circulating in the interferometer by placing a partially transmitting mirror between the laser and the beam splitter (the mirror is not shown in the figure). If the mirror is placed correctly and the transmission set to equal the losses in the interferometer, no light will be reflected toward the laser. The circulating power in the interferometer will be increased by the reciprocal of the losses. This technique, called power recycling, matches the laser to the interferometer without changing the spectral response of the instrument and is equivalent to using a higher power laser. The initial interferometer in LIGO (see Fig. 2) and the VIRGO projects will use this configuration.

To bring such an instrument into operation so that fundamental noise dominates the performance, several experimental techniques, first introduced into precision experiments by R. H. Dicke, are employed. In particular, the laser frequency, amplitude, and beam position, and the mirror positions and orientations are controlled and damped by low-noise servo systems to maintain the system at the proper operating point. An associated strategy is to impress high-frequency modulation on the important experimental variables to bring them into a spectral region above the ubiquitous $1/f$ noise.

The remaining noise can be classified into sensing noise—fluctuations in the optical phase independent of the motions of the mirrors—and stochastic-force noise—random forces on the mirrors that are not due to gravitational waves. Sensing noise has nonfundamental contributions from such phenomena as scattered optical fields derived from moving walls and gas molecules or excess amplitude noise in the light, which are controlled by good design, and in the case of the gas, by a vacuum system. The fundamental component is the intrinsic uncertainty of the optical phase and number of photons in the same quantum state of the laser light—referred to as shot noise in the literature. The phase noise varies as $1/\sqrt{P_{\text{splitter}}}$, the optical power at the symmetric port of the beam splitter. The increase in the noise at frequencies above the minima of all the detectors shown in Fig. 4 is due to the sensing-noise contribution.

The stochastic-force noise has both fundamental and nonfundamental components as well. A key feature seen in the low-frequency performance of the terrestrial interferometers in Fig. 4 is the sharp rise at the lowest frequencies below the minima. The noise is due to seismic accelerations not completely removed by the isolation stages and suspension systems. Seismic noise will yield to better engineering since it is a motion relative to the inertial frame and can be reduced by reference to this frame; it is not a fundamental noise. The Newtonian-gravitational gradients associated with density fluctuations of the ground that accompany the seismic waves (as well as density fluctuations of the atmosphere) cannot be shielded from the test masses and constitute a "fundamental" noise at low frequencies for terrestrial detectors; the extension of gravitational-wave observations to frequencies below a few Hz will require the operation of interferometers in space.

The most troublesome stochastic force in the current systems is thermal noise [Brownian motion; again Einstein, as it also is with the photon (Pais, 1982)] coming both from the center of mass motion of the test mass on the pendulum and through the thermal excitation of acoustic waves in the test mass causing random motions of the reflecting surface of the mirrors. The normal modes of the suspension as well as the internal modes of the test mass are chosen to lie outside the sensitive gravitational-wave frequency band. The off-resonance spectrum of the noise that falls in band depends on the dissipation mechanisms in the solid that couple to the thermal disorder (the fluctuation-dissipation theorem of statistical mechanics). The noise at the minima of the room-temperature, large-baseline terrestrial detectors in Fig. 4 is due to thermal excitation. Current strategies to deal with thermal noise use low-loss materials; future development may require selec-

Figure 2. Photograph of the Laser Interferometer Gravitational-wave Observatory (LIGO) site at the Hanford Reservation site in central Washington state. The LIGO is comprised of two sites, the other is in Livingston Parish, Louisiana, which are run in coincidence. The figure shows the central building housing offices and the vacuum and laser equipment area at the intersection of the two 4-km arms extending toward the top and left. The arms are 120-cm-diameter evacuated tubes fit with baffles to reduce the influence of scattered light. The tubes are enclosed in concrete arched covers to reduce the wind and acoustic forces on the tube as well as to avoid corrosion. At the Hanford site the initial interferometer configuration includes a 4-km- and a 2-km-long interferometer operating in the same evacuated tube. At Livingston there will initially be a single 4-km interferometer. The three interferometers will be operated in coincidence and the detection of a gravitational wave will require consistency in the data from the three. The first data runs are planned in November of 2001 at a sensitivity $h_{rms} \approx 10^{-21}$ around 100 Hz. The expectation is that the French/Italian VIRGO project, the German/Scotch GEO project, and the Japanese TAMA project will be operating at the same time.

tive refrigeration of normal modes by feedback or cryogenic operation of the test masses and flexures.

As with the acoustic detector but at a much lower level in the long-baseline systems, the combination of the sensing noise, varying as $1/\sqrt{P}$, and the stochastic forces associated with sensing, the fluctuating radiation pressure, varying as $\sqrt{P}$ on the test masses leads to the "naive" quantum limit. The physics is the same as the Heisenberg microscope we use to teach about the uncertainty relation. The electron has become the test mass, while the random recoil from the photon has been replaced by the beat between the zero-point vacuum fluctuations and the coherent laser light. The naive quantum limit for broadband detection, assuming a bandwidth equal to the frequency, is given by

$$h_{rms} = \frac{1}{2\pi L} \sqrt{\frac{4 h_{Planck}}{\pi m f}},$$

for example, $h_{rms} = 1 \times 10^{-23}$ at $f = 100$ Hz for a 100 kg mass placed in the 4-km arms of the LIGO.

Figure 4 shows several curves for the long-baseline detector. Enabling research is being carried out in many collaborating laboratories throughout the world to reduce the limiting noise sources to gain performance at the advanced detector level and ultimately to the gravity gradient and quantum limits.

Observations of low-frequency gravitational waves need the large baselines and low environmental perturbations afforded by operation in space. Searches for gravitational waves with periods of minutes to several hours have been executed using microwave Doppler ranging to interplanetary spacecraft. The strain levels shown in Fig. 4 are limited by the propagation fluctuations in the interplanetary solar plasma and can be reduced by operating at shorter wavelengths.

Currently there are efforts underway by both the European Space Agency (ESA) and NASA to study the Laser Interferometer Space Antenna (LISA). A concept for the project is shown in Fig. 3. Three spacecraft are placed in solar orbit at 1 a.u. trailing the Earth by 20 degrees. The spacecraft are located at the corners of an equilateral triangle with 5×10^6-km-long sides. Two arms of the triangle comprise a Michelson interferometer with vertices at the corners. The third arm permits another interferometric observable to be measured, which can determine a second polarization. The interferometers use the same 1 micron light as the terrestrial detectors but need only a single pass in the arms to gain the desired sensitivity. The end points of the interferometers are referenced to proof masses free-floating within and shielded by the spacecraft. The spacecraft is incorporated in a feedback loop with precision thrust control to follow the proof masses. This drag-free technology has a heritage in the military space program and will be further developed by the GPB program.

→

Figure 3. A schematic of the Laser Interferometer Space Antenna (LISA) being considered by ESA and NASA as a possible joint mission to observe gravitational waves at low frequencies between 10^{-5} and 1 Hz. This region of the gravitational wave spectrum includes several promising types of sources involving massive black holes at cosmological distances. It also includes the orbital frequencies of white dwarf and other types of binaries in our own Galaxy. The spectral region is precluded from terrestrial observations by gravity gradient fluctuations due to atmospheric and seismic density changes. The interferometric sensing is carried out by optical heterodyne using 1 micron, 1-watt lasers, and 30-cm-diameter optics. Current hopes are to launch LISA by about 2008.

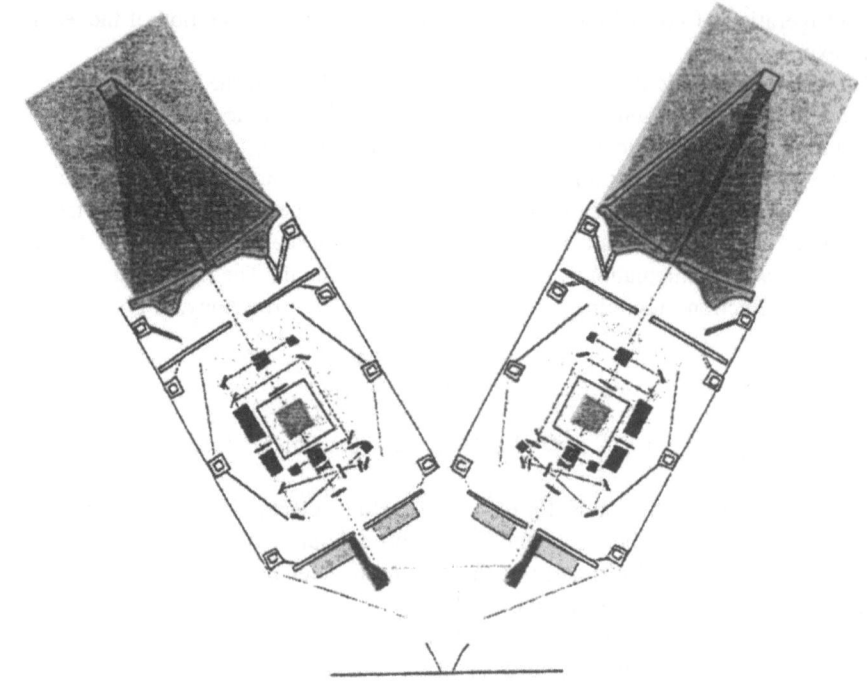

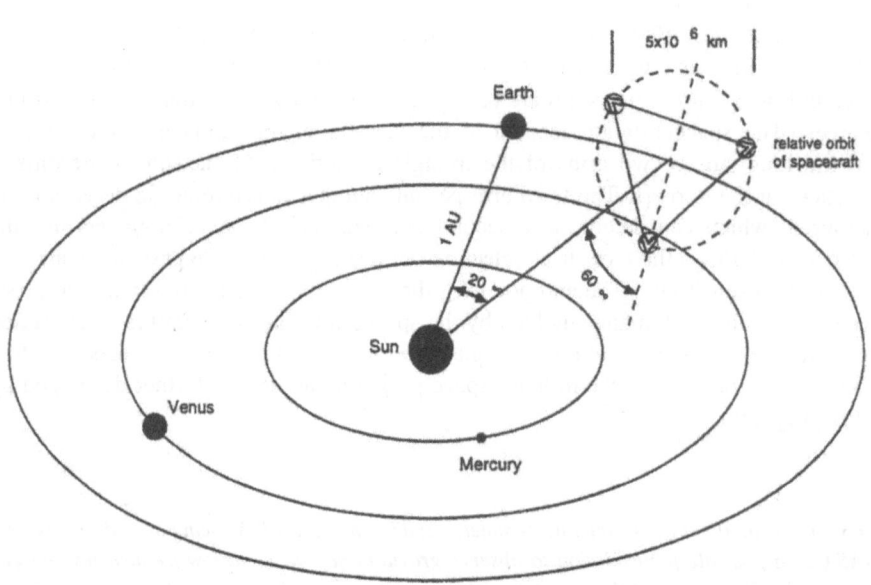

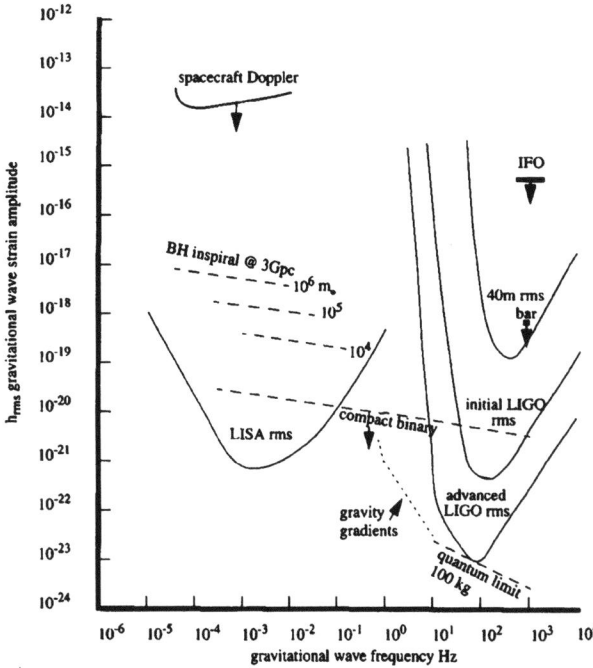

Figure 4. The rms gravitational-wave spectrum for impulsive (burst) sources using a bandwidth equal to the frequency. The figure shows currently established upper limits indicated by heavy lines associated with an arrow downward: "bars" is the rms limit of the LSU bar detector run in coincidence with the Rome group; "ifo" refers to the "100 hour" coincidence run made by the Glasgow and Max Planck groups; spacecraft Doppler ranging designates the limits achieved by the JPL group on the Galileo and other deep space missions. The curve labeled by "40 m" is the best spectrum attained in the LIGO 40-meter prototype at Caltech. The spectrum is not a measurement of an astrophysical limit. The curves labeled "initial LIGO" and "advanced LIGO" are projections for the rms sensitivity of the initial LIGO detector and a detector with improved seismic isolation, suspensions and optics to be placed in the LIGO facilities within a decade of the initial runs. A goal for the ultimate sensitivity of the terrestrial long-baseline detectors are the dotted lines labeled "gravity gradients" and "quantum limit." The curve designated by LISA is the rms noise projected for the space based LISA mission. The dashed lines are projections for a few burst sources. For chirp sources, to account for optimal filters in the detection, the strain amplitude is multiplied by the $\sqrt{n}$, where n is the number of cycles in the time series. The upper estimate curve labeled "compact binaries" is an optimistic one assuming 3NS/NS coalescences a year coming from all galaxies within 23 Mpc. The strength of the wave varies linearly as the reciprocal of the distance. In comparing the sources with the rms spectra of the detectors one needs to include a factor for the desired signal to noise and to take an average over all polarizations and directions of incidence. To gain a signal to noise of 5/1 and to take account of the possibility of nonideal orientation of the detector, it has become a standard practice to multiply the rms detector noise by about 11. The dashed lines at low frequencies "BH inspiral" are drawn for the chirp from BH/BH coalescence.

Figure 4 shows projections for LISA as a broadband burst detector. At frequencies below the minimum noise, the system is limited by a combination of stochastic forces acting on the proof masses. An important effect is from the fluctuating thermal radiation pressure on the proof mass. At frequencies above the minimum, the noise is mainly due to the sensing noise from the limited-light power. The limits follow from design choices and are far from the ultimate limits imposed by cosmic rays at low frequencies or by laser and optical technology at high frequencies.

Astrophysical Sources

Gravitational dynamics of self-gravitating objects has time scales derived from Newtonian arguments:

$$\tau \approx \frac{1}{\sqrt{G\rho}} \approx \sqrt{\left(\frac{R^3}{GM}\right)} \quad \text{(for black holes)} \rightarrow \frac{GM}{c^3}.$$

Spheroidal oscillations or orbits close to the surface of a neutron star (solar mass, 10 km radius, nuclear density) have periods around 1 msec. For black holes, the geometric relation between the mass and the radius of the horizon constrains the dynamics and the natural time scale becomes the light travel time around the horizon, about 0.1 msec for a solar mass black hole. Broadly, the terrestrial detectors will observe events at black holes in the range of $1-10^3$ solar masses, while space detectors can detect signals long before coalescence and observe black holes up to 10^8 solar masses.

Astrophysical sources have been classified by the gravitational-wave time series they generate as burst, chirp, periodic, and stochastic background sources. A comprehensive summary is presented by Thorne (1987). The new detectors will be able to detect all classes. The brief summary below begins with sources in the band of the terrestrial detectors.

The classical burst source with frequency components in the band of terrestrial detectors has been the supernova explosion for which the event rate per typical galaxy is once in 30 to 40 years. Although the rate is known, the energy radiated into gravitational waves is poorly estimated since the degree of nonsphericity in the stellar collapse is difficult to model. Systems with large specific angular momentum are expected to be strong radiators as are those which pass through a highly excited dense fluid state before becoming quiescent. A supernova losing 10^{-3} of its rest energy to gravitational waves in our own galaxy would produce an rms strain around 10^{-18} in the 100 Hz–1 kHz band. The sequence of events would be the detection of the gravitational-wave pulse followed shortly by the neutrinos and then hours later by the optical display. A high signal-to-noise detection with enough bandwidth to follow the motion of a close supernova could be one of the most interesting contributions of gravitational-wave observations to the knowledge of stellar dynamics. Even though such an event has a low probability, the long-baseline detectors are targeting almost continuous operation of at least a single interferometer for this eventuality.

The event rate of all classes of sources is increased by improving the sensitivity. For a given intrinsic source strength, the rate of events (once the sensitivity is enough not to be dominated by the local group of galaxies) will grow with the volume of space opened to the search, as the cube of the strain sensitivity. A benchmark for the field has been to bring the initial sensitivity to a level to include plausible sources at the distance of the Virgo cluster of galaxies, about 10^3 galaxies at a distance 10^3 times our galactic radius (10 Mpc). The supernova rate would then be about 3/year and the hypothetical supernova of the prior paragraph would provide a strain of 10^{-21}.

Black holes are sources with converse uncertainty, the event rate is uncertain but there is a reasonable estimate for the amplitude. The mass spectrum of black holes is not known although there is increasing evidence that most galaxies contain massive black holes in their cores; this, in part, has given impetus to place an interferometer in space. Even though the most energetic formations have not yet been successfully computer modeled, a reasonable estimate of the amplitude and time scales has been established from perturbation theory of black-hole normal modes. The radiating mechanism is the time dependence in the geometry of the event horizon as the hole changes mass; when matter falls into the hole, or when the hole forms initially. The horizon cannot readjust instantaneously to reflect the change, and a gravitational wave with periods determined by the local travel time of light around the event horizon is emitted. The radiation has a characteristic decaying exponential wave form of several cycles damped by the radiation itself. Currently, the only source for which a reasonably reliable rate and amplitude can be estimated is the neutron star/neutron star coalescence; the end point of a system like the Hulse-Taylor neutron star binary. In the final hours the stars orbit each other more and more rapidly until at an orbital frequency close to 1 kHz, the neutron stars collide. The collisions are possible candidate sources for the cosmological γ-ray bursts that have been observed since the mid-1970s. In the last $\frac{1}{4}$ hour before the collision, the system executes about 10^4 oscillations in a chirp extending over the sensitive band of the interferometric detectors. The wave form of the chirp can be modeled within a few milliseconds of the moment of contact and then new information concerning the equation of the state of the nuclear matter and conditions on the neutron star surface may become inscribed on the gravitational wave form.

The rate of coalescence is calculated from a number of such systems discovered in our own galaxy and from estimates of pulsar detection efficiencies. The expectation is that one needs to be able to look into the universe with a depth of 200 to 400 million light years to observe three coalescence chirps per year, at a strain integrated over the chirp of $h = 10^{-22}$. The chance of detecting these sources in the initial interferometer system is small but not vanishing. With improvements in the low-frequency performance, in particular, the thermal noise, the probability of detection improves significantly.

The neutron star/neutron star compact binary system is but one of several compact binary candidates; there should also be black hole/neutron star and black hole/black hole binaries. The stars in these systems will be more massive and stronger radiators. They may well be more interesting radiators since there will be new relativistic physics that can be studied in these systems. The Lense-Thirring, or "frame dragging" effect, should one of the compact objects be spinning, will cause new equations of motion and subtle modulations in the chirps. The detailed wave forms of the black-hole mergers are still not known and are the subject of extensive theoretical work. A major effort by the theoretical community in relativity is involved in calculating wave shapes to guide in the detection and to engender a comparison of theory with experiment as the field makes the transition into a real science.

A different class of sources are periodic or almost periodic systems in our galaxy that radiate extended wave trains. An example is a spinning neutron star with a time-dependent quadrupole moment due to a bump on its surface or an accretion-driven normal mode of oscillation. Such stars will radiate at twice the spin or oscillation frequency and at higher harmonics. They may show a small period derivative due to energy loss (possibly into gravitational waves) or spectral broadening from inhomogeneous excitation. The detection can take advantage of long integration times providing proper account is made of the frequency changes due to the motion of the detector relative to the source. The techniques required are similar to pulsar searches in radio astronomy. For a source at a specific location in the sky, it is possible to remove the Doppler shifts due to the Earth's rotation and orbit at all gravitational frequencies. The Fourier transform of the Doppler-corrected data is used as the narrow-band filter to search for periodicities. The concept is straightforward but the actual execution of a full-frequency/full-sky search poses a formidable computational challenge for extended integration times.

Periodic sources afford particularly attractive possibilities to test gravitational-wave kinematics from the amplitude modulation due to the rotation of the detector relative to the source. Such measurements would give information on the polarization state and propagation speed of the gravitational wave. The detection of the same periodicities in widely separated terrestrial detectors would provide strong confirmation, and would help in separating periodic signals with modulation at solar and sidereal days as well as to discriminate the artifacts from wandering local oscillators.

A stochastic background of gravitational waves may exist and is computationally one of the easier sources to search for. Such backgrounds could arise from the overlap of unresolved impulsive sources or the incomplete spectral resolution of many periodic sources. The most interesting source would be the random metric fluctuations associated with the primeval universe. These would constitute a gravitational equivalent to the cosmic microwave background radiation but come from a time much closer to the origin of the explosion, at an epoch inaccessible electromagnetically. The internal noise of the terrestrial detectors cannot be modeled well enough to establish a small excess due to a gravitational-wave noise. The detection of such a background requires the measurement of a small common noise in several detectors against a much larger uncorrelated component. The cross correlation depends on the gravitational-wave frequency and the separation of the detectors. For an isotropic background the correlation washes out at $f > (c/L_{\text{separation}})$. The detected correlation amplitude signal to noise grows slowly, only as $\frac{1}{4}$ power of the correlation time. The measurement of a stochastic background would benefit from the multiple correlations afforded by a network of detectors.

The sources to be studied by the LISA detector are quite different. There are binaries throughout our galaxy nearly certain to be observable at frequencies above 0.003 Hz. At lower frequencies, the spectrally unresolved high density of a white dwarf and other binary systems is anticipated to cause a background noise of gravitational radiation. The "gravitational confusion" will not compromise the main objective of the LISA to detect and study signals from massive black holes at cosmological distances.

The mass spectrum of black holes is not known, although there is increasing evidence that many galaxies contain massive black holes in their cores. One promising source for LISA is five to ten solar mass black holes orbiting and ultimately coalescing with the massive hole at the galactic center. The coalescence of massive galactic black holes during the merger of galaxies is another candidate as could be the metric perturbations during the initial formation of the massive holes themselves.

Detection Criteria

A signal needs to be above the noise experienced in the instrument and environment, however, this alone is insufficient to establish it as a gravitational wave in the terrestrial detectors. The most satisfying circumstance is that a gravitational-wave observation be made in a set of widely distributed detectors [the Gravitational-Wave Network (GWN)] and the recorded wave forms allow the solution for the polarization of the wave and the position of the source. Armed with this information, an electromagnetic (or neutrino) search could be attempted in the error circle of the gravitational wave detection; a time-honored approach bringing gravitational-wave observations into the main stream of astrophysics. The strategy would apply to all classes of sources: impulsive, chirps, quasiperiodic, and periodic.

The confident detection of impulsive sources is more difficult. While the periodic and quasiperiodic detections will have confidence limits based on quasi-stationary system noise (the signals last long enough to take a meaningful sample of the noise spectrum), the impulsive signals, especially if rare, will be particularly dependent on the non-Gaussian component of the noise; the noise most difficult to reduce and control in a single detector. The technique of multiple coincidence of several detectors is one of the best means to gain confidence. The coincidences must occur within a time window to permit a consistent solution for a location in the sky. If the general character of the source can be guessed in advance (for example, a binary coalescence chirp, or a black-hole normal-mode oscillation), the signal is filtered prior to the coincidence measurement to improve the sensitivity. The more detectors involved, the greater the confidence assigned to the detection.

There is still the possibility of coincidence due to environmental or anthropogenic causes. The various sites throughout the world are far enough apart that most environmental perturbations should not correlate between them. The acoustic noise, the seismic noise, and the power line (especially if the network includes detectors in different power grids and significantly different time zones) will be uncorrelated. There are correlations in the magnetic-field fluctuations (thunderstorms) and in radio frequency emissions. As part of the detection strategy a large number of environmental parameters will be measured along with the gravitational-wave signals at each site. One of the requirements for the authenticity of impulsive sources will be the lack of correlation with environmental perturbations and other ancillary internal signals developed to monitor the performance of the instruments.

The Future

As has been the rule rather than the exception in astrophysical observations, when new instrumentation offering a factor of 1000th improvement in sensitivity or bandwidth is applied to observing the universe; new phenomena are discovered. There is no reason to expect less for gravitation which involves looking at the universe in a new channel, going deep into the astrophysical processes to observe with no obscuration or scattering. The research has the two ingredients that make physics and astrophysics such a rewarding experience. There are the sharpshooter questions: the tests of the strong field, the confirmation of the wave kinematics, and the tests of astrophysical models; and there is also the buckshot part of the research with the high probability of discovering new and so far unthought-of processes—this gives an added romance to the new field.

Acknowledgments

Professor Anthony P. French of the MIT Physics Department helped locate the references of Newton and Einstein. I have learned a great deal from my colleagues in this adventure. In particular, I would like to acknowledge the efforts of Bruce Allen, Marcel Bardon, Barry Barish, David Berley, Peter Bender, Vladimir Braginsky, Alain Brillet, Mark Coles, Karsten Danzmann, Ronald Drever, Robert Eisenstein, Sam Finn, Peter Fritschel, Adelbert Giazotto, William Hamilton, Jim Hough, Richard Isaacson, Albert Lazzarini, Peter Michelson, Fred Raab, Albrecht Rudiger, Gary Sanders, Peter Saulson, Roland Schilling, Bernard Schutz, David Shoemaker, Kip Thorne, Rochus Vogt, Harry Ward, Stan Whitcomb, Walter Winkler, and Mike Zucker, some of whom have spent years in this business while others are relative newcomers, but all will be responsible for the revolution in astrophysics and physics that will hit us in the next decade.

References

Einstein, A., 1916, "Näherungsweise Integration der Feldgleichungen der Gravitation," Sitzungsberichte der Königlich Preussischen Akademie der Wissenschaften, Sitzung der physikalich-mathematischen Klasse, p. 688.

Einstein, A., 1918, "Über Gravitationswellen," Sitzungsberichte der Königlich Preussischen Akademie der Wissenschaften, Sitzung der physikalich-mathematischen Klasse, p. 154.

Hulse, R. A., and J. H. Taylor, 1974, "A high sensitivity pulsar survey," Astrophys. J. **191**, L59–L61.

Hulse, R. A., and J. H. Taylor, 1975, "Discovery of a pulsar in a binary system," Astrophys. J. **195**, L51–L53.

Landau, L. D., and E. M. Lifshitz, 1962, *The Classical Theory of Fields* (Addison-Wesley, Reading, MA).

Pais, A., 1982, *Subtle is the Lord* (Oxford University, New York).

Saulson, Peter R., 1994, *Fundamentals of Interferometric Gravitational Wave Detectors* (World Scientific, Singapore).

Smarr, Larry, 1979, Ed., *Sources of Gravitational Radiation* (Cambridge University, Cambridge, England).

Taylor, J. H., and J. M. Weisberg, 1982, "A new test of general relativity: Gravitational radiation and the binary pulsar PSR 1913+16," Astrophys. J. **253**, 908–920.

Taylor, J. H., and J. M. Weisberg, 1989, "Further experimental tests of relativistic gravity using the binary pulsar PSR 1913+16," Astrophys. J. **345**, 434–450.

Thorne, K. S., 1987, in *300 Years of Gravitation*, edited by S. W. Hawking and W. Israel (Cambridge University, Cambridge, England), Chap. 9.

Thorne, Kip, 1994, *Black Holes and Time Warps: Einstein's Outrageous Legacy* (Norton, New York).

Weber, J., 1961, *General Relativity and Gravitational Waves* (Interscience, New York).

Deciphering the Nature of Dark Matter
Bernard Sadoulet

The history of physics can be seen as the gradual discovery of structures invisible to the human eye and the instruments of the time. These structures are often hinted at first by indirect evidence, and new instrumentation has to be invented to fully establish their existence and to study their properties. The search for the dark matter in the universe represents an archetypal case study of such a process. It took nearly 60 years from the first evidence (Zwicki, 1933) for astronomers to reach a consensus that there is a dark component which dominates gravity but cannot be seen, as it neither emits nor absorbs light. The debate has now shifted to one about its nature, in particular whether dark matter is made of ordinary baryonic matter or whether new nonbaryonic components play a significant role. A number of innovative attempts to decipher this nature have been launched. The searches for baryonic forms of dark matter, including the evidence for massive halo compact objects (MACHOs), and for nonbaryonic forms are reviewed. The numerous attempts to detect weakly interactive, massive particles (WIMPs) are presented as an example of the novel instruments necessary to make progress in this new field of astrophysics. Because of space constraints, references are limited to recent reviews or representative works in each of the areas.

Dark Matter: Evidence and Nature

Dark matter

The existence of dark matter is now well established at a variety of scales (see, e.g., Trimble, 1987). In large spiral galaxies it is often possible to measure the rotation velocity of HII regions, atomic hydrogen clouds, or satellite galaxies out to large distances from the galactic centers. The constancy of these rotation velocities implies that the enclosed mass increases with radius well beyond the distance at which no more stars are observed. The effect is particularly spectacular for dwarf galaxies, which are totally dominated by dark matter. Similar evidence for dark matter is also observed in elliptical galaxies. The velocity dispersion of globular clusters and planetary nebulae, and the extended x-ray emission of the surrounding gas, show that most of the mass in outer parts of these galaxies is dark.

The dynamic effect of dark matter is even more pronounced in clusters of galaxies. It has been known for some time that dispersion velocities of the many hundreds of galaxies that constitute rich clusters are often in excess of 1500 km/s. Such large values indicate even deeper potential wells than for galaxies. In many clusters a large amount of gas is detected through its x-ray emission, and its high temperature (≈ 5 keV) implies similar dark masses. In the last few years, a third piece of evidence has been gathered that also points to a very large amount of dark matter in clusters. Galaxy clusters gravitationally lens the light emitted by quasars and field galaxies in the background. The mapping of the mass distribution through the many arclets seen in a number of clusters indicates potential wells qualitatively similar to those observed with the two other methods. These dark matter density estimates are confirmed by the combination of measurements of the gas mass fraction in clusters (typically 20%) and estimates of the baryon density from primordial nucleosynthesis (see, e.g., White et al., 1993).

At a larger scale, measurements of velocity flows and correlations hint at even larger amounts of dark matter. In this volume, Turner and Tyson summarize such measurements of the matter density in units of the critical density ρ_c as

$$\Omega_M = \frac{\rho_M}{\rho_c} = 0.35 \pm 0.07 \quad \text{with} \quad \rho_c = 1.88 \times 10^{-26} \, h^2 \, \text{kg m}^{-3},$$

where h is the Hubble expansion parameter in units of 100 km s^{-1} Mpc^{-1} ($h = 0.67 \pm 0.1$). Such a matter density is much greater than the visible matter density (less than 1% of the critical density).

While there is a broad consensus on the existence of such dark matter (unless Newton's laws are incorrect), there is still an intense debate on its nature. Can it be formed of ordinary baryons or is it something new?

Need for baryonic dark matter

An interesting element of this discussion is provided by the baryon density

$$\Omega_B = (0.02 \pm 0.002) h^{-2}$$

inferred from the observations of ^{4}He, D, ^{3}He, and ^{7}Li in the very successful standard scenario of homogeneous primordial nucleosynthesis (Schramm and Turner, 1998, and references therein). This is larger than the visible matter density, and we have to conclude that a component of the dark matter has to be baryonic. We need to understand where these dark baryons are hidden (Sec. III).

Need for nonbaryonic dark matter

It is clearly necessary to introduce a second type of dark matter to explain why measurements of Ω at large scales appear to be significantly higher than the baryonic density inferred from nucleosynthesis. Note that this argument is purely based on a set of converging observations, admittedly with large but different systematics, and not on inflation or the esthetic appeal of $\Omega = 1$. Homogeneous Big Bang nucleosynthesis may be wrong, but all attempts to produce significantly different results, for instance through inhomogeneities induced by a first-order quark hadron phase transition, have been unsuccessful.

A second argument for the nonbaryonic character of dark matter is that it provides the most natural explanation of the large-scale structure of the galaxies in the universe in terms of collapse of initial density fluctuations inferred from the COBE measurement of the temperature fluctuations of the cosmic microwave background. The deduced power spectrum of the (curvature) mass fluctuations at a very large scale connects rather smoothly with the galaxy power spectrum measured at lower scale, giving strong evidence for the formation of the observed structure by gravitational collapse. The observed spectral shape is natural with cold (that is, nonrelativistic) nonbaryonic dark matter but cannot be explained with baryons only; since they are locked in with the photons until recombination, they cannot by themselves grow large enough fluctuations to form the structure we see today.

A third general argument comes from the implausibility of hiding a large amount of baryons in the form of compact objects (routinely called MACHOs). For instance, if the ratio of the mass in gas and stars to the total mass in clusters is of the order of 20%, this would require 80% of the initial gas to have condensed into invisible MACHOs. This is very difficult to understand within the standard cooling and star formation scenarios. The same argument applies to galactic halos.

In conclusion, it seems very difficult to construct a self consistent cosmology without nonbaryonic dark matter. We therefore need at least two components of the matter in the universe. In addition, as explained by Turner and Tyson in this volume, there may be a third diffuse component, possibly with negative pressure, such as a cosmological constant. The fact that their densities are similar ($\Omega_B \approx 0.05$, $\Omega_{DM} \approx 0.3$, $\Omega_\Lambda \approx 0.65$) is somewhat disturbing, since they arise from *a priori* distinct physical phenomena and components with different equations of state evolve differently with time (e.g., if there is a sizable cosmological constant we live in a special time). This may indicate that our theoretical framework is incomplete. The task of the observer is clear however: to convincingly establish the existence of these three components and their equations of state. In addition to the confirmation of the recent indications for an accelerating universe provided by supernovae observations at high redshift, it is therefore important to solve the two dark matter problems: find the hidden baryon component and positively detect the nonbaryonic dark matter.

Searches for Baryonic Dark Matter

Where are the dark baryons? It is difficult to prevent baryons from emitting or absorbing light, and a large number of constraints obtained at various wavelengths considerably restrict the possibilities.

Gas

If the baryonic dark matter were today in the form of diffused nonionized gas, there would be a strong absorption of the light from the quasars, while if it were ionized gas, the x-ray background flux would be too large and the spectrum of the microwave background too much distorted by upward Compton scattering on the hot electrons. However, recent detailed measurements of the absorption lines in the spectrum of high redshift quasars (the so-called Lyman α forest) indicate that, at a redshift of three or so, the Lyman α gas clouds contain $(0.01-0.02)\ h^{-2}\ (h/0.67)^{1/2}$ of the critical density in ionized baryons, enough to account for all the baryons indicated by the primordial abundance of light elements.

The problem then shifts to explain what became of this ionized high redshift component. Two general answers are proposed:

(i) It can still be in the form of ionized gas with a temperature of approximately 1 keV. Such a component would be difficult to observe as it would be masked by the x-ray background from active galactic nuclei. This is the most natural solution, as it is difficult for ionized gas to cool off and clump significantly, as shown by hydrodynamical codes.
(ii) However, it has also been argued that our simulations are still too uncertain to believe these cooling arguments: this gas could have somehow condensed into poorly visible objects either in the numerous low surface brightness galaxies or in the halo of normal galaxies.

Atomic gas would be visible at 21 cm. Dust is excluded as it would strongly radiate in the infrared. Clumped molecular hydrogen regions are difficult to exclude but could in principle be detected as sources of gamma rays from cosmic-ray interactions. However, the most likely possibility, if this gas has been able to cool, is that it has formed compact objects. In particular, objects with masses below 0.08 solar masses, often called brown dwarfs, cannot start thermonuclear reactions and would naturally be dark. Black holes without companions would also qualify.

Massive halo compact objects

How do we detect such compact objects? Paczynski (1986) made the seminal suggestion of using gravitational lensing to detect such objects. Suppose that we observe a star, say in the Large Magellanic Cloud (LMC), a small galaxy in the halo of the Milky Way. If one MACHO assumed to be in the halo were to come close to the line of sight, it would gravitationally lens the light of the star and its intensity would increase. This object, however, cannot be static, lest it fall into the potential well. Therefore it will soon move out of the line of sight, and one would expect a temporary increase of the light intensity that, from the equivalence principle, should be totally achromatic. The duration of such a microlensing event is related to the mass m, distance x, and transverse velocity $v_\perp$ of the lens, and the distance L of the source by

$$\Delta t \propto \sqrt{mx(L-x)/v_\perp^2 L}.$$

The probability of lensing at a given time (the optical depth τ) is given by a weighted integral of the mass density $\rho(x)$ of MACHOs along the line of sight:

$$\tau \propto \int \rho(x) \frac{x(L-x)}{L} dx.$$

The maximum amplification unfortunately does not bring any additional information as it depends in addition on the random impact parameters.

To be sensitive enough, such a microlensing search for MACHOs in the halo should monitor at least a few million stars every night in the LMC. Following Alcock's observation that this was within the reach of modern instrumentation and computers, three groups (MACHO, EROS, OGLE) launched microlensing observations in 1992. Since then they have been joined by five other groups. The results of these five years can be summarized as follows:

(1) The observation of some 300 events towards the bulge of the galaxy has clearly established gravitational microlensing. The distribution of amplifications and the independence from the star population confirm this explanation. Microlensing has opened a new branch of astronomy which can now probe the mass distribution of condensed objects. It is even hoped that it will allow the detection of planets around lensing stars, as they would produce sharp amplification spikes.

(2) Probably the most important result of the microlensing experiments is that there is no evidence for short lensing events (corresponding to low-mass MACHOs) in the direction of the LMC. A combination of the EROS and MACHO results excludes (Fig. 1) the mass region between 10^{-7} and 10^{-1} solar masses (Alcock et al., 1998). Our halo is not made of brown dwarfs!

(3) However, a number of long-duration LMC events have been observed. EROS has detected two events and in five years the MACHO team has observed some 18 LMC lensing events of duration (defined as the full width at 1.5 amplification—the EROS group uses the half width) between 35 and 150 days. This cannot be explained in the standard picture of a rather thin Milky Way disk and a thin LMC. The main problem in interpreting this interesting result is that we usually do not know where the lenses are along the line of sight. As explained above, for each event we have only two

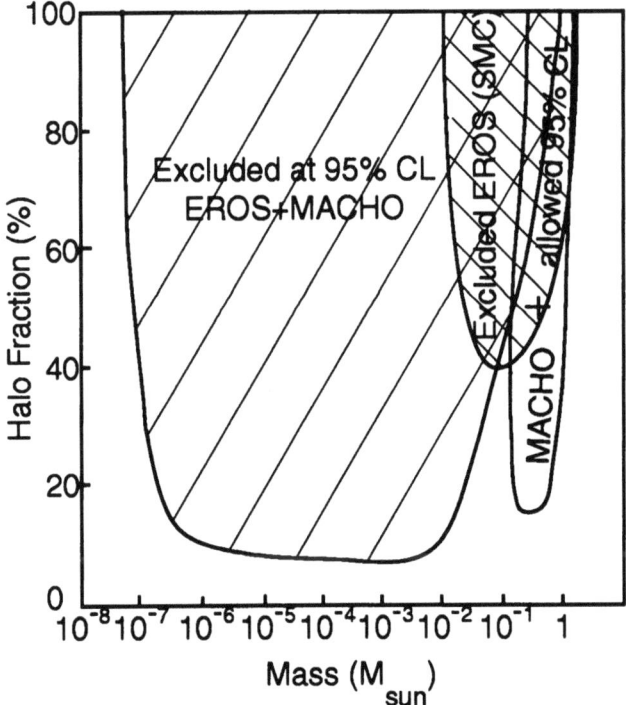

Figure 1. Excluded region (at 95% confidence level) of the halo fraction in MACHOs as a function of their mass in a standard halo model. The ellipse on the right is the 95% confidence level range allowed by the two year data of the MACHO collaboration.

experimental observables, the duration of the microlensing event and its probability, an insufficient amount of information to unravel the distance of the lens, its mass, and its transverse velocity. Only in specific events can we give the distance of the lens. One LMC event corresponds to a double lens that creates two amplification spikes at caustic crossing. This double lens is clearly in the LMC. One other event is produced by a disk star that we can see. For most of the observed events, the degeneracy between mass, distance, and velocity limits in a fundamental way our capability to interpret the results.

If we assume that the MACHOs are distributed in the same way as the galactic halo, they may represent a fraction of the halo density between 10 and 100% (Fig. 1). Although the compatibility with 100% may superficially indicate that the dark matter problem is solved, this interpretation encounters the serious difficulty that the mass of individual lenses would be typically one third of a solar mass. These objects are not brown dwarfs. They cannot be ordinary stars as this is incompatible with the Hubble Space Telescope surveys. The hypothesis that they could be very old white dwarfs requires an artificial initial mass function, an uncomfortable age of more than 18 billion years, and a totally unknown formation mechanism.

We are then led to question the assumed distance and velocity distributions. Four types of models have been proposed: an additional component of our galaxy such as a thick or warped disk, an extended spheroid, an intervening dwarf galaxy, or a tidally elongated LMC.

(4) The last model may be favored by the observations towards the Small Magellanic Cloud (SMC) that have so far detected two microlensing events. The first event is much longer (250 days) than all the LMC events and the absence of parallax due to the movement of the earth constrains it to be close to the SMC. While this result is unlikely for a halo-like distribution (some four events of duration similar to LMC events would have been expected), it is quite natural if the lenses are in the observed galaxies: the longer duration of the SMC event is due to the lower SMC dispersion velocity. The second SMC event is produced by a double lens clearly in the host galaxy. Although the SMC is known to be thicker than the LMC, these observations cast further doubt on a halo interpretation of the LMC events. A detailed mapping of the thickness of the LMC (for instance, by R. R. Lyrae's) is an important task to prove or disprove this self-lensing hypothesis. Note that lensing by a low-density extended component of the LMC would not produce a quadratic dependence on the lensing rate with respect to the star density on the sky. The apparent absence of such a dependence cannot be used as an argument against a self-lensing explanation. Note that the lack of events observed towards SMC excludes, as shown in Fig. 1, the possibility that MACHOs of mass smaller than a solar mass form a large fraction of the halo (Palanque Delabrouille et al., 1998).

It is clear that it is essential to break the degeneracy between mass, distance, and velocity. The data of double lenses, or the precise photometry of very long events, partially break this degeneracy. The different lines of sight such as the SMC or M31, which is beginning to be explored, are very important to test the assumption of a halo-like distribution. Unfortunately in the case of M31 one cannot see individual stars from the ground and one is limited to pixel lensing, in which interpretation depends on the good knowledge of the underlying star population. A satellite located one astronomical unit away would be a useful tool, as it may allow a parallax measurement as the lensing will be observed at a different time. The Space Interferometric Mission satellite to be launched in 2006 can also help break the degeneracy.

Dark matter black holes

Although black holes may not be initially formed by the collapse of baryonic objects and in any case have technically lost any information about their baryonic content, we summarize at the end of this baryonic section their possible contribution to dark matter.

Very low-mass black holes cannot form the bulk of dark matter, as they would evaporate through Hawking radiation and give rise to high energy gamma-ray flashes, which are not observed.

The quoted microlensing result exclusion of the mass range between 10^{-7} and 10^{-1} solar masses also applies to black holes. Note that primordial black holes of a solar mass or so could explain the MACHO observations towards the LMC and would otherwise behave as cold dark matter. One solar mass happens to be the mass inside the causal horizon at the quark hadron phase transition, and a strongly first-order transition may indeed induce density fluctuations large enough to produce these black holes. However, the needed abundance appears to require fine-tuning of parameters.

Very massive objects (VMOs), an early star population of at least a hundred solar masses, could have rapidly formed black holes without contaminating the interstellar medium with metals. However, we should now see the radiation of the progenitor stars in the far infrared and the diffuse infrared background experiment (DIRBE) severely constrains this possibility. Even more massive ones would disrupt galactic disks.

Searches for Nonbaryonic Dark Matter

The intrinsic degeneracy arising in the interpretation of microlensing observations prevents the fascinating MACHO results from seriously undermining the case for nonbaryonic dark matter. Moreover, if such a nonbaryonic component exists, as hinted by the cosmological arguments of Sec. II, it is difficult to prevent it from accreting (unless it is relativistic); even in the presence of MACHOs in the halos, it should constitute a significant portion of the halo and be present locally for detection. In fact, taking into account all kinematic information on the galaxy and the MACHO observations, the most likely density for a nonbaryonic component is close to the canonical 0.3 GeV/cm^3 inferred from the velocity curves of our galaxy.

A large number of candidates have been proposed over the years for such a nonbaryonic component. They range from shadow universes existing in some string models, strange quark nuggets formed at a first-order quark-hadron phase transition (Witten, 1984), charged massive particles (CHAMPs) (De Rujula, Glashow, and Sarid, 1990), and a long list of usually massive particles with very weak interactions. We should probably first search for particles that would also solve major questions in particle physics. According to this criterion, three candidates appear particularly well motivated.

Axions

Axions are an example of relic particles produced out of thermal equilibrium, a case in which we depend totally upon the specific model considered to predict their abundances. These particles have been postulated in order to dynamically prevent the violation of *CP* in strong interactions in the otherwise extremely successful theory of quantum chromodynamics. Of course there is no guarantee that such particles exist, but the present laboratory and astrophysical limits on their parameters are such that, if they exist, they would form a significant portion of cold dark matter (Turner, 1990). Such low-mass cosmological axions could be detected by interaction with a magnetic field that produces a faint microwave radiation detectable in a tunable cavity. The first two searches for cosmological axions performed a decade ago were missing a factor of 1000 in sensitivity. This is no longer the case; Livermore, MIT, Florida and Chicago are currently performing an experiment that has published preliminary limits (Hagmann et al., 1998). It will reach (Fig. 2) a cosmologically interesting sensitivity at least for one generic type of axion (the so-called hadronic model; see Turner, 1990). The collaboration hopes to improve their sensitivity down to the lowest couplings currently predicted (the DFZ model; see Turner, 1990). Matsuka and his collaborators in Kyoto are developing a more ambitious scheme using Rydberg atoms that are very sensitive photon detectors and should immediately reach the DFZ limit. Although these experiments are very impressive, it should be noted that the decade of frequency (and therefore of mass) that can be explored with the present method is only one out of three that is presently allowed.

Light massive neutrinos

Neutrinos of mass much smaller than 2 MeV/c fall in the generic category of particles that have been in thermal equilibrium in the early universe and decoupled when they were relativistic. Their current density is basically equal to that of the photons in the universe. The relic particle density is therefore directly related to its mass, and a neutrino species of 25 eV would give an Ω of the order of unity. Note that neutrinos alone cannot lead to the observed large-scale structure as fluctuations on scales greater than $40 h^{-1}$ Mpc are

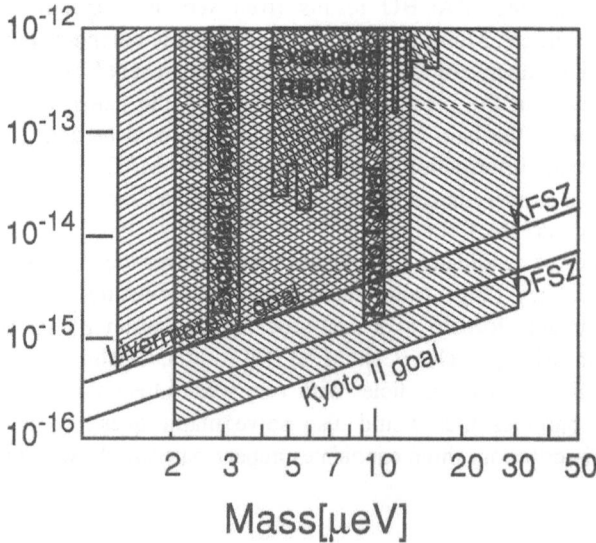

Figure 2. Expected sensitivity of the Livermore experiment. The lines labeled KSVZ and DFSZ refer to two generic species of axions. The shaded regions in the upper right are the previous experimental limits.

erased by relativistic neutrino streaming. They have to be mixed in with cold nonbaryonic dark matter (Klypin, Nolthenius, and Primack, 1997, and references therein) or seeded by topological defects. Moreover, because of phase space constraints, they cannot explain the dark matter halos observed around dwarf galaxies.

Unfortunately no good ideas have yet been put forward of possible ways to detect cosmological neutrinos (see, e.g., Smith and Lewin, 1990) and one can only rely on the mass measurements of neutrinos in the laboratory through the study of beta spectra, neutrinoless double beta decay, and oscillation experiments. One may summarize the situation (see accompanying review of Wolfenstein for details and references) as follows: The direct mass measurement of the electron neutrino gives limits of 5 eV. Model-dependent limits of the order of 1 eV on the mass of Majorana neutrinos are given by neutrinoless double beta decay searches (Heidelberg-Moscow). The claim by the LSND group for muon to electron neutrino oscillation with relatively large $\Delta m^2 \approx 6$ eV^2 oscillation is now challenged by the Karmen experiment.

The best indication that neutrinos have a nonzero mass comes from atmospheric and solar neutrinos. The SuperKamiokande group has recently presented statistically significant results demonstrating the disappearance of atmospheric muon neutrinos that points to an oscillation with Δm^2 of a few 10^{-3} eV^2 and a large mixing angle. The combination of the chlorine, water Cerenkov, and gallium experiments has been indicating for some time a depletion of solar neutrinos with respect to the standard solar model. The most natural explanation is a MSW (Mikheyev-Smirnov-Wolfenstein) or vacuum oscillation with Δm^2 of 10^{-6} eV^2 or 10^{-10} eV^2, respectively (Hata and Langacker, 1997).

Note, however, that these oscillation experiments do not give a direct measurement of the neutrino masses that may well be in the electron volt range (for nearly degenerate masses). It thus remains important for cosmology to improve the electron neutrino mass limit.

Weakly interactive massive particles

A generic class of candidates is constituted by particles that were in thermal equilibrium in the early universe and decoupled when they were nonrelativistic. In this case it can be shown that their present density is inversely proportional to their annihilation rate (Lee and Weinberg, 1977). For these particles to have the critical density, this rate has to be roughly the value expected from weak interactions (if they have masses in the GeV/c^2 to TeV/c^2 range). This may be a numerical coincidence, or a precious hint that physics at the W and Z^0 scale is important for the problem of dark matter. Inversely, physics at such a scale leads naturally to particles whose relic density is close to the critical density. In order to stabilize the mass of the vector-intermediate bosons, one is led to assume the existence of new families of particles, such as supersymmetry in the 100-GeV mass range. In particular, the lightest supersymmetric particle could well constitute the dark matter. We review in the next section the experimental challenge to detect them.

Searches for Weakly Interactive Particles

The most direct method to detect these WIMPs is by elastic scattering on a suitable target in the laboratory (Goodman and Witten, 1985; Primack, Seckel, and Sadoulet, 1988). WIMPs interaction with the nuclei in the target would produce a roughly exponential distribution of the recoil energy with a mean dependent on their mass; the hope is to identify such a contribution in the differential energy spectrum measured by an ultra-low background detector, or at least to exclude cross sections that would lead to differential rates larger than observation.

Experimental challenges

In specific models such as supersymmetry, the knowledge of the order of magnitude of the annihilation cross section allows an estimation of the WIMP elastic scattering, taking into account the coherence over the nucleus. Typically, if scalar (or "spin independent") couplings dominate, the interaction rate of WIMPs from the halo is expected to be of the order of a few events per kilogram of target per week for large nuclei like germanium. We display in Fig. 3, as the lower hatched region, the range of cross sections (rescaled to a proton target) expected (Jungman et al., 1996) in grand-unified-theory-inspired supersymmetric models, where scalar interactions usually dominate. The upper hatched regions summarize the current limits achieved with state-of-the-art techniques to achieve low radioactivity background. They barely skirt the supersymmetric region.

Unfortunately, the expected rates can be very small for specific combinations of parameters in which axial ("spin dependent") couplings dominate. In this case the interaction takes place with the spin of the nucleus, which limits the number of possible targets, and the current limits are very far above the supersymmetric region (Jungman et al., 1996).

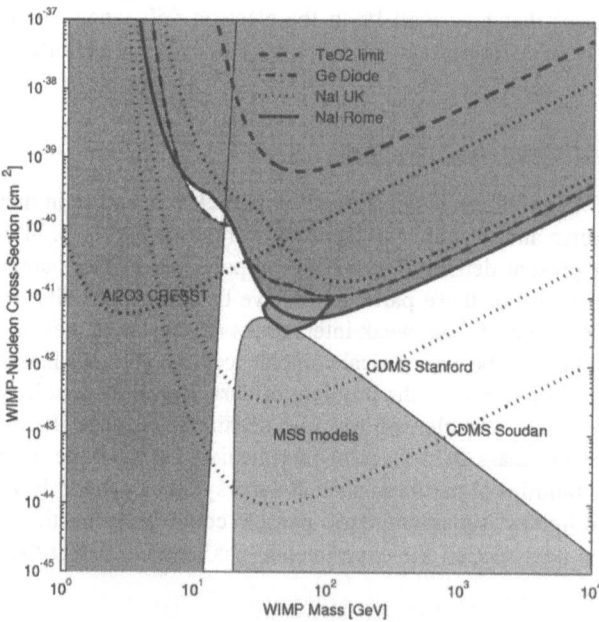

Figure 3. Current achieved limits for spin-independent couplings as a function of the WIMP mass. All the results have been converted to WIMP-nucleon cross sections assuming scalar interactions scaling as the square of the atomic number. The hatched region at the top is excluded by these experiments. The shaded regions at the bottom are the rates predicted by minimal supersymmetric models including the constraints from LEP and CDF experiments. The curves labeled CRESST and CDMS are goals of these experiments.

It is therefore essential to construct experiments with very low radioactive backgrounds and, if possible, with the instrumental capability to recognize nuclear recoils (only produced by WIMPs, if neutrons are eliminated) and actively reject the electron recoils produced by gamma rays and electrons from radioactivity. Note that, without this discrimination, the background is not measured independently of the signal. The experimental sensitivity to a small signal then ceases to improve with exposure, once the background level is measured with sufficient statistical accuracy. In contrast, with discrimination the combination of background rejection and subtraction of the remaining contamination allows a sensitivity increase as the square root of the target mass and the running time, until the subtraction becomes limited by systematics.

A second challenge faced by the experimentalist comes from the fact that the energy deposition is quite small, typically 10 keV for the mass range of interest. For detectors based only on ionization or scintillation light, this difficulty is compounded by the fact that the nuclear recoils are much less efficient in ionizing or giving light than electrons of the same energy. This increases the recoil energy threshold of such detectors, and one should be careful to distinguish between true and electron equivalent energy that may differ by a factor of 3 (Ge) to 12 (I).

A third challenge is to find convincing signatures linking detected events to particles in the halo of the galaxy. The best one would be the measurement of the direction of the scattered nucleus, a very difficult task. Short of that directionality signature, it is, in principle, possible to look for a change in the event rate and the spectrum of energy deposition with a change in the time of the year.

Prominent direct search strategies

In spite of these experimental challenges, low expected rates and low energy depositions, a number of experimental teams are actively attempting to directly detect WIMPs. The detection techniques are very diverse, ranging from mica, which integrates for billions of years over minute target masses, and superheated microdots, which should be only sensitive to nuclear recoil, to low pressure time projection chambers, which could give the directionality. However, we can identify three main experimental strategies.

(1) A first approach is to attempt to decrease the radioactive background as much as possible. Germanium is the detector of choice as it is very pure, and the first limits were obtained by decreasing the threshold of double-beta experiments. The most impressive results have been obtained by the Heidelberg-Moscow group (Baudis et al., 1998) with a background of 0.05 events/kg/day/(equivalent electron keV) around 20 keV (equivalent electron energy). The current combined exclusion plot is given in Fig. 3.

This strategy is pushed to the extreme by GENIUS, an ambitious proposal to immerse one ton of germanium detectors in an ultra-pure liquid nitrogen bath. However, this approach is fundamentally limited by the absence of discrimination against the radioactive background.

(2) A second approach has been to use large scintillators with pulse-shape discrimination of nuclear and electronic recoils, unfortunately with energy thresholds difficult to bring below 50 keV ($\approx$4 keV equivalent electron energy on iodine). The technique is simple and large masses can be assembled to search for modulation effects. The most impressive result so far has been obtained by NaI. The groups using NaI have published limits that are slightly better than those obtained with conventional germa-

nium detectors. The Rome group has recently announced (Bernabei et al., 1998) a close to 3σ detection of a signal using the annual modulation expected for a WIMP spectrum (heart-shaped region in Fig. 3). Note that because Na has a spin, these experiments so far give the best limits for spin-dependent couplings. It is too early to conclude, but it is unlikely that NaI could make significant additional progress as the small number of photoelectrons at the energies of interest and the lack of power of the pulse-shape discrimination make it highly susceptible to systematics.

(3) Thus more powerful discrimination methods need to be devised. Liquid xenon with simultaneous measurement of scintillation and ionization is a promising approach, albeit with relatively high thresholds, and not enough development so far to fully judge its potential. In contrast, the active development of novel "cryogenic" detectors based on the detection of phonons produced by particle interactions is beginning to bear fruit. In spite of the complexity of very low temperature operation, two large setups are currently being routinely operated (Milano: Alessandro et al., 1986; CDMS: Nam et al., in Cooper, 1997; CRESST: Sisti et al., in Cooper, 1997) with total detector mass ranging from 1 kg to 7 kg. For dark matter searches this technology appears to possess significant advantages.

To summarize, cryogenic detectors are making fast progress and appear currently to hold the most promise for exploring a significant portion of the supersymmetric WIMP space in the next few years.

Indirect detection methods

Let us note finally that several methods have been proposed for detecting WIMPs through their annihilation products (Primack, Seckel, and Sadoulet, 1988 and references therein). They of course assume dark matter exists in the form of both particles and antiparticles (or is self conjugate) as otherwise no annihilation would occur.

The detection of gamma-ray lines from their annihilation into two photons will require the resolution of the next generation of satellites and may be masked by the galactic background, especially if the dark matter density does not strongly peak at the galactic center. The first measurements of the energy spectra of antiprotons and antielectrons offered tantalizing hints of dark matter particle annihilations, but they turned out to be inaccurate. The interpretation of such spectra would in any case be very uncertain because of the uncertainty on the confinement time of these antiparticles in the halo of our galaxy.

A much more promising method is to search for high energy neutrinos coming from the centers of the earth and the sun. Since they can lose energy by elastic interactions, some dark matter particles would be captured by these objects, settle in their centers, and annihilate with each other producing, among other products, high energy neutrinos that can then be detected in underground detectors, especially through the muons produced by their interactions in the rock. The current generation of such detectors (Baksan, MACRO, and SuperKamiokande) of roughly 1000 m^2 area set a limit of the order of 10^{-14} muon cm^{-2} s^{-1} above 3 GeV. Such results exclude any charge-symmetric Dirac neutrino or scalar sneutrino and put limits on supersymmetric models that are generally in agreement but less restrictive than direct detection experiments. Fairly model-independent arguments (Kamionkowski et al., 1995) show that such an advantage of direct detection should be maintained for the next generation of detectors (cryogenic WIMP searches and 10^4 m^2 detectors such as AMANDA II), especially for scalar interactions. However, the very large neutrino detectors currently being studied (10^6 m^2) may be more sensitive than direct searches for large-mass WIMPs.

Conclusion

In the past decade astrophysicists have clearly confirmed the earlier indications that there is much more mass in the universe than we can see. This dark matter dominates gravity over a variety of scales, from dwarf galaxies to the larger structures and velocity flows that we can see. Representing more than 99% of the mass density, it is an essential component of any cosmology and appears responsible for the formation of structure, galaxies, stars, and planets. Ultimately, in spite of being totally inert, it may be an essential element for the appearance of life as planets would not exist without dark matter.

Elucidating the nature of this dark matter has therefore become a central question in astrophysics and probably one of the most fundamental and multidisciplinary quests in science today. Are we observing a new form of matter (and energy) in the universe? We have reviewed the large number of projects devoted to this question. They require long term efforts and highly sophisticated instrumentation, but after a decade of development, a number of searches are beginning to reach the necessary level of sensitivity. As often remarked, a positive answer would lead to another Copernican revolution; not only are we not the center of the universe, but we are not even made of what most of the universe is made of!

Acknowledgments

This work was supported by the Center for Particle Astrophysics, a National Science Foundation Science and Technology Center operated by the University of California under Cooperative Agreement no. AST-912005, and by the Department of Energy under the contract DE-AC03-76SF00098.

References

Ahlen, S. P., *et al.*, 1987, Phys. Lett. B **195**, 603.
Alcock, C., *et al.*, 1995b, Astrophys. J. **445**, 133.
Alcock, C., *et al.*, 1997, Astrophys. J. Lett. **491**, L11.
Alcock, C., *et al.*, 1998, Astrophys. J. Lett. **499**, L9.
Alessandrello, *et al.*, 1996, Nucl. Instrum. Methods Phys. Res. A **370**, 241.
Baudis, *et al.*, 1997, preprint hep-9811045.
Bernabei, R., *et al.*, 1998, preprint INFN/AE-98/20.
Cooper, S., 1997, in Proceedings of the VIIth International Workshop on Low Temperature Detectors, Munich, 1997 (Max Planck Institute of Physics, Munich), p. 237, and http://avmp01.mppmu.mpg.de/ltd7.
De Rujula, A., S. L. Glashow, and U. Sarid, 1990, Nucl. Phys. B **333**, 173.
Goodman, M. W., and E. Witten, 1985, Phys. Rev. D **31**, 3059.
Hagmann, C., *et al.*, 1998, Phys. Rev. Lett. **80**, 2043.
Hata, N., and P. Langacker, 1997, Phys. Rev. D **56**, 6107.
Jungman, G., M. Kamionkowski, and K. Griest, 1996, Phys. Rep. **267**, 195.
Kamionkowski, M., K. Griest, G. Jungman, and B. Sadoulet, 1995, Phys. Rev. Lett. **74**, 5174.

Klypin, A., R. Nolthenius, and J. Primack, 1997, Astrophys. J. **474**, 533.
Lee, B., and S. Weinberg, 1977, Phys. Rev. Lett. **39**, 165.
Palanque Delabrouille, N., *et al.*, 1998, preprint astro-ph/98-12173.
Primack, J. R., D. Seckel, and B. Sadoulet, 1988, Annu. Rev. Nucl. Part. Sci. **38**, 751.
Schramm, D. N., and M. S. Turner, 1998, Rev. Mod. Phys. **70**, 303.
Shutt, T., *et al.*, 1992, Phys. Rev. Lett. **29**, 3531.
Smith, P. F., and J. D. Lewin, 1990, Phys. Rep. **187**, 203.
Trimble, V., 1987, Annu. Rev. Astron. Astrophys. **25**, 425.
Turner, M. S., 1990, Phys. Rep. **197**, 67.
White, S. D. M., *et al.*, 1993, Nature (London) **366**, 429.
Witten, E., 1984, Phys. Rev. D **30**, 272.
Wolfenstein, L., 1979, Phys. Rev. D **20**, 2634.
Zwicky, F., 1933, Helv. Phys. Acta **6**, 110.

Nuclear Physics

Nuclear Physics at the End of the Century

E. M. Henley and J. P. Schiffer

As we are approaching the turn of the century we wish to review very briefly the status of nuclear physics as it has evolved in the course of our explorations of nature during this period. More first-hand details of this evolution can be found in the article by Hans A. Bethe in this issue (Bethe, 1999).

Nuclear physics was born over 100 years ago with the discovery of radioactivity by Becquerel and followed by the work of the Curies in identifying the sources of the new radiations. The nucleus as a small heavy center of the atom was only deduced in 1911 by Rutherford; its existence was a crucial feature of the Bohr atom—and thus central to the development of quantum mechanics. Until the discovery of the neutron in 1932 it was believed that the nucleus was made of electrons and protons. The nucleus of protons and neutrons, as we know it, can be said to date from that time.

It was quickly realized that to pursue such studies required more intense sources of charged particles. Several classes of accelerators, to provide such energetic particles, were invented and developed for the study of the nucleus in the 1930s, from the electrostatic generators of high voltage by Cockroft and Walton and Van de Graaff, to the cyclotron of Lawrence, to betatrons and their derivatives. Techniques for detecting particles also started in this period, from the early scintillating screens of Rutherford to the counters and cloud chambers of the 1920s and 30s. The early work on nuclear reactions quickly established the size of the nucleus and thus the range of nuclear forces. The electron spectrum of beta decay led Pauli to the supposition that there had to be an additional particle involved, the neutrino, and Fermi subsequently formulated the correct theory of the beta-decay process.

These early developments of nuclear physics provided a rich case study for the application of the new quantum theory.

One of the surprising features of the nucleus was the fact that nuclear forces seemed to have a very short range. In 1935 Yukawa postulated exchange forces and a "meson" as the carrier of the nuclear force that keeps the neutrons and protons bound inside the nucleus. Such a particle, of mass between the electron and proton that was thought to be the meson, was found in cosmic rays in 1937, but it was realized that its nuclear interactions were too weak. We now know that this was the muon, and the pion was not discovered until 1946. The 1930s also saw the first realization by Bethe of how nuclear reactions fuel the sun in converting hydrogen into the light elements.

Progress in nuclear physics was rapid after this start and the last 50 years have seen explosive growth in our understanding of the nucleus, its constituents, and its forces. In the present framework we cannot begin to do justice to many who have made important and often major contributions to the evolution of this part of science and, somewhat arbitrarily, we mention by name only Nobel laureates. The realization in the late 1940s by M. Goeppert-Mayer and J.H.D. Jensen that the structure of the nucleus can be understood in terms of a shell model, as in atomic physics, came as a great surprise because of the high density of the nucleus and the strength of nuclear forces, but the validity of this description has been verified by many phenomena. Its theoretical basis came later.

The large quadrupole deformation of some classes of nuclei led to a further understanding of new degrees of freedom, and the dynamical collective model of Bohr and Mottelson came out of this realization. These degrees of freedom, primarily collective rotations of nonspherical nuclei as well as vibrations, lead to a simplified description, particularly of deformed nuclei. When combined with the shell model in a deformed potential, this work led to a unified model of the nucleus (Bohr and Mottelson, 1969). An important degree of freedom in nuclei was found to be that of "pairing," the correlations between pairs of nucleons coupled to zero spin, and the theoretical understanding has a close analogy with the BCS theory of superconductivity involving the pairing of electrons. More recently, a very successful class of models has emerged, the so-called algebraic models of "dynamic symmetries," that has been given the name of the "interacting boson model." Here the degrees of freedom are those of a boson, primarily of zero spin (Arima and Iachello, 1981). The unified model and the mean-field theories also permitted descriptions of excited states and transition rates. They also led to an understanding of giant resonances seen in the excitation of a nucleus, be they monopole (breathing mode), dipole, or higher multipoles.

In parallel with the vast improvement in our understanding of the structure of nuclei was work on nuclear reactions. Early work showed pronounced resonances in nuclear reactions, particularly the absorption of slow neutrons, indicating long-lived intermediate states. This implied many degrees of freedom and apparently a complicated process. Considerable simplification resulted from models in which the description of the compound system through which a reaction proceeds is described in terms of the *average* properties of the resonances, without detailed consideration of their individual properties. Reactions can be considered in two limits, the first of which proceeds through long-lived intermediate resonances whose decay is independent of their mode of formation—the simple expression for a resonance was given by Breit and Wigner in the 1930s, followed by the detailed treatment in reaction formalisms, particularly of Wigner, in the 1940s and 50s. The other limit, developed in the 1960s, is that of direct, one-step reactions, proceeding in a time comparable to the passage of the projectile through the nuclear volume. In this limit, the reaction may be described by an effective "optical" potential with a real and imaginary part. Here the interaction can be described as a perturbation in a quasielastic scattering process with the incident wave modified by the interaction but continuing coherently in the outgoing channel. Formalisms for the descriptions of more complicated reactions between these two extremes were filled in more slowly.

The 1950s and 60s also saw major new developments in beta decay, the realm of the "weak" interactions. The ideas of Lee and Yang that parity need not be conserved in beta decay was quickly followed by the work of C.S. Wu showing that, indeed, parity conservation was not a valid symmetry in these processes. The study of nuclear beta decay has laid the foundations of the standard model of elementary particles.

The 1970s and 80s saw the consolidation of the description of the nucleus, with improved experimental techniques supporting the theoretical framework. The use of computers aided this endeavor greatly. The use of heavy-ion beams (accelerated heavy nuclei) was expanded and new features of nuclei were investigated. The use of electron beams for mapping of nuclear-charge distributions, pioneered by Hofstadter, was extended in precision. Distributions of charge and magnetization densities and transition probabilities were mapped out and this field was established as a powerful quantitative source of information about nuclei. More generally, detection techniques evolved considerably in resolution and in the capacity to handle complex information.

New accelerators using superconductivity were developed, both linear accelerators with superconducting rf cavities and cyclotrons with superconducting magnets. Also, the realization of the pion's role in the nucleus lead to the construction of "meson factories," where intense beams of pions were produced for studies of nuclear properties. Strange mesons produced at high-energy accelerators were used to produce "hypernuclei," a new class of nuclei in which a long-lived baryon with a "strange" quark is bound along with neutrons and protons.

Present perspective

In the current decade (1990s), nuclear physics continues to address the state of hadronic matter, which increasingly includes the structure of hadrons as well as the larger many-body aspects of nuclei. The field of nuclear physics at the end of the century encompasses a number of areas and in this article we will attempt to discuss briefly a few of the current thrusts and cite some review articles that provide more details.

The hadrons are the simplest entities of strongly interacting matter that can exist as free particles. Their properties are well established experimentally, but the way they are constituted out of quarks and gluons is still not well understood. Recent experimental results have shown that the spin of the nucleons, for instance, is not as simple as it seemed a few years ago, but has contributions from the polarization of the "sea" of quantum chromodynamics (QCD), arising from gluons and from the possible angular momentum of quarks and gluons. How protons and neutrons—the most stable hadrons—interact with each other to form simple nuclei has seen substantial progress. Evidence is now quite conclusive that simple two-body forces are insufficient to explain the properties of the simple nuclei and that many-body forces are important.

The understanding of the structure of nuclei in terms of the shell model and the various collective rotations, vibrations, and excitations that nuclei undergo has advanced in several directions. In particular, new detection techniques have helped unravel bands of states that correspond to shapes that are deformed from spherical symmetry much more drastically than previously observed—suggesting a region of stability with 2:1 (major to minor) axis ratios. These states appear to be rather pure and hardly mix at all with the more spherical normal states. Other advances in experimental capabilities have allowed physicists to explore the limits of nuclear stability, the so-called drip line.

One of the aspects of QCD that is not satisfactorily understood is the concept of *confinement*, the fact that the constituents of hadrons can never appear as free particles. At very high densities this confinement is expected to break down to the extent that quarks can travel freely within the region of high energy density. This is presumably a state that the universe passed through shortly after the big bang. There will soon be a new tool for investigating the state of matter at that time: a large collider of heavy (e.g., Au) ions is being constructed (RHIC) in which energy densities comparable to the big bang should be reached. A key symmetry that is broken in normal QCD, that of *chiral symmetry*, may well be restored in this regime of energy density. Both the experimental undertakings at this facility and the theoretical interpretations are a major challenge for the field in the coming decade.

The crucial role of nuclear physics in fueling the stars has been recognized since the early work of Bethe, who showed how stars are powered by fusion reactions, and later of Fowler and coworkers who developed the understanding of the processes responsible for the formation of elements. Specific nuclear properties play key roles in the big bang, in the energy production in our Sun and in other stars, and in all the nucleosynthetic processes of the universe. This intimate relationship is beautifully illustrated by the fact that the properties of the lightest neutrinos have been enormously clarified by the theoretical interpretation of experiments that searched for the nuclear reactions that these neutrinos induce on earth.

Finally, not only has the field depended critically on developing a large variety of experimental and theoretical techniques, but these techniques have in turn served society in a number of ways—nuclear medicine being a prominent example.

Hadron Physics

The smallest entities of accessible strongly interacting matter in the world are hadrons, either baryons that are aggregates of three quarks or mesons that are made from quark-antiquark pairs. The most stable baryons, the protons and neutrons, are the major constituents of atomic nuclei, and the lightest meson is the pion. Understanding the structure of hadrons and how the properties of these particles arise from QCD is a major interest of nuclear physics. This interest follows two paths. One concerns the properties of families of hadrons as they exist freely, to accurately characterize the members of the rich hadron spectrum in mass and decay properties and reflect the structure that arises from QCD. The other is to understand how these properties change when the hadrons are immersed in nuclei or nuclear matter.

Pions

Among the mesons, the lightest and most important one is certainly the pion. Thus, it is no accident that its properties, production, and interactions with nucleons and nuclei have received considerable attention in the past and again at the present time.

Because quantum chromodynamics (QCD) is the underlying theory of hadronic interactions, there have been many models built on one aspect or another of the theory. A particularly important symmetry of QCD, which is almost preserved at low energies, is chiral invariance or the symmetry between left and right handedness. (This differs from parity, which is invariance under mirror reflection.) Chiral symmetry is incorporated into the most recent treatments of few-body problems, with the use of low-energy effective theories, such as chiral perturbation theory, first introduced by Weinberg many years ago (Weinberg, 1979). Here an effective Lagrangian is constructed which incorporates all terms allowed by the symmetries of QCD. In QCD with massless up (u), down (d), and strange (s) quarks, the theory satisfies chiral invariance. This leads to both conserved-vector and axial-vector currents and to parity doublets. Since the axial current is not conserved in nature and parity doublets are not observed, one assumes that spontaneous symmetry breaking leads to the eight Goldstone (almost massless) pseudoscalar bosons. The finite quark masses also break the symmetry somewhat, and this leads to the nonvanishing pion and other light pseudoscalar-meson masses.

The approximate chiral invariance is incorporated in all low-energy effective theories. Chiral perturbation theory is a low-energy theory with a systematic expansion around the chiral limit in powers of m_π/Λ and p/Λ, where p is a typical momentum of order m_π or less and Λ is a QCD scale, of the order of 1 GeV. Because it is an effective theory, it needs to be renormalized at each order of the expansion. One introduces an effective operator in terms of the pion field. The most general Lagrangian density is then unique. When expanded in terms of m_π/Λ and p/Λ, the free-pion Lagrangian is obtained to lowest order and pion-pion scattering is found at the next order. Although the agreement with experiment is quite good, even at this order, it is improved by continuing the expansion through the inclusion of higher-order terms (Holstein, 1995).

Recently the photoproduction of pions has received considerable attention because it is a test of chiral perturbation theory. At threshold the production mechanism is dominated by the electric dipole amplitude which is given by gauge invariance (Ericson and Weise, 1988). In chiral perturbation theory the production amplitude is independent of the pion mass and the pion-nucleon coupling constant. At this order, the π^0 photoproduction from protons and neutrons vanishes. But at the next order, an expansion in terms of m_π^2/Λ^2 and $(p/\Lambda)^2$ gives a finite value which agrees quite well with recent experiments (Bernstein and Holstein, 1991, 1995; Holstein, 1995). Higher-order calculations do even better. We compare theory and experiment in Table I.

Nucleon structure

While we know that nucleons and mesons are composed of quarks and gluons, the transition from a description of nuclei in terms of nucleons and mesons to one in terms of quarks and gluons is still not understood. Nor do we fully understand the structure of the nucleons and mesons. Progress has been made by solving QCD numerically on a finite lattice rather than in continuous space-time.

In an effective theory the gluon degrees of freedom do not appear explicitly; in some models they are incorporated in the dressing of the quarks, which then are called constituent quarks. The constituent quarks (up and down) have masses close to one-third of the mass of the nucleon and thus have small binding energies. There are models of the nucleon made up of such quarks, often treated nonrelativistically and bound by harmonic or other simple forces. These models are amazingly successful in predicting the ratio of

Table 1. Magnitudes of the amplitude for photoproduction of pions in units of $10^{-3}/m_\pi^+$ to lowest ($n=1$) and higher ($n=2$ and 4) order in a chiral-perturbation-theory expansion compared to experiment (Holstein, 1995).

Amplitude	$n=1$	$n=2$	$n=4$	Experiment
$\gamma p \to \pi^+ n$	34.0	26.4		28.4 ± 0.6
$\gamma n \to \pi^- p$	-34.0	-31.5		-31.8 ± 1.2
$\gamma p \to \pi^0 p$	0	-3.58	-1.16	-1.31 ± 0.08
$\gamma n \to \pi^0 n$	0	0	-0.44	~ -0.4

the proton to neutron magnetic moments. However, a number of authors have pointed out that this result is not very model dependent. Most effective theories (e.g., chiral perturbation theory) use almost massless "current" quarks.

One of the original motivations for the study of nuclear structure was to gain an understanding of the strong interaction. Following this interest in the structure of hadronic matter, nuclear physicists have become more and more interested in a quantitative understanding of the structure of hadrons. QCD provides the framework for such understanding. For instance, QCD-based constituent quark models not only can reproduce accurately the masses of mesons with heavy quarks, but can also account for the main features of the masses and electromagnetic decays of baryons with light quarks. However, significant problems remain. In particular, some signs and magnitudes of strong decays of the higher-mass nucleon resonances are poorly understood, possibly because of the lack of a proper treatment of chiral symmetry. The lack of clear experimental evidence for particles that would correspond to excitations of the gluonic field or "hybrid" states of baryons that involve involve gluonic excitations is another outstanding puzzle.

In the last few years deep-inelastic polarized electron scattering on polarized H and D targets have provided insights into the spin structure of the nucleon and revealed that we do not yet fully understand it. For a given quark (q) species (u, d, or s) the fraction of the nucleon's spin that is carried by quark spins is defined as

$$\Delta q \equiv q\uparrow - q\downarrow + \bar{q}\uparrow - \bar{q}\downarrow, \tag{1}$$

and the total spin carried by the quarks is

$$\Delta \Sigma = \Delta u + \Delta d + \Delta s, \tag{2}$$

with

$$\Delta u - \Delta d = g_A, \tag{3}$$

where g_A is the weak axial-vector coupling constant, and arrows indicate spins parallel ($\uparrow$) and antiparallel ($\downarrow$) to the proton's spin. The nonrelativistic "naive" constituent quark model predicts

$$\Delta u = 4/3, \quad \Delta d = -1/3, \quad \Delta s = 0, \quad \Delta \Sigma = 1, \tag{4}$$

and

$$g_A = \Delta u - \Delta d = 5/3. \tag{5}$$

Equation (5) turns out to be far from the truth. Neutron beta decay yields $g_A = 1.26$ and the deep-inelastic scattering experiments show that only about 30% of the spin of the proton comes from the quarks. It is found that (Ashman *et al.*, 1989; 1997)

$$\Delta u = 0.84 \pm 0.04, \quad \Delta d = -0.42 \pm 0.04,$$

$$\Delta s = -0.09 \pm 0.04, \quad \Delta \Sigma = 0.33 \pm 0.08. \tag{6}$$

The experimental result came as a surprise and was called the "proton spin puzzle"; it implies that the major fraction of the proton spin comes from the gluons and possibly from orbital angular momentum. However, Δq is generally evaluated in the infinite-momentum frame and the nonrelativistic quark model is for a nucleon at rest. In terms of mesons, the angular momentum could come from pions coupled to quarks.

Another surprise was the relatively large (~10%) contribution to the spin from strange "sea" quarks. These are quarks and antiquarks in equal numbers that are in addition to those ("valence" quarks) that make up the charge of the nucleons. Again, in principle, this can be understood in terms of low-energy nuclear physics via the dissociation of protons into strange baryons and mesons such as Λ and K^+. Antiquarks play a role because gluons can split into $q\bar{q}$ pairs; the $\bar{q}$ can combine with valence quarks to form pions. To the extent that the QCD Fock space includes (nonperturbative) pions, one can understand the excess of $\bar{d}$ over $\bar{u}$ in a proton since a one-meson decomposition gives $p = p\pi^0$ or $n\pi^+$, with a $\pi^+ = u\bar{d}$. Indeed, there is evidence for an excess of $\bar{d}$ over $\bar{u}$ in the proton from inclusive hadronic reactions with lepton pair production (Drell-Yan processes) and from tests of the "Gottfried sum rule," which follows from the assumption of a flavorless sea of light quarks ($u\bar{u} = d\bar{d}$ in the sea).

Nuclear forces

The nucleon-nucleon force is basic to understanding nuclei and thus has been of great interest for many decades. Broadly speaking, potential representations of the force are either purely phenomenological, or based on meson exchange but with the parameters determined phenomenologically. All models have a one-pion exchange character at long range, which gives rise to a spin-spin central potential and a tensor term. The correctness of pion exchange and its dominance at large distances is clear from the nucleon-nucleon phase shifts at large angular momenta and from various properties of the deuteron ground state, such as the nonzero quadrupole moment. Indeed, the strong tensor component of the pion-exchange force is a unique feature that makes the solution of nuclear many-body problems particularly challenging.

Some potential models represent the shorter-range interaction by heavy-meson exchanges. The Reid and the Urbana-Argonne potentials are examples of more phenomenological models, while the Nijmegen, Paris, and Bonn potentials are based more on meson exchange (Ericson and Weise, 1988; see also Machleidt, 1989). In the last few years significant progress has been made in obtaining high-precision fits to the elastic-scattering data. The "Argonne V_{18}," the "CD Bonn," and several Nijmegen models all fit these data within the experimental accuracy.

These modern potentials, coupled with recent advances in nuclear many-body theory and in the capacity of computers, now makes it possible to understand the stability, structure, and reactions of light nuclei directly in terms of nucleons. Three- and four-nucleon systems are studied accurately in both bound and scattering states by Faddeev and hyperspherical harmonic methods. For nuclei with up to 7 nucleons, quantum Monte Carlo methods have been applied successfully. Ground states, stable against breakup into subclusters, are determined for ^{6}Li and ^{7}Li, and their binding and excited-state spectra agree reasonably. In this work three-nucleon forces are required; their strength is adjusted to reproduce the binding energy of ^{3}H and to give a reasonable saturation density for nuclear matter (Carlson and Schiavilla, 1998).

Chiral perturbative theories and other effective theories have also been applied to the nucleon-nucleon problem. Despite the difficulty caused by the large scattering length that characterizes the data for the $^1S_0$ state, and the bound $^3S_1$ states, these theories fit the phase shifts very well up to momenta of 300 MeV/c (see Fig. 1). The effective potentials or interactions include a contact term and pion exchange. Related techniques with effective chiral theories have been used by others (Meissner, 1992). It is interesting that chiral perturbation theory allows one to show that three-body forces are smaller than two-body ones by the ratio $(p/\Lambda)^2$, where p is a nucleon momentum and Λ, as before, is a QCD scale; for example, if the two-body potential has an average strength of 20 MeV, then the three-body one would have a strength of about 1 MeV. A nice feature of chiral perturbation theories is that they can rank the various classes of charge-independence and charge-symmetry breaking forces in powers of p/Λ.

Very sensitive tests of the nucleon-nucleon interaction in bound states are precise measurements of the radial distribution of nucleons in nuclei for comparison with *ab initio* calculations (Carlson, Hiller, and Holt, 1997). In many cases electron scattering can measure these distributions directly. However, in the case of the deuteron, which has unit spin, the orbital angular momentum 0 and 2 contributions can be separated using polarized electron-deuteron scattering in the tensor-polarization observable t_{20}. The current information on this quantity is illustrated in Fig. 2. A measurement that should provide such information to momentum transfer $Q > 6$ fm^{-1}, or about 0.15 fm in the distances in the radial distribution of the deuteron, is among the early experiments with the CEBAF electron accelerator.

While these measurements demonstrate the wide validity of the hadronic description of the deuteron, at the shortest distance scales the nucleon substructure does become important. To date, one nuclear reaction, the photodisintegration of the deuteron at large transverse momenta into a proton and a neutron, shows the behavior expected of coherently transferring the energy of the incident photon to the six constituent quarks in the deuteron. Indeed, this reaction seems to show "counting rule" behavior at considerably lower energies than was expected, suggesting a wider validity for quark descriptions.

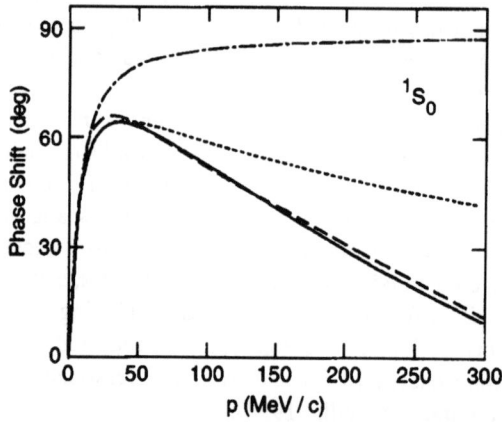

Figure 1. The phase shift δ for the $^1S_0$ channel. The dot-dash curve is a one-parameter fit in chiral perturbation theory at lowest order. The dotted and dashed curves are fits at the next order in the expansion; the dashed one corresponds to fitting the phase shift between $0 \leq p \leq 200$ MeV, whereas the dotted one is fit to the scattering length and effective range. The solid line corresponds to the phase shift obtained from a partial wave analysis carried out by the Nijmegen group (Kaplan, Savage, and Wise, 1998).

Nuclei

The understanding of the structure of nuclei is in terms of models, such as the shell model and the collective model, that necessarily involve approximations in the characterization of a finite many-body system with complex forces between the constituents.

The use of the Hartree approximation with simplified potentials representing the nucleon-nucleon interaction, or even a field theory with scalar and vector mesons ("quantum hadrodynamics") leads to a mean field in which the nucleons move. The theory can be extended to Hartree-Fock and to include deviations from the mean field; it can be compared to experiment. For instance, it does well in reproducing the charge density determined by electron scattering as is seen in Fig. 3.

Electrons are a great tool for precision studies of nuclei: their interaction is sufficiently weak that perturbation theory can be used, they cause little distortion of the system, and their wavelengths can be made sufficiently short to study both nuclei and nucleons in detail (Diepernick and de Witt Huberts, 1990). Electron scattering beautifully shows the single-particle structure of nuclei in a mean-field description by measuring the properties of individual shell-model orbitals and also by exploring the limitations of this description because of the correlation effects that arise from the short-range part of the nucleon-nucleon interaction (Pandharipande, Sick, and de Witt Huberts, 1997). This is clearly seen in the proton-knockout reactions illustrated in Fig. 4. The consequences of these correlations are manifold. They substantially renormalize the single-particle mean-field orbitals and appear to be the source of high-momentum nucleons that are important in "subthreshold" production of mesons and other particles (e.g., antiprotons). Such correlations can

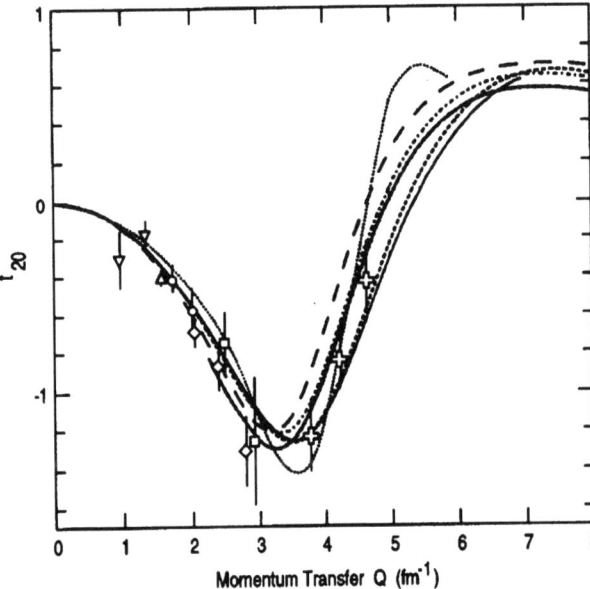

Figure 2. The results of polarization measurements from electron scattering from the deuteron. The quantity t_{20} from a variety of experiments is shown by different symbols, with predictions of different theoretical models of the nucleon-nucleon interactions drawn by lines. New measurements from CEBAF should help to better distinguish between the models.

have an effect on the mean-free paths of nucleons and other hadrons in nuclei and on mechanisms responsible for pion absorption in the nuclear medium. But we are only at the beginning of being able to separate the effects of short-range correlations from other many-body effects, both experimentally and theoretically.

Another perspective of the single-particle structure in nuclei with many nucleons comes from an entirely different dimension. While the successes of the single-particle description in heavy nuclei are remarkable, the description is tested almost entirely for the "valence" orbitals and not for the deeply bound states. Such tests become possible by introducing a different baryon into the nucleus that can settle into the lowest state without violating the exclusion principle. The most suitable baryon, because of its relative stability, is the Λ, and the structure of the deeply bound states of the so-called "hypernuclei" beautifully confirm the single-particle description with the mean field modified to account for the differences in the Λ-nucleon interaction (Chrien and Dover, 1989).

Major advances have been made recently in exploring the structure of nuclei in the limits of extreme conditions—at very high angular momentum, in approaching the limits of nuclear binding, and in temperature and energy density (discussed in Sec. V). We discuss the first two in the section below, as examples of recent developments in the field. The advances in approaching these limits have come in recent decades from novel and substantially improved experimental techniques coupled with new theoretical understanding.

Nuclei at high angular momentum

Accelerator developments have enormously expanded available beams. Beams of heavy nuclei, accelerated as projectiles, have made it possible to bring large amounts of angular momentum into nuclei (Diamond and Stephens, 1980). For instance, with a 200-MeV ^{48}Ca beam incident on a target of ^{120}Sn one forms a compound nucleus at high excitation energy, that first rapidly decays (in $\sim 10^{-21}$ sec) by emitting particles (neutrons), and then remains highly excited in a bound system, usually with high angular momentum. In a rotating reference frame, the excitation energy is not very high—it has to be measured

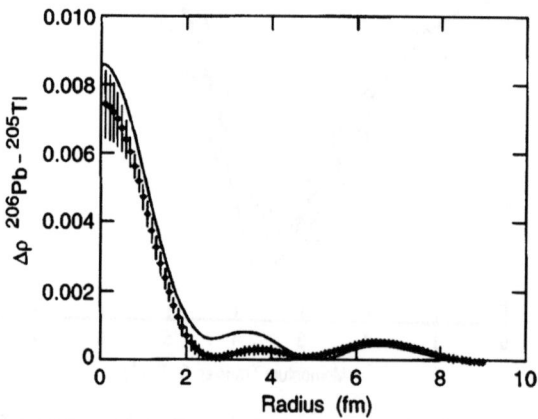

Figure 3. Difference in charge densities between ^{206}Pb and ^{205}Tl and oscillations due to the radial nodes in the wave function of the last proton (Pandharipande, Sick, and de Witt Huberts, 1997). The line is a Hartree-Fock calculation of the same difference.

with respect to the lowest energy that the nucleus can have at this angular momentum (the so-called "yrast line"). Under the influence of centrifugal forces the lowest configurations in the nucleus can be quite different from those at low angular momentum. The shell structure of nuclei becomes rearranged under the centrifugal effects of rotation and new pockets of relative stability, in the potential-energy surface of the nucleus, may develop as a function of quadrupole deformation. This then results in new classes of nuclear states, with high deformation, in which many of the nucleons have microscopic quantum numbers that are different from those of the ground state.

The exploration of nuclear structure at high angular momentum has shown that a few percent of the time fusion reactions populate states that decay by electromagnetic cascades which show characteristic rotational bands of remarkable simplicity as is shown

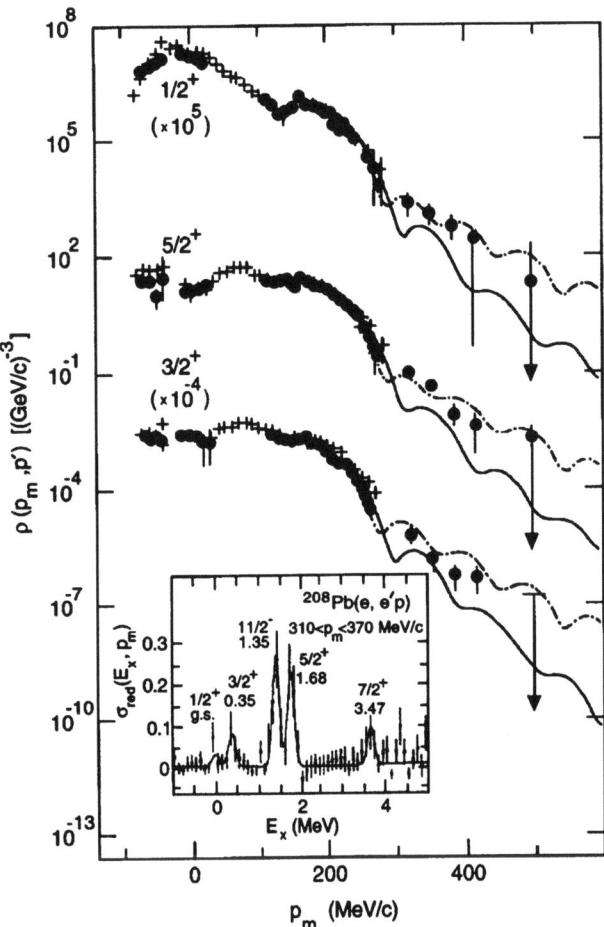

Figure 4. Demonstration of high-momentum components in the nuclear wave function. Transition densities from electron scattering knocking out a proton from the single-particle states in the doubly-magic nucleus ^{208}Pb. The inset shows the spectrum of hole states. The solid line represents the transition density calculated using a mean field for the protons, the dot-dashed curve includes the effects of short-range, high-momentum correlations (Pandharipande, Sick, and de Witt Huberts, 1997).

in Fig. 5. The energy spacings in these bands correspond to nuclei with much higher deformations (2:1 axis ratios) than those in the normal deformed bands that had been the basis of the collective model of Bohr and Mottelson (typically 1.3:1 axis ratios). The exploration of the properties of these *"superdeformed"* bands (Nolan and Twin, 1988; Janssens and Khoo, 1991) has uncovered a great deal of structural information. The precise energies in these, including the microscopic details of the small deviations from the pattern expected of perfect rotors, are reproduced with surprising accuracy in several nearby nuclei, giving rise to the *identical-band* phenomenon (Baktash, Haas, and Nazarewicz, 1995) as is illustrated in Fig. 6. It seems that these microscopic signatures carry over from one nucleus to another, without change, in a way that has not been seen elsewhere in nuclear structure and is not fully understood.

One of the most interesting features of superdeformed bands, mentioned above, is the fact that even when these states are well above the yrast line, they hardly mix at all into the higher-density states with more "normal" deformations. There appear to be two distinct classes of states corresponding to two minima in the potential-energy surface, with different deformations. The superdeformed states are closely related to a class of states that appeared in the 1960s in the study of delayed fission of very heavy nuclei. The mixing between these states leading to fission and the ordinary states is similarly inhibited.

The discovery and study of these phenomena in nuclei at high angular momentum have become possible through major advances in the detection and precision energy measurement of gamma rays with high-resolution germanium diodes. The size of the detectors, and their anticoincidence shields to suppress the Compton-scattering background, have now been developed to the point where complete spheres of detectors are used to search for multiple coincident gamma rays from a cascade. This instrumental advance culminated in detectors such as Gammasphere in the U.S. and Euroball in Europe. The new

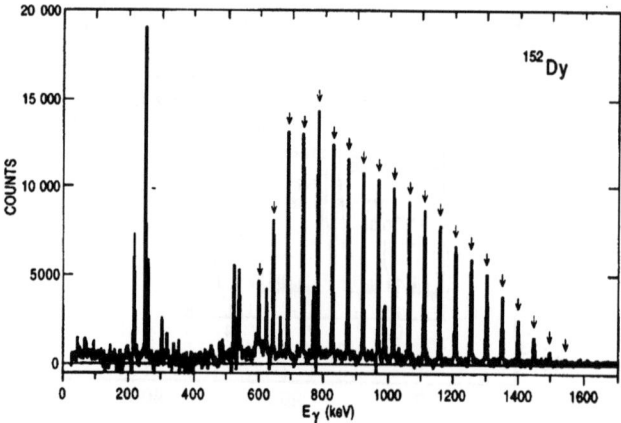

Figure 5. A *"superdeformed"* rotational band in ^{152}Dy showing the *gamma-ray transitions between members of the band. The constant increments in gamma-ray energies are characteristic of a band that follows the symmetry of a very good quantum-mechanical rotor. The variation in intensity of these gamma rays reflects the angular momentum distribution in the population of the band in the fusion-evaporation reaction that was used. The peaks that are not indicated by arrows correspond to states of lower energy of "normal" deformation, populated after the decay out of the super-deformed band (Twin et al., 1986).*

experimental information, in turn, has led to major advances in the theoretical insights into the structure of nuclei at the limits of large centrifugal stress.

Experimental work with these new instruments has also led to other new discoveries. One of the most interesting of these are bands of states connected by radiative transitions whose energies increase in small smooth increments, very much as those for rotational states (Baldsiefen et al., 1995). However, unlike the electric quadrupole radiations that are the earmark of transitions within rotational bands, the gamma-ray transitions in these sequences are magnetic dipole in character. This came as a dramatic surprise. The plausible explanation for these magnetic transitions is that there are two, rather stable, configurations, one for protons and one for neutrons, each of large angular momentum. The sequence of states with increasing angular momentum then correspond to states where the angular-momentum vectors of the two configurations simply are reoriented, as the closing blades of shears, to be more nearly parallel and thus give higher total angular momentum. This explains the large magnetic dipole transitions, but the smooth dependence to the sequence of energies is still very much a puzzle. The phenomenon suggests that some cooperative collective features are present, though theoretical understanding is still not complete.

Limits of binding

The 1950–1980 period saw the systematic exploration of the structure of nuclei in and near the valley of nuclear stability through a variety of techniques. The limits of nuclear

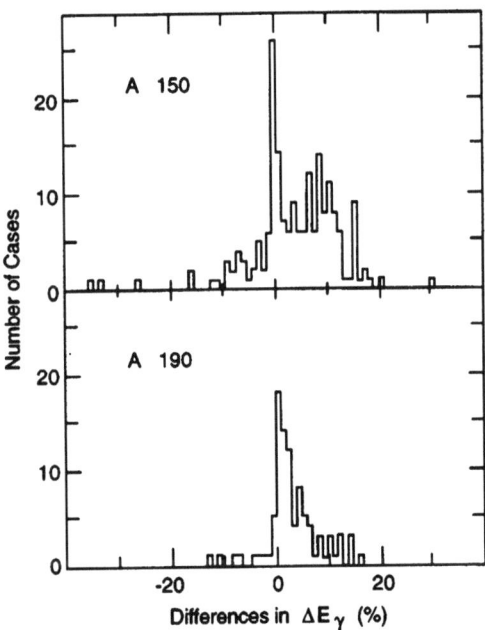

Figure 6. Similarities between superdeformed bands in nuclei in the same vicinity. The quantity plotted is the mean percentage difference in $E_\gamma(J+2 \to J) - E_\gamma(J \to J-2)$ between different bands. About sixty bands within the mass 150 region and fifty bands within the mass 190 region are compared. Note that in both regions there is a large excess of pairs of bands for which this quantity differs from zero by less than 2%—these are the "identical bands"—others differ by up to 20%. This identical reproduction of bands in different nuclei is not yet fully understood.

binding, where for a fixed number of protons no more neutrons can be bound (or for a fixed number of neutrons no more protons), the so-called "drip lines" were largely unknown, except for the lightest nuclei. The drip lines are of interest because nuclear properties might change, especially near the neutron drip line. They are also of particular interest in various stellar processes where, in a hot environment, a sequence of captures takes place rapidly. With recent advances the exploration of these limits has started in the 1990s.

One new phenomenon that occurs along the drip line is the observation of proton radioactivity, where the nuclei are literally dripping protons because their binding is insufficient but the Coulomb barrier retards their emission. At present, these have been identified in a number of elemental isotopic sequences, from Co to Bi, and the structure of these nuclei at the proton drip line is beginning to be explored (Woods and Davids, 1997).

Another result is in the limit of neutron binding where much less is known, because this regime is much more difficult to reach in laboratory experiments. Since neutrons see no Coulomb barrier, the density distributions of loosely bound neutrons can have long tails that reach far beyond the "normal" nuclear radii. These exponential tails fall off more and more slowly with decreasing binding energy. Thus the neutrons will reach far beyond the proton distributions in very neutron-rich nuclei. Such a separation between neutrons and protons, might in turn cause some qualitative changes in nuclear properties. For instance, it has been suggested that the spin-orbit term may be substantially reduced in such nuclei, and this would cause a change in shell structure that would be very interesting to observe experimentally. Such a change could also have serious consequences on the rapid neutron capture, r-process, in explosive stellar nucleosynthesis.

The best current example of a nucleus with diffuse neutron excess is in the very light nucleus ^{11}Li with three protons and eight neutrons. Here experiments show clearly how the last two, very loosely bound, neutrons form a diffuse tail, a "halo" around the protons. Thus the interaction radius of this nucleus is substantially larger than that of other Li isotopes as is shown in Fig. 7 and the structure of what would be the electric-dipole giant resonance in other nuclei is substantially different here, as is the momentum distribution (Hansen, Jensen, and Jonson, 1995). The further exploration of very neutron-rich nuclei, beyond the very light ones, requires major new advances in experimental techniques and facilities.

Another limit being explored is that of total mass, or nucleon number. Here ingenious improvements in experimental techniques permit the production of ever heavier new isotopes and elements, beginning to approach the region where calculations predict that, because of the stabilizing effects of shell structure, a new island of relatively stable "*superheavy*" nuclei should occur. A few nuclei of the new element with $Z=112$ have been produced—near one such possible island with $Z=114$, but further from another suggestion with $Z=126$. The results may in fact indicate a relatively stable bridge leading to a more stable island. Very heavy atoms are also of interest in QED, as they allow an exploration of vacuum polarization, and other relativistic effects under extreme conditions (Hofmann, 1996).

Hadrons in the nuclear medium

A crucial assumption in most many-body descriptions of nuclei is that nucleons and other hadrons do not change in the nuclear medium. Deep-inelastic scattering of electrons on nuclei has given evidence that nucleons and their quark structures are altered somewhat when they are placed in nuclei (Geesaman, Saito, and Thomas, 1995). This is called the EMC effect after the group that discovered it. Their finding has stimulated experimental and theoretical studies of these changes and, more generally, has raised important issues about how the properties of hadrons change in the hadronic medium of a nucleus. The changes can be investigated at the quark level in high-momentum-transfer reactions and at the hadron level in both electron scattering and heavy-ion collisions. Using electrons, the elastic form factors of the proton and neutron inside nuclei have been compared to those of free protons and neutrons in deuterium. Intriguing differences in the ratio of the magnetic to electric properties of the proton in light nuclei have been observed, and this is an active area of investigation.

On the other hand, there is clear evidence for changes in the effective nucleon-nucleon force in the nuclear medium. The challenge is to distinguish "normal" many-body effects from changes in the hadronic substructure. For instance, how do virtual pions from pion exchange manifest themselves in a change of the sea antiquark distribution of nucleons in nuclei? This is being examined at the hadronic level in looking for pions knocked out by electrons, in Drell-Yan processes that are directly sensitive to antiquarks, and in looking for pionic modes excited in proton scattering. If the structure and properties of the mesons change in the nuclear medium, there may be important implications for the effective internucleon forces.

These changes involve not only pions but also vector mesons. Theorists have suggested that the masses of the latter should decrease (Adami and Brown, 1993; Ko, Li, and Koch, 1997), but this remains controversial. The width of the ρ resonance is also likely to be affected. These alterations can be sought by searching for the leptonic decays of the vector mesons produced in heavy-ion collisions and studying their leptonic decays within the nucleus. Such experiments are being undertaken. Many of the changes of hadrons are not large, but the methodology seems to be at hand to seek them out and the consequences could have a profound impact throughout nuclear physics.

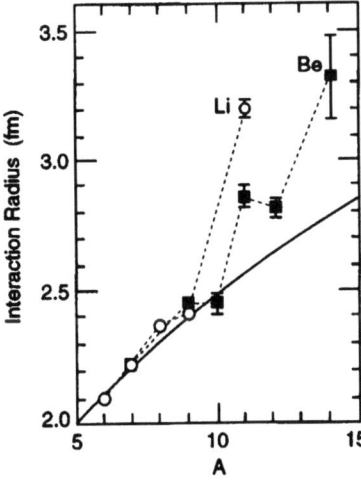

Figure 7. Interaction radii of Li (3 protons) and Be (4 protons) nuclei with carbon derived directly from total cross-section measurements. The line represents a smooth $A^{1/3}$ dependence of the radius (Tanihata, 1995).

When a colorless hadron is produced in a high-momentum-transfer reaction, the hadron is small at birth. Due to color screening of the quark components of this small hadron, its mean free path in the nucleus is large and its final-state interactions small. This phenomenon is called color transparency (Miller, 1994), and should be visible in quasielastic (e,ep) and (p,pp) reactions on a nuclear target. At present the evidence for nuclear transparency is ambivalent and further experiments are required to tie the effect down.

Nuclear Astrophysics

Nuclear physics plays a key role in the processes that take place in the universe, from the big bang on to energy production in stars. The synthesis of the chemical elements in the universe is the result of the various nuclear reactions that take place in different stellar environments. The Big Bang produced mostly protons and some of the lightest elements. When stars are formed from these remnants of the Big Bang, their matter is heated as the star contracts under gravity. In the hot star, nuclei run through cycles of nuclear reactions and hydrogen is gradually converted into helium, as in our sun. Somewhat hotter stars will form carbon, and as the carbon cycle described by Bethe becomes important more hydrogen is converted into helium. Further heating will cause captures of protons and alpha particles beyond the carbon cycle. Under appropriate conditions these reactions will produce elements up to about mass 56. To get further in mass, neutron capture is essential—and this can happen slowly in the "s-process" or explosively in the cataclysmic "r-process." All the elements heavier than iron in our world were produced in such stellar environments. Below we mention only a few of the key developments in our understanding of the recent past—following the major insights in this field by Bethe, Fowler, and others in identifying how nuclear processes determine the evolution of matter in the universe, in describing energy production in the Sun, and in explaining the formation of the elements from the Big Bang through the stages of stellar evolution.

Solar neutrinos

One of the intriguing developments of the past decades has been the study of neutrinos from the Sun. Other than the heat radiated, the neutrinos are the one accessible observable product from the chain of nuclear processes that take place in the interior. It was noted early that the number of neutrinos detected on earth was too small, given that the energy output of the sun is known accurately and thus the number of nuclear reactions leading to neutrinos is also known. The pioneering experiments of Davis and colleagues with a chlorine detector in the Homestake mine in South Dakota were sensitive primarily to the highest-energy neutrinos, those from the decay of ^{8}B. The observed neutrino flux was about a third of that expected, and in spite of extensive measurements and remeasurements of the nuclear parameters relevant to the solar processes and to the detection scheme, the discrepancy remains today as is shown in Fig. 8.

More recent experiments with gallium as the detecting material, in Europe and Russia are sensitive to much lower-energy neutrinos. They also find substantially fewer neutrinos than expected, as do experiments in which the high-energy neutrinos are detected more directly in large water detectors in Japan. The combined impact of these very different experiments has already been profound. The deficiency in the number of electron neutrinos might have its origins in the possibility that neutrinos have finite mass and that they oscillate between the originally emitted electron neutrinos and neutrinos of other flavors, with this oscillation enhanced by their passage through the dense matter of the Sun. Whether this is indeed the case will be tested in experiments to be carried out in the coming years (Haxton, 1995).

In a recent report of the Kamiokande experiment the signals from muon and electron neutrinos from cosmic rays were analyzed, rather than those from neutrinos coming from the sun. They appear to show neutrino oscillations, apparently $\nu_\mu \rightarrow \nu_\tau$, and thus also give tantalizing indications of neutrino mass and oscillations.

Supernovae

Among the most spectacular events in the universe are the supernova explosions that occur under the right circumstances during the evolution of sufficiently massive stars. The dynamics of such an explosion involves a complex interplay between nuclear properties, gravity, and the weak interaction. Supernovae play by far the dominant role in the synthesis of heavy elements. The decay in the light curve of a supernova, shown in Fig. 9, is governed predominantly by the decay of ^{56}Ni—the progenitor of ^{56}Fe, the most tightly bound nucleus and the most abundant element constituting the earth. The enormous flux of neutrinos from supernovae was evidenced by the dramatic detection of a neutrino pulse from the supernova SN1987A. Recent theoretical work has shown that the shock front of neutrinos interacting with nuclear matter plays a major role in the dynamics of supernova explosions.

Neutron stars

A typical neutron star has about 1.4 solar masses compressed in a sphere of 10 km radius. More than 500 neutron stars have been detected in our galaxy, most as radio pulsars and some at optical and x-ray wavelengths. A dozen neutron stars are found in close binary

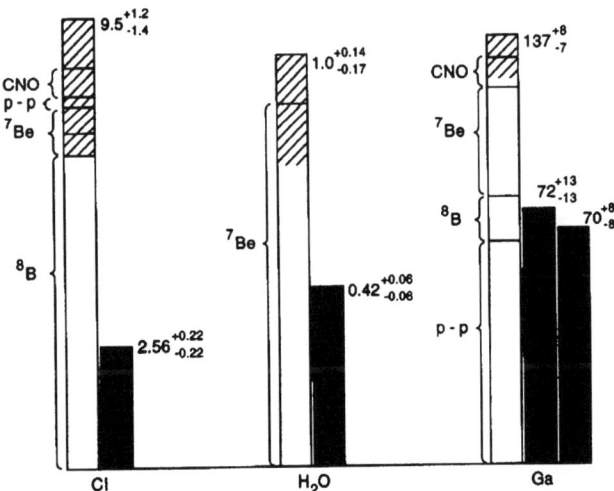

Figure 8. The solar neutrino signal expected with three different detection methods that are sensitive to different neutrino energies. The light bars indicate the expected neutrino yield, in appropriate units, from the various sources in the solar cycle. The solid bars represent the observed signals from the corresponding experiments. The shaded areas represent the uncertainties.

pairs, which has allowed their individual masses to be measured very accurately through orbital analysis. Binary neutron star coalescence events may be the source of observed extragalactic gamma-ray bursts, and our best hope for directly detecting gravitational radiation.

The structure of a neutron star is a consequence of the interplay between all interactions: the strong nuclear force, electroweak interactions, and gravity, with significant corrections from general relativity. The surface is metallic iron, which is the most stable form of matter at zero temperature and pressure. Underneath is a lattice of nuclei that grow progressively larger, more neutron-rich, and more tightly packed as the density increases with depth, since it is energetically favorable to capture electrons on protons to reduce the kinetic energy of the electron gas. At about 4×10^{11} g/cm^3 matter density, neutrons start to leak out of the nuclei, forming a low-density neutron superfluid in the intervening space. As the density increases further, around 10^{14} g/cm^3, various unusual shapes of nuclei may occur, with transitions from spheres to rods to sheets to tubes to bubbles. Eventually at a density around 2.7×10^{14} g/cm^3, or normal nuclear matter density, the nuclei dissolve into a uniform fluid which is over 90% neutrons, 5–10% protons and an equal number of electrons to preserve electrical neutrality, all in beta equilibrium. At successively higher densities muons, and perhaps pions and/or kaons may be present. At high enough densities, quark matter will probably appear, perhaps initially as bubbles in the nucleon fluid.

Progress in characterizing the nucleon-nucleon interaction has been key to our increased understanding of dense nucleon matter and consequently of neutron star structure. If neutrons were noninteracting, the maximum neutron star mass stable against gravitational collapse to a black hole would be 0.7 solar masses. However, many neutron stars have been observed with masses 1.3–1.6 times the solar mass, so the role of nuclear forces is clearly important. Recent work has set the minimum upper limit on neutron star mass at 2.9 solar masses, helping to further refine the observational boundary between neutron stars and black holes. Observations of quasi-periodic oscillations in binary x-ray

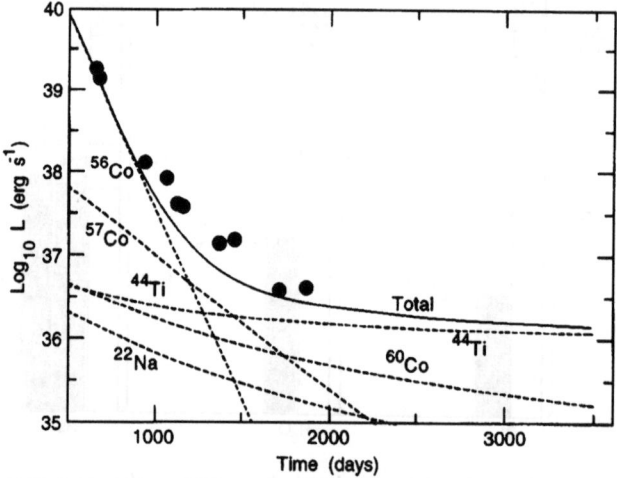

Figure 9. The logarithm of the light intensity from the supernova SN1987A as a function of time after the supernova explosion. Most of the intensity for the first two years comes from ^{56}Ni and its daughter ^{56}Co. The calculated contributions from other radioactive nuclei are also shown.

sources may also soon lead to limits on neutron star radii, which will give an even tighter constraint on the dense-matter equation of state.

Nucleosynthesis and reactions with unstable nuclei

The nuclei of the chemical elements are formed in the very hot environment inside stars. At higher stellar temperatures, these processes occur sufficiently fast that the nuclei involved are themselves short lived. New tools are being developed to determine properties of these short-lived nuclei, including the cross sections that are likely to be most important in astrophysical contexts. In addition, there are some expectations that the general features of nuclear structure may change near the limits of binding, where the path of explosive nucleosynthesis takes place.

The enormous improvements in the observational techniques of astronomy and astrophysics inevitably will require better quantitative understanding of the nuclear processes that yield energy in the universe and form an intriguing interface with nuclear physics.

Matter at High Energy Densities

One of the areas of intense interest is associated with the physics of very high densities, where the description of matter in terms of quarks and gluons contained in individual hadrons must break down. Calculations based on QCD suggest that when high densities and temperatures are reached in a volume large compared to that of a typical hadron, a transition to a state of matter will occur where the quarks are no longer confined to their individual hadrons but can move freely within the larger volume.

A schematic phase diagram is shown in Fig. 10 illustrating this transition and its relationship to the evolution of the universe following the Big Bang. To explore this state, experiments have been carried out, first at the Bevalac at Berkeley, then at increasingly higher energies at the AGS in Brookhaven and the SPS at CERN, and soon at a new collider, the RHIC (Relativistic Heavy-Ion Collider) facility at Brookhaven that is to come into operation shortly. Since the confinement of quarks in hadrons is one of the key features of the strong-interaction world, this deconfinement, to a volume larger than that of a hadron, is of intense interest. A key question is how such deconfinement might be observed unambiguously.

Collisions between two heavy nuclei at ultrarelativistic energies produce many thousands of fragments. Their detection, identification, and characterization is a formidable experimental challenge. The first question is whether large densities in energy and baryon number are indeed obtained in such collisions—whether the kinetic energy is absorbed or, whether the two nuclei primarily just pass through each other. This may be deduced from the measurements of the transverse momentum carried by the products of the interaction. Such "stopping" studies at the AGS, at laboratory energies up to 14 GeV per nucleon show clearly that the kinetic energy is absorbed, and that high temperature and high baryon density are indeed created. This remains true in work at CERN up to laboratory energies of 200 GeV per nucleon. As yet, there is no unequivocal evidence in these data for a phase transition. The energies at RHIC will be an order of magnitude higher—and it is expected that this will lead to the formation of a high-temperature low baryon-density system well past the expected transition point.

How the properties of this transitory state of matter can be deduced from the data is the subject of intense discussions between experimenters and theorists. Some of the possible "signatures" discussed and explored in current experiments are the suppression of the production of J/ψ mesons due to screening of the charm-anticharm interaction by the surrounding quarks and gluons, an increase in energetic leptons in radiation from the early stages of the transitory state, the modification of the width and decay modes of specific mesons, such as the ρ and ϕ, enhancement in the production of strangeness, the attenuation of high energy jets, etc. A change in the state of matter should first show up in a clear correlation between a number of such signatures.

Another possibility at high energy density is that the intrinsic chiral symmetry of QCD may be restored. This would show up through modifications in the masses of mesons. Possibly, nonstatistical fluctuations in the distributions of pions might signal the production of a so-called "disoriented chiral condensate." The major point is not the specific scenarios in such a complex environment, but that these energy densities will bring an unprecedented new regime of matter under experimental scrutiny where our current pictures will necessarily have to undergo radical changes (Harris and Mueller, 1996).

Tests of the Standard Model

Most of the tests of the standard model of the strong and electroweak interactions at nuclear energies have involved semileptonic interactions, particularly electrons and nucleons, and rare decay modes, primarily of kaons. Here we describe some of the semileptonic tests.

Beta decay

The study of superallowed Fermi beta decays (parent nucleus of spin/parity $J^\pi = 0^+ \rightarrow$ daughter 0^+) in the same family of isospin (isospin multiplet) permit an impor-

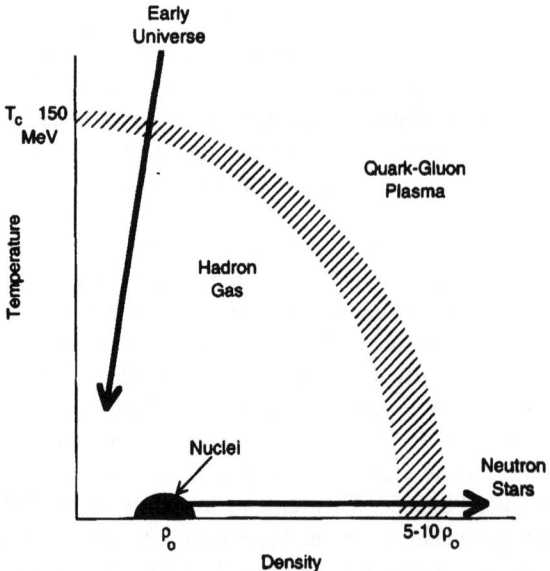

Figure 10. A qualitative phase diagram for hadronic matter showing the transition to a deconfined quark-gluon state at high temperature and high density. The paths followed in the early universe and in the interior of neutron stars is also indicated. ρ_0 is the density of normal nuclei. Collisions in the relativistic heavy-ion collider RHIC will explore this phase diagram in detail.

tant test of the standard model to be carried out via the unitarity of the matrix that describes the weak interaction connecting the various quarks: the Cabibbo-Kobayashi-Maskawa matrix (Towner and Hardy, 1995). The matrix element V_{ud} connecting up and down quarks is by far the largest one in the unitarity of

$$U \equiv |V_{ud}|^2 + |V_{us}|^2 + |V_{ub}|^2 = 1.$$

Here V_{us} and V_{ub} connect the up quark with the strange and bottom quarks, respectively. The precise measurements of superallowed transitions together with radiative corrections and removal of charge-dependent nuclear effects allow one to determine V_{ud} to better than 10^{-3}. In addition, these measurements, shown in Fig. 11, demonstrate that CVC (conserved vector current) holds to $\sim 4 \times 10^{-4}$. A straightforward analysis of the experiments, including a recent ^{10}C experiment, gives $V_{ud} = 0.9740 \pm 0.0006$. Together with the measurements of V_{us} and V_{ub}, one then obtains $U = 0.9972 \pm 0.0019$. There remain uncertainties, particularly in the charge-dependent nuclear effects; it has been proposed that these corrections can be approximated by a smooth Z dependence. In that case $U = 0.9980 \pm 0.0019$. However, better calculations of this charge dependence remain to be carried out.

Double beta decay

The decay $(A,Z) \rightarrow (A, Z+2) + e^- + e^- + \bar{\nu} + \bar{\nu}$ is expected in the standard model and has been seen in several nuclei (^{82}Se, ^{100}Mo, and ^{150}Nd) with half-lives of about 10^{20} y, consistent with the standard model (Moe and Vogel, 1994). Searches for the no-neutrino decay mode, important for determining whether ν's are massive and of the Majorana (neutrino and antineutrino are identical) type, are being continuously improved. The present lower limit on the half-life is 3×10^{24} y for ^{74}Ge.

Semileptonic parity-nonconservation studies

The surprising finding of strangeness in the nucleon led to considerable experimental and theoretical activity. Recently, in an ongoing experiment (SAMPLE) at MIT, the weak interaction is being used to investigate the contribution of strangeness to the proton's anomalous magnetic moment. As in all parity-violation experiments, it is the interference

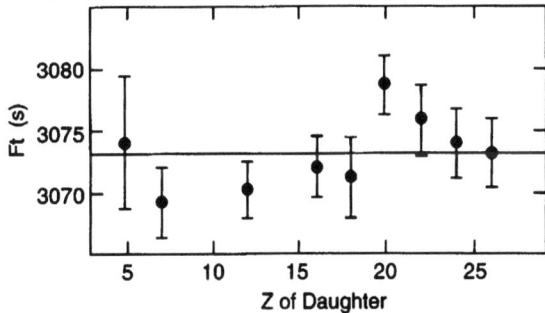

Figure 11. Beta-decay transition probabilities (Ft values) for nine superallowed Fermi β-decays and the best least-squares one-parameter fit, plotted as a function of the proton number of the final nucleus (Towner and Hardy, 1995).

of weak with electromagnetic amplitudes that is sought in a parity-odd signal such as $\langle \vec{j}\rangle \cdot \vec{p}$, where $\vec{p}$ is the incident momentum of the electron and $\langle \vec{j}\rangle$ is its polarization. The presence of strangeness in the nucleon can modify the momentum dependence of the form factors, but it can also add two new unconstrained ones, a vector magnetic form factor and an axial (isoscalar) form factor. The results to date (Mueller et al., 1997) are inconclusive, but do not indicate any strangeness (within large errors), as shown in Fig. 12. Further experiments are planned.

Other precision parity-violating (PV) studies of the weak interactions of electrons and nuclei have been carried out with atoms. Despite their being at lower momenta, where the PV effects are smaller, $\sim 10^{-11}$, these experiments have reached the incredible precision of 1/2% in the parity-violating asymmetry. At the present time theoretical errors are at the level of $\sim 1\%$. At this level of precision the atomic experiments provide meaningful tests of the standard model. The dominant weak-interaction term is $a_\mu V^\mu$ where a_μ is the axial current of the electron and V^μ is the hadronic vector current, which is coherent over the nucleus. The effective charge, the weak equivalent of the electrical charge in this case, is

$$Q_W = (1 - 4\sin^2\theta_W)Z - N, \qquad (7)$$

where θ_W is the Weinberg angle, $\sin^2\theta_W \approx 0.23$. Q_W is large for heavy atoms. The measurement on Cs, a one-valence-electron atom, at the 0.5% level, gives $Q_W = -72.35 \pm 0.27_{exp} \pm 0.54_{th}$ (Wood et al., 1997). This limits deviations from the standard model.

The term $v_\mu A^\mu$, where v_μ is the weak vector current of the electron and A^μ is the nuclear axial current, is much smaller than $a_\mu V^\mu$ because for the electron $v^\mu \propto (1 - 4\sin^2\theta_W) \sim 0.1$ and only a single nucleon contributes to $A^\mu \propto \langle \vec{\sigma}\rangle$, the nuclear spin. Thus the asymmetry is reduced by ≥ 500. The atomic measurements of this term make use of the hyperfine structure, which is due to the nuclear spin. This term has not yet been detected because it is hidden by the stronger nuclear "anapole" moment, a toroidal axial current of the nucleus coupling to photons. This is a weak parity-violating moment, which does not fit into the usual characterization of electromagnetic moments of nucleons. It is an effective axial vector coupling of the photon to the nucleus. The recent measurement in atomic Cs at the 1/2% level (Wood et al., 1997) has discovered the nuclear anapole

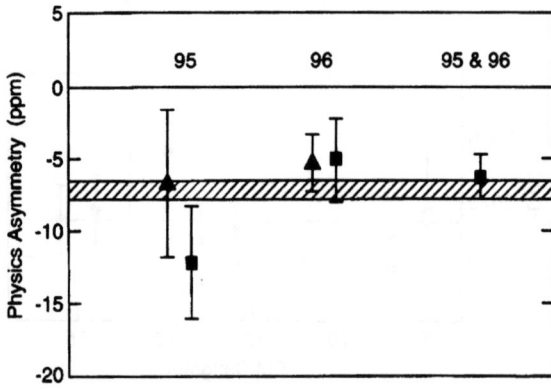

Figure 12. Results for the parity-violating asymmetry measured in the SAMPLE experiment in the 1995 and 1996 running periods. The hatched region is the asymmetry band (due to axial radiative corrections) for $F_2^s = 0$.

moment, the first anapole moment observed for such a microscopic system.

The experiment is of particular interest because it is sensitive to the weak internucleon force caused by neutral currents. This force has not been found in pure nuclear experiments and the upper limit found there (in ^{18}F) is at least a factor of three below that deduced from the anapole measurement, and also from a theoretical quark model. It remains to be seen whether this is an experimental or theoretical problem.

The nonleptonic weak interaction

The nonleptonic part of the weak interaction is of interest because it is the least well-understood one due to the strong interactions of all particles involved. The initial knowledge came from the weak decays of strange mesons and baryons. Even here, there remains the problem of fully understanding the ratio of parity-nonconserving to parity-conserving amplitudes in the decays of hyperons. Due to the change in flavor, only charged currents contribute to these decays in the electroweak theory. With the advent of precision nuclear experiments it became possible to study the weak interactions of nucleons by means of parity-violating asymmetries with polarized beams. These experiments do not probe the standard model as much as our understanding of the structure and weak forces of the nucleons. The asymmetry comes about from the interference of the weak and strong forces. Experiments in pp scattering and in light nuclei have provided the most reliable information (Adelberger and Haxton, 1985). The weak neutral currents have yet to be seen in nonleptonic weak interactions. They are the primary source of the isospin-changing $\Delta I = 1$ interaction, which arises from π exchange with one weak (f_π) and one strong pion coupling to nucleons. Since there is only an upper limit on the asymmetry in ^{18}F, which is determined by this mechanism, neutral-current effects have not yet appeared. Further measurements are planned for f_π, e.g., in low-energy polarized neutron capture by hydrogen, e.g., $n + p \rightarrow d + \gamma$.

Time-reversal invariance

Despite the finding of CP violation in 1964, over 30 years ago, we still do not have any definitive theory of time-reversal noninvariance. This is not due to a lack of effort. The only system where CP violation (and by implication T violation) has been found is in the K^0 and $\bar{K}^0$ system. To date, the most sensitive searches for time-reversal-invariance breaking are those for an atomic electric dipole moment of a neutron (Smith *et al.*, 1990; Alterev *et al.*, 1992) or atomic ^{199}Hg (Jacobs *et al.*, 1993; 1995); these tests are sensitive to simultaneous violations of parity and time-reversal invariance. No finite time-reversal violation effect has yet been published, but the upper limits keep decreasing and have already ruled out some models of CP violation.

Experimental facilities

The investigation of the properties of the nucleus has required the development of new tools with which to probe this realm of physics. These developments continue in a number of forms, an example is the new technologies based on superconductivity in both magnets and rf-accelerating structures. The pace of experimental exploration in nuclear physics is largely set by the rate at which the appropriate techniques become available.

While the early accelerators and other equipment were of relatively modest cost, and so could be duplicated at a number of university and other laboratories, the field has evolved in that latter part of the century, and it has increasingly required facilities that represent major investments. Though the scale of these has not been that of facilities in particle physics, it has required careful planning and priority choices, carried out primarily through the Nuclear Science Advisory Committee, with broad participation by the scientific community.

In the last two decades two major new facilities have emerged from this organized planning process (CEBAF and RHIC) along with a number of smaller ones. These will play a major role in determining the course of the science as we enter the next century.

Outlook

The study of the structure of nuclei, of hadronic matter, started less than 100 years ago. Enormous advances have been made in this time. But much work remains. Even the simplest building blocks of hadronic matter that we have in our world, the proton and neutron, are structures that are incompletely understood. So are the interactions between them, the quantitative features of the forces that hold nuclei together. The properties of nuclei as observed experimentally are understood in the framework of approximate models, but the more fundamental reasons for the validity of many of these models are not well understood—nor can we reliably extrapolate these properties to extreme conditions, whether in the limits of stability, or the limiting energy densities of matter.

Concerted efforts of nuclear physicists, theorists, and experimentalists is needed to pursue these areas of knowledge into the 21st century.

Acknowledgments

The authors would like to thank their colleagues G. Bertsch, D. Geesaman, W. Haxton, R. Janssens, G. Miller, and R. Wiringa for helpful discussions and advice in connection with the preparation of this article. This research was supported by the U.S. Department of Energy, Nuclear Physics Division, under Contract W-31-109-ENG-38 and Grant DE-FG06-90ER40561.

References

Adami, C., and G.E. Brown, 1993, Phys. Rep. **224**, 1.
Adelberger, E.G., and W.C. Haxton, 1985, Annu. Rev. Nucl. Part. Sci. **35**, 501.
Alterev, I.S., *et al.*, 1992, Phys. Lett. B **276**, 242.
Arima, A., and F. Iachello, 1981, Annu. Rev. Nucl. Part. Sci. **32**, 75.
Ashman, J., *et al.*, 1989, Nucl. Phys. B **328**, 1.
Ashman, J., 1997, Phys. Rev. D **56**, 5330.
Baktash, C., B. Haas, and W. Nazarewicz, 1995, Annu. Rev. Nucl. Part. Sci. **45**, 543.
Baldsiefen, G., *et al.*, 1995, Nucl. Phys. A **587**, 562.
Bernstein, A.M., and B.R. Holstein, 1991, Comments Nucl. Part. Phys. **20**, 197.
Bernstein, A.M., and B.R. Holstein, 1995, Int. J. Mod. Phys. E **4**, 193.
Bethe, H.A., 1999, Rev. Mod. Phys. **71**, 6; pp. 12–28 in this book.
Bohr, A., and B. Mottelson, 1969, *Nuclear Structure* (Benjamin, Inc., New York).

Bugg, D.V., 1985, Annu. Rev. Nucl. Part. Sci. **39**, 295.
Carlson, C.E., J.R. Hiller, and R.J. Holt, 1997, Annu. Rev. Nucl. Part. Sci. **47**, 395.
Carlson, J., and R. Schiavilla, 1998, Rev. Mod. Phys. **70**, 743.
Chrien, R., and C. Dover, 1989, Annu. Rev. Nucl. Part. Sci. **39**, 113.
Diamond, R.M., and F.S. Stephens, 1980, Annu. Rev. Nucl. Part. Sci. **30**, 85.
Dieperinck, A.E.L., and P.K.A. de Witt Huberts, 1990, Annu. Rev. Nucl. Part. Sci. **40**, 239.
Ericson, T., and W. Weise, 1988, *Pions and Nuclei* (Clarendon Press, Oxford).
Geesaman, D.F., K. Saito, and A.W. Thomas, 1995, Annu. Rev. Nucl. Part. Sci. **45**, 337.
Hansen, P.G., A.S. Jensen, and B. Jonson, 1995, Annu. Rev. Nucl. Part. Sci. **45**, 591.
Harris, J., and B. Mueller, 1996, Annu. Rev. Nucl. Part. Sci. **46**, 71.
Haxton, W.C., 1995, Annu. Rev. Astron. Astrophys. **33**, 459.
Hofmann, S., et al., 1996, Z. Phys. A **354**, 229.
Holstein, B.R., 1995, "Chiral Perturbation Theory: a Primer," hep-ph/9510344 (unpublished).
Jacobs, J.P., et al., 1993, Phys. Rev. Lett. **71**, 3782.
Jacobs, J.P., et al., 1995, Phys. Rev. A **52**, 3521.
Janssens, R.V.F., and T.L. Khoo, 1991, Annu. Rev. Nucl. Part. Sci. **41**, 321.
Kaplan, D.B., M. Savage, and M. Wise, 1998, Phys. Lett. B **424**, 390.
Ko, C.M., G. Li, and V. Koch, 1997, Annu. Rev. Nucl. Part. Sci. **47**, 505.
Machleidt, R., 1989, in *Advances in Nuclear Physics*, Vol. 19, edited by J.W. Negele and E. Vogt (Plenum, New York), p. 189.
Meissner, Ulf G., Ed., 1992, *Effective Theories of the Standard Model* (World Scientific, Singapore).
Miller, G.A., 1994, Annu. Rev. Nucl. Part. Sci. **45**, 501.
Moe, M., and S.P. Vogel, 1994, Annu. Rev. Nucl. Part. Sci. **44**, 247.
Mueller, B., et al., 1997, Phys. Rev. Lett. **78**, 3824.
Nolan, P.J., and P.J. Twin, 1988, Annu. Rev. Nucl. Part. Sci. **38**, 533.
Pandharipande, V.R., I. Sick, and P.K.A. de Witt Huberts, 1997, Rev. Mod. Phys. **69**, 981.
Smith, K.F., et al., 1990, Phys. Lett. B **234**, 191.
Tanihata, I., 1995, Prog. Part. Nucl. Phys. **3**, 505.
Towner, I.S., and J.C. Hardy, 1995, in *Symmetries and Fundamental Interactions in Nuclei*, edited by W.C. Haxton and E.M. Henley (World Scientific, Singapore), p. 183.
Twin, P.J., et al., 1986, Phys. Rev. Lett. **57**, 811.
Van Kolck, U., 1998, Sixth International Conference on Intersections of Particle and Nuclear Physics (AIP, New York), in press.
Weinberg, S., 1979, Physica A **96**, 327.
Wood, C.S., et al., 1997, Science **275**, 1759.
Woods, P.J., and C.N. Davids, 1997, Annu. Rev. Nucl. Part. Sci. **47**, 541.

Stellar Nucleosynthesis
Edwin E. Salpeter

Nuclear astrophysics started in the late 1930s, when Hans Bethe (1939) discussed both the proton-proton chain and the carbon-nitrogen cycle for turning hydrogen into helium. This conversion involves a series of nuclear reactions, but only two nuclei at a time, so the overall rate per proton is proportional to only one power of density. The next impetus to nuclear astrophysics came from the Kellogg Radiation Laboratory at Caltech, directed by Charles Lauritsen, in the late 1940s and the 1950s when William Fowler and his colleagues undertook experiments on energy releases, individual energy levels, and nuclear reaction rates. The first major impact for element production was a negative one when Fowler and his team showed that ^{8}Be is unstable, so that some triple collisions would be necessary to build up other elements from helium, and triple collisions would require very high densities. A major drama was provided by the precise position of resonant energy levels, since the typical space between levels is large in light nuclei but in the unlikely event that a level happens to lie in the appropriate narrow energy range it can increase thermonuclear reaction rates enormously. Such a level in ^{8}Be, measured by Fowler's team, was used by Salpeter in 1951 to show that the "triple-alpha reaction" can proceed fairly rapidly with ^{8}Be+^{4}He→^{12}C as the second step. In 1953 Fred Hoyle predicted at just what energy there should also be a resonance in ^{12}C to make the carbon production rate even faster, so that it could compete with the destruction rate of ^{12}C+^{4}He→^{16}O. This level was soon confirmed experimentally at the right energy. Nuclear experimental data have improved enormously since the 1950s, but it is ironic that one of the biggest remaining uncertainties in nuclear astrophysics today concerns the precise parameters of a pair of resonance levels in ^{16}O, just below the thermonuclear energy range.

The interior of the sun and of unevolved, main-sequence stars, where hydrogen is converted into helium, was understood much earlier, but the interior of red giant stars was a puzzle until the early 1950s. Numerical calculations by Hoyle, M. Schwarzchild, and others finally showed that, although the surface expands when a main-sequence star evolves into a red giant, the interior heats up and contracts. The central density in extreme red giant stars (the variable stars at the "tip of the red giant branch") is sufficiently large to initiate the triple-alpha conversion of helium into carbon and oxygen (plus smaller amounts of ^{20}Ne). In the big-bang model of cosmology high temperatures can also be reached in the early universe. Alpher, Herman, and Gamow first calculated in 1949 what elements could be built up from hydrogen. The density in the early universe, when its temperature was of order 10^8 K, was very much smaller than in red giant interiors, and little carbon and oxygen could be produced [modern primordial nucleogenesis is described in the Astrophysics and Cosmology section of this volume, while the history of controversies in the field of cosmology is treated in a book by Kragh (1996)].

The buildup of heavier elements in even more evolved stars, starting from ^{12}C, ^{16}O, and ^{20}Ne in red giant stars, consists of at least two stages: (i) The exothermic buildup of the iron region of nuclei, which have the largest binding energy per nucleon, and (ii) the much rarer, endothermic production of the very heavy nuclei, which requires some nonequilibrium catastrophic environment, such as a supernova. In reality, theorists had to consider more than two stages, with both fast and slow time scales, reprocessing in second-generation stars, etc. In spite of these complexities, the basic nuclear physics scheme for building up almost all the isotopes was already in place by the late 1950s. Two classic original papers by Burbridge, Burbridge, Fowler and Hoyle (1957) and by Cameron (1957) still serve as excellent reviews today.

While the nuclear reaction schemes *per se* were largely understood 40 years ago, embedding these schemes into quantitative stellar evolution calculations is a task that is still not fully completed. Moreover, while we know the present-day isotopic abundances in stellar atmospheres and in interstellar gas fairly well, we are only beginning to learn quantitatively what these abundances were when the universe was much younger. To illustrate the complexities with one surprising result: Although all stars start on the main sequence and produce helium in their interiors, most of the interstellar helium today was already there when the galaxy was formed, i.e., most of it is primordial and not from stars. One reason for this result is that there is little mixing from a star's center to its surface (and usually little mixing between stars and interstellar gas); another reason is that much of the interior helium is processed into heavier elements before a star dies. Furthermore, to predict today's interstellar abundances quantitatively we need to know how many stars of various masses were born and have already died, since only in old age (e.g., planetary nebulae) and death (supernovae) does the material from a star's interior reach interstellar space. This mass distribution, the "initial mass function," is still somewhat uncertain.

In spite of the pessimism of the last paragraph, much of stellar nucleosynthesis is now understood, and fortunately there are some excellent books and reviews which tell the story. We first give references to some of these (before returning to the uncertainties): The basic theory of stellar evolution is given in two textbooks (intelligible to physicists), Clayton (1968) and Kippenhahn and Weigert (1996), as well as in Iben (1991). Since most reviews and texts are written by theorists (like myself) the nuclear experimentalists often do not get enough credit, but luckily this is remedied by Rolfs and Rodney (1988). The whole present-day story of nucleosynthesis in stellar interiors and the chemical evolution of interstellar matter is given in two recent texts, Arnett (1996) and Pagel (1997), plus two collections of reviews by Prantzos *et al.* (1993) and Schramm and Woosley (1993). The following references give more details on two specialized subtopics: Clayton (1968) and Barnes *et al.* (1982) for nucleocosmochronology (ages and radioactivity); Simpson (1983) and Reeves (1994) for the very lightest isotopes and the effect of cosmic rays.

Theorists working on evolution of spherical stars used to make fun of colleagues in biophysics for inventing a spherical cow to make calculations easier. Overall deviations from sphericity in a star are still small percentagewise, but for nuclear reactions in a highly evolved star the deviations turn out to be just about as important as for a cow: Instabilities abound for both the hydrodynamics and the nuclear burning, and stellar evolution calculations now have to be carried out with three-dimensional codes. Such calculations are about as much of a challenge to modern supercomputers as one-dimensional calculations were to early electronic computers 40 years ago. One has to consider all types of nuclear burning—laminar flames, deflagration waves, and detonation waves—but also turbulence, convective overshot, Rayleigh-Taylor instabilities, etc. Such complications are particularly severe for the catastrophic explosions in supernovae (Thielemann et al., 1996). However, they are also present in somewhat less violent circumstances, such as intermediate-mass stars in the Asymptotic Giant Branch stage as they start to lose mass and evolve into planetary nebulae (Herwig et al., 1997). The uncertainties about the instabilities are exacerbated by the "onion-skin-layers" in highly evolved stars, e.g., a Si-core, next a C, O layer, then He and then H, etc., so that a nuclear flash in one layer may (or may not) trigger nuclear burning in others.

The hydro/nuclear instabilities are likely to remain the biggest challenge for stellar evolution theory for some time, but there are two additional challenges: Some attempts at including electron Coulomb screening corrections to thermonuclear reaction rates have been made for over 40 years, but the required nonequilibrium plasma physics is still not understood sufficiently well. In spite of a recent review article (Brown and Sawyer, 1997), the controversies continue and probably will have to be tackled with very extensive numerical calculations. What stellar evolution and nucleosynthesis were like when the universe and our Galaxy were young is now starting to be explored (see Pei and Fall, 1995). Since isotope abundance ratios are different for medium-mass and very massive stars, there is even a chance that we shall learn how the initial mass function has changed with time.

Nuclear Reactions in White Dwarf and Neutron Star Surfaces

Most of the medium-heavy and heavy elements in interstellar space were made in the deep interior of evolved stars and were then ejected from the surface, but some surface reactions are also of interest. Some fraction of stars have a close companion star that sheds mass, which then accretes onto the surface of the first star. The accreted material is mostly unprocessed, i.e., it consists mainly of hydrogen and helium which can burn if compressed to sufficiently high density and temperature. If the star is a white dwarf, its radius is small and its surface gravity is large, so the compression is achieved fairly rapidly; even more so on the surface of a neutron star, which has an even smaller radius. One should then expect some kind of relaxation oscillation: the hydrogen and helium build up until some kind of nuclear explosion burns up this fuel and gives a pulse of greatly increased luminosity, followed by another quiescent period while the hydrogen and helium build up again, etc. The accretion rate varies with time, and the flashes should be recurring but not accurately periodic.

The so-called "classical nova" is caused by such accretion onto a white dwarf star, followed by ignition of hydrogen burning by the C, N, O cycle plus a "nuclear runaway" (see Truran's report in Barnes et al., 1982). The white dwarf atmosphere, before the accretion, could in principle consist of helium but often has some kind of mixture of C, O, and Ne. The details of the nuclear flash depend on the atmospheric composition and on the white dwarf's mass (see, for example Nofar et al., 1991).

Rapidly rotating, magnetic, and isolated neutron stars often exhibit the pulsar phenomenon and a slight mass loss from the surface, but less extreme neutron stars that have a close companion star can accrete matter instead. This accretion can lead to nuclear flashes, as it does for an accreting white dwarf, but the details are different because of the smaller radius and even stronger gravitational field. For instance, the luminosity is emitted in x rays, and the steady luminosity between flashes, from the gravitational accretion energy, is larger than the averaged energy from the nuclear burning (Joss and Rappaport, 1984). The detailed shape of the x-ray bursts, produced by the thermonuclear flash, is particularly complex and depends strongly on the accretion rate. The complexity is partly due to the fact that both hydrogen and helium can be ignited, and the two burning shells interact with each other. Furthermore, the intrinsic burning time is shorter than the sound propagation time around the star's surface, and one can have localized ignition with the fire then spreading around the surface in some irregular manner (Bildsten, 1998). These x-ray bursts are also no more spherical than a cow and insight gained into instabilities from neutron star surfaces may well help to understand stellar evolution better.

This essay does not deal at all with neutron star interiors, nor with supernovae, but a few references may be useful: the book by Arnett (1996) deals with supernovae, the book by Tamagaki *et al.* (1992) with neutron stars in general. Two papers with references are Kalogera and Baym (1992) and Nomoto *et al.* (1997).

Summary

The main outline of nucleosynthesis inside stars is now understood and is an essential tool in interpreting observed chemical abundances in terms of the evolution of stellar populations in galaxies. However, electron Coulomb screening corrections to thermonuclear reaction rates are still uncertain. Beyond that are the instabilities which occur in nuclear burning stars like supernovae, asymptotic giant branch stars, and accreting neutron stars, which require three-dimensional dynamic simulations on supercomputers. Only such calculations can yield reliable results on chemical abundances that can be compared with observations.

References

Arnett, D., 1996, *Supernovae and Nucleosynthesis* (Princeton University, Princeton, NJ).
Barnes, C. A., D. D. Clayton, and D. N. Schramm, 1982, Eds., *Essays in Nuclear Astrophysics* (Cambridge University, Cambridge, England).
Bethe, H. A., 1939, Phys. Rev. **55**, 103 (letter) and 434 (full article).
Bildsten, L., 1998, in *The Many Faces of Neutron Stars*, edited by A. Alpar *et al.* (Kluwer, Dordrecht).
Brown, L. S., and R. F. Sawyer, 1997, Rev. Mod. Phys. **69**, 411.
Burbidge, E. M., G. R. Burbidge, W. A. Fowler, and F. Hoyle, 1957, Rev. Mod. Phys. **29**, 547.
Cameron, A. G. W., 1957, Atomic Energy of Canada Ltd., CRL-41; Publ. Astron. Soc. Pac. **69**, 201.
Clayton, D. D., 1968, 1984, *Principles of Stellar Evolution and Nucleosynthesis* (McGraw-Hill, New York, and University of Chicago, Chicago).
Herwig, F., T. Blöcker, D. Schönberner, and E. El Eid, 1997, Astron. Astrophys. **324**, L81.
Iben, I., Jr., 1991, Astrophys. J., Suppl. Ser. **76**, 55.
Joss, P. C., and S. A. Rappaport, 1984, Annu. Rev. Astron. Astrophys. **22**, 537.
Kalogera, V., and G. Baym, 1996, Astrophys. J. **470**, L61.
Kippenhahn, R., and A. Weigert, 1996, *Stellar Structure and Evolution* (Springer, New York).
Kragh, H., 1996, *Cosmology and Controversy* (Princeton University, Princeton, NJ).
Nofar, I., G. Shaviv, and S. Starrfied, 1991, Astrophys. J. **369**, 440.
Nomoto, K., K. Iwamoto, and N. Kishimoto, 1997, Science **276**, 1378.

Pagel, B. E. J., 1997, *Nucleosynthesis and Chemical Evolution of Galaxies* (Cambridge University, Cambridge, England).

Pei, Y. C., and S. M. Fall, 1995, Astrophys. J. **454**, 69.

Prantzos, N., E. Vangioni-Flam, and M. Cassé, 1993, Eds., *Origin and Evolution of the Elements* (Cambridge University, Cambridge, England).

Reeves, H., 1994, Rev. Mod. Phys. **66**, 193.

Rolfs, C. E., and W. S. Rodney, 1988, *Cauldrons in the Cosmos: Nuclear Astrophysics* (University of Chicago, Chicago).

Schramm, D. N., and S. E. Woosley, 1993, Phys. Rep. **227**, 1.

Simpson, J. A., 1983, Annu. Rev. Nucl. Part. Sci. **33**, 323.

Tamagaki, R., S. Tsuruta, and D. Pines, 1992, *The Structure and Evolution of Neutron Stars* (Addison-Wesley, Longman, Redwood City, CA).

Thielemann, F. K., K. Nomoto, and M. Hashimoto, 1996, Astrophys. J. **460**, 408.

Atomic, Molecular, and Optical Physics

Atomic Physics

*Sheldon Datz, G. W. F. Drake, T. F. Gallagher,
H. Kleinpoppen, and G. zu Putlitz*

Atomic physics[1] deals with any interactions in nature in which the electromagnetic force plays a dominant role and the other forces of nature play relatively minor roles, or no roles at all. Even so these "minor" roles can be employed to great advantage to study a wide variety of subjects, such as the determination of nuclear properties, and even delve into quantum electrodynamics and parity nonconservation in weak interactions. Data from atomic physics are necessary inputs for modeling phenomena in plasmas, condensed matter, neutral fluids, etc. Ionized and neutral atoms, positive and negative electrons, physical properties of molecules, as opposed to chemical and biological, and exotic systems such as muonium and positronium can be considered to be fair game for this enormous field. Atomic physics is periodically reinvigorated by new experimental and theoretical methods. It led the way to initial developments in quantum mechanics and remains extraordinarily vigorous to this day. It is remarkable for its diversity. For example, its energy domain extends from nanokelvin temperatures to relativistic energies. It is also the proving ground for studying the border area between quantum and classical mechanics.

Atomic physics has its own unique dynamics, continuing as it began with the study of the electromagnetic spectra of ions, atoms, and molecules, over a spectrum now extended to all frequencies from radio waves to x-rays and gamma-rays and continuing, as well, with the study of processes involving collisions between neutral and charged atomic and molecular systems. The two strands of research inspired two classic treatises, *The Theory of Atomic Spectra* by E. U. Condon and G. H. Shortly (1935), and *The Theory of Atomic Collisions* by N. F. Mott and H. S. W. Massey (1965). The two strands intertwine and both have been transformed by the appearance of new tools, especially the laser. Dramatic advances in the range, sensitivity, and precision of experiments have occurred, particularly in the past one or two decades.[2] Because of the rapid increase in computational power, problems once set aside can now be attacked directly. For example, simulations

---

[1]The best single sourcebooks for research in atomic physics are the biennial proceedings of the International Conference on Atomic Physics (ICAP), for example, 1997, University of Windsor, Windsor, Ontario, Canada.
[2]The best sourcebooks for an overview of activities in atomic collision are the proceedings of the biennial International Conference on the Physics of Electronic and Atomic Collisions (ICPEAC). See, for example, J. B. A. Mitchell *et al.*, Eds., 1996 ICPEAC XIX (American Institute of Physics, New York).

and visualizations can be constructed that reveal previously unseen aspects of the basic mechanisms that determine the outcome of atomic collisions. Details of atomic and molecular structural features can be explored by numerical experiments. Elaborate computer codes have been constructed that provide the data on atomic and molecular processes needed for the interpretation and prediction of the behavior of laboratory, fusion, terrestrial, and astrophysical plasmas.

Atomic physics is but a subclass of an even broader area of research, now canonized as a Division of the APS under the rubric Atomic, Molecular, and Optical ("AMO") physics. In this section we have chosen some representative examples of the scope of atomic physics in several areas. Many other important aspects of AMO physics are contained in other articles in this volume: molecular astrophysics—Herschbach; Bose-Einstein condensation and control of atoms by light—Wieman et al.; quantum optics and precision spectroscopy—Hänsch and Walther; the laser—Lamb et al. A short history of atomic, molecular, and optical physics is contained in the article by Kleppner.

Cold Collisions

T. F. Gallagher

Recent developments in the control of atoms by light (see Wieman et al., this issue) have led to the extension of atomic collision studies to the domain of ultracold atoms. Alkali atoms in a magneto-optical trap (MOT) typically have temperatures of hundreds of microdegrees Kelvin, and as a result collisions between them are qualitatively different from collisions between room-temperature or more energetic atoms. The collisions last for a long time, tens of nanoseconds rather than picoseconds. This time is longer than the radiative lifetime of the excited alkali atoms, so it is likely that excited atoms decay in mid collision, and it is possible to influence these collisions to some extent with near-resonant laser light. The collisions velocities are so low that the de Broglie wavelength λ_B is larger than the typical interaction length. Equivalently, the scattering is mainly s-wave scattering, and whether the interatomic interaction is attractive or repulsive determines the conditions under which a Bose-Einstein condensate can be created (see below).

Traditionally, one of the most useful ways of studying atomic collisions has been collisional line broadening (Gallagher, 1996). For cold collisions one form line broadening takes is photoassociation spectroscopy (Thorsheim et al., 1987). The essential notions of photoionization spectroscopy for Rb are shown in Fig. 1 (Heinzen, 1996). This shows the interatomic potential curves for two Rb atoms when both are in the ground state and when one is in ground and the other in the excited state. Only the attractive potential curve from the excited state is shown. (There is also a potential that is repulsive at long range, but we ignore it for the moment). In a MOT trap most atoms exist in the $s+s$ dissociation continuum, connected to two s states as the interatomic spacing $R \to \infty$. The atoms have low translational energy, <1 mK, and the squared amplitude of the wave function for the relative motion of two such ground-state atoms at relatively small separations is shown in Fig. 1. This pair of atoms can absorb a photon tuned to the red of the Rb $5s_{1/2}$-$5p_{1/2}$ atomic transition, making a transition to the more deeply bound molecular excited state. It is more deeply bound at long range due to the resonant dipole-dipole coupling of the molecular states $5s_{1/2}+5p_{1/2}$ and $5p_{1/2}+5s_{1/2}$, which are degenerate at infinite internuclear separation.

The temperature of the atoms in a MOT is less than 300 μK, which corresponds to 6 MHz. Therefore, instead of a continuous far wing of the resonance line, like that observed in a higher-temperature gas, well resolved transitions to high vibrational levels of the excited state are observed. The excited molecules can decay to bound vibrational levels of the ground state by path ω_{bb} of Fig. 1 or, more often, to the dissociative continuum states, by path ω_{bf} with enough kinetic energy for the atoms to escape from the trap. The loss of atoms from the trap is the usual method of detecting that a free-bound transition has been driven in the first place. Some of the atoms do decay to high vibrational levels of the ground state, forming translationally cold molecules, as has been observed by photoionizing the dimers and detecting the dimer ions using a time-of-flight technique.

Photoassociation spectroscopy gives direct insight into the scattering of the cold atoms. Since λ_B is roughly $400\,a_0$, it is far greater than the range of the interatomic potential between two ground-state atoms, as shown by Fig. 1, and their scattering is thus dominated by s-wave scattering. On a longer length scale than the interatomic separations shown in Fig. 1 we can represent the interatomic motion by a wave function that has a sinusoidal dependence on R, the interatomic distance. If there is no interaction between the two atoms the sine wave has a zero at $R=0$. If the interaction is attractive the nodes of the wave function are pulled towards $R=0$, and if the interaction is repulsive the nodes are pushed away from $R=0$. These phase shifts of the wave function are often described by the scattering length a, which is proportional to the tangent of the phase shift. If the interaction is attractive, $a<0$, while it is repulsive if $a>0$.

The scattering length for two ground-state atoms can be extracted from the photoassociation spectrum using the Franck-Condon principle. Briefly, each antinode of the ground-state vibrational wave function leads to strong transitions to vibrational states with

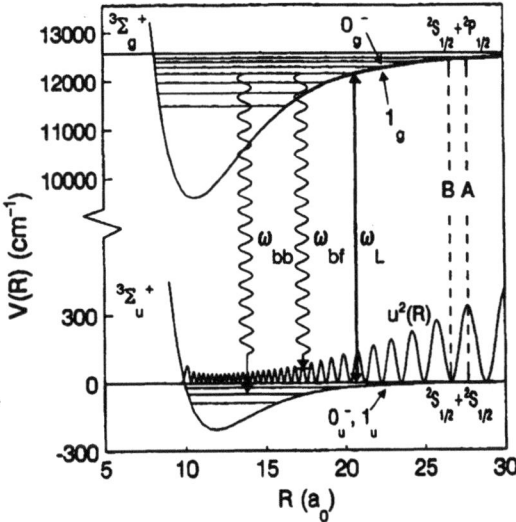

Figure 1. Cold atomic photoassociation. Colliding atoms incident on the ground-state potential are excited by a laser of frequency ω_L to bound excited states. At low temperature, the photon absorption rate exhibits a highly resolved peak when ω_L is tuned across a free-bound transition. The triplet states of Rb_2 are shown in this example. The solid oscillating curve shows the square of an approximate radial wave function $u(R)$ of the colliding atoms. From Miller, Cline, and Henzen (1993).

their outer turning points at the same R. Thus the transitions near the dashed line labeled A in Fig. 1 are strong. On the other hand, transitions at a slightly longer wavelength, resonant with transitions near the dashed line labeled B, are vanishingly weak since there is a node in the ground-state wave function at this interatomic spacing. In other words there is a slow modulation in the intensity of the lines in the photoassociation spectrum, which can be used to generate the wave function of the ground vibrational state and determine its phase shift relative to a sine wave starting from $R=0$. Measuring the phase shift, or equivalently, the scattering length tells whether or not it is possible to make a Bose-Einstein condensate in the system. It is possible if $a>0$, but if $a<0$ the atoms attract each other and BEC becomes more difficult, although it in fact has been achieved for Li, where $a<0$.

The collisions of cold atoms have shown initially surprising phenomena, and it has been possible to control them in a way not usually possible. It is likely that there will be many interesting future developments as well. One example is the use of cold atoms to make cold molecules. The first reports of the production of cold Cs_2 molecules came from Fioretti *et al.* (1998), and already a number of other cold-molecule experiments have been successful.

Accelerator-Based Atomic Collisions

Sheldon Datz

Accelerator-based atomic physics covers a huge number of phenomena ranging from multiple ionization events and electron capture from pair production, in very violent collisions, to charge transfer and electron-ion recombination in rather more delicate ones.[3] The increasing availability of high-energy accelerators and the development of sources that produce multicharged ions with large stored energy, coupled with greater sophistication in experimental techniques, has brought about a considerable increase in our understanding of these processes in recent years. Some examples of specific areas of recent accelerator-based atomic physics research are given below.

Cold-target recoil-ion momentum spectroscopy

An excellent example of experimental refinement is the cold-target recoil-ion momentum spectroscopy method for analysis of collision phenomena (Ulrich *et al.*, 1998). The technique employs a well-defined target in the form of a precooled supersonic gas jet that crosses the projectile beam. Collision products are extracted by an appropriate electric field and detected by position-sensitive detectors. The position and time-of-flight information are used to calculate the momentum of the collision products with high precision. This technique makes it possible to determine the full momentum vectors of all collision products yielding a kinematically complete experiment. It has been employed in the last ten years in a wide range of experiments including collisions of ions, electrons, photons, and exotic projectiles with atoms, molecules, and clusters. The technique yields detailed information about the collision, including the determination of impact parameter, scattering plane, energy loss, and scattering angles. In addition, it allows full detection at all solid angles of the emitted electrons. In two-body systems, as in the case of electron

---

[3]See, for example, *Atomic Physics; Accelerators* (Marton and Richard, 1980).

transfer, the momentum balance between projectile and target allows for a measurement of the energy gain, which is sufficiently precise to determine individual levels for transfer into a given n state.

Ultrarelativistic ion energies

Atomic collisions at ultrarelativistic energies (6.4-TeV S ions and 33.2-TeV Pb ions) have been studied using the SPS synchrotron at CERN (Vane et al., 1997). At these energies, the electric fields have Fourier components representing energies greater than twice the rest mass, $h\nu > 2m_0c^2$, and electrons can be promoted from the negative continuum to produce electron-positron pairs (via two virtual photons) or, alternatively, to promote an electron to a bound state of the projectile, thereby changing its charge. This process, dubbed capture from pair production, can be responsible for beam loss in relativistic colliders, for example.

Electron-ion recombination

At the other end of the energy scale, highly accelerated ions may be used to study collisions at very low center-of-mass energies ($\gtrsim 1$ meV), using "merging beams." An atomic ion can capture an electron and be stabilized by the emission of a photon (radiative recombination). In dielectronic recombination, the dominant process at high temperatures, an electron from the continuum can be captured into a bound state by the simultaneous excitation of a previously bound electron (inverse Auger effect) and the doubly excited state can be stabilized by the emission of a photon,

$$A^{q+}(n,\ell) + e \rightarrow A^{(q-1)+}(n'\ell', n''\ell'') \rightarrow A^{(q-1)+}(n\ell, n'\ell') + h\nu. \quad (1)$$

This is a resonant process that requires a specific collision energy to form a given doubly excited state.

At Oak Ridge National Laboratory, a tandem Van de Graaff accelerator was used to create an energetic multiply-charged ion beam which was merged with a collinear electron beam and the resultant reduced charge from the dielectronic recombination ion measured (Datz and Dittner, 1988). Why use an accelerator? High-charge states are easily obtained. The accelerated beams have a lower electron-capture cross section from background gas, and the higher-energy electron beams needed to achieve comparable velocities have higher space-charge limited currents. The system was used to measure $\Delta n = 1$ dielectronic recombination cross sections for a variety of ions from B^+ to S^{6+}. The resolution of ~ 4 eV was sufficient for integral measurements only. Much higher resolution was later achieved at Aarhus using a much improved merged electron beam ring. The advent of ion storage rings initiated a qualitative improvement in the study of electron-ion collisions (Larsson, 1995). A storage ring comprises a closed magnetic loop for the circulation of ions that have been injected from an ion source. Such ion storage rings are presently in operation in Sweden, Denmark Germany and Japan. The ions may be accelerated in the ring up to the limit of magnetic containment. In the case of the CRYRING in Stockholm, the strength is 1.44 Tm, equivalent to 96 MeV times the ion charge divided by the mass. During its storage time, the ion can relax from metastable states created in the source and, in the case of molecular ions, vibrational levels can also decay. In one of the straight sections of the ring, an electron beam is merged with the ion beam over a length of 1 m. The original purpose of this merged beam was to reduce the momentum spread in the stored ion beam ("cooling"). This is done by matching the ion and electron beam velocities and taking up the random motion within the ion beam by Coulomb scattering from "cold" electrons. The longitudinal energy spread of the electron beam is negligible and the transverse spread has now been reduced to 1 meV.

An intentional mismatch of velocities creates a variable-energy ion-electron collision target. The results obtained include highly precise dielectronic resonances which have led to more accurate determinations of the structure of doubly excited states. An unusual and, at this time, unexplained finding is a large increase in radiative recombination cross section above theory at very low collision energies.

In similar experiments, molecular ions can recombine and, upon regaining the ionization energy, dissociate into neutral fragments. Precision measurements of dissociative recombination of molecular ions in the energy range of 1 meV to 50 eV have been made. For diatomic molecular ions, the final states of the neutral atoms formed have been measured using ring techniques. The resonant structures and the fractionations observed present a challenge to current theory.

Fundamental Measurements in Atomic Collision Physics

Hans Kleinpoppen

Introductory remarks

Atomic collision physics can be broadly divided into cross-section measurements and analyses of fundamental quantum-mechanical processes. Such processes, particularly with regard to atomic and electron spin correlations and resonances, have only been accessible to detailed investigation since the second half of this century. Studies of this kind are particularly connected with "complete atomic scattering experiments" (see Sec. IV.B).

Approximately a hundred years ago the area of atomic collision physics was opened up when photoionization was discovered by H. Hertz in 1887 and electron collision cross sections were estimated by P. Lenard in 1903 from processes in Braun's classical electron tube. Lenard postulated that the electron impact cross section appeared to be much smaller than the atomic cross sections already known from chemical processes and kinetic theory. However, Lenard's experiments failed to reveal the presence of the massive positively charged nucleus of atoms discovered by E. Rutherford in 1911. While the detection of the inelastic energy loss of electrons in electron-impact excitation of atoms was discovered by J. Franck and G. Hertz in 1913, electron scattering from rare-gas atoms carried out by Ramsauer, Ramsauer and Kollath, and Townsend and Bailey in the 1920s revealed specific minima in the total electron-atom cross sections. In 1931 Bullard and Massey detected structure in the differential cross section which could theoretically be interpreted as interference effects in the partial waves of different orbital momenta. Early calculations by N. F. Mott (1929) predicted electron spin polarization through spin-orbit interaction in electron scattering by heavy atoms; however, this sensational prediction by Mott, in the 1920s, was only observed, much later, by Shull, Chase, and Myers (1943). A series of measurements to produce intense beams of polarized electrons and to apply collisions of them with polarized or unpolarized atoms only took place in the second half of this century. Since that time, further highlights in atomic collision physics were the scattering of electrons by polarized atoms and coincidence experiments between atomic particles and photons. Cross-section measurements were continuously stimulated by quantum-mechanical theories of atomic collisions by Born, Mott, Oppenheimer, Massey, and others [see, for example, Mott and Massey (1965) and more modern versions of theories by Bransden (1983), Joachain (1983), and Burke and Joachain (1995)]. One of the outstanding cross-section measurements in atomic collisions was connected with electron scattering on atomic hydrogen. The primary scientific aspect of such experiments was that they represented the simplest and most fundamental quantum-mechanical collision phenomenon in comparison to more complicated many-electron atoms. Experimentally, the production of atomic hydrogen targets was a very demanding and difficult problem, and, to a certain degree, it still remains a problem today. A pioneering experimental step forward in producing a sufficiently intense atomic hydrogen target was pioneered by Fite and Brackman (1958).

Another important electron impact excitation problem was related to the polarization of resonance radiation from atoms excited by electrons at or close to the threshold energy of the exciting process. There were old discrepancies between the theories of Oppenheimer (1927, 1929) and Penney (1932) and the experiments of Skinner and Appleyard (1927) which were only resolved much later by studying fine and hyperfine structure effects of threshold polarization in both theory and experiment. The fundamental quantum-mechanical theory of impact polarization, introduced by Percival and Seaton (1958), correctly took account of the fine and hyperfine splittings and the level widths of the excited states from which polarized radiation from the alkali resonance lines could be measured (Hafner et al., 1965). Such polarizations of impact line radiation can now be based on a proper theoretical quantum-mechanical theory, and numerous experimental comparisons are available confirming the theory of Percival and Seaton.

Cross-section measurements have been carried out since, approximately, the beginning of this century and they are important tests of atomic collision theories. In addition they are relevant to applications in plasma physics, astrophysics, and atmospheric physics; for examples, see reference books such as *Electronic and Ionic Impact Phenomena*, Vol. 1–4, by H. S. W. Massey (1979); *Atomic Collisions*, by E. W. McDaniel (1989); *Atomic and Electron Physics–Atomic Interactions*, by B. Bederson and W. L. Fite (1968), as well as the biennial series of proceedings of ICPEAC (footnote 2).

Complete atomic scattering experiments

Correlation and coincidence experiments

I shall now highlight developments (personally selected) that can be linked to correlation and coincidence measurements between atomic particles (including spin correlations) and to the so-called "complete (or perfect)" atomic scattering and photoionization experiments. Initial proposals for such types of experiments can be traced to the pioneering papers of U. Fano (1957) and B. Bederson (1969, 1970). The idea of "complete" experiments refers to the requirement that initially the collision process be described by a quantum-mechanical pure state vector. This means, for example, that particles A and B may be in quantum states $|n_A J_A m_A\rangle$ and $|n_B J_B m_B\rangle$ with the corresponding quantum numbers n, J, and m for the two particles. The interaction process of the collision with the particles in pure quantum states can, at least in principle, be described by a quantum-mechanical Hamilton operator H_{int}, which is determined by the interaction potential between the colliding partners. As a consequence of the linearity of the Schrödinger equation, the total system of the particles after the collision will also be in pure quantum states. In other words, we can represent atomic collision processes between atomic particles in pure quantum states as follows:

$$|\varphi_{in}\rangle = |A\rangle|B\rangle \xrightarrow[\text{linear operator } H_{int}]{} |\varphi_{out}\rangle = |C\rangle|D\rangle\cdots|K\rangle\cdots. \quad (2)$$

Before the collision, the colliding particles are in the joint quantum state $|\varphi_{in}\rangle$; after the collision the "collisional products" are in the state $|\varphi_{out}\rangle$.

If the state vector $|\varphi_{out}\rangle$ after the collision has been extracted from an appropriate experiment, it may be described by applying the quantum-mechanical superposition principle in the form

$$|\varphi_{out}\rangle = \sum_m f_m \varphi_m, \quad (3)$$

where φ_m are wave functions of possible substates of the state vector $|\varphi_{out}\rangle$ and f_m are complex amplitudes associated with the collision process. The extraction of the state vector $|\varphi_{out}\rangle$ represents the *maximum information and knowledge* that can be extracted from the experimental analysis of the collision process. Experiments that are successful in providing such maximum information are known as *complete experiments*. (Bederson initially used the expression "perfect experiments," but they are generally now called "complete experiments.")

Complete experiments on atomic collision processes require a high degree of experimental effort and special methods, which have only been successfully applied since the beginning of the 1970s. We cannot refer to primary citations because of space limitations; many review articles and books should serve instead, for example, those of Andersen *et al.* (1988, 1997).

Macek and Jaecks (1971) first developed a concise theory of the angular and polarization correlation between inelastically scattered electrons and polarized photons from excitation/deexcitation processes of atoms. With regard to the relatively simple singlet excitation process of $^1P_1$ states of helium, i.e.,

$$e(E) + \text{He}(1\,^1S_0) \rightarrow \text{He}(n\,^1P_1) + e(E - E_{thr}) \rightarrow \text{He}(1\,^1S_0) + e(E - E_{thr}) + h\nu, \quad (4)$$

two amplitudes describing coincidence signals between the inelastically scattered electron and the photon from the excitation process can be extracted from the coincidence experiment, namely, f_0 for the magnetic sublevel $m_1 = 0$ and f_1 for the sublevels $m_1 = \pm 1$ of the $^1P_1$ state. The helium excitation/deexcitation $1\,^1S_0 \rightarrow 3\,^1P_1 \rightarrow 2\,^1S_0$ was investigated by observing the photon emitted from the $3\,^1P_1$ state in coincidence with the inelastically scattered electron in the forward scattering direction; by measuring the photon linear polarization with reference to the incoming electrons, experimenters could confirm the selection rule $\Delta m = 0$ for the magnetic quantum number according to the Percival-Seaton (1958) theory of impact polarization of line radiation. They also could correct the previous discrepancy with theory of the noncoincidence threshold polarization measurements. Eminyan *et al.* (1973, 1974) extended the coincidence experiment by measuring a more general electron-photon angular correlation in the scattering place and determined a set of inelastic excitation amplitudes of the $2\,^1P_1$ state of helium for the first time. Standage and Kleinpoppen (1976) measured a full set of Stokes parameters (Born and Wolf, 1970) of the photon from the $3\,^1P_1 \rightarrow 2\,^1S_0$ excitation in coincidence with the scattered electron and obtained a complete amplitude analysis as well as the orientation vector $L_\perp$ (i.e., a finite expectation value of the angular momentum in the 3^1P_1 state) and an alignment (i.e., a nonisotropic distribution of the magnetic sublevels JM with expectation values $\langle M \rangle = \langle J^2 \rangle / 3$ of the excited $^1P_1$ state in accordance with the analysis of Fano and Macek, 1973). In addition to the amplitude and orientation/alignment analysis the "charge cloud" of the excited $^1P_1$ state of helium can also be determined (Andersen et al., 1984).

Figure 2 shows an example of a collisionally induced electron charge cloud distribution of a $^1P_1$ state detected by the coincidence of the scattered electron in the k_{out} direction and the photon detected in the z direction. The scattering plane is defined by the directions of the incoming k_{in} and the outgoing k_{out} relative momentum vectors of the electron. The atomic charge can be characterized by its relative length (l), width (w), and height (h), its alignment angle (γ) and its angular momentum ($L_\perp$).

Following the definitions of Born and Wolf (1970), the Stirling group (Standage and Kleinpoppen, 1976) could also prove that photon radiation from the electron impact excited $\text{He}(3^1P_1)$ state is completely polarized and the relevant degree of coherence for its excitation is approaching 100%. Many electron-photon coincidence experiments have been reported since the middle of the seventies [see, for example, Andersen et al. (1988)] and their results on coherence and correlation effects have opened up completely new research topics in atomic collision physics.

As already mentioned, electron impact excitation of atomic hydrogen is both one of the most fundamental and one of the simplest atomic collision processes. However, even for the excitation of the Lyman-α radiation, i.e., the $2^2P_{1/2,3/2}$ state decaying into $1^2S_{1/2}$ state, by electron impact there was a long-standing discrepancy between the theory and the experimental electron Lyman-α photon coincidences which has only recently been resolved (Yalim et al., 1997).

Many measurements on collisions between heavy atomic particles (neutral or ionized atoms) and atoms have been reported since the middle of the 1970s, for example, photon-particle (neutral or ionized atoms) coincidence experiments, in connection with charge-

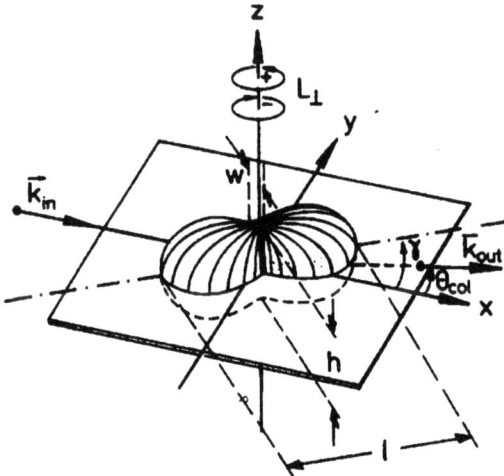

Figure 2. Schematic illustration of an electron charge cloud for a coherently excited P state induced by an incoming particle parallel to the x direction [after Andersen et al. (1984)].

exchange excitation of helium and molecular hydrogen with He^+ ions as projectiles. Angular correlation measurements from direct excitation of helium atoms have also been performed. Out of the many measurements of heavy-particle–photon excitation [see the review by Andersen et al. (1997)], we describe the alignment of the $2P$ state of atomic hydrogen produced in collisions between protons and atomic hydrogen as reported by the Hippler and Lutz group (Hippler et al., 1988).

The collision process considered is described by the following reaction with two outgoing results:

$$H^+ + H \begin{cases} H^+ + H(2P) \\ H(2P) + H^+ \end{cases} \quad (5)$$

i.e., a proton beam crosses an atomic hydrogen beam. At incident energies of a few keV the $H(2P)$ production results from a united-atom ($2p\sigma$-$2p\pi$) rotational coupling. Lyman-α photons from the decay of the $H(2P)$ state were detected perpendicular to the primary proton beam direction with a polarization-sensitive device. The degree of linear polarization $p = I_\parallel - I_\perp / I_\parallel + I_\perp$ of the Lyman-α radiation is defined with the light intensity components polarized parallel and perpendicular to the proton beam direction.

The polarization p can be related to the integral alignment A_{20} of the $H(2P)$ state by the following relations:

$$A_{20} = (Q_1 - Q_0)/Q(2P) = 6p/(p-3), \quad (6)$$

with the total excitation cross section $Q_1(=Q_{-1})$ and Q_0 for the magnetic sublevels m_1, m_0 and $Q(2P) = Q_0 + 2Q_1$ as the total cross section for the $2P$ state.

Figure 3 shows experimental data for the integral alignment A_{20} in comparison to several theoretical predictions. The theories fall into two categories in which either only atomic wave functions are used or, alternatively, molecular states or combinations of both atomic and molecular states are employed. Only theories within the second category include a quasimolecular rotational coupling mechanism. Good agreement between experiment and theory is obtained with the molecular orbital calculations. Such experiments are considered as a very important test case for theories of heavy-particle–atomic collisions.

Electron scattering by atoms in laser fields

Electron scattering by atoms in laser fields of various intensities has become of fundamental importance in understanding the interactions of the colliding electron and the electromagnetic field of the photon. Several outstanding contributions are listed here:

(a) Transfer of energy in multiples of the photon energy $h\nu$ to and from electrons while they are undergoing elastic scattering by atoms is described by the process

$$e(E) + A + nh\nu \rightarrow e(E \pm mh\nu) + A. \quad (7)$$

Such a process was first observed by Weingartshofer et al. (1977).

(b) The energy balance in electron impact excitation of atoms can be made up jointly by the electron energy E_0 and the photon energy in a photon and electron impact process such as $h\nu + e(E_0) + A \rightarrow A^* + e(E_0 + h\nu - E_{\text{thr}})$.

(c) As a mirror symmetric inverse process of the above electron-photon coincidences from electron impact excitation of atoms (reported in Sec. B.1) Hertel and Stoll (1974) studied electron scattering in conjunction with the resonance absorption of a laser photon energy of $h\nu = E_{thr} - E$ with E as the ground-state energy and E_{thr} as the threshold energy for the excited state. The analysis of the angular distribution related to the photon laser polarization again allows a description of the scattering process in terms of two amplitudes and their phase difference for $S \to P \to S$ transitions.

(d) By inducing a resonance absorption of atoms by photons, one can populate an excited state A_1^*. Electron-photon coincidences can then be detected from an even higher excited state A_2^*, i.e., $h\nu + e(E) + A \to A_1^* + e(E) \to A_2^* + h\nu_1 + e(E - E_{thr})$. Excitation amplitudes have been extracted from such "stepwise excitation electron-photon coincidences" for electron mercury scattering.

(e,2e) ionization processes of atoms

With regard to ionization of atoms by electrons, studies of the coincidence between the impinging scattered electron and the electron released from the ionization process, $(e,2e)$ experiments, were first reported by Amaldi et al. (1969) and by Ehrhardt et al. (1969). Since that time, many $(e,2e)$ and even $(e,3e)$ processes have been investigated, including theoretical approximations. As with the electron-photon coincidence experiments these represent more sensitive tests of a theory than the traditional total-cross-section measurements of ionization. As an example, we discuss here data on electron impact ionization of atomic hydrogen and refer the reader to other examples, e.g., to the review by McCarthy and Weigold (1991) and the monograph by Whelan and Walters (1997).

We first introduce the definition of the triply differential cross section: $d^3Q(E_0, E_a, E_b, \theta_a, \theta_b, \varphi_b)/dE_b d\Omega_a d\Omega_b$ with the polar scattering angles θ_a and θ_b for the

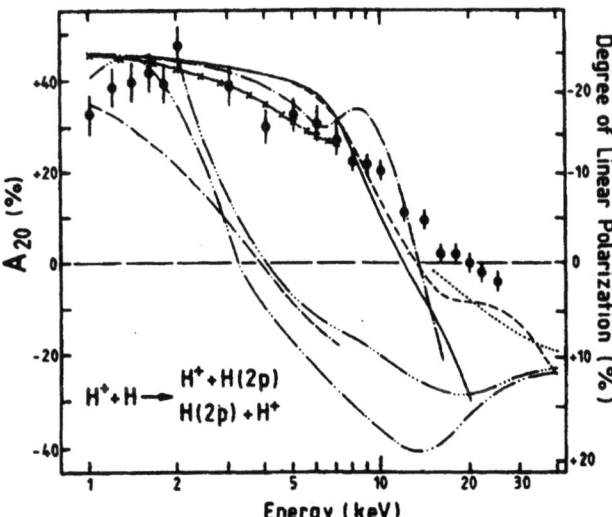

Figure 3. Integral alignment A^{20} (left-hand scale) and linear polarization (right-hand scale) of the Lyman-α radiation for H(2P) production of $H^+ + H$ collisions vs incident energy of the protons. Experimental data by Hippler et al. (1988) are compared to various calculations. Best agreement is obtained with calculations that include quasimolecular rotational coupling.

two electrons detected with an azimuthal angle $\varphi_b = 0$ (defining the scattering plane); θ_a and θ_b are finite, the energy of the scattered electron is E_a and the energy of the incoming electron is E_0 (E_b is then calculable from the energy balance); $d\Omega_a$ and $d\Omega_b$ are the solid angles for the coincident detection of the two electrons.

Figure 4 gives one example of theoretical and experimental data on ($e,2e$) processes for electron impact ionization of atomic hydrogen. As can be seen, the Born approximation essentially fails to reproduce the experimental data while the Coulomb correlation method appears to be in reasonable agreement. Direct coincidence measurements of electron momentum distributions of the ground state of atomic hydrogen and other atoms have also been performed.

Polarized-electron/polarized-photon interactions with atoms

This is the subfield where most of the exciting new approaches in experimental atomic collision physics are found at present, combining coincidence measurements with electron and atomic spin analysis. Early pioneering papers include the electron-polarized atom recoil experiment of Rubin et al. (1969) and the Mott spin polarization measurements of unpolarized electron scattered by partially polarized potassium atoms (Hils et al., 1972). Bederson, Rubin and collaborators applied a Stern-Gerlach hexapole magnet as a velocity selector and a polarizer to polarize potassium atoms. The electron impact process may result in a change of the polarization of the initially polarized atoms, which depend on Coulomb direct interaction (amplitudes f_0 and f_1 for the magnetic sublevels $m=0$ and $m=\pm 1$) and Coulomb exchange interaction (amplitudes g_0 and g_1). An E-H gradient analyzer acts as a "spin filter," which can be adjusted to pass only those atoms from the collision area to the detector which have changed their spin state into the antiparallel direction compared to the spin direction of the atoms leaving the Stern-Gerlach magnet.

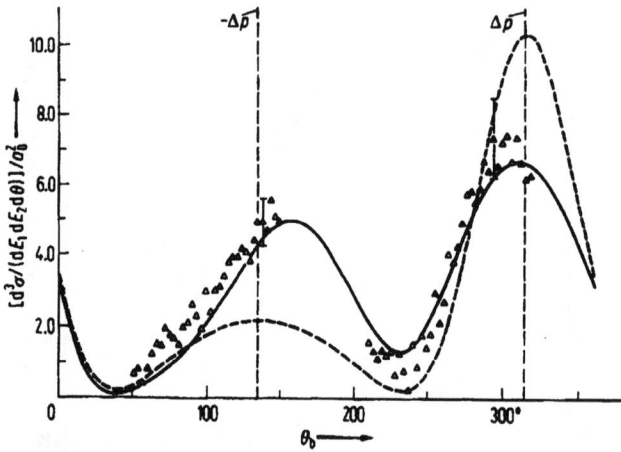

Figure 4. Triple differential cross section for the (e,2e) angular correlation of atomic hydrogen as a function of the scattering angle θ_b with fixed $\theta_a = 4°$, $E_b = 5\,eV$ and the primary incident electron energy $E_0 = 150\,eV$ [after Klar et al. (1987)]: dashed curve, Born approximation; solid curve, Coulomb-correlation method after Jetzke et al. (1989). The angle on the abscissa for θ_b is measured from the direction of the incident electron. The dotted vertical lines show the angle for scattering parallel (Δp) and antiparallel ($-\Delta p$) to the vector of momentum transfer $\Delta p = \lambda \Delta k$.

This experiment (undertaken approximately 30 years ago) could only be matched, some years later, by the JILA experimentalists S. J. Smith and collaborators (Hils et al., 1972), who introduced an alternative experiment in which a potassium atomic beam was polarized (and focused) by a commercial Hewlett-Packard magnetic hexapole and the spin polarization by the elastically scattered electrons was measured by a Mott detector. Neglecting spin-orbit interactions, they could then describe electron scattering by one-electron-atoms by the spin reactions shown in Table I, in which $e(\uparrow)$, $e(\downarrow)$, $A(\uparrow)$, and $A(\downarrow)$ are completely spin-polarized electrons or atoms and only Coulomb direct and Coulomb exchange interactions take part.

On the other hand, only partially polarized electrons and atoms (one-electron atoms) are so far available experimentally (i.e., spin-polarized electrons and spin-polarized atoms with their polarization degree P_e, P_e', P_a, and P_a' before and after scattering), which implies that combined measurements of spin reactions should be possible to determine both the modules of the amplitudes f and g and their relative phase difference. For example, the above experiment of Hils et al. (1972) gives information on the ratio $|f|^2/\sigma(E,\theta) = (1 - P_e' P_a)$, that is, the Coulomb direct cross section divided by the full differential cross section, which is determined by knowing the degree of polarization of the atomic target (P_a) and measuring the spin polarization (P_e'), of the scattered electrons.

The next step in polarized-electron/polarized-atom collisions was the detection of ion symmetries. The ion asymmetry in electron impact ionization can be seen in experiments in which polarized electrons collide with polarized atoms and ions are produced; by introducing the notations $N^{\uparrow\uparrow}(I)$ and $N^{\uparrow\downarrow}(I)$ for the number of ions produced, with the relevant spins of the incoming electrons and the target atoms either parallel or antiparallel to each other one can define the so-called ion asymmetry by the expression

$$A_{\text{ion}} = \frac{N^{\uparrow\uparrow}(I) - N^{\uparrow\downarrow}(I)}{N^{\uparrow\uparrow}(I) + N^{\uparrow\downarrow}(I)}. \tag{8}$$

The first experiments of this type were carried out with polarized atomic hydrogen (Alguard et al., 1977) and with polarized sodium atoms (Hils and Kleinpoppen, 1978).

In electron scattering by heavy atoms such as rubidium and cesium, spin-orbit interaction between the projectile electron and the target atom must be taken into account in addition to the direct Coulomb and exchange interactions. This situation is similar to that for the normal fine structure of excited atoms or the photoionization of heavy alkali atoms, in which spin-orbit interactions increase with increasing mass of the atoms involved. Six amplitudes are necessary for the description of elastic electron scattering on heavy alkali atoms, which means that 11 independent quantities, i.e., six moduli and five

Table 1. Electron scattering by one-electron atoms.

Scattering process	Interactions	Amplitudes	Cross sections
$e(\uparrow) + A(\downarrow) \to A(\downarrow) + e(\uparrow)$	Coulomb direct	f	$|f|^2$
$\to A(\uparrow) + e(\downarrow)$,	Coulomb exchange	g	$|g|^2$
$e(\uparrow) + A(\uparrow) \to A(\uparrow) + e(\uparrow)$,	Interference between direct and exchange interaction	$f-g$	$|f-g|^2$
Full differential cross section: $\sigma(E,\theta) = 1/2\{|f|^2 + |g|^2 + |f-g|^2\}$			

phase differences, have to be determined for complete analysis of the scattering process. A start in determining the six amplitudes has already been made with spin-polarized electrons scattered elastically by spin-polarized cesium atoms in the ground state (Brum *et al.*, 1997). By their measurements spin asymmetries could be detected including the one predicted by Farago (1974, 1976) as an interference effect between the spin-orbit interaction and the spin-exchange interaction.

The complication due to the large number of amplitudes is reduced by using target atoms without a resulting total electron spin ("spinless" atoms), as for example with rare-gas atoms or two-electron atoms with opposite spins. Two spin reactions can be defined for the scattering of polarized electrons on spinless atoms A:

$$(1) \quad e(\uparrow) + A \rightarrow A + e(\uparrow), \quad h, \quad |h|^2,$$

$$(2) \quad e(\uparrow) + A \rightarrow A + e(\downarrow), \quad k, \quad |k|^2. \tag{9}$$

We denote the first process, as before, as a direct process with amplitude h and the second one as a spin-flip process with amplitude k. The direct process can be superposed coherently with an electron exchange process; these processes cannot be individually observed due to their indistinguishability in the experiment. The spin polarization of the electrons after scattering determines the moduli $|h|$ and $|k|$ and their phase differences $\Delta\varphi = \gamma_1 - \gamma_2$.

Figure 5 shows an example for the moduli of the above amplitudes and their phase differences from elastic electron-xenon scattering. As one can see, the modulus $|h|$ of the direct scattering amplitude shows a distinctive diffraction structure, which is due to the superposition of several partial waves of scattered electrons with various angular momenta. This structure is determined by the dipole and exchange interactions, as previously described in connection with the Ramsauer-Townsend effect. The modulus of the spin-flip amplitude $|k|$, which originates from the spin-orbit interaction is considerably smaller than that of the direct amplitude $|h|$; the spin-flip amplitude is primarily determined by the ($\ell = 1$) partial wave of the scattered electrons, which has the result that the diffraction structure is hardly discernible.

Finally I should like to mention a new kind of complete experiment on photoionization of partially polarized atoms reported by Plotzke *et al.* (1996). By applying a hexapole magnet, they were able to polarize and focus atoms on the target area, where a small magnetic guiding field orients the atom spin or J-vector parallel or antiparallel to the direction of the incoming synchrotron radiation of BESSY (Berliner Elektronenspeicherring-Gesellschaft für Synchrotronstrahlung), which is linearly polarized. Photoionization experiments with randomly oriented targets yield two independent parameters, the cross section σ and the parameter β of the angular distribution of the emitted photoelectrons (Yang, 1948). However, this description of the photoionization process is restrictive and averages over the finer details. Klar and Kleinpoppen (1982) have shown that "complete" information on the photoionization process can be obtained by analyzing the photoelectron angular distributions from partially polarized atoms. This method is complementary to the spin analysis of photoelectrons from unpolarized atoms (see, for example, Heinzmann and Cherepkov, 1996).

A new approach to complete photoionization experiments, by means of coincidence measurement between autoionized electrons and polarized fluorescent photons, in the region of the $3p$-$3d$ resonance in calcium has recently been reported.

Summary

Atomic collision physics has developed to a high degree of sophistication particularly due to advanced experimental technology combined with ever-increasing computational capability. Recent emphasis of ICPEAC[1] (Aumayr and Winter, 1998) has been on fundamental quantum-mechanical aspects such as coherence, correlation, alignment, orientation, polarization and, lately, collisions of Bose-Einstein condensates. Recent experimental investigations that apply polarized electrons in electron-electron $(e,2e)$ and in electron-photon $(e,e\gamma)$ coincidence experiments have been reviewed by Hanne (1996a, 1996b).

Limited space does not permit discussion of the many other important recent accomplishments in fundamental atomic collision physics, for example, superelastic scattering by polarized excited atoms, synchrotron radiation experiments, spin asymmetries based on spin-orbit interactions in electron-atom and ion-atom collisions, resonances, and resonant interactions with radiation fields.

While we have concentrated here on fundamental types of experiments, it should be emphasized that many theoretical investigations were crucial for the interpretation of

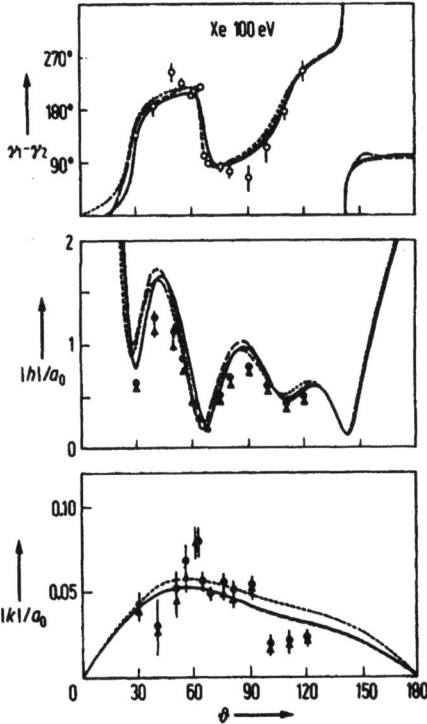

Figure 5. Moduli of amplitudes $|h|$ and $|k|$ and phase differences $\gamma_1 - \gamma_2$ between the two amplitudes for elastic scattering of polarized electrons on xenon atoms, as a function of the scattering angle at an energy of 100 eV. Experimental data points with error bars after Berger and Kessler (1986). The dotted lines and solid curves represent various theoretical predictions: after Haberland et al. (1986), McEachran and Stauffer (1986), Awe et al. (1983). The data $|h|$ and $|k|$ are given in units of the Bohr radius a_0 and are normalized to the measured differential cross section $\sigma = |h|^2 + |k|^2$.

"complete atomic collision experiments." See, for example, Karl Blum's book, *Density Matrix Theory and Applications*, in which the most relevant theoretical papers are discussed and listed. More basic treatments in connection with atomic and quantum collisions and scattering theories are covered by, for example, Massey, 1979; Bransden, 1983; Joachain, 1983; Merzbacher, 1998.

In a lecture given by Sir Harrie Massey approximately one year before his death he stated

We now have the opportunity for gaining deeper understanding of (atomic) collision mechanisms through providing information about the shapes and circulations of atomic wave functions in collision processes. The (experimental) techniques required are very elaborate. Indeed it is probably true that experiments in this field (of coherence and correlation) are among the most complicated in atomic physics today. They are very important for deepening our understanding of atomic collisions and it is essential that their very complexity should not be allowed to obscure their importance (Massey, 1979).

High-Precision Atomic Theory: The Few-Body Problem

G. W. F. Drake

Introduction

This section focuses on the remarkable advances that have taken place over the past fifteen years in the theory of atomic systems more complicated than hydrogen—specifically helium and other three-body systems. Since the time of Newton, the classic three-body problem has defied the best attempts of mathematicians to find exact analytic solutions. In the modern context, solutions to the three-body Schrödinger equation are equally difficult to obtain. However, once found to sufficient accuracy, they form the basis for studying the relativistic and quantum electrodynamic effects that must be included in order to account for the results of high-precision spectroscopic measurements. It is of course the advances in laser spectroscopy that provide the necessary accuracy to allow new and meaningful tests of quantum electrodynamics in systems more complicated than hydrogen, and possibly to find new physical effects. There is no guarantee that a consideration of pairwise interactions among the particles alone is sufficient or that there are no specifically three-body effects.

The nonrelativistic Schrödinger equation

The helium problem played an important role in the early history of quantum mechanics (Hylleraas, 1963). As pointed out by Max Born, the old Bohr-Sommerfeld quantization scheme worked perfectly well for hydrogen, and so helium became a crucial test of Schrödinger's (1926a) wave mechanics. Were it not for the e^2/r_{12} Coulomb repulsion between the two electrons, the three-body Schrödinger equation for helium would be separable, and therefore exactly soluble in terms of products of hydrogenic wave functions. Since the full equation is not separable, approximation methods must be applied, as described in the following sections.

The earliest tests involved the central-field approximation of Hartree (1928) in which each electron is assumed to move in the spherically averaged field of the other. Results for helium, and especially rubidium, yielded reasonable agreement with experiment and markedly different results from the classical orbital model. Hartree's result for the ionization energy of helium was 24.85 eV, as compared with the experimental value 24.60 eV. However, Hartree's central-field model, and its extension by Fock (1930) to include electron exchange, can never recover the remaining correlation energy of about 1 eV (relative to the Hartree-Fock ionization energy of 23.7 eV).

The Hartree-Fock approximation can be thought of as the best possible wave function that can be written in a separable product form. This has a clear physical meaning rooted in the shell model of an atom; but the exact wave function cannot be expressed in this form. The key innovation of Hylleraas in 1927 [with influences from Slater (1928), as well as Max Born, Eugene Wigner, and Hans Bethe] stems from the realization that the triangle formed by the nucleus and the two electrons of helium is determined by the three lengths r_1, r_2 and $\mathbf{r}_{12} = |\mathbf{r}_1 - \mathbf{r}_2|$, where $\mathbf{r}_1$ and $\mathbf{r}_2$ are the position vectors of the two electrons relative to the nucleus. Since the orientation of the triangle in space is not important, the essential dynamics of the system is contained in just these three variables out of the total of six (or equivalently $s = r_1 + r_2$, $t = r_1 - r_2$, $u = r_{12}$). With this in mind Hylleraas constructed a trial variational wave function for the ground state of helium consisting of a sum of terms of the form

$$A_{i,j,k} s^i t^j u^k \exp\left(-\frac{1}{2} Ks\right), \qquad (10)$$

where each term has different powers i,j,k, and the $A_{i,j,k}$ are linear variational coefficients determined by simultaneously diagonalizing the Hamiltonian and overlap integrals in this basis set of functions [see Drake (1998) for a review of variational methods]. Early numerical experiments by Hylleraas and others [see Bethe and Salpeter (1957), Sec. 32 for a review] showed that just a few terms involving powers of $u = r_{12}$ (especially the odd powers) were sufficient to recover nearly all the correlation energy.

Further calculations with basis sets of increasing size and sophistication, culminating with the work of Pekeris and co-workers in the 1960s (see Accad, Pekeris, and Schiff, 1971) showed that nonrelativistic energies accurate to a few parts in 10^9 could be obtained by this method, at least for the low-lying states of helium and He-like ions. However, these calculations also revealed two serious numerical problems. First, it is difficult to improve upon the accuracy of a few parts in 10^9 without using extremely large basis sets where roundoff error and numerical linear dependence become a problem. Second, as is typical of variational calculations, the accuracy is best for the lowest state of each symmetry, but rapidly deteriorates with increasing n.

Recent advances

Modern spectroscopic accuracies in the sub-MHz range require theoretical energies accurate to a few parts in 10^{12} to make meaningful comparisons with experiment. Over the past 15 years, both of the above limitations on accuracy have been resolved by "doubling" the basis set so that each combination of powers i,j,k is included twice with different exponential scale factors (Drake, 1993a, 1993b). A complete optimization with respect to all the nonlinear parameters leads to a natural partition of the basis set into two distinct distance scales—one appropriate to the long-range asymptotic behavior of the wave function, and one appropriate to the complex correlated motion near the nucleus. The greater flexibility in the available distance scales allows a much better physical description of the atomic wave function, especially for the higher-lying Rydberg states where two sets of distance scales are clearly important. However, the multiple distance scales also greatly improve the accuracy for the low-lying states.

For the classic example of the ground state of helium, the nonrelativistic energy is now known to be

$$E_{NR} = -2.903\,724\,377\,034\,119\,598\,13(23), \qquad (11)$$

obtained by extrapolation from a doubled basis set containing 2114 terms. The accuracy is about one part in 10^{19}. Other results approaching this accuracy have been obtained in recent years, using both Hylleraas and other basis sets (Bürgers et al., 1995; Goldman, 1998). The calculation of Baker et al. (1990) is significant for its accuracy, given that only 476 terms were used. The significant point is that all of these methods are evidently converging to the same numerical value.[4]

The same Hylleraas-type methods can in principle be applied to atoms more complex than helium, but the size of the basis set required for a given degree of accuracy and the demands on computer resources grow extremely rapidly with the number of particles. That is why alternative methods of more limited accuracy have been developed, such as multiconfiguration Hartree-Fock, configuration-interaction, many-body perturbation theory, finite-element, diffusion Monte Carlo, and variational Monte Carlo techniques. These techniques can readily be extended to arbitrarily complex systems, but the accuracy seldom exceeds one part in 10^6 for the energy. Hylleraas-type results with accuracies comparable to helium have been obtained only for lithium and similar four-body systems (Yan and Drake, 1998).

Asymptotic expansions

Results of similar accuracy are now available for all the higher-lying $1snl^{1,3}L$ Rydberg states of helium up to $n=10$ and $L=7$. One might object that these long strings of figures are just numerology with little physical content. However, with increasing L, one can give a full physical account of the variational results by means of a simple (in concept) core polarization model largely developed by Drachman (1993). An examination of the eigenvalues for Rydberg states reveals two significant features. First, with increasing L, the first several figures are accounted for by the screened hydrogenic energy

$$E_{SH} = -\frac{Z^2}{2} - \frac{(Z-1)^2}{2n^2} \qquad (12)$$

corresponding to the energy of the inner $1s$ electron with the full nuclear charge Z, and the outer nl electron with the screened nuclear charge $Z-1$. Second, the singlet-triplet splitting goes rapidly to zero with increasing L. This suggests that for sufficiently high L, one can treat the Rydberg electron as a distinguishable particle moving in the field of the polarizable core consisting of the nucleus and the tightly bound $1s$ electron. The various multipole moments of the core then give rise to an asymptotic potential of the form

---

[4]See the Atomic, Molecular, and Optical Physics Handbook (Drake, 1996), which includes a number of articles discussing computational techniques for the few- and many-body problems in detail, and which includes many important earlier references.

$$\Delta V(r) = \frac{c_4}{r^4} + \frac{c_6}{r^6} + \frac{c_7}{r^7} + \cdots, \tag{13}$$

where r is the coordinate of the Rydberg electron. In first order, the correction to the energy is then $\langle \Delta V(r) \rangle$, where the expectation value is with respect to the Rydberg electron. Since the core is a hydrogenic system, all the c_i coefficients and expectation values can be calculated analytically. For example, c_4 is related to the core polarizability $\alpha_1 = (9/32)a_0^3$ by $c_4 = -\alpha_1/2$ (a_0 is the Bohr radius), and c_6 is related to the quadrupole polarizability $\alpha_2 = (15/64)a_0^5$ and a nonadiabatic correction to the dipole polarizability $\beta_1 = (43/512)a_0^5$ by $c_6 = -\alpha_2/2 + 3\beta_1$. Detailed expressions for the higher-order terms up to c_{10} have been derived [see Drachman (1993) for further discussion]. Each term can be calculated analytically by repeated use of the perturbation methods of Dalgarno and Lewis (1956). However, the expansion must be terminated at $i = (L+1)$ because the expectation values $\langle r^{-i} \rangle$ diverge beyond this point. In this sense, the series must be regarded as an asymptotic expansion.

As an example, Table II shows that the terms up to c_{10}, together with a second-order perturbation correction, account for the variationally calculated energy of the $1s10k$ state to within an accuracy of only a few Hz. All the entries can be expressed analytically as rational fractions. For example, the $c_4 \langle r^{-4} \rangle$ contribution is exactly (in atomic units)

$$c_4 \langle r^{-4} \rangle = -\frac{3 \times 61}{2^{10} \times 5^6 \times 7 \times 13 \times 17} = -7.39334195 \cdots \times 10^{-9}. \tag{14}$$

Since the accuracy of the symptotic expansion rapidly gets even better with increasing L, there is clearly no need to perform numerical solutions to the Schrödinger equation for $L > 7$. The entire singly excited spectrum of helium is covered by a combination of high-precision variational solutions for small n and L, quantum defect extrapolations for high n, and asymptotic expansions based on the core polarization model for high L.

Asymptotic expansion methods have similarly been applied to the Rydberg states of lithium and compared with high-precision measurements (Bhatia and Drachman, 1997). This case is more difficult because the Li$^+$ core is a nonhydrogenic two-electron ion for

Table 2. *Asymptotic expansion for the energy of the 1s10k state of helium.*

Quantity	Value
$-Z^2/2$	$-2.000\ 000\ 000\ 000\ 000\ 00$
$-1/(2n^2)$	$-0.005\ 000\ 000\ 000\ 000\ 00$
$c_4 \langle r^{-4} \rangle$	$-0.000\ 000\ 007\ 393\ 341\ 95$
$c_6 \langle r^{-6} \rangle$	$0.000\ 000\ 000\ 004\ 980\ 47$
$c_7 \langle r^{-7} \rangle$	$0.000\ 000\ 000\ 000\ 278\ 95$
$c_8 \langle r^{-8} \rangle$	$-0.000\ 000\ 000\ 000\ 224\ 33$
$c_9 \langle r^{-9} \rangle$	$-0.000\ 000\ 000\ 000\ 002\ 25$
$c_{10} \langle r^{-10} \rangle$	$0.000\ 000\ 000\ 000\ 003\ 73$
Second order	$-0.000\ 000\ 000\ 000\ 070\ 91$
Total	$-2.005\ 000\ 007\ 388\ 376\ 30(74)$
Variational	$-2.005\ 000\ 007\ 388\ 375\ 8769(0)$
Difference	$-0.000\ 000\ 000\ 000\ 000\ 42(74)$

which the multipole moments cannot be calculated analytically, and variational basis-set methods must be used instead. However, the method is in principle capable of the same high accuracy as for helium.

Relativistic and QED effects

The accuracy of the foregoing results for helium exceeds that of the best measurements by a wide margin. However, numerous small corrections must be added before a meaningful comparison with experiment can be made. Many of these can also be calculated to high precision, leaving a finite residual piece due to higher-order relativistic and quantum-electrodynamic effects which lie at the frontier of current theory.

The two relevant parameters in calculating corrections to the nonrelativistic energy for infinite nuclear mass are μ/M and α, where $\mu = m_e M/(m_e + M)$ is the reduced electron mass and α is the fine-structure constant. Since $\mu/M \approx 1.3707 \times 10^{-4}$ for helium, and $\alpha^2 \approx 0.5328 \times 10^{-4}$, these terms are the same order of magnitude. The expansion then has the form (in units of e^2/a_μ)

$$E = E_0^0 + E_0^1(\mu/M) + E_0^2(\mu/M)^2 + E_2^0 \alpha^2 + E_2^1 \alpha^2(\mu/M) + E_3^0 \alpha^3 + \cdots . \quad (15)$$

The leading terms can be expressed as expectation values and accurately calculated. For example, $E_0^1 = -\langle \Delta_1 \cdot \Delta_2 \rangle$ is the specific mass shift due to the mass polarization operator, and E_0^2 is the second-order perturbation correction. The leading relativistic term E_2^0 is the expectation value of the well-known Breit operator (Bethe and Salpeter, 1957) for infinite nuclear mass, but the finite-mass correction E_2^1 contains new operators coming from a systematic reduction of the pairwise Breit interactions in the full three-body problem to center-of-mass plus relative coordinates along with mass scaling and mass polarization contributions. Although these terms become increasingly complicated, they can still be accurately calculated and subtracted from measured transition frequencies.

The leading QED term E_3^0 is the first term to present new computational challenges. It contains contributions coming from both the electron-nucleus interactions of leading order $\alpha^3 Z^4$ and the electron-electron interaction of leading order $\alpha^3 Z^3$. The general form of the electron-nucleus part $E_{3,Z}^0$ for helium is simply obtained from the corresponding hydrogenic case by inserting the correct electron density at the nucleus in place of the hydrogenic quantity $\langle \delta(\mathbf{r}) \rangle = Z^3/(\pi n^3)$. This part is easily done, but the Bethe logarithm $\beta(nLS)$, representing the emission and absorption of virtual photons, is much more difficult to calculate. It is defined in terms of a sum over virtual two-electron intermediate states,

$$\beta(nLS) = \frac{\sum_m |\langle 0|\mathbf{p}_1 + \mathbf{p}_2|m\rangle|^2 (E_m - E_0) \ln[2Z^{-2}(E_m - E_0)]}{\sum_m |\langle 0|\mathbf{p}_1 + \mathbf{p}_2|m\rangle|^2 (E_m - E_0)} . \quad (16)$$

The accurate calculation of $\beta(nLS)$ is one of the most challenging problems in atomic structure theory. The problem is that the sum in the numerator is very nearly divergent, and so the dominant contribution comes from states lying high in the scattering continuum (both one- and two-electron). In a monumental calculation based on earlier work by Schwartz (1961), Baker et al. (1993) have obtained accurate values of $\beta(nLS)$ for the low-lying S states of helium (1^1S, 2^1S, and 2^3S). These results have an important impact on bringing theory and experiment into agreement.

Although no other direct calculations of similar accuracy are available for other states, for sufficiently high L one can use instead the core polarization model. This picture shows that the dominant contribution to the change in $\beta(nLS)$ from the hydrogenic $\beta(1s)$ comes from the perturbing effect of the Rydberg electron on the $1s$ electron, rather than from the Rydberg electron itself. The dipole polarization result allows $\beta(nLS)$ to be expressed in terms of the known hydrogenic Bethe logarithms, plus a correction term proportional to $\langle r^{-4} \rangle_{nl}$ calculated with respect to the screened hydrogenic wave function of the Rydberg electron. This result is of pivotal importance because it allows the QED part of the D-state energies to be calculated to sufficient accuracy that these states can be taken as absolute points of reference in the interpretation of measured transition frequencies. In particular, the much larger S-state QED shift can then be extracted from measured nS-$n'D$ transition frequencies by subtraction of the other known terms.

Relativistic and QED terms of order α^4 a.u. and α^5 a.u. are also important in the comparison with experiment. The theory of these terms is incomplete. A complete treatment requires a systematic reduction of the Bethe-Salpeter equation in order to find equivalent nonrelativistic operators whose expectation values in terms of Schrödinger wave functions yield the correct coefficients for a given order of α. The result then represents an extension of the Breit interaction to higher order. To date, this ambitious program has been carried to completion only for the spin-dependent parts (see Sec. V.B.4). However, a comparison with experiment indicates that for S states, these higher-order terms are dominated by large QED contributions analogous to the corresponding terms in the one-electron Lamb shift. For example, the term $E_{4,Z}^0$ or order α^4 contributes to -771.1 MHz, -51.995 MHz, and -67.634 MHz, respectively, to the (positive) ionization energies of the helium $1s^2\,{}^1S$, $1s2s\,{}^1S$, and $1s2s\,{}^3S$ states, while the experimental uncertainties are more than an order of magnitude smaller. There would be large discrepancies between theory and experiment without this QED term. The comparison between theory and experiment [see Drake and Martin (1998)] shows that, for the ionization energy of the $1s^2\,{}^1S$ ground state of helium, the two agree at the ± 100 MHz level (1.7 parts in 10^8) out of a total ionization energy of $5\,945\,204\,226(100)$ MHz. The total QED contribution is $-41\,233(100)$ MHz. For the $1s2s\,{}^1S$ state, the agreement is spectacularly good. The difference between theory and experiment is only 1.1 MHz (1.2 parts in 10^9) out of a total ionization energy of $960\,332\,040$ MHz. Both of these results rely on the calculated ionization energies of the higher-lying P- and D-state energies as absolute points of reference.

Fine-structure splittings and the fine-structure constant

The helium $1s2p\,{}^3P$ manifold of states has three fine-structure levels labeled by the total angular momentum $J=0,-1$, and 2. If the largest $J=0\to 1$ interval of about 29 617 MHz could be measured to an accuracy of ± 1 kHz, this would determine the fine-structure constant α to an accuracy of ± 1.7 parts in 10^8, provided that the interval could be calculated to a similar degree of accuracy. This degree of accuracy would provide a significant test of other methods of measuring α, such as the ac Josephson effect and the quantum Hall effect, where the resulting values of α differ by 15 parts in 10^8 (Kinoshita and Yennie, 1990). Groups are now working toward the achievement of a ± 1 kHz measurement of the fine-structure interval.

Theory is also close to achieving the necessary accuracy. In lowest order, the dominant contribution of order α^2 a.u. comes from the spin-dependent part of the Breit interaction.

This part is known to an accuracy of better than 1 part in 10^9, and corrections of order α^3 a.u. and α^4 a.u. have similarly been calculated to the necessary accuracy. At each stage, the principal challenge is to find the equivalent nonrelativistic operators whose expectation value in terms of Schrödinger wave functions gives the correct coefficient of the corresponding power of α. This analysis has been completed for the next higher-order $\alpha^5 \ln \alpha$ and α^5 terms, and numerical results obtained for the former. A full evaluation of the remaining α^5 terms should be sufficient to reduce the theoretical uncertainty from the present ± 20 kHz to less than 1 kHz. Once both theory and experiment are in place to the necessary accuracy, a new value of α can be derived. At present, theory and experiment agree at the ± 20-kHz level (Storry and Hessels, 1998). This already represents a substantial advance in the accuracy that can be achieved for spin-dependent effects on helium.

Theory of few-electron ions

In the foregoing discussion of helium, the starting point was the nonrelativistic Schrödinger equation with relativistic and QED effects treated by successive orders of perturbation theory. However, measurements of transition frequencies are available for many He-like ions all the way up to two-electron uranium (U^{90+}). For such high-Z systems, relativistic effects predominate and a perturbation expansion in powers of αZ is no longer appropriate. On the other hand, electron correlation effects decrease in proportion to $1/Z$ relative to the one-electron energies due to the dominant Coulomb field of the nucleus. This suggests that one should start instead from the one-electron Coulomb-Dirac equation so that relativistic effects are included to all orders from the beginning, and treat electron correlation effects by perturbation theory. As a rough rule of thumb, the high-Z region begins when relativistic effects become larger than electron correlation effects, i.e., when $(\alpha Z)^2 > 1/Z$, or $Z > 26$.

The so-called Unified Method (Drake, 1988) provides a quick and simple way to merge $1/Z$ expansions for the nonrelativistic part of the energy with exact Dirac energies [i.e., $(\alpha Z)^2$ expansions summed to infinity] for the relativistic part. With allowance for QED corrections, the resulting energies are remarkably accurate over the entire range of nuclear charge. However, the accuracy of recent measurements for high-Z ions has now reached the point that higher-order contributions arising from the combined effects of relativity and electron correlation become important. The leading such term not included by the Unified Method is of order $(\alpha Z)^4$ a.u. and $\alpha^4 Z^2$ relative to the nonrelativistic energy. Terms of this order are automatically included in much more elaborate calculations based on the techniques of relativistic many-body perturbation theory and relativistic configuration interaction [see Sapirstein (1998) for a review]. Especially important are methods for evaluating all orders of perturbation theory at once. Although these methods are less accurate than the nonrelativistic Hylleraas-type calculations for the neutral helium and lithium atoms and their isoelectronic low-Z ions, they yield good agreement with experiment for intermediate- and high-Z ions. There is a broad range of Z between about 6 and 40 where both approaches yield results of useful accuracy and allow interesting comparisons between them.

Future prospects

The results described here indicate the high degree of understanding that has been achieved for few-electron systems over the entire range of nuclear charge from neutral atoms to highly ionized uranium. For the heliumlike systems, the Schrödinger equation has been solved and lowest-order relativistic corrections calculated to much better than spectroscopic accuracy. To a somewhat lesser extent, accurate solutions also exist for lithiumlike systems.

For highly ionized systems, many-body perturbation theory and relativistic CI provide powerful computational techniques. The residual discrepancies between theory and experiment determine the higher-order relativistic and QED (Lamb shift) contributions to nearly the same accuracy as in the corresponding hydrogenic systems. Interest therefore shifts to the calculation of these contributions, for which theory is far from complete for atoms more complicated than hydrogen. New theoretical formulations are needed, such as the simplifications recently discussed by Pachucki (1998). Each theoretical advance provides a motivation for parallel advances in the state of the art for high-precision measurement. The results obtained to date provide unique tests of both theory and experiment at the highest attainable levels of accuracy, and they point the way to applications to more complex atoms and molecules.

Exotic Atoms

G. zu Putlitz

"Exotic atoms" such as positronium and muonium are pure leptonic hydrogenlike systems perfectly suited for testing the electromagnetic interaction and possible tiny admixtures of other interactions like the weak and strong force or possible unknown forces beyond the standard model. The progress made in this field up to now has produced experimental values of such precision that the results and their interpretations compete with the experiments of elementary particle physics carried out at the highest energies available so far.

Of major importance are atoms with negatively charged particles in the atomic shell, particularly those with a negative muon.

The production of exotic atoms relies on the production of exotic particles with suitable intensity, phase space, and energy to be captured by a normal atom. For muonic atoms with a negative muon, formation initially involved a proton beam with energies of 0.5 to 2.0 GeV striking a target to produce negative pions. The π^- decay with a lifetime of $\tau_\pi = 2.6 \times 10^{-8}$ sec through the reaction $\pi^- \rightarrow \mu^- + \bar{\nu}_\mu$. The negative muons are subsequently decelerated by propagation through matter until they reach energies at which they are captured in states with rather high principal and orbital angular momentum quantum numbers. In the beginning of the capture process the energy is mostly released by Auger ionization of outer-shell electrons. Later on the muons reach lower states by emission of electromagnetic radiation with discrete energies. This part of the spectrum is particularly important not only for the investigation of the atomic potential but also for a measurement of the mass of the muon. In lower orbits the negative muons interact directly with the nucleus and annihilate through the weak interaction.

The electromagnetic spectrum emitted from a muon captured in an atom results in the emission of γ radiation where the energy levels E_n are

$$E_n = -\frac{\mu_{\text{red}} c^2}{2}\left(\frac{Z\alpha}{n}\right)^2. \tag{17}$$

The corresponding Bohr radius is

Table 3. Properties of nonstable leptons, mesons, hadrons and antiprotons (all with $Z=-1$).

Particle	Mass m [MeV]	Mean lifetime τ [s]	Bohr energy [KeV] $E_B = E_0 m/m_e z^2$	Bohr radius $a_B = m_e/m\, a_0/Z$ [fm]
μ^-	106	2.2×10^{-6}	$2.8 \times Z^2$	$255/Z$
τ^-	1784	2.9×10^{-13}	$47.5 \times Z^2$	$15/Z$
π^-	140	2.6×10^{-8}	$3.7 \times Z^2$	$193/Z$
K^-	494	1.2×10^{-8}	$13.1 \times Z^2$	$55/Z$
Σ^-	1197	1.5×10^{-10}	$31.8 \times Z^2$	$23/Z$
Ξ^-	1321	1.6×10^{-10}	$35.2 \times Z^2$	$21/Z$
Ω^-	1673	0.8×10^{-10}	$44.5 \times Z^2$	$16/Z$
$\bar{p}$	938	∞	$25.0 \times Z^2$	$29/Z$

$$\tau_n = \frac{\hbar^2}{\mu_{\text{red}} e^2} \frac{n^2}{Z}, \qquad (18)$$

where n is a principal quantum number of the state, μ_{red} is the reduced mass of the negative particle X in the shell $Z=$ nuclear charge number, and $\alpha = (e^2/\hbar c)$ = fine-structure constant $\approx \frac{1}{137}$. For a medium-mass nucleus the energy for the muonic Lyman-α line (10.2 eV for the electron in hydrogen) is in the MeV range. The corresponding orbit of the muon is located well inside the electronic shells for lower principal quantum numbers. The measured energy in this case is smaller than the point-charge value estimated above because of the different potential the μ^- experiences close to the nuclear surface. The Bohr radius and the nuclear radii become comparable.

Muonic x rays were first measured by Fitch and Rainwater in 1953. Over the decades the precision in muonic atom spectroscopy (as well as that for pionic, kaonic, and other spectra) has increased greatly due to increased beam intensities, the availability of solid-state detectors, and the construction of special crystal-diffraction spectrometers. Extensive data have been obtained on the energies of muonic spectra, which have led to the measurement of nuclear radii, of isotope and isotone displacements of muonic lines, and the change of the charge distribution of isomeric nuclei.

For larger nuclear charges, radiative corrections and corrections for effects of nuclear polarization are significant. Nevertheless, the spectra of pionic, kaonic, and hadronic atoms have played a very important role in the study of systematic trends of the charge and mass distribution in nuclei as well as nuclear deformations over large sequences of isotopes.

From the measurement of several transitions in a particular muonic atom at least two parameters of the charge distribution can be determined, the mean charge radius C and the skin thickness of the charge distribution t. In addition the width of the energy levels measured through the linewidth of the γ transition can be used to measure the rate of absorption processes of the exotic particles in the nucleus. For muonic hydrogen in the $1S$ state this rate is rather small compared to the natural decay rate of the muon of $R \approx (1/\tau_\mu) \approx 5 \times 10^5$ s^{-1} but has been measured rather precisely to be $R_c \approx 500$ s^{-1}. For lower nuclear charge numbers the capture probability for the muon by the nucleus increases with Z^4. For this reason even at relatively low Z capture processes dominate the linewidth of lower muonic transitions.

Muonic helium is an exotic atom with special features. It is produced by a capture reaction in which both of the electrons are emitted by the Auger effect. The remaining

$(He^{++}\mu^-)^+$ ion has a hydrogenlike spectrum and can capture an electron in the same way as does a proton. Its term energies have been recorded through soft γ-ray spectroscopy (in which the energy of the Lyman-α line=8.2 keV) and its fine structure was investigated in a laser experiment. A decade later neutral $(He^{++}\mu^-e^-)^0$ atoms were produced and the hyperfine structure was measured with high precision (Hughes, 1990). This atomic system is very peculiar because it constitutes a hydrogen atom within a hydrogen atom. Table III lists the negative exotic particles that are suitable for the production of exotic atoms. Tauonic atoms have not been observed so far because the lifetime of the τ lepton of 10^{-13} s is very short, and the probability of annihilation by nuclear absorption is very large.

Antiprotonic atoms are also formed by the capture of antiprotons into states with large quantum numbers. The spectra recorded for antiprotonic hydrogen, He, and other light atoms correspond, for large quantum numbers, to the theoretical predictions based on electromagnetic forces. However, for smaller quantum numbers the strong interaction starts to dominate and a strongly bound system, protonium $\bar{p}p$, is formed which annihilates very rapidly via the strong interaction into hadrons.

The study of antiprotonic atoms was stimulated by the discovery that some of the highly excited states in antiprotonic neutral helium $^4He^+\bar{p}^-e^-$ have a rather long lifetime (Yamazaki, 1992). This fact was utilized to investigate via laser spectroscopy transitions between states of quantum numbers around $n=35$ and the highest possible orbital angular momentum quantum numbers.

The investigation of neutral antimatter with high precision has been a target of scientific investigation for many decades, promoted by the synthesis of antideuterons some 30 years ago. With our present knowledge the strong electromagnetic and weak forces have to be considered as equal for matter and antimatter, but the question of the gravitational equality of matter and antimatter remains unsettled. For this reason, and also to achieve a more sensitive test of CPT invariance, the production of neutral antihydrogen ($\bar{H} = \bar{p}e^+$) is being actively pursued following three different pathways. These are (1) the storage of antiprotons and positrons in the same trap volume and obtaining $\bar{H}$ by three-body collisions; (2) laser-induced recombination of parallel $\bar{p}$ and e^+ beams of the same velocity from continuum states[5]; and (3) production of $\bar{p}$ and e^+ in the same phase space with simultaneous binding of oppositely charged particles. The latter method was successful in detecting antihydrogen at high energy. However, it is not well suited for precision measurements on antihydrogen. The forthcoming experiments with $\bar{H}$ in traps may clarify this interesting basic question about gravitation.

Positronium and muonium are prominent examples of the wealth of information which can be obtained from these exotic atoms. Positronium was discovered by M. Deutsch and collaborators (1951) and has been investigated with ever increasing precision with respect to its term energies and decay constants in the different angular momentum coupling states. Positronium is formed if positrons from a source (e.g., ^{22}Na, ^{64}Cu) are stopped in a gas or in a fine dispersed powder. Positron-electron capture results in two states, 1^1S_0 (parapositronium) and 1^3S_1 (orthopositronium). The decay of positronium is governed by C parity conservation, which requires that parapositronium decay collinearly into γ quanta with the energy of the rest mass of the electron $E_\gamma=511$ KeV and the lifetime $\tau(^1S_0)=1.25\times10^{-10}$ s. The $^3S_1$ state (orthopositronium) decays predominantly via 3γ quanta and has a lifetime of $\tau(^3S_1)=1.4\times10^{-7}$ s. The difference in decay modes and

---

[5]See also the discussion of Electron-Ion Recombination in an earlier section.

lifetimes makes possible high-precision spectroscopic measurements of the energy difference between $^1S_0$ and $^3S_1$. After the fast decay of the $^1S_0$ state formed initially, radio-frequency transitions into this state from the $^3S_1$ increase the number of two-quantum decays and thus provide a signal. Positronium is an ideal system in which to study quantum electrodynamics and radiative corrections. The splitting ΔE of the ground state can be written in lowest order as

$$\Delta E(1^3S_1 - 1^1S_0) = \left(\frac{4}{6} + \frac{3}{6}\right)\alpha^2 R_\infty, \qquad (19)$$

where R_∞ is the Rydberg constant.

The latest values for splitting are $\Delta E(1^3S_1 - 1^1S_0)_{\text{theor}} = 2.03380 \times 10^{11}$ Hz and $\Delta E(1^3S_1 - 1^1S_0)_{\text{exp}} = 2.03398(11) \times 10^{11}$ Hz, are in satisfactory agreement. The annihilation rates are a further test of the electromagnetic interaction. The corresponding results for the lifetime of the $^1S_0$ state are in adequate agreement, $\tau_{\text{theor}} = 0.79854(36) \times 10^{10}$ s^{-1} and $\tau_{\text{exp}} = 0.799(11) \times 10^{10}$ s^{-1}. However for the 1^3S_1 state, the results $\tau_{\text{exp}}^{-1} = 0.7262(15) \times 10^7$ s^{-1} and $\tau_{\text{theor}} = 0.72119(39) \times 10^7$ s^{-1} disagree substantially (Mills and Chu, 1990).

Like positronium, muonium is also a pure leptonic atom dominated by the electromagnetic interaction between the particles. Since pair annihilation cannot take place between two unequal leptons, muonium ($\mu^+ e^-$) is much more longer lived than positronium. Its lifetime is determined by the lifetime of the positive muon $\tau(\mu^+) = 2.19703(4) \times 10^{-6}$ s. This rather long lifetime makes muonium very suitable for high-precision experiments. The spectrum of muonium in its ground and first excited states (Fig. 6) shows which

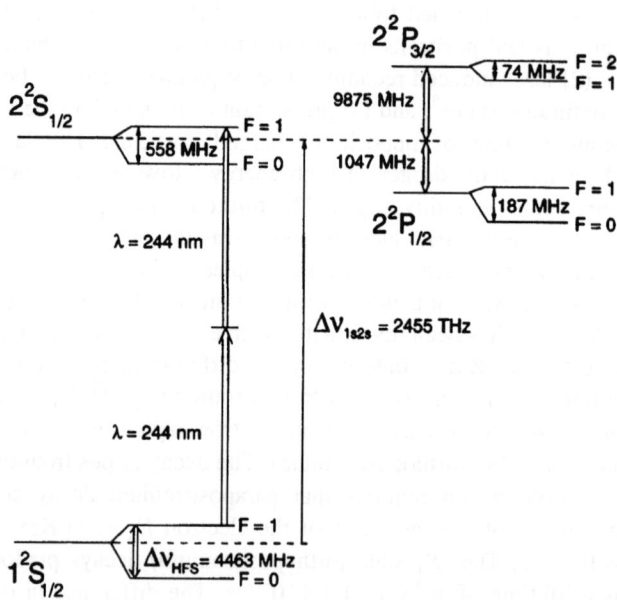

Figure 6. Level scheme of muonium for the two first main quantum numbers $n=1$ and $n=2$. Double arrows indicate those transitions which have been measured experimentally so far.

measurements are possible and desirable. The hyperfine-structure splitting of the ground state can be written in lowest order as

$$\Delta \nu_{hfs} = \left(\frac{16}{3}\alpha^2 cR_\infty \frac{\mu_\mu}{\mu_0}\right)\left(1 + \frac{m_e}{m_\mu}\right)^3. \quad (20)$$

Measurements in zero, weak, and strong magnetic fields can be utilized to extract the magnetic moment of the muon and the fine-structure constant.

Muonium was discovered by Hughes and collaborators (Hughes et al., 1960) when positive muons were stopped in highly purified argon gas. Muons are produced by the decay of pions in a two-body decay via the weak interaction. Consequently they exhibit spin polarization with respect to their momentum. Since the capture process of the electron does not destroy this polarization, muonium is polarized. The subsequent weak decay into positrons and neutrinos shows a $(1+\cos\theta)$-positron distribution with respect to the muon spin direction, which can be used to detect the μ^+ polarization at the time of the decay. Consequently spin depolarizing radio-frequency transitions can be detected through the change in the e^+ decay angular distribution (Hughes and zu Putlitz, 1990). After some 35 years of research the value of the hyperfine splitting in muonium $\Delta \nu_{hfssls}(M)$ has been improved from 4461.3 (2.2) MHz to 4463.302764 (54) MHz.

The mass of the muon can be determined from a measurement of the energy difference between the 1S and 2S levels. The method of collinear two-photon laser spectroscopy between the $1^2S_{1/2}-2^2S_{1/2}$ states at $\lambda = 244$ nm and the subsequent ionization by a third quantum of the same light field resulted in a best value for the splitting $\Delta \nu(1^2S_{1/2} - 2^2S_{1/2}) = 2\,455\,529\,002\,(33)\,(46)$ MHz. It should be mentioned also that muonium has been used to measure the Lamb shift in the $n=1$ and $n=2$ states and the fine-structure splitting $2^2P_{1/2}-2^2P_{3/2}$.

The formation and synthesis of the pionium atom $II = (\pi^+ e^-)$ have been observed. The pion has no spin and pionium no hyperfine structure, but the $n=1-n=2$ transition is attractive for an investigation of the pion form factor. An experiment to observe this transition can possibly be made in the future if larger intensities of such particles are available. Obviously high intensities of exotic particles open the way also to artificial bound systems containing two unstable particles like $\mu^+\mu^-$, $\pi^+\pi^-$, $K^+\pi^-$, and K^+K^-, as well as many other combinations. The $\pi^+\mu^-$ atom has been detected from the decay $K_L^0 \to \pi\mu\nu$ where π and μ were emitted with the same velocity and in a small solid angle (Coombes et al., 1973). Another five orders of magnitude in the flux of muons possibly available in a muon collider would result in unprecedented possibilities with low-energy high-intensity muon beams.

Exotic atoms have contributed to our understanding of exotic particles and their binding in an atom, as well as to our knowledge of the structure and deformation of atomic nuclei. Simple pure leptonic and hydrogenlike exotic atoms have provided some of the most precise tests of quantum electrodynamics in bound systems. In simple systems of bound elementary particles, fundamental symmetries in physics have been tested. Exotic atom spectroscopy will continue in the future to be a highly exciting field of research.

Acknowledgments

V. W. Hughes contributed significantly to the preparation of this article. Research support of G.W.F.D. by the Natural Sciences and Engineering Research Council of Canada is gratefully acknowledged.

References

Accad, Y., C. L. Pekeris, and B. Schiff, 1971, Phys. Rev. A **4**, 516, and earlier references therein.
Alguard, M. J., V. W. Hughes, M. S. Lubell, and P. F. Weinwright, 1977, Phys. Rev. Lett. **39**, 334.
Amaldi, U., R. Egidi, R. Marconero, and G. Pizzela, 1969, Rev. Sci. Instrum. **40**, 1001.
Andersen, N., I. V. Hertel, and H. Kleinpoppen, 1984, J. Phys. B **24**, L901.
Andersen, N., J. W. Gallagher, and I. V. Hertel, 1988, Phys. Rep. **165**, 1.
Andersen, N., J. T. Broad, E. E. B. Campbell, J. W. Gallagher, and I. V. Hertel, 1997, Phys. Rep. **278**, 107.
Atomic Physics 16: Proceedings of the 16th International Conference on Atomic Physics (ICAP), 1997 (University of Windsor, Windsor, Ontario, Canada).
Aumayr, F., and H. Winter, 1998, Eds., *Photonic, Electronic and Atomic Collisions: Invited Papers of the 20th International Conference*, Vienna, 1997 (World Scientific, Singapore).
Awe, B., F. Kemper, F. Rosicky, and R. Feder, 1983, J. Phys. B **16**, 603.
Baker, J. D., R. C. Forrey, J. D. Morgan, R. N. Hill, M. Jeziorska, and J. Schertzer, 1993, Bull. Am. Phys. Soc. **38**, 1127.
Baker, J. D., D. E. Freund, R. N. Hill, and J. D. Morgan III, 1990, Phys. Rev. A **41**, 1247.
Bederson, B., 1969, Comments At. Mol. Phys. **1**, 41 and 65.
Bederson, B., 1970, Comments At. Mol. Phys. **2**, 160.
Bederson, B., and W. L. Fite, 1968, Eds., *Atomic and Electron Physics: Atomic Interactions*, Methods of Experimental Physics No. 7 (Academic, New York).
Berger, O., and J. Kessler, 1986, J. Phys. B **19**, 3539.
Bethe, H. A., and E. E. Salpeter, 1957, *Quantum Mechanics of One- and Two-Electron Atoms* (Springer-Verlag, New York).
Bhatia, A. K., and R. J. Drachman, 1997, Phys. Rev. A **55**, 1842.
Blum, K., 1996, *Density Matrix Theory and Applications*, second edition (Plenum, New York/London).
Born, M., and E. Wolf, 1970, *Principles of Optics* (Pergamon, Oxford).
Bransden, B. H., 1983, *Atomic Collision Theory* (Benjamin/Cummings, Reading, MA).
Bürgers, A., D. Wintgen, and J.-M. Rost, 1995, J. Phys. B **28**, 3163.
Burke, P. G., 1965, Adv. Phys. **14**, 521.
Burke, P. G., and C. J. Joachain, 1995, *The Theory of Electron-Atom Collisions* (Plenum, New York/London).
Burke, P. G., and H. M. Schey, 1962, Phys. Rev. **126**, 147.
Condon, E. U., and G. H. Shortley, 1935, *The Theory of Atomic Spectra* (Cambridge University, Cambridge, England).
Coombes, R., *et al.*, 1973, Phys. Rev. Lett. **37**, 249.
Dalgarno, A., and J. T. Lewis, 1956, Proc. R. Soc. London, Ser. A **233**, 70.
Dalgarno, A., and A. L. Stewart, 1956, Proc. R. Soc. London, Ser. A **238**, 269.
Datz, S., and P. F. Dittner, 1988, Z. Phys. D **10**, 187.
Deutsch, M., 1951, Phys. Rev. **82**, 455.
Drachman, R. J., 1993, in *Long-Range Casimir Forces: Theory and Recent Experiments on Atomic Systems*, edited by F. S. Levin and David Micha (Plenum, New York), p. 219, and earlier references therein.

Drake, G. W. F., 1988, Can. J. Phys. **66**, 586.
Drake, G. W. F., 1993a, Adv. At., Mol., Opt. Phys. **31**, 1; 1994, *ibid.* **32**, 93.
Drake, G. W. F., 1993b, in *Long-Range Casimir Forces: Theory and Recent Experiments on Atomic Systems*, edited by F. S. Levin and D. A. Micha (Plenum, New York), p. 107.
Drake, G. W. F., 1996, Ed., *Atomic, Molecular, and Optical Physics Handbook* (AIP, New York).
Drake, G. W. F., 1998, in *Encyclopedia of Applied Physics*, edited by George L. Trigg (Wiley-VCH, Weinheim/New York), Vol. 23, p. 121.
Drake, G. W. F., and W. C. Martin, 1998, Can. J. Phys. **76**, 597.
Duké, L. J., J. B. A. Mitchell, J. W. McConkey, and C. E. Brion, 1996, Eds., *19th International Conference on the Physics of Electronic and Atomic Collisions*, Whistler, B.C., Canada, 1995, AIP Conf. Proc. No. 360 (AIP, New York).
Ehrhardt, H., M. Schulz, T. Tekaat, and K. Willmann, 1969, Phys. Rev. Lett. **22**, 89.
Eminyan, M., K. B. MacAdam, J. Slevin, and H. Kleinpoppen, 1973, Phys. Rev. Lett. **31**, 576.
Eminyan, M., K. B. MacAdam, J. Slevin, and H. Kleinpoppen, 1974, J. Phys. **12**, 1519.
Fano, U., 1957, Rev. Mod. Phys. **29**, 74.
Fano, U., 1973, Rev. Mod. Phys. **45**, 553.
Fano, U., and J. Macek, 1973, Rev. Mod. Phys. **45**, 553.
Farago, P., 1974, J. Phys. B **7**, L28.
Farago, P., 1976, *Electron and Photon Interactions with Atoms* (Plenum, New York/London), p. 235.
Fitch, V. L., and J. Rainwater, 1953, Phys. Rev. **92**, 789.
Fite, W. L., and R. T. Brackmann, 1958, Phys. Rev. **112**, 1151.
Fock, V., 1930, Z. Phys. **61**, 126.
Gallager, A., 1996, in *Atomic, Molecular, and Optical Physics Handbook*, edited by G. W. F. Drake (AIP, New York).
Goldman, S. P., 1998, Phys. Rev. A **57**, R677.
Haberland, R., L. Fritsche, and J. Noffke, 1986, Phys. Rev. A **33**, 2305.
Hafner, H., H. Kleinpoppen, and H. Krüger, 1965, Phys. Lett. **18**, 270.
Hanne, G. F., 1996a, Can. J. Phys. **74**, 811.
Hanne, G. F., 1996b, in *Selected Topics on Electron Physics*, edited by D. M. Campbell and H. Kleinpoppen (Plenum, New York/London), p. 57.
Hartree, D. R., 1928, Proc. Cambridge Philos. Soc. **24**, 89, 111.
Heinzen, D. J., 1996, in *Atomic Physics 14*, Proceedings of the 14th International Conference on Atomic Physics, 1995, edited by S. J. Smith, D. J. Wineland, and C. E. Wieman, AIP Conf. Proc. No. 323 (AIP, New York).
Heinzmann, U., and N. A. Cherepkov, 1996, in *VUV and Soft X-ray Photoionization*, edited by U. Becker and D. A. Shirley (Plenum, New York).
Hertel, I. V., and H. Stoll, 1974, J. Phys. B **7**, 570.
Hils, D., and H. Kleinpoppen, 1978, J. Phys. B **11**, L283.
Hils, D., V. McCusker, H. Kleinpoppen, and S. J. Smith, 1972, Phys. Rev. Lett. **29**, 398.
Hippler, R., H. Madeheim, W. Harbich, H. Kleinpoppen, and H. O. Lutz, 1988, Phys. Rev. A **38**, 1662.
Hughes, V. W., and G. zu Putlitz, 1990, in *Quantum Electrodynamics*, edited by T. Kinoshita (World Scientific, Singapore), p. 822.
Hughes, V. W., *et al.* 1960, Phys. Rev. Lett. **5**, 63.
Hylleraas, E. A., 1963, Rev. Mod. Phys. **35**, 421.
Jetzke, S., J. Zaremba, and F. H. M. Faisal, 1989, Z. Phys. D **11**, 63.
Joachain, C. J., 1983, in *Quantum Collision Theory* (North-Holland, Amsterdam).
Kinoshita, T., and D. R. Yennie, 1990, in *Quantum Electrodynamics*, edited by T. Kinoshita (World Scientific, Singapore), p. 1.
Klar, H., and H. Kleinpoppen, 1982, J. Phys. B **15**, 933.
Klar, H., A. C. Roy, P. Schlemmer, K. Jung, and H. Ehrhardt, 1987, J. Phys. B **20**, 821.

Kleinpoppen, H., and V. Raible, 1965, Phys. Lett. **18**, 24.
Krause, H. F., and S. Datz, 1996, in *Advances in Atomic, Molecular, and Optical Physics*, edited by B. Bederson and H. Walther (Academic, New York), Vol. 26, p. 139.
Larsson, M., 1995, Rep. Prog. Phys. **58**, 1267.
Macek, J., and D. H. Jaecks, 1971, Phys. Rev. A **4**, 1288.
Madden, R. P., and K. Codling, 1963, Phys. Rev. Lett. **10**, 516.
Marton, L. L., and Patrick Richard, 1980, Eds., *Atomic Physics; Accelerators*, Methods of Experimental Physics No. 17 (Academic, New York).
Massey, H. S. W., 1942, Proc. R. Soc. London, Ser. A **181**, 14.
Massey, H. S. W., 1979, *Atomic and Molecular Collisions* (Taylor and Francis, London).
McCarthy, I. E., and E. Weigold, 1991, Rep. Prog. Phys. **54**, 789.
McDaniel, E. W., 1989, *Atomic Collisions* (Wiley, New York).
McEachran, R. P., and A. D. Stauffer, 1986, J. Phys. B **19**, 3523.
Merzbacher, E., 1998, *Quantum Mechanics*, 3rd edition (Wiley, New York), Chap. 13.
Miller, J. D., R. A. Cline, and D. J. Henzen, 1993, Phys. Rev. Lett. **71**, 2204.
Mills, A. P., Jr., and S. Chu, 1990, in *Quantum Electrodynamics*, edited by T. Kinoshita (World Scientific, Singapore), p. 774.
Mott, N. F., 1929, Proc. R. Soc. London, Ser. A **124**, 425.
Mott, N. F., 1932, Proc. R. Soc. London, Ser. A **135**, 429.
Mott, N. F., and H. S. W. Massey, 1965, *The Theory of Atomic Collisions* (Clarendon, Oxford).
Oppenheimer, J. R., 1927, Fortschr. Phys. **43**, 27.
Oppenheimer, J. R., 1929, Phys. Rev. **32**, 361.
Pachucki, K., 1998, J. Phys. B **31**, 2489, 3547.
Penney, W. G., 1932, Proc. Natl. Acad. Sci. USA **18**, 231.
Percival, I. C., and M. J. Seaton, 1958, Philos. Trans. R. Soc. London, Ser. A **251**, 113.
Rubin, K., B. Bederson, M. Goldstein, and R. E. Collins, 1969, Phys. Rev. **182**, 201.
Sapirstein, J., 1998, Rev. Mod. Phys. **70**, 55.
Schrödinger, E., 1926a, Ann. Phys. (Leipzig) **79**, 361, 489, 734.
Schrödinger, E., 1926b, Ann. Phys. (Leipzig) **80**, 437.
Schrödinger, E., 1926c, Ann. Phys. (Leipzig) **81**, 109.
Schrödinger, E., 1926d, Phys. Rev. **28**, 1049.
Schulz, G. J., 1963, Phys. Rev. Lett. **10**, 104.
Schulz, G. J., 1973, Rev. Mod. Phys. **45**, 378.
Schwartz, C., 1961, Phys. Rev. **123**, 1700.
Shore, B. W., 1967, Rev. Mod. Phys. **39**, 439.
Shull, G. G., C. T. Chase, and F. E. Meyers, 1943, Phys. Rev. **135**, 29.
Skinner, H. W. B., and E. T. S. Appleyard, 1927, Proc. R. Soc. London, Ser. A **117**, 224.
Slater, J. C., 1928, Phys. Rev. **32**, 339, 349.
Smith, K., 1966, Rep. Prog. Phys. **29**, 373.
Standage, M. C., and H. Kleinpoppen, 1976, Phys. Rev. Lett. **36**, 577.
Storry, C. H., and E. A. Hessels, 1998, Phys. Rev. A **58**, R8.
Thorsheim, H. R., J. Weiner, and P. Julienne, 1987, Phys. Rev. Lett. **58**, 2420.
Ulrich, M., *et al.*, 1998, in *Photonic, Electronic, and Atomic Collisions*, edited by F. Aumayr and H.-P. Winter (World Scientific, Singapore), p. 421.
Vane, C. R., S. Datz, H. F. Krause, P. F. Dittner, E. F. Deveney, H. Knudson, P. Grafstroem, R. Schuck, H. Gao, and R. Hutton, 1997, Phys. Scr. **73**, 167.
Wannier, G., 1953, Phys. Rev. **90**, 817.
Weingartshofer, A., J. K. Holmes, G. Candle, E. M. Clarke, and H. Krüger, 1977, Phys. Rev. Lett. **39**, 269.
Whelan, C. T., and H. R. J. Walters, 1997, Eds., *Coincidence Studies of Electron and Photon Impact Ionization* (Plenum, New York/London).
Yalim, H. A., D. Crejanovic, and A. Crowe, 1997, Phys. Rev. Lett. **79**, 2951.

Yamazaki, T., 1992, in *Atomic Physics 13*, Proceedings of the Thirteenth Internation Conference, Munich, 1992, edited by T. W. Hänsch, B. Neizer, and H.-O. Walther (AIP, New York), p. 325.
Yan, Z.-C., and G. W. F. Drake, 1998, Phys. Rev. Lett. **81**, 774.
Yang, C. N., 1948, Phys. Rev. **54**, 93.

Laser Spectroscopy and Quantum Optics
T. W. Hänsch and H. Walther

Laser spectroscopy and quantum optics are both fields that have experienced a tremendous development in the last twenty years. Therefore the survey must be incomplete, and only a few highlights are touched on here.

Frontiers of Laser Spectroscopy

Although spectroscopy has taught us most of what we know about the physics of matter and light, it has repeatedly been overshadowed by seemingly more glamorous pursuits in physics, only to reemerge with unexpected vigor and innovative power. The advent of widely tunable and highly monochromatic dye lasers around 1970 ushered in a revolution in optical spectroscopy that has redefined the purpose and direction of the field. Intense highly monochromatic laser light has not only vastly increased the sensitivity and resolution of classical spectroscopic techniques, it has made possible many powerful new techniques of nonlinear spectroscopy. Laser spectroscopists are no longer just looking at light; they are using laser light as a tool to manipulate matter and even to create new states of matter.

Today, laser spectroscopy has found applications in most areas of science and technology. Voluminous monographs, textbooks, and conference proceedings are devoted to the subject (see Hänsch and Inguscio, 1984; Demtröder, 1996), tens of thousands of research papers have been written, and it has become impossible to do justice to the state, impact, and prospects of the field in an article of this scope. We can only discuss some areas, chosen by personal taste, where advances in laser spectroscopy appear particularly rapid, interesting, and promising. As a caveat we must keep in mind that any attempts at foretelling the future are almost certain to overestimate progress in the short term, only to severely underestimate advances in the more distant future, which will result from less predictable but important discoveries and new ideas.

Laser sources and other tools of the trade

In the past, advances in laser spectroscopy have often been driven by technological progress. On this technical side, we are now witnessing the emergence of such an abundance of new tunable sources and other sophisticated optoelectronic tools that we "old-timers" often feel that we may have entered the field too early. Aided by applications such as data storage, laser printing, telecommunications, and materials processing, researchers are able to devote large resources, for instance, to the development of diode lasers. Wavelength selection with integrated Bragg gratings or with tuning elements in an extended external cavity can turn such lasers into perfect tools for high-resolution spectroscopy. Tapered diode laser amplifiers or injection-seeded wide-stripe lasers yield substantial average power and can replace much more costly and complex dye-laser systems. Arrays of diode lasers operating at many different wavelengths, blue gallium nitride diode lasers, and quantum cascade lasers covering the important mid-infrared spectral region are further enhancing the arsenal of spectroscopic tools.

The generation of harmonics, sum or difference frequencies in nonlinear crystals, which can extend the wavelength range into the blue and ultraviolet or the infrared spectral region, is gaining much in conversion efficiency and versatility with quasi-phase-matched nonlinear crystals. This was first proposed in the sixties by N. Bloembergen and has now been realized with ferroelectric crystals such as periodically doped lithium niobate.

We are also witnessing dramatic progress with diode-pumped solid-state lasers and fiber lasers. Commercial frequency-doubled Nd:YAG or Nd:YLF lasers are now replacing power-hungry argon ion lasers as pump sources for tunable Ti:Sapphire lasers or dye lasers. With single-frequency output, such lasers have also pumped prototypes of continuous-wave optical parametric oscillators to generate widely tunable highly monochromatic infrared radiation with impressive efficiency.

Spectacular advances are coming from the frontier of femtosecond laser sources. Kerr-lens mode locking of Ti:Sapphire lasers, chirped pulse amplification, and linear or nonlinear pulse compression have created tabletop sources of intense ultrashort pulses with pulse lengths down to just a few optical cycles. Such lasers are revolutionizing the study of ultrashort phenomena in condensed-matter physics, molecular physics, and even in biological science. Nonlinear frequency conversion gives access to a very wide spectral range, from the submillimeter wavelengths of terahertz radiation to the soft x rays generated as high-order harmonics in gas jets. With direct diode pumping of laser crystals or fibers, shoebox-sized self-contained femtosecond laser systems are becoming a reality.

By greatly reducing the cost, size, and complexity of tunable laser systems, we can move sophisticated laser spectroscopic techniques from the research laboratory to the "real world." At the same time, we can now conceive new, more demanding research experiments, which may employ an entire "orchestra" of tunable lasers sources, directed and controlled in sophisticated ways by a personal computer.

Ultrasensitive spectroscopy

From its beginnings, laser spectroscopy has far surpassed classical spectroscopic techniques in sensitivity, resolution, and measurement accuracy, and we are witnessing unabated progress in these three directions. Advances in sensitivity have often led to unforeseen new applications. When laser-induced fluorescence of single sodium atoms was first observed at Stanford University in the early seventies, nobody anticipated that laser-induced fluorescence spectroscopy would be combined with scanning optical near-field microscopy to study single molecules in a solid matrix on a nanometer scale, that the fluorescence of a rotating molecule would reveal the superfluidity of small helium clusters, as recently demonstrated by J. P. Toennies in Göttingen, or that fluorescence spectroscopy of dye-labeled molecular fragments could greatly speed up the sequencing of DNA. Such examples provide persuasive arguments for curiosity-driven research. The observation of the laser-induced fluorescence of single trapped ions, first accomplished by P. Toschek and H. Dehmelt in Heidelberg around 1978, has opened a rich new regime of fascinating quantum physics to exploration. Extreme sensitivity is also offered by a growing number of other laser techniques, such as resonant photoionization, intracavity absorption spectroscopy, cavity ringdown spectroscopy, bolometric detection of energy deposition in a molecular beam, or sophisticated modulation and cross-modulation techniques, such as the resonator-enhanced conversion of frequency modulation to amplitude modulation, perfected by J. Hall in Boulder. Some obvious applications include spectroscopy of weak molecular vibrational overtones or the detection of trace impurities for industrial materials control, environmental research, or medical diagnostics.

Spectral resolution and accuracy—precision spectroscopy of atomic hydrogen

The progress in resolution and measurement accuracy is perhaps best illustrated with our own work on precision spectroscopy of atomic hydrogen, which began at Stanford in the early seventies and is continuing at Garching since 1986. The simple hydrogen atom has played a central role in the history of atomic physics. Its regular visible Balmer spectrum has been the key to deciphering the laws of quantum physics as we describe them today with quantum electrodynamics (QED), the most successful theory of physics. Spectroscopists have been working for more than a century to measure and compare the resonance frequencies of hydrogen with ever increasing resolution and accuracy in order to test basic laws of physics and to determine accurate values of the fundamental constants (Bassani, Inguscio, and Hänsch, 1989).

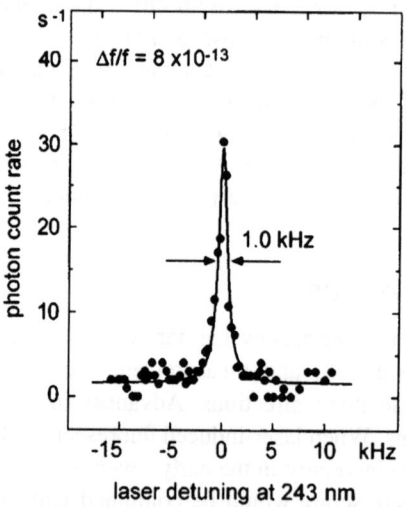

Figure 1. Doppler-free two-photon spectrum of the hydrogen 1S-2S resonance, observed by exciting a cold atomic beam with an optical standing wave at 243 nm. The ratio of linewidth to frequency is less than one part in a trillion.

One of the most intriguing resonances is the two-photon transition from the $1S$ ground state to the metastable $2S$ state with a natural linewidth of only 1.3 Hz, which has inspired advances in high-resolution spectroscopy and optical frequency metrology for almost three decades. By Doppler-free two-photon spectroscopy of a cold atomic beam, we are now observing this transition near 243 nm with a linewidth of 1 kHz, as illustrated in Fig. 1. The resolution of better than one part in 10^{12} corresponds to the thickness of a human hair compared to the circumference of the equator of the earth. Recently, we have measured the absolute optical frequency of this resonance to 13 decimal places. For the hyperfine centroid of the $1S$-$2S$ interval, we find a frequency $f = 2\,466\,061\,413\,187.34 \pm 0.84$ kHz, which is now one of the most accurately measured optical frequencies and the highest frequency that has been compared with the microwave frequency of a Cs atomic clock.

Figure 2 illustrates the progress in the accuracy of optical hydrogen spectroscopy over the past century. Classical spectroscopists were limited to accuracies of a part in 10^6, at best, by the large Doppler broadening due to the random thermal motion of the light atoms, which blurs and masks the intricate fine structure of the spectral lines. This barrier was overcome in the early seventies by the advent of highly monochromatic tunable dye lasers and nonlinear techniques of Doppler-free spectroscopy, such as saturation spectroscopy, as introduced by T. Hänsch and C. Bordé, following earlier work of W. Lamb, W. Bennett, and A. Javan, or polarization spectroscopy, first demonstrated by C. Wieman and T. Hänsch. In both methods, a gas sample is traversed by a strong saturating laser beam and a counterpropagating probe beam so that slow atoms can be selected. If the laser frequency is tuned to the center of a Doppler-broadened line, the two beams can interact with the same atoms, which can only be those at rest or moving sideways. V. Chebotaev in Novosibirsk was the first to realize that Doppler broadening in two-photon spectroscopy can be eliminated without any need to select slow atoms by exciting the

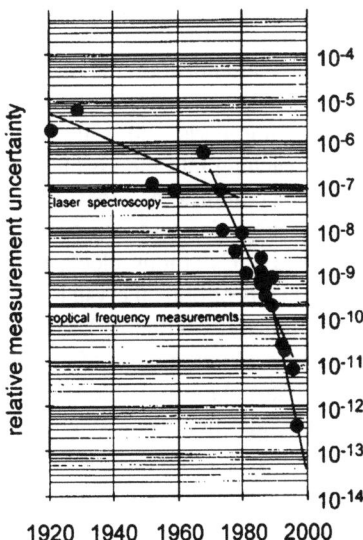

Figure 2. Advances in the accuracy of optical spectroscopy of atomic hydrogen during the 20th century.

atoms with two counterpropagating photons whose frequencies add to the atomic resonance frequency so that the Doppler shifts cancel to first order.

In the late eighties, laser spectroscopists reached another barrier, the limits of optical wavelength interferometry. Unavoidable geometric wave-front errors make it practically impossible to exceed an accuracy of one part in 10^{10}. In the early nineties, this problem was finally overcome in Garching and Paris with new precision experiments measuring the frequency of light with so-called frequency chains.

These new hydrogen experiments at Garching have already yielded a precise new value for the Lamb shift of the $1S$ ground state, which represents now the most stringent test of QED for an atom, a Rydberg constant that is now the most accurately measured fundamental constant, and a deuteron structure radius (from the H-D $1S$-$2S$ isotope shift) that disagrees with the measurements by electron scattering at large accelerators, but is about ten times more accurate and in good agreement with recent predictions of nuclear few-body theory. Future experiments with antihydrogen by two international collaborations at CERN may unveil conceivable small differences in the resonance frequencies or gravitational acceleration of matter and antimatter.

Optical frequency metrology

Extrapolation of the exponential advances shown in Fig. 2 suggests that laser spectroscopic experiments will soon reach the accuracy limit of a part in 10^{14}, imposed by the definition of the second in terms of 9-GHz hyperfine splitting of the Cs ground state. New microwave Cs atomic clocks using fountains of laser-cooled atoms, as first proposed by C. Wieman and S. Chu and now developed by C. Salomon in Paris and elsewhere, are pushing this limit to a few parts in 10^{15}. Much sharper resonances have been observed at optical frequencies with atomic hydrogen or other cold atoms and molecules or with laser-cooled trapped ions, as first proposed by H. Dehmelt and perfected in the laboratory of D. Wineland in Boulder and elsewhere. Accuracies of a part in 10^{17} or better should be achievable with future atomic clocks based on such optical transitions. These clocks will open a new era of precision metrology and fundamental physics tests, and they will likely find important applications in telecommunication, navigation, geological sciences, space research, and astronomy.

Only three laboratories (in Garching, Paris, and Braunschweig) have so far constructed harmonic laser frequency chains reaching into the visible spectrum. Traditional harmonic laser frequency chains are large and complex, and they are typically designed to measure just one particular optical frequency. In essence, such a chain synthesizes a high frequency, starting from a microwave Cs atomic clock, by repeated stages of harmonic generation in nonlinear circuit elements or optical crystals, boosting the feeble power after each step with a phase-locked "transfer oscillator." Many different and often delicate technologies are required to traverse a wide region of the electromagnetic spectrum.

New tools are now emerging that make it possible to design much more compact and versatile optical frequency counters and frequency synthesizers, using small and reliable solid-state components. In Garching, we have demonstrated a phase-locked chain of frequency interval dividers, which can reduce any frequency interval by cutting it repeatedly in half until it is small enough to be measured as a beat note with a fast photodetector. [A frequency interval divider stage receives two input laser frequencies f_1 and f_2 and it forces its own laser to oscillate at the precise midpoint, $f_3 = (f_1 + f_2)/2$. This is accomplished by generating the sum frequency $f_1 + f_2$ and the second harmonic $2f_3$ in nonlinear crystals and by observing a beat signal between these two frequencies. This beat frequency is then forced to zero or some chosen well-known local oscillator frequency with the help of a digital feedback circuit that controls the frequency and phase of the midpoint laser. All stages of the Garching divider chain are nearly identical and employ a small grating-tuned diode laser at visible or near-infrared wavelengths.]

To measure an absolute optical frequency, one may start with a laser frequency f and its second harmonic $2f$. After n bisections, the frequency difference f is reduced to $f/2n$. Once this interval is in the range of a few THz, it can be precisely measured with an optical comb generator, i.e., a fast electro-optic modulator in an optical cavity that produces a wide comb of modulation sidebands with a spacing precisely known from the driving modulation frequency. Even broader combs of precisely spaced longitudinal modes can be generated with a mode-locked femtosecond laser. In a direct comparison of a comb generator and an interval divider chain we have recently demonstrated that both these gears for optical clockworks of the future can work flawlessly over long periods, without losing even a single optical cycle. It is now worth dreaming about new experiments in high-resolution laser spectroscopy of atoms, ions, and molecules that will take advantage of the coming ability to measure any laser frequency quickly with extreme precision.

Manipulating matter with light

The ability to use intense laser light to manipulate matter has long fascinated laser spectroscopists. Even the preparation of spin-polarized gases by optical pumping, known from the work of A. Kastler at Paris in the fifties, has gained new importance for applications such as magnetic-resonance imaging of lungs and blood vessels with spin-polarized helium and xenon, as first demonstrated by W. Happer at Princeton. Atoms can be excited to high Rydberg states which approach macroscopic dimensions and have opened rich new fields of physics, such as the study of cavity QED or quantum chaos by S. Haroche in Paris, D. Kleppner in Cambridge, or H. Walther in Garching. Laser manipulation of molecules has advanced from the excitation of vibrational wave packets with ultrashort pulses to "coherent control" introduced by K. Wilson in San Diego and others, in which a specific quantum state is prepared with an elaborate sequence of radiation fields, similar to music creating a particular mood in our mind. The binding potential of a molecule can be drastically altered with an intense laser field once the resonant transition rate between two electronic states exceeds the vibrational frequency, and efforts are now underway to observe new man-made bound states in which the internuclear distance can be changed with the dressing laser wavelength. At even higher intensities, hot dense plasmas can be produced, with electrons reaching relativistic energies, and one can even hope to observe nonlinear optical phenomena such as photon-photon scattering in vacuum.

Laser cooling and trapping

On the other end of the energy scale, laser cooling has provided a tool to slow atoms to such low temperatures that the thermal De Broglie wavelength assumes macroscopic dimensions (Arimondo, Phillips, and Strumia, 1992; Adams and Riis, 1997). The original proposal for laser cooling of atomic gases by T. Hänsch and A. Schawlow in 1974 was motivated by the quest for higher resolution in hydrogen spectroscopy. Half a year later, H. Dehmelt and D. Wineland made a similar proposal for laser cooling of trapped ions.

If atoms are illuminated from all directions with laser light tuned below an atomic resonance, a moving atom sees oncoming light Doppler-shifted into resonance so that the radiation pressure exerts a strong viscous damping force. Such "optical molasses" was first realized by S. Chu in the eighties, and experiments of W. Phillips in Gaithersburg soon reached temperatures even lower than expected. J. Dalibard and S. Chu were the first to explain this violation of Murphy's law with an additional more subtle cooling mechanism involving light shifts and optical pumping between Zeeman sublevels that comes into play at very low temperatures. Even the remaining limit imposed by the recoil energy due to the emission of a single photon has since been overcome by C. Cohen-Tannoudji in Paris and S. Chu and M. Kasevich at Stanford by pumping the atoms into a velocity-selective dark state that no longer interacts with the light field.

Laser-cooled atoms can easily be trapped in the nodes or antinodes of an optical standing wave by the forces experienced by a light-induced electric dipole in an electric field gradient. Two- and three-dimensional atomic lattices bound by light, as first demonstrated by A. Hemmerich in Munich in 1992, represent an intriguing state of matter, combining the crystalline order of a solid with the density of a good vacuum. Unlike the periodic potential that electrons experience in a solid, the optical potential can be controlled and modulated from the outside at will, and phonons that obscure interesting coherent quantum phenomena in solid state physics are essentially absent. Such optical lattices are therefore intriguing quantum laboratories and provide a rich new playground for laser spectroscopy. High-resolution Raman spectroscopy reveals the vibrational levels of the atoms in their microscopic optical traps, and optical Bragg scattering has been used to probe the long-range order and to study the extension and dynamics of the trapped atomic wave packets. Other phenomena already observed in optical lattices include transport phenomena via Levy flights, Bloch oscillations, quantum tunneling, and quantum chaos. And the deposition of such laser-manipulated atoms on solid surfaces provides a promising tool for the creation of functional nanostructures.

One of the most fascinating phenomena created and studied with laser spectroscopic techniques is the Bose-Einstein condensation (BEC) of cold atomic gases, as discussed elsewhere in this volume by C. Wieman *et al.* If cold atoms are captured in a magnetic trap, and if fast energetic atoms are allowed to escape in "evaporative cooling," the phase-space density of the remaining thermalizing atoms can increase until the wave packets of neighboring atoms begin to overlap, and a large fraction of the atoms condense in the lowest vibrational state, thus losing their identity. Very recently, D. Kleppner has succeeded, after 20 years of effort at MIT, in producing large condensates of spin-polarized atomic hydrogen, which are observed by laser spectroscopy of the $1S-2S$ two-photon resonance.

BEC of atomic gases has already become a very active interdisciplinary field. Condensed-matter scientists are fascinated by the possibility of studying degenerate quantum gases at low densities with interactions dominated by simple two-body s-wave scattering. Such research will likely lead to new insights into correlation and damping effects or quantum transport phenomena such as superfluidity. Atomic spectroscopists, on the other hand, are intrigued by a far-reaching analogy between Bose-Einstein condensation and the creation of laser light. Their efforts may lead to atom lasers as intense sources of coherent matter waves, opening new opportunities in atom optics and atom interferometry.

From laser spectroscopy to quantum optics

With each new tool, we are extending the reach of laser spectroscopy. The field is in an excellent position to respond to societal pressures and put its emphasis on applied research with marketable products and human benefits as immediate goals. If such a policy had been pursued in the past, however, science and mankind would be much poorer today. Sometimes, seemingly useless and unrelated discoveries have been combined to create unexpected new science and powerful new technologies. And curiosity-driven research has greatly enriched human culture. In this wealthiest period that our planet has ever known, society must continue to make resources available to researchers dedicating their lives to such useless pursuits as the study of parity violation in atoms and molecules, tests of special and general relativity, the search for a dipole moment of the electron, for hypothetical exotic particles, or for conceivable slow variations of fundamental constants.

Atomic physicists have sometimes been pitied by their colleagues because their research is limited to studying a periodic system of just 92 elements. However, they are now becoming "quantum engineers," creating new systems that explore the boundaries between quantum physics and classical physics, which provide incredibly sensitive sensors for rotation or acceleration, or which can function as gates for cryptography or the still elusive quantum computers. The possibilities of such quantum engineering are limited only by human imagination and skill, and laser spectroscopy is sure to play a central role in this exciting endeavor.

Quantum Optics: Recent Advances and Trends

In this part of the paper we shall review recent advances and trends in the field of quantum optics. As for the laser spectroscopy section, it would be presumptuous to assume that complete coverage of the subject were possible. The field of quantum optics has developed tremendously under the influence of the technological progress in laser sources as well as in signal detection. Therefore the survey must remain incomplete and mainly influenced by the taste of the authors. The field of quantum optics covers quantum phenomena in the radiation-atom interaction in the broadest sense. It therefore provides interesting tools for testing basic quantum features and it is also an arena in which to illustrate and elucidate quantum effects which occasionally appear to be counterintuitive.

Introduction

The majority of processes in laser physics can be understood on the basis of semiclassical physics, where the atom is quantized but the radiation field treated classically. Therefore, looking back at the development of quantum optics during the late seventies or even the beginning of the eighties, we find that the only phenomena for which the quantization of the radiation field was important were almost exclusively phenomena related to spontaneous emission. In this way the linewidth of a laser and the related phase diffusion, which is caused by spontaneous emission processes in the laser medium, could only be calculated on the basis of a quantized field. Another related phenomenon was, of course, resonance fluorescence: the spectra of monochromatically driven atoms and the photon statistics of fluorescence radiation of a single atom show sub-Poissonian statistics and antibunching, which are both pure quantum features (Cresser *et al.*, 1982).

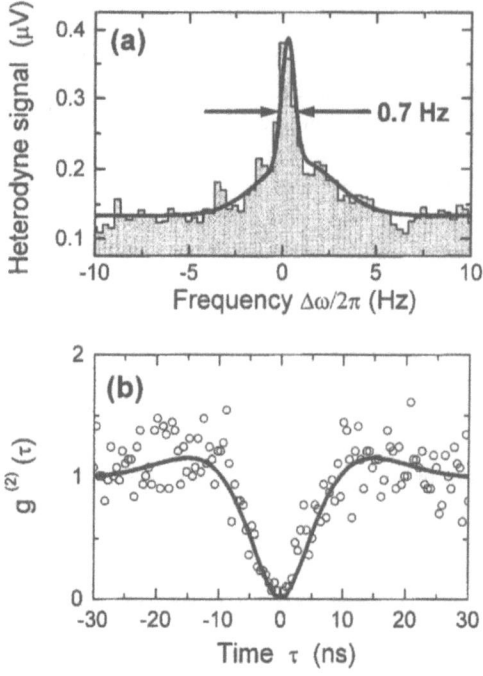

Figure 3. Resonance fluorescence of a single ion. Part (a) of the figure shows the spectrum measured in a heterodyne experiment. The reemitted fluorescence radiation at low intensities is monochromatic. The linewidth is limited by phase fluctuations of the light beam in the laboratory air. The same radiation was investigated in a Hanbury-Brown and Twiss experiment (b). This setup measures the intensity correlation or the probability that a second photon follows at time τ after the first one. The measurement shows antibunching (anticorrelation of the photon detection events). The statistics of the photons is sub-Poissonian, both being nonclassical properties. Depending on the observation, the radiation shows "wave" (a) or "particle" (b) character. The result is thus a nice demonstration of complementarity. See Höffges et al. (1997) for details.

With their experiments on intensity or photon correlations in the mid fifties Hanbury-Brown and Twiss directed the attention of physicists to the question of photon statistics. It was found that photon statistics of normal thermal light show photon bunching, referring to the fact that photons are counted with statistical fluctuations greater than would be expected on the basis of purely random (that is, Poisson) statistics. Photon bunching arises from the Bose-Einstein distribution. A coherent photon field, as represented by laser light, shows Poisson statistics, indicating that the photons arrive randomly. A qualitative classical explanation of photon bunching is sometimes made by saying that light from any natural source arises from broadband multimode photon emission by many independent atoms. There are naturally random periods of constructive and destructive interference among the modes, giving rise to large intensity "spikes," or "bunches" of photons, in the light beam. Unbunched light comes from a coherently regulated collection of atoms, such as from a well-stabilized single-mode laser. From this point of view, unbunched coherent light is optimally ordered and corresponds to a classical coherent wave.

However, as mentioned above, in a quantum treatment the statistics of a light beam can also show photon antibunching. For "antibunched" light, photons arrive with lower statistical fluctuations than predicted from a purely coherent beam and represent therefore a nonclassical light. The first observation of an antibunched beam was accomplished by L. Mandel, H.-J. Kimble and others in 1977 in connection with the above-mentioned resonance fluorescence of a single atom (Cohen-Tannoudji, 1977; Cresser et al., 1982). Antibunching occurs in such light for a very simple reason. A single two-level atom "regulates" the occurrence of pairs of emitted photons very severely, even more so than the photons are regulated in a single-mode laser. A second fluorescent photon cannot be emitted by the same two-level atom until it has been reexcited to its upper level by the absorption of a photon from the main radiation mode. The significance of photon statistics and photon counting techniques in quantum optics and in physics is clear from this discussion. They permit a direct examination of some of the fundamental distinctions between the quantum-mechanical and classical concepts of radiation (see Fig. 3).

The experimental situation with respect to the observation of quantum phenomena changed drastically during the eighties. Owing to the progress of experimental techniques it became possible to realize many of the "gedanken" experiments that were previously only found in quantum mechanics textbooks or discussed in respective courses. Furthermore during the last decade many new ideas were developed, allowing the control of fundamental quantum phenomena. This changed the picture of quantum optics completely. Some of the important developments will be summarized in the following sections; owing to the lack of space, only a few main topics can be touched upon.

Cavity quantum electrodynamics, micromasers, and microlasers

The simplest and most fundamental system for studying radiation-matter coupling is a single two-level atom interacting with a single mode of an electromagnetic field in a cavity. This system received a great deal of attention shortly after the maser was invented, but at that time, the problem was of purely academic interest as the matrix elements describing the radiation-atom interaction are so small that the field of a single photon is not sufficient to lead to an atom-field evolution time shorter than other characteristic times of the system, such as the excited-state lifetime, the time of flight of the atom through the cavity, and the cavity mode damping time. It was therefore not possible to test experimentally the fundamental theories of radiation-matter interaction such as the Jaynes-Cummings model predicting amongst other effects (Meystre, 1992; Walther, 1992) (a) a modification of the spontaneous emission rate of a single atom in a resonant cavity; (b) oscillatory energy exchange between a single atom and the cavity mode; and (c) the disappearance and quantum revival of optical (Rabi) nutation induced in a single atom by a resonant field.

The situation changed drastically when tunable laser light allowed the excitation of highly excited atomic states, called Rydberg states. Such excited atoms are very suitable for observing quantum effects in radiation-atom coupling for three reasons. First, the states are very strongly coupled to the radiation field (the induced transition rates between neighboring levels scale as n^4); second, transitions are in the millimeter wave region, so that low-order mode cavities can be made large enough to allow rather long interaction times; finally, Rydberg states have relatively long lifetimes with respect to spontaneous decay.

The strong coupling of Rydberg states to radiation resonant with transitions between neighboring levels can be understood in terms of the correspondence principle: with increasing n the classical evolution frequency of the highly excited electron becomes identical with the transition frequency to the neighboring level, and the atom corresponds to a large dipole oscillating at the resonance frequency. (The dipole moment is very large since the atomic radius scales with n^2).

In order to understand the modification of the spontaneous emission rate in an external cavity, we have to remember that in quantum electrodynamics this rate is determined by the density of modes of the electromagnetic field at the atomic transition frequency ω_0, which in turn depends on the square of the frequency. If the atom is not in free space, but in a resonant cavity, the continuum of modes is changed into a spectrum of discrete modes, one of which may be in resonance with the atom. The spontaneous decay rate of the atom in the cavity γ_c will then be enhanced in relation to that in free space γ_f by a factor given by the ratio of the corresponding mode densities (Haroche and Kleppner, 1989):

$$\frac{\gamma_c}{\gamma_f} = \frac{\rho_c(\omega_0)}{\rho_f(\omega_0)} = \frac{2\pi Q}{V_c \omega_0^3} = \frac{Q\lambda_0^3}{4\pi^2 V_c},$$

where V_c is the volume of the cavity and Q is a quality factor of the cavity which expresses the sharpness of the mode. For low-order cavities in the microwave region $V_c \simeq \lambda_0^3$ this means that the spontaneous emission rate is increased by roughly a factor of Q. However, if the cavity is detuned, the decay rate will decrease. In this case the atom cannot emit a photon, as the cavity is not able to accept it, and therefore the energy will stay with the atom.

Many experiments have been performed with Rydberg atoms to demonstrate this enhancement and inhibition of spontaneous decay in external cavities or cavitylike structures. More subtle effects due to the change of the mode density can also be expected: radiation corrections such as the Lamb shift and the anomalous magnetic dipole moment of the electron are modified with respect to the free-space value, although changes are of the same order as present day experiments allow us to measure. Roughly speaking, one can say that such effects are determined by a change of virtual transitions and not by real transitions as in the case of spontaneous decay (see the articles in Berman, 1994).

If the rate of atoms crossing a cavity exceeds the cavity damping rate ω/Q, the photon released by each atom is stored long enough to interact with the next atom. The atom-field coupling becomes stronger and stronger as the field builds up and evolves into a steady state. The system is a new kind of maser, which operates with exceedingly small numbers of atoms and photons. Atomic fluxes as small as 100 atoms per second have generated maser action, as could be demonstrated by H. Walther *et al.* in 1985. For such a low flux there is never more than a single atom in the resonator—in fact, most of the time the cavity is empty. It should also be mentioned that a single resonant photon is sufficient to saturate the maser transition.

A scheme of this one-atom maser or micromaser is shown in Fig. 4. With this device the complex dynamics of a single atom in a quantized field predicted by the Jaynes-Cummings model could be verified. Some of the features are explicitly a consequence of the quantum nature of the electromagnetic field: the statistical and discrete nature of the photon field leads to a new characteristic dynamics such as collapse and revivals in the Rabi nutation.

The steady-state field of the micromaser shows sub-Poisson statistics. This is in contrast with regular masers and lasers where coherent fields (Poisson statistics) are observed. The reason that nonclassical radiation is produced is due to the fixed interaction

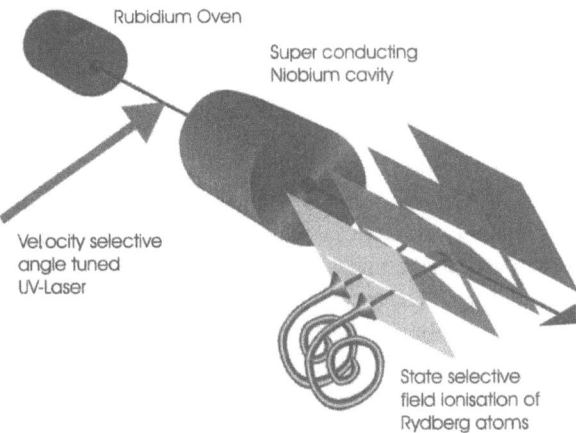

Figure 4. Micromaser setup of Rb Rydberg atoms. The velocity of the atoms is controlled by exciting a velocity subgroup of atoms in the atomic beam. The atoms in the upper and lower maser levels are selectively detected by field ionization.

time of the atoms leading to a careful control of the atom-field interaction dynamics (Meystre, 1992; Berman, 1994; see also the article on lasers by W. E. Lamb *et al.* on pp. 442–459 in this book where also a micromaser with ultracold atoms is discussed).

The micromaser also led to interferometry experiments and inspired many gedanken experiments on using two cavities for Ramsey-type interferometry (Briegel *et al.*, 1997).

Recently S. Haroche, J. M. Raimond, *et al.* succeeded in realizing a Schrödinger cat state in a cavity (Haroche, 1998). Studies on the decoherence of this state could be carried out. The experiments are another demonstration of the boundaries between the quantum and the classical world.

There is an interesting equivalence between an atom interacting with a single-mode field and a quantum particle in a harmonic potential, as was first pointed out by D. F. Walls and H. Risken; this connection results from the fact that the radiation field is quantized on the basis of the harmonic oscillator. Therefore the Jaynes-Cummings dynamics can also be observed with trapped ions, as demonstrated recently in a series of beautiful experiments by D. Wineland *et al.* They also produced a Schrödinger cat state by preparing a single trapped ion in a superposition of two spatially separated wave packets, which are formed by coupling different vibrational quantum states in the excitation process (see the contribution by C. E. Wieman *et al.* on pp. 426–441 of this book).

Besides the experiments in the microwave region, a single-atom laser emitting in the visible range has also been realized by M. Feld *et al.* Furthermore, cavity quantum electrodynamic effects have been studied in the optical spectral region by J. L. Kimble *et al.* (Berman, 1994).

Today's technology in the microfabrication of semiconductor diode structure allows the realization of low-order cavity structures for diode lasers. In these systems the spontaneous emission is controlled in the same way as in the micromaser. Since spontaneous decay is a source of strong losses, the control of this phenomenon leads to highly efficient laser systems. The quantum control of spontaneous decay has thus important consequences for technical applications (Yamamoto and Slusher, 1993; see also the contribution by R. S. Slusher on pp. 798–811 in this book).

Quantum interference phenomena

Interference occurs in classical optics when two or more wave amplitudes are added with different phases. As a classical particle does not have a phase, only waves can give rise to interference in classical physics. By contrast, interference is a general feature in quantum mechanics which is not limited to waves, but shows up whenever the outcome of a measurement can be arrived at via several indistinguishable paths, the probability amplitudes of which must be added to calculate the result of the measurement. Therefore both particle and wave aspects manifest themselves in the quantum-mechanical description of interference. The particle aspect of a photon is apparent in that it cannot be detected at two separate positions at the same time, i.e., the detection of a photon at one point eliminates the probability of detecting the photon at any other point; therefore classical optics, which treats light only in terms of waves, cannot explain some photon interference effects.

In order that photon interference experiments can be performed, photon pairs have to be generated. This has been done using spontaneous parametric down conversion or parametric fluorescence. In this process an ultraviolet "pump" photon decays inside the crystal into two red photons, called the signal and idler, which are highly correlated. Interferences are observed when both photons can reach a first and a second detector and when the different paths are indistinguishable. The setup introduced by L. Mandel and co-workers has been used to investigate a series of quantum phenomena including photon tunneling, the quantum eraser, the electron-paramagnetic-resonance paradox, and Bell inequalities (Chiao, Kwiat, and Steinberg, 1994; Chiao and Steinberg, 1998; Shih, 1998). With type-II parametric down conversion entangled pairs of photons can also be generated. They are the basis for the experiments performed in connection with quantum information to be discussed later in this paper (see also the contribution of A. Zeilinger in this issue).

Quantum nondemolition measurements

In quantum systems the process of measurement of an observable introduces noise; therefore successive measurements of the same observable yield different results in general. A typical example from the field of quantum optics that is useful to illustrate this phenomenon is the detection of the field variables by photon counting techniques, which are, of course, field destructive, i.e., the field is modified as a consequence of the measurement. For general applications it is desirable to find schemes that avoid back action during a measurement. A possible way to perform such a quantum nondemolition measurement is to hide the noise introduced in the measurement in a conjugate observable which is not of particular interest in that case. The original idea of quantum nondemolition was introduced theoretically in connection with gravitational wave detection by Braginski in 1974 and Thorne in 1978 (Braginsky and Khalili, 1992); the first implementation was performed in quantum optics. Several schemes were implemented in the eighties and early nineties. The first experiment to measure a light field was performed by I. Imoto, H. A. Haus, and Y. Yamamoto. It is based on the optical Kerr effect whereby the change of the refractive index depends on the intensity of a transmitted beam. A weak probe wave therefore experiences a phase shift which is proportional to the intensity of the signal wave.

Many other schemes along the same lines were implemented. Two very special ones should be mentioned here: single photons in cavities were measured by S. Haroche *et al.* (Berman, 1994) employing the dispersive level shift of probing atoms; furthermore, the nonlinearity of cold trapped atoms was used by Grangier *et al.*, leading to extremely small excess noise.

Related to the topic of quantum nondemolition measurements is the topic of squeezed radiation, both fields actually developing in parallel and cross-fertilizing each other.

Squeezed radiation

A classical electromagnetic field consists of waves with well-defined amplitude and phase. This is not the case in a quantum treatment: fluctuations are associated with both conjugate variables. The case of a coherent state that most nearly describes a classical electromagnetic field has an equal amount of uncertainty in the two variables (when normalized to the field of a single photon). Equivalently the field can be described in two conjugate quadrature components and the uncertainties in the two conjugate variables satisfy the Heisenberg uncertainty principle. The coherent state represents a minimum uncertainty state with equal uncertainties in the two quadrature components. This case is usually called the shot-noise limit.

In a quantum treatment of radiation it is also possible to generate states that are not present in the classical limit. They can show fluctuations reduced below the classical limit in one of the quadrature components, while the canonically conjugate quadrature component must display enhanced fluctuations in order to fulfill the Heisenberg uncertainty principle. Those states are called "squeezed states." An electromagnetic field with fluctuations below the standard quantum limit in one of the quadrature components has in principle many attractive applications, e.g., in optical communication, in precision and sensitive measurements such as gravitational wave detection, or in noise-free amplification. Therefore there has been great interest in generating squeezed radiation. Optical measurements have three characteristics that enable them to reach the quantum noise level more readily than in other fields of physics: (1) optical signals are naturally immune to external sources of noise; (2) thermal noise at room temperature is negligible in the optical domain; (3) the outstanding equalities of the optical sources and detectors allow a very-low-level instrumental noise. The first observation of squeezing was achieved by R. S. Slusher *et al.* in 1987 in an experiment of parametric generation involving four-wave mixing in sodium vapor.

Later many other systems were proposed and realized also involving laser oscillators. The field is still strongly progressing, yet technical applications in any of the above-mentioned fields have not yet evolved (for a review, see Berman, 1994).

Measurement of quantum states

In quantum theory all information on a quantum system is contained in the wave function. It is only in recent years that the theoretical knowledge and the experimental technology were acquired to prepare a specific system such as light or atoms in a particular quantum state. The preparation of squeezed states discussed above is one example. The ability to "engineer" quantum states offers fascinating possibilities for testing elementary predictions of quantum mechanics. Furthermore there are practical implications involved; specifically designed laser fields (e.g., squeezed light) may allow high-precision interferometry owing to a possible reduction of quantum noise. The preparation of particular molecular states could lead to a detailed control of chemical reactions. Engineering of states also requires a successful measurement. The wave function of a quantum state cannot, of course, be determined in a single measurement. Therefore the initial conditions in a measurement have to be reproduced repeatedly in order to sample the data. There has been impressive progress in theoretical as well as in experimental work in developing strategies to reconstruct the various quasiprobability distributions and other representations of a quantum state. Since pointlike sampling in quantum-mechanical phase space is forbidden by the uncertainty principle, a tomographic method is used, circumventing this problem by investigating thin slices, circular discs, or infinitesimal rings of the quasiprobability distribution in phase space. In this way quasiprobability distributions (the Wigner function or Q function) are obtained as a representation of the quantum state.

The first reconstruction of a Wigner function was demonstrated for a light field by M. Raymer *et al.* Since the quantization of the radiation field is performed in analogy to the harmonic oscillator, the dynamics of a single mode of light correspond to the dynamics of a quantum particle in a harmonic potential. In particular, the amplitudes of the magnetic and electric fields are the conjugate variables corresponding to position and momentum. Hence phase-space considerations can immediately be transferred to light. The actual analysis of the light field is performed by a homodyne detector. The unknown light is mixed at a beam splitter with a strong coherent laser field. The signal in the two arms is analyzed when the phase of the laser light is shifted, leading to the reconstruction of the Wigner function. More recently the group of D. Wineland at NIST in Boulder managed to determine the vibrational quantum state of a single ion stored in a Paul trap.

Quantum state measurements is a new and fascinating area of research that opens an important new window on the quantum world. It has now become possible to extract complete information about an elementary quantum system such as a single mode of light, a molecule, or a single trapped particle and to determine the wave function of such a system.

Quantum information

Quantum physics provides means for processing and transmitting information that differ fundamentally from classical physics. It turns out that information theory and quantum mechanics fit together very well. The entanglement of quantum objects and the inherent quantum nonlocality as the basis of the Einstein-Podolsky-Rosen paradox and the Bell inequalities forms the essential new ingredient that distinguishes quantum from classical information theory. The new disciplines can essentially be subdivided into three overlapping areas: quantum computation, quantum cryptography, and quantum communication (Zeilinger *et al.*, 1998).

Quantum computers as introduced by D. Deutsch in 1985 make use of the special properties of quantum theory. The binary information is stored in a quantum object. In principle, any two-state system can be used as a quantum bit (qu bit). Examples used in experiments include the polarizations of photons, the orientations of electron and nuclear spins, and energy levels of atoms or quantum dots. For simplicity we shall limit the discussion here to a two-level atom. Apart from the information that the atom is in either an upper or lower state, superpositions of the states are also possible. The reason why quantum computers are faster is their ability to process quantum superpositions of many numbers in one computational step, each computational step being a unitary transformation of quantum registers. The ability of particles to be in a superposition of more than one quantum state naturally introduces a form of parallelism that can, in principle, perform some traditional computing tasks faster than is possible with classical computers. In recent years a new quantum theory of computation has been developed and it has been shown that the computer power grows exponentially with the size of the quantum register. An open question is whether it would ever be possible or practical to build physical devices to perform such computations or whether they would forever remain theoretical curiosities. Quantum computers require a coherent, controlled evolution of the quantum system and a certain period of time to complete the computations. Quantum states are notoriously delicate, and it is not clear at the moment whether the system could be isolated sufficiently from environmental influences to prevent the original states from being destroyed (Williams and Clearwater, 1997; Steane, 1998).

Quantum logic gates perform unitary operations on qu bits, and in order to implement them it is necessary to perform a unitary transformation on one physical subsystem conditioned upon the quantum state of another one. A basic operation of this sort is the quantum-controlled NOT gate. The special feature of this operation is that it transforms superpositions into entanglements. This transformation can be reversed by applying the same controlled NOT operation, again being equivalent to a Bell measurement. Controlled NOT operation has been demonstrated for a single trapped ion by D. Wineland *et al.*, and for an atom in an optical cavity by H. J. Kimble *et al.*

To bring atoms into an entangled state, strong coupling between them is necessary. This can be realized in experiments described in the chapter on cavity quantum electrodynamics in Englert *et al.* (1998). These experiments achieve an entanglement between cavity field and atoms through the strong coupling between both. Therefore entanglement between subsequent atoms via the field is also possible. Another way to entangle atoms is to use ion traps. The ions are harmonically bound to their position of rest. If they are in the quantum limit of their motion, an entanglement between vibrational and electronic excitation may be generated with the help of laser pulses. It is even possible to entangle different ions when a collective vibration is excited. A linear ion chain can thus be an

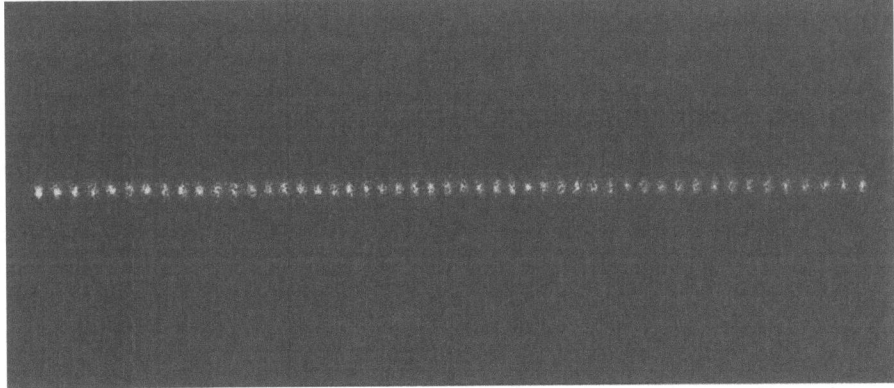

Figure 5. Linear chain of ions in a trap. The position of the ions is marked through the fluorescence radiation. They are constantly emitting when illuminated by a laser beam. The laser beam is used simultaneously to cool the ions so that they stay in an ordered configuration. Such an ion chain has been proposed as the register of a quantum computer. (See Color Plate 5.)

ideal quantum register. The common center-of-mass oscillation of the ion chain along the axis is used to entangle different ions (see Fig. 5).

While the experimental realization of a quantum computer is still lagging behind, the realization of elements of quantum cryptography and quantum communication is much closer. In quantum cryptography, or quantum key distribution, the secret key is sent encoded in single-photon pulses, so that eavesdroppers will leave their mark behind due to the Heisenberg uncertainty principle when they try to read the information, which makes them immediately noticeable. The theory of quantum cryptography has been worked out in quite some detail (Bennett, 1992), and practical quantum channels have been realized by the use of optical fibers and low-intensity light pulses containing, on average, fewer than one photon. Current state-of-the-art techniques provide kilobit-per-second transmission rates with errors at the level of 1%, this being sufficiently low for practical purposes. The main problem, however, is the noise in the avalanche photodiodes used as simple photon detectors today, with the photodiodes still showing very bad performance, especially in the region of 1.55 μm where the optical fibers have their smallest losses.

Quantum information, another widely discussed new field, also offers very interesting new possibilities. We shall mention just two: quantum dense coding and quantum teleportation. The term dense coding describes the possibility of transmitting more than one bit by manipulating only a single two-state particle. This is possible when sender and receiver share a pair of entangled photons. The information that can be coded in one of the photons is twice as large as in the case of a classical information channel. An entangled photon pair also plays the decisive role in the case of teleportation, experimentally demonstrated by the groups of F. de Martini and A. Zeilinger and recently also by H. J. Kimble (see also the contribution by A. Zeilinger in this issue). Teleportation enables information to be sent on a particular quantum state, e.g., of a photon, by sending classical information containing the result of a joint Bell state measurement performed on the photon to be teleported and one auxiliary photon provided by the entangled photon pair. This measurement projects the other photon of the pair into a quantum state uniquely related to the original one owing to entanglement. With the result of this measurement transmitted classically to the receiver, a transformation of the auxiliary state is performed to reproduce the original quantum state.

Conclusion—quantum optics

The examples given in this review demonstrate that the field of quantum optics has been branching out tremendously in recent years. Whereas the pure quantum phenomena observed in the late seventies were very few and mostly connected to spontaneous emission processes, the field has now broadened to include many basic phenomena of quantum physics, for example, the measurement process and the preparation of quantum states leading to a deeper understanding of quantum physics and the peculiarities of the quantum world. On the other side it is also obvious that important new technical applications may evolve. It is still too early to make predictions about applications like quantum computing, but it is already clear that quantum cryptography will be a useful tool. Another development is the microlaser which resulted from the studies of cavity quantum electrodynamics. It has proven to be an efficient laser system, which is already in use in optical communication.

References

Adams, C. S., and E. Riis, 1997, Prog. Quantum Electron. **21**, 1.
Arimondo, E., W. D. Phillips, and F. Strumia, 1992, Eds., *Laser Manipulation of Atoms and Ions, Proceedings of the International School of Physics Enrico Fermi, Course CXVIII* (North-Holland, Amsterdam).
Bassani, G. F., M. Inguscio, and T. W. Hänsch, 1989, Eds., *The Hydrogen Atom* (Springer, Berlin).
Bennett, C. H., 1992, Science **257**, 737.
Berman, P., 1994, Ed., *Cavity Quantum Electrodynamics* (Academic, New York).
Georgia Southern University, 1998, Bose-Einstein Condensation Homepage, http://amo.phy.gasou.edu/bec.html/
Braginsky, V. B., and F. Ya. Khalili, 1992, *Quantum Measurement* (Cambridge University, Cambridge, England).
Briegel, H., B.-G. Englert, M. O. Scully, and H. Walther, 1997, in *Atom-Interferometry*, edited by P. Berman (Academic, New York), p. 217.
Chiao, R. Y., P. G. Kwiat, and A. M. Steinberg, 1994, in *Advances in Atomic, Molecular, and Optical Physics*, edited by Benjamin Bederson and Herbert Walther (Academic, New York), Vol. 34, p. 35.
Chiao, R. Y., and A. M. Steinberg, 1998, in *Progress in Optics*, edited by E. Wolf (North-Holland, Amsterdam), Vol. 37, p. 345.
Cohen-Tannoudji, C., 1977, in *Frontiers in Laser Spectroscopy*, edited by R. Balian, S. Haroche, and S. Libeman (North-Holland, Amsterdam), Vol. 1, p. 3.
Cresser, J. D., J. Häger, G. Leuchs, M. Rateike, and H. Walther, 1982, in *Dissipative Systems in Quantum Optics*, edited by R. Bonifacio, Topics in Current Physics No. 27 (Springer, Berlin), p. 21.
Demtröder, W., 1996, *Laser Spectroscopy: Instrumentation and Basic Concepts*, 2nd edition (Springer, Heidelberg).
Englert, B. G., M. Löffler, O. Benson, B. Varcoe, M. Weidinger, and H. Walther, 1998, Fortschr. Phys. **46**, 6.
Hänsch, T. W., and M. Inguscio, 1994, Eds., *Frontiers in Laser Spectroscopy, Proceedings of the International School of Physics Enrico Fermi, Course CXX* (North-Holland, Amsterdam).
Haroche, S., 1998, Phys. Today **51**, 36.
Haroche, S., and D. Kleppner, 1989, Phys. Today **42**, 24.
Höffges, J. T., H. W. Baldauf, W. Lange, and H. Walther, 1997, J. Mod. Opt. **44**, 1999.
Leonhardt, U., 1997, *Measuring the Quantum State of Light* (Cambridge University, Cambridge, England).
Meystre, P., 1992, in *Progress in Optics*, edited by E. Wolf (North-Holland, Amsterdam), Vol. 30, p. 261.
Reynaud, S., H. Heidmann, E. Giacobino, and C. Fabre, 1992, in *Progress in Optics*, edited by E. Wolf (North-Holland, Amsterdam), Vol. 30, p. 1.
Schleich, W. P., and M. Raymer, 1997, Eds., Quantum State Preparation and Measurement, Special Issue, J. Mod. Opt. **44**.
Shih, Y., 1998, in *Advances in Atomic, Molecular, and Optical Physics*, edited by Benjamin Bederson and Herbert Walther (Academic, New York), Vol. 41, p. 1.
Steane, A., 1998, Rep. Prog. Phys. **61**, 117.
Walther, H., 1992, Phys. Rep. **219**, 263.
Williams, C. P., and S. H. Clearwater, 1997, *Explorations in Quantum Computing* (Springer, Berlin).
Yamamoto, Y., and R. S. Slusher, 1993, Phys. Today **46**, 66.
Zeilinger, A., D. Deutsch, A. Ekert, W. Tittel, G. Ribordy, N. Gisin, D. DiVincenzo, and B. Terhal, 1998, Phys. World **11**, 33.

Atom Cooling, Trapping, and Quantum Manipulation

Carl E. Wieman, David E. Pritchard, and David J. Wineland

Modern atomic, molecular, and optical physics has advanced primarily by using known physics to devise innovative techniques to better isolate and control the atomic system, and then exploiting this nearly ideal system to achieve higher precision and discover new physical phenomena. The striking advances along these lines have been recognized by awards of Nobel prizes to 21 individuals in this area; most recently, the 1997 Nobel prize was given for laser cooling and trapping of neutral atoms (Phys. Today, 1997). In the first half of the twentieth century, the Stern-Gerlach magnet, and later optical pumping, allowed the preparation and analysis of internal quantum numbers. Resonance techniques allowed the quantum state to be changed controllably, and methods such as Ramsey's separated oscillatory fields, and spin echos created and exploited coherent superposition of internal quantum states. This control of internal states ultimately led to the invention of the maser and the laser. For a brief discussion of what might be called "Rabi physics," see the article by Kleppner in this volume.

This paper discusses the extension of this pattern of control and study to the *external* degrees of freedom (position and velocity) that has occurred in the last few decades. The strong forces of electric and magnetic fields on ions allow them to be trapped with high kinetic energy, and once trapped they can be cooled in various ways. The forces available to trap neutral atoms are much weaker. In order to trap them, they must first be cooled below 1 K by radiation pressure that cannot exceed 1 meV/cm for a strong resonant transition. For cold atoms, trapping has been achieved using resonant radiation pressure and/or forces from field gradients acting on either the atoms' magnetic moments or their induced electric dipole moments. The latter force is produced by the electric field of a near-resonant, tightly focused laser beam. All of these traps have maximum depths, expressed in terms of temperature, on the order of 1 K for practical situations.

Traps, together with cooling methods that have achieved kinetic temperatures as low as nanokelvins, have now created the ultimate physical systems thus far for precision spectroscopy, frequency standards, and tests of fundamental physics. Atomic collisions are qualitatively different, and the cooling has produced new states of matter: ion liquids, crystals, and (neutral) atomic Bose-Einstein condensates. Atoms have been placed in the lowest quantum state of the confining trap potentials, and coherent superpositions of translational states have been created, often entangled with the internal quantum states. These provide novel test beds for quantum mechanics and manipulation of quantum information.

Manipulating Position and Velocity

Cooling

Radiation pressure arises from the transfer of momentum when an atom scatters a photon. Radiation pressure cooling uses the Doppler shift with light tuned just below the atomic resonance frequency (Wineland and Itano, 1987; Wieman and Chu, 1989). Atoms that are moving towards the light will see the light Doppler shifted nearer to resonance, and hence will scatter more photons than slower atoms. This slows the faster atoms and compresses the velocity distribution (i.e., cooling the atom sample). A single laser beam is sufficient to cool a sample of trapped atoms or ions; however, free atoms must be irradiated with laser beams from all directions. For atoms with velocities that cause Doppler shifts comparable to the natural transition width (typically several meters per second), this "optical molasses" takes just microseconds to cool to the "Doppler limit." This limit is somewhat under 1 mK for a typical, allowed, electric dipole transition.

A variety of methods have been found for cooling isolated atoms and ions to lower temperatures (Wineland and Itano, 1987; Wieman and Chu, 1989; Cohen-Tannoudji and Phillips, 1990). These include sub-Doppler laser cooling, evaporative cooling, and laser "sideband" cooling. Sub-Doppler laser cooling uses standing-wave laser beams that give rise to potential energy hills and valleys due to the spatial variations in the atom's ac Stark shift. As an atom moves up a hill, it loses kinetic energy. Near the top of a hill, optical excitations tend to reorient the atom relative to the local field so that it then sees that location as a potential valley. This efficiently transfers kinetic energy into photon energy, and can cool atoms into the microkelvin range, a few times the recoil energy/temperature gained by the scattering of a single photon. Even lower temperatures can be achieved by evaporative cooling of trapped atoms. The process is analogous to the way a cup of hot coffee cools down by giving off the most energetic molecules as steam. As the energetic atoms are removed from the trap, collisions readjust the remaining atoms into a lower temperature thermal distribution. Trapped atoms have been evaporatively cooled to 50 nK by precisely controlling the removal of the energetic atoms and making traps with very good thermal isolation. Cooling with resolved sidebands (Wineland and Itano, 1987) is a straightforward realization of the principle that a laser transition can simultaneously change the internal and motional quantum states of a trapped atom. For example, if it is sufficiently tightly bound in a harmonic potential, the atom's optical spectrum has resolvable Doppler-effect-generated frequency-modulation sidebands. Absorption on a lower sideband reduces the atom's motional state energy; if the atom's recoil energy is smaller than the motional quanta, overall cooling occurs when the photon is reemitted. This has been used to cool small numbers of trapped ions and atoms to the ground state of the confining potential with high efficiency.

Atom optics

Given an ensemble of atoms localized in phase space, a growing cadre of techniques, collectively called "atom optics," have been developed for manipulating atoms with full retention of their quantum coherence (Pritchard, 1991; Adams, Sigel, and Mlynek, 1994). The most salient feature of atom optics is the small size of atomic de Broglie wavelengths relative to optical wavelengths—an order of magnitude smaller for atoms with submillikelvin temperatures, and four orders of magnitude smaller for room-temperature atoms.

Preserving atomic coherence for such wavelengths is a major experimental challenge. However, these short wavelengths also suggest that atom optics offers possibilities for precise measurements and subnanometer fabrication. The principal applications of atom optics have been in the creation and use of atom interferometers (Berman, 1997), and for atom lithography—the deposition of precise patterns of atoms on surfaces (Thywissen et al., 1997).

The principal tools of atom optics have been light forces and nanofabricated mechanical structures, and the major technique has been diffraction. If a highly collimated atom wave crosses a standing wave of near-resonant light at right angles, the spatially periodic variation of the light-atom interaction potential energy causes a corresponding variation in the local de Broglie wave number. This diffracts the atom wave like a phase grating in classical optics, where the diffraction orders correspond to the successive absorption and stimulated emission of a pair of photons traveling in opposite directions. The resultant momentum transfer to the atoms is twice the photon momentum. When the atom waves travel through a thick standing wave, one diffracted order predominates. For a thin standing wave, the diffraction pattern is spread over many orders. Blazed gratings have also been demonstrated, as well as Raman and adiabatic-dark-state gratings in which the diffraction is accompanied by a specific change in the internal state of the atoms.

Mechanical diffractive structures (Keith et al., 1988) differ qualitatively from light gratings. They are purely amplitude gratings (with concomitant loss of intensity), species independent (do not depend on internal structure), and can have periods several times smaller than light. They can be made with arbitrary patterns using electron-beam lithography. Examples include spherical and cylindrical zone plates; a combination of lens and hologram (Morinaga et al., 1996), which produced an atom image with 10^4 resolution elements; and a sieve for "sizing" molecules (Luo et al., 1996).

There are no achromatic partially reflecting mirrors for atoms, hence diffraction gratings have been pressed into service as beam splitters and combiners for atom waves. These have been used to make a variety of atom (and even molecule) interferometers since 1991. Nearly all use "white fringe" designs that compensate for the dependence of diffraction angle on the wavelength of the individual atoms. A majority have used the three-grating configuration (Fig. 1) in which the first grating splits the incident beam, the second reverses the differential momenta given by the first, and the third recombines the two beams at the location where they overlap. Both mechanical and all types of light gratings have been successfully used. Interference fringes have been read out using both detectors sensitive to atom position and detectors of the atom's internal state.

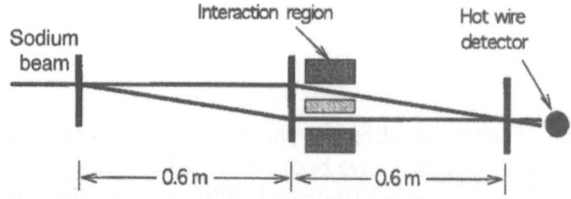

Figure 1. An example of a three-grating atom interferometer. The sodium beam is split up and then later recombined after the interaction region where the atoms in one arm can be perturbed in various ways. In this example, a hot wire detector detects the fringe pattern.

Traps

Trapping atoms can be as simple as putting a gas of atoms in a storage vessel that has walls that inhibit sticking. However, electromagnetic fields can also be configured to confine atoms with much less perturbation to their internal structure and minimal heating from the surrounding environment (Wieman and Chu, 1989; Ghosh, 1995; Newbury and Wieman, 1996). Although Maxwell's equations put severe constraints on how this can be done (for example, Earnshaw's theorem and its optical analog), numerous clever designs have been found, a few of which are presented here. A very useful trap for neutral atoms is the magnetic bottle. Magnetic substates that are attracted to regions of lower field can be trapped if $|\mathbf{B}|$, and thus the trap potential $U(\mathbf{r})$ have a local minimum. One choice of $\mathbf{B}$ that provides such a potential is a quadrupole field, formed by an "anti-Helmholtz" coil (currents in opposite directions in the coils of a Helmholtz pair). This configuration is effective but has the problem that at the center of the coil, the fields vanish. For atoms passing near the center of the coil, the field becomes so small that the magnetic moment alignment is lost with respect to the field direction ("Majorana transitions"), transferring the atoms into untrapped magnetic substates. A popular choice to overcome this problem has been a trap composed of a linear magnetic quadrupole along whose axis is superimposed an axial field with maxima that "close" the ends (the Ioffe-Pritchard configuration). This gives a local, but nonzero minimum in $|\mathbf{B}|$, and hence eliminates the leak in the center.

A trap that works for both neutral and charged atoms relies on the time-averaged force produced by a rapidly oscillating inhomogeneous field. For example, if a charged particle is placed in the center of an oscillating spherical quadrupole field, it is initially pushed outward, taking it to a region of higher field where it will experience a larger push inward when the field has reversed. Thus its micromotion at the field oscillation frequency will cause a nonzero average confining force (the "ponderomotive force") that can be described by a pseudopotential. The Paul trap confines atomic ions using this principle. High-energy particle accelerators and storage rings are another form of the ponderomotive force trap, where the oscillating fields arise from the particles traversing inhomogeneous static fields. Induced dipole-moment optical traps rely on the same pondermotive principle. Another trap for charged particles is the Penning ion trap, which uses static electric and magnetic fields. The magnetic field provides confinement normal to this field, while the electric field confines the particles axially along the magnetic field. It is particularly useful for producing large cold samples (see next section on One-Component Plasmas).

The workhorse of cold neutral atom research is the magneto-optical trap (MOT), because of both its simplicity of construction and its depth (Raab et al., 1987). Radiation pressure from laser beams converging on a center provides the trapping force, but a weak inhomogeneous magnetic field acts as a spatially dependent control on this force. It shifts the magnetic sublevels so that the atoms preferentially absorb the polarized light going toward the trap center. The magnetic field is a spherical quadrupole configuration with gradients of several gauss per centimeter about the zero of the magnetic field, which is the center of the trap. This field has a linear gradient in all three directions, permitting the use of three mutually perpendicular pairs of oppositely circularly polarized laser beams that are detuned a few natural linewidths below a strong atomic transition. Therefore in addition to three-dimensional confinement, the light also provides Doppler and sub-Doppler cooling of the atoms. With only milliwatts of laser power, a typical MOT is 1 K deep, sufficient to capture atoms out of a room-temperature vapor cell.

New Physics from Cold Atoms

One-component plasmas

A collection of trapped ions can be viewed as a "one-component" plasma. By cooling such a plasma to very low temperatures, novel liquid and crystalline plasma states have been created (Fig. 2). These crystals can be regarded as the classical limit of Wigner crystallization, where the wave functions of the ions do not overlap and quantum statistics do not play a role. Instead, the crystallization arises entirely from the balance of the trapping fields and the strong long-range Coulomb repulsion between the ions. The struc-

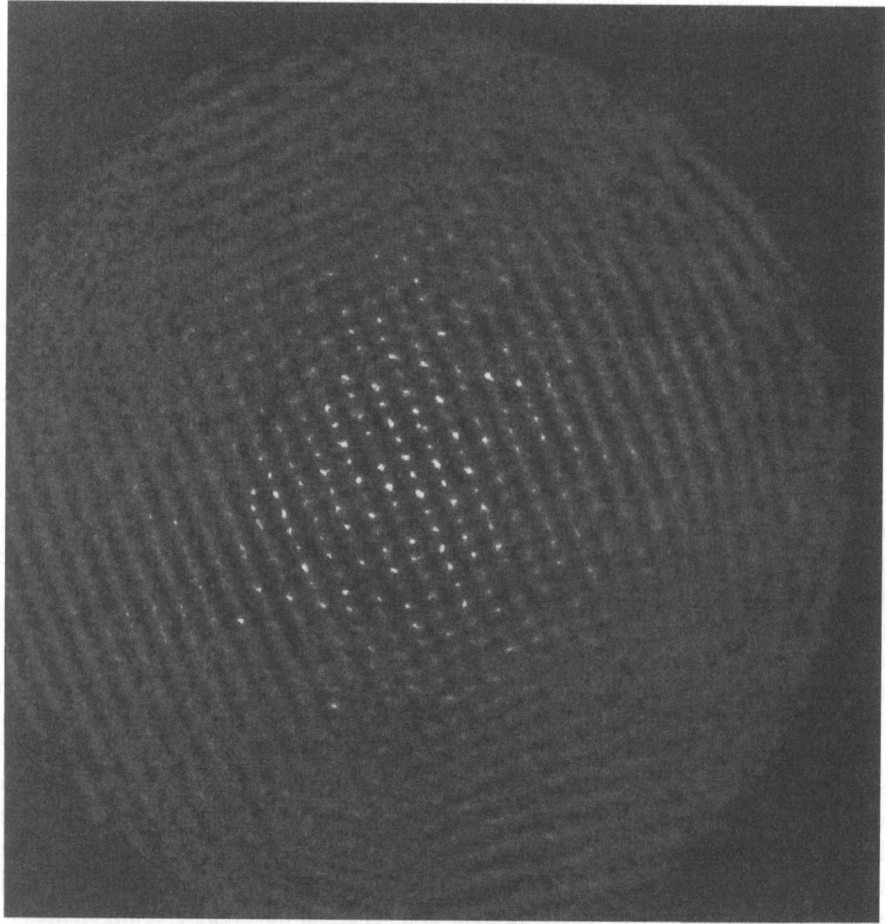

Figure 2. Photograph of a large ion crystal. When atomic ions are trapped and cooled, they form crystals whose minimum energy configurations are determined by a balance between the trap potentials and the ions' mutual Coulomb repulsion. In this photograph, about 2000 laser-cooled beryllium ions are confined in a Penning trap. The ion crystal has a BCC configuration with about a 15 μm spacing between ions. As in all Penning traps, the ions rotate about the trap axis (normal to the center of photo) necessitating stroboscopic imaging. (Image courtesy of John Bollinger, NIST.) (See Color Plate 6.)

ture and formation of these crystals have been studied using Bragg scattering (Itano et al., 1998) and direct imaging of the ions (Walther, 1993; Mitchell et al., 1998).

Bose-Einstein condensation in dilute gases

Perhaps the most exciting physics outcome of cooling and trapping techniques has been the creation of a novel macroscopic quantum system, the Bose-Einstein condensate (BEC), in a dilute gas. Bose-Einstein condensate in a gas was first predicted in 1924 (Einstein, 1924, 1925); as a phase transition it is unique because it is driven only by statistics rather than energetics. Superfluid helium, superconductivity, and certain excitation behavior are all manifestations of BEC in various systems. The condensation in a dilute atomic gas was first achieved (Anderson et al., 1995) by cooling a cloud of trapped rubidium atoms so that they were sufficiently cold (~200 nK) and dense enough that their de Broglie wave packets began to overlap. A large fraction of the atoms then condensed into the ground state of the trapping potential. Bose-Einstein condensate in a gas is proving to be a fascinating new macroscopic quantum system because of the experimental capabilities to manipulate and study it in great detail. Moreover, it is quite amenable to theoretical analysis because the interatomic interactions are relatively weak.

The achievement of BEC (Anderson et al., 1995; Davis et al., 1995) required the combination of many of the techniques for cooling and trapping neutral atoms that had been developed over the previous two decades. It also built on much of the understanding of basic atomic processes at very low temperatures that had been obtained using these techniques. A Bose-Einstein condensate was created (Fig. 3) by first collecting a cloud of laser-cooled atoms in a MOT and cooling them by sub-Doppler laser cooling. At ~10 μK, these were much too hot and dilute for BEC, however. These laser-trapped and cooled atoms were then transferred to a magnetic bottle and evaporatively cooled to

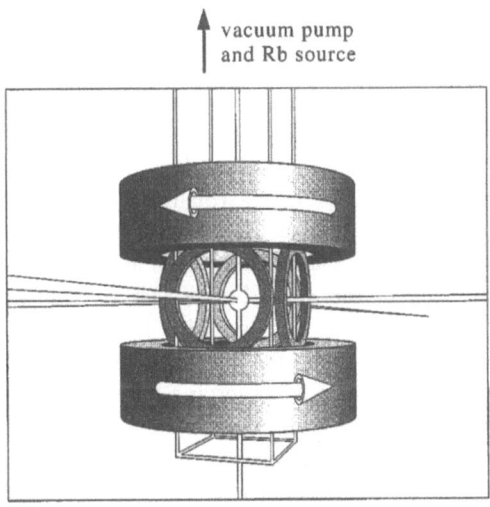

Figure 3. Schematic of the first apparatus used to create BEC in a dilute gas (Anderson et al., 1995). A room-temperature rectangular glass cell 2.5 cm square by 10 cm high is attached to a vacuum pump and rubidium reservoir (not shown). Light from diode lasers comes from all six directions to form a MOT in the middle of the cell. Running current through the magnetic-field coils shown surrounding the cell creates the magnetic trap.

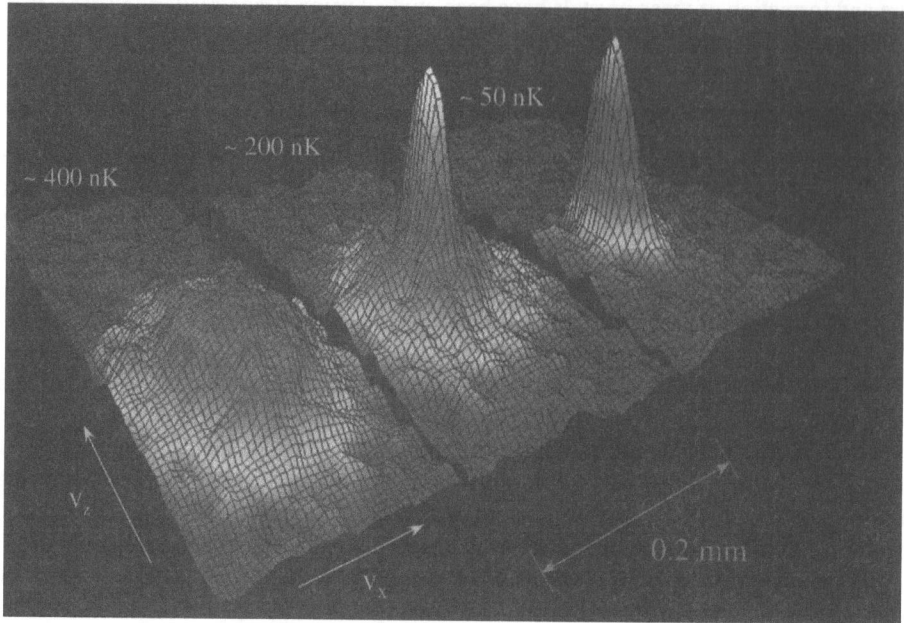

Figure 4. Two-dimensional velocity distributions of cold atomic clouds showing BEC. These are for three experimental runs with different amounts of cooling. The axes are the x and z velocities and the number density of atoms. The distribution on the left shows a gentle hill that corresponds to a temperature of about 400 nK. The middle picture is at 200 nK, and shows the asymmetric condensate spire in the center of the hill. The picture on the right is at about 50 nK and about 90% of the atoms are in the condensate. (Courtesy of M. Matthews, University of Colorado.) (See Color Plate 7.)

below the condensation temperature. Evaporative cooling of magnetically trapped atoms was developed for the pioneering efforts to achieve BEC in gaseous hydrogen (Greytak, 1995). However, it ultimately turned out that laser "precooled" alkali atoms had more favorable collision properties for evaporative cooling (Ketterle and van Druten, 1996). Relative to hydrogen, for every "bad" inelastic collision that causes atoms to be lost from the magnetic trap, there are more "good" thermalizing elastic collisions.[1]

The same convenient optical transitions that provide laser cooling also make it easy to use light scattering to image the cooled alkali clouds and thereby study the condensate. The macroscopic occupation of the ground state that is BEC has been seen both in momentum space, as a peak at zero velocity (Fig. 4), and in real space, as a sudden increase in the density of the atoms in the center of the trap. The condensate images provide a unique opportunity for directly observing the shape of a quantum wave function. An advantage of the inhomogeneous trapping potential is that there is spatial separation of condensed and noncondensed portions of the cloud. This makes it possible to distinguish, manipulate, and study the condensed and noncondensed portions of the cloud separately, as well as to create samples of nearly pure condensate. Adding optical fields to the magnetic traps is proving to be a particularly convenient technique to manipulate

---

[1] After the completion of this article, BEC was reported in a gas of spin-polarized hydrogen (D. Kleppner, private communication).

the condensates and the shape of the confining potential in useful ways (Stamper-Kurn et al., 1998).

There has been an explosion of experimental and theoretical activity in the study of condensates. Initial work considered basic aspects such as the shape of the condensate wave function and how it was distorted by interactions between the atoms. The experimental observations were found to be well described by solutions of the Gross-Pitaevski equation (Gross, 1961, 1963; Pitaevski, 1961) where the self-interaction of the condensate is characterized by a single parameter, the S-wave scattering length. (This is independently measured in cold atom collision experiments.) The validity of this equation has been confirmed over a wide range of interaction strengths by varying the number of atoms in the condensate, and for both positive (repulsive interaction) and negative (attractive interaction) scattering lengths. The fraction of atoms in the condensate and the specific heat, as a function of temperature, have similarly been found to agree very well (within a few percent) with theory.

The dynamical behavior of condensates, including the effects of interactions, has also received considerable study. By modulating the magnetic confining potential, phononlike collective modes of the condensate have been excited. The frequency and damping of these excitations have been studied over a wide range of conditions. For very low temperatures, it was found that the measured resonant frequencies, and their dependencies on the self interaction, agree very precisely with those predicted by the Gross-Pitaevski equation. However, the temperature dependence of the damping and resonant frequencies has proven to be much more difficult to explain theoretically (Jin et al., 1997). These effects, particularly the shifts in frequency, were much larger than simple intuitive models would have suggested, and to explain them requires a more sophisticated treatment of the coupling of condensate and noncondensate phases. This is an area of considerable theoretical and experimental activity.

The various coherence properties of the condensate wave function have been examined in several different ways. The most dramatic was the observation of first-order coherence that occurred when two independent condensates were allowed to pass through each other (Andrews et al., 1997), and the fringes formed in the density distribution as the two-condensate wave functions interfered with each other (Andrews et al., 1997). Third-order coherence (the probability of three-condensate atoms being in the same place) was measured by looking at the rate of three-body recombination in the condensate. This process, in which two atoms bind to form a diatomic molecule and the third atom carries off energy and momentum, is found to be the dominant process by which atoms are lost from the condensate. It was predicted and then confirmed by experiment that the loss rate would be six times lower in a condensate sample than a noncondensate sample of the same density because of the higher-order coherence of the condensate (a lack of spatial fluctuations in the density).

Among the many other areas of study currently underway or planned, studies of multicomponent condensates appear to be among the richest. The static and dynamical behavior of these interacting quantum fluids can be explored by observing both density and phase of the wave functions. A variety of techniques have been demonstrated for creating multicomponent condensates, and the dynamical evolution of the spatial structure and the phases of the wave functions have been studied in these condensates. These techniques include forming two separate condensates of the same type in a double-well trapping potential, and creating condensates in coherent superpositions of different spin states using radio frequency and microwave magnetic fields.

Quantum measurements on single atoms

A single trapped atom affords an opportunity to make repeated measurements on a single quantum system. This provides a display of how an atom will absorb and emit light and exhibit behavior not predicted by the density-matrix formalism which describes ensemble averages. A simple example is an ion with two excited states: one, denoted $|s\rangle$, with strong coupling to the ground state; the other, denoted $|w\rangle$, with weak coupling and a correspondingly long lifetime. If a laser excites the strong transition, strong fluorescence will be observed until the ion is somehow (for example, by spontaneous decay or the action of another laser tuned to the weak transition) transferred to state $|w\rangle$. Then the fluorescence will stop for a time characteristic of the decay time of this state. This behavior is called atom "shelving" and provides a way to tell when the ion is in the $|w\rangle$ state without disturbing it if it is in that state (Fig. 5) (Blatt and Zoller, 1988). Even a single driven two-state atom exhibits interesting correlations: if a fluorescence photon is observed, another photon cannot be emitted for a time on the order of the excited-state lifetime while the excited-state amplitude builds up again. This is called "photon antibunching" and is a purely quantum effect.

Quantum-state engineering

Atom manipulation techniques provide a means to synthesize arbitrary and, in general, entangled quantum states from initially unentangled quantum systems (Monroe and Bollinger, 1997). A simple example is an optical beamsplitter in atom interferometry. When a two-level atom (states g and e) is excited by a laser beam directed perpendicular to its motion, a $\pi/2$ pulse creates an entangled state of the form $2^{-1/2}[\Psi(g,0) + \Psi(e,\hbar k)]$ where the second argument denotes the momentum state along the laser beam direction. Thus the atom has states where the momentum is entangled with the internal state. For an atom confined in a harmonic trap, we can create an entangled state of the form $2^{-1/2}[\Psi(g)\Psi(\alpha) + \Psi(e)\Psi(\alpha')]$, where $\Psi(\alpha)$ and $\Psi(\alpha')$ are coherent states—states that are most nearly classical in that they correspond to Gaussian wave packets that oscillate in the trap without changing shape. This state has been created by first optically pumping and laser-cooling a single ion to its internal and motional ground states. Next, the ion's internal state is placed into a coherent superposition state

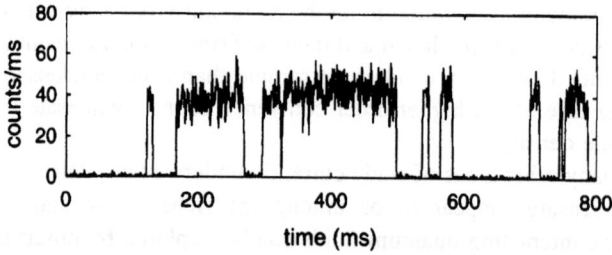

Figure 5. Quantum jumps. The number of fluorescent photons detected in 0.5 ms sampling times from a trapped and cooled mercury ion. The fluorescence corresponds to laser scattering on the transition between the $6s\,^2S_{1/2}$ ground state and $6p\,^2P_{1/2}$ state (level s in the text). With 10^{-7} probability, the $^2P_{1/2}$ state decays to the lower lying ^2D level (level w in text) and fluorescence stops until the ^2D levels decay back to the $^2S_{1/2}$ ground state. By discriminating between fluorescence levels, it is possible to detect the "w" state with nearly 100% efficiency.

$2^{-1/2}[\Psi(g)+\Psi(e)]$ with resonant radiation. Optical dipole forces that are modulated at the ion's trap oscillation frequency create coherent states of motion. These forces can be applied with two different laser beams whose polarizations are chosen to selectively excite first the state $\Psi(g)$ and then the state $\Psi(e)$ with different modulation phases, with the result that each internal state is associated with a different coherent motional state. When $\Psi(\alpha)$ and $\Psi(\alpha')$ correspond to well-separated, localized spatial wave packets (e.g., oscillations out of phase), this state is called a "Schrödinger-cat" state because a classical-like property (the position) is entangled with a quantum property (the internal state) (Monroe and Bollinger, 1997). Schemes exist to generate arbitrary entangled states between many internal and motional states of an atom. These entangled states can be completely characterized by a family of operations that selectively map different parts of the wave function onto a particular internal state that is then detected—so-called tomographic techniques (Leibfried et al., 1998).

Quantum-state engineering methods can be extended from a single atom to entangled states of many atoms, if a suitable coupling mechanism can be found. The strong Coulomb interaction between cold trapped ions provides one such mechanism. Two ions in the sample can be entangled by first entangling the internal state of one ion with a collective mode of the ions' motion using operations similar to the creation of the Schrödinger cat. This mode, which is *shared* among all ions, could be the center-of-mass mode of motion where all ions oscillate together at the frequency of a single ion as described in the previous paragraph. This mode of motion can then be entangled with a second ion, thereby entangling the internal states of the two selected ions. This can be done in such a way that, at the end, the motional state factors out of the wave function, leaving only the ions' internal states entangled (Cirac and Zoller, 1995).

The ideas of quantum computation have provided a useful framework in which to cast these methods. This is because a general computation can be broken down into a series of elementary operations involving single ion ("qubit") internal state rotations and a single type of entangling operation between two ions, and because a "computation" can always be devised to create an arbitrary entangled state. Quantum computation algorithms have recently been shown to be capable of solving certain problems that are intractable on a classical computer, such as factorization of large numbers (Ekert and Jozsa, 1996; Steane, 1998). These algorithms may remain technically unfeasible in the near future because of the fragility of the entangled states; however, more modest algorithms, such as one for efficiently measuring atomic spectral lines (Bollinger et al., 1996), appear to be within reach. Independent of the outcome of quantum computation, quantum state engineering is allowing detailed studies of the ideas of coherence and decoherence in quantum mechanics (as represented by the fragility of entangled states) and, correspondingly, the capabilities and limits of quantum measurement.

Precision Measurements

Spectroscopy and clocks

Trapping, combined with very low temperatures, can lead to very long observation times and suppression of Doppler effects (including time dilation). This leads to very accurate high-resolution spectroscopy. A classic example is provided by the single electrons (and positrons) trapped in Penning traps, by the group of Dehmelt and van Dyck (Dehmelt, 1995; Ghosh, 1995). These experiments effectively measured the ratio of the electron's spin-flip frequency to its cyclotron resonance frequency in the same magnetic field, thereby determining the g factor of the electron to an inaccuracy of less than 5 parts in 10^{12}. The comparison of this measurement with the value predicted by quantum electrodynamics is the most accurate comparison of the experimental value of a quantity to its theoretical value in all of physics.

Another application of spectroscopy is to atomic clocks, where accurate time intervals are realized by counting cycles of radiation that is exactly in resonance with an atomic transition (Bergquist, 1996). The fundamental limit to the measurement resolution is the duration of the observation time. Major advances have been possible using trapped ions and very cold neutral atoms in "atomic fountains." Now, the world's most precise cesium clocks (which define the second) are based on a cesium fountain. In this device, a sample of 10^6 cold (≈ 1 μK) cesium atoms is launched upward. The atoms go up 1 m and then fall back toward the source. At both the beginning and the end of their trajectory (1 s apart), the atoms pass through a microwave cavity. Ramsey's method of separated oscillatory fields is used to drive the hyperfine "clock" transition (≈ 9.2 GHz), achieving a linewidth of about 1 Hz. This narrow linewidth, coupled with the relatively small perturbations on the atoms in free flight, has led to a measurement inaccuracy of only 2 parts in 10^{15}, currently the most precise direct measurement of any physical quantity (Simon *et al.*, 1997).

Since the fields in ion traps act on the ions' overall charge and do not significantly perturb their internal structure, trapped ions can also provide very accurate high-resolution clocks (Fisk, 1997). A linewidth of less than 0.001 Hz has been obtained in a trapped-ion clock (Fig. 6). The advantages of this very narrow linewidth are offset by the fact that trapped-ion experiments typically must use relatively few atoms because of attendant higher velocities (and Doppler shifts) associated with trapping large numbers. Therefore, the performance of the most accurate ion clocks is currently about equal to that of the best cesium clocks (Berkeland *et al.*, 1998).

The most accurate mass spectroscopy is now performed with ions in Penning ion traps (Ghosh, 1995). Ion mass ratios have been determined with an inaccuracy of about 1 part in 10^{10} by measuring the ratio of the cyclotron frequencies for different mass ions in the

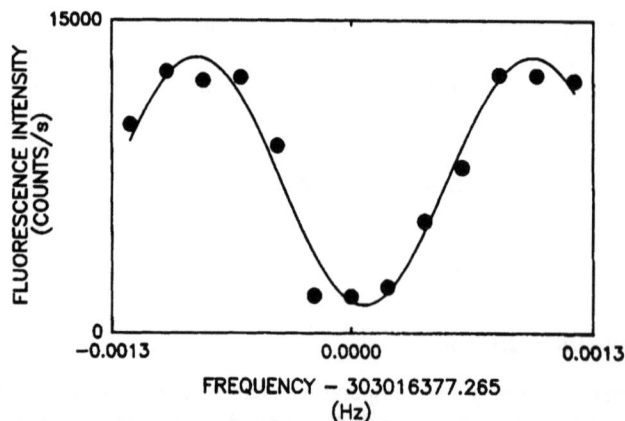

Figure 6. Atom traps enable long observation times and thus high resolution for atomic clocks. The figure shows a resonance curve for a ground-state hyperfine transition (303 MHz) in laser-cooled beryllium ions that is less than 0.001 Hz wide.

same trap. Proton and antiproton masses have been measured to be equal at this level, providing the most stringent test of *CPT* invariance for baryons (Gabrielse et al., 1998).

Inertial measurements with atom interferometers

Phase shifts arise when matter-wave interferometers accelerate, and atom interferometers have proved to be sensitive inertial sensors because such phase shifts generally vary with the mass and inversely with the velocity of the interfering particle (Adams et al., 1994). Because of their high potential sensitivity, atom interferometers are being developed for use as accelerometers, rotation sensors, gravimeters, and gradiometers.

The freely propagating matter waves in an interferometer form fringes with respect to an inertial reference frame. Hence, if the interferometer moves noninertially while the atoms are in transit, the fringes are shifted from the location where they would have been if the interferometer were stationary. For example, if the interferometer in Fig. 1 has acceleration upward, atom waves that pass through the (accelerating) middle grating at the same grating location shown in the figure will form fringes that will be observed a distance $D = a\tau^2$ below the centerline (where $\tau = L/v$ is the time for an atom moving with velocity v to travel the distance L between adjacent gratings). This displacement is observed as a phase shift

$$\varphi_{\text{acceleration}} = 2\pi\left(\frac{-D}{d_g}\right) = -\frac{2\pi}{d_g}\left(\frac{L}{v}\right)^2 a = -\frac{2\pi m^2 \lambda_{\text{dB}} A}{h^2} a, \qquad (1)$$

where d_g is the period of the gratings, $\lambda_{\text{dB}} = h/mv$ is the de Broglie wavelength for an atom with mass m and velocity v, and $A = L^2(\lambda_{\text{dB}}/d_g)$ is the area enclosed by the paths of the interferometer. If the interferometer rotates with angular rate Ω, the resultant Coriolis acceleration $\vec{a} = 2\vec{v} \times \vec{\Omega}$ gives rise to a rotational phase shift,

$$\varphi_{\text{rotation}} = \left[\frac{2\pi}{d_g}\left(\frac{L}{v}\right)^2 2v\right]\Omega = \left[4\pi\frac{mA}{h}\right]\Omega. \qquad (2)$$

The second expression is the usual Sagnac phase shift. This phase shift exceeds that of a light interferometer with the same enclosed area by the ratio, $mc^2/\hbar\omega$, which can exceed 10^{10}.

Atom interferometers have already made dramatic improvements in measurements of phase shifts due to rotation and gravitation relative to earlier measurements with neutron and electron interferometers. Gravitational measurements with an accuracy approaching 10^{-9} g have been made using a laser-cooled-atom interferometer (Kasevich and Chu, 1991). Interferometers using an uncooled atomic beam have measured rotations with a sensitivity of 4 millidegree per hour in a 1 s measurement (Gustavson et al., 1997; Lenef et al., 1997), about 3 orders of magnitude better than measurements with neutron or electron interferometers.

Interferometric measurements of gravity confirm the weak equivalence principle. Orbiting atom interferometers might improve on this important null test. Similarly, orbiting atom rotation sensors should have the sensitivity to test the frame drag predictions of general relativity.

Measurements of atomic and molecular properties

Better determination of atomic properties is another one of the payoffs of the new techniques for atom manipulation. This has led to a more precise spectroscopic determination of many atomic energy levels. In addition, atom interferometers allow sensitive absolute measurements of the perturbations applied to atoms or molecules in one of the two arms. Examples are the determination of the electric polarizability of Na in its ground state (Fig. 7), and the measurement of the phase shift associated with passage of Na and Na_2 waves through a gaseous medium, essentially measuring the matter-wave index of refraction.

Neutral atom traps with their dense submillikelvin samples have revolutionized free-to-bound spectroscopy (also called photoassociative spectroscopy), in which an unbound atom pair is excited to a bound molecular state (Walker and Feng, 1994). The resolution, limited by the thermal energy of the free atoms, has been reduced from hundreds of inverse centimeters to 0.001 cm^{-1}, while the angular momentum of the colliding atoms, which determines the complexity of the molecular rotational spectra, has been reduced from hundreds of $\hbar$ to one or two. Consequently, extremely high-resolution free-bound

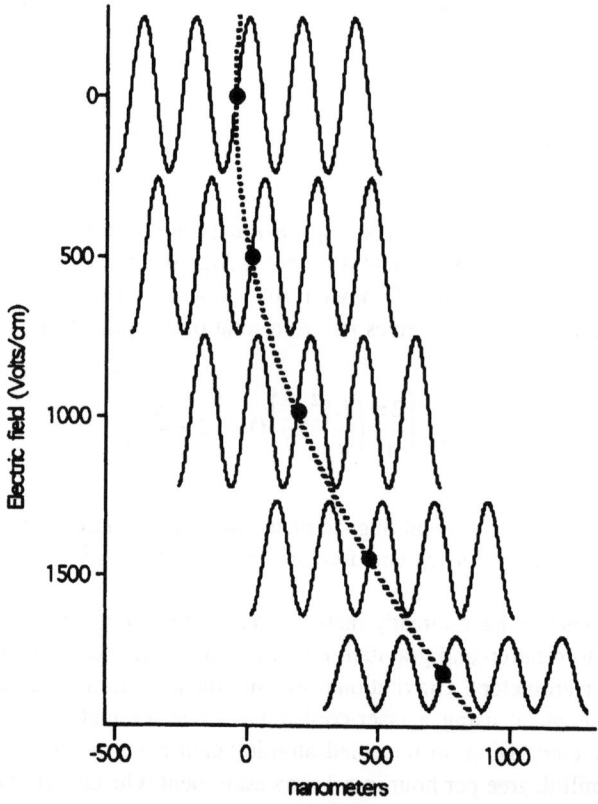

Figure 7. The fringes in a separated-beam interferometer (shown in Fig. 1) shift in proportion to the Stark shift, $V_{Stark} = -1/2\alpha E^2$, where α is the ground-state electric polarizability and E is the electric field applied to the sodium atoms in one side of the interferometer (Ekstrom et al., 1995). The decrease of amplitude at larger phase shift reflects the finite coherence length of the beam. The quadratic fit (dashed line) allows 0.3% absolute determination of α. This exceeds the accuracy of the best theoretical calculations.

spectra have been obtained with resolved hyperfine and rovibrational structure. For the first time it has become possible to study excited states very near to dissociation with corresponding internuclear separations of 2.5 to 10 nm. This has allowed the observation of pure long-range molecules—excited-state molecules that have shallow (e.g., 1 K) well depths with inner potential barriers beyond 2.5 nm. The structure of these molecules is well described by theories based entirely on the properties of the separated atoms. Remarkably, free-bound spectroscopy has allowed determination of oscillator strengths for the important resonance transitions of Na, K, and Rb to 0.1%, better than they can be determined by any other technique. It has also provided improved determinations of the interatomic potentials, and corresponding improvements in the calculations of many atomic collision processes that depend on these potentials.

Extremely weakly bound molecules have been studied using nanofabricated structures. A diffraction grating was used to analyze a collimated beam from a supersonic expansion of cold He gas, showing the existence $^4He_3$ trimers and higher n-mers and conclusively demonstrating the existence of the $^4He_2$ dimer (Schollkopf and Toennies, 1996), which was previously thought not to be bound (Fig. 8). Subsequent measurements of the attenuation of $^4He_2$ by a nanofabricated sieve showed it to have an internuclear separation of 6.5 nm, making it by far the largest ground-state diatomic molecule and also the most weakly bound, with a dissociation energy of only 10^{-7} eV.

Conclusion

We have presented a brief summary of some of the novel techniques for manipulating atoms, and the measurements that these techniques have made possible. Probably the most exciting thing about this field is that many of these techniques have matured from research projects to useful tools only in the past few years. Thus the next decade promises

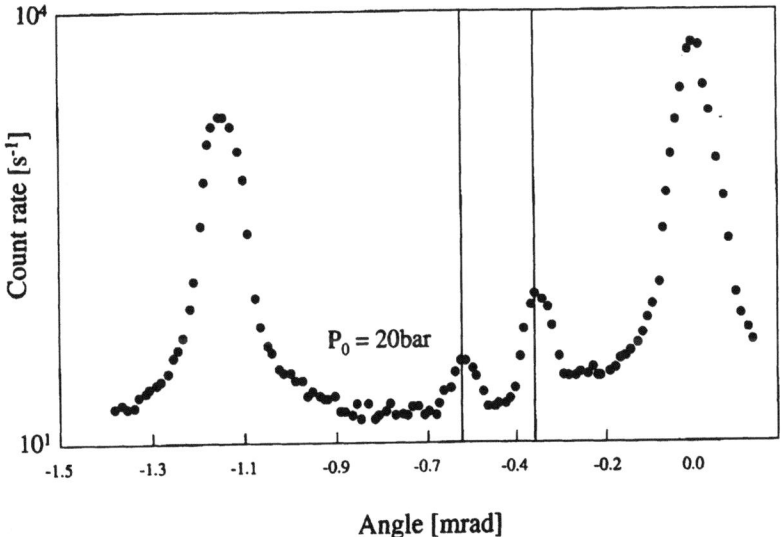

Figure 8. Diffraction pattern for a helium cluster beam after passing through a transmission grating. The vertical lines indicate the peaks of the dimer (left) and trimer (right).

to see an explosive growth of the manipulation and study of individual atoms at the quantum wave-packet level. This will allow many basic issues of quantum mechanics to be explored, and is also likely to give rise to a new generation of practical measurement devices that use the basic external and internal quantum properties of atoms.

References

Adams, C., M. Sigel, and J. Mlynek, 1994, Phys. Rep. **240**, 143; provides a review of atom optics and atom interferometry.
Anderson, M. H., J. R. Ensher, M. R. Matthews, C. E. Wieman, and E. A. Cornell, 1995, Science **269**, 198; a bibliography of about 600 papers on BEC in dilute gases can be found online at http://amo.phy.gasou.edu/bec.html/bibliography.html.
Andrews, M. R., C. G. Townsend, H.-J. Miesner, D. S. Durfee, D. M. Kurn, and W. Ketterle, 1997, Science **275**, 637.
Atomic Physics, 1980–1996, Vols. 7–15, *Proceedings of the International Conference on Atomic Physics*; has many articles covering the topics discussed in this article.
Bergquist, J., 1996, Ed., *Proceedings of the Fifth Symposium on Frequency Standards and Metrology*, Woods Hole, Massachusetts, 1995 (World Scientific, Singapore).
Berkeland, D. J., J. D. Miller, J. C. Bergquist, W. M. Itano, and D. J. Wineland, 1998, Phys. Rev. Lett. **80**, 2089.
Berman, P. R., 1997, *Atom Interferometry*, Suppl. 3 Adv. At. Mol. Phys. (Academic, San Diego, CA, 1997).
Blatt, R., and P. Zoller, 1988, Eur. J. Phys. **9**, 250.
Bollinger, J. J., W. M. Itano, D. J. Wineland, and D. J. Heinzen, 1996, Phys. Rev. A **54**, R4649.
Cirac, J. I., and P. Zoller, 1995, Phys. Rev. Lett. **74**, 4091.
Cohen-Tannoudji, C., and W. D. Phillips, 1990, Phys. Today **43**, 33.
Davis, K. B., *et al.*, 1995, Phys. Rev. Lett. **75**, 3969.
Dehmelt, H., 1995, Science **247**, 539.
Einstein, A., 1924, Sitzungsber. K. Preuss. Akad. Wiss., Phys. Math. Kl. **1924**, 261.
Einstein, A., 1925, Sitzungsber. K. Preuss. Akad. Wiss., Phys. Math. Kl. **1925**, 3.
Ekert, A., and R. Jozsa, 1996, Rev. Mod. Phys. **68**, 733.
Ekstrom, C. J., J. Schmiedmayer, M. S. Chapman, T. D. Hammond, and D. E. Pritchard, 1995, Phys. Rev. A **51**, 3883.
Fisk, P. T. H., 1997, Rep. Prog. Phys. **60**, 761.
Gabrielse, G., D. Phillips, W. Quint, H. Kalinowsky, G. Rouleau, and W. Jhe, 1995, Phys. Rev. Lett. **74**, 3544.
Ghosh, P. K., 1995, *Ion Traps* (Clarendon, Oxford); Bergström, I., C. Carlberg, and R. Schuch, Eds., 1995, *Proceedings of the Nobel Symposium on Trapped Charged Particles and Related Fundamental Physics*, Lysekil, Sweden, 1994, Physica Scr. **59**, both provide extensive discussions of ion traps and work that has been done with them.
Greytak, T., 1995, in *Bose-Einstein Condensation*, edited by A. Griffin, D. Snoke, and S. Stringari (Cambridge University, Cambridge, England).
Gross, E. P., 1961, Nuovo Cimento **20**, 454.
Gross, E. P., 1963, J. Math. Phys. **4**, 195.
Gustavson, T. L., P. Bouyer, and M. A. Kasevich, 1997, Phys. Rev. Lett. **78**, 2046.
Itano, W. M., *et al.*, 1998, Science **279**, 686.
Jin, D., C. E. Wieman, and E. A. Cornell, 1997, Phys. Rev. Lett. **78**, 764.
Kasevich, M., and S. Chu, 1991, Phys. Rev. Lett. **67**, 181.
Kasevich, M., and S. Chu, 1992, Appl. Phys. B **54**, 321.
Keith, D. W., M. L. Shattenburg, H. I. Smith, and D. E. Pritchard, 1988, Phys. Rev. Lett. **61**, 1580.
Ketterle, W., and N. J. van Druten, 1996, Adv. At., Mol., Opt. Phys. **37**, 181.

Leibfried, D., T. Pfau, and C. Monroe, 1998, Phys. Today **51**, 22.
Lenef, A., T. D. Hammond, E. T. Smith, M. S. Chapman, R. A. Rubenstein, and D. E. Pritchard, 1997, Phys. Rev. Lett. **78**, 760.
Luo, F., C. F. Giese, W. R. Gentry, 1996, J. Chem. Phys. **104**, 1151.
Metcalf, H., and P. van der Straten, 1994, Phys. Rep. **244**, 203; a review of laser cooling and trapping of neutral atoms.
Meystre, P., and S. Stenholm, Eds. 1985, J. Opt. Soc. Am. B **2**, 1706; a special issue on the mechanical effects of light.
Mitchell, T. B., J. J. Bollinger, D. H. E. Dubin, X.-P. Huang, W. M. Itano, and R. H. Baughman, 1998, Science **282**, 1290.
Monroe, C., W. Swann, H. Robinson, and C. Wieman, 1990, Phys. Rev. Lett. **65**, 1571.
Monroe, C., and J. Bollinger, 1997, Phys. World **10**, 37.
Morinaga, M., M. Yasuda, T. Kishimoto, and F. Shimizu, 1996, Phys. Rev. Lett. **77**, 802.
Newbury, N. R., and C. E. Wieman, 1996, Am. J. Phys. **64**, 18, is an extensive bibliography on neutral atom trapping.
Phys. Today 1997, **50**, 17.
Pitaevskii, L. P., 1961, Sov. Phys. JETP **13**, 451.
Pritchard, D. E., 1991, in *Atomic Physics 12*, Proceedings of the 12th International Conference on Atomic Physics, edited by Jens C. Zorn and Robert L. Lewis (AIP, New York) p. 165.
Raab, E. L., M. Prentiss, A. Cable, S. Chu, and D. E. Pritchard, 1987, Phys. Rev. Lett. **59**, 2631.
Schollkopf, W., and J. Toennies, 1996, J. Chem. Phys. **104**, 1155.
Simon, E., P. Laurent, C. Mandache, and A. Clairon, 1997, in Proceedings of the *11th European Frequency and Time Forum* (Swiss Foundation for Research in Microtechnology, Neuchâtel, Switzerland), p. 43.
Stamper-Kurn, D. M., M. R. Andrews, A. P. Chikkatar, S. Inouye, H.-J. Miesner, J. Stenger, and W. Ketterle, 1998, Phys. Rev. Lett. **580**, 2027.
Steane, A., 1998, Rep. Prog. Phys. **61**, 117.
Thywissen, J. H., K. S. Johnson, R. Younkin, N. H. Dekker, K. K. Berggren, A. P. Chu, M. Prentiss, and S. A. Lee, 1997, J. Vac. Sci. Technol. B **15**, 2093.
Walker, T., and P. Feng, 1994, Adv. At., Mol., Opt. Phys. **34**, review ultralow-temperature collision studies.
Walther, H., 1993, Adv. At., Mol., Opt. Phys. **31**, 137.
Wieman, C., and S. Chu, Eds. 1989, J. Opt. Soc. Am. B **6**, 2020; a special issue with many articles on laser cooling.
Wineland, D. J., and W. M. Itano, 1987, Phys. Today **40**, 34.

Laser Physics: Quantum Controversy in Action

W. E. Lamb, W. P. Schleich, M. O. Scully, and C. H. Townes

Laser physics is a subject whose roots go back to the very source of quantum thought, namely, blackbody radiation and the Planck distribution. But the road to the first maser device showing radiation amplification by stimulated emission was rocky indeed. Likewise, the struggle to produce the first laser was hampered by much false "wisdom." Furthermore, the understanding of the laser, its extraordinary radiation, and its limits of operation is a great example of how our understanding of quantum physics has developed through and is illuminated by vigorous argument and debate, which extends even unto today.

For example, the very nature of our understanding of light (photon-wave) is a matter of much discussion and misunderstanding. We shall endeavor to present maser/laser physics from the prospective of quantum controversy; looking at the pros and cons of various issues, false starts, and key clues. This, after all, is the way physics develops, moving forward in the real-life give-and-take of debate, and often, intense controversy.

In the first section we briefly review the premaser days and highlight some stepping stones towards the maser and laser. In particular, we focus on the basic underpinning of laser physics: stimulated emission. The next two sections present the development of maser/laser science and the semiclassical theory of its operation. However, the many subtle features of the laser come to light only within the quantum theory of the laser. This is the topic of the next section. We conclude by presenting a few exotic laser concepts.

We emphasize that there are many more exciting developments in laser physics than we can cover in this review. For example, space does not allow us to discuss in detail the field of semiconductor lasers and its theory. Nor can we but mention the fields of excimer or x-ray lasers. Femto-second pulses have opened the new field of wave-packet dynamics and femto-chemistry, which represents another frontier of laser science that we can only allude to in this article. Nevertheless, these examples illustrate in a vivid way that laser physics is alive and well, both as a scientific and engineering discipline as we enter the 21st century.

Maser Prehistory

Conventional wisdom associates the beginning of the maser/laser with Einstein's discovery of the phenomenon of stimulated emission; even the acronym of microwave/

light amplification by stimulated emission radiation suggests this. However, in the spirit of the title of this paper, Quantum Controversy, we argue that the path leading to these devices started even earlier.

The search for the perfect oscillator

When God said "let there be light" he surely must have meant perfectly coherent light, that is, a perfect oscillator. But how to create such a perfect oscillator? Start from a source of dc energy for an oscillator; then by some trick, change the dc into ac: the result is a self-sustained oscillator.

The most elementary example is a grandfather clock. Here, we turn a knob, and raise a weight. Its potential energy is converted into periodic swinging of the pendulum bob. Moreover, a ratchet serves as a control mechanism between the super-hot reservoir—that is, the weight in the gravitational field—and the bob system.

Another example, from the realm of acoustical oscillators, is the aeolian harp known since the time of King David. In a paper reprinted in his collected works, Lord Rayleigh reports about an experiment in which he has a steady stream of wind blowing across the strings of a harp, creating sound. This is similar to the "Galloping Gertie," which is the oscillating Tacoma Narrows Bridge that in the 1940s was swinging for days and finally collapsed. These are three examples in which a steady source of energy creates a self-sustained oscillator.

Why not mention in the domain of electronics the triode vacuum-tube radio-frequency oscillator developed by L. De Forest in 1912? This was, in fact, the first electrical oscillator made by man. In 1921 E.V. Appleton and B. Van der Pol gave a classical theory of this device. It introduced for the first time the concept of "negative resistance." We can think of the negative resistance as a source of dc energy that can drive oscillations. For example, two-level atoms with a population inversion are a source of energy that can drive "laser" oscillations.

Later, in his book *Principles of Electricity and Magnetism*, Gaylord P. Harnwell (1938) gave a simple model for a de Forrest triode oscillator. He showed that a nonlinear negative resistance in an RLC circuit can generate self-sustained oscillations. If the applied voltage is zero, and the circuit has a little noise, the oscillations build up from noise to a steady state. Here, the effective resistance becomes zero. For large currents, the effective resistance becomes positive instead of negative. This leads to saturation. This is very similar to the laser.

In the late 1930s, two brothers, Russell and Sigurd Varian[1], invented the klystron oscillator at Stanford. Here, a dc electron beam passes through two cavities. The first cavity is called buncher, and the second cavity is called catcher, which is where the electromagnetic oscillations build up. There is feedback from the catcher to the buncher.

Another electromagnetic oscillator is the magnetron, developed in 1920 by Albert W. Hull. It is like a triode oscillator or a klystron. It is fully classical in operation, except it is also made of electrons and atoms, which need quantum mechanics. Magnetrons and

---

[1] Russell Varian was a physicist and considered a genius. However, he did not hold a Ph.D. although he was awarded honorary degrees later on. His brother Sigurd Varian was a professional airplane pilot and was therefore interested in short-wave radio communications. Eventually, the brothers founded a company, Varian Associates, that produced klystron devices and many instruments for scientific and commercial purposes.

klystrons have played a central role in war research related to radar. It was this radar work that laid the path for the maser as we discuss in the next section.

"Negative absorption"

Intimately related to the concept of "negative resistance," discussed in the preceding section, are the ideas of "negative absorption" and "population inversion." They occur already in various early publications on atomic spectroscopy and constitute important stepping stones on the path to the maser and laser. Since space does not allow us to discuss all these papers in detail we shall highlight only a few.

In 1924 R. Tolman wrote in a long article on an excited gaseous medium:

> The possibility arises, however, that molecules in the upper quantum state may return to the lower quantum state in such a way as to reinforce the primary beam by "negative absorption"...

However, he concluded that "... for absorption experiments as usually performed, the amount of 'negative absorption' can be neglected."

Two years later an interesting episode took place. The chemist G. N. Lewis in Berkeley proposed in a paper entitled "The conservation of photons" a mechanism for chemical bonding. The particle that achieves this he called *photon*. He certainly meant something completely different from Einstein's light quantum. The word "photon" caught on but not his meaning (Lamb, 1995).

Rudolf Walther Ladenburg played an important role in the history of the maser and laser. He came very close to discovering amplification by stimulated emission. Indeed, in the 1920s, while he was still in Germany, Ladenburg performed experiments with his co-workers (most prominently Hans Kopfermann), on the dispersion of gaseous neon near the red emission lines. Neon was excited in a tube by means of an electric discharge and the dispersion was studied as a function of the discharge current density. As Ladenburg summarized his experiments in 1933:

> ...the experiments prove the influence of the negative term in the dispersion formula. This "negative dispersion" corresponds to the negative absorption of the theory of radiation...

The theme of negative absorption reoccurs in the context of the fine structure of the hydrogen atom (Lamb and Retherford, 1950). The authors show that if the state "... $2^2P_{3/2}$ is more highly populated, there will be a net induced emission (negative absorption!)."

We conclude this discussion of the prehistory of the maser by briefly mentioning the work of V. A. Fabrikant in the former Soviet Union. In his thesis in 1940 he also discussed the consequences of population inversion.

Stimulated emission: Einstein and Dirac versus Maxwell

In his derivation of the Planck radiation formula in 1917 Albert Einstein introduced the A coefficient for the rate of spontaneous emission by atoms and the B coefficient for their absorption of radiation. He also introduced the new process of stimulated emission of radiation and found that the B coefficient determined its rate. Ten years later the quantum electrodynamics (QED) of P. A. M. Dirac provided the deeper foundation.

However, Einstein's result is perfectly natural when we disregard, for a moment, Maxwell's electromagnetic theory and, instead, believe in the 1905 concept of photons and in the Bohr orbits. Then it is natural to have spontaneous emission and absorption of the light particles, and the new feature is, indeed, stimulated emission. However, we emphasize that Maxwell's theory also predicts these phenomena.

To bring this out most clearly, we consider a charged particle oscillating back and forth in an electromagnetic wave. We recall that a particle of charge q moving with velocity v in an electric field E, gains or loses energy depending on the algebraic sign of the product qEv. An increase of the energy of the charge implies a loss of energy in the field. This is equivalent to the process of absorption of radiation. Likewise, if the charge is losing energy, the electromagnetic field must be gaining energy. This is equivalent to stimulated emission of radiation. The relative direction of the velocity and the electric-field vectors determines the direction of the energy flow between field and matter (Lamb, 1960). Moreover, the fact that an accelerated charge radiates corresponds to the process of spontaneous emission.

How does this translate into the language of QED? To answer this question, we consider the change of the electromagnetic field due to the transition of an excited atom into its ground state. We assume that initially only one mode is occupied by n-quanta and all the other modes are empty. The atomic transition creates one quantum of field excitation in any field mode. However, due to the property

$$\hat{a}^\dagger |n\rangle = \sqrt{n+1}\,|n+1\rangle$$

of the creation operator, the mode with n-quanta already present has a higher probability compared to the vacuum modes where $n=0$. Hence the amplification, which is stimulated emission, is preferentially in the mode of the incident radiation.

But how can we use Maxwell's theory to explain the directionality of the emitted radiation, which is so obvious in the QED formulation? On first sight this seems to be impossible: A dipole does not radiate in the direction in which it is driven. However, when we calculate the energy flow, that is, the Poynting vector of the total field consisting of the incident and the radiated electromagnetic field, the interference term between the two provides the directionality. Indeed, this term is rapidly oscillating in space except along a narrow cone along the axis of propagation of the incident radiation (Sargent *et al.*, 1974).

We conclude this section by briefly alluding to one more feature of stimulated emission. Stimulated emission is said to be in phase with the incident radiation. We can understand this feature when we recall that the induced dipole is a driven oscillator. Therefore it is in phase and has the same frequency as the incident light—there is no way to see this easily from QED!

The Maser—How It Came to Be

In the present section we briefly follow the path from the early work on microwave absorption in water vapor to the conception and realization of the ammonia maser. We also speculate why the maser was not discovered earlier.

From water vapor and radar to microwave spectroscopy

The absorption of microwave radiation in water vapor was an important question during W.W. II. Indeed, it was recognized by J. H. Van Vleck and V. F. Weisskopf (1945) that the shortest waves (K band) might be absorbed in water vapor. If correct, this would have drastically reduced the use of radar in the South Pacific, an area of high rainfall. Therefore Isidor I. Rabi got authorization to study this question at the Columbia Radiation Laboratory at Columbia University.

Willis E. Lamb was involved in this research directed by J. M. Kellogg. Water vapor was inside a 8 ft.×8 ft.×8 ft. resonator made of copper sheet. Large rotating copper fans "mixed up" the mode structure of the 1 cm microwaves in this resonator. The room was heated with steam radiators, to simulate tropical conditions. The experiments (Becker and Autler, 1946) indicated that there was some microwave absorption, but not bad enough to give up on the K-band completely (Lamb, 1946). As Lamb recalls those years in the labs:

I learned something from this work, although not enough to invent microwave ovens. But I might have been the first to warm up cold hamburgers or coffee using centimeter microwave radiation. That came in handy for lunch.

During the war Charles H. Townes worked at Bell labs on radar bombing and navigation systems. He also studied Van Vleck's work, and argued strongly that water vapor would have a disastrous effect on K-band radar in the Pacific arena. But after contacting several people about it, high-level officials simply told him the decision had been made to proceed and could not be changed.

In the process of examination and argument, Townes recognized that microwave absorption by such molecules at low pressures could provide a new kind of high-resolution spectroscopy (Townes, 1946). The new field largely used components of the K-band radar, which were in surplus because that wavelength was indeed relatively useless for radar. About ten years earlier, Claude E. Cleeton and Neil H. Williams (1934) at the University of Michigan had studied absorption of microwaves by the inversion of ammonia molecules, an absorption which had been predicted by David Dennison (1932) and colleagues at their university. This was a striking demonstration of the inversion of ammonia, a resonance which was later to be the basis for the first masers. However, they used ammonia at atmospheric pressure which gave absorption over a range almost as large as the frequency itself, not the high-resolution-at-low-pressure characteristic of postwar microwave spectroscopy.

The idea on the park bench

As the field of microwave spectroscopy progressed, Townes grew increasingly eager to extend it into the millimeter and submillimeter wavelengths, where absorption would be still stronger. He worked on several possible methods to achieve this, including Cerenkov radiation, magnetron harmonics, and electron-spin resonances. Knowing of his interest, the Office of Naval Research asked him to head a national committee to explore how shorter-wavelength oscillators might be achieved.

After about 18 months of committee discussions and visits to laboratories in the field, a final meeting was held in Washington in the spring of 1951. Townes was frustrated that no great ideas had turned up and, waking up early before the committee meeting, left his hotel and sat down on a park bench. In musing over the problem and his frustrations with it, he suddenly recognized that molecules could produce much more than thermal radiation intensities if they were not thermally distributed but had more molecules or atoms in an upper than in a lower state. Within about ten minutes he had invented such a system using a beam of ammonia and a cavity, and calculated that it seemed practical to get enough molecules to cross the threshold of oscillation. This meant that molecular-stimulated emission at a given radiation intensity would be greater than energy loss in the walls of the cavity.

He did not know that an eventual outcome would be fantastic communication by optical fibers, nor did he know at the time that he was sitting next to the building where Alexander Graham Bell had worked for a long time trying to successfully communicate with light waves. However, the fact that the new idea occurred right next to Bell's old laboratory may now seem a bit mystical.

The birth of the maser

Townes's initial plan was to use ammonia rotational levels and work in the far-infrared or submillimeter region. But the K-band region, or centimeter wavelengths and the ammonia inversion spectrum, seemed an easier start. So, after a few months, and carefully checking the coherence of stimulated emission using his notes on the quantum theory of radiation from his student days, Townes got together with an excellent student and postdoc, James Gordon and Herbert Zeiger, and set to work on the new type of oscillator.

However, as Townes (1999) recalls, his maser team at the Columbia Radiation Lab did not get much encouragement:

One day, after we had been at it for about two years, Rabi and Kusch, the former and current chairmen of the department, both of them Nobel Laureates for their work with atomic and molecular beams and with a lot of weight behind their opinions, came into my office and sat down. They were worried. Their research depended on support from the same source as did mine. "Look," they said, "you should stop the work you are doing. You're wasting money. Just stop!"

At this moment Townes was indeed thankful that he came to Columbia with tenure already.

One of the problems that almost prevented the birth of the maser was the worry about too much radiation leaking through the entrance and the exit holes in the cavity for the molecular beam. Various metal rings designed to keep the microwaves inside of the resonator were tried. No success! Only when Jim Gordon, skipping a seminar, opened the ends almost completely did the device work (Gordon *et al.*, 1954, 1955). As Townes recalls:

The first maser had been born. This was about three months after Poly Kusch had insisted it would not work. But when it worked, he was gracious about it, commenting that he should have realized I probably knew more about what I was doing than he did.

Why not earlier?

Probably the reason why quantum electronics, that is, masers and lasers, did not develop sooner is that some aspects of the maser are most easily envisioned from a classical point of view and stimulated by an interest in electronics, while others require an understanding of quantum mechanics. In the early days of the field, it was clear that electrical engineers intuitively understood some characteristics of the maser oscillator much better than many physicists, who usually thought in terms of photons. However, at that time few engineers understood stimulated emission.

That a background in microwave spectroscopy, with a combination of engineering and quantum mechanics, was important to putting together appropriate concepts of useful amplification by stimulated emission has empirical support. The general idea appeared at approximately the same time to three groups: Joseph Weber of the University of Maryland (1953), Nickolai Basov and Alexander Prokhorov of the Lebedev Institute (1955), and Charles Townes at Columbia (Gordon *et al.*, 1954), all of whom were working on microwave spectroscopy.

On the Road towards the Laser

In the present section we discuss the development of the laser. In particular, we review the basic obstacles lying in the path of this development and discuss their solutions.

Early maser research

Interest in masers grew rapidly, partly because of their value as frequency standards, and partly because as amplifiers, they could be one or two orders of magnitude more sensitive than other types available. The spectral width of maser oscillators in the presence of thermal noise was derived by Gordon *et al.* (1955). A more quantum-mechanical approach to maser oscillation was developed by Shimoda *et al.* (1956), which gave spectral widths when dominated by quantum emission. A quantum-mechanical theory of noise in maser (or laser) amplifiers was given by Shimoda *et al.* (1957). For amplification, electron-spin resonances in solids (Combrisson *et al.*, 1956) and particularly 3-level systems (Bloembergen, 1956) were especially attractive since they could be fairly broadband and tunable. Application of the uncertainty principle to complementary uncertainties in the phase of a wave and the number of quanta $\Delta n \Delta \theta \geq 1/2$ was also soon discussed (Serber and Townes, 1960), along with the possibility of measuring the phase more accurately than the normal precision, which, with $\Delta n = \sqrt{\bar{n}}$, is $\Delta \theta \geq 1/2\sqrt{\bar{n}}$.

Why lasers will not work

With a working maser in hand, thoughts naturally turned to the possibility of an optical maser. Townes recalls that many colleagues argued that the maser would not work at shorter wavelengths. An illustration of this skepticism is a 1958 report for the U.S. Air Force on technology which might be of importance in the following 25 years. Townes was a member of a committee to write the report in the summer of 1957. He persuaded the group to mention not only the development of masers in the microwave region, but also the possibility of extending them to shorter wavelengths. The report was not issued in 1957, and a further study was made to complete it in the summer of 1958 by the same group except that Townes was absent. That was the same summer Schawlow and Townes (1958) finished and began to circulate their paper on "Optical Masers." But the committee, which had not seen the paper, did not accept Townes's previous recommendation: The committee decided to remove any mention of the possibility of extending masers to shorter wavelengths, and included only microwave masers as the technology of the next 25 years.

One of the many reasons for this early pessimism was that the rate of energy radiated spontaneously from a molecule increases as the fourth power of the frequency, assuming other characteristics of the molecule remain generally the same. Townes's original goal was to amplify at a wavelength of, say, one tenth of a millimeter instead of one centimeter. Hence to keep the electrons or molecules excited in a regime, this would require an increase in the pumping power by many orders of magnitude.

Another "problem" was that for gas molecules or atoms, Doppler effects increasingly broaden the emission spectrum as the frequency goes up. This leaves less amplification per molecule to drive any specific resonant frequency. For a description of this battle with the Doppler effect, see Lamb (1984).

Finally, there are no nice "cavities" for light. Indeed, the maser operating at wavelengths of the order of a centimeter can oscillate in a cavity whose length is equal to the wavelength. However, it is difficult to make similar cavities much smaller than a millimeter and virtually impossible to make them as small as an optical wavelength. A cavity related to the Fabry-Perot etalon or interferometer, whose length is many times the wavelength of resonance is the solution of the optical-resonator problem. This was pointed out by Schawlow and Townes (1958); a summarizing discussion is given by Siegman (1986).

Nothing stops "naysayers" like a working device. Theodore Maiman (1960) gave the first working laser to us—the pulsed ruby system. This was soon followed by other types of lasers, including those using gaseous discharges (Javan *et al.*, 1961) and semiconductors (Basov *et al.*, 1959 and Hall *et al.*, 1962).

Lamb dip

Motivated by the experiments on the ammonia maser and building on his theoretical work on water-vapor absorption, Lamb, during the years 1954–1956, developed a theory of the maser (see Lamb, 1960). Later, he worked out a complete semiclassical theory of laser action (Lamb, 1964a, 1964b). It is based on a self-consistent treatment of the polarization of the masing/lasing medium as it drives the electric-field oscillator in the laser cavity. This semiclassical description of laser operation shown in Fig. 1 has been a touchstone of laser physics over the years.

We conclude by returning for a moment to the Doppler broadening problem. It turned out to be a blessing in disguise. As shown by the semiclassical theory, there is a dip in the laser power as a function of cavity detuning. This so-called Lamb dip, shown in Fig. 2, was verified experimentally by McFarlane *et al.* (1963) and Szöke and Javan (1963) and has proven to be very useful in building ultrastable lasers: It provides a narrow resonance allowing to lock lasers to the center of the dip.

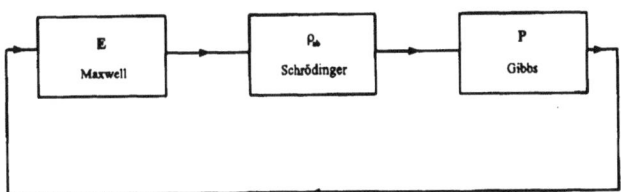

Figure 1. Semiclassical theory of the laser as a self-consistent analysis. We start with a nonvanishing seed electric field E in the laser cavity. This field induces microscopic dipole moments in the active medium according to the laws of quantum mechanics. These moments represented by off-diagonal elements ρ_{ab} of the density operator sum up statistically through a Gibbs ensemble to a macroscopic polarization P. This generates via Maxwell's equations a new electric field E'. The condition of self-consistency requires that the seed field is equal to the generated field, that is, $E = E'$.

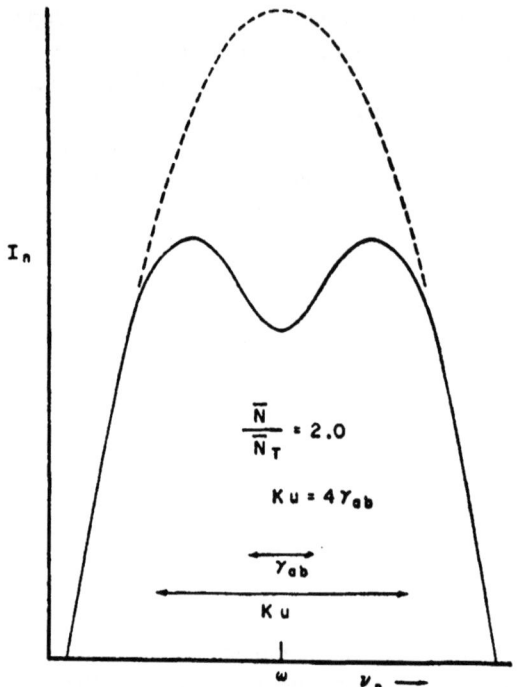

Figure 2. Lamb dip. The laser intensity shows a characteristic dip as a function of cavity detuning. [From the original paper by Lamb (1964b).]

Quantum Effects in Masers and Lasers

Laser theory has come a long way from the early approaches using birth and death equations (Shimoda et al., 1957) via the semiclassical theory of the laser to the fully quantized version. The three approaches towards the quantum theory of the laser[2] are the Fokker-Planck method (Haken and Risken), noise operators (Lax and Louisell), and density matrix techniques (Scully and Lamb). In the present section we now turn to the quantum aspects of laser light.

Linewidth

In order to study the laser linewidth (Schawlow and Townes, 1958) we represent the laser field by a vector in the complex plane as shown in Fig. 3. This vector describes the amplitude and the phase of the electromagnetic field. Strictly speaking, both are operators. However, in the present discussion it suffices to take them as fluctuating classical quantities. We emphasize that this replacement is not a trivial step and the question of the appropriate definition of a Hermitian phase operator is still a research topic (Schleich and Barnett, 1993).

Due to the spontaneous emission of the atom, the electric field experiences small changes and undergoes a random walk. Here, we assume that the small change takes place

---

[2]For the expositions of the various schools of laser theory, see Haken (1970), Louisell (1973), and Sargent et al. (1974).

on a far shorter time scale than the overall evolution of the field. Moreover, we consider a situation in which the laser is operating sufficiently far above threshold so that the amplitude fluctuations can be ignored. The field $E(t)$ can then be written as

$$E(t) = \sqrt{\bar{n}} e^{i\theta(t)} e^{-i\nu_0 t},$$

where $\bar{n}$ is the mean number of photons of frequency ν_0.

Since we ignore amplitude fluctuations, the phase θ performs a one-dimensional random walk around a circle, as indicated by Fig. 3. The quantum theory of the laser shows that the probability $P(\theta)$ of finding the phase θ obeys a phase diffusion equation of the Fokker-Planck type (Risken, 1984) with the diffusion constant $D = A/(4\bar{n})$. Here A denotes the linear gain. This diffusion causes a decaying average electric field

$$\langle E(t) \rangle = \int d\theta \, P(\theta, t) E(\theta) = E_0 e^{-Dt},$$

which gives rise to the laser linewidth.

This situation is analogous to ferromagnetism, where the magnetization of an open system (magnet) experiences the same kind of decay, albeit on a geological time scale. Measured in terms of the atomic and cavity lifetimes, the decay of $\langle E(t) \rangle$ is also very slow, typically estimated to take many minutes. Consequently, the laser/maser linewidth is much narrower than that of the atoms or molecules that drive it.

To many physicists steeped in the uncertainty principle, the maser's performance, at first blush, made no sense at all: Molecules spend so little time in the cavity of a maser, about one-ten-thousandth of a second, that it seemed impossible for the frequency of the radiation to also be narrowly confined. Townes recalls a conversation with Niels Bohr on a sidewalk in Denmark in which Bohr emphasized this very argument. After Townes persisted, he said: "Oh, well, yes, maybe you are right." But, the impression was that he was simply trying to be polite to a younger physicist.

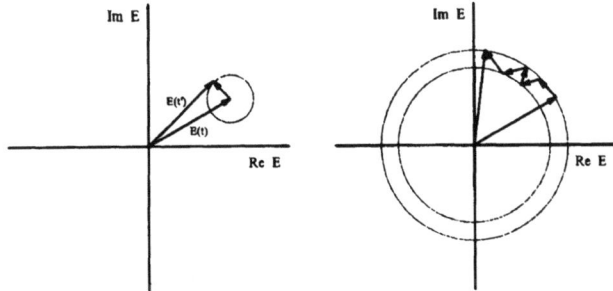

Figure 3. Phase diffusion in a laser as a random walk in complex space. We represent the electric field of the laser by a vector in the complex plane. Its length is the amplitude of the field, whereas, its angle with the horizontal axis of the coordinate system is its phase. The spontaneous emission of an atom changes the electric field (left). The amplitude of the emitted field is small compared to the original field and its phase is completely uncorrelated. Due to many spontaneous emission events, the electric field undergoes Brownian motion in complex space (right). The saturation of the laser stabilizes the amplitude of the electric field and forces the phase to undergo diffusion in a band.

Squeezing and the correlated spontaneous emission laser

Interferometers have great potential for high-precision measurements and tests of the foundations of physics. Examples include optical rotation sensors, such as fiber-optic or ring-laser gyroscopes (Chow et al., 1985), and probes of gravitomagnetic fields à la Lense-Thirring, and the search for gravitational waves (Schleich and Scully, 1984). In all of these examples the signal is encoded in the phase difference between two electromagnetic fields. Hence the fluctuations in the phases of the two fields determine the ultimate sensitivity of these devices.

In a passive interferometer, such as the fiber-optic laser gyro or a gravitational-wave detector, the laser is outside of the interferometer. The phase noise in the two arms of the interferometer, originating from spontaneous emission, is correlated and cancels since it comes from the same source. The only remaining noise is then due to the quantum nature of the radiation. It is the uncertainty $\Delta\theta$ of the laser field in phase as governed[3] by $\Delta n \Delta \theta \geq 1/2$. Since for a coherent state we have $\Delta n = \sqrt{\bar{n}}$ we find $\Delta\theta \cong 1/2\sqrt{\bar{n}}$. However, when we use appropriately "squeezed light" we can achieve a situation such that $\Delta\theta \cong 1/\bar{n}$ as shown by Caves (1981). In the mid-1980s the pioneering, squeezing experiments by the groups of R. Slusher and H. J. Kimble created a new and thrilling field. For a review of the activities in this arena, see Kimble and Walls (1987) and Grangier et al. (1995). A different method for obtaining $\Delta\theta = 1/(2\bar{n})$ has been discussed earlier by Serber and Townes (1960).

In an active device, such as the laser gyro, the laser medium is inside the ring cavity. Hence spontaneous emission of the atoms represents the ultimate limitation. Can we overcome this noise? The correlated spontaneous emission laser (CEL) (Scully, 1985), shown in Fig. 4, provides the definite answer, Yes!

The correlated spontaneous emission laser relies on a specially prepared lasing medium such that the noise in the relative phase angle of the emitted radiation is eliminated. For this purpose we use a three-level atom wherein the two transitions, from two excited states to a common ground state, drive a doubly resonant cavity. They are strongly correlated by preparing the two upper levels in a coherent superposition. This is similar to quantum-beat experiments in which a strong external microwave signal produces coherent mixing and correlates the spontaneously emitted fields.

The correlated-spontaneous-emission-laser noise reduction was observed in a series of beautiful experiments (Winters et al., 1990, and Steiner and Toschek, 1995). However, the CEL concept is somewhat tricky, as expressed by a senior researcher in a discussion with Marlan O. Scully:

The CEL is vastly more general than you think and vastly more trivial. It is simply due to the fact that the two lasers are locked together.

It is true that the CEL involves phase-locked lasers. However, there is more: The spontaneous emission events in the two transitions are correlated. This leads to dramatically different equations of motion for the phase difference: Whereas, in a phase-locked laser, the spontaneous emission noise is additive, in the CEL it enters in a multiplicative way (Schleich and Scully, 1988).

---

[3]According to Serber and Townes (1960) the photon number and phase uncertainty relation reads $\Delta n \Delta \theta \geq 1/2$ rather than $\Delta n \Delta \theta = 1$. This is illustrated by a specific example given by these authors, where $\Delta n \Delta \theta = 1/2\sqrt{(\bar{n}+1)/\bar{n}}$. For large $\bar{n}$ the right-hand side indeed reduces to 1/2.

Photon statistics

As with the linewidth, so with the photon statistics; confusion abounded in the early days of the laser. Some people said that since "photons" are Bose-Einstein (BE) "particles" they must obey BE statistics. However, when we recall that the BE distribution applies to a system in thermal equilibrium and recognize that a laser is a system far away from thermodynamic equilibrium, we discover that this argument is false.

Indeed, the quantum theory of the laser predicts (Scully and Lamb, 1967) that the photon statistics of a laser, shown in Fig. 5, are substantially different from the photon statistics

$$P_n = (1+\bar{n})^{-1} \left(\frac{\bar{n}}{1+\bar{n}} \right)^n$$

of a thermal state or even from the Poissonian distribution (Glauber, 1964)

$$P_n = \frac{\bar{n}^n}{n!} e^{-\bar{n}},$$

of a coherent light beam. This feature has been verified by the groups of Tito Arecchi, Werner Martienssen, and Roy Pike in the early days of laser physics (Mandel and Wolf, 1970).

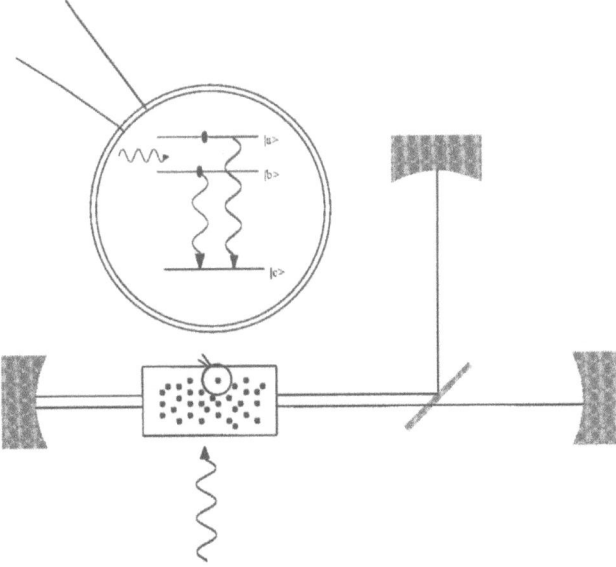

Figure 4. Schematic setup of the correlated spontaneous emission laser (CEL) in the form of the quantum-beat laser. A microwave drives the two upper levels of a three-level atom. The two cavities, resonant with the two transitions to the ground state, are aligned in an orthogonal way as to be sensitive to gravitational waves or to rotation. Due to the coherence of the two upper states, the spontaneous emission in the two transitions is correlated. This leads to a dramatic suppression of the laser phase noise between the two fields and thus to a small Schawlow-Townes linewidth.

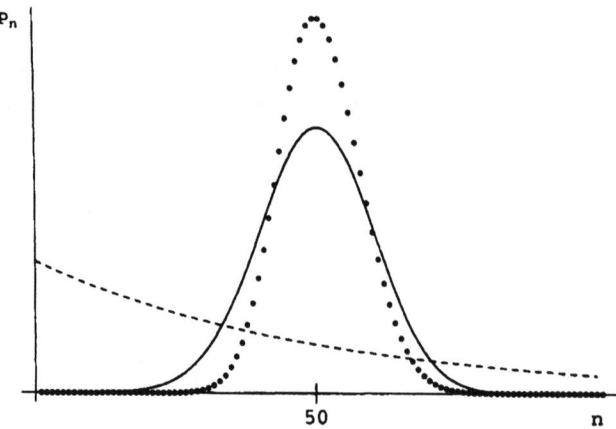

Figure 5. Comparison between the photon statistics of a thermal state (dashed curve), coherent state (dotted curve), and a laser field (solid curve). In all three cases we have taken the average photon number $\bar{n} = 50$. Whereas, the thermal distribution has its maximum at $n=0$ the Poissonian distribution of the coherent state and the laser photon statistics have their maximum at $n=\bar{n}$. We note however, that the photon statistics of the laser is broader than that of the coherent state.

One-atom masers/lasers and the mazer

The one-atom maser[4] or "micromaser" is an ideal testing ground for the quantum theory of the laser and extensions thereof. In the micromaser a stream of two-level atoms passes through a superconducting cavity with a high-quality factor. The injection rate can be such that only one atom is present inside the resonator at any time. Due to the high quality factor of the cavity, the radiation decay time is much larger than the characteristic time of the atom-field interaction. A field builds up inside the cavity when the mean time between the atoms injected into the cavity is shorter than the cavity decay time. Therefore a micromaser allows sustained oscillations with less than one atom on the average in the cavity.

The enormous progress in producing mirrors with almost 100 percent reflectivity has opened a new era in cavity quantum electrodynamics. Experiments, previously performed only in the microwave domain, have now been extended to the optical domain (Kimble et al., 1996). Moreover, an optical version of the one-atom maser, the so-called one atom laser (An et al., 1994), microlasers based on molecules (De Martini et al., 1993), and condensed matter systems (Yamamoto and Slusher, 1993), have been realized experimentally. For a more detailed discussion of laser technology we refer to the article by R. Slusher in this issue.

An interesting "spin-off" of the micromaser is its extension into the microwave amplification via z-motion-induced emission of radiation (mazer) regime (Scully et al., 1996) when the kinetic energy is comparable to the interaction energy. Indeed, in this regime very slow (laser-cooled) atoms can be reflected from or tunnel through the cavity. In the process they undergo a new kind of induced emission, which is different from the stimu-

---

[4]For a review of the work of the Garching and Paris groups see Walther (1992) and Raithel et al. (1994), and Haroche and Raimond (1994) and Haroche (1998), respectively. For the theoretical work, see Meystre (1992) and Scully and Zubairy (1997).

lated emission of the ordinary micromaser regime with fast, that is, thermal atoms.

The difference between the classical and the quantum treatment of the center-of-mass motion comes out most clearly in the probability that an excited atom launched into an empty cavity will emit a photon. Figure 6 shows that when an integral number of half de Broglie wavelengths of the atomic center-mass-motion equals the cavity length, the atoms are transmitted through the cavity and new type of induced emission comes into play.

The laser phase transition analogy

We conclude this section by noting that in many ways the physics of the laser is analogous to that of a cooperative system near a phase transition. The root of the analogy is the fact that both the laser and the ferromagnet (or superconductor) are, to a conceptual idealization, mean-field or self-consistent field theories. In the many-body ferromagnet or superconductor problem, any given element of the ensemble sees all the other elements through the self-consistent mean field. For example, in the magnetic problem each spin communicates with all other spins through the average magnetization. Similarly, the atoms in the laser contribute to the total electric field by means of their induced dipole. This dipole is in turn induced by the mean electric field as contributed by all of the other atoms in the ensemble.

This analogy comes to light when we expand the density operator

$$\rho = \int P(E)|E\rangle\langle E| d^2 E$$

of the laser field in terms of coherent states $|E\rangle$. The Glauber-Sudarshan P-distribution (Sargent et al., 1974) then reads (DiGiorgio and Scully, 1970; Graham and Haken, 1970)

$$P(E) = \frac{1}{Z} e^{-G(E)/k_L \sigma},$$

where

$$G(E) \equiv \frac{A}{2}(\sigma - \sigma_t)|E|^2 + \frac{B}{4}\sigma|E|^4.$$

Here A and B are linear and nonlinear gain parameters, and σ and σ_t denote the inversion and the threshold inversion. The spontaneous emission rate $4k_L$ per atom plays the role

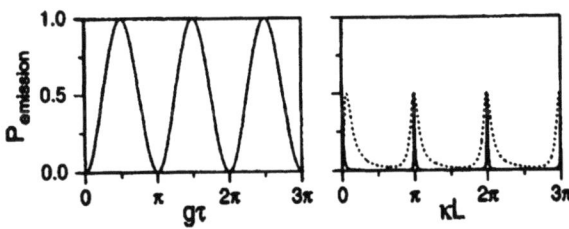

Figure 6. *Maser and mazer compared and contrasted using the emission probability as a function of interaction time (left) or interaction length (right). A two-level atom of mass M enters in its excited state a cavity which is initially in the vacuum state. On the left, the kinetic energy $(\hbar k)^2/2M$ is much larger than the interaction energy $\hbar g \equiv (\hbar \kappa)^2/2M$, that is, $k/\kappa = 10$. On the right, it is smaller (dotted curve for $k/\kappa = 0.1$) or much smaller (solid curve for $k/\kappa = 0.01$) than $\hbar g$.*

of the Boltzmann constant of equilibrium statistical mechanics.

This is to be compared to a ferromagnet. Here, in the Ginzburg-Landau approximation to the second-order phase transition, the free energy $F(M)$ reads

$$F(M) = C(T - T_c)M^2 + DTM^4,$$

where C and D are constants and T_c is the Curie temperature.

Exotic Lasers and Outlook

In the following subsections, we briefly discuss three exotic laser systems—the free-electron laser, the laser without inversion, and the so-called atom laser—illustrating in an impressive way that laser physics is still an evolving field. We do not concern ourselves with the details, but rather bring out the underlying physics of these devices. In the spirit of quantum controversy, we focus on the quantum aspects.

The free-electron laser

The free-electron laser (FEL) is an interesting device because it is largely free from the constraints imposed by the atomic medium of an ordinary laser. In a free-electron laser, a relativistic electron beam interacts with a periodic static magnetic field called a wiggler (Brau, 1990). The electrons convert their kinetic energy into laser light. Indeed, the individual electrons passing through the magnetic field undergo transverse oscillations. Thus the electrons periodically accelerate and decelerate while they absorb and emit radiation. Depending on the wiggler length and the velocity of the electrons, we therefore have net gain or net loss.

We note, however, that the physical principle governing the FEL was initially strongly debated. It was first thought that FEL operation takes place only because of the quantum nature of the electron and photon. It was later shown that the FEL is a classical device (Hopf *et al.*, 1976). The free-electron laser is therefore an excellent example of how the "photon-picture" can obscure the physics, which is electron bunching in a ponderomotive potential. The lasing process in a FEL may thus be compared with amplification and oscillation in electron tubes, which were characteristically used for radio frequencies before quantum electronics appeared on the scene.

Lasers without inversion

A laser seems to require population inversion in order to overcome absorption. Can we engineer a laser medium in which we cancel absorption but keep stimulated emission? Can we then lase without inversion? The answer is "Yes"!

To understand this we again focus on a three-level atom. However, in contrast to the discussion of the quantum-beat laser in the earlier section on Squeezing and the Correlated Spontaneous Emission Laser, we now consider transitions between a single upper level and two lower levels. The absorption probability is then the *coherent* sum of the *probability amplitudes* of the two transitions. They can therefore cancel each other. This feature of vanishing absorption in this particular three-level configuration is yet another manifestation of quantum coherence. For a comprehensive discussion of these ideas we refer to Kocharovskaya and Khanin (1988), Harris (1989), and Scully *et al.* (1989). In contrast, in the emission process we end up in two different states; consequently, the two corresponding *probabilities* add up. This asymmetry between absorption and emission is the basis for the phenomenon of lasing without inversion.

We conclude by noting that many experiments have verified this and other related coherence effects, for example, electromagnetically induced transparency. However, this is not the place to go deeper into this exciting field. For an introduction and overview, we refer to Arimondo (1996) and Harris (1997).

The atom laser

Bose-Einstein condensation (BEC) of dilute He^4 in a porous jell has been observed (Reppy, 1992). In addition, the pioneering experiments reporting BEC in laser-cooled atoms have ushered in a new era in the study of BEC in dilute low-temperature gases. In particular, the striking experimental demonstration of interference between two condensates and the time evolution of a condensate provide incontrovertible evidence for BEC. For a summary of these experiments we refer to the article by Wieman *et al.* in the present issue.

They have clearly demonstrated that bosonic-stimulated emission is important in the process. This has suggested a comparison with the laser and coined the name "atom laser" for certain kinds of coherent atom beams. The MIT group has already realized experimentally such an atom laser. In this context it is interesting to note that an analog of the semiclassical theory of the optical laser can describe the essential features of this device (Kneer *et al.*, 1998). This is only one of many applications of laser concepts to BEC, which will continue to be a low-temperature "hot topic."

Acknowledgments

We thank M. Hug, S. Meneghini, P. Meystre, and H. Walther for their comments on and help with this manuscript. W.E.L. expresses his appreciation to the Humboldt Stiftung. The work of W.P.S. was partially supported by the Deutsche Forschungsgemeinschaft, and that of M.O.S. by the Office of Naval Research, the National Science Foundation, and the Welch Foundation.

References

Arimondo, E., 1996, in *Progress in Optics*, edited by E. Wolf (North-Holland, Amsterdam), Vol. 35, p. 257.
An, K., J.J. Childs, R.R. Desari, and M.S. Feld, 1994, Phys. Rev. Lett. **73**, 3375.
Basov, N.G., and A.M. Prokhorov, 1955, Dokl. Akad. Nauk **101**, 47.
Basov, N.G., B.M. Vul, and Yu.M. Popov, 1959, Zh. Eksp. Teor. Fiz., **37**, 587 [Sov. Phys. JETP **10**, 416].
Becker, G.E., and S.H. Autler, 1946, Phys. Rev. **70**, 300.
Bloembergen, N., 1956, Phys. Rev. **104**, 324.
Brau, C.A., 1990, *Free-Electron Lasers* (Academic Press, Orlando).
Caves, C.M., 1981, Phys. Rev. D **23**, 1693.
Chow, W.W., J. Gea-Banacloche, L. Pedrotti, V. Sanders, W. Schleich, and M.O. Scully, 1985, Rev. Mod. Phys. **57**, 61.
Cleeton, C.E., and N.H. Williams, 1934, Phys. Rev. **45**, 234.
Combrisson, J., A. Honig, and C.H. Townes, 1956, Phys. Rev. **242**, 2451.

De Martini, F., M. Marrocco, P. Mataloni, and D. Murra, 1993, J. Opt. Soc. Am. B **10**(2), 360.
Dennison, D.M., and G.E. Uhlenbeck, 1932, Phys. Rev. **41**, 313.
DiGiorgio, V., and M.O. Scully, 1970, Phys. Rev. A **2**, 1170.
Glauber, R., 1964, in *Quantum Optics and Electronics*, edited by C. DeWitt, A. Blandin, and C. Cohen-Tannoudji (Gordon and Breach, New York).
Gordon, J.P., H.J. Zeiger, and C.H. Townes, 1954, Phys. Rev. **95**, 282L.
Gordon, J.P., H.J. Zeiger, and C.H. Townes, 1955, Phys. Rev. **99**, 1264.
Graham, R., and H. Haken, 1970, Z. Phys. **237**, 31.
Grangier, P., E. Giacobino, and P. Tombesi, 1995, Eds. *Nonclassical Light*, J. Quantum Semiclass. Opt. **7**, 499.
Haken, H., 1970, *Laser Theory* in *Encyclopedia of Physics* Vol. XXV/2c, edited by S. Flügge (Springer, Berlin).
Hall, R.N., G.E. Fenner, J.D. Kingsley, T.J. Soltys, and R.O. Carlson, 1962, Phys. Rev. Lett. **9**, 366.
Harnwell, G.P., 1938, *Principles of Electricity and Electromagnetism* (McGraw-Hill, New York).
Haroche, S., 1998, Phys. Today **51**(7), 36.
Haroche, S., and J.-M. Raimond, 1994, in *Cavity Quantum Electrodynamics*, Adv. At. Mol., and Opt. Phys., Suppl. 2, edited by P.R. Berman (Academic Press, New York), p. 123.
Harris, S.E., 1989, Phys. Rev. Lett. **62**, 1033.
Harris, S.E., 1997, Phys. Today **50**(7), 36.
Hopf, F.A., P. Meystre, M.O. Scully, and W.H. Louisell, 1976, Phys. Rev. Lett. **37**, 1342.
Javan, A., W.R. Bennett, and D.R. Herriott, 1961, Phys. Rev. Lett. **6**, 106.
Kimble, H.J., Q.A. Turchette, N.Ph. Georgiades, C.J. Hood, W. Lange, H. Mabuchi, E.S. Polzik, and D.W. Vernooy, 1996, in *Coherence and Quantum Optics VII*, edited by J.H. Eberly, L. Mandel, and E. Wolf (Plenum Press, New York), p. 203.
Kimble, H.J., and D.F. Walls, 1987, Eds., *Squeezed States of the Electromagnetic Field*, J. Opt. Soc. Am. B **4**, 1450.
Kneer, B., T. Wong, K. Vogel, W.P. Schleich, and D.F. Walls, 1998, Phys. Rev. A **58**, 4841.
Kocharovskaya, O., and Ya. I. Khanin, 1988, Pis'ma Zh. Éksp. Teor. Fiz. **48**, 581 [JETP Lett. **48**, 630].
Ladenburg, R., 1933, Rev. Mod. Phys. **5**, 243.
Lamb, W.E., 1946, Phys. Rev. **70**, 308.
Lamb, W.E., 1960, in *Lectures in Theoretical Physics*, edited by W.E. Brittin and B.W. Downs (Interscience Publishers, New York), Vol. 2.
Lamb, W.E., 1964a, in *Quantum Optics and Electronics*, edited by C. DeWitt, A. Blandin, and C. Cohen-Tannoudji (Gordon and Breach, New York).
Lamb, W.E., 1964b, Phys. Rev. **134**, A1429.
Lamb, W.E., 1984, IEEE J. Quantum Electron. **20**, 551.
Lamb, W.E., 1995, Appl. Phys. B: Lasers Opt. **60**, 77.
Lamb, W.E., and R.C. Retherford, 1950, Phys. Rev. **79**, 549.
Lax, M., 1965, in *Proceedings of the International Conference on Quantum Electronics, Puerto Rico, 1965*, edited by P.L. Kelley, B. Lax, and P. Tannenwald (McGraw Hill, New York).
Lewis, G.N., 1926, Nature (London) **118**, 874.
Louisell, W.H., 1973, *Quantum Statistical Properties of Radiation* (Wiley, New York).
Maiman, T.H., 1960, Nature (London) **187**, 493.
Mandel, L., and E. Wolf, 1970, *Selected Papers on Coherence and Fluctuations of Light* (Dover, New York).
McFarlane, R.A., W.R. Bennett, and W.E. Lamb, 1963, Appl. Phys. Lett. **2**, 189.
Meystre, P., 1992, in *Progress in Optics*, edited by E. Wolf (North-Holland, Amsterdam), Vol. 30, p. 263.
Raithel, G., C. Wagner, H. Walther, L. Narducci, and M.O. Scully, 1994, in *Cavity Quantum Electrodynamics*, Adv. At. Mol., and Opt. Phys., Suppl. 2, edited by P.R. Berman (Academic Press, New York), p. 57.

Reppy, J.D., 1992, J. Low Temp. Phys. **87**, 205.
Risken, H., 1965, Z. Phys. **186**, 85.
Risken, H., 1984, *The Fokker Planck Equation* (Springer, Heidelberg).
Sargent, M., M.O. Scully, and W.E. Lamb, 1974, *Laser Physics* (Addison-Wesley, Reading).
Schawlow, A.L., and C.H. Townes, 1958, Phys. Rev. **112**, 1940.
Schleich, W., and S.M. Barnett, 1993, *Quantum Phase and Phase Dependent Measurements*, Phys. Scr. **T48**, 3.
Schleich, W., and M.O. Scully, 1984, in *Modern Trends in Atomic Physics*, edited by G. Grynberg and R. Stora (North-Holland, Amsterdam).
Schleich, W., and M.O. Scully, 1988, Phys. Rev. A **37**, 1261.
Scully, M.O., W.E. Lamb, and M.J. Stephen, 1965, in *Proceedings of the International Conference on Quantum Electronics, Puerto Rico, 1965*, edited by P.L. Kelley, B. Lax, and P. Tannenwald (McGraw Hill, New York).
Scully, M.O., and W.E. Lamb, 1967, Phys. Rev. **159**, 208.
Scully, M.O., 1985, Phys. Rev. Lett. **55**, 2802.
Scully, M.O., G.M. Meyer, and H. Walther, 1996, Phys. Rev. Lett. **76**, 4144.
Scully, M.O., S.-Y. Zhu, and A. Gavrielides, 1989, Phys. Rev. Lett. **62**, 2813.
Scully, M.O., and M.S. Zubairy, 1997, *Quantum Optics* (Cambridge University Press, Cambridge).
Serber, R., and C. H. Townes, 1960, *Quantum Electronics*, edited by C.H. Townes (Columbia University, New York).
Shimoda, K., T.C. Wang, and C.H. Townes, 1956, Phys. Rev. **102**, 1308.
Shimoda, K., H. Takahasi, and C.H. Townes, 1957, J. Phys. Soc. Jpn. **12**, 686.
Siegman, A.E., 1986, *Lasers* (University Science Books, Mill Valley).
Steiner, I., and P.E. Toschek, 1995, Phys. Rev. Lett. **74**, 4639.
Szöke, A., and A. Javan, 1963, Phys. Rev. Lett. **10**, 521.
Tolman, R.C., 1924, Phys. Rev. **23**, 693.
Townes, C.H., 1946, Phys. Rev. **70**, 665.
Townes, C.H., 1999, *How the Laser Happened; Adventures of a Scientist* (Oxford University, Oxford).
Van Vleck, J.H., and V.F. Weisskopf, 1945, Rev. Mod. Phys. **17**, 227.
Walther, H., 1992, Phys. Rep. **219**, 263.
Weber, J., 1953, IRE Trans. Electron Devices **3**, 1.
Winters, M.P., J.L. Hall, and P.E. Toschek, 1990, Phys. Rev. Lett. **65**, 3116.
Yamamoto, Y., and R.E. Slusher, 1993, Phys. Today **46**(6), 66.

Quantum Effects in One-Photon and Two-Photon Interference
L. Mandel

Although interference is intrinsically a classical wave phenomenon, the superposition principle which underlies all interference is also at the heart of quantum mechanics. Feynman has referred to interference as really "the only mystery" of quantum mechanics. Furthermore, in some interference experiments we encounter the idea of quantum entanglement, which has also been described as really the only quantum mystery. Clearly interference confronts us with some quite basic questions of interpretation. Despite its long history, going back to Thomas Young at the beginning of the 19th century, optical interference still challenges our understanding, and the last word on the subject probably has not yet been written. With the development of experimental techniques for fast and sensitive measurements of light, it has become possible to carry out many of the Gedanken experiments whose interpretation was widely debated in the 1920s and 1930s in the course of the development of quantum mechanics. Although this article focuses entirely on experiments with light, interference has also been observed with many kinds of material particles like electrons, neutrons, and atoms. We particularly draw the reader's attention to the beautiful experiments with neutron beams by Rauch and co-workers and others (see, for example, Badurek et al., 1988). Quantum optical interference effects are key topics of a recent book (Greenstein and Zajonc, 1997), an extended rather thorough review (Buzek and Knight, 1995) and an article in *Physics Today* (Greenberger et al., 1993).

The essential feature of any optical interference experiment is that the light from several (not necessarily primary) sources like S_A and S_B (see Fig. 1) is allowed to come together and mix, and the resulting light intensity is measured at various positions. We characterize interference by the dependence of the resulting light intensities on the optical path length or phase shift, but we need to make a distinction between the measurement of a single realization of the optical field and the average over an ensemble of realizations or over a long time. A single realization may exhibit interference, whereas an ensemble average may not. We shall refer to the former as transient interference, because a single realization usually exists only for a short time. Transient interference effects have been observed in several optical experiments in the 1950s and 1960s (Forrester et al., 1955; Magyar and Mandel, 1963; Pfleegor and Mandel, 1967, 1968).

We now turn to interference effects that are defined in terms of an ensemble average. Let us start by distinguishing between second-order or one-photon, and fourth-order or two-photon interference experiments. In the simplest and most familiar type of experiment, one photodetector, say D_1, is used repeatedly to measure the probability $P_1(x_1)$ of detecting a photon in some short time interval as a function of position x_1 [see Fig. 1(a)]. Interference is characterized by the (often, but not necessarily, periodic) dependence of $P_1(x_1)$ on the optical path lengths $S_A D_1$ and $S_B D_1$ or on the corresponding phase shifts ϕ_{A1} and ϕ_{B1}. Because $P_1(x_1)$ depends on the second power of the optical field and on the detection of one photon at a time, we refer to this as second-order, or one-photon, interference. Sometimes two photodetectors D_1 and D_2 located at x_1 and x_2 are used in coincidence repeatedly to measure the joint probability $P_2(x_1,x_2)$ of detecting one photon at x_1 and one at x_2 within a short time [see Fig. 1(b)]. Because $P_2(x_1,x_2)$ depends on the fourth power of the field, we refer to this as fourth-order, or two-photon, interference. For the purpose of this article, a photon is any eigenstate of the total number operator belonging to the eigenvalue 1. That means that a photon can be in the form of an infinite plane wave or a strongly localized wave packet. Because most photodetectors function by photon absorption, the appropriate dynamical variable for describing the measurement is the photon annihilation operator. If we make a Fourier decomposition of the total-field operator $\hat{E}(x)$ at the detector into its positive- and negative-frequency parts $\hat{E}^{(+)}(x)$ and $\hat{E}^{(-)}(x)$, then these play the roles of photon annihilation and creation operators in configuration space. Let $\hat{E}^{(+)}(x_1)$, $\hat{E}^{(+)}(x_2)$ be the positive-frequency parts of the optical field at the two detectors. Then $P_1(x_1)$ and $P_2(x_1,x_2)$ are given by the expectations in normal order:

$$P_1(x_1) = \alpha_1 \langle \hat{E}^{(-)}(x_1) \hat{E}^{(+)}(x_1) \rangle, \tag{1}$$

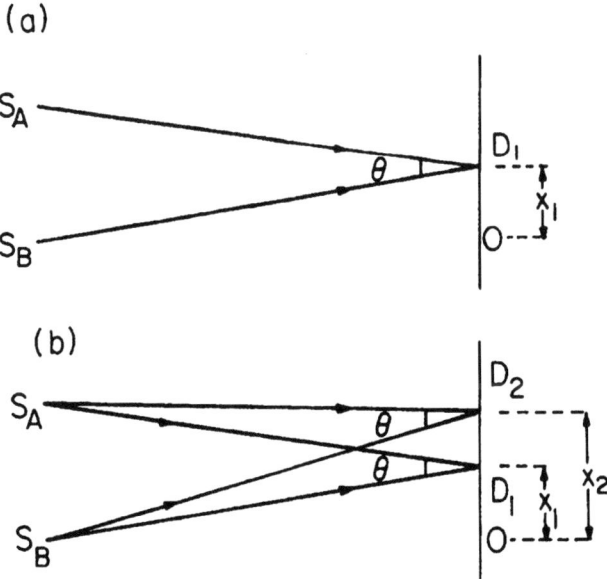

Figure 1. Principle of photon interference: (a) one-photon or second-order interference; (b) two-photon or fourth-order interference. S_A and S_B are sources. D_1 and D_2 are photodetectors.

$$P_2(x_1,x_2) = \alpha_1\alpha_2 \langle \hat{E}^{(-)}(x_1)\hat{E}^{(-)}(x_2)\hat{E}^{(+)}(x_2)\hat{E}^{(+)}(x_1)\rangle, \qquad (2)$$

where α_1, α_2 are constants characteristic of the detectors and the measurement times.

Second-Order Interference

Let us decompose $\hat{E}^{(+)}(x_1)$ and $\hat{E}^{(+)}(x_2)$ into two normal modes A and B, such that $\hat{a}_A, \hat{a}_B$ are the annihilation operators for the fields produced by the two sources S_A and S_B, respectively. Then we may put $\hat{E}^{(+)}(x_1) = f_A e^{i\phi_{A1}}\hat{a}_A + f_B e^{i\phi_{B1}}\hat{a}_B$, where f_A, f_B are complex parameters, and similarly for $\hat{E}^{(+)}(x_2)$. From Eq. (1) we then find

$$P_1(x_1) = \alpha_1[|f_A|^2 \langle \hat{n}_A \rangle + |f_B|^2 \langle \hat{n}_B \rangle + f_A^* f_B e^{i(\phi_{B1}-\phi_{A1})}\langle \hat{a}_A^\dagger \hat{a}_B \rangle + \text{c.c.}]. \qquad (3)$$

If second-order interference is characterized by the dependence of $P_1(x_1)$ on the optical path lengths or on the phase difference $\phi_{B1} - \phi_{A1}$, then clearly the condition for the system to exhibit second-order interference is that $\langle \hat{a}_A^\dagger \hat{a}_B \rangle \neq 0$. This is usually achieved most easily if the fields from the two sources S_A and S_B are at least partly correlated or mutually coherent. We define the degree of second-order mutual coherence by the normalized correlation ratio

$$|\gamma_{AB}^{(1,1)}| \equiv |\langle \hat{a}_A^\dagger \hat{a}_B \rangle|/(\langle \hat{a}_A^\dagger \hat{a}_A \rangle \langle \hat{a}_B^\dagger \hat{a}_B \rangle)^{1/2}, \qquad (4)$$

so that, by definition, $|\gamma_{AB}^{(1,1)}|$ lies between 0 and 1. But such correlation is not necessary for interference. Even with two independent sources it is apparent from Eq. (3) that interference can occur if $\langle \hat{a}_A \rangle \neq 0 \neq \langle \hat{a}_B \rangle$. An example would be the two-mode coherent state $|v_A\rangle_A |v_B\rangle_B$, with complex eigenvalues v_A, v_B, for which $\langle \hat{a}_A^\dagger \hat{a}_B \rangle = v_A^* v_B$, which is nonzero because of the definite complex amplitude of the field in a coherent state. The field of a single-mode laser is often well approximated by a coherent state for a short time. On the other hand the corresponding expectations vanish for a field in a Fock (photon number) state $|n_A\rangle_A |n_B\rangle_B$, for which $\langle \hat{a}_A^\dagger \hat{a}_B \rangle = 0$. Therefore there is no second-order interference in this case. Needless to say, this situation has no obvious counterpart in classical optics.

In order to understand why interference effects occur in some cases and not in others, we need to recall that interference is the physical manifestation of the intrinsic indistinguishability of the sources or of the photon paths. If the different possible photon paths from source to detector are indistinguishable, then we have to add the corresponding probability amplitudes before squaring to obtain the probability. This results in interference terms as in Eq. (3). On the other hand, if there is some way, even in principle, of distinguishing between the possible photon paths, then the corresponding probabilities have to be added and there is no interference.

Let us see how this argument works when each source consists of a single two-level atom. When the atom is in the fully excited state (an energy eigenstate), its energy can be measured, in principle, without disturbing the atom. Suppose that both sources are initially in the fully excited state and that the energy of each atom is measured immediately after the detection of a photon by D_1. If source A is found to be in the ground state whereas source B is found to be still excited, then, obviously, S_A can be identified as the source of the photon detected by D_1. Therefore there is no second-order interference in this case, and this conclusion holds regardless of whether the energy measurement is actually carried out. In this case, the optical field is in a one-photon Fock state $|1\rangle_A |0\rangle_B$

for which $\langle \hat{a}_A^\dagger \hat{a}_B \rangle = 0$. On the other hand, if the atoms are in a superposition of upper and lower states initially, then the atomic energy has no well-defined initial value and it cannot be measured without disturbing the atom. The source of the detected photon therefore cannot be identified by measuring the atomic energy, or in any other way, and, as a result, second-order interference is observed. This argument can be made more quantitative in that the degree of second-order coherence $|\gamma_{AB}^{(1,1)}|$ in Eq. (4) can be shown to equal the degree of path indistinguishability (Mandel, 1991).

It should be clear from the foregoing that in these experiments one photon does not interfere with another one; only the two probability amplitudes of the same photon interfere with each other. This has been confirmed more explicitly in interference experiments with a single photon (Grangier et al., 1986) and in experiments with two independent laser beams, in which interference was observed even when the light was so weak that one photon passed through the interferometer and was absorbed by the detector long before the next photon came along (Pfleegor and Mandel, 1967, 1968).

Fourth-Order Interference

We now turn to the situation illustrated in Fig. 1(b), in which two photodetectors are used in coincidence to measure the joint probability $P_2(x_1, x_2)$ of detecting one photon at x_1 and one at x_2. Fourth-order interference occurs when $P_2(x_1, x_2)$ depends on the phase differences $\phi_{A1} - \phi_{B2}$, and this happens when the different paths of the photon pair from the sources to the detectors are indistinguishable. Then we again have to add the corresponding (this time two-photon) probability amplitudes before squaring to obtain the probability. From Eq. (2) one can show that

$$\begin{aligned}P_2(x_1,x_2) = \alpha_1\alpha_2\{&|f_A|^4\langle:\hat{n}_A^2:\rangle + |f_B|^4\langle:\hat{n}_B^2:\rangle + 2|f_A|^2|f_B|^2\langle\hat{n}_A\rangle\langle\hat{n}_B\rangle[1+\cos(\phi_{B2}-\phi_{A2}\\&+\phi_{A1}-\phi_{B1})] + f_A^{*2}f_B^2\langle\hat{a}_A^{\dagger 2}\hat{a}_B^2\rangle e^{i(\phi_{B2}-\phi_{A2}+\phi_{B1}-\phi_{A1})} + \text{c.c.}\\&+|f_A|^2 f_A^* f_B \langle\hat{a}_A^{\dagger 2}\hat{a}_A\hat{a}_B\rangle[e^{i(\phi_{B1}-\phi_{A1})}+e^{i(\phi_{B2}-\phi_{A2})}]+\text{c.c.}\\&+|f_B|^2 f_B^* f_A \langle\hat{a}_B^{\dagger 2}\hat{a}_B\hat{a}_A\rangle[e^{i(\phi_{A1}-\phi_{B1})}+e^{i(\phi_{A2}-\phi_{B2})}]+\text{c.c.}\},\end{aligned} \quad (5)$$

where $\langle:\hat{n}^r:\rangle$ denotes the rth normally ordered moment of $\hat{n}$.

For illustration, let us focus once again on the special case in which each source consists of a single excited two-level atom. We have seen that in this case there is no second-order interference, because the source of each detected photon is identifiable in principle. But the same is not true for fourth-order interference of the photon pair. This time there are two indistinguishable two-photon paths, viz., (a) the photon from S_A is detected by D_1 and the photon from S_B is detected by D_2 and (b) the photon from S_A is detected by D_2 and the photon from S_B is detected by D_1. Because cases (a) and (b) are indistinguishable, we have to add the corresponding two-photon amplitudes before squaring to obtain the probability, and this generates interference terms. In this case most terms on the right of Eq. (5) vanish, and we immediately find the result given by (see Box A)

> Box A: Comparison of quantum and classical fourth-order interference.
>
> For the case in which exactly one photon is emitted by S_A and one photon by S_B, we have
>
> $\langle :\hat{n}_A^2: \rangle = \langle \hat{n}_A(\hat{n}_A-1) \rangle = 0, \quad \langle :\hat{n}_B^2: \rangle = \langle \hat{n}_B(\hat{n}_B-1) \rangle = 0,$
>
> $\langle \hat{a}_A^{\dagger 2} \hat{a}_B^2 \rangle = 0, \quad \langle \hat{a}_A^{\dagger 2} \hat{a}_A \hat{a}_B \rangle = 0, \quad \langle \hat{a}_B^{\dagger 2} \hat{a}_B \hat{a}_A \rangle = 0.$
>
> Therefore of all the terms on the right hand side of Eq. (5) only the third survives, and we obtain finally,
>
> $$P_2(x_1,x_2) = \alpha_1 \alpha_2 2|f_A|^2|f_B|^2[1+\cos(\phi_{B2}-\phi_{A2}+\phi_{A1}-\phi_{B1})]. \qquad (6)$$
>
> This result exhibits two-photon interference with 100% visibility.
> Let us contrast this conclusion for a two-photon state with the result given by the same Eq. (5) for a classical state of the incoming field, when the two sources are completely independent. In this case we have to treat $\hat{a}_A, \hat{a}_B$ as complex c-number amplitudes, $\langle \hat{n}_A \rangle$ becomes the mean light intensity $\langle I_A \rangle$, and $\langle :\hat{n}_A^2: \rangle$ becomes $\langle I_A^2 \rangle$. Because of the phase independence, all terms on the right hand side of Eq. (5) beyond the first three vanish again, but this time terms one and two are nonzero, and we have the result
>
> $$P_2(x_1,x_2) = \alpha_1 \alpha_2 [\langle (|f_A|^2 I_A + |f_B|^2 I_B)^2 \rangle + 2|f_A|^2|f_B|^2 \langle I_A \rangle \langle I_B \rangle \cos(\phi_{B2}-\phi_{A2}+\phi_{A1}-\phi_{B1})].$$
>
> It is not difficult to prove that in this classical field case the visibility of the interference has an upper bound of 1/2, compared with the value 1 given by Eq. (6) for the case of a quantum field (Richter, 1979).

$$P_2(x_1,x_2) = \alpha_1 \alpha_2 2|f_A|^2|f_B|^2[1+\cos(\phi_{B2}-\phi_{A2}+\phi_{A1}-\phi_{B1})]. \qquad (6)$$

Despite the fact that the two sources are independent, they exhibit two-photon interference with 100 percent visibility.

Two-photon interference exhibits some striking nonlocal features. For example, $P_2(x_1,x_2)$ given by Eq. (6) can be shown to violate one or more of the Bell inequalities that a system obeying local realism must satisfy. This violation of locality, which is discussed more fully in the article by Zeilinger in this issue, has been demonstrated experimentally. [See, for example, Mandel and Wolf, 1995.]

Interference Experiments with a Parametric Downconverter Source

The first two-photon interference experiment of the type illustrated in Fig. 1(b), in which each source delivers exactly one photon simultaneously, was probably the one reported by Ghosh and Mandel in 1987. They made use of the signal and idler photons emitted in the splitting of a pump photon in the process of spontaneous parametric downconversion in a nonlinear crystal of $LiIO_3$. The crystal was optically pumped by the 351.1-nm uv beam from an argon-ion laser and from time to time it gave rise to two simultaneous signal and idler photons at wavelengths near 700 nm. A modified and slightly improved version of the experiment was later described by Ou and Mandel (1989). The signal (s) and idler (i) photons were incident from opposite sides on a 50%:50% beam splitter that mixed them at a small angle $\theta \approx 1$ mrad, and the two mixed beams then fell on detectors D_1 and D_2, each of which carried a 0.1-mm-wide aperture. The photons counted by each detector separately and by the two detectors in coincidence in a total time of a few minutes were

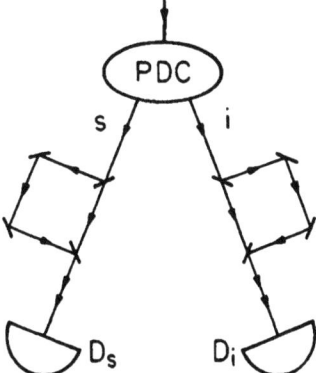

Figure 2. Principle of the Franson (1989) two-photon interference experiment in which signal and idler photons never mix. PDC is the parametric downconverter. D_s and D_i are photodetectors.

registered for various positions of the detectors. Because of the two-photon state, no second-order interference is expected from quantum mechanics, as we have seen, and none was observed. However, the two-photon coincidence rate exhibited the expected interference in the form of a periodic variation of the rate with detector position, because the photon pair detected by D_1 and D_2 could have originated as signal and idler, respectively, or vice versa.

An ingenious variation on the same theme of two-photon interference was proposed by Franson (1989) and is illustrated in Fig. 2. Signal and idler photons emitted simultaneously in two slightly different directions from a parametric downconverter (PDC) fall on two detectors D_s and D_i, respectively. The two beams never mix. On the way to the detector each photon encounters a beam splitter leading to an alternative time-delayed path, as shown, and each photon is free to follow either the shorter direct or the longer delayed path. If the time difference T_D between the long and short paths is much longer than the coherence time T_C of the downconverted light, and much longer than the coincidence resolving time T_R, no second-order interference is to be expected, and at first glance it might seem that no fourth-order interference would occur either. But the signal and idler photons are emitted simultaneously, and, within the coincidence resolving time, they are detected simultaneously. Therefore in every coincidence both photons must have followed the short path or both photons must have followed the long path, but we cannot tell which. When $T_C \ll T_D \ll T_R$ two more path combinations are possible. With continuous pumping of the parametric downconverter the emission time is random and unknown, and there is no way to distinguish between the light paths. We therefore have to add the corresponding probability amplitudes, which leads to the prediction of fourth-order interference as the path difference in one arm is varied. This has been confirmed experimentally. A different outcome may be encountered with pulsed rather than continuous excitation of the parametric downconverter.

Interference Experiments With Two Parametric Downconverters

Next let us consider the experiment illustrated in Fig. 3, which allows both one-photon and two-photon interference to be investigated at the same time. Two similar nonlinear

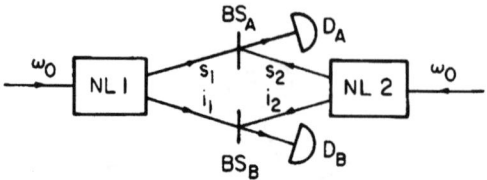

Figure 3. Principle of the interference experiment with two downconverters in which both one-photon and two-photon interference can be investigated (after Ou et al., 1990).

crystals NL1 and NL2, which both function as parametric downconverters, are optically pumped simultaneously by mutually coherent pump beams that we shall treat classically and represent by the complex field amplitudes V_1 and V_2. As a result downconversion can occur at NL1, with the simultaneous emission of a signal s_1 and an idler i_1 photon in two slightly different directions, or downconversion can occur at NL2, with the simultaneous emission of an s_2 and an i_2 photon, as shown. The question we wish to address is whether, in view of the mutual coherence of the two pump beams, the s_1 and s_2 beams from the two downconverters are mutually coherent and exhibit interference when they are mixed, and similarly for the i_1 and i_2 beams. In order to answer the question the experiment illustrated in Fig. 3 is carried out. s_1 and s_2 are allowed to come together; they are mixed at the 50%:50% signal beam splitter BS_A, and the combined beam emerging from BS_A falls on the photon detector D_A. If s_1 and s_2 are mutually coherent, then the photon counting rate of D_A varies sinusoidally as the phase difference between the two pump beams V_1 and V_2 is slowly increased. Similarly for the two idlers i_1 and i_2, which are mixed by BS_B and detected by D_B.

In order to treat this problem theoretically we represent the quantum state of the signal and idler photon pair from each crystal by the entangled state $|\Psi_j\rangle = M_j |\text{vac}\rangle_{s_j, i_j} + \eta V_j |1\rangle_{s_j} |1\rangle_{i_j}$ ($j=1,2$). The combined state is then the direct product state $|\Psi\rangle = |\Psi_1\rangle \times |\Psi_2\rangle$, because the two downconversions proceed independently. V_1 and V_2 are the c-number complex amplitudes of the pump fields. η represents the coupling between pump modes and the downconverted signal and idler modes, such that $\langle |\eta V_j|^2 \rangle$ ($j=1,2$) is the small probability of downconversion in a short measurement time. M_1 and M_2 are numerical coefficients that ensure the normalization of $|\Psi_1\rangle$ and $|\Psi_2\rangle$, which we take to be real for simplicity. Because $\langle |\eta V_j|^2 \rangle \ll 1$ it follows that M_1 and M_2 are very close to unity. We shall retain the coefficients M_1 and M_2 nevertheless, because they provide us with useful insight into the role played by the vacuum. Of course the downconverted light usually has a very large bandwidth, and treating each signal and idler as occupying one monochromatic mode is a gross oversimplification. However, a more exact multimode treatment leads to very similar conclusions about the interference.

The positive-frequency parts of the signal and idler fields at the two detectors can be given the two-mode expansions $\hat{E}_s^{(+)} = \hat{a}_{s1} e^{i\theta_{s1}} + i\hat{a}_{s2} e^{i\theta_{s2}}$ and $\hat{E}_i^{(+)} = \hat{a}_{i1} e^{i\theta_{i1}} + i\hat{a}_{i2} e^{i\theta_{i2}}$, where $\theta_{s1}, \theta_{s2}, \theta_{i1}, \theta_{i2}$ are phase shifts corresponding to the propagation from one of the two sources NL1, NL2 to one of the two detectors D_A, D_B. Then the expectations of the number of photons detected by D_A and by D_B are $\langle \Psi | \hat{E}_s^{(-)} \hat{E}_s^{(+)} | \Psi \rangle$ and $\langle \Psi | \hat{E}_i^{(-)} \hat{E}_i^{(+)} | \psi \rangle$, and for the quantum state $|\Psi\rangle = |\Psi_1\rangle \times |\Psi_2\rangle$ we obtain immediately

$$\langle \Psi | \hat{E}_s^{(-)} \hat{E}_s^{(+)} | \Psi \rangle = |\eta|^2 (\langle |V_1|^2 \rangle + \langle |V_2|^2 \rangle) = \langle \Psi | \hat{E}_i^{(-)} \hat{E}_i^{(+)} | \Psi \rangle. \tag{7}$$

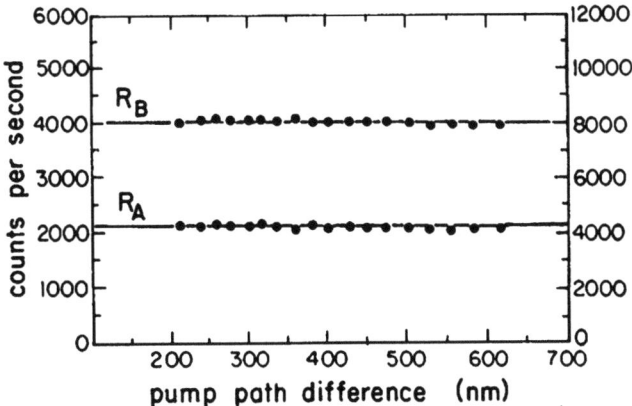

Figure 4. Results of measurements of the photon counting rate by D_A and D_B in Fig. 3 as a function of path difference, showing the absence of one-photon interference.

These averages are independent of the interferometric path lengths and of the phases of the two pump beams, showing that there is no interference and no mutual coherence between the two signals s_1, s_2 or between the two idlers i_1, i_2. These conclusions are confirmed by the experimental results presented in Fig. 4, which exhibit no sign of second-order or one-photon interference.

Next let us look at the possibility of fourth-order or two-photon interference, by measuring the joint probability of detecting a signal photon and an idler photon with both detectors in coincidence. This probability is proportional to $P_{12} = \langle \Psi | \hat{E}_s^{(-)} \hat{E}_i^{(-)} \hat{E}_i^{(+)} \hat{E}_s^{(+)} | \Psi \rangle$, and it is readily evaluated. If $|V_1|^2 = I = |V_2|^2$ and $|\eta|^2 I \ll 1$, so that terms of order $|\eta|^4 I^2$ can be neglected, we find

$$P_{12} = 2|\eta|^2 \langle I \rangle [1 - M_1 M_2 |\gamma_{12}^{(1,1)}| \cos \Theta], \tag{8}$$

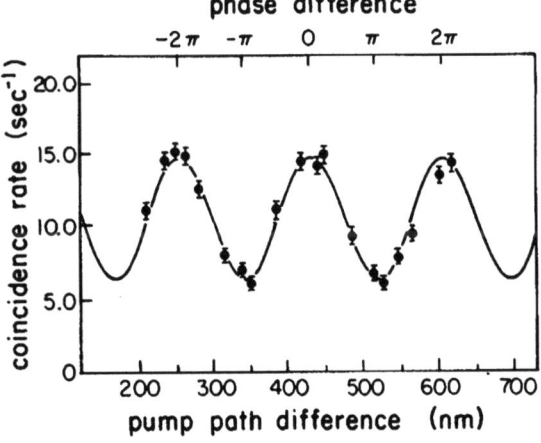

Figure 5. Results of coincidence measurements by D_A and D_B in Fig. 3 as a function of path difference, showing two-photon interference. The continuous curve is theoretical.

where $\Theta \equiv \theta_{s2} + \theta_{i2} - \theta_{s1} - \theta_{i1} + \arg(\gamma_{12}^{(1,1)})$ and $\gamma_{12}^{(1,1)} \equiv \langle V_1^* V_2 \rangle / \langle I \rangle$ is the complex degree of coherence of the two classical pump beams. A two-photon coincidence measurement with both detectors D_A and D_B is therefore expected to exhibit interference as the optical path difference or the pump phase difference is varied. This is confirmed by the experimental results shown in Fig. 5. It is interesting to note that the vacuum contribution to the state plays an essential role, because of the presence of the $M_1 M_2$ coefficients in Eq. (8).

Finally, we would like to understand in physical terms why no second-order interference is registered by detectors D_A and D_B separately, but fourth-order interference is registered by the two together. Here it is helpful to recall the relationship between interference and indistinguishability. From the coincidence measurement in Fig. 5 it is impossible to determine whether the detected photon pair originates in NL1 or in NL2, and this indistinguishability is manifest as a fourth-order interference pattern. However, if we are interested only in the interference of, say, the signal photons registered by D_A, we can use the detection of the idlers as an auxiliary source of information, to determine where each detected signal photon originated. This destroys the indistinguishability of the two sources and kills the interference of the signal photons, whether or not the auxiliary measurement is actually carried out.

Figure 6 illustrates a one-photon interference experiment with two downconverters that exhibits interesting nonclassical features (Zou et al., 1991). NL1 and NL2 are two similar nonlinear crystals of LiIO$_3$ functioning as parametric downconverters. They are both optically pumped by the mutually coherent uv light beams from an argon-ion laser oscillating on the 351.1-nm spectral line. As a result, downconversion can occur at NL1 with the simultaneous emission of a signal s_1 and an idler i_1 photon at wavelengths near 700 nm, or it can occur at NL2 with the simultaneous emission of an s_2 and i_2 photon. Simultaneous downconversions at both crystals is very improbable. NL1 and NL2 are aligned so as to make i_1 and i_2 collinear and overlapping, as shown, so that a photon detected in the i_2 beam could have come from NL1 or NL2. At the same time the s_1 and s_2 signal beams come together and are mixed at beam splitter BS$_0$. The question to be explored is whether, in view of the mutual coherence of the two pump beams, s_1 and s_2 are also mutually coherent and exhibit interference, under the conditions when the downconversions at NL1 and NL2 are spontaneous and random. More explicitly, if BS$_0$ is translated in a direction normal to its face. Will the photon counting rate of detector D_s vary sinusoidally, thereby indicating that interference fringes are passing across the photocathode?

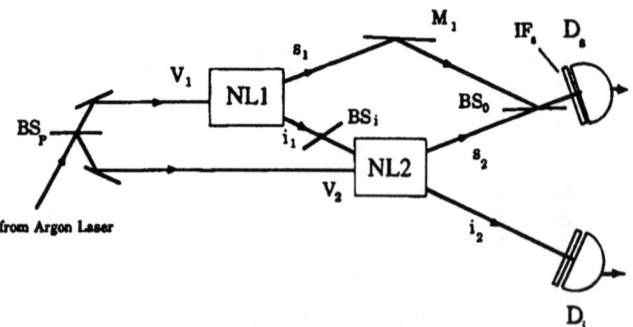

Figure 6. Outline of the one-photon interference experiment with two downconverters (Zou et al., 1991). See text for description.

With the experiment in Fig. 3 in mind, one might not expect to see one-photon interference at D_s, but, as shown in Fig. 7 (curve A), interference fringes were actually observed so long as i_1 and i_2 were well aligned and overlapped. The relatively small visibility of the interference is largely due to the incomplete overlap of the two idlers. However, after deliberate misalignment of i_1 and i_2, or if i_1 was blocked from reaching NL2, all interference disappeared, as shown by curve B in Fig. 7. Yet the average rate of photon emission from NL2 was unaffected by blocking i_1 or by misalignment. In the absence of induced emission from NL2, how can this be understood?

Here it is instructive again to invoke the relationship between interference and indistinguishability. Let us suppose that an auxiliary perfect photodetector D_i is placed in the path of the i_2 beam equidistant with D_s from NL2, as shown in Fig. 6. Now the insertion of D_i in the path of i_2 does not in any way disturb the interference experiment involving the s_1 and s_2 beams. However, when i_1 is blocked, D_i provides information about the source of the signal photon detected by D_s. For example, if the detection of a signal photon by D_s is accompanied by the simultaneous detection of an idler photon by D_i, a glance at Fig. 6 shows immediately that the signal photon (and the idler) must have come from NL2. On the other hand, if the detection of a signal photon by D_s is not accompanied by the simultaneous detection of an idler by D_i, then the signal photon cannot have come from NL2 and must have originated in NL1. With the help of the auxiliary detector D_i we can therefore identify the source of each detected signal photon, whenever i_1 is blocked, and this distinguishability wipes out all interference between s_1 and s_2. A similar conclusion applies when the two idlers i_1 and i_2 do not overlap, so that they can be measured separately. However, when i_1 is unblocked and the two idlers overlap, this source identification is no longer possible, and s_1 and s_2 exhibit interference. Needless to say, it is not necessary actually to carry out the auxiliary measurement with D_i; the mere possibility, in principle, that such a measurement could determine the source of the signal photon is sufficient to kill the interference of s_1 and s_2.

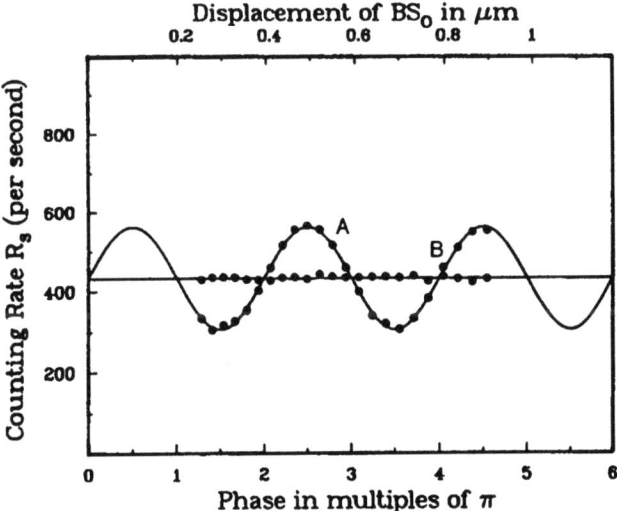

Figure 7. Results of the one-photon interference experiment shown in Fig. 6: A, data with i_I unblocked; B, data with i_I blocked.

This kind of argument leads to an important conclusion about the quantum state of a system: in an experiment the state reflects not what is actually known about the system, but rather what is knowable, in principle, with the help of auxiliary measurements that do not disturb the original experiment. By focusing on what is knowable in principle, and treating what is known as largely irrelevant, one completely avoids the anthropomorphism and any reference to consciousness that some physicists have tried to inject into quantum mechanics. We emphasize here that the act of blocking the path of i_1 between NL1 and NL2 kills the interference between s_1 and s_2 not because it introduces a large uncontrollable disturbance. After all, the signal photons s_1 and s_2 are emitted spontaneously and the spontaneous emissions are not really disturbed at all by the act of blocking i_1. In this experiment the disturbance introduced by blocking i_1 is of a more subtle kind: it is only the possibility of obtaining information about the source of the signal photon which is disturbed by blocking i_1.

If, instead of blocking i_1 completely from reaching NL2, one merely attenuates i_1 with some sort of optical filter of complex transmissivity T, then the degree of coherence and the visibility of the interference pattern formed by s_1 and s_2 are reduced by the factor $|T|$ (Zou et al., 1991). This provides us with a convenient means for controlling the degree of coherence of two lightbeams s_1 and s_2 with a variable filter acting on i_1, without affecting the light intensities of s_1 and s_2. Finally, insofar as i_1 falling on NL2 may be said to induce coherence between s_1 and s_2 from the two sources NL1 and NL2, we have here an example of induced coherence without induced emission.

Measurement of the Time Interval between Two Photons by Interference

The same fourth-order two-photon interference effect has been used to measure the time separation between two photons with time resolution millions of times shorter than the resolution of the detectors and the electronics (Hong et al., 1987). Let us consider the experiment illustrated in Fig. 8. Here the signal and idler photons emitted from a uv-pumped crystal of potassium dihydrogen phosphate (KDP) serving as parametric downconverter are sent in opposite directions through a symmetric 50%:50% beam splitter (BS) that mixes them. The emerging photon pair is allowed to impinge on two similar photon detectors D_1 and D_2, whose output pulses are counted both separately and

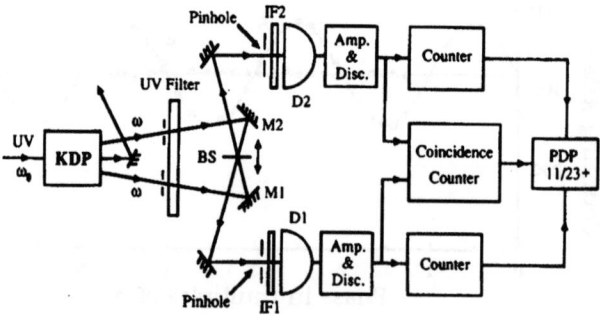

Figure 8. Outline of the two-photon interference experiment to measure the time separation between signal and idler photons (Hong et al., 1987). See text for description.

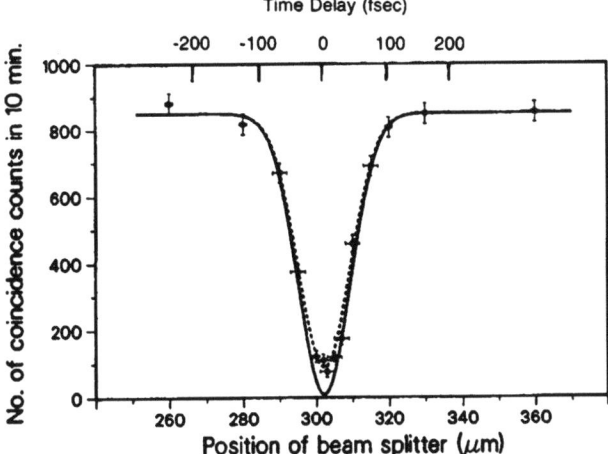

Figure 9. Results of the two-photon interference experiment shown in Fig. 8. The measured coincidence rate is plotted as a function of beam-splitter displacement in μm or differential time delay in fsec. The continuous curve is theoretical.

in coincidence as the beam splitter is translated in the direction shown though a distance of a few wavelengths. The coherence time T_c of the downconverted light is made about 10^{-13} sec with the help of the interference filters IF_1, IF_2.

Let us consider the quantum state $|\psi\rangle$ of the photon pair emerging from the beam splitter. With two photons impinging on BS from opposite sides there are really only three possibilities for the light leaving BS: (a) one photon emerges from each of the outputs 1 and 2; (b) two photons emerge from output 1 and none emerges from output 2; (c) two photons emerge from output 2 and none emerges from output 1. The quantum state of the beam-splitter output is actually a linear superposition of all three possibilities in the form

$$|\psi\rangle = (|\mathcal{R}|^2 - |\mathcal{T}|^2)|1\rangle_1|1\rangle_2 + \sqrt{2}i|\mathcal{RT}|[|2\rangle_1|0\rangle_2 + |0\rangle_1|2\rangle_2], \qquad (9)$$

where $\mathcal{R}$ and $\mathcal{T}$ are the complex beam-splitter reflectivity and transmissivity. When $|\mathcal{R}| = 1/\sqrt{2} = |\mathcal{T}|$, the first term on the right vanishes, which implies the destructive interference of the photon pair in arms 1 and 2, and both photons emerge together either in arm 1 or in arm 2. Therefore no coincidence counts (other than accidentals) between detectors D_1 and D_2 are registered. The reason for this can be understood by reference to Fig. 8. A coincidence detection between D_1 and D_2 can occur only if the two incoming signal and idler photons are either both reflected from the beam splitter or are both transmitted through the beam splitter. Because these two possible photon paths are indistinguishable, we have to add the corresponding two-photon probability amplitudes before squaring to obtain the probability. But because of the phase change that occurs on reflection from the beam splitter, as compared with that on transmission, the two-photon probability amplitude for two reflections from BS is 180° out of phase with the corresponding two-photon probability amplitude for two transmissions through BS. When these two amplitudes are added they give zero.

Needless to say, this perfect destructive interference of the photon pair requires two identical incident photons, and their description goes well beyond our oversimplified two-mode treatment. If we think of the incoming entangled photon pair as two identical wave packets that overlap completely in time, then it should be obvious that if one wave packet is delayed even slightly relative to the other, perfect destructive interference is no longer possible, and the apparatus in Fig. 8 no longer yields zero coincidences. The greater the relative time delay τ_D, the greater is the two-photon coincidence rate R_c, and by the time the delay τ_D exceeds the time duration of the wave packet, the coincidence rate R_c becomes constant and independent of the time delay τ_D between the wave packets. For wave packets of Gaussian shape and bandwidth $\Delta\omega$, and with a 50%:50% beam splitter, one finds that R_c is given by (see Box B)

$$R_c \propto K[1 - e^{-\tau_D^2(\Delta\omega)^2}]. \tag{10}$$

The two-photon coincidence rate R_c is therefore expected to vary with the time delay τ_D as in Fig. 9. This has indeed been observed in an experiment in which the differential time delay τ_D was introduced artificially by translating the beam splitter BS in Fig. 8 (Hong et al., 1987). It is worth noting that the measurement achieved a time resolution of a few femtoseconds, which is a million times shorter than the time resolution of the photon detectors and the associated electronics. This is possible because the measurement was really based on optical interference. In some later experiments the resolution time was

Box B: Measuring the time separation between two photons.

The photon pair s,i from the downconverter is mixed at a symmetric 50%:50% beam splitter after a known differential time delay τ_D is introduced between the signal (s) and idler (i). The joint probability for detecting two photons at the two beam-splitter outputs 1 and 2 at time t and $t+\tau$ is proportional to

$$P_{12}(t,t+\tau) = \langle \hat{E}_1^{(-)}(t)\hat{E}_2^{(-)}(t+\tau)\hat{E}_2^{(+)}(t+\tau)\hat{E}_1^{(+)}(t)\rangle.$$

and when this is evaluated, the result is expressible in the form.

$$P_{12}(t,t+\tau) = K[g^2(\tau_D+\tau) + g^2(\tau_D-\tau) - 2g(\tau_D-\tau)g(\tau_D+\tau)].$$

Here $g(\tau)$ is the Fourier transform of the spectral function that characterizes the downconverted field, and $g(\tau)$ is real, symmetric, and normalized so that $g(0)=1$. The rate R_c at which photon pairs in beams 1 and 2 are detected in coincidence is given by

$$R_c = \int_{-T_R/2}^{T_R/2} P_{12}(t,t+\tau)d\tau,$$

where T_R is the resolving time of the coincidence detector. When T_R greatly exceeds the reciprocal bandwidth $\Delta\omega$ of the downconverted light, the limits on the integral effectively become $\pm\infty$, with the result that

$$R_c = 2K\int_{-\infty}^{\infty} d\tau g^2(\tau)\left\{1 - \int_{-\infty}^{\infty} d\tau g(\tau_D-\tau)g(\tau_D+\tau)\bigg/\int_{-\infty}^{\infty}d\tau g^2(\tau)\right\}.$$

In the special case in which $g(\tau)$ is Gaussian and of the form

$$g(\tau) = e^{-(\tau\Delta\omega)^2/2},$$

we obtain

$$R_c = (2\sqrt{\pi}/\Delta\omega)K[1 - e^{-\tau_D^2(\Delta\omega)^2}].$$

even shorter than the period of the light. The same principle has been used by Chiao and co-workers to measure photon tunneling times through a barrier.

Conclusions

We have seen that quantum effects can show up in both one-photon and two-photon interference. The analysis of some interference experiments confronts us with fundamental questions of interpretation and brings out that the quantum state reflects not what we know about the system, but rather what is knowable in principle. This avoids any reference to consciousness in the interpretation of the state. Finally, quite apart from their fundamental interest, quantum interference effects have led to some valuable practical applications, such as the new method for measuring the time separation between two photons on a femtosecond time scale, and new techniques of controlling the degree of coherence of two light beams without change of intensity.

Acknowledgments

This article is based on research resulting from collaboration with numerous former graduate students and postdocs whose help is gratefully acknowledged. The research was supported by the National Science Foundation and by the U.S. Office of Naval Research.

References

Badurek, G., H. Rauch, and A. Zeilinger, 1988, Eds., *Matter Wave Interferometry* (North-Holland, Amsterdam).
Buzek, V., and P. L. Knight, 1995, in *Progress in Optics XXXIV*, edited by E. Wolf (Elsevier, Amsterdam), p. 1.
Forrester, A. T., R. A. Gudmundson, and P. O. Johnson, 1955, Phys. Rev. **99**, 1691.
Franson, J. D., 1989, Phys. Rev. Lett. **62**, 2205.
Ghosh, R., and L. Mandel, 1987, Phys. Rev. Lett. **59**, 1903.
Grangier, P., G. Roger, and A. Aspect, 1986, Europhys. Lett. **1**, 173.
Greenberger, D. M., M. A. Horne, and A. Zeilinger, 1993, Phys. Today **46**, 22.
Greenstein, G., and A. G. Zajonc, 1997, *The Quantum Challenge* (Jones and Bartlett, Sudbury, MA).
Hong, C. K., Z. Y. Ou, and L. Mandel, 1987, Phys. Rev. Lett. **59**, 2044.
Kwiat, P. G., W. A. Vareka, C. K. Hong, H. Nathel, and R. Y. Chiao, 1990, Phys. Rev. A **41**, 2910.
Magyar, G., and L. Mandel, 1963, Nature (London) **198**, 255.
Mandel, L., 1991, Opt. Lett. **16**, 1882.
Mandel, L., and E. Wolf, 1995, *Optical Coherence and Quantum Optics*, 1st Ed. (Cambridge University Press, Cambridge, UK).
Ou, Z. Y., and L. Mandel, 1989, Phys. Rev. Lett. **62**, 2941.
Ou, Z. Y., L. J. Wang, X. Y. Zou, and L. Mandel, 1990, Phys. Rev. A **41**, 566.
Pfleegor, R. L., and L. Mandel, 1967, Phys. Rev. **159**, 1034.
Pfleegor, R. L., and L. Mandel, 1968, J. Opt. Soc. Am. **58**, 946.
Richter, Th., 1979, Ann. Phys. (Leipzig) **36**, 266.
Zou, X. Y., L. J. Wang, and L. Mandel, 1991, Phys. Rev. Lett. **67**, 318.

From Nanosecond to Femtosecond Science

N. Bloembergen

The measurement of short time intervals during the first half of this century was limited to intervals longer than one nanosecond. Before 1900 it was already known that electrical sparks and Kerr-cell shutters could have response times as short as 10^{-8} s. Abraham and Lemoine (1899) used a discharge from condenser plates which simultaneously activated a spark gap and a Kerr-cell shutter. The condenser plates were placed in a cell containing carbon disulfide between crossed polarizer and analyzer. The light path between the spark source and the Kerr cell was variable. If the delay path was longer than four meters, no light was transmitted. This proved that the Kerr-cell response was faster than 10^{-8} s. We now know that the response of the Kerr effect in CS_2 is about two picoseconds. Thus Abraham and Lemoine really measured the duration of light emitted by the spark. This experiment used the essential feature of a variation in light path length for the measurement of short time intervals. In most picosecond and femtosecond time-resolved experiments, the delay between a pump and a probe pulse is varied by changes in optical path length. Norrish and Porter (1950) used flash lamps to excite photochemical reactions and probed them spectroscopically with varying delays in the millisecond to microsecond range. This work in "microchemistry" was recognized by the Nobel prize for chemistry in 1967.

The advent of lasers in 1960 revolutionized the field of time-resolved spectroscopy. Shapiro (1977) has presented a historical overview and edited an early volume dedicated to ultrashort light pulses. During the past two decades the field has grown explosively and new scientific subfields, denoted by femtochemistry and femtobiology, have been created.

In Sec. II, the highlights in the experimental development are surveyed, leading from nanosecond to femtosecond light pulse generation and detection. In Sec. III, some paradigms of transient excitations, involving nuclear displacements and vibrations in molecules and in condensed matter, are presented.

A final section draws attention to the fact that the concentration of light in space and time, as achieved in focused femtosecond laser pulses, can reach power flux densities exceeding 10^{18} watts/cm^2. Some ultrahigh-field phenomena that are currently under intense investigation include ultrafast phase transitions in electronic structure and the creation of high-density, high-temperature plasmas. Other phenomena include above-threshold ionization of atoms, high-order harmonic generation, and the acceleration of relativistic electrons.

The vastness of the new orders of magnitude in time and in radiative density that have been opened up to experimental investigation by short light pulses may be illustrated by the following two observations: The minute is the geometric mean between the lifetime of the universe and the duration of a 10-fs pulse. The radiative density of blackbody radiation at ten million degrees, corresponding to star interiors, is 10^{17} watts/cm^2.

Highlights in Short-Pulse Generation and Detection

The first laser, realized by Maiman (1960), was based on a ruby crystal, pumped by a xenon flash discharge. It created a laser pulse of fluctuating intensity lasting between a microsecond and a millisecond. Hellwarth (1961a, 1961b) proposed the concept of Q switching at the second conference on quantum electronics, held in Berkeley, California, in March 1961. He reported Q switching of a ruby laser by means of a Kerr-cell shutter at the APS Chicago meeting in November 1961. A "giant" pulse with a duration of about 10 nanoseconds was reported. Others achieved Q switching by means of turbine-driven rapidly rotating reflecting prisms. The higher intensities available in Q-switched pulses aided in the detection and study of numerous nonlinear optical phenomena.

Mode locking of a large number of longitudinal modes all activated by the gain profile of the lasing medium was proposed to obtain shorter pulses. Active mode locking was first demonstrated for a helium-neon laser by Hargrove, Fork, and Pollack (1964). By modulating the index of refraction acoustically at the period of the round-trip time of the light in the laser cavity, they obtained an output from the He-Ne laser consisting of a series of narrow pulses spaced by the cavity round-trip transit time. Since the gain profile of the active Ne atoms was narrow, the pulse duration remained well above one nanosecond.

Passive mode locking of a ruby laser by means of a saturable absorber was first demonstrated by Mocker and Collins (1965). Pulses shorter than a nanosecond were obtained by DeMaria et al. (1966) by passive mode locking of a Nd-glass laser, which has a broader gain profile. The generation of the short pulse may be qualitatively understood as follows: A stochastic peak in the amplified spontaneous emission output will bleach the dye absorber more and will consequently be preferentially amplified. Thus the emitted power will be concentrated in a short pulse with a duration determined by the gain bandwidth product of the laser. The broadband response of Nd-glass lasers, and later the Nd-Yag (yttrium aluminum garnet) lasers, made passive mode locking by a thin film of saturable absorbing dye in the laser cavity very effective. Pulses of 10-ps duration were readily obtained with such systems and utilized in many laboratories.

The first pulses shorter than 1 ps were obtained by Shank and Ippen (1974) with tunable broad-gain dye laser media in combination with a saturable dye absorber. An analysis by New (1972) showed that it is possible to obtain pulses shorter than the characteristic response time of either the absorbing or the emitting dye, as the pulse duration is determined by the balance of saturable gain and saturable absorption. Further developments include the use of counterpropagating pulses in a ring dye laser system. The two pulses cross each other in a thin film of saturable dye, with a thickness of a few μm. With compensation of group velocity dispersion by a configuration of glass prisms in the laser cavity, a repetitive train of six femtosecond pulses was reported by Fork et al. (1987). These systems, with pulse durations between 10 and 100 fs, required delicate adjustments, but were used by many investigators during the decade of the eighties.

In the current decade they are rapidly being replaced by Ti-sapphire lasers. Spence, Kean, and Sibbett (1991) discovered that these lasers may show "spontaneous" dynamic mode locking without the use of a saturable absorber. The effect is based on the occurrence of self-focusing by the intensity-dependent index of refraction in the Ti-sapphire crystal. In combination with a suitable aperture, the round-trip gain is increased at higher intensities, as more light is focused to pass through the aperture inside the optical resonator. Again, compensation of group velocity dispersion is essential to obtain very short pulses. Zhou et al. (1994) obtained a pulse of 8-fs duration from a Ti-sapphire laser. An all-solid-state system based on pumping the Ti-sapphire with a semiconductor array to emit a continuous train of short pulses is feasible. Thus femtosecond pulse generators are rapidly becoming a standard laboratory tool.

The measurement of the duration of picosecond pulses was first carried out by fast photoelectronic streak cameras, but soon autocorrelation and cross-correlation techniques were introduced. In a typical arrangement one divides a pulse by a beam splitter. The two pulses are recombined with a variable delay in one path. The state of polarization may also be changed. When the two pulses travel in opposite directions in a fluorescent medium, two-photon-induced fluorescence may be observed at right angles. This fluorescence is enhanced in a narrow strip where the two pulses overlap. The two pulses may also be recombined traveling in different directions in a thin sliver of a nonlinear material. Second-harmonic generation in the phase-matched directions occurs only when the two pulses are simultaneously present. Note that a picosecond differential corresponds to a delay in optical path length of 0.3 mm, and a femtosecond differential to 0.3 μm. A picosecond traveling-wave Kerr-cell shutter activated by an intense pump pulse traveling in the same direction was realized rather early. The change in index of refraction or birefringence is proportional to the intensity of the pump pulse. These detection methods are all based on the nonlinear response of an optical medium. The temporal response of the amplitude or the intensity of the probe pulse is readily measured.

Important information about the temporal behavior of the phase of the pulsed field may be obtained by spectrally analyzing the second-harmonic or sum frequency signal produced by the two pulses as a function of delay time. This technique was first introduced by Trebino and Kane (1993) and is called FROG (frequency-resolved optical gating). The time variations in the observed spectrum of the combination signal give the temporal variations in phase. Thus a complete picture of the pulsed field was obtained for a Ti-sapphire laser by Taft et al. (1995).

Conversely, it is also possible to generate a pulse with prescribed amplitude and phase. To achieve complete pulse shaping one obtains a spectrum of the collimated pulse from a grating and a lens combination. In the focal plane one has separated the various Fourier components spatially. In this focal plane one can insert an amplitude and phase filter. The various frequency components are then recombined by a second lens and grating. The results of this pulse-shaping technique may be verified by FROG analysis.

A detailed pedagogical survey of short-pulse techniques by Glezer (1997) has recently been published. Ippen (1994) has written a review of theory and experiment of passive mode locking. This section has extracted much information from these papers.

Some Highlights of Ultrafast Spectroscopy

Transient Raman scattering was one of the phenomena extensively investigated early on with picosecond pulses. Carman et al. (1970) presented a detailed description of how stimulated Raman scattering changes its characteristics when the pulse duration t_p is shorter than the characteristic damping times of the Raman-active vibration. A paradigm experiment was carried out by Laubereau, von der Linde, and Kaiser (1972). A picosecond laser pulse was partially converted to a second-harmonic pulse. The two pulses were separated by a dichroic mirror. The strong fundamental pulse triggered a coherent vibrational excitation, by stimulated Raman scattering. The vibration continued to ring after the exciting pulse had passed. The second-harmonic pulse probed the vibrational excitation with a variable time delay.

Stokes and anti-Stokes components in the forward direction of the second-harmonic pulse are created as the sum and difference frequencies of the incident light and the coherent vibration. These signals decay with a characteristic phase coherence relaxation time T_2 of the vibrational excitation. Spontaneous emission of anti-Stokes light in arbitrary directions measures the population in the excited vibrational state. This light decays with a different characteristic time T_1 for vibrational energy relaxation. Many molecular liquids and mixtures have been analyzed by this system. Heavily damped rotational excitations can also be investigated by stimulated Rayleigh wing scattering with short pulses. Time-resolved and frequency-resolved spectroscopic measurements are complementary. Time-resolved observation is especially useful to detect very fast phenomena hidden in the weak far wings of frequency-resolved spectra.

An entirely new regime occurs when the pulse duration t_p is short compared to the vibrational period itself, $\omega_{\text{vib}} t_p \ll 1$. In this case the vibrational mode is impulsively excited. The pulse contains many pairs of Fourier components with a difference in frequency equal to the vibrational resonance. In experiments by Ruhman, Joly, and Nelson (1988) the very short pulse is split into two parts of equal intensity which are recombined spatially to form a diffraction pattern in a Raman-active liquid. Thus a grating of impulsively excited vibrations is established. This grating is probed by the diffraction of a weak pulse with a variable delay. Alternatively one may detect the impulsively excited coherent vibration by sending a probe pulse along the same path as the first pulse. If the delay of the probe is an integral number of vibrational periods, the probe will enhance the excitation. As it loses energy from Fourier components of higher energy to those of lower energy, a redshift occurs. When the delay is an odd number of half vibrational periods, a blueshift is detected, as the vibrational energy is shifted back from the vibrational excitation to the probe in an anti-Stokes scattering process.

Impulsive Raman scattering has also been observed in a single crystal of germanium by Kutt, Albrecht, and Kurz (1992). The Raman-active vibration can be probed in reflection, as it produces a small modulation in the effective index of refraction.

Impulsive excitations must be sharply distinguished from displacive excitations, which can be induced by the absorption of short pulses in many crystalline materials, including semiconductors, high-temperature superconductors, and metals. A sudden change in carrier density or in electron temperature causes a change in equilibrium internuclear separations. Zeiger et al. (1992) describe the excitation of totally symmetric vibrations in Bi, Sb, Ti_2O_3, and high-temperature superconductors. Kutt et al. (1992) observe excitations of longitudinal phonons in GaAs, as the sudden creation of carriers changes the internal space-charge field near the surface. The excitation is detected by probing the oscillations in reflectivity which occur following the pump pulse.

A displacive type of vibrational excitation occurs during an optical transition in a molecule. The Franck-Condon principle states that the internuclear distance coordinates do not change during the transition to an electronically excited state. The equilibrium nuclear distances in this excited state are different from those in the electronic ground-state configuration. A short optical pulse, absorbed in a molecular gas, excites a rotational vibrational wave packet. The time evolution of this wave packet has been observed by Dantus, Bowman, and Zewail (1990) in an I_2 molecule by probing the excited configuration with a second absorption process to a still higher configuration from which fluorescence can be served. Oscillations in the fluorescent intensity as a function of pump-probe delay demonstrate the temporal evolution of the ro-vibrational wave packet in a time-resolved manner.

A paradigm of femtochemistry is the excitation of the NaI molecule to a predissociative state. The evolution of the wave packet following femtosecond optical excitation shows periodic Landau-Zener tunneling to separated atoms. This time dependence can be observed by resonant fluorescence of the Na atom induced by a probe pulse. Zewail (1993) has presented an overview of the rich field of femtochemistry.

As a final example, the *cis-trans* configurational change of the 11-*cis* retinal molecule should be mentioned. The absorption of a photon to an excited electronic state induces this transition. It is the first step in the vision process. The *trans* configuration has a different absorption spectrum. Thus the temporal evolution of this configurational change may again be probed by a femtosecond pump-probe technique. Matthies *et al.* (1994) established that the configurational change takes place in less than 200 fs. It is a prototypical example of the new field of femtobiology.

Phenomena at High Flux Densities

The development of femtosecond pulses has led to very high instantaneous power levels attainable with a relatively small table-top laser configuration. Diffraction-limited laser pulses may be focused to a spot size of an optical wavelength. The concentration of light in space and time has opened up new regimes of high-intensity radiation to experimental investigation.

A pulse of 1 μJ of 100-fs duration is readily available from a Ti-sapphire laser system. When such a pulse is focused onto a surface of an absorbing medium with a spot size of 10^{-5} cm^2, the fluence is 0.1 J/cm^2. A large number of electron-hole pairs exceeding 10^{22} cm^3 is created in the absorption depth. This carrier density, created in 10^{-13} s, will change the band structure of gallium arsenide. Huang *et al.* (1998) have shown that the effective indirect band gap decreases to zero by measuring the complex dielectric function by reflection of a time-delayed white-light probe pulse. When more than about 10 percent of the valence-band electrons have been promoted to the conduction band, the tetrahedral lattice structure becomes unstable and the second-order nonlinear susceptibility, observed by reflected second-harmonic light in reflection, vanishes within 100 fs, before significant lattice heating occurs.

On time scales longer than several picoseconds the energy transfer to the lattice causes melting and evaporation, both of which may be verified by post-pulse inspection. A well-defined damage threshold for the 001 surface of a gallium arsenide crystal irradiated by a 70-fs pulse at 635-nm wavelength is 0.1 J/cm^2.

The peak power flux density of a 1-μJ pulse of 100-fs duration, focused onto an area of 10^{-5} cm^2 is one terawatt/cm^2. In a transparent medium, dielectric breakdown is initiated by multiphoton absorption and enhanced by avalanche ionization. Glezer *et al.* (1996) have proposed the use of damage spots with submicron dimensions inside transparent materials, such as glass, silica, or sapphire, for three-dimensional optical storage.

Pulsed irradiation of metallic targets at fluences exceeding 1 J/cm^2 creates very hot, high-density plasmas, which can serve as pulsed x-ray sources.

A special technique, first introduced by Strickland and Mourou (1985) is required to amplify short pulses to attain higher flux densities. An unfocused 10-fs beam of 1 cm^2 cross section with 1 μJ energy has a power flux density of 10^8 watts/cm^2. At this power level the beam becomes susceptible to self-focusing and filamentation in solid-state amplifier media with good energy storage characteristics such as Ti-sapphire, Nd-Yag, and alexandrite. The technique of chirped pulse amplification permits amplification by a factor of 10 000 or more. For this purpose the pulse is first stretched in time. Different frequencies have different optical path lengths, if diffracted between a pair of antiparallel gratings. A stretched pulse, chirped in frequency, is amplified and then recompressed by a matched combination of gratings. A pulse is also stretched by group velocity dispersion in an optical medium, but the recompression by a combination of gratings or optical prisms is more difficult in this case. Champaret et al. (1996) have amplified a 10-fs pulse with subnanojoule energy to a level near one joule, after stretching by a factor of 10^5 and recompression close to the initial pulse duration.

When a one-joule 10-fs pulse is focused, power flux densities in the range of 10^{18} to 10^{20} watts/cm^2 are attained. Previously such power levels were only available in an assembly of many Nd-glass laser beams of large cross sections, as operated at the Lawrence Livermore National Laboratories and a few other large installations. These pulses with energies of 1–100 kJ have been used in the study of inertially confined fusion plasmas.

New physical regimes of exploration have been opened up by the use of one-joule femtosecond pulses. The Coulomb field responsible for the binding of valence electrons in atoms and molecules is on the order of e/a_0^2, where a_0 is the Bohr radius. According to Poynting's formula, the light field amplitude equals this Coulomb field of about 10^9 volts/cm at a power level of about 10^{15} watts/cm^2. In this regime the light field can no longer be considered as a small perturbation. At this power level harmonics of more than one hundred times the laser frequency are created. The phenomenon of above-threshold ionization is related to the quiver energy of a free electron in an intense oscillating field.

The quiver motion of the electron becomes dominant at higher power levels. Here one should first solve for the motion of the free electron in the strong pulse, and the Coulomb atomic field becomes a perturbation.

A low-energy free electron will start to oscillate parallel to the transverse electron field. The Lorentz force due to the transverse magnetic field will cause an oscillation at twice the light frequency in the longitudinal direction. Classically and nonrelativistically the quiver energy $e^2E^2/2m\omega^2$ at 10^{19} watts/cm^2 at 1 μm wavelength would be three times m_0c^2, where m_0 is the rest mass of the electron. The electron reaches relativistic velocities and the Lorentz force attains the same magnitude as the electric force. A Lorentz-invariant formulation is indicated and detailed solutions have recently been given by Startsev and McKinstrie (1997).

In short, focused pulses with very large spatial and temporal gradients in intensity give rise to ponderomotive forces, related to the gradient in quiver energy. In a plasma the positive heavy ions will not be displaced much, but the electrons may oscillate with large amplitude. Huge internal longitudinal electric fields may be generated in the wake of a short light pulse. Tajima and Dawson (1979) have proposed the acceleration of electrons traveling in synchrony with the wake field of a light pulse. Recent papers by Umstadter, Kim, and Dodd (1996) and by Siders *et al.* (1996) provide many references to the extensive literature. "Table top" electron accelerators using femtosecond pulses may become a reality.

The scattering of femtosecond pulses with highly relativistic electrons, produced by a LINAC accelerator, has been studied. Burke *et al.* (1997) have observed nonlinear Compton scattering and electron-positron pair production. The field of ultrahigh-intensity laser physics has recently been reviewed by Mourou, Barty, and Perry (1998).

In conclusion, ultrashort laser pulses have opened up not only the field of femtosecond time-resolved spectroscopy, but also the study of relativistic electrons and plasmas at ultrahigh intensities.

References

Abraham, H., and T. Lemoine, 1899, C. R. Acad. Sci. **129**, 206.
Burke, D. L., *et al.*, 1997, Phys. Rev. Lett. **79**, 1626.
Carman, R. L., F. Shimizu, C. S. Wang, and N. Bloembergen, 1970, Phys. Rev. A **2**, 60.
Champaret, J. P., C. LeBlanc, G. Cheriaux, P. Curley, C. Darpentigny, and P. Rousseau, 1996, Opt. Lett. **21**, 1921.
Dantus, M., R. M. Bowman, and A. H. Zewail, 1990, Nature (London) **343**, 73.
DeMaria, A. J., D. A. Stetser, and H. Heinan, 1966, Appl. Phys. Lett. **8**, 174.
Fork, R. L., C. H. Brito Cruz, P. C. Becker, and C. V. Shank, 1987, Opt. Lett. **12**, 483.
Glezer, E. N., M. Milosavljevie, L. Huang, R. J. Finlag, T. H. Her, J. P. Callau, and E. Mazur, 1996, Opt. Lett. **21**, 2023.
Glezer, E. N., 1997, in *Spectroscopy and Dynamics of Collective Excitations in Solids*, edited by B. DiBartolo (Plenum, New York), p. 375.
Hargrove, L. E., R. L. Fork, and M. A. Pollack, 1964, Appl. Phys. Lett. **5**, 4.
Hellwarth, R. W., 1961a, in *Advances in Quantum Electronics*, edited by J. R. Singer (Columbia University, New York), p. 334.
Hellwarth, R. W., 1961b, Bull. Am. Phys. Soc. **6**, 414.
Huang, L., J. P. Callan, E. N. Glezer, and E. Mazur, 1998, Phys. Rev. Lett. **80**, 185.
Ippen, E. P., 1994, Appl. Phys. B: Lasers Opt. **58**, 159.
Kutt, W. A., W. Albrecht, and H. Kurz, 1992, IEEE J. Quantum Electron. **28**, 2434.
Laubereau, A., D. von der Linde, and W. Kaiser, 1972, Phys. Rev. Lett. **28**, 1162.
Maiman, T. H., 1960, Nature (London) **187**, 493.
Matthies, R. A., R. W. Schoenlein, L. A. Peteanu, Q. Wang, and C. V. Shank, 1994, in *Femtosecond Reaction Dynamics*, edited by D. A. Wiersman (North-Holland, Amsterdam), p. 229.
Mocker, H. W., and R. J. Collins, 1965, Appl. Phys. Lett. **7**, 270.
Mourou, G. A., C. P. J. Barty, and M. D. Perry, 1998, Phys. Today, January 22.
New, G. H. C., 1972, Opt. Commun. **6**, 188.
Porter, G., 1950, Proc. Phys. Soc. London, Sect. A **200**, 284.
Ruhman, S., A. G. Joly, and K. A. Nelson, 1988, IEEE J. Quantum Electron. **24**, 460.
Shank, C. V., and E. P. Ippen, 1974, Appl. Phys. Lett. **24**, 373.
Shapiro, S. L., 1977, in *Ultrashort Light Pulses*, edited by S. L. Shapiro (Springer, Berlin), p. 1.
Siders, C. W., S. P. LeBlanc, D. Fisher, T. Tajima, M. C. Downer, A. Babine, A. Stephanov, and A. Sergeev, 1996, Phys. Rev. Lett. **76**, 3570.
Spence, D. E., P. N. Kean, and W. Sibbett, 1991, Opt. Lett. **16**, 42.
Startsev, E. A., and C. J. McKinstrie, 1997, Phys. Rev. E **55**, 7527.
Strickland, D., and G. Mourou, 1985, Opt. Commun. **56**, 219.
Taft, G., A. Rundquist, M. M. Murnane, and H. C. Kapteyn, 1995, Opt. Lett. **20**, 743.

Tajima, T., and J. M. Dawson, 1979, Phys. Rev. Lett. **43**, 267.
Trebino, R., and D. J. Kane, 1993, J. Opt. Soc. Am. A **10**, 1101.
Umstadter, D., J. K. Kim, and E. Dodd, 1996, Phys. Rev. Lett. **76**, 2073.
Zeiger, H. J., J. Vidal, T. K. Cheng, E. P. Ippen, G. Dresselhaus, and M. S. Dresselhaus, 1992, Phys. Rev. B **45**, 768.
Zewail, A. H., 1993, J. Phys. Chem. **97**, 12427.
Zhou, J. P., G. Taft, C. P. Huang, M. M. Murnane, and H. C. Kaptyn, 1994, Opt. Lett. **19**, 1149.

Experiment and the Foundations of Quantum Physics

Anton Zeilinger

Quantum physics, a child of the early 20th century, is probably the most successful description of nature ever invented by man. The range of phenomena it has been applied to is enormous. It covers phenomena from the elementary-particle level all the way to the physics of the early universe. Many modern technologies would be impossible without quantum physics—witness, for example, that all information technologies are based on a quantum understanding of solids, particularly of semiconductors, or that the operation of lasers is based on a quantum understanding of atomic and molecular phenomena.

So, where is the problem? The problem arises when one realizes that quantum physics implies a number of very counterintuitive concepts and notions. This has led, for example, R. P. Feynman to remark, "I think I can safely say that nobody today understands quantum physics," or Roger Penrose (1986) to comment that the theory "makes absolutely no sense."

From the beginning, gedanken (thought) experiments were used to discuss fundamental issues in quantum physics. At that time, Heisenberg invented his gedanken gamma-ray microscope to demonstrate the uncertainty principle while Niels Bohr and Albert Einstein in their famous dialogue on epistemological problems in what was then called atomic physics made extensive use of gedanken experiments to make their points.

Now, at the end of the 20th century, the situation has changed dramatically. Real experiments on the foundations of quantum physics abound. This has not only given dramatic support to the early views, it has also helped to sharpen our intuition with respect to quantum phenomena. Most recently, experimentation is already applying some of the fundamental phenomena in completely novel ways. For example, quantum cryptography is a direct application of quantum uncertainty and both quantum teleportation and quantum computation are direct applications of quantum entanglement, the concept underlying quantum nonlocality (Schrödinger, 1935).

I will discuss a number of fundamental concepts in quantum physics with direct reference to experiments. For the sake of the consistency of the discussion and because I know them best I will mainly present experiments performed by my group. In view of the limited space available my aim can neither be completeness, nor a historical overview. Rather, I will focus on those issues I consider most fundamental.

A Double Slit and One Particle

Feynman (1965) has said that the double-slit "has in it the heart of quantum mechanics. In reality, it contains the only mystery." As we shall see, entangled states of two or more particles imply that there are further mysteries (Silverman, 1995). Nevertheless, the two-slit experiment merits our attention, and we show the results of a typical two-slit experiment done with neutrons in Fig. 1 (Zeilinger et al., 1988). The measured distribution of the neutrons has two remarkable features. First, the observed interference pattern showing the expected fringes agrees perfectly well with theoretical prediction (solid line), taking into account all features of the experimental setup. Assuming symmetric illumination the neutron state at the double slit can be symbolized as

$$|\psi\rangle = \frac{1}{\sqrt{2}}(|\text{passage through slit } a\rangle + |\text{passage through slit } b\rangle). \tag{1}$$

The interference pattern is then obtained as the superposition of two probability amplitudes. The particle could have arrived at a given observation point $\vec{r}$ either via slit 1 with probability amplitude $a(\vec{r})$ or via slit 2 with probability amplitude $b(\vec{r})$. The total probability density to find the particle at point $\vec{r}$ is then simply given as

$$p(\vec{r}) = |a(\vec{r}) + b(\vec{r})|^2. \tag{2}$$

This picture suggests that the pattern be interpreted as a wave phenomenon.

Yet, second, we note that the maximum observed intensity is of the order of one neutron every two seconds. This means that, while one neutron is being registered, the

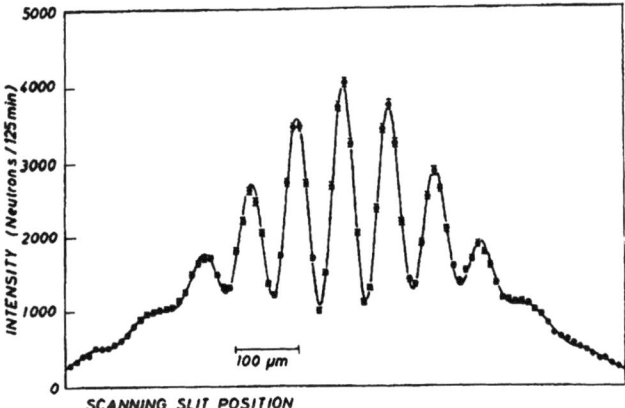

Figure 1. A double-slit diffraction pattern measured with very cold neutrons with a wavelength of 2 nm corresponding to a velocity of 200 ms^{-1}. The two slits were 22 μm and 23 μm wide, respectively, separated by a distance of 104 μm. The resulting diffraction angles were only of the order of 10 μrad, hence the observation plane was located 5 m downstream from the double slit in order to resolve the interference pattern. (For experimental details see Zeilinger et al., 1988.) The solid line represents first-principles prediction from quantum mechanics, including all features of the experimental apparatus. For example, the fact that the modulation of the interference pattern was not perfect can fully be understood on the basis that a broad wavelength band had to be used for intensity reasons and the experiment was not operated in the Fraunhofer regime.

next one to be registered usually is still confined to its uranium nucleus inside the nuclear reactor, waiting for nuclear fission to release it to freedom!

This feature of very low-intensity interference is shared by all existing neutron interferometer experiments (Rauch and Werner, in press). These pioneering matter-wave experiments led to the realization of a number of very basic experiments in quantum mechanics including the change of the sign of a spinor under a full rotation, the effect of gravity on the phase of a neutron wave, a number of experiments related to quantum complementarity, and many others.

Thus the interference pattern is really collected one by one and this suggests the particle nature. Then the famous question can be posed: through which of the two slits did the particle actually pass on its way from source to detector? The well-known answer according to standard quantum physics is that such a question only makes sense when the experiment is such that the path taken can actually be determined for each particle. In other words, the superposition of amplitudes in Eq. (1) is only valid if there is no way to know, even in principle, which path the particle took. It is important to realize that this does not imply that an observer actually takes note of what happens. It is sufficient to destroy the interference pattern, if the path information is accessible in principle from the experiment or even if it is dispersed in the environment and beyond any technical possibility to be recovered, but in principle still "out there." The absence of any such information is *the essential criterion* for quantum interference to appear. For a parallel discussion, see the accompanying article by Mandel (1999) in this volume.

To emphasize this point, let us consider now a gedanken experiment where a second, probe, particle is scattered by the neutron while it passes through the double slit. Then the state will be

$$|\psi\rangle = \frac{1}{\sqrt{2}} \frac{(|\text{passage through slit } a\rangle_1 |\text{scattered in region } a\rangle_2}{+ |\text{passage through slit } b\rangle_1 |\text{scattered in region } b\rangle_2)}. \quad (3)$$

There the subscripts 1 and 2 refer to the neutron and the probe particle, respectively. The state (3) is entangled and if the two states for particle 2 are orthogonal, no interference for particle 1 can arise. Yet, if particle 2 is measured such that this measurement is not able, *even in principle*, to reveal any information about the slit particle 1 passes, then particle 1 will show interference. Obviously, there is a continuous transition between these two extreme situations.

We thus have seen that one can either observe a wavelike feature (the interference pattern) or a particle feature (the path a particle takes through the apparatus) depending on which experiment one chooses. Yet one could still have a naive picture in one's mind essentially assuming waves propagating through the apparatus which can only be observed in quanta. That such a picture is not possible is demonstrated by two-particle interferences, as we will discuss now.

A Double Slit and Two Particles

The situation is strikingly illustrated if one employs pairs of particles which are strongly correlated ("entangled") such that either particle carries information about the other (Horne and Zeilinger, 1985; Greenberger, Horne, and Zeilinger, 1993). Consider a setup where a source emits two particles with antiparallel momenta (Fig. 2). Then, whenever

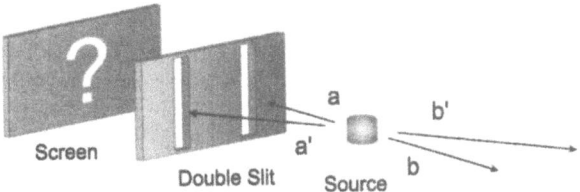

Figure 2. A source emits pairs of particles with total zero momentum. Particle 1 is either emitted into beams a or a' and particle 2 into beams b or b' with perfect correlations between a and b and a' and b', respectively. The beams of particle 1 then pass a double-slit assembly. Because of the perfect correlation between the two particles, particle 2 can serve to find out which slit particle 1 passed and therefore no interference pattern arises.

particle 1 is found in beam a, particle 2 is found in beam b and whenever particle 1 is found in beam a', particle 2 is found in beam b'. The quantum state is

$$|\psi\rangle = \frac{1}{\sqrt{2}}(|a\rangle_1|b\rangle_2 + |a'\rangle_1|b'\rangle_2). \qquad (4)$$

Will we now observe an interference pattern for particle 1 behind its double slit? The answer has again to be negative because by simply placing detectors in the beams b and b' of particle 2 we can determine which path particle 1 took. Formally speaking, the states $|a\rangle_1$ and $|a'\rangle_1$ again cannot be coherently superposed because they are entangled with the two orthogonal states $|b\rangle_2$ and $|b'\rangle_2$.

Obviously, the interference pattern can be obtained if one applies a so-called quantum eraser which completely erases the path information carried by particle 2. That is, one has to measure particle 2 in such a way that it is not possible, even in principle, to know from the measurement which path it took, a' or b'.

A recent experiment (Dopfer, 1998) used the so-called process of parametric down conversion to create entangled pairs of photons (Fig. 3) where a UV beam entering a nonlinear optical crystal spontaneously creates pairs of photons such that the sum of their linear momenta is constant. In type-I parametric down conversion, the two photons carry equal polarization. Parametric down conversion is discussed in somewhat more detail below. Although the experimental situations are different, conceptually this is equivalent to the case discussed above. In this experiment, photon 2 passes a double slit while the other, photon 1, can be observed by a detector placed at various distances behind the Heisenberg lens which plays exactly the same role as the lens in the gamma-ray microscope discussed by Heisenberg (1927) and extended by Weizsäcker (1931). If the detector is placed at the focal plane of the lens, then registration of a photon there provides information about its direction, i.e., momentum, before entering the lens. Thus, because of the strict momentum correlation, the momentum of the other photon incident on the double slit and registered in coincidence is also well defined. A momentum eigenstate cannot carry any position information, i.e., no information about which slit the particle passes through. Therefore, a double-slit interference pattern for photon 2 is registered conditioned on registration of photon 1 in the focal plane of the lens. It is important to note that it is actually necessary to register photon 1 at the focal plane because without registration one could always, at least in principle, reconstruct the state in front of the lens. Most strikingly, therefore, one can find out the slit photon 2 passed by placing the

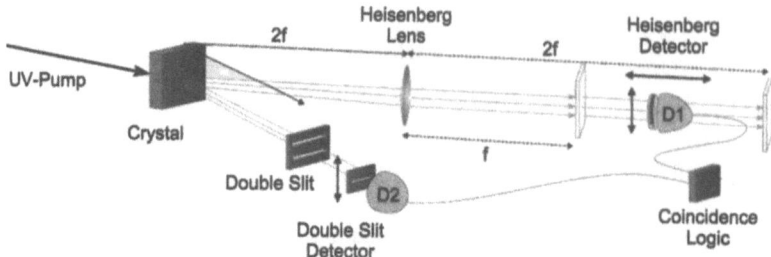

Figure 3. Two photons and one double slit. A pair of momentum-entangled photons is created by type-I parametric down conversion. Photon 2 enters a double-slit assembly and photon 1 is registered by the Heisenberg detector arranged behind the Heisenberg lens. If the Heisenberg detector is placed in the focal plane of the lens, it projects the state of the second photon into a momentum eigenstate which cannot reveal any position information and hence no information about slit passage. Therefore, in coincidence with a registration of photon 1 in the focal plane, photon 2 exhibits an interference pattern. On the other hand, if the Heisenberg detector is placed in the imaging plane at 2 f, it can reveal the path the second photon takes through the slit assembly which therefore connot show the interference pattern (Dopfer, 1998).

detector for photon 1 into the imaging plane of the lens. The imaging plane is simply obtained by taking the object distance as the sum of the distances from the lens to the crystal and from the crystal to the double slit. Then, as has also been demonstrated in the experiment, a one-to-one relationship exists between positions in the plane of the double slit and in the imaging plane and thus, the slit particle 2 passes through can readily be determined by observing photon 1 in the imaging plane. Only after registration of photon 1 in the focal plane of the lens is any possibility to obtain any path information from photon 1 irrecoverably destroyed.

We note that the distribution of photons behind the double slit without registration of the other photon is just an incoherent sum of probabilities having passed through either slit and, as shown in the experiment, no interference pattern arises if one does not look at the other photon. This is again a result of the fact that, indeed, path information is still present and can easily be extracted by placing the detector of photon 1 into the imaging plane of the lens.

Likewise, registration of photon 2 behind its double slit destroys any path information it may carry and thus, by symmetry, a Fraunhofer double-slit pattern is obtained for the distribution of photon 1 in the focal plane behind its lens, even though that photon never passed a double slit (Fig. 4)! This experiment can be understood intuitively if we carefully analyze what registration of a photon behind a double slit implies. It simply means that the state incident on the double slit is collapsed into a wave packet with the appropriate momentum distribution such that the wave packet peaks at both slits. By virtue of the strong momentum entanglement at the source, the other wave packet then has a related momentum distribution which actually is, according to an argument put forward by Klyshko (1988), the time reversal of the other wave packet. Thus, photon 1 appears to originate backwards from the double slit assembly and is then considered to be reflected by the wave fronts of the pump beam into the beam towards the lens which then simply realizes the standard Fraunhofer observation conditions.

One might still be tempted to assume a picture that the source emits a statistical mixture of pairwise correlated waves where measurement of one photon just selects a certain,

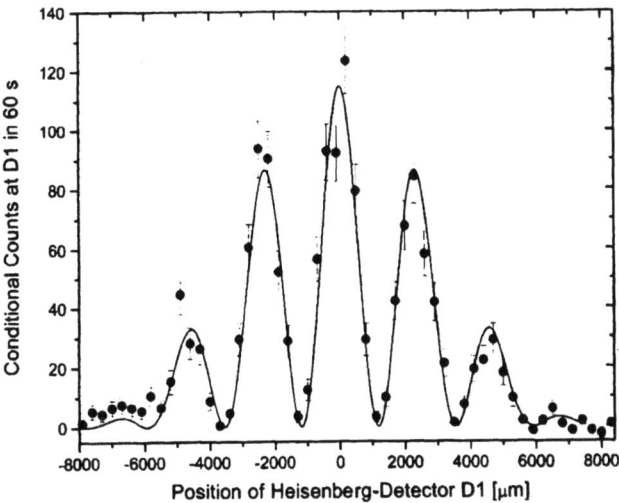

Figure 4. Double-slit pattern registered by the Heisenberg detector of photon 1 (Fig. 3). The graph shows the counts registered by that detector as a function of its lateral position, if that detector is arranged in the focal plane of the lens. The counts are conditioned on registration of the second photon behind its double slit. Note that the photons registered in detector D1 exhibit a double-slit pattern even though they never pass through a double-slit assembly. Note also the low intensity which indicates that the interference pattern is collected photon by photon.

already existing, wavelet for the other photon. It is easy to see that any such picture cannot lead to the perfect interference modulation observed. The most sensible position, according to quantum mechanics, is to assume that no such waves preexist before any measurement.

Quantum Complementarity

The observation that particle path and interference pattern mutually exclude each other is one specific manifestation of the general concept of complementary in quantum physics. Other examples are position and linear momentum as highlighted in Heisenberg's uncertainty relation, or the different components of angular momentum. It is often said that complementarity is due to an unavoidable disturbance during observation. This is suggested if, as in our earlier example, we consider determining the path a particle takes through the double-slit assembly by scattering some other particle from it. That this is too limited a view is brought out by the experiment discussed in the preceding section.

The absence of the interference pattern for photon 2 if no measurement is performed on photon 1, is not due to it being disturbed by observation; rather, it can be understood if we consider the complete set of possible statements which can be made about the experiment as a whole (Bohr, 1935) including the other photon.

As long as no observation whatsoever is made on the complete quantum system comprised of both photons our description of the situation has to encompass all possible experimental results. The quantum state is exactly that representation of our knowledge of the complete situation which enables the maximal set of (probabilistic) predictions for any possible future observation. What comes new in quantum mechanics is that, instead of just listing the various experimental possibilities with the individual probabilities, we have to represent our knowledge of the situation by the quantum state using complex amplitudes. If we accept that the quantum state is no more than a representation of the information we have, then the spontaneous change of the state upon observation, the so-called collapse or reduction of the wave packet, is just a very natural consequence of the fact that, upon observation, our information changes and therefore we have to change our representation of the information, that is, the quantum state. From that position, the so-called measurement problem (Wigner, 1970) is not a problem but a consequence of the more fundamental role information plays in quantum physics as compared to classical physics (Zeilinger, 1999).

Quantum complementarity then is simply an expression of the fact that in order to measure two complementary quantities, we would have to use apparatuses which mutually exclude each other. In the example of our experiment, interference pattern and path information for photon 2 are mutually exclusive, i.e., complementary, because it is not possible to position the detector for photon 1 simultaneously in the focal plane and in the image plane of the lens. Yet the complete quantum state encompasses both possible experiments.

We finally note two corollaries of our analysis. First, it is clearly possible to have a concept of continuous complementarity. In our case, placing the detector of photon 1 somewhere in between the two extreme positions mentioned will reveal partial path information and thus an interference pattern of reduced visibility. And second, the choice whether or not path information or the interference pattern become manifest for photon 2 can be delayed to arbitrary times after that photon has been registered. In the experiment discussed, the choice where detector D_1 is placed can be delayed until after photon 2 has been detected behind its double slit. While we note that in the experiment, the lens was already arranged at a larger distance from the crystal than the double slit, a future experiment will actually employ a rapidly switched mirror sending photon 1 either to a detector placed in the focal plane of the lens or to a detector placed in the imaging plane.

This possibility of deciding long after registration of the photon whether a wave feature or a particle feature manifests itself is another warning that one should not have any realistic pictures in one's mind when considering a quantum phenomenon. Any detailed picture of what goes on in a specific individual observation of one photon has to take into account the whole experimental apparatus of the complete quantum system consisting of both photons and it can only make sense after the fact, i.e., after all information concerning complementary variables has irrecoverably been erased.

Einstein-Podolsky-Rosen and Bell's Inequality

In 1935 Einstein, Podolsky, and Rosen (EPR) studied entangled states of the general type used in the two-photon experiment discussed above. They realized that in many such states, when measuring either linear momentum or position of one of the two particles, one can infer precisely either momentum or position of the other. As the two particles might be widely separated, it is natural to assume validity of the locality condition suggested by EPR: "Since at the time of measurement the two systems no longer interact, no real change can take place in the second system in consequence of anything that may be done to the first system." Then, whether or not momentum or position can be assigned to particle (system) 2 must be independent of what measurement is performed on particle 1 or even whether any measurement is performed on it at all. The question therefore arises whether the specific results obtained for either particle can be understood without reference to which measurement is actually performed on the other particle. Such a picture would imply a theory, underlying quantum physics, which provides a more detailed account of individual measurements. Specifically, following Bell, it might explain "why events happen" (Bell, 1990; Gottfried, 1991).

In the sixties, two different developments started, which nicely complement each other. First, it was initially argued by Specker (1960) for Hibbert spaces of dimension larger than two that quantum mechanics cannot be supplemented by additional variables. Later it was shown by Kochen and Specker (1967) and by Bell (1966; for a review see Mermin, 1993), that for the specific case of a spin-1 particle, it is not possible to assign in a consistent way measurement values to the squares of any three orthogonal spin projections, despite the fact that the three measurements commute with each other. This is a purely geometric argument which only makes use of some very basic geometric considerations. The conclusion here is very important. The quantum system cannot be assigned properties independent of the context of the complete experimental arrangement. This is just in the spirit of Bohr's interpretation. This so-called contextuality of quantum physics is another central and subtle feature of quantum mechanics.

Second, a most important development was due to John Bell (1964) who continued the EPR line of reasoning and demonstrated that a contradiction arises between the EPR assumptions and quantum physics. The most essential assumptions are realism and locality. This contradiction is called Bell's theorem.

To be specific, and in anticipation of experiments we will discuss below, let us assume we have a pair of photons in the state:

$$|\psi\rangle = \frac{1}{\sqrt{2}}(|H\rangle_1|V\rangle_2 - |V\rangle_1|H\rangle_2). \tag{5}$$

This polarization-entangled state implies that whenever (Fig. 5) photon 1 is measured and found to have horizontal (H) polarization, the polarization of photon 2 will be vertical (V) and vice versa. Actually, the state of Eq. (5) has the same form in any basis. This means whichever state photon 1 will be found in, photon 2 can definitely be predicted to be found in the orthogonal state if measured.

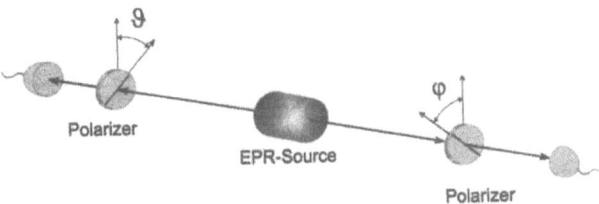

Figure 5. Typical experimental arrangement to test Bell's inequality. A source emits, say, polarization-entangled pairs of photons. Each photon is sent through a polarizer whose orientation can be varied. Finally behind each polarizer, the transmitted photons are registered. Quantum mechanics predicts a sinusoidal variation of the coincidence count rate as a function of the relative angular orientation of the polarizers. Any such variation violates local realism as expressed in Bell's inequality.

Following EPR one can apply their famous reality criterion, "If, without in any way disturbing a system, we can predict with certainty (i.e., with probability equal to unity) the value of a physical quantity, then there exists an element of physical reality corresponding to this physical quantity." This would imply that to any possible polarization measurement on any one of our photons we can assign such an element of physical reality on the basis of a corresponding measurement on the other photon of any given pair.

The next step then is to assume the two photons (systems) to be widely separated so that we can invoke EPR's locality assumption as given above. Within this line of reasoning, whether or not we can assign an element of reality to a specific polarization of one of the systems must be independent of which measurement we actually perform on the other system and even independent of whether we care to perform any measurement at all on that system. To put it dramatically, one experiment could be performed here on earth and the other on a planet of another star a couple of light years away. It is this very independence of a measurement result on one side from what may be done on the other side, as assumed by EPR, which is at variance with quantum mechanics. Indeed, this assumption implies that certain combinations of expectation values have definite bounds. The mathematical expression of that bound is called Bell's inequality, of which many variants exist. For example, a version given by Clauser, Horne, Shimony, and Holt (1969) is

$$|E(\alpha,\beta)-E(\alpha',\beta)|+|E(\alpha,\beta')+E(\alpha',\beta')|\leq 2, \qquad (6)$$

where

$$E(\alpha,\beta)=\frac{1}{N}[C_{++}(\alpha,\beta)+C_{--}(\alpha,\beta)-C_{+-}(\alpha,\beta)-C_{-+}(\alpha,\beta)]. \qquad (7)$$

Here we assume that each photon is subject to a measurement of linear polarization with a two-channel polarizer whose outputs are + and −. Then, e.g., $C_{++}(\alpha,\beta)$ is the number of coincidences between the + output port of the polarizer measuring photon 1 along α and the + output port of the polarizer measuring photon 2 along β. Maximal violation occurs for $\alpha=0°$, $\beta=22.5°$, $\alpha'=45°$, $\beta'=67.5°$. Then the left-hand side of Eq. (6) will be $2\sqrt{2}$ in clear violation of the inequality. Thus Bell discovered that the assumption of local realism is in conflict with quantum physics itself and it became a matter of experiment to find out which of the two world views is correct.

Interestingly, at the time of Bell's discovery no experimental evidence existed which was able to decide between quantum physics and local realism as defined in Bell's derivation. An earlier experiment by Wu and Shaknov (1950) had demonstrated the existence of spatially separated entangled states, yet failed to give data for nonorthogonal measurement directions. After the realization that the polarization entangled state of photons emitted in atomic cascades can be used to test Bell's inequalities, the first experiment was performed by Freedman and Clauser in 1972 (Fig. 6). By now, there exists a large number of such experiments. The ones showing the largest violation of a Bell-type inequality have for a long time been the experiments by Aspect, Grangier, and Roger (1981, 1982) in the early eighties. Aside from two early experiments, all agreed with the predictions of quantum mechanics and violated inequalities derived from Bell's original version using certain additional assumptions. Actually, while the experimental evidence strongly favors quantum mechanics, there remained two possible mechanisms for which a local realistic view could still be maintained.

One problem in all experimental situations thus far is due to technical insufficiencies, namely that only a small fraction of all pairs emitted by the source is registered. This is a standard problem in experimental work and experimentalists take great care to ensure that it is reasonable to assume that the detected pairs are a faithful representative of all pairs emitted. Yet, at least in principle, it is certainly thinkable that this is not the case and that, should we once be able to detect all pairs, a violation of quantum mechanics and data in agreement with local realism would be observed. While this is in principle possible, I would agree with Bell's judgment (1981) that "although there is an escape route there, it is hard for me to believe that quantum mechanics works so nicely for inefficient practical set-ups, and is yet going to fail badly when sufficient refinements are made. Of more importance, in my opinion, is the complete absence of the vital *time* factor in existing experiments. The analyzers are not rotated during the flight of the particles. Even if one is obliged to admit some long-range influence, it need not travel faster than light—and so would be much less indigestible." Until recently, there has been only one experiment where the time factor played a role. In that experiment (Aspect, Dalibard, and Roger, 1982) each of the two photons could be switched between two different polarizers on a time scale which was small compared to the flight time of the photons. Due to technical limitations at the time of the experiment and because this switching back and forth between two different polarizations was periodic, the experiment does not completely fulfill Bell's desideratum, but it is an important step.

Experimental development in the last decade is marked by two new features. First, it was realized initially by Horne and Zeilinger (1985, 1988) for momentum and position, and then by Franson (1989) for time and energy, that situations can arise where Bell's inequality is violated not just for internal variables, like spin, but also for external ones. This observation put Bell's theorem in a much broader perspective than before. Second, a new type of source was employed (Burnham and Weinberg, 1970), based on the process of spontaneous parametric down conversion. The first to use such a source in a Bell-inequality experiment were Alley and Shih in 1986. In such experiments, a nonlinear

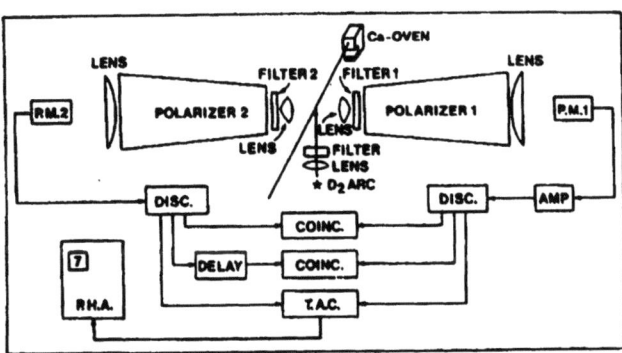

Figure 6. Sketch of the experimental setup used in the first experiment demonstrating a violation of Bell's inequality (Freedman and Clauser, 1972). The two photons emitted in an atomic cascade in Ca are collected with lenses and, after passage through adjustable polarizers, coincidences are registered using photomultiplier detectors and suitable discriminators and coincidence logic. The observed coincidence counts violate an inequality derived from Bell's inequality under the fair sampling assumption.

optical crystal is pumped by a sufficiently strong laser beam. Then, with a certain very small probability, a photon in the laser beam can spontaneously decay into two photons. The propagation directions of the photons and the polarization are determined by the dispersion surfaces inside the medium. The so-called phase-matching conditions of quantum optics, which for sufficiently large crystals are practically equivalent to energy and momentum conservation, imply that the momenta and the energies of the two created photons have to sum up to the corresponding value of the original pump photon inside the crystal. In effect, a very rich entangled state results. The two emerging photons are entangled both in energy and in momentum. In type-I down conversion, these two photons have the same polarization while in type-II down conversion, they have different polarization.

A recent experiment utilized type-II down conversion (Figs. 7 and 8) such that the two emerging photons having orthogonal polarizations effectively emerge in a polarization-entangled state as discussed above [see Eq. (5)]. In the experiment (Weihs et al., 1998), the photons were coupled into long glass fibers and the polarization correlations over a distance of the order of 400 m was measured. The important feature of that experiment is that the polarization of the photons could be rotated in the last instant, thus effectively realizing the rotatable polarizers suggested by Bell. The decision whether or not to rotate the polarization was made by a physical random-number generator on a time scale short compared to the flight time of the photons. Figure 9 shows the principle of the experimental setup. Due to technological progress it is possible now in such experiments to violate Bell's inequality by many standard deviations in a very short time: in this experiment by about 100 standard deviations in measurement times of the order of a minute. In a related experiment (Tittel et al., 1998), entanglement could be demonstrated over distances of more than 10 km but without random switching.

A few points deserve consideration regarding future experiments. On the one hand experiments must be improved to high enough pair-collection efficiencies in order to finally prove that the fair sampling hypothesis used in all existing experiments was justified. On the other hand, and more interesting from a fundamental point of view, one could still assume that both random-number generators are influenced by joint events in their common past. This suggests a final experiment in which two experimenters exercise their free will and choose independently the measurement directions. Such an experiment

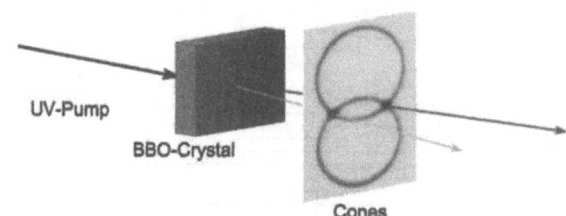

Figure 7. Principle of type-II parametric down conversion to produce directed beams of polarization entangled photons (Kwiat et al., 1995). An incident pump photon can spontaneously decay into two photons which are entangled in momentum and energy. Each photon can be emitted along a cone in such a way that two photons of a pair are found opposite to each other on the respective cones. The two photons are orthogonally polarized. Along the directions where the two cones overlap, one obtains polarization-entangled pairs. In the figure, it is assumed that a filter already selects those photons which have exactly half the energy of a pump photon.

would require distances of order of a few light seconds and thus can only be performed in outer space.

Another future direction of research will certainly be directed at quantum entanglement employing more systems, or at larger, specifically more massive, systems. A first experiment in Paris was able to demonstrate entanglement between atoms (Hagley *et al.*, 1997).

Quantum Information and Entanglement

While most work on the foundations of quantum physics was initially motivated by curiosity and even by philosophical considerations, this has recently led to the emergence

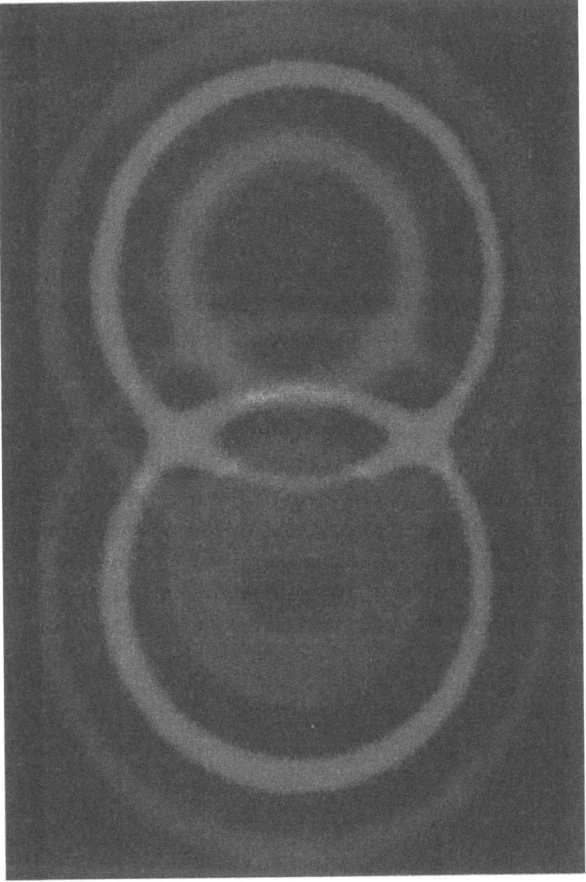

Figure 8. A more complete representation of the radiation produced in type-II parametric downconversion (photo: Paul Kwiat and Michael Reck). Three photographs taken with different color filters have been superposed here. The colors are actually false colors for clarity of presentation. The photons emitted from the source are momentum and energy entangled in such a way that each photon can be emitted with a variety of different momenta and frequencies, each frequency defining a cone of emission for each photon. The whole quantum state is then a superposition of many different pairs. For example, if measurement reveals a photon to be found somewhere on the red small circle in the figure, its brother photon is found exactly opposite on the blue small circle. The green circles represent the case where the two colors are identical. (See Color Plate 8.)

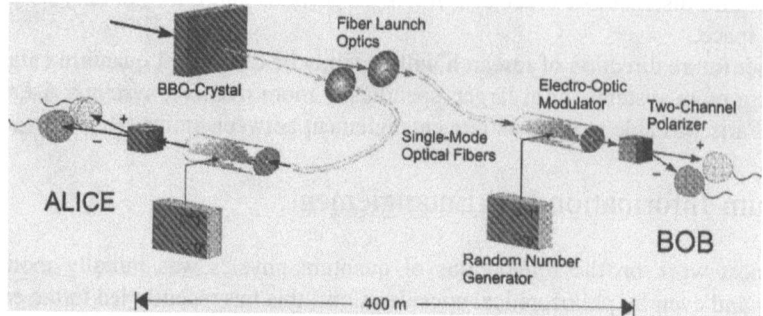

Figure 9. Long-distance Bell-inequality experiment with independent observers (Weihs et al., 1998). The two entangled photons are individually launched into optical fibers and sent to the measurement stations of the experimenters Alice and Bob which are separated from each other by a distance of 400 m. At each of the measurement stations, an independent, very fast, random-number generator decides, while the photons are really in flight, the direction along which the polarization will be measured. Finally, events are registered independently on both sides and coincidences are only identified long after the experiment is finished.

of novel ideas in information science. A significant result is already a new perspective on information itself. Eventually, applications might include quantum communication, quantum cryptography, possibly even quantum computation.

Some of the basic novel features are contained in quantum teleportation involving two distant experimenters, conventionally called Alice and Bob (Fig. 10). Here, Alice initially has a single particle in the quantum state $|\psi\rangle$ (the "teleportee"). The state may be unknown to her or possibly even undefined. The aim is that the distant experimenter Bob obtains an exact replica of that particle. It is evident that no measurement whatsoever Alice might perform on the particle could reveal all necessary information to enable Bob to reconstruct its state. The quantum teleportation protocol (Bennett et al., 1993) proceeds by Alice and Bob agreeing to share initially an entangled pair of "ancillary" photons. Alice then performs a joint Bell-state measurement on the teleportee and her ancillary

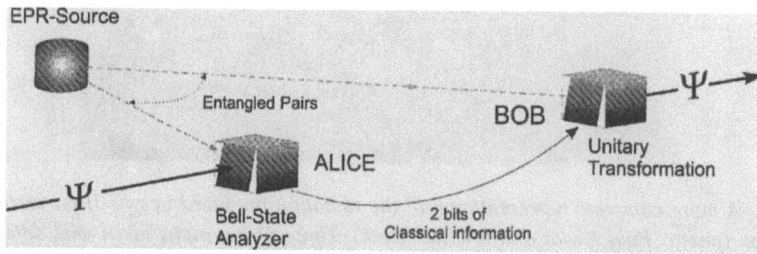

Figure 10. Principle of quantum teleportation (Bennett et al., 1993). In order to teleport her, possibly even unknown, quantum state $|\psi\rangle$ to Bob, Alice shares with him initially an entangled pair. She then performs a Bell-state analysis and, after receipt of Alice's measurement result, Bob can turn his member of the entangled pair into the original state by applying a unitary transformation which only depends on the specific Bell state result obtained by Alice and is independent of any properties of the teleportee state $|\psi\rangle$.

photon, and obtains one of the four possible Bell results. The four possible Bell states (Braunstein et al., 1992) are

$$|\psi^{\pm}\rangle = \frac{1}{\sqrt{2}}(|H\rangle_1|V\rangle_2 \pm |V\rangle_1|H\rangle_2),$$

$$|\phi^{\pm}\rangle = \frac{1}{\sqrt{2}}(|H\rangle_1|H\rangle_2 \pm |V\rangle_1|V\rangle_2). \tag{8}$$

They form a maximally entangled basis for the two-photon four-dimensional spin Hilbert space. These Bell states are essential in many quantum information scenarios. Alice's measurement also projects Bob's ancillary photon into a well-defined quantum state. Alice then transmits her result as a classical two-bit message to Bob, who performs one of four unitary operations, independent of the state $|\psi\rangle$, to obtain the original state. In the experiment (Bouwmeester et al., 1997), femtosecond pulse technology had to be used in order to obtain the necessary nontrivial coherence conditions for the Bell-state measurements.

While teleportation presently might sound like a strange name conjuring up futuristic images, it is appropriate. The reader should be reminded of the strange connotations of the notion of magnetism before its clear definition by physicists. Quantum teleportation actually demonstrates some of the salient features of entanglement and quantum information. It also raises deep questions about the nature of reality in the quantum world.

Most important for the understanding of the quantum teleportation scheme is the realization that maximally entangled states such as the Bell basis are characterized by the fact that none of the individual members of the entangled state, in our case, the two photons, carries any information on its own. All information is only encoded in joint properties. Thus, an entangled state is a representation of the relations between two possible measurements on the two members of the entangled pair. In the most simple case, the state $|\psi^-\rangle$ is a representation of the prediction that in any basis whatsoever, the two photons will be found to have orthogonal states with none of the photons having any well-defined state before measurement. The teleportation scheme then simply means that Alice's Bell-state measurement results in a well-defined relational statement between the original $|\psi\rangle$ and one of the two photons emerging from the EPR source. The specific entangled state emitted by the source then implies another relational statement with Bob's photon, and thus, by this line of reasoning, we have a clear relational statement connecting his photon with Alice's original. That statement is independent of the properties of $|\psi\rangle$, and Bob just has to apply the proper unitary transformation defined by the specific one of the four Bell states Alice happened to obtain randomly. In the most simple case, suppose Alice's Bell-state measurement happens to give the same result as the state emitted by the source. Then, Bob's particle is immediately identical to the original, and his unitary transformation is the identity. Even more striking is the possibility to teleport a quantum state which itself is entangled to another particle. Then, the teleported state is not just unknown but undefined. This possibility results in entanglement swapping (Zukowski et al., 1993; Pan et al., 1998), that is, in entangling two particles which were created completely independently and which never interacted.

The essential feature in all these schemes is again entanglement. Information can be shared by two photons in a way where none of the individuals carries any information on its own.

As a most striking example consider entangled superpositions of three quanta, e.g.,

$$|\psi\rangle = \frac{1}{\sqrt{2}}(|H\rangle|H\rangle|H\rangle + |V\rangle|V\rangle|V\rangle). \tag{9}$$

Such states, usually called Greenberger-Horne-Zeilinger states (Greenberger *et al.*, 1989; Greenberger *et al.*, 1990), exhibit very rich perfect correlations. For such states, these perfect correlations lead to a dramatic conflict with local realism on an event-by-event basis and not just on a statistical basis as in experiments testing Bell's inequality. Such states and their multiquanta generalizations are essential ingredients in many quantum communication and quantum computation schemes (*Physics World*, 1998).

Final Remarks and Outlook

I hope that the reader can sympathize now with my viewpoint that quantum physics goes beyond Wittgenstein, who starts his Tractatus Logico-Philosophicus with the sentence, "The world is everything that is the case." This is a classical viewpoint, a quantum state goes beyond. It represents all possibilities of everything that could be the case.

In any case, it will be interesting in the future to see more and more quantum experiments realized with increasingly larger objects. Another very promising future avenue of development is to realize entanglements of increasing complexity, either by entangling more and more systems with each other, or by entangling systems with a larger number of degrees of freedom. Eventually, all these developments will push the realm of quantum physics well into the macroscopic world. I expect that they will further elucidate Bohr's viewpoint that over a very large range the classical-quantum boundary is at the whim of the experimenter. Which parts we can talk about using our classical language and which parts are the quantum system depends on the specific experimental setup.

In the present brief overview I avoided all discussion of various alternative interpretations of quantum physics. I also did not venture into analyzing possible suggested alternatives to quantum mechanics. All these topics are quite important, interesting and in lively development. I hope my omissions are justified by the lack of space. It is my personal expectation that new insight and any progress in the interpretive discussion of quantum mechanics will bring along fundamentally new assessment of our humble role in the Universe.

Acknowledgments

I would like to thank all my colleagues and collaborators who over many years shared with me the joy of working in the foundations of quantum physics. I am deeply indebted to Ben Bederson and Kurt Gottfried for careful reading of my initial manuscript and for detailed comments. I would also like to thank those who supported my work financially, particularly the Austrian Science Foundation FWF and the U.S. National Science Foundation NSF.

References

As impossible as it is to do justice to all the work done in the field, it is equally impossible to attempt to quote even a minute fraction only of the relevant papers published. The general references below should help the reader to delve deeper into the subject:

Bell, J. S., 1987, *Speakable and Unspeakable in Quantum Mechanics* (Cambridge University, Cambridge).
Feynman, R. P., R. B. Leighton, and M. Sands, 1965, *The Feynman Lectures of Physics*, Vol. III, Quantum Mechanics (Addison-Wesley, Reading).
Greenberger, D. M., and A. Zeilinger, 1995, Eds., *Fundamental Problems in Quantum Theory* (Annals of the New York Academy of Sciences, Vol. 755, New York).
Greenstein, G. and A. G. Zajonc, 1997, *The Quantum Challenge: Modern Research on the Foundations of Quantum Mechanics* (Jones and Bartlett, New York).
Klyshko D., 1988, *Photons and Nonlinear Optics* (Gordon and Breech, New York).
Quantum Information, 1998, Phys. World **11** (3).
Rauch, H., and S. A. Werner, *Neutron Interferometry. Lessons in Experimental Quantum Mechanics* (Oxford University, Oxford) (in press).
Silverman, M. P., 1995, *More than One Mystery: Explorations in Quantum Interference* (Springer, Berlin).
Wheeler J. A., and W. H. Zurek, 1983, Eds. *Quantum Theory and Measurement* (Princeton University Princeton).

Specific papers quoted in the present paper are:
Alley, C. O., and Y. H. Shih, 1986, in *Proceedings of the Second International Symposium on Foundations of Quantum Mechanics in the Light of New Technology*, Tokyo, 1986, edited by M. Namiki *et al.* (Physical Society of Japan), p. 47.
Aspect, A., J. Dalibard, and G. Roger, 1982, Phys. Rev. Lett. **49**, 1804.
Aspect, A., P. Grangier, and G. Roger, 1981, Phys. Rev. Lett. **47**, 460.
Aspect, A., P. Grangier, and G. Roger, 1982, Phys. Rev. Lett. **49**, 91.
Bell, J. S., 1964, Physics (Long Island City, N.Y.) **1**, reprinted in J. S. Bell, 1987, *Speakable and Unspeakable in Quantum Mechanics* (Cambridge University, Cambridge).
Bell, J. S., 1966, Rev. Mod. Phys. **38**, 447.
Bell, J. S., 1981, J. Phys. C2 **42**, 41, reprinted in J. S. Bell, 1987, *Speakable and Unspeakable in Quantum Mechanics* (Cambridge University, Cambridge).
Bell, J. S., 1990, Phys. World 3(August), 33.
Bennett, C. H., G. Brassard, C. Crépeau, R. Josza, A. Peres, and W. K. Wootters, 1993, Phys. Rev. Lett. **70**, 1895.
Bohr, N., 1935, Phys. Rev. **48**, 696.
Bouwmeester, D., J. W. Pan, K. Mattle, M. Eibl, H. Weinfurter, and A. Zeilinger, 1997, Nature (London) **390**, 575.
Braunstein, S. L., A. Mann, and M. Revzen, 1992, Phys. Rev. Lett. **68**, 3259.
Burnham, D. C., and D. L. Weinberg, 1970, Phys. Rev. Lett. **25**, 84.
Clauser, J. F., M. A. Horne, A. Shimony, and R. A. Holt, 1969, Phys. Rev. Lett. **23**, 880.
Dopfer, B., 1998, Ph.D. thesis (University of Innsbruck).
Einstein, A., B. Podolsky, and N. Rosen, 1935, Phys. Rev. **47**, 777.
Franson, J. D., 1989, Phys. Rev. Lett. **62**, 2205.
Freedman, S. J., and J. S. Clauser, 1972, Phys. Rev. Lett. **28**, 938.
Gottfried, K., 1991, Phys. World 4(October), 41.
Greenberger, D. M., M. A. Horne, and A. Zeilinger, 1993, Phys. Today (8), 22.
Greenberger, D. M., M. A. Horne, and A. Zeilinger, 1989, in *Bell's Theorem, Quantum Theory, and Conceptions of the Universe*, edited by M. Kafatos (Kluwer, Dordrecht), p. 74.
Greenberger, D. M., M. A. Horne, A. Shimony, and A. Zeilinger, 1990, Am. J. Phys. **58**, 1131.

Hagley, E., X. Maître, G. Nogues, C. Wunderlich, M. Brune, J. M. Raimond, and S. Haroche, 1997, Phys. Rev. Lett. **79**.
Heisenberg, W., 1927, Z. Phys. (Leipzig) **43**, 172, reprinted in English in Wheeler and Zurek (above).
Horne, M. A., and A. Zeilinger, 1985, in *Proceedings of the Symposium Foundations of Modern Physics*, edited by P. Lahti and P. Mittelstaedt (World Scientific, Singapore), p. 435.
Horne, M. A., and A. Zeilinger, 1988, in *Microphysical Reality and Quantum Formalism*, edited by A. van der Merwe, F. Seller, and G. Tarozzi (Kluwer, Dordrecht), p. 401.
Kochen, S. and E. Specker, 1967, J. Math. Mech. **17**, 59.
Kwiat, P., H. Weinfurter, T. Herzog, and A. Zeilinger, 1995, Phys. Rev. Lett. **74**, 4763.
Mandel, L., 1999, Rev. Mod. Phys. **71**, 274; pp. 460–473 in this book.
Mermin, N. D., 1993, Rev. Mod. Phys. **65**, 803.
Pan, J. W., D. Bouwmeester, H. Weinfurter, and A. Zeilinger, 1998, Phys. Rev. Lett. **80**, 3891.
Penrose, R., 1986, in *Quantum Concepts in Space and Time*, edited by R. Penrose and C. J. Isham (Clarendon Press, Oxford), p. 139.
Schrödinger, E., 1935, Naturwissenschaften **23**, English translation in *Proceedings of the American Philosophical Society*, 124 (1980) reprinted in Wheeler & Zurek, above.
Specker, F., 1960, Dialectica **14**, 239.
Tittel, W., J. Brendel, B. Gisin, T. Herzog, H. Zbinden, and N. Gisin, 1998, Phys. Rev. A **57**, 3229.
Weihs, G., T. Jenewein, C. Simon, H. Weinfurter, and A. Zeilinger, 1998, Phys. Rev. Lett **81**, 5039.
Weizsächer, K. F., 1931, Z. Phys. **40**, 114.
Wigner, E. P., 1970, Am. J. Phys. **38**, 1005.
Wu, C. S., and I. Shaknov, 1950, Phys. Rev. **77**, 136.
Zeilinger, A., 1999, Found. Phys. (in press).
Zeilinger, A., R. Gähler, C. G. Shull, W. Treimer, and W. Hampe, 1988, Rev. Mod. Phys. **60**, 1067.
Zukowski, M., A. Zeilinger, M. A. Horne, and A. K. Ekert, 1993, Phys. Rev. Lett. **71**, 4287.

MIX
Papier aus verantwortungsvollen Quellen
Paper from responsible sources
FSC® C105338

If you have any concerns about our products,
you can contact us on
ProductSafety@springernature.com

In case Publisher is established outside the EU,
the EU authorized representative is:
Springer Nature Customer Service Center GmbH
Europaplatz 3, 69115 Heidelberg, Germany

Printed by Libri Plureos GmbH
in Hamburg, Germany

More Things in Heaven and Earth

A Celebration of Physics at the Millennium

Springer-Verlag Berlin Heidelberg GmbH

BENJAMIN BEDERSON
EDITOR

More Things in Heaven and Earth

A Celebration of Physics at the Millennium

 Springer

APS
The American Physical Society

Benjamin Bederson
Department of Physics
New York University
New York, NY 10003
USA

Library of Congress Cataloging-in-Publication Data
More things in heaven and earth : a celebration of physics
at the millennium / edited by Ben Bederson.
   p. cm.
   Includes bibliographical references.
   ISBN 978-1-4612-7174-1   ISBN 978-1-4612-1512-7 (eBook)
   DOI 10.1007/978-1-4612-1512-7
   1. Physics—History—20th century. 2. Astrophysics—History—20th century.
   I. Bederson, Benjamin. II. Title: A celebration of physics at the millennium.
   QC7.M568   1999   09-53116
   530'.09'04—dc21   CIP

Printed on acid-free paper.

©1999 Springer-Verlag Berlin Heidelberg
Originally published by Springer-Verlag Berlin Heidelberg New York in 1999

All rights reserved. This work may not be translated or copied in whole or in part without the written permission of The American Physical Society, One Physics Ellipse, College Park, MD 20740-3844, USA, except for brief excerpts in connection with reviews or scholarly analysis. Use in connection with any form of information storage and retrieval, electronic adaptation, computer software, or by similar or dissimilar methodology now known or hereafter developed is forbidden.
The use of general descriptive names, trade names, trademarks, etc., in this publication, even if the former are not especially identified, is not to be taken as a sign that such names, as understood by the Trade Marks and Merchandise Marks Act, may accordingly be used freely by anyone.

Production managed by MaryAnn Cottone; manufacturing supervised by Nancy Wu.
Photocomposed pages prepared by American Institute of Physics, Woodbury, NY.

9 8 7 6 5 4 3 2 1

ISBN 978-1-4612-7174-1   SPIN 10697621

Contents

- ix   Contributors
- xv   Foreword
  Martin Blume
- xvii   Preface
  Benjamin Bederson
- xxi   Introduction
  Hans A. Bethe

Historic Perspectives–Personal Essays on Historic Developments

- 3   Quantum Theory
  Hans A. Bethe
- 12   Nuclear Physics
  Hans A. Bethe
- 29   Theoretical Particle Physics
  A. Pais
- 43   Elementary Particle Physics: The Origins
  Val L. Fitch
- 56   Astrophysics
  George Field
- 69   A Century of Relativity
  Irwin I. Shapiro
- 89   From Radar to Nuclear Magnetic Resonance
  Robert V. Pound
- 96   An Essay on Condensed Matter Physics in the Twentieth Century
  W. Kohn
- 129   A Short History of Atomic Physics in the Twentieth Century
  Daniel Kleppner

Particle Physics and Related Topics

- 143   Quantum Field Theory
  Frank Wilczek
- 161   The Standard Model of Particle Physics
  Mary K. Gaillard, Paul D. Grannis, and Frank J. Sciulli

188 String Theory, Supersymmetry, Unification, and All That
John H. Schwarz and Nathan Seiberg

203 Accelerators and Detectors
W. K. H. Panofsky and M. Breidenbach

223 Anomalous *g* Values of the Electron and Muon
V. W. Hughes and T. Kinoshita

234 Neutrino Physics
L. Wolfenstein

Astrophysics

245 Cosmology at The Millennium
Michael S. Turner and J. Anthony Tyson

278 Cosmic Rays: The Most Energetic Particles in the Universe
James W. Cronin

291 Cosmic Microwave Background Radiation
Lyman Page and David Wilkinson

302 Black Holes
Gary T. Horowitz and Saul A. Teukolsky

313 Gravitational Radiation
Rainer Weiss

329 Deciphering the Nature of Dark Matter
Bernard Sadoulet

Nuclear Physics

345 Nuclear Physics at the End of the Century
E. M. Henley and J. P. Schiffer

370 Stellar Nucleosynthesis
Edwin E. Salpeter

Atomic, Molecular, and Optical Physics

377 Atomic Physics
Sheldon Datz, G. W. F. Drake, T. F. Gallagher, H. Kleinpoppen, and G. zu Putlitz

408 Laser Spectroscopy and Quantum Optics
T. W. Hänsch and H. Walther

426 Atom Cooling, Trapping, and Quantum Manipulation
Carl E. Wieman, David E. Pritchard, and David J. Wineland

442 Laser Physics: Quantum Controversy in Action
W. E. Lamb, W. P. Schleich, M. O. Scully, and C. H. Townes

460 Quantum Effects in One-Photon and Two-Photon Interference
     L. Mandel

474 From Nanosecond to Femtosecond Science
     N. Bloembergen

482 Experiment and Foundations of Quantum Physics
     Anton Zeilinger

Condensed Matter Physics

501 The Fractional Quantum Hall Effect
     Horst L. Stormer, Daniel C. Tsui, and Arthur C. Gossard

515 Conductance Viewed as Transmission
     Yoseph Imry and Rolf Landauer

526 Superconductivity
     J. R. Schrieffer and M. Tinkham

534 Superfluidity
     A. J. Leggett

543 In Touch with Atoms
     G. Binnig and H. Rohrer

555 Materials Physics
     P. Chaudhari and M. S. Dresselhaus

563 The Invention of the Transistor
     Michael Riordan, Lillian Hoddeson, and Conyers Herring

Statistical Physics and Fluids

581 Statistical Mechanics: A Selective Review of Two Central Issues
     Joel L. Lebowitz

601 Scaling, Universality, and Renormalization: Three Pillars of Modern Critical Phenomena
     H. Eugene Stanley

617 Insights from Soft Condensed Matter
     Thomas A. Witten

629 Granular Matter: A Tentative View
     P. G. de Gennes

644 Fluid Turbulence
     Katepalli R. Sreenivasan

665 Pattern Formation in Nonequilibrium Physics
     J. P. Gollub and J. S. Langer

Plasma Physics

679 The Collisionless Nature of High-Temperature Plasmas
T. M. O'Neil and F. V. Coroniti

Chemical Physics and Biological Physics

693 Chemical Physics: Molecular Clouds, Clusters, and Corrals
Dudley Herschbach

706 Biological Physics
Hans Frauenfelder, Peter G. Wolynes, and Robert H. Austin

726 Brain, Neural Networks, and Computation
J. J. Hopfield

Computational Physics

741 Microscopic Simulations in Physics
D. M. Ceperley

Applications of Physics to Other Areas

755 Physics and Applications of Medical Imaging
William R. Hendee

767 Nuclear Fission Reactors
Charles E. Till

775 Nuclear Power—Fusion
T. Kenneth Fowler

781 Physics and U.S. National Security
Sidney D. Drell

798 Laser Technology
R. E. Slusher

812 Physics and the Communications Industry
W. F. Brinkman and D. V. Lang

827 Index

Contributors

Robert H. Austin
Department of Physics
Princeton University
Princeton, New Jersey 08544

Hans A. Bethe
Floyd R. Newman Laboratory
  of Nuclear Studies
Cornell University
Ithaca, New York 14853

Gerd K. Binnig
IBM Research Division
Zurich Research Laboratory
Säumerstrasse 4
8803 Rüschlikon, Switzerland

Nicolaas Bloembergen
Department of Physics
Harvard University
Cambridge, Massachusetts 02138

Martin Breidenbach
Stanford Linear Accelerator Center
Stanford University
Stanford, California 94305

William F. Brinkman
Physical Sciences Research
AT&T Bell Laboratories
Lucent Technologies
Murray Hill, New Jersey 07974

David M. Ceperley
National Center for Superconducting
  Applications
University of Illinois
Urbana, Illinois 61801

Praveen Chaudhari
IBM Thomas J. Watson Research Center
Yorktown Heights, New York 10598

Ferdinand V. Coroniti
Department of Physics and Astronomy
University of California
Los Angeles, California 90095

James W. Cronin
Enrico Fermi Instiute
University of Chicago
Chicago, Illinois 60637

Sheldon Datz
Atomic Physics Division
Oak Ridge National Laboratory
Oak Ridge, Tennessee 37831

Pierre Giles de Gennes
College de France
Physique de la Matiere Condensee
75231 Paris Cedex 05
France

Gordon W. F. Drake
Department of Physics
University of Windsor
Windsor, Ontario
Canada

Sidney D. Drell
Stanford Linear Accelerator Center
Stanford University
Stanford, California 94309

Mildred S. Dresselhaus
Massachusetts Institute of Technology
Cambridge, Massachusetts 02139

George B. Field
Harvard-Smithsonian Center for Astrophysics
Cambridge, Massachusetts 02138

Val L. Fitch
Department of Physics
Princeton University
Princeton, New Jersey 08544

T. Kenneth Fowler
Nuclear Engineering
University of California
4155 Etcheverry Hall
Berkeley, California 94720

Hans Frauenfelder
Center for Nonlinear Studies
Los Alamos National Laboratory
Los Alamos, New Mexico 87545

Mary K. Gaillard
Physics Department
Lawrence Berkeley Laboratory
University of California
Berkeley, California 94720

Thomas F. Gallagher
Department of Physics
University of Virginia
Charlottesville, Virginia 22901

Jerry P. Gollub
Physics Department
Haverford College
Haverford, Pennsylvania 19041

Arthur C. Gossard
Materials Department
University of California
Santa Barbara, California 93106

Paul D. Grannis
Department of Physics
State University of New York
Stony Brook, New York 11794

Theodore W. Hänsch
Max-Planck-Insitut fur Quantenoptik
D-85740 Garching
Germany

William R. Hendee
Medical College of Wisconsin
8701 Watertown Plank Road
Milwaukee, Wisconsin 53226

Ernest M. Henley
Department of Physics
University of Washington
Seattle, Washington 98195

Conyers Herring
Department of Applied Physics
Stanford University
Stanford, California 94305

Dudley R. Herschbach
Chemistry and Biochemistry Departments
Harvard University
Cambridge, Massachusetts 02138

Lillian Hoddeson
Department of Physics
University of Illinois
Urbana, Illinois 61801

John J. Hopfield
Department of Molecular Biology
Princeton University
Princeton, New Jersey 08544

Gary T. Horowitz
Department of Physics
University of California
Santa Barbara, California 93106

Vernon W. Hughes
Department of Physics
Yale University
New Haven, Connecticut 06520

Yoseph Imry
Department of Physics
Weizmann Institute of Science
Rehovot 76 100
Israel

Thomas Kinoshita
Department of Physics
Cornell University
Ithaca, New York 14853

Hans Kleinpoppen
Department of Physics
University of Stirling
Stirling FK9 4LA
Scotland

Contributors

Daniel Kleppner
Department of Physics
Massachusetts Institute of Technology
Cambridge, Massachusetts 02139

Walter Kohn
Department of Physics
University of California
Santa Barbara, California 93106

Willis E. Lamb
Optical Sciences Center
University of Arizona
Tucson, Arizona 85721

Rolf W. Landauer
IBM Thomas J. Watson Research Center
Yorktown Heights, New York 10598

David V. Lang
AT&T Bell Laboratories
Lucent Technologies
Murray Hill, New Jersey 07974

James S. Langer
Department of Physics
University of California
Santa Barbara, California 93106

Joel L. Lebowitz
Department of Mathematics and Physics
Rutgers University
Piscataway, New Jersey 08854

Anthony J. Leggett
Department of Physics
University of Illinois
Urbana, Illinois 61801

Leonard Mandel
Department of Physics and Astronomy
University of Rochester
Rochester, New York 14627

Thomas M. O'Neil
Department of Physics
University of California at San Diego
La Jolla, California 92093

Lyman A. Page
Department of Physics
Princeton University
Princeton, New Jersey 08544

Abram Pais
Department of Physics
Rockefeller University
New York, New York 10021

Wolfgang K. H. (Pief) Panofsky
Stanford Linear Accelerator Center
Stanford University
Stanford, California 94309

Robert V. Pound
Department of Physics
Harvard Univerity
Cambridge, Massachusetts 02138

David E. Pritchard
Department of Physics
Massachusetts Institute of Technology
Cambridge, Massachusetts 02140

Michael Riordan
4532 Cherryvale
Soquel, California 95073

Heinrich Rohrer
IBM Research Division
Zurich Research Laboratory
8805 Richterswil
Zurich, Switzerland

Bernard Sadoulet
Center for Particle Astrophysics
Lawrence Berkeley National Laboratory and
  Physics Department
University of California
Berkeley, California 94720

Edwin E. Salpeter
Center for Radiophysics and Space Research
Cornell University
Ithaca, New York 14853

John P. Schiffer
Physics Division
Argonne National Laboratory
Argonne, Illinois 60439

Wolfgang P. Schleich
Abteilung fur Quantenphysik
Universität Ulm
D-89069 Ulm
Germany

J. Robert Schrieffer
National High Magnetic Field Laboratory
Florida State University
Tallahassee, Florida 32310

John H. Schwarz
Department of Theoretical Physics
California Institute of Technology
Pasadena, California 91125

Frank J. Sciulli
Professor of Physics
Columbia University
New York, New York 10027

Marlan O. Scully
Department of Physics
Texas A&M University
College Station, Texas 77843 and
Max-Planck-Institut fur Quantenoptik
Hans-Kopfermann Strasse 1
D-85748 Garching
Germany

Nathan Seiberg
Institute for Advanced Study
Princeton, New Jersey 08854

Irwin I. Shapiro
Harvard-Smithsonian Center for Astrophysics
Harvard University
Cambridge, Massachusetts 02138

Richard E. Slusher
AT&T Bell Laboratories
Lucent Technologies
Murray Hill, New Jersey 07974

Katepalli R. Sreenivasan
Department of Mechanical Engineering
Yale University
New Haven, Connecticut 06520

H. Eugene Stanley
Department of Physics
Boston University
Boston, Massachusetts 02215

Horst L. Stormer
Departments of Physics and of Applied
  Physics
Columbia University
New York, New York 10027 and
AT&T Bell Laboratories
Lucent Technologies
Murray, Hill New Jersey 07974

Saul A. Teukolsky
Floyd R. Newman Laboratory of Nuclear
  Studies
Cornell University
Ithaca, New York 14853

Charles E. Till
Argonne National Laboratory
Argonne, Illinois 60439

Michael Tinkham
Department of Physics
Harvard University
Cambridge, Massachusetts 02138

Charles H. Townes
Department of Physics
University of California
Berkeley, California 94720

Daniel C. Tsui
Department of Electrical Engineering
Princeton University
Princeton, New Jersey 08544

Michael S. Turner
Departments of Physics and of Astronomy &
  Astrophysics
Enrico Fermi Institute
University of Chicago
Chicago, Illinois 60637

J. Anthony Tyson
AT&T Bell Laboratories
Lucent Technologies
Murray Hill, New Jersey 07974

Herbert Walther
Max-Planck-Institut fur Quantenoptik
D-85740 Garching
Germany

Rainer Weiss
Department of Physics
Massachusetts Institute of Technology
Cambridge, Massachusetts 02139

Carl E. Wieman
Joint Institute for Laboratory Astrophysics
  and Department of Physics
University of Colorado
Boulder, Colorado 80309

Frank A. Wilczek
Institute for Advanced Study
School of Natural Sciences
Princeton, New Jersey 08540

David Todd Wilkinson
Department of Physics
Princeton University
Princeton, New Jersey 08544

David Wineland
National Institute of Standards and
  Technology
Time and Frequency Division
Boulder, Colorado 80309

Thomas A. Witten
James Franck Institute
University of Chicago
Chicago, Illinois 60637

Lincoln Wolfenstein
Department of Physics
Carnegie-Mellon University
Pittsburgh, Pennsylvania 15213

Peter G. Wolynes
Department of Chemistry
University of Illinois
Urbana, Illinois 61801

Anton Zeilinger
Institut für Experimentalphysik
University of Vienna
Boltzmanngasse 5
A-1000 Vienna
Austria

Gisbert zu Putlitz
Physikalisches Institut der Universität
  Heidelberg
D-6900 Heidelberg
Germany

Foreword

It is most appropriate that the Centennial of the American Physical Society be marked by a special issue of *Reviews of Modern Physics*, here published in a hardcover edition. RMP was begun when the American Physical Society was thirty years old, as the Society began its rise to importance on the international scene of physics. Many of the splendid review articles that appeared in RMP have chronicled the development of physics since that time, and several of the distinguished contributors to this volume have authored some of those articles. The forward march of physics continues unabated. There has been some hesitation, however, on looking ahead for the next hundred years—the example of predictions made at the end of the last century should be sufficient to give pause to all but the most determined prognosticators. I will only say that startling new ideas and phenomena will continue to appear, and I hope and expect that the most novel of these will continue to be revealed in *Physical Review* and summarized in *Reviews of Modern Physics*. We are all deeply indebted to Ben Bederson, my predecessor as Editor-in-Chief, for taking on the job of special editor for this issue. He deserves our thanks and gratitude for this splendid volume.

Martin Blume
Editor-in-Chief
The American Physical Society

Preface

Inspired by the 100th anniversary of the founding of the American Physical Society, we undertook this project in order to display the vast canvas that encompasses a century of magnificent accomplishment in physics as we enter the new millennium. Our intent was to present a contemporary portrait of physics, based on its past achievements and its current vitality. With regard to its future, maybe the best statement we can offer is contained in the last sentence of Hans Bethe's Introduction: *Looking at the predictions of 100 years ago, it would be foolish to make predictions for the next 100 years.* But, even so, scattered throughout the volume will be found somewhat more modest extrapolations, on a more limited time scale.

The scientific content of this volume is the same as the special Centenary issue of *Reviews of Modern Physics*. In a sense it can be considered as a companion to the collection edited by H. Henry Stroke, *The Physical Review: The First Hundred Years*, a book and CD-ROM which contains about 1000 articles from the *Physical Review* and *Physical Review Letters*, published in 1995 by the American Institute of Physics (with an upgraded CD to be issued by Springer-Verlag in 1999). That volume commemorated the slightly earlier centenary of the *Physical Review*, whose birth preceded that of the APS by three years.

We were all too painfully aware that a single volume, however ambitious its scope, could not hope to present a full portrait of all of physics—an understatement of the limitations that this project would surely be up against. As a result, we set ourselves a somewhat more limited goal. While we invited articles in the major subfields of physics, we asked authors to deal with these in a relatively informal, even personal way. We encouraged them not to try to include everything, and where appropriate to discuss subjects that were of particular interest to themselves. In some cases we chose to deal with specific aspects of a subject. These should be considered as "case studies." As a result readers will find many favorite topics omitted. Emphatically, it does not pretend to be comprehensive and readers should not be surprised or chagrined to discover significant omissions. This is the price we had to pay to accomplish our limited goals.

We asked authors to write their articles at the level of a good departmental colloquium. How successful our authors have been in abiding by this guideline is for the reader to determine.

The first order of business was to form an advisory group, a special "Editorial Board." The group, all of whom readily agreed to serve, consisted of six very special individuals: Kurt Gottfried (Cornell University), representing theoretical particle, nuclear physics, and general theoretical physics; Walter Kohn (University of California Santa Barbara), for theoretical condensed matter physics; Eugene Merzbacher (University of North Carolina), a past APS president, for atomic, molecular, and optical physics; Myriam Sarachik (City College of New York), for experimental condensed matter physics; David Schramm (University of Chicago), representing astrophysics and cosmology; and Andrew Sessler (University of California Berkeley), for experimental high energy physics and a variety of other subfields and, incidentally, President of APS in 1998. Sadly David Schramm perished in an airplane accident about halfway through the project, although he played a major role during its initial phase in developing the outline and in identifying appropriate authors. We were very fortunate to have George Field, Harvard Smithsonian Center for Astrophysics, then agree to serve in his stead. He has played an equally invaluable role in overseeing the astrophysics and cosmology aspects of the volume.

Inevitably, in reviewing past achievements, the history of physics, as a subject in itself, had to play a major role. In order to present this history in the same spirit as for the subject matter of the volume we decided to have a special history section, prepared by what one might characterize as "eyewitnesses"— though the contributors to this section are by no means as old as the APS. Peter Galison, Professor of the History of Science at Harvard University, has served as coordinator of this section and has participated in the reviewing of these articles. Of course the history of physics cannot be totally separated from physics itself, so history appears interwoven with scientific presentations throughout the volume, although with a different emphasis in the main body of the volume.

Physics is a living subject—reshaping, developing, evolving even as you look at it. Just within the brief period during which we have been receiving manuscripts major developments have occurred in many fields which warrant mention, even without our expressed goal of not being all-inclusive. Our authors have continued to revise their articles up to the very last possible minute.

To simplify the difficult task facing our authors we have encouraged them to keep citations to original research articles to a minimum, referring instead where possible to review articles and books. This necessarily results in omissions of worthy citations. Please bear this in mind, especially if you happen to be one of those omitted. However, authors did not uniformly abide by this rule, and we did not attempt to enforce it rigorously; this explains the inconsistencies with regard to references in some of the articles. It should not surprise anyone that there is an enormous variety of styles, range of material covered, and historical material included. This reflects, we believe, the enormous range of styles and personal taste of physicists themselves. On the other hand all articles were reviewed by at least one expert, either an editorial board member or an outside referee, although primarily for scientific content rather than uniformity of presentation. We are also aware that the volume possesses a daunting size, much larger than that which was originally intended. The editor must bear the brunt of possible criticism in this regard—in the last analysis he was unwilling to insist on Draconian measures to cut articles down to previously allotted lengths.

The time constraint—eighteen months from inception to completion—has doubtlessly resulted in the appearance of some scientific and possibly historic errors. We have of course worked diligently to keep these to a minimum; still we cannot claim that they do not exist.

Acknowledgments

It is literally true that there would have been no volume at all were it not for the continuing and generous advice, criticism, and encouragement of the extraordinary *ad hoc* Editorial Board, whose members are listed above. They deserve very special thanks.

Obviously such a project could not have come to fruition without the full cooperation, assistance, and encouragement of a large number of people, beyond the members of the

Editorial Board and, of course, the authors. Among these are the many other physicists whose advice we sought and received and anonymous reviewers of submitted articles.

I wish to acknowledge the following individuals and groups for their specific roles in producing this volume:

First, the authors. On very short notice, and with very severe constraints in time and space, our authors responded magnificently to this challenge. They good-naturedly acquiesced to the editor's haggling and imposed deadlines; they generally agreed to what we recognize as outrageous requests to cut out mention of important work, reduce the number of references to a painful few, generally making their articles possibly less authoritative and comprehensive than they would have been with more reasonable space allotments. We like to believe that their extraordinary efforts have been worth their while.

The APS Publications Oversight Committee, and the APS Executive Board, for their support and encouragement, especially Marty Blume, APS Editor-in-Chief, and Tom McIlrath, APS Treasurer, for their enthusiastic support, and George Bertsch, the Editor of *Reviews of Modern Physics*, for his willingness to relinquish his oversight authority for what appears within his journal for this single supplementary issue.

Karie Friedman, Assistant Editor of RMP, for playing the dominant role in editing final copy, performing for us what she has done with distinction for the regular issues of the journal over many years, continuing to earn the thanks and appreciation of our authors.

Maria Taylor, Executive Editor of Springer-Verlag, New York, for her invaluable assistance in arranging for and in producing this companion hardcover version of the volume.

We received enormous help from both the American Institute of Physics and Springer-Verlag, New York, in implementing remarkably tight production schedules—a crucial element in satisfying our requirement of having the supplement available for mail distribution to subscribers before the Centennial meeting and the hardcover volume available for distribution at the meeting itself.

Finally, I must state that as essential as was the advice and assistance received from all the groups and individuals noted above, the final responsibility for what appears in this volume rests with the Editor.

Benjamin Bederson
Professor of Physics Emeritus, New York University
Editor-in-Chief Emeritus, The American Physical Society

Introduction

A hundred years ago, some of the great physicists in England and Continental Europe predicted that physics was at an end. We know what actually happened, and we are proud of the contribution that The American Physical Society and *Reviews of Modern Physics*, where the material appearing in this volume first appeared, have made to it.

This volume tries to summarize some of the important areas of twentieth century physics. We have not covered all areas, and undoubtedly the essays on individual areas are not complete. But we hope some of them may be useful.

Looking at the predictions of 100 years ago, it would be foolish to make predictions for the next 100 years.

Hans A. Bethe
Cornell University

Condensed Matter Physics

The Fractional Quantum Hall Effect
Horst L. Stormer, Daniel C. Tsui, and Arthur C. Gossard

The fractional quantum Hall effect is an example of the new physics that has emerged from the enormous progress made during the past few decades in material synthesis and device processing. Driven by the increasing demands of the electronic and photonic industry for material control, ultrathin semiconductor layers of exceptional purity and smoothness are now being fabricated routinely. This technology has made it possible to realize two-dimensional (2D) electron systems of unprecedentedly low disorder, which have become an ideal laboratory in which to study many-particle physics in lower dimensions. In particular for 2D electrons in a magnetic field the discoveries have been stunning: An abundance of new energy gaps exists where one expected none. Hall resistances are quantized to exact rational fractions of the resistance quantum, while the magnetoresistance is vanishing. Huge external magnetic fields are apparently eliminated, and new particles with ballistic trajectories emerge. Theory has constructed a powerful and elegant model to account for these strange observations: Electrons condense into novel quantum liquids leading to rational fractional quantum numbers and to carriers with exactly fractional charge. Electrons absorb magnetic flux quanta, seemingly eliminating external magnetic fields. Particle statistics are altered from fermionic to bosonic and back to fermionic. And a novel, field-induced particle pairing mechanism is foreseen.

All these fascinating properties arise not from single electrons interacting with an external field, but rather from the strongly correlated motion of many electrons: It is not electron fission that leads to fractionally charged quasiparticles, but the interplay of some 10^{11} electrons per cm^2 that collaborate and create these bizarre objects.

The following pages are intended to provide a brief survey of the physics of strongly correlated 2D electron systems in the presence of a high magnetic field. By no means can it be considered an exhaustive review. For the sake of brevity many exciting topics and important contributions to the field have not been incorporated. The report focuses on the fractional quantum Hall effect and closely related phenomena. Rather than to instruct the expert, the aim is to sketch for the general, scientifically knowledgeable reader the strangeness of the experimental observations, reveal the simple beauty of the extracted concepts, and communicate the elegance of the still-evolving many-particle theory.

Background

Nearly ideal two-dimensional electron systems can be realized by quantum-mechanically confining charge carriers in thin potential wells. They are fabricated by epitaxially growing high-quality single-crystal films of selected semiconductors with different energy, 1979). The carriers in such structures are free to move along the 2D plane, but their motion perpendicular to the planes is quantized. As a result, a low-density 2D metal of high perfection emerges ($n \sim 0.2-4 \times 10^{11}$ cm^{-2}). In the cleanest case of GaAs/AlGaAs heterostructures, the 2D carriers show low-temperature mean free paths as long as 0.1 mm (!).

Application of a magnetic field normal to the plane further quantizes the in-plane motion into Landau levels at energies $E_i = (i + 1/2)\hbar \omega_c$, where $\omega_c = eB/m^*$ represents the cyclotron frequency, B the magnetic field, and m^* the effective mass of electrons having charge e. The number of available states in each Landau level, $d = 2eB/h$, is linearly proportional to B. The electron spin can further split the Landau level into two, each holding eB/h states per unit area. Thus the energy spectrum of the 2D electron system in a magnetic field is a series of discrete levels, each having a degeneracy of eB/h (Ando et al., 1983).

At low temperature ($T \ll$ Landau/spin splitting) and in a B field, the electron population of the 2D system is given simply by the Landau-level filling factor $\nu = n/d = n/(eB/h)$. As it turns out, ν is a parameter of central importance to 2D electron physics in high magnetic fields. Since $h/e = \phi_0$ is the magnetic-flux quantum, ν denotes the ratio of electron density to magnetic-flux density, or more succinctly, the number of electrons per flux quantum. Much of the physics of 2D electrons in a B field can be cast in terms of this filling factor.

Most of the experiments performed on 2D electron systems are electrical resistance measurements, although in recent years several more sophisticated experimental tools have been successfully employed. In electrical measurements, two characteristic voltages are measured as a function of B, which, when divided by the applied current, yield the magnetoresistance R_{xx} and the Hall resistance R_{xy} (see insert Fig. 1). While the former, measured along the current path, reduces to the regular resistance at zero field, the latter, measured across the current path, vanishes at $B = 0$ and, in an ordinary conductor, increases linearly with increasing B. This Hall voltage is a simple consequence of the Lorentz force's acting on the moving carriers, deflecting them into the direction normal to current and magnetic field. According to this classical model, the Hall resistance is $R_{xy} = B/ne$, which has made it, traditionally, a convenient measure of n.

It is evident that in a B field current and voltage are no longer collinear. Therefore the resistivity $\hat{\rho}$ which is simply derived from R_{xx} and R_{xy} by taking into account geometrical factors and symmetry, is no longer a number but a tensor. Accordingly, conductivity $\hat{\sigma}$ and resistivity are no longer simply inverse to each other, but obey a tensor relationship $\hat{\sigma} = \hat{\rho}^{-1}$. As a consequence, for all cases of relevance to this review, the Hall conductance is indeed the *inverse* of the Hall resistance, but the magnetoconductance is under most conditions *proportional* to the magnetoresistance. Therefore, at vanishing resistance ($\rho \to 0$), the system behaves like an insulator ($\sigma \to 0$) rather than like an ideal conductor. We hasten to add that this relationship, although counterintuitive, is a simple consequence of the Lorentz force's acting on the electrons and is not at the origin of any of the phenomena to be reviewed.

Figure 1 shows a classical example of the characteristic resistances of a 2D electron system as a function of an intense magnetic field at a temperature of 85 mK. The striking observation, peculiar to 2D, is the appearance of steps in the Hall resistance R_{xy} and exceptionally strong modulations of the magnetoresistance R_{xx}, dropping to vanishing values. These are the hallmarks of the quantum Hall effects.

The Integral Quantum Hall Effect

Integer numbers in Fig. 1 indicate the position of the integral quantum Hall effect (IQHE) (Von Klitzing, et al., 1980). The associated features are the result of the discretization of the energy spectrum due to confinement to two dimensions plus Landau/spin quantization.

At specific magnetic fields B_i, when the filling factor $\nu = n/(eB/h) = i$ is an integer, an exact number of these levels is filled, and the Fermi level resides within one of the energy gaps. There are no states available in the vicinity of the Fermi energy. Therefore, at these singular positions in the magnetic field, the electron system is rendered incompressible, and its transport parameters (R_{xx}, R_{xy}) assume quantized values (Laughlin, 1981). Localized states in the tails of each Landau/spin level, which are a result of residual disorder in the 2D system, extend the range of quantized transport from a set of precise points in B to finite ranges of B, leading at integer filling factors to the observed plateaus in the Hall resistance and stretches of vanishing magnetoresistance (Prange and Girvin, 1990; Chakraborty and Pietiläinen, 1995).

In essence, the transport features are the result of transitions between alternating metallic and insulating behavior, i.e., from E_f within a Landau/spin band to E_f in a gap

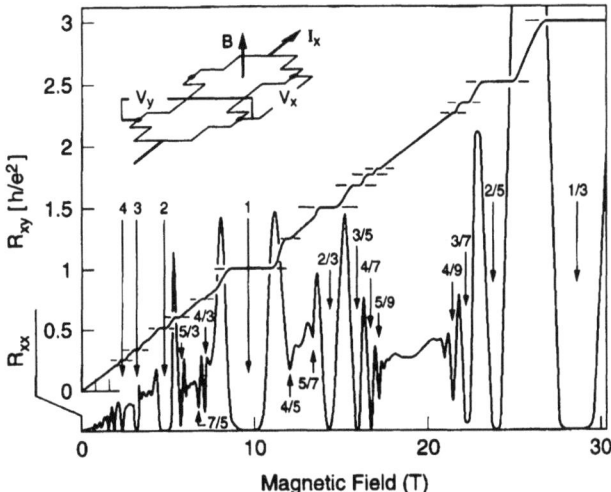

Figure 1. Composite view showing the Hall resistance $R_{xy} = V_y/I_x$ and the magnetoresistance $R_{xx} = V_x/I_x$ of a two-dimensional electron system of density $n = 2.33 \times 10^{11}$ cm^{-2} at a temperature of 85 mK, vs magnetic field. Numbers identify the filling factor ν, which indicates the degree to which the sequence of Landau levels is filled with electrons. Instead of rising strictly linearly with magnetic field, R_{xy} exhibits plateaus, quantized to $h/(\nu e^2)$ concomitant with minima of vanishing R_{xx}. These are the hallmarks of the integral ($\nu = i$ = integer) quantum Hall effect (IQHE) and fractional ($\nu = p/q$) quantum Hall effect (FQHE). While the features of the IQHE are the results of the quantization conditions for individual electrons in a magnetic field, the FQHE is of many-particle origin. The insert shows the measurement geometry. B = magnetic field, I_x = current, V_x = longitudinal voltage, and V_y = transverse or Hall voltage. From Eisenstein and Störmer, 1990.

between Landau/spin bands. These IQHE states occur at integer filling factor i and display quantization of the Hall resistance to $h/(ie^2)$, as indicated in Fig. 1. They identify and exhaust all single-particle energy gaps. The IQHE is the result of the quantization conditions for noninteracting 2D electrons in a magnetic field.

The Fractional Quantum Hall Effect

Different from the IQHE, the fractional quantum Hall effect (FQHE; Tsui, et al., 1982) occurs at fractional level filling and its quantum numbers are not integers but rational fractions p/q (see Fig. 1). Features at these fractional fillings cannot be explained in terms of single-electron physics. They occur when the Fermi energy resides *within* a highly degenerate Landau or spin level and imply the existence of energy gaps of many-particle origin.

The fractional quantum Hall effect is the result of the highly correlated motion of many electrons in 2D exposed to a magnetic field. Its driving force is the reduction of Coulomb interaction between the like-charged electrons. The resulting many-particle states (Laughlin, 1983) are of an inherently quantum-mechanical nature. Fractional quantum numbers and exactly fractionally charged quasiparticles are probably the most spectacular of its implications (Chakraborty and Pietilainen, 1995).

Today, the attachment of magnetic vortices to electrons has become the unifying principle underlying the multiple many-particle states of the FQHE (Read, 1994; DasSarma and Pinczuk, 1997). Laughlin's wave function describing the $\nu = 1/3$ state is the prime example for this principle at work.

The presence of the magnetic field requires the many-electron wave function to assume as many zeroes within a unit area as there are magnetic flux quanta penetrating it. Each zero "heals" on the scale of a magnetic length ($l_0 = \sqrt{\hbar/(eB)}$) and, limiting ourselves to the lowest Landau level, each such "hole" in the electron sheet represents an overall charge deficit of νe. Since the magnetic field also imparts a 2π phase twist to the wave function at the position of each such zero, these objects are termed vortices. In a certain sense, vortices are the embodiment of flux quanta in an electron system. A tiny coil threaded through the plane of the electrons and energized to generate just one magnetic flux quantum through its core would create one such vortex (Laughlin, 1984). Therefore, loosely speaking, vortices are often equated with flux quanta.

Just like electrons, vortices are delocalized in the plane. However, since electrons represent a charge accumulation and vortices a charge deficit, they attract each other. Considerable Coulomb energy can be gained by placing vortices onto electrons. At $\nu = 1/3$ there exist three times as many vortices as there are electrons, each vortex representing a local charge deficit of $\frac{1}{3}e$.

Each electron must carry at least one vortex equivalent to one zero in the wave function to satisfy the Pauli principle. Additional vortex attachment is "optional," driven by Coulomb gain. Vortex attachment to an electron, representing a local depletion of companion electrons, is always energetically beneficial. The situation is somewhat reminiscent of the screening cloud around an electron in a regular metal, although in the case of the FQHE, such "screening" is very rigid and quantized in units of vortex charge.

The attachment of exactly three vortices to each electron is at the origin of the prominent $\nu = \frac{1}{3}$ FQHE state expressed by Laughlin's wave function as

$$\psi_{1/3} = \prod_{i<j}^{n} (Z_i - Z_j)^3 \exp\left(-\frac{1}{4} \sum_{k}^{n} |Z_k|^2\right).$$

The $Z_{i,j,k}$'s represent the coordinates of n electrons in a complex 2D plane, which renders the wave function more compact. Normalization and magnetic length are set to unity. All electron-electron correlations derive from the first term, which is a product over all complex pair distances between electrons. The exponent 3 in each factor expresses in mathematical terms the attachment of three vortices exactly to the position of each electron. More generally, states at $\nu = 1/q$ (q = odd) consist of electrons dressed by q vortices, and their wave function differs from the above only by the exponent, which changes from 3 to q. Only odd q are allowed, since only they guarantee antisymmetry of this electron wave function.

Any deviation from such a commensurate electron-vortex ratio comes at a considerable energetic cost. Creation of additional vortices, as induced by an increase in magnetic field, requires a finite amount of energy. This is the origin of the energy gap at $\nu = 1/3$ and of the incompressibility of the electronic state at this filling factor. Unbound vortices become quasiholes in the sea of electrons, each carrying exactly $+1/3$ of an electronic charge (Laughlin, 1983). Gap formation, charged quasiparticles, and their localization at residual potential fluctuations are the ingredients required to account for the transport features in Fig. 1.

Returning to the representation of vortices as flux quanta, one can also regard the $\nu = 1/3$ state as consisting of new objects: electrons to which three *vortex-generating* flux quanta have been attached (Kivelson et al., 1992). This viewpoint has interesting conceptual consequences. First, since the total external magnetic field consists of exactly three flux quanta per electron, it appears that the entire magnetic field has been attached to electrons, reducing the magnetic field felt by these composite objects at $\nu = 1/3$ to zero. Second, the flux carried by such objects has a dramatic effect on their statistics. Exchange of two such "magnetized" particles introduces an Aharanov-Bohm phase which turns them into bosons for an odd number of attached flux quanta while reverting them back to fermions for an even number of attached flux quanta. Thus the $\nu = 1/3$ state consists of composite bosons created by the attachment of three flux quanta to each electron, which Bose condense in the apparent absence of an external magnetic field. This view, developed in conjunction with the composite fermion model (see below), links the physics of the $1/q$ FQHE state directly to the statistics of peculiar new particles.

Today, fractional charge and energy gap have been observed independently in several experiments (Clark et al., 1988; Simmons et al., 1989; Chang and Cunningham, 1990; Kukushkin et al., 1992; Pinczuk et al., 1993; Goldman and Su, 1995; de-Picciotto et al., 1997; Saminadayar et al., 1997). The states at primary filling factor $\nu = 1/q$, as well as at $\nu = 1 - 1/q$ (electron-hole symmetry), are well understood. Their description in terms of a practically exact and remarkably succinct electronic wave function is a triumph of many-particle physics.

Composite Fermions in the FQHE

Composite fermions (CFs) are a new concept that provides us with a concise way of accounting for the so-called higher-order FQHE states at odd-denominator fractional filling factors, different from the primary states at $\nu = 1/q$. They have also been used in an interpretation of the long-time enigmatic states at even-denominator fractional filling such as those at $\nu=1/2$. These will be addressed in the next section.

Higher-order FQHE states (e.g., $\nu=2/5$, 5/9, or 5/7) are abundant in Fig. 1, and the features they generate are very similar to those at $\nu=1/q$. This is particularly apparent for the states at $\nu=p/(2p\pm1)$ which converge towards half filling. Of course, the same principle of Coulomb energy optimization, that governs the physics at $\nu=1/q$, is expected to be at work. However, such higher-order states are much more difficult to describe in terms of many-electron wave functions than the states at $\nu=1/q$. An early hierarchical scheme (Haldane, 1983; Halperin, 1984; Laughlin, 1984), starting from the primary fractions, was able to rationalize the abundance of FQHE features and the odd-denominator rule: Just as the original electrons correlate to form primary states at $\nu=1/q$, at sufficient deviation from such filling factors, the so-created, charged quasielectrons or quasiholes can correlate and form new quantum liquids of quasiparticles at neighboring rational filling factors. This process can be repeated *ad infinitum*, eventually covering all odd-denominator fractions. However, wave functions proposed for such states at $\nu=p/q$ lack the simplicity of the Laughlin wave function at $\nu=1/q$, and they do not contain a simple underlying principle to generate a sequence of wave functions for the higher-order states.

Building on the notion of the transmutability of statistics in 2D (Wilczek, 1982; Halperin, 1984; Arovas *et al.*, 1985; Girvin and MacDonald, 1987; Laughlin, 1988; Lopez and Fradkin, 1991; Moore and Read, 1991; Zhang *et al.*, 1989), a new model for the generation of higher-order FQHE states was introduced by Jain (1989, 1990). It is best described for the concrete set of states at $\nu=p/(2p+1)$, starting at $\nu=1/3$ and converging towards $\nu=1/2$ (see Fig. 1). Again, vortex attachment to electrons plays a pivotal role. As elucidated in the previous section, the $\nu=1/3$ state consists of electrons to which three vortices have been attached by virtue of the electron-electron interaction and is expressed by Laughlin's wave function (exponent "3"). Formally, this wave function can be factorized into

$$\psi_{1/3} = \prod_{i<j}^{n} (Z_i - Z_j)^2 \prod_{i<j}^{n} (Z_i - Z_j)^1 \exp\left(-\frac{1}{4}\sum_k^n |Z_k|^2\right) = \prod_{i<j}^{n} (Z_i - Z_j)^2 \psi_1.$$

The second factor turns out to represent exactly one fully filled Landau level, Ψ_1, in which each electron is provided with one vortex, the minimum needed to satisfy the Pauli principle (spin neglected). Hence, at least formally, the FQHE state at $\nu=\frac{1}{3}$ can also be viewed as an IQHE state at $\nu=1$. However, now each electron carries *two* attached vortices. In the flux-quanta attachment language, two out of three of the flux quanta per electron have been incorporated into the new particle, reducing the external magnetic field to an *effective magnetic field* of only one flux quantum per composite, which is equivalent to the field at $\nu=1$. These composites, carrying an even number of flux quanta, behave as fermions and fill exactly one Landau level.

While being of largely philosophical nature at $\nu=1/3$, generalization of this concept to the filling of more than one composite fermion Landau level has important formal implications. Such a generalized composite fermion model generates very good many-particle wave functions for the higher-order states at $\nu=p/(2p+1)$, as deduced from comparison with exact, few-particle numerical calculations (Dev and Jain, 1992). Furthermore, it has provided a rationale for the dominance of this particular sequence of states in experiment (see Fig. 1).

Exploiting the analogies between the FQHE at $\nu=p/(2p+1)$ and an IQHE of composite fermions at Landau level filling factor p, we can postulate an analogy between composite fermion gap energies (i.e., FQHE gaps of higher-order fractions, $\nu=p/q$) and Landau gaps in the electron case (Halperin *et al.*, 1993): Similar to electrons, for which the Landau gap opens linearly with the magnetic field, the FQHE gap energies are conjectured to open practically linearly with effective magnetic field B_{eff}. However, the mass value m^*, derived from such composite fermion cyclotron gaps ($\hbar e B_{\text{eff}}/m^*$), is unrelated to the cyclotron mass of the electron. The composite fermion mass is exclusively a consequence of electron-electron interaction and, hence, for a given fraction, exclusively a function of electron density. This generation of mass solely from many-particle interactions is very peculiar.

Experimentally, such a quasilinear relationship between the energy gap and the effective magnetic field is indeed borne out (Du *et al.*, 1993; Leadley *et al.*, 1994; Manoharan *et al.*, 1994; Coleridge *et al.*, 1995). As an example, Fig. 2 shows measurements of the activation energies of the sequence of higher-order FQHE states in the vicinity of filling

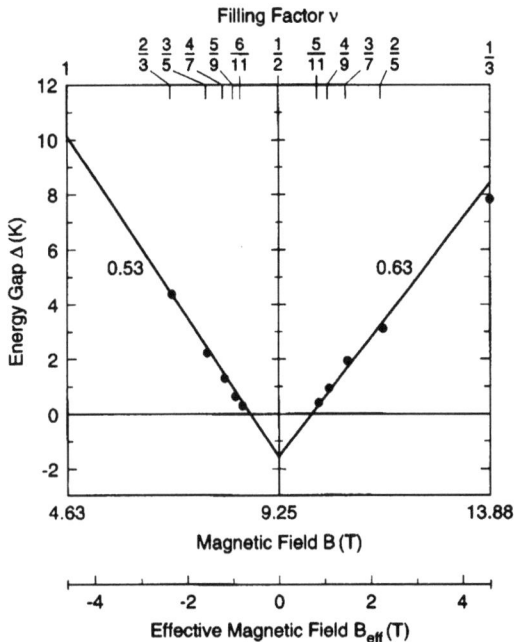

Figure 2. Magnetic-field dependence of the energy gaps of fractional quantum Hall states at filling factors $\nu=p/(2p\pm1)$ around $\nu=1/2$, as determined from thermal activation energy measurements on a sample of density $n=0.83\times10^{11}$ cm^{-2}. Lines are a guide to the eye. Gaps are clearly opening roughly linearly with an effective magnetic field, $B_{\text{eff}}=B-B(\nu=1/2)$, whose origin is at $\nu=1/2$. It is reminiscent of the opening of Landau gaps for electrons around $B=0$. Therefore the FQHE states around $\nu=1/2$ can be viewed as arising from the Landau quantization of new particles, so-called composite fermions (CFs), which consist of electrons to which two magnetic flux quanta have been bound by virtue of the electron-electron interaction. Their mass, $m^*\sim0.53-0.63m_0$, determined from $\hbar e B_{\text{eff}}/m^*$, is unrelated to the electron mass and of purely many-particle origin. The negative intercept at exactly $\nu=1/2$ is believed to be a result of level broadening, which reduces the gaps by a fixed amount. From Du et al., 1993.

factor $\nu = 1/2$. The derived gap energies vary practically linearly with effective magnetic field, and masses of $\sim 0.6 m_0$ are obtained. The measured masses are about ten times larger than the electron mass in GaAs and increase in specimen with higher electron density n. The spirit of the composite fermion model has been most extensively tested for fractions in the vicinity of $\nu = 1/2$. However, in principle, the model is expected to hold for all odd-denominator rational filling factors, giving rise to multiple self-similarities in the experimental data. Studies of other such fractions are still in their infancy.

The composite fermion model has provided us with a rationale for the existence of higher-order FQHE states, the relative size of their energy gaps, and the appearance of particular sequences. It has also established a procedure by which to generate excellent many-particle wave functions as well as their excitations. How then are the FQHE and the IQHE related? Is the physics of the FQHE the same as the physics of the IQHE? Certainly not. The composite fermion model reveals a similarity in the mathematics between the sequence of IQHE states in terms of electrons and the sequence of FQHE states in terms of composite fermions. However, it also reveals a new class of strange particles, consisting of electrons dressed by vortices, which display a peculiar mass of purely many-particle origin. Excitations from such states carry an exactly fractional charge $e/3$, $e/5$, $e/7$, etc., and the Hall resistance exhibits odd rational fractional quantum numbers, such as 2/5, 3/13, and 5/23. These are truly remarkable characteristics of the fractional quantum Hall effect, having been brought into existence through the intricate cooperation of many electrons.

Even-Denominator Filling

What is the nature of the states at even-denominator filling? Compared to the odd-denominator FQHE states, even-denominator states appear uninspiring, showing hardly any variation in temperature-dependent transport. Yet, as it turns out, they are just as fascinating as their odd-denominator counterparts.

Early on, subtle features in the resistivity at $\nu = 1/2$ hinted at unanticipated interactions at half filling (Jiang et al., 1989). However, it took a new probe, surface acoustic waves, to detect a distinctively different behavior at even-denominator filling as compared to odd-denominator filling.

As their name implies, surface acoustic waves are acoustic waves, typically at hundreds of MHz, that travel along the surface of a specimen. In the long-wavelength limit, velocity and attenuation of a surface acoustic wave are uniquely determined by material parameters of the semiconductor and the dc conductivity σ of the 2D electron gas. Actual surface acoustic wave data diverge markedly from those calculated from σ at $\nu = 1/2$ (see Fig. 3; Willett et al., 1990). The origin of the discrepancy is found in the finite wavelength of the surface acoustic wave probe as compared to the dc transport measurements. Today we interpret this discrepancy as evidence for the existence of a novel Fermi system at $\nu = 1/2$ with a well-defined Fermi wave vector. Unlike the incompressible FQHE states away from half filling, such a Fermi system, having no energy gap, allows for infinitesimal excitations, and their specific wave vector dependence is qualitatively reflected in the surface acoustic wave data.

During the past few years, several experiments have unequivocally demonstrated the existence of a Fermi wave vector k_f at $\nu = 1/2$ (Kang et al., 1993; Willett et al., 1993; Goldman et al., 1994; Smet et al., 1996). They rely on the commensuration resonance between the classical cyclotron radius $r_c = \hbar k_f/(eB_{eff})$ at filling factors somewhat off $\nu = 1/2$ and an externally imposed length. Remarkably, the relevant magnetic field, $B_{eff} = B - B(\nu = 1/2)$, is the deviation of B from $\nu = 1/2$. The experiments reveal, in a very pictorial manner, the ballistic trajectories of the current-carrying objects. These particles, under the influence of B_{eff}, execute large classical cyclotron orbits, which degenerate into straight lines at exactly half filling. For them, the magnetic field seems to have vanished at $\nu = 1/2$. How can a degenerate Fermi system emerge and what causes the field to vanish? Once again, an analogy between composite fermions and electrons offers an explanation.

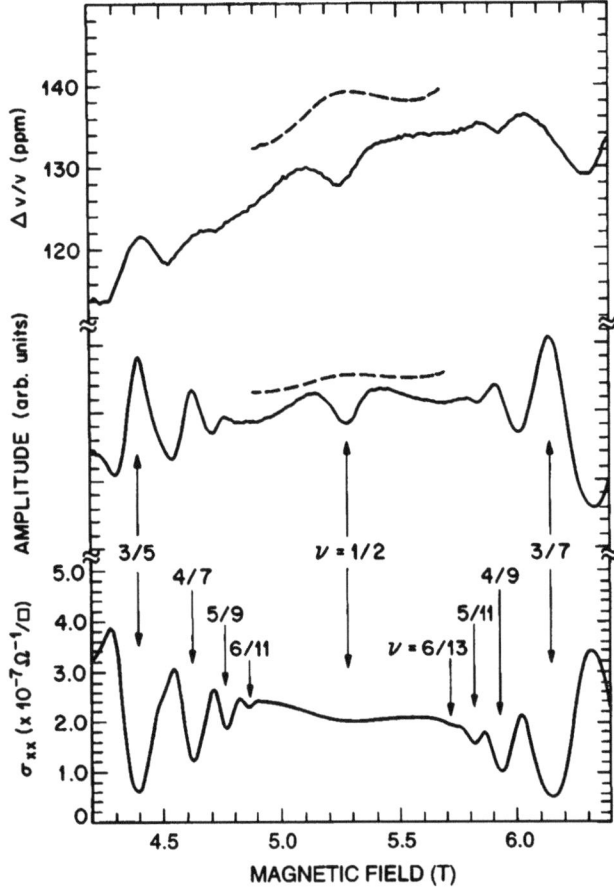

Figure 3. Changes in amplitude and velocity ($\Delta v/v$) of surface acoustic waves through a 2D electron system as a function of magnetic field compared with its electrical conductivity σ_{xx}. Fractions ν indicate the Landau-level filling factor. Amplitude and $\Delta v/v$ agree closely (not shown) with the behavior calculated from σ_{xx} in most regions of the magnetic field, but deviate strongly at exactly $\nu = 1/2$. This is indicated by the dashed curves, which are calculated in this regime from σ_{xx} and superimposed on the surface acoustic wave traces. These data provided evidence for exceptional properties at half filling. Today, we interpret this finding as being due to the formation of a novel Fermi system of composite fermions at exactly even-denominator fractional filling. The external magnetic field has been incorporated into the particles, and they move in the apparent absence of a magnetic field (see also Fig. 2). From Willett et al., 1990.

As the filling factor converges towards $\nu = 1/2$, composite fermions are populating more and more Landau levels. The effective magnetic field they are experiencing is the true external field reduced by two flux quanta per electron on account of the two vortices bound to each of them. At $\nu = 1/2$ the effective field for composite fermions has dropped to zero, and the limit of infinitely many-Landau-level occupation has been reached. For electrons of fixed density, increasing Landau-level occupation is synonymous with convergence towards the electron Fermi sea at $B = 0$. Hence one may conjecture that the state at $\nu = 1/2$ represents a Fermi sea of composite fermions at $B_{\text{eff}} = 0$, i.e., in the apparent absence of a magnetic field (Kalmeyer and Chang, 1992; Halperin et al., 1993). This model can account for many of the experimental observations, in particular the surface acoustic wave resonance data, the particle trajectories, and the geometrical resonances. Properties related to energy, such as particle mass at $\nu = 1/2$, scattering rate, and a quantitative interpretation of the surface acoustic wave data, require further exploration, experimentally as well as theoretically. In fact, the Fermi system at $\nu = 1/2$ has been conjectured to be of the marginal kind, exhibiting various singularities at E_f (Halperin et al., 1993).

In the spirit of the above reasoning, a trial wave function for the state at $\nu = 1/2$ has been presented (Rezayi and Read, 1994):

$$\psi_{1/2} = P \prod_{i<j}^{n} (Z_i - Z_j)^2 \times \psi_\infty,$$

where ψ_∞ denotes the electron Fermi sea at $B = 0$. Composite fermions carry two vortices each (exponent 2) and are filling up a Fermi sea. The projection operator P ensures that the many-particle wave function resides wholly within the lowest Landau level, as required by the actual physical situation. What is the nature of this state, and what are its particles?

Just as at $\nu = 1/3$, so also at $\nu = 1/2$ vortex attachment to electrons reduces Coulomb interaction. The simplest and energetically most beneficial way to achieve this at $\nu = 1/2$ is to place both vortices exactly onto the electron's position. Such a configuration represents a straight generalization of Laughlin's electronic wave function for the $\nu = 1/3$ FQHE state. However, such a state violates the antisymmetry requirement for an electron wave function.

Barring exact vortex attachment, vortex proximity still remains beneficial. The electron system can maintain antisymmetry while reaping considerable Coulomb gain by making a slight adjustment. Of the two vortices per electron at $\nu = 1/2$, one vortex is directly placed on each electron satisfying the requirements of the Pauli principle. The second vortex is being kept as close by as possible while obeying the antisymmetry requirement for the overall electron wave function. Successively increasing this separation in electron-vortex pairs, we find that a maximum distance r_f is reached which corresponds to k_f of the Fermi liquid. The resulting particles are no longer monopoles as they are at $\nu = 1/3$ due to exact vortex attachment, but electron-vortex dipoles. Both "charges" of the dipole extend approximately one magnetic length and overlap strongly. This overlap considerably reduces their Coulomb energy.

Many of the properties of the state at $\nu = 1/2$ can be visualized through such a simple picture, which regards the composite fermions at this filling factor as electrical dipoles created from electrons and vortices. Even a mass can be derived from such a classical,

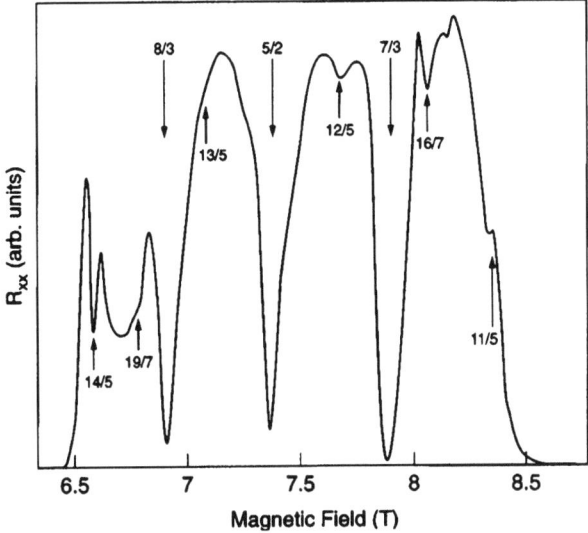

Figure 4. Magnetoresistance data from the second Landau level (between $\nu = 2$ and $\nu = 3$) of a very high quality 2D electron system of density $n = 4.45 \times 10^{11}$ cm^{-2} at 32 mK. Several FQHE states are visible. Most of them represent replicas, $\nu = 2 + p/q$, of FQHE states in the lowest Landau level at $\nu = p/q$. However, the state at $\nu = 5/2$ is very unusual, showing the deep minimum in R_{xx} and plateau in R_{xy} (not shown) of a FQHE state while occurring at an even-denominator fraction. The origin of this state remains unclear. One possible explanation is the formation of composite fermions, as expected at even-denominator filling, which subsequently organize into pairs and condense into a state akin to superconductivity. From Du et al., unpublished.

albeit vortex-based, model. Again, the mass depends only on electron density and is unrelated to the mass of the electron.

A much more sophisticated reasoning than has been presented here for considering composite fermions as dipoles is currently being advanced (Shankar and Murthy, 1997; Lee, 1998; Pasquier and Haldane, 1998). It remains unclear to what degree it is valid and what all its implications are, in particular for such properties as the density of states in the vicinity of the Fermi energy.

The existence of such fascinating new Fermi systems is not limited to filling factor $\nu = 1/2$. In fact, equivalent physics is expected to occur at all even-denominator fractions. Their study is just beginning, and data around $\nu = 3/2, 1/4, 3/4$ reveal the anticipated composite fermion behavior. However, we already know of a major exception to the rule.

Composite Fermion Pairs?

At filling factor $\nu = 5/2$, the lowest Landau level is totally filled and the lowest spin level of the second Landau level is half filled. The state at $\nu = 5/2$ has always been puzzling. It has all the characteristics of a FQHE state, including energy gap and quantized Hall resistance, in spite of its even-denominator classification (see Fig. 4). Though it was discovered more than a decade ago (Willett et al., 1987), its origin has never been clearly resolved (Rezayi and Haldane, 1988).

With the advent of the composite fermion model, the $\nu=5/2$ state has recently been revisited. An earlier proposal for the states at half filling (Moore and Read, 1991; Greiter, et al., 1991) is now being examined as a contender for the state at $\nu=5/2$. This so-called "Pfaffian state," which has been characterized as a state of paired composite fermions, represents a very attractive option. If applicable, the state at $\nu=5/2$ can be thought of as arising in a two-step process. First, electrons in the second Landau level at half filling would form the "familiar" Fermi sea of composite fermions. These would then pair in a BCS-like fashion and condense into a novel "superconducting" ground state of composite fermions. Although this last analogy applies only in a very loose sense, the concept is truly exciting, and one should remain hopeful that further investigations into the state at $\nu=5/2$ may reveal an extraordinary new phase of 2D electrons (Morf, 1998).

Conclusion

Two-dimensional electrons in high magnetic fields have revealed to us totally new many-particle physics. Confined to a plane and exposed to a magnetic field, such electrons display an enormously diverse spectrum of fascinating new properties—fractional charge and fractional quantum numbers, new particles obeying either Bose or Fermi statistics, absorption of exceedingly high magnetic fields, apparent microscopic dipoles, and possibly a very unusual kind of particle pairing. They are just electrons—although lots of them. Indeed, "More is different" (Anderson, 1972).

Acknowledgments

We should like to thank our numerous experimental collaborators for having been part of this quest, and we are grateful to many of our theorist colleagues for not giving up lecturing us on many-particle physics. In particular, we are indebted to L. N. Pfeiffer, K. W. West, and K. W. Baldwin, who created and processed most of the magnificent samples.

References

Anderson, P. W., 1972, Science **177**, 393.
Ando, T., A. B. Fowler, and F. Stern, 1983, Rev. Mod. Phys. **54**, 437.
Arovas, D. P., J. R. Schrieffer, F. Wilczek, and A. Zee, 1985, Nucl. Phys. B **251**, 117.
Chakraborty, T., and P. Pietilainen, 1995, *The Quantum Hall Effects*, Springer Series in Solid State Sciences No. 85 (Springer, New York).
Chang, A. M., and J. E. Cunningham, 1990, Surf. Sci. **229**, 216.
Clark, R. G., J. R. Mallett, S. R. Haynes, J. J. Harris, and C. T. Foxon, 1988, Phys. Rev. Lett. **60**, 1747.
Coleridge, P. T., Z. W. Wasilewski, P. Zawadzki, A. S. Sachrajda, and H. A. Carmona, 1995, Phys. Rev. B **52**, 11 603.
DasSarma, S., and A. Pinczuk, 1997, Eds., *Perspectives in Quantum Hall Effects* (Wiley, New York).
de-Picciotto, R., M. Reznikov, M. Heiblum, V. Umansky, G. Bunin, and D. Mahalu, 1997, Nature (London) **389**, 162.
Dev, G., and J. K. Jain, 1992, Phys. Rev. Lett. **69**, 2843.

Du, R.-R., H. L. Stormer, D. C. Tsui, L. N. Pfeiffer, and K. W. West, 1993, Phys. Rev. Lett. **70**, 2944.
Du, R.-R., A. S. Yeh, H. L. Stormer, D. C. Tsui, L. N. Pfeiffer, and K. W. West, 1998 (unpublished).
Eisenstein, J. P., and H. L. Stormer, 1990, Science **248**, 1461.
Girvin, S. M., and A. H. MacDonald, 1987, Phys. Rev. Lett. **58**, 1252.
Goldman, V. J., and B. Su, 1995, Science **267**, 1010.
Goldman, V. J., B. Su, and J. K. Jain, 1994, Phys. Rev. Lett. **72**, 2065.
Greiter, M., X.-G. Wen, and F. Wilczek, 1991, Phys. Rev. Lett. **66**, 3205.
Haldane, F. D. M., 1983, Phys. Rev. Lett. **51**, 605.
Halperin, B. I., 1984, Phys. Rev. Lett. **52**, 1983; **52**, 2390(E).
Halperin, B. I., P. A. Lee, and N. Read, 1993, Phys. Rev. B **47**, 7312.
Jain, J. K., 1989, Phys. Rev. Lett. **63**, 199.
Jain, J. K., 1990, Phys. Rev. B **41**, 7653.
Jiang, H. W., H. L. Stormer, D. C. Tsui, L. N. Pfeiffer, and K. W. West, 1989, Phys. Rev. B **40**, 12 013.
Kalmeyer, V., and S. I. Chang, 1992, Phys. Rev. B **46**, 9889.
Kang, W., H. L. Stormer, L. N. Pfeiffer, K. W. Baldwin, and K. W. West, 1993, Phys. Rev. Lett. **71**, 3850.
Kivelson, S., D.-H. Lee, and S. C. Zhang, 1992, Phys. Rev. B **48**, 2223.
Kukushkin, I. V., N. J. Pulsford, K. von Klitzing, R. J. Haug, S. Koch, and V. B. Timofeev, 1992, Europhys. Lett. **18**, 62.
Laughlin, R. B., 1981, Phys. Rev. B **22**, 5632.
Laughlin, R. B., 1983, Phys. Rev. Lett. **50**, 1395.
Laughlin, R. B., 1984, in *Two-Dimensional Systems, Heterostructures, and Superlattices*, edited by G. Bauer, F. Kuchar, and H. Heinrich, Springer Series in Solid State Sciences, No. 53 (Springer, Berlin/New York).
Laughlin, R. B., 1988, Phys. Rev. Lett. **60**, 2677.
Leadley, D. R., R. J. Nicholas, C. T. Foxon, and J. J. Harris, 1994, Phys. Rev. Lett. **72**, 1906.
Lee, D.-H., 1998, Phys. Rev. Lett. **80**, 4745.
Lopez, A., and E. Fradkin, 1991, Phys. Rev. B **46**, 5246.
Manoharan, H. C., M. Shayegan, and S. J. Klepper, 1994, Phys. Rev. Lett. **73**, 3270.
Moore, G., and N. Read, 1991, Nucl. Phys. B **360**, 362.
Morf, R., 1998, Phys. Rev. Lett. **80**, 1505.
Pasquier, V., and F. D. M. Haldane, 1998, Nucl. Phys. B **515**, (F), in press.
Pinczuk, A., B. S. Dennis, L. N. Pfeiffer, and K. W. West, 1993, Phys. Rev. Lett. **70**, 3983.
Prange, R. E., and S. M. Girvin, 1990, Eds., *The Quantum Hall Effect*, 2nd Ed. (Springer, New York).
Read, N., 1994, Semicond. Sci. Technol. **9**, 1859.
Rezayi, E., and F. D. M. Haldane, 1988, Phys. Rev. Lett. **60**, 956.
Rezayi, E., and N. Read, 1994, Phys. Rev. Lett. **71**, 900.
Saminadayar, L., D. C. Glattli, Y. Jin, and B. Etienne, 1997, Phys. Rev. Lett. **79**, 2526.
Shankar, R., and G. Murthy, 1997, Phys. Rev. Lett. **79**, 4437.
Simmons, J. A., H. P. Wei, L. W. Engel, D. C. Tsui, and M. Shayegan, 1989, Phys. Rev. Lett. **63**, 1731.
Smet, J. H., D. Weiss, R. H. Blick, G. Lutjering, K. von Klitzing, R. Fleischmann, R. Ketzmerick, T. Geisel, and G. Weimann, 1996, Phys. Rev. Lett. **77**, 2272.
Stormer, H. L., R. Dingle, A. C. Gossard, W. Wiegmann, and M. D. Sturge, 1979, Solid State Commun. **29**, 705.
Tsui, D. C., H. L. Stormer, and A. C. Gossard, 1982, Phys. Rev. Lett. **48**, 1559.
von Klitzing, K., G. Dorda, and M. Pepper, 1980, Phys. Rev. Lett. **45**, 494.
Wilczek, F., 1982, Phys. Rev. Lett. **48**, 1144.

Willett, R. L., J. P. Eisenstein, H. L. Stormer, D. C. Tsui, A. C. Gossard, and J. H. English, 1987, Phys. Rev. Lett. **59**, 1776.
Willett, R. L., M. A. Paalanen, R. R. Ruel, K. W. West, L. N. Pfeiffer, and D. J. Bishop, 1990, Phys. Rev. Lett. **54**, 112.
Willett, R. L., R. R. Ruel, K. W. West, and L. N. Pfeiffer, 1993, Phys. Rev. Lett. **71**, 3846.
Zhang, S. C., H. Hanson, and S. A. Kivelson, 1989, Phys. Rev. Lett. **62**, 82.

Conductance Viewed as Transmission
Yoseph Imry and Rolf Landauer

Early quantum theories of electrical conduction were semiclassical. Electrons were accelerated according to Bloch's theorem; this was balanced by back scattering due to phonons and lattice defects. Cross sections for scattering, and band structures, were calculated quantum-mechanically, but the balancing process allowed only for occupation probabilities, not permitting a totally coherent process. Also, in most instances, scatterers at separate locations were presumed to act incoherently. Totally quantum-mechanical theories stem from the 1950s, and have diverse sources. Particularly intense concern with the need for more quantum mechanical approaches was manifested in Japan, and Kubo's formulation became the most widely accepted version. Quantum theory, as described by the Schrödinger equation, is a theory of conservative systems, and does not allow for dissipation. The Schrödinger equation readily allows us to calculate polarizability for atoms, molecules, or other isolated systems that do not permit electrons to enter or leave. Kubo's linear-response theory is essentially an extended theory of polarizability. Some supplementary handwaving is needed to calculate a dissipative effect such as conductance, for a sample with boundaries where electrons enter and leave (Anderson, 1997). After all, no theory that ignores the interfaces of a sample to the rest of its circuit can possibly calculate the resistance of such a sample of limited extent. Modern microelectronics has provided the techniques for fabricating very small samples. These permit us to study conductance in cases where the carriers have a totally quantum mechanically coherent history within the sample, making it essential to take the interfaces into account. Mesoscopic physics, concerned with samples that are intermediate in size between the atomic scale and the macroscopic one, can now demonstrate in manufactured structures much of the quantum mechanics we associate with atoms and molecules.

When scattering by a randomly placed set of point defects was under consideration, it quickly became customary in resistance calculations to evaluate the resistance after averaging over an ensemble of all possible defect placements. This removed the effects of quantum-mechanically coherent multiple scattering, which depends on the distance between the scatterers. This approach also made the unwarranted assumption that the variation of resistance between ensemble members was small. The approach made it impossible to ask about spatial variations of field and current within the sample. Unfortunately, as a result, the very existence of such questions, which distinguish between the ensemble average and the behavior of a particular sample, was ignored.

Electron transport theory has typically viewed the electric field as a cause and the current flow as a response. Circuit theory has had a broader approach, treating voltage sources and current sources on an equal footing. The approach to be emphasized in the following discussion is a generalization of the circuit theory alternative: Transport is a result of the carrier flow incident on the sample boundaries. The voltage distribution within the sample results from the self-consistent pileup of carriers.

The viewpoint stressed in this short note has been explained in much more detail in books and review papers. We can cite only a few: Beenakker and van Houten (1991), Datta (1995), Ferry and Goodnick (1997), Imry (1997). There are also many conference proceedings and special theme volumes related to this subject, e.g., Sohn et al., 1997; Datta, 1998.

It seems obvious that the ease with which carriers penetrate through a sample should be closely related to its conductance. But this is a viewpoint that, with the exception of some highly specialized limiting cases, found slow acceptance. That the conductance of a single localized tunneling barrier, with a very small transmission probability, is proportional to that probability was understood in the early 1930s (Landauer, 1994). In the case of a simple tunneling barrier it has always been apparent that the potential drop across the barrier is localized to the immediate vicinity of the barrier and not distributed over a region of the order of the mean free path in the surrounding medium. The implicit acceptance of a highly localized voltage drop across a single barrier did not, however, readily lead to a broader appreciation of spatially inhomogeneous transport fields in the presence of other types of scattering. The localized voltage drop across a barrier has been demonstrated with modern scanning tunneling microscopy (STM) probing methods (Briner et al., 1996). We return to spatial variations in the next section.

Figure 1 shows an ideal conducting channel with no irregularities or scattering mechanisms along its length. A long perfect tube is tied to two large reservoirs via adiabatically tapered nonreflecting connectors. Carriers approaching a reservoir pass into that reservoir with certainty. The reservoirs are the electronic equivalent of a radiative blackbody; the electrons coming out of a reservoir are occupied according to the Fermi distribution that characterizes the deep interior of that reservoir. Assume, initially, that the tube is narrow enough so that only the lowest of the transverse eigenstates in the channel has its energy below the Fermi level. That makes the channel effectively one-dimensional. Take the zero temperature case and let the left reservoir be filled to up to level μ_1, higher than that of the right-hand reservoir, μ_2. Then in the range between μ_1 and μ_2 we have fully occupied states pouring from left to right. Thus the current is

$$j = -(\mu_1 - \mu_2)ev(dn/d\mu), \quad (1.1)$$

where $dn/d\mu$ is the density of states (allowing for spin degeneracy) and v is the velocity

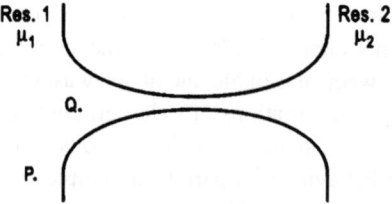

Figure 1. Two reservoirs on each side of a perfect tube, at different electrochemical potentials μ_1 and μ_2. P is well inside reservoir 1; Q is in its entrance.

component along the tube at the Fermi surface. Now $(\mu_1-\mu_2)=-e(V_1-V_2)$, where V is a voltage and e the magnitude of the electronic charge. Furthermore $dn/d\mu = 1/\pi\hbar v$. Therefore the net current flow is given by $-(e/\pi\hbar)(\mu_1-\mu_2)$. The resulting conductance is

$$G = \frac{j}{V_1-V_2} = e^2/\pi\hbar. \tag{1.2}$$

This is the conductance of an ideal one-dimensional conductor. The conditions along the uniform part of the channel are the same; there is no potential drop there. The potential drop associated with the resistance specified in Eq. (1.2) occurs at the connections to the reservoir (Imry, 1986). Consider the left reservoir. Deep inside that reservoir there is a thermal equilibrium population. In the 1D channel only the right-moving electrons are present. The effective Fermi-level, or effective electrochemical potential, measures the level to which electrons are occupied. At point P in Fig. 1, deep inside the left-hand reservoir, the electron distribution is that characteristic of thermal equilibrium with the Fermi level μ_1. At point Q in Fig. 1, in the tapered part of the connection, the electron population shows some effect of the lowered density of electrons which have come out of the right hand reservoir with electrochemical potential μ_2. Thus there is a potential difference between P and Q. Along the ideal one-dimensional channel the electron population is equally controlled by both reservoirs, and the electrochemical potential there must be $\frac{1}{2}(\mu_1+\mu_2)$. Therefore the voltage drop specified by Eq. (1.2) is divided equally between the two tapered connectors. The physics we have just discussed is essential. Conductance can only be calculated after specifying the location where the potential is determined. The voltage specification deep inside the reservoir and the geometrical spreading, are essential aspects of the derivation of Eq. (1.2). Unfortunately, supposed derivations that ignore these geometrical aspects are common in the literature.

If we insert an obstacle into the channel, which transmits with probability T, the current will be reduced accordingly, and we find

$$G = (e^2/\pi\hbar)T. \tag{1.3}$$

Note that it does not matter whether the T in Eq. (1.3) is determined by a single highly localized barrier or by a more extended and complex potential profile. Expressions for the conductance with this same current, but with the potential measured within the narrow channel, on the two sides of the obstacle, also exist (Sec. 2.2 of Datta, 1995; Sec. 1.2.1 of Ferry and Goodnick, 1997; Chap. 5 of Imry, 1997). If, in that case, the potential is averaged over a region long enough to remove interference oscillations, then T in Eq. (1.3) is replaced by $T/(1-T)$.

The preceding discussion can easily be generalized to a channel that involves more than one transverse eigenstate with energy below the Fermi level (Imry, 1986). In that case we utilize the transmission matrix t of the scattering obstacle, which specifies the transmitted wave functions relative to the incident wave, utilizing the transverse eigenstates of the channel as a basis. This yields

$$G = (e^2/\pi\hbar)Tr(tt^\dagger). \tag{1.4}$$

In the particular case where we have N perfectly transmitting channels this becomes

$$G = N(e^2/\pi\hbar). \tag{1.5}$$

One of the earliest and most significant experimental verifications of this approach came from celebrated studies of quantum point contacts (QPC). These are narrow two-dimensional channels connecting wide reservoirs. The channel width can be controlled by externally applied gate voltages. As the conducting channel is widened, the number of transverse eigenstates below the Fermi level increases. Conductance steps corresponding to increasing values of N in Eq. (1.5) are clearly observed. The original 1988 experiments were carried out at Cambridge University and by a Delft-Philips collaboration (van Houten and Beenakker, 1996).

The material of this section has been extended in many directions; we list only a few. Büttiker (1986) describes the widely used results when more than two reservoirs are involved. Among a number of very diverse attempts to describe ac behavior we cite only one (Büttiker, 1993). In the ac case, however, the method of applying excitation to the sample matters. Moving the Fermi level of reservoirs up and down is one possibility; applying an electric field through an incident electromagnetic field is another. The discussion of systems that consist of incoherent semiclassical scatterers occurs repeatedly; we cite one with device relevance [Datta, Assad, and Lundstrom, in Datta (1998)]. The extension to nonvanishing temperatures is contained in many of our broader citations. Nonlinearity has been treated repeatedly. The correction for reservoirs of limited lateral extent has been described by Landauer (1989). It must be stressed that the severe restrictions needed for the derivation of Eq. (1.4), i.e., the existence of ideal conducting tubes on both sides of the sample joined smoothly to the reservoirs, are only conditions for that particular expression. Transmission between reservoirs can be calculated under many other circumstances. Equation (1.4) has been applied to a wide variety of geometries. Many of the early experiments emphasized analogies to waveguide propagation. Transmission through cavities with classical chaotic motion has been studied extensively. Systems with superconducting interfaces and Andreev reflections have been examined; see Chap. 7 of Imry (1997). Three-dimensional narrow wires, resulting either from an STM geometry or from mechanically pulling wires, to or past their breaking point, have received extensive attention (Serena and Garcia, 1997).

The preceding discussion assumes that we can ascribe a transmission coefficient to electrons whose interactions while in the reservoir are neglected. That does not prevent a Hartree approximation Coulomb interaction along the conductor. Electron-electron interactions of almost any kind can exist *within* a sample, but that still permits us to discuss the transmission of uncorrelated electrons *through* that sample.

Feeding current from reservoirs, with the carriers coming from each side characterized by a thermal equilibrium distribution, is only one possible way of driving a sample. The exact distribution of arriving carriers, both in real space and in momentum, matters. A sample does not really have a unique resistance, independent of the way we attach to it. Wide reservoirs, connected to a narrower sample, and emitting a thermal equilibrium distribution, are a good approximation to many real experimental configurations.

Equation (1.4) describes conduction as a function of quantum mechanically coherent transmission. Current flow in the presence of a limited conductance is a dissipative process. Where are the dissipation and the irreversibility (Sec. 2.3 of Datta, 1995)? They are in the reservoirs; carriers returning to them from the sample eventually suffer inelastic collisions. These inelastic collisions give the carriers, when they later again reach the transmissive sample, the occupation probability characteristic of the reservoir. The inelastic collisions in the reservoir also serve to eliminate any phase memory of the carrier's earlier history. Thus the sample determines the size of the conductance, even though the irreversible process takes place elsewhere. For a narrow conductor, attached to reservoirs which can serve as effective heat sinks, this means that the energy is released where it can easily be carried away and allows surprisingly large currents. Frank *et al.* (1998) pass current through a carbon nanotube, which would heat it to 20 000 K if the dissipation occurred along the tube. Such large currents and the accompanying changes in the wave functions of the binding electrons may induce temporary atomic displacements (Sec. 14 of Sorbello, 1997).

Spatial Variation, Conductance Fluctuations, Localization

We have already emphasized that ensemble members differ and that transport fields are spatially inhomogeneous. Spatial variations of current and field exist for two reasons. First of all geometry and preparation can impose obvious patterns in space, as in a transistor or scanning tunneling microscope. But a random arrangement of point scatterers can also provide inhomogeneity with easy and hard paths through the sample. Why are spatial variations of interest? Calculating conductance from Eq. (1.4) does not require an understanding of the spatial variations within the sample. But spatial variations are vital in other contexts. We can actually probe spatial distributions (Briner *et al.*, 1996; Eriksson *et al.*, 1996). The notion, common in the middle 1980s, that transport could only be examined by very invasive extra conducting leads has been replaced by the awareness that there is a growing set of minimally disturbing probes, including, for example, electro-optic effects. Spatial variations also matter in nonlinear transport. In that case the transport field itself is part of the field that determines transmission through the sample, and a self-consistent analysis invoking Poisson's equation is needed. Datta *et al.* (1997), in an analysis of conduction through an organic molecule caught in an STM configuration, illustrate this. Spatial variations matter in high-frequency behavior. In that case there can be capacitive shorting across the resistively hard parts of the sample. Spatial variations matter for electromigration (Sorbello, 1997). Electromigration is the motion of lattice defects induced by electron transport. The moving defects probe their local environment, not a volume average. In that analysis care must be taken to include the spatial variation induced by that defect.

We have alluded to the localized voltage drop across a planar barrier. If, instead, we introduce a point scatterer, in the presence of a constant current flow, we can also expect an increase in transport field related to the defect's location. In fact, a planar barrier can be built from an array of point defects, and the two cases must show related behavior. In the planar case the localized voltage drop arises from the pileup of incident electrons on one side of the barrier and their removal from the other side, allowing for self-consistent screening of these piled-up charges. A similar pileup at a point defect will generate a dipole field, called the residual resistivity dipole (RRD), which has been studied for over four decades (Sorbello, 1997; Zwerger, 1997; also see Ref. 19 of Landauer's introductory chapter in Serena and Garcia, 1997). A volume with incoherent point scatterers will generate a set of dipole fields, one dipole per scatterer. The resulting space-average field is that given by other elementary semiclassical theories; there is no new result for the resistivity. We have only emphasized the strong spatial variation of the transport field. The current flow pattern is also spatially nonuniform (Zwerger, 1997), representing the fact that the incident carrier flux, scattered by a localized defect, has to be carried around that defect much as a current has to be carried around a macroscopic cavity.

A review should not be confined to progress, but can also list questions. The spatial variation of the field in the presence of randomly placed point scatterers, providing *coherent* multiple scattering, is not understood. A set of dipole fields can still be expected, but the size of each dipole can no longer depend only on the scattering action of a particular defect. The striking nonuniformity of the potential drop along a coherent disordered one-dimensional array has been demonstrated (Maschke and Schreiber, 1994).

An array of randomly placed point scatterers, acting incoherently, will allow for some variation in transmission depending on the carrier's path. In the presence of coherence this variation is sensitive not only to density fluctuations among the obstacles, but also to the exact relative phasing of scattered waves. At one extreme the random placement includes ensemble members that give a periodic, or almost periodic, arrangement and cause the electron to see an allowed band, giving excellent transmission. But the ensemble will also include members that in all or part of space simulate a forbidden gap, yielding exponentially small transmission. The relative phasing of scattered waves can be altered not only by changing atomic placement, but also by changing parameters, such as the Fermi energy or a magnetic field. The resulting variations of conductance have been widely observed (Washburn and Webb, 1992; Sec. 5.2 of Ferry and Goodnick, 1997) and are called universal conductance fluctuations (UCF). Relative phasing of waves, taking alternative paths, also plays a critical role in the observed oscillations in the conductance in a solid-state analog of the Aharonov-Bohm effect (Washburn and Webb, 1992).

Consider a one-dimensional ensemble of a fixed number of identical localized scatterers, allowing for all possible relative phases between adjacent scatterers at the Fermi wavelength. (The total length of the chains cannot be held fixed.) This is a particularly simple case of disorder. As already stated, this includes ensemble members that, over portions of the chain, simulate a forbidden band, resulting in an exponential decay of the wave function. (The forbidden band can be associated with a superlattice formed from the scatterers and in any case represents only a physically suggestive way of pointing to constructive interference in the buildup of reflections.) For an electron incident, say from the left, there will be no regions providing a compensating exponential increase to the right. An ensemble average of the resulting resistance, rather than conductance, weights the high-resistance ensemble members and can be shown to increase exponentially with the number of obstacles (Sec. 5.3 of Imry, 1997). The problem of treating such a highly dispersive ensemble (Azbel, 1983) was solved by Anderson *et al.* (1980), who emphasized $\ln(1+g^{-1})$, where $g = G\pi\hbar/e^2$, which behaves like a typical extensive quantity. The ensemble average of this quantity is proportional to the number of obstacles, and this quantity also has a mean-squared deviation which scales linearly with length. The exponential decay, with length, of transmission through a disordered array is a particularly simple example of *localization*; electrons cannot propagate as effectively as classical diffusion would suggest. In two or three dimensions a carrier can detour around a poorly transmissive region. As a result localization in higher dimensions is not as pronounced and is more complex.

Equation (1.4) tells us that the conductance can be considered to be a sum of contributions over the eigenvalues of $tt^\dagger$. These represent, effectively, channels that transmit independently. The relative phase of what is incident in different channels does not matter. The variation of transmission, which would depend on the exact choice of path in a semiclassical discussion, is now represented by the distribution of eigenvalues of $tt^\dagger$.

For a sample long enough so that conduction is controlled by many random elastic-scattering events, producing diffusive carrier motion, but not long enough to exhibit localization, this distribution is bimodal (Beenakker, 1997). Most eigenvalues are very small, corresponding to channels that transmit very poorly. There is, however, a cluster of highly conducting channels, which transmit most of the current. Oakeshott and MacKinnon (1994) have modeled the striking nonuniformity of current flow, showing filamentary behavior, in a disordered block. This is a demonstration of the fact that the bimodal distribution is related to the distribution in real space. The bimodal distribution, the strong variation between ensemble members, and the geometrical nonuniformity have also been persistent themes in the work of Pendry (e.g., Pendry et al., 1992), who stresses the importance of necklaces i.e., chains of sites that permit tunneling from one to the next, in the limit where localization matters.

Electron Interactions

Both the thermodynamics and the transport properties of independent electrons propagating in a finite system with random scatterers are relatively well understood. It is also well known that for the infinite *homogeneous* electron gas the Hartree term for the interactions cancels the ionic background. The remaining exchange-correlation contributions still play an important role, especially in soluble 1D models (Emery, 1979), but a noninteracting model allows considerable progress in most higher-dimensional situations. A partial justification for this, with modified parameters, is provided by the Landau Fermi-liquid picture. In this description, the low-energy excitations of the interacting system are Fermion quasiparticles with a renormalized dispersion relation and a finite lifetime due to collisions. This is valid as long as the quasiparticle width is much smaller than its excitation energy, which is the case for homogenous systems at low enough excitation energies. However, when strong inhomogeneities exist, a rich variety of new phenomena opens up. Here we briefly consider the effects of disorder and finite size.

Disorder turns out to enhance the effects of the interactions, as explained by Altshuler and Aronov (1985). This enhancement is not only in the Hartree term, but also in greatly modified exchange and correlation contributions. These effects become very strong for low dimensions or strong disorder.

Singular behavior was found in the single-particle density of states (DOS) near the Fermi energy. For disordered 3D systems, the magnitude of this singularity (Altshuler and Aronov, 1985) is determined by the ratio of Fermi wavelength to mean free path. That is small for weak disorder, but increases markedly for stronger disorder. The situation is much more interesting for effectively 2D thin films and 1D narrow wires. There, when carried to low excitation energies, these corrections diverge, respectively, like the log and the inverse square root of $E - E_F$. Thus a more complete treatment, which is still lacking, is needed. These "zero-bias" DOS corrections should be ubiquitous. They have been observed experimentally and agree semiquantitatively with the theory, as long as they are not too large. Direct interaction-induced corrections to the conductivity in the metallic regime were also predicted by Altshuler and Aronov and confirmed by numerous experiments. The 2D case, realized in semiconductor heterojunctions and in thin metal films, is of special interest. The theory for noninteracting electrons (Abrahams et al., 1979) predicts an insulating behavior as the temperature $T \rightarrow 0$. The effect of electron-electron interactions is hard to treat fully. The calculations by Finkelstein (1983) show a window of possible metallic behavior characterized by a strong sensitivity to a parallel magnetic field. Several experimental studies suggest a similar metallic behavior in 2D (Lubkin, 1997), whose origin is still under debate.

Schmid (1974) found the interaction-induced lifetime broadening strongly enhanced by the disorder, varying in 3D as $(E-E_F)^{3/2}$ at $T=0$ instead of the usual ballistic $(E-E_F)^2$ Landau result. These changes are stronger in 1D and 2D, where the expressions obtained for the disorder-dominated $e\text{-}e$ scattering rate *diverge* at nonvanishing T. However, physically meaningful scattering rates are finite. For example, Altshuler *et al.* (1982) found that the rate of dephasing of the relative phase between two different paths is regular and goes as T (with a small logarithmic prefactor) and $T^{2/3}$ in 2D and 1D respectively. These results are crucial for numerous mesoscopic interference situations and agree quantitatively with experiments at temperatures that are not too low.

A particular case of strong inhomogeneity occurs when the electrons are confined to a small spatial range, such as a lattice site or a small grain or "quantum dot" (a two-dimensionally confined region of electrons). The deviation from charge neutrality is accompanied by an energy cost $e^2/2C$, where C is an effective capacitance. This energy, when it is large compared to the thermal energy, can prohibit double electronic occupancy for a hydrogenic impurity or exclude the electron transfer into or through a quantum dot. The latter phenomenon has been dubbed "the Coulomb blockade," and it is relevant to many experimental situations, including the optimistically named *single-electron transistor* (Chap. 4 of Ferry and Goodnick, 1997). Most interestingly, the correlation embodied by this strong inhibition of electrons to populate certain locations often results in subtle and dramatic phenomena. Those include a *Fermi-edge singularity* and the Kondo effect, both appearing (like the Altshuler-Aronov singularities) at low energies, near the Fermi level.

The Fermi-edge singularity is well known from x-ray absorption in metals. The attraction between the core hole left by the photoexcitation and the conduction electrons causes the absorption to diverge at its edge. Matveev and Larkin (1992) suggested that an analogous effect should exist for tunneling through a resonant impurity (or a quantum dot) state in a small tunnel junction. The hole left in that state by an electron tunneling out plays the same role as the above core hole. The interaction between this hole and the conduction electrons increases the transmission amplitude near E_F. The logarithmic divergence near threshold obtained from the simplest low-order perturbation theory is replaced in the full theory by a power-law singularity. Geim *et al.* (1994) observed this effect in impurity-assisted tunneling through small resonant tunneling diodes.

Another interesting near-Fermi-level structure is the Kondo resonance, due to the repulsion between the two opposite-spin electrons on the same impurity (or quantum dot) state and their hybridization with the conduction electrons. Again, for a quantum dot connected to two electrodes via tunnel junctions, this leads to a resonance in the zero-bias transmission. The splitting of this peak by a bias and its magnetic-field dependence were predicted by Meir *et al.* (1993). Very recently (Goldhaber-Gordon *et al.*, 1998) this effect was observed with a high-quality quantum dot.

Electron Coherence; Persistent Currents; Other Features

It was pointed out earlier that, in the absence of inelastic scattering in an intervening sample, the conductance between reservoirs is determined by the elastic scattering in that sample. In this case the ultimate source of irreversibility is in the inelastic scattering in the reservoirs. What happens if we eliminate the reservoirs and their inelastic scattering? We can do that by tying the leads to the sample to each other and creating a loop, considering the response of this quantum-mechanical system to an external magnetic flux through the ring. This will be done later in this section.

Inelastic scattering of the electron by other degrees of freedom of the ring acts like distributed coupling to external reservoirs. Inelastic scattering causes the electron waves to lose phase coherence; effects due to the interference between electron waves following alternative paths are eliminated (Chap. 3 of Imry, 1997; Chap. 6 of Ferry and Goodnick, 1997). Because the eigenstates of a closed system are determined by periodicity and boundary conditions, the energy levels of a bounded system with some inelastic scattering are no longer sharp and gradually lose their dependence on the periodicity condition when the inelastic scattering increases. For our ring, once the inelastic-scattering length exceeds the circumference of the ring, the waves can respond to the total set of boundary conditions. The ensuing flux sensitivity of the energy levels and their associated "persistent currents" are discussed below. The significant difference between the effects of elastic and inelastic scattering has been highlighted by recent research in disordered and mesoscopic systems. Equilibrium and transport experiments have determined the length the electron can propagate without losing phase coherence, often in good agreement with theory.

That the π electrons on a benzene-type ring molecule have a large orbital magnetic response has been known for more than half a century. The explanation in terms of "ring currents" was advanced by Pauling in 1936. Ordinary metals exhibit a small but measurable orbital diamagnetism, a purely quantum-mechanical phenomenon. However, the prediction that a metallic ring structure in the usual mesoscopic size range can support an equilibrium circulating current in response to an external flux was greeted with skepticism 15 years ago. This applied particularly to the diffusive regime with repeated elastic defect scattering for a carrier traversing the ring. When most of the flux is through the ring's opening, rather than the conductor, the response is periodic in the flux, with period $\Phi_0 = h/e$. This follows because such a flux Φ causes a phase change of $2\pi\Phi/\Phi_0$ in the phase of the eigenstates, upon taking an electron around the ring. Therefore the energy levels depend periodically on the flux. This leads to a dependence of the thermodynamic potential on the flux and hence, by thermodynamics, to an equilibrium current. Contrary to classical intuition, elastic scattering alone does not cause current decay.

There are now a number of experiments confirming the existence of persistent currents in single mesoscopic rings and also in rather large ensembles of rings, as summarized in Chap. 4, Sec. 2 of Imry (1997). In the latter case, the experiment measures the persistent current averaged over many rings. These rings differ through varying realizations of the random impurity potentials; an averaging over these differing realizations is effectively performed. The results of the single-ring experiments agree roughly with the theory for noninteracting electrons, and it can be shown (see p. 75 of Imry, 1997) that the interactions do not change the order of magnitude of the single-ring, sample-specific current. The situation is very different for the ensemble-averaged current. The periodicity in the flux was experimentally found to be $h/2e$, rather than h/e, in accordance with the theory for the ensemble-averaged persistent current for noninteracting electrons. However, these theories underestimate the magnitude of the measured current by more than two orders of magnitude. Introducing electron-electron interactions perturbatively gives a result with the required period $h/(2e)$, but still smaller than the experiment by a factor of 5 to 10.

This clearly goes in the right direction, but there is still no definitive understanding of the magnitude of the measured ensemble-averaged persistent currents.

An interesting case in which persistent normal currents *may* exist on the millimeter length scale is provided by recent experiments on the magnetic response in a proximity-effect system. Very-low-temperature measurements of that response for a superconducting cylinder with a normal-metal coating (Mota *et al.*, 1994), revealed an unexpected strong paramagnetic moment in addition to the usual Meissner effect induced by the superconductor in the proximity layer. This paramagnetic moment is comparable to a diamagnetic moment. Whispering gallery modes of the normal electrons, bouncing around the outer perimeter of the normal layer, qualitatively and speculatively explain this magnetic moment as due to unusually large normal persistent currents flowing near that surface (Bruder and Imry, 1998).

Our subject has many more facets than we can discuss, or even list, in this short paper. We allude to only one subtopic. Noise measurements have developed into a surprisingly accurate probing method. Noise is more sensitive than the dc conductance to electron correlations. Schoelkopf *et al.* (1997) have studied the frequency dependence of noise in a current-carrying metallic conductor, with enough elastic scattering to give diffusive carrier motion, and find remarkable agreement with the simplest independent-electron models. Reznikov *et al.*, in Datta (1998), discuss recent measurements at the Weizmann Institute and by a CEA/Saclay-CNRS/Bagneux collaboration, using shot-noise measurements to demonstrate the effective $e/3$ charge of the tunneling entity in a fractional-quantum-Hall-effect experiment.

Many alternative views of quantum transport have been developed, e.g., the Keldysh formulation, adaptable to inelastic processes as treated in Chap. 8 of Datta (1995). Especially powerful is a block-scaling picture due to Thouless (p. 21 of Imry, 1997), which generalized 1D localization to finite cross-section wires. Later it was broadened into an intuitive and successful theory at higher dimension (Abrahams *et al.*, 1979).

Only the most settled aspects of electron-electron interaction have been discussed, slighting a number of fashionable efforts. The approach emphasized in this paper should permit some generalization to the case in which carrier interactions in the reservoir are critical. For a given potential difference between two reservoirs there is a maximum current that can be passed through a smooth and long laterally constricted connection. [The lateral dimension(s) of the connecting pipe can be less than the range of the electron interactions.] In the presence of irregularities only a portion of that maximum will pass.

Acknowledgments

The research of Y.I. was supported by grants from the German-Israeli Foundation (GIF) and the Israel Science Foundation, Jerusalem.

References

Abrahams, E., P. W. Anderson, D. C. Licciardello, and T. V. Ramakrishnan, 1979, Phys. Rev. Lett. **42**, 673.

Altshuler, B. L., and A. G. Aronov, 1985, in *Electron-Electron Interactions in Disordered Systems*, edited by A. L. Efros and M. Pollak (North-Holland, Amsterdam), p. 1.

Altshuler, B. L., A. G. Aronov, and D. E. Khmelnitskii, 1982, J. Phys. C **15**, 7367.

Anderson, P. W., 1997, *The Theory of Superconductivity in the High-T_C Cuprates* (Princeton University, Princeton); see p. 158 and index.

Anderson, P. W., D. J. Thouless, E. Abrahams, and D. S. Fisher, 1980, Phys. Rev. B **22**, 3519.
Azbel, M. Ya., 1983, Solid State Commun. **45**, 327.
Beenakker, C. W. J., 1997, Rev. Mod. Phys. **69**, 731.
Beenakker, C. W. J., and H. van Houten, 1991, in *Quantum Transport in Semiconductor Nanostructures*, Solid State Physics **44**, edited by H. Ehrenreich and D. Turnbull (Academic, New York), p. 1.
Briner, B. G., R. M. Feenstra, T. P. Chin, and J. M. Woodall, 1996, Phys. Rev. B **54**, R5283.
Bruder, C., and Y. Imry, 1998, Phys. Rev. Lett. **80**, 5782.
Büttiker, M., 1986, Phys. Rev. Lett. **57**, 1761.
Büttiker, M., 1993, J. Phys.: Condens. Matter **5**, 9361.
Datta, S., 1995, *Electronic Transport in Mesoscopic Systems* (Cambridge University Press, Cambridge).
Datta, S., 1998, Ed., Superlattices and Microstruct., Vol. 23, Nos. 3/4, pp. 385–980.
Datta, S., W. Tian, S. Hong, R. G. Reifenberger, J. I. Henderson, and C. P. Kubiak, 1997, Phys. Rev. Lett. **79**, 2530.
Emery, V. J., 1979, in *Highly Conducting One-Dimensional Solids*, edited by J. T. Devreese, Roger P. Evrad, and Victor E. Van Doren (Plenum, New York), p. 247.
Eriksson, M. A., R. G. Beck, M. Topinka, J. A. Katine, R. M. Westervelt, K. L. Campman, and A. C. Gossard, 1996, Appl. Phys. Lett. **69**, 671.
Ferry, D. K., and S. M. Goodnick, 1997, *Transport in Nanostructures* (Cambridge University, New York).
Finkelstein, A. M., 1983, Zh. Eksp. Teor. Fiz. **84**, 168 [Sov. Phys. JETP **57**, 97 (1983)].
Frank, S., P. Poncharal, Z. L. Wang, and W. A. de Heer, 1998, Science **280**, 1744.
Geim, A. K., P. C. Main, N. La Scala Jr., L. Eaves, T. J. Foster, P. H. Beton, J. W. Sakai, F. W. Sheard, M. Henini, G. Hill, and M. A. Pate, 1994, Phys. Rev. Lett. **72**, 2061.
Goldhaber-Gordon, D., H. Shtrikman, D. Mahalu, D. Abusch-Magder, U. Meirav, and M. A. Kastner, 1998, Nature (London) **391**, 156.
Imry, Y., 1986, in *Directions in Condensed Matter Physics*, edited by G. Grinstein and G. Mazenko (World Scientific, Singapore), p. 101.
Imry, Y., 1997, *Introduction to Mesoscopic Physics* (Oxford University, New York).
Landauer, R., 1989, J. Phys.: Condens. Matter **1**, 8099.
Landauer, R., 1994, in *Coulomb and Interference Effects in Small Electronic Structures*, edited by D. Glattli, M. Sanquer, and J. Trân Thanh Vân (Editions Frontières, Gif-sur-Yvette), p. 1.
Landauer, R., 1998, in *Nanowires*, edited by P. A. Serena and N. Garcia (Kluwer, Dordrecht), p. 1.
Lubkin, G. B., 1997, Phys. Today **50**, 19.
Maschke, K., and M. Schreiber, 1994, Phys. Rev. B **49**, 2295.
Matveev, K. A., and A. I. Larkin, 1992, Phys. Rev. B **46**, 15 337.
Meir, Y., N. S. Wingreen, and P. A. Lee, 1993, Phys. Rev. Lett. **70**, 2601.
Mota, A. C., P. Visani, A. Pollini, and K. Aupke, 1994, Physica B **197**, 95.
Oakeshott, R. B. S., and A. MacKinnon, 1994, J. Phys.: Condens. Matter **6**, 1513.
Pendry, J. B., A. MacKinnon, and P. J. Roberts, 1992, Proc. R. Soc. London, Ser. A **437**, 67.
Schmid, A., 1974, Physics **271**, 251.
Schoelkopf, R. J., P. J. Burke, A. A. Kozhevnikov, D. E. Prober, and M. J. Rooks, 1997, Phys. Rev. Lett. **78**, 3370.
Serena, P. A., and N. Garcia, 1997, Eds., *Nanowires* (Kluwer, Dordrecht).
Sohn, L., L. Kouwenhoven, and G. Schön, 1997, Eds., *Mesoscopic Electron Transport* (Kluwer, Dordrecht).
Sorbello, R. S., 1997, in Solid State Physics: Advances in Research and Applications No. 51, edited by H. Ehrenreich and F. Spaepen (Academic, Boston).
van Houten, H., and C. Beenakker, 1996, Phys. Today **49**, 22.
Washburn, S., and R. A. Webb, 1992, Rep. Prog. Phys. **55**, 1311.
Zwerger, W., 1997, Phys. Rev. Lett. **79**, 5270.

Superconductivity
J. R. Schrieffer and M. Tinkham

Superconductivity was discovered in 1911 by H. Kamerlingh Onnes (1911) in Leiden just three years after he first liquified helium, which made sufficiently low temperatures available. What he found was that the electrical resistance of some metals, such as lead, mercury, tin, and aluminum, disappeared completely in a narrow temperature range at a critical temperature T_c (typically a few Kelvin) specific to each metal. Twenty-two years later, Meissner and Ochsenfeld (1933) discovered that these superconductors were perfectly diamagnetic (the "Meissner effect") as well as perfectly conducting. These remarkable properties were neatly described by the phenomenological theory of F. and H. London (1935). Their model postulated a density of "superconducting electrons" n_s per unit volume, whose response to electromagnetic fields could be described by

$$\mathbf{J}_s = -(c/4\pi\lambda_L^2)\mathbf{A} \tag{1}$$

(with a specific "London gauge" choice for the vector potential). The time derivative of Eq. (1) implies that the superconducting electrons respond to an electric field **E** essentially as Drude free electrons with an infinitely long relaxation time. Combined with the Maxwell equations, this leads to a frequency-independent skin depth, called the London penetration depth:

$$\lambda_L = (mc^2/4\pi n_s e^2)^{1/2}. \tag{2}$$

The curl of Eq. (1) (with Maxwell's equations) implies the static flux expulsion of the Meissner effect, which cannot be interpreted in a classical way. Since λ_L was found experimentally to diverge at T_c roughly as $[1-(T/T_c)^4]^{-1/2}$, n_s was presumed to go continuously to zero at T_c, as in a second-order phase transition.

Ginzburg and Landau (1950) extended the London phenomenology in a brilliant stroke based on Landau's theory of second-order phase transitions. They introduced as an order parameter, a complex "wave function of the superconducting electrons," $\psi(\mathbf{r})=|\psi(\mathbf{r})|e^{i\varphi(r)}$, such that $n_s \propto |\psi(\mathbf{r})|^2$. Their theory reproduced Eq. (1) in a gauge-invariant form,

$$\mathbf{J} = 2e|\psi(\mathbf{r})|^2 \mathbf{v}_s, \tag{3}$$

where

$$m^* \mathbf{v}_s = \hbar \nabla \varphi - 2e\mathbf{A}/c, \tag{3a}$$

with the effective mass m^* usually taken to be $2m$. Moreover, this $\psi(\mathbf{r})$ was shown to be governed by a nonlinear differential equation, so that it could vary with position and field strength, in addition to the temperature dependence of n_s in the London picture. For example, it provided a natural description for the interface between normal and superconducting phases in the presence of a critical magnetic field H_c. This theory was later shown by Gor'kov (1959) to be a limiting case of the BCS (Bardeen, Cooper, and Schrieffer, 1957a, 1957b) theory and remains today as the standard initial approach to problems with a spatially varying superconducting state.

Aided by wartime developments in high-frequency technology, Pippard was able to make very precise measurements of λ_L to compare with Eq. (2), using parameters determined from similar measurements of the skin depth in the normal state. He found that, even at $T \approx 0$, the fitted value of n_s was less than the density of conduction electrons in the normal state, by a ratio that was larger for low-T_c materials like Al($T_c \approx 1$ K) than for metals like Pb ($T_c \approx 7$ K). Building on Chambers' equation for the anomalous skin effect in normal metals, Pippard (1953) was able to explain this reduced value of n_s by introducing a "coherence length"

$$\xi_0 = a\hbar v_F / k_B T_c \qquad (4)$$

into the London electrodynamics, where the coefficient a was of order unity. In a review published in 1956, shortly before the discovery of the BCS microscopic theory of superconductivity, Bardeen (1956) was able to show that just such a "nonlocal" electrodynamics would be a consequence of an energy gap Δ in the electronic spectrum, if the energy gap was proportional to T_c. And, indeed, when the BCS theory was created, it predicted the nonlocal electrodynamics, with $\xi_0 = \hbar v_F / \pi \Delta(0)$, in agreement with Pippard's brilliant conjecture.

This was the state of our understanding of the electrodynamics of classic superconductors in the mid 1950s—a very satisfactory phenomenology, but no "explanation" in microscopic terms. What was the nature of the superconducting state that made it have these remarkable properties? This question was answered in one stroke by the classic paper of Bardeen, Cooper, and Schrieffer (1957a), which is the subject of the next section of this article.

The BCS Microscopic Theory

The discovery of the isotope effect by Maxwell (1950) and Reynolds et al. (1950), namely, that $T_c \propto M^{-\alpha}$ where M is the ionic mass and $\alpha \approx 1/2$, gave strong support to the view that superconductivity is the result of the electron-phonon interaction. Prior to this discovery, Fröhlich (1950) had worked out a model based on this interaction but ran into formal difficulties and the approach did not describe the properties of a superconductor. In fact, Shafroth (1958) proved that the Meissner effect could only be obtained by going beyond perturbation theory in treating the effective interaction between electrons.

In 1955 Bardeen considered attacking the problem using the techniques of quantum field theory and invited Cooper to join the effort since Cooper's background was in particle physics. It soon became clear that since the existing field-theoretic methods were based on perturbation theory, another scheme would have to be devised.

Bardeen stressed the importance of an energy gap in the excitation spectrum and that superconductivity is due to a condensation in momentum space of a coherent superposition of normal-state configurations. A major difficulty existed in that the correlation energy in the normal phase is of order 1 eV per electron, while the energy distinguishing the normal and super phases is of order 10^{-6} eV per electron. Fortunately, Landau's theory of a Fermi liquid provided the necessary basis for treating the normal-state excitations in one-to-one correspondence with the free-electron gas so that the small condensation energy between the super and normal phases could be isolated.

Cooper (1956) studied the problem of two electrons interacting via an attractive effective potential above a frozen Fermi sea. He found that the normal state is unstable regardless of how weak the attraction is. Bardeen, Cooper, and Schrieffer (1957a, 1957b) then studied a reduced Hamiltonian which included interactions involving only paired states,

$$H_{\text{red}} = \sum_{ks} \varepsilon_k n_{ks} + \sum_{kk'} V_{kk'} b_k^+ b_k, \tag{5}$$

where b_k^+ creates an electron pair in $(k\uparrow, -k\downarrow)$, and ε_k is the normal-state quasiparticle energy measured relative to the chemical potential.

Bardeen argued on the basis of the uncertainty principle that the overlap of pair wave functions is extremely large because of the large ratio of the Fermi and critical temperatures. Thus one cannot think of the pairs as bosons since the Pauli principle plays a crucial role in the problem.

Schrieffer constructed a variational trial function in analogy with the Tomonaga (1947) approach to the pion nucleon problem,

$$\Psi = \prod_k (u_k + v_k b_k^+)|0\rangle, \tag{6}$$

where $u_k^2 + v_k^2 = 1$ for normalization, and the parameters v_k are to be chosen to minimize the energy. This prescription describes pairing in a spin singlet and orbital s-wave state. One finds that the energy minimization leads to a self-consistency condition

$$\Delta_k = -\sum_{k'} V_{kk'} \Delta_{k'}/2E_{k'} \tag{7}$$

with

$$v_k^2 = \frac{1}{2}\left(1 - \frac{\varepsilon_k}{E_k}\right) \tag{8}$$

and

$$E_k = (\varepsilon_k^2 + \Delta_k^2)^{1/2}. \tag{9}$$

The excitation spectrum based on this state exhibits quasiparticles of energy E_k with an energy gap Δ_k. For k far above the Fermi surface the excitations are electronlike, and far below k_F they are holelike, while at k_F they are an equal mixture of electron and hole, having charge zero but spin one-half. This is an example of charge-spin separation since the charge of an injected electron at the Fermi surface shifts the mean number of pairs by one-half with the spin remaining with the quasiparticle.

Since the spectrum exhibits a gap, it follows as Bardeen had argued, that the theory predicts a Meissner effect. The electrodynamics is nonlocal, involving a coherence length of a form [Eq. (4)] proposed by Pippard (1953).

The theory predicts a second-order phase transition at a temperature given by

$$k_B T_c \approx \hbar \bar{\omega}_0 e^{-1/N(0)V}, \tag{10}$$

with the gap vanishing at T_c as $(T_c - T)^{1/2}$. Here $\hbar \bar{\omega}_0$ is the mean phonon energy and V is the pair interaction. For weak-coupling superconductors, the ratio of the zero-temperature gap $2\Delta(0)$ and the transition temperature is predicted to be 3.52.

Magnetic flux trapped in a superconducting ring is predicted to be in units of $\Phi_0 = hc/2e$, reflecting the fact that the condensate is formed by electron pairs. This was observed experimentally by Deaver and Fairbank (1961) and by Doll and Näbauer (1961).

Gor'kov (1958) suggested the quantum field formulation of the BCS theory by making use of Δ_k as the "off-diagonal" long-range order parameter. By including spatial variation of the gap function $\Delta(r)$, he succeeded (Gor'kov, 1959) in deriving the Ginzburg-Landau phenomenological theory from the BCS theory.

Strong-coupling effects were explained by Eliashberg (1960) by extending the Gor'kov equations to include retardation effects in the pairing interaction and damping of the quasiparticles arising from phonon emission.

Shortly after the pairing theory was advanced, it was proposed (Bohr et al., 1958) that the theory also described many features of atomic nuclei, such as the even vs odd effects on adding one nucleon to the nucleus. Moreover, the deviation of the moment of inertia from the rigid moment is the analog of the Meissner effect. ^{3}He is another Fermi liquid, which was discovered by Osheroff et al. (1972) to undergo a transition to a superfluid state in which the pairing is in a spin-triplet state with an orbital angular momentum one.

Experimental Confirmation of the BCS Energy Gap and Coherence Factors

The BCS theory described a radically new vision of the nature of the superconducting state, which had eluded theorists for 46 years. Yet it was accepted by the great majority of physicists almost immediately. Why was that? For one thing, its predictions of the low-frequency electrodynamics essentially reproduced the results of the London and Pippard phenomenological theories, which were known to describe in detail the experimental data for the penetration depth. More decisive support for the new theory was provided by other experiments, which tested *new* predictions of the theory that went well beyond the general two-fluid models which had been available earlier.

One such prediction was the existence of an energy gap $2\Delta(T)$ for the creation of a pair of quasiparticle excitations. For weak-coupling superconductors, the theory predicted that $2\Delta = 3.52\, kT_c$ at $T=0$, falling continuously to zero at the second-order transition to the normal state at T_c. Such an energy gap was consistent with the exponential temperature dependence found in the latest specific-heat measurements (Corak et al., 1954). It was supported more decisively by spectroscopic microwave absorption measurements (Biondi et al., 1956) and spectroscopic far-infrared transmission experiments (Glover and Tinkham, 1956), the latter extending to frequencies well above the energy gap even at $T=0$. This allowed a quantitative test of the predictions of the BCS theory for the frequency-dependent complex conductivity $\sigma_1(\omega) - i\sigma_2(\omega)$ near the energy-gap frequency in the superconducting state. After the gap width was scaled up from 3.52 to ~$4.2kT_c$ for lead, which is not a weak-coupling superconductor, the transmission curve $T(\omega)$ predicted by the theoretical $\sigma_1(\omega)-i\sigma_2(\omega)$ was in excellent agreement with the experimental data, including the size and shape of a nontrivial *peak* in transmission near the energy-gap frequency, where both $\sigma_1(\omega)$ and $\sigma_2(\omega)$ are relatively small. In an elegant experiment, Hebel and Slichter (1957) observed a coherence peak in the NMR spectrum that probed details of the paired state and its excitations, in agreement with the BCS theory. The energy gap $\Delta(T)$ in the superconducting density of states was subsequently measured directly in an important pioneering experiment by Giaever (1960a, 1960b), in which he measured the minimum energy in eV required to insert an electron into a superconductor by a tunneling process.

A particularly distinctive prediction of the BCS theory is the existence of coherence factors in the transition probabilities which distinguish processes, like ultrasonic absorption, that are even under time reversal from those, like nuclear relaxation, that are odd. This difference in coherence factors was predicted to cause the ultrasonic attenuation to drop very sharply on cooling through T_c, as confirmed by Morse (1959), while the nuclear relaxation rate was predicted to rise to a maximum above the normal-state value just below T_c, before dropping exponentially at lower temperatures, as was confirmed by Hebel and Slichter (1957). Since both of these processes depend on the density of quasiparticles, which correspond to the "normal electrons" of a two-fluid picture, the fact that the nuclear relaxation rate goes up while the ultrasonic attenuation rate goes down on cooling below T_c is inexplicable without the coherence factors, which are a unique and specific feature of the BCS theory.

Type-II Superconductors

In 1957, the same year as the BCS theory, Abrikosov (1957) also published a groundbreaking paper, based on the Ginzburg-Landau theory, in which he explored theoretically what would happen if the inequality $\lambda < \xi$ typical of superconductors like tin and lead were reversed. He found that when the ratio $\kappa = \lambda/\xi$ exceeded $1/\sqrt{2}$, the magnetic properties were completely different from the classic superconductors; he called these high-κ materials "type-II superconductors." Instead of showing a first-order transition from superconducting flux exclusion (Meissner effect) to the normal state at a critical field H_c like the classic, or type-I, superconductors, type-II superconductors above a lower critical field H_{c1} were predicted to allow magnetic flux to penetrate in a regular array of quantum units of $\Phi_0 = hc/2e$, each flux tube being confined by a circulating vortex of current. These materials were predicted to remain superconducting until a second-order transition at an upper critical field $H_{c2} = \Phi_0/2\pi\xi^2 = \sqrt{2}\kappa H_c > H_c$. (Here H_c is the thermodynamic critical field such that $H_c^2/8\pi$ equals the free-energy difference between superconducting and normal states of the metal.) Since for "dirty" metals, with short mean free path ℓ, the BCS theory shows that $\xi^2 \approx \xi_0 \ell \approx \ell \hbar v_F/kT_c$, this H_{c2} can be very high ($>10^5$ Oe) if ℓ is small and/or T_c is high. These type-II materials thus made possible the fabrication of high-field superconducting magnets, which play an important role both in the laboratory and in large-scale applications of superconductivity.

Superconducting materials research was rejuvenated by the discovery by Bednorz and Müller (1986) of new classes of oxide-based high-temperature superconductors, some of which have T_c in excess of 100 K and extremely high values of H_{c2}. The detailed origin of superconductivity in these materials is still unclear, but there is considerable evidence indicating that the pairing has a predominantly d-wave symmetry as opposed to the s-wave symmetry of conventional BCS superconductivity. This field remains one of vigorous research activity at the time of this writing.

The Josephson Effect

In 1962, Josephson (1962) made the remarkable prediction that a zero-voltage supercurrent of magnitude

$$I_s = I_c \sin(\Delta\varphi) \qquad (11)$$

should flow between two superconducting electrodes separated by a thin tunnel barrier. Here $\Delta\varphi$ is the difference in the phase of the Ginzburg-Landau ψ in the two electrodes. Although it was startling at the time, in retrospect this relation is now recognized as a general property of "weak links" between superconductors, and it can be derived as a discrete form of Eq. (3) for a short superconducting constriction. Josephson also predicted that if a voltage difference V were maintained across the junction, the phase difference would evolve as

$$d(\Delta\varphi)/dt = 2eV/\hbar. \qquad (12)$$

Thus the current would be ac current of amplitude I_c and frequency

$$f = 2eV/h, \qquad (12a)$$

consistent with the Planck-Einstein relation $E = hf$ relating frequency to the energy change associated with transfer of a Cooper pair from one electrode to the other. This fundamental relation is now used to define the standard volt in terms of a precise frequency.

In the presence of a magnetic field, $\Delta\varphi$ in these expressions must be generalized to a gauge-invariant phase difference, consistent with the general expression (3a). The resulting sensitivity of the Josephson current to magnetic fields stems from the fact that a single quantum of flux $\Phi_0 = hc/2e$ enclosed in a superconducting circuit shifts $\Delta\varphi$ by a full 2π. This has made possible the development of SQUID (Superconducting QUantum Interference Device) magnetometers of extreme sensitivity $\sim 10^{-6}\,\Phi_0$, which are approaching the ultimate limit set by the quantum-mechanical uncertainty principle.

Phase and Number Variables

In its canonical form (6), the BCS ground-state wave function is a superposition of states with many different numbers of pairs in a grand canonical ensemble. In reality, because the electrons carry a charge, there is a Coulomb energy $(\delta N)^2 E_c$ associated with any imbalance (δN) between the number of electrons and the number of positive nuclear charges in the sample. Here $E_c = e^2/2C$ is the charging energy associated with a single electronic net charge on a system with self-capacitance C. Since the capacitance scales with physical size, E_c is small for macroscopic superconductors, and this energy term can usually be neglected. However, in mesoscopic superconductors E_c can become the dominant energy term, and the electron number must be precisely fixed in the ground state of the system. As pointed out by Anderson (1967), this can be accomplished by associating a Ginzburg-Landau-like phase variable with each pair in the BCS ground state and then projecting out the part with a definite number of pairs. More explicitly, if we generalize Eq. (6) to the form

$$\Psi_\varphi = \prod_k (u_k + v_k e^{i\varphi} b_k^+)|0\rangle, \tag{13}$$

where u_k and v_k are taken to be real, then we can obtain an eigenfunction containing N electrons ($N/2$ pairs) by writing

$$\Psi_N = \int_0^{2\pi} e^{-iN\varphi/2} \Psi_\varphi d\varphi. \tag{14}$$

This Fourier transform relation between eigenfunctions of phase and number has the same form as that between eigenfunctions of position and momentum for a particle. Accordingly, it also implies an uncertainty relation between phase and number of the form

$$\Delta N \Delta \varphi \geq 1. \tag{15}$$

In dealing with macroscopic superconductors, for example in the Josephson effect, it is more appropriate to use eigenfunctions of the form of Eq. (13), in which the phase variable φ is well defined and identified with the phase variable in the Ginzburg-Landau equations. However, for describing small isolated superconducting particles, the Ψ_N of Eq. (14) is more appropriate.

An interesting illustration of the use of superconducting eigenstates of number rather than of phase is offered by the superconducting single-electron tunneling transistor. This device consists of a nanoscale superconducting island connected to two leads by high-resistance, low-capacitance tunnel junctions and capacitively coupled to a gate electrode. If the tunnel resistance is greater than $R_Q \sim h/e^2$, the number of electrons on the island is a good quantum number, and if the capacitance is small enough that $E_c = e^2/2C \gg kT$, a unique choice of electron number is energetically favored. If one measures the current through the device for a fixed small-bias voltage between the leads while sweeping the charge $C_g V_g$ induced by a voltage V_g on the gate, one finds periodic current peaks spaced $2e$ apart in gate charge (Tuominen et al., 1992). These peaks occur at values of V_g at which states with successive integer numbers of pairs on the island are degenerate, allowing pairs to be transferred without an energy barrier. This phenomenon provides a rather direct demonstration of the paired nature of the superconducting ground state.

In conclusion, we point out that the above discussion is necessarily incomplete, due to length limitations. Instead of attempting a brief review of the entire field, we have focused on the development of the pairing theory, together with some key points in the prehistory and later consequences of the theory.

References

Abrikosov, A. A., 1957, Zh. Éksp. Teor. Fiz. **32**, 1442 [Sov. Phys. JETP **5**, 1174].
Anderson, P. W., 1967, "The Josephson Effect and Quantum Coherence Measurements in Superconductors and Superfluids," in *Progress in Low Temperature Physics*, edited by C. J. Gorter (Wiley, New York), Vol. 5, p. 5.

Bardeen, J., 1956, "Theory of Superconductivity," in *Handbuch der Physik*, edited by S. Flügge (Springer, Berlin), Vol. XV, p. 274.
Bardeen, J., L. N. Cooper, and J. R. Schrieffer, 1957a, Phys. Rev. **106**, 162.
Bardeen, J., L. N. Cooper, and J. R. Schrieffer, 1957b, Phys. Rev. **108**, 1175.
Bednorz, G., and K. A. Müller, 1986, Z. Phys. B **64**, 189.
Biondi, M. A., M. P. Garfunkel, and A. O. McCoubrey, 1956, Phys. Rev. **102**, 1427.
Bohr, A., B. R. Mottelson, and D. Pines, 1958, Phys. Rev. **110**, 932.
Cooper, L. N., 1956, Phys. Rev. **104**, 1189.
Corak, W. S., B. B. Goodman, C. B. Satterthwaite, and A. Wexler, 1954, Phys. Rev. **96**, 1442.
Deaver, B. S., and W. M. Fairbank, 1961, Phys. Rev. Lett. **7**, 43.
Doll, R., and M. Näbauer, 1961, Phys. Rev. Lett. **7**, 51.
Eliashberg, 1960, Zh. Éksp. Teor. Fiz. **38**, 966 [Sov. Phys. JETP **11**, 696].
Fröhlich, H., 1950, Phys. Rev. **79**, 845.
Giaever, Ivar, 1960a, Phys. Rev. Lett. **5**, 147.
Giaever, Ivar, 1960b, Phys. Rev. Lett. **5**, 464.
Ginzburg, V. L., and L. D. Landau, 1950, Zh. Eksp. Teor. Fiz. **20**, 1064.
Glover, III, R. E., and M. Tinkham, 1956, Phys. Rev. **104**, 844.
Gor'kov, L. P., 1958, Zh. Eksp. Teor. Fiz. **36**, 1918 [Sov. Phys. JETP **7**, 505].
Gor'kov, L. P., 1959, Zh. Eksp. Teor. Fiz. **34**, 735 [Sov. Phys. JETP **9**, 1364].
Hebel, L. C., and C. P. Slichter, 1957, Phys. Rev. **107**, 901.
Josephson, B. D., 1962, Phys. Lett. **1**, 251.
London, F., and H. London, 1935, Proc. R. Soc. London, Ser. A **149**, 71.
Maxwell, E., 1950, Phys. Rev. **78**, 477.
Meissner, W., and R. Ochsenfeld, 1933, Naturwissenschaften **21**, 787.
Morse, R. W., 1959, "Ultrasonic attenuation in metals at low temperatures," in *Progress in Cryogenics*, edited by K. Mendelssohn (Heywood, London), Vol. I, p. 219.
Onnes, H. K., 1911, Commun. Phys. Lab. Univ. Leiden **120b,122b,124c**, .
Osheroff, D. D., R. C. Richardson, and D. M. Lee, 1972, Phys. Rev. Lett. **28**, 885.
Pippard, A. B., 1953, Proc. R. Soc. London, Ser. A **216**, 547.
Reynolds, C. A., B. Serin, W. H. Wright, and L. B. Nesbitt, 1950, Phys. Rev. **78**, 487.
Shafroth, M. R., 1958, Phys. Rev. **111**, 72.
Tomonaga, S., 1947, Prog. Theor. Phys. **27**, 1697.
Tuominen, M. T., J. M. Hergenrother, T. S. Tighe, and M. Tinkham, 1992, Phys. Rev. Lett. **69**, 1997.

Superfluidity
A. J. Leggett

The original observation of the phenomenon, or more precisely the complex of phenomena, known as "superfluidity" was made simultaneously in liquid 4-He in 1938 by two groups, Kapitza in Moscow and Allen and Misener in Cambridge. It had been known for some years previously that liquid helium (which, until the early 1950s when the light isotope 3-He began to be produced in experimentally useful quantities from nuclear reactors, was synonymous with liquid 4-He) did not freeze under its own vapor pressure down to the lowest attainable temperatures, and during the early- and mid-1930s it had become clear that some peculiar things happened at and below a characteristic temperature (~2.17 K), which became known as the "lambda temperature." Stimulated by measurements that seemed to show that below the lambda temperature the heat flow was not simply proportional to the temperature gradient, Allen and Misener, and simultaneously Kapitza, decided to measure the resistance to the flow of liquid helium clamped in narrow channels and subjected to a pressure drop. They found that while the so-called He-I phase, i.e., helium above the lambda temperature, showed a behavior that could be described in terms of a conventional viscosity, below the lambda point (in the so-called He-II phase) the liquid flowed so easily that if the concept of viscosity was applicable at all, it would have to be at least a factor of 1500 smaller than in the He-I phase. It was this anomalous behavior for which Kapitza coined the term "superfluidity." Actually, as we shall see below, this "ability to flow without apparent friction" in the kind of geometry employed in the Moscow and Oxford experiments, while spectacular, is not the conceptually simplest manifestation of superfluidity.

Within a few months of the experimental observation Fritz London came up with a qualitative explanation that has stood the test of time. The He atom is composed of an even number of elementary particles (2 protons, 2 neutrons, and 2 electrons) and thus according to the general precepts of quantum field theory, the many-body wave function of the system should be symmetric under the exchange of any two atoms; in technical language, the system should obey "Bose statistics." Fourteen years earlier Albert Einstein had studied the thermodynamic behavior of a gas of noninteracting atoms of this type, and had shown that below a characteristic temperature, which depends on the mass and density, it should manifest a peculiar behavior, which is nowadays known as Bose-Einstein condensation (BEC); a finite fraction of all the atoms (and at zero temperature, all of them) should occupy a *single* one-particle state. At the time Einstein made this suggestion this behavior was widely suspected of being a pathology of the noninteracting gas, which would disappear as soon as the interatomic interactions were taken into account. However, London now resurrected it and, noting that for a noninteracting gas with the mass and density of 4-He, the BEC phenomenon would occur at 3.3 K, suggested that this was exactly what was going on at the observed lambda transition (2.17 K). Very soon thereafter Laszlo Tisza pushed the idea further by suggesting that the anomalous flow behavior seen in the He-II phase could be qualitatively understood in terms of a "two-fluid" model in which the "condensate" (that is, those atoms which occupy the "special" one-particle state) behaves completely without friction, while the rest behave qualitatively like an ordinary liquid. One striking prediction that he was able to make on this basis was of a new type of collective excitation in which the two components—the condensate and the rest—oscillate out of phase.

A major landmark in the history of superfluidity was the appearance in 1941 of a paper by Lev Landau in which he developed in a quantitative way the "two-fluid" description of liquid He-II. (It seems likely that because of wartime conditions, Landau was unaware of Tisza's earlier, more qualitative work.) It is interesting that in this paper Landau never explicitly introduced the idea of BEC (indeed, he seems to have been opposed to it, regarding it as a pathology of the noninteracting gas), but rather posited, on intuitive grounds, various properties of the "ground state" of a Bose liquid, which with hindsight can in fact be seen to be natural consequences of the BEC phenomenon (see below). This paper marks the first explicit introduction into condensed-matter physics of the seminal notion of a "quasiparticle," that is, an excitation of the system from the ground state, which is characterized by a definite energy and momentum, and such that, at least at sufficiently low temperatures, the total energy, momentum, etc., of the system can be regarded as the sum of that carried by the quasiparticles. Landau identified the quasiparticles of a Bose liquid as of two types: quantized sound waves or phonons, with an energy ε, which depends on momentum p as $\varepsilon = cp$ (c = speed of sound), and "rotons," which he regarded as corresponding to quantized rotational motion and to which he originally assigned an energy spectrum $\varepsilon(p) = \Delta + p^2/2\mu$ (later modified, see below). An immediate prediction of this ansatz was that in the limit of low temperatures ($T \ll \Delta$) the rotons give negligible contribution to the specific heat, which in this regime is entirely due to the phonons and is proportional to T^3 (just as in an ordinary insulating crystalline solid).

To construct a quantitative theory of the flow properties of He-II, Landau postulated that it consisted of two components: the "superfluid" component, which he identified, in an intuitive way, with the part of the liquid that remained in its ground state, and a "normal" component, which corresponded to the quasiparticles. The superfluid component was conceived as carrying zero entropy and flowing irrotationally (i.e., its velocity v_s satisfied the condition curl $v_s = 0$); by contrast, the normal component behaved like any other viscous liquid. From these apparently minimal postulates Landau was able to derive a complete, quantitative theory of two-fluid hydrodynamics. It made, in particular, three remarkable predictions: (1) If the liquid (or more precisely the superfluid component of it) flows relative to the walls of the vessel containing it at a velocity smaller than velocity v_c (nowadays known as the Landau critical velocity) given by the minimum value of $\varepsilon(p)/p$ (usually this is the speed of sound c), then it may be able to do so without dissipation; otherwise the flow will be unstable against creation of quasiparticles. (2) If the boundary conditions rotate slowly (as, for example, in a rotating bucket), then only the fraction ρ_n of the liquid which corresponds to the normal component will rotate with them; Landau gave a formula for ρ_n in terms of the excitation spectrum. (3) It should be possible (as had also been suggested by Tisza) to set up an oscillation (nowadays known as "second sound") in which the normal and superfluid components oscillate out of phase; we now know (though Landau originally did not) that in liquid helium such a wave corresponds to substantial oscillations in temperature but only a very slight variation in pressure. Predictions (2) and (3) were verified within a few years in experiments carried out in the Soviet Union, by Andronikashvili and by Peshkov, respectively; prediction (3), though of fundamental importance conceptually, proved much more difficult to verify explicitly, and it is only comparatively recently that a direct measurement of the Landau critical velocity has been made, with the flow in question being relative not to the walls of the vessel but to ions moving through it (arguably the only case to which Landau's argument actually applies in its original form without a string of caveats).

While Landau's two-fluid hydrodynamics provides a conceptual basis for superfluidity, which still stands today, it is phenomenological in the sense that both the properties of the superfluid and the nature of the excitation spectrum are postulated in an intuitive way rather than being explicitly demonstrated to be a consequence of the Bose statistics obeyed by the atoms. This lacuna was partially filled in 1946 in a paper by N. N. Bogoliubov, which may for practical purposes be taken as ushering in the area of research known today as the "many-body problem." Bogoliubov considered a dilute gas of atoms obeying Bose statistics and interacting via an interatomic interaction, which is weakly repulsive. He *assumed* that such a system, like the completely free Bose gas, would undergo the phenomenon of BEC, and then, using a series of controlled approximations, was able to show that while the energy spectrum for large momentum p corresponds approximately to the simple excitation of free atoms from the condensate [$\epsilon(p) = p^2/2m$], at smaller momenta it has precisely the phonon-like form $\epsilon(p) = cp$ postulated by Landau, where the velocity of sound c is derived from the bulk compressibility in the standard way. (However, in Bogoliubov's work there is no obvious trace of the second, "roton" branch of the excitation spectrum postulated by Landau.) This work was subsequently refined and extended by Lee, Huang, Yang, Girardeau, and others, and actually turns out to be applicable in more or less its original form to the recently stabilized BEC alkali gases (see below).

While Bogoliubov's results were extremely suggestive, they referred to a dilute system, which is rather far from real-life liquid He-II (where the atoms are so closely packed as to be sampling both the attractive and the repulsive parts of the van der Waals interaction virtually all of the time). Thus a number of attempts were made to treat the realistic helium problem by variational or related methods; a particularly successful attack on the problem was made in 1956 by Feynman and Cohen on the basis of Feynman's earlier work. Among other things, this work predicted that the excitation spectrum of real liquid He-II should go over from the "phonon-like" behavior $\epsilon(p) = cp$ at small momenta predicted by Bogoliubov to a "roton-like" form $\epsilon(p) = \Delta + (p - p_0)^2/2m$, at larger values of the momentum. (This revision of his original hypothesis had actually been advanced a few years earlier by Landau himself, on the basis of experimental measurements of the temperature-dependence of the second-sound velocity.) Actually, in the early 1950s the use of reactor sources permitted for the first time experiments on the scattering of neutrons from various materials including liquid 4-He. The neutrons essentially measure the energy distribution of a particular kind of excitation, namely the density fluctuations, which have given momentum p; what is seen is that for a given p the energy is indeed approximately unique (so that the "quasiparticle" hypothesis indeed seems to be valid), and furthermore, that the spectrum has exactly the general form predicted by the Landau-Feynman-Cohen ansatz.

Rather than reviewing further in historical sequence the important advances made throughout the 50s, 60s, and 70s in the study of superfluid 4-He, it may be useful at this point to stand back and try to give a brief overview of our current understanding of the subject, bringing in the relevant experiments to illustrate them as we go. This understanding is, in some sense, a coherent amalgam of the ideas of London on the one hand, and Landau on the other, as refined and amplified by many subsequent workers. It should be remarked that these ideas developed in parallel with similar considerations concerning superconductivity, and indeed from a modern point of view superconductivity *is* nothing but superfluidity occurring in a charged system (or vice versa)—an idea which was extensively exploited by Fritz London in his 1950 two-volume book *Superfluids*, which covers both subjects.

The fundamental assumption that underlies the modern theory of superfluidity in a simple Bose system such as liquid 4-He is that the superfluid phase is characterized by what one might call "generalized BEC." By this I mean the following: we assume that at any given time t it is possible to find a complete orthonormal basis (which may itself depend on time) of single-particle states such that *one and only one* of these states is occupied by a finite fraction of all the particles, while the number of particles in any other single-particle state is of order 1 or less. (In technical language: at any given time the one-particle density matrix has exactly one eigenvalue N_0 which is of order N, while all the other eigenvalues are of order unity or less.) The corresponding single-particle wave function $\chi_0(r,t)$ is then called the "condensate wave function," and the N_0 particles occupying it, the "condensate." It is not necessary that the number N_0 be equal to the total number of particles N in the system, even at zero temperature, and indeed it seems almost certain that in real-life liquid 4-He, the T=0 condensate fraction N_0/N is only in the region of 10% (see below).

Just why this state of affairs should be realized is quite a subtle question. First, why should there be macroscopic [$O(N)$] occupation of *any* single-particle state? The only case for which a totally rigorous argument can be given (at least to my knowledge) is the one originally considered by Einstein, namely a completely noninteracting gas in thermal equilibrium. While it can be shown that a calculation that starts from the BEC state of the noninteracting gas and does perturbation theory in the interatomic interactions leads to a finite value (generally less than 100%) of the condensate fraction in thermal equilibrium, there is no general proof that an arbitrary system of Bose particles must show BEC at T=0, and indeed the existence of the solid phase of 4-He is a clear counterexample to this hypothesis. Whether the crystalline solid and the Bose-condensed liquid exhaust the possible T=0 phases of such a system is, as far as I know, an open question. For nonequilibrium states the situation is even less clear.

An even trickier question is why, given that macroscopic occupation occurs, it occurs only in a *single* one-particle state. A relatively straightforward argument shows that, at least within the Hartree-Fock approximation, macroscopic occupation of more than one state is always energetically unfavorable *provided* the effective low-energy interaction is repulsive, as is believed to be the case for 4-He. For the case of an attractive interaction the problem is complicated by the fact that in the thermodynamic limit, as usually understood, ($N\to\infty, V\to\infty; N/V\to$const) the system is unstable against a collapse in real space; for the finite geometries which are of interest in the case of the alkali gases, the issue is, at this time, controversial. Also, even in the repulsive case, it is not entirely obvious that one can exclude "multiple condensates" in certain nonequilibrium conditions.

Given that BEC occurs in the sense defined above, i.e., that at any given time there exists *one and only one* single-particle state $\chi_0(r,t)$ that is macroscopically occupied, the conceptual basis for superfluidity is quite simple. We write $\chi_0(r,t)$

$=|\chi_0(r,t)|\exp i\phi(r,t)$, and define the *superfluid velocity* $v_s(r,t)$ by the prescription

$$v_s(r,t) \equiv \frac{\hbar}{m}\nabla\phi(r,t). \tag{1}$$

This immediately leads to the result $\nabla\times v_s = 0$, i.e., the "superfluid" flow is irrotational. Moreover, we observe that no "ignorance" is associated with the single state χ_0, and thus the entropy must be carried entirely by the "normal" component, i.e., the particles occupying single-particle states other than χ_0. (Obviously, this argument can be made more precise.) These two observations provide the basis for Landau's phenomenological two-fluid hydrodynamics. However, it should be emphasized that the "superfluid density" ρ_s, which occurs in the latter is, in general, *not* simply given by N_0/V, where N_0 is the number of particles condensed into χ_0; indeed, in the case of liquid 4-He, it is believed that as T→0, ρ_s tends to the total density N/V, while N_0 remains only about 10% of N.

For a simply connected region of space in which $|\chi_0|$ is everywhere nonzero, the application of Stokes' theorem to the curl of Eq. (1) leads at once to the conclusion that the integral of v_s around any closed curve is zero. A more interesting application of Eq. (1) is to the case in which there is a line, or more generally, a region infinite in one dimension, on which $|\chi_0(r,t)|$ vanishes. This may happen either because the liquid is physically excluded from this region, as in the example considered below, or because, while atoms are present in the region in question, the particular single-particle state into which BEC has taken place happens to have a nodal line there. In either case we can consider the integral of Eq. (1) around a circuit that encloses the one-dimensional region in question, while we are no longer entitled to use Stokes' theorem to conclude that this integral is zero, the fact that the phase of the wave function χ_0 must be single-valued modulo 2π leads to the *Onsager-Feynman quantization condition*

$$\oint v_s \cdot dl = nh/m. \tag{2}$$

In a region of space which is, from a purely geometrical point of view, simply connected, Eq. (2) can be satisfied by a "vortex," that is, a pattern of flow in which $v_s \sim 1/r$, where r is the perpendicular distance from the "core"; the singularity which formally appears at the core is physically irrelevant because by hypothesis $|\chi_0|$ vanishes there and thus v_s is not defined. The statics and dynamics of vortices is, of course, a subject that has been extensively studied in *classical* hydrodynamics; but in that case the circulation, while independent of path, can take any value and, in addition, vortices tend to be stable only under nonequilibrium conditions. By contrast, in a superfluid system the circulation is quantized according to Eq. (2) (it is actually found that the only values of n of interest are $n = \pm 1$, since vortices with higher values of n are unstable against decay into these), and in addition, for reasons we shall see, vortices can be metastable, even under equilibrium conditions, for essentially astronomical times.

The most interesting application of Eq. (2), and the most clear-cut definition of the various phenomena which together constitute what we call superfluidity, occurs in a literally multiply connected geometry, let us say for definiteness the annular region between two concentric cylinders. In the following I consider such a geometry, with the mean radius of the annulus denoted R and its thickness d taken small compared to R; I neglect corrections of relative order d/R. The superfluid velocity $v_s(rt)$ is not itself a directly observable quantity, and in practice we are interested in the value of the mass current $J(rt)$. With Landau we argue that in (stable or metastable) equilibrium this quantity should be given by an expression of the form

$$J(r,t) = \rho_s v_s(rt) + \rho_n v_n(rt), \qquad (3)$$

where the "superfluid" and "normal" densities ρ_s and $\rho_n \equiv \rho - \rho_s$ are functions only of temperature, and where the "normal velocity" $v_n(rt)$ is assumed to behave just like the velocity field of a normal (nonsuperfluid) liquid; in particular, in equilibrium v_n should be zero in the frame of reference in which the walls of the vessel are at rest. In the following I mean by the scalar quantities v, v_s, and J, the tangential (circumferential) components of the respective vectors.

Consider two different thought-experiments, in each of which the cylinders are rotated synchronously with angular velocity omega; we note from Eq. (2) that a natural unit in which to measure omega is the angular velocity corresponding to $n=1$, that is $\omega_c = \hbar/mR^2$. In the first experiment, we start with the liquid above the lambda-temperature T_λ, rotate the cylinders with some *small* angular velocity ω and wait for thermal equilibrium to be established. Since for $T > T_\lambda$ the helium behaves like any other ("normal") liquid, e.g., H_2O, we see that in the rotating equilibrium the fluid velocity will be simply ωR and the total angular momentum $I_{cl}\omega$, where the classical moment of inertia I_{cl} is just NmR^2. Now, while continuing to rotate the container, we cool the liquid through T_λ. Below T_λ, the "superfluid fraction" is finite and moves, according to Eq. (3), with the superfluid velocity v_s. However, v_s is constrained by the quantization condition and in general *cannot* be taken equal to ωR. In fact, a simple statistical-mechanical argument shows that the lowest free energy is obtained when n takes the value closest to ω/ω_s; for $\omega \ll \omega_c$ this is obviously zero. Consequently, in Eq. (3) the superfluid component no longer contributes to the circulating current. Meanwhile, the quantity v_n is still given by ωR, and consequently the total angular momentum is reduced by a factor $\rho_n(T)/\rho$. Thus by ramping the temperature up and down below T_λ, the angular momentum can be *reversibly* increased or decreased; in particular, for $T \to 0$ it tends to zero in the laboratory frame (or more accurately in frame of the fixed stars) even though the vessel is still rotating. At larger values of $\omega (> \omega_c/2)$ the superfluid will contribute to the angular momentum an amount $\sim n\omega_c$, where n is the nearest integer to ω/ω_c; thus for $\omega = 0.75\omega_c$, for example, the apparent velocity of the liquid may *exceed* that of the container. This remarkable effect, which turns out to be a close analog of the Meissner effect in superconductors, was originally predicted by F. London and eventually observed (in effect) by Hess and Fairbank in 1967; it is essential to appreciate that it is a manifestation of the *equilibrium* behavior of the system and has nothing to do with long relaxation times.

A second experiment, which is at first sight, closely related to the above but is conceptually quite different, goes as follows: we again start above T_λ, but now with the liquid rotating at a much higher angular velocity $\omega \gg \omega_c$, so that, as above, velocity v is ωR. We next cool, still rotating, through T_λ; according to the prescription given above, the superfluid component will take the quantized value of circulation which makes n closest to ω/ω_c; but since ω/ω_c is very large this means that the fractional change is proportional to ω_c/ω and in practice unobservably small, and the angular momentum is to all intents and purposes $I_{cl}\omega$. Finally, still keeping the temperature below T_λ, we stop the rotation of the container. What happens?

It should be strongly emphasized that in contrast to the "Hess-Fairbank" experiment discussed above, the present problem does not concern the nature of the thermodynamic equilibrium state under the new (final) conditions; the latter rather obviously corresponds to zero circulating current. Rather, the question concerns the *degree of metastability* of the circulating-current state. In practice we find that when we stop the rotation, the contribution of the normal component to Eq. (3) rapidly relaxes to zero, but the superfluid contribution persists for a time, which, except under very special conditions, is effectively infinite, and moreover can be reversibly increased or decreased by sweeping the temperature up and down (but never allowing it to exceed T_λ). In other words, the system preserves the value of the superfluid circulation [Eq. (2)] that it originally had, even though it is clearly not the equilibrium one. This is the phenomenon of *metastable superflow*, which should be carefully distinguished from the (equilibrium) Hess-Fairbank effect. Unfortunately, the term "persistent currents," frequently used in the literature, is ambiguous and tends to confuse these two conceptually very different effects. It is amusing that the phenomenon of "frictionless flow" originally discovered by Kapitza, Allen, and Misener may, depending on the parameters, be a manifestation of either of these effects.

Unlike the Hess-Fairbank effect, which can be understood at least qualitatively in terms of the behavior of a single atom under the same conditions, a viable explanation of the phenomenon of metastable superflow requires explicit consideration of the effects of the interatomic interactions; indeed, it is believed that a noninteracting Bose gas, even in the BEC state, would *not* display this behavior. Crudely speaking, the argument goes as follows: to go continuously from a state in which a macroscopic number N_0 of atoms occupies the state corresponding to a finite value of n, say n_0, in Eq. (2) to one in which the same N_0 atoms occupy the state $n=0$, we must do one of two things: either we scatter particles one by one out of the state $n=n_0$ and into $n=0$, thereby creating, at intermediate times, a state in which *two* single-particle states are simultaneously macroscopically occupied, or we keep N_0 particles in a single one-particle wave function but modify the latter so as to go continuously from χ_{n_0} at $t=-\infty$ to χ_0 at $t=+\infty$. *Provided there is no extra "internal" quantum number* and the low-energy effective interatomic interaction is repulsive (as is the case for 4-He), it is straightforward to show that for not too large values of n_0 both of these "paths" involve surmounting a free-energy barrier, which except for T extremely close to T_λ, is so enormous that the chance of doing so is negligible even on astronomical timescales. When T is extremely close to T_λ, this energy barrier (which scales as ρ_s and hence vanishes in the limit $T \to T_\lambda$) becomes surmountable with difficulty, and indeed it is found experimentally that there is a measurable relaxation of superflow in this regime.

Thus a theory based on Eqs. (1)–(3) and the considerations of the last paragraph can account not only qualitatively but, as it turns out, quantitatively for the main phenomena of superfluidity in 4-He. (In addition, it predicts other characteristic phenomena, such as the Josephson effect, which have been searched for and found, but there is no space to discuss this topic here.) However, there is one feature of this whole scenario that might leave one with a feeling of slight disquiet: in the sixty years since London's original proposal, while there has been almost universal belief that the key to superfluidity is indeed the onset of BEC at the lambda-temperature it has proved very difficult, if not impossible, to verify the existence of the latter phenomenon directly. The main evidence for it comes from high-energy neutron scattering and, very recently, from the spectrum of atoms evaporated from the surface of the liquid, and while both are certainly consistent with the existence of a condensate fraction of approximately 10%, neither can be said to establish it beyond all possible doubt.

All the above refers to our best-known superfluid, liquid 4-He below the lambda-temperature. However, that is not the end of the story. In 1972 it was discovered that the light isotope of helium, 3-He (which is also liquid under its own vapor pressure down to the lowest temperatures) possesses, below the much lower temperature of 3 mK, not one but three anomalous phases, each of which appears to display most of the properties expected of a superfluid, so that these new phases are usually referred to collectively as "superfluid 3-He." In this case, since the 3-He atom obeys Fermi rather than Bose statistics, the mechanism of superfluidity cannot be simple BEC as in 4-He. Rather, it is believed that, just as in metallic superconductors, the fermions pair up to form "Cooper pairs"—a sort of giant diatomic quasimolecule whose characteristic "radius" is very much larger than the typical interatomic distance—and that these molecules, being composed of two fermions, effectively obey Bose statistics and can thus undergo BEC. However, it should be emphasized that, at least within the context of the traditional theory, the formation of the Cooper pairs and the process of BEC are not two independent phenomena, rather they occur simultaneously and are intimately connected. A microscopic theory that is a generalization of the BCS theory of superconductivity can be constructed for these new phases, and in fact, over the last 25 years has had a remarkable degree of quantitative as well as qualitative success in explaining their properties, to the extent that we can now claim an understanding of these materials which is more quantitative than that which we at present have of the apparently simpler system 4-He.

Although not all the phenomena that accompany the onset of superfluidity in 4-He have been explicitly demonstrated in the low-temperature phases of 3-He, the general pattern is sufficiently similar that there is a fair degree of confidence that the underlying scenario is parallel in the two cases, with the role of the condensate wave function in 4-He being played by the center-of-mass wave function of the Cooper pairs in 3-He. However, there is one very important difference: as well as their center-of-mass degree of freedom, the pairs in 3-He turn out to have also *internal* degrees of freedom; if one thinks of them as like diatomic molecules, they turn out, crudely speaking, to possess total spin $S=1$ and also "intrinsic" orbital angular momentum $L=1$, and the corresponding vectors can be oriented, *prima facie*, in arbitrary directions. (By contrast, the Cooper pairs in traditional superconductors have $L=S=0$ and thus do not possess any interesting internal degrees of freedom.) A crucial aspect of BEC in such a system is that the "condensed" pairs should not only all possess the same center-of-mass wave function, *they should also all behave identically as regards their internal degrees of freedom.*

Now, one might at first sight think that the arguments given regarding the Hess-Fairbank effect and the metastability of superflow which, *prima facie*, refer only to the center-of-mass behavior, would be qualitatively unaffected by the presence or absence of internal degrees of freedom. This is indeed so with regard to the Hess-Fairbank effect, and in one of the three phases (the B phase) it is also true in the context of metastability of superflow. However, with regard to the other two phases, the situation is more intriguing: it turns out that the nature of the internal degree of freedom in these phases is such that once it is taken into account, at least within the simplest approximation, the argument that any attempt to deform the condensate wave function so as to pass continuously from $n = n_0$ to $n = 0$ *no longer holds*, so that within such an approximation superflow is no longer stable for $|n| > 1$. (For $n = \pm 1$, for a subtle reason, it is still metastable.) In real life superflow does appear to be metastable in all the new phases, but both the experimental and the theoretical situation is considerably more complicated than in 4-He.

Finally, it may be remarked that there now exists a third electrically neutral laboratory system, which is generally expected to show behavior characteristic of a superfluid, namely various monatomic alkali gases (87-Rb, 23-Na, 7-Li, and also, very recently, 1-H) at ultralow temperatures. These atoms possess an odd number of electrons, thus an even total number of fermions, and so should obey Bose statistics, and under appropriately extreme conditions, display the phenomenon of BEC and the resulting superfluid behavior. However, because of the nature of the "confinement" of these systems (usually by magnetic or laser traps) the situation with regard to BEC and superfluidity is reversed with respect to 4-He: The onset of BEC should be spectacular in the form of a dramatic change in the density profile, while that of superfluidity should be much more subtle and difficult to observe. Indeed, since June 1995, many experiments have seen such a change of profile, or closely related effects, in these systems at μK or nK temperatures, and their low-temperature states universally believed to exhibit BEC; but, at least at this time, the evidence for superfluidity is still quite circumstantial.

References

Feynman, R. P., 1955, *Applications of Quantum Mechanics to Liquid Helium*, Progress in Low Temperature Physics, Vol. I, edited by C. J. Gorter and D. F. Brewer (North-Holland, Amsterdam), p. 17.
Gorter, C. J., and D. F. Brewer, Eds., 1955, *Progress in Low Temperature Physics* (North-Holland, Amsterdam).
Keesom, W. H., 1942, *Helium* (Elsevier, New York/Amsterdam).
London, F., 1950, *Superfluids I: Macroscopic Theory of Superconductivity* (Wiley, New York).
London, F., 1954, *Superfluids II: Macroscopic Theory of Superfluid Helium* (Wiley, New York).
Vollhardt, D., and P. Wölfle, 1990, *The Superfluid Phases of Helium 3* (Taylor & Francis, London).
Wilks, J., 1967, *The Properties of Liquid and Solid Helium* (Clarendon, Oxford).

In Touch with Atoms
G. Binnig and H. Rohrer

Quantum mechanics has dramatically changed our perception of atoms, molecules, and condensed matter and established the central role of electronic states for electronic, chemical, and mechanical properties. Electronics, understood broadly as the motion of electrons and the deformation of their arrangements, has become the basis of our high-tech world, including "electronics," computer science, and communications. Mechanics, on the other hand, understood as the motion of the mass of atomic cores and the deformation of their arrangements, played a lesser role, at best that of the guardian of the electron. Quantum mechanics has become, for many, synonymous with electronic states and electronics, whereas mechanics is considered the Stone Age. In this respect, the "mechanical" scanning tunneling microscope (STM) came as a surprise. The STM is a mechanically positioned, electrically sensitive kind of nanofinger for sensing, addressing, and handling individually selected atoms, molecules, and other tiny objects and for modifying condensed matter on an atomic scale (Sarid, 1991; Güntherodt and Wiesendanger, 1992; Chen, 1993; Stroscio and Kaiser, 1993; Hamers, Weaver, Weimer, and Weiss, 1996). And like with finger tips, it is the "touch" that makes the difference (see Fig. 1). Back to the future of mechanics: Nanomechanics, a new era.

The STM emerged as a response to an issue in technology. [For a historical review of STM see Binnig and Rohrer (1987a, 1987b).] Inhomogeneities on the nanometer scale had become increasingly important as the miniaturization of electronic devices progressed. Condensed matter physics, on the other hand, was occupied predominantly with periodic structures in solids and on surfaces and thus had developed very successfully momentum-space methods and concepts for the nanometer scale. Inspired by the specific problem of inhomogeneities in thin insulating layers — a central challenge to our colleagues working on the development of a computer based on Josephson tunnel junctions — and realizing the general scientific significance associated with it, we started to think in terms of local phenomena. Tunneling appeared a natural and promising solution. This was the beginning of a new approach to the nanometer scale, the local-probe methods.

Local probes are small-sized objects, usually the very end of a sharp tip, whose interactions with a sample or a field can be sensed at selected positions. Proximity to or contact with the sample is required for good resolution. This is in principle an old concept, the medical doctor's stethoscope being a well-known example. "Small sized" in this case means small compared to the wavelength of the sound to be heard and comparable to the distance from the sound source. The local-probe concept even appeared sporadically in the scientific literature in context with electromagnetic radiation (Synge, 1928, 1932; O'Keefe, 1956; Ash and Nicolls, 1972), but met with little interest and was not pursued. Nanoprobes require atomically stable tips and high-precision nanodrives. The latter are based on mechanical deformations of springlike structures by given forces—piezoelectric, mechanical, electrostatic, or magnetic—to ensure continuous and reproducible displacements with precision down to the picometer level. They also require very good vibration isolation. Furthermore, the concept of contact—electrical or mechanical—blurs at the nanometer scale. In the case of electrical contact, no sharp boundaries exist because of the penetration of electronic wave functions into the potential barriers of finite height, giving rise to electron tunneling (Bardeen, 1961; Güntherodt and Wiesendanger, 1992; Chen, 1993; Stroscio and Kaiser, 1993). On the other hand, interference and quantum effects can lead to discontinuities like in quantum conduction (Imry and Landauer, 1998).

The resolution f of local-probe methods is given mainly by an effective probe size r, its distance from the object d, and the decay of the interaction. The latter can also be considered to create an effective aperture, e.g., by selecting a small feature of the overall geometry of the probe tip, which then corresponds to the effective probe. If the decay in the distance range of interest can be approximated by an exponential behavior, $\exp(-x/l)$, with an effective decay length l, a good approximation of the resolution is $f = A\sqrt{(r+d)l}$, where A is of order unity [e.g., $A \simeq 3$ for a spherical STM tip of radius r and electronic s-wave functions (Tersoff and Hamann, 1983)]. Atomic resolution

Figure 1. Principle of a local probe: The gentle touch of a nanofinger. If the interaction between tip and sample decays sufficiently rapidly on the atomic scale, only the two atoms that are closest to each other are able to "feel" each other.

therefore requires probe size, proximity, and decay length, respectively, of atomic dimension.

In STM, the interaction can be described as the wave-function overlap of empty and filled states of a tip and sample, respectively, or vice versa, which leads to a tunnel current when a voltage is applied (Bardeen, 1961). The interaction and, thus the tunneling current, decay exponentially with increasing separation with a decay length l (nm) $\simeq 0.1/\sqrt{\phi_{eff}}$ for free electrons, where ϕ_{eff} is the effective tunnel barrier. For electrons at the Fermi energy with momentum perpendicular to the tunnel barrier, ϕ_{eff} is the average of sample and tip work functions $\bar{\phi}$. For most tip and sample materials, $\bar{\phi}$ is 4 to 6 eV and thus $l \simeq 0.05$ nm.

This short decay length ensures that the tunnel current is carried mainly by the frontmost atom of the tip, which thus represents a local probe of atomic dimensions as depicted in Fig. 1. For a tunnel current in the nanoampere to picoampere range, the distance has to be less than 1 nm. This leads in a natural way to atomic resolution, provided that tips and samples are mechanically and chemically stable. In other words, once tunneling was chosen, atomic resolution was inevitable. Fast fluctuations owing to thermal excitations such as phonons or the diffusion of atoms are largely averaged out. Therefore STM can be operated at elevated temperatures or in ambient or liquid environments with an acceptable signal-to-noise ratio.

The STM is an electronic-mechanical hybrid. The probe positioning is mechanics, whereas the interaction is sensed by the tunneling current, which is of quantum-mechanical origin. The most common imaging mode is the constant interaction or the "mechanical" mode in which a feedback loop adjusts the probe position with respect to the sample, say in the z direction, to a given tunneling current while scanning in the x-y direction over the surface. The x-y-z positions of the probe, i.e., the image, represent a contour of constant tunnel current or of whatever the tunnel current can be related to, e.g., in many cases a contour of constant local density of electronic states. On smooth surfaces, faster imaging can often be achieved by measuring the tunneling current while scanning on a given, smooth x-y-z contour, e.g., a plane parallel to an average surface portion, which is then called constant-height mode. For very weak interaction, i.e., for tunneling currents at or below 1 pA, the imaging speed is, however, limited by the current measuring bandwidth, not by the mechanical system response.

In view of its conceptional simplicity, the fact that no new theoretical insight and concepts nor new types of materials or components were required, and the prospect of a fundamentally new approach to the nanometer scale, it is remarkable that the STM—or local probes in general for the nanometer scale—was not invented much earlier. And when it was, it did not happen in one of the obvious communities. And after it was, it took several years and an atomically resolved, real-space image of the magical Si(111) 7×7 reconstruction (Binnig, Rohrer, Gerber, and Weibel, 1983) to overcome the reservations of a skeptical and sometimes rather conservative scientific community. There were certain exceptions, though.

Colorful Touch

The "touch" of a local probe with the nano-object is essentially given by the *type* of interaction, which addresses a distinct property, process, or function, by the *strength* of the interaction, which can make a tool out of the probe, and by the *medium*, such as liquid, ambient or vacuum, that provides the specific local "atmosphere," the "nanosphere." A colorful touch, indeed, with a rainbow of possibilities. However, to have a large variety of interactions at one's disposal is one thing, to differentiate among them is another. Ideally, one would like to control the position of the probe and guide it over a specific, easily understandable contour, such as the object's topography, by means of a control interaction and to work with other interactions, the working interactions, for addressing or changing the properties, processes, and functions of interest. Admittedly, the topography is a fuzzy concept on the nanometer scale but it nevertheless might be a useful one in many cases. But even if an appropriate control interaction can be found, one must still separate the working interactions because the measuring signal includes the effect of an entire class of interactions. In STM, the contribution of the electronic states to the tunneling current depends on their energy, momentum, symmetry, and density as well as on the tunneling barrier height and width and the tunneling process. Or the response of a force sensor is the composite action of different forces. Separating the interactions for imaging and working is the challenge and the art, the "touch," of working with local-probe methods.

Tunneling spectroscopy is the major technique to separate the contributions of the various electronic states and various tunneling processes to the tunneling current in order to associate specific image features with a characteristic surface property or process [see the chapters on spectroscopy in Güntherodt and Wiesendanger (1992), Chen (1993), and Stroscio and Kaiser (1993)]. The tunneling current due to those electronic states that are homogeneous on the surface and reflect in many cases the total density of states is usually used as the control interaction. In this case, the contour traced can be regarded as the surface topography. At each measuring point on the topography, the electronic states of interest are extracted by the tunneling current in the appropriate voltage, i.e., energy window. This is done in practice with various techniques [see Hamers, Weaver, Weimer, and Weiss (1996)]. There are many other ways to extract information from a local tunneling experiment. Ballistic electron-emission microscopy (BEEM) tests buried potential barriers; distance-current characteristics yield tunnel barrier heights and decay lengths of the electronic wave functions. Emitted photons owing to inelastic tunneling processes (Lambe and McCarthy, 1976; Gimzewski, 1995) are characteristic of local excitations such as surface plasmons, of local densities of states, of energy levels of adsorbed molecules, of electron-hole pair formation and recombination, and of spin polarization. Very powerful light emission from a Cu surface covered with polyaromatic molecules equipped with molecular spacers has recently been observed within an STM configuration (Gimzewski, 1998). The estimated conversion rate is as high as 30% or 10^8 photons/sec from a volume of several cubic nanometers, although inelastic-tunneling processes are otherwise lower than the elastic ones by four to six orders of magnitude.

STM was followed by the scanning near-field optical microscope (SNOM) [Pohl, Denk, and Lanz, 1984) and the atomic force microscope (AFM) (Binnig, Gerber, and Quate, 1986; Sarid, 1991) with all its different force derivatives. In the SNOM a photon current is a measure of the interaction. Although it extended the resolution of optical microscopy far beyond the diffraction limit and offers the power of optical spectroscopy, it did not arouse widespread attention. The atomic-resolution capability of STM appeared to be a serious handicap for SNOM, just like the STM once had to overcome the bias for established surface-science methods. Fortunately, this has changed and SNOM has found its champions and proper place.

A major extension of local-probe methods was brought about by the invention of the AFM. It allows nanometer–resolution, in special cases even atomic–resolution, imaging of conducting and nonconducting objects and local force detection below the picoNewton level. The various forces that are mainly used for imaging are repulsive interatomic, electrostatic, magnetic, van der Waals (all of electronic origin), and lateral (friction) forces (Mate, McClelland, Erlandson, and Chiang, 1987; Güntherodt and Wiesendanger, 1992; Chen, 1993; Stroscio and Kaiser, 1993; see also Hamers, Weaver, Weimer, and Weiss, 1996). The AFM also uses a sharp tip as local probe, but, unlike STM where the tunnel current is a measure of the interaction, a force between tip and sample is detected via the deformation of a spring, generally the bending of a cantilever beam carrying the tip at one end. In the static mode, the excursion of the beam determines the force, in the dynamic mode it is the amplitude and frequency responses of the oscillating cantilever, e.g., a shift of resonance frequency or damping, that are measured and can be used to control the lever position. The force interaction is first transformed into mechanics before being measured. The AFM, therefore, is of an even more mechanical nature than the STM. Today, a large number of deflection sensors yield subangstrom sensitivity; some are electrical and integrated into the lever, others are external. For sensitivity, the beam has to be soft, for vibration protection and to achieve an acceptable imaging speed, its eigenfrequency has to be high. Both requirements can be satisfied by miniaturization in all dimensions because both compliance and resonant frequency increase linearly with decreasing dimension. Microfabricated cantilevers with resonance frequencies above 1 MHz and spring constants below 1 N/m are in use today. Designs are flexible for applications requiring either higher frequencies or lower spring constants.

Shortly after the introduction of the AFM, atomic periodicities—easily confused with atomically resolved structures—were observed. It took a few years, however, before true atomic resolution could be achieved (Ohnesorge and Binnig, 1993). Repulsive forces of the order of only 10^{-10} N between the frontmost atom of the tip and the closest sample atom can deform even a hard sample and tip such that they adapt their shapes to each other. The resolution then is no longer given by the frontmost atom of the tip but rather by its overall radius of curvature. For sharp tips there are nevertheless only a small number of tip atoms in contact with the sample, and periodicities are not completely averaged out. This then simulates atomic resolution, however, with defects either smeared out or not visible at all.

Most scientists operate the AFM in air. In contrast to STM, atomic resolution in air is hardly possible with an AFM. There will be always some humidity present, and therefore the tip and sample will be covered with a water film. As a result, capillary forces will drive the tip with a relatively strong force against the sample. In principle this force can be counterbalanced by pulling the lever away from the sample and prebending it this way. Unfortunately the capillary forces and the maximum tolerable loading forces differ by so many orders of magnitude that a counterbalancing is spoiled by tiny variations in these forces during the scan. Operating the cantilever in vacuum solves the problem. Operating it in water or an aqueous solution also solves this problem and another one: van der Waals forces do not decay as rapidly as tunneling currents and therefore a background attraction of the tip (and not just its very front) is present. In liquids, e.g., aqueous solution, this background attraction can be counterbalanced completely by the van der Waals forces that act on the solution by pulling it into the gap between tip and sample (Garcia and Binh, 1992).

Of prime interest in AFM are the topography, the type of contact, and the local mechanical properties. The dynamic mode is used to address local elastic constants, whereas force-distance curves in and out of contact provide information about contact, intermolecular forces, and binding. Adding a known Coulomb force allows one to separate Coulomb, van der Waals, and magnetic forces. These methods have their counterparts in spectroscopy in STM.

Following the scanning near-field optical microscope and the AFM, a profusion of local-probe techniques using various interactions appeared, each geared to solve a specific class of problems in a given environment. They include Maxwell stress microscopy, ion conductance microscopy, scanning electrochemical microscopy, higher-harmonics generation of microwaves and optical photons, and many others, and more are still appearing. Their adaptability to different types of interactions and working environments is one of the greatest assets of local-probe methods.

Another one is the ease of making a nanometer-scale tool out of a probe (see Fig. 2). Probe or tool is a matter of the strength of the interaction and of the local sensitivity to it. Changing the distance between probe and object by a fraction of a nanometer can change the interaction strength by several orders of magnitude. Alternatively, applying a few volts can result in electric fields of the order of intramolecular fields, which are sufficient to break individual chemical bonds or to initiate a local chemical reaction. A wide variety of local manipulation and modification possibilities are in use, ranging from gentle atom and molecule displacements to their individually selected removal and deposition, to local chemical changes, to brute-force nanometer-sized scratching and chiseling (Güntherodt

Figure 2. STM image of a quantum corral for electrons built with 48 iron atoms on copper. The same tip is used to position the iron atoms into a 12.4-nm-diameter ring and to image them and the wave-structure interior caused by the confined surface-state copper electrons. Courtesy D. Eigler, IBM Research Center, Almaden, CA. (See Color Plate 9.)

and Wiesendanger, 1992; Chen, 1993; Stroscio and Kaiser, 1993; Hamers, Weaver, Weimer, and Weiss, 1996).

Change and Challenge

Since the advent of local-probe methods, atoms, molecules, and other nanometer-sized objects are no longer "untouchables." They forsook their anonymity as indistinguishable members of a statistical ensemble and became individuals. We have established a casual relationship with them, and quite generally with the nanometer scale. Casual, however, does not mean easy. They are fragile individuals, whose properties and functions depend strongly on their context and which are usually quite different from those in the isolated state. Interfacing them to the nanoscopic, microscopic, and macroscopic worlds in a controlled way is one of the central challenges of nanotechnology. Imaging them or communicating with them is the easiest of these tasks, although not always trivial. Besides the effects of immobilization, even weak-electric-contact probes like STM tips are not strictly noninvasive because of the forces present at tunneling distances, in particular when weak contrast or weak electric signals require extreme proximity. Adhesive and electrostatic forces, both from applied voltage and contact potential, can lead to reversible local deformations, in many cases even to irreversible displacements. The latter, undesirable in imaging, has become the basis for atom manipulation (Crommie, Lutz, and Eigler, 1993). The measuring process on the nanoscale is somewhere between an intricate quantum-mechanical one and a straightforward macroscopic one. Generally speaking, the smaller the object or the higher the required resolution, the more delicate the measuring process; and the stronger the required interactions, e.g., for controlling a function or process, the more demanding their control.

The real-space, nano- to atomic-scale resolution of local probes changed our way of thinking and working in many areas and created new ones with distinct properties and behavior on the nanometer scale such as nanotribology, nanoelectrochemistry, and nanomechanics.

In surface science, most of the more complex surface structures and reconstructions could be resolved by STM, often together with other surface-science techniques, and are understood reasonably well. The real-space imaging capability proved to be crucial to unravel the structure of the enlarged unit cell of reconstructions [for an example of the richness of reconstructions see Xue, Hashizume, and Sakurai (1997)]. This is even more so for the study of more local phenomena such as surface structures coexisting on short length scales, nucleation and growth phenomena, heterogeneous catalysis, phase transitions, and surface chemistry. Changes always occur and propagate locally. Still awaited is a general nanoscopic chemical-analysis method.

In electrochemistry, local probes brought in a new era by advancing *in situ* resolution from at best that of optical microscopy for observation and macroscopic for processes to the atomic and nanometer scale, respectively [for a review, see Siegenthaler (1998)]. The significance of working in a liquid environment, however, extends far beyond electrochemistry. The liquid-solid interface is, in our opinion, the interface of the future, at least on equal footing with the solid-vacuum interface of classical surface science. Liquids provide a very adaptive environment for protection, process control, and modification of surfaces, they carry ionic charges and atomic and molecular species, and they remove many of the "traffic restrictions" typical for a two-dimensional solid surface.

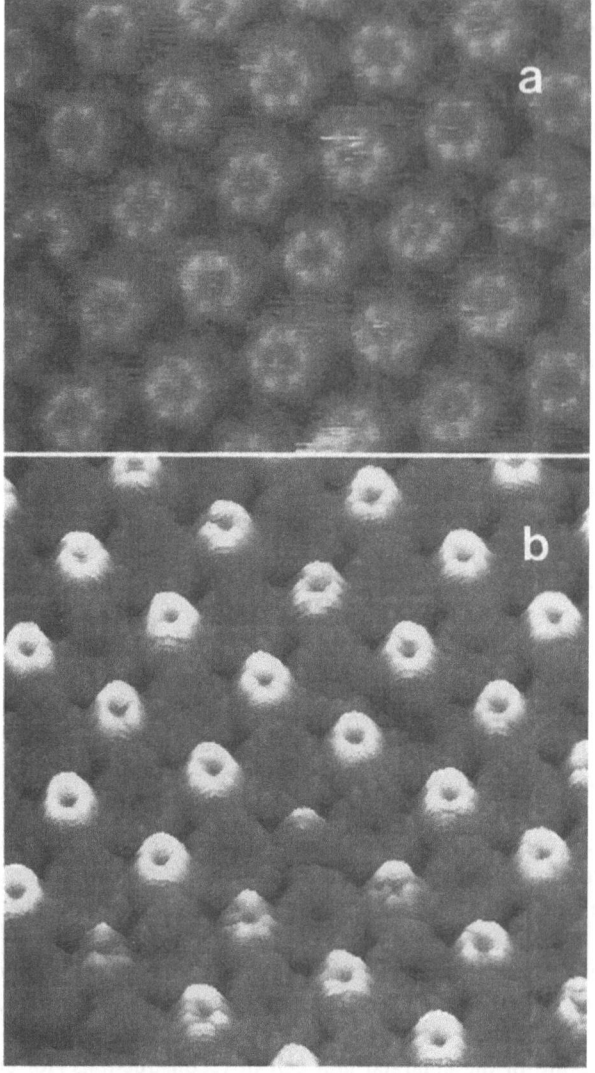

Figure 3. (a) The cytoplasmic surface of the hexagonally packed intermediate (HPI) layer is an essential part of the cell envelope of Deinococcus radiodurans. It is supposed to have a protective function and to act as a molecular sieve. The pores seen in the protruding cores are probably the channels of this sieve, and as shown by AFM for the first time, the channels exhibit two conformations that change dynamically. The unit cell size is 18 nm, and the brightness range corresponds to 3 nm (Müller, Baumeister, and Engel, 1996). (b) Two-dimensional crystals of bacteriophage F 29 head-tail connectors recorded with the AFM in buffer solution. The connectors are packed in up-and-down originations, exposing their narrow ends that connect to the tail and their wide ends that connect to the head. The 12-fold symmetry and vorticity of this complex is clearly demonstrated by this topograph. The unit cell size is 16.5 nm, whereas the brightness range corresponds to 4 nm (Müller et al., 1997). Courtesy A. Engel, Univ. Basel. (See Color Plate 10.)

Ambient environment and liquids are also a key for *in situ* and *in vivo* local-probe methods for macromolecules and biomaterial (Drake *et al.*, 1989). STM and AFM imaging have made good progress, both in problem areas not accessible to other methods as well as complementary to electron microscope imaging (Engel and Gaub, 1997; and references therein) Fig. 3; breakthroughs such as decoding DNA still lie ahead.

In the technology domain, local-probe imaging and measurements in vacuum, at ambient and in liquids, have begun to be applied routinely in the surface analytical sector, where instrumentation is predominantly of the AFM type. The long-range perspective of local probes in general, however, is their use as local sensors, flexible and adaptable tools, and in massive parallel operating devices. Cantilever probes have a special status. Besides their great force and strain sensitivity, they are fast, yielding, and robust. They ensure soft contact in the microNewton to the nanoNewton range. They are, therefore, especially suited for cantilever array applications where fine control of each individual cantilever might be too cumbersome, impractical, or infeasible. Even though the individual local experiments in a specific application might still require μsec to milliseconds, massive parallel operation of cantilevers in batch-fabricated arrays opens new possibilities. Lithography applications (Minne *et al.*, 1998; Wilder *et al.*, 1998) take advantage of very fast "chemics" on the nanometer scale, as diffusion times scale with the square of linear dimension.

An illustrative example of an array application with mechanics is the "Millipede," a mechanical-electronic pocket-sized terabit storage device (Binnig, Rohrer, and Vettiger, 1997; Lutwyche *et al.*, 1998). It consists essentially of a two-dimensional array of smart AFM cantilever beams with integrated read, write, and actuation capabilities, which are addressed and controlled via a multiplex scheme. With feasible bit-space requirements of 30×30 nm^2, e.g., indentations in a polymer storage medium, a million cantilevers serve one terabit on 3×3 cm^2. At realistic read and write speeds of 10 to 1000 kbit/sec, the data-transfer rate is limited by the multiplex speed rather than by mechanics. The architecture of the Millipede solves two basic issues in miniaturization to the nanometer scale, namely the effective-space requirement and the deterioration of signal strength. The degree of miniaturization is determined by the active part and the periphery, which is necessary to build a functional element. The periphery often becomes the space-limiting requirement for nanometer-scale active parts. In the Millipede, the periphery, i.e., the smart cantilever, is of the same size as the megabit it addresses. The effective miniaturization is, therefore, given by the bit size. Secondly the read/write signal can be prepared during the multiplex cycle. In spite of the enormous data-transfer rate, the signal deterioration due to both decreasing bit size and increasing read/write speed is greatly reduced. Most exciting, however, is the prospect of new approaches for combining data storage and *in situ* processing.

The prime activity in local-probe methods focused initially on super-resolution and the understanding of imaging, manipulation, and modification processes. Interest is now expanding to high-sensitivity measuring processes, which often require a tradeoff between resolution and sensitivity/precision, to local-probe systems, and generally to include more complexity in local probes and systems of them such as the cantilever arrays mentioned above. Combining spin-resonance techniques with magnetic force microscopy introduces the possibility of unprecedentedly high-resolution spin-resonance imaging (Sidles, 1991; Rugar *et al.*, 1994). Often, however, imaging merely serves to determine the appropriate position for the experiment or is not used at all. Studies performed predominantly in STM configurations include electron transfer through individual molecules and other nano-objects, frequency mixing using nonlinear tunnel characteristics, multiprobe systems for correlation and local resistivity measurements, and quantum transport through and mechanical properties of metallic nanoconstrictions—the latter in an AFM configuration with conducting tip. Functionalized cantilever-type force sensors are used to detect forces in the picoNewton to femtoNewton range as well as ultrasmall strains produced on the beam itself. Examples are molecular recognition via the binding behavior between two selected molecules, one of which is attached to the tip, measurement of reaction heat in the femtojoule to picojoule range on a functionalized bimorph (a double-layer lever of silicon and aluminium with very different thermal expansion) cantilever, or detection of dilution in the (10^{-18} mol) range due to the strain induced by adsorbed molecules. Smallness comes to bear in three ways: picometer deflection detection brings the extreme sensitivity, and the small dimensions of the cantilever yield short response times and allow nearly noninvasive local sensing. There are still many other uses of cantilevers, e.g., the water meniscus which can form at ambient conditions between tip and surface and which is undesirable in imaging, can serve attomol chemistry in modification processes (Garcia, Calleja, and Perez-Murano, 1998), or the nonlinear coupling of cantilevers can be used for mechanical processing. It is amazing how much can be done and how much potential lies in a primitive cantilever, when it is small enough and properly functionalized.

Local probes play a crucial role in our understanding of how to create an interface to molecular and biofunctional units and, quite generally, they pave the way to building problem-specific nanosystems. In many cases, they might not be the final word, but act merely as a midwife for new experimental approaches and novel technological devices.

Nature's Way

Problem-specific nanosystems allow us to work on the same scale as nature does. Nature has built life on nanofunctionality, the ultimate purpose of nanotechnology. Sensing, processing, actuation, and growth take place on the nanometer scale and are joined in intricate ways to macroscopic properties, processes, and functions.

Nature uses mechanics abundantly and generally does not even separate it from electronics. Nanomechanics has many attractive features: energies required to produce the deformations useful for sensing and actuation are in the thermal energy (kT) range, strains obtained from bending scale with thickness, mechanical eigenfrequencies reach megahertz to gigahertz values and can be adapted to the problem, e.g., low attempt frequencies for transitions, and diffusion times come down to μsec to picoseconds. Nature's nanomechanics rests predominantly on deformation and on the transport of atoms, molecules, small entities, and ionic charges, in contrast to translation and rotation in macromechanics. Simple deformations on the nanometer scale can be synthesized to create complex macromotions. Finally, the small energies required for local activation, sensing, and processing can be provided by distributed chemical-energy reservoirs.

"Distributed" seems to be Nature's general approach to solving so many tasks much more elegantly, efficiently, and successfully than we can do or even attempt to do with present-day macroinstrumentation, central processing, and computation. The nanometer-scale elements allow all kinds of densely interwoven, distributed storage, programming and processing; software is built into the hardware. The same should become true for lifting disciplinary boundaries. The nanoscale is the bifurcation point where materials develop their properties and the science and engineering disciplines their particularities in thinking, working, and terminology. Coming down from the macro and micro scales, the nanometer scale should become the merging point. This then could also be the starting point for the human bottom-up approach to functionality—Nature's way.

References

Appl. Phys. A **66**.
Ash, E.A., and G. Nicolls, 1972, Nature (London) **237**, 510.
Bardeen, J., 1961, Phys. Rev. Lett. **6**, 57.
Binnig, G., and H. Rohrer, 1987a, "Scanning Tunneling Microscopy—from Birth to Adolescence" (Nobel Lecture), in *Les Prix Nobel 1986* (The Nobel Foundation, Stockholm), p. 85. Reprinted in Rev. Mod. Phys. **59**, 615 (1987).
Binnig G., and H. Rohrer, 1987b, Angew. Chem. Int. Ed. Engl. **26**, 606.
Binnig G., Ch. Gerber, and C.F. Quate, 1986, Phys. Rev. Lett. **56**, 930.
Binnig, G.K., H. Rohrer, and P. Vettiger, 1997, "Mass-Storage Applications of Local Probe Arrays," Patent No. WO97/05610 (February 13, 1997).
Binnig, G., H. Rohrer, Ch. Gerber, and E. Weibel, 1983, Phys. Rev. Lett. **50**, 120.
Chen, C. J., 1993, *Introduction to Scanning Tunneling Microscopy* (Oxford University Press, New York).
Crommie, M.F., L.P. Lutz, and D.M. Eigler, 1993, Science **262**, 218.
Drake, B., C.B. Prater, A.L. Weisenhorn, S.A.C. Gould, T.R. Albrecht, C.F. Quate, D.S. Cannell, H.G. Hansma, and P.K. Hansma, 1989, Science **243**, 1586.
Engel, A., and H.E. Gaub, 1997, Eds. "Imaging and Manipulating Biological Structures with Scanning Probe Microscopies," J. Struct. Biol. **119**, 83.
Garcia, N., and V.T. Binh, 1992, Phys. Rev. B **46**, 7946.
Garcia, R., M. Calleja, and F. Perez-Murano, 1998, Appl. Phys. Lett. **72**, 2295.
Gimzewski, J., 1995, *Photons and Local Probes*, NATO ASI Series E: Applied Sciences, edited by O. Marti, and R. Möller (Kluwer, Dordrecht), Vol. 300, p. 189.
Gimzewski, J.K., 1998, private communication.
Güntherodt, H.-J., and R. Wiesendanger, 1992, Eds., *Scanning Tunneling Microscopy*, Vols. I-III (Springer, Berlin).
Hamers, R., M. Weaver, M. Weimer, and P. Weiss, 1996, Eds., "Papers from the Eighth International Conference on Scanning Tunneling Microscopy/Spectroscopy and Related Techniques," J. Vac. Sci. Tech. **14**, 787.
Imry, Y., and R. Landauer, 1999, Rev. Mod. Phys. **71**, 306; pp. 515–525 in this book.
Lambe, J., and S.L. McCarthy, 1976, Phys. Rev. Lett. **37**, 923.
Lutwyche, M., C. Andreoli, G. Binnig, J. Brugger, U. Drechsler, W. Haeberle, H. Rohrer, H. Rothuizen, and P. Vettiger, 1998, Sensors and Actuators A, in press.
Mate, C.M., G.M. McClelland, R. Erlandson, and S. Chiang, 1987, Phys. Rev. Lett. **59**, 1942.
Minne, S.C., G. Yaralioglu, S.R. Manalis, J.D. Adams, A. Atalar, and C.F. Quate, 1998, Appl. Phys. Lett. **72**, 2340.
Müller, D.J., W. Baumeister, and A. Engel, 1996, J. Bacteriol. **178**, 3025.
Müller, D.J., A. Engel, J.L. Carrascosa, and M. V'elez, 1997, EMBO J., **16**, 2547.
Ohnesorge, F., and G. Binnig, 1993, Science **260**, 1451.
O'Keefe, J.A., 1956, J. Opt. Soc. Am. **46**, 359.
Pohl, D.W., W. Denk, and M. Lanz, 1984, Appl. Phys. Lett. **44**, 651.
Rugar, D., O. Züger, S. Hoen, C.S. Yannoni, H.-M. Veith, and R.D. Kendrick, 1994, Science **264**, 1560.
Sarid, D., 1991, *Scanning Force Microscopy* (Oxford University Press, New York).
Sidles, J.A., 1991, Appl. Phys. Lett. **58**, 2854.

Siegenthaler, H., 1998, in *Forum on Nanoscience and Technology*, NATO ASI Series E: Applied Sciences, edited by N. Garcia (Kluwer, Dordrecht), in press.
Stroscio, J.A., and W.J. Kaiser, 1993, Eds., *Scanning Tunneling Microscopy, Methods of Experimental Physics*, Vol. 27 (Academic, New York).
Synge, E.H., 1928, Philos. Mag. **6**, 356.
Synge, E.H., 1932, Philos. Mag. **13**, 297.
Tersoff, J., and D.R. Hamann, 1983, Phys. Rev. Lett. **50**, 1998.
Wilder, K., B. Singh, D.F. Kyser, C.F. Quate, 1998, J. Vac. Sci. Tech. B **16** (in press).
Xue, Q., T. Hashizume, and T. Sakurai, 1997, Prog. Surf. Sci. **56**, 1.

Color Plates

Color Plate 1 (see page 92)

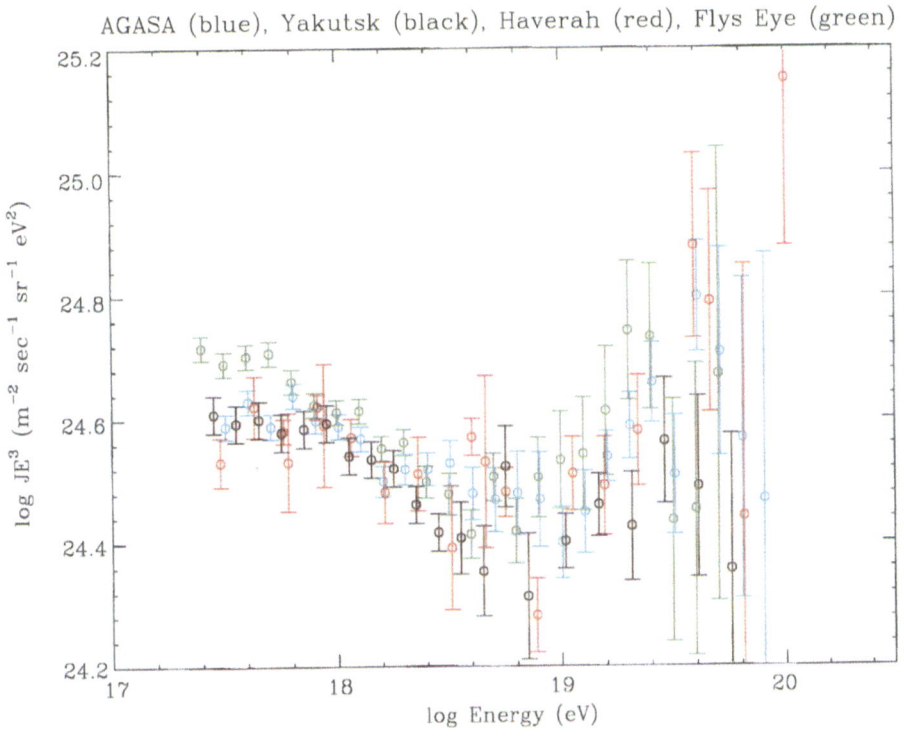

Color Plate 2 (see page 283)

Color Plates

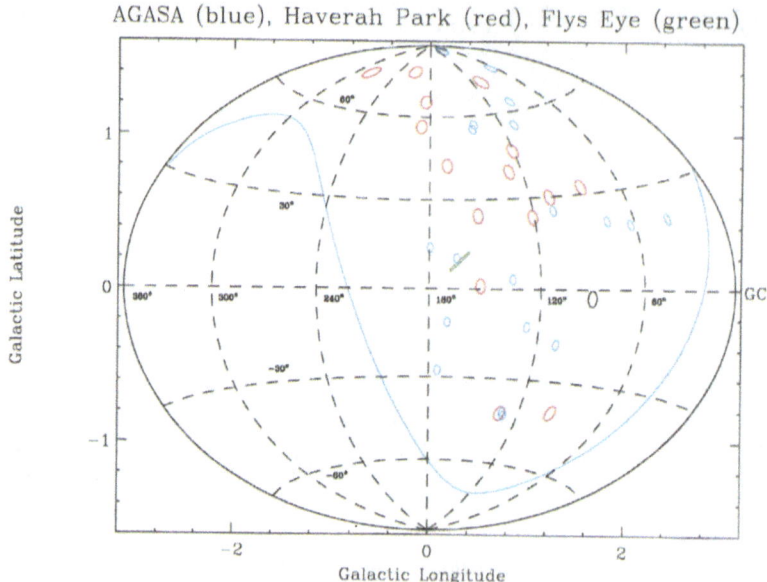

Color Plate 3 (see page 287)

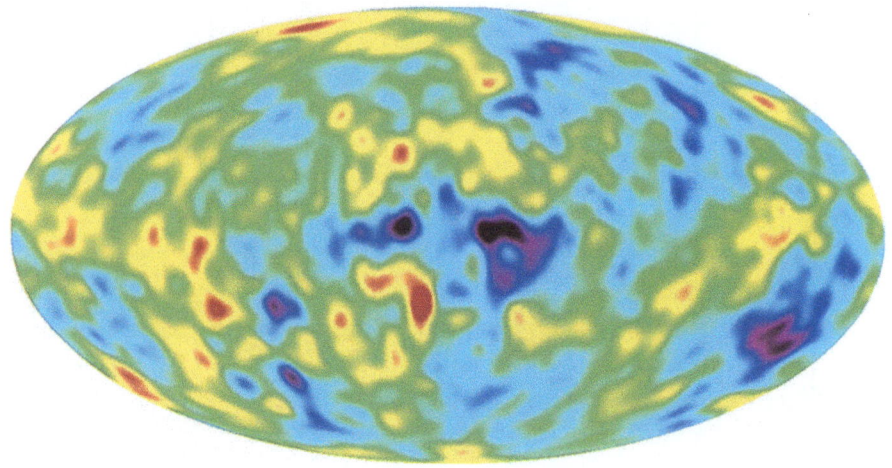

Color Plate 4 (see page 295)

Color Plates

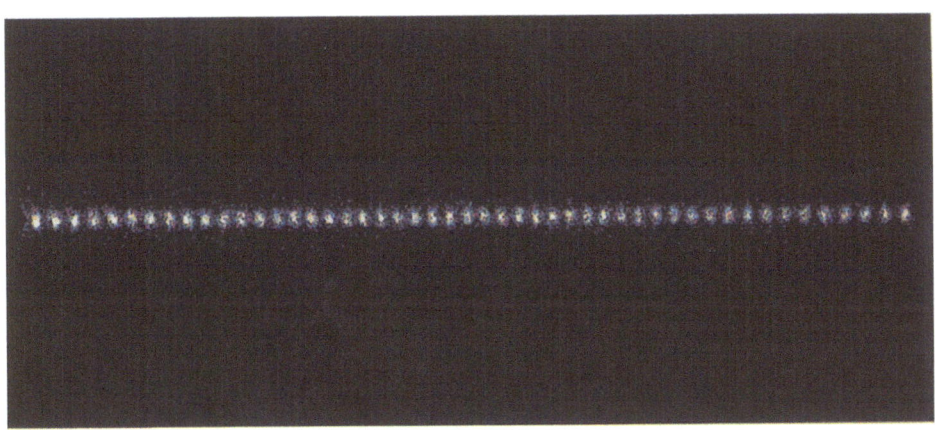

Color Plate 5 (see page 423)

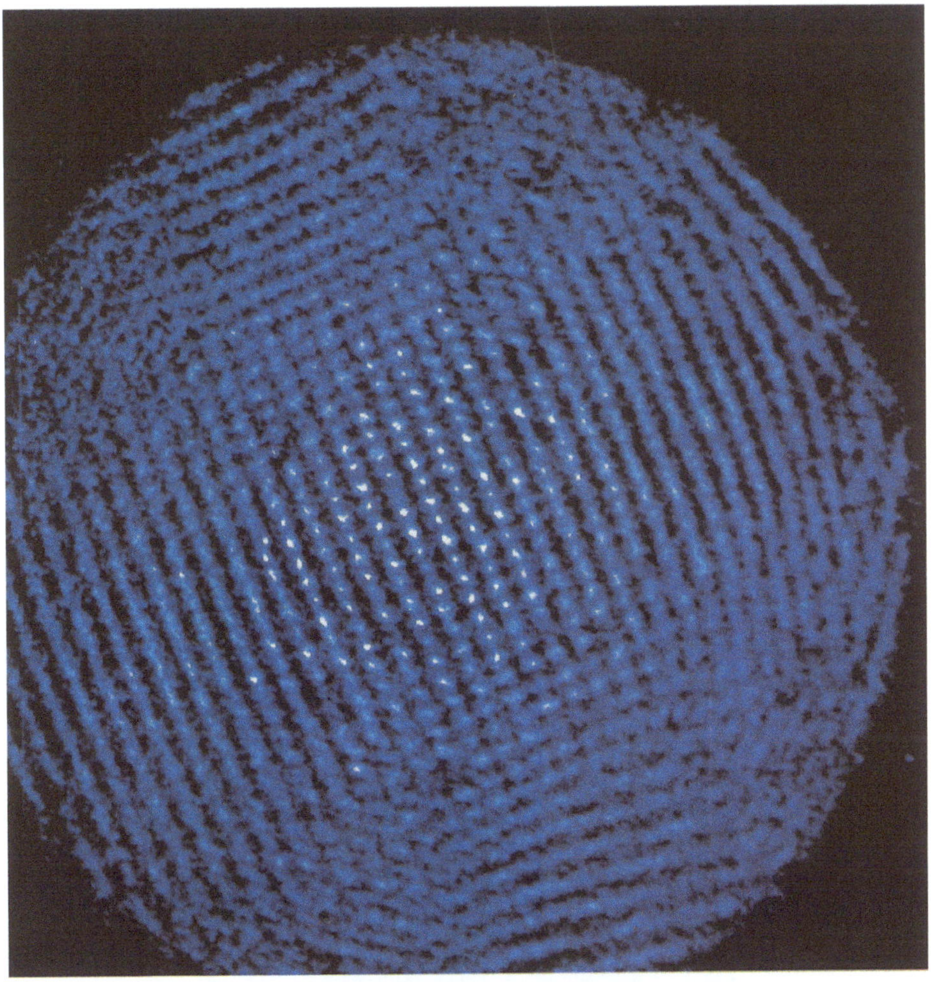

Color Plate 6 (see page 430)

Color Plates

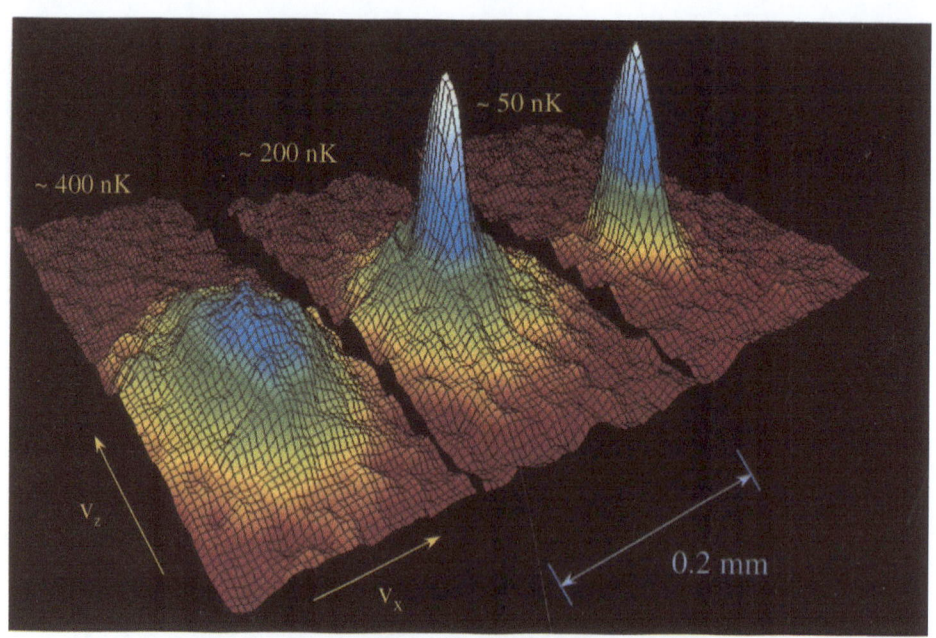

Color Plate 7 (see page 432)

Color Plate 8 (see page 493)

Color Plates

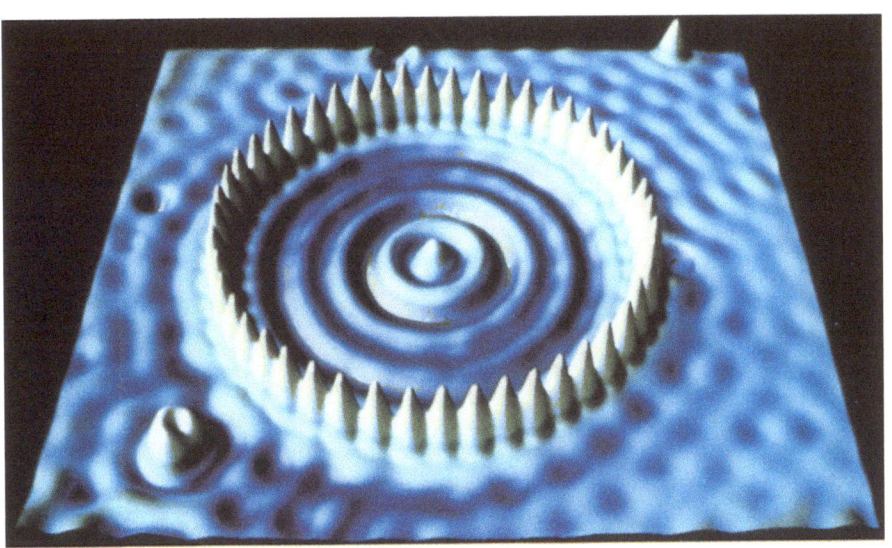

Color Plate 9 (see page 548)

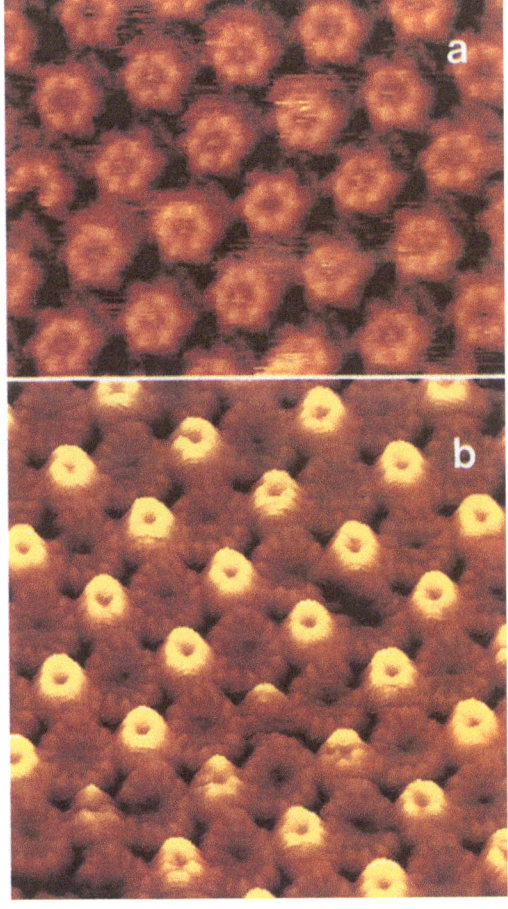

Color Plate 10 (see page 550)

Color Plates

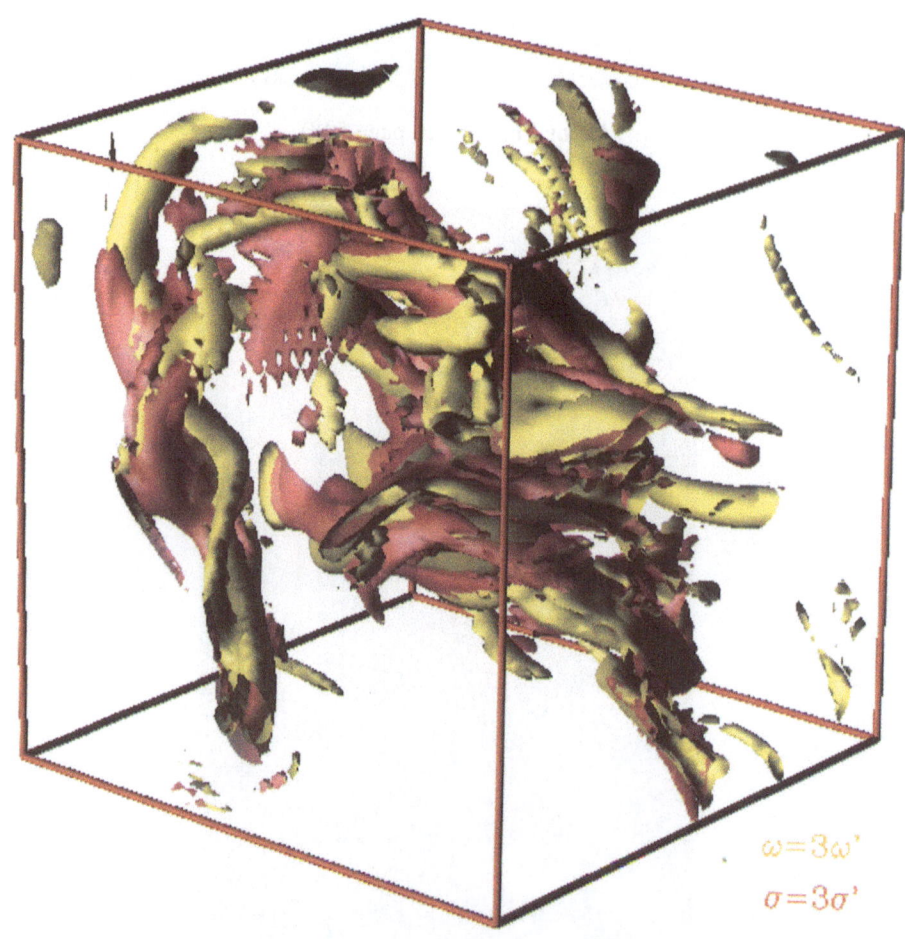

Color Plate 11 (see page 658)

Color Plates

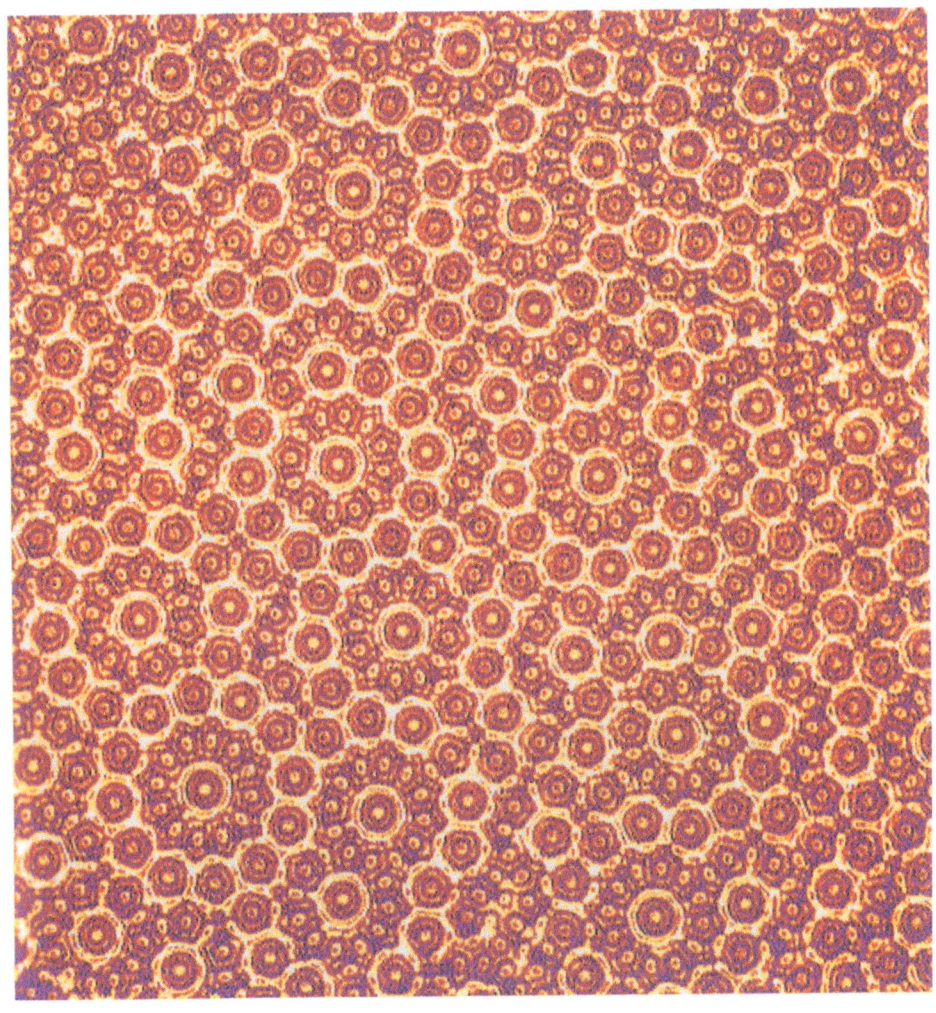

Color Plate 12 (see page 670)

Color Plates

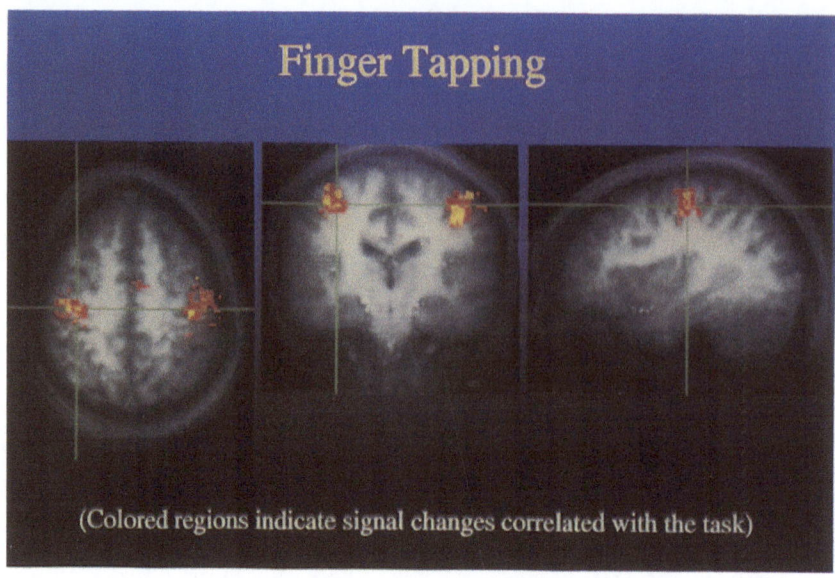

Color Plate 13 (see page 764)

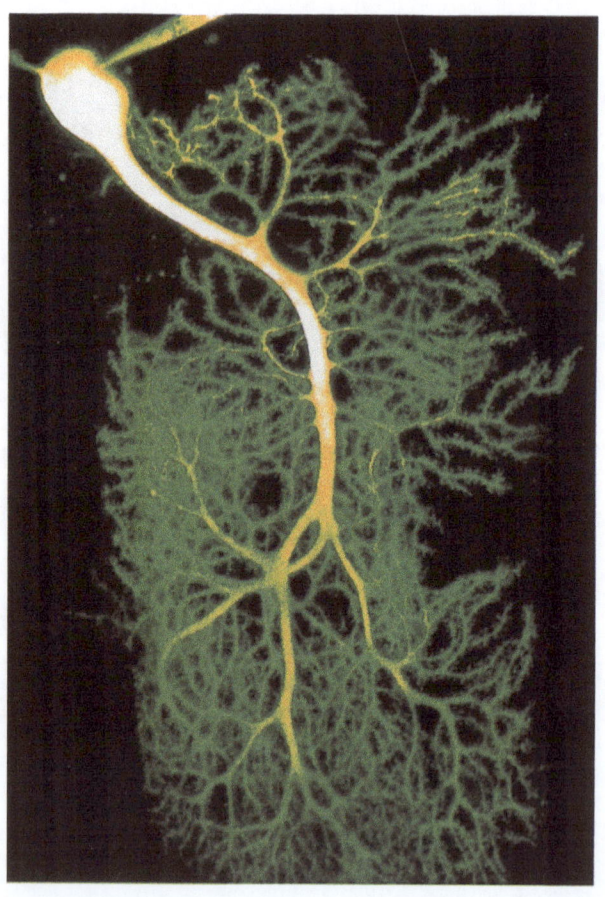

Color Plate 14 (see page 808)

Materials Physics
P. Chaudhari and M. S. Dresselhaus

Over the last one hundred years there have been stunning advances in materials research. At the turn of the last century we did not know what the atomic structure of a material was. Today, not only do we know the structure, but we routinely make artificial structures that require placement of atoms at specified locations, that mix atoms to create properties not found in naturally occurring materials, that have the functionality needed by today's technology, and that adjust their properties to a changing environment (smart materials). Over the last few years we have begun to manipulate individual atoms to form structures that enable us to explore scientific issues, but that will surely lead to profound technological and social consequences; for example, the manipulation of nucleotides in a DNA molecule, which is then correlated with the functioning of, say, a gene and with its expression in the control of disease.

A hundred years ago there were no electronic devices, and today there is hardly any electrical appliance without them. It is anticipated that in the near future there will be a microprocessor embedded in almost all electrical appliances and not just in those used for computation or information storage. These devices, inconceivable a century ago, could without exception not be made without the knowledge gained from materials research on insulators, semiconductors, metals, ceramics, and polymers. At the end of this century, we have begun a debate on how far the present devices can continue to develop, given the limits imposed by the speed of light and the discrete nature of atoms, a debate that would have been incomprehensible to scientists and technologists of a century ago and a debate in which we now discuss the possibilities of using single electrons to switch a device on or off.

Our ability to measure temporal phenomena was limited to fractions-of-a-second resolution a hundred years ago. Today we can measure changes in properties with a few femtoseconds' resolution. Strobelike probes enable us to measure phenomena ranging over time scales covering more than ten orders of magnitude. We can, for example, study the relaxation of electrons in a semiconductor on a femtosecond time scale, the visible motion of bacteria in a petri dish, or the slow motion of a sand dollar on a beach.

Materials research spans the range from basic science, through engineering, to the factory floor. This has not changed over the last hundred years or, for that matter, throughout the history of human civilization. Materials research came out of the practical needs of mankind. Eras of civilization were named after materials, so central has been their role in achieving mankind's mastery over nature. The field of materials research can trace its roots to alchemy, metallurgy, natural philosophy, and even art, as practiced over many centuries. However, the field of modern materials research, as represented by materials physics, is only about sixty years old.

Shifts in materials usage from one type to another are usually gradual. This is due to the very large investments associated with products in the materials-related industries, complex relationships between reliability and functionality, environmental issues, and energy demands. However, measured over time, these shifts become quite perceptible. For example, in automobiles the ratio of plastic components to iron-based alloys has changed from less than 3% to more than 15% over the last two decades. Although the percentage change appears to be modest, the actual volume of material is large; over 40 million tons of structural materials are used annually in cars.

The advances of materials research in this century, which far exceed those of all prior centuries put together, can be illustrated by three examples: structural, polymeric, and electronic materials. Our choice of these three is somewhat, but not completely, arbitrary. It is out of structural materials, particularly from the fields of metallurgy and metal physics, that modern materials physics has evolved. From the crude weaponry of our forefathers to our mastery of air travel, space flight, surface transportation, and housing, structural materials play a role that is unequivocally important. Nature uses the second category of materials, polymers, in amazing ways, to perform very complex functions. We humans are an example of that. Over the last hundred years, we have begun to understand and develop polymers for uses in food packaging, fabrics, and structural applications. We anticipate that polymer research will play an increasingly important role in biomaterials of the future. The third category, electronic materials, was not conceived until quantum mechanics was discovered in this century. Today we cannot imagine a world without telecommunication, computers, radio, and television. These and future devices that will make information available instantly are only possible because of advances in the control of materials structure and processing to achieve a desired functionality.

Structural Materials

At the turn of the last century, mankind's use of structural materials was limited primarily to metals, particularly iron and its alloys, ceramics (most notably Portland cement), and polymers, which were limited to naturally occurring rubbers, glues, and fibers. Composites, as a concept were nonexistent even though wood and animals, each composed of different materials, were used in a variety of ways. However, the uses of alloying to enhance the strength of lightweight materials, such as pewter, or copper additions to aluminum, were established techniques, known well before this century. This knowledge was used to build the first dirigibles. The useful nature of a material was often understood through serendipity and not through an understanding of its structure or the relation between structure and properties. We still cannot predict in any quantitative way the evolution of structure with deformation or processing of a material. However, we have come a very long way from the situation that existed a hundred years ago, thanks to the contributions of twentieth century science to our understanding of atomic arrangement and its determination in a material. Our classification of materials by symmetry considerations came into existence once atomic arrangement became known. To the seven crystal systems and amorphous structures, typified by the glasses and liquids, we can now add quasicrystals and molecular phases, such as fullerenes and nanotubes, in a crystalline solid.

The crystal systems define perfect crystals. At finite temperatures, the crystals are no longer perfect but contain defects. It is now understood that these defects are responsible for atomic transport in solids. In fact, the structural properties of materials are not only a function of the inherent strength of a material but also of the defects that may be present. We know that aluminum is soft because crystallographic defects, called dislocations, can be readily generated and moved in this metal. In contrast, in alumina (Al_2O_3), dislocation generation and motion are difficult; hence alumina can be strong but brittle at room temperature. The addition of copper or manganese to aluminum creates second-phase precipitates, which inhibit the motion of dislocations, thus enhancing its strength-to-weight ratio. Our ability to improve the strength-to-weight ratio in materials has increased more than tenfold during the twentieth century. This is to be compared with a change of less than ten over the last twenty centuries. Much of the increase in this century has come from an understanding of the relationship between the processing of materials and their structure. The highest strength-to-weight ratios have been achieved in materials in the form of fibers and nanotubes. In these structures, dislocations either do not exist or do not move.

Most structural materials are not single crystals. In fact, they consist of a large number of crystals joined at interfaces, which in single-phase materials are called grain boundaries. These interfaces can, for example, influence the mechanical and electrical properties of materials. At temperatures where the grain boundary diffusion rate is low, a small grain size enhances the strength of a material. However, when the grain boundary diffusion rate is high, the material can exhibit very large elongation under a tensile load (superplastic behavior), or can exhibit high creep rates under moderate or small conditions of loading. In demanding high-temperature environments, such as the engine of a modern aircraft, grain boundaries are eliminated so that a complex part, such as a turbine blade, made of a nickel alloy, is a single crystal. Thus the use of materials for structural purposes requires an understanding of the behavior of defects in solids. This is true for metallic, ceramic, and glassy materials.

Both ceramics and glasses were known to ancient civilizations. Ceramics were used extensively in pottery and art. The widespread use of ceramics for structural purposes is largely limited by their brittle behavior. This is now well understood, and schemes have been proposed to overcome brittleness by controlling the propagation of cracks. In metals, dislocations provide the microscopic mechanism that carries energy away from the tip of a crack, thereby blunting it. In ceramics, the use of phase transformations induced at the tip of a propagating crack is one analog of dislocations in metals. Other schemes involve the use of bridging elements across cracks so as to inhibit their opening and hence their propagation. Still another scheme is to use the frictional dissipation of a sliding fiber embedded in a matrix not only to dissipate the energy of crack propagation, but also, if the crack propagates through the material, to provide structural integrity. Use of these so-called fault-tolerant materials requires both an understanding of mechanical properties and control over the properties of interfaces to enable some sliding between the fiber and the matrix without loss of adhesion between them. Such schemes rely either on composite materials or on microstructures that are very well controlled.

The widespread use of silicate glasses, ranging from windows to laptop displays, is only possible through the elimination of flaws, which are introduced, for example, by inhomogeneous cooling. These flaws, which are minute cracks, are eliminated during processing by controlling the cooling conditions, as in a tempered glass, and also by introducing compressive strains through composition modulations.

There are a number of fibers that are available for use with ceramics, polymers, and metals to form composite materials with specific applications; these include carbon fibers, well known for their use in golf clubs and fishing rods, and silicon carbide or nitride fibers. Optical fibers, which are replacing copper wires in communication technologies, owe their widespread use not only to their optical transparency, but also to improvements in their structural properties. Fibers must withstand mechanical strains introduced during their installation and operation.

The use of composite materials in today's civilization is quite widespread, and we expect it to continue as new applications and "smart" materials are developed. An outstanding example of a functional composite product comes from the electronics industry. This is a substrate, called a package, which carries electronic devices. Substrates are complicated three-dimensionally designed structures, consisting of ceramics, polymers, metals, semiconductors, and insulators. These packages must satisfy not only structural needs but also electrical requirements.

Although we have made great progress over the last hundred years in materials physics, our microscopic understanding of the physics of deformation (particularly in noncrystalline solids), fracture, wear, and the role of internal interfaces is still far from complete. There has been considerable progress in computer simulation of some of these issues. For example, there is now a concerted effort to model the motion of dislocations, during deformation, in simulations of simple metallic systems. We anticipate that within the next decade, as computational power continues to increase, many of these problems will become tractable. The ultimate goal is to design a structural component for a set of specified environmental conditions and for a predictable lifetime.

Polymers

Polymers, also known as macromolecules, are long-chain molecules in which a molecular unit repeats itself along the length of the chain. The word polymer was coined approximately 165 years ago (from the Greek *polys*, meaning many, and *meros*, parts). However, the verification of the structure of polymers, by diffraction and other methods, had to wait, approximately, another 100 years. We now know that the DNA molecule, proteins, cellulose, silk, wool, and rubber are some of the naturally occurring polymers. Synthetic polymers, derived mainly from petroleum and natural gas, became a commodity starting approximately 50 years ago. Polymers became widely known to the public when nylon was introduced as a substitute for silk and, later, when Teflon-coated pans became commercially available. Polymers are now widely used in numerous household applications. Their industrial use is even more widespread.

Most of the applications associated with polymers have been as structural materials. Since the 1970s it was realized that with suitable doping of the polymers, a wide variety of physical properties could be achieved, resulting in products ranging from photosensitive materials to superconductors. The field of materials physics of polymers has grown rapidly from this period onwards.

Polymers are a remarkably flexible class of materials, whose chemical and physical properties can be modified by molecular design. By substitution of atoms, by adding side groups, or by combining (blending) different polymers, chemists have created a myriad of materials with remarkable, wide ranging, and useful properties. This research is largely driven by the potential applications of these materials in many diverse areas, ranging from cosmetics to electronics. Compared to most other materials, polymers offer vast degrees of freedom through blending and are generally inexpensive to fabricate in large volumes. They are light weight and can have very good strength-to-weight ratios.

Polymers have traditionally been divided into five classes:

(1) Plastics are materials that are molded and shaped by heat and pressure to produce low-density, transparent, and often tough products, for uses ranging from beverage bottles to shatterproof windows.
(2) Elastomers are chemically cross-linked or entangled polymers in which the chains form irregular coils that straighten out during strain (above their glass transition temperatures), thus providing large elongations, as in natural and synthetic rubbers.
(3) Fibers, which are spun and woven, are used primarily in fabrics. About fifty million tons of fibers are produced annually for uses ranging from clothing to drapes. Apart from naturally occurring fibers such as silk and wool, there are regenerated fibers made from cellulose polymers that make up wood (rayon) and synthetic fibers, comprising molecules not found in nature (nylon).
(4) Organic adhesives have been known since antiquity. However, with demanding environments and performance requirements, synthetic adhesives and glues have largely replaced natural ones. The microscopic mechanisms of adhesion and the toughness of joints are still debated. There is an increasing trend to use UV radiation to promote polymerization in adhesives and, more generally, as a method of polymerization and cross linking in polymers.
(5) Finally, polymers, frequently with additives, are used as protective films, such as those found in paints or varnishes.

Physicists have played a significant role in explaining the physical properties of polymeric materials. However, the interest of physicists in polymers accelerated when it was discovered that polyacetylene could be made conductive by doping. This development was noteworthy for it opened the possibility of deliberately controlling conductivity in materials that are generally regarded as good insulators. The structure of all conjugated polymers, as these materials are known, is characterized by a relatively easily delocalized π bond, which, with suitable doping, results in effective charge motion by solitons, polarons, or bipolarons. Since the discovery that polymers could be electrical conductors, active research areas have developed on the physics of polymer superconductors, ferro- and ferri-magnets, piezoelectrics, ferroelectrics, and pyroelectrics. Within the field of doped polymers, devices have been built to demonstrate light-emitting diodes, photovoltaic cells, and transistors.

Conjugated polymers have also been investigated extensively for their large nonlinear, third-order polarizability, which is of interest to the field of nonlinear optics. Large nonlinearities are associated with the strong polarizability of the individual molecules that make up the building blocks of the polymer. Furthermore, the flexibility of polymer chemistry has allowed the optical response of polymers to be tailored by controlling their molecular structure, through the selective addition of photoactive molecules. Hence these materials have been widely investigated by physicists and engineers for optical applications, such as in holographic displays (dichromated gelatin), diffraction gratings, optocouplers, and wave guides.

Polymers have long interested physicists for their conformational and topological properties. This interest has shifted from the conformational behavior of individual molecules to that of a macromolecular assembly, phase behavior, and a search for universal classes. Block copolymers, consisting of two or more polymers, can give rise to nanoscale phases, which may, for example, be present as spheres, rods, or parallel lamallae. The distribution of these phases and their topologies are of current theoretical and practical interest. Block copolymer morphologies are also being used as nanoscale templates for production of ceramics of unique properties having the same morphology.

Block copolymers are also of interest as biomaterials. Proteins are an example of block copolymers, in which the two phases form helical coils and sheets. Attempts to mimic the hierarchical structure present in natural polymers have only been partly successful. The principal difficulty has been to control the length of the polymer chains to the precision that Nature demands. Significant progress has been made in controlling polymer morphologies with the use of new catalysts. For example, metallocenes have been used as catalysts to control branched polymers and organonickel initiators to suppress chain transfer and termination, so that polypeptides with well-defined sequences and with potential for applications in tissue engineering could be made. The growth of well-controlled polymer chains is an example of "living" polymerization.

The static and dynamic arrangement of atoms on the surfaces and interfaces of polymers is another area of active investigation. For example, thin films of polymers, in which the chain lengths are long compared to the thickness of a film, show unusual physical properties: the glass transition temperature for a thin-film polymer decreases significantly, but between solid surfaces polymer liquids solidify.

Even though we have some way to go in making tailored proteinlike structures, polymer research has played a significant role in the class of materials called biomaterials. Polymers have been used, for example, to produce artificial skin, for dental fillings that are polymerized *in situ* by a portable UV lamp, and for high-density polyethylene used in knee prostheses. Physicists play a significant role in these developments, not only for their interest in the materials, but also because of their familiarity with physical processes that can be used to tailor the properties of polymers. A particularly good example of this interplay is the recent and rapidly growing use of excimer laser radiation to correct corneal abnormalities; using a technology developed from studies of the ablation of polymeric materials for applications in the electronics industry, physicists realized that the small, yet precisely controlled, ablation of a polymeric surface might be useful in shaping the surface of an eye.

Electronic Materials

The roots of the electronic materials field can be traced back to Europe in the 1920s, with the advent of quantum mechanics and its application to periodic structures like those occurring in crystals. The early experimental focus was on alkali halides, because these materials could be prepared in a controlled way from both a structural and a compositional standpoint. The creation of a strong academic program in solid-state physics at the University of Illinois in the 1930s had an important impact on the early history of the electronic materials field in the United States. This knowledgeable human resource played a significant role in mobilizing the national materials program during World War II, especially in the development of semiconducting materials with enhanced purity, suitable for use in diode detectors at microwave frequencies for communications applications. The availability of these new semiconducting materials in purified, crystalline form soon led to the discovery of the transistor, which ushered in the modern era of electronics, computers, and communications, which is now simply called the "information age."

Semiconductors have been a central focus for electronic materials. Quantum-mechanical treatments of a periodic lattice were successful in laying the groundwork for describing the electronic band structure, which could account for electrical conduction by electrons and holes, carrier transport under the action of forces and fields, and the behavior of early electronic devices. Because of the interest of industrial laboratories and the Defense Department in the newly emerging field of semiconductor electronics, semiconductor physics developed rapidly, and this focus soon led to the development of the integrated circuit and the semiconductor laser.

The strong interplay between technological advances and basic scientific discovery has greatly energized semiconductor physics, by raising challenging fundamental questions and by providing new, better materials and devices, which in turn opened up new research areas. For example, the development of molecular-beam epitaxy in the 1960s and 1970s led to the ability to control layer-by-layer growth of semiconductor quantum wells and superlattices. The use of modulation doping of the quantum wells, whereby the dopants are introduced only in the barrier regions, led to the possibility of preparing semiconductors with low-temperature carrier mobilities, orders of magnitude greater than in the best bulk semiconductors. These technological advances soon led to the discovery of the quantum Hall effect, the fractional quantum Hall effect, and a host of new phenomena, such as Wigner crystallization, which continue to challenge experimentalists and theorists. Lithographic and patterning technologies developed for the semiconductor industry have led to the discovery of the quantized conductance for one-dimensional semiconductors and to the fabrication of specially designed semiconductor devices, in which the transport of a single electron can be controlled and studied. The ever decreasing size of electronic devices (now less than 0.2 microns in the semiconductor industry) is greatly stimulating the study of mesoscopic physics, in which carriers can be transported ballistically without scattering and the effect of the electrical leads must be considered as part of the electronic system. New materials, such as carbon nanotubes with diameters of 1 nm, have recently been discovered, and junctions between such nanotubes are being considered for possible future electronics applications on the nm scale, utilizing their unique one-dimensional characteristics.

The electronic materials field today is highly focused on the development of new materials with special properties to meet specific needs. Advances in condensed-matter physics offer the possibility of new materials properties. In photonics, new materials are providing increased spectral range for light-emitting diodes, smaller and more functional semiconducting lasers, new and improved display materials. The new field of photonic band-gap crystals, based on structures with periodic variations in the dielectric constant, is just now emerging. Research on optoelectronic materials has been greatly stimulated by the optical communications industry, which was launched by the development of low-loss optical fibers, amplifiers, and lasers.

Ferroelectrics have become important for use as capacitors and actuators, which are needed in modern robotics applications, as are also piezoelectric materials, which are critical to the operation of scanning tunneling probes that provide information at the atomic level on structure, stoichiometry, and electronic structure. The technological development of microelectromechanical systems (MEMS), based on silicon and other materials, is making possible the use of miniature motors and actuators at the micrometer level of integrated circuits. Some of these have already found applications, such as the triggering mechanism for the release of airbags in automobiles. Such developments are not only important to the electronics industry, but are also having great impact on fields such as astronomy and space science, which are dependent on small, light-weight instruments with enhanced capabilities to gather signals at ever increasing data rates and from ever increasing distances from Earth. The developments in new materials and low-dimensional fabrication techniques have recently rejuvenated the field of thermoelectricity, where there is now renewed hope for enhanced thermoelectric performance over a wider temperature range.

Research on magnetic materials has been strongly influenced by applications ranging from the development of soft magnetic materials (by the utilization of rapid solidification techniques) to hard magnetic materials such as neodymium-iron-boron for use in permanent magnets. In the 1980s efforts focused on the development of small magnetic particles for magnetic memory storage applications. New magnetic materials, especially magnetic nanostructures, are now an extremely active research field, where the discovery of new phenomena such as giant magnetoresistance and colossal magnetoresistance are now being developed for computer memory applications.

The strong interplay between fundamental materials physics and applications is also evident in the area of superconducting materials. Early use of superconducting materials was in the fabrication of superconducting magnets, which in turn promoted understanding of type-II superconductors, flux dynamics, and flux pinning phenomena. The discovery of the Josephson tunneling effect led to the development of the SQUID (superconducting quantum interference device), which has become a standard laboratory tool for materials characterization and for the sensitive measurement of extremely small magnetic fields, such as the fields associated with brain stimuli. The discovery of high-T_c superconductivity in 1986 has revolutionized this field, with much effort being devoted to studies of the mechanism for high-T_c superconductivity, along with efforts to discover materials with yet higher T_c and critical current values, to improve synthesis methods for the cuprate superconductors, and to develop applications for these materials to electronics, energy storage, and high-magnetic-field generation.

When viewed from the perspective of time, the developments in electronic materials have been truly remarkable. They have generated businesses that approach a trillion dollars, have provided employment to millions of workers, either directly and indirectly associated with these industries, and have enabled us, as humans, to extend our abilities, for example, in information gathering, communication, and computational capabilities. Science has been the key to these marvelous developments, and in turn these developments have enabled us, as scientists, to explore and understand the subtleties of nature.

Summary

In this very brief note, we have only touched on some of the advances made in structural, polymeric, and electronic materials over the last century, showing how materials physics has played a central role in connecting science to technology and, in the process, revolutionized our lives.

The Invention of the Transistor

Michael Riordan, Lillian Hoddeson, and Conyers Herring

Arguably the most important invention of the past century, the transistor is often cited as the exemplar of how scientific research can lead to useful commercial products. Emerging in 1947 from a Bell Telephone Laboratories program of basic research on the physics of solids, it began to replace vacuum tubes in the 1950s and eventually spawned the integrated circuit and microprocessor—the heart of a semiconductor industry now generating annual sales of more than $150 billion. These solid-state electronic devices are what have put computers in our laps and on desktops and permitted them to communicate with each other over telephone networks around the globe. The transistor has aptly been called the "nerve cell" of the Information Age.

Actually the history of this invention is far more involved and interesting than given by this "linear" account, which overlooks the intricate interplay of scientific, technological, social, and personal interests and developments. These and many other factors contributed to the invention of not one but two distinctly different transistors—the point-contact transistor by John Bardeen and Walter Brattain in December 1947, and the junction transistor by William Shockley a month later.[1] The point-contact transistor saw only limited production and never achieved commercial success. Instead, it was the junction transistor that made the modern semiconductor industry possible, contributing crucially to the rise of companies such as Texas Instruments, SONY, and Fairchild Semiconductor.

Given the tremendous impact of the transistor, it is surprising how little scholarship has been devoted to its history.[2] We have tried to fill this gap in recent publications (Herring, 1992; Riordan and Hoddeson, 1997a, 1997b). Here we present a review of its invention, emphasizing the crucial role played by the postwar understanding of solid-state physics. We conclude with an analysis of the impact of this breakthrough upon the discipline itself.

Preliminary Investigations

The quantum theory of solids was fairly well established by the mid-1930s, when semiconductors began to be of interest to industrial scientists seeking solid-state alternatives to

---

[1] This paper is based in large part on Riordan and Hoddeson (1997a). The best scholarly historical account of the point-contact transistor is that of Hoddeson (1981); on the invention of the junction transistor, see Shockley (1976).
[2] In addition to the above references, see Bardeen (1957), Brattain (1968), Shockley (1973, 1976), Weiner (1973), Holonyak (1992), Riordan and Hoddeson (1997b), Ross (1998), and Seitz and Einspruch (1998b). Scholarly books that cover the topic well include those of Braun and MacDonald (1978) and Seitz and Einspruch (1998a).

vacuum-tube amplifiers and electromechanical relays. Based on the work of Felix Bloch, Rudolf Peierls, and Alan Wilson, there was an established understanding of the band structure of electron energies in ideal crystals (Hoddeson, Baym, and Eckert, 1987; Hoddeson et al., 1992). This theory was then applied to calculations of the energy bands in real substances by groups of graduate students working with Eugene Wigner at Princeton and John Slater at MIT. Bardeen and Frederick Seitz, for example, wrote dissertations under Wigner, calculating the work function and band structure of sodium; studying with Slater, Shockley determined the band structure of sodium chloride (Bardeen, 1936; Shockley, 1936; Herring, 1992). By the mid-1930s the behavior of semiconductors was widely recognized to be due to impurities in crystals, although this was more a qualitative than quantitative understanding. The twin distinctions of "excess" and "defect" semiconductors could be found in the literature; their different behavior was thought to be the result of electrons added to the conduction band or removed from the valence band by impurity atoms lodged in the crystal lattice (Wilson, 1931; Mott and Jones, 1936).

There were a few solid-state electronic devices in use by the mid-1930s, most notably the copper-oxide rectifier, on which Brattain worked extensively at Bell Labs during that period (Brattain, 1951). Made by growing an oxide layer on copper, these rectifiers were used in AC-to-DC converters, in photometers and as "varistors" in telephone circuitry made for the Bell System. But the true nature of this rectification, thought to occur at the interface between the copper and copper-oxide layers, was poorly understood until the work of Nevill Mott (1939) and Walther Schottky (1939) showed the phenomenon to be due to the establishment of an asymmetric potential barrier at this interface. In late 1939 and early 1940, Shockley and Brattain tried to fabricate a solid-state amplifier by using a third electrode to modulate this barrier layer, but their primitive attempts failed completely.

One of the principal problems with this research during the 1930s was that the substances generally considered to be semiconductors were messy compounds such as copper oxide, lead sulfide, and cadmium sulfide. In addition to any impurities present, there could be slight differences from the exact stoichiometric ratios of the elements involved; these were extremely difficult, if not impossible, to determine and control at the required levels. Semiconductor research therefore remained more art than science until World War II intervened.

During the War, silicon and germanium rose to prominence as the preferred semiconductors largely through the need for crystal rectifiers that could operate at the gigahertz frequencies required for radar receivers. Driven by this requirement, the technology of these two semiconductor materials advanced along a broad front (Torrey and Whitmer, 1948). Where before the War it was difficult to obtain silicon with impurity levels less than one percent, afterwards the DuPont Company was turning out 99.999 percent pure silicon (Seitz, 1994, 1995; Seitz and Einspruch, 1998a). The technology of doping silicon and germanium with elements from the third and fifth columns of the periodic table (such as boron and phosphorus) to produce p-type and n-type semiconductor materials had become well understood. In addition, the p-n junction had been discovered in 1940 at Bell Labs by Russell Ohl—although its behavior was not well understood, nor was it employed in devices by War's end (Scaff and Ohl, 1947; Scaff, 1970; Riordan and Hoddeson, 1997a, 1997c).

There was also extensive research on semiconductors in the Soviet Union during the same period, but this work does not seem to have had much impact in the rest of Europe and the United States (Herring, Riordan, and Hoddeson, n.d.). Of course, contributions of well-known theorists, such as Igor Tamm on surface-bound electron levels and Yakov Frenkel on his theory of excitons, attracted wide interest (Tamm, 1932; Frenkel, 1933, 1936); published in German and English, they were quickly incorporated into the corpus of accepted knowledge.

But the work of Boris Davydov on rectifying characteristics of semiconductors seems to have eluded notice until after the War, even though it was available in English-language publications (Davydov, 1938). Working at the Ioffe Physico-Technical Institute in Leningrad, he came up with a model of rectification in copper oxide in 1938 that foreshadowed Shockley's work on p-n junctions more than a decade later. His idea involved the existence of a p-n junction in the oxide, with adjacent layers of excess and deficit semiconductor forming spontaneously due to an excess or deficit of copper relative to oxygen in the crystal lattice. Nonequilibrium concentrations of electrons and holes—positively charged quantum-mechanical vacancies in the valence band—could survive briefly in each other's presence before recombining. Using this model, Davydov successfully derived the current-voltage characteristics of copper-oxide rectifiers; his formula was essentially the same as the one that Shockley would derive a decade later for p-n junctions (Shockley, 1949). But his cumbersome mathematics and assumptions may have obscured the importance of his physical ideas to later workers. Bardeen, for example, was aware of Davydov's publications by 1947 but does not seem to have recognized their significance until a few years later.

The Invention of the Point-Contact Transistor

Both the point-contact transistor and the junction transistor emerged from a program of basic research on solid-state physics that Mervin Kelly, then Bell Labs Executive Vice President, initiated in 1945. He recognized that the great wartime advances in semiconductor technology set the stage for electronic advances that could dramatically improve telephone service. In particular, he was seeking solid-state devices to replace the vacuum tubes and electromechanical relays that served as amplifiers and switches in the Bell Telephone System (Hoddeson, 1981; Riordan and Hoddeson, 1997a). He had learned valuable lessons from the wartime efforts at Los Alamos and the MIT Radiation Laboratory, where multidisciplinary teams of scientists and engineers had developed atomic bombs and radar systems in what seemed a technological blink of an eye (Hoddeson et al., 1993).

Kelly perceived that the new quantum-mechanical understanding of solids could be brought to bear on semiconductor technology to solve certain problems confronting his company. "Employing the new theoretical methods of solid state quantum physics and the corresponding advances in experimental techniques, a unified approach to all of our solid state problems offers great promise," he wrote that January. "Hence, all of the research activity in the area of solids is now being consolidated . . . ." (Riordan and Hoddeson, 1997a, pp. 116–117.) At the helm of this Solid State Physics Group he put Shockley and chemist Stanley Morgan. Soon Brattain and Bardeen joined a semiconductor subgroup within it headed by Shockley.

While planning the new solid-state group in April 1945, Shockley proposed a device now called the "field-effect" transistor (Shockley, 1976; Hoddeson, 1981). Here an externally applied transverse electric field is arranged so that it can increase or decrease the number of charge carriers in a thin film of silicon or germanium, thus altering its conductivity and regulating the current flowing through it. By applying suitable voltages to two circuit loops passing through this semiconductor material, Shockley predicted that an input signal applied to one loop could yield an amplified signal in the other. But several attempts to fabricate such a field-effect device in silicon failed. So did Shockley's theoretical attempt to explain why, on the basis of Mott and Schottky's rectification theory, his conceptual field-effect device did not work as predicted (Hoddeson, 1981, pp. 62–63).

In October 1945 Shockley asked Bardeen, who had just joined the group, to check the calculations that he had made in an attempt to account for the failure of his field-effect idea. By March 1946 Bardeen had an answer. He explained the lack of significant modulation of the conductivity using a creative heuristic model, based on the idea of "surface states" (Bardeen, 1947). In this model, electrons drawn to the semiconductor surface by the applied field become trapped in these localized states and are thus unable to act as charge carriers.[3] As Shockley (1976, p. 605) later recalled, the surface states "blocked the external field at the surface and . . . shielded the interior of the semiconductor from the influence of the positively charged control plate."

But were these postulated states real? If so, how did they generally behave? These questions became intensely interesting to the Bell Labs semiconductor group, which in the following months responded to Bardeen's surface-state idea with an intensive research program to explore this phenomenon. Bardeen worked closely on the problem with the group's experimental physicists, Brattain and Gerald Pearson.[4]

On 17 November 1947, Brattain made an important discovery. Drawing on a suggestion by Robert Gibney, a physical chemist in the group, he found that he could neutralize the field-blocking effect of the surface states by immersing a silicon semiconductor in an electrolyte (Brattain, 1947b, pp. 142–151; 1968). "This new finding was electrifying," observed Shockley (1976, p. 608); "At long last, Brattain and Gibney had overcome the blocking effect of the surface states." Their discovery set in motion events that would culminate one month later in the first transistor.

Four days after this discovery, Bardeen and Brattain tried to use the results to build a field-effect amplifier. Their approach was based on Bardeen's suggestion to use a point-contact electrode pressed against a specially prepared silicon surface. Rather than the thin films employed in the 1945 experiments by Shockley and his collaborators, Bardeen proposed the use of an n-type "inversion layer" a few microns thick that had been chemically produced on the originally uniform surface of p-type silicon. Because charge carriers—in this case, electrons—would have higher mobility in such an inversion layer than they had in vapor-deposited films, Bardeen believed that this approach would work better in a field-effect amplifier (Bardeen, 1957). In particular, this layer would act as a shallow channel in which the population of charge carriers could be easily modulated by an applied external field. The device tested on 21 November used a drop of electrolyte on the surface as one contact and the metal point as the other; Bardeen and Brattain obtained a small but significant power amplification, but the device's frequency response was poor (Bardeen, 1946, pp. 61–70).

---

[3]Previous work on surface states had been done by Tamm (1932) and Shockley (1939). Bardeen, however, was the one who applied these ideas to understanding the surface behavior of semiconductors (Bardeen, 1946, pp. 38–57; 1947).
[4]The research program is described in the laboratory notebooks of Bardeen (1946), Brattain (1947b, 1947c), Pearson (1947), and Shockley (1945); it is summarized by Hoddeson (1981). The sequence of steps to the point-contact transistor detailed here largely follows the account in Hoddeson (1981) and Riordan and Hoddeson (1997a).

The next crucial step occurred on 8 December. At Bardeen's suggestion, Brattain replaced the silicon with an available slab of n-type, "high-back-voltage" germanium, a material developed during the wartime radar program by a research group at Purdue directed by Karl Lark-Horovitz (Henriksen, 1987). They obtained a power gain of 330—but with a negative potential applied to the droplet instead of positive, as they had expected. Although the slab had not been specially prepared, Bardeen proposed that an inversion layer was being induced electrically, by the strong fields under the droplet. "Bardeen suggests that the surface field is so strong that one is actually getting P type conduction near the surface," wrote Brattain (1947b, pp. 175–176) that day, "and the negative potential on the grid is increasing the P type or hole conduction."[5] This was a crucial perception on Bardeen's part, that holes were acting as charge carriers within a slab of n-type germanium.

Later that week Brattain evaporated a gold plate onto a specially prepared germanium slab that already had an inversion layer. In an attempt to improve the frequency response by eliminating the sluggish droplet, he employed instead a thin germanium-oxide layer grown on the semiconductor surface. He thought the gold would be insulated from the germanium by this layer, but unknown to him the layer had somehow been washed away, and the plate was now directly in contact with germanium. This serendipitous turn of events proved to be a critical step toward the point-contact transistor (Hoddeson, 1981).

The following Monday, 15 December, Bardeen and Brattain were surprised to discover that they could still modulate the output voltage and current at a point contact positioned close to the gold plate, but only when the plate was biased *positively*—the opposite of what they had expected![6] "An increase in positive bias *increased* rather than decreased the reverse current to the point contact," wrote Bardeen (1957) ten years later. This finding suggested "that holes were flowing into the germanium surface from the gold spot and that the holes introduced in this way flowed into the point contact to enhance the reverse current. This was the first indication of the transistor effect."

Although Brattain and Bardeen failed to observe power amplification with this configuration, Bardeen suggested that it would occur if two narrow contacts could be spaced only a few thousandths of an inch apart. Brattain (1947b, pp. 192–93) achieved the exacting specifications by wrapping a piece of gold foil around one edge of a triangular polystyrene wedge and slitting the foil carefully along that edge. He then pressed the wedge—and the two closely spaced gold contacts—down into the surface of the germanium using a makeshift spring (see Figs. 1 and 2). In their first tests, made on 16 December, the device worked as expected. It achieved both voltage and power gains at frequencies up to 1000 Hz. The transistor had finally been born. A week after that, on 23 December 1947, the device was officially demonstrated to Bell Labs executives in a circuit that allowed them to hear amplified speech in a pair of headphones (Brattain, 1947c, pp. 6–8; Hoddeson, 1981).

---

[5]This was the first recorded instance we can find in which Bardeen and Brattain recognized the possibility that holes were acting as charge carriers. Note that Bardeen still proposed that the flow occurred within a shallow inversion layer at the semiconductor surface.

[6]Hoddeson (1981, p. 72) states that this event occurred on Thursday, 11 December. A closer examination of Brattain (1947b, pp. 183–92) indicates that there was a period of confusion followed by the actual breakthrough on 15 December. See Riordan and Hoddeson (1997a), Chapter 7, for a more complete discussion of this sequence of events.

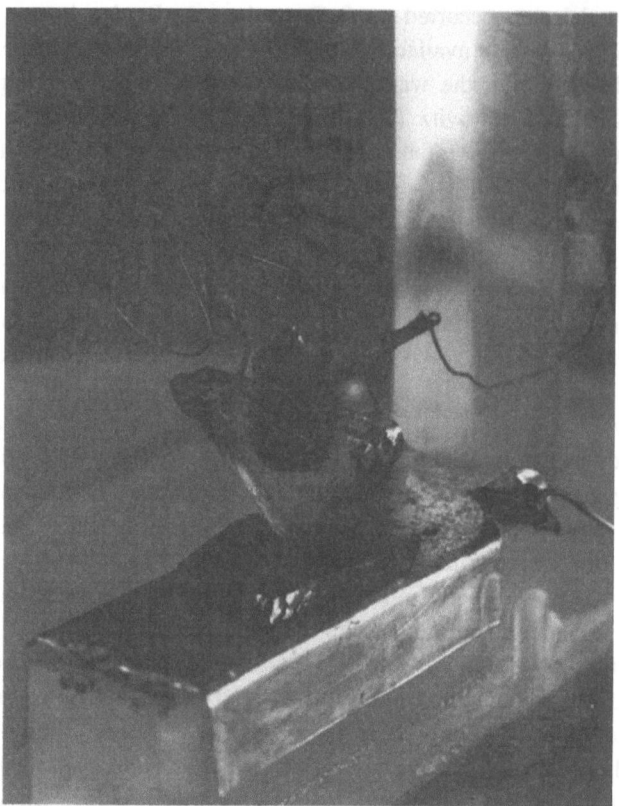

Figure 1. Photograph of the point-contact transistor invented by Bardeen and Brattain in December 1947. A strip of gold foil slit along one edge is pressed down into the surface of a germanium slab by a polystyrene wedge, forming two closely spaced contacts to this surface. (Reprinted by permission of AT&T Archives.)

The Flow of Charge Carriers

An important issue that has engendered much recent debate is how Bardeen and Brattain conceptualized the flow of charge carriers while they were developing the first transistor. Memory is imperfect, and later accounts are often subject to what is called "retrospective realism,"[7] a process whereby conjectures become imbued with an aura of certainty, or embellished with details that became known only at a later time. Fortunately, we have available several telling entries that Bardeen, Brattain and Shockley made in their laboratory notebooks during those pivotal weeks before and after Christmas 1947.[8]

On 19 December, three days after the first successful test of their device, Brattain (1947c, p. 3) wrote: "It would appear then that the modulation obtained when the grid point is bias+is due to the grid furnishing holes to the plate point." By grid point and plate point, he was referring to what we now call the emitter and collector: he was

---

[7]This phrase and concept is due to Pickering (1984).
[8]Some of the entries in Brattain's notebooks during those critical weeks in December 1947 are written in Bardeen's handwriting. The two obviously were working side by side in the laboratory.

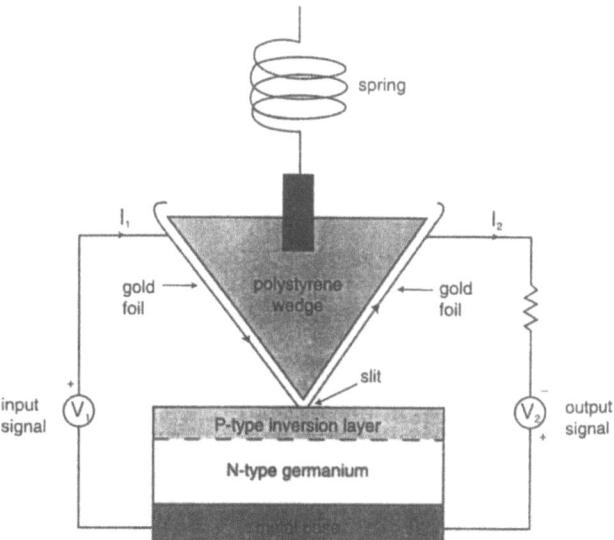

Figure 2. Schematic diagram of the first transistor (Fig. 1). The signal current I_1 flows through the input circuit, generating holes in a p-type inversion layer that modulate the flow of current I_2 in an output circuit. (Reprinted from M. Riordan and L. Hoddeson, Crystal Fire.)

obviously using a familiar vacuum-tube analogy. Although we cannot determine from this passage exactly how he conceived the details of their flow, we can be sure he understood that holes were responsible for modulation.

Bardeen gave a more detailed explanation in a notebook entry on 24 December, the day after the team made its official demonstration. After describing their setup, which used a slab of *n*-type germanium specially prepared to produce a shallow inversion layer of *p*-type conductivity near its surface (see Fig. 3), he portrayed the phenomenon as follows (Bardeen, 1946, p. 72):

When A is positive, holes are emitted into the semi-conductor. These spread out into the thin P-type layer. Those which come in the vicinity of B are attracted and enter the electrode. Thus A acts as a cathode and B as a plate in the analogous vacuum tube circuit.

Again it is clear that Bardeen also attributed the transistor action to the holes, but he went a step farther and stated that the flow of these holes occurs within the inversion layer.

This emerging theory of the transistor based on the flow of holes at or near the surface of the germanium developed further during the following months, the period in which Bell Labs kept the discovery of the transistor "laboratory secret," while patent applications were being drawn up. A drawing found in Bardeen and Brattain's patent application of 17 June 1948 (revised from a version submitted on 25 February) suggests that although the flow of charge carriers was thought to occur largely within the *p*-type inversion layer, they were by this time allowing that some holes might diffuse through the body of the *n*-type germanium. They state (Bardeen and Brattain, 1948a):

... potential probe measurements on the surface of the block, made with the collector disconnected, indicate that *the major part* of the emitter current travels on or close to the surface of the block, substantially laterally in all directions away from the emitter ....

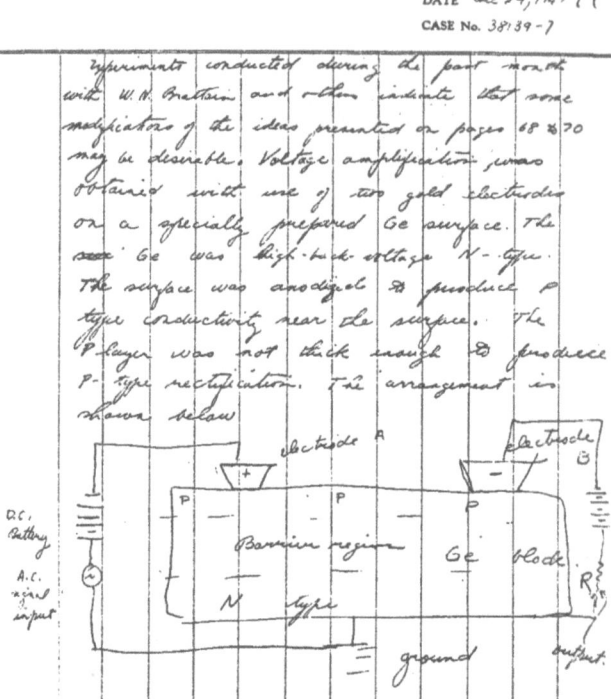

Figure 3. Entry in Bardeen's lab notebook dated 24 December 1947, giving his conception of how the point-contact transistor functions. (Reprinted by permission of AT&T Archives.)

In a famous letter submitted to the *Physical Review* on 25 June 1948, they wrote (Bardeen and Brattain, 1948b) that as a result of the existence of the shallow p-type inversion layer next to the germanium surface, "the current in the forward direction with respect to the block is composed in large part of holes, i.e., of carriers of sign opposite to those normally in excess in the body of the block." In a subtle shift from their earlier conception, they envisioned that holes flow predominantly in the p-type inversion layer, but with a portion that can also flow through the n-type layer beneath it.

It is not clear from these entries just how and why this shift occurred. But both the revised patent application and the *Physical Review* letter are dated well after Shockley's conception of the junction transistor in late January and a crucial mid-February experiment (discussed below) by John Shive.

The Conception of the Junction Transistor

During the weeks that followed the invention of the point-contact transistor, Shockley was torn by conflicting emotions. Although he recognized that Bardeen and Brattain's invention had been a "magnificent Christmas present" to Bell Labs, he was chagrined that he had not had a direct role to play in this obviously crucial breakthrough. "My elation with the group's success was tempered by not being one of the inventors," he recalled a quarter century later (Shockley, 1976). "I experienced frustration that my personal efforts, started more than eight years before, had not resulted in a significant inventive contribution of my own."

Since the failure of his field-effect idea more than two years earlier, Shockley had paid only passing attention to semiconductor research. During the months before the invention, he had mainly been working on the theory of dislocations in solids. He had, however, thought about the physics of p-n junctions and their use in such practical devices as lightning arrestors and high-speed thermistors (Shockley, 1945, pp. 71, 76–78, 80, 88–89).

Brattain and Gibney's discovery in November 1947 stimulated Shockley's thinking. A few days after that he suggested fabricating an amplifier using a drop of electrolyte deposited across a p-n junction in silicon or germanium; this approach worked when Brattain (1947b, pp. 169–70) and Pearson (1947, p. 75) tried it. On 8 December 1947, more than a week before the point-contact transistor was invented. Shockley (1945, p. 91) outlined an idea in his laboratory notebook for an n-p-n sandwich that had current flowing laterally in the interior p-layer and with the n-layers around it acting as control electrodes.

The 16 December invention of the point-contact transistor and Bardeen's interpretation of its action in terms of the flow of holes galvanized Shockley into action. Bardeen's above-quoted analogy with the operation of a vacuum tube—in which the current carriers were holes instead of electrons—was in fact due to Shockley,[9] who applied it in his first attempt at a junction transistor, written in a room in Chicago's Hotel Bismarck on New Year's Eve of 1947. In this first stab at a junction transistor, one can see a clear analogy with a vacuum tube; its "control" electrode acts as a grid to control the flow of holes from a "source" to a "plate" (Shockley, 1945, pp. 110–13). On this disclosure of a p-n-p device, however, Shockley (1976) admitted that he had "failed to recognize the possibility of minority carrier injection into a base layer.... What is conspicuously lacking [in these pages] is any suggestion of the possibility that holes might be injected into the n-type material of the strip itself, thereby becoming minority carriers in the presence of electrons."

A little more than three weeks later, this time working at his home on the morning of 23 January, 1948, Shockley conceived another design in which n-type and p-type layers were reversed and electrons rather than holes were the current carriers (see Fig. 4). Applying a positive voltage to the interior p-layer should lower its potential for electrons; this he realized would "increase the flow of electrons over the barrier exponentially" (Shockley, 1945, p. 129). As Shockley (1976) observed nearly thirty years later, this n-p-n sandwich device finally contained the crucial concept of "exponentially increasing minority carrier injection across the emitter junction." Minority carriers, in this case the electrons, had to flow in the presence of the dominant majority carriers—the holes of the p-type layer.

A Crucial Experiment

Almost another month passed before Shockley revealed his breakthrough idea to anyone in his group other than physicist J. Richard Haynes, who witnessed the entry in his

---

[9]Bardeen (1946) credits Shockley with this suggestion in his notebook on p. 72.

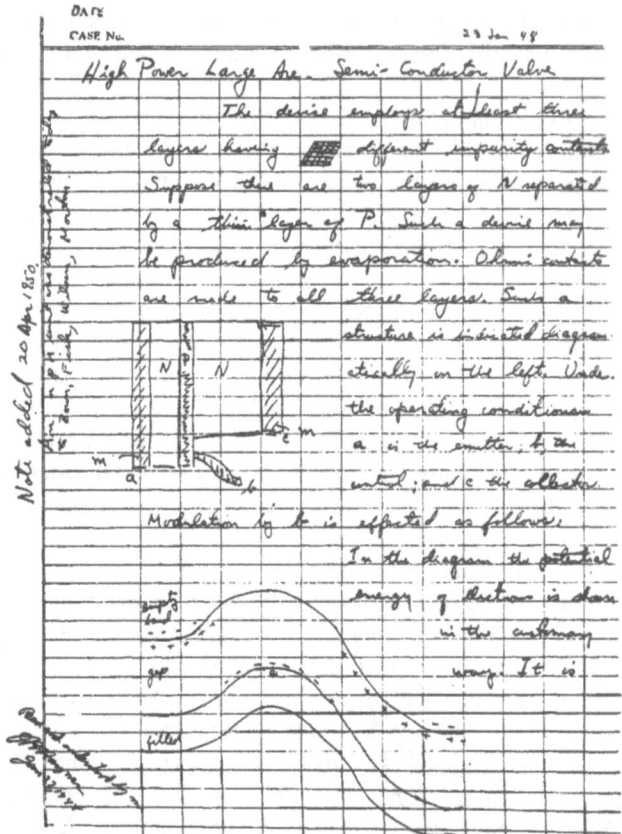

Figure 4. Entry in Shockley's lab notebook dated 23 January 1948 recording his conception of the junction transistor. He wrote this page at home on a piece of paper, which he later pasted into his notebook. (Reprinted by permission of AT&T Archives.)

logbook. Why did Shockley keep the information to himself? Did he recognize that he had made a major conceptual advance but decide to keep it quiet to give himself more time to follow up its theoretical and practical ramifications? Was he afraid that Bardeen and Brattain were so close to making a similar discovery themselves that knowledge of his idea would push them to publish before him?[10] Or was he simply so unsure of the idea that he avoided discussing it with them until he could think about it further? We do not know.

In order to function, Shockley's n-p-n device required additional physics beyond that involved in the point-contact transistor. It was crucial to understand that minority carriers are able to diffuse *through* the base layer in the presence of majority carriers. Bardeen

---

[10]Nick Holonyak gave a reasonable argument that once the point-contact transistor and the notion of minority carrier flow were on hand, the p-n junction transistor was "bound to follow." He recalled a dinner in the mid-1980s in Urbana at which Bardeen stated that he and Brattain had planned to move on to that device as soon as they completed their time-consuming work of preparing patents for the original transistor, only to find that Shockley had already tied up this area of work with the Bell Labs patent attorney (who, in John's words, "was in Shockley's pocket"). Holonyak interview by L. Hoddeson, 10 January 1992.

may have in fact had such an understanding, but it is not obvious from his logbook entries at the time. And at the time, he and Brattain were preoccupied with preparing patent documents dealing with their point-contact device. They still apparently believed that nearly all the hole flow occurred in a micron-deep *p*-type layer at the semiconductor surface.

Evidence for the required diffusion of the minority carriers into the bulk material was not long in coming. In a closed meeting at Bell Labs on 18 February 1948, physicist John Shive revealed that he had just tested a successful point-contact transistor using a very thin wedge of *n*-type germanium, but with the emitter and the collector placed on the *opposite* faces of the wedge (Shive, 1948, pp. 30–35). At the position where the two contacts touched it, the wedge was only 0.01 cm thick, while the distance between these points along the germanium surface was much larger. Shockley immediately recognized what this revelation meant. In this geometry, the holes had to flow by diffusion in the presence of the majority carriers, the electrons in the *n*-type germanium, through the bulk of the semiconductor; they were not confined to an inversion layer on the surface, as Bardeen and Brattain had been suggesting occurred in their contraption. "As soon as I had heard Shive's report," Shockley (1976) recalled, "I presented the ideas of my junction transistor disclosure and used them to interpret Shive's observation."

This experiment may have injected a heady dose of urgency into the Bell Labs solid-state physics group. On 26 February, the company applied for four patents on semiconductor amplifiers, including Bardeen and Brattain's original application on the point-contact transistor. Their two landmark papers, "The Transistor, a Semi-Conductor Triode" (Bardeen and Brattain, 1948b) and "Nature of the Forward Current in Germanium Point Contacts" (Brattain and Bardeen, 1948), were sent to the *Physical Review* four months later, on 25 June. One day later, Bell applied for Shockley's patent on the junction transistor, and on 30 June it announced the invention of the transistor in a press conference.

In July 1948 Shockley proved that hole "injection" (as he dubbed the flow of minority carriers in transistor action) was indeed occurring in *n*-type germanium. Working with Haynes, he showed that the charge carriers traveling from the emitter to the collector were in fact "positive particles with a mobility of about 1.2×10^3 cm^2/volt-sec" (Haynes and Shockley, 1949). Their paper was published in early 1949 together with Shive's article (Shive, 1949) on the two-sided transistor. Much of this research was discussed in detail in *Electrons and Holes in Semiconductors, with Applications to Transistor Electronics* (Shockley, 1950), which became the bible of the new discipline.

Shockley had another blind spot to overcome in his thinking about minority carriers before it finally became possible to fabricate working junction transistors. One of the problems behind the failure of his field-effect transistor had been how slowly charge carriers diffused through the polycrystalline silicon and germanium films used in the early experiments. Gordon Teal, a Bell Labs physical chemist, recognized the merits of using single crystals of germanium and silicon (Teal, 1976; Goldstein, 1993). He realized that in polycrystalline films minority carriers cannot survive long enough to make it from emitter to collector in sufficient numbers, but that they would have lifetimes 20 to 100 times longer in single crystals. Teal tried to convince Shockley of this critical advantage, but Shockley ignored his suggestion.

Fortunately Jack Morton, an engineer who headed the Bell Labs efforts to develop the point-contact transistor into a commercially viable product, took Teal seriously and in late 1949 gave him a small amount of support to pursue this avenue. Working with physical chemist Morgan Sparks, Teal modified the crystal-growing machine that he and a colleague had developed for pulling single crystals out of molten germanium (Goldstein, 1993). This alteration allowed them to dope the germanium in a controlled manner and thereby fabricate the first practical n-p-n junction transistor in April 1950. On the date of its demonstration, 20 April 1950, Shockley (1945, p. 128) penned a note in the margin of his 23 January 1948 entry (see Fig. 4): "An n-p-n unit was demonstrated today to Bown, Fisk, Wilson, Morton."[11]

Conclusions

On 10 December 1956 Shockley, Bardeen, and Brattain (in that order) were awarded the Nobel prize in physics for their "investigations on semi-conductors and the discovery of the transistor effect."[12] Taken together, their physical insights into the flow of electrons and holes in the intimate presence of one another were what made the invention of the transistor possible. Following Brattain's initial experiment indicating that the surface states could be overcome, Bardeen recognized in early December 1947 that holes could flow *as minority carriers* in a surface layer on a slab of n-type germanium; they employed this understanding to invent the point-contact transistor. But the possibility of minority-carrier *injection* into the bulk of the semiconductor, which made the junction transistor feasible, apparently occurred first to Shockley. In 1980 Bardeen reflected on the two interpretations of transistor action:

The difference between ourselves [Bardeen and Brattain] and Shockley came in the picture of how the holes flow from the emitter to the collector. They could flow predominantly through the inversion layer at the surface, which does contain holes. And the collector would be draining out the holes from the inversion layer. They could also flow through the bulk of the semiconductor, with their charge compensated by the increased number of electrons in the bulk . . . .[13]

It was this detailed understanding of semiconductor physics, which emerged in the course of a basic research program at Bell Labs, that overcame the barriers that had foiled all previous attempts to invent a solid-state amplifier.

It is important to recognize, however, that this physical insight was applied to a new technological base that had emerged from World War II. The very meaning of the word "semiconductor" changed markedly during that global confrontation. Where before the War, scientists commonly used the word to refer to compounds such as copper oxide, lead sulfide (or galena), and cadmium sulfide, afterwards it meant silicon and germanium doped with small amounts of highly controllable impurities. These crucial technological advances were mainly due to the work of physical chemists and electrochemists working in relative obscurity (Scaff and Ohl, 1947; Scaff, 1970; Seitz, 1995; Seitz and Einspruch, 1998a). Thus the "linear model" of technological development—wherein scientific

---

[11]This work was published (Shockley, Sparks, and Teal, 1951) in *Physical Review* over a year later, after a microwatt junction transistor operating at 10 kHz had been announced to the press. For the full story of the invention and development of the junction transistor, see Riordan and Hoddeson (1997a), Chapter 8.
[12]Quoted from Felix Belair, Jr., "Nobel Physics Prize Goes to 3 Americans; 2 Chemists Honored," *The New York Times*, 2 November 1956, p. 1.
[13]Interview of Bardeen by L. Hoddeson, 13 February 1980 (AIP Niels Bohr Library archives, College Park, MD), p. 2.

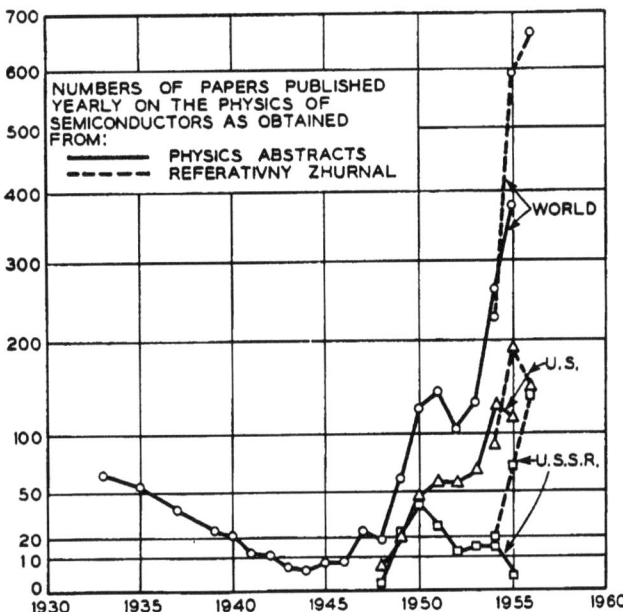

Figure 5. Variation with time in the annual numbers of papers on semiconductor physics listed in Physics Abstracts and (for 1954–56) in the Russian publication Refarativny Zhurnal. Symbols represent the number of papers from the United States, the Soviet Union, and the entire world, as indicated. The correspondence to actual publication rates is only rough, as the abstract journals fluctuate in the breadth of their coverage and in the time lag from publication of the papers to appearance of the abstracts.

research precedes technological development, from which useful products emerge—does not encompass very well what happened in the case of the transistor.

This new technology and the invention of the transistor have influenced the progress of science in many ways through the revolutionary impact of computers and electronic information processing. A more immediate impact was the stimulus on the field of solid-state physics that came in the few years after the breakthrough; a rough measure of this stimulus is given by the publication statistics plotted in Fig. 5 (Herring, 1957). The publication rate for research papers in all fields of physics showed a sizable decline during the combat years followed by a postwar recovery to a level above the pre-war rate—as one might expect due to lessened monetary and manpower resources during the War, followed by eventual return to a slowly expanding peacetime rate. In contrast, the publication rate in semiconductor physics suffered a gradual decline even in the pre-war years and almost disappeared during the War, but it recovered to a nearly level value in the period 1950–53 and then rose again spectacularly. We can reasonably attribute this great burst of activity to an increase in the number of people working in the field, and in the industrial and governmental support for such research.[14]

---

[14] Note that a similar rise appears in the world totals of papers on semiconductor physics, as listed by the Soviet abstract publication *Referativny Zhurnal*. But when one compares the curves for papers published in the United States with those in the Soviet Union, the rise in the latter seems to begin about a year later, a modest delay in view of the poor communication between scientists of the two countries during the Stalin years.

The research that led to the transistor had a psychological and intellectual impact that not only accelerated the growth of semiconductor research but also stimulated work in other areas of solid-state physics. Like semiconductors, most of these areas had seemed "dirty" to many physicists because relevant measurements were sensitive to factors such as the purity of materials, cleanliness of surfaces, and perfection of crystals, which made the phenomena too complicated to be understood in terms of simple theories. The research involved in the invention and development of the transistor showed that materials and experimental conditions could indeed be controlled, after all, and that many phenomena, such as the behavior of p-n junctions could be interpreted quantitatively using soundly based theories. Awareness of these advances was probably a major factor in the enthusiasm and resulting wave of publication that swept through the solid-state community in the early 1950s.[15]

The transistor discovery has clearly had enormous impact, both intellectually and in a commercial sense, upon our lives and work. A major vein in the corpus of condensed-matter physics quite literally owes its existence to this breakthrough. It also led to the microminiaturization of electronics, which has permitted us to have powerful computers on our desktops that communicate easily with each other via the Internet. The resulting globalization of science, technology, and culture is now transforming the ways we think and interact.

Acknowledgments

We thank William Brinkman, Nick Holonyak, Howard Huff, and Frederick Seitz for helpful discussions. This work was supported in part by grants from the Alfred P. Sloan Foundation and the Richard Lounsbery Foundatin.

---

[15]There were, of course, other favorable influences, such as the new availability of microwave tools. And the Cold War confrontation of the United States and Soviet Union probably also played a part. Another indication of the change in perspective can be seen in some of the statistics on ten-minute papers presented at meetings of the American Physical Society on non-semiconductor solid-state work done at governmental or industrial laboratories (other than Bell Labs): in 1949–50, about a sixth of such papers were theoretical; in 1956, about a third.

References

Bardeen, J., 1936, "Theory of the work function II: the surface double layer," Phys. Rev. **49**, 653-663.
Bardeen, J., 1946, Bell Labs Notebook No. 20780 (AT&T Archives, Warren, NJ).
Bardeen, J., 1947, "Surface states and rectification at a metal-semi-conductor contact," Phys. Rev. **71**, 717-727.
Bardeen, J., 1957, "Semiconductor research leading to the point-contact transistor," in *Les Prix Nobel en 1956*, edited by K. M. Siegbahn *et al.* (P. A. Nordstet & Sons, Stockholm), pp. 77-99; edited and reprinted in Science **126**, 105-112.
Bardeen, J., and W. H. Brattain, 1948a, "Three-electrode circuit element utilizing semiconductive materials," US Patent No. 2,524,035 (Washington, DC).
Bardeen, J., and W. H. Brattain, 1948b, "The transistor, a semi-conductor triode," Phys. Rev. **74**, 230-231.

Bardeen, J., and W. H. Brattain, 1949, "Physical principles involved in transistor action," Phys. Rev. **75**, 1208-1225.
Brattain, W. H., 1947a, "Evidence for surface states on semiconductors from change in contact potential on illumination," Phys. Rev. **72**, 345.
Brattain, W. H., 1947b, Bell Labs Notebook No. 18194 (AT&T Archives, Warren, NJ).
Brattain, W. H., 1947c, Bell Labs Notebook No. 21780 (AT&T Archives, Warren, NJ).
Brattain, W. H., 1951, "The copper oxide rectifier," Rev. Mod. Phys. **23**, 203-212.
Brattain, W. H., 1968, "Genesis of the transistor," Phys. Teach. **6**, 109-114.
Brattain, W. H., and J. Bardeen, 1948, "Nature of the forward current in germanium point contacts," Phys. Rev. **74**, 231-232.
Braun, E., and S. MacDonald, 1978, *Revolution in Miniature: The History and Impact of Semiconductor Electronics* (Cambridge University Press, Cambridge).
Davydov, B., 1938, "On the rectification of current at the boundary between two semi-conductors," C. R. (Dokl.) Acad. Sci. URSS **20**, 279-282; "On the theory of solid rectifiers," 1938, **20**, 283-285.
Frenkel, J., 1933, "Conduction in poor electronic conductors," Nature (London) **132**, 312-313.
Frenkel, J., 1936, "On the absorption of light and the trapping of electrons and positive holes in crystalline dielectrics," Phys. Z. Sowjetunion **9**, 158-186.
Goldstein, A., 1993, "Finding the right material: Gordon Teal as inventor and manager," in *Sparks of Genius: Portraits of Electrical Engineering*, edited by F. Nebeker (IEEE Press, New York), pp. 93–126.
Guerlac, H., 1987, *Radar in World War II* (AIP Press, New York).
Haynes, J. R., and W. Shockley, 1949, "Investigation of hole injection in transistor action," Phys. Rev. **75**, 691.
Henriksen, P. W., 1987, "Solid state physics research at Purdue," Osiris **2:3**, 237-260.
Herring, C., 1957, "The significance of the transistor discovery for physics," paper presented at a Bell Labs symposium on the Nobel prize (unpublished).
Herring, C., 1992, "Recollections from the early years of solid-state physics," Phys. Today **45**(4), 26-33.
Herring, C., M. Riordan, and L. Hoddeson, "Boris Davydov's theoretical work on minority carriers," n.d. (unpublished).
Hoddeson, L., 1981, "The discovery of the point-contact transistor," Hist. Stud. Phys. Sci. **12**, 41-76.
Hoddeson, L., G. Baym, and M. Eckert, 1987, "The development of the quantum mechanical theory of metals," Rev. Mod. Phys. **59**, 287-327.
Hoddeson, L., *et al.*, 1992, *Out of the Crystal Maze: Chapters in the History of Solid State Physics* (Oxford University Press, New York).
Hoddeson, L., *et al.*, 1993, *Critical Assembly: A Technical History of Los Alamos during the Oppenheimer Years, 1943–45* (Cambridge University Press, New York).
Holonyak, N., 1992, "John Bardeen and the point-contact transistor," Phys. Today **45**(4), 36-43.
Mott, N. F., 1939, "The theory of crystal rectifiers," Proc. R. Soc. London, Ser. A **171**, 27-38.
Mott, N. F., and H. Jones, 1936, *Theory of the Properties of Metals and Alloys* (Oxford University Press, Oxford).
Pearson, G., 1947, Bell Labs Notebook No. 20912 (AT&T Archives, Warren, NJ).
Pickering, A., 1984, *Constructing Quarks: A Sociological History of Particle Physics* (Edinburgh University Press, Edinburgh).
Riordan, M., and L. Hoddeson, 1997a, *Crystal Fire: The Birth of the Information Age* (W. W. Norton, New York).
Riordan, M., and L. Hoddeson, 1997b, "Minority carriers and the first two transistors," in *Facets: New Perspectives on the History of Semiconductors*, edited by A. Goldstein and W. Aspray (IEEE Center for the History of Electrical Engineering, New Brunswick, NJ), pp. 1–33.
Riordan, M., and L. Hoddeson, 1997c, "The origins of the pn junction," IEEE Spectr. **34**(6), 46-51.

Ross, I., 1998, "The invention of the transistor," Proc. IEEE **86**, 7-28.
Scaff, J., 1970, "The role of metallurgy in the technology of electronic materials," Metall. Trans. A **1**, 561-573.
Scaff, J., and R. S. Ohl, 1947, "The development of silicon crystal rectifiers for microwave radar receivers," Bell Syst. Tech. J. **26**, 1-30.
Schottky, W., 1939, "Zur halbleitertheorie der sperrschict- und spitzengleichrichter," Z. Phys. **113**, 367-414.
Seitz, F., 1994, *On the Frontier: My Life in Science* (AIP Press, New York).
Seitz, F., 1995, "Research on silicon and germanium in World War II," Phys. Today **48**(1), 22-27.
Seitz, F., and N. Einspruch, 1998a, *Electronic Genie: The Tangled History of Silicon* (University of Illinois Press, Urbana).
Seitz, F., and N. Einspruch, 1998b, "The tangled history of silicon and electronics," in *Semiconductor Silicon/1998*, edited by H. R. Huff, U. Gösele, and H. Tsuya (Electrochemical Society, Pennington, NJ), pp. 69–98.
Shive, J. N., 1948, Bell Labs Notebook No. 21869 (AT&T Archives, Warren, NJ).
Shive, J. N., 1949, "The double-surface transistor," Phys. Rev. **75**, 689-690.
Shockley, W., 1936, "Electronic energy bands in sodium chloride," Phys. Rev. **50**, 754-759.
Shockley, W., 1939, "On the surface states associated with a periodic potential," Phys. Rev. **56**, 317-323.
Shockley, W., 1945, Bell Labs Notebook No. 20455 (AT&T Archives, Warren, NJ).
Shockley, W., 1949, "The theory of p-n junctions in semiconductors and p-n junction transistors," Bell Syst. Tech. J. **28**, 435-489.
Shockley, W., 1950, *Electrons and Holes in Semiconductors, with Applications to Transistor Electronics* (Van Nostrand, New York).
Shockley, W., 1973, in *Proceedings of the Second European Solid State Device Research Conference* (Institute of Physics, London), pp. 55–75.
Shockley, W., 1976, "The path to the conception of the junction transistor," IEEE Trans. Electron Devices **ED-23**, 597-620.
Shockley, W., M. Sparks, and G. Teal, 1951, "P-N junction transistors," Phys. Rev. **83**, 151-162.
Tamm, I., 1932, "Uber eine mögliche art der elektronenbind-ung an kristalloberflächen," Phys. Z. Sowjetunion **1**, 733-746.
Teal, G., 1976, "Single crystals of germanium and silicon—basic to the transistor and integrated circuit," IEEE Trans. Electron Devices **ED-23**, 621-639.
Torrey, H. C., and C. A. Whitmer, 1948, *Crystal Rectifiers* (McGraw-Hill, New York; republished in 1964 by Boston Technical Publishers, Boston).
Weiner, C., 1973, "How the transistor emerged," IEEE Spectr. **10**(1), 24-33.
Wilson, A. H., 1931, "The theory of electronic semi-conductors," Proc. R. Soc. London, Ser. A **133**, 458-491; 1931, "The theory of electronic semi-conductors—II," **134**, 277-287.

Statistical Physics and Fluids

Statistical Mechanics: A Selective Review of Two Central Issues

Joel L. Lebowitz

Nature has a hierarchical structure, with time, length, and energy scales ranging from the submicroscopic to the supergalactic. Surprisingly, it is possible, and in many cases essential, to discuss these levels independently—quarks are irrelevant for understanding protein folding and atoms are a distraction when studying ocean currents. Nevertheless, it is a central lesson of science, very successful in the past three-hundred years, that there are no new fundamental laws, only new phenomena, as one goes up the hierarchy. Thus arrows of explanations between different levels always point from smaller to larger scales, although the origin of higher-level phenomena in the more fundamental lower-level laws is often very far from transparent. (In addition some of the dualities recently discovered in string theory suggest possible arrows from the highest to the lowest level, closing the loop.)

Statistical mechanics provides a framework for describing how well-defined higher-level patterns or behavior may result from the nondirected activity of a multitude of interacting lower-level individual entities. The subject was developed for, and has had its greatest success so far in, relating mesoscopic and macroscopic thermal phenomena to the microscopic world of atoms and molecules. Fortunately, many important properties of objects containing very many atoms—such as the boiling and freezing of water—can be obtained from simplified models of the structure of atoms and the laws governing their interactions. Statistical mechanics therefore often takes as its lowest-level starting point—and so will I in this article—Feynman's description of atoms (Feynman, Leighton, and Sands, 1963) as "little particles that move around in perpetual motion, attracting each other when they are a little-distance apart, but repelling upon being squeezed into one another." Why this crude classical picture (a refined version of that held by some ancient Greek philosophers) gives predictions that are not only qualitatively correct but in many cases also highly accurate, is certainly far from clear to me—but that is another story or article.

Statistical mechanics explains how macroscopic phenomena originate in the cooperative behavior of these "little particles." Some of the phenomena are simple additive effects of the actions of individual atoms, e.g., the pressure exerted by a gas on the walls of its container, while others are paradigms of emergent behavior, having no direct counterpart in the properties or dynamics of individual atoms. Particularly fascinating and important examples of such emergent phenomena are the irreversible approach to equilibrium and phase transitions in equilibrium. Both of these would (or should) be astonishing if they were not so familiar. Their microscopic derivation and analysis forms the core of statistical mechanics.

For a more general survey of statistical mechanics in the past hundred years, the reader is referred to the other chapters in this section as well as to my article in the special volume celebrating the first-hundred years of the *Physical Review* (Lebowitz, 1995a) where there are also reprints of some of the original papers as well as references to others. For some very recent reviews of specific topics see Fisher (1998) and Brydges and Martin (1999).

Microscopic Origins of Irreversible Macroscopic Behavior

There are many conceptual and technical problems encountered in going from a time-symmetric description of the dynamics of atoms to a time-asymmetric description of the evolution of macroscopic systems. This involves a change from Hamiltonian (or Schrödinger) equations to hydrodynamical ones, e.g., the diffusion equation. The problem of reconciling the latter with the former became a central issue in physics during the last part of the nineteenth century. It was also, in my opinion, essentially resolved at that time, at least in the framework of nonrelativistic classical mechanics. To quote from Thomson's (later Lord Kelvin) 1874 article (Thomson, 1874), "The essence of Joule's discovery is the subjection of physical phenomena to dynamical law. If, then, the motion of every particle of matter in the universe were precisely reversed at any instant, the course of nature would be simply reversed for ever after. The bursting bubble of foam at the foot of a waterfall would reunite and descend into the water.... Physical processes, on the other hand, are irreversible: for example, the friction of solids, conduction of heat, and diffusion. Nevertheless, the principle of dissipation of [organized] energy is compatible with a molecular theory in which each particle is subject to the laws of abstract dynamics." Unfortunately there is still much confusion about this issue among some scientists which is the reason for my discussing it here.[1]

Formally the problem considered by Thomson is as follows: The complete microscopic (or micro) state of an isolated classical system of N particles is represented by a point X in its phase space Γ, $X = (\mathbf{r}_1, \mathbf{p}_1, \mathbf{r}_2, \mathbf{p}_2, \ldots, \mathbf{r}_N, \mathbf{p}_N)$, $\mathbf{r}_i$ and $\mathbf{p}_i$ being the position and momentum of the ith particle. The evolution is governed by Hamiltonian dynamics, which connects a microstate at some time t_0, $X(t_0)$, to the microstate $X(t)$ at all other times t, $-\infty < t < \infty$. Let $X(t_0)$ and $X(t_0 + \tau)$, with τ positive, be two such microstates. Reversing (physically or mathematically) all velocities at time $t_0 + \tau$, we obtain a new microstate. If we now follow the evolution for another interval τ we find that the new

---

[1]This issue was the subject of a "round table" at the 20th IUPAP International Conference on Statistical Physics held in Paris, July 20–25, 1998. The panel consisted of M. Klein, who gave a historical overview, myself, who presented the Boltzmannian point of view described in the text which follows, I. Prigogine, who disagreed strongly with this point of view, claiming that the explanation lies in some (to me abstruse) new mathematical formalism developed by his group, and D. Ruelle, who presented some recent developments in the dynamical systems approach to far from equilibrium stationary states. The proceedings of that conference, which contain the presentations of the panel as well as some of the latest developments in statistical mechanics, will appear in *Physica A*. (See also Lebowitz, 1993a; 1993b; 1994; 1995b.) For a clear defense of Boltzmann's views against some recent attacks see Bricmont (1996). This article first appeared in the publication of the Belgian Physical Society, Physicalia Magazine **17**, 159 in 1995, where it is followed by an exchange between Prigogine and Bricmont.

microstate at time $t_0 + 2\tau$ is just $RX(t_0)$, the microstate $X(t_0)$ with all velocities reserved; $RX = (\mathbf{r}_1, -\mathbf{p}_1, \mathbf{r}_2, -\mathbf{p}_2, \ldots, \mathbf{r}_N, -\mathbf{p}_N)$. Hence, if there is an evolution [i.e., a trajectory $X(t)$] of a system in which some property of the system described by some function $f(X) = f(RX)$, which increases as t increases, e.g., particle densities get more uniform by diffusion, then there is also one in which the density profile evolves in the opposite direction, since the density is the same for X and RX. So why is one direction, identified with "entropy" increase by the second "law," common and the other never seen?

The explanation of this apparent paradox, due to Thomson, Maxwell, and Boltzmann, which I will now describe, shows that not only is there no conflict between reversible microscopic laws and irreversible macroscopic behavior, but, as clearly pointed out by Boltzmann in his later writings,[2] there are extremely strong, albeit subtle, reasons to expect the latter from the former. These involve several interrelated ingredients which together provide the sharp distinction between microscopic and macroscopic variables required for the emergence of definite time-asymmetric behavior in the evolution of the latter despite the total absence of such asymmetry in the dynamics of individual atoms. They are: (a) the great disparity between microscopic and macroscopic scales, (b) the fact that events are, as put by Boltzmann, determined not only by differential equations, but also by initial conditions, and (c) the use of probabilistic reasoning: it is not every microscopic state of a macroscopic system that will evolve in accordance with the second law, but only the "majority" of cases—a majority which however becomes so overwhelming when the number of atoms in the system becomes very large that irreversible behavior becomes a near certainty. (The characterization of the set whose "majority" we are describing will be discussed later.)

To see how the explanation works let us denote by M the macrostate of a macroscopic system. For a system containing N atoms in a box V, the microstate X is a point in the $6N$-dimensional phase space Γ while M is a much cruder description, e.g., the specification, to within a given accuracy, of the energy of the system and of the number of particles in each half of the box. [A more refined (hydrodynamical) description would divide V into K cells, where K is large, but still $K \ll N$, and specify the number of particles and energy in each cell, again with some tolerance.] Thus, while M is determined by X, there are many X which correspond to the same M. We will call Γ_M the region in Γ consisting of all microstates X corresponding to a given macrostate M and take as a measure of the "number" of microstates corresponding to a subset A of Γ_M to be equal to the $6N$-dimensional Liouville volume of A normalized by the volume of Γ_M, denoted by $|\Gamma_M|$: $|\Gamma_M| = \int_{\Gamma_M} \Pi_{i=1}^N d\mathbf{r}_i d\mathbf{p}_i$. (This corresponds to the classical limit of "counting" states in quantum mechanics.)

Consider now a situation in which there is initially a wall confining a dilute gas of N atoms to the left half of the box V. When the wall is removed at time t_a, the phase-space volume available to the system is fantastically enlarged, roughly by a factor of 2^N. (If the system contains one mole of gas in a container then the volume ratio of the unconstrained region to the constrained region is of order $10^{10^{20}}$.) This region will contain new macrostates with phase-space volumes very large compared to the initial phase-space volume available to the system. We can then expect (in the absence of any obstruction, such as a hidden conservation law) that as the phase point X evolves under the uncon-

---

[2]Boltzmann's early writings on the subject are sometimes unclear, wrong, and even contradictory. His later writings, however, are superbly clear and right on the money (even if a bit verbose for Maxwell's taste). I strongly recommend the references cited at the end.

strained dynamics it will with very high "probability" enter the newly available regions of phase space and thus find itself in a succession of new macrostates M for which $|\Gamma_M|$ is increasing. This will continue until the system reaches its unconstrained macroscopic equilibrium state, M_{eq}, that is, until $X(t)$ reaches $\Gamma_{M_{eq}}$, corresponding to approximately half the particles in each half of the box, say within an interval ($\frac{1}{2}-\epsilon, \frac{1}{2}+\epsilon$), $\epsilon \ll 1$, since in fact $|\Gamma_{M_{eq}}|/|\Sigma_E| \simeq 1$, where $|\Sigma_E|$ is the total phase-space volume available under the energy constraint. After that time we can expect only small fluctuations about the value $\frac{1}{2}$, well within the precision ϵ, typical fluctuations being of the order of the square root of the number of particles involved.

To extend the above observation to more general situations, Boltzmann associated with each microscopic state X of a macroscopic system, be it gas, fluid, or solid, a number S_B, given, up to multiplicative and additive constants (in particular we set Boltzmann's constant, k_B, equal to unity), by

$$S_B(X) = \log|\Gamma_{M(X)}|. \qquad (2.1)$$

A crucial observation made by Boltzmann was that when $X \in M_{eq}$ then $S_B(X)$ agrees (up to terms negligible in the size of the system) with the thermodynamic entropy of Clausius and thus provides a microscopic definition of this macroscopically defined, operationally measurable (à la Carnot), extensive property of macroscopic systems in *equilibrium*. Having made this connection Boltzmann found it natural also to use Eq. (2.1) to define the entropy for a macroscopic system not in equilibrium and thus to explain (in agreement with the ideas of Maxwell and Thomson) the observation, embodied in the second law of thermodynamics, that when a constraint is lifted, an isolated macroscopic system will evolve toward a state with greater entropy,[3] i.e., that S_B will *typically* increase in a way which *explains* and describes qualitatively the evolution towards equilibrium of macroscopic systems.

Typical, as used here, means that the set of microstates corresponding to a given macrostate M for which the evolution leads to a macroscopic decrease in the Boltzmann entropy during some fixed time period τ, occupies a subset of Γ_M whose Liouville volume is a fraction of $|\Gamma_M|$ which goes very rapidly (exponentially) to zero as the number of atoms in the system increases.

It is this very large number of degrees of freedom involved in the specification of macroscopic properties that distinguishes macroscopic irreversibility from the weak approach to equilibrium of ensembles for systems with good ergodic properties (Lebowitz, 1993a, 1993b, 1994, 1995b). While the former is manifested in a typical evolution of a single macroscopic system, the latter, which is also present in *chaotic* systems with but a few degrees of freedom, e.g., two hard spheres in a box, does not correspond to any appearance of time asymmetry in the evolution of an individual system. On the other hand, because of the exponential increase of the phase-space volume, even a system with only a few hundred particles (commonly used in molecular-dynamics computer simula-

---

[3]When M specifies a state of local equilibrium, $S_B(X)$ agrees up to negligible terms, with the "hydrodynamic entropy." For systems far from equilibrium the appropriate definition of M and thus of S_B is more problematical. For a dilute gas in which M is specified by the density $f(\mathbf{r},\mathbf{v})$ of atoms in the six-dimensional position and velocity space $S_B(X) = -\int\int f(\mathbf{r},\mathbf{v})\log f(\mathbf{r},\mathbf{v})d\mathbf{r}d\mathbf{v}$. This identification is, however, invalid when the potential energy is not negligible; cf., Jaynes (1971). Following Penrose (1970), we shall call $S_B(X)$ the Boltzmann entropy of the macrostate $M = M(X)$.

tions) will, when started in a nonequilibrium "macrostate" M, with "random" $X \in \Gamma_M$, appear to behave like a macroscopic system.[4] This will be so even when integer arithmetic is used in the simulations so that the system behaves as a truly isolated one; when its velocities are reversed the system retraces its steps until it comes back to the initial state (with reversed velocities), after which it again proceeds (up to very long Poincare recurrence times) in the typical way (Levesque and Verlet, 1993; see also Nadiga, Broadwell, and Sturtevant, 1989).

Maxwell makes clear the importance of the scale separation when he writes (Maxwell, 1878): "the second law is drawn from our experience of bodies consisting of an immense number of molecules. ... it is continually being violated, ..., in any sufficiently small group of molecules .... As the number ... is increased ... the probability of a measurable variation ... may be regarded as practically an impossibility." We might take as a summary of the discussions in the late part of the last century the statement by Gibbs (Gibbs, 1875), quoted by Boltzmann (in a German translation) on the cover of his book *Lectures on Gas Theory II*: "In other words, the impossibility of an uncompensated decrease of entropy seems to be reduced to an improbability."

As already noted, typical here refers to a measure which assigns (at least approximately) equal weights to the different microstates consistent with the "initial" macrostate M. (This is also what was meant earlier by the "random" choice of an initial $X \in \Gamma_M$ in the computer simulations.) In fact, any meaningful statement about probable or improbable behavior of a physical system has to refer to some agreed upon measure (probability distribution). It is, however, this use of probabilities (whose justification is beyond the reach of mathematical theorems) and particularly of the notion of typicality for explaining the origin of the apparently deterministic second law which was most difficult for many of Boltzmann's contemporaries, and even for some people today, to accept (Lebowtiz, 1993a, 1993b, 1994, 1995b; Bricmont, 1996). This was clearly faced by Boltzmann when he wrote, in his second reply to Zermelo in 1897 (Boltzmann, 1897): "The applicability of probability theory to a particular case cannot of course be proved rigorously. ... Despite this, every insurance company relies on probability theory. ... It is even more valid [here], on account of the huge number of molecules in a cubic millimetre.... The assumption that these rare cases are not observed in nature is not strictly provable (nor is the entire mechanical picture itself) but in view of what has been said it is so natural and obvious, and so much in agreement with all experience with probabilities...[that]...[it] is completely incomprehensible to me how anyone can see a refutation of the applicability of probability theory in the fact that some other argument shows that exceptions must occur now and then over a period of eons of time; for probability theory itself teaches just the same thing."

It should be noted here that an important ingredient in the above analysis is the constancy in time, of the Liouville volume of sets in the phase space Γ as they evolve under the Hamiltonian dynamics (Liouville's Theorem). Without this invariance the connection between phase-space volume and probability would be impossible or at least very problematic. We also note that, in contrast to $S_B(X)$, the Gibbs entropy $S_G(\mu)$,

$$S_G(\mu) = -\int \mu \log \mu \, dX, \qquad (2.2)$$

---

[4]After all the likelihood of hitting, in the course of say one- thousand tries, on something which has probability of order 2^{-N} is, for all practical purposes, the same, whether N is a hundred or 10^{23}.

is defined not for individual microstates but for statistical ensembles or probability distributions μ. For equilibrium ensembles $S_G(\mu_{eq}) \sim \log|\Sigma_E| \sim S_B(X)$, for $X \in M_{eq}$, up to terms negligible in the size of the system. However, unlike S_B, S_G does not change in time even for time-dependent ensembles describing (isolated) systems not in equilibrium. Hence the relevant entropy for understanding the time evolution of macroscopic systems is S_B and not S_G.

Initial conditions

Once we accept the statistical explanation of why macroscopic systems evolve in a manner that makes S_B increase with time, there remains the nagging problem (of which Boltzmann was well aware) of what we mean by "with time": since the microscopic dynamical laws are symmetric, the two directions of the time variable are *a priori* equivalent and thus must remain so *a posteriori*.

Put another way: why can we use phase-space arguments (or time-asymmetric diffusion-type equations) to predict the behavior of an *isolated* system in a nonequilibrium macrostate M_b at some time t_b, e.g., a metal bar with a nonuniform temperature, in the future, i.e., for $t > t_b$, but not in the past, i.e. for $t < t_b$? After all, if the macrostate M is invariant under velocity reversal of all the atoms, then the analysis would appear to apply equally to $t_b + \tau$ and $t_b - \tau$. A plausible answer to this question is to assume that the nonequilibrium state of the metal bar M_b had its origin in an even more nonuniform macrostate M_a, prepared by some experimentalist at some earlier time $t_a < t_b$ and that for states thus prepared we can apply our (approximately) equal *a priori* probability of microstates argument, i.e., we can assume its validity at time t_a. But what about events on the sun or in a supernova explosion where there are no experimentalists? And what, for that matter, is so special about the status of the experimentalist? Isn't he or she part of the physical universe?

Before trying to answer the last set of "big" questions let us consider whether the assignment of equal probabilities for $X \in \Gamma_{M_a}$ at t_a permits the use of an equal probability distribution of $X \in \Gamma_{M_b}$ at time t_b for predicting *future* macrostates: in a situation where the system is isolated for $t > t_a$. Note that the microstates in Γ_{M_b}, which have come from Γ_{M_a} through the time evolution during the time interval from t_a to t_b, make up only a very small fraction of the volume of Γ_{M_b}, call it Γ_{ab}. Thus we have to show that the overwhelming majority of points in Γ_{ab} (with respect to Liouville measure on Γ_{ab}, which is the same as Liouville measure on Γ_{M_a}) have *future* macrostates like those typical of Γ_b—while still being very special and unrepresentative of Γ_{M_b} as far as their *past* macrostates are concerned.[5] This property is explicitly proven by Lanford (1981) in his derivation of the Boltzmann equation (for short times), and is part of the derivation of hydrodynamic equations (Lebowitz, Presutti, and Spohn, 1988; De Masi and Presutti, 1991; Spohn, 1991; Esposito and Marra, 1994; Landim and Kipnis, 1998; see also Lebowitz and Spohn, 1983).

To see intuitively the origin of this property we note that for systems with realistic interactions the domain Γ_{ab} will be so convoluted as to *appear* uniformly smeared out in Γ_{M_b}. It is therefore reasonable that the future behavior of the system, as far as macrostates

---

[5]We are considering here the case where the macrostate $M(t)$, at time t, determines $M(t')$ for $t' > t$. There are of course situations where $M(t')$ depends also (weakly or even strongly) on the history of $M(t)$ in some time interval prior to t', e.g., in materials with memory.

go, will be unaffected by their past history. It would of course be nice to prove this in all cases, e.g., justifying (for practical purposes) the factorization or "Stosszahlansatz" assumed by Boltzmann in deriving his dilute-gas kinetic equation for all times $t > t_a$, not only for the short times proven by Lanford. Our mathematical abilities are, however, equal to this task only in very simple situations as we shall see below. This should, however, be enough to convince a "reasonable" person.

The large number of atoms present in a macroscopic system plus the chaotic nature of the dynamics also explains why it is so difficult, essentially impossible (except in some special cases such as experiments of the spin-echo type, and then only for a limited time), for a clever experimentalist to deliberately put such a system in a microstate which will lead it to evolve contrary to the second law. Such microstates certainly exist—just start with a nonuniform temperature, let it evolve for a while, then reverse all velocities. In fact, they are readily created in the computer simulations with no roundoff errors as discussed earlier (Levesque and Verlet, 1993; see also Nadiga, Broadwell, and Sturtevant, 1989). To quote again from Thomson's article (Thomson, 1984): "If we allowed this equalization to proceed for a certain time, and then reversed the motions of all the molecules, we would observe a disequalization. However, if the number of molecules is very large, as it is in a gas, any slight deviation from absolute precision in the reversal will greatly shorten the time during which disequalization occurs." *In addition*, the effect of unavoidable small outside influences, which are unimportant for the evolution of macrostates in which $|\Gamma_M|$ is increasing, will greatly destabilize evolution in the opposite direction when the trajectory has to be *aimed* at a very small region of the phase space (Lebowitz, 1993a, 1993b, 1994, 1995b).

Let us return now to the big question posed earlier: what is special about t_a compared to t_b in a world with symmetric laws? Put differently, where ultimately do initial conditions such as those assumed at t_a come from? In thinking about this we are led more or less inevitably to cosmological considerations and to postulate an initial "macrostate of the universe" having a very small Boltzmann entropy at some time t_0. To again quote Boltzmann (1896): "That in nature the transition from a probable to an improbable state does not take place as often as the converse, can be explained by assuming a very improbable [small S_B] initial [macro]state of the entire universe surrounding us. This is a reasonable assumption to make, since it enables us to explain the facts of experience, and one should not expect to be able to deduce it from anything more fundamental." We do not, however, have to assume a very special initial microstate X, and this is a very important aspect of our considerations. As Boltzmann further writes: "we do not have to assume a special type of initial condition in order to give a mechanical proof of the second law, if we are willing to accept a statistical viewpoint ... if the initial state is chosen at random ... entropy is almost certain to increase." All that is necessary to assume is a far from equilibrium initial macrostate and this is in accord with all cosmological and other independent evidence.

Feynman (1967) clearly agrees with this when he says, "it is necessary to add to the physical laws the hypothesis that in the past the universe was more ordered, in the technical sense, than it is today ... to make an understanding of the irreversibility." More recently the same point was made very clearly by Penrose (1990) in connection with the "big-bang" cosmology. Penrose, unlike Boltzmann, believes that we should search for a more fundamental theory that will also account for the initial conditions. Meanwhile he takes for the initial macrostate of the universe the smooth energy-density state prevalent soon after the big bang. Whether this is the appropriate initial state or not, it captures an essential fact about our universe. Gravity, being purely attractive and long range, is unlike any of the other natural forces. When there is enough matter/energy around, it completely overcomes the tendency towards uniformization observed in ordinary objects at high energy densities or temperatures. Hence, in a universe dominated, like ours, by gravity, a uniform density corresponds to a state of very low entropy, or phase-space volume, for a given total energy.

The local "order" or low entropy we see around us (and elsewhere)—from complex molecules to trees to the brains of experimentalists preparing macrostates—is perfectly consistent with (and possibly even a consequence of) the initial macrostate of the universe. The value of S_B of the present clumpy macrostate of the universe, consisting of planets, stars, galaxies, and black holes, is much much larger than what it was in the initial state and also quite far away from its equilibrium value. The "natural" or "equilibrium" state of the universe is, according to Penrose, one with all matter and energy collapsed into one big black hole which would have a phase-space volume some $10^{10^{120}}$ times that of the initial macrostate. (So we may still have a long way to go.)

Quantitative considerations

Let me now describe briefly the very interesting work, still in progress, in which one rigorously derives time-asymmetric hydrodynamic equations from reversible microscopic laws (Lebowitz, Presutti, and Spohn, 1988; De Masi and Presutti, 1991; Spohn, 1991; Esposito and Marra, 1994; Landim and Kipnis, 1998). While many qualitative features of irreversible macroscopic behavior depend very little on the positivity of Lyapunov exponents, ergodicity, or mixing properties of the microscopic dynamics, such properties are important for the quantitative description of the macroscopic evolution, i.e., for the derivation of time-asymmetric autonomous equations of hydrodynamic type. The existence and form of such equations depend on the instabilities of microscopic trajectories induced by chaotic dynamics. When the chaoticity can be proven to be strong enough (and of the right form) such equations can be derived rigorously from the reversible microscopic dynamics by taking limits in which the ratio of macroscopic to microscopic scales goes to infinity. Using the law of large numbers one shows that these equations describe the behavior of almost all individual systems in the ensemble, not just that of ensemble averages, i.e., that the dispersion goes to zero in the scaling limit. The equations also hold, to a high accuracy, when the macro/micro ratio is finite but very large.

A simple example in which this can be worked out in detail is the periodic Lorentz gas (or Sinai billiard). This consists of a *macroscopic number of noninteracting particles* moving among a periodic array of fixed convex scatterers, arranged in the plane in such a way that there is a maximum distance a particle can travel between collisions. The chaotic nature of the microscopic dynamics, which leads to an approximately isotropic local distribution of velocities, is directly responsible for the existence of a simple autonomous deterministic description, via a diffusion equation, for the macroscopic particle profiles of this system. A second example is a system of hard spheres at very low densities for which the Boltzmann equation has been shown to describe the evolution of the density in the six-dimensional position and velocity space (at least for short times) (Lanford, 1981). I use these examples, despite their highly idealized nature, because here all the mathematical i's have been dotted. They thus show *ipso facto*, in a way that should convince even (as Mark Kac put it) an "unreasonable" person, not only that there is no conflict between reversible microscopic and irreversible macroscopic behavior but also that, *for essentially all initial microscopic states consistent with a given nonequilibrium macroscopic state*, the latter follows from the former—in complete accord with Boltzmann's ideas. Yet the debate goes on.

Phase Transitions in Equilibrium Transitions in Equilibrium Systems

Information about the equilibrium phases of a homogeneous macroscopic system is conveniently encoded in its phase diagram. Phase diagrams can be very complicated but their essence is already present in the familiar, simplified two-dimensional diagram for a one-component system like water or argon. This has axes marked by the temperature T and pressure p, and gives the decomposition of this thermodynamic parameter space into different regions: the blank regions generally correspond to parameter values in which there is a unique pure phase, gas, liquid, or solid, while the lines between these regions represent values of the parameters at which two pure phases can exist. At the triple point, the system can exist in any of three pure phases.

In general, a macroscopic system with a given Hamiltonian is said to *undergo* or *be at* a first-order phase transition when the temperature and pressure or, more generally, the temperature and chemical potentials do not uniquely specify its homogeneous equilibrium state. The different properties of the pure phases coexisting at such a transition manifest themselves as discontinuities in certain observables, e.g., a discontinuity in the density as a function of temperature at the boiling point. On the other hand, when one moves between two points in the thermodynamic parameter space along a path which does not intersect any coexistence line the properties of the system change smoothly.

I will now sketch a mathematically precise formulation of what is meant by coexistence of phases, and give some rigorous results about phase diagrams. This is a beautiful part of the developments in statistical mechanics during this century, it is also one which is essential to a full understanding of the singular behavior of macroscopic systems at phase transitions, e.g., the discontinuity in the density mentioned earlier. These singularities can only be captured precisely through the infinite volume or *thermodynamic limit*; a formal mathematical procedure in which the size of the system becomes infinite while the number of particles and energy per unit volume (or the chemical potential and temperature) stay fixed. While at first sight entirely unrealistic, such a limit represents an idealization of a macroscopic physical system whose spatial extension, although finite, is very large on the microscopic scale of interparticle distances or interactions. The advantage of this idealization is that boundary and finite-size effects present in real systems, which are frequently irrelevant to the phenomena of interest, are eliminated in the thermodynamic limit. As Robert Griffiths once put it, every experimentalist implicitly takes such a limit when he or she reports the results of a measurement, like the magnetic susceptibility, without giving the size and shape of the sample.

My starting point here is the Gibbs formalism for calculating equilibrium properties of macroscopic systems as ensemble averages of *functions* of the microscopic state of the system. While the use of ensembles was anticipated by Boltzmann (Boltzmann, 1884; Broda, 1973; Klein, 1973; Flamn, 1973) and independently discovered by Einstein, it was Gibbs who, by his brilliant systematic treatment of statistical ensembles, i.e., probability measures on the phase space, developed statistical mechanics into a useful elegant tool for relating, not only typical but also fluctuating behavior in equilibrium systems, to microscopic Hamiltonians. In a really remarkable way the formalism has survived essentially intact the transition to quantum mechanics. Here, however, I restrict myself to classical mechanics.[6]

As in the previous section the microscopic state of a system of N particles in a spatial domain V is given by a point X in the phase space, $X=(\mathbf{r}_1,\mathbf{p}_1,\ldots,\mathbf{r}_N,\mathbf{p}_N)$. We are generally interested in the values of suitable *sum functions* of X: those which can be written as a sum of terms involving only a fixed finite number of particles, e.g., $F_{(1)}(X)=\Sigma f_1(\mathbf{r}_i,\mathbf{p}_i)$, $F_{(2)}(X)=\Sigma_{i,j}f_2(\mathbf{r}_i,\mathbf{p}_i,\mathbf{r}_j,\mathbf{p}_j)$ (with $f_2(\mathbf{r}_i,\mathbf{p}_i,\mathbf{r}_j,\mathbf{p}_j)\to 0$ when $|\mathbf{r}_i-\mathbf{r}_j|\to\infty$), etc. (Familiar examples are the kinetic and potential energies of the system.) Typical macroscopic properties then correspond to sum functions which, when divided by the volume $|V|$, are essentially constant on the energy surface Σ_E of a macroscopic system. Consequently, if we take the thermodynamic limit, defined by letting $N\to\infty$, $E\to\infty$, and $|V|\to\infty$ in such a way that $N/|V|\to\rho$ and $E/|V|\to e$, then these properties assume deterministic values, i.e., their variances go to zero. They also become (within limits) independent of the shape of V and the nature of the boundaries of V. (As a less familiar concrete example, let $f_1(\mathbf{r}_i,\mathbf{p}_i)=[(1/2m)\mathbf{p}_i^2]^2$, the square of the kinetic energy of the ith particle. Then, in the thermodynamic limit, $|V|^{-1}F_1(X)\to\frac{9}{4}\rho T^2(e,\rho)$ for *typical* X, with T the temperature of the system given by $[(\partial/\partial e)s(e,\rho)]^{-1}$, with $s(e,\rho)$ the thermodynamic limit of $|V|^{-1}\log|\Sigma_E|$.)

It is this property of sum functions which makes meaningful the use of ensembles to describe the behavior of individual macroscopic systems as in the previous section. In particular it assures the "equivalence" of ensembles: microcanonical, canonical, grand canonical, pressure, etc. for computing equilibrium properties. The use of the thermodynamic limit actually extends this equivalence, in that part of the phase diagram where the system has a unique phase, to the probability distribution of fluctuating quantities, e.g., the correlation functions. These are translation invariant and independent of boundary conditions in the thermodynamic limit. (See later for what happens on coexistence lines.)

To actually obtain the phase diagram of a system with a given Hamiltonian is a formidable mathematical task. It has still not been solved even for such simple continuum systems as particles interacting via a Lennard-Jones pair potential. I will therefore postpone further discussion of continuum systems until later and switch now to lattice systems for which such results are available. These come from a variety of techniques some of which, I shall not be able to discuss here at all.

Lattice systems

Lattice systems can be considered approximations to the continuum particle systems (the cell theory of fluids) or as representations of spins in magnetic systems (Lee and Yang, 1952; Yang and Lee, 1952). They also arise as models of a variety of nonthermal physical phenomena (Liggett, 1985; Vicsek, 1989; Meakin, 1998). I shall consider for simplicity the simple cubic lattice Z^d, in d dimensions. At each site $\mathbf{x}\in Z^d$ there is a spin variable $S(\mathbf{x})$ which can take k discrete values, $S(\mathbf{x})=\xi_1,\ldots,\xi_k$. The configuration of the system

---

[6]It is clearly impossible to cite here all or even a significant fraction of all the good reviews and textbooks on the subject. The reader would do well however to browse among the original works (Thomson, 1874; Penrose, 1970) and in particular read Gibbs (1960) beautiful book. A partial list of books and reviews with a mathematical treatment of Gibbs measures and phase transitions which contain the results presented follows: Fisher (1964); Ruelle (1969); Griffiths (1972); Baxter (1982); Sinai (1982); Georgii (1988); Fernández, Fröhlich, and Sokal (1992); Simon (1993).

in a region $V \subset Z^d$ containing $|V|$ sites, is denoted by $\mathbf{S}_V$, it is one of the $k^{|V|}$ points in the set $\Omega = \{\xi_1, \ldots, \xi_k\}^V$. There is an interaction energy U which is a sum of *internal* interactions assumed to be translation invariant and *boundary terms*.

To be specific, consider the Ising model, $S(\mathbf{x}) = \pm 1$, with uniform magnetic field h and pair interactions $u(\mathbf{r})$. The energy of a configuration $\mathbf{S}_V$ is given by

$$U(\mathbf{S}_V | \bar{S}_{V^c}) = -h \sum_{\mathbf{x} \in V} S(\mathbf{x}) - \frac{1}{2} \sum_{\mathbf{x}, \mathbf{y} \in V} \sum u(\mathbf{x}-\mathbf{y}) S(\mathbf{x}) S(\mathbf{y})$$

$$- \sum_{\mathbf{x} \in V} \left\{ \sum_{\mathbf{y} \in V^c} u(\mathbf{x}-\mathbf{y}) \bar{S}(\mathbf{y}) \right\} S(\mathbf{x}). \tag{3.1}$$

In Eq. (3.1) $\bar{S}(\mathbf{y})$ denotes the *preassigned* value of the spin variables at sites $\mathbf{y}$ in V^c, the complement (or outside) of V, which act as boundary conditions (BC). They contribute, through the last sum in Eq. (3.1), an energy term which is proportional to the surface area of V whenever the interactions have finite range or decay fast enough to be summable, e.g., $u(\mathbf{r})$ decays faster than $|\mathbf{r}|^{-(d+\epsilon)}$, $\epsilon > 0$. We can also consider periodic or free BC: the latter corresponds to dropping the last term in Eq. (3.1). We will indicate all possible boundary conditions by the letter b; sometimes setting $b = p$ or $b = f$ for periodic or free BC.

When the system is in equilibrium at temperature T, the probability of finding the configuration $\mathbf{S}_V$ is given by the Gibbs formula (see footnote 6)

$$\mu_V(\mathbf{S}_V | b) = \frac{1}{Z(\mathbf{J}; b, V)} \exp[-\beta U(\mathbf{S}_V | b)], \tag{3.2}$$

where $\beta^{-1} = T$, and Z is the partition function,

$$Z(\mathbf{J}; b, V) = \sum_{\mathbf{S}_V} \exp[-\beta U(\mathbf{S}_V | b)]. \tag{3.3}$$

The sum in Eq. (3.3) is over all possible microscopic configurations of the system in V and we have used $\mathbf{J}$ to refer to all the parameters entering Z through the interactions (including β) while b represents the BC specified by $\bar{S}_{V^c}$, p, or f. The Gibbs free-energy density of the finite system is given by

$$\Psi(\mathbf{J}; b, V) \equiv |V|^{-1} \log Z(\mathbf{J}; b, V). \tag{3.4}$$

To get an *intrinsic* free energy, which determines the bulk properties of a macroscopic system, one needs to let the size of V become infinite while keeping $\mathbf{J}$ fixed in such a way that the ratio of surface area to volume goes to zero, i.e., to take the thermodynamic limit, $V \nearrow Z^d$, in Eq. (3.4).

It is one of the most important rigorous results of statistical mechanics, to whose proof many have contributed (see Fisher, 1964; Ruelle, 1969; Griffiths, 1972) that when the interactions decay in a summable way, the limit $V \nearrow Z^d$ of Eq. (3.5) in fact exists and is independent of the boundary condition b:

$$\Psi(\mathbf{J}; b, V) \to \Psi(\mathbf{J}). \tag{3.5}$$

We shall call $\Psi(\mathbf{J})$ the thermodynamic free-energy density. It has all the convexity properties of the free energy *postulated* by macroscopic thermodynamics as a stability requirement on the equilibrium state. (For Coulomb interactions see below and Brydges and Martin, 1998.)

We now note that as long as V is finite, $Z(\mathbf{J};b,V)$ is a finite sum of positive terms and so $\Psi(\mathbf{J};b,V)$ is a smooth function of the parameters $\mathbf{J}$ (including β and h) entering the interaction. This is also true for the probabilities of the spin configuration in a set $A \subset V$, $\mu_V(\mathbf{S}_A|b)$ obtained from the Gibbs measure Eq. (3.2) or equivalently the correlation functions. In other words, once b is specified, all equilibrium properties of the finite system vary smoothly with the parameters $\mathbf{J}$. The only way to get nonsmooth behavior of the free energy or nonuniqueness of the measure is to take the thermodynamic limit. In that limit the b-independent $\Psi(\mathbf{J})$ can indeed have singularities. Similarly, the measure defined by a specification of the probabilities in *all* fixed regions $A \subset \mathbb{Z}^d$, $\hat{\mu}(\mathbf{S}_A|\hat{b})$, can depend on the way in which the thermodynamic limit was taken and in particular on the boundary conditions at "infinity," here denoted symbolically by $\hat{b}$ (Fisher, 1964; Ruelle, 1969; Griffiths, 1972; Baxter, 1982; Sinai, 1982; Georgii, 1988; Fernández, Fröhlich, and Sokal, 1992; Simon, 1993).

To see this explicitly, let us specialize even further and consider isotropic nearest neighbor (NN) interactions:

$$u(\mathbf{r}) = \begin{cases} J, & \text{for } |\mathbf{r}| = 1 \\ 0, & \text{otherwise} \end{cases} \quad (3.6)$$

with J constant. For this model the effect of the spins outside V, $\bar{S}_{V^c}$, is just to produce an additional magnetic fields $h_b(\mathbf{x})$, for $\mathbf{x}$ on the inner boundary of V. The finite-volume free energy $\Psi(J_1,J_2;b,V)$, where $\beta h = J_1$ and $\beta J = J_2$, is then clearly real analytic for all $J_1, J_2 \in (-\infty, \infty)$. The phase diagram of this system after taking the thermodynamic limit is given in Fig. 1 where we have used axes labeled by $h/|J|$ and J_2^{-1}. Note that $J_2 > 0$ ($J_2 < 0$) corresponds to ferromagnetic (antiferromagnetic) interactions.

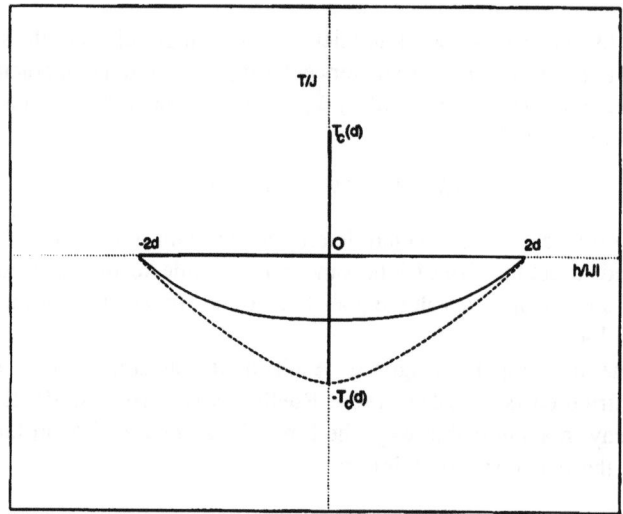

Figure 1. Schematic phase diagram of the nearest-neighbor Ising model on a simple cubic lattice in dimensions $d \geq 2$. The ground states of the antiferromagnetic system are degenerate for $|h| \leq 2|J|d$. For $d=1$, $T_c=0$.

For the ferromagnetic Ising model, corresponding to the upper half of this figure, *almost* everything is known rigorously. In the region where the magnetic field h is not zero, both $\Psi(J_1,J_2)$ and the infinite-volume Gibbs measure, i.e., the $\hat{\mu}(S_A|b)$, are independent of the BC and are real analytic in J_1 and J_2. The analyticity results follow from the remarkable Lee-Yang theorem (Lee and Yang, 1952; Yang and Lee, 1952) which states that for $J_2 \geq 0$ fixed, and $b=p$ or f, the only singularities of $\Psi(J_1,J_2:b,V)$ (corresponding to zeros of the partition function) in the complex J_1 plane occur on the line $\mathrm{Re}\,J_1=0$. Uniqueness of $\hat{\mu}$ follows (Lebowitz and Martin-Löf, 1972) from an argument combining the Lee-Yang theorem with the equally remarkable Fortuin, Kasteleyn, and Ginibre inequalities (Fortuin, Kasteleyn, and Ginibre, 1971).

Furthermore, for small values of $|J_2|$, Ψ is analytic in both J_1 and J_2 and the measure $\hat{\mu}$ is unique. This fact, which holds for general interactions at high temperatures, follows either from the existence of a convergent high-temperature expansion for Ψ and for the correlation functions in powers of β or from the Dobrushin-Shlosman uniqueness criterion (Dobrushin and Shlosman, 1985a, 1985b). On the other hand, for $J_1=0$ and J_2 large enough there is the ingenious argument due to Peierls (1936), made fully rigorous by Dobrushin and by Griffiths (Dobrushin, 1968; Griffiths, 1972), which proves that in dimension $d \geq 2$, the probability that the spin $S(\mathbf{x})$ has value $+1$ is different for "$b=+$" and "$b=-$," corresponding to BC for which $\bar{S}(\mathbf{y})=+1$, or $\bar{S}(\mathbf{y})=-1$, respectively, for all $\mathbf{y}$ outside V. The crucial point of the Peierls argument is that this difference persists *no matter how large V is*: the probability being greater (less) than $\frac{1}{2}$ for $+\,(-)$ BC. This implies that the average value of the magnetization is positive at low temperatures for $+$ BC, even when $h=0$, independent of V. By symmetry the opposite is true for $-$ BC. Thus for $J_1=0$ and J_2 large, the limiting Gibbs measures $\hat{\mu}_+$ and $\hat{\mu}_-$ (obtained with $+$ or $-$ BC), which can be shown to exist, are different. *It is this nonuniqueness of the Gibbs measure $\hat{\mu}$, for specified $\mathbf{J}$, which corresponds to the coexistence of phases in macroscopic systems.*

The expected value of $S(\mathbf{x})$ in the "$+$ state," denoted by $m^*(\beta)$, is independent of $\mathbf{x}$ and is equal to the value of the average of the magnetization in all of V obtained when one lets $h \to 0$ from the positive side after taking the thermodynamic limit. (Remember that $\hat{\mu}$ and hence the magnetization, $m(\beta,h)$, is independent of BC for $h \neq 0$.) It can be further shown, using the second Griffiths inequality that $m^*(\beta)$ is monotone increasing in β (Griffiths, 1967; Kelley and Sherman, 1969). Hence there is, for a given $J_2 > 0$, a unique critical temperature T_c, such that for $h=0$ and $T<T_c$, $m^*(\beta)>0$ while for $T>T_c$, $m^*(\beta)=0$. T_c depends on the dimension d, $T_c(d)>0$ for $d \geq 2$, $T_c(1)=0$.

There is a unique infinite-volume Gibbs measure for $T \geq T_c$ and (essentially) only two, $\hat{\mu}_+$ ands $\hat{\mu}_-$, *extremal, translation-invariant* Gibbs measures for $T<T_c$. The latter statement means that every infinite-volume translation-invariant Gibbs measure $\hat{\mu}_b$ is a convex combination of $\hat{\mu}_+$ and $\hat{\mu}_-$, i.e.,

$$\hat{\mu}(S_A|b) = \alpha\hat{\mu}_+(S_A) + (1-\alpha)\hat{\mu}_-(S_A), \tag{3.7}$$

for some α, $0 \leq \alpha \leq 1$. For periodic or free BC $\alpha = \frac{1}{2}$ by symmetry, so that $\hat{\mu}_p = \hat{\mu}_f = \frac{1}{2}(\hat{\mu}_+ + \hat{\mu}_-)$. This means physically that when V is large the system with "symmetric" BC will, with equal probability, be found in *either* the "$+$ state" *or* in the opposite "$-$ state." Of course as long as the system is finite it will "fluctuate" between these two pure phases, but the "relaxation times" for such fluctuations grows (for any reasonable dynamics) exponentially in $|V|$, so the *either/or* description correctly captures the behavior of macroscopic systems. This phenomena is the paradigm of *spontaneous symmetry breaking* which occurs in many physical situations.

The fact that free bc lead to translation-invariant measures is a consequence of the Griffiths inequalities (Griffiths, 1967; Kelley and Sherman, 1969). There also exist nontranslation-invariant $\hat{\mu}$ for temperatures below the "roughening" temperature $T_R \leq T_c$. These are obtained as the thermodynamic limit of systems with "Dobrushin BC" favoring an interface between the $+$ and $-$ phase. Dobrushin (1972) proved that $T_R > 0$ in $d \geq 3$ while Aizenman (1979; 1980) showed that long-wavelength fluctuations destroy these states in two dimensions at all $T > 0$, i.e., $T_R = 0$ in $d = 2$. Using inequalities van Beijeren (1975; 1977) showed that $T_R(d) \geq T_c(d-1)$.

We also know that at T_c, $m^*(\beta_c) = 0$ in $d = 2$ and for $d \geq 4$; the former from Onsager's (1944) exact solution (see also Fröhlich and Spencer, 1981) and the latter from general results about mean field-like behavior for $d > 4$ (Aizenman, 1981) (with logarithmic corrections for $d = 4$). Of course one expects continuity of $m^*(\beta)$ for this system also in $d = 3$, but this is not yet proven. As the temperature is lowered, the $+$ and $-$ states come to resemble the two ground states corresponding to all spins up or all spins down, and there is a convergent low-temperature "cluster expansion," in which the low-order terms correspond to excitations consisting of small isolated domains of down (up) spins in the $+$ ($-$) phase.

The absence of any homogeneous pure phases other than $\hat{\mu}_+$ and $\hat{\mu}_-$, i.e., the validity of Eq. (3.7) for all translation-invariant $\hat{\mu}$, is only proven subject to the condition that the average energy is a continuous function of the temperature (Lebowitz, 1976). This is known in $d = 2$ from Onsager's solution which also gives the exact value of T_c. In $d > 2$ the continuity of the energy is known to hold at low temperatures (where the cluster expansion is valid) and at *almost all* temperatures otherwise. There is, however, much numerical and analytic evidence that $\Psi(J_1 = 0, J_2)$ is real analytic in J_2 everywhere away from the critical temperature. The story is similar for the decay of correlations. This is known to be exponential for $h \neq 0$ at high temperatures and at low temperatures in the $+$ and $-$ phases. Similar behavior is expected at all $T \neq T_c$, but this is only proven for $d = 2$ (and in $d = 1$ where $T_c = 0$). Note that for mixed states, when $\alpha \neq 0$ or 1 in Eq. (3.7), there is no decay of correlations.

Essentially everything said above for the ferromagnetic Ising model with NN interactions holds also for more general ferromagnetic pair interactions, $u(\mathbf{r}) \geq 0$ in Eq. (3.1) with $u(\mathbf{r})$ of finite range or decaying faster than $r^{-(d+1+\epsilon)}$. (An exception is the decay of correlations, which is never faster than the decay of the interactions.) It follows in fact from the Griffiths-Kelley-Sherman inequalities (Griffiths, 1967; Kelley and Sherman, 1969) that adding ferromagnetic pair or multispin interactions to an already ferromagnetic Ising system (with $h \geq 0$) can only increase the magnetization. A particular consequence of this is that the critical temperature for the nearest-neighbor Ising model cannot decrease with dimension: going from d to $d+1$ can be viewed of as adding ferromagnetic couplings. This argument works also when we increase the "thickness" of a d-dimensional system, e.g., adding layers to a $d = 2$ Ising model. To show that T_c actually increases, not just stays fixed, is more difficult. In fact, going from $d = 1$ to a strip of finite width (and infinite length) does not increase T_c from zero, its value for $d = 1$.

An interesting situation occurs in $d = 1$ when the ferromagnetic pair interaction decays like $r^{-\gamma}$, $1 < \gamma \leq 2$, so the thermodynamic limit of Ψ still exists. The $d = 1$ system then has a $T_c > 0$ with the spontaneous magnetization *discontinuous* at T_c (coexistence of phases) for the borderline case $J(r) \sim r^{-2}$. A result of this type was first found by

Anderson and Yuval (1969; 1971), then proven for the hierarchical model by Dyson (1969; 1971) and for the Ising model by Aizenman *et al.* (1988).

For general lattice systems we still have the existence of the thermodynamic limit of the free energy, independent of the BC as well as the general connection between pure phases and extremal translation-invariant or periodic Gibbs states. (Any periodic Gibbs state can be made translation invariant by enlarging the "unit cell" of the lattice.) We know, however, much less about the phase diagram, except at very low temperatures. Here the Pirogov-Sinai theory and its extensions (Pirogov and Sinai, 1975a-1975b; Dobrushin and Zahradnik, 1986; Kotecky and Preiss, 1986; Dinaburg and Sinai, 1988) show how the existence of different periodic ground states, corresponding to a "ground-states" diagram in the space of interactions $\mathbf{J}$, at $T=0$, gives rise to a similar phase diagram of the pure phases at sufficiently low temperatures. The great advantage of this theory, compared to arguments of the Peierls type, is that there is no requirement of symmetry between the phases—the existence of which was of crucial importance for the ferromagnetic examples discussed earlier.

This can be seen already for the NN Ising model with antiferromagnetic interaction, $J<0$ Eq. in (3.6). For $h=0$ this system can be mapped into the ferromagnetic one by changing $S(\mathbf{x})$ into $-S(\mathbf{x})$ on the odd sublattice—but what about $h \neq 0$? If we look at this system at $T=0$ we find two periodic ground states for $|h|<2d|J|$ corresponding to $S(\mathbf{x})=-1$ on the even (odd) and $S(\mathbf{x})=1$ on the odd (even) sublattice. For $|h|>2d|J|$ there is a unique ground state: all up for $h>2d|J|$, all down for $h<-2d|J|$. At $|h|=2d|J|$ there are an infinite number of ground states with positive entropy per site (in violation of the third law). The existence of two periodic phases for sufficiently low temperatures at $|h|<2d|J|$, and of a unique translation-invariant phase for $|h|>2d|J|$ then follows from Pirogov-Sinai theory. Of course for $h=0$ we know, from the isomorphism with the ferromagnetic system, that there are two periodic states for all $T<T_c$. This, however, doesn't strictly (i.e., rigorously) tell us anything about $h \neq 0$. I am not aware of any argument which proves that the boundary of the curve enclosing the coexistence region in the antiferromagnetic part of Fig. 1 has to touch the point corresponding to $h=0$, $T=T_c$. There is also for this system, a generalization of the Peierls argument, due to Dobrushin (1968), which exploits the symmetry of this system and is therefore simpler than Pirogov-Sinai theory. This proves the existence of the two periodic states in a portion of the phase diagram (indicated by the solid curve in Fig. 1).

Unfortunately Pirogov-Sinai theory does not say anything about the immediate neighborhood of the points $|h|=2d|J|$ where the system has such a high degeneracy. It does follow, however, from the general Dobrushin-Shlosman uniqueness criteria (Dobrushin, Kolafa, and Shlosman, 1985; Dobrushin and Shlosman, 1985a-1985b; Radulescu and Styer, 1987), implemented by computer enumerations, that, for $d=2$, the boundary of the coexistence region at $4|J|$ has to curve to the left, hence we expect that for $|h|=4|J|$ there is a unique phase for all $T>0$.

Continuum systems

The existence of the thermodynamic limit of the free energy, independent of b as in Eq. (3.5), also holds for continuum systems (classical or quantum) with Hamiltonians, having the form

$$H(X) = (2m)^{-1} \sum \mathbf{p}_i^2 + \sum_{i \neq j} u(\mathbf{r}_i - \mathbf{r}_j), \tag{3.8}$$

and satisfying certain conditions (Fisher, 1964; Ruelle, 1969; Griffiths, 1972; Sinai, 1982. These conditions are readily shown to hold for systems with Lennard-Jones type potentials. For Coulomb systems, where u contains explicitly terms of the form $e_{\alpha_i} e_{\alpha_j} |\mathbf{r}_i - \mathbf{r}_j|^{-1}$ in $d=3$ (logarithmic ones in $d=2$, etc.) it is required that the system be overall charge neutral. For classical systems it is further required that there be some cutoff preventing arbitrarily large negative "binding" energies between positive and negative charges, e.g., a hard-core exclusion. For quantum systems it is sufficient if either the positive or negative charges obey Fermi statistics—as electrons indeed do, (see Dyson and Lenard, 1967; Lenard and Dyson, 1968; Lebowitz and Lieb, 1969; Lieb and Lebowitz, 1972; Lieb, 1976; Brydges and Martin, 1998; and references therein).

Remarkably enough it is possible to prove (subject to some assumptions) that a system of protons and electrons will, in certain regimes of sufficiently low temperatures and densities, consist mostly of a gas of atoms or molecules in their ground states (Brydges and Martin, 1998). This may be the *beginning* of a theory which would justify, from first principles, the use of effective potentials, e.g., Feynman's "little particles" (Feynman, Leighton, and Sands, 1963) for obtaining properties, such as phase transitions, of macroscopic systems (Fisher, 1988).

I return now to the theme of this section with a discussion of first-order phase transitions in continuum systems, a subject of much curent interest to me. While the general theory concerning infinite-volume Gibbs measures readily extends to such systems, the techniques used for proving existence of phase transitions in lattice systems are harder to generalize. The ground states of even the simplest model continuum systems are difficult to characterize; they are presumed to be periodic or quasiperiodic configurations which depend in some complicated way on the interparticle forces. This is however far from proven, and hence the analysis of the fluctuations that appear when we increase the temperature above zero is correspondingly harder, indeed very much harder, to study than in lattice systems. Moreover, key inequalities are no longer available. These problems have been overcome for some multicomponent systems with special features. In particular Ruelle (1971) proved that the symmetric two-component Widom-Rowlinson (1970) model has a demixing phase transition in $d \geq 2$. There are also later proofs of phase transitions in $d \geq 2$ for generalizations of this model as well as for $d=1$ continuum systems with interactions which decay very slowly (Felderhof and Fisher, 1970; Johansson, 1995).

The first proof of a liquid-vapor transition in a one-component continuum systems with finite-range interactions and no symmetries was given only very recently (Lebowitz, Mazel, and Presutti; 1998a; 1998b). The basic idea there is to study perturbations not of the ground state but of the mean field state which describes systems with infinite-range interactions. These interactions are parametrized by their range γ^{-1} and the perturbation is about $\gamma = 0$. The proof of mean-field behavior, in the limit $\gamma \to 0$, was first given by Kac, Uhlenbeck, and Hemmer (1963a; 1963b, 1964) for $d=1$. These results were later generalized (Lebowitz and Penrose, 1966) to d-dimensional systems with suitable short-range repulsive interactions and general Kac potentials of the form

$$\phi_\gamma(q_i, q_j) = -\alpha \gamma^d J(\gamma |q_i - q_j|), \tag{3.9}$$

with $\int_{R^d} J(r) dr = 1$, $J(r) > 0$. In the thermodynamic limit, followed by the limit $\gamma \to 0$, the Helmholtz free energy a takes a mean-field form:

$$\lim_{\gamma \to 0} a(\rho, \gamma) = CE\left\{ a_0(\rho) - \frac{1}{2} \alpha \rho^2 \right\}. \tag{3.10}$$

Here ρ is the particle density and a_0 is the free-energy density of the reference system, i.e., the system with no Kac potential. a_0 is convex in ρ (by general theorems) and $CE\{f(x)\}$ is the largest convex lower bound of f. For α large enough the term in the curly brackets in Eq. (3.10) has a double-well shape and the CE corresponds to the Gibbs double-tangent construction. This is equivalent to Maxwell's equal-area rule applied to a van der Waals'-type equation of state where it gives the coexistence of liquid and vapor phases. In this limit, $\gamma \to 0$, the correlation functions in the pure phases are those of the reference system at the corresponding densities.

The assumption of strongly repulsive short-range interactions by Kac, Uhlenbeck, and Hemmer (1963a; 1963b; 1964) and Lebowitz and Penrose (1966) in addition to the long-range attractive Kac-type interactions, was dictated not only by realism but also by the need to insure stabilization against collapse, which would be induced by a purely attractive pair potential. The approach by Lebowitz, Mazel, and Presutti (1998a, 1998b), however, which proves a liquid-vapor phase coexistence for $\gamma > 0$, needs a cluster expansion for the unperturbed reference system (i.e., without the Kac interaction) at values of the chemical potential or density for which it is not proven to hold in systems with strong short-range interactions. Stability is therefore produced by a positive four-body potential of the same range as the attractive two-body one. The reference system is then the free, ideal gas for which the cluster expansion holds trivially. The proof of the existence of phase transitions in fluids with Lennard-Jones-type potentials is therefore still an open problem. Hopefully we will not have to wait another century for its resolution.

Acknowledgments

I would like to thank S. Goldstein and E. Speer for very many, very helpful comments. Thanks are also due to the anonymous referee for many suggestions, and to D. Chelst for a careful reading of the manuscript. The research was supported in part by NSF Grant No. 95-23266, and AFOSR Grant No. 95-0159.

References

Aizenman, M., 1979, Phys. Rev. Lett. **43**, 407.
Aizenman, M., 1980, Commun. Math. Phys. **73**, 83.
Aizenman, M., 1981, Phys. Rev. Lett. **47**, 1.
Aizenman, M., J.T. Chayes, L. Chayes, and C.M. Newman, 1988, J. Stat. Phys. **50**, 1.
Anderson, P.W., and G. Yuval, 1969, Phys. Rev. Lett. **23**, 89.
Anderson, P.W., and G. Yuval, 1971, J. Phys. C **4**, 607.
Baxter, R., 1982, *Exactly Solvable Models in Statistical Mechanics* (Academic, New York).
Boltzmann, L., 1884 in *Wssenschaftliche Abhandluncen*, 3, 122, (reprinted by Chelsea, New York, 1968).
Boltzmann, L., 1896, 1898, *Vorlesungen uber Gastheorie,* 2 vols. (Barth, Leipzig). [Translated into English By S.G. Brush, 1964, *Lectures on Gas Theory* (Cambridge University, London).]

Boltzmann, L., 1897, Ann. Phys. (Leipzig) **60**, 392.
Boltzmann, L., 1896, Ann. Phys. (Leipzig) **57**, 773.
Bricmont, J., 1996, Ann. (N.Y.) Acad. Sci. **79**, 131.
Broda, E., 1973, in *The Boltzmann Equation, Theory and Application*, edited by E.G.D. Cohen and W. Thirring (Springer-Verlag).
Broda, E., 1983, *Ludwig Boltzman, Man-Physicist-Philosopher* (Ox Bow, Woodbridge, CT).
Brush, S.G., 1966, *Kinetic Theory* (Pergamon, Oxford).
Brush, S.G., 1976, *The Kind of Motion We Call Heat, Studies in Statistical Mechanics*, Vol. VI, edited by E.W. Montroll and J.L. Lebowitz (North Holland, Amsterdam).
Brydges, D., and Ph. A. Martin, 1998, Rev. Mod. Phys. (in press).
De Masi, A., and E. Presutti, 1991, *Mathematical Methods for Hydrodynamic Limits, Lecture Notes in Math 1501* (Springer-Verlag, Berlin).
Dinaburg, E.I., and Ya.G. Sinai, 1988, Sel. Math. Sov. **7**, 291.
Dobrushin, R.L., 1968, Funct. Anal. Appl. **2**, 31.
Dobrushin, R.L., 1972, Theor. Probab. Appl. **17**, 382.
Dobrushin, R.L., J. Kolafa, and S.B. Shlosman, 1985, Commun. Math. Phys. **102**, 89.
Dobrushin, R.L., and S.B. Shlosman, 1985a, in *Statistical Physics and Dynamical Systems*, edited by J. Fritz, A. Jaffe, and D. Szasz (Birkenhauser, New York), p. 347.
Dobrushin, R.L., and S.B. Shlosman, 1985b, in *Statistical Physics and Dynamical Systems*, edited by J. Fritz, A. Jaffe, and D. Szasz (Birkenhauser, New York), p. 371.
Dobrushin, R. L., and M. Zahradnik, 1986, in *Mathematical Problems of Statistical Mechanics and Dynamics*, edited by R.L. Dobrushin (Dordrecht, Boston), p. 1.
Dyson, F.J., 1969, Commun. Math. Phys. **12**, 91.
Dyson, F.J., 1971, Commun. Math. Phys. **21**, 269.
Dyson, F.J., and A. Lenard, 1967, J. Math. Phys. **8**, 423.
Esposito, R., and R. Marra, 1994, J. Stat. Phys. **74**, 981.
Felderhof, B.U., and M.E. Fisher, 1970, Ann. Phys. (Paris) **58** N1, 176 (Part IA); 217 (Part IB); 268 (Part II).
Fernandez, R., J. Frolich, and D.A. Sokal, 1992, *Random Walks, Critical Phenomena, and Triviality in Quantum Field Theory* (Springer, Berlin).
Feynman, R.P., 1967, *The Character of Physical Law* (MIT, Cambridge, MA), Chap. 5.
Feynman, R.P., R.B. Leighton, and M. Sands, 1963, *The Feynman Lectures on Physics*, Section 1-2 (Addison-Wesley, Reading, MA).
Fisher, M.E., 1964, Arch. Ration. Mech. Anal. **17**, 377.
Fisher, M.E., 1988, in *Niels Bohr: Physics and the World*, edited by H. Feshbach, T. Matsui, and A. Oleson (Harwood Academic, Chur), p. 65.
Fisher, M.E., 1998, Rev. Mod. Phys. **70**, 653.
Flamn, L., 1973, in *The Boltzmann Equation, Theory and Application*, edited by E.G.D. Cohen and W. Thirring (Springer-Verlag, Berlin).
Fortuin, C.M., P.W. Kasteleyn, and J. Ginibre, 1971, Commun. Math. Phys. **22**, 89.
Fröhlic, J., and T. Spencer, 1981, Phys. Rev. Lett. **46**, 1006.
Georgii, H.O., 1988, *Gibbs Measures and Phase Transitions* (W. De Gruyter, Berlin).
Gibbs, J.W., 1875, Connecticut Academy Transaction **3**, 229. Reprinted in *The Scientific Papers* (Dover, New York, 1961), Vol. 1, p. 167.
Gibbs, J.W., 1960, *Elementary Principles in Statistical Mechanics* (Dover, New York) (reprint of 1902 edition).
Griffiths, R.B., 1967, J. Math. Phys. **8**, 478.
Griffiths, R.B., 1972, in *Phase Transitions and Critical Phenomena*, Vol. 1, edited by C. Domb and H.S. Green (Academic, New York).
Jaynes, E.T., 1971, Phys. Rev. A **4**, 747.
Johansson, K., 1995, Commun. Math. Phys. **169**, 521.
Kac, M., G. Uhlenbeck, and P.C. Hemmer, 1963a, J. Math. Phys. **4**, 216.

Kac, M., G. Uhlenbeck, and P.C. Hemmer, 1963b, J. Math. Phys. **4**, 229.
Kac, M., G. Uhlenbeck, and P.C. Hemmer, 1964, J. Math. Phys. **5**, 60.
Kelley, D.G., and S. Sherman, 1969, J. Math. Phys. **9**, 466.
Klein, M., 1973, in *The Boltzmann Equation, Theory and Application*, edited by E.G.D. Cohen and W. Thirring (Springer-Verlag, Berlin).
Kotecky, R., and D. Preiss, 1986, Commun. Math. Phys. **103**, 491.
Landim, C., and C. Kipnis, 1998, *Scaling Limits for Interacting Particle Systems* (Springer-Verlag, Heidelberg).
Lanford, O., 1981, Physica A **106**, 70.
Lebowitz, J.L., 1976, J. Stat. Phys. **16**, 463.
Lebowitz, J.L., 1993a, Physica A **194**, 1.
Lebowitz, J.L., 1993b, Phys. Today **46**, 32.
Lebowitz, J.L., 1994, Phys. Today **47**, 113.
Lebowitz, J.L., 1995a, Introduction to Chapter 6, Phys. Rev. Centennial, in *The Physical Review: The First Hundred Years*, edited by H. Stroke (AIP, New York), p. 369.
Lebowitz, J.L., 1995b, in *25 Years of Non-equilibrium Statistical Mechanics*, Proceedings of the Sitges Conference, Barcelona, Spain, 1994, in Lecture Notes in Physics, edited by J. J. Brey, J. Marro, J.M. Rubi, and M. San Miguel (Springer, New York).
Lebowitz, J.L., A. Mazel, and E. Presutti, 1998a, Phys. Rev. Lett. **80**, 4701.
Lebowitz, J.L., A. Mazel, and E. Presutti, 1998b, J. Stat. Phys. (in press).
Lebowitz, J.L., and A. Martin-Lof, 1972, Commun. Math. Phys. **25**, 276.
Lebowitz, J.L., and E.H. Lieb, 1989, Phys. Rev. Lett. **22**, 631.
Lebowitz, J.L., and H. Spohn, 1983, Commun. Pure Appl. Math. **36**, 595; in particular section 6(i).
Lebowitz, J.L., and O. Penrose, 1966, J. Math. Phys. **7**, 98.
Lebowitz, J.L., E. Presutti, and H. Spohn, 1988, J. Stat. Phys. **51**, 841.
Lee, T.D., and C.N. Yang, 1952, Phys. Rev. **87**, 410.
Lenard, A., and F.J. Dyson, 1968, J. Math. Phys. **9**, 689.
Levesque, D., and L. Verlet, 1993, J. Stat. Phys. **72**, 519.
Lieb, E.H., 1976, Rev. Mod. Phys. **48**, 553.
Lieb, E.H., and J.L. Lebowitz, 1972, Adv. Math. **9**, 316.
Liggettt, T.M., 1985, *Interacting Particle Systems* (Springer, New York).
Maxwell, J.C., 1878, Nature (London) **17**, 257. Quoted in M. Klein, 1973, in *The Boltzmann Equation, Theory, and Application*, edited by E.G.D. Cohen and W. Thirring (Springer-Verlag, Berlin).
Meakin, P., 1998, *Fractals, Scaling, and Growth Far from Equilibrium* (Cambridge University, Cambridge).
Nadiga, B.T., J.E. Broadwell, and B. Sturtevant, 1989, *Rarified Gas Dynamics: Theoretical and Computational Techniques*, Progress in Astronautics and Aeronautics, Vol. 118, edited by E.P. Muntz, D.P. Weaver, and D.H. Campbell (AIAA, Washington, DC).
Onsager, L., 1944, Phys. Rev. **65**, 117.
Peierls, R., 1936, Proc. Cambridge Philos. Soc. **32**, 477.
Penrose, O., 1970, *Foundations of statistical Mechanics* (Pergamon, Elmsford, New York), Chap. 5.
Penrose, R., 1990, *The Emperor's New Mind* (Oxford University Press, New York), Chap. 7.
Pirogov, S.A., and Ya.G. Sinai, 1975a, Theor. Math. Phys. **25**, 358.
Pirogov, S.A., and Ya.G. Sinai, 1975b, Theor. Math. Phys. **25**, 1185.
Radulescu, D.C., and D.F. Styer, 1987, J. Stat. Phys. **49**, 281.
Ruelle, D., 1969, *Statistical Mechanics* (Benjamin, New York).
Ruelle, D., 1971, Phys. Rev. Lett. **27**, 1040.
Simon, B, 1993, *The Statistical Mechanics of Lattice Gases* (Princeton University, Princeton, NJ).
Sinai, Ya.G., 1982, *Theory of Phase Transitions: Rigorous Results* (Pergamon, New York).
Spohn, H.M., 1991, *Large Scale Dynamics of Interacting Particles* (Springer-Verlag, New York).

Thomson, W., 1874, Proc. R. Soc. Edinburgh **8**, 325.
van Beijeren, H., 1975, Commun. Math. Phys. **40**, 1.
van Beijeren, H., 1977, Phys. Rev. Lett. **38**, 993.
Vicsek, T., 1989, *Fractal Growth Phenomena* (World Scientific, Singapore).
Widom, B., and J.S. Rowlinson, 1970, J. Chem. Phys. **52**, 1670.
Yang, C.N., and T.D. Lee, 1952, Phys. Rev. **87**, 404.

Scaling, Universality, and Renormalization: Three Pillars of Modern Critical Phenomena

H. Eugene Stanley

The First Question: "What Are Critical Phenomena?"

Suppose we have a simple bar magnet. We know it is a ferromagnet because it is capable of picking up thumbtacks, the number of which is called the order parameter M. As we heat this system, M decreases and eventually, at a certain critical temperature T_c, it reaches zero: no more thumbtacks remain! In fact, the transition is remarkably sharp, since M approaches zero at T_c with infinite slope. Such singular behavior is an example of a "critical phenomenon."

Critical phenomena are by no means limited to the order parameter. For example, the response-functions constant-field specific heat C_H and isothermal susceptibility χ_T both become infinite at the critical point.

The Second Question: "Why Do We Care?"

One reason for interest in any field is that, simply put, we do not fully understand the basic phenomena. For example, for even the simplest three-dimensional system we cannot make exact predictions of all the relevant quantities from any realistic *microscopic* model at our disposal. Of the models that can be solved in closed form, most make the same predictions for behavior near the critical point as the classical mean-field model, in which one assumes that each magnetic moment interacts with all other magnetic moments in the entire system with equal strength (see, e.g., the review of Domb, 1996). The mean-field model predicts that both M^2 and χ_T^{-1} approach zero *linearly* as $T \to T_c$, and that C_H does not diverge at all. In fact, the mean-field theory cannot locate the value of T_c to better than typically about 40%.

A second reason for our interest is the striking similarity in behavior near the critical point among systems that are otherwise quite different in nature. A celebrated example is the "lattice-gas" analogy between the behavior of a single-axis ferromagnet and a simple fluid, near their respective critical points (Lee and Yang, 1952). Even the numerical values of the critical-point exponents describing the quantitative nature of the singularities are identical for large groups of apparently diverse physical systems.

A third reason is awe. We wonder how it is that spins "know" to align so suddenly as $T \to T_c^+$. How can the spins propagate their correlations so extensively throughout the entire system that $M \neq 0$ and $\chi_T \to \infty$?

The Third Question: "What Do We Do?"

The answer to this question will occupy the remainder of this brief overview. The recent past of the field of critical phenomena has been characterized by several important conceptual advances, three of which are scaling, universality, and renormalization.

Scaling

The scaling hypothesis was independently developed by several workers, including Widom, Domb and Hunter, Kadanoff, Patashinskii and Pokrovskii, and Fisher (authoritative reviews include Fisher, 1967 and Kadanoff, 1967). The scaling hypothesis has two categories of predictions, both of which have been remarkably well verified by a wealth of experimental data on diverse systems. The first category is a set of relations, called *scaling laws*, that serve to relate the various critical-point exponents. For example, the exponents α, 2β, and γ describing the three functions C_H, M^2, and χ_T are related by the simple scaling law $\alpha + 2\beta + \gamma = 2$. Here the exponents are defined by $C_H \sim \epsilon^{-\alpha}$, $M^2 \sim \epsilon^{2\beta}$, and $\chi_T \sim \epsilon^{-\gamma}$, where $\epsilon \equiv (T - T_c)/T_c$ is the reduced temperature.

The second category is a sort of *data collapse*, which is perhaps best explained in terms of our simple example of a uniaxial ferromagnet. We may write the equation of state as a functional relationship of the form $M = M(H, \epsilon)$, where M is the order parameter and H is the magnetic field. Since $M(H, \epsilon)$ is a function of two variables, it can be represented graphically as M vs ϵ for a sequence of different values of H. The scaling hypothesis predicts that all the curves of this family can be "collapsed" onto a single curve provided one plots not M vs ϵ but rather a *scaled* M (M divided by H to some power) vs a *scaled* ϵ (ϵ divided by H to some different power).

The predictions of the scaling hypothesis are supported by a wide range of experimental work, and also by numerous calculations on model systems such as the n-vector model. Moreover, the general principles of scale invariance used here have proved useful in interpreting a number of other phenomena, ranging from elementary-particle physics (Jackiw, 1972) to galaxy structure (Peebles, 1980).

Universality

The second theme goes by the rather pretentious name "universality." It was found empirically that one could form an analog of the Mendeleev table if one partitions all critical systems into "universality classes." The concept of universality classes of critical behavior was first clearly put forth by Kadanoff, at the 1970 Enrico Fermi Summer School, based on earlier work of a large number of workers including Griffiths, Jasnow and Wortis, Fisher, Stanley, and others.

Consider, e.g., experimental M-H-T data on five diverse magnetic materials near their respective critical points (Fig. 1). The fact that data for each collapse onto a scaling function supports the scaling hypotheses, while the fact that the scaling function is the

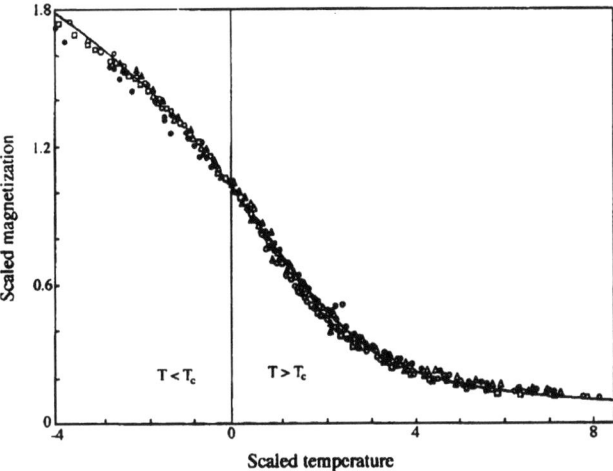

Figure 1. Experimental MHT data on five different magnetic materials plotted in scaled form. The five materials are $CrBr_3$, EuO, Ni, YIG, and Pd_3Fe. None of these materials is an idealized ferromagnet: $CrBr_3$ has considerable lattice anisotropy, EuO has significant second-neighbor interactions. Ni is an itinerant-electron ferromagnet, YIG is a ferrimagnet, and Pd_3Fe is a ferromagnetic alloy. Nonetheless, the data for all materials collapse onto a single scaling function, which is that calculated for the $d=3$ Heisenberg model [after Milošević and Stanley (1976)].

same (apart from two material-dependent scale factors) for all five diverse materials is truly remarkable. This apparent universality of critical behavior motivates the following question: *"Which features of this microscopic interparticle force are important for determining critical-point exponents and scaling functions, and which are unimportant?"*

Two systems with the same values of critical-point exponents and scaling functions are said to belong to the same universality class. Thus the fact that the exponents and scaling functions in Fig. 1 are the same for all five materials implies they all belong to the same universality class.

Renormalization

The third theme stems from Wilson's essential idea that the critical point can be mapped onto a fixed point of a suitably chosen transformation on the system's Hamiltonian (see the recent reviews: Goldenfeld, 1994; Cardy, 1996; Lesne, 1998). This resulting "renormalization group" description has (i) provided a foundation for understanding the themes of scaling and universality, (ii) provided a calculational tool permitting one to obtain numerical estimates for the various critical-point exponents, and (iii) provided us with altogether new concepts not anticipated previously.

One altogether new concept that has emerged from renormalization is the idea of upper and lower marginal dimensionalities d_+ and d_- (see the review of Als-Nielsen and Birgeneau, 1977). For $d > d_+$, the classical theory provides an adequate description of critical-point exponents and scaling functions, whereas for $d < d_+$, the classical theory breaks down in the immediate vicinity of the critical point because statistical fluctuations neglected in the classical theory become important. The case $d = d_+$ must be treated with great care; usually, the classical theory "almost" holds, and the modifications take the form of weakly singular corrections.

For $d<d_-$, fluctuations are so strong that the system cannot sustain long-range order for any $T>0$. For $d_-<d<d_+$, we do not know exactly the properties of systems (in most cases) except when n approaches infinity, where n will be introduced below as the spin dimension. One can, however, develop expansions in terms of the parameters (d_+-d), $(d-d_-)$, and $1/n$ (see, e.g., the reviews of Fisher, 1974; and Brézin and Wadia, 1993).

In the remainder of this brief overview, we shall attempt to define somewhat more precisely the concepts underlying the three themes of scaling, universality, and renormalization without sacrificing the stated purpose, that of a colloquium-level presentation.

What is Scaling?

I offer here a very brief introduction to the spirit and scope of the scaling approach to phase transitions and critical phenomena using, for the sake of concreteness, a simple system: the Ising magnet. Further, we discuss only the simplest static property, the order parameter, and the two response functions C_H and χ_T. The rich subject of dynamic scaling is beyond our scope here (see, e.g., the authoritative review of Hohenberg and Halperin, 1977).

The scaling hypothesis

The scaling hypothesis for thermodynamic functions is made in the form of a statement about one particular thermodynamic potential, generally chosen to be the Gibbs potential per spin, $G(H,T)=G(H,\epsilon)$. One form of the hypothesis is the statement (see, e.g., Stanley, 1971) that asymptotically close to the critical point, $G_s(H,\epsilon)$, the singular part of $G(H,\epsilon)$, is a generalized homogeneous function (GHF). Thus the scaling hypothesis may be expressed as a relatively compact statement that asymptotically close to the critical point, there exist two numbers, a_H and a_T (termed the field and temperature scaling powers) such that for all positive λ, $G_s(H,\epsilon)$ obeys the functional equation:

$$G_s(\lambda^{a_H}H,\lambda^{a_T}\epsilon)=\lambda G_s(H,\epsilon). \tag{1}$$

Exponent relations: The scaling laws

The predictions of the scaling hypothesis are simply the properties of GHFs: (i) Legendre transforms of GHFs are also GHFs, so all thermodynamic potentials are GHFs. (ii) Derivatives of GHFs are also GHFs. Since *every* thermodynamic function is expressible as some derivative of some thermodynamic potential, it follows that the singular part of every thermodynamic function is asymptotically a GHF.

Two useful facts are worth noting:

(a) The critical-point exponent for any function is simply given by the ratio of the scaling power of the function to the scaling power of the path variable along which the critical point is approached:

$$\text{arbitrary exponent}=\frac{a_{\text{function}}}{a_{\text{path}}}. \tag{2}$$

Thus one can "write down by inspection" expressions for any critical-point exponent. Equation (2) holds generally, and proves useful in practice. For the special case of a uniaxial ferromagnet, we have

$$a_{\text{path}} = \begin{cases} a_H & \text{strong path } [T=T_c, H\to 0], \\ a_T & \text{weak path } [H=0^\pm, T\to T_c^\pm]. \end{cases} \quad (3)$$

From property (ii), it follows that

$$a_{\text{function}} = \begin{cases} 1-a_H & \text{for } \bar{M} \propto (\partial G/\partial H)_T, \\ 1-a_T & \text{for } \bar{S} \propto (\partial G/\partial T)_H. \end{cases} \quad (4a)$$

Similarly, from the definitions for the susceptibility and specific heat, we have

$$a_{\text{function}} = \begin{cases} 1-2a_H & \text{for } \bar{\chi}_T \propto (\partial^2 G/\partial H^2)_T, \\ 1-2a_T & \text{for } \bar{C}_H \propto (\partial^2 G/\partial T^2)_H. \end{cases} \quad (4b)$$

(b) Since each critical-point exponent is directly expressible in terms of a_H and a_T, it follows that one can eliminate these two unknown scaling powers from the expressions for three different exponents, and thereby obtain a family of equalities called *scaling laws*.

To illustrate the utility of facts (a) and (b), we note from Eqs. (3) and (4b) that

$$-\alpha' = \frac{1-2a_T}{a_T}, \quad (5a)$$

$$\beta = \frac{1-a_H}{a_T}, \quad (5b)$$

and

$$-\gamma' = \frac{1-2a_H}{a_T}. \quad (5c)$$

We thus have three equations and two unknowns. Eliminating a_H and a_T, we find

$$\alpha' + 2\beta + \gamma' = 2, \quad (6)$$

which is the Rushbrooke inequality $\alpha' + 2\beta + \gamma \geq 2$ in the form of an equality. Defining δ through $M \sim H^\delta$, it follows that

$$\delta^{-1} = \frac{a_M}{a_H} = \frac{1-a_H}{a_H}. \quad (7)$$

Eliminating a_H and a_T from Eqs. (5a), (5b), and (7), we obtain the Griffiths equality

$$\alpha' + \beta(\delta+1) = 2. \quad (8)$$

Similarly, Eqs. (5b), (5c), and (7) give the Widom equality

$$\gamma' = \beta(\delta-1). \quad (9)$$

Thus one hallmark of the scaling approach is a family of three-exponent equalities—called *scaling laws*—of which Eqs. (6), (8), and (9) are but examples. In general, it suffices to determine two exponents since these will in general fix the scaling powers a_H and a_T, which in turn may be used to obtain the exponents for *any* thermodynamic function.

Equation of state and scaling functions

Next we discuss a second hallmark of the scaling approach, the equation of state. The scaling hypothesis of Eq. (1) constrains the form of a thermodynamic potential, near the critical point, so this constraint must have implications for quantities derived from that potential, such as the equation of state.

Consider, for example, the $M(H,T)$ equation of state of a uniaxial ferromagnet near the critical point $[H=0, T=T_c]$. On differentiating Eq. (1) with respect to H, we find

$$M(\lambda^{a_H}H, \lambda^{a_T}\epsilon) = \lambda^{1-a_H}M(H,\epsilon). \tag{10}$$

Since Eq. (10) is valid for all positive values of λ, it must certainly hold for the particular choice $\lambda = H^{-1/a_H}$. Hence

$$M_H = M(1, \epsilon_H) = \mathcal{F}^{(1)}(\epsilon_H), \tag{11a}$$

where

$$M_H \equiv \frac{M}{H^{(1-a_H)/a_H}} = \frac{M}{H^{1/\delta}}, \tag{11b}$$

and

$$\epsilon_H \equiv \frac{\epsilon}{H^{a_T/a_H}} = \frac{\epsilon}{H^{1/\Delta}} \tag{11c}$$

are termed the *scaled magnetization* and *scaled temperature*, while the function $\mathcal{F}^{(1)}(x) = M(1,x)$ defined in Eq. (11a) is called a *scaling function*.

In Fig. 1, the scaled magnetization M_H is plotted against the scaled temperature ϵ_H, and the entire family of $M(H=\text{const}, T)$ curves "collapse" onto a single function. This scaling function $\mathcal{F}^{(1)}(H) = M(1, \epsilon_H)$ evidently is the magnetization function in fixed nonzero magnetic field.

What is Universality?

Empirically, one finds that all systems in nature belong to one of a comparatively small number of such universality classes. Two specific microscopic interaction Hamiltonians appear almost sufficient to encompass the universality classes necessary for static critical phenomena.

The first of these is the Q-state Potts model (Potts, 1952; Wu, 1982). One assumes that each spin i can be in one of Q possible discrete orientations ζ_i ($\zeta_i = 1, 2, \ldots, Q$). If two neighboring spins i and j are in the same orientation, then they contribute an amount $-J$ to the total energy of a configuration. If i and j are in different orientations, they contribute nothing. Thus the interaction Hamiltonian is [Fig. 2(a)]

$$\mathcal{H}(d,s) = -J\sum_{\langle ij \rangle} \delta(\zeta_i, \zeta_j), \tag{12a}$$

where $\delta(\zeta_i, \zeta_j) = 1$ if $\zeta_i = \zeta_j$, and is zero otherwise. The angular brackets in Eq. (12a) indicate that the summation is over all pairs of nearest-neighbor sites $\langle ij \rangle$. The interaction energy of a pair of neighboring parallel spins is $-J$, so that if $J > 0$, the system should order ferromagnetically at $T = 0$.

The second such model is the n-vector model (Stanley, 1968), characterized by spins capable of taking on a continuum of states [Fig. 2(b)]. The Hamiltonian for the n-vector model is

$$\mathcal{H}(d,n) = -J\sum_{\langle ij \rangle} \vec{S}_i \cdot \vec{S}_j. \tag{12b}$$

Here, the spin $\vec{S}_i \equiv (S_{i1}, S_{i2}, \ldots, S_{in})$ is an n-dimensional unit vector with $\sum_{\alpha=1}^{n} S_{i\alpha}^2 = 1$, and $\vec{S}_i$ interacts isotropically with spin $\vec{S}_j$ localized on site j. Two parameters in the n-vector model are the system dimensionality d and the spin dimensionality n. The parameter n is sometimes called the order-parameter symmetry number; both d and n determine the universality class of a system for static exponents.

Both the Potts and n-vector hierarchies are generalization of the simple Ising model of a uniaxial ferromagnet. This is indicated schematically in Fig. 2(c), in which the Potts hierarchy is depicted as a north-south "Metro line," while the n-vector hierarchy appears as an east-west line. The various stops along the respective Metro lines are labeled by the appropriate value of s and n. The two Metro lines have a *correspondence* at the Ising model, where $Q = 2$ and $n = 1$.

Along the north-south Metro line (the Q-state hierarchy), Kasteleyn and Fortuin showed that the limit $Q = 1$ reduces to the random percolation problem, which may be relevant to the onset of gelation (Stauffer and Aharony, 1992; Bunde and Havlin, 1996). Stephen demonstrated that the limit $Q = 0$ corresponds to a type of treelike percolation, while Aharony and Müller showed that the case $Q = 3$ has been demonstrated to be of relevance in interpreting experimental data on structural phase transitions and on absorbed monolayer systems.

The east-west Metro line, though newer, has probably been studied more extensively than the north-south line; hence we shall discuss the east-west line first. For $n = 1$, the spins S_i are one-dimensional unit vectors which take on the values ± 1. Equation (12b), $\mathcal{H}(d,1)$, is the Ising Hamiltonian, which has proved extremely useful in interpreting the properties of the liquid-gas critical point (Levelt Sengers *et al.*, 1977). This case also corresponds to the uniaxial ferromagnet introduced previously.

Other values of n correspond to other systems of interest. For example, the case $n = 2$ describes a set of isotropically interacting classical spins whose motion is confined to a plane. The Hamiltonian $\mathcal{H}(d,2)$ is sometimes called the plane-rotator model or the XY model. It is relevant to the description of a magnet with an easy plane of anisotropy such that the moments prefer to lie in a given plane. The case $n = 2$ is also useful in interpreting experimental data on the λ-transition in ^{4}He.

For the case $n = 3$, the spins are isotropically interacting unit vectors free to point anywhere in three-dimensional space. Indeed, $\mathcal{H}(d,3)$ is the classical Heisenberg model, which has been used for some time to interpret the properties of many isotropic magnetic materials near their critical points.

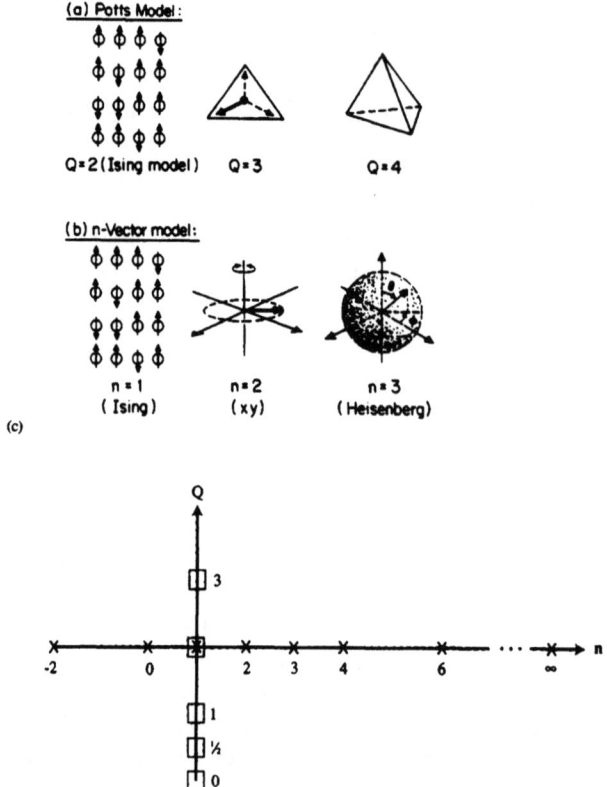

Figure 2. Schematic illustrations of the possible orientations of the spins in (a) the s-state Potts model, and (b) the n-vector model. Note that the two models coincide when $Q=2$ and $n=1$. (c) North-south and east-west "Metro lines."

Two particular "Metro stops" are more difficult to see yet nevertheless have played important roles in the development of current understanding of phase transitions and critical phenomena. The first of these is the limiting case $n \to \infty$, which Stanley showed (in a paper reprinted as Chapter 1 of Brézin and Wadia, 1993) corresponds to the Berlin-Kac spherical model of magnetism, and is in the same universality class as the ideal Bose gas. The second limiting case $n=0$ de Gennes showed has the same statistics of a d-dimensional self-avoiding random walk, which in turn models a system of dilute polymer molecules (see, e.g., de Gennes, 1979 and references therein). The case $n=-2$ corresponds, as Balian and Toulouse demonstrated, to random walks, while Mukamel and co-workers showed that the cases $n=4,6,8,\ldots$ may correspond to certain antiferromagnetic orderings.

What is Renormalization?

This is the second most-often-asked question. In one sense this question is easier to answer than "what is scaling," because to some degree renormalization concepts lead to a well-defined prescription for obtaining numerical values of critical exponents, unlike the scaling hypothesis which leads only to relations among exponents. Answering the question can involve considerable mathematics, so we concentrate here not on momentum-space renormalization but rather on the simpler position space. Instead of treating thermal phenomena we treat a different class of critical phenomena, the purely geometric connectivity phenomena generally called "percolation." The example we give requires such simple mathematics that one could imagine that renormalization could have been invented by the Greek geometers.

The percolation problem

We begin by defining the percolation problem. This is a phase-transition model that was formulated only in comparatively recent times. Recent reviews describing the wealth of current research on percolation include Stauffer and Aharony (1992), and Bunde and Havlin (1996).

Suppose a fraction p of the sites of an infinite d-dimensional lattice are occupied. For p small, most of the occupied sites are surrounded by vacant neighboring sites. However as p increases, many of the neighboring sites become occupied, and the sites are said to form clusters (sites i and j belong to the same cluster if there exists a path joining nearest-neighbor pairs of occupied sites leading from site i to site j). One can describe the clusters by various functions, such as their characteristic linear dimension $\xi(p)$. As p increases, $\xi(p)$ increases monotonically, and at a critical value of p—denoted p_c—it diverges:

$$\xi(p) \sim |p - p_c|^{-\nu}. \tag{13}$$

For $p \geq p_c$ there appears, in addition to the finite clusters, a cluster that is infinite in extent.

The number p_c is referred to as the *connectivity threshold* because of the fact that for $p < p_c$ the connectivity is not sufficient to give rise to an infinite cluster, while for $p > p_c$ it is. Indeed, we shall see that the role of $\epsilon \equiv (T - T_c)/T_c$ is played by $(p_c - p)/p_c$. The numerical value of p_c depends upon both the dimensionality d of the lattice and on the lattice type; however percolation exponents depend only on d.

Kadanoff cells and the renormalization transformation

Percolation functions can be calculated in closed form for $d=1$ by Reynolds and co-workers (see, e.g, the review Stanley, 1982). In particular, one finds that $p_c = 1$, and that $\nu = 1$. It is instructive to illustrate some aspects of the position-space renormalization approach on this exactly-soluble system (Stanley, 1982). The treatment presented below is intended to illustrate—in terms of a simple example—some of the features of the position-space renormalization approach.

The starting point of our illustrative example is the Kadanoff-cell transformation (Kadanoff, 1967). This is illustrated for one-dimensional percolation in Fig. 3(a), which shows b^d-site Kadanoff cells with $b=2$ and $d=1$. Just as each *site* in the lattice is described by a parameter p, its probability of being occupied, so each *cell* is described by a parameter p', which we may regard as being the "cell occupation probability" [Fig. 3(b)]. The essential step in the renormalization-group approach is the construction of a functional relation between the original parameter p and the "renormalized" parameter p',

$$p' = R_b(p). \tag{14}$$

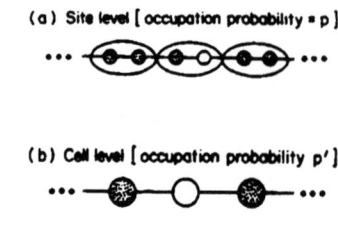
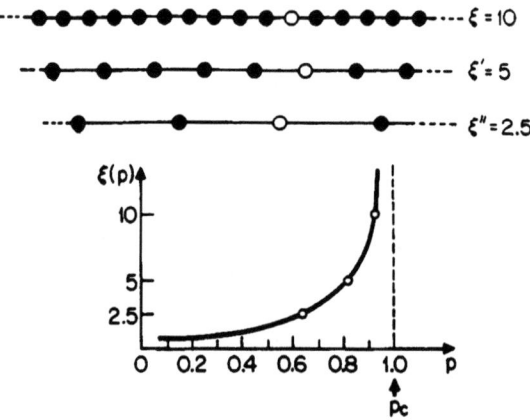

Figure 3. The Kadanoff-cell transformation applied to the example of one-dimensional percolation. The site level in (a) is characterized by a single parameter p—the probability of a site being occupied. The cell level in (b) is characterized by the parameter p'—the probability of a cell being occupied. The relation between the two parameters, p and p', is given by the renormalization transformation R(p) of Eqs. (14) and (15). Also shown are successive Kadanoff-cell transformations. After each transformation, the correlation length ξ(p) is halved. The corresponding value of occupation probability is reduced to $p' = p^b = p^2$, thus taking the system "farther away" from the critical point $p = p_c = 1$. Occupied sites and cells are shown solid, while empty sites and cells are open.

The function $R_b(p)$ is termed a renormalization transformation.

The transformation $R_b(p)$ is particularly simple for one-dimension percolation. Since the percolation threshold is a connectivity phase transition, it is reasonable to say that a cell is "occupied" only if all the sites in the cell are occupied (for if a single site were empty, then the connectivity would be lost). If the probability of a single site being occupied is p, then the probability of all b sites in the cell being occupied is p^b. Hence $R_b(p) = p^b$, and Eq. (14) becomes

$$p' = p^b. \tag{15}$$

Fixed points of the renormalization transformation

The actual choice of the function $R_b(p)$ varies, of course, from one problem to the other. However the remaining steps to be followed after selecting a suitable $R_b(p)$ are essentially the same for all problems. First, we note [Fig. 3(c)] that on carrying out the

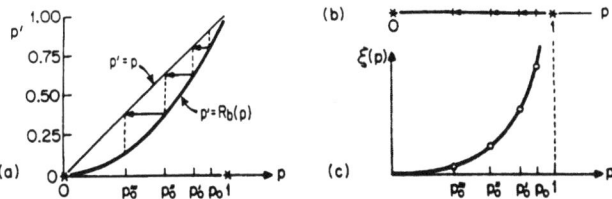

Figure 4. Generic idea of a flow diagram, illustrated here for the pedagogical example of one dimension. (a) Two curves, $p'=p$ and $p'=R_2(p)=p^2$. The fixed points $p^*=0,1$ are given by the intersection of these two curves; the "thermal" scaling power a_T is related to the slope of $R_b(p)$ at the unstable fixed point $p^*=1$. Also shown is the effect of successive Kadanoff-cell transformations, Eq. (15), on a system whose initial value of the parameter p is $p_0=0.9$. This information is capsulized in the one-dimensional flow diagram of part (b), which illustrates the result of Eq. (16)—that each renormalization serves to halve the correlation length ξ.

renormalization transformation, the new correlation length $\xi'(p')$ is smaller than the original correlation length $\xi(p)$ by a factor of b:

$$\xi'(p')=b^{-1}\xi(p). \tag{16}$$

Next we consider the effect of carrying out *successive* Kadanoff-cell transformations with our one-dimensional example. Suppose the system starts out at an initial parameter value $p=p_0=0.9$, as shown schematically in Fig. 4. After a single renormalization transformation, the value of p becomes $p_0'=R_b(p_0)=0.81$ by Eq. (15). The transformed system is *farther* from the critical point, and hence $\xi'(p')$ is smaller—just as we noted in Fig. 3(c). If we now perform a renormalization transformation on the transformed system, we have $p_0''=R_b[R_b(p_0)]=(p_0')^2=0.64$. The doubly-transformed system is now farther still from the critical point.

Thus the effect of successive Kadanoff-cell transformations for the example at hand is to take the system *away* from its critical point. An important exception to this statement is the following: if a system is initially *at* its critical point (e.g., if $p_0=p_c=1$), then $\xi=\infty$ and hence ξ', by Eq. (16), is also infinite. A necessary but not sufficient condition that this occur is for p' to equal p. The values of p for which $p'=p$ are termed the *fixed points* p^* of the transformation $R_b(p)$,

$$R_b(p^*)=p^*. \tag{17}$$

Thus, by obtaining all the fixed points of a given renormalization transformation $R_b(p)$, we should be able to obtain the critical point. For the example of one-dimension percolation, $R_b(p)=p^b$ and there are two fixed points. One is $p^*=0$ and the other is $p^*=1$. Indeed, we recognize the critical point, $p_c=1$, as one of the two fixed points.

Now if the system is initially at a value $p=p_0$, which is close to the $p^*=1$ fixed point, then under the renormalization transformation it is carried to a value of p_0', which is farther from that fixed point. We may say a fixed point is *unstable* for the "relevant" scaling field $u=(p-p_c)$. Conversely, if p_0 is close to the $p^*=0$ fixed point, then it is carried to a value p_0' that is still closer to that fixed point; we term such a fixed point *stable*. Thus for the example at hand, there is one unstable fixed point, $p^*=1$, and one stable fixed point, $p^*=0$.

We often indicate the results of successive renormalization transformations schematically by means of a simple *flow* diagram, as is shown in Fig. 4(b). The arrows in the flow diagram indicate the effect of successive renormalization on the system's parameters. Note that the "flow" under successive transformations is from the unstable fixed point toward the stable fixed point. In the example treated here, there is only one parameter p and hence the flow diagram is one dimensional; in general, there can be many parameters, and the flow diagram is multidimensional.

Calculations of the "thermal" scaling power

We can also obtain numerical values for the scaling powers once we have a renormalization transformation. The "thermal" scaling power can be calculated for the basic reason that knowledge of $R_b(p)$ near p^* provides information on how $\xi(p)$ behaves for p near p^*. Perhaps the simplest and most straightforward fashion of demonstrating this fact is to expand $R_b(p)$ about $p = p^*$:

$$R_b(p) = R_b(p^*) + \lambda_T(b)(p - p^*) + \mathcal{O}(p - p^*)^2. \tag{18}$$

Here we use the symbol $\lambda_T(b)$ to denote the first derivative of the renormalization function evaluated at the fixed point p^*. From Eq. (15), we find

$$\lambda_T(b) = \left(\frac{dR_b}{dp}\right)_{p=p^*} = b. \tag{19}$$

If we now substitute Eqs. (14) and (17) into (18), and if we neglect terms of order $(p - p^*)^2$, then we obtain simply

$$p' - p^* = \lambda_T(b)(p - p^*). \tag{20}$$

Equation (20) expresses the deviation of p' from the fixed point in the transformed system in terms of the deviation of p from the fixed-point value in the original system.

As we noted above, the effect of the renormalization transformation on $\xi(p)$ is given by Eq. (16). If we regard Eq. (16) as a functional equation valid for all values of p, p', and b, then we can set $b = 1$ and conclude that

$$\xi'(p) = \xi(p), \tag{21}$$

where the equality $p' = p$ follows from Eqs. (19) and (20). Thus ξ' and ξ are the same functions, so that if $\xi(p)$ has a power-law dependence near the critical point—given by Eq. (13)—then it follows from Eq. (21) that

$$\xi'(p') = |p' - p_c|^{-\nu}. \tag{22a}$$

Substituting Eqs. (13) and (22a) into Eq. (16), we have

$$|p' - p_c|^{-\nu} \sim b^{-1}|p - p_c|^{-\nu}. \tag{22b}$$

Since p_c is the value of p at which ξ diverges, we set $p^* = p_c$ in Eq. (20). Hence

$$|p' - p_c|^{-\nu} = [\lambda_T(b)]^{-\nu}|p - p_c|^{-\nu}. \tag{22c}$$

Comparing Eqs. (22b) and (22c), we can express ν in terms of the scale change b and the "derivative" $\lambda_T(b)$,

$$\nu = \frac{\ln b}{\ln \lambda_T(b)}. \qquad (22d)$$

The argument thus far is valid generally. Returning to the example of one-dimensional percolation, we note from Eq. (19) that $\lambda_T(b) = b$. Hence from Eq. (22d) $\nu = 1$, which is the exact result.

The renormalization approach to critical phenomena leads to scaling (see, e.g., the discussion in Nelson and Fisher, 1975 and Fisher, 1998). As a result of scaling, knowledge of ν is sufficient to determine the value of a_T, the "thermal" scaling power for the weak direction, since

$$a_T = \frac{1}{d\nu}. \qquad (23a)$$

It is becoming customary to normalize scaling powers by a factor of d, the system dimensionality. Thus one defines $y_T \equiv d a_T$ and finds from Eqs. (22d) and (23a) that

$$y_T = \frac{\ln \lambda_T(b)}{\ln b}. \qquad (23b)$$

Do We Understand the Critical Point?

About half of the physicists I know feel the critical point is not understood, while the other half seem to feel that it is. It all depends on what we mean by the word "understood." For some, the term means that one can solve a model in closed form and calculate all the exponents. Then the situation is like Schubert's unfinished symphony—albeit perhaps not finished, it is nonetheless very beautiful. And, like Schubert's symphony, what is not finished will never be since even the "simple" Ising model is believed hopelessly insoluble except for the case of $d = 1, 2$. Even the $d = 2$ case is hopeless to solve in nonzero magnetic field, so do not expect exact calculations of scaling functions and all the field-dependent exponents. In three dimensions, no models are solved in closed form, with a few notable exceptions such as the $n \to \infty$ limit of the n-vector model, and some initial terms for the $1/n$ expansion (Brézin and Wadia, 1993).

If we relax our standards of rigor and consider the scaling hypothesis, then we can make some concrete predictions for all dimensions, but not for the exponent values or the threshold values. While not rigorous, the various "handwaving" arguments to justify scaling and renormalization are sufficient to convince a reasonable person—but not a stubborn one (to paraphrase the critical-phenomena pioneer Marc Kac). But even the handwaving arguments do not explain why in some systems scaling holds for only 1–2 % away from the critical point and in other systems it holds for 30–40 % away. Moreover, no modern theory makes exact predictions for experimentally interesting critical parameters such as T_c, which varies from one material to the next by as much as six orders of magnitude.

Despite this "unfinished" situation, the conceptual framework of critical phenomena is increasingly finding application in other fields, ranging from chemistry and biology on the one hand to econophysics (Mantegna and Stanley, 1999) and even liquid water (Stanley et al., 1997; Mishima and Stanley, 1998). Why is this? One possible answer concerns the way in which correlations spread throughout a system comprised of subunits. Like the economy, "everything depends on everything else." But how can these interdependencies give rise not to exponential functions, but rather to the power laws characteristic of critical phenomena?

The paradox is simply stated. The probability that a spin at the origin 0 is aligned with a spin a distance r away, $(1+\langle s_0 s_r\rangle)/2$, is unity only at $T=0$. For $T>0$, our intuition tells us that the spin correlation function $C(r) \equiv \langle s_0 s_r\rangle - \langle s_0\rangle \langle s_r\rangle$ must decay exponentially with r—for the same reason that the value of money stored in a mattress decays exponentially with time (each year it loses a constant fraction of its worth). Thus we might expect that $C(r) \sim e^{-r/\xi}$, where ξ, the correlation length, is the characteristic length scale above which the correlation function is negligibly small. Experiments and also calculations on mathematical models confirm that correlations do indeed decay exponentially, but if the system is at its critical point, then the rapid exponential decay magically turns into a long-range power-law decay of the form $C(r) \sim 1/r^{d-2+\eta}$.

So then how can correlations actually propagate an infinite distance, without requiring a series of amplification stations all along the way? We can understand such "infinite-range propagation" as arising from the huge multiplicity of interaction paths that connect two spins if $d>1$ (if $d=1$, there is no multiplicity of interaction paths, and spins order only at $T=0$). Enumeration algorithms take into account exactly the contributions of such interaction paths of length ℓ—up to a maximum length that depends on the strength of the computer used. Remarkably accurate quantitative results are obtained if this hierarchy of exact results for successive finite values of ℓ is then extrapolated to $\ell=\infty$.

For any $T>T_c$, the correlation between two spins along each of the interaction paths that connect them *decreases* exponentially with the length of the path. On the other hand, the number of such interaction paths *increases* exponentially, with a characteristic length that is temperature independent, depending primarily on the lattice dimension. This exponential increase is multiplied by a "gently decaying" power law that is negligible except for one special circumstance which we will come to.

Consider a fixed temperature T_1 far above the critical point, so that ξ is small, and consider two spins separated by a distance r which is larger than ξ. The exponentially decaying correlations along each interaction path connecting these two spins is so severe that it cannot be overcome by the exponentially growing number of interaction paths between the two spins. Hence at T_1 the exponential decrease in correlation along each path wins the competition between the two exponentials, and we anticipate that $\langle s_0 s_r\rangle$ falls off exponentially with the distance r. Consider now the same two spins at a fixed temperature T_2 far below the critical point. Now the exponentially decaying correlation along each interaction path connecting these two spins is insufficiently severe to overcome the exponentially growing number of interaction paths between the two spins. Thus at T_2 the exponential increase in the number of interaction paths wins the competition. Clearly there must exist some intermediate temperature in between T_1 and T_2 where the the two exponentials just balance, and this temperature is the critical temperature T_c. Right at the critical point, the gently decaying power-law correction factor in the number of interaction paths, previously negligible, emerges as the victor in this stand-off between the two warring exponential effects. As a result, two spins are well correlated even at arbitrarily large separation.

Acknowledgments

I conclude by thanking those under whose tutelage I learned what little I understand of this subject. These include my thesis advisors T. A. Kaplan and J. H. Van Vleck, the students, postdocs, and faculty visitors to our research group over the past 30 years with whom I have enjoyed the pleasure of scientific collaboration, as well as many of the genuine pioneers of critical phenomena from whom I have learned so much: G. Ahlers, G. B. Benedek, K. Binder, R. J. Birgeneau, A. Coniglio, H. Z. Cummins, C. Domb, M. E. Fisher, M. Fixman, P. G. de Gennes, R. J. Glauber, R. B. Griffiths, B. I. Halperin, P. C. Hohenberg, L. P. Kadanoff, K. Kawasaki, J. L. Lebowitz, A. Levelt-Sengers, E. H. Lieb, D. R. Nelson, J. V. Sengers, G. Stell, H. L. Swinney, B. Widom, and F. Y. Wu. Any of these individuals could have prepared this brief overview more authoritatively than I. Finally, I acknowledge my debt to the late M. Kac, W. Marshall, E. W. Montroll, and G. S. Rushbrooke, to whose memory I dedicate this "minireview." My critical phenomena research would not have been possible without the financial support of NSF, ONR, and the Guggenheim Foundation.

References

Als-Nielsen, J., and R. J. Birgeneau, 1977, Am. J. Phys. **45**, 554.
Brézin, E., and S. R. Wadia, 1993, *The Large N Expansion in Quantum Field Theory and Statistical Physics: From Spin Systems to 2-dimensional Gravity* (World Scientific, Singapore).
Bunde, A., and S. Havlin, Eds., 1996, *Fractals and Disordered Systems, Second Edition* (Springer-Verlag, Berlin).
Cardy, J. L., 1996, *Scaling and Renormalization in Statistical Physics* (Cambridge University Press, Cambridge, England).
de Gennes, P.-G., 1979, *Scaling Concepts in Polymer Physics* (Cornell University, Ithaca).
Domb, C., 1996, *The Critical Point: A Historical Introduction to the Modern Theory of Critical Phenomena* (Taylor & Francis, London).
Fisher, M. E., 1967, Rep. Prog. Phys. **30**, 615.
Fisher, M. E., 1974, Rev. Mod. Phys. **46**, 597.
Fisher, M. E., 1998, Rev. Mod. Phys. **70**, 653.
Goldenfeld, N., 1994, *Renormalization Group in Critical Phenomena* (Addison-Wesley, Reading).
Hohenberg, P. C., and B. I. Halperin, 1977, Rev. Mod. Phys. **49**, 435.
Jackiw, R., 1972, Phys. Today **25**, 23.
Kadanoff, L. P., *et al.*, 1967, Rev. Mod. Phys. **39**, 395.
Lee, T. D., and C. N. Yang, 1952, Phys. Rev. **87**, 410.
Lesne, A., 1998, *Renormalization Methods: Critical Phenomena, Chaos, Fractal Structure* (Wiley, New York).
Levelt Sengers, J. M. H., R. Hocken, and J. V. Sengers, 1977, Phys. Today **30**, 42.
Mantegna, R. N., and H. E. Stanley, 1999 *Econophysics: An Introduction* (Cambridge University Press, Cambridge, England).
Milošević, S., and H. E. Stanley, 1976, in *Local Properties at Phase Transitions*, Proceedings of Course 59, Enrico Fermi School of Physics, edited by K. A. Müller and A. Rigamonti (North-Holland, Amsterdam), pp. 773–784.
Mishima, O., and H. E. Stanley, 1998, Nature (London) **396**, 329.
Nelson, D. R., and M. E. Fisher, 1975, Ann. Phys. (N.Y.) **91**, 226.
Peebles, P. J. E., 1980, *The Large-Scale Structure of the Universe* (Princeton University Press, Princeton, NJ).

Potts, R. B., 1952, Proc. Cambridge Philos. Soc. **48**, 106.
Stanley, H. E., 1968, Phys. Rev. Lett. **20**, 589.
Stanley, H. E., 1971, *Introduction to Phase Transitions and Critical Phenomena* (Oxford University Press, London).
Stanley, H. E., P. Reynolds, S. Redner, and F. Family, 1982, in *Real-Space Renormalization*, edited by T. W. Burkhardt and J. M. J. van Leeuwen (Springer-Verlag, Berlin).
Stanley, H. E., L. Cruz, S. T. Harrington, P. H. Poole, S. Sastry, F. Sciortino, F. W. Starr, and R. Zhang, 1997, Physica A **236**, 19.
Stauffer, D., and A. Aharony, 1992, *Introduction to Percolation Theory* (Taylor & Francis, Philadelphia).
Wu, F. Y., 1982, Rev. Mod. Phys. **54**, 235.

Insights from Soft Condensed Matter
Thomas A. Witten

The physicist's method of reflection about the world has, especially in this century, opened our eyes to the world as have few activities in history. Relativity has shown a profound link between space and time, unveiling a startling unity of mass and energy. Quantum mechanics has revealed that motion entails spatial and temporal periodicity, leading to phenomena of beauty, precision, and utility that are inconceivable in terms of everyday experience. Perhaps this is why ours has been called the century of physics.

Though the implications of relativity and quantum mechanics have dominated physics in the twentieth century, these are still just particular examples of the physics enterprise. As these examples have developed, other insights from completely new directions have appeared. One such insight is the symmetry under magnification that many condensed-matter systems show. This symmetry dominates the behavior of matter undergoing a continuous phase transition, for example. This magnification symmetry has arisen not from relativistic, quantum, or other properties of the matter's ultimate constituents, but rather from the workings of mundane forces and random fluctuations.

Another such domain—soft condensed matter—is the subject of this essay. Soft matter occupies a middle ground between two extremes: the fluid state and the ideal solid state. The mobile molecules of a simple fluid may freely exchange positions, so that their new positions are permutations of their old ones. By contrast, the molecules of an ideal solid are fixed in position and may not readily permute in this way. Soft condensed matter is a fluid in which large groups of the elementary molecules have been constrained so that the permutation freedom within the group is lost. For example, thousands of small molecules may be fastened together to form a rigid rod or a flexible chain. The new behavior of soft matter emerges because these groups of fastened-together molecules are large. It emerges because the thermal fluctuations that dominate the fluid state coexist with the stringent constraints characteristic of the solid state.

The best-known soft-matter structures are polymers—hydrocarbon molecules in which many repeating subunits are connected to form a flexible chain. Colloids, emulsions, and foams are fluids containing compact grains or droplets or bubbles of matter, each large enough to constitute a distinct phase, but small enough that thermal fluctuations are important for their properties. Surfactants are small molecules that contain within themselves two strongly immiscible parts. A surfactant molecule in a liquid such as water has one part that is strongly immiscible with the water. The immiscible parts spontaneously aggregate into spheres, flexible rods, bilayers, and more elaborate structures. These structures may grow to arbitrary size. Their variety increases when a second liquid, such as oil, is added. The surfactant aggregates then become receptacles for the oil. Such a liquid, formed from an equilibrium dispersion of two immiscible fluids plus a surfactant is called a microemulsion.

The field of soft condensed matter has developed many aspects (see, for example, Chaikin and Lubensky, 1995). In surveying these below, I emphasize ways that these materials have brought us qualitative new notions about how matter can behave. A central theme of this field is self-organization: the spontaneous creation of regular structure that strongly constrains the spatial arrangement of the system over and above the constraints on the polyatomic constituents. I shall discuss new types of periodic structure, as well as dilation-symmetric, fractal structure. But self-organized structure is only part of the new behavior of soft condensed matter. Another part involves new kinds of organized motion. Soft matter can move in unprecedented ways and thereby confer unprecedented flow properties on fluids.

These distinctive behaviors of soft matter offer potential impact on other fields: the world of electronic structure and optical excitation, the world of chemical transport and reactions, and the world of living organisms. I shall close by mentioning some of these potential impacts.

Principles

Pick up a cube of Jello[1] in one hand and a cube of ice in the other. The contrast between soft condensed matter and conventional matter is immediately apparent. The rigidity and strength of the ice arises from its atomic composition. The molecules lie packed adjacent to one another; compressing the ice between the fingers pushes these packed molecules into one another. The compression creates virtual atomic excitation and requires energy—an energy of the order of electron volts per atom. With only a fraction of a percent compression, the compressional energy approaches the binding energy of the atoms. Then the ice is susceptible to brittle fracture. The Jello by contrast is much more resilient: it sustains a factor-of-two compression with no apparent damage, though it is much weaker than the ice.

The resiliency of Jello arises from its weak connectivity. Its atoms are not packed together. Most of them are in the form of liquid water. The restoring force of the compressed Jello cube arises from the long, randomly coiling gelatin polymers that have been dispersed in the water and have then intertwined to form a network as the water cooled. Compressing the cube distorts the network and pulls on the ends of the gelatin molecules. Because it is contorted, each molecule can elongate easily, just as a spring or a loose wad of wire can. Its molecular bonds can rotate and allow elongation with virtually none of the atomic distortion seen in the ice. A second important feature of the gelatin molecules is their low density. There is plenty of room for the polymer coils to distort without crowding one another the way ice atoms do. Surfactant rods and sheets, aggregated colloidal particles, and the droplet interfaces of emulsions have this same low density. It is this low density and weak connectivity that make soft matter soft.

A second feature of Jello is needed in order to account for its easy deformability. The gelatin molecules are not frozen into a single configuration but are constantly fluctuating

---

[1]Trade name for a common food product made by the General Foods Company of White Plains, New York, consisting of water, gelatin, fruit syrup, and sugar.

from one random configuration to another. The resulting random-walk shape has a preferred size and end-to-end distance. If this distance is distorted by an imposed compression, there are necessarily fewer configurations accessible. Thus the distortion reduces the entropy of the molecule and requires work. This work is the origin of the tension in a gelatin molecule. A loose string buffeted by a random barrage of ping-pong balls would feel an analogous tension. The work is far smaller than the electron-volt-per-atom scale of an ordinary solid. It is of the order of a thermal energy kT per crosslinked section of the gelatin. The thermal energy itself is only 1/40 of an electron volt; moreover, the number of crosslinked segments in the Jello is far fewer than the number of atoms.

Self-organization

Soft matter's most striking property is its ability to self-organize—to create a coherent order throughout spatial regions indefinitely larger than the soft-matter constituents. The most familiar type of order is the breaking of rotational or translational symmetry to form spatially oriented or periodic states. Soft matter, like ordinary matter, shows a wealth of such states: the liquid crystals (de Gennes and Prost, 1993). Soft matter has revealed another form of organization that involves another symmetry: spatial dilation.

Dilational order

When a flexible polymer such as a gelatin molecule is put in solution, each successive segment extends in a random direction relative to its predecessor: the molecule has the form of a random walk. This in itself is a primitive form of organization. Though the correlations of steps of a random walk are structureless, the distribution of steps in space is not. The density of steps near an arbitrary segment falls off inversely with distance instead of being uniform. This length-scale-dependent density reflects a type of spatial order that is not connected to rotations or translations but to dilation. Dilation is the transformation to a magnified coordinate system. A structureless, uniform material looks the same when magnified, provided the magnification is too weak to see its molecular constituents.

A section of a very long random walk also looks the same when magnified. A set of configurations viewed at one magnification is indistinguishable from a set viewed at another, provided the two magnifications are not too strong or too weak. The dilation symmetry may be characterized quantitatively in terms of the falloff of density of steps with distance. In d-dimensional space the density falls off as the $2-d$ power of the distance. This scaling exponent (and related ones) characterizes the dilation symmetry in the same way that linear momentum characterizes translational symmetry and angular momentum characterizes rotation symmetry.

In the last three decades, dilation symmetry has emerged as a powerful form of self-organization in soft matter. This era has brought the means of conceiving of this symmetry, of framing it in mathematical terms, and even of predicting the power laws that define it. It has also brought a wealth of new realizations, which have expanded the range of dilation-symmetric phenomena to a startling degree. The first inkling of this breadth came with the realization that real polymers do not have the size in solution expected for random walks: they are too big. Empirically, their size grows as the 0.6 power of their length, not the 0.5 power expected for random walks. Somehow the polymer expands because its subunits prefer the solvent to one another. In effect, the subunits repel one another. Though *ad hoc* ways of accounting for this swelling date from the '50s (see Chap. XII of Flory, 1971), it was not until 1971 that the meaning of this swelling emerged (de Gennes, 1972). The solvent effect produces not just a global swelling but an expansion at all length scales. As such, it destroys the dilation symmetry of the random walk and replaces it with a new representation of dilation symmetry, characterized by a new set of scaling exponents.

To understand this shift of symmetry required a major reframing of our description of many-body matter. There is no simple explanation of the exponents as there is for a random walk. Here, the -1 power law arose straightforwardly from the two spatial derivatives in the diffusion equation describing random walks. If the power is to shift to a fractional power, must this second derivative somehow change to a fractional derivative? Clearly, a major generalization of the mathematical description is required. In 1971, de Gennes (1972) recognized that the generalization needed for the self-repelling polymer was the same as that needed to describe continuous phase transitions in matter such as a critically opalescent fluid or a magnet at its Curie temperature. He realized that the 0.6 power was in essence a critical exponent. Meanwhile, Wilson and Fisher (1972) showed that the needed mathematical language for all such systems was that of renormalized field theory.

Self-organization within a molecule

Some forms of self-organization occur at the level of a single polyatomic constituent, such as a polymer molecule. Others are collective phenomena involving many polyatomic constituents. I turn first to organization within a molecule.

The classical examples above show the power of self-organization that soft matter can generate. The forms of self-organization can also show a dazzling variety, as the following examples illustrate. Some forms occur within a single molecule, such as DNA, that resists twisting and bending. If such a molecule's ends are joined to form a loop, the preferred shape is circular. But if the loop is twisted, the preferred shape is no longer planar. Instead the molecule twists around itself in a shape called a supercoil (Marko and Siggia, 1995). A mechanical twisting at one point in the molecule can thus cause a global change in the molecule's conformation. Organisms may well use this phenomenon as a means of coiling and uncoiling DNA and otherwise manipulating it in the process of cell replication or gene expression.

A more subtle level of single-molecule self-structuring occurs when the "molecule" is in the form of a two-dimensional membrane, such as graphite oxide (Spector *et al.*, 1994) or a red blood cell membrane (Schmidt *et al.*, 1993) or a sheet of paper. Such membranes resist bending or stretching. When confined into a small space, they develop the familiar but mysterious structure of the crumpled state (Kantor and Nelson, 1987; Kramer and Witten, 1997). Alternatively, they may be made sufficiently flexible that thermal fluctuations bend them strongly on a local scale. These thermal fluctuations might be expected to leave the membrane in a strongly contorted state, as they do for the polymers discussed above. Remarkably, the reverse happens: the local undulations lead to a global flattening. This surprising reversal happens because of the two-dimensional connectivity of the membrane. Such a membrane may bend easily in one direction. But it cannot bend in both directions at once (producing Gaussian curvature) without also stretching: corrugating a sheet in one direction makes it stiffer in the perpendicular direction. The membrane's resistance to stretching thus makes the sheet bend less than it would otherwise.

Two possibilities are open to the membrane. On the one hand, it might break its directional symmetry and choose to contort in one direction while remaining flat in the perpendicular direction. A moderate amount of local bending anisotropy can tip the balance in this direction, according to recent calculations (Bowick *et al.*, 1997; Radzihovsky and Toner, 1998). On the other hand, the membrane might retain its directional isotropy and simply moderate its bending so as to avoid excessive stretching. This possibility implies a progressive flattening at large length scales. The local bending means that the membrane has an effective thickness, like the corrugated wall of a cardboard box. This thickness makes the membrane stiff, and limits similar corrugations at larger scales. The result is that both the bending and stretching stiffness change progressively when measured on larger and larger length scales. According to recent simulations, the bending modulus grows roughly as the 0.6 power of the linear size, while the stretching modulus decreases roughly as the -0.5 power (Bowick *et al.*, 1996). This dependence is consistent with analytical estimates (Aronovitz and Lubensky, 1988) and experiments (Spector *et al.*, 1994). This subtle form of scaling is a startling consequence of the interplay of elasticity with thermal fluctuations.

Self-organization from entropic interactions

The thermal fluctuations of chain molecules or molecular sheets give rise to mutual self-organizing effects on the ensemble of constituents in the fluid. One of the most dramatic forms of spatial organization results from these fluctuations in a molten polymer liquid. Two immiscible polymer chains, A and B, can be joined together at the ends to form a "diblock copolymer." In a liquid of such copolymers the A species must segregate from the B species. But this phase separation cannot proceed normally since the A and B chains are connected. The result is a microscopic phase separation into small regions containing A or B chains. The size of these domains is limited, since the A-B junctions are necessarily at the boundaries of a domain, and the chains extending from the boundary must fill the interior. Increasing the domain size reduces the interfacial area between A and B regions and thus reduces the energetic cost, as in any phase-separating fluid. But this increased size ultimately forces the chains to stretch, reducing their entropy.

The domains must find the most favorable balance between these opposing tendencies. The domains typically contain hundreds of kT of interfacial energy. Thus there is a large energetic incentive to attain an optimal domain structure. The result is a strongly periodic pattern, whose shape depends on the relative length of the A and B chains used. The intricate bicontinuous structure of Fig. 1 is one example. Since the domain energy depends only on the random-walk nature of the chains and their interfacial energy, the optimal structure is independent of chemical details. Thus the structures should be accurately predictable using generic numerical methods (Helfand and Wasserman, 1978; Scheutjens and Fleer, 1979; Noolandi and Hong, 1982). The variety of known patterns is increasing rapidly, as copolymers of different block sequences and varying degrees of elastic entropy are used (Sakurai *et al.*, 1998). These microdomains provide a means of arranging material in space whose power is only beginning to be exploited.

These copolymers are only one example of how polymer entropy may create spatial organization. Polymers grafted to colloidal particles can create large enough forces to separate these particles and induce periodic order in solution (McConnell *et al.*, 1994).

Polymers can create strong forces to separate macroscopic surfaces (Klein *et al.*, 1994) or surfactant bilayers (Warriner *et al.*, 1996).

Two-dimensional structures can also create order via their entropic interactions. The simplest example occurs when certain lipid surfactants are dissolved in oil. These surfactants spontaneously form bilayers, with the oil-loving ends facing outward. These bilayers are not elastic membranes; they have no shear modulus. Thus thermal fluctuations make them undulate readily and give the bilayers entropy. If anything confines such a bilayer to a gap of fixed width, this entropy is reduced and a repulsive force varying as the inverse cube of the thickness results (Helfrich, 1978). The same repulsion acts between a bilayer and its neighbors. The result of the bilayer repulsion is a long-range smectic order in the bilayers, even when the proportion of surfactant in the fluid is small. As with polymers, the addition of further features leads to a barely explored wealth of strongly selected structures. Polymers that live in the oil, that graft to the bilayers, or that bridge between bilayers create some of these variations (Warriner *et al.*, 1996).

A further form of entropic interaction appears when electrically charged species are introduced. It is entropy that causes oppositely charged pairs of ions to dissociate in the liquid. Strong structure formation arises when macroions are introduced. Macroions are colloids, polymers, or surfactant interfaces with many dissociating charges. For typical charged colloids the number of charges is in the hundreds. The dissociation creates an electrostatic energy that is comparable to the entropic free energy kT times the entropy of the dissociated ions. This energy amounts to several hundred times the thermal energy kT. Thus the expected interaction energy between macroions is large. Not surprisingly, the macroions take on a strong crystalline order even when separated by many volumes of solvent. This order is readily disrupted by flow, but immediately reasserts itself. Charged lipid bilayers show similar interaction; it enhances the repulsions between bilayers and thus enhances their smectic order. The expected interaction between like-charged macroions is repulsive. But there is increasing evidence that significant attraction can occur (Kepler and Fraden, 1994; Crocker and Grier, 1996). Our understanding of such

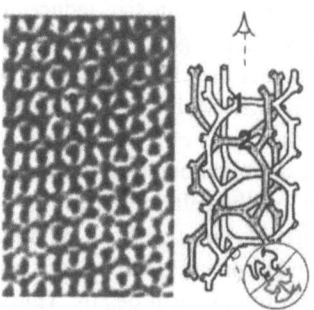

Figure 1. Transmission electron micrograph of the bicontinuous copolymer domain structure, reproduced from Hasegawa et al., 1987. Repeat distance is about 100 nm. The minority species in black forms two disjoint domains, labeled 1 and 2. The sketch at right, after Matsen, 1998, shows the three-dimensional structure inferred from such micrographs and from x-ray diffraction studies; the vertical dashed line gives the line of sight for the micrograph. The thickness of the two minority domains is reduced to make the illustration clearer. Domains 1 and 2 are shown with different shadings to distinguish them. The magnified sketch at bottom shows the orientation of individual polymers in the domains.

attractions has not yet reached the point of consensus. Still, this puzzle shows that there is more to the interaction of macroions than their obvious Coulomb repulsion.

When the macroion is a polymer with its own internal entropy, the interplay with counterion entropy leads to strong elongation of the polymer coils. It also leads to strong short-range order in solutions of charged polymers and to anomalous flow properties that are far from being well understood (Micka and Kremer, 1997). The interaction between oppositely charged macroions, such as a polymer and a colloidal sphere, are likely to lead to further new forms of self-organization. These should be strong effects because of the large energies at stake (Pincus, 1998).

Kinetically driven self-organization

The self-organized structures discussed above all lie within the compass of equilibrium statistical mechanics. Further undreamed-of structures arise during irreversible, kinetic processes. A striking example is the remarkable structure of colloidal aggregates. Colloidal particles have a mutual interaction energy that grows with their size. Thus particles much larger than the solvent molecules have a nominal interaction strength that may be many times the thermal energy kT. Since the basic energy scale is large, a modest change in the fluid composition can change the mutual interaction energy from being repulsive to being strongly attractive. Now, instead of forming a stable dispersion, the particles flocculate irreversibly. The attraction can easily be so strong that the particles stick together on contact without being able to migrate over each other's surfaces.

This extreme aggregation leads to irregular shapes as particles form small clusters, and these aggregate into progressively larger clusters (Vold, 1959, 1963; Sutherland, 1967; Witten, 1987). Each pair of clusters that collides is frozen in the configuration it had at collision. Each collision causes a growth of the cluster mass to the sum of the constituent masses. Each also causes a growth of the radius by a finite factor that has no fixed relation to the mass factor. The result is a fractional power law relating the radius to the mass. Naturally, the same power governs each subcluster of a large cluster, as well as its own subclusters, and so forth. Thus the cluster has the same structure at many length scales; it is dilation symmetric. Like polymers, these clusters have an average density near an arbitrary particle which falls off as a power of the distance. The overall average density decreases as a power of their mass. They (like polymers) are "fractal objects" (Mandelbrot, 1982).

The main geometric, elastic, and hydrodynamic properties of these colloidal aggregates have now been well characterized and shown to be independent of materials and conditions over wide ranges (Lin et al., 1989). These aggregates are a distinctive organization of matter that have emerged through the study of soft matter.

Flow

The kinetic effects of flow exert a major restructuring influence on structured fluids. The conventional shear flow of two surfaces sliding past each other can align smectic layers (Fredrickson and Bates, 1996), elongate polymers (Bird et al., 1987), and induce melting or ordering in a colloidal dispersion (Stevens et al., 1991). But beyond these effects, flow can create qualitative new structures whose origin remains a mystery. A striking example occurs when a smectically ordered solution of bilayers is subjected to shear flow of the right magnitude. The lamellae reorganize to form a closely packed, regular array of onionlike structures, each with dozens of concentric bilayer vesicles (Diat et al., 1993).

The size and number of layers may be varied by changing the shear rate, even after the onions have initially been formed. Yet when the shear flow is stopped, the onions are stable over many months, even if solvent is added. To understand how shear induces the transition among the smectic and the various onion states is a great challenge.

Other challenging flow effects occur when a solution of suspended particles (Hu et al., 1994) or DNA (Isambert et al., 1997) is subjected to an oscillating electric field. Field of the proper amplitude and frequency leads to islands of many circulating molecules and to a herringbone arrangement of these islands (Fig. 2). These are just two of a variety of unstable, pattern-forming flows involving vesicles, surfaces, or fluid-fluid interfaces (Bacri et al., 1991; Bar-Ziv and Moses, 1994). Since kinetic effects can lead to such startling and inexplicable reorganization, we must look forward to even further surprises.

Rheology

The last section showed how flow can have strong effects on structure. The effects of structure on flow are just as dramatic. In soft matter the sharp distinction between solids and liquids blurs. We may characterize a simple liquid by its viscosity, the proportionality between the stress and the rate of strain. We characterize the mechanics of an ordinary solid by the modulus, the proportionality between stress and strain. In a soft material like Jello, we must speak of a modulus that depends on the strain amplitude and frequency. Often it depends on history of the strain, since this history may have organized and altered the structure within the sample.

In Jello the rheological properties follow readily from the known properties of its polymeric subunits and their hydrodynamic interaction with the solvent. In other polymer

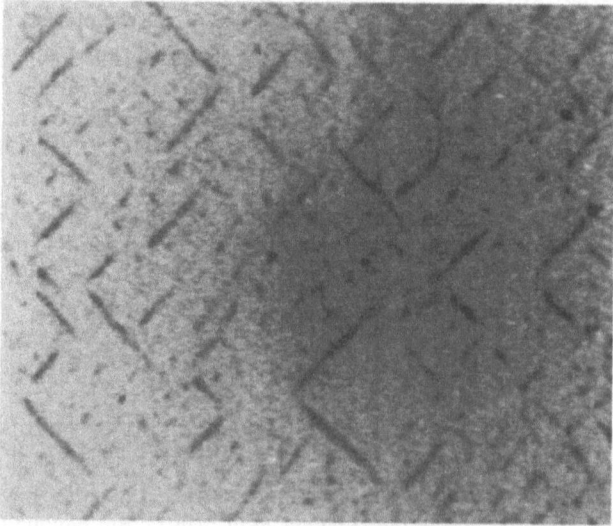

Figure 2. Precipitation patterns in a DNA solution induced by an oscillating field, reproduced from Isambert et al., 1997, by permission. Horizontal electric field had an amplitude of 300 V/cm, a frequency of 2 Hz and was applied for two minutes. Sample volume is a horizontal slab 10 microns deep. Black streaks are regions of concentrated DNA about 10 microns wide.

fluids (and fluids of wormlike micelles), flow is strongly inhibited by the mutual *entanglement* of the molecules. To understand the flow requires an additional fundamental insight about the molecular motions that can lead to disentanglement. Such an insight came in the 1970s with the work of de Gennes, Edwards, and Doi (see Doi and Edwards, 1986). The theory supposes that the important motion for disentanglement is the Brownian motion of each molecule along its own contour, a motion known as "reptation." The reptation hypothesis has grown into a powerful and successful kinetic theory of disentanglement rheology—a "standard model" to which experiments and other theories are compared. By addressing the phenomenon of polymer flow, the reptation hypothesis has brought the murky notion of disentanglement to a level approaching conceptual mastery and confident prediction.

Many other aspects of soft-matter rheology still baffle us. The most perplexing is the phenomenon of turbulent drag reduction (de Gennes, 1992). A very few large polymers in a liquid are able to alter significantly the way the liquid flows through a pipe. A few parts per million of polymer can reduce the drag during turbulent flow by nearly a factor of two. The polymers seem to forestall the large dissipation that occurs near the pipe walls during turbulent flow and to affect the flow by undergoing a "coil-stretch transition," in which elongational flow stretches the chain from a contorted random walk to nearly full elongation. But beyond these elements, little is understood or confirmed.

An even more fundamental phenomenon occurs when a few large particles are suspended in a flowing fluid. The flow can easily be made fast enough so that convective motion of the particles completely dominates any diffusive Brownian motion. At the same time the flow can be slow enough that the inertial effects leading to turbulence are completely negligible. And yet, when one observes the particles through a microscope, their motion, except for its glacial pace, resembles a turbulent flow (Segrè *et al.*, 1997). Instead of moving in a coherent fashion, the particles collide, swirl, and move at greatly different speeds. The basic effect responsible is the hydrodynamic interaction, the alteration of one particle's motion owing to the perturbed flow near another particle. This chaotic flow appears to be a motion as fundamental as turbulence, though conceptually simpler. But our understanding of it is much more primitive than even our meager understanding of turbulence. No analog of the Reynolds number has been proposed to quantify the passage from coherent motion to this chaotic state, nor has any asymptotic scaling hypothesis been convincingly formulated.

Soft glasses

Other strange flow phenomena occur in dense emulsions like mayonnaise. Mayonnaise holds its shape because its many micron-sized drops of oil are packed together, with only a small amount of water between the drops. Any external deformation distorts the drops further and increases their interfacial area; this increase costs energy and thus creates a stress opposing the deformation. In this way a collection of liquid droplets becomes a disordered elastic solid—a glass.

Under strong enough stress an emulsion or foam flows. This flow is unlike the yielding of ordinary solids made of stiff atomic particles. There strain becomes concentrated in fracture cracks or plastic flow zones. In a typical emulsion the stress relaxes instead by means of discrete, localized yield events, each involving a handful of droplets (Durian *et al.*, 1991). The yielding events do not seem to persist at preferred places, but occur throughout the material. Thus a yielding event at one point seems to shift the nearby droplets enough that all regions are equivalent after sufficient time. The overall stress in such materials shows suggestive dependence on frequency and strain rate; often this dependence resembles a power law (Kahn et al., 1988; Ketz et al., 1988). The basic nature of all simple emulsions is the same; one differs from another only in volume fraction and droplet size distribution. Thus it should be possible to come to a common understanding of how their yielding differs from that of conventional solids and how the apparent power-law responses arise. These questions are being actively pursued (Langer and Liu, 1997; Sollich et al., 1997).

Bridges to Other Fields

Electronic condensed matter

In the sections above, we have seen a wealth of behavior emerge from simply defined systems. It is only natural that soft-matter phenomena should prove useful for understanding and manipulating phenomena in other domains. The domain of electronic motion is a prime example. Progress in our control of electron motion and ordering in recent decades has hinged in large part on our ability to create interfaces between metals and insulators, or between one semiconductor and another. The self-organizing abilities of soft matter provide a natural avenue to make such interfaces. For example, the copolymers discussed above can be readily designed so as to make domains of desired shapes and sizes. Since the domains arise from a small number of polymer properties, there is great flexibility to make them with a range of polymer types having a range of desired electronic properties. Such structures are being developed to study magnetic incommensurability effects (Harrison et al., 1997).

The organized "self-assembled surfactant monolayers" that form at a metal or semiconductor surface offer further potential for electronic control. The charged surfactant heads can create a controllable electron depletion in the substrate. Such a system can be operated as a field-effect transistor whose conductance is controlled by the state of the monolayer, with its sensitivity to solvent or flow effects (Gartsman et al., 1998).

Chemistry

Soft-matter structures also allow one to manipulate chemical reactions. Often molecule-sized pores are used as the site of organic reactions, in order to limit the transport of reactants and products selectively according to their size and shape. Up to now the claylike zeolite minerals have been the porous materials used. Their range of pore structures is limited by their bonding configurations. Recently it has proved possible to use the microdomain structures of surfactant-water mixtures to create inert, mineralized pore structures. The soft-matter microdomains were "transcribed" into hard, mineral form (Beck et al., 1992).

In the domain of materials science, colloidal structures have long been used to create materials. Composite materials of polymers with colloidal aggregates make tough rubber compounds (Witten et al., 1993). Likewise, particulate minerals form the green bodies which are heat treated to make strong, precisely shaped ceramic parts. The emerging knowledge of soft matter must increase the control and understanding of these strong materials.

Biology

Living cells are full of polymer chains, lipid vesicles, macroions, and self-assembled rods and tubes. Still, it is not clear whether the distinctive behavior of these entities is of the essence in understanding the machinery of life. On the one hand physical effects like microphase separation, entropic interaction, and membrane stress propagation may be crucial. Such effects are attractive candidates to explain such important processes as protein folding, the formation of ordered complexes in the cell, or organized motility in cells. On the other hand, the important process may be better thought of via the enzyme paradigm of explicit molecular reactions being catalyzed in an inter-related sequence. In either case, understanding how these physical processes influence the cell must be helpful.

Conclusion

Physics may be viewed as a space of concepts. By combining these concepts, we may understand a broad range of phenomena. But some phenomena cannot be understood in this way; they lie beyond the compass of established concepts. These phenomena, if they are simple and broad, expand our space of concepts into new dimensions. The physics of soft condensed matter is valuable in large part because it has been a source of such new dimensions. The field has shown us how to conceive nature in ways formerly beyond our imagining. Soft-matter physics has brought much understanding, but is continually uncovering basic mysteries. There is no sign that this creative process is abating.

Acknowledgments

I am grateful to Michael Cates, David Grier, Tom Lubensky, Scott Milner, Adrian Parsegian, Didier Roux, Alexei Tkachenko, and Molly Romer Witten for their helpful comments. This work was supported in part by the National Science Foundation under Award Number DMR-95 28957.

References

Space restraints do not permit citations of detailed references. A complete list can be obtained at: http://mrsec.uchicago.edu/~tten/Bederson.pdf

Chaikin, P. M., and T. C. Lubensky, 1995, *Principles of Condensed Matter Physics* (Cambridge University, Cambridge, England).
de Gennes, P. G., 1972, Phys. Lett. A **38**, 339.
de Gennes, P. G., 1992, Rev. Mod. Phys. **64**, 645.
de Gennes, P. G., and J. Prost, 1993, *The Physics of Liquid Crystals*, 2nd ed. (Oxford University, New York).
Doi, M., and S. F. Edwards, 1986, *The Theory of Polymer Dynamics* (Oxford University, Oxford).
Flory, P., 1971, *Principles of Polymer Chemistry* (Cornell University, Ithaca).
Hasegawa, H., H. Tanaka, K. Yamazaki, and T. Hashimoto, 1987, Macromolecules **20**, 1651.
Langer, S. A., and A. J. Liu, 1997, J. Phys. Chem. B **101**, 8667.
Mandelbrot, B. B., 1982, *The Fractal Geometry of Nature* (Freeman, San Francisco).
Matsen, M. W., 1998, Phys. Rev. Lett. **80**, 4470.
Sollich, P., F. Lequeux, P. Hebraud, and M. E. Cates, 1997, Phys. Rev. Lett. **78**, 2020.

Vold, M. J., 1963, J. Colloid Sci. **18**, 684.
Wilson, K. G., and M. E. Fisher, 1972, Phys. Rev. Lett. **28**, 240.
Witten, T. A., 1987, in *Chance and Matter, 46th Les Houches Summer School*, edited by J. Souletie, J. Vannimenus, and R. Stora (North Holland, Amsterdam).
Witten, T. A., M. Rubinstein, and R. H. Colby, 1993, J. Phys. II **3**, 367.

Granular Matter: A Tentative View
P. G. de Gennes

Granular matter refers to particle systems in which the size d is larger than one micron. Below one micron, thermal agitation is important, and Brownian motion can be seen. Above one micron, thermal agitation is negligible. We are interested here in many-particle systems, at zero temperature, occupying a large variety of metastable states: if we pour sand on a table, it would like to go to a ground state, with a monolayer of grains giving the lowest gravitational energy. But in reality the sand remains as a heap; the shape of the heap and the stress distribution inside depend critically on how the heap was made. Hence come many difficulties.

We cited sand as an example: a desert like the Sahara provides us with a gigantic laboratory model. The grains are silica (rounded by collisions) of ~100 microns in size. They form ripples and dunes. These deserts have fascinated a number of great men—Lawrence and Thesinger in Arabia, Monod in the western Sahara, and R. Bagnold in the Libyan desert. Bagnold knew physics and fluid mechanics: he had very much the style of G. I. Taylor. He made precise observations in the desert, then returned to England, built a cheap but efficient wind tunnel (with plywood, etc.), and determined with it the basic laws for the transport of sand. His book *Physics of Blown Sand and Sand Dunes*, published in 1941, remains a basic reference sixty years later (Bagnold, 1941). We shall give an "idealized summary" of his views in the section on Macroscopic Stress Fields.

Of course, there are many other important granular systems in nature: snow is an obvious example; but snow is frightfully complex, because water can show up in all its natural states, and the resulting phase transitions imply deep macroscopic consequences. In the present text, we shall try to concentrate on dry systems. This may be sand, but it may also be mustard seed (the latter being very convenient for certain nuclear resonance studies).

Many industrial products are powders:

- "clinkers" (the starting point of cement) are complex mixtures of silicoaluminates, calcium silicates, etc.
- "builders" are an important part of a commercial detergent: they are based on inorganic particles such as calcium carbonate.
- most pharmaceutical products are derived from powders, obtained by precipitation, crystallization, or prilling (prilling is based on a molten thread of material, which breaks into droplets via the Rayleigh instability; the droplets then reach a cool region where they freeze, giving grains with a very-well-defined size).

If we measure it by tons, the material most manipulated by man is water; the second-most-manipulated is granular matter. But in our supposedly sophisticated 20th century, the manipulation of powders still involves some very clumsy and/or dangerous operations.

(1) Milling is slow, inefficient, and generates a very broad distribution of final sizes.
(2) The smaller-size component of these distributions is often toxic.
(3) Many powders, when dispersed in air, achieve a composition that is ideal for strong detonations. Certain workshops or silos explode unexpectedly. One of the main reasons for this is electrostatic: many grains, when manipulated, hit each other or hit a wall, generating triboelectric charges, which ultimately end up in sparks. To understand this, a new type of mass spectrometry is now set up, in which the particles are grains rather than molecules. The grains are studied after a sequence of wall collisions; here, the interest is more in the charge than in the weight.
(4) When feeding, for instance, a glass furnace with a mixture of oxides, one finds that the corresponding flow of oxide in the hoppers can lead to *segregation*—thus creating dangerous inhomogeneities in the final glass: the manipulation of mixtures is delicate.

Certain other operations are quite successful, although their basic principles are only partly understood: for instance, by injecting a gas at the bottom of a large column filled with catalytic particles, one can transform them into a *fluidized bed*. This is crucial for many processes, such as the production of polyethylene. But the dynamics of these beds is still not fully understood.

We see, at this level, the importance of fundamental research in granular matter. This was appreciated very early in mechanical and chemical engineering; physicists have joined in more recently. For them, granular matter is a new type of condensed matter, as fundamental as a liquid or a solid and showing in fact two states: one fluidlike, one solidlike. But there is as yet no consensus on the description of these two states! Granular matter, in 1998, is at the level of solid-state physics in 1930.

There are some excellent reviews (e.g., Jaeger *et al.*, 1996) but very few textbooks—apart from Bagnold (1941) and Brown and Richards (1970). The most recent one is (at the moment) published only in French (Duran, 1997).

In the present short survey, we shall talk only about the *statics* of heaps and silos. The dynamics will be presented elsewhere.

"We fill a glass column with sand." This innocent statement hides many subtleties. Did we fill it from a jet of sand near the axis, or did we sprinkle the sand over the whole section? Did we shake the object after filling?

A first, obvious problem is *compaction*. Bernal (1964) and Scott (1962) measured the average density of containers filled with ball bearings. They were in fact concerned with models for amorphous systems at the atomic level, but their results are of wider utility. Computer simulations (Finney, 1970) indicate that the maximum volume fraction achieved in a random packing of spheres is $\phi_{rp} = 0.64$—significantly smaller than the face-centered-cubic (or hexagonal) compact packing $\phi_{max} = 0.74$. Compaction is favored by the weight of the grains themselves. Immersing the grains in a fluid of matched density (Onoda *et al.*, 1990), one can study weaker compactions and more or less reach the connectivity limit or, as it is called, the random loose-stacked limit, which for spheres is around $\phi_{min} = 0.56$.

When a powder is gently shaken, it densifies. In fact, a useful method for characterization of a new granular material is based on tapping a vertical column (see, for instance, Selig and Ladd, 1973). Powders that compact fast are expected to flow easily, while powders that compact slowly, more or less refuse to flow. Fundamental studies on the compaction of noncohesive grains have been performed by the Chicago group (Knight et al., 1995; Nowak et al., 1998). The density plots ϕ_n (after n taps) depend on the amplitude of the taps. At small amplitudes, they follow a logarithmic law:

$$\phi_n = \frac{a}{\ln(n)+b}. \tag{1}$$

Many frustrated, frozen systems are expected to show similar forms of creep (Coniglio and Hermann, 1996; Nicodemi et al., 1997). The simplest interpretation of Eq. (1) is based on free-volume models (Knight et al., 1995; Boutreux et al., 1997), which are familiar from the physics of glasses. The case of strong tapping is more complex (Nowak et al., 1998), but some relevant simulations and modelizations have been performed (Barker and Mehta, 1991, 1992, 1993).

Even if we do not perform any tapping, we must specify how the grains were brought in: there is a critical moment, where the grains stop and adopt a *frozen conformation*. For instance, if we build a heap of sand from an axial jet falling on the center, we create avalanches from the center towards the edges; the freezing process takes place via grains that roll and stop.

The distinction between rolling and frozen grains is crucial. It is reminiscent of a phase transition. If we accept it, we may describe the later evolution of the frozen phrase by a *displacement field* $\sim u(x,y,z,t)$. This is defined by the following gedanken experiment. We focus our attention on one rolling grain and watch when it stops, at a certain point x,y,z. This will define the origin of its displacements. Later, with other grains added and loading the system, our grain will move by an amount $\sim u(x,y,z,t)$. Its position will thus depend on the whole history of loading. The resulting displacement field is continuous. Inside the frozen phase, we may define deformations ∇u. We may also define a (coarse-grained average) stress field $\sigma_{\alpha\beta}$ and relate it by some empirical relation to the deformations.

This procedure is essentially what has been used in mechanics departments: see, for instance, the review by Biarez and Gourves (1989). But the precise definition of $\sim u$ is not always stated, and thus the very notion of a displacement field has been questioned by a number of physicists (for a recent summary, see Cates et al., 1998a, 1998b).

The present author's belief is that $\sim u$ is well defined, provided that there is a sharp distinction between fluid particles and frozen particles.[1] We shall come back to this discussion later in the section on Macroscopic Stress Fields.

Another important point is the role of *boundary conditions*, on the frozen piece:

(a) At the free surface: a heap, for instance, shrinks under its own weight, and this renormalizes the relation between deformations and displacements.

(b) At the interface between the grains and a solid wall, the normal displacements must, of course, be continuous. The delicate part is the description of friction, i.e., of tangential stresses σ_t at the surface. The natural scheme is as follows:
  (i) If the tangential component of $\sim u (\sim u_t)$ has grown monotonically and is large

---

[1]This may exclude certain complex problems such as tapping.

enough, the reaction σ_t from the wall is opposed to $\sim u_t$. For a cohesionless interface, we may write the classical relation (Amontons' law; see, for instance, Bowden and Tabor, 1973)

$$\sigma_t = \mu_f \sigma_n,$$

where σ_n is the normal stress and μ_f is a friction coefficient. We call this regime "fully mobilized friction."

(ii) If the tangential displacement $|\sim u_t|$ is smaller than a certain microscopic length Δ, the friction is only partly mobilized. We call Δ the "anchoring length" (de Gennes, 1997). It is usually related to the size of microscopic roughness. For macroscopic solids in contact, Δ is of order one micron.

(iii) If we reverse the displacements (as may happen in experiments where weight and thermal expansions are in conflict) the friction force will reverse fully, only if we move backwards by more than 2Δ.

Thus the state of friction may be influenced by minute displacements of the grains (of order Δ) with respect to the container walls. In a recent experiment on columns (Vanel et al., 1998), the apparent weight at the bottom was found to vary cyclically between day and night: as pointed out by the authors, this is probably due to thermal expansion, inducing some (very small) relative displacements between the grains and the lateral walls, and changing drastically the mobilization of friction.

To summarize: the definition of an initial state, in an experiment on granular matter, requires great care. Many theories and some experiments suffer from a lack of precise definitions.

Macroscopic Stress Fields

The general problem

For more than a hundred years, departments of applied mechanics, geotechnical engineering, and chemical engineering have analyzed the static distribution of stresses in granular samples. What is usually done is to determine the relations between stress and strain on model samples, using the so-called triaxial tests. Then, these data are integrated into the problem at hand, with the material divided into finite elements (see, for instance, Schofield and Wroth, 1968).

In a number of cases, the problem can be simplified, assuming that the sample has not experienced any dangerous stress since the moment when the grains "froze" together: this leads to a *quasielastic description*, which is simple. I shall try to make these statements more concrete by choosing one example: a silo filled with grain.

The Janssen picture for a silo

The filled silo is shown in Fig. 1. The central observation is that stresses, measured with gauges at the bottom, are generally much smaller than the hydrostatic pressure $\rho g H$ which we would have in a liquid (here ρ is the density, g is the gravitational acceleration, and H is the column height). A first modelization for this was given long ago by Janssen and Vereins (1895) and Lord Rayleigh (1906a, 1906b, 1906c, 1906d).

(a) Janssen assumes that the horizontal stresses in the granular medium (σ_{xx}, σ_{yy}) are proportional to the vertical stresses:

$$\sigma_{xx} = \sigma_{yy} = k_j \sigma_{zz} = -k_j p(z), \tag{2}$$

where k_j is a phenomenological coefficient and $p = -\sigma_{zz}$ is a pressure.

(b) An important item is the friction between the grains and the vertical walls. The walls endure a stress σ_{rz}. The equilibrium condition for a horizontal slice of grain (area πR^2, height dz) gives

$$-\rho g + \frac{\partial p}{\partial z} = \frac{2}{R} \sigma_{rz}\bigg|_{r=R}, \tag{3}$$

where r is a radical coordinate and z is measured positive towards the bottom.

Janssen assumes that, everywhere on the walls, the friction force has reached its maximum allowed value—given by the celebrated law of L. da Vinci and Amontons (Bowden and Tabor, 1973):

$$\sigma_{rz} = -\mu_f \sigma_{rr} = -\mu_f k_j p, \tag{4}$$

where μ_f is the coefficient of friction between grains and wall.

Accepting Eqs. (2) and (4), and incorporating them into Eq. (3), Janssen arrives at

$$\frac{\partial p}{\partial z} + \frac{2\mu_f}{R} k_j p = \rho g. \tag{5}$$

This introduces a characteristic length

$$\lambda = \frac{R}{2\mu_f k_j} \tag{6}$$

and leads to pressure profiles of the form

$$p(z) = p_\infty [1 - \exp k(-z/\lambda)], \tag{7}$$

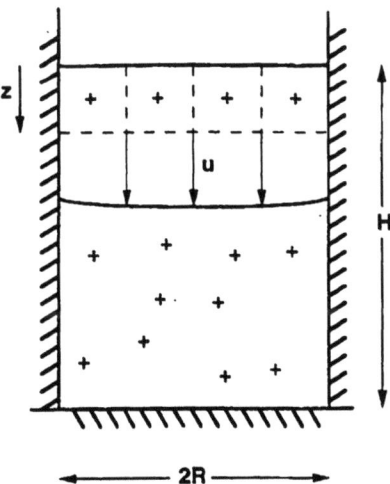

Figure 1. A silo filled with granular material: the material falls slightly under its own weight, by an amount u. The width of the silo has been exaggerated to display the expected profile of u in a quasielastic model.

with $p_\infty = \rho g \lambda$. Near the free surface ($z < \lambda$) the pressure is hydrostatic ($p \sim \rho g z$). But at larger depths ($z > \lambda$) $p \to p_\infty$: all the weight is carried by the walls.

Critique of the Janssen model

This picture is simple and does give the gross features of stress distributions in silos. But the two assumptions are open to some doubt.

(a) If we take an (excellent) book describing the problem as seen from the point of view of the mechanics department (Nedermann, 1992), we find that Eq. (1) is criticized: a constitutive relation of this sort might be acceptable if x,y,z were the principal axes of the stress tensor, but in fact, in the Janssen model, we also need nonvanishing off-diagonal components σ_{xz}, σ_{yz}.

(b) For the contact with the wall, it is entirely arbitrary to assume full mobilization of the friction, as in Eq. (4). In fact, any value σ_{rz}/σ_{rr} below threshold would be acceptable. Some tutorial examples of this condition and of its mechanical consequences are presented in Duran's book (1997). I discussed some related ambiguities in a recent note (de Gennes, 1997) emphasizing the role of the anchoring length.

Quasielastic model

When a granular sample is prepared, we start from grains in motion, and each grain freezes at a certain moment. This defines our reference state: (i) the origin of the grain displacements is the freezing point; (ii) the reference density (for defining deformations) is the density achieved immediately upon freezing.

If we fill a silo from the center, we have continuous avalanches running towards the walls, which stop and leave us with a certain slope.

Recent theoretical studies on avalanches (Boutreux et al., 1998) suggest that this final slope, in a "closed-cell" geometry like the silo, should always be below critical: we do not expect to be close to an instability in shear, and the material is under compression everywhere. In situations like this, we may try to describe the granular medium as a *quasielastic medium*. The use of "quasi" must be explained at this point.

When we have a granular system in a certain state of compaction, it will show a resistance to compression, measured by a macroscopic bulk modulus K. But the forces are mediated by small contact regions between two adjacent grains, and the contact areas increase with pressure. The result is that $K(p)$ increases with p. For spheroidal objects and purely Hertzian contacts, one would expect $K \sim p^{1/3}$, while most experiments are closer to $K \sim p^{1/2}$ (Duffy and Mindlin, 1957). Various interpretations of the $p^{1/2}$ law have been proposed (Goddard et al., 1990; de Gennes, 1996).

Evesque and the present author (1998) recently used the quasielastic picture to describe displacements and stresses in a silo. The displacements are vertical and correspond to a slight collapse of the column under its own weight. They increase during filling: their description involves the whole sample history. (The displacements are also slightly smaller near the walls than in the center. This creates the shear stresses that worried Nedermann.)

The result is a Janssen relation of the form of Eq. (2), with a value of k_j that depends only on the Poisson ratio σ_p of the material:

$$k_j = \frac{\sigma_p}{1-\sigma_p}. \tag{8}$$

Although the elastic moduli do depend on pressure, it may be that σ_p and k_j are pressure independent. Then the Janssen pressure profile should hold, provided that mobilization of the wall friction is complete. For long columns ($H \gg \lambda$) the maximum displacement is achieved at mid-height and is

$$|u|_{\max} = \frac{\lambda^2}{\lambda_c}, \qquad (9)$$

where $\lambda_c = E/\rho g$ is what we call the compaction length (E = the Young modulus; ρ = the density). Mobilization is indeed complete if $|u|_{\max} \gg \Delta$ (the anchoring length), or equivalently $\lambda > H^*$, where

$$H^* = (\Delta \lambda_c)^{1/2}. \qquad (10)$$

In this formula, Δ is very small, but λ_c is very large. Typical values of H^* depend on E, but may be centimetric. Thus, if the quasielastic model makes sense, the Janssen picture should hold for silos ($\lambda \cong$ meters, $\lambda > H^*$) but not necessarily for laboratory columns ($\lambda \cong 1$ cm).

Stress distribution in a heap

Below a heap of sand, the distribution of normal pressures on the floor is not easy to guess. In some cases, the pressure is not a maximum at the center point! This has led to a vast number of physical conjectures, describing "arches" in the structure (Bouchaud et al., 1995; Edwards and Mounfield, 1996). In their most recent form (Wittmer et al., 1997), what is assumed is that, in a heap, the principal axes of the stress are fixed by the deposition procedure. Near the free surface, following the pioneering work of Coulomb, it is usually assumed that (for a material of zero cohesion) the shear and normal components of the stress (τ and σ_n) are related by the condition

$$\tau = \sigma_n \mu_i = \sigma_n \tan \theta_{\max}, \qquad (11)$$

where μ_i is an interval friction coefficient and $\tan \theta_{\max}$ is the resulting slope. Equation (11) should hold for a dry system with no cohesion between grains. In a two-dimensional geometry, this corresponds to a principal axis that is at an angle $2\theta_{\max}$ from the horizontal (Nedermann, 1992). The assumption of Wittmer et al. is that this orientation is retained in the left-hand side of the heap (plus a mirror symmetry for the right-hand side). Once this is accepted, the equilibrium conditions incorporating gravity naturally lead to a "channeling of forces" along the principal axis, and to a distribution of loads on the bottom that has two peaks. More generally, in the description of Bouchaud et al., the transmission of stresses is described by *hyperbolic* equations, leading to certain preferred directions. In the classical approach from continuum mechanics, the transmission is ruled by *elliptic* equations. In the first picture, the entire heap is pictured as being in some sort of critical state. In the second picture, we are far from criticality, and the heap is not dramatically different from a conventional solid—although the sample history is important for a clear definition of deformations.

The "critical" view has been challenged by S. Savage (1997a, 1997b) and by J. D. Goddard (1998). Savage gives a detailed review of the experimental and theoretical literature. He makes the following claims:

(a) For two-dimensional heaps ("wedges") with a rigid support plane, there is no dip in the experiments.
(b) If the support is (very slightly) deformable, the stress field changes deeply, and a dip occurs. This is another example of the role of minute displacements, which was already emphasized in Sec. III.E.
(c) For the 3D case ("cones"), the results are extremely sensitive to the details of the deposition procedure.

The most recent data on cones are by Brockbank *et al.* (1997). They use an accurate optical measurement of the local load under a conical heap of steel balls. The balls in the bottom layer deform the support, which is made of a transparent rubber film (~2 mm in thickness) lying over a glass surface. They do find a dip with steel, and also with glass heads of diameter 0.18 mm. But, when going to larger glass beads (~0.6 mm), the dip disappears!

Savage also describes finite element calculations, where one imposes the Mohr-Coulomb conditions (to which we come back in Sec. III) at the free surface of a wedge. If we had assumed a quasielastic description inside, we would have found an inconsistency: there is a region, just below the surface, which becomes unstable towards shear and slippage. Thus Savage uses Mohr-Coulomb conditions in a finite sheet near the surface, plus elastic laws in the inner part. With a rigid support he finds no dip, but with a deformable support he gets a dip.

The Savage methodology is similar in spirit to the quasielastic method although the details of the boundary conditions could possibly be altered. For instance, there may exist an extra simplification—which I already announced in connection with the silos. If we look at the formation of the heap, we find that the slope angle upon deposition should be slightly lower than the critical angle θ_{max}. Thus our system is prepared under noncritical conditions: all of the sample may then be described as quasielastic. This, in fact, should not produce very different results from those of Savage.

But there is a certain doubt, formulated by M. Cates and others: if the grains were glued together by microscopic glue patches at the contact point, indeed we might define displacements and deformations and use the Savage picture. But there is no glue! Certain grains might then be under tension (even if we are under a global compressive load): mechanical integrity is not granted!

In reply to this, the present author proposes three observations, which tend to support the classical view from mechanics.

(i) *Shear tests:* Under compressive load (in conditions without fracture) the stress strain relations are clearly history dependent, but do not display (as far as we can tell) any singular power laws.
(ii) *Lack of criticality:* If we examine the local density in a horizontal bed of sand, or the volume fraction ϕ as a function of depth, we find that ϕ is nearly constant and significantly larger than the critical value ϕ_{min} mentioned in Sec. II.[2] For these practical ϕ values (as we shall see in Sec. IV) the few indications available on correlation lengths ξ suggest that ξ is not large (at most of order 5 to 10 grain diameters). The singularities linked with arches, with tensile microcracks, should thus be confined to very small scales $\Delta x < \xi$.

---

[2]Note that although ϕ is nearly constant, in a bed of sand elastic moduli increase dramatically with depth. This is the basis of the "quasielastic" model.

(iii) *Texture:* One of the features that the physicists really wanted to incorporate is the possible importance of an internal *texture*. If we look at the contacts $(1,2,\ldots i,\ldots p)$ of a grain in the structure, we can form two characteristic tensors: one is purely geometrical and defines preferred directions of contact. It is

$$Q_{\alpha\beta} = i \sum x_\alpha^{(i)} x_\beta^{(i)}, \tag{12}$$

where x_α are the distances measured from the center of gravity of the grain. $Q_{\alpha\beta}$ is also called the "fabric tensor" (Oda, 1972, 1993; Oda and Sudoo, 1989). It is related to the "ellipsoid of contacts" introduced by Biarez and Wiendick (1963). The other tensor is the static stress:

$$\sigma_{\alpha\beta} = \frac{1}{2} i \sum (x_\alpha^{(i)} F_\beta^{(i)} + x_\beta F_\alpha^{(i)}), \tag{13}$$

where $\sim F^i$ is the force transmitted at contact (i). There is no reason for the axes of these two tensors to coincide. For instance, in an ideal hexagonal crystal, one major axis of the Q tensor is the hexagonal axis, while the stresses can have any set of principal axes. In the heap problem, I am personally inclined to believe that the deposition process freezes a certain structure for the Q tensor, but not for the stress tensor. However, this is still open to discussion! Recent arguments defending the opposite viewpoint have been given by Cates *et al.* (1998b).

The presence of a nontrivial Q tensor (or "texture") can modify the quasielastic model: instead of using an isotropic medium, we may need an anisotropic medium. In its simplest version, we would assume that the coarse-grained average $Q_{\alpha\beta}$ had two degenerate eigenvalues and a third eigenvalue, along a certain unit vector (the director) $\sim n(\sim r)$. Thus a complete discussion of static problems (in the absence of strong shear bands) would involve an extra field $\sim n$ defined by the construction of the sample. This refinement may modify the load distribution under a heap. But, conceptually, it is, in my opinion, minor. Texture effects should not alter deeply the quasielastic picture.

Strong deformations

Sophisticated tools have been designed for measuring the yield stress τ_y of granular materials in simple shear (Jenike, 1961; for a review, see for instance Brown and Richards, 1970). There is an elastic response at low shears, followed by yield at a certain value of the stress τ_y:

$$\tau_y = C + \mu p_n, \tag{14}$$

where p_n is the normal pressure. The constant C represents adhesive interactions between grains, and μ is a friction coefficient. An important feature of these strongly sheared systems—emphasized long ago by Reynolds (1885) is *dilatancy*: when the material was originally rather compact and is forced to yield, it increases in volume. This can be qualitatively understood by thinking of two compact layers of spheres sliding over each other.

In some cases, these strong deformations, with dilatancy, are present over large volumes. In other cases, they may be concentrated on *slip bands* (see, for instance, Desrues, 1991; Tillemans and Herrmann, 1995). For instance, if we remove sand with a bulldozer, slip bands will start from the bottom edge of the moving plate. Sometimes, the size of these slip bands is large and depends on the imposed boundary conditions (on the sharpness of the plate edge). But there seems to be a minimal thickness for a slip band: for spheroidal grains, without cohesion, it may be of order 5 to 10 grain diameters. We shall come back to this thickness when discussing microscopic properties.

Microscopic Features

Correlation lengths

We have talked about macroscopic stresses σ_{ij}: they must represent some coarse-grained averages over a certain volume. The implicit assumption here is that, indeed, a granular medium can be considered as homogeneous at large scales. This is not obvious: if we were talking about noncompacted material, with a density close to the lower limit $\phi_{min} = 0.56$, we might have a structure of weakly connected clusters (similar to percolation clusters). Exactly at threshold ($\phi = \phi_{min}$) a structure like this would probably be self-similar and not homogeneous at all. However, in real life, we always operate on systems with $\phi > \phi_{min}$, and we can expect that, at scales larger than a certain correlation length $\xi(\phi)$, our system may be treated as homogeneous.

Various experiments (Liu et al., 1995) and simulations (Moreau, 1994; Ouageni and Roux, 1995; Zhuang et al., 1995; Radjai et al., 1996) have investigated the local distribution of forces between grains. The central conclusion is that there are *force channels*, which build up a certain mesh with a characteristic size ξ. For spherical objects and ϕ values in the usual range, this ξ is somewhat larger than the grain diameter d ($\xi/d \sim 5$ to 10).

The network is obviously sensitive to variations in size among the grains. This "polydispersity" is always present and plays an important role in the actual value of ξ.

It may well be that the minimum thickness of a slip band (as introduced earlier) is equal (within coefficients) to the correlation length ξ. Thus we have at least two empirical ways of estimating ξ for a given system.

Fluctuations of the local load

It is also of interest to probe the local distribution of forces on all grains in contact with a supporting (horizontal) plate. This has been done in experiments by the Chicago group (Liu et al., 1995; Mueth et al., 1998), together with some simulations. Their trick is to lay the granular sample on a sequence carbon paper/white paper/solid plate. There is an empirical relation between the size of the dots printed by each grain on the white paper, and the force (w) with which it presses the ground. What Liu et al. found was a distribution of w, of the form

$$p(w) = \frac{w^2}{2\bar{w}^3} e^{-2w/\bar{w}}. \tag{15}$$

Liu et al. (1995) constructed a simple model for this statistical behavior, ignoring the vector character of the forces. They stipulated that each grain receive a load (w) from three neighbors above it:

$$w = q_1 w_1 + q_2 w_2 + q_3 w_3, \tag{16}$$

where w_1, w_2, w_3 are the loads on the "parents," and q_1, q_2, q_3 are three coupling factors statistically distributed between 0 and 1, and independent. Conversely, each parent sends some of its weight on to three "children" with fractions q'_1, q'_2, q'_3, and these fractions satisfy the sum rule $\Sigma q'_i = 1$. But apart from this constraint, all the q'_s are independent.

The law (15) can be understood as follows:

(a) For $w \gg \bar{w}$, we must have $q_1, q_2, q_3 \sim 1$, and we can then factorize $p(w) \sim p(w_1)p(w_2)p(w_3)$, with $w = w_1 + w_2 + w_3$. As pointed out by T. Witten, this condition is similar to the problem of a Boltzmann distribution of energies in thermal physics, and the solution is exponential $p(w) \sim \exp(-\alpha w)$.

(b) For $w \ll \bar{w}$, the weights carried by the three parents are much larger than w, and the probability $p(w)$ is essentially proportional to the phase space available in (q_1, q_2, q_3) where the q_s are linked by Eq. (16). This corresponds to a triangle of edges, $(w_1/w, 0, 0)$, $(0, w/w_2, 0)$, $(0, 0, w/w_3)$ in the (q_1, q_2, q_3) space, with an area $\sim w^2$. [However, on the experimental side, the more recent data of Mueth et al. (1998) give a different law!]

To summarize: (i) the fluctuations of w are comparable to the average ($\bar{w}$); (ii) the tail of the distribution at large w is exponential. The probabilities q_1, for very small loads ($w \to 0$), are still open to discussion.

A subsidiary question is: what are the correlations $\langle w(\sim x)w(\sim y)\rangle$ between grains at different locations (x, y) on the ground? The natural guess is that the range of these correlations is the correlation length ξ.

Of course, the model should be refined by introducing the vector character of the forces. The vectorial features are crucial when the *average* load is variable from point to point on the bottom plate. Consider, for instance, a horizontal slab of grains, with a thickness H and a very large aspect ratio. Impose a weak localized force F downwards, at the center of the upper surface ($x = y = 0$; see Fig. 2).

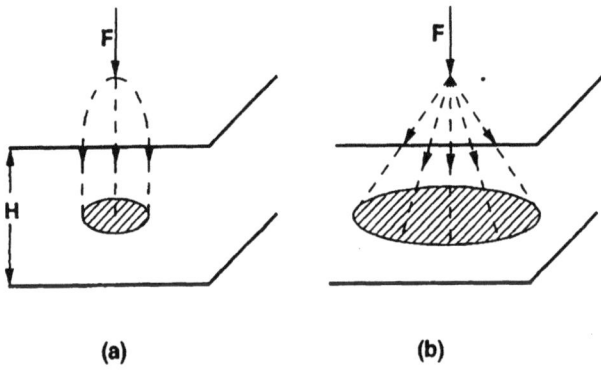

Figure 2. A crucial experiment, which to the author's knowledge has not yet been performed in a completely conclusive way. A bed of sand is deposited uniformly on a large flat surface and fills a height H. A small local force F is applied vertically at one point of the top surface. What are the resulting extra loads on the bottom plate? (a) In the "elliptic" models, used in soil mechanics, the load is spread over a region of size $\sim H$. (b) In the "hyperbolic" models of Bouchaud et al. the load is distributed over an annulus.

(a) The scalar model of (Liu et al., 1995) would give an average load profile on the bottom plate with a peak at the center and a width $\Delta x \sim \Delta y \sim \sqrt{dH}$ (where d is the grain diameter).
(b) With a tensorial stress field and a quasielastic model, we expect $\Delta x \sim \Delta y \sim H$.
(c) With "singular" models that predict transmission of the weight only in special directions (e.g., Bouchaud et al., 1995), the load would be concentrated in a ring, and disorder would make this ring slightly diffuse.

Concluding Remarks

The science of granular materials started with outstanding pioneers: Coulomb, Reynolds, Bagnold . . . . In recent years, it has benefited from the impact of very novel techniques—e.g., nuclear imaging of grains at rest or in motion (Nakagawa et al., 1993). A strong stimulus has also come from computer simulations—which have not been adequately described in the present text, because of the author's inexperience. It is clear that virtual experiments with controlled, simplified interactions between grains can have a major impact. A review of the tools, and of certain difficulties, can be found in Duran (1997). Recent advances are described in the proceedings of the Cargèse Workshop (Hermann, 1997).

However, in spite of these powerful tools, and even for the simplest "dry" systems, the statistical physics of grains is still in its infancy. Some basic notions may emerge: (a) the sharp distinction between a fluid phase and a frozen phase, with the resulting possibility of defining a displacement field to describe the evolution of the frozen phase; (b) a displacement field containing a memory of all the sample history; (c) the possibility of describing surface flows with equations coupling the two phases and reduced to a simplicity reminiscent of the Landau-Ginsburg picture of phase transitions.

But we are still left with strong disputes, and large sectors of unraveled complexity.

Two fundamentally different pictures of the static behavior of heaps are facing each other: one represents the material as a deformable solid, the other assumes a completely singular state of matter, with stress fields transmitted along special directions and with microscopic instabilities (earthquakes) occurring all the time (see, for instance, Miller et al., 1996).

We have to know more! Here are some examples:

(a) The problem raised in Fig. 2: If we press *gently* at the free surface of a large, flat bed of sand, are the stresses below widely spread (as expected from a quasielastic solid) or are they localized on a cone (as expected in "singular" models)? The word "gently" is important here: if we go to strong, local loads, we shall of course, generate shear bands.
(b) Acoustic propagation in a granular bed: It is mainly controlled by the (nonlinear) quasielastic features plus mild effects of disorder. Or is it qualitatively different, because a sound wave, even at small amplitudes, starts some sort of earthquake?
(c) Decompaction: If we open the bottom of a vertical column, we see pieces of solidlike matter which separate from each other. Can we think of this as propagation of fractures in a quasielastic solid, or is it completely different?
(d) Similarly, when we perform a sequence of taps on a column, as mentioned in Sec. II, should we visualize the grains during the tap as a solid with microcracks or as a liquid (if the amplitudes are high enough)?

We mentioned some current uncertainties for the solid phase. There are uncertainties of comparable magnitude for the fluid phases. Think, for instance, of fluidized beds: an intelligent literature (describing both transport and macroscopic instabilities) has been-built up, but we are still looking for a unified vision. The link between mechanics, tribology, statistical physics, surface chemistry, ... remains to be built.

Acknowledgments

My education on the physics of grains is due mainly to J. P. Bouchaud, J. Duran, P. Evesque, H. Herrmann, and T. Witten. I am deeply thankful for their patient explanations. However, in many instances, the perspectives that are proposed here do not coincide with their own views. They should not be held responsible for my (naive) attempts at unification.

I also greatly benefited from discussions with and/or messages from R. Behringer, J. Biarez, D. Bideau, T. Boutreux, M. Cates, S. Coppersmith, S. Fauve, H. Jaeger, J. Kakalios, L. Limat, H. Makse, J. J. Moreau, S. Nagel, E. Raphael, J. Rajchenbach, P. Rognon, and S. Roux.

References

Ahn, A., C. Brennen, and R. Sabersky, 1988, in *Micromechanics of Granular Materials*, edited by M. Satake and J. Jenkins (Elsevier, New York).
Bagnold, E. R., 1941, *Physics of Blown Sand and Sand Dunes* (Chapman and Hall, London).
Barker, G. C., and A. Mehta, 1991, Phys. Rev. Lett. **67**, 394.
Barker, G. C., and A. Mehta, 1992, Phys. Rev. A. **65**, 3935.
Barker, G. C., and A. Mehta, 1993, Phys. Rev. E. **47**, 184.
Bernal, J. D., 1964, Proc. R. Soc. London, Ser. A **280**, 299.
Bernu, B., and R. Mazichi, 1990, J. Phys. A **23**, 5745.
Biarez, J., and M. Gourves, 1989, Eds., *Powders and Grains* (Balkema, Rotterdam).
Biarez, J., and K. Wiendick, 1963, C. R. Acad. Sci. (Paris) **256**, 1217.
Bouchaud, J. P., P. Claudin, M. Cates, and M. Cates, 1997, in *Physics of Dry Granular Media*, Proceedings of the Cargèse Workshop, edited by H. Hermann, P. Hovi, and S. Ludwig (Kluwer Academic, Dordrecht), p. 97.
Bouchaud, J. P., M. Cates, and P. Claudin, 1995, J. Phys. II (Paris), 639.
Bouchaud, J. P., M. Cates, R. Prakash, and S. F. Edwards, 1994, J. Phys. II **4**, 1383.
Boutreux, T. et al., 1997, Physica A **266**, 59.
Boutreux, T. et al., 1998, Phys. Rev. E **58**, 4692.
Bowden, P., and D. Tabor, 1973, *Friction: An Introduction* (Doubleday, New York).
Brockbank, R., J. M. Huntley, and R. C. Ball, 1997, J. Phys. II **7**, 1521.
Brown, R., and J. C. Richards, 1970, *Principles of Powder Mechanics* (Pergamon, Oxford).
Cates, M., J. P. Wittmer, J. P., Bouchaud, and P. Claudin, 1998a, Phys. Rev. Lett. **81**, 1861.
Cates, M., J. P. Wittmer, J. P. Bouchaud, and P. Claudin, 1998b, Philos. Trans. A R. Soc. London, in press.
Clément, E., S. Luding, A. Blumen, J. Rajchenbach, and J. Duran, 1993, Int. J. Mod. Phys. **9**, 1807.
Coniglio, A., and H. Hermann, 1996, Physica A **225**, 1.
de Gennes, P. G., 1996, Europhys. Lett. **35**, 145.
de Gennes, P. G., 1997, C. R. Acad. Sci. (Paris) **325** II, 9.
de Gennes, P. G., 1998, in *Powders and Grains* (Balkema, Rotterdam), p. 4.

Desrues, J., 1991, in *Physics of Granular Media*, Les Houches Series, edited by D. Bideau and J. Dodds (Nova Sci.), p. 127.
Drake, J. J., 1991, J. Fluid Mech. **121**, 225.
Duffy A., and M. Mendlin, 1957, J. Appl. Mech. (ASME), **24**, 585.
Duran, J., 1997, *Sables, Poudres et Grains* (Eyrolles, Paris).
Edwards, S. F., and C. C. Mounfield, 1996, Physica A **226**, 1,12,25.
Evesque, P., 1991, Phys. Rev. A **43**, 2720.
Evesque, P., and P. G. de Gennes, 1998, C. R. Acad. Sci. (Paris), Ser. II, **326**, 761.
Finney, J. L., 1970, Proc. R. Soc. London, Ser. A **319**, 479.
Goddard, J., 1990, Proc. R. Soc. London, Ser. A **430**, 105.
Goddard, J., 1998, in *Physics of Granular Media, Proceedings of NATO Institute*, edited by H. Herrmann, J. P. Hovi, and S. Luding (Kluwer Academic, Dordrecht).
Hagen, G., 1852, Monatsber. Dtsch. Akad. Wiss. Berlin, p. 35.
Hanes, D., and D. Inman, 1985, Proc. R. Soc. London, Ser. A **150**, 357.
Herrmann, H., 1998, in *Physics of Granular Media, Proceedings of NATO Institute*, edited by H. Herrmann, J. P. Hovi, and S. Luding (Kluwer Academic, Dordrecht), p. 319.
Jaeger, H., C. Liu, and S. Nagel, 1988, Phys. Rev. Lett. **62**, 40.
Jaeger, H., S. Nagel, and R. Behringer, 1996, Rev. Mod. Phys. **68**, 1250.
Janssen, H. A., and Z. Vereins, 1895, Dtsch. Eng. **39(25)**, 1045.
Jenike, A. W., 1961, Bulletin 108, Utah Eng. Exp. Station (University of Utah).
Knight, J. B., C. Fandrich, C. Lau, H. Jaeger, and S. Nagel, 1995, Phys. Rev. E **51**, 3957.
Levine, D., 1997, Phys. World **30**, 26.
Liu, C. H., S. Nagel, D. Shechter, S. Coppersmith, S. Majumdar, O. Narayan, and T. Witten, 1995, Science **269**, 513.
Miller, B., C. O'Hern, and R. P. Behringer, 1996, Phys. Rev. Lett. **77**, 3110.
Moreau, J. J., 1994, Eur. J. Mech. A/Solids **13**, 93.
Mueth, D., H. Jaeger, and S. Nagel, 1998, Phys. Rev. E **57**, 3164.
Nakagawa, H. *et al.*, 1993, Exp. Fluids **16**, 56.
Nedermann, R., 1992, *Statics and Kinematics of Granular Materials* (Cambridge University Press, Cambridge, England).
Nedermann, R., U. Tuzun, S. Savage, and G. Houlsby, 1982, Chem. Eng. Sci. **37**, 1597.
Nicodemi, M., A. Coniglio, and H. Herrmann, 1997, Phys. Rev. E **55**, 3962.
Nowak, E. R., J. B. Knight, E. Ben Haim, H. Jaeger, and S. Nagel, 1998, Phys. Rev. E, in press.
Oda, M., 1972, *Soils and Foundations*, Vol. 2, p. 1.
Oda, M., 1993, in *Powders and Grains*, edited by Thornton (Balkema, Rotterdam), p. 161.
Oda, M., and T. Sudoo, 1989, in *Powders and Grains*, edited by J. Biarez and M. Gourves (Balkema, Rotterdam), p. 156.
Onoda, G., and E. Lininger, 1990, Phys. Rev. Lett. **64**, 2727.
Ouaguenouni, S., and J. N. Roux, 1995, Europhys. Lett. **32**, 449.
Radjai, F., M. Jean, J. J. Moreau, and S. Roux, 1996, Phys. Rev. Lett. **77**, 274.
Rayleigh, Lord (John William Strutt), 1906, Philos. Mag. **36**, 11.
Rayleigh, Lord (John William Strutt), 1906, Philos. Mag. **36**, 61.
Rayleigh, Lord (John William Strutt), 1906, Philos. Mag. **36**, 129.
Rayleigh, Lord (John William Strutt), 1906, Philos. Mag. **36**, 206.
Reynolds, O., 1885, Philos. Mag. **5**, 469.
Savage, S. B., 1997a, *Disorder and Granular Media* (North-Holland, Amsterdam).
Savage, S. B., 1997b, in *Powders and Grains*, edited by R. Behringer and Jenkins (Balkema, Rotterdam), p. 185.
Schofield, A. N., and C. P. Wroth, 1968, *Critical State of Soil Mechanics* (McGraw-Hill, New York).
Scott, G. D., 1962, Nature (London) **194**, 956.

Selig, E. T., and R. Ladd, 1973, American Society for Testing of Materials, Publication No. 523 (Baltimore, Maryland).

Thompson, P. A., and G. S. Grest, 1991, Phys. Rev. Lett. **67**, 1751.

Tillemans, H. J., and H. J. Herrmann, 1995, Physica A **217**, 261.

Umbanhowar, P., F. Melo, and H. Swinney, 1996, Nature (London) **382**, 793.

Vanel, L., E. Clément, J. Lanuza, and J. Duran, 1998, in *Physics of Dry Granular Media*, edited by H. Herrmann, J. P. Hovi, and S. Luding (Kluwer Academic, Dordrecht), p. 249.

Wittmer, J. P., M. Cates, and P. Claudin, 1997, J. Phys. I **7**, 39.

Zenz, F. A., 1997, in *Handbook of Powder Science and Technology*, edited by M. El Fayed and L. Otten (Chapman and Hall, London), Chap. 11.

Zhuang, X., A. Didwanis, and J. Goddard, 1995, J. Comput. Phys. **121**, 331.

Fluid Turbulence
Katepalli R. Sreenivasan

The fascinating complexity of turbulence has attracted the attention of naturalists, philosophers, and poets alike for centuries, and ubiquitous allusions have been made to the turbulence of agitated minds and disturbed dreams, of furious rivers and stormy seas. Perhaps the earliest sketches of turbulent flows, capturing details with some degree of realism, are those of Leonardo da Vinci. A serious scientific study has been in progress for more than a hundred years, but the problem has not yet yielded to our efforts. As Liepmann (1979) has pointed out, the outlook, optimism, and progress have waxed and waned over time.

Few would dispute the importance of turbulence. Without it, the mixing of air and fuel in an automobile engine would not occur on useful time scales; the transport and dispersion of heat, pollutants, and momentum in the atmosphere and the oceans would be far weaker; in short, life as we know would not be possible on the earth. Unfortunately, turbulence also has undesirable consequences: it enhances energy consumption of pipe lines, aircraft and ships, and automobiles; it is an element to be reckoned with in air-travel safety; it distorts the propagation of electromagnetic signals; and so forth. A major goal of a turbulence practitioner is the prediction of the effects of turbulence and control them—suppress or enhance them, as circumstances dictate—in various applications such as industrial mixers and burners, nuclear reactors, aircraft and ships, and rocket nozzles.

Less well appreciated is the intellectual richness of the subject and the central place it occupies in modern physics. Looking into the problem, we are immediately faced with an apparent paradox. Even with the smoothest and most symmetric boundaries possible, flowing fluids—except when their speed is very low—assume the irregular state of turbulence. This feature, though not fully understood, is now known to bear *some* connection with the occurrence of dynamical chaos in nonlinear systems. Turbulence has constantly challenged and expanded the horizons of modern dynamics, the theory of differential equations, scaling theory, multifractals, large-scale computing, fluid mechanical measurement techniques, and the like.

Until the 1960s, turbulence was *the* paradigm system in which the excitation of many length scales was recognized as important. The powerful notions of scaling and universality, which matured when renormalization group theory was applied to critical phenomena, had already manifested in turbulence a couple of decades earlier. Turbulence and critical phenomenon share the feature that a continuous range of scales is excited in both; however, they are different in that the fluctuations in turbulence are strong and there exists no small parameter. Thus, turbulence is a paradigm in non-equilibrium statistical physics, in which fluctuations and macroscopic space-time structure coexist. It is an example like no other of spatially extended dissipative systems.

An excellent case can thus be made that turbulence is central to flow technology as well as modern statistical and nonlinear physics. The reader wishing to learn about the subject should begin with Monin and Yaglom (1971, 1975), and move on, for different specialized perspectives, to the books of Batchelor (1953), Townsend (1956), Bradshaw (1971), Leslie (1972), Lesieur (1990), McComb (1990), Chorin (1994), Frisch (1995), and Holmes et al. (1998). There are many useful review articles, each emphasizing a different aspect. Some examples are Corrsin (1963), Saffman (1968), Roshko (1976), Cantwell (1981), Narasimha (1983), Hussain (1983), Frisch and Orszag (1990), Lumley (1990), Sreenivasan (1991), Nelkin (1994), Siggia (1994), L'vov and Procaccia (1996), Sreenivasan and Antonia (1997), Zhou and Speziale (1998), Smith and Woodruff (1998), and Canuto and Christensen-Dalsgaard (1998). The two volumes of Monin and Yaglom, covering the subject only until the early seventies, contain more than 1600 pages. Several hundred papers have appeared on the subject since then. Discussing this vast subject in any depth and completeness would be a herculean task. This article makes no such pretensions; instead, it makes a few isolated and qualitative observations to suggest the nature of progress made: slow, multi-faceted, useful, insightful—but often soft. While the importance of turbulence has long made its study imperative, all the tools needed for such a complex undertaking are not fully in place. In this sense, despite its age, turbulence is a frontier subject.

The Phenomenon and the Goal

Water flowing from a slightly open faucet is smooth and steady, or laminar. As the faucet opens up more, the flow becomes erratic. Figure 1 illustrates that a seemingly erratic turbulent flow is actually a labyrinth of order and chaos. Swirling flow structures—or patterns—of various sizes are intertwined with fluid mass of indifferent shape. Being static, however, the picture does no justice to the dynamical interaction among the constituent scales of the flow. Casual observations suggest that the patterns get stretched, folded and tilted as they evolve, losing shape by agglomeration or breakup—all in a manner that does not repeat itself in detail. Unlike patterns in equilibrium systems, which are associated with phase transitions, those in fluid flows are intimately related to transport processes. The patterns in fluid systems exhibit varying sensitivity to initial and boundary conditions, and are rich in morphology (see, e.g., Cross and Hohenberg, 1994).

The key to the onset of turbulence has long been believed to be the successive loss of stability that occurs with ever increasing rapidity as a typical control parameter in a flow problem is increased (e.g., Landau and Lifshitz, 1959). The most familiar control parameter is the Reynolds number[1] Re, which expresses the balance between the nonlinear and dissipative properties of the flow. This scenario is thought to be relevant especially for flows whose vorticity attains a maximum in the interior, instead of at the boundary. Linear and nonlinear stability theories have been successful in describing the initial stages of the transition to turbulence (e.g., Drazin and Reid, 1981), but the later stages seem quite abrupt (e.g., Gollub and Swinney 1975), and not amenable to stability analysis. This

---

[1] Reynolds number is the dimensionless parameter UL/ν, where U and L are the characteristic velocity and length scales of a turbulent flow and ν is the fluid viscosity. Depending on the purpose, different velocity and length scales become relevant.

Figure 1. A turbulent jet of water emerging from a circular orifice into a tank of still water. The fluid from the orifice is made visible by mixing small amounts of a fluorescing dye and illuminating it with a thin light sheet. The picture illustrates swirling structures of various sizes amidst an avalanche of complexity. The boundary between the turbulent flow and the ambient is usually rather sharp and convoluted on many scales. The object of study is often an ensemble average of many such realizations. Such averages obliterate most of the interesting aspects seen here, and produce a smooth object that grows linearly with distance downstream. Even in such smooth objects, the averages vary along the length and width of the flow, these variations being a measure of the spatial inhomogeneity of turbulence. The inhomogeneity is typically stronger along the smaller dimension (the "width") of the flow. The fluid velocity measured at any point in the flow is an irregular function of time. The degree of order is not as apparent in time traces as in spatial cuts, and a range of intermediate scales behaves like fractional Brownian motion.

abruptness is qualitatively in the spirit of the modern theory of deterministic chaos (Ruelle and Takens 1971), and is especially characteristic of boundary layers[2] (Emmons, 1951).

---

[2]The boundary layer is the thin region close to a solid body moving relative to the fluid. Processes in this thin layer are the source, among other things, of fluid dynamical resistance and aerodynamic lift.

While this situation is reminiscent of second-order phase transitions in condensed matter, it is unclear if the analogy is helpful in a serious way.

In any case, at high enough Reynolds numbers, nonlinear interactions produce finer and finer scales, and the scale range in developed turbulence is of $O(Re^{9/4})$. The Reynolds number could be several million in the earth's atmosphere a few meters above the ground or in the boundary layer of an aircraft fuselage. Clearly, in such instances, only a statistical description of turbulence and the prediction of its consequences—such as increased mixing, transport, and energy loss—are of practical value. The discovery of an efficient procedure to do this is the principal and outstanding challenge of the subject.

The goal just mentioned is no different from that of statistical thermodynamics. The statistical assumptions made there possess vast applicability and powerful predictive capability. Unfortunately, those made in turbulence have enjoyed far less success, even though much about the behavior of turbulence has been learned in the process of their application. The era in which the statistical approach was the norm—one in which developments in turbulence occurred, on the whole, in the context of fluid dynamics—is called here the "classical era." Because of the continuing awareness of the limitations of statistical theories, one has more recently begun to ask whether this basic approach needs to be augmented by a different outlook. This outlook has the common element that it focuses on mechanisms rather than flows, and is influenced by developments in neighboring fields such as bifurcations, chaos, multifractals, and modern field theory. The intent is often to acquire qualitative understanding of fluid turbulence through model nonlinear equations. This era, which we shall loosely call "modern," has benefitted tremendously by the availability of powerful computers and the qualitative theory of differential equations (e.g., the study of space-time singularities).

The Classical Era

Before Osborne Reynolds

Unlike many other problems in condensed matter physics, the equations governing turbulence—the Navier-Stokes equations—have been known for some 150 years. All available evidence suggests that the phenomenon of turbulence is consistent with these equations, and that the molecular structure makes little difference (except for their role in prescribing gross parameters such as the viscosity coefficient). The Navier-Stokes equations and the use of proper boundary conditions are the result of the cumulative work of heroes such as J. R. d'Alembert, L. Euler, L. M. H. Navier, A. L. Cauchy, S. D. Poisson, J.-C. B. Saint-Venant, and G. G. Stokes. Even as the equations were being refined, controlled experiments were discovering, or rediscovering, that fluid motion occurs in two states—laminar and turbulent—and that a transition from the former to the latter occurs in distinctive ways. It was realized that turbulent flows transport heat, matter, and momentum far better than laminar flows. The concept of "eddy viscosity," attesting to this enhancement of transport, was discussed by Saint-Venant and J. Boussinesq. From observations in water canals, the latter deduced that an apparent analogy exists between gas molecules and turbulent eddies as they carry and exchange momentum.

Contributions of Osborne Reynolds

It was Reynolds (1883, 1894) who heralded a new beginning of the study of turbulence: he visualized laminar and turbulent motions in pipe flows; identified the criterion for the onset of turbulence in terms of the nondimensional parameter that now bears his name; showed that the onset is in the form of intensely choatic "flashes" in the midst of otherwise laminar motion; introduced statistical methods by splitting the fluid motion into mean and fluctuating parts ("Reynolds decomposition"); and identified that nonlinear terms in the Navier-Stokes equations yield additional stresses ("Reynolds stresses" or "turbulent stresses") when the equations are recast for the mean part. A *tour de force* indeed! Reynolds' equations for the mean velocity demonstrated the so-called "closure problem" in turbulence: if one generates from the Navier-Stokes equations an auxiliary equation for a low-order moment such as the mean value, that equation contains higher-order moments, so that, at any level in the hierarchy of moments, there is always one unknown more than the available equations. High-order moments are not related to low-order moments as (for example) in a Gaussian process. Thus, even though the Navier-Stokes equations are themselves closed, some additional assumptions are required to close the set of auxiliary equations at any finite level. This feature has defined the framework for much of the turbulence research that has followed.

From Reynolds until the 1960s

Closure models

Although the closure problem was apparent in Reynolds' work, its fundamentals seem to have been spelled out first by Keller and Friedmann (1924). They derived the general dynamical equations for two-point velocity moments and showed that the equations for each moment also contain high-order moments. Since there is no apparent small parameter in the problem, there is no rational procedure for closing the system of equations at any finite level. The moment equations have been closed by invoking various statistical hypotheses. The simplest of them is Boussinseq's pedagogical analogy—already mentioned—between gas molecules and turbulent eddies. Taylor (1915, 1932), Prandtl (1925), and von Kármán (1930) postulated various relations between turbulent stresses and the gradient of mean velocity (the so-called mixing length models) and closed the equations. Truncated expansions, cumulant discards, infinite partial summations, etc., have all been attempted (see, e.g., Monin and Yaglom, 1975; Narasimha, 1990). Another interesting idea (Malkus, 1956) is that the mean velocity distribution is maintained in a kind of marginally stable state, the turbulence being self-regulated by the transport it produces.

In a paper less known than it deserves, Kolmogorov (1942) augmented the mean velocity equation by two *differential equations* for turbulent energy and (effectively) the energy dissipation, thus anticipating the so-called two-equation models of turbulence; this is a common practice even today in turbulence modeling (although its development was essentially independent of Kolmogorov's original proposal). Other schemes of varying sophistication and complexity have been developed (see, e.g., Reynolds, 1976; Lesieur, 1990).

Similarity arguments

Given that the equations governing turbulence dynamics have been known for so long, the paucity of results that follow from them exactly is astonishing (for an exception under certain conditions, see Kolmogorov 1941a). This situation speaks for the complexity of the equations. Much effort has thus been expended on dimensional and similarity arguments,[3] as well as asymptotics, to arrive at various scaling relations. This type of work continues unabated and with varying degrees of success (see, e.g., Townsend, 1956; Tennekes and Lumley, 1972; Narasimha 1983). For instance, a result from similarity arguments is that the *average* growth of turbulent jets, of the sort shown in Fig. 1, is linear with downstream distance, with the proportionality constant independent of the detailed initial conditions at the jet orifice. Likewise, the energy dissipation on the jet axis away from the orifice depends solely on the ratio U_o^3/D, with the coefficient of proportionality of order unity. Here, D and U_o are the orifice diameter and the velocity at its exit, respectively. These (and similar) scaling results seem to be correct to first order, and so have been used routinely in practice. However, they are working approximations at best: the conditions under which similarity arguments hold are not strictly understood, and the constants of proportionality cannot be extracted from dynamical equations in any case. One should therefore not be too surprised if such relations do not work in every instance (Wygnanski et al., 1986): there is some reason or another to hesitate about the bedrock accuracy of almost every such relation used in the literature. Yet, this should not detract us from appreciating that such results are extremely useful for solving practical problems.

An important relation obtained by asymptotic arguments and supplementary assumptions concerns the distribution of mean velocity in boundary layers, pipes, and flow between parallel plates (e.g., Millikan, 1939). The result is that, in an intermediate region not too close to the surface nor too close to the pipe axis or the boundary layer edge, the mean velocity is proportional to the logarithm of the distance from the wall. This so-called log-law has for a long time enjoyed a preeminent status in turbulence theory (see, however, later section on A Brief Assessment of the Classical Era). Again, the additive and proportionality constants in the log-law are known only from empirical data.

Homogeneous and isotropic turbulence

In another important turn of events, a considerable simplification of the general dynamical problem of turbulence was achieved by Taylor (1935) with the introduction of the concept of homogeneous and isotropic turbulence, that is, turbulence that is statistically invariant under translation, rotation and reflection of coordinate axes. Experimentally, nearly homogeneous and isotropic turbulence (see Fig. 2) was developed in the late 1930's using uniform grids of bars in a wind tunnel (e.g., Comte-Bellot and Corrsin 1966). The use of tensors in isotropic turbulence was introduced by von Kármán (1937) who also studied the dynamical consequences of isotropy (von Kármán and Howarth, 1938). Taylor (1938) derived an equation for turbulent vorticity and, almost simultaneously, initiated the use of Fourier transform and spectral representation. Since that time, isotropic turbulence has been the testing ground for most of the analytical theories of turbulence.

Local isotropy and universality of small scales: The Kolmogorov turbulence

The reality is that no turbulent flow is homogeneous and isotropic. Further, there are many types of turbulence—depending on boundary conditions, body forces, and other auxiliary

---

[3] A similarity transformation is an affine transformation that reduces a set of partial differential equations to an ordinary differential equation. For a turbulent jet far away from the orifice, it takes the form that mean velocity distribution preserves its shape when scaled on *local* velocity and length scales. By demanding that the coefficients of the resulting ordinary differential equation be constants, one obtains power-laws for the variation of these velocity and length scales along the jet axis. However, only the power-law exponent can be determined in this way, not prefactors.

parameters: incompressible, compressible, homogeneously sheared, inhomogeneous, stratified, magnetohydrodynamic, superfluid turbulence, and so forth. They are all similar in some respects (e.g., they are highly dissipative), but also different in some respects (e.g., the topology of the large structure is different). This situation is somewhat similar to that in chemistry: while all compounds have the same essential elements, they are also different from each other. It is therefore useful to ask whether "turbulence"—when divorced from a specific context—has a meaningful existence at all.

Enter Kolmogorov (1941b) and his revolutionary postulate that small scales of turbulence are *statistically* isotropic—no matter how the turbulence is produced. This

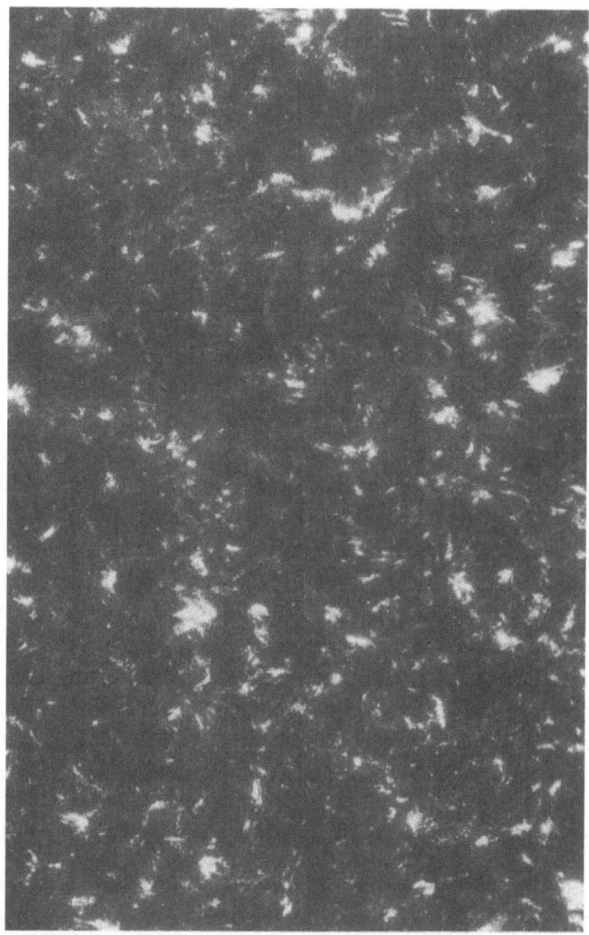

Figure 2. This picture depicts homogeneous and isotropic turbulence produced by sweeping a grid of bars at a uniform speed through a tank of still water. Unlike the jet turbulence of Fig. 1, turbulence here does not have a preferred direction or orientation. On the average, it does not possess significant spatial inhomogeneities or anisotropies. The strength of the structures, such as they are, is weak in comparison with such structures in Fig. 1. Homogeneous and isotropic turbulence offers considerable theoretical simplifications, and is the object of many studies.

postulate,[4] coupled with Kolmogorov's other hypothesis—known for short in the jargon as K41—has allowed several detailed predictions to be made with regard to the scaling properties of "small-scale" turbulence. The spirit of K41, a major fore-runner for which are Richardson's (1922) qualitative ideas of self-similar distribution of turbulent eddies, is to assume that the "small" scales of turbulence are universal, even though the "large" scales are specific to a given flow—or class of flows with the same boundary conditions. While a full understanding of a turbulent flow requires attention to large as well as small scales (whose mix varies from flow to flow), K41 presupposes that the small scales can be understoood independent of the specifics that determine the large scales. In particular, towards the upper end of the small-scale range (the so-called inertial subrange), K41 shows that the energy spectral density $\phi(\kappa)$ varies with the wave number κ according to $\phi(\kappa) = c_\kappa \varepsilon^{2/3} \kappa^{-5/3}$. Here ε is the rate at which energy is dissipated by the low end of the small scales, and c_κ is an unknown but universal constant. Embedded in K41 is the notion that the large scales—at which the energy is injected—transfer it to the small scales—where it is dissipated—through a series of steps, each of which is dissipationless and involves the interaction of only neighboring scales (instead of all possible triads of wave numbers allowed by the Navier-Stokes equations). The transfer is supposed to occur with ever-increasing rapidity as one approaches increasingly smaller scales. This process of energy transfer, which is at best a good abstraction of a more complex reality, is picturesquely known as energy cascade (Onsager, 1945). Besides Onsager, the other early workers who independently contributed to the understanding of the inertial subrange are von Weizsäcker (1948) and Heisenberg (1948). It is worth stressing that K41 makes no direct connection to the Navier-Stokes equations.

We shall not discuss here Kolmogorov's form for the dissipative scales, but refer to Monin and Yaglom (1975) and Frisch (1995). We shall also say nothing about the consequences of K41 for turbulent diffusion except to note that Richardson's (1926) law for the diffusion of particle pairs can be recovered from its application.

A first-order verification of K41 in a tidal channel at very high Reynolds numbers (Grant *et al.* 1962) is a milestone in the history of turbulence. This rough experimental confirmation and its alluring simplicity have made K41 a staple of turbulence research. However, we shall presently see that K41 is not correct in detail.

Experimental tools

Until the late 1920s, the types of turbulence measurements that could be made were limited to time-average properties such as mean velocity and pressures differences. It was not possible to measure fluctuations faithfully because of the demands of spatial and temporal resolution: the spatial resolution required is $O(Re^{-9/4})$ and the temporal resolution $O(Re^{-1/2})$. The technique commonly used for the study of turbulent characteristics was the visualization of flow by injecting a dye or a tracer. This type of work led to valuable insights in the hands of stalwarts such as Prandtl (see Prandtl and Tietjens, 1934). Since the 1950's, which is when thermal anemometry came into being in a robust form, the technique has been the workhorse of turbulence research. Briefly, a fine wire of low thermal capacity is heated to a certain temperature above the ambient, and the change

---

[4] A second important postulate, already mentioned in the specific context of the turbulent jet, is that the rate of energy dissipation at high Reynolds numbers far away from solid boundaries—although mediated by fluid viscosity—is independent of it. Experiments support the postulate on balance, but the evidence leaves much to be desired.

in resistance encountered by it, as a fluid with fluctuating velocity flows around it, is measured. This change is related to the flow velocity through a calibration. Late in the period being considered here, optical techniques such as laser Doppler velocimetry began to make inroads, but hotwires are still the probes of choice in a number of situations.

A brief assessment of the classical era

As already mentioned, the statistical principles used for closing the moment equations have enjoyed only transient success. The eddy viscosity and mixing length principles, despite the initial triumphs (see Schlichting, 1956), proved to be flawed (though this has not prevented their use—with varying levels of discernment). Similarly, some closure models (e.g., the so-called quasinormal approximation) often violate the *realizability* condition, namely the positivity of probabilities (or other related results that follow). Kraichnan (1959, and later) has emphasized the need for dealing with this issue directly, and devised models that ensure realizability. These models have certain consistency properties that conventional closure schemes may not. Realizability constraints have now become a standard test in turbulence modeling (Speziale, 1991), especially in modern computing efforts. Further, the general scaling results such as for the overall growth of turbulent flows and energy dissipation (see earlier section on Similarity Arguments) seem to derive rough experimental support, but reveal many open problems upon close scrutiny. Even the log-law, long regarded as a crowning achievement in turbulence, has been questioned vigorously in recent years (Barenblatt et al., 1997). The issue on hand is not simply whether the log-law or an alternative power-law fits the data better. At stake is the validity of the underlying principles of similarity that each argument employs.

The lack of successful closure models on the one hand, and the apparent success of K41 in describing low-order statistics of the small-scale on the other, have led to an excessive tendency to regard turbulence as a single, unified phenomenon. This development has not always been healthy.

One cannot escape the feeling that much of the work has a tentative character to it. This is not the norm in mechanics or other branches of classical physics.

The Modern Era

Large-scale coherent structures

Even casual observations of turbulent flows reveal well-organized motions on scales comparable to the flow width (see the splendid collection of pictures by Van Dyke, 1982); indeed, experimentally measured correlation functions had occasionally pointed to the existence of organized large scales (Liepmann, 1952; Favre et al., 1962). Yet, this aspect was not the central theme of turbulence research in the classical era. On hindsight, many aspects contributed to this neglect: the realization that statistical description was inevitable, preoccupation with isotropic turbulence where the spatial organization is minimal, the absence of historical precedents of physical systems in which order and chaos coexist, and so forth. The important role of large-scale organized motions for transport processes has since been emphasized (Kline et al., 1967; Brown and Roshko, 1974; Head and Bandyopadhyay, 1981), leading to a resurgence of interest in them.

It is a nontrivial matter that the large scales can maintain their coherence in the presence of a superimposed incoherent activity. The origin of the large structure has often been sought in terms of the instability of the (hypothetical) mean velocity distribution, or something even simpler, but there are conspicuous gaps in the arguments employed. It is worth recalling that complicated, nonlinear, systems with many degrees of freedom do sometimes develop organized structures such as solitons (Zabusky and Kruskal, 1965). If solitons have anything to do with coherent structures in turbulence, that connection remains obscure.

Taking for granted the importance of the large scales, the question is how to identify them objectively. An experimentally useful tool is the so-called conditional averaging (e.g., Kovasznay et al., 1970), in which one averages over preselected members of an ensemble. Suitable wavelets have sometimes been used as templates for the large scale. The difficult question is how to describe them analytically and construct usefully approximate dynamical systems, preferably of low dimensions. This is not a simple task, but some success has been attained in special cases via the so-called Karhunen-Lóeve procedure (e.g., Sirovich, 1987; Holmes et al., 1998).

A hope in the work on coherent structures has been that they could lead to efficient methods for predicting overall features of turbulent flows. The verdict on this effort is still unclear (e.g., Hussain, 1983). Another quest has been to control, or manage, turbulent flows via large-scale coherent structures. The verdict on this line of inquiry is mixed (e.g., Gad-el-Hak et al., 1998).

Small-scale turbulence: Repercussions of Kolmogorov's "refinement"

It has been hinted already that Kolmogorov's arguments of local isotropy and small-scale universality have pervaded all aspects of turbulence research (e.g., Monin and Yaglom, 1975; Frisch, 1995). Deeper exploration has revealed that strong departures from the K41 universality exist, and that they are due to less benign interactions between large and small scales than was visualized in K41. Following a remark of Landau (see Frisch, 1995), Kolmogorov (1962) himself provided a "refinement" of his earlier hypotheses. In reality, this refinement is a vital revision (Kraichnan, 1974), and its repercussions are being felt even today (e.g., Chorin, 1994; Stolovitzky and Sreenivasan, 1994). One of its manifestations is that the various scaling exponents characterizing small-scale statistics are anomalous (that is, the exponent for each order of the moment has to be determined individually in a nontrivial manner, and cannot be guessed from dimensional arguments). Although the anomaly is still *essentially* an empirical fact, and its existence has yet to be established beyond blemish due to various experimental ambiguities,[5] it seems unlikely that we will return to K41 universality. Even the nature of anomaly seems to depend on the particular class of flows. However, these subtle differences might arise from finite Reynolds number effects, large-scale anisotropies, and so forth; without quantitative

---

[5]There are several of them. First, measured time traces of turbulent quantities are interpreted as spatial cuts by assuming that turbulence gets convected by the mean velocity without distortion. This is the so-called Taylor's hypothesis. Second, one cannot often measure the quantity of theoretical interest in its entirety, but only a part of it. The practice of replacing one quantity by a similar one is called surrogacy. Surrogacy is often a necessary evil in turbulence work, and makes the interpretation of measurements ambiguous (e.g., Chen et al., 1993). Finally, the scaling region depends on some power of the Reynolds number, and also on the nature of large-scale forcing. The scaling range available in most accessible flows—especially in numerical simulations where the first two issues are not relevant—is small because the Reynolds numbers are not large enough.

ability to calculate these effects, one will always have lingering doubts about the true nature of anomaly and of scaling itself (e.g., Barenblatt and Goldenfeld, 1995). These issues are being constantly investigated with increasing precision (e.g., Anselmet *et al.*, 1983; Benzi *et al.*, 1993; L'vov and Procaccia, 1995; Arneodo *et al.*, 1996; Cao *et al.*, 1996; Tabeling *et al.*, 1996; Sreenivasan and Dhruva, 1998).

The anomaly of scaling exponents is related to small-scale intermittency. Roughly speaking, intermittency means that extreme events are far more probable than can be expected from Gaussian statistics and that the probability density functions of increasingly smaller scales are increasingly non-Gaussian (Fig. 3). This is a statistical consequence of uneven spatial distribution of the small-scale (Fig. 4), and can be modeled by multifractals (Mandelbrot, 1974; Parisi and Frisch, 1985; Meneveau and Sreenivasan, 1991). Most nonlinear systems are intermittent in time, space, or both, and the study of intermittency in turbulence is useful in a broad range of circumstances (e.g., Halsey *et al.*, 1986).

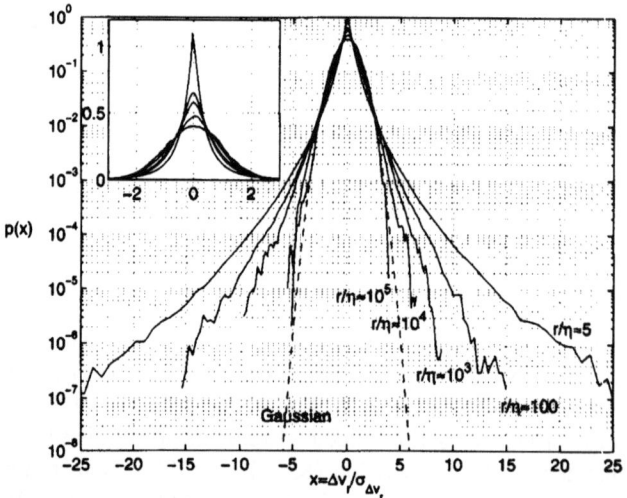

Figure 3. *The probability density functions, of differences of velocity fluctuations, obtained in atmospheric turbulence about 30 m above the ground. The ordinate is logarithmic in the main figure and linear in the inset. Each curve is for a different separation distance (using Taylor's hypothesis). The separation distance is transverse to the direction of the velocity component. The smallest separation distance (about 2.5 mm) is only five times the Kolmogorov scale η, denoting the smallest scale of fluctuations, while the largest (about 50 m) is comparable to the height of the measurement point. For small separation distances, very large excursions (even as large as 25 standard deviations) occur with nontrivial frequency; they are far more frequent than is given by a Gaussian distribution (shown by the full line), which is approached only for large separation distances. Extended tails over a wide range of scales is related to the phenomenon of small-scale intermittency (that is, uneven distribution in space of the small scales). These probability density functions are nonskewed. If the separation distance is in the direction of the velocity component measured, the probability density functions possess a definite skewness, as shown by Kolmogorov (1941a). This skewness is related to the energy transfer from large to small scales. In contrast to velocity increments, velocity fluctuations themselves have a nearly Gaussian character at this height above the ground. The shape of the probability density function depends on the flow and the spatial position in an inhomogeneous flow. For isotropic and homogeneous turbulence, it is marginally sub-Gaussian for high fluctuation amplitudes.*

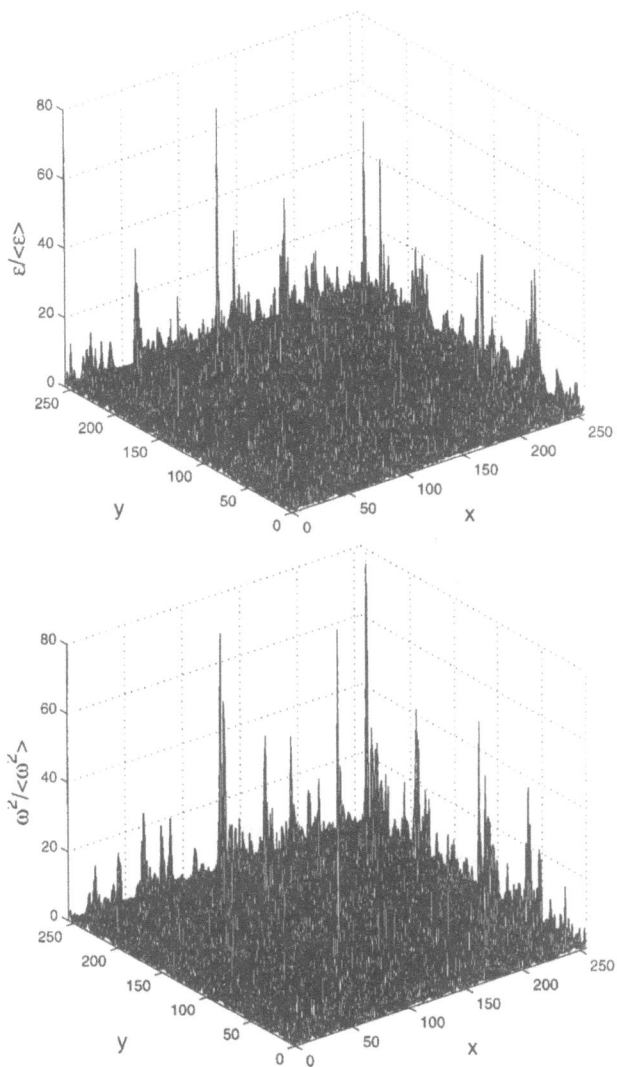

Figure 4. Planar cuts of the three-dimensional fields of (a) energy dissipation and (b) squared vorticity in a box of homogeneous and isotropic turbulence. The data are obtained by solving the Navier-Stokes equations on a computer. Not uncommon are amplitudes much larger than the mean; these large events become stronger with increasing Reynolds number. Such quantities are not governed by the central limit theorem. The statistics of large deviations are relevant here, as in many other broad contexts of modern interest. Kolmogorov (1962) proposed log-normal distribution to model energy dissipation (and, by inference, squared vorticity), but there seems to be a general agreement that lognormality is in principle incorrect (e.g., Mandelbrot, 1974; Narasimha, 1990; Novikov, 1990; Frisch 1995). Both these quantities have been modeled successfully by multifractals (Meneveau and Sreenivasan, 1991). A promising alternative is the log-Poisson model (She and Leveque, 1994).

Taken together, a major thrust of theoretical efforts has been the understanding of intermittency, multifractality, and the anomaly of scaling exponents. Many pedagogically illuminating models have been invented (see Sreenivasan and Antonia, 1997, for a summary), and a few rigorous inequalities are known (e.g., Constantin and Fefferman, 1994).

Some recent efforts

Theoretical issues

As a guide to further discussion, it is helpful to recall the mathematical problems associated with the Navier-Stokes equations. First, as already emphasized, there is no obvious small parameter on which to base a systematic perturbation theory. Second, the equations are nonlinear. The effects include energy redistribution among the constituent scales, as well as the so-called sweeping effect, which represents the manner in which the small scales are swept by the large. Third, the equations are dissipative even when the fluid viscosity is infinitesimally small ($Re \to \infty$). Fourth, there are dominant nonlocal effects arising from pressure.

The desire to understand qualitative aspects of each of these effects has led to different approaches. For instance, the inadequacy of perturbation methods have led to the exploration of nonperturbative alternatives. (An incomplete list of references in this regard, not necessarily alike in philosophy or detail, are Kraichnan, 1959; Martin et al., 1973; Forster et al., 1977; Yakhot and Orszag, 1986; McComb, 1990; Avellaneda and Majda, 1994; Eyink, 1994; Mou and Weichman, 1995; L'vov and Procaccia, 1996.) To understand nonlinear effects in forced systems, researchers have explored various alternatives such as Burgers equation with stochastic forcing (e.g., Cheklov and Yakhot, 1995; Polyakov, 1995), and shell models (e.g., Jensen et al., 1992) or their variants (Grossmann and Lohse, 1994).[6] Some attention has been paid to possible depletion of nonlinearity in parts of the real space (e.g., Frisch and Orszag, 1990). For passive scalars, the anomaly of scaling exponents is being explored via the rapidly-varying-velocity model for passive scalars (e.g., Kraichnan, 1994; Frisch et al., 1998). The interest in the small viscosity limit in the problem has led to serious studies of the singularities of the governing equations (Caferelli et al., 1982), especially of the inviscid counterpart—namely, the Euler equations (see, e.g., Beale et al., 1989). The multifractal analysis of dissipation fits in this broad picture. There is substantial interest in the physics of vortex dynamics (e.g., Saffman, 1992), particularly vortex reconnections (e.g., Kida and Takaoka, 1994). It is not always clear how centrally these studies bear on developed turbulence.

Advanced experimental methods

Traditional turbulence measurements are made at a single spatial position or at a few positions as functions of time, and yield time traces of velocity, temperature, or other quantities. These are treated as spatial cuts through the flow by invoking Taylor's hypoth-

---

[6]Burger's equation is the one-dimensional version of the Navier-Stokes equation, but without the pressure term; it possesses no chaotic solutions without forcing. Shell models are severe truncations of the Navier-Stokes equations, retaining only a few representative Fourier modes in any wavenumber band. Only nearest, or the next nearest, couplings are allowed. The models retain several symmetry properties of the Navier-Stokes equations.

esis, whose limitations are not fully understood (Lumley, 1965). A major accomplishment in recent years is the direct measurement of spatio-temporal fields of turbulence, obviating the need for this plausible but uncertain assumption. The techniques are typically the laser-induced fluorescence for passive scalars (e.g., Dahm *et al.*, 1991) and particle image velocimetry for flow velocity (e.g., Adrian, 1991). Unfortunately, available technology restricts true spatio-temporal measurements to low Reynolds numbers.

An experimental goal is to produce high Reynolds number turbulence and measure all the desired properties with adequate resolution in space and time. To obtain high Re, one may use the high speeds of fluid (but one is then limited by compressibility effects for gases and cavitation problems for liquids), a large-scale apparatus (which is limited by cost and available space), or use fluids of low viscosity (such as air at very high pressures or cryogenic fluids such as He I). For He I, the exquisite control on viscosity allows one to obtain, in an apparatus of a fixed size, a large *range* of Reynolds numbers than is possible by varying flow speed alone. This advantage has been exploited adroitly in a few instances (e.g., Castaing *et al.*, 1989; Tabeling *et al.*, 1996). In these instances, one has been forced to limit oneself to single-point data; the challenge is to develop instrumentation for obtaining spatial data, especially resolving small scales (for an account of some progress, see, e.g., Donnelly, 1991).[7]

Computational efforts

Another major advance is the use of powerful computers to solve Navier-Stokes equations exactly to produce turbulent solutions (e.g., Chorin, 1967; Orszag and Patterson, 1972). These are called direct numerical simulations (DNS). The DNS data are in some respects superior to experimental data because one can study experimentally inaccessible quantities such as tensorial invariants or pressure fluctuations at an interior point in the flow. The DNS data have allowed us to visualize details of small-scale vorticity and other similar features. For instance, they show that intense vorticity is often concentrated in tubes[8] (She *et al.*, 1990; Jimenez *et al.*, 1993); see Fig. 5. Yet, available computer memory and speed limit calculations to Reynolds numbers of the order of a few thousand. This limit is at present slightly better than the experimental range (see previous subsection).

It thus becomes necessary to adopt different strategies for computing high-Reynolds-number flows (e.g., Leonard, 1985; Lesieur and Metais, 1996; Moin, 1996; Moin and Kim, 1997). A fruitful avenue is the so-called Large Eddy Simulation method, in which one resolves what is possible, and suitably models the unresolved part. The modeling schemes vary in nature from an *a priori* prescription of the properties of the unresolved scales to computing their effect as part of the calculation scheme itself; the latter makes use of the known scaling properties of small-scale motion such as the locality of wavenumber interaction or spectral-scale similarity. As a computational tool for *practical* applications, the Large Eddy Simulation method has much promise. Increasing its versatility and adaptability near a solid surface is a major area of current research.

---

[7]For a fixed Re, a far smaller apparatus suffices when He is used instead of say, air, which makes the smallest scale that much smaller: recall that the ratio of the smallest scale to the flow apparatus is $O(Re^{-3/4})$.

[8]Experimental demonstration that vortex tubes can often be as long as the large-scale of turbulence can be found in Bonn *et al.* (1993).

We shall not remark here at length on engineering models of turbulence. They range from modified mixing length theories to those based on supplementary differential equations (e.g., Reynolds, 1976; Lumley, 1990; Lesieur, 1990) to the adaptation of the renormalization group methods (e.g., Yakhot and Orszag, 1986). These models cleverly exploit symmetries, conservation properties, realizability constraints, and other general principles to make headway in practical problem solving. Their short-term importance cannot be exaggerated.

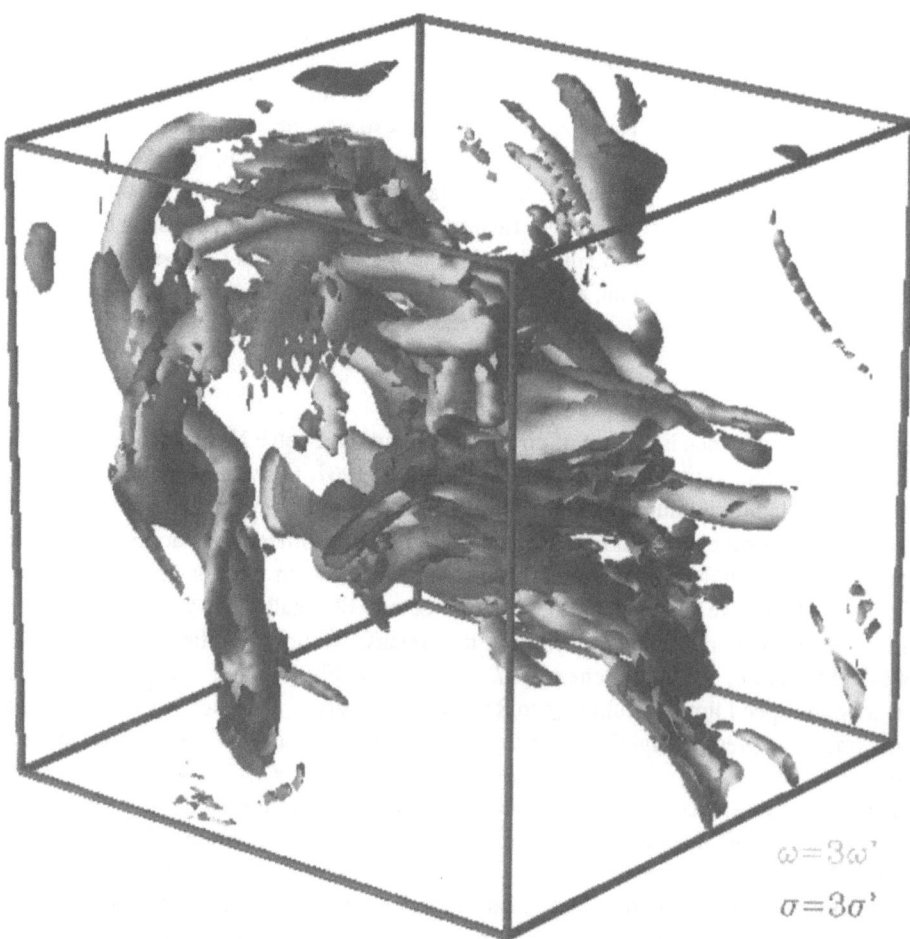

Figure 5. Demonstration that vorticity at large amplitudes, say greater than 3 standard deviations, organizes itself in the form of tubes (shown in yellow), even though the turbulence is globally homogeneous and isotropic. Large-amplitude dissipation (shown in red) is not as organized, and seems to surround regions of high vorticity. Smaller amplitudes do not possess such structure even for vorticity. In principle, the multifractal description of the spiky signals of Fig. 4 is capable of discerning geometric structures such as sheets and tubes, but no particular shape plays a central role in that description. The dynamical reason for this organization of large-amplitude vorticity is unclear. The ubiquitous presence of vortex tubes raises a number of interesting questions, some of which are mentioned in the text. At present, elementary properties of these tubes, such as their mean length and scaling of their thickness with Reynolds number, have not been quantified satisfactorily; nor has their dynamical significance. (See Color Plate 11.)

Prospects for the Near Future

It is useful to reiterate that turbulence research spans a wide spectrum from practical applications to fundamental physics. At one end of this spectrum are problems such as the prediction of fluctuating pressure field on the skin of an aircraft wing, or the hydrodynamic noise emitted by a submarine. Interactions with complexities such as combustion, rotation, and stratification pose a plethora of further questions: for instance, what is the amount of heat transported by the outer convective motion in the Sun? These problems involve nonlinearly complex interactions of the many parts of which they are composed, and so will necessarily remain too specific to expect general solutions. A sensible goal in such instances will always be to obtain reliable working approximations.

At the other end of the spectrum are deep physics issues arising from the nonperturbative nature of the turbulence problem. How may one understand precisely this many-scale problem with strong coupling among its constituent scales? It is natural to seek clues to this question in the analytic structure of the Navier-Stokes equations, but this task has so far proved hopelessly difficult. Therefore, one often seeks guidance via simpler problems of the same class, even if some essential elements are lost along the way.

Between the two ends is a wide middle, consisting of a study of carefully chosen idealized configurations. Typical problems follow: How much mixing occurs between two parallel streams in a well-conditioned flow apparatus? What is the net force exerted on a flat plate parallel to a smooth stream? What is the best way to parametrize the flow near smooth boundaries where viscosity affects all scales of turbulence? Such problems are approached by several complementary methods, but their broad content is the splitting of the overall motion into large and small scales—the former may well be the mean motion—and mastering the latter by combining phenomenology with aspects of universality. A sensible goal here is to put this practice on firmer physical principles.

Since these physical principles are still unclear, the task has an iterative character to it; thus, each generation of students of the subject has lived through them in different forms and made incremental progress. Progress has demanded that this grand problem (often hailed as the last such problem in classical physics) be split into various sub-problems—some closer to basic physics and some to working practice. Some in either variety may ultimately prove inessential to the overall purpose, but there can be no room for impatience or prejudice.

Listing all useful sub-problems without trivializing them is itself a challenge. We will unfortunately not rise to the occasion here, but list a few illustrative ones—making no effort to describe the progress being made. With respect to small scales, one interesting question is the dynamical importance of the highly anisotropic vortex tubes, and whether their existence is consistent with the universal (albeit anomalous) scaling presumed to exist in high-Reynolds-number turbulence (Moffatt, 1994; Moffatt et al., 1994): What is the connection between scaling (which emphasizes the sameness at various scales) and structure (which becomes better defined and topologically more anisotropic at larger fluctuation amplitudes)? In some problems of condensed matter physics—for example, anisotropic ferromagnets near the critical point—the critical indices are oblivious to the magnitude of anisotropy. However, this is not always the case. As Mandelbrot (1982) has emphasized in several contexts, this type of question necessarily forces the marriage of geometry with analysis (for some progress, see, Constantin, 1994); a particular case of this bigger picture is the stochastic geometry of turbulent/nonturbulent interfaces and of isoscalar surfaces (e.g., Constantin et al., 1991). A second question is the understanding of the effect of finite Reynolds number and of finite shear and anisotropy, comparable in scope, say, to that of finite-size effects in typical scaling problems in critical phenomena. This is a crucial undertaking for all issues related to scaling. Third, shifting focus from scaling exponents to scaling functions, and from the tails of probability density functions to the entire distribution, would be a useful relief. Fourth, a study of objects more complex than two-point structure functions would be highly informative (e.g., L'vov and Procaccia, 1996; Chertkov et al., 1998). Fifth, while the overall flux of energy from the large to the small scale is unidirectional on the average, the instantaneous flux is in both directions; is the overall average flux a small difference between the forward and reverse fluxes, or only a small fraction of the average? The answer to this question changes our perception of the degree of non-equilibrium present in the energy cascade, and influences the development of sound Large Eddy Simulation models. Sixth, one may usefully focus attention on other problems where violations of the K41 universality are first-order in importance—e.g., the problem of passive admixtures (Sreenivasan, 1991; Shraiman and Siggia, 1995), of pressure, and of acceleration statistics (e.g., Nelkin, 1994). As far as the large structures are concerned, the outstanding question is the determination of their origin, topology, frequency, and relation to small scales (e.g., Roshko, 1976; Hussain, 1983). Finally, an overarching issue is the abstraction of the small-scale influence on the small scales.

Some degree of progress has occurred on all these fronts, and has accelerated in recent years. Much of it is due to a powerful combination of experimental methods, computer simulations, and analytical advances in neighboring fields. Our hope lies in this synergism, whose importance cannot be exaggerated. It is trite but true to say that advancing experimental methods will imporve our understanding of turbulence significantly. (Recall the motto of Kamerlingh Onnes, the father of low temperature physics: "through measurement to knowledge".) In this regard, the key lies in measurements at high Reynolds numbers. How high a Reynolds number is "high enough" depends on the context and purpose. Yet, without a proper knowledge of Reynolds-number-scaling, one can be lured into false certainty by focusing exclusively on low Reynolds numbers. Presently, one obtains high-Reynolds-number small-scale data either in atmospheric flows or specialized facilities. Among the latter are facilities meant for testing large-scale aeronautical and navy vehicles, or those that use helium (e.g., Castaing et al., 1989; Tabeling et al., 1996), or use compressed air at very high pressures (Zagarola and Smits, 1996). Atmospheric flows are not controlled and stationary over long intervals of time, and only a few probes can be used at a given time. Among the specialized facilities, the large ones are very expensive to operate and, to a first approximation, unavailable for basic research. The smaller specialized flows allow, because of instrumentation limitations, only a small number of quantities to be measured with limited resolution. These shortcomings have been alleviated to some degree by computer simulation of the equations of motion, and a great deal can indeed be learned by combining such simulations at moderate Reynolds numbers with experiments at high Reynolds numbers. It is clear that the next generation of simulations, now already in progress, will produce data at high enough Reynolds numbers to begin to close the existing gap.

Concluding Remarks

From Osborne Reynolds at the turn of the last century to the present day, much qualitative understanding has been acquired about various aspects of turbulence. This progress has been undoubtedly useful in practice, despite large gaps that exist in our understanding. As a problem in physics or mechanics—contrasted, for example, against the rigor with which potential theory is understood—the problem is still in its infancy.

It has already been remarked that viewing turbulence as one grand problem may be debilitating. The large and diverse clientele it enjoys—such as astrophysicists, atmospheric physicists, aeronautical, mechanical, and chemical engineers—has different needs and approaches the problem with correspondingly different emphases. This makes it difficult to mount a focused frontal attack on a single aspect of the problem. It is therefore intriguing to ask: how may one recognize that the "turbulence problem" has been solved? It would be a great advance for an engineer to determine from fluid equations the pressure needed to push a certain volume of fluid through a circular tube. Even if this particular problem, or another like it, were to be solved, might it be deemed too special unless the effort paved the way for attacking similar problems?

There are two possible scenarios. Our computing abilities may improve so much that any conceivable turbulent problem can be "computed away" with adequate accuracy, so the problem disappears in the face of this formidable weaponry. One may still fret that computing is not understanding, but the issue assumes a more benign complexion. The other scenario—which is common in physics—is that a particular special problem that is sufficiently realistic and close enough to turbulence, will be solved in detail and understood fully. After all, no one can compute the detailed structure of the nitrogen atom from quantum mechanics, yet there is full confidence in the fundamentals of that subject. Unfortunately, the appropriate "hydrogen atom" or the "Ising model" for turbulence remains elusive.

In summary, there is a well-developed body of knowledge in turbulence that is generally self-consistent and useful for problem solving. However, there are lingering uncertainties at almost all levels. Extrapolating from experience so far, future progress will take a zigzag path, and further order will be slow to emerge. What is clear is that progress will depend on controlled measurements and computer simulations at high Reynolds numbers, and the ability to see in them the answers to the right theoretical questions. There is ground for optimism, and a meaningful interaction among theory, experiment, and computations must be able to take us far. It is a matter of time and persistence.

Acknowledgments

I am grateful to Rahul Prasad, Günter Galz, Brindesh Dhruva, Inigo Sangil and Shiyi Chen for their help with the figures, and to Steve Davis for comments on a draft version. The work was supported by the National Science Foundation grant DMR-95-29609.

References

Adrian, R.J., 1991, Annu. Rev. Fluid Mech. **23**, 261.
Arneodo, A., *et al.*, 1996, Europhys. Lett. **34**, 411.
Anselmet, F., Y. Gagne, E.J. Hopfinger, and R.A. Antonia, 1983, J. Fluid Mech. **140**, 63.
Avellaneda, M., and A.J. Majda, 1994, Philos. Trans. R. Soc. London, Ser. A **346**, 205.
Barenblatt, G.I., A.J. Chorin, and V.M. Prostokishin, 1997, Appl. Mech. Rev. **50**, 413
Barenblatt, G.I., and N. Goldenfeld, 1995, Phys. Fluids **7**, 3078.

Batchelor, G.K., 1953, *The Theory of Homogeneous Turbulence* (Cambridge University Press, England).
Beale, J.T., T. Kato, and A.J. Majda, 1989, Commun. Math. Phys. **94**, 61.
Benzi, R., S. Ciliberto, R. Tripiccione, C. Baudet, F. Massaioli, and S. Succi, 1993, Phys. Rev. E **48**, R29.
Bonn, D., Y. Couder, P.H.J. van Damm, and S. Douady, 1993, Phys. Rev. E **47**, R28.
Bradshaw, P., 1971, *An Introduction to Turbulence and Its Measurement* (Pergamon, New York).
Brown, G.L., and A. Roshko, 1974, J. Fluid Mech. **64**, 775.
Cafarelli, L., R. Kohn, and L. Nirenberg, 1982, Commun. Pure Apppl. Math. **35**, 771.
Cantwell, B.J., 1981, Annu. Rev. Fluid Mech. **13**, 457.
Canuto, V.M., and J. Christensen-Dalsgaard, 1998, Annu. Rev. Fluid Mech. **30**, 167.
Cao, N., S. Chen, and Z.-S. She, 1996, Phys. Rev. Lett. **76**, 3714.
Castaing, B., *et al.*, 1989, J. Fluid Mech. **204**, 1.
Cheklov, A., and V. Yakhot, 1995, Phys. Rev. E **51**, R2739.
Chen, S., G.D. Doolen, R.H. Kraichnan, and Z.-S. She, 1993, Phys. Fluids A **5**, 458.
Chertkov, M., A. Pumir, and B.I. Shraiman, 1998, Phys. Fluids (submitted).
Chorin, A.J., 1967, J. Comput. Phys. **2**, 1.
Chorin, A.J., 1994, *Vorticity and Turbulence* (Springer-Verlag, New York).
Comte-Bellot, G., and S. Corrsin, 1966, J. Fluid Mech. **25**, 657.
Constantin, P., 1994, SIAM (Soc. Ind. Appl. Math.) Rev. **36**, 73.
Constantin, P., and C. Fefferman, 1994, Nonlinearity **7**, 41.
Constantin, P., I. Procaccia, and K.R. Sreenivasan, 1991, Phys. Rev. Lett. **67**, 1739.
Corrsin, S., 1963, in *Handbuch der Physik, Fluid Dynamics II*, edited by S. Flugge and C. Truesdell (Springer-Verlag, Berlin), p. 524.
Cross, M.C., and P.C. Hohenberg, 1994, Rev. Mod. Phys. **65**, 851.
Dahm, W., K.B. Southerland, and K.A. Buch, 1991, Phys. Fluids A **3**, 1115.
Donnelly, R.J., 1991, Ed., *High Reynolds Number Flows Using Liquid and Gaseous Helium* (Springer-Verlag, New York).
Drazin, P.G., and W.H. Reid, 1981, *Hydrodynamic Stability* (Cambridge University Press, England).
Emmons, H.W., 1951, J. Aeronaut. Soc. **18**, 490.
Eyink, G., 1994, Phys. Fluids **6**, 3063.
Favre, A., J. Gaviglio, and R. Dumas, 1958, J. Fluid Mech. **3**, 344.
Forster, D., D.R. Nelson, and M.J. Stephen, 1977, Phys. Rev. A **16**, 732.
Frisch, U., 1995, *Turbulence: The Legacy of A.N. Kolmogorov* (Cambridge University Press, England)
Frisch, U., and S.A. Orszag, 1990, Phase Transit. **43**(1), 24.
Frisch, V., A. Mazzino, and M. Vergassola, 1998, Phys. Rev. Lett. **80**, 5532.
Gad-el-Hak, M., A. Pollard, and J.-P. Bonnet, 1998, Eds., *Flow Control: Fundamentals and Practices* (Springer-Verlag, New York).
Gollub, J.P., and H.L. Swinney, 1975, Phys. Rev. Lett. **35**, 927.
Grant, H.L., R.W. Stewart, and A. Moilliet, 1962, J. Fluid Mech. **12**, 241.
Grossmann, S., and D. Lohse, 1994, Phys. Rev. E **50**, 2784.
Halsey, T.C., M.H. Jensen, L.P. Kadanoff, I. Procaccia, and B.I. Shraiman, 1986, Phys. Rev. A **33**, 1141.
Head, M.R., and P. Bandyopadhyay, 1981, J. Fluid Mech. **107**, 297.
Heisenberg, W., 1948, Z. Phys. **124**, 628.
Holmes, P., J.L. Lumley, and G. Berkooz, 1998, *Turbulence, Coherent Structures, Dynamical Systems and Symmetry* (Cambridge University Press, England).
Hussain, A.K.M.F., 1983, Phys. Fluids **26**, 2816.
Jensen, M.H., G. Paladin, and A. Vulpiani, 1991, Phys. Rev. A **43**, 798.
Jimenez, J., A.A. Wray, P.G. Saffman, and R.S Rogallo, 1993, J. Fluid Mech. **255**, 65.

Keller, L.V., and A. Friedmann, 1924, in *Proceedings of the First International Congress on Applied Mechanics*, edited by C.B. Biezeno and J.M. Burgers (Technische Boekhandel en drukkerij, J. Waltman, Jr., Delft), p. 395.
Kida, S., and M. Takaoka, 1994, Annu. Rev. Fluid Mech. **26**, 169.
Kline, S.J., W.C. Reynolds, F.A. Schraub, and P.W. Runstadler, 1967, J. Fluid Mech. **30**, 741.
Kolmogorov, A.N., 1941a, Dokl. Akad. Nauk SSSR **32**, 19.
Kolmogorov, A.N., 1941b, Dokl. Akad. Nauk SSSR **30**, 299.
Kolmogorov, A.N., 1942, Izv. Akad. Nauk SSSR, Ser. Fiz. **VI (1-2)**, 56.
Kolmogorov, A.N., 1962, J. Fluid Mech. **13**, 82.
Kovasznay, L.S.G., V. Kibens, and R.F. Blackwelder, 1970, J. Fluid Mech. **41**, 283.
Kraichnan, R.H., 1959, J. Fluid Mech. **5**, 497.
Kraichnan, R.H., 1974, J. Fluid Mech. **62**, 305.
Kraichnan, R.H., 1994, Phys. Rev. Lett. **72**, 1016.
Landau, L.D., and E.M. Lifshitz, 1959, *Fluid Mechanics* (Pergamon, Oxford).
Leonard, A., 1985, Annu. Rev. Fluid Mech. **17**, 523.
Lesieur, M., 1990, *Turbulence in Fluids* (Kluwer, Dordrecht).
Lesieur, M., and O. Metais, 1996, Annu. Rev. Fluid Mech. **28**, 45.
Leslie, D.C. 1972, *Developments in the Theory of Turbulence* (Clarendon, Oxford).
Liepmann, H.W., 1952, Z. Angew. Math. Phys. **3**, 321.
Liepmann, H.W., 1979, Am. Sci. **67**, 221.
Lumley, J.L., 1965, Phys. Fluids **8**, 1056.
Lumley, J.L., 1990, Ed., *Whither Turbulence? Turbulence at the Crossroads* (Springer-Verlag, New York).
L'vov, V., and I. Procaccia, 1995, Phys. Rev. Lett. **74**, 2690.
L'vov, V., and I. Procaccia, 1996, Phys. World **9**, 35.
Majda, A., 1993, J. Stat. Phys. **73**, 515.
Malkus, W.V.R., 1956, J. Fluid Mech. **1**, 521.
Mandelbrot, B.B., 1974, J. Fluid Mech. **62**, 331.
Mandelbrot, B.B., 1982, *The Fractal Geometry of Nature* (Freeman, San Francisco).
Martin, P.C., E.D. Siggia, and H.A. Rose, 1973, Phys. Rev. A **8**, 423.
McComb, W.D., 1990, *The Physics of Fluid Turbulence* (Oxford University Press, England).
Meneveau, C., and K.R. Sreenivasan, 1991, J. Fluid Mech. **224**, 429.
Millikan, C.B., 1939, *Proceedings of the Fifth International Congress on Applied Mechanics*, edited by J.P. Den Hartog and H. Peters (John Wiley, New York), p. 386.
Moffatt, H.K., 1994, J. Fluid Mech. **275**, 406.
Moffatt, H.K., S. Kida, and K. Ohkitani, 1994, J. Fluid Mech. **259**, 241.
Moin, P., 1996, in *Research Trends in Fluid Mechanics*, edited by J.L. Lumley, A. Acrivos, L.G. Leal, and S. Leibovich (AIP Press, New York), p. 188.
Moin, P., and J. Kim, 1997, Sci. Am. **276**, 62.
Monin, A.S., and A.M. Yaglom, 1971, *Statistical Fluid Mechanics* (M.I.T. Press, Cambridge, MA), Vol. 1.
Monin, A.S., A.M. Yaglom, 1975, *Statistical Fluid Mechanics* (MIT Press, Cambridge, MA), Vol. 2.
Mou, C.-Y., and P. Weichman, 1995, Phys. Rev. E **52**, 3738.
Narasimha, R., 1983, J. Indian Inst. Sci. **64**, 1.
Narasimha, R., 1990, in *Whither Turbulence? Turbulence at Cross Roads*, edited by J.L. Lumley (Springer-Verlag, New York), p. 1.
Nelkin, M., 1994, Adv. Phys. **43**, 143.
Novikov, E.A., 1990, Phys. Fluids A **2**, 814.
Onsager, L., 1945, Phys. Rev. **68**, 286.
Orszag, S.A., and G.S. Patterson, 1972, in *Statistical Models and Turbulence*, Lecture Notes in Physics, edited by M. Rosenblatt and C.W. Van Atta (Spring-Verlag, Berlin), Vol. 12, p. 127.

Parisi, G., and U. Frisch, 1985, in *Turbulence and Predictability in Geophysical Fluid Dynamics*, edited by M. Ghil, R. Benzi, and G. Parisi (North Holland, Amsterdam), p. 84.
Polyakov, A., 1995, Phys. Rev. E **52**, 6183.
Prandtl, L., 1925, Z. Angew. Math. Mech. **5**, 136.
Prandtl, L., and O.G. Tietjens, 1934, *Applied Hydro- and Aeromechanics* (Dover, New York).
Reynolds, O., 1883, Philos. Trans. R. Soc. London **174**, 935.
Reynolds, O., 1894, Philos. Trans. R. Soc. London, Ser. A **186**, 123.
Reynolds, W.C., 1976, Annu. Rev. Fluid Mech. **8**, 183.
Richardson, L.F., 1922, *Weather Prediction by Numerical Process* (Cambridge University Press, England).
Richardson, L.F., 1926, Proc. R. Soc. London, Ser. A **A110**, 709.
Roshko, A., 1976, AIAA J. **14**, 1349.
Ruelle, D., and F. Takens, 1971, Commum. Math. Phys. **20**, 167.
Saffman, P.G., 1968, in *Topics in Nonlinear Physics*, edited by N.J. Zabusky (Springer, Berlin), p. 485.
Saffman, P.G., 1992, *Vortex Dynamics* (Cambridge Univeristy Press, England)
Schlichting, H., 1956, *Boundary Layer Theory* (McGraw Hill, New York).
She, Z.-S., E. Jackson, and S.A. Orszag, 1990, Nature (London) **344**, 226.
She, Z.-S., and E. Leveque, 1994. Phys. Rev. Lett. **72**, 336.
Shraiman, B., and E.D. Siggia, 1995, C.R. Acad. Sci. Ser. IIb: Mec., Phys., Chim., Astron. **321**, 279.
Siggia, E.D., 1994, Annu. Rev. Fluid Mech. **26**, 137.
Sirovich, L., 1987, Quart. Appl. Math. **45**, 561.
Smith, L.M., and S.L. Woodruff, Annu. Rev. Fluid Mech. **30**, 275.
Speziale, C.G., 1991, Annu. Rev. Fluid Mech. **23**, 107.
Sreenivasan, K.R., 1991, Proc. R. Soc. London, Ser. A **434**, 165.
Sreenivasan, K.R., and R.A. Antonia, 1997, Annu. Rev. Fluid Mech. **29**, 435.
Sreenivasan, K.R., and B. Dhruva, 1998, Prog. Theor. Phys. Suppl. **130**, 103.
Stolovitzky, G., and K.R. Sreenivasan, 1994, Rev. Mod. Phys. **66**, 229.
Tabeling, P., G. Zocchi, F. Belin, J. Maurer, and J. Williame, 1996, Phys. Rev. E **53**, 1613.
Taylor, G.I., 1915, Philos. Trans. R. Soc. London, Ser. A **215**, 1.
Taylor, G.I., 1932, Proc. R. Soc. London, Ser. A **135**, 685.
Taylor, G.I., 1935, Proc. R. Soc. London, Ser. A **151**, 421.
Taylor, G.I., 1938, Proc. R. Soc. London, Ser. A **164**, 15,476.
Tennekes, H., and J.L. Lumley, 1972, *A First Course in Turbulence* (MIT Press, Cambridge, MA).
Townsend, A.A., 1956, *The Structure of Turbulent Shear Flow* (Cambridge University Press, England).
Van Dyke, M., 1982, *An Album of Fluid Motion* (Parabolic Press, Stanford).
von Kármán, T., 1930, Nachr. Ges. Wiss. Goettingen, Math. Phys. **K1**, 58.
von Kármán, T., 1937, J. Aeronaut. Sci. **4**, 131.
von Kármán, T., and L. Howarth, 1938, Proc. R. Soc. London, Ser. A **164**, 192.
von Weizsäcker, C.F., 1948, Z. Phys. **124**, 614.
Wygnanski, I.J., F.H. Champagne, and B. Marasli, 1986, J. Fluid Mech. **168**, 31.
Yakhot, V., and S.A. Orszag, 1986, Phys. Rev. Lett. **57**, 1722.
Zabusky, N.J., and M. D. Kruskal, 1965, Phys. Rev. Lett. **15**, 240.
Zagaraola, M., and A.J. Smits, 1996, Phys. Rev. Lett. **78**, 239.
Zhou, Y., and C.G. Speziale, 1998, Appl. Mech. Rev. **1**, 267.

Pattern Formation in Nonequilibrium Physics

J. P. Gollub and J. S. Langer

The complex patterns that appear everywhere in nature have been cause for wonder and fascination throughout human history. People have long been puzzled, for example, about how intricate snowflakes can form, literally, out of thin air; and our minds boggle at the elegance of even the simplest living systems. As physicists, we have learned much about natural pattern formation in recent years; we have discovered how rich this subject can be, and how very much remains to be understood. Our growing understanding of the physics of pattern formation has led us to speculate—so far with only limited success—about a more general science of complexity, and to pose deep questions about our ability to predict and control natural phenomena.

Although pattern formation—i.e., morphogenesis—has always been a central theme in natural philosophy, it has reemerged in mainstream nonequilibrium physics only in the last quarter of the 20th Century. This has happened, in part, as an outgrowth of physicists' and materials scientists' interest in phase transitions. Many of the most familiar examples of pattern formation occur in situations in which a system is changing from one phase to another—from a liquid to a geometrically patterned solid, for example, or from a uniform mixture of chemical constituents to a phase-separated pattern of precipitates. As scientists have learned more about the equilibrium aspects of phase transitions, many have become interested in the non-equilibrium processes that accompany them. This line of investigation has led directly to questions of pattern formation.

Another direction from which physicists have approached the study of pattern formation has been the theory of nonlinear dynamical systems. Mechanical systems that can be described by ordinary differential equations often undergo changes from simple to complex behavior in response to changes in their control parameters. For example, the periodically forced and damped pendulum shows chaotic motion for certain intervals of the forcing amplitude, as well as periodic windows within the chaotic domains—a temporal "pattern" with considerable complexity. This is a simple case, however, with only a few degrees of freedom. More relevant for the present purposes are spatially extended dynamical systems with many degrees of freedom, for which partial differential equations are needed. Corresponding physical systems include fluids subjected to heating or rotation, which exhibit sequences of increasingly complex spatiotemporal patterns as the driving forces change. These are all purely deterministic pattern-forming systems. Understanding how they behave has been a crucial step toward understanding deterministic chaos—one of the most intriguing and profound scientific concepts to emerge in this century.

At the center of our modern understanding of pattern formation is the concept of instability. It is interesting to note that the mathematical description of instabilities is strikingly similar to the phenomenological theory of phase transitions first given by Landau (Landau and Lifshitz, 1969). We now know that complex spatial or temporal patterns emerge when relatively simple systems are driven into unstable states, that is, into states that will deform by large amounts in response to infinitesimally small perturbations. For example, solar heating of the earth's surface can drive Rayleigh-Bénard-like convective instabilities in the lower layer of the atmosphere, and the resulting flow patterns produce fairly regular arrays of clouds. At stronger driving forces, the convection patterns become unstable and turbulence increases. Another familiar example is the roughness of fracture surfaces produced by rapidly moving cracks in brittle solids. When we look in detail, we see that a straight crack, driven to high enough speeds, becomes unstable in such a way that it bends, sends out sidebranching cracks, and produces damage in the neighboring material. In this case, the physics of the instability that leads to these irregular patterns is not yet known.

After an instability has produced a growing disturbance in a spatially uniform system, the crucial next step in the pattern-forming process must be some intrinsically nonlinear mechanism by which the system moves toward a new state. That state may resemble the unstable deformation of the original state—the convective rolls in the atmosphere have roughly the same spacing as the wavelength of the initial instability. However, in many other cases, such as the growth of snowflakes, the new patterns look nothing like the linearly unstable deformations from which they started. The system evolves in entirely new directions as determined by nonlinear dynamics. We now understand that it is here, in the nonlinear phase of the process, that the greatest scientific challenges arise.

The inherent difficulty of the pattern-selection problem is a direct consequence of the underlying (linear or nonlinear) instabilities of the systems in which these phenomena occur. A system that is linearly unstable is one for which some response function diverges. This means that pattern-forming behavior is likely to be extremely sensitive to small perturbations or small changes in system parameters. For example, many patterns that we see in nature, such as snowflake-like dendrites in solidifying alloys, are generated by selective amplification of atomic-scale thermal noise. The shapes and speeds of growing dendrites are also exquisitely sensitive to tiny crystalline anisotropies of surface energies.

Some important questions, therefore, are: Which perturbations and parameters are the sensitively controlling ones? What are the mechanisms by which those small effects govern the dynamics of pattern formation? What are the interrelations between physics at different length scales in pattern-forming systems? When and how do atomic-scale mechanisms control macroscopic phenomena? At present, we have no general strategy for answering these questions. The best we have been able to do is to treat each case separately and—because of the remarkable complexity that has emerged in many of these problems—with great care.

In the next several sections of this article we discuss the connection between pattern formation and nonlinear dynamics, and then describe just a few specific examples that illustrate the roles of instability and sensitivity in nonequilibrium pattern formation. Our examples are drawn from fluid dynamics, granular materials, and crystal growth, which are topics that we happen to know well. We conclude with some brief remarks, mostly in the form of questions, about universality, predictability, and long-term prospects for this field of research.

Pattern Formation and Dynamical Systems

Our understanding of pattern formation has been dramatically affected by developments in mathematics. Deterministic pattern-forming systems are generally described by nonlinear partial differential equations, for example, the Navier-Stokes equations for fluids, or reaction-diffusion equations for chemical systems. It is characteristic of such nonlinear equations that they can have multiple steady solutions for a single set of control parameters such as external driving forces or boundary conditions. These solutions might be homogeneous, or patterned, or even more complex. As the control parameters change, the solutions appear and disappear and change their stabilities. In mathematical models of spatially extended systems, different steady solutions can coexist in contact with each other, separated by lines of defects or moving fronts.

The best way to visualize the solutions of such equations is to think of them as points in a multidimensional mathematical space spanned by the dynamical variables, that is, a "phase space." The rules that determine how these points move in the phase space constitute what we call a "dynamical system." One of the most important developments in this field has been the recognition that dynamical systems with infinitely many degrees of freedom can often be described by a finite number of relevant variables, that is, in finite-dimensional phase spaces. For example, the flow field for Rayleigh-Bénard convection not too far from threshold can be described accurately by just a few time-dependent Fourier amplitudes. If we think of the partial differential equations as being equivalent to finite sets of coupled ordinary differential equations, then we can bring powerful mathematical concepts to bear on the analysis of their solutions.

As we shall emphasize in the next several sections of this article, dynamical-systems theory provides at best a qualitative framework on which to build physical models of pattern formation. Nevertheless, it has produced valuable insights and, in some cases, has even led to prediction of novel effects. It will be useful, therefore, to summarize some of these general concepts before looking in more detail at specific examples. An introductory discussion of the role of dynamical systems theory in fluid mechanics has been given by Aref and Gollub (1996).

In dynamical-systems theory, the stable steady solutions of the equations of motion are known as "stable fixed points" or "attractors," and the set of points in the phase space from which trajectories flow to a given fixed point is its "basin of attraction." As the control parameters are varied, the system typically passes through "bifurcations" in which a fixed point loses its stability and, at the same time, one or more new stable attractors appear. An especially simple example is the "pitchfork" bifurcation at which a stable fixed point representing a steady fluid flow, for example, gives rise to two symmetry-related fixed points describing cellular flows with opposite polarity. Many other types of bifurcation have been identified in simple models and also have been seen in experiments.

The theory of bifurcations in dynamical systems helps us understand why it is sometimes reasonable to describe a system with infinitely many degrees of freedom using only a finite (or even relatively small) number of dynamical variables. An important mathematical result known as the "center manifold theorem" (Guckenheimer and Holmes,

1983) indicates that, when a bifurcation occurs, the associated unstable trajectories typically move away from the originally stable fixed point only within a low-dimensional subspace of the full phase space. The subspace is "attracting" in the sense that trajectories starting elsewhere converge to it, so that the degrees of freedom outside the attracting subspace are effectively irrelevant. It is for this reason that we may need only a low-dimensional space of dynamical variables to describe some pattern-formation problems near their thresholds of instability—a remarkable physical result.

Time-varying states, such as oscillatory or turbulent flows, are more complex than simple fixed points. Here, some insight also has been gained from considering dynamical systems. Oscillatory behavior is generally described as a flow on a limit cycle (or closed loop) in phase space, and chaotic states may be represented by more complex sets called "strange attractors." The most characteristic feature of the latter may be understood in terms of the Lyapunov exponents that give the local exponential divergence or convergence rates between two nearby trajectories, in the different directions along and transverse to those trajectories. If at least one of these exponents, when averaged over time, is positive, then nearby orbits will separate from each other exponentially in time. Provided that the entire attracting set is bounded, the only possibility is for the set to be fractal.

In the early 1970s, strange attractors were thought by some to be useful models for turbulent fluids. In fact, the phase-space paradigm can give only a caricature of the real physics because of the large number of relevant degrees of freedom involved in most turbulent flows. While much less than $6N$ (the number of degrees of freedom for a system of N molecules), that number still grows in proportion to $R^{3/4}$, where R is the Reynolds number. We do not yet know whether weakly turbulent states ("spatiotemporal chaos") that sometimes occur near the onset of instability may be viewed usefully using the concepts of dynamical systems.

Patterns and Spatiotemporal Chaos in Fluids: Nonlinear Waves

Pattern formation has been investigated in an immense variety of hydrodynamic systems. Examples include convection in pure fluids and mixtures; rotating fluids, sometimes in combination with thermal transport; nonlinear surface waves at interfaces; liquid crystals driven either thermally or by electromagnetic fields; chemically reacting fluids; and falling droplets. Some similar phenomena occur in nonlinear optics. Many of these cases have been reviewed by Cross and Hohenberg (1993); there also have been a host of more recent developments. Since it is not possible in a brief space to discuss this wide range of phenomena, we focus here on an example that poses interesting questions about the nature of pattern formation: waves on the surfaces of fluids. We shall also make briefer remarks about other fluid systems that have revealed strikingly novel phenomena.

The surface of a fluid is an extended dynamical system for which the natural variables are the amplitudes and phases of the wavelike deformations. These waves were at one time regarded as being essentially linear at small amplitudes. However, even weak nonlinear effects cause interactions between waves with different wave vectors and can be important in determining wave patterns. When the wave amplitudes are large, the nonlinear effects lead to chaotic dynamics in which many degrees of freedom are active.

A convenient way of exciting nonlinear waves in a manner that does not directly break any spatial symmetry is to subject the fluid container to a small-amplitude vertical excitation. This leads to standing waves at half the driving frequency, via an instability and an associated bifurcation first demonstrated by Faraday (1831), 68 years before the American Physical Society was founded. The characteristic periodicity of the resulting patterns is approximately the same as the wavelength of the most rapidly growing linear instability determined by the dispersion relation for capillary-gravity waves, but the wave patterns themselves are far more complex and interesting.

All of the regular patterns that can tile the plane have been found in this system, including hexagons, squares, and stripes (Kudrolli and Gollub, 1996). In addition, various types of defects that are analogous to crystalline defects occur: grain boundaries, dislocations, and the like. Which patterns are stable depends on parameters: the fluid viscosity, the driving frequency, and the acceleration. Significant domains of coexistence between different patterns are also known, where patterns with different symmetry are simultaneously stable.

These phenomena have resisted quantitative explanation for a number of reasons: the difficulty of dealing with boundary conditions at the moving surface of the fluid; the nonlinearity of the hydrodynamic equations; and the complex effects of viscosity. However, a suitable mathematical description, consistent with the general framework of dynamical-systems theory, is now available (Chen and Viñals, 1997), and it leads to a satisfactory explanation of these pattern-forming phenomena. The basic idea is to regard the surface as a superposition of interacting waves propagating in different directions. Coupled evolution equations can be written for the various wave amplitudes. The coupling coefficients depend on the angles between the wave vectors, and these coupling functions depend in turn on the imposed parameters (such as wave frequency). The entire problem is variational, but only near the threshold of wave formation. That is, the preferred pattern near the onset of instability is the one that minimizes a certain functional of the wave amplitudes, in much the same way that the preferred state of a crystal is the one that minimizes its free energy. Away from threshold, on the other hand, no such variational principle exists, and the variety of behaviors is correspondingly richer.

These results raise the question of whether other types of regular patterns can be formed that are not spatially periodic but do have rotational symmetry, i.e., quasicrystalline patterns. In fact they do occur (Christiansen *et al.*, 1992; Edwards and Fauve, 1994), just as they do in ordinary crystals. The way in which these different patterns become stable or unstable as the parameters are varied has now been worked out in some detail and appears to be in accord with experiment (Binks and van de Water, 1997). An example of a quasicrystalline pattern is shown in Fig. 1.

When the wave amplitudes are raised sufficiently, transitions to spatially and temporally disordered states occur (Kudrolli and Gollub, 1996 and references therein). In the language of dynamical systems, some of these new states might be called "strange attractors," although they are certainly not low-dimensional objects. They are much less well understood than the standing-wave states, and the ways in which they form appears to depend on the ordered states from which they emerge. For example, the hexagonal lattice appears to melt continuously, while the striped phase breaks down inhomogeneously in regions where the stripes are most strongly curved. The resulting states of spatiotemporal chaos are not completely disordered; there can be regions of local order. Furthermore, if the fluid is not too viscous, so that the correlation length of the pattern is relatively long, then the symmetry imposed by the boundaries can be recovered by averaging over a large number of individually fluctuating patterns (Gluckman *et al.*, 1995). A case of strongly turbulent capillary waves has also been studied experimentally (Wright *et al.*, 1997).

Certain other fluid systems have chaotic states that occur closer to the linear threshold of the primary pattern. In these cases, quantitative comparison with theory is sometimes possible. An example is the behavior of Rayleigh-Bénard convection in the presence of rotation about a vertical axis (Hu *et al.*, 1997). This problem is relevant to atmospheric dynamics. Though the basic pattern consists of rolls, as shown in Fig. 2, they are unstable. Patches of rolls at different angles invade each other as time proceeds, and the pattern remains time dependent indefinitely. This phenomenon has been discussed theoretically (Tu and Cross, 1992) using a two-dimensional nonlinear partial differential equation known as the complex Ginzburg-Landau equation. Similar models have been used successfully for treating a variety of nonchaotic pattern-forming phenomena. In this case, the model is able to reproduce the qualitative behavior of the experiments, but does not

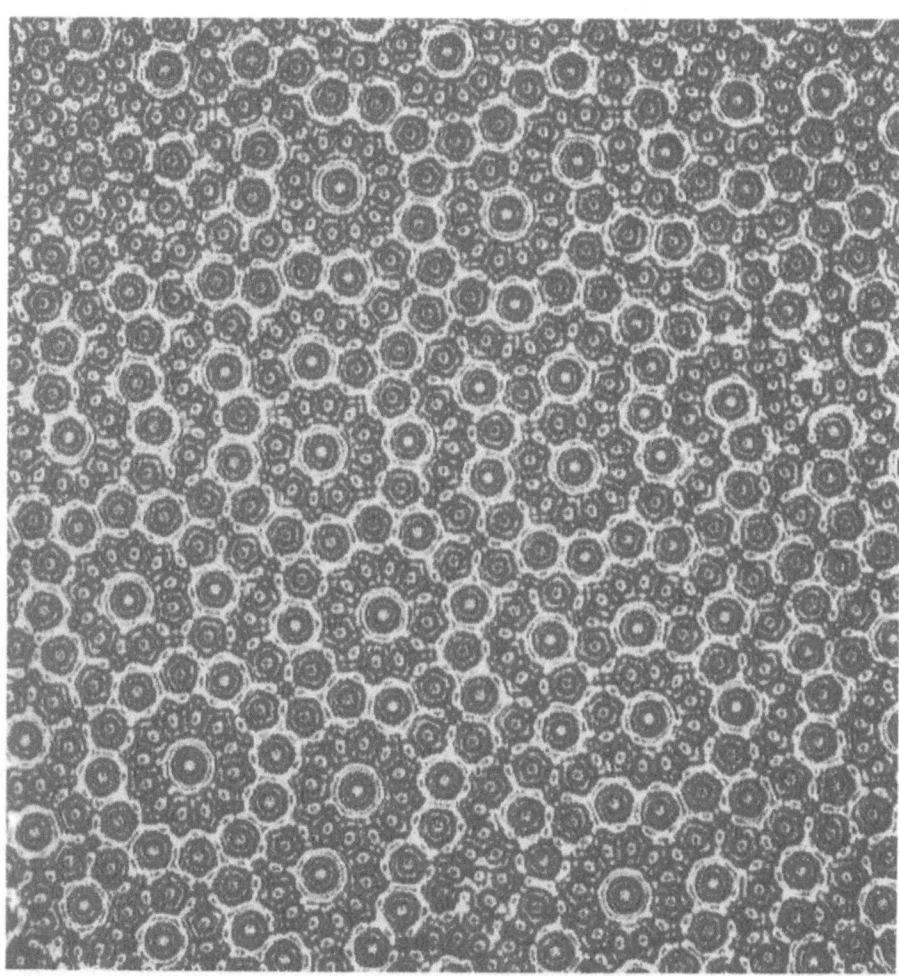

Figure 1. A quasicrystalline wave pattern with 12-fold rotational symmetry. This standing-wave pattern was produced by forcing a layer of silicone oil simultaneously at two frequencies, using a method invented by Edwards and Fauve. The brightest regions are locally horizontal, whereas darker colors indicate inclined regions. From work done at Haverford for an undergraduate thesis by B. Pier. (See Color Plate 12.)

Figure 2. Spatiotemporal chaos in rotating Rayleigh-Bénard convection shown at two different times. Patches of rolls at different angles invade each other as time proceeds. Courtesy of G. Ahlers.

successfully describe the divergence of the correlation length of the patchy chaotic fluctuations as the transition is approached. Thus the goal of understanding spatiotemporal chaos has remained elusive.

There is one area where the macroscopic treatment of pattern-forming instabilities connects directly to microscopic physics: the effects of thermal noise. Macroscopic patterns often emerge from the amplification of noise by instabilities. Therefore, fluctuations induced by thermal noise (as distinguished from chaotic fluctuations produced by nonlinearity) should be observable near the threshold of instability. This remarkable effect has been demonstrated quantitatively in several fluid systems, for example, ordinary Rayleigh-Bénard convection (Wu et al., 1995). As we shall see later in the section on Growth at Interfaces: Dendritic Solidification, very similar amplification of thermal noise occurs in dendritic crystal growth.

Patterns in Granular Materials

Patterns quite similar to the interfacial waves described in the previous section occur when the fluid is replaced by a layer of granular matter such as sand or, in well-controlled recent experiments, uniform metallic or glassy spheres (Melo et al., 1995). Depending on the frequency and amplitude of the oscillation of the container, the upper surface of the grains can arrange itself into arrays of stripes or hexagons, as shown in Fig. 3. Lines dividing regions differing in their phase of oscillation, and disordered patterns, are also evident.

In addition, granular materials can exhibit localized solitary excitations known as "oscillons" (Umbanhowar et al., 1996). These can in turn organize themselves into clusters, as shown in Fig. 3(d). This striking discovery has given rise to a number of competing theories and has been immensely provocative. It is interesting to note that localized excitations are also found in fluids. For example, they have been detected in instabilities induced by electric fields applied across a layer of nematic liquid crystal (Dennin et al., 1996). All of these localized states are intrinsically nonlinear phenomena, whether they occur in granular materials or in ordinary fluids, and do not resemble any known linear instability of the uniform system.

Granular materials have been studied empirically for centuries in civil engineering and geology. Nevertheless, we still have no fundamental physical understanding of their nonequilibrium properties. In fact, to a modern physicist, granular materials look like a novel state of matter. For a review of this field, see Jaeger *et al.* (1996), and references therein.

There are several clear distinctions between granular materials and other, superficially comparable, many-body systems such as fluids. Because they have huge numbers of degrees of freedom, they can only be understood in statistical terms. However, individual grains of sand are enormously more massive than atoms or even macromolecules; thus thermal kinetic energy is irrelevant to them. On the other hand, each individual grain has an effectively infinite number of internal degrees of freedom; thus the grains are generally inelastic in their interactions with each other or with boundaries. They also may have irregular shapes; arrays of such grains may achieve mechanical equilibrium in a variety of configurations and packings. It seems possible, therefore, that concepts like "temperature" and "entropy" might be useful for understanding the behavior of these materials (for example, see Campbell, 1990).

In the oscillating-granular-layer experiments, some of the simpler transitions can be explained in terms of temporal symmetry breaking resulting from the the low-dimensional dynamics of the particles as they bounce off the oscillating container surface. The onset of temporal period doubling in the particle dynamics coincides with the spatial transition from stripes to hexagons. Both the particle trajectories and the spatial patterns are then different on successive cycles of the driver. The analogous hexagonal state for Faraday waves in fluids can be induced either by temporal symmetry breaking of the external forcing, or by the frequency and viscosity dependence of the coupling between different traveling-wave components. Both of these mechanisms are quite different microscopically from the single-particle dynamics that generates the hexagons in granular materials.

On the other hand, there are substantial similarities between the granular and fluid behaviors. The granular material expands or dilates as a result of excitation. Roughly speaking, dilation of the granular layer reduces the geometrical constraints that limit flow.

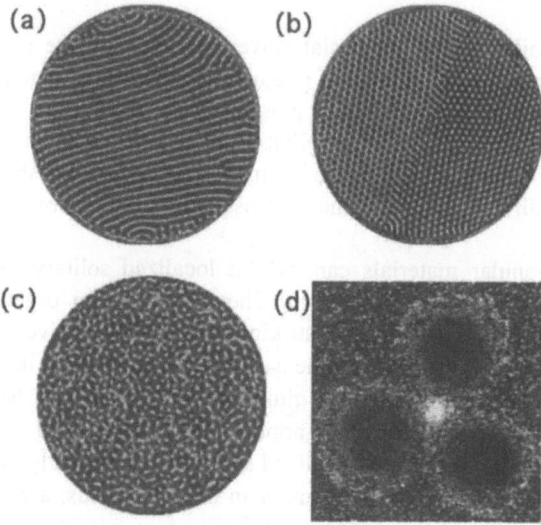

Figure 3. Standing-wave patterns in a vibrating layer of granular material. (a) Stripes; (b) hexagons and defects; (c) disordered waves; (d) clusters of localized "oscillons." Courtesy of P. Umbanhowar and H. Swinney.

This correponds to lower viscosity of the conventional fluid. Dilation accounts in physical terms for the fact that the striped phase that occurs at high fluid viscosity may be found at low acceleration of the granular material.

The differences, however, seem to be emerging dramatically as new experiments and numerical simulations probe more deeply into granular phenomena. For example, stresses in nearly static granular materials are highly inhomogeneous, forming localized stress chains that are quite unlike anything seen in ordinary liquids and solids. Even in situations involving flow, the behavior of granular materials often seems to be governed by their tendency to "jam," that is, to get themselves into local configurations from which they are temporarily unable to escape. This happens during a part of each cycle in the oscillating-layer experiments. (The concept of "jammed" systems was the topic of a Fall 1997 program at the Institute for Theoretical Physics in Santa Barbara. For more information, consult the ITP web site: http://www.itp.ucsb.edu/online/jamming2/schedule.html.)

Growth at Interfaces: Dendritic Solidification

We turn finally to the topic of dendritic pattern formation. It is here that some of the deepest questions in this field—the mathematical subtlety of the selection problem and the sensitivity to small perturbations—have emerged most clearly in recent research.

Dendritic solidification, that is, the "snowflake problem," is one of the most thoroughly investigated topics in the general area of nonequilibrium pattern formation. It is only in the last few years, however, that we finally have learned how these elegant dendritic crystals are formed in the atmosphere, and why they occur with such diversity that no two of them ever seem to be exactly alike. Nevertheless, our present understanding is still far from good enough for many practical purposes, for example, for predicting the microstructures of multicomponent cast alloys.

Much of the research on dendritic crystal growth has been driven, not only by our natural curiosity about such phenomena, but also by the need to understand and control metallurgical microstructures. (For example, see Kurz and Fisher, 1989) The interior of a grain of a freshly solidified alloy, when viewed under a microscope, often looks like an interlocking network of highly developed snowflakes. Each grain is formed by a dendritic, i.e., treelike, process in which a crystal of the primary composition grows out rapidly in a cascade of branches and sidebranches, leaving solute-rich melt to solidify more slowly in the interstices. The speed at which the dendrites grow and the regularity and spacing of their sidebranches determine the observed microstructure which, in turn, governs many of the properties of the solidified material such as its mechanical strength and its response to heating and deformation.

The starting point for investigations of metallurgical microstructures or snowflakes is the study of single, isolated, freely growing dendrites. Remarkable progress has been made on understanding this phenomenon recently. The free-dendrite problem is most easily defined by reference to the xenon dendrite shown in Fig. 4. Here, we are looking at a pure single crystal growing into its liquid phase. The speed at which the tip is advancing, the radius of curvature of the tip, and the way in which the sidebranches emerge behind the tip, all are determined uniquely by the degree of undercooling, i.e., by the degree to which the liquid is colder than its freezing temperature. The question is: How?

In the most common situations, dendritic growth is controlled by diffusion—either the diffusion of latent heat away from the growing solidification front or the diffusion of chemical constituents toward and away from that front. These diffusion effects very often lead to shape instabilities; small bumps grow out into fingers because, like lightning rods, they concentrate the diffusive fluxes ahead of them and therefore grow out more rapidly than a flat surface. This instability, generally known as the "Mullins-Sekerka instability," is the trigger for pattern formation in solidification.

Today's prevailing theory of free dendrites is generally known as the "solvability theory" because it relates the determination of dendritic behavior to the question of whether or not there exists a sensible solution for a certain diffusion-related equation that contains a singular perturbation. The term "singular" means that the perturbation, in this case the surface tension at the solidification front, completely changes the mathematical nature of the problem whenever it appears, no matter how infinitesimally weak it might be. In the language of dynamical systems, the perturbation controls whether or not there exists a stable fixed point. Similar situations occur in fluid dynamics, for example, in the "viscous fingering" problem, where a mechanism similar to the Mullin-Sekerka instability destabilizes a moving interface between fluids of different viscosities, and a solvability mechanism determines the resulting fingerlike pattern. (See Langer, 1987, for a pedagogical introduction to solvability theory, and Langer, 1989, for an overview including the viscous fingering problem.)

The solvability theory has been worked out in detail for many relevant situations such as the xenon dendrite shown in Fig. 4. (See Bisang and Bilgram, 1996, for an account of the xenon experiments, and also for references to recent theoretical work by Brener and colleagues.) The theory predicts how pattern selection is determined, not just by the surface tension (itself a very small correction in the diffusion equations), but by the crystalline anisotropy of the surface tension—an even weaker perturbation in this case. It further predicts that the sidebranches are produced by secondary instabilities near the tip that are triggered by thermal noise and amplified in special ways as they grow out along the sides of the primary dendrite. The latter prediction is especially remarkable because it

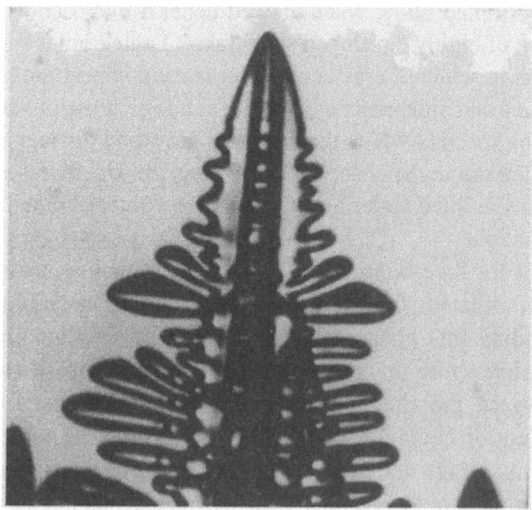

Figure 4. Dendritic xenon crystal growing in a supercooled melt. Courtesy of J. Bilgram.

relates macroscopic features—sidebranches with spacings of order tens of microns—to molecular fluctuations whose characteristic sizes are of order nanometers.

Each of those predictions has been tested in the xenon experiment, quantitatively and with no adjustable fitting parameters. They have also been checked in less detail in experiments using other metallurgical analog materials. In addition, the theory has been checked in numerical studies that have probed its nontrivial mathematical aspects (Karma and Rappel, 1996). As a result, although we know that there must be other cases (competing thermal and chemical effects, for example, or cases where the anisotropy is large enough that it induces faceting), we now have reason for confidence that we understand at least some of the basic principles correctly.

Reflections and Conclusions

We have illustrated pattern-forming phenomena through a few selected examples, each of which has given rise to a large literature. Many others could be cited. As we have seen, the inherent sensitivity of pattern-forming mechanisms to small perturbations means that research in this field must take into account physical phenomena across extraordinarily wide ranges of length and time scales. Moreover, studies of pattern formation increasingly are being extended to materials that are more complex than isotropic classical fluids and homogeneous solids. The case of granular matter described here is one example of this trend. An important example for the future is pattern formation in biological systems, where the interplay between physical effects and genetic coding leads to striking diversity.

The expanding complexity and importance of this field brings urgency to a set of deep questions about theories of pattern formation and, more generally, about the foundations of nonequilibrium statistical physics. What does sensitivity to noise and delicate perturbations imply about the apparent similarities between different systems? Are there, for example, deep connections between dendritic sidebranching and fracture, or are the apparent similarities superficial and unimportant? What about the apparent similarities between the patterns seen in fluids and granular materials? In short, are there useful "universality classes" for which detailed underlying mechanisms are less important than, for example, more general symmetries or conservation laws? Might we discover some practical guidelines to tell us how to construct predictive models of pattern-forming systems, or shall we have to start from the beginning in considering each problem?

These are not purely philosophical questions. Essentially all processes for manufacturing industrial materials are nonequilibrium phenomena. Most involve, at one stage or another, some version of pattern formation. The degree to which we can develop quantitative, predictive models of these phenomena will determine the degree to which we can control them and perhaps develop entirely new technologies. Will we be able, for example, to write computer programs to predict and control the microstructures that form during the casting of high-performance alloys? Can we hope to predict, long in advance, mechanical failure of complex structural materials? Will we ever be able to predict earthquakes? Or, conversely, might we discover that the complexity of many systems imposes intrinsic limits to our ability to predict their behavior? That too would be an interesting and very important outcome of research in this field.

Acknowledgments

J.S.L. acknowledges support from DOE Grant DE-FG03-84ER45108. J.P.G. acknowledges support from NSF Grant DMR-9704301.

References

Aref, H., and J.P. Gollub, 1996, in *Research Trends in Fluid Dynamics: Report From the United States National Committee on Theoretical and Applied Mechanics*, edited by J.L. Lumley (American Institute of Physics, Woodbury, NY), p. 15.
Binks, D., and W. van de Water, 1997, Phys. Rev. Lett. **78**, 4043.
Bisang, U., and J.H. Bilgram, 1996, Phys. Rev. E **54**, 5309.
Campbell, C.S., 1990, Annu. Rev. Fluid Mech. **22**, 57.
Chen, P., and J. Viñals, 1997, Phys. Rev. Lett. **79**, 2670.
Christiansen, B., P. Alstrom, and M.T. Levinsen, 1992, Phys. Rev. Lett. **68**, 2157.
Cross, M.C., and P.C. Hohenberg, 1993, Rev. Mod. Phys. **65**, 851.
Dennin, M., G. Ahlers, and D.S. Cannell, 1996, Phys. Rev. Lett. **77**, 2475.
Edwards, S., and S. Fauve, 1994, J. Fluid Mech. **278**, 123.
Faraday, M., 1831, Philos. Trans. R. Soc. London **121**, 299.
Gluckman, B.J., C.B. Arnold, and J.P. Gollub, 1995, Phys. Rev. E **51**, 1128.
Guckenheimer, J., and P. Holmes, 1983, *Nonlinear Oscillations, Dynamical Systems, and Bifurcations of Vector Fields* (Springer, New York).
Hu, Y., R.E. Ecke, and G. Ahlers, 1997, Phys. Rev. E **55**, 6928.
Jaeger, H.M., S.R. Nagel, and R.P Behringer, 1996, Rev. Mod. Phys. **68**, 1259.
Karma, A., and J. Rappel, 1996, Phys. Rev. Lett. **77**, 4050.
Kudrolli, A., and J.P. Gollub, 1996, Physica D **97**, 133.
Kurz, W., and D.J. Fisher, 1989, *Fundamentals of Solidification* (Trans Tech Publications, Brookfield, VT).
Landau, L.D., and E.M. Lifshitz, 1969, *Statistical Physics* (Pergamon, Oxford), Chap. XIV.
Langer, J.S., 1987, in *Chance and Matter, Proceedings of the Les Houches Summer School, Session XLVI*, edited by J. Souletie, J. Vannimenus, and R. Stora (North Holland, Amsterdam), p. 629.
Langer, J.S., 1989, Science **243**, 1150.
Melo, F., P.B. Umbanhowar, and H.L. Swinney, 1995, Phys. Rev. Lett. **75**, 3838.
Miles, J., and D. Henderson, 1990, Annu. Rev. Fluid Mech. **22**, 143.
Tu, Y., and M. Cross, 1992, Phys. Rev. Lett. **69**, 2515.
Umbanhowar, P.B., F. Melo, and H.L. Swinney, 1996, Nature (London) **382**, 793.
Wright, W.B., R. Budakian, D.J. Pine, and S.J. Putterman, 1997, Science **278**, 1609.
Wu, M., G. Ahlers, and D.S. Cannell, 1995, Phys. Rev. Lett. **75**, 1743.

Plasma Physics

The Collisionless Nature of High-Temperature Plasmas

T. M. O'Neil and F. V. Coroniti

An important property that distinguishes high temperature plasmas from normal fluids, even from conducting fluids such as liquid metals, is that the plasmas are to a first approximation collisionless. In a laboratory plasma, the mean-free-path between collisions can be much larger than the dimensions of the plasma. In space and astrophysical plasmas, the mean free path can easily exceed the dimensions of the structures of interest. The collisionless nature necessitates a kinetic treatment and introduces a variety of subtle new phenomena. For example, Landau damping (or growth) results from the resonant interaction of a wave with free streaming particles, a resonance that would be spoiled by collisions in a normal fluid. Also, the collisionless nature challenges us to find new descriptions for familiar phenomena. For example, what is the nature of a shock wave in a collisionless plasma?

This review provides a brief introduction to the collisionless nature of plasmas. Taking the resonant wave-particle interaction as characteristic, we follow Landau's idea through the years as it is tested experimentally, extended nonlinearly, and applied. We then describe the Earth's bow shock, which is an important example of a collisionless shock wave. Finally, we touch on cosmic ray acceleration by supernova shocks.

Wave-Particle Interactions

Linear theory

The first proper treatment of modes in a collisionless plasma was provided by Landau (1946). Using the collisionless Boltzmann equation and Poisson's equation he obtained the dispersion relation for electron plasma oscillations (Langmuir oscillations). Tonks and Langmuir (1929) had described these simple electrostatic modes many years earlier using fluid equations. In fact, it was in this early paper that Tonks and Langmuir coined the name "plasma." Landau's main correction to the earlier fluid description was that the modes experience a collisionless damping (or growth). The electric potential for a mode that is characterized by wave number $\mathbf{k} = \hat{z}k$ damps (or grows) temporally at the rate

$$\gamma(k) = \frac{\pi}{2}\omega(k)\frac{\omega_p^2}{k^2}\frac{\partial f_0}{\partial v_z}\bigg|_{\omega(k)/k}, \tag{1}$$

where $\omega(k)$ is the mode frequency and $\omega_p = (4\pi n e^2/m)^{1/2}$, is the electron plasma frequency. Here, e and m are the electron charge and mass and n is the density. The function $f_0(v_z) = \int dv_x dv_y F_0(v_x, v_y, v_z)$ is the distribution of electron velocities parallel to the direction of propagation, where $F_0(v_x, v_y, v_z)$ is the distribution over all three velocity components. The subscript zero indicates that the distribution refers to the unperturbed equilibrium state, which is assumed to be spatially homogeneous. Langmuir waves are special in that the electrons dominate the dynamics; more generally, both electrons and ions contribute to $\gamma(k)$. Clearly, the damping (or growth) is associated with electrons that satisfy the relation $kv_z = \omega(k)$ [or equivalently, $\mathbf{k} \cdot \mathbf{v} = \omega(k)$]; these electrons maintain a constant phase relative to the wave and resonantly exchanging energy with the wave. For a Maxwellian velocity distribution, the derivative $\partial f_0/\partial v_z|_{\omega(k)/k}$ is negative so $\gamma(k)$ is negative and the mode damps. However, for a non-Maxwellian distribution, corresponding, say, to the case where a small warm beam drifts through the plasma, $\partial f_0/\partial v_z|_{\omega(k)/k}$ can be positive implying wave growth. Two caveats should be noted here. The first is that expression (1) is an approximate form for $\gamma(k)$ that is valid when the damping is weak (i.e., $|\gamma/\omega| \ll 1$). The second more important caveat is that Landau linearized the collisionless Boltzmann equation neglecting a term that is second order in the mode amplitude.

Landau's work served as a model for the theoretical description of many kinds of plasma modes, both electrostatic and electromagnetic, and the Landau resonance for an unmagnetized plasma was generalized to the cyclotron resonance for a magnetized plasma (Stix, 1962, 1992). By the late 1950s, a large body of theory had been developed for the kinetic description of plasma modes. However, there was a concern that the theory had not been tested adequately, so several small scale laboratory experiments were developed to isolate and test the basic elements of the theory. Happily, scientific opportunity and availability of funds converged to make this a "golden era" for such small scale experiments.

Figure 1 shows the results of an experiment that investigated Landau damping for the simple case of Langmuir waves (Malmberg and Wharton, 1966). The plasma was a steady-state (continuously produced and lost) 2-m-long column that was immersed in an axial magnetic field. A Langmuir wave was transmitted continuously by a probe (wire)

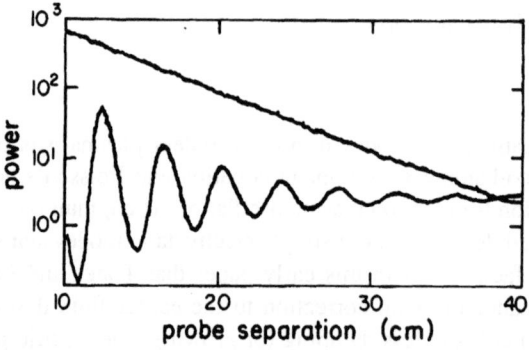

Figure 1. Measurement of spatial Landau damping. The upper curve is the logarithm of the power measured by a receiver probe plotted as a function of the distance from the transmitter probe. The lower curve is the instantaneous wave form, obtained by operating the two-probe system as an interferometer. From Malmberg and Wharton, 1966.

that was inserted into the plasma and made to oscillate in potential. Waves propagated axially in both directions away from the transmitter damping spatially as they propagated. Landau also considered the case of spatial damping, predicting the spatial damping rate (imaginary wave number) $k_i(\omega) = \gamma[k(\omega)]/v_g$, where $\gamma(k)$ is the temporal damping rate and $v_g = d\omega/dk$ is the group velocity. The finite radial size of the column and the large axial magnetic field produce slight changes in the form of $\gamma(k)$ and of $\omega(k)$ relative to the results for an unmagnetized homogeneous plasma, but the changes are technical details, not matters of principle. The upper curve in Fig. 1 is the logarithm of the power measured by a receiver probe, plotted as a function of the distance from the transmitter probe. The oscillatory curve was obtained by operating the two-probe system as an interferometer. The nearly straight line dependence of the upper curve demonstrates that the damping was exponential, and the slope is twice the spatial damping decrement $k_i(\omega)$. The factor of 2 enters because power is proportional to the square of the wave amplitude. The measured decrement was far too large to be accounted for by collisional processes, but was in good agreement with the predictions of Landau's theory. A particularly convincing demonstration was that the damping ceased when the velocity distribution was manipulated to remove the resonant electrons. Also, Landau growth was observed when a warm beam was injected to make $\partial f_0/\partial v_z$ positive at the resonant velocity. In other early experiments, Landau's theory was tested using ion acoustic waves (Chen, 1984). By now, predictions of damping and growth due to the Landau and cyclotron resonances have been verified for many modes in a wide range of experimental settings.

Nonlinear theory

We will consider three nonlinear extensions of Landau damping, discussing in each case the original work from the 1960s and then a modern incarnation (or application) of that work. The first two extensions illustrate complementary physical interpretations of Landau damping: damping as a result of energy exchange with electrons that "surf" on the wave field and damping of the wave field because of phase mixing in velocity space.

Trapped particle oscillations and the plasma wave accelerator

Consider a resonant electron that is trapped in the trough of a large amplitude Langmuir wave and is accelerated forward as it slides down one side of the wave trough. The electron gains energy by "surfing" on the wave. However, when the electron reaches the bottom of the trough and decelerates as it moves up the other side, it loses energy. Thus, we expect the wave damping decrement, $\gamma(t)$, to oscillate in time at the frequency of oscillation of an electron that is trapped in the trough of the wave, $\omega_{osc} = (ek^2\delta\phi/m)^{1/2}$. Here, $\delta\phi$ is the amplitude of the wave potential. Formally, one can check that Landau's linearization procedure fails after a time ω_{osc}^{-1}. For a small amplitude wave (i.e., $\omega_{osc} \ll |\gamma|$), the wave damps away long before the trapped electrons can complete an oscillation, so Landau's theory is valid. In the opposite limit of a very large amplitude wave (i.e., $\omega_{osc} \gg |\gamma|$), the trapping oscillations stop the damping before the wave amplitude can change by a significant amount (Mazitov, 1965; O'Neil, 1965). In general trapped particle oscillations have been found to dominate the nonlinear wave-particle interaction in many situations. Most importantly, the Landau growth of a single wave (or narrow spectrum of waves) saturates nonlinearly when the amplitude is large enough that $\omega_{osc} \sim |\gamma|$ (Drummond et al., 1970; Onishchenko et al., 1970).

Figure 2 shows measurements of spatial Landau damping when the transmitter power was turned up until trapped particle oscillations dominated the evolution (curve C) (Malmberg and Wharton, 1967). For spatial damping, the oscillations occur spatially and are characterized by the wave number $k_{osc} = \omega_{osc}/v_{ph}$, since $v_{ph} = \omega/k$ is the speed of the resonant electrons. The measured value of k_{osc} scaled with wave amplitude as $\sqrt{\delta\phi}$, as expected.

A modern version of this experiment is the plasma wave accelerator (Tajima and Dawson, 1979). Conventional accelerators are limited to acceleration rates of 100 MeV/m, the limit where radio frequency breakdown occurs. The longitudinal electric field of a plasma wave can be much larger than this, and the phase velocity can be relativistic ($v_{ph} \simeq c$). In principle, trapped bunches of electrons can be accelerated resonantly to high energy in a relatively short distance, so there is the promise of a compact accelerator. In recent experiments (Everett et al., 1995), a large-amplitude, relativistically propagating plasma wave was generated by beats between two co-propagating laser fields of slightly different frequency (i.e., $\Delta\omega \simeq \omega_p$). Trapping of externally injected electrons and an acceleration rate of 2.8 GeV/m were demonstrated. For these small scale experiments, the interaction length was only a cm, so the energy gain was modest (28 MeV).

Plasma wave echoes and beam echoes

The plasma wave echo (Gould, O'Neil, and Malmberg, 1967) is another nonlinear extension of Landau's theory. The echo explicitly demonstrates that the free energy associated with the wave is not dissipated in collisionless damping, but is stored in the distribution function and can reappear later as a wave electric field. The plasma wave echo is closely related to other echo phenomena such as the spin echo. Landau's analysis shows that macroscopic quantities such as the electric field or charge density damp away, but that the perturbation in the distribution function, $\delta f(z, v_z, t)$, oscillates indefinitely. Since the perturbed electron density is given by $\delta n(z,t) = \int dv_z \delta f(z, v_z, t)$, one may think of Landau damping as a phase mixing of different parts of the distribution function. When an electric field of spatial dependence $\exp(-ik_1 z)$ is excited and then Landau damps away, it modulates the distribution function leaving a perturbation of the form $\delta f = f_1(v_z)\exp[-ik_1 z + ik_1 v_z t]$. This perturbation propagates at the local streaming velocity in phase space, v_z. For large t, there is no electric field associated with the perturbation since a velocity integral over the perturbation phase mixes to zero. If after a time Δt an electric field of spatial dependence $\exp[ik_2 z]$ is excited and then damps away, it moderates

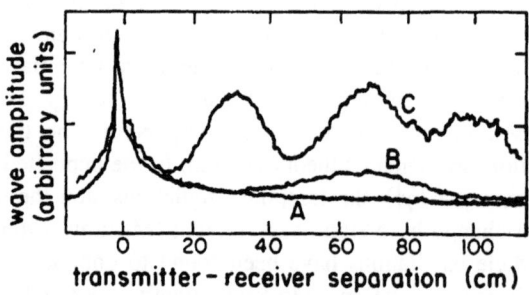

Figure 2. Wave amplitude vs position. The transmitter voltage was 0.9, 2.85, and 9 V for curves A, B, and C, respectively. From Malmberg and Wharton, 1967.

the unperturbed part of the distribution leaving a first-order term $f_2(v_z)\exp[ik_2z-ik_2v_z(t-\Delta t)]$. However, it also modulates the perturbation due to the first field leaving a second-order perturbation of the form

$$\delta f^{(2)} = f_1(v_z)f_2(v_z)\exp[i(k_2-k_1)z+ik_2v_z\Delta t-i(k_2-k_1)v_zt]. \qquad (2)$$

The coefficient of velocity in this exponential vanishes when $t = \Delta t k_2/(k_2-k_1)$, so at this time a velocity integral over this second-order perturbation does not phase mix to zero and an electric field (the echo) reappears in the plasma. This is a temporal echo, but there are also spatial echoes, where the wave fields damp spatially, and the echo is separated spatially.

Soon after they were predicted, spatial echoes were observed experimentally using both Langmuir waves and ion-acoustic waves (Chen, 1984). Also, echoes were used to make very sensitive measurements of small angle scattering due to Coulomb collisions (and to stochastic fields). The echo depends on very fine scale structure in the phase space distribution, which is easily smoothed out by small angle scattering. For example, the plasma column that was used to make the measurements shown in Figs. 1 and 2 was only 2 m long, but Langmuir wave echoes in this plasma were used to measure an effective mean free path of 2 km.

A modern version of this experiment was used recently to measure the energy diffusion rate (due to intra-beam Coulomb collisions) of a coasting antiproton beam in the Fermilab Antiproton Accumulator (Spentzouris, Ostiguy, and Colestock, 1996). Figure 3 shows the signal for a temporal echo on the beam. These echoes are the same as plasma echoes except that the beam particles are relativistic and that collective fields can be ignored in the dynamics. By measuring the decay of the echo amplitude as a function of time to the echo, an intra-beam collision frequency of $(3.0\pm0.8)\times10^{-4}$ Hz was obtained. More recently, an effective collision frequency of 10^{-13} Hz was measured for a higher-energy coasting proton beam at CERN (Brüning et al., 1997). Clearly, the echo provides an exquisitely sensitive measure of small angle scattering.

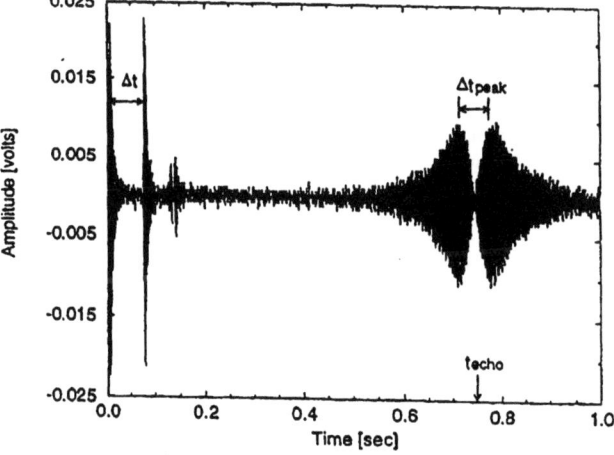

Figure 3. Temporal echo observed on a stored, coasting antiproton beam in the Fermilab Antiproton Accumulator. The two excitations and the echo are shown. From Spentzouris et al., 1996.

Quasilinear theory and current drive

The most widely used of the nonlinear extensions is quasilinear theory (Drummond and Pines, 1962; Vedenov et al., 1962). This physically appealing theory provides a simplified description of the nonlinear wave-particle interaction for the case of a broad spectrum of randomly phased waves. The auto-correlation time for the field as seen by a resonant particle is assumed to be short compared to the time for a trapped particle oscillation. The sign of the field experienced by a resonant particle then undergoes rapid random changes, and the particle experiences a kind of Brownian diffusion in velocity space. For the simple case where all of the waves propagate in the z direction (or a strong magnetic field constrains the particle motion to the z direction), quasilinear theory predicts that the equilibrium velocity distribution evolves according to the diffusion equation

$$\frac{\partial f_0}{\partial t}(v_z,t) = \frac{\partial}{\partial v_z} D(v_z,t) \frac{\partial f_0}{\partial v_z}(v_z,t). \tag{3}$$

In the resonant region, the diffusion coefficient $D(v_z,t)$ is proportional to the energy in the waves that are characterized by phase velocity $\omega/k = v_z$; these waves satisfy a Landau resonance with particles at velocity v_z. Equations (1) and (3) govern the evolution of the wave energy [or, equivalently, of $D(v_z,t)$] and of $f_0(v_z,t)$. This truncated description conserves particle number, momentum, and energy, and is in reasonably good agreement with experiment (Roberson and Gentle, 1971; Hartmann et al., 1995). Quasilinear theory has generated a vast literature, including many applications and many attempts to place the theory on a stronger theoretical foundation. Efforts to understand quasilinear theory from a first principles dynamical perspective helped to motivate early work on the dynamical origins of chaos.

A modern application of the wave-particle interaction, where the quasilinear diffusion equation is used to describe the theory, is rf current drive in tokamaks (Fisch, 1984). As described by Fowler (see article in this volume), a tokamak is a toroidal magnetic confinement device for high temperature plasmas, and is the leading contender to be a fusion reactor. Confinement in a tokamak requires that the plasma carry a toroidal electric current, and in a conventional tokamak this current is driven by an inductive electric field that is directed toroidally. The plasma is the secondary in a transformer circuit where the primary passes through the hole in the torus. Since the magnetic flux in the primary is finite, a substantial inductive electric field can be maintained only for a finite time (about an hour for a reactor scale tokamak). However, there would be technological advantage in the steady-state operation of a tokamak reactor.

In recent years, experiments on tokamaks have demonstrated that the required steady state current can be driven with the wave-particle interaction. Mega-amps of current have been driven by this method in large tokamaks. A phased array of wave guides is used to launch lower hybrid plasma waves so that they propagate in a particular direction around the torus. The waves Landau damp on the tail of the electron distribution, transferring wave momentum to the resonant electrons. In this way, electrons are pulled out from the Maxwellian to produce a high velocity tail (or plateau) in the direction of wave propagation. From a quasilinear perspective, the spectrum of waves diffusively sweeps particles down hill on the Maxwellian distribution forming the high velocity plateau. The current resides in this high velocity plateau. Of course, the steady-state shape of the plateau is determined by a balance between momentum deposition by the waves and collisional drag on the ions and slow electrons. Incidentally, it is advantageous to deposit the momentum in fast electrons since the collisional drag on these electrons is less than that on thermal electrons. When research on current drive was beginning, critics worried that the high rf power levels and the high velocity electrons would produce anomalous (collective) processes and that these would confuse the theory and spoil the efficiency of current drive. However, this has not been the case; traditional theory (e.g. quasilinear theory and the classical collision operator) provides a good description of experimental results over a wide parameter range.

Collisionless Shocks in Space Plasmas

In the late 1950s, the question of whether shocks exist in collisionless plasmas posed a great challenge to the developing discipline of plasma physics. Gas dynamic shocks and magnetohydrodynamic fast and slow shocks form as the steepened limit of a nonlinear compression wave in which the thickness of the shock layer (shock front) is determined by the characteristic dissipation length associated with the particle collision mean free path. Since early laboratory experiments with plasma shocks were partially collisional, the collisionless shock challenge was first met by the satellite study of space plasmas.

The Earth's dipole magnetic field is immersed in the supersonic, super-Alfvenic solar wind whose low density (5 cm^{-3}) and moderate temperature (2×10^5 K) result in a 1 AU mean free path. Nevertheless, plasma physicists speculated that a bow shock would stand in the solar wind upstream of the Earth's magnetosphere. In late 1964, using the magnetometer measurements from NASA's IMP 1 satellite, Ness et al. (1964) unambiguously identified the magnetic compression signature of a thin magnetohydrodynamic fast mode collisionless bow shock; the IMP 1 plasma measurements subsequently confirmed the shock heating and the slowing of the solar wind. Today high Mach number bow shocks have been detected at all the planets that have been visited by spacecraft and around three comets. Although tantalizing evidence of slow magnetohydrodynamic shocks was found in solar wind magnetic structures, the unambiguous detection of a magnetohydrodynamic collisionless slow shock did not come until Feldman et al. (1984) used plasma and magnetic field measurements from NASA's ISEE 3 spacecraft to verify the slow shock Rankine-Hugoniot relations at the plasma sheet boundary in the distant geomagnetic tail.

Early collisionless shock models

For collisionless shocks, the critical question is, What dissipation mechanisms replace particle-particle collisions? Viewed broadly, the answer is the wave-particle interaction discovered by Landau, although a wide variety of different wave modes with different dispersive properties are involved even for a single type of shock. In the late 1950s, Adlam and Allen (1958) and R. Z. Sagdeev (reported in Sagdeev, 1966) showed that the thickness of a steepening nonlinear fast magnetohydrodynamic wave that propagates perpendicular to the magnetic field would be limited by the finite inertia of the plasma electrons at the dispersive scale length c/ω_p, the so-called collisionless skin depth. The balance between nonlinear steepening and dispersion results in a steady compressive soliton that propagates in the ideal fluid plasma without dissipation. With the addition of resistivity, however, the soliton converts into a sharp leading front, in which the magnetic field strength rises to above the downstream Rankine-Hugoniot value, followed by a train of trailing wave oscillations that resistively damp to the downstream Rankine-Hugoniot state. Recall that the Rankine-Hugoniot relations, which are robust consequences of conservation theorems applied across the shock, determine the downstream state in terms of the upstream state independent of the details of the dissipation mechanism in the shock interior. Sagdeev argued that, for a collisionless shock, the resistivity would be provided by ion-acoustic wave turbulence which is self-consistently excited by the strong cross-field current drift of the electrons in the magnetic field ramp at the leading edge of the shock; in quasilinear theory, ion acoustic waves elastically scatter the current-carrying electrons and transfer electron momentum to ions. For oblique fast shocks, ion inertial dispersion speeds up the fast wave, so that the soliton is a rarefaction pulse; with the addition of resistivity, the soliton becomes a leading whistler wave train that propagates into the upstream region ahead of the magnetic ramp, and is spatially damped with increasing upstream distance.

A quite different approach to the collisionless shock dissipation mechanism was proposed by E. Parker and H. E. Petschek. Parker (1961) was interested in high-β plasmas (β is the ratio of the plasma to magnetic pressure). He argued that the shock would consist of interpenetrating upstream and downstream ion beams that would excite the beam-firehose instability, an instability driven by velocity space anisotropy. A similar quasi-parallel shock theory was proposed by Kennel and Sagdeev (1967) in which the shock compression creates a plasma distribution with a higher parallel than perpendicular temperature; this anisotropy excites magnetosonic Alfvén turbulence via the temperature anisotropy version of the firehose instability. They developed this model into a full quasilinear theory of weak, high-β quasiparallel shocks in which the scattering of the upstream ions by the fluctuating wave magnetic fields provides the shock dissipation.

Petschek (1965) was interested in high-Mach-number (supercritical) quasiperpendicular shocks for which resistivity alone (as in the Sagdeev model) cannot provide sufficient dissipation to satisfy the Rankine-Hugoniot relations. In Petschek's shock model, the increased entropy associated with the whistler turbulence replaces the thermal heating of the plasma; plasma heating occurs as the Alfvén waves are gradually absorbed by Landau damping on spatial scales that are much greater than the shock thickness.

Fast shocks

After the discovery of the Earth's bow shock, subsequent NASA spacecraft, especially OGO 5, with its high time resolution magnetometer and the first electrostatic wave measurements, compiled an observational database on collisionless fast shock structure for a wide range of Mach numbers, propagation angles, and plasma betas. In addition, a powerful new diagnostic—the plasma numerical simulation of collisionless shocks—was pioneered at Los Alamos by D. Forslund, and developed into a technique that permitted a detailed comparison of theory and spacecraft observations. Today, although many details remain, a broad outline of collisionless shock dissipation mechanisms has been established.

The observed structure of low-Alfvén-Mach-number ($M_A < 2$) oblique fast shocks consists of the predicted leading whistler wave train. However, the source of the anomalous resistivity that converts the rarefaction soliton into the shock wave train is still uncertain. In the solar wind the electron to ion temperature ratio is too low for the electron drifts associated with the whistler's magnetic field oscillations to destabilize the ion acoustic wave. Other modes, such as the lower hybrid drift wave, could be excited for lower electron drift speeds, or some other process may be responsible for introducing irreversibility into whistler wave trains.

At higher Alfvén Mach numbers, the ISEE 1 and 2 spacecraft typically observed the following: a large magnetic overshoot of a factor of 2 or so above the downstream field strength, very little evidence of an upstream whistler wave train, and a double-peaked downstream ion distribution. These observations were beautifully explained by the plasma simulations of supercritical shocks (Leroy *et al.*, 1982). In addition to the magnetic ramp, supercritical shocks have a net electrostatic potential jump across the shock front. Roughly, due to their large inertia, the ions plough through the sharp magnetic ramp whereas the electrons remain attached to the field lines; thus a charge separation electric field develops to slow the incoming ions. The combined $\mathbf{v} \times \mathbf{B}$ and normal electric field force reflects a fraction of the incoming ions back upstream; the reflected ions then gain energy by drifting parallel to the tangential shock electric field and penetrate through the potential barrier into the downstream region on their next gyro-encounter with the shock front. Downstream, the reflected ion population and the still-unshocked upstream ions form a plasma distribution with a velocity space gyrating beam, which appear to spacecraft as a double-peaked distribution. Subsequent 2D and 3D simulation studies have shown that the beam ions ballistically mix and eventually thermalize with the directly transmitted upstream ions via turbulent interactions with electromagnetic ion cyclotron waves; the reflected ions create a distribution that is more perpendicular than parallel and a temperature that destabilizes the ion cyclotron waves. In the 2D and 3D simulations, ion reflection produces a shock that is steady only in an average sense; the number of reflected ions, the size of the magnetic overshoot, and the thickness of the magnetic ramp actually vary in time and in position along the shock front. Observations at the bow shocks of the outer planets and some simulation studies suggest that very high Mach number ion reflection shocks may be intrinsically and violently unsteady.

For quasiparallel shocks, since the downstream flow velocity is less than the speed of sound, the shocked ions can readily travel back upstream along the magnetic field. The interpenetrating inflowing and backstreaming ions create a beam-firehose distribution (as envisioned by Parker), but the most unstable waves are actually fast magnetosonic modes (Krauss-Varban and Omidi, 1993). For quasiparallel terrestrial bow shocks, the amplitudes of the observed magnetic field fluctuations can be comparable to the DC field, an example of order-one magnetic turbulence.

Cometary bow shocks

Near perihelion cometary nuclei emit large fluxes of neutral hydrogen and water group molecules that are then ionized by the solar UV, forming a large halo around the comet. Unless the solar wind velocity is exactly parallel to the interplanetary magnetic field, a newborn ion is first accelerated by the solar wind $\mathbf{v} \times \mathbf{B}$ electric field, thereby forming a cold gyrating beam in velocity space. The energy and momentum of this ion pick-up process must come at the expense of the solar wind flow. Thus the pick-up ions inertially load and slow the solar wind, thereby forcing a bow shock to form around the comet. Since the interaction is collisionless, the coupling between the pick-up ions and the solar wind is mitigated by the unstable excitation of low-frequency hydromagnetic waves that scatter the pick-up ions in pitch angle and energy and extract momentum from the solar wind. The spacecraft that encountered comets Giacobini-Zinner, Halley, and Grigg-Skjellerup observed large-amplitude magnetic turbulence both upstream and downstream of the cometary bow shocks.

Cosmic ray acceleration by supernova shocks

The galactic component of cosmic rays extends from about 1 GeV to at least 10^6 GeV with a momentum distribution that decreases as a simple power law. The maintenance of the cosmic rays against escape losses from the galaxy requires an energy input of roughly 10^{41} ergs/sec whose only known source is blast waves from supernova explosions. The problem of how shock energy can be efficiently converted into very high energy particles was solved independently by G. F. Krimskii (1977), I. W. Axford *et al.* (1977), and R. D. Blandford and J. Ostriker (1978). They recognized that a quasiparallel collisionless shock could establish the physical conditions for a particle to undergo type-I Fermi acceleration.

In a quasiparallel shock, suprathermal particles (seed cosmic rays) can rather freely travel upstream along the magnetic field. As a streaming or beamlike distribution, these particles excite parallel propagating magnetosonic waves that scatter the streaming particles in pitch angle. In the shock frame, the unstable waves, which travel at the Alfvén speed relative to the plasma at rest, are convected into the shock by the upstream flow. Once the streaming particles are scattered through 90° in pitch angle, they also travel back to the shock and into the downstream flow region where they can again be scattered in pitch angle. As a net result of these wave-particle scattering interactions, the particles diffuse back and forth across the shock. Since the downstream flow speed is less (typically 1/4 for a strong shock) than the upstream flow speed, to the particles, the wave scattering centers appear to converge; thus the particles are effectively trapped between converging mirrors and experience type-I Fermi acceleration. Rather amazingly, the steady-state momentum distribution that the particles acquire from the shock acceleration process is a power law whose spectral index is independent of the pitch angle or spatial diffusion coefficient (i.e., the intensity of the unstable waves), and only depends on the shock compression ratio. The predicted power spectral index closely agrees with the observed cosmic-ray spectrum from 1 GeV to about 10^6 GeV.

Since the original model was proposed, the theory of cosmic-ray accelerating shocks has been greatly elaborated and refined. In particular, the energy density of the accelerated cosmic rays can evolve to become comparable to the flow energy associated with the shock; thus the cosmic rays become part of the overall shock structure, so that supernova shocks actually extend over vast distances. The acceleration theory was extended to solar flare blast waves traveling in the interplanetary medium by M. A. Lee (1983), and was thoroughly and successfully tested by Kennel *et al.* (1984) in the spacecraft study of a particle-accelerating interplanetary shock. Finally, shock acceleration may also explain the production of relativistic electrons in synchrotron extragalactic radio jets. Today, the shock acceleration of cosmic rays stands as, perhaps, the one major successful application of collisionless plasma physics to astrophysics.

References

Adlam, J. H., and J. E. Allen, 1958, Philos. Mag. **3**, 448.

Axford, W. I., E. Leer, and G. Skadron, 1977, in Proceedings of the 15th International Conference on Cosmic Rays, Plovidiv, Bulgaria, Vol. 11, p. 132.

Blandford, R. D. and J. P. Ostiker, 1978, Astrophys. J. **221**, 129.

Brüning, O., T. Linnecar, F. Ruggio, W. Scandale, and E. Shaposhnikova, 1997, *Nonlinear and Collective Phenomena in Beams Physics*, edited by S. Chattopadhyay, M. Cornacchia, and C. Pellegrini (AIP, New York), p. 155.

Chen, F. F., 1984, *Introduction to Plasma Physics and Controlled Fusion*, Second Ed. (Plenum Press, New York).
Drummond, W. E., J. H. Malmberg, T. M. O'Neil, and J. R. Thompson, 1970, Phys. Fluids **13**, 2422.
Drummond, W. E., and D. Pines, 1962, Nucl. Fusion Suppl. **3**, 1049.
Everett, M., *et al.*, 1994, Nature (London) **368**, 527; see also Madena, A., *et al.*, 1995, Nature (London) **377**, 606.
Feldman, W. C., S. J. Schwartz, S. J. Bame, D. N. Baker, J. Birn, J. T. Gosling, E. W. Hones, D. J. McComas, J. A. Slavin, E. J. Smith, and R. D. Zwickl, 1984, Geophys. Res. Lett. **11**, 599.
Fisch, N.J., 1984, Rev. Mod. Phys. **59**, 175.
Gould, R. W., T. M. O'Neil, and J. H. Malmberg, 1967, Phys. Rev. Lett. **19**, 219.
Hartmann, D. A., C. F. Driscoll, T. M. O'Neil, and V. Shapiro, 1995, Phys. Plasmas **2**, 654.
Kennel, C. F., *et al.*, 1984, J. Geophys. Res. **89**, 5419.
Kennel, C. F., and R. Z. Sagdeev, 1967, J. Geophys. Res. **72**, 3303.
Krauss-Varban, D., and N. Omidi, 1993, Geophys. Res. Lett. **20**, 1007.
Krimskii, G. F., 1977, Dokl. Akad. Nauk SSSR **234**, 1306.
Landau, L., 1946, J. Phys. (U.S.S.R.) **10**, 45.
Lee, M. A., 1983, J. Geophys. Res. **88**, 6109.
Leroy, M. M., D. Winske, C. C. Goodrich, C. S. Wu, and K. Papadopoulos, 1982, J. Geophys. Res. **87**, 5081.
Malmberg, J. H., and C. B. Wharton, 1966, Phys. Rev. Lett. **17**, 175.
Malmberg, J. H., and C. B. Wharton, 1967, Phys. Rev. Lett. **19**, 775.
Mazitov, R. K., 1965, Zh. Prikl. Mekhan. Tekh. Fiz. **1**, 27.
Ness, N. F., C. S. Scearce, and J. B. Seek, 1964, J. Geophys. Res. **69**, 3531.
O'Neil, T. M., 1965, Phys. Fluids **8**, 2255.
Onishchenko, I. N., A. R. Lineskii, M. G. Matsiboroko, V. D. Shapiro, and V. I. Shevchenko, 1970, JETP Lett. **12**, 218.
Parker, E. N., 1961, J. Nucl. Energy, Part C **2**, 146.
Petschek, H. E., 1965, *Plasma Physics* (Intl. Atomic Energy Agency, Vienna), p. 567.
Roberson, C., and K. W. Gentle, 1971, Phys. Fluids **14**, 2462.
Sagdeev, R. Z., 1966, Rev. Plasma Phys. **4**, 23.
Spentzouris, L. K., J. F. Ostiguy, and P. L. Colestock, 1996, Phys. Rev. Lett. **76**, 620.
Stix, T., 1962, *The Theory of Plasma Waves* (McGraw-Hill, New York).
Stix, T., 1992, *Waves in Plasmas* (AIP, New York).
Tajima, T., and J. M. Dawson, 1979, Phys. Rev. Lett. **43**, 267.
Tonks, L., and I. Langmuir, 1929, Phys. Rev. **33**, 195.
Vedenov, A. A., E. P. Velikhov, and R. Z. Sagdeev, 1962, Nucl. Fusion Suppl. **2**, 465.

*Chemical Physics and
Biological Physics*

Chemical Physics: Molecular Clouds, Clusters, and Corrals

Dudley Herschbach

In its modern incarnation, chemical physics as a field is generally regarded as having been born in 1933, along with the *The Journal of Chemical Physics*. Its first editor, H. C. Urey, declared that "the boundary...has been completely bridged...chemists and physicists have become equally serious students of atoms and molecules" (Urey, 1933). Among other evangelical founders were P. Debye, H. Eyring, G. B. Kistiakowsky, I. Langmuir, G. N. Lewis, L. Pauling, K. S. Pitzer, J. C. Slater, J. H. Van Vleck, and E. B. Wilson, Jr. Actually, Urey's bridge was still rickety and had to stretch over a wide cultural gulf (Nye, 1993).[1] A major impetus for the new journal was the fact that *The Journal of Physical Chemistry* refused to accept any purely theoretical paper (and continued to do so for another two decades). By 1939, however, Slater had published his *Introduction to Chemical Physics*, and by 1942 Wilson and Van Vleck had established at Harvard the first Ph.D. program in chemical physics. Over the next 50 years, means of elucidating molecular structure and dynamics developed enormously, by virtue of pervasive applications of quantum theory and experimental tools provided by physics, especially myriad spectroscopic methods.

Chemical physics and physical chemistry no longer differ appreciably, either in research journals or in academic programs, now that computers and lasers have become ubiquitous. Yet creative tension persists at the interface of chemistry and physics. The chemist wants above all to understand why one substance behaves differently from another; the physicist wants to find disembodied principles that transcend the specific substances. A chemical physicist thus sometimes stands awkwardly astride a widening intellectual abyss. However, often the duality of outlook provokes invigorating perspectives. This article takes a brief look at three frontier areas that offer such perspectives. Necessarily, these vignettes are idiosyncratic and impressionistic; only a few leading references and reviews can be cited, and many other fruitful areas are left out altogether. The chief aim is to exemplify the characteristic eclectic style of chemical physics,

---

[1] Nye concludes that chemistry, as a discipline, preceded and aided the establishment of physics as an academic and laboratory discipline. She points out that the term "chemical physics" often appears in the titles and chapters of textbooks in the latter half of the 19th century. These treated heat, light, and electricity as chemical agents, topics regarded as prefatory to the core of chemistry dealing with properties and reactions of inorganic and organic substances.

coupling theory and experiment, probing structural and dynamical aspects, and ranging from *ab initio* rigor to heuristic extrapolation.

Chemistry in Interstellar Molecular Clouds

Over the past 30 years, radioastronomy has revealed a rich variety of molecular species in the interstellar medium of our galaxy and even others. Well over 100 molecules have now been identified in the interstellar gas or in circumstellar shells (Thaddeus et al., 1998). These include H_2, OH, H_2O, NH_3, and a few other small inorganic species, but most are organic molecules, many with sizable carbon chains involving double or triple bonds. To appreciate how surprising this proliferation of organic molecules is, we need to review some aspects of the interstellar environment.

Uniform radiation, multiform chemistry

As early as 1941, optical absorption lines of the CN molecule were observed in an interstellar cloud that fronted a bright star, which served as the light source. The intensity ratio of lines originating from the ground level and first excited rotational levels provided the first evidence for the 3 °K cosmic background radiation, although not recognized as such until 25 years later (Thaddeus, 1972). This background contains roughly 99% of the electromagnetic energy in the known universe. It is now established as isotropic, blackbody radiation and attributed to a frigid whimper of radiation, still in thermal equilibrium, from a primordial inferno, the Big Bang.

The cosmic abundance of the elements is drastically nonuniform. Hydrogen comprises over 92%, helium over 6%; next come oxygen at 0.07%, carbon at 0.04%, and nitrogen at 0.009%. The average density within our galaxy is only about one H atom per cubic centimeter. Yet the density is about a hundredfold higher in what are called diffuse interstellar clouds and up to a millionfold higher in dark clouds. In diffuse clouds, such as those in which the CN rotational states were found to be in thermal equilibrium with the 3 °K background radiation, a molecule collides with another (H_2 or He) only once every two months or so. This led, in the early days of radioastronomy, to the expectation that emission spectra from rotational levels of any polar molecules in the interstellar medium would be unobservable, because collisions were much too infrequent to alter the rotational temperature.

Such pessimistic anticipation was dispelled by the discovery of molecular rotational emissions from dark clouds (Rank et al., 1971). This discovery showed that the gas density in many interstellar regions was actually high enough to enable the population of rotational levels to be governed more strongly by collisions than by the background radiation. However, the observed molecular abundances departed enormously from estimates derived by assuming chemical equilibrium. For instance, next to H_2, carbon monoxide is the most abundant interstellar molecule (although typically down by a factor of 10^{-4} or more). But thermodynamic calculations predict that under typical dark cloud conditions (20 °K, density of $H_2 \sim 10^5$ cm^{-3}) at chemical equilibrium there would be fewer than one CO molecule in the volume (10^{84} cm^3) of the observable universe. Likewise, the prevalence of organic molecules containing many carbon atoms and relatively little hydrogen is inexplicable by thermodynamics.

Synthesis of interstellar molecules

This situation led Klemperer (1995, 1997) to propose a nonequilibrium kinetic scheme for the synthesis of interstellar molecules, to show how "chemistry can, in the absence of biological direction, achieve complexity and specificity." The scheme invokes sequences of exoergic, bimolecular ion-molecule reactions. Extensive laboratory experiments have shown that these processes are typically quite facile and uninhibited by activation energy barriers, unlike most gas-phase chemical reactions not involving ions. Uninhibited reactions in two-body collisions are the only plausible candidates for gas-phase chemistry at the low density and temperature of an interstellar cloud. The clouds also contain dust particles of unknown composition. Formation of hydrogen molecules from atoms is probably catalyzed on the surface of dust particles, but the host of other molecules seem more likely to be produced by nonequilibrium gas-phase kinetics.

The dark clouds where most interstellar molecules have been seen are immense, typically comprised of hydrogen and helium with a million times the mass of our Sun. In our galaxy such clouds loom as huge dark blotches obscuring regions of the Milky Way. Ionization by the pervasive flux of 100-MeV cosmic rays seeds the clouds with a little H_2^+ and He^+ (about one ion per 500 cm^3), from which sprout many reaction sequences.

The H_2^+ rapidly reacts with H_2 to form H_3^+ which, as known from laboratory studies, itself readily transfers a proton to many other molecular species. Most of the H_3^+ is converted to HCO^+, a very stable species. This prediction was a triumph for Klemperer's model. Soon thereafter interstellar emission from a species dubbed Xogen, which had not yet been seen on earth, was shown to come from the HCO^+ ion. It has proved to be the most abundant ion in dark clouds and has even been observed in several distant galaxies.

Much else offers support for the kinetic model. For instance, proton transfer from H_3^+ to nonpolar molecules such as N_2 and CO_2 converts them to polar species HN_2^+ and $HOCO^+$, which are capable of emitting rotational spectra. Again, laboratory observation of these spectra (Saykally and Woods, 1981) confirmed the detection of interstellar emissions from these species.

Most striking are offspring of the He^+ ions, which exemplify how chemical kinetics can produce paradoxical results. The extraction by He^+ of a hydrogen atom from H_2, the most abundant molecule in interstellar clouds, would be very exoergic. Yet, for reasons described below, that reaction does not occur. Instead, He^+ reacts with CO, the second most abundant molecule, to form C^+ and O. The ionization of helium is almost quantitatively transferred to C^+, enhancing its concentration a thousandfold (by the He/CO abundance ratio). In turn, the C^+ ion reacts only feebly with H_2 (via radiative association), but reacts avidly with methane, CH_4, and acetylene, C_2H_2, to launch sequences that build up many organic compounds, including chains punctuated with double and triple bonds. The paradoxical irony is that the mutual distaste of the simplest inorganic species, He^+ and H_2, gives rise to the proliferation of complex organic molecules in the cold interstellar clouds.

Electronic structure and reaction specificity

The three-electron system involving only helium and two hydrogen atoms offers a prototypical example for interpretation of chemical dynamics in terms of electronic structure (Mahan, 1975). As shown in ion-beam scattering experiments, the reaction $He + H_2^+ \rightarrow HeH^+ + H$ is endoergic by 0.8 eV, but occurs readily if at least that amount of energy is supplied, either as relative kinetic energy of the collision partners or as vibrational excitation of H_2^+. In contrast, the reaction $He^+ + H_2 \rightarrow HeH^+ + H$ is exoergic by 8.3 eV, but appears not to occur at all; the less exoergic pathway to form $He + H^+ + H$ has been observed, but its reaction rate is four orders of magnitude smaller than for comparable exoergic ion-molecule reactions.

Figure 1 provides an explanation, due to Mahan (1975), for the drastic difference in reactivity of $He + H_2^+$ and $He^+ + H_2$. Plotted are diatomic potential-energy curves for the reactants and products; these represent cuts through the triatomic potential-energy surfaces in the asymptotic entrance and exit channels. Consider first the lowest-lying trio of separated atoms, $He + H^+ + H$. Since both the reactants $He + H_2^+$ and the products $HeH^+ + H$ correlate adiabatically to $He + H^+ + H$, the reaction can be expected to proceed on a single triatomic potential-energy surface.

However, for the upper trio of atoms, $He^+ + H + H$, this does not hold. The ground-state H_2 diatomic potential curve; $^1\Sigma_g^+$, which arises from bringing together two H atoms with antiparallel spins, represents a cut in the asymptotic reactant region through the potential surface for $He^+ + H_2$ collisions. The corresponding cut in the product region, generated by bringing together $He^+ + H$, yields an excited singlet state that is totally repulsive, according to electronic structure calculations. Likewise, the accompanying excited triplet state with parallel spins is at best only very weakly bound. Accordingly, colliding $He^+ + H_2$ is very unlikely to form a stable HeH^+ molecule.

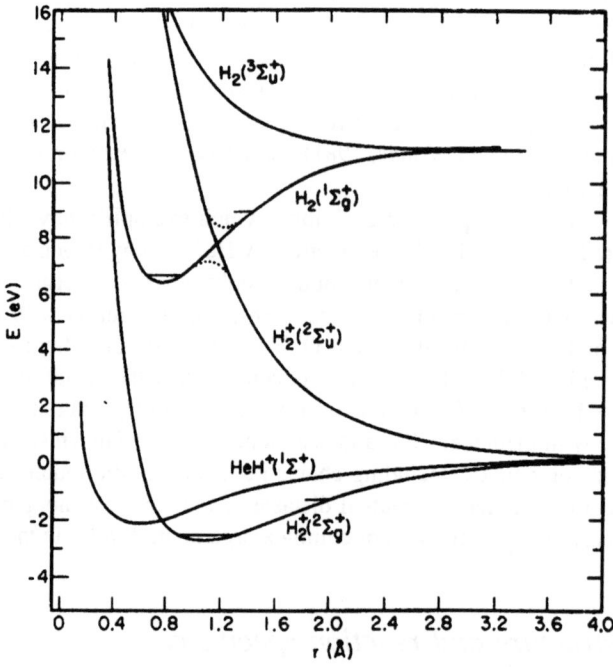

Figure 1. Potential-energy curves for the diatoms in the asymptotic reactant and product regions of the $(He-H_2)^+$ system. Since the energies of He and He^+ are included, here the ground $^1\Sigma_g^+$ state of H_2 lies above the states of H_2^+. Note the crossing of the curves for H_2 ($^1\Sigma_g^+$) and H_2^+ ($^2\Sigma_u^+$), which occurs in the reactant region of $He^+ + H_2$, but which becomes an avoided intersection (indicated by dashes) when all three atoms are close to each other. From Mahan, 1975.

As can be seen in Fig. 1, in the asymptotic reactant region, the ground-state H_2 ($^1\Sigma_g{}^+$) potential curve crosses at about 1.1 Å that for H_2^+ ($^2\Sigma_u{}^+$), a strongly repulsive state. When He^+ approaches H_2, however, the resulting interaction induces these states to mix, as both then acquire the same symmetry (A' under the C_S point group). The crossing thus evolves into an avoided intersection (indicated schematically by dashes). If electron transfer occurs, the adiabatically formed products initially are $He+H_2^+$ ($^2\Sigma_u{}^+$), which dissociate directly to $He+H^++H$.

Exemplary of chemical physics, the profusion of organic molecules tumbling in the heavens is linked to devilish details that govern an electron hopping between helium and hydrogen.

Molecular Clusters, Superstrong or Superfluid

The properties and interactions of molecules are often much influenced by the company they keep. Molecular clusters, generated by supersonic expansion of gas into a vacuum apparatus, have now become a favorite medium for the study of reactions and spectra. Such clusters, composed of from two to up to a billion molecules, offer means for interpolating between gaseous and condensed phases or solution chemistry (Castleman and Bowen, 1996). Before considering two examples from this cornucopian field, we describe the key experimental tool that made it possible, the supersonic nozzle.

The versatile supersonic beam

The canonical physics literature on molecular beams, going back to Otto Stern and I. I. Rabi, stressed that the pressure within the source chamber should be kept low enough so that molecules, as they emerged from the exit orifice, did not collide with each other. In this realm of effusive or molecular flow, the emergent beam provides a true random sample of the gas within the source, undistorted by collisions. Chemical physicists, in desperate need of intensity for studies of reactions in crossed beams, violated the canonical ideal by using much higher source pressures. Collisions within the orifice then produced hydrodynamic, supersonic flow. This realm, avidly explored by chemical engineers (Fenn, 1996), proved to offer many advantages.

When a gas expands isentropically into a vacuum through a pinhole nozzle, the pressure and temperature both drop abruptly. The nozzle imposes collisional communication that brings the gas molecules to nearly the same direction and velocity. It also efficiently relaxes thermal excitation of molecular rotation and (less so) vibration. Thus not only is the intensity of a supersonic beam far higher than that from an effusive source, but the spreads in velocity and rotational states are markedly narrowed. The effective temperature for relative motion of molecules within such a beam is typically only a few °K. Moreover, by seeding heavy molecules in a large excess of light diluent gas, one can accelerate the heavy molecules to the exit velocity of the light gas. Translational energies much higher than are feasible with an effusive source can thereby be obtained, up to a few eV.

Within the markedly nonequilibrium environment of a supersonic expansion, chemical interactions are liberated from thermodynamic constraints. In effect, in the free energy, $\Delta H - T\Delta S$, the entropy term ΔS is suppressed by the low internal temperature within the beam. Even a weakly favorable enthalpy term ΔH can then suffice to produce large yields of molecular clusters.

Balls and tubes of carbon

The discovery of carbon-60 and kindred fullerene molecules ranks among the most important achievements of chemical physics (Dresselhaus et al., 1996; Baum, 1997). It also affirms the value of fostering eclectic collaborations and the playful pursuit of curious observations. The crucial ingredient was a technique, devised by Richard Smalley, to generate clusters from solid samples. This procedure uses a laser to vaporize material, enabling it to be entrained in a supersonic gas flow. In the early 1980s, several laboratories had adopted this technique, chiefly to study clusters of metals or semiconductor materials, of interest for catalysis or microelectronics. Among many curious results were features of a mass spectrum of carbon clusters from laser-vaporized graphite, published by a group at the Exxon laboratory as part of an Edisonian survey (Rohlfing et al., 1984). For C_{40} and larger clusters, only those with an even number of carbon atoms appeared, in a broad distribution extending above C_{100}. Especially prominent in the mass spectrum was the C_{60} peak, about twice as tall as its neighbors.

What is now justly regarded as the discovery of C_{60} did not come until nearly a year later. Smalley's group at Rice University was visited by Harry Kroto from Sussex, who had long pursued work on carbon-containing interstellar molecules. Kroto wanted to examine carbon clusters and their reactions with other molecules, in hopes of identifying candidates for unassigned interstellar spectra. Smalley was reluctant to interrupt other work, particularly since vaporizing carbon would make the apparatus very dirty. Fortunately, hospitality and willing graduate students prevailed. On repeating the Exxon work, the Rice group found that, when conditions were varied, the C_{60} peak became far more prominent. That result led them to play with models and propose as an explanation the celebrated soccer-ball structure, dubbed Buckminsterfullerene. It contains 12 pentagonal and 20 hexagonal carbon rings, with all 60 atoms symmetrically equivalent and linked to three neighbors by two single bonds and one double bond. Soon other fullerene cage molecules were recognized, differing from C_{60} by the addition or subtraction of hexagonal rings, in accord with a theorem proved in the 18th century by Euler.

These elegant structures, postulated to account for cluster mass spectra, remained unconfirmed for five years. Then, in 1990, it was not chemists but astrophysicists who found a way to extract C_{60} in quantity from soot produced in an electric arc discharge. As well as enabling structural proofs, that discovery opened up to synthetic chemistry and materials science a vast new domain of molecular structures, built with a form of carbon that has 60 valences rather than just four. It is striking, however, that despite the great stability of C_{60} and its self-assembly after laser ablation or arc discharge of graphite, as yet all efforts to synthesize C_{60} by conventional chemical means have failed. Such means, which operate under thermodynamic equilibrium conditions, evidently cannot access facile reaction pathways.

Among the burgeoning families akin to fullerene molecules are carbon nanotubes, first discovered by Sumio Iijima at NEC Fundamental Research Laboratories in Tsukuba, Japan. Particularly intriguing is a single-walled nanotube (designated 10,10) of the same diameter as C_{60} (7.1 Å). In principle, its chicken-wire pattern of hexagons can be extended indefinitely; in practice, nanotubes of this kind have now been made that contain millions of carbon atoms in a single molecule. The electrical conductivity of this hollow carbon tube is comparable to copper, and it forms fibers 100 times stronger than steel but with only one-sixth the weight. Already carbon nanotubes have provided much enhanced performance as probe tips in atomic force microscopy. Chemically modifying the nanotube tips has even been shown to create the capability of chemical and biological discrimination at the molecular level, in effect directly reading molecular braille (Wong *et al.*, 1998). A host of other applications is in prospect.

Reactions and spectra in clusters

Much current work examines the effect of solvation on reaction dynamics or spectra by depositing solute reactants or "guests" on a cluster of solvent or "host" molecules, bound by van der Waals forces or by hydrogen bonds (Mestdagh *et al.*, 1997). Often photoinduced reactions, particularly those involving electron or proton transfer, are studied in this way, as are processes involving ion-molecule reactions within clusters (Castleman and Bowen, 1996), including "cage" effects due to the solvent. Among many variants is work in which high-velocity clusters are made to collide with metal or crystal surfaces. Such collisions can induce even guest species that are ordinarily inhibited by a high activation barrier to react (Raz and Levine, 1995). Here we consider a quite different special realm, employing spectra of a guest molecule to study clusters that are finite quantum fluids.

In the prototype experiment, shown in Fig. 2, a supersonic expansion generates large He or Ar clusters, each with 10^3 to 10^5 atoms. In flight these clusters pick up one or more small guest molecules while passing through a gas cell, without suffering appreciable attenuation or deflection. The cluster beam is probed downstream by a laser, coaxial or transverse to the flight path, and spectroscopic transitions of the guest molecule are detected by laser-induced fluorescence or by beam depletion. This pickup technique, originally developed by Giacinto Scoles (Lehmann and Scoles, 1998), is well suited to the study of unstable or highly reactive chemical species. Indeed, these may be synthesized *in*

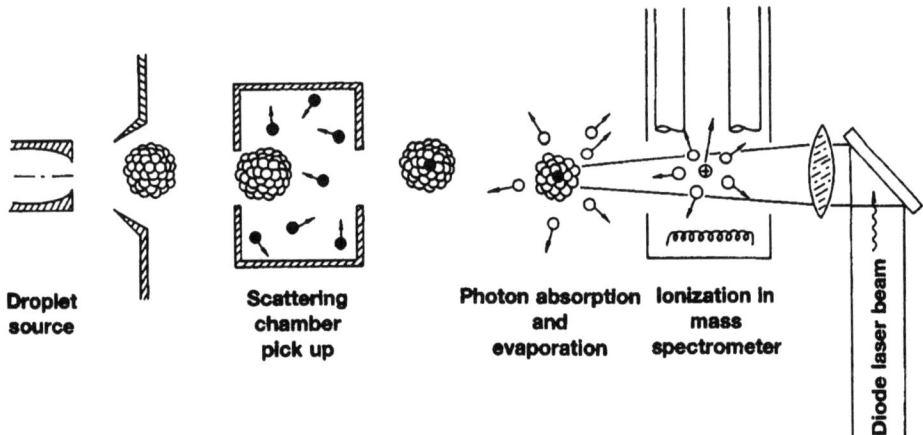

Figure 2. Schematic diagram (omitting vacuum pumps) of molecular-beam apparatus for depletion spectroscopy of molecules embedded in helium clusters. The large clusters or droplets of He, formed in a supersonic nozzle, pick up a guest molecule while passing thorough a scattering chamber. En route to the mass spectrometric detector, the cluster is irradiated by a tunable, coaxial laser. When the laser is in resonance with the guest molecule, the absorbed energy induces evaporation from the cluster and a depletion in the mass spectrometer signal. From Hartmann et al., 1996.

situ by using more than one pickup cell, or by introducing into the cell species generated in a discharge or pyrolytic decomposition. Instead, a stable guest molecule can serve to probe the environment within its solvent cluster. In this way, the group of Peter Toennies at Göttingen (Hartmann *et al.*, 1996; Grebenev *et al.*, 1998) has recently obtained striking results for superfluid helium clusters.

These ^{4}He clusters are produced by a strong supersonic expansion (e.g., He at 5-bar pressure and 6.6 °K behind a 5-μm nozzle), which drops the internal temperature to about 0.4 °K, well below the transition temperature for superfluidity ($T_\lambda = 2.12$ °K). When a hot guest molecule comes aboard (from the pickup cell at 10^{-5} mbar and $\sim$300 °K), the cluster rapidly evaporates away a few hundred He atoms, thereby cooling itself and the guest to the original internal temperature (in about 10^{-6} sec). In the process, the guest molecule also migrates from the surface to the center of the cluster (as deduced from mass spectroscopic experiments and predicted by theory). For a variety of guest species, among them the linear triatomic molecule OCS, the Göttingen group found spectra with well-resolved rotational structure. This indicates the guest molecules are rotating freely within the clusters, although with an effective moment of inertia larger (by a factor of 2.7 for OCS) than that for an isolated, gas-phase molecule.

In normal liquids, rotational structure in spectroscopic transitions is destroyed by diffusional and librational processes; free rotation is seen only for light, weakly interacting molecules such as H_2 or CH_4. Free rotation within the ^{4}He clusters thus can plausibly be attributed to superfluidity, but it might instead result from the exceptionally cool and feeble guest-host interactions, further blurred by the large zero-point oscillations of the helium atoms.

As a diagnostic test for the role of superfluidity, the experiments were repeated using ^{3}He clusters. These have lower density, so are more weakly interacting and somewhat colder (about 0.15 °K), although far above the superfluid range ($T_\lambda = 3 \times 10^{-3}$ °K). Indeed, in ^{3}He clusters the OCS spectrum showed no rotational structure, but rather had the broad, featureless form typical for heavy molecules in normal liquids. A further elegant test was obtained in a series of runs made with increasing amounts of ^{4}He added to the OCS vapor in the pickup cell. Because of their high diffusivity and lower zero-point energy, the ^{4}He atoms entering a ^{3}He cluster gathered around the guest molecule. The average number of such friendly ^{4}He atoms that were picked up was estimated from a Poisson distribution. When this number reached about 60, the rotational structure in the OCS spectrum had again grown in, just as sharp as for pure ^{4}He clusters.

Thus, in the pure, nonsuperfluid ^{3}He clusters, the guest molecule does not rotate freely, but it does so in pure, superfluid ^{4}He clusters or when surrounded by about 60 atoms of ^{4}He, enough to form about two shells around the OCS molecule. Rough estimates suggest the increase in effective moment of inertia may be due chiefly to dragging along the vestigial normal-fluid component of these shells. The Göttingen experiments offer strong evidence that a sharp guest rotational spectrum is diagnostic of superfluidity and it can occur even in ^{4}He clusters with as few as 60 atoms.

Corralling Molecules, Controlling Reactions

Under ordinary experimental conditions, gas molecules careen about in all directions with a broad range of thermal velocities and also tumble erratically with random spatial orientations. Taming that molecular wildness has been a major odyssey of chemical physics, still unfolding. It is part of a modern alchemical quest to exploit molecular dynamics in developing means to control the outcome of chemical reactions. We review some recent advances, including efforts towards achieving spatial trapping of molecules and utilizing lasers as the philosopher's stone.

Pendular orientation and alignment

Although supersonic beams have long served to subdue the translational wildness of molecules, until a few years ago there was no generally applicable technique for constraining the spatial orientation of a molecular axis. Without that capability, major directional features of collisional interactions were averaged out by molecular rotation. The only previous method for producing beams of oriented molecules, developed in the late 1960s, used inhomogeneous electric focusing fields to select intrinsically oriented rotational states in which the molecular axis precessed rather than tumbled. This is an excellent method; it has made possible incisive studies of "head vs tail" reaction probabilities in collisions with both gas molecules and surfaces. However, the method requires an elaborate apparatus and is only applicable to low rotational states of symmetric top molecules (or equivalent) that exhibit a first-order Stark effect.

A different method, much wider in chemical scope and far simpler to implement, exploits the low rotational temperatures attainable in supersonic beams. This method, introduced in 1990, uses a strong homogeneous electric or magnetic field to create oriented or aligned states of polar or paramagnetic molecules (Loesch, 1995; Friedrich and Herschbach, 1996). In the presence of the field, the eigenstates become coherent linear superpositions or hybrids of the field-free rotational states. These hybrids coincide with the familiar Stark or Zeeman states when the dipole and/or the moment of inertia is small or the field is weak; then the molecule continues to tumble like a pinwheel. When the interaction is sufficiently strong, however, the hybrids become librational; then the molecule swings to and fro about the field direction like a pendulum. Such pendular states can be produced for linear or asymmetric rotors as well as for symmetric tops. The magnetic version produces alignment rather than orientation,[2] but is applicable to many molecules not accessible to the electric version; this includes paramagnetic nonpolar molecules and molecular ions (which would just crash into an electrode if subjected to an electric field). Either version requires that the interaction of the molecular dipole with the external field exceed the kinetic energy of tumbling; hence the key role of drastic rotational cooling by a supersonic expansion.

The experimental simplification is major because a focusing field (typically a meter long and expensive to fabricate) is not needed. Instead, the molecular beam is merely sent between the plates of a small condenser (usually about 1 cm^2 in area and a few mm apart) or between the pole pieces of a compact magnet. The uniform field which creates the hybrid eigenstates need only extend over the small region in which the beam actually interacts with its target.

A kindred variety of pendular states can be produced by utilizing the induced dipole moment created by nonresonant interaction of intense laser radiation with the molecular polarizability (Friedrich and Herschbach, 1995a, 1995b). This can produce alignment whether the molecule is polar, paramagnetic, or neither, as long as the polarizability is

---

[2] As usual, here axial anisotropy is designated *orientation* if it behaves like a single-headed arrow and *alignment* if it behaves like a double-headed arrow.

anisotropic. That is generally the case; for example, for a linear molecule the polarizability is typically about twice as large along the axis as transverse to it. Although the electric field of the laser rapidly switches direction, since the interaction with the induced dipole is governed by the square of the field strength, the direction of the aligning force experienced by the molecule remains the same. Experimental evidence for strong alignment arising from the polarizability interaction has been found in nonlinear Raman spectra (Kim and Felker, 1997).

Pendular states make accessible many stereodynamical properties. Studies of steric effects in inelastic collisions or chemical reactions are a chief application (Loesch, 1995). The ability to turn the molecular orientation on or off makes possible modulation of angular distributions and other collision properties, thereby revealing anisotropic interactions not otherwise observable. In photodissociation of oriented molecules (Wu et al., 1994), pendular hybridization renders the laboratory photofragment distributions much more informative. For all applications, the spectroscopy of pendular states has an important role, as the field dependence of suitable transitions reveals the extent of molecular orientation or alignment. Other features arising from the hybrid character of pendular states also prove valuable in spectroscopy, including the ability to tune transitions over a wide frequency range and to access states forbidden by the field-free selection rules.

Towards trapping molecules

The advent of powerful methods for cooling, trapping, and manipulating neutral atoms has led to dramatic achievements, including Bose-Einstein condensation of atomic vapor, an atom laser, atom interferometry, and atom lithography. Molecular physics yearns to follow suit. However, many optical manipulation methods that are effective for atoms fail for molecules because of the complexity of the energy-level structure, with its myriad vibrational and rotational components. Here we merely note some promising approaches to manipulating or trapping molecules, most not yet demonstrated experimentally.

In addition to its utility for molecular alignment, the polarizability interaction with an intense, directional laser field provides a lensing effect acting on the translational motion of molecules. Seideman (1996, 1997) has given a theoretical analysis showing how this arises. The interaction with the field produces molecular states, "high-field seekers," whose energy levels decrease as the field increases. This generates a force that moves the molecule towards the spatial region of highest laser intensity. Thereby focusing occurs, subject to dynamics and dependent on the ratio between the molecular translational kinetic energy and the maximum attractive field-induced potential. In the strong-interaction limit, where that ratio becomes much less than unity, the molecules can become trapped in the laser field (Friedrich and Herschbach, 1995a, 1995b).

An electrostatic storage ring for polar molecules has also been proposed (Katz, 1997), modeled on a neutron storage ring. This would employ an inhomogeneous hexapolar toroidal field, within which molecules in "low-field seeking" states would be confined and follow orbits determined by their rotational state and translational velocity. Design calculations limited to practical parameters indicate that storage lifetimes of the order of 10^3-10^4 s can be expected. However, since the molecular trajectories must bend to stay in the ring, only molecules with low translational kinetic energy can be stored.

Whatever means are used to create an attractive potential region, molecular trapping requires a way to remove enough kinetic energy so that the molecule cannot escape from that region. Since collisions between trapped molecules can redistribute rotational or vibrational excitation into translation, those internal modes need to be quenched also. Photoassociation of trapped atoms can produce trapped diatomic molecules, but must contend with small yields and a strong propensity for vibrational excitation in forming the molecules. Collisional relaxation by means of a cold buffer gas (Doyle et al., 1995) has worked well for loading atoms into a magnetic trap. Recently this technique has achieved a large yield (about 10^8) of trapped CaH molecules (Weinstein et al., 1998). The best buffer gas is ^{3}He; it can be maintained by a dilution refrigerator at about 0.24 °K, where its vapor density is 5×10^{15} cm^{-3}, ample for collisional quenching. Since the helium interaction with any trap potential will be negligibly weak, it can be pumped away after cooling down the molecules. A disadvantage is that the mean lifetime of a molecule in the trap is much shorter than the time required to remove the buffer gas.

A dizzying proposal for disposing of kinetic energy involves mounting a supersonic nozzle on a high-speed rotor, in order to cancel the velocity of the emerging molecules (Herschbach, 1998a). Design calculations indicate that centrifuge action should enhance markedly the supersonic beam quality, thereby shrinking the velocity width and lowering the equivalent temperature for both relative translation and rotation to a few hundred m °K. Preliminary experiments, as yet at modest rotor speeds, confirm that gas can indeed be introduced along the rotor axis and emerge from a whirling arm in a supersonic beam, with the expected velocity subtraction. This approach, if it proves feasible, would avoid cryogenic technology and provide an intense source of molecules deprived of kinetic energy.

If a sufficient density of molecules can be confined, the temperature of the trapped ensemble can be lowered much further by evaporative cooling (Doyle et al., 1991). The aim is to get to the range of a few millidegrees Kelvin or below, where the de Broglie wavelength exceeds the size of the molecule (e.g., $\Lambda_{de\,B} = 20$ Å for Cl_2 at 10 mK). This would give access to an exotic regime for chemical reactivity, governed by quantum tunneling and resonances (Herschbach, 1998b; Forrey et al., 1998). It should be remarked that "trapping," although now a firmly established term, is ridiculously inappropriate. The tightest traps in prospect have linear dimensions at least 10^3 times larger than molecules; that is not a cage, but a roomy corral.

Laser control of reaction pathways

In "real world" chemistry, reaction pathways and yields have to be cajoled or conjured, by adjusting macroscopic conditions (temperature, concentrations, pH) or catalysts. Chemical physicists have sought genuine control of molecular pathways (Zare, 1998). For bimolecular reactions, this has been done by varying the collision energy, orientation, or vibrational excitation of reactant molecules or by selecting particular alignments of excited electronic orbitals of atoms. For unimolecular processes, which we consider briefly, control has been achieved by utilizing the coherence of laser light (Gordon and Rice, 1997). This enables the outcome to be governed by quantum interference arising from the phase difference between alternate routes or by the temporal shape and spectral content of ultrashort light pulses.

The method employing phase control, first proposed by Brumer and Shapiro (1986, 1997), offers a molecular analog of Young's two-slit experiment. An upper state of a molecule is simultaneously excited by two lasers, of frequencies ω_n and ω_m, absorbing n photons of one color and m of the other, with $n\omega_m = m\omega_n$. This prepares a superposition of continuum eigenstates $\Psi_n + \Psi_m$ that correlate asymptotically with different product channels. The coefficients of the components of this superposition state are determined by the relative phases and amplitudes of the two lasers. Since the wavelengths differ markedly, the light waves do not interfere, but the wave functions produced by them strongly interfere. The cross term in $|\Psi_n + \Psi_m|^2$ thus governs the product distribution, and the outcome can be controlled by varying the relative phases and amplitudes of the lasers. This scheme has received several experimental demonstrations in which product yields from two competing channels exhibit large modulations, 180° out of phase.

Another method, operating in the time domain, was introduced by Tannor and Rice (1985). This employs a sequence of ultrashort light pulses to create a wave packet of molecular eigenstates. The frequencies, amplitudes, and phases of the pulse sequence are tailored to favor a particular outcome as the wave packet evolves in time. By means of optimal control theory, the pulse shape and spectral content can be adjusted to maximize the yield of a desired product (Kosloff et al., 1989; Peirce et al., 1990). Moreover, with iterative feedback, the experimenter can be taught by the molecule how best to tailor the light pulses (Judson and Rabitz, 1992). The first reports of such molecular instruction, aided by a computer-controlled pulse shaper, have recently appeared (Bardeen et al., 1997; Assion et al., 1998). These may portend an era in which chemical physicists, like chess masters, learn of many subtle moves beyond their imagination.

Benediction

This quick glance at a bulging family album has pointed to just a few snapshots. Anyone who looks at *Advances in Chemical Physics* or *Annual Reviews of Physical Chemistry* or the *Faraday Discussions* or many other journals will quickly learn of remarkable work in reaction dynamics, femtosecond chemistry, single-molecule spectroscopy, surface chemistry, phase transitions, protein dynamics, electronic structure theory, and a host of other vigorous domains of chemical physics. Having left unmentioned or implicit so much landmark work, I can only hope for a holographic effect, wherein even fragments convey something of the enterprising spirit of the field. In my own experience, it has been exhilarating over nearly five decades to witness what has happened on Urey's bridge. Another sentence from his preface of 65 years ago (Urey, 1993) seems apt as an abiding creed or benediction for chemical physics: "New and effective methods, experimental and theoretical, for the study of these units from which massive matter is composed, have developed largely from physical discoveries which at the time did not appear to have the importance to centuries-old chemical problems that they have since assumed."

References

Assion, A., T. Baumert, M. Bergt, T. Brixner, B. Kiefer, V. Seyfried, M. Strehle, and G. Gerber, 1998, Science **282**, 919.
Bardeen, C. J., V. V. Yakovlev, K. R. Wilson, S. D. Carpenter, P. M. Weber, and W. S. Warren, 1997, Chem. Phys. Lett. **289**, 151.
Baum, R., 1997, Chem. Eng. News **75**, 29.
Brumer, P., and M. Shapiro, 1986, Chem. Phys. Lett. **126**, 541.
Brumer, P., and M. Shapiro, 1997, J. Chem. Soc., Faraday Trans. **93**, 1263.
Castleman, A. W., and K. H. Bowen, 1996, J. Phys. Chem. **100**, 12911.
Doyle, J. M., J. C. Sandberg, I. A. Yu, C. L. Cesar, D. Kleppner, and T. J. Greytak, 1991, Physica B **194**, 13.

Doyle, J. M., B. Friedrich, J. Kim, and D. Patterson, 1995, Phys. Rev. A **52**, R2525.
Dresselhaus, M. S., G. Dresselhaus, and P. C. Eklund, 1996, *Science of Fullerenes and Carbon Nanotubes* (Academic, New York).
Fenn, J. B., 1996, Annu. Rev. Phys. Chem. **47**, 1.
Forrey, R. C., N. Balakrishnan, V. Kharchenko, and A. Dalgarno, 1998, Phys. Rev. A **58**, R2645.
Friedrich, B., and D. Herschbach, 1995a, Phys. Rev. Lett. **74**, 4623.
Friedrich, B., and D. Herschbach, 1995b, J. Phys. Chem. **99**, 15686.
Friedrich, B., and D. Herschbach, 1996, Int. Rev. Phys. Chem. **15**, 325.
Gordon, R. J., and S. A. Rice, 1997, Annu. Rev. Phys. Chem. **48**, 601.
Grebenev, S., J. P. Toennies, and A. F. Vilesov, 1998, Science **279**, 2083.
Hartmann, M., R. E. Miller, J. P. Toennies, and A. F. Vilesov, 1996, Science **272**, 1631.
Herschbach, D., 1998a, Nucleus (Boston) **76**(10), 37.
Herschbach, D., 1998b, in *Chemical Research—2000 and Beyond: Challenges and Visions*, edited by P. Barkan (Oxford University, New York).
Judson, R. S., and H. Rabitz, 1992, Phys. Rev. Lett. **68**, 1500.
Katz, D. P., 1997, J. Chem. Phys. **107**, 8491.
Kim, W., and P. M. Felker, 1997, J. Chem. Phys. **107**, 2193.
Klemperer, W., 1995, Annu. Rev. Phys. Chem. **46**, 1.
Klemperer, W., 1997, Proc. R. Institution 209.
Kosloff, R., S. A. Rice, P. Gaspard, S. Tersigni, and D. J. Tannor, 1989, Chem. Phys. **139**, 201.
Lehmann, K. K., and G. Scoles, 1998, Science **279**, 2065.
Loesch, H. J., 1995, Annu. Rev. Phys. Chem. **46**, 555.
Mahan, B. H., 1975, Acc. Chem. Res. **8**, 55.
Mestdagh, J. M., M. A. Gaveau, C. Gee, O. Sublemontier, and J. P. Visticot, 1997, Int. Rev. Phys. Chem. **16**, 215.
Nye, M. J., 1993, From Chemical Philosophy for Theoretical Chemistry: Dynamics of Matter and Dynamics of Disciplines, 1800–1950 (University of California, Berkeley).
Peirce, A. P., M. A. Dahleh, and H. Rabitz, 1988, Phys. Rev. A **37**, 4950.
Peirce, A. P., M. A. Dahleh, and H. Rabitz, 1990, Phys. Rev. A **42**, 1065.
Rank, D. M., C. H. Townes, and W. J. Welch, 1971, Science **174**, 1083.
Raz, T., and R. D. Levine, 1995, J. Phys. Chem. **99**, 7495.
Rohlfing, E. A., D. M. Cox, and A. Kaldor, 1984, J. Chem. Phys. **81**, 3322.
Saykally, R. J., and R. C. Woods, 1981, Annu. Rev. Phys. Chem. **32**, 403.
Seideman, T., 1996, J. Chem. Phys. **106**, 2881.
Seideman, T., 1997, J. Chem. Phys. **107**, 10420.
Tannor, D. J., and S. A. Rice, 1985, J. Chem. Phys. **83**, 5013.
Thaddeus, P., 1972, Annu. Rev. Astron. Astrophys. **10**, 303.
Thaddeus, P., M. C. McCarthy, M. J. Travers, C. A. Gottlieb, and W. Chen, 1998, Faraday Discuss. **109**, 1.
Urey, H. C., 1933, J. Chem. Phys. **1**, 1.
Weinstein, J. D., R. deCarvalho, T. Guillet, B. Friedrich, and J. M. Doyle, 1998, Nature (London) **395**, 148.
Wong, S. S., E. Joselevich, A. T. Woolley, C. L. Cheung, and C. M. Lieber, 1998, Nature (London) **394**, 52.
Wu, M., R. J. Bemish, and R. E. Miller, 1994, J. Chem. Phys. **101**, 9447.
Zare, R. N., 1998, Science **279**, 1875.

Biological Physics
Hans Frauenfelder, Peter G. Wolynes, and Robert H. Austin

Physics and biology have interacted at least since Galvani and physicists have always been intrigued by biological problems. Erwin Schrödinger's book (1944) led many physicists to study biology. Despite its inspirational character and its stressing the importance of biomolecules, many of the detailed ideas in the book proved to be wrong and have had a limited impact on mainstream molecular biology. The connection between physics and biology has also been treated quite early in the *Reviews of Modern Physics*. In 1940, Loofbourow (1940) described the application of physical methods. Oncley *et al.* (1959) edited a study program on biophysical science; the report written by many outstanding scientists is still worth reading.

In the foreword to his article, Loofbourow wrote: "*I was tempted to use the title 'Biophysics' for this review as more succinctly delimiting the field discussed. But despite the obviously increasing interest in biophysical problems, there does not seem to be clear agreement, even among biophysicists, as to what the term biophysics means.*" This confusion still exists. Here we use Stan Ulam's remark ("*Ask not what physics can do for biology, ask what biology can do for physics*") and define biological physics as the field where one extracts interesting physics from biological systems. Much like the terms physical chemistry and chemical physics, the terminological differences represent only psychological style and current attitude; the same person at different times could be thinking as a biophysicist or as a biological physicist.

The connection between biology and physics is a two-way street. However, the heavy traffic has gone one way. Many tools from physics have been adopted by researchers in the biological sciences. The return traffic, where biological ideas motivate physical considerations, has been less visible, but the study of biological systems has already led to some interesting results, particularly concerning the physics of complexity and of disordered systems. Here we focus on biological physics at the molecular level. However, biological physics is much broader. Organismal physiology has inspired much work, for example, the study of neural networks (J. Hertz *et al.*, 1991) and immunology (Perelson and Weisbuch, 1997). The mathematics of evolution and population biology has attracted much attention by theoretical physicists who have pioneered a mutually beneficial connection with computational and statistical physics.

Biophysics and biological physics cover an enormously broad field, and offer an exciting future. Unlike Janus, the Roman god who could see both to the past and the future, the present authors only know what has happened and even then have a limited view of that terrain. For the present review, in the spirit of the centennial celebration, we have looked mostly to the past and what has been successful. More than fifty years ago,

Loofbourow (1940) cited 1203 references in biological physics. Since then, the number of papers has increased nearly exponentially with a time constant of about 15 years. We had to make a biased selection in order to present a coherent story and so a great deal of important work is not mentioned. We cite reviews rather than original papers wherever possible but urge the reader to consult other books (Flyvbjerg et al., 1997; Peliti, 1991) which present a broader picture than we can in these short pages.

The Structures of Biological Systems

In nearly every field of physics, experimental study of the structure of a system has been an essential first step leading to models and theories. Structural studies have also been crucial in biology.

From Röntgen to synchrotrons and NMR

The most important contribution of physics to molecular biology has been x-ray structure determination. Wilhelm Conrad Röntgen discovered x rays in 1895 and his discovery has affected all scientific fields (Haase et al., 1997). Max von Laue introduced x-ray diffraction and W. L. Bragg determined the first crystal structures. Laue believed that the structure of biomolecules would never be solved. He was wrong. In 1953, James Watson and Francis Crick deduced the exquisite structure of the DNA double helix. In 1958 John Kendrew determined the structure of myoglobin. Shortly afterwards, Max Perutz solved the structure of the much larger hemoglobin. These structure determinations were heroic efforts and took years of work, but they showed how DNA and proteins are built and laid the foundation for an understanding of the connection between structure and function (Branden and Tooze, 1991).

Computers, synchrotron radiation, and improved detectors have changed the field radically. By the year 2000 about 25 000 structures will have been deposited in the Protein Data Bank. Moreover, cryogenic experiments permit the determination of nonequilibrium states produced for instance by photodissociation (Schlichting et al., 1994). The x-ray diffraction technique has, however, two limitations: (1) X-ray diffraction requires good crystals, but not all proteins can be crystallized easily. This problem is particularly severe for the large and important class of membrane-bound proteins which require a heterogeneous environment. Even for those proteins which are soluble in water it is not always clear if the protein in a crystal has the same structure as in solution. (2) Water molecules, crucial for the function of biomolecules, are difficult to see with x rays; their positions must be inferred from the positions of the heavier atoms. The first limitation was overcome by another technique from physics, NMR (Wüthrich, 1986; Clore and Gronenborn, 1991). The main geometric information used in the NMR structure determination resides in short interproton-distance restraints derived from the observation of nuclear Overhauser effects. The second limitation is overcome by neutron diffraction (Schoenborn and Knott, 1996) which can locate hydrogen atoms directly because of their large scattering cross section and can distinguish hydrogen and deuterium, thus making labeling of exchangeable protons possible.

Biological systems

Living things can be viewed hierarchically: Genes (DNA)↔proteins↔organelles↔cells↔tissues ↔organs↔organisms. Explanations in terms of cause must ultimately deal with the last level of description (an extreme view is the dictum: "Nothing in biology makes sense but in the light of the theory of evolution"). But for biological physics the first steps are intriguing enough to provide much inspiration and challenge. The genetic information is coded in the genes in the form of three-letter words on a linear unbranched DNA molecule. Organisms that have chromosomes have the DNA molecule wound around protein molecules (histones) for compact storage and access. Without this compactification, the enormous 3-m length of the DNA molecules in a human cell (with 1 billion basepairs) would not be able to fit as a random Gaussian coil within the 5-μm-diameter nucleus. The information for the construction of a particular protein is read and transcribed onto an RNA molecule. The RNA molecule is by itself also quite interesting because unlike DNA it is conformationally flexible due to its ability to basepair intramolecularly. Like proteins some RNA molecules can fold into three-dimensional catalytically active structures called ribozymes. Those RNA molecules destined to code for proteins are edited to decrease the error rate and this process also leads to interesting physics (Hopfield, 1978). The RNA molecule is then transported to a ribosome, where the protein assembly takes place. The protein is also built as a linear chain, but the building blocks of nucleic acids and proteins are different: Nucleic acids are built from four different nucleotides, proteins from twenty different amino acids. The RNA instructs the ribosome in which order the amino acids must be assembled to form the primary sequence of the protein. When the primary sequence emerges from the ribosome, it folds into the functionally active three-dimensional structure. Sometimes chaperone proteins are involved, but their role now seems to be one of correcting errors rather than being instructive.

Myoglobin, the hydrogen atom of biology

As an example of a typical protein, we discuss myoglobin (Mb). Myoglobin stores oxygen (O_2), facilitates oxygen diffusion, and mediates oxidative phosphorylation in muscles (Wittenberg and Wittenberg, 1990). In phosphorylation, the free-energy donor molecule ATP is formed as a result of the transfer of electrons from NADH to oxygen. This process clearly involves physics. Myoglobin also binds carbon monoxide (CO). The reversible binding processes

$$Mb + O_2 \leftrightarrow MbO_2,$$

$$Mb + CO \leftrightarrow MbCO \tag{1}$$

can be used to study reaction theory, protein dynamics, and protein function. CO and O_2, and other molecules that are bound are called ligands. Myoglobin consists of 153 amino acids, also called residues. Its secondary structure consists of eight alpha helices. These fold into a boxlike tertiary structure, shown in Fig. 1(a), with approximate dimensions $2 \times 3 \times 4$ nm^3 that encloses a heme group (protoporphyrin IX) with an iron atom at its center. The ligand, O_2 or CO, binds at the iron. At first glance the structure is complicated, but it does contain some elements of a rough symmetry—a polyhedral arrangement of cylindrical helices.

The Complexity of Biological Systems

Biological systems, in particular biomolecules, are ideal systems to study complex phenomena. Biological systems are reproducible and thus offer many experimental advantages over other seemingly complex systems such as disordered magnetic alloys.

Distributions

A "simple" system such as an atom or a small molecule has a unique ground state and its properties have, within the limits of the uncertainty relation, sharp values. At finite

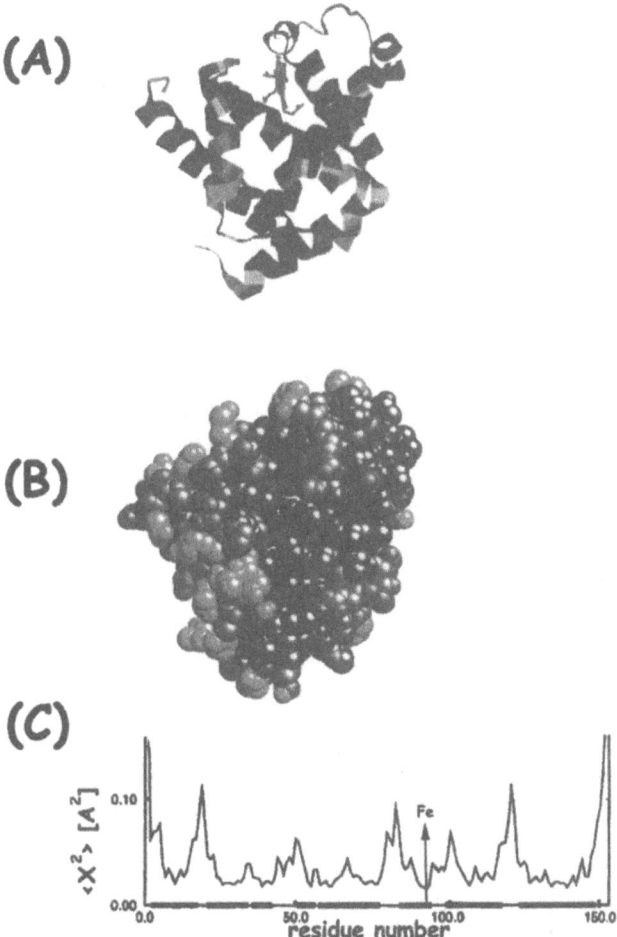

Figure 1. Myoglobin. (a) Skeleton, showing the protein backbone. The Debye-Waller factors averaged over the amino acid group are gray-scale coded, a lighter gray equals more movement. (b) Computer-produced space-filling view of myoglobin, again gray-scale coded in terms of Debye-Waller factors. (c) Debye-Waller factors. The mean-square deviations, plotted as a function of the amino acid number for deoxyMb. The data for parts (a) and (b) are taken from the Brookhaven Protein Data bank entry IMBC (Kuriyan, Wilz, Karplus, and Petsko). The data for part (c) are from F. Parak, personal communication.

temperatures small molecular systems usually sample configurations in the vicinity of this ground state. The transformations of small molecules then satisfy very simple phenomenological laws familiar in elementary chemistry. The time dependence of simple unimolecular reactions is usually given by a single exponential and the temperature dependence follows an Arrhenius law,

$$k(H,T) = A \exp(-H/RT), \qquad (2)$$

where H is a barrier height, R the gas constant, and A the preexponential factor. In complex systems, such as, for instance, glasses, the behavior is different (Richert and Blumen, 1993). The time dependence of relaxation phenomena is usually nonexponential in time and can often be described by a power law or a stretched exponential:

$$N(t) = N(0) \exp\{-[k(T)t]^\beta\}, \qquad (3)$$

where β is less than 1. The rate coefficient, $k(T)$, often does not follow the Arrhenius relation Eq. (2), but can be approximated by, for instance, the Ferry relation (Fig. 2):

$$k(H^*, T) \approx A \exp[-(H^*/RT)^2]. \qquad (4)$$

For many years, reactions observed in proteins were assumed always to be simple and were described by Eqs. (1) and (2). As in the chemistry of small molecules, deviations from these laws were ascribed to mechanisms involving the concatenation of a few elementary steps. The impression that proteins were simple was fortified by the structures inferred from x-ray diffraction. These showed each atom in a unique position, but actually this is the result of the model usually used in data reduction. A study of the reaction Eq. (1) at low temperatures changed the picture (Austin et al., 1975). The binding of CO or O_2 was not exponential in time; between 40 K and about 200 K it could be described by a distribution of barrier heights H:

$$N(t) = N(0) \int g(H) \exp(-H/RT) dH. \qquad (5)$$

Here $g(H)dH$ gives the probability of finding a barrier between H and $H+dH$.

The appearance of a distribution rather than a single value for H in MbCO is not an exception. Both at low temperatures and at room temperature at short times, protein properties must be described by distributions. There are two different explanations: Either all myoglobin molecules are identical, but processes are intrinsically nonexponential in time (homogeneous case), or different myoglobin molecules are different (heterogeneous case). The experiment gives a clear answer: Each myoglobin molecule rebinds exponentially, but different molecules are different. This conclusion is supported by many experiments (Nienhaus and Young, 1996). Particularly convincing are spectral hole-burning experiments. If protein molecules with identical primary sequence indeed differ in tertiary structure, spectral lines should be inhomogeneously broadened and it should be possible to use a sharp laser line to burn a hole into the band. This phenomenon is indeed observed; it proves the inhomogeneity and permits the study of many protein characteristics (Friedrich, 1995).

The energy landscape of biomolecules

Why do distributions occur? A possible answer was implicitly contained in a visionary talk by Cyrus Levinthal (1969) who asked if the final conformation after folding necessarily has to be the one of lowest free energy. He concluded that it did not have to be the case, but that it must be a metastable state in a sufficiently deep well to survive possible perturbations. If the lowest state is not reached, the observation of a distribution of activation barriers, Eq. (5), can be explained by saying that the protein can assume a very large number of related, but different conformational substates that are only potentially related to the ground-state structure. It must be described by an energy landscape (Frauenfelder, Sligar, and Wolynes, 1991; Frauenfelder et al., 1997). To completely describe an energy landscape, the energy of the protein should be given as function of the $3N$-6 (>1000) coordinates of all atoms. It is not enough to exhibit the energy function. The organization of the hyperspace that results from the energy function must be understood. Each substate is a valley in this hyperspace. The activation barriers in the different substates are different and the observed $g(H)$ is explained. A one-dimensional schematic of an energy landscape is given in Fig. 2. The energy barriers between different valleys (different conformational substates) range from about 0.2 kJ/mol to about 70 kJ/mol. The kinetic observations suggest that the energy landscape might have a hierarchical structure, arranged in a number of tiers, with different tiers having widely separated average barrier

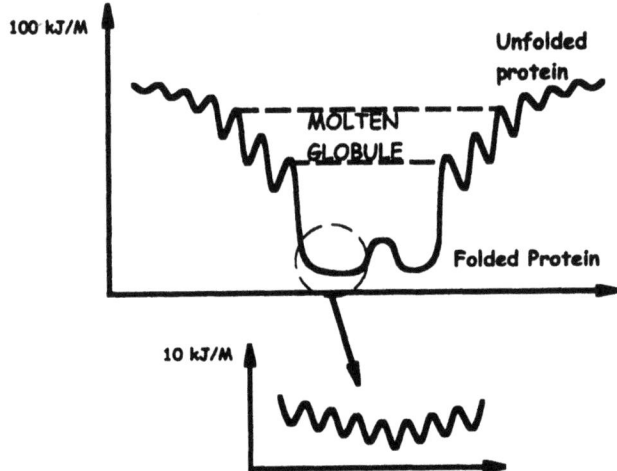

Figure 2. The energy landscape. The main figure shows the funnel in the energy landscape that leads towards the folded protein. The vertical axis is the difference in enthalpy between the folded and the unfolded state; the difference between the unfolded and folded states is of the order of 100 kJ/mol. The horizontal axis is simply a crude one-dimensional representation of a many-dimensional space of coordinates for the amino acids. The bistable minima in the figure are meant to represent the possibility of the protein existing in more than one metastable conformation. The lower figure shows the energy landscape magnified by roughly 10 for the folded protein; it gives H^*, Eq. (4), as a function of one conformation coordinate. Note that H^* characterizes the roughness of the energy landscape and not the height of an Arrhenius barrier.

heights. A strictly hierarchical energy landscape arises in other complex systems such as spin glasses. Understanding the nature of such a hierarchy remains a hot topic.

The Debye-Waller factor

Can the existence of an energy landscape and conformational substates be reconciled with the apparently unique structure that emerges from the x-ray diffraction? Yes! Debye and Waller proved that for a harmonic oscillator with mean-square deviation, $\langle x^2 \rangle$, the intensity of an x-ray-diffraction spot, with wavelength λ at the scattering angle θ decreases by a factor (Willis and Pryor, 1975):

$$f_{DW} = \exp(-16\pi^2 \langle x^2 \rangle \sin^2 \theta/\lambda^2). \qquad (6)$$

If proteins have substates, it should show up in the Debye-Waller factor for the individual atoms. This effect is indeed observed. In Fig. 1(c) the $\langle x^2 \rangle$ for deoxy Mb are plotted versus the amino acid number. The figure shows that the $\langle x^2 \rangle$ in different parts of the proteins differ. The same plot for a crystalline solid would show a uniform and smaller $\langle x^2 \rangle$ that would vanish at $T=0$. At present, $\langle x^2 \rangle$ are routinely determined in x-ray structures of proteins (Rejto and Freer, 1996).

Dynamics

Motions are essential in biology, from the transport of oxygen by hemoglobin to muscle action. Study of the motions of biomolecules is a central part of biological physics.

Fluctuations and relaxations

Fluctuations at equilibrium and relaxations from nonequilibrium states are essential for the function of proteins. The processes can be described in terms of the energy landscape (Fig. 2). Fluctuations correspond to equilibrium transitions among the conformational substates in the folded protein; relaxation processes correspond to transitions towards equilibrium from an out-of-equilibrium state. The folding of the nascent polypeptide chain is the prototype of a relaxation process. Near equilibrium, fluctuations and relaxations are directly connected, but this is not always the case if the system is strongly perturbed.

One goal of the studies of relaxation and fluctuation phenomena is to connect them to structural entities within the protein. Another is to determine their functional importance. Not all motions of a biomolecule are directly relevant to its biological role. Piecing together the information from experiments, theory, and computation (McCammon and Harvey, 1987; Brooks *et al.*, 1988; Nienhaus and Young, 1996) yields striking results. Differential scanning calorimetry and flash-photolysis studies indicate, for instance, that the distribution of relaxation times in myoglobin is extremely broad. Additional information comes from inelastic neutron scattering, spectral hole burning, optical spectroscopy, NMR, the Mössbauer effect, and other techniques. A synthesis that connects structure, energy landscape, dynamics, and function has not yet been achieved.

Proteins and glasses

Proteins and glasses share one fundamental property, the existence a large number of nearly isoenergetic minima and some of the terminology from glasses is now also used for proteins (Frauenfelder *et al.*, 1991). Significant protein motions involve transitions between substates. The harmonic vibrations occur within the substates and are possibly too fast to be directly involved in most physiology, although the role of harmonic and anharmonic effects in biological processes remains an active area of research (Christiansen and Scott, 1990). As the temperature is lowered, the transitions become slower. An arbitrary glass temperature T_g can be defined as the temperature where the transition rate is 10^{-5} s^{-1}. In the simplest view, the protein moves above T_g, and is metastable below T_g. The existence of a hierarchy of conformational substates complicates the situation, because motions in different tiers freeze out at different temperatures. Some motions occur even at 100 mK.

The glass transition in proteins is more involved than in an ordinary structural glass. If a protein is embedded in a glass-forming solvent, T_g for large-scale protein motions is very similar to T_g of the surrounding glass. The motions are slaved to the solvent. Protein and surrounding must consequently be treated together and the environment can control protein motions. This fact may be used by biological systems for control.

Protein folding

The complexity of biomolecules ultimately derives from the information contained in the sequence of nucleotide bases in DNA. To a strict reductionist, the task of biological physics is to decode this message or at least to describe how the phenomena of biology at different levels emerge from this kernel of information. Remarkably, the first few steps of information flow from the DNA to biological behavior, while physically complex, can be algorithmically understood in a simple way. With high but not perfect reliability, because of RNA editing, the sequence of a protein can be inferred from the DNA, which encodes it. This simplicity seems to derive ultimately from both having a complex biological apparatus to transfer the information from DNA into protein, including error-correction machinery. But the simplicity also comes from the simple structure of DNA, which can accommodate extraordinarily different sequences in the same structural format—just as many different words can all be written in a book using the same typeface. The next stage of information flow—how the protein molecule obtains a three-dimensional shape which allows it to function in the ways previously described—has been particularly inspiring to the current generation of biological physicists. There are several reasons for this. First, this self-organization can proceed without additional biological machinery and thus can be studied in detail in the test tube. The spontaneous act of folding is quite remarkable in that the complex motion of the protein transfers the information contained in a one-dimensional sequence of data into a three-dimensional object: sculpture by Brownian motion. Folding resembles a phase transition like crystallization, but is much more complex, since there are so many different shapes a protein can have. Since folding must usually occur before any further functioning but is also directly related to the genetics, the understanding of folding intellectually intersects the study of molecular evolution and the origin of life. Finally, there is an important practical motivation for the study of folding. Sequence data are cheap while structural data are still expensive. Even the frightening rate of experimental determination of protein structures cannot keep up with the more horrifying rate of acquisition of the DNA sequence data, which ultimately encodes it. Understanding protein folding can improve the capability of predicting protein structure from sequence. This engineering goal of making structure predictions pinpoints a place where theory can be of direct economic value.

Appreciating that protein folding is sculpture by Brownian motion has led to a view which unites the study of the folding process with the investigation of protein motions that occur in the folded protein. Understanding folding, however, requires a broader picture of the energy landscape that includes not only the states that are excited during function but also those far from equilibrium in which the protein is found in, early after the molecule is synthesized. Directly upon synthesis, the protein molecule is nearly a random coil much like many artificial polymers in solution. The molecule condenses into a more compact, but still highly fluctuating, structure and finally chooses to organize itself into the much smaller set of structures that are involved in function and whose average is obtained by x-ray diffraction and NMR. The protein starts out soft and squishy like molecular spaghetti and becomes harder and more organized: an aperiodic crystal in Schrödinger's language. Many of these folding transformations can be described using the language of phase transitions in small systems. The mesoscopic size of proteins, along with their reproducibility, makes the study of the folding process an excellent test bed for statistical physics of small systems, much as the study of nuclei was for quantum mechanics in the 1940s. It is the variety of possible structures, along with the relative specificity of the structures actually formed, that brings a truly novel element into the physics of folding. The complexity of the protein sequence might suggest that the sequence could be treated as specific but random. Thus, phase transitions in a folding protein would resemble those seen in disordered magnetic alloys or spin glasses, which are known to have complex energy landscapes. The analysis of spin glasslike models of folding has, therefore, been very useful. However, an important consequence of the analysis of the folding of random heteropolymers is the realization that achieving organized structures probably requires a preselection of protein sequences so that the energy landscape of a protein is, in some respects, simplified from the worst case of a highly disordered system. The spin-glass landscape has many alternate basins, statistically similar but in detail different. Likewise, for a random heteropolymer, each alternate basin could act as a trap for the configurational motions of the molecule impeding the folding process. Some specially chosen sequences have landscapes that eliminate these traps leaving primarily only one dominant basin in which the minima required for function can coexist. Thus the complexity of the landscape is partially removed by selective evolution. This aspect of landscapes needed for efficient folding is known as the "principle of minimal frustration" (Bryngelson and Wolynes, 1987). Using the theory of spin glasses and polymer theory, the principal of minimal frustration can be translated into a quantitative statement about the statistical characteristics of the energy landscape, namely the depth of the main basin must exceed the amount by which the energy varies from configuration to configuration, the "ruggedness" of the landscape. These statistical characteristics are related to underlying thermodynamic transition temperatures, one being a glass transition driven by the ruggedness, the other equilibrium folding temperature related to the mean basin depth. The simplified energy landscape, which now allows rapid folding, can be described as a funnel (see Fig. 2). A single basin with many minima dominates although within it there are side basins, which can act transiently as traps. In a funnel landscape, sculpture by Brownian motion becomes easy and a folded state is nearly inevitable at a low enough temperature.

The energy-landscape description inherent in this funnel picture has influenced the experimental work on folding. Levinthal's argument that pointed out the difficulty of finding a folded state had inspired an experimental program of searching for specific paths to the folded state and emphasized finding slow folders with intermediates. In contrast, the energy-landscape picture suggested that many of these studies actually were exploring traps that weren't helping the folding process. Landscape ideas shifted the emphasis to the faster folding proteins whose kinetics do not show intermediates. Studying these molecules using protein engineering (Fersht, 1997) and fast spectroscopic methods (Eaton *et al.*, 1998) has yielded much information about the structure of the landscape. The new emphasis in experiment is on characterizing the ensembles of different structures, not trying to find individual specific ones that occur in the folding process.

A fruitful question has been to find out how evolution was able to select sequences that would obey the minimal frustration principle and lead to funnel-like landscapes. This question reminds us of the "Hoyle paradox" named after the astronomer who whimsically argued that the difficulty of biological design buttresses the case for an extraterrestrial origin of life in a steady-state universe (Hoyle, 1957). Statistical physics shows that this "design problem" is intrinsically easier than the search problem faced by a random sequence with its complex energy landscape. This has been made clear by the development of many algorithms that "design proteins" on lattices that fold readily in a computer simulation. The emerging ability of chemists to design laboratory proteins from scratch reinforces this lesson.

One way in which a funneled landscape can be achieved is for the folded structure to be particularly symmetric. This property has led to a search for "magic numbers" in the database of known protein structures, much like the earlier search for magic numbers in nuclear abundance. The occurrence of certain super families of protein structures seem to be explicable on the grounds that some structures are particularly appropriate for funneled landscapes. The quantitative nature of the energy-landscape approach to protein folding allows statistical physics to be used as a new tool in creating protein-structure prediction algorithms. The statistical examination of the simplified energy landscapes of proteins used for computational structure prediction allows one to assess which models are better and which are worse as prediction schemes. Using optimization strategies to find energy functions that lead to minimally frustrated landscapes for known sequences with known structures also provides a route to approximate energy functions for use in structure prediction (Onuchic *et al.*, 1997). Physically based algorithms that come from this approach are now competitive with others that use the evolutionary trees of proteins to predict their structure by analogy to related proteins of known structure.

Reaction Theory

The complex motions of proteins and their self-organization into three-dimensional structures have been inspirational for physicists. But even for investigating simple chemical transformations, biomolecules, in particular proteins, have proved to be excellent laboratories.

Dynamic effects in chemical reactions

Biological reactions occur in condensed phases. For many years, textbooks used the transition-state theory to describe such reactions. This theory, however, does not take the effect of the surrounding into account and it cannot, therefore, adequately describe biological reactions. Fortunately, a better theory exists, created by Kramers in 1940

(Frauenfelder and Wolynes, 1985; Hänggi et al., 1990). The interaction of the reacting system with the environment is characterized by a friction coefficient which, by Stokes law, is approximately proportional to the viscosity η. At small viscosities, the system must make a Brownian walk in energy to move over the barrier and the rate coefficient k is proportional to η. At high viscosities, the system moves like through molasses and must diffuse over the barrier; k then is inversely proportional to η. While many studies of this rate theory concentrated on small molecule reactions in liquids, the biomolecular problems were an important proving ground for the theory.

Experimentally, the viscosity effect has, for instance, been studied in the case of the binding of CO to myoglobin [Eq. (1)]. Binding involves a series of steps. The CO enters the protein from the solvent and moves into a cavity near the heme iron (Fig. 1). Once there, it can establish a covalent bond with the iron atom at the center of the heme group. The viscosity dependence of the different steps has been examined. The results support the Kramers equation, but with modifications. Both the viscosity of the solvent and of the protein must be considered. As the action moves deeper into the protein, the effect of the environment is attenuated, and even in the high-friction range, the rate coefficient is not proportional to $1/\eta$, but to $\eta^{-\kappa}$, with $\kappa < 1$.

Tunnel effects

The Arrhenius and the Kramers equations contain the factor $\exp(-H/RT)$ and therefore predict that the reaction rate should vanish in the limit $T \to 0$. It has, however, been known since Hund's work in 1927 that quantum tunneling takes over at low temperatures (Goldanskii et al., 1989). The theory has been worked out in detail (Hänggi et al., 1990), but for some insight, a simple expression suffices. The rate coefficient k_t for tunneling of a particle with mass M through a barrier of height H and width d can be approximated by

$$k_t \approx A \exp[-\pi^2 d(2ME)^{1/2}/h], \tag{7}$$

where h is Planck's constant. This relation shows that tunneling is essentially temperature independent and decreases exponentially with increasing distance and mass. Indeed, electrons tunnel easily, and electron tun-neling is crucial in photosynthesis. Protons are also known to tunnel, but for heavier particles, the barrier H and the distance d must be very small for tunneling to be measurable. Here proteins again provide an excellent laboratory. The last step in the binding of CO to heme proteins is the bond formation Fe-CO. The low-temperature data (Austin et al., 1975) show that the barrier H for this step can be as small as a few kJ/mol. Distances d for bond formation are of the order of 1 nm. Equation (7) then implies that tunneling of CO should be observable below, say, 20 K. The rate coefficient for the CO binding indeed follows an Arrhenius law down to about 20 K, but then becomes essentially temperature independent, implying tunneling. Since Eq. (7) indicates that k_t depends on mass, it is also possible to study the isotope effect in tunneling.

Gated reactions

The treatment so far assumed static barriers. In a fluctuating protein, however, the barriers themselves fluctuate. An example is the entrance of ligands into myoglobin. It is known that even isonitriles, molecules much larger than CO, can enter myoglobin and bind. The x-ray structure of myoglobin shows, however, no channel where molecules could enter and leave. Thus fluctuations must open channels. This opening must involve large-scale motions of the protein that are coupled to the solvent and hence depend on solvent viscosity.

The theory of fluctuating barriers involves two cases. The fluctuations can either be energetic or geometric. The rate coefficient for passage through a gate depends on the rate coefficient k_f of the fluctuations that open the gate and the rate coefficient k_p for passage of the ligand through the open gate. The calculations yield a fractional-power dependence on viscosity. A resonance occurs when $k_p \sim k_f$ (Gammaitoni et al., 1998). This stochastic resonance between fluctuation (noise) and transition leads to an enhancement of the effective passage rate. The enhancement has been observed in many systems, but it is not yet clear if proteins take advantage of it.

In chemical reactions, both nuclei and electrons move. The discussion so far has only considered nuclear motions, assuming that the electrons adjust to the nuclear position. There are, however, situations where this assumption fails. Consider the binding of CO to the heme iron. Before binding the Fe-CO system is in a quintuplet (q) state, in the bound state it is in a singlet (s) state. If the matrix element connecting s and q, $\Delta = V_{sq}$, vanishes, the free CO cannot bind; it will remain on the diabatic curve q. If Δ is very large, the system will change from q to s in the transition region, the reaction will be adiabatic, and the CO can bind. The condition for adiabaticity and the probability of changing from q to s have been calculated by Zener (Zener, 1932), and by Stueckelberg (Frauenfelder and Wolynes, 1985). In the intermediate case, the transition $q \rightarrow s$ depends on Δ. Here is another case where biomolecules may use quantum mechanics to regulate a reaction (Redi et al., 1981).

Bioenergetics and Physics

Bioenergetics, accounting for how energy flows in biological systems, has been a major inspiration for physicists. We often forget that Mayer, a physician, who was contemplating why sailors' blood in the tropics was of a different color than in temperate climates, took the first steps toward the law of conservation of energy in thermodynamics. The myoglobin example derives ultimately also from understanding respiration, the first step of energy transformation for animals. Later steps in energy transduction involve setting up transmembrane potential gradients ultimately caused by the transfer of electrons and protons between different biomolecules. For plants, the transformation of light into chemical and electrical energy has led also to much good physics. Both charge transport and light energy transduction very early forced biological physicists to face the quantum. In both areas, the progress of biology would have been impossible without the contributions of physical scientists.

Charge transport

Charge transport in molecular systems is generally different from the free flow studied in simple metals. An isolated charge in a biomolecule strongly perturbs its environment and actually acts much like a polaron. Distortions of the molecular framework must accompany the motion of the charge, dramatically affecting the rate of charge-transfer processes. The study of biological electron transfer, however, brought new surprises. The environmental distortions accompanying charge transport in proteins at low temperature can occur by quantum-mechanical tunneling and involve nonadiabatic effects.

Biological electron transfer was one of the first tunneling processes observed in complex molecular systems. Electron transfer in biology occurs over large physical distances. In ordinary electrochemical reactions, the molecules transferring charge come nearly into contact, but the big proteins separate the small prosthetic groups in which the labile electrons reside often by tens of angstroms. While physicists were familiar with such large-distance electron transfer processes as occurring by electron tunneling in metal-oxide metal junctions, this idea was controversial among biologists and chemists for quite some time. Hopfield pointed out the analogy and suggested that charge transfer could be mediated and controlled by tunneling through the intervening protein medium. At first glance, an exponential decay of the tunneling probability with characteristic length of approximately 1 Å fits many experimental data. Crucial experiments that used protein engineering to place the electron donor and acceptor sites in well-defined locations confirmed the outline of this electron tunneling picture but showed there was a still deeper aspect that involved the structure of the pro-tein. Theorists finally showed that charge is transported in biomolecules through quantum-mechanical tunneling of holes, mostly along the covalent backbone of the molecules (Onuchic et al., 1992). Tunneling rates for electrons can be predicted by finding the tubes along which these holes are transported. It seems likely that evolution has made use of the details of the tunneling process in modulating charge transport in many systems.

Light transduction in biology

Both the processes of vision in animals and photosynthesis in plants have brought forth new biomolecular physics. Our vision is sensitive at the single-photon level. The process starts with a photoinduced isomerization, which is one of the fastest processes in biology occurring in <1 ps. This speed makes it comparable to many of the processes of vibrational-energy flow in small molecules. While much is known about the process, it remains a controversial area which attracts laser physicists with powerful new ultrafast techniques and theoreticians developing new computational methods for quantum dynamics.

The study of photosynthesis has a longer history and has been even more fruitful. One of the first steps in photosynthesis is the capture of light energy by chlorophyll molecules in a so-called antenna system. In the 1940s, J. R. Oppenheimer reasoned that the transfer of energy within the photosynthesis apparatus could occur by a process analogous to internal conversion in nuclear physics. This process was independently described later by Förster in greater detail and is now known as Förster transfer—a general mechanism for energy flow between electronically excited molecules. Recent experiments suggest that this transfer is not quite so simple as Förster imagined. The transfer occurs so rapidly that quantum- mechanical coherence is not completely lost. Recent structural characterization of the light-harvesting apparatus has allowed Schulten and his co-workers to give a more complex quantum-mechanical description of this process (Hu and Schulten, 1997).

In photosynthesis, light energy is ultimately transduced into chemical and electronic energy through the apparatus of the photosynthetic reaction center. Here the excitation of a chlorophyll molecule by the photon's energy initiates a series of charge-transfer processes. Again, the first steps are so fast that the quantum dynamics of the nuclear motion needs to be accounted for as well as the electron tunneling *per se*. Theorists have brought to bear much of the heavy machinery of quantum dynamics to address this problem, ranging from large-scale molecular dynamics coupled with polaron theory to real time-dependent quantum Monte Carlo methods. In many of these processes, the precise tuning of energy levels of the molecules, probably largely through electrostatics seems to play a crucial role. This fine-tuning represents a puzzle that needs to be reconciled with the disorder intrinsic in the energy landscape of proteins, suggesting that there are still mysteries to be resolved.

Forces

One of the grand themes of subatomic physics has been the exploration of the forces through which elementary particles interact. For biological physics, an understanding of the forces between biomolecules, organelles, and cells is of equal importance, but there is a difference. In subatomic physics, the forces and the underlying entities were unknown. In biological physics the force is well known, it is the electromagnetic interaction. In principle the Schrödinger equation with a suitable potential should describe all phenomena. The difficulty comes from the complexity of the systems that interact which leads to a description in terms of effective "forces" that are really not fundamental, but correspond to suitable approximations for the interactions between larger objects. The situation is like that of deducing the properties of nuclei directly from QCD.

The building blocks in the primary sequence of proteins and nucleic acids are held together by covalent bonds. These bonds are quite strong with binding potentials on the order of 1 to 2 eV. Rupture of these bonds, done by enzymes in biology, are "violent" events and the subject of a great deal of work. Such events which involve moving atoms apart to the point of dissociation are highly nonlinear and have attracted a great deal of interest within the theoretical physics community. Davidov (Davidov, 1987) proposed that not only was the catalytic event the result of nonlinear force-displacement relationships but that the transport of the energy used in catalysis was due to the movement of a solitonic elastic wave propagating down the backbone of the protein. A soliton, in an over-simplified view, is a nonlinear wave which moves in a highly dispersive medium where the phase velocity of the wave is a strong function of the frequency of the wave. When the amplitude of the wave in the medium is of the appropriate size the nonlinear modulation of the phase velocity exactly cancels the dispersion and the wave travels without spreading. Solitons exist and are very important but the relevance to biological systems is still very much in doubt (Christiansen and Scott, 1990) and await new experiments. The critical event in enzyme catalysis, the breakage of a bond, still remains the province of the chemist. Perhaps in the future aspects of nonlinear dynamics and energy flow will help us obtain insight into general aspects of this complex event.

At a lower-energy scale we consider the weaker forces that determine how biological polymers self-interact as they bend and twist and approach other molecules. Electrostatic, van der Waals, entropic, and undulation (elasticity) forces determine the three-dimensional structure of biopolymers and the interactions between biological entities up to the cell. Rather old discussions of the first two forces are still relevant (Gabler, 1978), but the entropic and undulation forces are not yet universally appreciated.

Entropic "forces" are due to phase-space considerations and seek disorder, fighting enthalpic forces which want to bring objects together (Leikin, Parsegian, and Rau, 1993). An interesting aspect of biological polymers (and solvents like water) is the relatively large magnitude of both the entropic (S) and the enthalpic (H) energies due to the large number of atoms which are linked together by the covalent backbone. Elementary thermodynamic arguments show that the fractional occupation of a state B which lies higher in free energy Δg above a ground state A is given by

$$\frac{N_B}{N_A+N_B}=[1-\tanh(\Delta G/kT)]/2, \tag{8}$$

where $\Delta G = \Delta H - T\Delta S$. Since ΔH and ΔS can be large for complex biomolecules, the temperature dependence of the populations of the two states can be steep. The midpoint temperature T_M is given by $\Delta H/\Delta S$ and the width of the transition ΔT is given by $k/\Delta S$. Thus the entropic contribution to the transitions is critical and for large molecules dominating.

The undulation forces are due to physical strains in the surface of biomembranes, and bring into play enthalpic considerations. The deformation of a cell, under complex cytoskeleton control, is directly concerned with the undulation force. The idea behind the undulation force is simple (Albersdorfer et al., 1997). A biological object has a complex surface containing elements that interact with other elements through one or more of the first three forces. The movement of these elements towards or away from each other strains the connecting parts, adding an elastic energy to the interaction term. Consider a biological membrane of thickness d and Young's modulus E (the elastic modulus is a general concept which can be used to characterize any material). The bending modulus κ_M of such a membrane is

$$\kappa_M = \frac{Ed^3}{12(1-\nu^2)}, \tag{9}$$

where ν is Poisson's ratio. The bending energy H_{bend} stored in a membrane then is

$$H_{bend} = \frac{1}{2}\kappa_M \int_{surface} dA \left(\frac{\partial^2 u}{\partial x^2} + \frac{\partial^2 u}{\partial y^2} - C_0\right)^2, \tag{10}$$

where u is the magnitude of the membrane normal vector and C_0 is the spontaneous curvature of the membrane due to asymmetrical sides (Gruner, 1994). Variations on this theme, done in a far more sophisticated way than we have outlined here, can be applied at many different length scales to understand the deformation of biological polymers, membranes and organelles.

The undulation force is mechanical and thus can have a long range, just as when you pull on a rope the tension is transmitted over a long distance. If the object through which the force is transmitted is heterogeneous in composition due either to local Young's modulus or through local variations in the entropy density there is a complex dependence of the force with distance as the strained medium responds. The strain dependence of the force can be highly nonlinear. The strain dependence of the bending energy can be viewed as another form of "energy landscape." However, now the energy landscape is that of the bond itself. Thus it is not only the affinity of two ligands for each other, but also the landscape of the binding surface, that can make a great deal of biological difference. Implications are only recently being explored, and are amenable to the analytical tools of the physicist.

One example of this basic idea is the use of the DNA-sequence-dependent Young's modulus of DNA to predict the binding coefficient of a protein repressor (Hogan and Austin, 1987) which was known to strain noncontacted regions of the double helix. Through the use of elastic-energy considerations it is possible to predict quantitatively the dependence of the binding of a protein which induces helical strain on a basepair sequence.

Single-Molecule Experiments

Just as physicists have recently learned how to image single atoms and hold single electrons in confining traps, biological physicists are learning to study single biomolecules (Moerner, 1996; Nie and Zare, 1997). Unlike sodium atoms or electrons, biomolecules are individuals. New insight into biomolecular dynamics and function may result if the distributions discussed earlier are observed on single molecules. We give as an example of single-molecule experiments the stretching of DNA molecules. DNA is particularly easy to study at the single-molecule level because it is incredibly long and has a large value for its persistence length. The challenge of studying individual protein molecules is still very much in its infancy. Different approaches permit the study of the reaction kinetics and thermodynamics of individual proteins. The key is to use extreme dilution so that only a single biomolecule is in the reaction volume. Reactions or excitations are then induced in the same biomolecule many times and observed, for instance, through fluorescence. Such studies can provide additional information on the energy landscape and explore for instance the role of intermittency (Zeldovich *et al.*, 1990).

Nature not only knows chemistry and physics well, she uses them as an excellent engineer. She has built sophisticated linear and rotary motors even at the molecular level (Kreis and Vale, 1993). Linear motors, powered by the splitting of the fuel molecule ATP, actively transport molecules and organelles along the cytoskeleton from one part of the cell to another. Rotary motor proteins are powered by a flux of ions between the cytoplasm and the periplasmic lumen, transforming the ion flux into a rotary motion to drive bacterial flagella (Schuster and Khan, 1994). Motor proteins are exciting for biological physics both because they can be studied in single-molecule experiments and because they have given rise to sophisticated theoretical work (Jülicher *et al.*, 1997). Some of the ideas underlying the protein motors which involve the phenomena of rectified Brownian motion had already been discussed by Feynman (1963).

Single-DNA-molecule experiments are well advanced. A *single* DNA molecule is a polymer with a diameter of a few nm, but can have a length up to about 50 mm (Austin *et al.*, 1997). The mechanical property of individual DNA molecules can be studied for instance by attaching one end to a glass plate and the other end to a magnetic bead. A different technique uses microfabricated arrays to measure the static and dynamic properties of DNA (Bakaijn *et al.*, 1998). Such experiments permit a comparison of the properties of individual DNA molecules with theory.

A long thin polymer molecule such as DNA is entropically extensible like a rubber band. The origin of this entropic elasticity is connected to the mechanical rigidity of the polymer. The mechanical rigidity κ_P of the polymer is a function of the modulus of elasticity E and the cross-sectional shape of the (long axis is the z axis) polymer in the xy plane:

$$\kappa_P = E \int x^2 dx\, dy. \quad (11)$$

While the rigidity tries to keep the DNA straight, the thermal forces buffeting the molecule act to bend it in random directions. The molecule is constantly in motion. The interplay between Brownian agitation and rigidity, then, determines the *persistence length* P of the DNA—the length scale on which the directionality of the polymer is maintained,

$$P = \frac{\kappa_P}{k_B T}. \tag{12}$$

Zooming in to scales shorter than P, the molecule appears straight. But looking from a distance, the molecule appears to be randomly coiled. For DNA in normal physiological conditions, $P \approx 50$ nm which is considerably longer than the molecular diameter of 12 nm but much smaller then the length of the total molecule. When the length L of the polymer is much greater than the persistence length the polymer acts as a linear hookean spring with effective spring coefficient $k_S \sim (3k_B T)/(2PL)$. However, as the strain increases at some force $F_{stretch}$ the polymer no longer responds in a linear manner to applied stress:

$$F_{stretch} \approx \frac{k_B T}{P}. \tag{13}$$

At room temperature $F_{stretch} \approx 0.1$ pN. This value is surprisingly small, weaker than the typical force generated by individual motor proteins and similar in magnitude to the typical drag forces acting on micron-sized objects as they are transported in the cell. When the applied force is stronger than $F_{stretch}$, the elasticity becomes nonlinear. It becomes harder and harder to stretch the DNA as it straightens out and the end-to-end separation approaches the contour length L. By pulling hard on the ends of a DNA molecule, it is possible to "wring out" all of the entropy and straighten the polymer. By pulling harder yet, might one stretch the molecular backbone, just as one can stretch a nylon thread? If so, the elasticity would correspond to the straining of chemical bonds along the DNA axis, and would therefore be of enthalpic, rather than entropic, origin.

There has been an explosion of work on single DNA molecules based upon the rather simple physical modeling that makes this molecule so accessible. The work basically has divided into two parts: studies on the fundamental statistical mechanics of long thin polymers (Perkins *et al.*, 1997; Bakajin *et al.*, 1998) and biological applications involving the influence of supercoiling and overstretching on the DNA and the onset of nonlinearities (Cluzel *et al.*, 1996; Smith *et al.*, 1996). The beauty of this work is the smooth junction between physics and biology, theory and experiment.

The Future

It took a long time until the energy levels of atoms, small molecules, nuclei, and particles were well enough known so that fundamental theories could be constructed. Since organisms are made of biological molecules, it might have been thought that fundamental theories for biology could now be built up easily. The great complexity of these systems, however, has required a repetition of that earlier development now at a new level. Our present understanding of the energy landscapes and motions of biomolecules is probably no further along than the theory of the Bohr atom or the early shell model of the nucleus, despite the heavy mathematics already being used. It has been amazing how some simple ideas have emerged only recently and already proved unexpectedly useful. Today the physicist interested in biology is in a good position to provide such pictures for the future of biology both at the molecular and higher levels. We have emphasized the molecular aspects of biological physics in this brief review. The far greater problems of the brain loom ahead for those physicists that are brave (Hopfield, 1986).

Many scientists believe that each biological situation is unique, the results of unpredictable quirks of evolution. If so, the quest of biological physics to search for generalizations is quixotic. However, just the last few years of progress suggest that there is plenty of room to find new general concepts and principles through the study of biological systems. Therefore we have no doubt that the study of biological systems will continue to inspire the development of new physics. Ultimately, however, physics must transcend biology. The principles gleaned from biological physics should be extended to other systems of the same complexity as natural organisms. Barring the discovery of life on other planets, these more general objects of study will have to be constructed by us. Perhaps they already have been (Langton, 1988). One hundred years from now, the *Reviews of Modern Physics* will certainly contain discussions of what has been learned in biological physics. The only question is whether its authors will be carbon-based life forms like us.

Acknowledgments

We thank our friends from both sides of the divide for many inspiring discussions and for many collaborative efforts. The work of H.F. was performed under the auspices of the U.S. Department of Energy. P.G.W. has been supported by the National Science Foundation and the National Institutes of Health. R.H.A. has been supported in part by the National Institutes of Health and the Office of Naval Research.

References

Albersdorfer, A., T. Feder, and E. Sackmann, 1997, "Adhesion-induced domain formation by interplay of long-range repulsion and short-range attraction force: a model membrane study," Biophys. J. **73**, 245–257.

Austin, R. H., K. W. Beeson, L. Eisenstein, H. Frauenfelder, and I. C. Gunsalus, 1975, "Dynamics of ligand binding to myoglobin," Biochemistry **13**, 5355–5373.

Austin, R. H., J. P. Brody, E. C. Cox, T. Duke, and W. Volkmuth, 1997, "Stretch genes," Phys. Today **50**(2), 32–38.

Bakajin, O. B., T. A. J. Duke, C. F. Chou, S. S. Chan, R. H. Austin, and E. C. Cox, 1998, "Electrohydrodynamic stretching of DNA in confined environments," Phys. Rev. Lett. **80**, 2737–2740.

Branden, C., and J. Tooze, 1991, *Introduction to Protein Structure* (Garland, New York).

Brooks III, C. L., M. Karplus, and B. M. Pettitt, 1988, *Proteins—A Theoretical Perspective of Dynamics, Structure, and Thermodynamics* (Wiley, New York).

Bryngelson, J. D., and P. G. Wolynes, 1987, "Spin glasses and the statistical mechanics of protein folding," Proc. Natl. Acad. Sci. USA **84**, 7524–7528.

Christiansen, P. L., and A. C. Scott, 1990, Eds., *Davydov's Soliton Revisited* (Plenum, New York).

Clore, G. M., and A. M. Gronenborn, 1991, "Two-, three-, and four-dimensional NMR methods for obtaining larger and more precise three-dimensional structures of proteins in solution," Annu. Rev. Biophys. Biophys. Chem. **20**, 29–63.

Cluzel, P., A. Lebrun, C. Heller, R. Lavery, J.-L. Viovy, D. Chatenay, and F. Caron, 1996, "DNA: an extensible molecule," Science **271**, 792–794.

Davidov, A. S., 1987, *Solitons in Molecular Systems* (D. Reidel, Boston).

Eaton, W. A., J. Hofrichter, V. Munoz, E. R. Henry, and P. A. Thompson, 1998, "Kinetics and dynamics of loops, α-helices, β-hairpins, and fast-folding proteins," Accounts Chemical Research, 745–753.

Fersht, A. R., 1997, "Nucleation Mechanisms in Protein-Folding," *Current Opinion in Structural Biology* **7**(1), 3–9.

Feynman, R. P., R. B. Leighton, and M. Sands, 1963, *The Feynman Lectures on Physics* (Addison-Wesley, Reading, MA), Vol. 1, Chap. 46.

Flyvbjerg, H., J. Hertz, M. H. Jensen, O. G. Mouritsen, and K. Sneppen, 1997, Eds., *Physics of Biological Systems* (Springer, Berlin).

Frauenfelder, H., and P. G. Wolynes, 1985, "Rate theories and puzzles of hemeprotein kinetics," Science **229**, 337–345.

Frauenfelder, H., S. G. Sligar, and P. G. Wolynes, 1991, "The Energy Landscapes and Motions of Proteins," Science **254**, 1598–1603.

Frauenfelder, H., A. R. Bishop, A. Garcia, A. Perelson, P. Schuster, D. Sherrington, and P. J. Swart, 1997, Eds., *Landscape Paradigms in Physics and Biology* (Elsevier, Amsterdam); reprinted from Physica D **107**, 117–435 (1997).

Friedrich, J., 1995, "Hole burning spectroscopy and physics of proteins," Methods Enzymol. **246**, 226–259.

Gabler, R., 1978, *Electrical Interactions in Molecular Biophysics* (Academic, New York).

Gammaitoni, L., P. Hänggi, P. Jung, and F. Marchesoni, 1998, "Stochastic resonance," Rev. Mod. Phys. **70**, 223–287.

Goldanskii, V. I., L. I. Trakhtenberg, and V. N. Fleurov, 1989, *Tunneling Phenomena in Chemical Physics* (Gordon and Breach, New York).

Gruner, S. M., 1994, "Coupling between bilayer curvature activity and membrane protein activity," in *ACS Advances in Chemistry Series No. 235, Biomembrane Electrochemistry*, edited by M. Blank and I. Vodanoy (ACS Books, Washington, DC).

Haase, A., G. Landwehr, and E. Umbach, 1997, Eds., *Röntgen Centennial—X-rays in Natural and Life Sciences* (World Scientific, Singapore).

Hänggi, P., P. Talkner, and M. Borkovec, 1990, "Reaction-rate theory: fifty years after Kramers," Rev. Mod. Phys. **62**, 251–341.

Hertz, J., A. Krogh, and R. G. Palmer, 1991, *Introduction to the Theory of Neural Computation* (Addison-Wesly, Reading, MA).

Hogan, M. E., and R. H. Austin, 1987, "The importance of DNA stiffness in protein-DNA binding specificity," Nature (London) **329**, 263–265.

Hopfield, J. J., 1978, "Origin of the genetic code: A testable hypothesis based on tRNA structure, sequence, and kinetic proofreading," Proc. Natl. Acad. Sci. USA **75**, 4334–4338.

Hopfield, J. J., 1986, in *Lessons of Quantum Theory. Niels Bohr Centenary Symposium* (North-Holland, Amsterdam), p. 295–314.

Hoyle, F., 1957, *The Black Cloud* (Harper, New York).

Hu, X., and K. Schulten, 1997, "How nature harvests sunlight," Phys. Today **50**(8), 28–34.

Jülicher, F., A. Ajdari, and J. Prost, 1997, "Modeling molecular motors," Rev. Mod. Phys. **69**, 1269–1281.

Kreis, T., and R. Vale, 1993, *Cytoskeletal and Motor Proteins* (Oxford University Press, New York).

Langton, C. G., 1988, *Artificial Life: Proceedings of an Interdisciplinary Workshop on the Synthesis and Simulation of Living Systems* (Addison-Wesley, Reading, MA).

Leikan, S., V. A. Parsegian, and D. C. Rau, 1993, "Hydration Forces," Annu. Rev. Phys. Chem. **44**, 369–395.

Levinthal, C., 1969, "How to fold graciously," in *Mössbauer Spectroscopy in Biological Systems*, Vol. 22-24, edited by P. Debrunner, J. C. M. Tsibris, and E. Münck (University of Illinois Engineering, Urbana).

Loofbourow, J. R., 1940, "Borderland problems in biology and physics," Rev. Mod. Phys. **12**, 267–358.

McCammon, J. A., and S. C. Harvey, 1987, *Dynamics of Proteins and Nucleic Acids* (Cambridge University Press, New York).

Moerner, W. E., 1996, Ed., "Single molecules and atoms," Acc. Chem. Res. **29**, 561–613.

Nie, S., and R. N. Zare, 1997, "Optical detection of single molecules," Annu. Rev. Biophys. Biomol. Struct. **26**, 567–596.

Nienhaus, G. U., and R. D. Young, 1996, "Protein Dynamics," Encyclopedia Appl. Phys. **15**, 163–184.

Oncley, J. L., F. O. Schmitt, R. C. Williams, M. D. Rosenberg, and R. H. Bolt, 1959, "Biophysical Science," Rev. Mod. Phys. **31**, 1–1072.

Onuchic, J. N., D. N. Beratan, J. R. Winkler, and H. B. Gray, 1992, "Pathway analysis of proton-electron transfer reactions," Annu. Rev. Biophys. Biomol. Struct. **21**, 349–377.

Onuchic, J. N., Z. Luthey-Schulten, and P. G. Wolynes, 1997, "Theory of protein folding: The energy landscape perspective," Annu. Rev. Phys. Chem. **48**, 545–600.

Parak, F., private communication.

Peliti, L., 1991, Ed., *Biologically Inspired Physics* (Plenum, New York).

Perelson, A. S., and G. Weisbuch, 1997, "Immunology for Physicists," Rev. Mod. Phys. **69**, 1219–1267.

Perkins, T. T., D. E. Smith, and S. Chu, 1994, "Direct observation of tube-like motion of a single polymer chain," Science **264**, 819–822.

Perkins, T. T., D. E. Smith, and S. Chu, 1997, "Single polymer dynamics in an elongational flow," Science **276**, 2016–2021.

Redi, M. J., J. Hopfield, and B. Gerstman, 1981, "Hemoglobin-carbon monoxide binding rate. Low temperature magneto-optical detection of spin-tunneling," Biophys. J. **35**, 471–484.

Rejto, P. A., and S. T. Freer, 1996, "Protein conformational substrates from x-ray crystallography," Prog. Biophys. Mol. Biol. **66**, 167–196.

Richert R., and A. Blumen, 1994, Eds., *Disorder Effects on Relaxation Processes* (Springer, Berlin).

Schlichting, I., J. Berendzen, G. N. Phillips, Jr., and R. M. Sweet, 1994, "Crystal structure of photolysed carbonmonoxy-myoglobin," Nature (London) **371**, 808–812.

Schoenborn, B. P., and R. B. Knott, 1996, Eds., *Neutrons in Biology* (Plenum, New York).

Schrödinger, E., 1944, *What is Life?* (Cambridge University Press, Cambridge, England).

Schuster, S. C., and S. Khan, 1994, "The bacterial flagellar motor," Annu. Rev. Biophys. Biomol. Struct. **23**, 509–539.

Smith, S. B., Y. Cui, and C. Bustamante, 1996, "Overstretching B-DNA: The elastic response of individual double-stranded and single-stranded DNA molecules," Science **271**, 795–799.

Wang, J., and P. Wolynes, 1993, "Passage through fluctuating geometrical bottlenecks. The general Gaussian fluctuating case," Chem. Phys. Lett. **212**, 427–433.

Willis, B. T. M., and A. W. Pryor, 1975, *Thermal Vibrations in Crystallography* (Cambridge University Press, London).

Wittenberg, J. B., and B. A. Wittenberg, 1990, "Mechanisms of cytoplasmic hemoglobin and myoglobin function," Annu. Rev. Biophys. Biophys. Chem. **19**, 217–241.

Wüthrich, K., 1986, *NMR of Proteins and Nucleic Acids* (Wiley, New York).

Zeldovich, Ya. B., A. A. Ruzmaikin, and D. D. Sokoloff, 1990, *The Almighty Chance* (World Scientific, Singapore).

Zener, C., 1932, "Non-adiabatic crossing of energy levels," Proc. R. Soc. London, Ser. A **137**, 696–702.

Brain, Neural Networks, and Computation

J. J. Hopfield

The method by which brain produces mind has for centuries been discussed in terms of the most complex engineering and science metaphors of the day. Descartes described mind in terms of interacting vortices. Psychologists have metaphorized memory in terms of paths or traces worn in a landscape, a geological record of our experiences. To McCulloch and Pitts (1943) and von Neumann (1958), the appropriate metaphor was the digital computer, then in its infancy. The field of "neural networks" is the study of the computational properties and behavior of networks of "neuronlike" elements. It lies somewhere between a model of neurobiology and a metaphor for how the brain computes. It is inspired by two goals: to understand how neurobiology works, and to understand how to solve problems which neurobiology solves rapidly and effortlessly and which are very hard on present digital machines.

Most physicists will find it obvious that understanding biology might help in engineering. The obverse engineering-toward-biological link can be made by testing a circuit of "model neurons" on a difficult real-world problem such as oral word recognition. If the "neural circuit" with some particular biological feature is capable of solving a real problem which circuits without that feature solve poorly, the plausibility that the biological feature selected is computationally useful in biology is bolstered. If not, then it is more plausible that the feature can be dispensed with in modeling biology. These are not strong arguments, but they do provide an approach to finding out what, of the myriad of details in neurobiology, is truly important and what is merely true. The study of a 1950 digital computer, in the spirit of neurobiology, would have a strong commitment to studying BaO, then the material of vacuum tube cathodes. The study of the digital computer in 1998 would have a strong commitment to SiO_2, the essential insulating material below each gate. Yet the computing structure of the two machines could be identical, hidden amongst the lowest levels of detail. The study of "artificial neural networks" in the spirit of biology will relate to aspects of how neurobiology computes in the same sense that understanding the computer of 1998 relates to understanding the computer of 1950.

Brain as a Computer

A digital machine can be programmed to compare a present image with a three-dimensional representation of a person, and thus the problem of recognizing a friend can be solved by a computation. Similarly, how to drive the actuators of a robot for a desired motion is a problem in classical mechanics that can be solved on a computer. While we may not know how to write efficient algorithms for these tasks, such examples do illustrate that what the nervous system does might be described as computation.

For present purposes, a computer can be viewed as an input-output device, with input and output signals that are in the same format (Hopfield, 1994). Thus in a very simple digital computer, the input is a string of bits (in time), and the output is another string of bits. A million axons carry electrochemical pulses from the eye to the brain. Similar signaling pulses are used to drive the muscles of the vocal tract. When we look at a person and say, "Hello, Jessica," our brain is producing a complicated transformation from one (parallel) input pulse sequence coming from the eye to another (parallel) output pulse sequence which results in sound waves being generated. The idea of *composition* is important in this definition. The output of one computer can be used as the input for another computer of the same general type, for they are compatible signals. Within this definition, a digital chip is a computer, and large computers are built as composites of smaller ones. Each neuron is a simple computer according to this definition, and the brain is a large composite computer.

Computers as Dynamical Systems

The operation of a digital machine is most simply illustrated for batch-mode computation. The computer has N storage registers, each storing a single binary bit. The logical state of the machine at a particular time is specified by a vector $10010110000\ldots$ of N bits. The state changes each clock cycle. The transition map, describing which state follows which, is implicitly built into the machine by its design. The computer can thus be described as a dynamical system that changes its discrete state in discrete time, and the computation is carried out by following a path in state space.

The user of the machine has no control over the dynamics, which is determined by the state transition map. The user's program, data, and a standard initialization procedure prescribe the starting state of the machine. In a batch-mode computation, the answer is found when a stable point of the discrete dynamical system is reached, a state from which there are no transitions. A particular subset of the state bits (e.g., the contents of a particular machine register) will then describe the desired answer.

Batch-mode analog computation can be similarly described by using continuous variables and continuous time. The idea of computation as a process carried out by a dynamical system in moving from an initial state to a final state is the same in both cases. In the analog case, the possible motions in state space describe a flow field as in Fig. 1, and computation done by moving with this flow from start to end. (Real "digital" machines contain only analog components; the digital description is a representation in fewer variables which contains the essence of the continuous dynamics.)

Dynamical Model of Neural Activity

The anatomy of a "typical" neuron in a mammalian brain is sketched in Fig. 2 (Kandel, Schwartz, and Jessell, 1991). It has three major regions: dendrites, a cell body, and an axon. Each cell is connected by structures called synapses with approximately 1000 other cells. Inputs to a cell are made at synapses on its dendrites. The output of that cell is through synapses made by its axon onto the dendrites of other cells. The interior of the neuron is surrounded by a membrane of high resistivity and is filled with a conducting ionic solution. Ion-specific pumps transport ions across the membrane, maintaining an electrical potential difference between the inside and the outside of the cell. Ion-specific channels whose electrical conductivity is voltage dependent and dynamic play a key role in the evolution of the "state" of a neuron.

A simple "integrate and fire" model captures much of the mathematics of what a compact nerve cell does (Junge, 1981). Figure 3 shows the time-dependent voltage difference between the inside and the outside of a simple functioning neuron. The electrical potential is generally slowly changing, but occasionally there is a stereotype voltage spike of about two milliseconds duration. Such a spike is produced every time the interior potential of this cell rises above a threshold, u_{thresh}, of about 53 millivolts. The voltage then resets to a u_{reset} of about -70 millivolts. This "action potential" spike is caused by the dynamics of voltage-dependent ionic conductivities in the cell membrane. If an electrical current is injected into the cell, then except for the action potentials, the interior potential approximately obeys

$$C \, du/dt = -(u - u_{rest})/R + i(t), \tag{1}$$

where R is the resistance of the cell membrane, C the capacitance of the cell membrane, and u_{rest} is the potential to which the cell tends to drift. If $i(t)$ is a constant i_c, then the cell potential will change in an almost linear fashion between u_{rest} and u_{thresh}. An action potential will be generated each time u_{thresh} is reached, resetting u to u_{reset} similar to what is seen in Fig. 3. The time P between the equally spaced action potentials when R is very large is

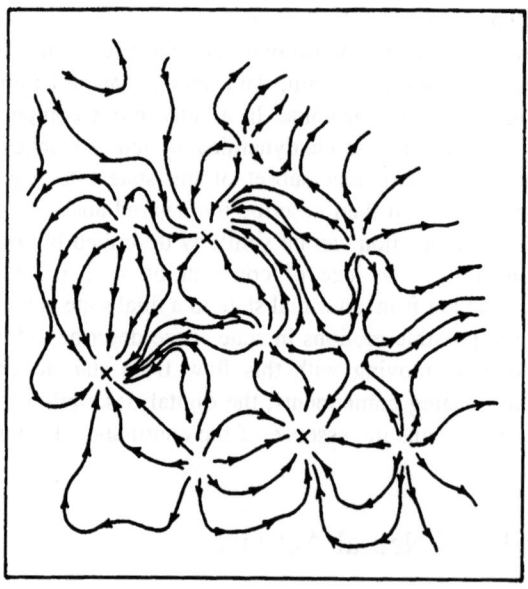

Figure 1. The flow field of a simple analog computer. The stable points of the flow, marked by x's, are possible answers. To initiate the computation, the initial location in state space must be given. A complex analog computer would have such a flow field in a very large number of dimensions.

$$P = C(u_{thresh} - u_{rest})/ic \quad \text{or firing rate} \quad 1/P \sim i_c. \tag{2}$$

If i_c is negative, no action potentials will be produced. The firing rate $1/P$ of a more realistic cell is not simply linear in i_c, but asymptotes to a maximum value of about 500

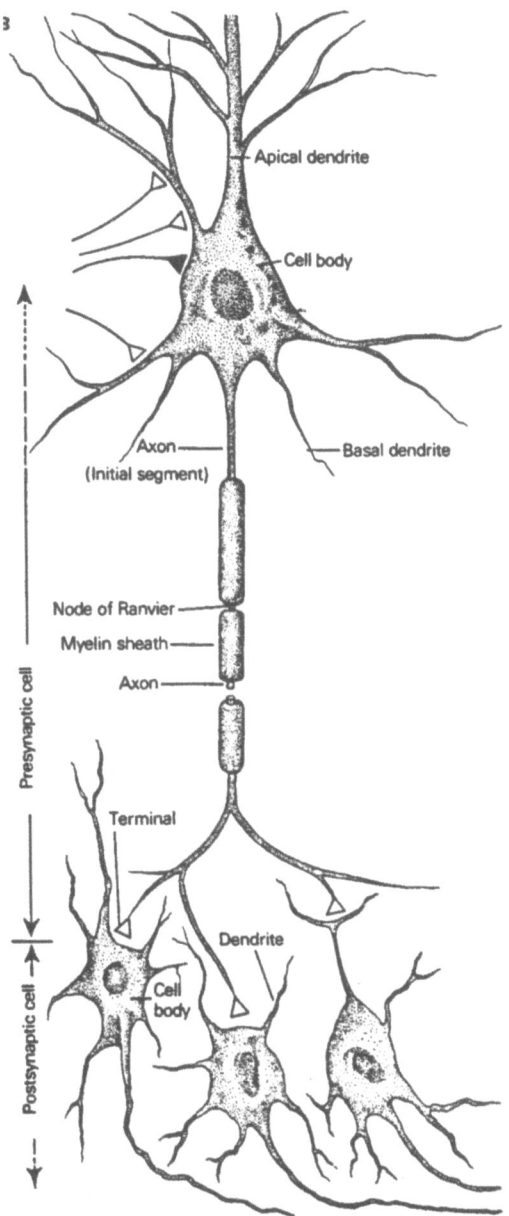

Figure 2. A sketch of a neuron and its style of interconnections. Axons may be as long as many centimeters, though most are on the scale of a millimeter. Typical cell bodies are a few microns in diameter.

per second (due to the finite time duration of action potentials). It may also have a nonzero threshold current due to leakage currents (of either sign) in the membrane.

Action potentials will be taken to be δ functions, lasting a negligible time. They propagate at constant velocity along axons. When an action potential arrives at a synaptic terminal, it releases a neurotransmitter which activates specific ionic conductivity channels in the postsynaptic dendrite to which it makes contact (Kandel, Schwartz, and Jessell, 1991). This short conductivity pulse can be modeled by

$$\sigma(t) = 0 \quad t < t_0$$
$$= s_{kj} \exp[-(t-t_0)/\tau] \quad t > t_0. \quad (3)$$

The maximum conductivity of the postsynaptic membrane in response to the action potential is s. The ion-specific current which flows is equal to the chemical potential difference V_{ion} times $\sigma(t)$. Thus at a synapse from cell j to cell k, an action potential arriving on axon j at time t_0 results in a current

$$i_{kj}(t) = 0 \quad t < t_0$$
$$= S_{kj} \exp[-(t-t_0)/\tau] \quad t > t_0 \quad (4)$$

flows into the cell k. The parameter $S_{kj} = V_{\text{ion}} s_{kj}$ can take either sign. Define the instantaneous firing rate of neuron k, which generates action potentials at times t_n^k, as

$$f_k(t) = \sum_n \delta(t - t_n^k). \quad (5)$$

By differentiation and substitution

$$di_k/dt = -i_k/\tau + \sum_j S_{kj} * f_j(t) + \text{another term if a sensory cell}. \quad (6)$$

This equation, though exact, is awkward to deal with because the times at which the action potentials occur are only implicitly given.

The usual approximation relies on there being many contributions to the sum in Eq. (6) during a reasonable time interval due to the high connectivity. It should then be permissible to ignore the spiky nature of $fj(t)$, replacing it by a convolution with a smoothing

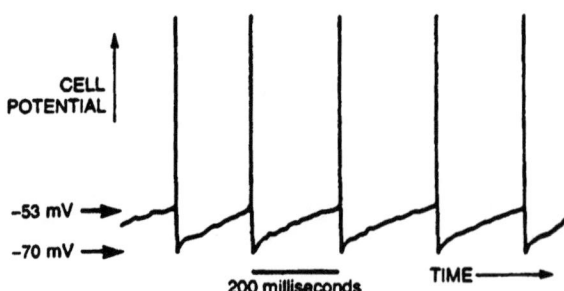

Figure 3. The internal potential of a neuron, driven with a constant current, as a function of time. In response to a steady current, many neurons (those which do not adapt) generate stereotype action potentials at a regular rate.

function. In addition, the functional form of $V(i_c)$ is presumed to hold when i_c is slowly varying in time. What results is like Eq. (6), but with $f_j(t)$ now a smooth function given by $f_j(t) = V[i_j(t) = V_j(t)]$:

$$di_k/dt = -i_k/\tau + \sum_j S_{kj} * V[i_j(t)] + I_k \quad \text{(last term only if a sensory cell).} \quad (7)$$

The main effect of the approximation, which assumes no strong correlations between spikes of different neurons, is to neglect shot noise. (Electrical circuits using continuous variables are based on a similar approximation.) While equations of this structure are in common use, they have somewhat evolved, and do not have a sharp original reference.

The Dynamics of Synapses

The second dynamical equation for neuronal dynamics describes the changes in the synaptic connections. In neurobiology, a synapse can modify its strength or its temporary behavior in the following ways:

(1) As a result of the activity of its presynaptic neuron, independent of the activity of the postsynaptic neuron.
(2) As a result of the activity of its postsynaptic neuron, independent of the activity of the presynaptic neuron.
(3) As a result of the coordinated activity of the pre- and postsynaptic neurons.
(4) As a result of the regional release of a neuromodulator. Neuromodulators also can modulate processes 1, 2, and 3.

The most interesting of these is (3) in which the synapse strength S_{kj} changes as a result of the roughly simultaneous activity of cells k and j. This kind of change is needed to represent information about the association between two events. A synapse whose change algorithm involves only the simultaneous activity of the pre- and postsynaptic neurons and no other detailed information is now called a Hebbian synapse (Hebb, 1949). A simple version of such dynamics (using firing rates rather than detailed times of individual action potentials) might be written

$$dS_{kj}/dt = \alpha * i_k * f_j(t) - \text{decay terms.} \quad (8)$$

Decay terms, perhaps involving i_k and $f(i_j)$, are essential to forget old information. A nonlinearity or control process is important to keep synapse strength from increasing without bound. The learning rate α might also be varied by neuromodulator molecules which control the overall learning process. The details of synaptic modification biophysics are not completely established, and Eq. (8) is only qualitative. A somewhat better approximation replaces α by a kernel over time and involves a more complicated form in i and f. Long-term potentiation (LTP) is the most significant paradigm of neurobiological synapse modification (Kandel, Schwartz, and Jessell, 1991). Synapse change rules of a Hebbian type have been found to reproduce results of a variety of experiments on the development of the eye dominance and orientation selectivity of cells in the visual cortex of the cat (Bear, Cooper, and Ebner, 1987).

The tacit assumption is often made that synapse change is involved in learning and development, and that the dynamics of neural activity is what performs a computation.

However, the dynamics of synapse modification should not be ignored as a possible tool for doing computation.

Programming Languages for Artificial Neural Networks (ANN)

Let batch-mode computation, simple (point) attractor dynamics, and fixed connections be our initial "neural network" computing paradigm. The connections need to be chosen so that for any input ("data") the network activity will go to a stable state, and so that the state achieved from a given input is the desired "answer." Is there a programming language?

The simplest approaches to this issue involve establishing an *overall architecture* or "anatomy" for the network which *guarantees going to a stable state*. Within this given architecture, the connections can be arbitrarily chosen. "Programming" involves the "inverse problem" of finding the set of connections for which the dynamics will carry out the desired task.

Feed-forward networks

The simplest two styles of networks for computation are shown in Fig. 4. The feed-forward network is mathematically like a set of connected nonlinear amplifiers without feedback paths, and is trivially stable.

This fact allows us to evaluate how much computation must be done to find the stable point. It is:

(1 multiply + 1 add) (number of connections) + (number of "neurons") (1 look-up).

This evaluation requires no dynamics and involves a trivial amount of computation. How is it then that feed-forward ANN's, which have almost no computing power, are very useful even when implemented inefficiently on digital machines? The answer is that their utility comes chiefly from the immense computational work necessary to find an appropriate set of connections for a problem which is implicitly described by a large data base. The resulting network is a compact representation of the data, which allows it to be used with much less computational effort than would otherwise be necessary.

The output of the network is merely a *function* of its input. In this case the problem of finding the best set of connections reduces to finding the set of connections that minimizes output error.

When the inputs of all amplifiers are connected by a network to the external inputs and the outputs of the same amplifiers are used as the desired outputs, a feed-forward network is said to have "no hidden units." If the amplifiers have a continuous input-output relation, a best set of connections can be found by starting with a random set of connections and doing gradient descent on the error function. For most problems, the terrain is relatively smooth, and there is little difficulty of being trapped in poor local minima by doing gradient descent. When the input-output relation is a step function, as it was chosen to be in the perceptron (Rosenblatt, 1962), the problem is somewhat more difficult, but satisfactory algorithms can still be found. An interesting "statistical mechanics" of their capabilities in random problems has been described (Gardiner, 1988).

Unfortunately, networks with a single layer of weights are severely limited in the functions they can represent. The detailed description of that limitation by Minsky and

Pappert (1969) and "our view that the extension [to multiple layers] is sterile" had a great deal to do with destroying a budding perceptron enthusiasm in the 1960s. It was even then clear that networks with hidden units are much more powerful, but the "failure to produce an interesting learning theorem for the multilayered machine" was chilling.

For the analog feed-forward ANN with hidden units, the problem of finding the best fit to a desired input-output relation is relatively simple since the output can be explicitly written in terms of the inputs and connections. Gradient descent on the error surface in "weight space" can be carried out, beginning from small random initial connections, to find a locally optimal solution to the problem. This elegant simple point was noted by Werbos (1974), but had no impact at the time, and was independently rediscovered at least twice in the 1980s. A variety of more complex ways to find good sets of connections have since been explored.

Why was the Werbos suggestion not followed up and subsequently lost? Several factors were involved. First, the landscape of the function on which gradient descent is being done is very rugged; local minima abound, and whether a useful network can be found is a computational issue, not a question which can be demonstrated from mathematics. There was little understanding of such landscapes at the time. Worse, the demon-

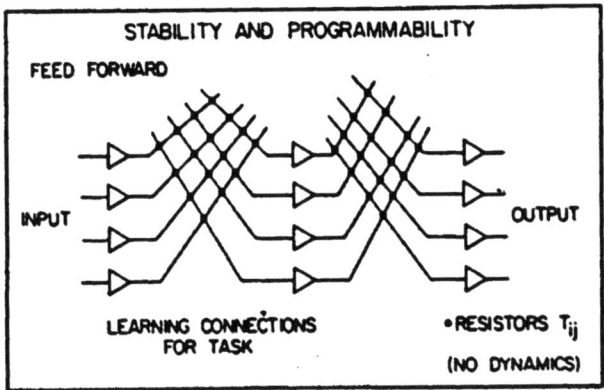

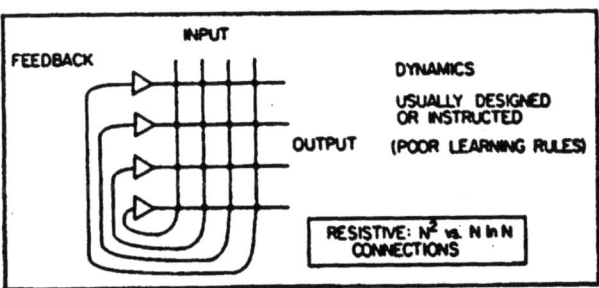

Figure 4. Two extreme forms of neural networks with good stability properties but very different complexities of dynamics and learning. The feedback network can be proved stable if it has symmetric connections. Scaling of variables generates a broad class of networks which are equivalent to symmetric networks, and surprisingly, the feed-forward network can be obtained from a symmetric network by scaling.

strations that such a simple procedure would actually work consumed an immense amount of computer cycles even in its time (1983-5) and would have been impossibly costly on the computers of 1973. Artificial intelligence was still in full bloom, and no one that was interested in pattern recognition would waste machine cycles on searches in spaces having hundreds of dimensions (parameters) when sheer logic and rules seemed all that was necessary.

And finally, the procedure looks absurd. Consider as a physicist, being told to fit a 200-parameter, highly nonlinear model to 500 data points. (And sometimes, the authors would be fitting 200 parameters to 150 data points!) We were all brought up on the Wignerism "if you give me two free parameters, I can describe an elephant. If you give me three, I can make him viggle his tail." We all knew that the parameters would be meaningless. And so they are. Two tries from initially different random starting sets of connections usually wind up with entirely different parameters. For most problems, the connection strengths seem to have little meaning. What is useful in this case, however, is *not* the connection strengths, but the *quality of fit* to the data. The situation is entirely different from the usual scientific "fits" to data, normally designed chiefly to derive meaningful parameters.

Feed-forward networks with hidden units have been successfully applied to evaluating loan applications, pap smear classification, optical character recognition, protein folding prediction, adjusting telescope mirrors, and playing backgammon. The nature of the features must be carefully chosen. The choice of network size and structure is important to success (Bishop, 1995) particularly when generalization is the important aspect of the problem (i.e., responding appropriately to a new input pattern that is not one on which the network was trained).

Feedback networks

There is immense feedback in brain connectivity. For example, axons carry signals from the retina to the LGN (lateral geniculate nucleus). Axons originating in the LGN carry signals to cortical processing area $V1$. But there are more axons carrying signals from $V1$ back to LGN than in the "forward" direction. The axons from LGN make synapses on cells in layer IV of area $V1$. However, most of the synaptic inputs within layer IV come from other cells within $V1$. Such facts lead to strong interest in neural circuits with feedback.

The style of feedback circuit whose mathematics is most simply understood has symmetric connection, i.e., $S_{kj} = S_{jk}$. In this case, there is a Lyapunov or "energy" function for Eq. (8), and the quantity f_i

$$E = -\tfrac{1}{2}\sum S_{ij}V_iV_j - \sum I_iV_i + 1/\tau \sum \int V^{-1}(f')df' \qquad (9)$$

always decreases in time (Hopfield, 1982, 1994). The dynamics then is described by a flow to an attractor where the motion ceases.

In the high-gain limit, where the input-output relationship is a step between two asymptotic values, the system has a direct relationship to physics. It can be stated most simply when the asymptotic values are scaled to $+-1$. The stable points of the dynamic system then have each $V_i = +-1$, and the stable states of the dynamical system are the stable points of an Ising magnet with exchange parameters $J_{ij} = S_{ij}$.

The existence of this energy function provides a programming tool (Hopfield and Tank, 1985; Takefuji, 1991). Many difficult computational problems can be posed as optimization problems. If the quantity to be optimized can be mapped onto the form Eq. (8), it defines the connections and the "program" to solve the optimization problem.

The trivial generalization of the Ising system to finite temperature generates a statistical mechanics. However, a "learning rule" can then be found for this system, even in the presence of hidden units. This was the first successful learning rule used for networks with hidden units (Hinton and Sejnowski, 1983). Because it is computationally intensive, practical applications have chiefly used analog "neurons" and the faster "back-propagation" learning rule when applicable. The relationship with statistical mechanics and entropic information measures, however, give the Boltzmann machine continuing interest.

Associative memories are thought of as a set of linked features $f1, f2$, etc. The activity of a particular neuron signifies the presence of the feature represented by that neuron. A memory is a state in which the cells representing the features of that memory are simultaneously active. The relationship between features is symmetric in that each implies the other and is expressed in a symmetric network. An elegant analysis of the capacity of such memories for random patterns is related to the spin glass (Amit, 1989).

Many nonsymmetric networks can be mapped onto networks with related Lyaupanov functions. Thus, while symmetric networks are exceptional in biology, the study of networks with Lyapunov functions is a useful approach to understanding biological networks. Line attractors have been used in connection with keeping the eye gaze stable at any set position (Seung, 1996).

Networks which have feedback may oscillate. The olfactory bulb is an example of a circuit with a strong excitatory-inhibitory feedback loop. In mammals, the olfactory bulb bursts into 30–50 Hz oscillations with every sniff (Freeman and Skarda, 1985).

Development and Synapse Plasticity

For simple animals such as the *C. elegans* (a round worm) the nervous system is essentially determined. Each genetically identical *C. elegans* has the same number of nerve cells, each cell identifiable in morphology and position. The synaptic connections between such "identical" animals are 90% identical. Mammals, at the other end of the spectrum, have identifiable cell types, identifiable brain structures and regions, but no cells in 1:1 correspondence between different individuals. The "wiring" between cells clearly has rules, and also a strong random element arising from development. How, then, can we have the system of fine-tuned connections between neurons which produces visual acuity sharper than the size of a retinal photoreceptor, or coordinates the two eyes so that we have stereoscopic vision? The answer to this puzzle lies in the synapses change due to coordinated activity during development. Coordinated activity of neurons arises from the correlated nature of the visual world and is carried through to higher level neurons. The importance of neuronal activity patterns and external input is dramatically illustrated in depth perception. If a "wandering eye" through muscular miscoordination, is corrected in the first six months, a child develops normal binocular stereopsis. Corrected after two years, the two eyes are used in a coordinate fashion and seem completely normal, but the child will never develop stereoscopic vision.

When multiple input patterns are present, the dynamics generates a cellular competition for the representation of these patterns. The idealized mathematics is that of a symmetry breaking. Once symmetry is broken, the competition continues to refine the connections (Linsker, 1986). This mathematics was originally used to describe the development of connections between the retina and the optic tectum of the frog. It describes well the generation of orientation-selective cells in the cat visual cortex. A hierarchy of such symmetry breakings has been used to describe the selectivity of cells in the mammalian visual pathway. This analysis is simple only in cases where the details of the biology have been maximally suppressed, but such models are slowly being given more detailed connections to biology (Miller, 1994).

There is an ongoing debate in such areas about "instructionism" versus "selectionism," and on the role of genetics versus environmental influences. "Nature" versus "nurture" has been an issue in psychology and brain science for decades and is seen at its most elementary level in trying to understand how the functional wiring of an adult brain is generated.

Action Potential Timing

The detailed timing in a train of action potentials carries information beyond that described by the short-term firing rate. When several presynaptic neurons fire action potentials simultaneously, the event can have a saliency for driving a cell that would not occur if the events were more spread out in time. These facts suggest that for some neural computations, Eq. (7) may lose the essence of Eq. (6). Theoretically, information can be encoded in action potential timing and computed efficiently and rapidly (Hopfield, 1995).

Direct observations also suggest the importance of action potential timing. Experiments in cats indicate that the synchrony of action potentials between different cells might represent the "objectness" of an extended visual object (Gray and Singer, 1989). Synchronization effects are seen in insect olfaction (Stopfer et al., 1997). Azimuthal sound localization by birds effectively involves coincidences between action potentials arriving via right- and left-ear pathways. A neuron in rat hippocampus which is firing at a low rate carries information about the spatial location of the rat in its phase of firing with respect to the theta rhythm (Burgess, O'Keefe, and Recce, 1993). Action potentials in low-firing-rate frontal cortex seem to have unusual temporal correlation. Action potentials propagate back into some dendrites of pyramidal cells, and their synapses have implicit information both from when the presynaptic cell fired *and when the postsynaptic cell fired*, potentially important in a synapse-change process.

The Future

The field now known as "computational neurobiology" has been based on an explosion in our knowledge of the electrical signals of cells during significant processing events and on its relationship to theory including understanding simple neural circuits, the attractor model of neural computation, the role of activity in development, and the information-theoretic view of neural coding. The short-term future will exploit the new ways to visualize neural *activity*, involving multi-electrode recording, optical signals from cells (voltage-dependent dyes, ion-binding fluorophores, and intrinsic signals) functional magnetic resonance imaging, magnetoencephalography, patch clamp techniques, confocal microscopy, and microelectrode arrays. Molecular biology tools have now also begun to be significant for computational neurobiology. On the modeling side it will involve understanding more of the computational power of biological systems by using additional biological features.

The study of silicon very large scale integrated circuits (VLSI's) for analog "neural" circuits (Mead, 1989) has yielded one relevant general principle. When the physics of a device can be used in an algorithm, the device is highly effective in computation compared to its effectiveness in general purpose use. Evolution will have exploited the biophysical molecular and circuit devices available. For any particular behavior, some facts of neurobiology will be very significant because they are used in the algorithm, and others will be able to be subsumed in a model which is far simpler than the actual biophysics of the system. It is important to make such separations, for neurobiology is so filled with details that we will never understand the neurobiological basis of perception, cognition, and psychology merely by accumulating facts and doing ever more detailed simulations. Linear systems are simple to characterize completely. Computational systems are highly nonlinear, and a complete characterization by brute force requires a number of experiments which grows exponentially with the size of the system. When only a limited number of experiments is performed, the behavior of the system is not fully characterized, and to a considerable extent the experimental design builds in the answers that will be found. For working at higher computational levels, experiments on anaesthetized animals, or in highly simplified, overlearned artificial situations, are not going to be enough. Nor will the characterization of the behavior of a very small number of cells during a behavior be adequate to understand how or why the behavior is being generated. Thus it will be necessary to build a better bridge between lower animals, which can be more completely studied, and higher animals, whose rich mental behavior is the ultimate goal of computational neurobiology.

References

Amit, D., 1989, *Modeling Brain Function* (Cambridge University, Cambridge).
Bear, M. F., L. N. Cooper, and F. F. Ebner, 1987, Science **237**, 42.
Bishop, C. M., 1995, *Neural Networks for Pattern Recognition* (Oxford University, Oxford).
Burgess, N., J. M. O'Keefe, and M. Recce, 1993, in *Advances in Neural Information Processing Systems*, Vol. 5, edited by S. Hanson, C. L. Giles, and J. D. Cowan (Morgan Kaufman, San Mateo, CA), p. 929.
Freeman, W. F., and C. A. Skarda, 1985, Brain Res. Rev. **10**, 147.
Gardiner, E., 1988, J. Phys. A **21**, 257.
Gray, C. M., and W. Singer, 1989, Proc. Natl. Acad. Sci. USA **86**, 1698.
Hebb, D. O., 1949, *The Organization of Behavior* (Wiley, New York).
Hinton, G. E., and T. J. Sejnowski, 1983, *Proceedings of the IEEE Conference on Computer Vision and Pattern Recognition* (IEEE, Washington, D.C.), p. 488.
Hopfield, J. J., 1982, Proc. Natl. Acad. Sci. USA **79**, 2554.
Hopfield, J. J. and D. G. Tank, 1985, Biol. Cybern. **52**, 141.
Hopfield, J. J., 1994, Phys. Today **47**(2), 40.
Hopfield, J. J., 1995, Nature (London) **376**, 33.
Junge, D., 1981, *Nerve and Muscle Excitation* (Sinauer Assoc., Sunderland, MA).
Kandel, E. R., J. H. Schwartz, and T. M. Jessell, 1991, *Principles of Neural Science* (Elsevier, New York).
Linsker, R., 1986, Proc. Natl. Acad. Sci. USA **83**, 8390, 8779.
McCulloch, W. W., and W. Pitts, 1943, Bull. Math. Biophys. **5**, 115.

Mead, C. A., 1989, *Analog VLSI and Neural Systems* (Addison Wesley, Reading, MA).
Miller, K. D., 1994, Prog. Brain Res. **102**, 303.
Minsky, M., and S. Papert, 1969, *Perceptrons: An Introduction to Computational Geometry* (MIT, Cambridge), p. 232.
Rosenblatt, F., 1962, *Principles of Perceptrons* (Spartan, Washington, D.C.).
Seung, S. H., 1996, Proc. Natl. Acad. Sci. USA **93**, 13339.
Stopfer, M. S., S. Bhagavan, B. H. Smith, and G. Laurent, 1997, Nature (London) **390**, 70.
Takefuji, Y., 1991, *Neural Network Parallel Computing* (Kluwer, Boston).
von Neumann, J., 1958, *The Computer and the Brain* (Yale University, New Haven).
Werbos, P. J., 1974, Ph.D. thesis (Harvard University).

Computational Physics

Microscopic Simulations in Physics
D. M. Ceperley

In 1820, Laplace speculated on determining the consequences of physical laws:

An intelligent being who, at a given moment, knows all the forces that cause nature to move and the positions of the objects that it is made from, if also it is powerful enough to analyze this data, would have described in the same formula the movements of the largest bodies of the universe and those of the lightest atoms. Although scientific research steadily approaches the abilities of this intelligent being, complete prediction will always remain infinitely far away.

His intuition about complete predictability has been borne out: in general, dynamics is chaotic, thus making long-range forecasts unreliable because of their sensitivity to initial conditions.

The question remains whether average properties such as those that arise in statistical mechanics and thermodynamics may be predictable from first principles. Shortly after the formulation of quantum mechanics Dirac (1929) recognized

The general theory of quantum mechanics is now almost complete. The underlying physical laws necessary for the mathematical theory of a large part of physics and the whole of chemistry are thus completely known, and the difficulty is only that the exact application of these laws leads to equations much too complicated to be soluble.

Today, we might add the disciplines of biology and materials science to physics and chemistry as fundamentally based on the principles of the Maxwell, Boltzmann, and Schrödinger theories. The complication in solving the equations has always been in the many-body nature of most problems.

Rather than trying to encapsulate the result in a formula as Laplace and Dirac would have done, in the last half century we have turned to computer simulations as a very powerful way of providing detailed and essentially exact information about many-body problems, enabling one to go directly from a microscopic Hamiltonian to the macroscopic properties measured in experiments. Because of the power of the methods, they are used in most areas of pure and applied science; an appreciable fraction of total scientific computer usage is taken up by simulations of one sort or another.

Computational physics has been said to constitute a third way of doing physics, comparable to theory and experiment. What is the role of theory or simulation in physics today? How can simulations aid in providing understanding of a physical system? Why not just measure properties in the laboratory? One answer is that simulation can give reliable predictions when experiments are not possible, very difficult, or simply expensive. Some examples of such questions are: What is the behavior of hydrogen and other elements under conditions equivalent to the interior of a planet or star? How do phase transitions change in going from two to three to four dimensions? Is the standard model of QCD correct?

Part of the reason for the pervasiveness of simulations is that they can scale up with the increase of computer power; computer speed and memory have been growing geometrically over the last 5 decades with a doubling time of roughly 18 months (Moore's law). The earliest simulations involved 32 particles; now one can do hundreds of millions of particles. The increase in hardware speed will continue for at least another decade, and improvements in algorithms will hopefully sustain the growth for far longer than that. The discipline of computer simulation is built around an instrument, the computer, as other fields are built around telescopes and microscopes. The difference between the computer and those instruments is evident both in the computer's pervasive use in society and in its mathematical, logical nature.

Simulations are easy to do, even for very complex systems; often their complexity is no worse than the complexity of the physical description. In contrast, other theoretical approaches typically are applicable only to simplified models; methods for many-body problems involve approximations with a limited range of validity. To make theoretical progress, one needs a method to test out or benchmark approximate methods to find this range of application. Simulations are also a good educational tool; one does not have to master a particular theory to understand the input and output of a simulation.

Two different sorts of simulation are often encountered. In the first approach, one assumes the Hamiltonian is given. As Dirac said above, it is just a question of working out the details—a problem for an applied mathematician. This implies that the exactness of simulation is very important. But what properties of a many-body system can we calculate without making any uncontrolled approximations and thereby answer Laplace's and Dirac's speculations? Today, we are far from solving typical problems in quantum physics from this viewpoint. Even in classical physics, it takes a great deal of physical knowledge and intuition to figure out which simulations to do, which properties to measure, whether to trust the results, and so forth. Nevertheless, significant progress has been made.

The second approach is that of a modeler; one is allowed to invent new models and algorithms to describe some physical system. One can invent a fictitious model with rules that are easy to carry out on a computer and then systematically study the properties of the model. Which precise equations they satisfy are secondary. Later, one might investigate whether some physical system is described by the model. Clearly, this approach is warranted in such fields as economics, ecology, and biology since the "correct" underlying description has not always been worked out. But it is also common in physics, and occasionally it is extremely successful, as for example, in the lattice gas description of hydrodynamics and models for self-organized criticality. Clearly, the methodology for this kind of activity is different from that of the applied mathematics problem mentioned above, since one is testing the correctness both of the model and of the numerical implementation.

Lattice models, such as the well-known Ising, Heisenberg, and Hubbard models of magnetism, are intermediate between these two approaches. It is less important that they precisely describe some particular experiment than that they have the right "physics." What one loses in application to real experiments, one gains in simulation speed. Lattice models have played a key role in understanding the generic properties of phase transitions and in modeling aspects of the oxide superconductors. Since, necessarily, this review will just hit a few highlights, I shall concentrate on the first type of simulation and the road to precise predictions of the microscopic world.

Classical Simulations

The introduction of the two most common algorithms, molecular dynamics (Alder and Wainright, 1957) and Monte Carlo (Metropolis *et al.*, 1953), occurred shortly after the dawn of the computer age. The basic algorithms have hardly changed in the intervening years, although much progress has been made in elaborating them (Binder 1978, 1984, 1995; Ciccotti and Hoover, 1986; Binder and Ciccotti, 1996; Ferguson *et al.*, 1998). The mathematical problem is to calculate equilibrium and/or dynamical properties with respect to the configurational Boltzmann distribution:

$$\langle \mathcal{O} \rangle = \int dr_1 \cdots dr_N \mathcal{O}(R) e^{-\beta V(R)} \bigg/ \int dr_1 \cdots dr_N e^{-\beta V(R)}, \tag{2.1}$$

where $\mathcal{O}(R)$ is some function of the coordinates $R = (\mathbf{r}_1, \mathbf{r}_2, \ldots, \mathbf{r}_N)$ and $V(R)$ is the potential-energy function.

Part of the appeal of these simulations is that both methods are very easy to describe. Molecular dynamics is simply the numerical solution of Newton's equation of motion; thermal equilibrium is established by ergodicity. Monte Carlo (Metropolis or Markov Chain) is a random walk through phase space using rejections to achieve detailed balance and thereby sample the Boltzmann distribution. Molecular dynamics can be used to calculate classical dynamics; Monte Carlo only calculates static properties, unless you accept that a random walk is an interesting dynamical model. The two methods are not completely different; for example, there exist hybrid methods in which molecular dynamics are used for awhile, after which the velocities are randomized.

What is not always appreciated is that one does not do a brute force integration with Monte Carlo because the integrand of Eq. (2.1) is very sharply peaked in many dimensions. By doing a random walk rather than a direct sampling, one stays where the integrand is large. But this advantage is also a curse because it is not obvious whether any given walk will converge to its equilibrium distribution in the time available; this is the ergodic problem. This aspect of simulation is experimental; there are no useful theorems, only lots of controlled tests, the lore of the practitioners, and occasional clean comparisons with experimental data. Other subtleties of these methods are how to pick the initial and boundary conditions, determine error bars on the results, compute long-range potentials quickly, and determine physical properties (Allen and Tildesley, 1988).

An important reason why certain algorithms become more important over time lies in their scaling with respect to the number of variables: the complexity. To be precise, if we want to achieve a given error for a given property, we need to know how the computer time scales with the degrees of freedom, say the number of particles. The computer time will depend on the problem and property, but the exponents might be "universal." For the algorithms that scale the best, computer power increases with a low power of the number of degrees of freedom. Order (N) is the best and is achievable on the simplest classical problems. For those systems, as already noted, the number of particles used in simulations has gone from 224 hard spheres in 1953 to hundreds of millions of realistically interacting atoms today. On the other hand, while a very accurate quantum scattering calculation could be done for two scattering particles in the 1950s, only four particles can be done with comparable accuracy today. During this time, the price of the fastest computer has remained in the range of $20 million. This difference in scaling arises from the exponential complexity of quantum scattering calculations.

Applying even the best order (N) scaling to a macroscopic system from the microscopic scale is sobering. The number of arithmetic operations per year on the largest machine is approximately 10^{19} today. Let us determine the largest classical calculation we can consider performing using that machine for an entire year. Suppose the number of operations per neighbor of a particle is about 10 and that each atom has about 10 neighbors. Then the number of particles N times the number of time steps T achievable in one year is $NT \approx 10^{17}$. For a physical application of such a large system, at the very minimum one has to propagate the system long enough for sound to reach the other side so that $T \gg L$ where L is the number of particles along one edge. Taking $T = 10L$ for simplicity, one finds that even on the fastest computer today we can have a cube roughly 10^4 atoms on a side (10^{12} atoms altogether) for roughly $T = 10^5$ times steps. Putting in some rough numbers for silicon, that gives a cube 2 μm on a side for 10 ps. Although Moore's law is some help, clearly we need more clever algorithms to treat truly macroscopic phenomena! (Because spacetime is four dimensional, the doubling time for lengths scales will be six years.)

It has not escaped notice that many of the techniques developed to model atoms have applications in other areas such as economics. Although 10^{12} atoms is small in physical terms, it is much larger than the number of humans alive today. Using today's computers we can already simulate the world's economy down to the level of an individual throughout a lifetime (assuming the interactions are local and as simple as those between atoms).

A key early problem was the simulation of simple liquids. A discovery within the first few years of the computer era was that even a hundred particles could be used to predict things like the liquid-solid phase transition and the dynamics and hydrodynamics of simple liquids for relatively simple, homogeneous systems. Later on, Meiburg (1986) was able to see vortex shedding and related hydrodynamic phenomena in a molecular dynamics simulation of 40000 hard spheres moving past a plate. Much work has been performed on particles interacting with hard-sphere, Coulombic, and Lennard-Jones systems (Hansen and McDonald, 1986; Allen and Tildesley, 1988). Many difficult problems remain even in these simple systems. Among the unsolved problems are how hard disks melt, how polymers move, how proteins fold, and what makes a glass special.

Another important set of early problems was the Ising model and other lattice models. These played a crucial role in the theory of phase transitions, as elaborated in the scaling and renormalization theory along with other computational (e.g., series expansions) and theoretical approaches. There has been steady progress in calculating exponents and other sophisticated properties of lattice spin models (Binder, 1984) which, because of universality, are relevant to any physical system near the critical point. An important development was the discovery of cluster algorithms (Swendsen and Wang, 1975). These are special Monte Carlo sampling methods, which easily move through the phase space even near a phase transition, where any local algorithm will become very sluggish. A key challenge is to generalize these methods so that continuum models can be efficiently simulated near phase boundaries.

Quantum Simulations

A central difficulty for the practical use of classical simulation methods is that the forces are determined by an interacting quantum system: the electrons. Semiempirical pair potentials, which work reasonably well for the noble gases, are woefully inadequate for most materials. Much more elaborate potentials than Lennard-Jones (6-12) are needed for problems in biophysics and in semiconductor systems. There are not enough high-quality experimental data to use to parametrize these potentials even if the functional form of the interaction were known. In addition, some of the most important and interesting modern physics phenomena, such as superfluidity and superconductivity, are intrinsically nonclassical. For progress to be made in treating microscopic phenomena from first principles, simulations have to deal with quantum mechanics.

The basis for most quantum simulations is imaginary-time path integrals (Feynman, 1953) or a related quantum Monte Carlo method. In the simplest example, the quantum statistical mechanics of bosons is related to a purely classical problem, but one that has more degrees of freedom. Suppose one is dealing with a quantum system of N particles interacting with the standard two-body Hamiltonian:

$$\mathcal{H} = -\sum_{i=1}^{N} \frac{\hbar^2}{2m_i} \nabla_i^2 + V(R). \qquad (3.1)$$

Path integrals can calculate the thermal matrix elements: $\langle R|e^{-\beta \mathcal{H}}|R'\rangle$. We still want to perform integrations as in Eq. (2.1), except now we have an operator to sample instead of a simple function of coordinates. This is done by expanding into a path average:

$$\langle R|e^{-\beta \mathcal{H}}|R'\rangle = \int dR_1 \cdots \int dR_M \exp[-S(R_1, \ldots, R_M)]. \qquad (3.2)$$

In the limit that $\tau = \beta/M \to 0$, the action S has the explicit form

$$S(R_1, \ldots, R_M) = \sum_{i=1}^{M} \left(\left[\sum_{k=1}^{N} \frac{m_k (\mathbf{r}_{k,i} - \mathbf{r}_{k,i-1})^2}{2\hbar^2 \tau} \right] - \tau V(R_i) \right). \qquad (3.3)$$

The action is real, so the integrand is non-negative, and thus one can use molecular dynamics or Monte Carlo as discussed in the previous section to evaluate the integral. Doing the trace in Eq. (3.2) means the paths close on themselves; there is a beautiful analogy between quantum mechanics and the statistical mechanics of ring polymers (Feynman, 1953). Figure 1(a) shows a picture of a typical path for a small sample of "normal" liquid ^{4}He.

Most quantum many-body systems involve Fermi or Bose statistics which cause only a seemingly minor modification: one must allow the paths to close on themselves with a permutation of particle labels so that the paths go from R to PR as in Fig. 1(b), with P a permutation of particle labels. In a superfluid system, exchange loops form that have a macroscopic number of atoms connected on a single path stretching across the sample. Superfluidity is equivalent to a problem of percolating classical polymers. It is practical to perform a simulation of thousands of helium atoms for temperatures both above and below the transition temperature (Ceperley, 1995). The simulation method gives considerable insight into the nature of superfluids. Using this method, we have recently

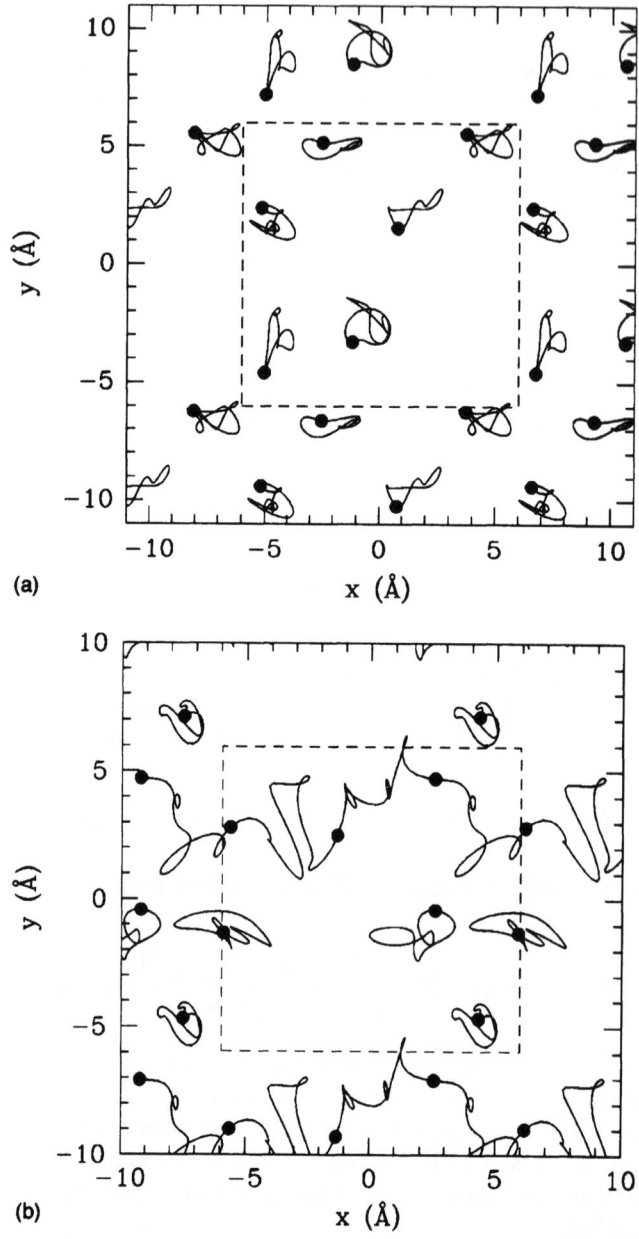

Figure 1. Typical "paths" of six helium atoms in 2D. The filled circles are markers for the (arbitrary) beginning of the path. Paths that exit on one side of the square reenter on the other side. The paths show only the lowest 11 Fourier components. (a) shows normal ^{4}He at 2 K, (b) superfluid ^{4}He at 0.75 K.

predicted that H_2 placed on a silver surface salted with alkali-metal atoms will become superfluid below 1 K (Gordillo and Ceperley, 1997). Krauth (1996) did simulations comparable to the actual number of atoms in a Bose-Einstein condensation trap (10^4).

Unfortunately, Fermi statistics are not so straightforward: one has to place a minus sign in the integrand for odd permutations and subtract the contribution of negative permutations from that of the positive permutations. This usually causes the signal/noise ratio to approach zero rapidly so that the computer time needed to achieve a given accuracy will grow exponentially with the system size, in general as $\exp[2(F_f - F_b)/(k_B T)]$, where F_f is the total free energy of the fermion system, F_b the free energy of the equivalent Bose system, and T the temperature (Ceperley, 1996). The difference in free energy is proportional to the number of particles, so the complexity grows exponentially. The methods are exact, but they become inefficient when one tries to use them on large systems or at low temperatures. But we do not know how to convert a fermion system into a path integral with a non-negative integrand to avoid this "sign" problem.

A related area where these methods are extensively used is the lattice gauge theory of quantum chromodynamics. Because one is simulating a field theory in four dimensions, those calculations are considerably more time consuming than an equivalent problem in nonrelativistic, first-quantized representation. Special-purpose processors have been built just to treat those models. Quantum Monte Carlo methods are also used to make precise predictions of nuclear structure; those problems are difficult because of the complicated nuclear interaction, which has spin and isospin operators. Currently up to seven nucleons can be treated accurately with direct quantum Monte Carlo simulation (Carlson and Schiavilla, 1998).

All known general exact quantum simulation methods have an exponential complexity in the number of quantum degrees of freedom (assuming one is using a classical computer). However, there are specific exceptions (solved problems) including thermodynamics of bosons, fermions in one dimension, the half-filled Hubbard model (Hirsch, 1985), and certain other lattice spin systems.

Approaches with good scaling make approximations of one form or another (Schmidt and Kalos, 1984). The most popular is called the fixed-node method. If the places where the wave function or density matrix changes sign (the nodes) are known, then one can forbid negative permutations (and matching positive permutations) without changing any local properties. This eliminates the minus signs and makes the complexity polynomial in the number of fermions. For systems in magnetic fields, or in states of fixed angular momentum, one can fix the phase of the many-body density matrix and use quantum Monte Carlo for the modulus (Ortiz et al., 1993). Unfortunately, except in a few special cases, such as one-dimensional problems where the nodes are fixed by symmetry, the nodal locations or phases must be approximated. Even with this approximation, the fixed-node approach gives accurate results for strongly correlated many-body systems.

Even at the boson level, many interesting problems cannot be solved. For example, there are serious problems in calculating the dynamical properties of quantum systems. Feynman (1982) made the following general argument that quantum dynamics is a very hard computational problem: If, to make a reasonable representation of the initial wave function for a single particle involves giving b complex numbers, then N particles will take on the order of b^N numbers. Just specifying the initial conditions gets out of hand very rapidly once b and N get reasonably large. (Remember that Laplace's classical initial conditions only required $6N$ numbers.) One may reduce this somewhat by using symmetry, mainly permutational symmetry, but on close analysis that does not help nearly enough. The only way around this argument is to give up the possibility of simulating general quantum dynamics and to stick to what is experimentally measurable; arbitrary initial conditions cannot be realized in the laboratory anyway. If quantum computers ever become available, they would at least be able to handle the quantum dynamics, once the initial conditions are set (Lloyd, 1996).

Mixed Quantum and Classical Simulations

A major development on the road to simulating real materials in the last decade has been the merging of quantum and classical simulations. In simulations of systems at room temperature or below, electrons are to a good approximation at zero temperature, and in most cases the nuclei are classical. In those cases, there exists an effective potential between the nuclei due to the electrons. Knowing this potential, one could solve most problems of chemical structure with simulation. But it needs to be computed very accurately because the natural electronic energy scale is the Hartree or Rydberg ($me^4/\hbar^2$), and chemical energies are needed to better than $k_B T$. At room temperature this requires an accuracy of one part in 10^3 for a hydrogen atom. Higher relative accuracy is needed for heavier atoms since the energy scales as the square of the nuclear charge.

Car and Parrinello (1985) showed that it is feasible to combine classical molecular dynamics with the simultaneous evaluation of the force using density-functional theory. Earlier, Hohenberg and Kohn (1964) had shown that the electronic energy is a functional only of the electronic density. In the simplest approximation to that functional, one assumes that the exchange and correlation energy depend only on local electron density (the local-density approximation). This approximation works remarkably well when used to calculate minimum-energy structures. The idea of Car and Parrinello was to evolve the electronic wave function with a fictitious dynamics as the ions are moving using classical molecular dynamics. A molecular dynamics simulation of the enlarged system of the ions plus the electronic wave functions is performed (Payne et al., 1992). Because one does not have to choose the intermolecular potential, one has the computer doing what the human was previously responsible for. The method has taken the field by storm in the last decade. There is even third-party software for performing these simulations, a situation rare in physics and an indication that there is an economic interest in microscopic simulation. Some recent applications are to systems such as nanotubes, water, liquid silicon, and carbon. Why was the combination of electronic structure and molecular dynamics made in 1985 and not before? Partly, because only in 1985 were computers powerful enough that such a combined treatment was feasible for a reasonably large system. We can anticipate more such examples as computer power grows and we go towards *ab initio* predictions. For example, Tuckerman et al. (1997) performed simulations of small water clusters using path-integral molecular dynamics for the nucleus and density-functional calculations for the electronic wave functions.

Today, the combined molecular dynamics method is too slow for many important problems; in practice one can only treat hundreds of electrons. Also, even though the LDA method is much more accurate than empirical potentials, on many systems it is not accurate enough. It has particular problems when the electronic band gap is small and the ions are away from equilibrium such as when bonds are breaking. Much work is being devoted to finding more accurate density-functional approximations (Dreizler and Gross, 1990).

Prospects

As computer capabilities grow, simulations of many-body systems will be able to treat more complex physical systems to higher levels of accuracy. The ultimate impact for an extremely wide range of scientific and engineering applications will undoubtedly be profound. The dream that simulation can be a partner with the experimentalist in designing new molecules and materials has been suggested many times in the last thirty years. Although there have been some successes, at the moment, more reliable methods for calculations energy differences to 0.01 eV (or 100 K) accuracy are needed. One does not yet have the same confidence in materials calculations that Laplace would have had in his calculations of planetary orbits. Because of the quantum nature of the microscopic world, progress in computers does not translate linearly into progress in materials simulations. Thus there are many opportunities for progress in the basic methodology.

It is unlikely that any speedup in computers will allow direct simulation, even at the classical level, of a truly macroscopic sample, not to speak of macroscopic quantum simulations. A general research trend is to develop multiscale methods, in which a simulation at a fine scale is directly connected to one at a coarser scale, thus allowing one to treat problems in which length scales differ. Historically this has been done by calculating parameters, typically linear response coefficients, which are then used at a higher level. For example, one can use the viscosity coefficient calculated with molecular dynamics in a hydrodynamics calculation. Abraham *et al.* (1998) describe a single calculation which goes from the quantum regime (tight-binding formulation), to the atomic level (classical molecular dynamics) to the continuum level (finite element). They apply this methodology to the propagation of a crack in solid silicon. Quantum mechanics and the detailed movement of individual atoms are important for the description of the bond breaking at the crack, but away from the crack, a description of the solid in terms of the displacement field suffices. While the idea of connecting scales is easy to state, the challenge is to carry it out in a very accurate and automatic fashion. This requires one to recognize which variables are needed in the continuum description and to take particular care to have consistency in the matching region.

On a technical level, a large fraction of the growth in computer power will occur for programs that can use computer processors in parallel, a major focus of researchers in the last decade or so, particularly, in areas requiring very large computer resources. Parallelization of algorithms ranges from the trivial to the difficult, depending on the underlying algorithm.

Computational physics and simulations in particular have both theoretical and experimental aspects, although from a strict point of view they are simply another tool for understanding experiment. Simulations are a distinct way of doing theoretical physics since, properly directed, they can far exceed the capabilities of pencil and paper calculations. Because simulations can have many of the complications of a real system, unexpected things can happen as they can in experiments. Sadly, the lore of experimental and theoretical physics has not yet fully penetrated into computational physics. Before the field can advance, certain standards, which are commonplace in other technical areas, need to be adopted so that people and codes can work together. Today, simulations are rarely described sufficiently well that they can be duplicated by others. Simulations that use unique, irreproducible and undocumented codes are similar to uncontrolled experiments. A requirement that all publications include links to all relevant source codes, inputs, and outputs would be a good first step to raising the general scientific level.

To conclude, the field is in a state of rapid growth, driven by the advance of computer technology. Better algorithms, infrastructure, standards, and education would allow the field to grow even faster and growth to continue when the inevitable slowing down of computer technology happens. Laplace's and Dirac's dream of perfect predictability may not be so far off if we can crack the quantum "nut."

Acknowledgments

The writing of this article was supported by NSF Grant DMR94-224-96, by the Department of Physics at the University of Illinois Urbana-Champaign, and by the National Center for Supercomputing Applications.

References

Abraham, F. F., J. Q. Broughton, N. Bernstein, and E. Kaxiras, 1998, Comput. Phys. (in press).
Alder, B. J., and T. E. Wainwright, 1957, J. Chem. Phys. **27**, 1208.
Allen, M. P., and D. J. Tildesley, 1988, *Computer Simulation of Liquids* (Oxford University Press, New York/London).
Binder, K., 1978, Ed., *Monte Carlo Methods in Statistical Physics*, Topics in Current Physics No. 7 (Springer, New York).
Binder, K., 1984, Ed., *Applications of the Monte Carlo Method in Statistical Physics*, Topics in Current Physics No. 36 (Springer, New York).
Binder, K., 1995, Ed., *The Monte Carlo Method in Condensed Matter Physics*, Topics in Applied Physics No. 71 (Springer, New York).
Binder, K., and G. Ciccotti, 1996, Eds., *Monte Carlo and Molecular Dynamics of Condensed Matter Systems* (Italian Physical Society, Bologna, Italy).
Car, R., and M. Parrinello, 1985, Phys. Rev. Lett. **55**, 2471.
Carlson, J., and R. Schiavilla, 1998, Rev. Mod. Phys. **70**, 743.
Ceperley, D. M., 1995, Rev. Mod. Phys. **67**, 279.
Ceperley, D. M., 1996, in *Monte Carlo and Molecular Dynamics of Condensed Matter Systems*, edited by K. Binder and G. Ciccotti (Italian Physical Society, Bologna, Italy), p. 443.
Ciccotti, G., and W. G. Hoover, 1986, *Molecular-Dynamics Simulation of Statistical-Mechanical Systems, Proceedings of the International School of Physics "Enrico Fermi," Course XCVII, 1985* (North-Holland, Amsterdam).
Dirac, P. A. M., 1929, Proc. R. Soc. London, Ser. A , **123**, 714.
Dreizler, R. M., and E. K. U. Gross, 1990, *Density Functional Theory* (Springer, Berlin).
D. Ferguson, J. L. Siepmann, and D. J. Truhlar, 1998, *Monte Carlo Methods in Chemical Physics*, Advances in Chemical Physics (Wiley, New York).
Feynman, R. P., 1953, Phys. Rev. **90**, 1116.
Feynman, R. P., 1982, Int. J. Theor. Phys. **21**, 467.
Gordillo, M. C., and D. M. Ceperley, 1997, Phys. Rev. Lett. **79**, 3010.
Hansen, J. P., and I. R. McDonald, 1986, *Theory of Simple Liquids* 2nd edition (Academic, New York).
Hirsch, J. E., 1985, Phys. Rev. B **31**, 4403.
Hohenberg, P., and W. Kohn, 1964, Phys. Rev. **136**, B864.
Kalia, R. K., 1997, Phys. Rev. Lett. **78**, 2144.
Krauth, W., 1996, Phys. Rev. Lett. **77**, 3695.
Laplace, P.-S., 1820, *Theorie Analytique des Probabilités* (Courcier, Paris), Sec. 27, p. 720.
Lloyd, S., 1996, Science **273**, 1073.
Meiburg, E., 1986, Phys. Fluids **29**, 3107.

Metropolis, N., A. W. Rosenbluth, M. N. Rosenbluth, A. H. Teller, and E. Teller, 1953, J. Chem. Phys. **21**, 1087.
Ortiz, G., D. M. Ceperley, and R. M. Martin, 1993, Phys. Rev. Lett. **71**, 2777.
Payne, M. C., M. P. Teter, D. C. Allan, T. A. Arias, J. D. Joannopoulos, 1992, Rev. Mod. Phys. **64**, 1045.
Rahman, X., 1964, Phys. Rev. **136**, A 405.
Schmidt, K. E., and M. H. Kalos, 1984, in *Applications of the Monte Carlo Method in Statistical Physics*, edited by K. Binder, Topics in Applied Physics No. 36 (Springer, New York), p. 125.
Swendsen, R. H., and J.-S. Wang, 1987, Phys. Rev. Lett. **58**, 86.
Tuckerman, M. E., D. Marx, M. L. Klein, and M. Parrinello, 1997, Science **275**, 817.
Wood, W. W., and F. R. Parker, 1957, J. Chem. Phys. **27**, 720.

Applications of Physics to Other Areas

Physics and Applications of Medical Imaging

William R. Hendee

The discovery of x rays by Wilhelm Röntgen in 1895 opened new pathways for the detection and diagnosis of disease in humans. Before this discovery, most diagnoses were made from a verbal description of the patient's history and symptoms, combined with the use of the physician's senses to detect peculiar odors, unusual visual signs, abnormal sounds, unnatural physical sensations, and, occasionally, odd tastes. X rays provided a novel approach to patient examination whereby the physician could study the internal anatomy and physiology of the patient. Today, x rays are used in hospitals, clinics, offices, and emergency facilities worldwide, and contribute essential information for the detection and diagnosis of a wide spectrum of illnesses and injuries in millions of patients each year.

Projection Radiography

A "projection image" is formed by x rays transmitted through a region of the body following their release from an x-ray tube. Each point in the projection image reveals the intensity of x rays directed towards the point, modulated by differences in density and atomic number of various tissue constituents in the path of the x-ray beam. The projection image is a two-dimensional depiction of a three-dimensional distribution of tissue constituents, with the third (depth) dimension of the body represented as overlapping shadows in the image plane. Mentally reconstructing the third dimension of the image is one of the challenges of learning radiology, the science and art of image interpretation in medicine.

Analog x-ray image receptors

The projection image formed by transmitted x rays is usually captured on photographic film. Film alone may be used as the receptor when images of exquisite spatial resolution are desired, such as for detection of hairline fractures in bones of the extremities. However, the film's thin layer of photographic emulsion is a relatively inefficient x-ray absorber. Film alone as a receptor requires long exposure times and, for most applications, yields unacceptable image blurring caused by voluntary and involuntary patient motion. In most cases, the film is sandwiched between fluorescent "intensifying" screens that absorb x rays with up to 50% efficiency, and in turn emit visible light that exposes the photographic emulsion on the film.

The first intensifying screens were developed by Edison and contained calcium tungstate ($CaWO_4$) as the light-emitting ingredient. In the 1970s, these were replaced by screens containing a rare-earth element such as gadolinium, lanthanum, or yttrium complexed with oxysulfide or oxybromide crystals embedded in a plastic matrix. Compared with $CaWO_4$, rare-earth screens are better absorbers of x rays, and some emit more light for each x ray absorbed. Today, most x-ray images are captured on x-ray film sandwiched between rare-earth intensifying screens in an imaging cassette that is tightly sealed to provide intimate contact and reduce geometric blurring between the film and screens. Cassettes are available for film of different sizes ranging from about $10 \times 10 \, cm^2$ to approximately $35 \times 45 \, cm^2$ used principally for chest imaging. Once the film has been exposed to x rays or light from the intensifying screens, it contains a "latent image" that can be made visible by chemical processing. In the final image, darker areas represent anatomic regions penetrated by a greater number of x rays, whereas lighter areas depict regions where fewer x rays have been transmitted. Often, the lighter areas reveal bony structures or regions where a contrast agent is present, because the higher density and atomic number of the bone or contrast agent causes greater x-ray absorption.

X-ray fluoroscopy

Projection radiography yields exquisite two-dimensional images that present a "snapshot" of the patient's anatomy at a particular moment in time. Although these images are often sufficient to detect an abnormal condition and lead to a diagnosis of the cause of the abnormality, they are limited in their ability to depict rapid changes in the anatomy caused by underlying physiologic processes. For this purpose, a continuous x-ray image is required. The technique that yields such a continuous image is termed "fluoroscopy" because it captures the image instantaneously on a fluorescent screen and displays the image in real time to the viewer.

Early applications of fluoroscopy employed a fluorescent screen that was viewed directly by the physician. The image was so dim that it could be seen only after the observer's vision had been "dark adapted," and even then only gross features in the image could be distinguished. Fluoroscopy was improved in the 1950s with invention of the image intensifier. In this device, transmitted x rays are captured on a CsI intensifying screen and converted instantaneously into a two-dimensional distribution of electrons ejected from a photocathode juxtaposed to the CsI screen. The electrons are accelerated through 25–35 kV onto an output screen that emits a small image in response to the impinging electrons. The gain in image brightness (on the order of $50\,000\times$) in the image intensifier alleviates the need for dark adaptation. The image on the output screen may be viewed directly through an optical system of lenses and mirrors, or captured by a television camera and transmitted electronically to a remote television monitor for viewing. This process converts the light image into an electronic signal, which can then be digitized for integration into an imaging network.

Fluoroscopy is an important part of x-ray imaging, especially for studies of the gastrointestinal tract and for angiographic studies of the central and peripheral circulatory system. It is used for image-guided therapeutic approaches such as interventional radiology and minimally invasive surgery that are growing in popularity because of their reduced patient morbidity and, usually, lower cost compared with alternative therapeutic approaches.

Digital x-ray image receptors

The combination of intensifying screens and film has many advantages as an image receptor for projection radiography. It is simple, portable, inexpensive, and yields images with excellent anatomic detail. The method has some disadvantages as well. Its use is restricted to a narrow range of exposures, and sometimes repeat examinations are necessary because films are over- or under-exposed. Film images are bulky to store and easy to misplace, and they must be physically transported from one location to another. Replacing film-based x-ray image receptors with digital x-ray detectors reduces these problems, and offers other advantages as well such as (1) image processing to improve contrast, sharpen edges, and reduce image noise, (2) integration of x-ray images with those from other digital imaging methods such as computed tomography, nuclear medicine, and magnetic-resonance imaging, (3) electronic transmission of x-ray images within the institution to provide immediate access for individuals caring for patients, (4) electronic transmission of images to and from distant locations to improve the care of patients in remote areas (teleradiology), and (5) more effective use of algorithms for computer-assisted diagnosis.

Three approaches to digital x-ray imaging are currently available commercially. One approach is to capture the projection image with an image intensifier, and to digitize the resulting signal from a video camera optically coupled to the intensifier's output screen. This approach yields a continuous digital image, but has limited spatial resolution compared with images on x-ray film. Another approach is to use a phosphor screen that forms a latent image by trapping electrons excited by the absorption of the incident x rays. By subsequently illuminating the phosphor with a scanning laser beam, the trapped electrons escape and release blue light that is captured by a photomultiplier tube to yield an electronic signal that can be digitized. This approach provides single images like those obtained with x-ray film and intensifying screens, except with reduced spatial resolution. The third commercial approach uses an amorphous selenium (aSe) photoconductive screen to convert incident x rays directly into a distribution of charge carriers on the screen's surface. This technique employs the photoconductor (aSe) widely used in earlier times for photocopying, and resembles xeroradiography, a method used in the 1970s for breast imaging. In the present case, however, the charge distribution is converted into a digital readout. The commercial version of the aSe approach uses a large rotating drum that is too large to be housed in most existing examination rooms (Neitzel, Maack, and Guenther-Kohlfahl, 1994).

Although still experimental, flat-panel display technologies are rapidly improving and show considerable promise as digital x-ray receptors capable of providing immediate images for radiography and fluoroscopy with exquisite spatial resolution. One approach is the use of an x-ray sensitive, light-emitting phosphor that is optically coupled to an array of photodetectors (e.g., aSi:H) to yield digitizable electronic signals (Fig. 1). A more direct approach is the use of a photoconductor such as aSe combined with an array of flat-panel thin-film transistors (termed an active matrix array) to yield a "direct conversion" image receptor in which the projection x-ray image is converted directly into a digitizable charge distribution over a large-area plate. The five requirements of an ideal x-ray photoconductor are well met by aSe: (1) the relatively high atomic number Z of aSe provides efficient x-ray absorption; (2) only a small amount of energy is required to create an electron-hole pair in aSe; (3) aSe has negligible dark current; (4) the charge carriers can migrate a considerable distance along the applied electric field in aSe without being trapped; and (5) since most of the impinging x-ray energy is absorbed in aSe, images are produced at relatively low doses (Rowlands and Kasap, 1997). Other possible photoconductors include lead iodide, thallium bromide, and cadmium-zinc telluride. Although the relatively low Z of organic photoconductors limits their x-ray absorption efficiency, it is conceivable that they could be used in an organic binder containing high-Z x-ray absorbing particles (Wang and Herron, 1996).

A major advantage of the direct-conversion approach is its structural flexibility that allows fabrication of large-area detectors for medical x-ray imaging. This approach to digital x-ray imaging has yielded spatial resolutions on the order of 150 μm, and has the potential to achieve resolutions as fine as 50 μm coincident with the most exacting requirements of x-ray imaging. With continued research and demonstration of efficacy through clinical applications, direct-conversion, flat-panel receptors could be the pathway over which x-ray projection imaging progresses into the digital era.

X-Ray Tomography

Introduction

In projection radiography, the third (depth) dimension of tissue is represented as overlapping shadows in a two-dimensional image. As a result, an anatomic structure of interest is frequently obscured by shadows of objects above or below it in the patient. Removing these shadows often improves the delineation of the shape and composition of the structure. Two approaches, analog tomography and computed tomography, can be used to remove the shadows. Over the past two decades, computed tomography has grown widely in acceptance, and analog tomography has declined in popularity.

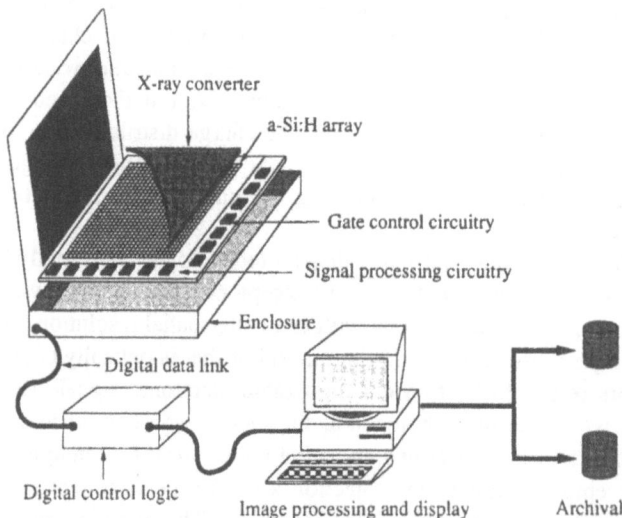

Figure 1. A flat-panel phosphor/photodiode detector for digital x-ray projection imaging (from Antonuk et al., 1995).

Analog tomography

In analog tomography, anatomic structures in a specific image plane (actually an image section of specific thickness, often a few mm) in the patient are kept in focus in the image, while structures above and below the image plane are blurred. The blurring is accomplished by moving the x-ray tube and the image receptor in synchrony during exposure around a pivot (fulcrum) in the image plane. The depth of the image plane in the patient can be altered by moving the patient up or down with respect to the fulcrum. Analog tomography is limited by the presence of image artifacts (tomographic ghosts) that interfere with image clarity, and by low image contrast caused by scattered radiation and the inability of conventional x-ray image receptors to detect differences in x-ray intensity of less than a few percent. Because of its low cost and simplicity, analog tomography is still used today in certain applications such as the acquisition of supplemental information about suspicious areas seen in chest radiographs. However, most applications of analog tomography have been superseded by computed tomography.

Computed tomography

The first commercial computed-tomography (CT) unit was announced in 1972, and reflected the pioneering work of the Austrian mathematician Radon, the South African physicist Cormack, and the English engineer Hounsfield. The latter two individuals shared the 1979 Nobel Prize in medicine for their contributions to CT. The first unit employed a narrow beam of x rays scanned across the patient in synchrony with a scintillation detector moving on the patient's opposite side (Hounsfield, 1973). The intensity I of x rays measured by the detector is

$$I = I_0 \exp[-\Sigma \mu_i x_i],$$

where μ_i represents the linear attenuation coefficient of each of "i" structures in the path of the narrow x-ray beam, and x_i represents the thickness of each of the "i" structures.

With a single measurement of x-ray transmission, the separate attenuation coefficients cannot be determined. However, these coefficients can be distinguished if enough transmission measurements are obtained at different orientations through the patient, with the aid of calculations using some type of back-projection algorithm. The result of such calculations is a two-dimensional map of linear attenuation coefficients distributed across the imaging section with a thickness defined by the width of the scanning x-ray beam. These coefficients can be converted into CT numbers with the formula

$$\text{CT number} = 1000[\mu - \mu_w]/\mu_w,$$

where μ is the linear attenuation coefficient at a specific location in the section, and μ_w is the coefficient of water for the x-ray energy employed for CT scanning. The distribution of CT numbers across the section can be displayed in various shades of gray to yield an image of the distribution of tissues in the section, each with its own CT number. This image is referred to as a CT image.

In early CT units, the x-ray tube and scintillation detector scanned the patient along a linear path perpendicular to the axis of rotation of tube and detector. Transmission data were subjected to an iterative method for computing attenuation coefficients. The process of acquiring transmission data and producing a gray-scale image was too time consuming for a busy clinical environment. The computational problem was solved by developing faster algorithms using a convolution (filtered back projection) model for image reconstruction. This approach subjects the transmission data to a Fourier transform into frequency space, permitting use of ramp and cutoff-frequency filters to improve image quality and enhance subtle features in the image. The acquisition time for x-ray transmission data was shortened markedly by development of purely rotational CT units that complete the entire scanning process in a few seconds. Combining this motion with simultaneous movement of the patient along the axis of rotation permits accumulation of many cross-sectional images during one relatively short examination period. This process, known as "spiral" or "helical" scanning, has significantly expanded the applications of CT, especially in the thorax and abdomen. Spiral scanning yields a three-dimensional array of CT numbers, and images parallel ("sagittal" and "coronal" slices), perpendicular ("transaxial" slices), or at any angle to the long axis of the patient, by compiling arrays of attenuation coefficients across the corresponding planes. The three-dimensional database can be configured to yield images that appear to be three-dimensional, and windowed to provide images of specific ranges of CT numbers corresponding to selected tissues.

Although rotational CT units can produce images in a few seconds, they are unable to acquire data quickly enough (<0.1 s) to capture images of the heart and other blood-perfused organs without significant blurring caused by motion. For these images, a way is needed to acquire x-ray transmission data from various angles without mechanical motion of the scanner. A scanner designed for this purpose employs an electron gun that scans a stationary metal annulus to generate x-ray beams along different projections. The resulting examination times are as short as 50–100 ms (Hendee and Ritenour, 1992).

Emission Tomography

In nuclear imaging, a small amount of a radioactive pharmaceutical is administered to the patient. The pharmaceutical carries the radioactivity to different organs or tissues according to its biokinetic properties. As the radioactive "tag" decays, it emits γ rays or, in the case of positron imaging, annihilation photons produced during annihilation of positrons released by the tag. As the emitted radiation escapes from the body, it is detected by one or more scintillation detectors positioned near the patient. In conventional nuclear imaging, the signals from the detectors are processed to yield a two-dimensional planar image of the three-dimensional distribution of radioactivity inside the patient. In emission tomography, two-dimensional cross-sectional images are reconstructed from multiple projections obtained at different angles around the patient. Emission tomography with radioactive pharmaceuticals emitting γ rays is referred to as "single-photon emission computed tomography (SPECT)." When annihilation photons are imaged following positron decay of radioactive pharmaceuticals, the technique is termed "positron tomography (PET)." Until recently, PET had the advantage of using coincidence measurement of annihilation photons to yield much higher radiation-detection efficiencies compared with SPECT.

An exciting recent development in nuclear imaging is operation of multiple-detector SPECT cameras in coincidence mode to yield images of positron-emitting radioactive pharmaceuticals. This potential is reinforced by regional supplier networks for ^{18}F-labeled deoxyglucose that obviate the need for an on-site cyclotron to produce positron-emitting ^{18}F. These developments are enhanced by growing recognition of the usefulness of positron imaging for detecting and staging cancer in a variety of anatomic sites, including brain, breast, and lung. It is conceivable that patients at high genetic risk for cancer will someday be administered ^{18}F-labeled deoxyglucose and scanned at periodic intervals for early detection of cancer.

Advances in molecular biology and genetics are yielding new knowledge at an astonishing rate about the molecular and genetic infrastructure underlying human health and disease. New knowledge about receptor sites, metabolic pathways, and "antisense" molecular technologies promises to yield increasingly specific agents that can be tagged with radioactive markers to permit visualization of normal and abnormal tissue structure and function at microscopic levels. These possibilities, referred to collectively as "molecular medicine," have the potential to enhance the contributions of nuclear imaging to clinical medicine.

Magnetic Resonance

Introduction

Damadian used nuclear magnetic resonance (NMR) in 1971 in an effort to distinguish normal from cancerous tissue in rats, and extended these studies to humans in 1973 (Damadian, 1973). That same year, Lauterbur published the first magnetic-resonance images (Lauterbur, 1973). The first human images were acquired in 1977 (Hinshaw, Bottomley, and Holland, 1977). The first magnetic-resonance image of the human brain was demonstrated in 1980 (Holland, Moore, and Hawkes, 1980). Unlike x-ray imaging, magnetic-resonance imaging does not depend on the transmission through tissue of radiation from an external source. Instead, the tissue itself is the source of imaging signals that arise from the macroscopic spin magnetization **M** of polarized water protons in tissue. Far less frequently, the signals may originate from other nuclei such as phosphorus or sodium. The motion of the magnetization vector **M** of uncoupled spins of water protons is given by the Block equation

$$d\mathbf{M}/dt = \gamma \mathbf{M} \times \mathbf{H} - M_{xy}/T_2 - (M_z - M_0)/T_1$$

where γ is the gyromagnetic ratio of hydrogen, **H** is the effective magnetic field, T_1 is the spin-lattice relaxation time defined as the time constant for the longitudinal magnetization M_z to return to its equilibrium value M_0 following receipt of a radio-frequency (rf) pulse that orients it at 90°, and T_2 is the spin-spin relaxation time defined as the time constant for decay of the coherent magnetization M_{xy} in the transverse plane to the equilibrium value $M_{xy} = 0$. Both T_1 and T_2 are affected by interactions of water with tissue molecules. By depicting these effects in magnetic-resonance images through selected sequences of rf pulses, T_1 and T_2 can be made to contribute independently by varying degrees to the contrast among different tissues in magnetic-resonance images.

Magnetic-resonance imaging

At the heart of a magnetic-resonance-imaging (MRI) system is a magnet that provides a highly stable and uniform magnetic field for nuclear polarization. Although resistive and permanent magnets have been used, most MRI units employ superconducting magnets. These magnets provide field strengths typically between 0.3 and 2 tesla, and the NbTi alloy conductors must be cooled to superconducting temperature (~4 K) with liquid helium. A natural next step in the evolution of MRI is the use of high-temperature superconductors such as bismuth-strontium-calcium-copper oxide/silver (BSCCO/Ag) that can accommodate greater currents at higher temperatures (National Research Council, 1996).

A radio-frequency coil serves two purposes in MRI. First, by sending a brief rf pulse into a region of tissue, it misaligns the magnetization vector of water protons with respect to the applied magnetic field. Second, it receives the weak rf signal emitted by the tissue as the magnetization vector realigns with the field according to the relaxation times T_1 and T_2. The design of rf coils has been largely experimental and focused primarily on the goals of improving signal quality and data-acquisition rate. Future developments may include improved computational design, cooled or superconducting rf coils to reduce noise, and multiple coils operated in parallel to improve data-acquisition efficiency.

Spatial localization of the rf signals emitted from tissue is accomplished by spatially encoding the signals through use of gradient magnetic coils that impose linear magnetic gradients along three-dimensional orthogonal axes in the tissue. This gradient approach is also used to encode dynamic information related to studies of blood flow and diffusion. Factors to be considered in the design of gradient coils include geometry, size, anatomic region of interest, desired gradient strength, efficiency, inductance, eddy currents, gradient uniformity, forces and torques on the coils, heat dissipation, and nerve stimulation in tissue (National Research Council, 1996).

Although relatively new, MRI is an exceptionally powerful imaging technique in clinical medicine. To date it has been used principally for anatomic imaging (Fig. 2), although flow and diffusion images are growing in popularity. Functional MRI (fMRI) is a rapidly developing area with significant clinical potential. This technique exploits the paramagnetic behavior of deoxyhemoglobin in red blood cells as an intrinsic intravascular contrast agent. When in a magnetic field, a blood vessel containing deoxyhemoglobin distorts the field in its immediate environs, with the degree of distortion increasing with the concentration of deoxyhemoglobin. This distortion affects the behavior of water protons in the environs and, consequently, the magnetic-resonance signal arising from these protons.

Neural activation of a region of the brain stimulates increased arterial flow of oxygenated blood, thereby decreasing the concentration of deoxyhemoglobin in the region. This decrease affects the immediate magnetic field and changes the intensity of the magnetic-resonance signals from the region. Changes in the magnetic-resonance signal can be detected and displayed as functional-MRI images (Fig. 3). These images, termed BOLD (blood-oxygen-level dependent) images, are useful in mapping functional neural activities onto the cerebral cortex and studying such activities in response to various somatosensory and cognitive tasks. Functional MRI is an exciting and fast-evolving technology that could benefit immeasurably from improvements in areas such as (1) motion detection and compensation, (2) characterization of temporal response patterns, (3) characterization of physiological noise and its effects of functional MRI, and (4) display of volumetric functional-MRI images, especially in real time.[1]

---

[1] See the chapter by Kleppner on pp. 129–140 of this book for a discussion of the use of optical pumping of He^3 to enhance MRI imaging in lung diagnostics.

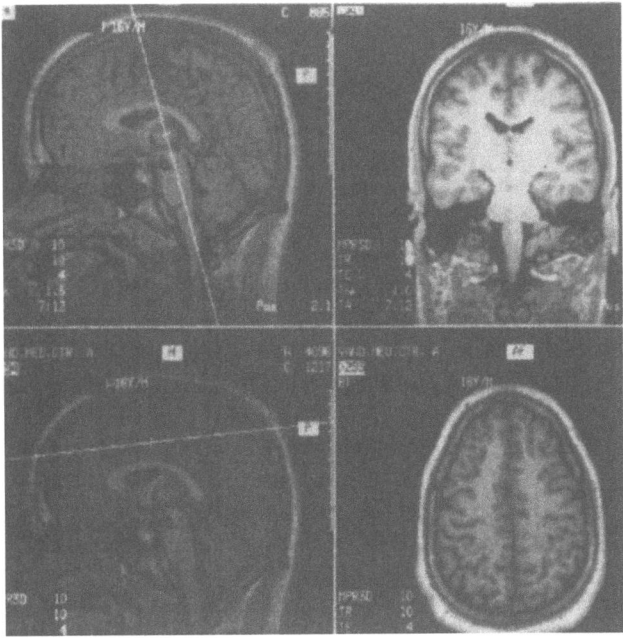

Figure 2. MRI anatomic images through different planes from a 128×256×256 three-dimensional dataset (from Price, 1995).

Magnetic-resonance spectroscopy

Combining MRI and magnetic-resonance spectroscopy permits noninvasive acquisition of unique *in vivo* information about the chemical composition of human tissues. Through the technique of chemical-shift imaging (CSI), spatial and temporal changes in tissue function can be studied by examining the magnetic-resonance spectra usually of ^{1}H, but occasionally of other nuclei such as ^{13}C, ^{14}N, ^{15}N, ^{19}F, ^{23}Na, ^{31}P, ^{39}K, ^{35}Cl, and ^{37}Cl. Compared with ^{1}H, these other nuclei occur less frequently in tissue and yield much weaker magnetic-resonance signals that are difficult to separate from noise. Magnetic-resonance spectroscopy is being investigated especially for its ability to distinguish benign from malignant tumors and recurrent tumors from scar tissue caused by earlier radiation therapy. It also is promising as a method to monitor tumor regression during radiation therapy and chemotherapy.

Image Networking

Most medical imaging systems (CT, MRI, ultrasound, nuclear medicine) present imaging data in digital form. The major exceptions to this rule are projection radiography and fluoroscopy. As described earlier, these applications are also becoming digital. Several advantages can be achieved by linking digital imaging systems electronically in a "picture archiving and communications system (PACS)" [referred to as an "image management, archiving and communication system (IMACS)" when integrated with other information networks such as hospital and clinic information systems]. These

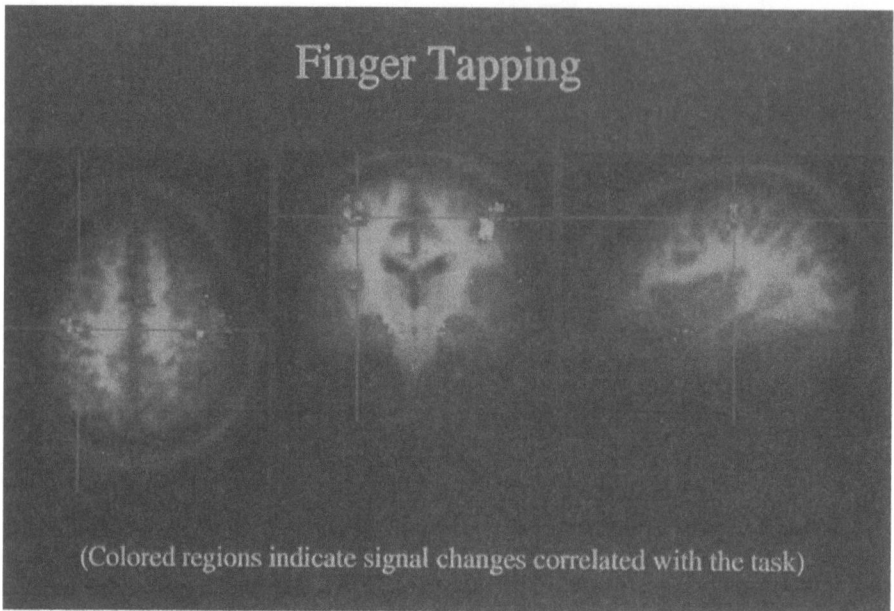

Figure 3. Orthogonal fMRI images of the human brain averaged over five subjects showing neural activation of the motor cortex (light areas in the image) associated with finger tapping (Birn, 1998). (See Color Plate 13.)

advantages include the capability to store images from many units electronically in one location, retrieve and assemble images for comparative studies from different techniques used to examine the same patient, and transmit images to remote locations for viewing by specialists without loss of the central data file. A PACS has the potential to expedite diagnoses, improve diagnostic accuracy, enhance information transmission to other physicians, and eliminate misplaced and lost films. Many imaging specialists believe that a fully integrated PACS will be essential to operation of a radiology service in the future. Others are convinced that the cost (several million dollars for a typical imaging service) will prevent many departments from converting completely to PACS, at least in the near future. They believe that conversion will most frequently occur incrementally, with film remaining as the preferred image receptor for certain applications for some time to come.

Medical Images in Radiation Treatment Planning

Medical images are used to detect, diagnose, and stage many types of cancer. They also are used to design, guide, and monitor the treatment of cancer, and to follow the patient after treatment to detect possible recurrence of the disease. Cancer is a disease characterized by the uncontrolled proliferation of cells. Cancer is treated by removing the cancerous cells, either through surgical extraction or by killing them with poisons (chemotherapy) or ionizing radiation (radiation therapy). Although medical images are important to all three approaches, they are employed most widely in radiation therapy. Their application to radiation therapy is discussed here.

The successful treatment of cancer with radiation requires that the cancer is localized to a specific region of tissue, and that the cancer and its microscopic extensions into normal tissue receive a dose of radiation sufficient to kill the tumor cells, while keeping the dose to nearby normal tissues low enough to avoid serious complications. Achieving this balance between radiation doses to tumor and normal tissues demands careful planning and precise treatments, including accurate delineation of the margins of the cancer and identification of the location of nearby radiosensitive normal tissues. These needs frequently are met by incorporating CT and MRI into the treatment planning process. CT and MRI images provide not only a cross-sectional picture of the patient's anatomy, including the cancer and surrounding tissues, but also an accurate representation of the body contour and organs that are especially sensitive to radiation. The digital data from these imaging units can be entered directly into the treatment-planning computer, and proposed treatment plans can be superimposed onto the cross-sectional images. Images acquired over the course of treatment can be examined to monitor the regression of the cancer and to make adjustments in the treatment plan in response to changes in the patient's anatomy.

In most radiation-therapy services, commercial imaging units are used to generate cross-sectional information for treatment planning. To improve the alignment of the cross-sectional information with the actual geometry encountered during treatment, some physicists have built a CT scanner on a gantry identical to that supporting the linear accelerator used for radiation therapy. Although these CT units yield spatial resolution inferior to that of commercial CT units, they duplicate the treatment geometry and provide images that are good enough for treatment planning.

Many x-ray treatments consist of multiple fixed radiation fields converging on the cancer from different directions. This approach concentrates the radiation dose in the cancer while delivering much lower doses to surrounding normal tissues. In some cases an even better dose distribution can be achieved by rotating the treatment machine partially or completely around the patient during treatment so that only the cancer is always in the path of the radiation beam. Since cancers are asymmetrical, the size of the x-ray beam should be expanded and contracted continually during treatment in order to restrict the dose to normal tissues to the lowest possible level. Often the dose distribution can be improved even more by varying the dose rate during rotation. This approach is referred to as "conformal therapy." Its applications require detailed three-dimensional knowledge of the anatomy of the irradiated tissue through accumulation of medical images, together with exquisite computerized control of the treatment unit and patient couch (Mageras *et al.*, 1994).

A further advance in conformal therapy would be the convergence of CT imaging and x-ray therapy into a single gantry, so that tomographic images could be used to monitor treatment alignment and dose distribution continually as the treatment progresses. This hybrid approach is being pursued by a few medical physicists (Convery and Rosenbloom, 1995; Mackie *et al.*, 1993); it presents formidable technical challenges as well as promises for improved radiation therapy.

Conclusions

Medical imaging celebrated its centennial anniversary in 1995, and today it continues to push the frontiers of research and clinical applications forward. It is an excellent example of what can be done through a multidisciplinary effort, in this case involving physicists, engineers, and physicians. Many research opportunities are available, as much remains to be done in further improving the applications of medical imaging to reducing human disease and disability.

References

Antonuk, L., *et al.*, 1995, Radiographics **15**, 999.
Birn, R., 1998, Medical College of Wisconsin (unpublished).
Convery, D., and M. Rosenbloom, 1995, Phys. Med. Biol. **40**, 979.
Damadian, R., 1973, Ann. (N.Y.) Acad. Sci. **222**, 1048.
Hendee, W. and E. Ritenour, 1992, *Medical Imaging Physics* (Mosby-YearBook, St. Louis).
Hinshaw, W., P. Bottomley, and G. Holland, 1977, Nature (London) **270**, 722.
Holland, G., W. Moore, and R. Hawkes, 1980, J. Comput. Assist. Tomogr. **4**, 1.
Hounsfield, G., 1973, Br. J. Radiol. **46**, 1016.
Lauterbur, P., 1973, Nature (London) **242**, 190.
Mackie, T., T. Holmes, S. Swerdloff, P. Reckwerdt, J. Deasy, J. Yang, B. Paliwal, and T. Kinsella, 1993, Med. Phys. **20**, 1709.
Mageras, G., Z. Fuks, J. O'Brien, L. Brewster, C. Burman, C. Chui, S. Leibel, C. Ling, M. Masterson, and R. Mohan, 1994, Int. J. Radiat. Oncol., Biol., Phys. **30**, 971.
National Research Council, 1996, Mathematics and Physics of Emerging Biomedical Imaging (National Academy Press, Washington, DC).
Neitzel, U., I. Maack, and S. Guenther-Kohlfahl, 1994, Med. Phys. **21**, 509.
Price, R., 1995, Radiographics **15**, 175.
Rowlands, J., and S. Kasap, 1997, Phys. Today **50**, 24.
Wang, Y., and N. Herron, 1996, Science **273**, 632.

Nuclear Fission Reactors
Charles E. Till

The story of fission is a story of drama and emotion, and it is a story that demonstrates possibly better than any other the overwhelming importance of modern physics to the events of this past century. Its discovery, or more properly, its identification, was delayed for a surprisingly long time after the phenomenon had in fact been induced in several of the first-rank laboratories in the world. Over four years passed without its recognition, but when recognition did come, it came at the most dramatic possible time. It was just before the war that engulfed the world, and knowledgeable physicists in every country were in a position to guess at its possible military significance.

The delay in the identification of the phenomenon of fission was caused by misdirection, in that the experimenters expected another result, based on the understanding of the day. When this was combined with measurement imprecision inherent in the techniques used in the experiments, proper interpretation of the results was obscured. For the discovery of fission rested on delicate radiochemical separations of unknown elements created in tiny amounts by neutron bombardment of uranium. The systematics of similar experiments in lighter elements suggested that new transuranic elements, heavier than uranium, were to be expected. The results therefore were interpreted in this way, but with increasing puzzlement as time passed. Nevertheless, confidence in this interpretation caused even the evidence provided by large fission pulses actually seen in a few ion-chamber measurements to be attributed to instrument malfunction. Recognition of fission was therefore delayed from 1934 to 1938, and it was very late in 1938 when the phenomenon was recognized unequivocally for what it was.

Once the fact of fission was recognized, however, its implications, military and otherwise, were very quickly grasped. With World War II rapidly coming on, the most interesting and important physics of the day appeared briefly, then disappeared from the technical journals of the time, first through a form of secrecy self-imposed by the scientists involved, and then by the heavy formalized security of wartime. Behind this wall of secrecy, the key technical issues were quickly settled—the presence of neutrons from fission itself in sufficient number to sustain a chain reaction ($2\frac{1}{2}$ neutrons in uranium) and the adequacy of two different materials, heavy water and graphite,[1] the latter readily available, to moderate the energy of the neutrons from fission to energies such that there could be sufficient neutron capture in uranium to make a self-sustaining fission reaction possible.

---

[1] The advantage of heavy water (D_2O) over light water (H_2O) lies in the far smaller neutron absorption cross section of D compared to H.

The principles underlying the first successful reactor demonstration were quite simple. The principal obstacle to overcome was the fact that the fuel had to be uranium that contained only the naturally occurring trace concentration (0.7%) of the actual fissile element U-235. No enrichment facilities to increase the concentration of U-235 had yet been contemplated. Absorption of neutrons in the fuel itself therefore needed to be maximized, useless neutron absorption in nonfuel materials minimized, the surface where neutrons could leave the system, also uselessly, minimized with respect to the volume where useful absorption could take place, and finally, absorption of neutrons in the fuel had to take place at the (low) energies where the probability of fission was high. Only then could criticality be achieved and a chain reaction begin to proceed at steady power. The probability of fission is highest at thermal energies. An efficient moderating material was needed to moderate the energy of neutrons released in fission all the way to thermal energies without the moderator itself absorbing an inordinate number of neutrons. Both graphite and heavy water would work, if free of impurities, and as we shall see, both were in fact used exclusively in the early programs, but the ready availability of graphite led to its choice for the world's first nuclear reactor.

Graphite as the moderator and natural uranium as the fuel became the basis for the world's first reactor, Chicago Pile #1, or CP-1. Assembled by Enrico Fermi and his group at the University of Chicago, following measurements by the same group indicating feasibility both in number of neutrons and in moderator efficiency, it went critical on December 2, 1942, under cover of complete secrecy. And in total secrecy research facilities, production reactors, processing plants, and ancillary facilities were quickly constructed at places such as Los Alamos, Hanford, and Oak Ridge. The scale of the effort was extraordinary, urgency was felt at every step, and it led in just two and a half years from the successful CP-1 demonstration to the successful development of the "atomic bomb."

Secrecy came to an abrupt end with the explosions in August of 1945 over Hiroshima and Nagasaki. The use of the "atomic bomb" to end World War II, so controversial later, was greeted with overwhelming relief and approval by the public of the Allied Nations at the time. Such feeling is captured in the reaction of a young Australian woman, her husband an American captain serving in the Pacific. "At that moment," she said, "I knew God was on our side."

Such strong emotion, for and against, has been the hallmark of nuclear development throughout its history. It provided the driving force for the huge national programs after World War II, on the one hand, and on the other hand it fueled the intense organized opposition to further development seen in recent years.

In the months following World War II, while the world grappled with the stupendous implications of the existence of fission, those nations in a position to do so turned quickly to exploiting the fission process. Driven in part by the desire to attain, or in the case of the U.S., to maintain, military advantage, the national programs that were put in place nevertheless explored on a very large scale the science, the technology, and the potential, civilian as well as military, of fission-based nuclear energy. In particular, programs that would lead to rapid development of commercial nuclear electric power were soon in place in six nations of the world: the U.S., the U.K., Canada, France, the U.S.S.R., and Sweden. All other nations that deployed nuclear power transferred the necessary technology from these six. United States technology, and to a much lesser extent, Canadian technology, eventually predominated.

At the end of the war the United States had an impressive lead, for the wartime projects provided a very broad technological base—of knowledge, processes, and facilities—for further nuclear power development. The passage of the Atomic Energy Act of 1946 established the Atomic Energy Commission, and the principle of civilian control of nuclear activities; it provided the framework for private industrial involvement and it began the system of great national laboratories created specifically for the development of nuclear energy. Argonne National Laboratory, established on July 1, 1946, was the first, and in 1948 it became the designated center for reactor development for the nation. Progress was rapid. On December 20, 1951, the first electricity produced by nuclear energy was generated by the Argonne group in the Experimental Breeder Reactor #1, at the newly commissioned National Reactor Test Station on the Great Basin Desert in Idaho.

Principal among the myriad problems faced by the early reactor inventors and designers were problems with materials. They needed materials that first of all were suitable in nuclear properties, but that also had adequate corrosion resistance, radiation resistance, and fabricability. The characteristics of any reactor are very largely set by the choice of fuel material, moderator materials, and coolant material. Insightful design can ameliorate disadvantages and emphasize corresponding advantages, but once the main material choices are made the principal characteristics of that reactor type are set. Because the U.S. had uranium isotopic enrichment facilities from the wartime project, and, in fact, for a considerable time had a monopoly on such facilities, the constraint put on material choice by the need to use only natural uranium did not exist for the U.S. As a result an astonishing array of material choices and combinations were exploited in the first demonstration phase of reactor prototypes in the U.S. This remarkably diverse set of utility-operated reactors had powers generally in the tens of MWe range,[2] too small to be economic, but large enough to firmly establish technical feasibility, and they were deployed by the late 1950s and early 1960s.

The goal of that first U.S. demonstration phase was an admirable one. It aimed at establishing as much information as possible concerning the strength and weaknesses of various reactor possibilities, at a time when the actual need for early deployment of nuclear power was not pressing. Alternatives could then be weighed and the best possible choices made. However, events of the day quickly overtook this studied approach. The head start given by the Naval Reactors program to their choice of light-water reactor technology—a single-minded choice that meant rapid development of that technology for Navy purposes—and the political imperative of the day to show a successful reactor type in operation combined to settle the question of civilian reactor technology choice. The pressing need to show successful reactor commercialization arose from the U.S.-led Atoms for Peace initiative of 1955. The promise of successful civilian power technology to other nations was felt to be necessary, in return for nonproliferation commitments. Light-water reactors were the natural choice, as the easiest, the fastest, and the most established technology. It is important to note today that the considerations of the time did not give the slightest weight to whether this technology would be the right long-term choice for large-scale permanent deployment. In fact, it was assumed that it certainly would not be the best long-term choice, but that it would probably make a perfectly

---

[2]MWe is the abbreviation for megawatts of electrical power and makes the distinction between electrical power output and total (or thermal) power output, denoted by MWt. The difference is the waste heat.

suitable choice for a first generation of nuclear power.

It was in this way that the light-water reactor came to dominate the U.S. market, as it eventually came to dominate that of the entire world. The distinction of being the first commercial reactor in the U.S. is credited to the 620-MWe General Electric Boiling-Water Reactor ordered in 1963 by Jersey Central Power and Light. With it, a commercialization rush began. In 1966, 20 plants were ordered, 30 more in 1967. By 1974, the last big year for reactor orders in the U.S., well over a hundred large nuclear power plants were on the books of U.S. nuclear vendors. In later years some were canceled, as overcapacity, cost overruns, and a new, much more skeptical environment for nuclear power diminished the early optimism. Some reactors have now reached the end of their useful life, and some, after many years of operation, have been closed down simply as an economic business decision. Nevertheless, today the U.S. has 108 nuclear power plants in operation, supplying 22% of the nation's electricity.

The pattern in other parts of the world initially differed from the U.S., principally because of the U.S. monopoly on enrichment capacity. In other nations natural uranium with its low (0.7%) fraction of the fissile element U-235 had to be utilized, at least initially. Essentially, each nation had to start with its equivalent of CP-1, in most cases using graphite as the moderator. The principal design consideration was correct arrangement of the graphite and uranium to make the array as reactive as possible, as this combination of materials gives little margin in reactivity, but CP-1 had shown the way. Two nations, Canada and Sweden, had sufficient heavy water, a much superior moderator, for their prototypes. Because heavy water gave a satisfactory system using natural uranium, Canada was able to base its successful development of the commercial CANDU reactor on the technology of its earliest prototypes. The same was not true of the nations forced to use graphite-moderated systems. Eventually all such national programs were displaced by U.S. light-water technology.

In each nation such a natural-uranium-based reactor was assembled within a year or two of the war's end, and its technology became the basis for the early power reactor programs. In the U.S.S.R., the U.K., and France, the programs were graphite-moderator based because of the inexpensive availability of graphite and the simplicity that was possible in the initial designs. But the simplicity, even crudity, of the early graphite-based designs led directly to two of the three major reactor accidents that the world has seen: The Windscale production reactor accident in 1957 in the U.K. and the 1986 accident at Chernobyl, both of which spread radioactive contamination beyond, and, as the world knows, in the Chernobyl case, far beyond, the boundary of the reactor site itself.[3] Both accidents resulted from poor operator understanding of the characteristics of a primitive reactor whose design made little allowance for operator misjudgments.

The successful Canadian reactor, CANDU, still based on natural uranium and heavy water, needing neither enrichment nor reprocessing facilities, has been deployed in several countries of the world. Several percent of the world's nuclear electrical capacity is based on this reactor, most of it, of course, in Canada itself.

In France, after a relatively short graphite-based phase, a very successful program began, based on the Westinghouse PWR, that transformed the nation's electrical system largely to nuclear generation. It led to an electrical system in France that is now 77% nuclear, with nuclear electricity now being exported to neighboring countries, and a very

---

[3]The other was the TMI-2 event in 1979, where a commercial LWR, lacking cooling, melted a substantial fraction of its core.

stable domestic electrical system with by far the highest nuclear portion of any of the large industrialized countries.

The U.K. stayed with a relatively successful graphite-moderated, gas-cooled program for a considerable time, but it too, in its recent reactor additions, has turned to the light-water reactor.

In the countries of the former Soviet Union, the graphite-based design RBMK continued to be deployed, even after the light-water reactor technology was phased in and, as mentioned, it was the former reactor type that was involved in the Chernobyl event. Although some RBMK's remain in operation, the former Soviet Union nations now largely use light-water technology as well.

Around the globe, the world's nuclear capacity is about 350 GWe, about 18% of the total world electricity generation, as other nations have adopted the technology of the pioneers. Essentially all capacity is based on light-water technology. It is widely known that the light-water reactor cannot be the sole long-term choice for nuclear deployment, as it uses the uranium resource very inefficiently, but there is little incentive, or, indeed, will, at present to do more. The R&D programs, once so large, have shrunk in all countries, and in many have been eliminated entirely. Any change in this picture will require a change in today's attitudes towards the place of nuclear power in the energy mix worldwide. For nuclear technology is capable of more, much more.

Thus it is of the greatest importance to examine what is required of nuclear technology to make it a large, permanent solution to world's energy needs. Clearly, technological improvements are needed, and these we shall examine here. Though all reactors function in basically the same way, in that the fuel material is very gradually consumed by fission, they can differ appreciably in characteristics, depending on the choice of materials used. The fuel material, always, is basically uranium—it may actually be uranium, or it may be substances, principally plutonium, derived from uranium. Whatever the material, it is usually fabricated in pencil-sized rods and enclosed in a sleeve of some ordinary metal like steel. Several thousand of these rods make up the core of any nuclear reactor, assembled in a fairly close-packed array several feet in diameter and several feet high. Circulating around the rods is some kind of cooling fluid. When the fuel mass is sufficient, the fission process becomes self-sustaining and the reactor is said to be "critical."

The fission process in a critical reactor steadily produces heat, and it can quickly and easily be adjusted to produce steady heat at any rate that is desired. But the right rate is just the rate at which heat can get out of the fuel and be carried away by the coolant. This rate varies in the different kinds of reactors made of different materials. Fuels, for example, can be metals or ceramics; coolants can be liquids—water or liquid metals are used routinely—or gas. Both helium and carbon dioxide have been used. Whatever the coolant, its heat is used to produce steam which in turn generates electricity. Now cooled, it is circulated back to carry heat away from the reactor core again.

The essential point is that the heat generation in the fuel must match the rate at which the heat can be removed by the coolant. This is not difficult to do, and reactors tend to run very stably and evenly. But in the matching of heating to cooling lies the basis for understanding reactor safety. All of the serious accidents that can happen to a nuclear reactor involve upsetting this balance. There is really no nuclear accident unless radioactivity is released. As long as the sleeve around the fuel, the "fuel cladding," remains intact, radioactivity is contained in the fuel and no-one, worker or public, can be affected. The only way to get widespread cladding failure is by melting. Melting occurs in only one way: from an imbalance between heat generation in the fuel and heat carried away by the coolant. If the fuel generates more heat than the coolant can carry away, the temperatures of both—fuel and coolant—rise. Unless corrective action is taken, temperatures can rise to a point at which the cladding melts and radioactivity is released from the fuel (which itself may also be molten). The initial release of radioactivity is to the coolant. To reach the outside atmosphere, the radioactivity must first get out of the vessel and piping containing the coolant and then out of the containment building, which itself is constructed expressly to prevent contamination of the outside atmosphere in just such situations. These multiple barriers to the escape of radioactivity have generally been very effective, but clearly, it is much better if the first barrier, that is, the fuel cladding, never fails. If that is achieved, then there can be no radioactive release, no "meltdown," in fact, no effect on the public at all.

These potentially dangerous imbalances between heat generation and heat removal can happen due to mechanical failure or simply to mistakes, as in any other machine. The corrective action to prevent cladding failure is always the same: the heat production—the power that is being generated—must be reduced quickly to match the heat removal capability that remains after the mechanical failure. (Typical mechanical failures are coolant pumps failing, pipes breaking, valves not working, and so on—all the things to be expected occasionally in a large hydraulic system.)

Normally, power is reduced (but never entirely shut off) by inserting neutron-absorbing safety rods into the reactor to stop the fission process. This, too, relies on mechanical devices' functioning properly—sensors, motors, springs, relays, etc. Such safety systems are always designed to be redundant, so if one device fails there is backup. But reactors can also be developed that shut themselves down harmlessly under these conditions without the need for mechanical devices' functioning properly. These reactor concepts are said to be "passively safe" or "inherently safe." In essence, they automatically match heat production to heat removal. Thus under conditions in which heat removal is failing, the balance is maintained without the need for mechanical devices like safety rods to reduce the heat production.

The principle underlying passively safe behavior is quite simple: choice of reactor materials to give a strongly negative temperature coefficient of reactivity overall, with little contribution to the coefficient from the fuel itself. Under these conditions, if heat removal begins to become inadequate for the power being generated, the resulting temperature rise is immediately offset by reactivity reduction and consequent power reduction due to the overall negative temperature coefficient. This power reduction drops the fuel temperature, and in the absence of a significant contribution to the coefficient by the fuel itself—in particular, no significant positive contribution from the dropping fuel temperature—the power simply adjusts to the level of cooling available. The reactor remains critical, the coolant temperatures remain at operating levels, the fuel temperatures are reduced, and the power level steadies at the new equilibrium level. An entire class of possible serious accidents can be avoided in this way.

In essence, this is how reactors work; this is what can go wrong with them; and this is the basis for improved safety in advanced reactors.

World consumption of coal, oil, and gas now generates 90% of the world's energy, about 300 quads/yr, and growing.[4] The ability of nuclear energy to displace a significant fraction of this fossil fuel usage depends on the amount of uranium needed to produce a given amount of energy. World uranium resources, while large by present standards, are

---

[4] 1 quad = 1 quadrillion BTU = 1.055 exajoule = 1.055×10^{18} J.

not large when set against future needs for energy. It is important to understand that energy production for a given amount of uranium varies very greatly with the type of reactor used. In present-day light-water reactors uranium is used very inefficiently. Less than 1% is actually used, and all the rest is waste. In fact this makes up a large part of the nuclear waste that eventually has to be placed in a repository. No state wants to host such a repository, although Nevada has been selected at the present time. Ninety-nine percent of the uranium that goes into light-water reactors therefore is part of a problem—the nuclear waste problem—and not part of any energy solution.

By present estimates the total world's known uranium resources, if used in today's reactors, would produce electricity equivalent to about 40% of present world energy generation for 30 years. Even if the estimates of uranium resources are wrong by a large amount, even by a factor of 2, the conclusion would be the same: This amount of energy is useful in the overall picture but is not a permanent solution to the world's needs for energy.

If reactors other than the light-water reactor of today were to be developed, reactors that have the property of breeding, thus making much better utilization of the uranium resource, nuclear power could be a large part of a permanent solution to the world's long-term needs for energy. Breeding causes the 99% of the uranium that is now a part of the waste to be gradually converted to a new fuel material—plutonium—that can be burned just as the fraction of 1% of the uranium that is burned today. In this way, all, or essentially all, uranium is burnable fuel and the entire uranium resource can be used.

The breeder reactor is the only reactor that can substitute for a large part of fossil fuel usage for a long period of time and thus impact in a significant way greenhouse gas increase. With uranium utilization at least a factor of 70 greater than today's reactors, the breeder reactor could theoretically displace all fossil fuel consumption for at least 1000 yr. If electricity generation could be made half the total world energy generation, the current economically recoverable uranium reserves would last for 2000 yr. Nuclear power then becomes a significant part of a permanent solution.

Breeding today is controversial, because of its focus on the conversion of nonfissile uranium to a fissionable plutonium fuel, and the concomitant concern about the use of such plutonium to make nuclear weapons. Such concerns, however, eventually must be reconciled with energy needs. If nuclear energy is to contribute substantively to meeting mankind's future energy needs and maintain the present atmospheric environment, as it is capable of doing, breeding is a fundamental requirement. The concomitant proliferation problem is one that needs to be solved by the world community, a problem that is beyond the scientific and engineering efforts discussed in this article.

There are other implications for the long term. Assume, for illustration, that just half the present rate of world energy usage is electric and its generation is all nuclear. About 1500 1000-MW reactors would be needed today, increasing each year with further world energy growth. In the present reactors a plant performance goal of 10^{-4} reactor yr^{-1} for a large-scale core melt accident is agreed to be a reasonable standard. That is, the probability of widespread melting of fuel and cladding should not be greater for any single reactor than 0.01% a year. Accepting this as a goal, we can see that 1500 reactors imply a serious accident every seven years and, of course, with growth, even more. Such an accident rate is unacceptably high for long-term widely deployed reactor usage, and the adoption of passive safety features may be wise in the long run.

There are implications for the nuclear waste as well. From a global perspective, the present reactor system converts the world's supply of uranium to a form that is a cancer risk to man for tens of thousands of years, orders of magnitude worse than the ore it was mined from. This can be defended on the basis of our need for energy, but it is useful to remember that in the present reactors this will have been done without significantly affecting the long-term consumption of carbon fuels or the long-term outlook for global climate change, and the opportunity for doing better will have passed.

Just as the type of nuclear reactor dictates the magnitude of its energy contribution, the reactor type and the choice of the fuel cycle can have a radical effect on the nuclear waste, too. It is unlikely that nuclear power will develop in an orderly way on the scale required without a solution to the nuclear waste problem. Many people do not have confidence that wastes can be isolated for the millions of years that they are harmful. There is a consensus in the scientific community, by no means unanimous, that this can be done, but the public is skeptical. Present policy is to bury the fuel coming out of light-water reactors. Burying this spent fuel, as it is called, buries much of the uranium's energy potential and commits the burial ground—the repository—to 1 000 000-yr isolation.

But just as there are different types of reactors, so too are there different types of fuel cycles. The fuel cycle is the way the spent reactor fuel is treated and, in the case of the breeder reactor, cycled back to the reactor. Fuel cycle alternatives can have great impact on the waste issue. The actinide elements, uranium and the derivatives of uranium created while the fuel is in the reactor, cause the 1 000 000-yr problem. Practically everything else that is radioactive in the waste will become relatively harmless in a few hundred years. A fuel cycle that removes the actinides from the waste changes the isolation period necessary for it from millions of years to hundreds of years. Structures to isolate waste from the environment can be designed with confidence for such periods. As for the actinide materials themselves, in the right breeder reactor type they can be destroyed by fissioning them. Such fissioning in turn creates energy rather than a 1 000 000-yr waste commitment. If all actinide elements, not just plutonium, are recycled, the usefulness of such material for weapons is substantially lessened.

The point in all of this, therefore, is that improved reactor types are needed and that from a global perspective, the characteristics they should have are clear—in breeding, in safety, and in waste. New technologies, new processes, new ways of exploiting basic properties of reactor materials can fundamentally alter the role nuclear power can and will be allowed to play in our future. Nuclear power may even be essential to a stable environment with a climate as we now know it. Of the technologies known to be feasible today, it alone is capable of the magnitudes required to supply future energy needs. Advanced reactor development that promises to radically improve the outlook for large-scale nuclear power may be one of the most important challenges presented to man. But undertaken successfully, nuclear fission could well have beneficial effects on the next centuries even greater than its effect in this century of its discovery.

Nuclear Power—Fusion
T. Kenneth Fowler

The earliest speculations about nuclear power—first, about nuclear fusion—followed soon after the publication of Albert Einstein's special theory of relativity. A story related by Edward Teller tells of the young George Gamow's being offered, in 1929, the nightly use of the full electric power grid of Leningrad if he would undertake to create in the laboratory the fusion energy that Atkinson and Houtermans were claiming to be sufficient to explain stars (Teller, 1981). Then, when fission was discovered in 1939, fusion took a back seat as the more readily exploitable fission process forged ahead, culminating in the first fission power reactors in the 1950s.

While the early success of fission reactors came at dazzling speed, the story of fusion power—still in the research stage—is one of persistent determination driven on the one hand by the alluring goal of virtually unlimited and environmentally attractive nuclear power, and on the other by the intellectual appeal of unprecedented technical and scientific challenges that have created the field of modern plasma physics.

Both the allure and the challenges of fusion arise from the nature of the fusion process. Fusion fuel is abundant and cheap, the most easily exploitable fuels being deuterium, occurring naturally in all water, and tritium, which can easily be manufactured inside the fusion reactor by the neutron bombardment of lithium, also abundant in nature. And fusion does not produce nuclear waste directly, though tritium is mildly radioactive and neutron activation of the reactor chamber dictates which structural materials are most useful to minimize waste disposal of components discarded in maintenance or the entire reactor assembly at the end of its life. However, whereas fission occurs at normal temperatures, fusion occurs only at the extreme temperatures characteristic of stars (aside from muon catalysis, which does not require high temperatures but thus far poses other unsolved problems). The fuel with the lowest kindling point is a mixture of deuterium and tritium that ignites at temperatures around 50 keV or 50 million degrees Kelvin. At such high temperatures, the fuel becomes a fully ionized gas, or plasma, hence the prominence of plasma physics in fusion research.

Two approaches to obtaining high-temperature plasmas have dominated the field. One is the confinement of the fuel at moderate pressure by means of magnetic fields, and the other—called inertial-confinement fusion (ICF)—utilizes solid DT targets heated by intense laser beams or ion beams. Impressive progress has been made. While self-sustaining ignition has not yet been achieved, experiments in tokamak magnetic fusion devices have approached "breakeven"—power out equal to the power in—at fusion power levels exceeding 10 MW (Strachan *et al.*, 1994; JET Team, 1997), and great progress has also been made with ICF laser experiments (Lindl, 1995). Based on these results, the physics requirements for achieving ignition are now fairly clear, for both approaches (Fowler, 1997).

In this article, we shall focus on the scientific developments that led to these achievements. Largely through the impetus of fusion research, plasma physics has reached a level of sophistication comparable to older fields of applied science, such as solid-state physics and fluid mechanics. Mastery of plasma physics at a level adequate for understanding fusion plasmas requires a complete synthesis of classical physics. A resurgence of interest in this fundamental discipline has benefited astrophysics, space physics, and applied mathematics and has trained many scientists and engineers who have made outstanding contributions in industry and academia.

Creating Magnetic Fusion Science

Magnetic fusion research began in the 1950s, initially in secret but soon declassified, in 1958, in recognition of the fact that the research would benefit greatly from a concerted world effort and had little connection with nuclear weapons technology or weapons proliferation. Research on the ICF approach began about a decade later and remained classified for a longer time, especially in the U.S., but it too is now largely declassified. The value of international cooperation in fusion research cannot be overstated, in terms both of science and of its contributions to East-West communication during the Cold War. A famous event in fusion history, which heralded the dominant role of the Russian tokamak in magnetic fusion research, was the "airlift" to Moscow, in 1969, of a British research team using their own equipment to verify Russian claims that they had achieved new records of plasma confinement and the then-unprecedented temperature of 10 million degrees Kelvin. This event, soon followed by confirming experiments in the U.S. and Europe, paved the way for the large tokamak facilities constructed in aftermath of the oil crises of the 1970s—the facilities that have now achieved near-breakeven in the 1990s.

The development of plasma physics for magnetic fusion research has been strongly influenced by the requirements for achieving ignition. It is useful to think of ignition requirements in two steps—first, the creation of a stable magnetic configuration to confine the fuel plasma, and, second, doing so at a critical size large enough so that the fusion reactions heat the fuel faster than the heat can leak away. Examining the history of the tokamak in light of these two requirements will serve to illustrate how and why fusion science developed as it did. A more thorough discussion of tokamaks and other magnetic configurations can be found in Teller (1981) and Sheffield (1994) and a discussion of fusion nuclear engineering in Holdren et al. (1988), which compares safety and nuclear waste characteristics of fusion and fission reactors.

The starting point is a magnetic configuration to confine the plasma in a state of equilibrium between magnetic forces and pressure forces. Whereas gravitational forces are symmetrical, so that stars are spheres, the magnetic force is two-dimensional, acting only perpendicular to a current, so that a magnetically confined plasma is a cylinder. The tokamak, invented by Igor Tamm and Andrei Sakharov in the Soviet Union, is descended from the linear "pinch," a plasma column carrying currents along its length whose mutual attraction constricts the plasma away from the walls of the tube that contains it, as discovered by Willard Bennett in 1934. Early linear pinch experiments at Los Alamos and elsewhere proved to be unstable, and heat leaked out the ends, defects remedied in the tokamak by bending the cylinder into a closed ring or torus, stabilized by a strong field generated by a solenoid wrapped around the toroidally shaped vacuum vessel. Bending the current channel into a circle requires an additional "vertical" field perpendicular to the plane of the torus. Thus the tokamak solves the requirement of stable confinement using three sources of magnetic field—the vertical field to confine the current, the current to confine the plasma, and the solenoid to stabilize the current channel.

The stability of the tokamak follows from its magnetic geometry, in which the field lines produced by the toroidal solenoid are given a helical twist by the current. Ideally, these twisting field lines trace out symmetric, closed toroidal surfaces—called flux surfaces—nested one inside the other. A similar concept, not requiring currents in the plasma, is the stellarator, invented by Lyman Spitzer in the early 1950s. A major theoretical achievement, published by Bernstein, Frieman, Kruskal, and Kulsrud (1958), is the energy principle, whereby the stability of tokamaks or any other magnetic configuration can be determined exactly within the constraints of magnetohydrodynamic (MHD) theory borrowed from astrophysics, in which the plasma is treated as a fluid represented by averaging the equations of motion over all particles in the plasma.

By the late 1960s, fusion scientists had repaid their debt to astrophysics by their own extensions of the theory including the effects of resistivity due to Coulomb collisions between electrons and ions. In fusion devices carrying current, the magnetic structure can be disrupted by breaking or "tearing" of the field lines due to resistivity—an example of magnetic reconnection prevalent in many astrophysical phenomena and in planetary magnetic fields, and an early example of "chaos" in which the current channel breaks up into filaments that create islands in the field structure or field lines wandering out of the machine. For tokamaks, the study of tearing was motivated by occasional violent disruptions of the current channel, which must be understood and controlled to avoid severe damage to the machine. In other magnetic confinement geometries, discussed below, tearing can actually serve the useful purpose of self-organization of plasma currents into a stable configuration. Thus we see how, in concentrating on the creation of stable magnetic configurations to meet a basic ignition requirement, fusion science has evolved a fundamental understanding of magnetized plasmas that has simultaneously contributed to solving a practical problem in tokamak design, shed light on phenomena ubiquitous in nature, contributed to the development of chaos theory in applied mathematics and many fields of physics, and stimulated new inventions in fusion research.

Of even greater impact on plasma physics, and on our understanding of turbulence in fluids, has been the extensive body of experimental and theoretical work aimed at the second ignition requirement, to determine the critical size at which fusion power production exceeds heat transport out of the plasma. Given a stable magnetic structure, heat can still be transported by entropy generation associated with processes not included in the energy principle, which assumes perfect conductivity along field lines. One such process, already mentioned, is resistivity, for which the entropy generation rate can be calculated accurately. "Classical" resistive transport, due to Coulomb collisions, is relatively weak and diminishes greatly at high temperatures. More important but more difficult to calculate is transport due to microscopic turbulence associated with the buildup of weak electric fields parallel to the magnetic field **B**. Though usually too weak to affect the resistivity in fusion plasmas, these weak parallel electric fields also imply components perpendicular to **B** that cause plasma particles to execute cycloidal orbits drifting between flux surfaces. Small perturbations grow into turbulence if the drifting motion is amplified, as can be true in tokamaks for perturbation wavelengths a few times the orbital radius of ions spinning in the magnetic field. At this time, the main information about turbulent transport is obtained empirically, by fitting formulas to the results of numerous tokamak experiments, guided in part by dimensional analysis to suggest scaling laws appropriate for particular physical processes. Computer codes, called particle-in-cell (PIC) codes, have been used to simulate drift motion turbulence (drift waves) by following the detailed motion of thousands of particles representing charge clouds generated by the turbulence. Other codes, focusing on magnetic turbulence, follow the nonlinear evolution of "tearing" modes. Calibration of code results with experimental data shows promise, though machine designers still must rely heavily on the empirical approach.

Theoretically, the potential for microturbulence is studied by examining the stability properties of the Vlasov equation, in which ions and electrons are represented by distribution functions $f(\mathbf{x},\mathbf{v},t)$ in the phase space of position $\mathbf{x}$ and velocity $\mathbf{v}$. The Vlasov equation is just the continuity equation in this phase space—a Liouville equation with Hamiltonian forces, coupled to Maxwell's equations in which the charge and current densities are obtained by velocity averages of the Vlasov distribution function. The MHD fluid equations can be derived as velocity moments of the Vlasov equation. Stability is studied by searching for growing eigenmodes of the Vlasov equation linearized around an equilibrium distribution. It can also be shown that, as for the MHD equations, there must exist a corresponding "energy principle" for the linearized Vlasov equation; correspondingly, the nonlinear theory should possess a generalized entropy and associated "free energy" from which transport could be derived. Though useful conceptually, this approach has not yet yielded many calculational results (Fowler, 1968).

At the time this article appears, a promising new direction—already being exploited experimentally—is the reduction of transport by the deliberate introduction of sheared flows and "reversed" magnetic shear that break up the collective motions produced by turbulence. Initially discovered experimentally in the 1980s as the "H mode" of operation with reduced transport at the plasma edge, with theoretical guidance this technique has now been extended throughout the plasma volume, resulting in heat transport associated with the ions at the minimum rates allowed by Coulomb collisions, though electron-related transport still appears to be governed by turbulence.

As a final example of fusion-inspired plasma physics, we return to the tokamak and its requirement for a strong current circulating around the torus. In existing tokamaks, the current is induced by the changing flux of a transformer, the plasma ring itself acting as the secondary winding, but already methods have been demonstrated that can drive a steady current (energetic atomic beams, microwaves, etc.). All such methods would be too inefficient, producing unwanted heating, were it not for the fact that, miraculously, the tokamak can generate most of its own current, called the "bootstrap" current. Again the reason lies in the equation of motion, now having to do with the nonuniformity of the magnetic field in a tokamak, which causes additional oscillatory drift motion due to changes in the orbital radius as particles spin around field lines. The enhanced transport of particles due to collisions among these magnetically drifting orbits, called "neoclassical" transport, drives a dynamo-like response as the conducting plasma flows across magnetic field lines, and this dynamo drives the bootstrap current. It is neoclassical heat transport by ions that gives the minimum possible rate of entropy generation and the minimum possible heat transport in a tokamak.

A different mechanism of current generation is the magnetic dynamo, now arising from the statistical average of the $\mathbf{v} \times \mathbf{B}$ force in a magnetic field undergoing turbulent fluctuations due to tearing and reconnection. Whereas the collisional bootstrap current creates magnetic flux, this magnetic dynamo mainly reorganizes the field due to the approximate conservation of a quantity called "helicity," given by an integral of the scalar product of **B** and the vector potential **A**. Though of limited importance in tokamaks, in which tearing is largely suppressed by the strong toroidal field, in other concepts, such as the reversed-field pinch (RFP), the plasma generates its own toroidal field as it relaxes toward a state of minimum magnetic energy at fixed helicity—known as "Taylor relaxation" (Taylor, 1986). An open question at this time is the extent to which turbulent relaxation creates unacceptable heat transport as the field continually readjusts to compensate for resistive decay near the plasma boundary where the temperature is lowest and the resistivity is highest.

Inertial-Confinement Fusion

Turning briefly to the ICF approach, we find entirely different physics issues, reflecting an entirely different solution to the basic requirements of a stable assembly of fuel and the critical size to achieve ignition. Though heating the solid target also forms a plasma, its density is so high that plasma turbulence is irrelevant in calculating the critical size. However, achieving a useful critical size requires that the fuel first be compressed to densities many hundreds of times that of ordinary materials. This is accomplished by heating the spherical target uniformly from all sides, whereby the intense heating of the surface creates an inward implosion of the fuel as the surface layer, called the ablator, explodes outward. At the intensity of giant lasers now available, implosion pressures of millions of atmospheres are created, compressing the fuel to 100 times liquid density, and a 1000-fold or more compression should be possible. The physics issues concern mainly the uniformity of illumination and target design requirements to suppress hydrodynamic instability that amplifies imperfections in the surface finish.

Plasma physics enters mainly in ensuring efficient absorption of the laser energy before the beams are reflected at the "cutoff" density at which the laser frequency matches the "plasma frequency" (the same condition as that for the reflection of light from an ordinary mirror). Two methods are employed, the direct-drive approach, in which laser beams shine directly on the target, and indirect drive, in which the target is mounted inside a tiny metal cylinder, called a hohlraum, which converts laser light to x rays that in turn irradiate the target. For both approaches, efficient absorption—either in a plasma cloud surrounding the ablator for direct drive, or in the metallic plasma formed where laser beams strike the hohlraum wall for indirect drive—requires the use of ultraviolet light to penetrate to densities where collisional absorption dominates over collective "laser-plasma interactions" (Lindl, 1995). The invention at the University of Rochester in the late 1970s of efficient methods to convert the infrared light produced by glass lasers into ultraviolet light was an important milestone in ICF research.

Future Directions

Magnetic fusion research is now focused on an international effort to achieve ignition in a tokamak and on improvements in the concept, including other means for creating the nested toroidal flux surfaces so successfully utilized to confine plasmas in tokamaks. A central issue is to what extent one should rely on internal currents, as the tokamak does. As noted earlier, the stellarator avoids internal currents altogether by creating closed toroidal flux surfaces. Its external helical coils impart a twist to the field lines as current does. At the opposite extreme are the reversed-field pinch devices, with only a weak external toroidal field, and a very compact device called the spheromak, which has no external toroidal field and relies totally on Taylor relaxation to create the desired field configuration (Taylor, 1986). This is more than an intellectual exercise, since the size and cost of toroidal confinement devices tends to increase as they rely more heavily on externally generated toroidal fields, requiring large coils looping through the plasma torus. The spherical tokamak is an innovative exception designed to minimize this difficulty (Peng and Strickler, 1986).

For ICF, the immediate goal is ignition, in the National Ignition Facility now under construction in the U.S. Application of the concept to electric power production will require new laser technology, or perhaps ion beams, capable of rapid repetition—several times per second—and greater efficiency than existing glass lasers. An innovative means for reducing the laser energy required for compression is the "fast ignitor," using a small but very-high-power laser to ignite the target after it has been compressed Tabak *et al.* (1994).

References

Bernstein, I. B., E. A. Frieman, M. D. Kruskal, and R. M. Kulsrud, Proc. Soc. London, Ser. A **244**, 17 (1958).
Fowler, T. K., 1968, in *Advances in Plasma Physics*, edited by A. Simon and W. B. Thompson (Interscience, New York), Vol. 1, p. 201.
Fowler, T. K., 1997, *The Fusion Quest* (Johns Hopkins University, Baltimore, MD).
Holdren, J. P., D. H. Berwald, R. J. Budnitz, J. G. Crocker, J. G. Delene, R. D. Endicott, M. S. Kazimi, R. A. Krakowski, B. G. Logan, and K. R. Schutty, 1988, Fusion Technol. **13**, 7.
JET Team, 1997, Plasma Phys. Controlled Fusion (Suppl.) **2B**, 1.
Lindl, John, 1995, Phys. Plasmas **2**, 3933.
Peng, Y.-K., and D. J. Strickler, 1986, Nucl. Fusion **26**, 769.
Sheffield, John, 1994, Rev. Mod. Phys. **66**, 1015.
Strachen, J. D., and TFTR Team, 1994, Phys. Rev. Lett. **72**, 3526.
Tabak, Max, J. Hammer, M. E. Glinsky, W. L. Kruer, S. C. Wilks, J. Woodworth, E. M. Campbell, M. D. Perry, and R. J. Mason, 1994, Phys. Plasmas **1**, 1626.
Taylor, J. B., 1986, Rev. Mod. Phys. **58**, 741.
Teller, E., 1981, Ed., *Fusion* (Academic, New York), Vol. I, Parts A, B.

Physics and U.S. National Security
Sidney D. Drell

Throughout history scientists and engineers have contributed to the military strength and ultimate security of their societies through the development of new technologies for warfare. And throughout history the military and the governmental leaders have called on scientists and engineers to help devise the means to counter or neutralize the technologies developed by adversaries as a threat to their national security.

Looking back to the third century B.C., one recalls the legend of Archimedes designing the great catapult to help thwart the Romans besieging Syracuse. That was but one example of a variety of fortifications and instruments of war he contributed. Perhaps best known of the great military scientists throughout history is Leonardo da Vinci, of whom Lord Zuckerman wrote in his 1982 book, *Nuclear Illusion and Reality* (Zuckerman, 1982):

The letter which [Leonardo] wrote to Ludovico Sforza, the ruler of the principality of Milan, offering to provide any instruments of war which he could desire—military bridges, mortars, mines, chariots, catapults, and other "machines of marvelous efficacy not in common use"—was that of an arms salesman, the sort of offer which a later generation might have regarded as emanating from a "merchant of death."

And later Michaelangelo spent time as the engineer-in-chief of the fortifications in Florence. Based on their expertise, scientists and engineers can contribute to developing new military technology. Equally importantly, their understanding of the laws of nature helps them define the limits of what one can expect from technology—existing and prospective—which must be understood when governments formulate military plans and national security policy. Nature cannot be coerced to meet unrealistic military goals.

During World War II, in most of the combatant countries there was a total mobilization of scientists into the war effort. In the United States and Britain they tackled many technical problems, from rockets and antisubmarine warfare to operations research (Jones, 1978). Physicists played an especially important role in collaboration with the military in developing microwave radar (Buderi, 1996; see the chapter by Pound on pp. 89–95 in this book) and the atomic bomb (Rhodes, 1986). And the decisive role of these weapons has been widely chronicled. This collaboration and its achievements formed the foundation for expanded cooperation following W.W. II.

A new circumstance emerged in the 1950s with the development of the hydrogen bomb. With its greatly enhanced energy release from a second, or fusion, stage, the hydrogen bomb meant that science had now created a weapon of such enormous devastating potential that, if used in large numbers in a future conflict, it could threaten the very existence of civilization as we know it.

This new circumstance, and the growing danger of renewed conflict in the developing Cold War, greatly enhanced the importance of cooperation and understanding between physicists and the military and national policy leaders. A whole raft of new, serious issues had to be explored and understood—not only hydrogen bombs, but additional challenges including worldwide radioactive fallout from nuclear weapon tests above ground, the global effects of large-scale nuclear war, the leap into space with missiles and rockets, and the role of anti–ballistic missile (ABM) systems. It was also important to communicate with Soviet and other international scientific colleagues to develop a mutual understanding of these issues.

Upon becoming involved with technical issues of national security, scientists must also deal with the inhibiting requirements of secrecy. There is a natural tension between the openness that we scientists value so highly in our research and the secrecy that surrounds so much of the technical work for national security. This is clearly evident in the domain of technical intelligence and in issues pertaining to nuclear weapons, which were born secretly during W.W. II and have remained so to this date. The need for secrecy is understandable, but it is important to recognize that there are serious costs when the walls of secrecy are too encompassing or too high for too long. One cost is the loss of critical analyses of assumptions and decisions by highly qualified peers, a process of proven importance to scientific progress. The barrier of secrecy also makes it difficult to develop an informed citizenry, beyond the policy specialists, whose inputs can be important for making enlightened policy choices. Scientists are in a good position to judge which matters are readily open to discovery by first-class minds anywhere, and therefore futile to guard by secrecy. We have a special role to help establish a proper balance between secrecy and openness.

In the following, I shall discuss three general topics of importance raised by the new circumstance. First I shall describe some of the scientific initiatives by physicists that were designed to help meet the new challenges starting shortly after W.W. II. In this I shall be selective on two counts: I shall only discuss efforts of which I have firsthand knowledge and about which sufficient information can be publicly analyzed to make fully informed judgments. There are many others. Then I shall discuss briefly a mechanism for ensuring that needed scientific advice is available to the President on important issues of national security. Finally I shall discuss the responsibility of the scientific community in helping the U.S. and, more generally, all societies meet these new challenges that we have created and that leave us precious little margin for error.

Photoreconnaissance from Space

Aerial reconnaissance for tactical purposes emerged as a major asset for battle planning and fighting during W.W. II. It provided important targeting information on enemy facilities and deployments, as well as post-action damage assessment as a guide for future missions. As the Cold War confrontations intensified, the U.S. military and national security leaders realized the enormous potential of reconnaissance from space for strategic purposes, especially against a society so obsessed with secrecy as the Soviet Union was. Strategic reconnaissance from space could alert us to threats that might be developing, as well as dispel some that we incorrectly assumed to exist. This was especially critical in an era of nuclear weapons and intercontinental-range delivery systems no more than 30 minutes away with their almost unimaginable destructive potential.

This concern stimulated a major effort by the U.S. to develop means to penetrate the Iron Curtain in peacetime from high above ground and into space. The means would have to be sufficiently nonintrusive and nonmilitarily threatening to avoid triggering conflict, and also be essentially invulnerable to being intercepted and destroyed (Hall, 1996; this article contains many further references on this subject).

The original vehicles for strategic reconnaissance were high-flying aircraft, notably the U2, which began to overfly the Soviet Union in 1956. It could fly at altitudes above 60 000 feet, sufficiently high to avoid being shot down by the interceptors or surface-to-air missiles (SAMs) that existed prior to 1960. It was also quiet and invisible from the ground by the naked eye, and thereby its intrusion of sovereign airspace was not politically embarrassing to Soviet leaders. Aside from the challenge of building the U2, there were the technical challenges of developing an accurate high-resolution camera system operating near its diffraction limit, with compensation for image motion during the filming, among other major operational requirements. It was also realized that it was only a matter of time before the U2 could be shot down as technology improved. This led to a major effort to develop photoreconnaissance satellites circling the earth at sufficiently high altitudes (above 100 miles) that they could stay aloft for days to weeks, and eventually years, without having to carry prohibitively heavy fuel loads to compensate for atmospheric friction.

The strategic value of overhead photographic intelligence—from aircraft and subsequently from satellites—has been recognized for many years. It is best illustrated by several examples. During the 1950s, before the U.S. could rely on effective aerial reconnaissance, we understandably made worst-case assumptions and spent huge sums on a very large and expensive air defense system against what was erroneously believed to be a major Soviet bomber threat, but which in reality was nonexistent. Furthermore a fictitious missile gap played center stage in the Presidential election of 1960. No one alive at the time will ever forget the intensely frightening drama of the Cuban missile crisis in October of 1962. It was the U2 overflights of Cuba by the Central Intelligence Agency that alerted the U.S. to the introduction there of nuclear-capable missiles and bombers by the Soviet Union early enough to enable a peaceful resolution before matters got out of hand, possibly triggering nuclear conflict. In the 1970s and 1980s the "national technical means" of surveillance, as photoreconnaissance satellites were called, gave sufficient transparency through the Soviet Union's Iron Curtain that the U.S. could enter into strategic arms negotiations to control and eventually start to reduce the threat of nuclear weapons mounted on large delivery systems of intercontinental ranges. The number of such missiles that were actually deployed could be counted from space, so that compliance with treaty limitations could be verified. Verification, understandably, was a *sine qua non* of arms control agreements. Until recently the nonintrusive means of inspection from space were the primary means of acquiring the necessary information to ensure compliance. Photoreconnaissance satellites were the first big step toward achieving the Open Skies that President Eisenhower had first called for in 1955, and opened the door to arms control negotiations.

With the declassification in 1995 of the first generation of photoreconnaissance satellites, known as Corona, it is now possible to give a technical description of the remarkable scientific and engineering achievements in developing that system (Wheelon, 1997; this article contains references to a number of more detailed documents; Day *et al.*, 1998).

Over a twelve-year period from August 12, 1960, the date of its first successful mission, until its last one on May 25, 1972, more than two million feet of film were recovered from 145 satellites successfully placed in low polar orbits, above the atmosphere. During this period more than 800 000 photographs of earth targets were taken with a camera system that ultimately achieved a ground resolution of six feet.

A large number of American scientists and engineers contributed to this remarkable feat. I will mention four physicists with whom I collaborated extensively through the years on this subject, who made major technical and leadership contributions to the development and the continued technical advances of aerial and space reconnaissance: Edwin Land, founder of Polaroid; Edward Purcell from Harvard; Richard Garwin from IBM; and Albert D. Wheelon, an MIT physics Ph.D. who was the CIA's Deputy Director for Science and Technology from 1962 to 1966, in which capacity he headed the U2 program, development of the Mach 3 SR71 airplane that was its successor, and the Corona program, plus follow-on satellite systems.

To give a sense of the magnitude of what was accomplished, consider the following numbers. An optical system with a 1-m aperture, operating under ideal conditions, can resolve a separation of approximately 5 inches from a distance of 100 miles. This is the theoretical limit of performance. What one actually achieves from a camera moving in earth orbit depends on how well system vibrations are damped; how accurately the image motion is compensated from a satellite moving at a speed of 5 miles per second in orbit; and the limiting resolution of the photographic film. In addition there is the unavoidable image degradation as a consequence of atmospheric scattering and turbulence and their effects on image contrast and blurring.

The Corona satellite used a panoramic camera that panned 70° transverse to the orbit direction. For most of its life this system had two counter-rotating cameras with 7-inch-diameter lenses, one tipped 15° forward of the vertical and the other 15° to the rear, for stereoscopic coverage. This provided important information on the vertical dimension of the targets on the ground. With a 70-mm-wide film format and a focal length of 24 inches, the camera panned a ground swath width approximately 120 miles long cross-track and 10 miles wide along orbit. With a pan being made every two seconds, successive ground swaths were contiguous with one another. On successive orbital passes new swaths of denied territory were photographed as the earth rotated under the 90-minute retrograde polar orbits (with $\approx$100-mile perigees and $\approx$240-mile apogees). During each pan the film was held stationary on a cylindrical platen as the rotating lens assembly ("telescope") scanned a slit image over it. The film then moved forward and the process was repeated when the telescope reached the starting position.

A very thin but strong acetate-base film was developed so that the camera could carry a large load for maximum photographic coverage. In its ultimate design 16 000 feet of film were carried aloft for each Corona mission, 8000 feet for each of the stereoscopic pair. This film was three-thousandths of an inch thick and had the capacity to resolve 170 lines per millimeter at 2 to 1 contrast ratio. For reference, the best film used in aerial reconnaissance during W.W. II resolved 50 lines per millimeter. With this resolution and the 24-inch focal length for its optics, Corona could theoretically achieve, and eventually closely approximated in practice, 6-foot ground resolution from orbit.

On this scale of resolution, the degradation of image quality due to random inhomogeneities in the atmosphere's refractive index arising from turbulence plays a minor role under normal viewing conditions. As was known from extensive data from ground-based telescopes, as well as from theoretical studies of atmospheric properties, a beam of light vertically traversing the atmosphere is typically spread over 1 to 2 arc seconds of angle. Since most of the turbulent atmosphere lies below 40 000-feet altitude, this translates into a blur circle with a radius of 2 to 4 inches for the image of a point target as seen from high altitudes by the satellites.

The development of the Corona system also had to surmount many additional operational problems ranging from the powerful rocket stages needed to insert it into the proper near-polar orbits, to the design of the film capsules to survive reentry into the earth's atmosphere, where they would be caught by trailing hooks from a C119 aircraft before sinking into the ocean.

Corona—a truly wondrous achievement—was just the first of several generations of satellites that have advanced space reconnaissance to higher resolutions and provided real-time return of valuable intelligence information when solid-state detectors replaced film. With this technical advance and others, including satellite communication and data relays, both tactical battlefield intelligence and strategic intelligence have been achieved, as made evident by recent experience in the Gulf War and Bosnia. The full story of impressive achievements in surveillance from space, utilizing a broad range of the electromagnetic spectrum, is yet to be told.

ABM Systems

The enormous destructive potential of nuclear weapons has greatly changed the balance between offense and defense. The British won the Battle of Britain in World War II with an aerial defense that shot down approximately 1 in 10 attacking aircraft. This meant that the attacking units of the German Air Force were reduced by close to two-thirds after ten raids, but London still stood, although battered. Not much of London or its population would survive today, however, if but one modern thermonuclear warhead arrived and exploded. Death and destruction would be very extensive within five miles of a one-megaton air burst at an altitude of 6–8 thousand feet. A fire storm could be ignited, further extending the range of destruction. Clearly a defense would have to be essentially perfect to provide effective protection against nuclear-tipped ballistic missiles arriving at speeds of ≈ 7 km/sec on trajectories above the atmosphere from across the oceans.

This new circumstance triggered extensive and intense debates. Would an effective nationwide anti–ballistic missile (ABM) defense be feasible against a massive attack, and what impact would efforts to develop and deploy such an ABM system have on strategic stability? Or would it be more realistic and conducive to arms control to deploy a survivably based retaliatory force of missiles and long-range bombers, and rely on mutual assured destruction to deter a would-be attacker and maintain a stable strategic balance? This debate has lasted for more than three decades, with periodic crescendos that have often been driven more by political agendas than by technical realities.

On an issue as important and complex as this, it is critical to have the technical factors right before drawing conclusions or making decisions. There are no simple or obvious answers when all the relevant strategic and economic factors are included with the technical ones. For example, if one starts with the assumption that the threat is known and severely limited and will not change or grow over the many years required to build a defensive system, many, if not most, scientists would agree that an effective defense could be constructed, given the necessary resources. After all, we met the challenge to put a man on the moon and that was indeed a major technical challenge. However, in the real world, the offense would not remain frozen in time. There is a whole repertoire of countermeasures that a determined opponent could rely on to overpower a growing defense: decoys, maneuvering reentry bodies, reentry bodies with reduced radar cross sections, piling more multiple independently targetable reentry vehicles (MIRVs) atop the missiles, or attacking the defensive system. In a competition of countermeasures and counter-countermeasures, what are the costs of a defensive system designed to maintain a desired level of effectiveness against the offense? Can a defensive system, whether ground-based, space-based, or both be made invulnerable to being blinded or destroyed if it, itself, were attacked as the initial target? Answers to such questions depend on how far one can reach in developing a reliable and effective system with newly developed technologies, as well as on the overall scale and scope of the planned defensive and offensive systems.

No less important than these technical and economic questions is the strategic one: what will be the effect, over the long run, for strategic stability and for future prospects of reducing the nuclear threat of a continuing offense-defense competition? When the Soviet Union started building an ABM system ringing Moscow, which was perceived as the first step in a possible nationwide ABM deployment, the U.S. responded by deploying many MIRVs, on both land- and sea-based ballistic missiles. This was the surest and cheapest way to overpower the emerging defense with more warheads than it could engage and thus to maintain our nuclear deterrent. The net result was that the nuclear danger, measured in terms of total number of threatening warheads and worldwide consequences of a total nuclear war, increased greatly. With very accurate and highly MIRVd missiles sitting in silos (like the Soviet SS18 and the U.S. MX) there was also a threat to strategic stability in a perceived advantage of launching first to take advantage of the MIRV multiplier that allows one or two warheads to destroy up to 10 warheads per missile in each silo.

Another important factor fueling the ABM debate was a very human and emotional one. Throughout history, protecting our families and defending our homes has been one of the most basic human instincts. Did we now have to accept, as inescapable, the conclusion that defense in the nuclear age was no longer possible and that we would have to settle for deterrence? The technical realities lie at the core of a responsible answer to this question. Clearly this was an issue in which physicists would play a central role, together with other scientists and engineers and the military. Whatever we may prefer as the goal of our policy, realities consistent with the laws of nature cannot be denied.

The first serious commitment by a U.S. President to deploy an ABM system was a "thin" area defense against China proposed by President Lyndon Johnson in 1967. President Richard Nixon mutated the mission of this system in 1969 and proposed deploying it as a "hard-point" defense of Minutemen ICBM silos. It acquired the system name of "Safeguard." At that time the technology at hand consisted of phased-array radars for long-range target acquisition and more sophisticated ones that managed the battle by launching and guiding nuclear-tipped defensive missiles to intercept the incoming warheads. The interceptors were of two kinds. For defending hardened targets like Minutemen silos one could engage the incoming warhead well below the top of the atmosphere, thereby using its friction to strip away decoys. In this case the interceptor was relatively small, had very high acceleration, and was armed with an enhanced neutron warhead for relatively long-range kill of the incoming warhead. For defending soft targets like cities, the interceptor was designed for long-range flyout to keep the engagement far from the target, and it was intended to engage an incoming warhead, or any decoys accompanying it on identical Newtonian trajectories, above the atmosphere relying on long-range x-ray kill.

The problems and limitations of an anti–ballistic missile defense based on the then-available technology were described by Hans Bethe and Richard Garwin (Garwin and Bethe, 1968) in *Scientific American*. This was the first comprehensive technical analysis to be published in the unclassified literature and played an important role in informing and framing the public debate. So did extensive congressional testimony on the technical and strategic issues raised by the proposed Safeguard deployment. The debate involved many scientists in the first comprehensive public hearings on a proposed major new weapon system. In the end Safeguard was deployed at one field of Minuteman ICBM silos, at Grand Forks, North Dakota, consistent with the provisions of the 1972 ABM Treaty; and soon thereafter it was decommissioned and abandoned because of cost ineffectiveness.

Following the negotiation of the 1972 ABM Treaty by the United States and the Soviet Union, technology continued to advance swiftly and new possibilities such as directed energy weapons and space-based sensors came to the fore for ABMs. In his famous Star Wars speech in March, 1983, President Ronald Reagan sought to rely on the new and emerging technologies to build a nationwide defense, which some supporters claimed would create an "astrodome" or impenetrable defense of the entire nation. In the absence of a careful analysis of practical possibilities and limitations, fanciful claims preceded more measured judgments, and a largely political and highly acrimonious debate ensued. Many physicists contributed to the careful analyses that led eventually to more realistic goals for a much more modest potential ABM system (cf. Carter and Schwartz, 1984; Drell, Farley, and Holloway, 1985; Office of Technology Assessment, 1985).

A very important contribution by physicists to the ultimate resolution of that debate was the report prepared by the American Physical Society Study Group on "Science and Technology of Directed Energy Weapons" co-chaired by N. Bloembergen of Harvard and C. K. Patel, then of Bell Labs. It was published in the *Reviews of Modern Physics*, July 1987, four years after President Reagan's speech. This was a definitive analysis of the new and prospective technologies along with the relevant operational issues. Laser and particle beams; beam control and delivery; atmospheric effects; beam-material interactions and lethality; sensor technology for target acquisition, discrimination, and tracking; systems integration including computing power needs and testing; survivability; and system deployment were all analyzed carefully, as were some countermeasures. Aspects of boost-phase, midcourse, and terminal intercepts that were all parts of the Star Wars concept of a layered "defense in depth" were included in their comprehensive analysis.

The sober findings of the APS Directed Energy Weapons Study are summarized in part as follows:

Although substantial progress has been made in many technologies of DEW over the last two decades, the Study Group finds significant gaps in the scientific and engineering understanding of many issues associated with the development of these technologies. Successful resolution of these issues is critical for the extrapolation to performance levels that would be required in an effective ballistic missile defense system. At present, there is insufficient information to decide whether the required extrapolations can or cannot be achieved. Most crucial elements required for a DEW system need improvements of several orders of magnitude. Because the elements are inter-related, the improvements must be achieved in a mutually consistent manner. We estimate that even in the best of circumstances, a decade or more of intensive research would be required to provide the technical knowledge needed for an informed decision about the potential effectiveness and survivability of directed energy weapon systems. In addition, the important issues of overall system integration and effectiveness depend critically upon information that, to our knowledge, does not yet exist . . . . Since a long time will be required to develop and deploy an effective ballistic missile defense, it follows that a considerable time will be available for responses by the offense. Any defense will have to be designed to handle a variety of responses since a specific threat can not be predicted accurately in advance of deployment.

Physicists and, more generally, scientists and engineers, played quite different roles vis-à-vis the military and government in the ABM debates relative to the development of space-based reconnaissance. In the latter case we faced what was exclusively a technical challenge, with physicists in the forefront pushing and accelerating technical advances to open new possibilities for enhanced and more timely coverage. The work was done totally in secret with no political or policy debates beyond bureaucratic wars (some very intense) for control and budgets. In contrast, the ABM debates were and remain very public and political, and many of us found ourselves arguing on technical grounds for more realism in making claims for what could be achieved, as opposed to what would be pie-in-the-sky against a determined opponent. The point is that, as described above, ABM defenses presented not only a technical challenge, as did space reconnaissance, but also major strategic and economic challenges, as countermeasures and counter-countermeasures were developed. Fundamentally it came down to man and resources against the same. Putting man on the moon was indeed a great technical challenge and a glorious success. But the moon did not object to being landed on. It could not and did not, for example, maneuver away, or turn out its lights, or deploy decoys, or destroy the invading lander.

Physicists in the ABM debate also played a very significant role in the education of a public constituency for arms control. The public discussion of the competition between offense and defense, and their countermeasures and counter-countermeasures, went beyond purely technical matters. Out of the many exchange calculations of casualties due to blast, burns, and radioactive fallout, with or without various assumed ABM systems, two conclusions became indelibly etched in the public awareness:

1. In any large-scale nuclear conflict, the unknowns far exceeded what could be predicted. In any event, the level of casualties and destruction would be almost unimaginably high.
2. The sizes of nuclear arsenals of the U.S. and the Soviet Union were large beyond all reason or purpose. What does one do with 10–20 thousand warheads when Hiroshima and Nagasaki revealed to us the enormous destruction caused by the mere trigger of one such modern bomb?

An awareness of these facts that was sharpened by the ABM debates added to the cogency of the public support for arms reductions. As a strong believer in the power of an informed public constituency, I see this as a very important achievement by scientists in the ABM debates, and more generally in efforts to reduce nuclear danger.

The continuing debate on ABMs in the post-Cold War world of 1998 is concerned with two issues. The first is the deployment of regional ABM systems in areas of potential conflict to provide a defense of societies and combatants against short-range ballistic missiles carrying non-nuclear warheads. During the Gulf War, upgraded air defenses provided political comfort, though they were not effective for physical protection when used in Israel and Saudi Arabia against short-range SCUD missiles launched by Iraq. This was not surprising since the so-called Patriot defense system had not been designed as a ballistic missile defense. After the Gulf War there were very heated debates as grossly exaggerated claims of Patriot effectiveness were made and then used in a political effort to sell the case for nationwide ballistic missile defenses against strategic systems. Those claims were effectively debunked after extensive analysis (Lewis and Postol, 1993) and current U.S. programs are focused primarily on the limited mission of protection in regional conflict against tactical, or relatively slow and short-range, ballistic missiles. Technical analyses and political discussions with Russian officials are addressing the important issue of establishing an appropriate demarcation between permitted activities and deployments, and deployments of nationwide strategic defenses that are severely constrained by the ABM Treaty of 1972.

The second issue that is currently being addressed by the United States is the potential deployment of long-range ballistic missile threats against our society by other so-called "third-world" nations, such as North Korea, who do not at present pose such a threat, but whose activities can be perceived as an effort to develop one. Of course such nations who wish to develop threats to the U.S. mainland can rely on much less demanding technologies, such as, for example, ship-launched cruise or ballistic missiles whose range is measured in hundreds, rather than many thousands, of kilometers; or covert delivery into harbors. We see here yet again the critical value of strategic intelligence, including space reconnaissance to keep us informed of what, if any, threat is emerging, and to do so in a timely way, so that we will have ample opportunity to respond appropriately, if need be.[1]

Nuclear Testing

When President Clinton signed the Comprehensive Test Ban Treaty (CTBT) at the United Nations on September 26, 1996, he said that the CTBT was "The longest sought, hardest fought prize in the history of arms control." The effort to end all nuclear tests commenced four decades earlier. Upon leaving office President Eisenhower commented that not achieving a nuclear test ban "would have to be classed as the greatest disappointment of any administration—of any decade—of any time and of any party . . . ."

A decisive political and strategic reason for the United States and the other four declared nuclear powers—China, France, Russia, and the United Kingdom[2]—to sign a ban on all nuclear testing in 1996 was the importance of such a treaty for accomplishing broadly shared nonproliferation goals. This was made clear in the debate at the United Nations in May 1995 by 181 nations when they signed on to the indefinite extension of the Non-Proliferation Treaty (NPT) at its fifth and final scheduled five-year review. A commitment by the nuclear powers to cease testing and developing new nuclear weapons was a condition for many of the non-nuclear nations when they signed on to the Treaty. Not only will the CTBT help limit the spread of nuclear weapons through the nonproliferation regime, particularly if current negotiations succeed in strengthening the provisions

---

[1] In this connection see "Intelligence analysis of the long-range missile threat to the United States: hearing before the Select Committee on Intelligence of the United States Senate, One Hundred Fourth Congress, second session . . . Wednesday, December 4, 1996." (U.S. Government Printing Office). See also the report of the "Commission to Assess the Ballistic Missile Threat to the United States," the so-called Rumsfield Commission, whose unclassified Executive Summary was issued July 15, 1998.
[2] In May 1998 the two so-called "threshold" or undeclared nuclear powers, India and Pakistan, joined the nuclear club with a short series of underground nuclear tests. Since then there have been no further test explosions.

for verifying that treaty and appropriate sanctions are applied for noncompliance. It will also dampen the competition among nations who already have nuclear warheads, but who now will be unable to develop and deploy with confidence more advanced ones at either the high or the low end of destructive power. The CTBT would also force rogue states seeking a nuclear capability to place confidence in untested bombs.

Notwithstanding a strong case for the CTBT, the United States, if it is to be a signatory of this treaty, must be confident of a positive answer to the following question. Under a ban on all nuclear explosions, will it be possible to retain the currently high confidence in the reliability of our nuclear arsenal over the long term, as the weapons age and the numbers are reduced through arms control negotiations? A study was organized in 1995 to address the scientific and technical challenge of answering this question. It was sponsored by the Department of Energy (DOE) and done under the auspices of JASON, an independent group of predominantly academic scientists who work as consultants for the government on issues of national importance. The participants were academic research physicists plus leading weapons designers with long and distinguished careers at the three weapons labs.

We analyzed[3] in great detail the experimental and theoretical basis for understanding the performance of each of the weapon types that is currently planned to remain in the U.S.'s enduring stockpile. This understanding has been gained from 50 years of experience and analysis of data from more than 1000 nuclear tests, including the results of approximately 150 nuclear tests of modern weapon types in the past 25 years. We found that this experience does, indeed, provide a solid basis for the U.S. to place high confidence for today and the near-term future in the safety, reliability, and performance of the nuclear weapons that are designated to remain in the enduring stockpile.

We also studied in detail the full range of activities that would enable us to extend our present confidence in the stockpile decades into the future under a CTBT. This greater challenge can be addressed with a more comprehensive science-based understanding than now exists of the processes occurring at each stage during the explosion of a modern thermonuclear warhead. What is required to gain this understanding is enhanced surveillance and forensic studies of the aging stockpile, coupled with improved diagnostic data against which to benchmark full-physics three-dimensional codes of material behavior in conditions representative of those occurring in a nuclear explosion [Office of Defense Programs, October 1997; see also JASON Report JSR-94-345 (November, 1994); Peña, 1997]. It will then be possible to simulate accurately, with greatly enhanced computer speed and power, the effects of aging on the performance of a warhead. These studies will alert us to when remedial actions will be needed to avoid significant performance degradations, and will also provide confidence in retaining weapons that exhibit no such need. Finally facilities are required for undertaking necessary remanufacturing or refurbishing of components in a timely fashion as may be needed. These are identified as the necessary components of a science-based stockpile stewardship program being implemented by the DOE as a substitute for continued nuclear testing. This program will permit the United States to preserve the integrity of our enduring nuclear stockpile. It is fully consistent with the spirit and intent of the CTBT: in the absence of tests the U.S. will not be able to develop and deploy with confidence new, improved warheads.

---

[3]For an unclassified summary of the JASON study on "Nuclear Testing" (JSR-95-320) see the Congressional Record—Senate: S-11368 (August 4, 1995).

To see what is involved, here is a brief schematic review of the successive stages of a modern warhead. The first step is to ignite the layer of chemical high explosive that surrounds the primary assembly, which has a central core called the "pit." The "pit," which contains the fissile material, Pu^{239} or highly enriched U^{235}, is driven into a highly compressed mass at the center of the primary assembly by the imploding chemical shock. A technique called "boosting" is used to achieve higher explosive yields from relatively small primaries. Boosting is accomplished by injecting a mixture of D and T gases, stored separately in high-pressure reservoirs, into the pit just before it starts imploding. With the onset of fission in the compressed pit, the D-T gas mixture is heated to the point of initiating D-T fusion, with the subsequent production of large numbers of fast neutrons via the nuclear reaction $D+T \Rightarrow He^4 + n + 17.6$ MeV. These neutrons produce many more fission reactions, thereby boosting the yield of the primary sufficiently to drive the secondary assembly, or main stage, which contains lithium deuteride and other materials. A large fraction of the bomb's energy release comes from the secondary, as $D+D$ and $D+T$ neutrons convert Li^6 to He^4 plus T, which in turn undergoes fusion with the D.

The operation of the primary of a nuclear weapon is critical to its performance: if the primary does not work nothing nuclear happens, and if its yield is too low it will not succeed in igniting the secondary, or main stage, as expected. Age-related changes that can affect a nuclear weapon and that must be understood and evaluated include the following:

(i) Structural or chemical degradation of the high explosive leading to a change in performance during implosion,
(ii) Changes in plutonium properties as impurities build up due to radioactive decay,
(iii) Corrosion along interfaces, joints, and welds,
(iv) Chemical or physical degradation of other materials or components.

An intensified stockpile surveillance program that looks for cracks, component failures, or other signs of deterioration, and that develops quantitative measures to determine when these unacceptably affect the performance of the primary, will be crucial for the short-term confidence in the stockpile over the coming decade. There are very many other non-nuclear components of a weapon system that are crucial to its successful operation, including arming and firing systems, neutron generators, explosive actuators, safing components, permissive action link coded control, radar components, batteries, and aerodynamic surfaces. All of these are critical to mission success, but testing of these non-nuclear components and making improvements as may be indicated are not restricted by a CTBT.

To ensure performance over the longer term, there will be a need for new facilities to do more detailed measurements of the behavior of the bomb components right up to the initiation of fission, including greatly increased computer power for analyzing effects of aging on bomb performance to the required accuracy.

A hydrotest is the closest non-nuclear simulation of the operation of a primary. In these experiments, the fissile material is replaced by another material, e.g., depleted uranium, tantalum, or lead. The behavior of an imploding pit that has been modified by this substitution is then studied very close to, but not beyond, the point where a real weapon would become critical; i.e., the nuclear chain reaction would be ignited. These experiments are consistent with the CTBT and can be done above ground. Properly designed, they can address issues of aging as well as safety by providing dynamic radiography of the imploding pit, and therefore they can detect any possible changes in the energy or symmetry of the imploding primary assembly due to aging that could unacceptably alter performance. "Core punching" is the technique of dynamic radiography that allows one to study properties of the pit at the late stages up to what would be ignition in a real primary with fissile material. The idea is quite simple. The x-ray source is an accelerator producing precisely timed bursts of 10–30-MeV electrons with pulse widths of perhaps 60 ns and spot sizes of approximately 1 mm that impinge on a high-Z target to yield a burst of gamma rays. These intense beams of photons (currently comprising a dose of about 300 roentgens at a meter) have a broad energy spectrum, with a mean energy of several MeV, and can penetrate the imploding pit from one side and be detected on the other side to produce an image. A high-resolution system with a small beam spot and an efficient, high-speed solid-state detector can produce an accurate image of x rays penetrating some 100 g/cm^2 of heavy metal. The ultimate resolution depends not only upon the spot size of the electron beam but also on the efficiency with which the transmitted gamma rays are detected.

In order to get several looks from two different directions for better diagnostic details, a Dual-Axis Radiographic Hydrodynamic Test Facility (DARHT) that produces two beams from two electron linear accelerators at right angles, with multipulsing in one arm to provide successive snapshots, is now under construction at Los Alamos. Beyond this, the value of obtaining tomographic movies of the late stages of an imploding pit using a variety of different look angles and time intervals is under study, including the relative merits of electron/photon and proton beams.

Beyond hydrotesting there are still important nuclear aspects that need to be carefully measured and analyzed in order to develop a deeper science-based understanding of the performance of the enduring stockpile. These include the behavior of plutonium at high temperatures and pressures, the nature of the ejecta and spall from the surface of the imploding plutonium, and a better knowledge of its equation of state at pressures and temperatures created in a nuclear explosion. Important information on such properties is being obtained from a series of subcritical underground experiments. The meaning of "subcritical" is that fewer nuclei fission in each successive generation of the chain reaction after it has been initiated, perhaps by a high explosive shock or by injection of a burst of neutrons. This contrasts with a sustained and steady rate of chain reaction, as in a reactor, or a positive coefficient of exponential growth, as in a bomb. Such subcritical experiments are consistent with the CTBT as generally interpreted.

In addition, the National Ignition Facility (NIF) is being built at Livermore to study inertial confinement fusion. It is designed to deliver a high-energy laser pulse of about 1.8 MJ, which is divided into 192 beamlets to excite a hohlraum to temperatures of 300 eV and higher, in which to symmetrically implode and initiate fusion in millimeter-size pellets, as well as for the study of other phenomena. Its relevance to the weapons program includes the study of the physics in the bomb's secondary stage during an explosion and measurements of opacity, hydrodynamic behavior, and equations of state of constituents during the primary explosion. It will also be possible to benchmark advanced explosion codes by comparing their predictions with NIF data, including analysis of changes due to aging. Many scientists also see NIF as an important facility for the study of inertial confinement fusion and the processes that control its efficiency and prospects for energy production.

Finally supercomputer capacity and codes are being rapidly expanded by factors of greater than 10^3 in the so-called Accelerated Strategic Computing Initiative (ASCI) in order to make effective use of all this data from the new facilities for realistic bomb performance calculations. Beyond their weapons-related activities, these advanced facilities will be of great interest to physicists in better understanding extreme temperature and pressure conditions in burning stars, as well as advancing our understanding of inertial confinement fusion.

No stockpile stewardship program can be better than the quality of its scientists and engineers. An important consequence of this multifaceted program that the DOE has developed is that it will generate a large body of valuable new data and challenging opportunities capable of attracting and retaining experienced nuclear weapons scientists and engineers.

To summarize the JASON conclusions, with a strong science-based stockpile stewardship and management program, equipped with advanced diagnostic equipment and led by first-class scientists and engineers at the national weapons laboratories, there is no need to continue nuclear testing at any level of yield. Instead the U.S. will rely on enhanced surveillance and diagnostic information and far more accurate and reliable simulations, to deepen our understanding of the physical processes in a nuclear explosion. We shall fill substantial gaps in that understanding, gaps that we were formerly willing to accept as long as we could monitor the performance of our bombs by testing. This will provide the necessary scientific basis for retaining confidence in our ability to hear whatever warning bells may ring, however unanticipated they may be, alerting us to the deterioration of an aging stockpile. We shall also maintain facilities to provide for warhead refurbishing or remanufacture in response to identified needs. This program, as emphasized earlier, is consistent with the spirit, as well as the letter of the CTBT: without testing, the U.S. will not be able to develop and deploy with confidence more advanced weapons.

This conclusion was endorsed by the weapons laboratories and proved to be persuasive in Washington. It provided the technical base for President Clinton's decision, announced in September, 1996, for the United States to support and seek a true zero-yield Comprehensive Test Ban Treaty. The scientific challenge of developing, successfully accomplishing, and correctly interpreting the findings of such a program is a major one for the weapons laboratories and for all physicists involved in the process.

Science Advice

In World War II, the American scientific community—from university and industrial research laboratories and including many refugees from persecution in Europe—was recruited to large projects focused on developing the latest scientific advances in support of the military effort of the U.S. and its Allies. The Radiation Laboratory organized at MIT under the leadership of Lee duBridge developed microwave radar into instruments that proved decisive in the aerial defense of England and in the ultimate defeat of German U-boat raiders against the lifeline of convoys crossing the Atlantic Ocean. Out of the latest developments in nuclear physics and the theory of fission, a successful nuclear chain reaction was achieved by Enrico Fermi and his collaborators at the University of Chicago's Metallurgical Laboratory on December 2, 1942. This led to the production of plutonium and eventually to the construction of the first atomic bomb at Los Alamos under J. Robert Oppenheimer's leadership.

These are the two best-known examples of a focused massive civilian effort by scientists to create new weapons of war that played critically important roles in the outcome of a military conflict. They served as models of continuing close scientific-military relations and large secret projects at national laboratories through the Cold War and up to the present. In all the major industrial powers, such laboratories have continued to develop new technologies that have had great impact on policy options for their governments. As we discussed, American Presidents were presented with important choices for this nation's security as a result of some of these developments: What should we do about ABM defenses? Are advanced diagnostics and simulations of the behavior of nuclear weapons adequate for us to maintain confidence in our nuclear deterrent under a CTBT? The rapid and, in some instances, revolutionary advances in military technology have created a growing gap between science and government leaders. This circumstance led former British Prime Minister Harold Macmillan to lament, in his 1972 book *Pointing The Way* (MacMillan, 1972) that

In all these affairs Prime Ministers, Ministers of Defense, and Cabinets are under a great handicap. The technicalities and uncertainties of the sophisticated weapons which they have to authorize are out of the range of normal experience. There is today a far greater gap between their own knowledge and the expert advice which they receive than there has ever been in the history of war.

President Eisenhower understood very well the importance of closing this gap. Following the Soviet launch of Sputnik and development of long-range missiles as a potential threat to the U.S. in 1957 he created the position of a full-time Science Advisor in the White House and also established the President's Science Advisory Committee. This mechanism was his resource for direct, in-depth analyses and advice as to what to expect from science and technology, both current and in prospect, in establishing realistic national policy goals. Members of the PSAC and consultants who served on its hard-working panels were selected apolitically and solely on the grounds of demonstrated achievements in science and engineering. Two things set the PSAC apart from the existing governmental line organizations and cabinet departments with operational responsibilities, as well as from nongovernmental organizations engaged in policy research. First of all, they had White House backing and the requisite security clearances to gain access to all the relevant information for their studies on highly classified national security issues. Secondly, the individual scientists were independent and presumably, therefore, immune from having their judgments affected by operational and institutional responsibilities. Therein lay their unique value (Golden, 1988).

Unfortunately, the advisory mechanism that served the White House and the nation well when it was created eroded in the late 1960s under the political strains and public discord of the Viet Nam conflict. Although a scientific presence in the White House has been recreated in various forms since then, it has not been reenergized effectively to bridge the gap on issues of military and national security importance. Thus the very influential White House advisory mechanism, which was so effective in advancing the development of space-based photoreconnaissance, was notably absent in 1983 at the time of the Star Wars decision, with unfortunate consequences as we discussed earlier. It was simply fortuitous that the JASON study on nuclear testing was completed and used to brief senior officials in time to influence the 1995 policy choice by the Clinton Administration to support a true zero-yield CTBT. Fortuitous timing, however, is a very poor and unreliable substitute for a formal, nonpartisan, high-level scientific presence in the White House to help ensure that the President has the technical input and advice needed when he faces major policy and strategic decisions.

The importance of recreating such a mechanism cannot be overemphasized. The President will continue to face decisions on issues vital to U.S. national security with major technical components. In the nuclear area we still have a long way to go, and major decisions to make, in reducing nuclear danger, and not just by reducing the sizes of the arsenals. There is need, and technologies offer new opportunities, to strengthen safeguards against the accidental launch of nuclear weapons due to faulty indicators that they are under attack, and against unauthorized launch. There is an urgent need to reduce the danger of rogue leaders or terrorists acquiring "loose nukes," i.e., nuclear weapons or their fissile material.

More broadly we must prepare to deal with emerging threats posed by new weapons capable of large-scale indiscriminate destruction. In particular, biological weapons present a rapidly growing—indeed already imminent—global danger as a result of advances in biotechnology, spurred by recent discoveries in molecular biology and genetics (Office of Technology Assessment, December 1993; Stimson Center, January 1998). A growing number of countries are capable of biological warfare. Reports have identified a dozen or more countries that already have, or may be developing, biological weapons. These include some of the smallest and poorest countries, with relatively primitive technical infrastructures and led by reactionary and unstable regimes.

Such weapons may be of limited value for tactical military purposes due to the unpredictability in the spread of microbial pathogens and the incubation period of days to weeks between infection and the appearance of their debilitating effects in humans. However, their devastating potential against society if used by terrorists is terrifying and presents a threat that can no longer be ignored. Modern biotechnology has increased the accessibility of virulent pathogens, and the prospect of terrorists using bugs grows even more frightening with anticipated development of genetically engineered pathogens that are easier to manufacture, propagate, and deliver, as well as being more toxic, more difficult to detect, and harder to counter.

The Ethical Dilemma of Scientists

I have always felt that the scientific community has a special responsibility to be alert to the implications and practical uses of our progress. We bear an obligation to assist society, in its political deliberations, to understand the potential benefits and risks and to shape in beneficial ways the applications of scientific progress for which we are responsible. Though it need not be fulfilled by each individual scientist, this is a moral obligation of the community as a whole, including scientists engaged in basic research and in applied industrial and weapons research and development.

The moral dilemma is particularly sharp for individual scientists when facing decisions as to whether or how to involve themselves in work on nuclear weapons. With their scientific training, they can contribute to public understanding of the devastating effects of nuclear explosions and of the importance of arms control efforts to achieve truly major reductions in the size of today's bloated arsenals. There are scientists and citizens alike who believe that getting rid of all nuclear weapons is more than a distant vision, but is a realistic possibility. Some scientists believe that the prospect of achieving such a goal is improved if all work on nuclear weapons ceases or is reduced to a minimal custodial role attending to their safety and security. Others see the prospects for reducing nuclear danger to be better served by contributing to a strong science-based stockpile stewardship program, as described earlier, as a necessary basis for adhering to a CTBT.

Which of these two, or other possible, choices to make is a decision individuals must answer for themselves. This issue is highlighted in ongoing discussions of the DOE's Stockpile Stewardship Program and of worldwide calls to get rid of all nuclear weapons sooner rather than later. I am reminded of the saga of Andrei Sakharov, who in 1948 was drawn to work on the development of the Soviet hydrogen bomb by his judgment that the world would be safer with a socialist bomb to balance the capitalist bomb (Sakharov, 1990). But by the 1960s, after seeing his work help fuel a worldwide arms race of tens of thousands of nuclear bombs, Sakharov felt increasing concern about the dangers to mankind of thermonuclear war. Disillusioned when Soviet leaders rejected his advice not to resume atmospheric testing in 1961 after a three year moratorium, Sakharov turned into an energetic, outspoken, courageous dissident and opponent of a continuing nuclear arms buildup of mindless proportions. Can we or should we make a judgment that Sakharov was wrong in 1948 and right in the 1960s?

Sakharov's saga is but one example of the ethical dilemmas that scientists must resolve when entering into work that can seriously impact the human condition. Decisions will and should be affected by political circumstances that, in contrast to the immutable and rational laws of nature, can and frequently do change unpredictably. The best a scientist can do is to carefully weigh the ethical dimensions, in addition to the political and technical ones, before making a decision as to whether or not, or how, to become involved. One cannot ask more of one's colleagues, nor should one expect less from them or from oneself, than to make the best informed and objective technical judgments, to try seriously to understand the often murky and confusing political issues, and ultimately to anchor one's actions solidly in one's true principles.

Acknowledgment

I wish to thank R. L. Garwin, K. Gottfried, W. K. H. Panofsky, R. L. Peurifoy, and A. D. Wheelon, who read earlier versions of this article, for valuable suggestions.

References

Bloembergen, N., C. K. Patel, et al. (APS Study Group on Science and Technology of Directed Energy Weapons), 1987, Rev. Mod. Phys.
Buderi, R., 1996, *The Invention that Changed the World* (Simon and Schuster, New York).
Congressional Record, 1995, Senate: S-11368 (August 4, 1995) [Summary of the JASON Study, "Nuclear Testing" (JSR-95-320)].
Carter, A. B., and D. N. Schwartz, 1984, Eds., *Ballistic Missile Defense* (The Brookings Institution, Washington, D.C.).
Day, D. A., J. M. Logsdon, and B. Latell, 1998, Eds., *Eye in the Sky: the Story of the Corona Spy Satellites* (Smithsonian Institution Press, Washington, D.C.).
Drell, S. D., P. J. Farley, and D. Holloway, 1985, *The Reagan Strategic Defense Initiative: A Technical, Political, and Arms Control Assessment* (Ballinger Publishing, Cambridge, MA).
Garwin, R. L., and H. A. Bethe, 1968, "Antiballistic-Missile systems," Sci. Am. March, pp. 21–31.
Golden, W. T., 1988, Editor, *Science and Technology, Advice to the President, Congress, and Judiciary* (Pergamon, New York).
Hall, R. Cargill, 1996, Quarterly of the National Archives and Records Administration **28**, 107.
JASON Report JSR-94-345, November 1990, "Science Based Stockpile Stewardship."
Jones, R. V., 1978, *The Wizard War* (Coward, McCann, and Geoghegan, New York).

Lewis, G. N., and T. A. Postol, 1993, "Video Evidence on the Effectiveness of Patriot during the 1991 Gulf War," Science and Global Security, Vol. 4, pp. 1–63.

MacMillan, H., 1972, *Pointing the Way* (Macmillan, London).

Office of Defense Programs, October 1997, *Stockpile Stewardship Program: Overview and Progress* (Department of Energy, Office of Defense Programs).

Office of Technology Assessment, 1985, *Ballistic Missile Defense Technologies*, report (Office of Technology Assessment, Washington, D.C.).

Office of Technology Assessment, December 1993, "Technologies Underlying Weapons of Mass Destruction," U.S. Congress OTA-BP-ISC-115 (U.S. Government Printing Office, Washington, D.C.).

Peña, F., 1997, testimony before the Senate Energy and Water Development Appropriations Subcommittee, October 29, 1997.

Rhodes, R., 1986, *The Making of the Atomic Bomb* (Simon and Schuster, New York).

Sakharov, A. D., 1990, *Memoirs* (Knopf, New York).

Select Committee on Intelligence, U.S. Senate, 1996, "Intelligence analysis of the long range missile threat to the United States: hearing before the Select Committee . . . One hundred fourth Congress, second session . . . Wednesday, December 4 (U.S. Government Printing Office, Washington, D.C.).

Stimson, H. L., Center, January 1998, Report No. 24 "Biological Weapons Proliferation: Reasons for Concern, Courses of Action."

Wheelon, A. D., 1997, Phys. Today **50**, 24, and references therein.

Zuckerman, L., 1982, *Nuclear Illusion and Reality* (Viking, New York).

Laser Technology
R. E. Slusher

Light has always played a central role in the study of physics, chemistry, and biology. Light is key to both the evolution of the Universe and to the evolution of life on Earth. This century a new form of light, laser light, has been discovered on our small planet and is already facilitating a global information transformation as well as providing important contributions to medicine, industrial material processing, data storage, printing, and defense. This review will trace the developments in science and technology that led to the invention of the laser and give a few examples of how lasers are contributing to both technological applications and progress in basic science. There are many other excellent sources that cover various aspects of the lasers and laser technology including articles from the 25th anniversary of the laser (Ausubell and Langford, 1987) and textbooks (e.g., Siegman, 1986; Agrawal and Dutta, 1993; and Ready, 1997).

Light amplification by stimulated emission of radiation (LASER) is achieved by exciting the electronic, vibrational, rotational, or cooperative modes of a material into a nonequilibrium state so that photons propagating through the system are amplified coherently by stimulated emission. Excitation of this optical gain medium can be accomplished by using optical radiation, electrical current and discharges, or chemical reactions. The amplifying medium is placed in an optical resonator structure, for example between two high reflectivity mirrors in a Fabry-Perot interferometer configuration. When the gain in photon number for an optical mode of the cavity resonator exceeds the cavity loss, as well as loss from nonradiative and absorption processes, the coherent state amplitude of the mode increases to a level where the mean photon number in the mode is larger than one. At pump levels above this threshold condition, the system is lasing and stimulated emission dominates spontaneous emission. A laser beam is typically coupled out of the resonator by a partially transmitting mirror. The wonderfully useful properties of laser radiation include spatial coherence, narrow spectral emission, high power, and well-defined spatial modes so that the beam can be focused to a diffraction-limited spot size in order to achieve very high intensity. The high efficiency of laser light generation is important in many applications that require low power input and a minimum of heat generation.

When a coherent state laser beam is detected using photon-counting techniques, the photon count distribution in time is Poissonian. For example, an audio output from a high efficiency photomultiplier detecting a laser field sounds like rain in a steady downpour. This laser noise can be modified in special cases, e.g., by constant current pumping of a diode laser to obtain a squeezed number state where the detected photons sound more like a machine gun than rain.

An optical amplifier is achieved if the gain medium is not in a resonant cavity. Optical amplifiers can achieve very high gain and low noise. In fact they presently have noise figures within a few dB of the 3 dB quantum noise limit for a phase-insensitive linear amplifier, i.e., they add little more than a factor of two to the noise power of an input signal. Optical parametric amplifiers (OPAs), where signal gain is achieved by nonlinear coupling of a pump field with signal modes, can be configured to add less than 3 dB of noise to an input signal. In an OPA the noise added to the input signal can be dominated by pump noise and the noise contributed by a laser pump beam can be negligibly small compared to the large amplitude of the pump field.

History

Einstein (1917) provided the first essential idea for the laser, stimulated emission. Why wasn't the laser invented earlier in the century? Much of the early work on stimulated emission concentrates on systems near equilibrium, and the laser is a highly nonequilibrium system. In retrospect the laser could easily have been conceived and demonstrated using a gas discharge during the period of intense spectroscopic studies from 1925 to 1940. However, it took the microwave technology developed during World War II to create the atmosphere for the laser concept. Charles Townes and his group at Columbia conceived the maser (microwave amplification by stimulated emission of radiation) idea, based on their background in microwave technology and their interest in high-resolution microwave spectroscopy. Similar maser ideas evolved in Moscow (Basov and Prokhorov, 1954) and at the University of Maryland (Weber,1953). The first experimentally demonstrated maser at Columbia University (Gordon *et al.*, 1954, 1955) was based on an ammonia molecular beam. Bloembergen's ideas for gain in three level systems resulted in the first practical maser amplifiers in the ruby system. These devices have noise figures very close to the quantum limit and were used by Penzias and Wilson in the discovery of the cosmic background radiation.

Townes was confident that the maser concept could be extended to the optical region (Townes, 1995). The laser idea was born (Schawlow and Townes, 1958) when he discussed the idea with Arthur Schawlow, who understood that the resonator modes of a Fabry-Perot interferometer could reduce the number of modes interacting with the gain material in order to achieve high gain for an individual mode. The first laser was demonstrated in a flash lamp pumped ruby crystal by Ted Maiman at Hughes Research Laboratories (Maiman, 1960). Shortly after the demonstration of pulsed crystal lasers, a continuous wave (CW) He:Ne gas discharge laser was demonstrated at Bell Laboratories (Javan *et al.*, 1961), first at 1.13 μm and later at the red 632.8 nm wavelength lasing transition. An excellent article on the birth of the laser is published in a special issue of Physics Today (Bromberg, 1988).

The maser and laser initiated the field of quantum electronics that spans the disciplines of physics and electrical engineering. For physicists who thought primarily in terms of photons, some laser concepts were difficult to understand without the coherent wave concepts familiar in the electrical engineering community. For example, the laser linewidth can be much narrower than the limit that one might think to be imposed by the laser transition spontaneous lifetime. Charles Townes won a bottle of scotch over this point from a colleague at Columbia. The laser and maser also beautifully demonstrate the interchange of ideas and impetus between industry, government, and university research.

Initially, during the period from 1961 to 1975 there were few applications for the laser. It was a solution looking for a problem. Since the mid-1970s there has been an explosive growth of laser technology for industrial applications. As a result of this technology growth, a new generation of lasers including semiconductor diode lasers, dye lasers, ultrafast mode-locked Ti:sapphire lasers, optical parameter oscillators, and parametric amplifiers is presently facilitating new research breakthroughs in physics, chemistry, and biology.

Lasers at the Turn of the Century

Schawlow's "law" states that everything lases if pumped hard enough. Indeed thousands of materials have been demonstrated as lasers and optical amplifiers resulting in a large range of laser sizes, wavelengths, pulse lengths, and powers. Laser wavelengths range from the far infrared to the x-ray region. Laser light pulses as short as a few femtoseconds are available for research on materials dynamics. Peak powers in the petawatt range are now being achieved by amplification of femtosecond pulses. When these power levels are focused into a diffraction-limited spot, the intensities approach 10^{23} W/cm^2. Electrons in these intense fields are accelerated into the relativistic range during a single optical cycle, and interesting quantum electrodynamic effects can be studied. The physics of ultrashort laser pulses is reviewed in this centennial series (Bloembergen, 1999).

A recent example of a large, powerful laser is the chemical laser based on an iodine transition at a wavelength of 1.3 μm that is envisioned as a defensive weapon (Forden, 1997). It could be mounted in a Boeing 747 aircraft and would produce average powers of 3 megawatts, equivalent to 30 acetylene torches. New advances in high quality dielectric mirrors and deformable mirrors allow this intense beam to be focused reliably on a small missile carrying biological or chemical agents and destroy it from distances of up to 100 km. This "star wars" attack can be accomplished during the launch phase of the target missile so that portions of the destroyed missile would fall back on its launcher, quite a good deterrent for these evil weapons. Captain Kirk and the starship Enterprise may be using this one on the Klingons!

At the opposite end of the laser size range are microlasers so small that only a few optical modes are contained in a resonator with a volume in the femtoliter range. These resonators can take the form of rings or disks only a few microns in diameter that use total internal reflection instead of conventional dielectric stack mirrors in order to obtain high reflectivity. Fabry-Perot cavities only a fraction of a micron in length are used for VCSELs (vertical cavity surface emitting lasers) that generate high quality optical beams that can be efficiently coupled to optical fibers (Choquette and Hou, 1997). VCSELs may find widespread application in optical data links.

Worldwide laser sales in the primary commercial markets for 1997 (Anderson, 1998; Steele, 1998) are shown schematically in Fig. 1. Total laser sales have reached 3.2 billion dollars and at a yearly growth rate of nearly 27% will exceed 5 billion dollars by the year 2000. The global distribution of laser sales is 60% in the U.S., 20% in Europe, and 20% in the Pacific. Semiconductor diode lasers account for nearly 57% of the 1997 laser market. Diode lasers in telecommunications alone account for 30% of the total market.

Materials processing is the second largest market with applications such as welding, soldering, patterning, and cutting of fabrics. CO_2 lasers with average powers in the 100 W range account for a large fraction of the revenues in this category. High power diode lasers with power output levels between 1 and 20 W and wavelengths in the 750 to 980 nm range are now finding a wide variety of applications in materials processing as well as ophthalmic and surgical applications, instrumentation, and sensing.

Growth in medical laser applications is largely due to cosmetic laser procedures such as skin resurfacing and hair removal. A large fraction of medical lasers are still used in ophthalmological and general surgical applications. Frequency-doubled Nd:YAG lasers and diode laser systems are replacing argon-ion lasers in ophthalmology. New lasers, including the erbium-doped YAG laser, are being widely used in dermatology, dentistry, and ophthalmology.

Optical storage accounts for 10% of the market where one finds the lasers used in the compact disk (CD) players for both the entertainment and computer markets. The GaAs semiconductor laser at 800 nm wavelengths for these applications are manufactured so efficiently today that the laser costs are down to nearly $1 each. Over 200 million diode lasers, with wavelengths in the 750 to 980 nm range and powers of a few milliwatts, were sold for optical storage in 1997. The advent of digital video disks (DVDs) with 4.7 Gbytes of storage capacity and blue diode lasers (DenBaars, 1997) will lead to further growth in this field.

Image recording laser applications include desktop computer printers, fax machines, copiers, and commercial printing (Gibbs, 1998). Low power, single-mode diode lasers emitting at 780 to 670 nm wavelengths are being used in image recorders used to produce color-separation films with high sensitivity in this wavelength range. This laser-based color printing technology has combined with desktop publishing software to allow high quality page designs. Computer-to-plate technology is another important development in printing. A printing plate surface is directly imaged by exposing it with a laser beam instead of using film-based color separations. For example, photopolymer-coated plates can be exposed with frequency-doubled diode pumped Nd:YAG lasers at a wavelength of 532 nm. Most recently, thermally sensitive plates have been developed for use with near infrared patterning lasers.

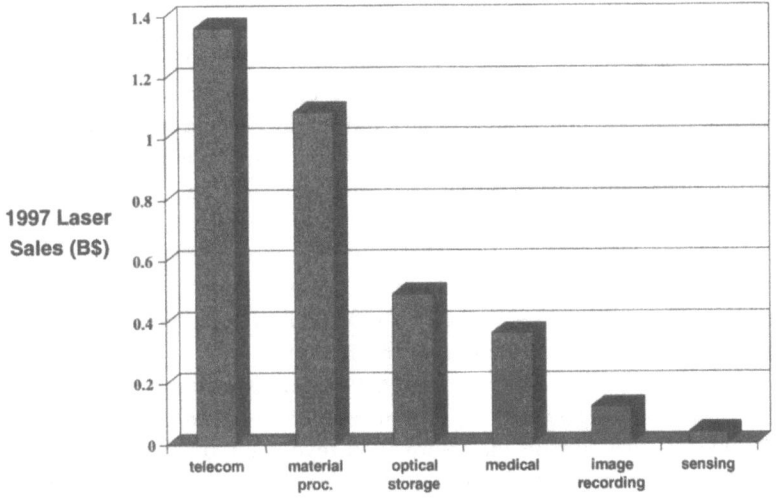

Figure 1. Worldwide laser sales in billions of dollars for laser markets in 1997.

Remote sensing laser markets include automotive collision avoidance, atmospheric chemical detectors, and air movement detection. Laser ranging is providing detailed elevation maps of the Earth including land mass movements, biomass, cloud and haze coverage, and ice cap evolution. Laser ranging from satellites can achieve subcentimeter resolution of elevation features and land mass movement on Earth. The Moon, Mars, and other planets are also being mapped by laser ranging. For the planets the measurement precision ranges between meters and centimeters. Detailed features of the ice cap on Mars as well as clouds near the edge of the ice cap have recently been mapped.

Laser applications in research, barcode scanning, inspection, art, and entertainment are small but significant markets. Lasers sold for basic research in 1997 accounted for 132 million dollars in revenues. Low power consumption, frequency-doubled diode sources emitting in the green at power levels near 10 W are being used as pump lasers for frequency tunable lasers like the Ti:sapphire laser and optical parametric amplifiers. Even a tabletop research laser can reach the petawatt peak power regime with large-volume optical amplifiers. These highly tunable, ultra short pulses are leading to advances in many research fields.

Lasers in Communications

Laser light sources have revolutionized the communications industry. Voice communications increased the demand for information transmission capacity at a steady pace until the mid-1970s. The doubling time for transmission capacity during this period was approximately 8 years. The basic data rate was in the range between 10 and 80 kHz based on audio transmissions. During this period first copper wires and then microwaves were the primary communications technologies. Then in the 1980s an explosive information rate increase began, with data, fax, and images added to the information stream. The new technology of optical fiber communications using laser light sources was developed to keep pace with this new demand. The advent of the global Internet resulted in an even more surprising explosion in capacity demand. At the data source, computer terminals are used to access the Internet in homes and businesses around the world, resulting in data rates that are increasing exponentially. As workstation computer rates approach 1000 MIPS, fiber communication links to the computer in the 1000 Mb/sec range will be required. Note the coincidence of these rates and that both are increasing exponentially. It is clear that there will continue to be an exponentially increasing demand for information transmission capacity. In response to this demand, the information capacity on a single optical fiber during the past four years, between 1994 and 1998, has increased 160 fold in commercial systems from 2.5 Gbits/sec to 400 Gbits/sec. This amazing increase has been achieved by using up to 100 different laser wavelengths (dense wavelength division multiplexing, DWDM) on each fiber. The data rates at a single wavelength have increased from tens of Mbits/sec in the 1970s to 10 Gbits/sec at present, and 40 Gbits/sec will probably be in use before the turn of the century.

This information revolution is reshaping the global community just as strongly as the printing press revolution and the industrial revolution reshaped their worlds. Two of the basic technologies that support the information revolution are the semiconductor diode laser and the erbium-doped fiber optical amplifier. The low noise, high intensity, and narrow linewidths associated with laser oscillators and amplifiers are absolutely essential to optical fiber communications systems. Wider bandwidth incoherent sources like light emitting diodes or thermal sources fall short of the needed intensities and spectral linewidths by many orders of magnitude.

Semiconductor laser diodes were first demonstrated in 1962 at GE, IBM, and Lincoln Laboratories as homojunction devices based on III-V materials. A history of these early diode lasers and references can be found in Agrawal and Dutta (1993). When the first heterojunction GaAs/AlGaAs room temperature, continuous wave diode lasers were operated in 1970 by Hayashi and Panish (Hayashi *et al.*, 1970) at Bell Labs and Alferov (Alferov *et al.*, 1970) in Russia, their lifetimes were measured in minutes. Diode laser reliabilities have increased dramatically since that time. Diode laser lifetimes at present are estimated to be hundreds of years, and the wavelength stabilities are greater than 0.1 nm over a period of 25 years. These amazing stabilities are necessary for the new DWDM systems with over 100 wavelength channels spanning 100 nm wavelength ranges. As the optimum wavelength for low-loss in silica fiber increased in wavelength from 800 nm to 1500 nm during the 1970s, diode laser wavelengths followed by evolving from GaAs to the InGaAsP system. During the late 1980s and early 1990s, quantum wells replaced the bulk semiconductor in the active optical gain region in order to enhance the laser operating characteristics. A schematic diagram of a present-day telecommunications diode laser integrated with an electro-absorption modulator is shown in Fig. 2. The overall dimensions are less than 1 mm. An elevated refractive index region and buried distributed feedback (DFB) grating, below the active quantum wells, defines the laser optical cavity and laser wavelength, respectively.

Fiber optic communication systems also rely strongly on the erbium-doped fiber amplifier developed in the late 1980s (Urquhart, 1988). These amplifiers have high gain, typically near 25 dB, and low noise figures near the 3 dB quantum noise limit for a linear phase-insensitive amplifier. The gain in these amplifiers can be equalized over bandwidths of up to 100 nm, covering nearly a quarter of the low-loss silica fiber window between 1.2 and 1.6 μm wavelengths. Optical fiber systems can be made "transparent" over thousands of kilometers using erbium-doped fiber amplifiers spaced at distances of approximately 80 km, where fiber losses approach 20 dB.

As the century closes we are rapidly approaching fundamental physical limits for lasers, optical amplifiers, and silica fibers. Laser linewidths are in the 10 MHz range,

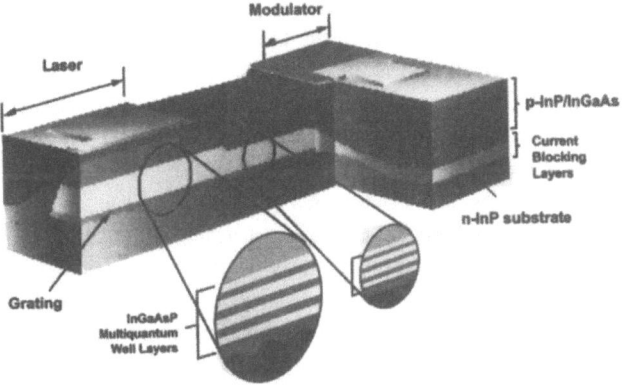

Figure 2. A schematic diagram of a semiconductor laser diode with an electro-absorption modulator used in optical communications systems. (Courtesy of R. L. Hartman, Lucent Technologies.)

limited by fundamental spontaneous emission fluctuations and gain-index coupling in semiconductor materials. The number of photons in a detected bit of information is approaching the fundamental limit of approximately 60 photons required when using coherent-state laser light fields in order to maintain an error rate of less than 1 part in 10^9. A bandwidth utilization efficiency of 1 bit/sec/Hz has recently been demonstrated. Optical amplifier bandwidths do not yet span the 400 nm width of the low-loss fiber window, but they are expanding rapidly. Fundamental limits imposed by nonlinear and dispersive distortions in silica fibers make transmission at data rates over 40 Gbits/sec very difficult over long distances. Optical solitons can be used to balance these distortions, but even with solitons fundamental limits remain for high bit rate, multiwavelength systems. The channel capacity limits imposed by information theory are on the horizon. It is clearly a challenge for the next centuries to find even more information transmission capacity for the ever-expanding desire to communicate.

Materials Processing and Lithography

High power CO_2 and Nd:YAG lasers are used for a wide variety of engraving, cutting, welding, soldering, and 3D prototyping applications. rf-excited, sealed off CO_2 lasers are commercially available that have output powers in the 10 to 600 W range and have lifetimes of over 10 000 hours. Laser cutting applications include sailclothes, parachutes, textiles, airbags, and lace. The cutting is very quick, accurate, there is no edge discoloration, and a clean fused edge is obtained that eliminates fraying of the material. Complex designs are engraved in wood, glass, acrylic, rubber stamps, printing plates, plexiglass, signs, gaskets, and paper. Three-dimensional models are quickly made from plastic or wood using a CAD (computer-aided design) computer file.

Fiber lasers (Rossi, 1997) are a recent addition to the materials processing field. The first fiber lasers were demonstrated at Bell Laboratories using crystal fibers in an effort to develop lasers for undersea lightwave communications. Doped fused silica fiber lasers were soon developed. During the late 1980s researchers at Polaroid Corp. and at the University of Southampton invented cladding-pumped fiber lasers. The glass surrounding the guiding core in these lasers serves both to guide the light in the single mode core and as a multimode conduit for pump light whose propagation is confined to the inner cladding by a low-refractive index outer polymer cladding. Typical operation schemes at present use a multimode 20 W diode laser bar that couples efficiently into the large diameter inner cladding region and is absorbed by the doped core region over its entire length (typically 50 m). The dopants in the core of the fiber that provide the gain can be erbium for the 1.5 μm wavelength region or ytterbium for the 1.1 μm region. High quality cavity mirrors are deposited directly on the ends of the fiber. These fiber lasers are extremely efficient, with overall efficiencies as high as 60%. The beam quality and delivery efficiency is excellent since the output is formed as the single mode output of the fiber. These lasers now have output powers in the 10 to 40 W range and lifetimes of nearly 5000 hours. Current applications of these lasers include annealing micromechanical components, cutting of 25 to 50 μm thick stainless steel parts, selective soldering and welding of intricate mechanical parts, marking plastic and metal components, and printing applications.

Excimer lasers are beginning to play a key role in photolithography used to fabricate VLSI (very large scale integrated circuit) chips. As the IC (integrated circuit) design rules decrease from 0.35 μm (1995) to 0.13 μm (2002), the wavelength of the light source used for photolithographic patterning must correspondingly decrease from 400 nm to below 200 nm. During the early 1990s mercury arc radiation produced enough power at sufficiently short wavelengths of 436 nm and 365 nm for high production rates of IC devices patterned to 0.5 μm and 0.35 μm design rules respectively. As the century closes excimer laser sources with average output powers in the 200 W range are replacing the mercury arcs. The excimer laser linewidths are broad enough to prevent speckle pattern formation, yet narrow enough, less than 2 nm wavelength width, to avoid major problems with dispersion in optical imaging. The krypton fluoride (KF) excimer laser radiation at 248 nm wavelength supports 0.25 μm design rules and the ArF laser transition at 193 nm will probably be used beginning with 0.18 μm design rules. At even smaller design rules, down to 0.1 μm by 2008, the F_2 excimer laser wavelength at 157 nm is a possible candidate, although there are no photoresists developed for this wavelength at present. Higher harmonics of solid-state lasers are also possibilities as high power UV sources. At even shorter wavelengths it is very difficult for optical elements and photoresists to meet the requirements in the lithographic systems. Electron beams, x rays, and synchrotron radiation are still being considered for the 70 nm design rules anticipated for 2010 and beyond.

Lasers in Medicine

Lasers with wavelengths from the infrared through the UV are being used in medicine for both diagnostic and therapeutic applications (Deutsch, 1997). Lasers interact with inhomogeneous tissues through absorption and scattering. Absorbers include melanin skin pigment, hemoglobin in the blood, and proteins. At wavelengths longer than 1 μm the primary absorber is water. Dyes can also be introduced into tissue for selective absorption. For example, in photodynamic therapy hematoporphyrin dye photosensitizers that absorb in the 630 nm to 650 nm wavelength range can be introduced into the system and used to treat cancer tumors by local laser irradiation in the urinary tract or esophagus. Scattering in tissue limits the penetration of radiation; for example, at a wavelength of 1 μm scattering limits the penetration depths to a few millimeters. Scattering processes are being studied in the hope of obtaining high-resolution images for breast cancer screening. Laser interaction with tissue depends on whether the laser is pulsed or CW. Short laser pulses where no thermal diffusion occurs during the pulse can be used to confine the depth of laser effects. This phenomenon along with selective tuning of the laser wavelength is used in dermatology for treatment of skin lesions and in the removal of spider veins, tattoos, and hair. Nonlinear interactions also play an important role. For example, laser-induced breakdown is used for fragmentation of kidney and gallbladder stones.

Since the interior of the eye is easily accessible with light, ophthalmic applications were the first widespread uses of lasers in medicine. Argon lasers have now been used for many years to treat retinal detachment and bleeding from retinal vessels. The widespread availability of the CO_2 and Nd:YAG lasers that cut tissue while simultaneously coagulating the blood vessels led to their early use in general surgery. The Er:YAG laser has recently been introduced for dental applications with the promise of dramatic reduction in pain, certainly a welcome contribution from laser technology.

Diagnostic procedures using the laser are proliferating rapidly. Some techniques are widely used in clinical practice. For example the flow cytometer uses two focused laser beams to sequentially excite fluorescence of cellular particles or molecules flowing in a liquid through a nozzle. The measured fluorescent signals can be used for cell sorting or analysis. Routine clinical applications of flow cytometry include immunophenotyping and DNA content measurement. Flow cytometers are used to physically separate large numbers of human chromosomes. The sorted chromosomes provide DNA templates for the construction of recombinant DNA libraries for each of the human chromosomes. These libraries are an important component of genetic engineering.

A new laser based medical imaging technique (Guillermo et al., 1997) based on laser technology called optical coherence tomography (OCT) is achieving spatial resolution of tissues in the 10 μm range. Ultrasound and magnetic resonance imaging (MRI) resolutions are limited to the 100 μm to 1 mm range. The new high-resolution OCT technique is sensitive enough to detect abnormalities associated with cancer and atherosclerosis at early stages. The OCT technique is similar to ultrasound, but it makes use of a bright, broad spectral bandwidth infrared light source with a coherence length near 10 μm, resulting in at least an order of magnitude improvement in resolution over acoustic and MRI techniques. The source can be a super luminescent diode, Cr:forsterite laser, or a mode-locked Ti:sapphire laser. OCT performs optical ranging in tissue by using a fiber optic Michelson interferometer. Since interference is observed only when the optical path lengths of the sample and the reference arms of the interferometer match to within the coherence length of the source, precision distance measurements are obtained. The amplitude of the reflected/scattered signal as a function of depth is obtained by varying the length of the reference arm of the interferometer. A cross-sectional image is produced when sequential axial reflection/scattering profiles are recorded while the beam position is scanned across the sample. Recent studies have shown that OCT can image architectural morphology in highly scattering tissues such as the retina, skin, the vascular system, the gastrointestinal tract, and developing embryos. An image of a rabbit trachea obtained using this technique coupled with a catheterendoscope is shown in Fig. 3. OCT is already being used clinically for diagnosis of a wide range of retinal macular diseases.

An elegant and novel optical technique using spin-polarized gases (Mittleman et al., 1995) is being explored to enhance MRI images of the lungs and brain. Nuclear spins in Xe and ^{3}He gases are aligned using circularly polarized laser radiation. These aligned nuclei have magnetizations nearly 10^5 times that for protons normally used for MRI imaging. Xenon is used as a brain probe since it is soluble in lipids. In regions like the lungs, that do not contain sufficient water for high-contrast MRI images, ^{3}He provides the high-contrast images. One can even watch ^{3}He flow in the lungs for functional diagnostics.

Lasers in Biology

Laser applications in biology can be illustrated with two examples, laser tweezers and two-photon microscopy. When collimated laser light is focused near or inside a small dielectric body like a biological cell, refraction of the light in the cell causes a lensing effect. A force is imparted to the cell by transfer of momentum from the bending light beam. Arthur Ashkin at Bell Laboratories (Ashkin, 1997) found that by varying the shape and position of the focal volume in a microscopic arrangement, a cell can be easily moved or trapped with these "laser tweezer" forces using light intensities near 10 W/cm^2. At these light levels and wavelengths in the near infrared, there is no significant damage or heating of cell constituents. Laser tweezers are now being used to move subcellular bodies like mitochondria within a cell (Sheetz, 1998). Tweezer techniques can also be used to stretch DNA strands into linear configurations for detailed studies. Two laser beams can be used to stabilize a cell and then a third laser beam at a different wavelength, can be used for spectroscopic or dynamic studies. Pulsed lasers are being used as "scissors" to make specific modifications in cell structures or to make small holes in cell membranes so that molecules or genetic materials can be selectively introduced into the cell.

Scanning confocal and two-photon optical microscopy are excellent examples of the contribution of laser technology to biology. Three-dimensional imaging of nerve cells nearly 200 μm into functioning brains and developing embryos is now a reality. Practical confocal microscopes came into wide use in the late 1980s as a result of reliable laser light sources. The resolution of the lens in a confocal microscope is used both to focus the light to a diffraction-limited spot and then again to image primarily the signal photons, i.e., those that are not strongly scattered by the sample, onto an aperture. Even though high-resolution 3D images are obtained, this single-photon scheme is a wasteful use of the illuminating light since a major fraction is scattered away from the aperture or is absorbed by the sample. In fluorescent microscopy, photodamage to the fluorophore is an especially limiting factor for single-photon confocal microscopy.

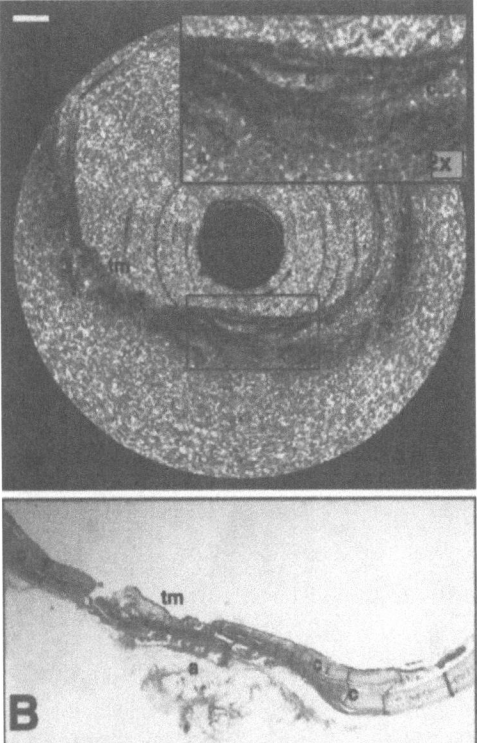

Figure 3. Optical coherence tomography images of a rabbit trachea in vivo. (Top) This image allows visualization of distinct architechtual layers, including the epithelium (e), the mucosal stroma (m), cartilage (c), and adipose tissue (a). The trachealis muscle (tm) can be easily identified. (Bottom) Corresponding histology. Bar, 500 μm.

Multiphoton scanning confocal microscopy was introduced in 1990 and solves many of the problems of single-photon techniques. A typical two-photon microscope uses short 100 fs pulses from a Ti:sapphire mode-locked laser at average power levels near 10 mW. The high intensity at the peak of each pulse causes strong two-photon absorption and fluorescence only within the small focal volume, and all the fluorescent radiation can be collected for high efficiency. The exciting light is chosen for minimal single-photon absorption and damage, so that the two-photon technique has very high resolution, low damage, and deep penetration.

A beautiful two-photon fluorescent image of a living Purkenji cell in a brain slice is shown in Fig. 4 (Denk and Svoboda 1997). Neocortical pyrimidal neurons in layers 2 and 3 of the rat somatosensory cortex have been imaged at depths of 200 μm below the brain surface. Even more impressive are motion pictures of embryo development. Embryo microscopy is particularly sensitive to photodamage and the two-photon technique is opening new vistas in this field.

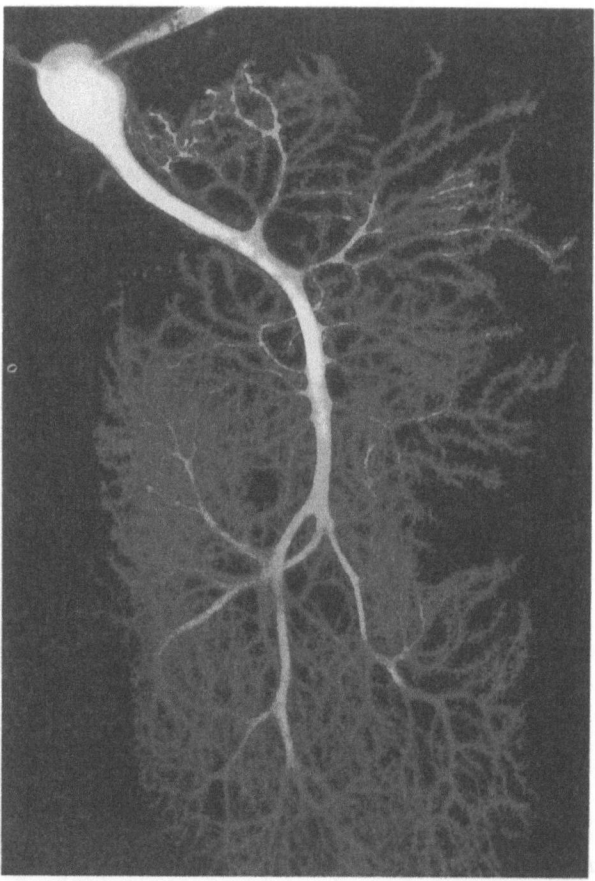

Figure 4. Two-photon confocal microscope fluorescent image of a living Purkenji cell in a brain slice. The cell dimensions are of the order of 100 µm. (See Color Plate 14.)

Lasers in Physics

Laser technology has stimulated a renaissance in spectroscopies throughout the electromagnetic spectrum. The narrow laser linewidth, large powers, short pulses, and broad range of wavelengths has allowed new dynamic and spectral studies of gases, plasmas, glasses, crystals, and liquids. For example, Raman scattering studies of phonons, magnons, plasmons, rotons, and excitations in 2D electron gases have flourished since the invention of the laser. Nonlinear laser spectroscopies have resulted in great increases in precision measurement as described in an article in this volume (Hänsch and Walther, 1999).

Frequency-stabilized dye lasers and diode lasers precisely tuned to atomic transitions have resulted in ultracold atoms and Bose-Einstein condensates, also described in this volume (Wieman et al., 1999). Atomic-state control and measurements of atomic parity nonconservation have reached a precision that allows tests of the standard model in particle physics as well as crucial searches for new physics beyond the standard model. In recent parity nonconservation experiments (Wood et al., 1997) Ce atoms are prepared in specific electronic states as they pass through two red diode laser beams. These prepared atoms then enter an optical cavity resonator where the atoms are excited to a higher energy level by high-intensity green light injected into the cavity from a frequency-stabilized dye laser. Applied electric and magnetic fields in this excitation region can be reversed to create a mirrored environment for the atoms. After the atom exits the excitation region, the atom excitation rate is measured by a third red diode laser. Very small changes in this excitation rate with a mirroring of the applied electric and magnetic fields indicate parity nonconservation. The accuracy of the parity nonconservation measurement has evolved over several decades to a level of 0.35%. This measurement accuracy corresponds to the first definitive isolation of nuclear-spin-dependent atomic parity violation. At this accuracy level it is clear that a component of the electron-nuclear interaction is due to a nuclear anapole moment, a magnetic moment that can be visualized as being produced by toroidal current distributions in the nucleus.

Lasers are also contributing to the field of astrophysics. A Nd:YAG laser at 10.6 μm wavelength will be used in the first experiments to attempt detecting gravitational waves from sources like supernovas and orbiting neutron stars. These experiments use interferometers that should be capable of measuring a change in length between the two interferometer arms to a precision of one part in 10^{22}. A space warp of this magnitude is predicted for gravitational radiation from astrophysical sources. The terrestrial experiments are called LIGO (Light Interferometer Gravitational Wave Observatory) in the U.S. and GEO in Europe. A space-based experiment called LISA (Light Interferometer Space Antenna) is also in progress. The LIGO interferometer arms are each 4 km long. A frequency-stable, low noise, high-spatial-beam-quality laser at a power level of 10 W is required for the light source. Cavity mirrors form resonators in each interferometer arm that increase the power in the cavities to nearly 1 kW. Four Nd:YAG rods, each side pumped by two 20 W diode bars, amplify the single frequency output of a nonplanar ring oscillator from 700 mW to at least 10 W. Achieving the required sensitivity for detecting gravitational waves means resolving each interferometer fringe to one part in 10^{11}, a formidable, but hopefully achievable goal.

Future Laser Technologies

The free-electron laser and laser accelerators are examples of developing laser technologies that may have a large impact in the next century. The free-electron laser (FEL) is based on optical gain from a relativistic electron beam undulating in a periodic magnetic field (Sessler and Vaugnan, 1987). Electron beam accelerators based on superconducting microwave cavities are being developed at a new FEL center at Jefferson Laboratories. These accelerating cavities generate high fields in the 10 to 20 MeV/m range and allow very efficient generation of FEL light that can be tuned from the infrared to the deep ultraviolet with average power levels in the kilowatt range (Kelley et al., 1996). At present a 1 kW average power infrared FEL is near completion and an upgrade to a powerful, deep-UV FEL is being planned. At these immense powers, a number of new technologies may be commercially interesting. Short, intense FEL pulses may allow rapid thermal annealing and cleaning of metal surfaces. Pulsed laser annealing may result in nearly an order of magnitude increase in hardness for machine tools. The high average FEL powers may be sufficient to make commercial production of laser-enhanced tools a reality. Another large market that requires high powers for processing of large volumes is polymer wraps and cloth. In this case intense FEL pulses can induce a wide range of modified polymer properties including antibacterial polymer surfaces that could be used for food wrappings and clothing with pleasing textures and improved durability. High average powers and wavelength tunability are also important for patterning of large area micromaching tools used to imprint patterns in plastic sheets.

Petawatt-class lasers may provide the basis for a new generation of particle accelerators. The frequency of microwave field accelerators being used at present will probably be limited by self-generated wakes to less than 100 GHz where the accelerating fields reach the 100 MeV/m range. Intense laser beams are being used to generate much higher fields in the 100 GeV/m range (Madena et al., 1995). For example, one technique uses two laser beams whose difference frequency is tuned to the plasma frequency of a gas ionized by the laser. Accelerating fields as high as 160 GeV/m can be generated between the periodic space charge regions of the plasma wave. The propagation velocities of these gigantic fields can be engineered to match the relativistic velocities of the accelerated particles. Much work remains in order to achieve practical accelerators but proof of principle has already been achieved.

Developing laser technologies and their contributions to science are too numerous to cover adequately in this brief review. Laser communications between satellite networks, laser propelled spacecraft, and laser fusion are additional examples of developing laser technologies. In the basic sciences there are many new experiments that are being enabled by laser technology including correction for atmospheric distortions in astronomy using laser reflections from the sodium layer in the upper atmosphere and studies of quantum electrodynamics using ultra-intense laser beams. Just as it was hard to envision the potential of laser technologies in the 1960s and 1970s, it seems clear that we cannot now envision the many new developments in lasers and their applications in the next century we will see. Our new laser light source is sure to touch us all, both in our ordinary lives and in the world of science.

References

Agrawal, G. P., and N. K. Dutta, 1993, *Semiconductor Lasers* (Van Nostrand Reinfield, New York).
Alferov, Zh. I., V. M. Andreev, D. Z. Garbuzov, Yu. V. Zhilyaev, E. P. Morozov, E. L. Portnoi, and V. G. Trofim, 1971, Sov. Phys. Semicond. **4**, 1573.
Anderson, S. G., 1998, Laser Focus World **34**, 78.
Ashkin, A., 1997, Proc. Natl. Acad. Sci. USA **94**, 4853.
Ausubel, J. H., and H. D. Langford, 1987, Eds., *Lasers: Invention to Application* (National Academy, Washington, DC).
Basov, N. G., and A. M. Prokhorov, 1954, J. Exp. Theor. Phys. **27**, 431.
Bloembergen, N., 1999, Rev. Mod. Phys. **71**, 283; pp. 474–481 in this book.
Bromberg, J. L., 1988, Phys. Today **41**, 26.
Choquette, K. D., and H. Q. Hou, 1997, Proc. IEEE **85**, 1730.
Deutsch, T. F., 1997, Proc. IEEE **85**, 1797.
DenBaars, S. P., 1997, Proc. IEEE **85**, 1740.
Denk, W., and K. Svoboda, 1997, Neuron **18**, 351.
Einstein, J., 1917, Phys. Z. **18**, 121.
Forden, G. E., 1997, IEEE Spectr. **34**(9), 40.
Gibbs, R., 1998, Laser Focus World **34**, 135.
Gordon, J. P., H. J. Zeigler, and C. H. Townes, 1954, Phys. Rev. **95**, 282.
Gordon, J. P., H. J. Zeigler, and C. H. Townes, 1955, Phys. Rev. **99**, 1264.
Guillermo, J. T., M. E. Brezinski, B. E. Bouma, S. A. Boppart, C. Pitris, J. F. Southern, and J. G. Fujimoto, 1997, Science **276**, 2037.
Hänsch, T., and H. Walther, 1999, Rev. Mod. Phys. **71**, 242; pp. 408–425 in this book.
Hayashi, I., M. B. Panish, P. W. Foy, and S. Sumski, 1970, Appl. Phys. Lett. **17**, 109.
Javan, A., W. R. Bennett, Jr., and D. R. Herriott, 1961, Phys. Rev. Lett. **6**, 106.
Kelley, M. J., H. F. Dylla, G. R. Neil, L. J. Brillson, D. P. Henkel, and H. Helvajian, 1996, SPIE **2703**, 15.
Maiman, T. H., 1960, Nature (London) **187**, 493.
Mittleman, H., R. D. Black, B. Saam, G. D. Cates, G. P. Cofer, R. Guenther, W. Happer, L. W. Hedlund, G. A. Johnson, K. Juvan, and J. Scwartz, 1995, Magn. Reson. Med. **33**, 271.
Modena, A., Z. Najmudin, A. E. Dangor, C. E. Clayton, K. A. Marsh, C. Joshi, V. Malka, C. B. Darrow, C. Danson, D. Neely, and F. N. Walsh, 1995, Nature (London) **377**, 606.
Ready, J., 1997, *Industrial Applications of Lasers* (Academic, San Diego).
Rossi, B., 1997, Laser Focus World **33**, 78.
Schawlow, A. L., and C. H. Townes, 1958, Phys. Rev. **112**, 1940.
Sesseler, A. M., and D. Vaughan, 1987, Am. Sci. **75**, 34.
Sheetz, M. P., 1998, Ed., *Laser Tweezers in Cell Biology*, in *Methods in Cell Biology* (Academic, San Diego), Vol. 55.
Siegman, A. E., 1986, *Lasers* (University Science Books, Mill Valley, CA).
Steele, R., 1998, Laser Focus World **34**, 72.
Townes, C. H., 1995, *Making Waves* (AIP, New York).
Urquhart, P., 1988, Proc. IEEE **135**, 385.
Weber, J., 1953, IRE Trans. Prof. Group on Electron Devices **3**, 1.
Wieman, C. E., D. E. Pritchard, and D. J. Wineland, 1999, Rev. Mod. Phys **71**, 253; pp. 426–441 in this book.
Wood, C. S., S. C. Bennett, D. Cho, B. P. Masterson, J. L. Roberts, C. E. Tanner, and C. E. Wieman, 1997, Science **275**, 1759.

Physics and the Communications Industry

W. F. Brinkman and D. V. Lang

Today's communications industry is a leading force in the world's economy. Our lives would be vastly different without the telephone, fax, cell-phone, and the Internet. The commonplace and ubiquitous nature of this technology, which has been evolving over a period of 125 years,[1] tends to overshadow the dominant role that physics and physicists have played in its development.[2] The purpose of this review is to explore this coupling and to show that the communications industry has not only made use of the results of academic physics research but has also contributed significantly to our present understanding of fundamental physics. Due to limitations of space, we shall primarily use examples from the Bell System and Bell Laboratories.

The foundations of communications technology lay in the discoveries of the great physicists of the early 19th century: Oersted, Ampere, Faraday, and Henry. The telegraph was invented only seventeen years after the discovery of electromagnetism by H. C. Oersted in 1820. In spite of this connection, however, much of the early work on communications was done by inventors, such as Morse, Bell, and Edison, who had no formal scientific background. The telegraph was a fairly simple electromechanical system which did not require the development of new scientific principles to flourish commercially in the mid 19th century. Closer coupling between physics and communications occurred shortly after the invention of the telephone by Alexander Graham Bell in 1876. The telephone concept immediately captured the imagination of the scientific community, where the "hot physics" of the period was electromagnetism and wave propagation.

Over the next century, industrial physics research on communications improved the technology, as well as spawning fundamental results of interest to the broad physics community. The types of physics specifically devoted to communications have varied continuously with the evolution of the technology and discontinuously with major physics discoveries over the past 125 years. We can identify four broad eras of physics that have impacted communications: (1) the era of electromagnetism (starting in 1820); (2) the era

---

[1] Much of this review is based on the telecommunications histories complied by Fagen (1975) and Millman (1983, 1984).
[2] The solid-state physics history is based on Hoddeson et al. (1992) and Riordan and Hoddeson (1997). The laser and optical communications history is based on Whinnery (1987), Agrawal and Dutta (1993), and Kaminow and Koch (1997).

of the electron (starting in 1897); (3) the era of quantum mechanics (starting in the 1920s); and (4) the era of quantum optics (starting in 1958).

The Era of Electromagnetism

This era dates from 1820, when Oersted discovered that an electric current generates a magnetic field. The first electromagnet was built in 1825, and in 1831 Faraday and Henry independently discovered that electric currents can be induced in wires moving in a magnetic field. The concept of the electromagnet was exploited in two independently invented telegraph systems in 1837—an analog system by Cooke and Wheatstone in Britain and a digital system by Samuel Morse in the U.S., using a dot-dash code that he also invented for the purpose. Because of its simple and robust design, the latter system achieved widespread commercial use in less than fifteen years. In 1861 Western Union had completed the first transcontinental telegraph line across the U.S. The British physicist William Thomson (later Lord Kelvin) was largely responsible for the construction of the first successful transatlantic telegraph cable in 1866 and developed the mirror galvanometer needed to detect the extremely weak signals.

One of the many young inventors excited by telegraph technology was Alexander Graham Bell, who was also interested in teaching the deaf to speak. Bell was working on devices that would enable deaf people to visualize the sounds they could make but not hear. He constructed a mechanical strip chart recorder, known as a phonautograph, which used human ear bones to couple the vibrations of a diaphragm to a stylus that traced the voice oscillations on a moving glass slide covered with lampblack. He was interested in making an electromagnetic analog for "electric speech" and as early as 1874 had the concept of vibrating a small magnet with sound waves and inducing a speech current in an electromagnet. In 1875, during his concurrent work on the multiplexed "harmonic telegraph," he accidentally discovered that useful audio-frequency currents could indeed be induced by a vibrating magnetic reed over an electromagnet and could be transmitted over such a system. By coupling the vibrating magnet to a diaphragm, as in hisphonautograph, he could transmit speech sounds. Figure 1 shows this first crude telephone, which

Figure 1. Alexander Graham Bell's first telephone. Voice sounds were transmitted for the first time on June 3, 1875, over this gallows-shaped instrument. From Fagen, 1975.

Bell patented in 1876. For the next several years he empirically optimized the telephone while he and his financial backers started various corporations to exploit his invention.

Bell was not the only person exploring speech transmission. The idea of transmitting voice using "harmonic telegraph" technology also occurred to Elisha Gray of the Western Union Company. Working with Thomas Edison, who invented the carbon microphone in 1877, Gray developed and patented a telephone design that was technically superior to Bell's. The sound transmission efficiency of the transmitter-receiver pair in Bell's original telephone was roughly -60 dB. Edison's granular carbon microphone transmitter increased this efficiency by 30 dB. As a result, Western Union had 50 000 of Gray's telephones in service by 1881 when various patent lawsuits were settled giving the American Bell Telephone Company complete control of the technology. Note that the rapid commercialization of the telephone occurred within only five years of its invention—a time scale usually associated with contemporary computer technology.

While the telephone entrepreneurs were busy deploying the new technology, academic physicists were laying the groundwork that would be necessary to create a long-distance telephone network. James Clerk Maxwell developed the unified equations governing electromagnetism in 1864. Electromagnetic wave propagation, predicted by Maxwell, was observed in the 1880s by Heinrich Hertz. Also in the 1880s, the British physicist Oliver Heaviside applied Maxwell's theory to show that the propagation of speech currents over wires in telephone systems needed to be understood on the basis of wave propagation, not simple currents. In 1884, Lord Rayleigh showed that such speech currents would be exponentially attenuated in a telegraph cable, calculating that a 600-Hz signal would be reduced by a factor of 0.135 over 20 miles in the transatlantic telegraph cable (0.27 dB/km). His paper was very pessimistic about the prospects for use of telephone technology for long-distance communication, compared to the well-established telegraph.

Results such as Rayleigh's, as well as the poor sound quality of the early telephone system, created the first real opportunity for physics in the infant telephone industry. In 1885, the managers of the engineering department of the American Bell Telephone Company, which Bell had left in 1881 and which later became AT&T, realized that it would not be easy to improve the transmission distance and quality on long-distance lines by trial and error. Therefore they formed a research department in Boston specifically focused on the physics of electromagnetic propagation on long-distance telephone lines. Hammond Hayes, one of Harvard's first physics Ph.D.s, organized the department and over the next twenty years hired other physicists from Harvard, MIT, Yale, Chicago, and Johns Hopkins to explore this new area of applied physics. This marked the beginning of industrial research in applied physics for communications.

The problem faced by early telephone researchers can be described rather simply from the vantage point of today's understanding. The original telephone transmission system was based on telegraph technology and used a single iron wire for each circuit with a return path through the ground. The attenuation A (in dB per unit length) of such a line at telephone frequencies ($\sim$kHz) is given approximately by

$$A \sim R\sqrt{C/L} + G\sqrt{L/C} \tag{1}$$

where R is the series resistance, L the series inductance, C the shunt capacitance, and G the shunt conductance, all per unit length. At telegraph frequencies ($\sim$10 Hz) such a line is almost purely resistive. For a multiple-wire telephone cable, which has much higher capacitance and lower inductance, the attenuation is 10 to 25 times greater than Eq. (1) and can be approximated by

$$A \sim \sqrt{RCf} \qquad (2)$$

where f is the frequency. The first solution to the telephone attenuation problem was to reduce R by an order of magnitude by replacing the iron wire with copper. A second problem was that the interference from outside sources picked up by the single, unshielded wire was an order of magnitude greater at telephone frequencies than for the telegraph. This was solved by adding a second wire to make a so-called "metallic circuit." However, such solutions dramatically increased costs. First, the copper wire had to be of large enough gauge to be self-supporting, and second, twice as much copper was needed in a metallic circuit. The situation in long-distance cables was even worse, where the higher intrinsic attenuation of Eq. (2) could only be solved by using even heavier-gauge copper (the longest cables used one-tenth-inch-diameter wire). Thus cables were only cost effective within cities to solve the congestion and weather problems illustrated in Fig. 2.

By the turn of the century, an understanding of the physics of transmission lines had produced a dramatic solution to these problems. In 1887, Heaviside developed the transmission line theory which we now understand as Eqs. (1) and (2). He pointed out that the attenuation could be reduced by increasing the series inductance per unit length, which is obvious from Eq. (1) when the first term dominates, as is usually the case. In 1899, George Campbell, at AT&T's Boston laboratory, and Michael Pupin, at Columbia University, almost simultaneously concluded that discrete inductors could simulate the continuous inductance of Eq. (1) as long as the spacing was not larger than one-tenth of a wavelength. For telephone frequencies this corresponded to a spacing of eight miles in open lines and one mile in cables. The effect of these so-called "loading coils" was dramatic. The maximum transmission distance of open lines nearly doubled, thus allowing the long-distance network to extend from New York to Denver by 1911. The effect on telephone cables was even more dramatic. The attenuation was decreased by a

Figure 2. Boston central telephone station at 40 Pearl Street after the blizzard of 1881. Inset shows the installation of underground cables which solved the weather and congestion problems of thousands of open wires. From Fagen, 1975.

factor of four and the frequency distortion of Eq. (2) was greatly reduced. Since cables were primarily used for relatively short distances, these gains were traded off against the series resistance of the conductor and allowed smaller-gauge wires to be used, thus saving an order of magnitude in the cost of copper for the same cable length.

In addition to the transmission enhancements of loaded lines, there was a strong desire to develop some kind of "repeater" that would strengthen the weakened signals and retransmit them. Electromechanical repeaters were commonly used in the telegraph system for many years and were the reason that the telegraph was quickly extended coast to coast. The early attempts to invent telephone repeaters were electromechanical analogs of the telegraph systems. A number of inventors patented telephone repeaters that were essentially a telephone receiver placed next to a carbon microphone. The carbon microphone modulated a large current by means of speech-induced vibrations of the weakly touching particles, and the resulting gain formed the basis of an amplifier. An improved version was developed at AT&T's Boston laboratory and used commercially in the New York to Chicago route in 1905. Such electromechanical amplifiers had considerable distortion and narrow dynamic range, but their limited success served to focus research energies on finding a more useful type of amplifier. This leads us to the next era in the relationship between physics and the communications industry, when such an amplifier was indeed invented.

The Era of the Electron

At about the same time that the telephone was invented, various electrical experiments in gas-discharge tubes were laying the physics groundwork for the next phase of communications technology. In about 1878, Sir William Crookes developed a specially designed tube to study the mysterious phenomenon of cathode rays which caused gas-discharge tubes to glow. In 1897, the cathode-ray studies of J. J. Thomson, Cavendish Professor of Physics at Cambridge, led to the discovery of the electron. The first thermionic vacuum-tube diode was invented by Sir John Fleming in 1904, following Edison's 1883 observation of current flow between the filament of a light bulb and a nearby electrode. Fleming's device was an excellent detector of wireless telegraph signals, which had been invented by Marconi in 1896. In 1907, Lee de Forest invented the vacuum-tube triode, which for five years remained almost the exclusive province of wireless entrepreneurs. It was a much better rf detector than Fleming's diode but was never used as a power amplifier.

Meanwhile, pressure to develop a telephone amplifier intensified. In 1909, Theodore Vail, the new president of AT&T, who was rapidly acquiring the small telephone companies that had sprung up after Bell's patents expired, set forth the vision that would define the Bell System for many years to come—"One Policy, One System, Universal Service." He probably did not know that his grand vision of coast-to-coast service was not possible without an effective telephone amplifier. In about 1910, Frank B. Jewett, who was in charge of transmission engineering at AT&T's Western Electric subsidiary, guessed that an improved amplifier might be possible with the "inertialess moving parts" of the electron beams that his friend Robert Millikan at the University of Chicago had been studying. Jewett asked Millikan to recommend one of his top students trained in the "new physics" whom he could hire to do amplifier research. The new hire, H. D. Arnold, started in 1911 and within two years achieved the goal of the useful vacuum-tube amplifier that Vail's vision required. In 1913, Arnold's vacuum-tube amplifier, shown in Fig. 3, was being used in commercial service. As a result, the Panama-Pacific International Exposition in San Francisco opened to great fanfare in 1915 with the dedication of the first transcontinental telephone circuit and a conversation between President Woodrow Wilson in the White House, Alexander Graham Bell in New York, and Bell's original assistant Thomas Watson in San Francisco. This unqualified success of physics convinced AT&T officials that paying topnotch Ph.D. physicists to do communications research was good business.

Arnold's vacuum-tube triumph was facilitated by de Forest's triode demonstration to Bell officials in 1912. Even though the demonstration failed under the output power levels required for telephone applications, Arnold immediately saw how to improve the tube. He made three improvements that were critical: a higher vacuum, an oxide-coated cathode, and a better grid position. Irving Langmuir of General Electric had independently pursued a high-vacuum design for improving the triode in 1913. These vacuum-tube amplifiers not only dramatically increased the transmission distance of telephone lines, but also made long-distance wireless telephony possible. In 1915, using high-power vacuum-tube amplifiers, Bell System scientists transmitted speech for the first time by wireless telephony from Arlington, Virginia, to both Paris and Honolulu. Wireless telephony was not widely used for two-way communications, however, until the development of microwaves during W.W. II and cellular wireless technology in the latter part of the 20th century.

Vacuum tubes also made possible dramatic increases in the capacity of the long-distance telephone system, which lowered the cost per call. This was done by multiplexing a number of calls in parallel over the same pair of wires using independently modulated carriers of different frequencies, a technique sometimes called "wired wireless." The idea of multiplexing was originally developed for the telegraph, using electromechanical resonators to generate and separate the various carrier tones. Indeed, Bell had been experimenting with just such a "harmonic telegraph" when he invented the telephone. However, to apply the same principle to voice-frequency signals required carriers of much higher frequency, ~5–25 kHz. This was not possible without vacuum tubes for oscillators, modulators, and amplifiers. In addition, technology from the era of electromagnetism was needed for the bandpass filters to demultiplex the signals at the

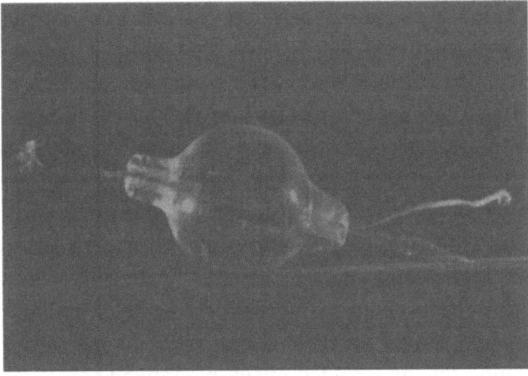

Figure 3. Arnold's high-vacuum tube, first used as a telephone repeater at Philadelphia on a New York to Washington cable circuit in October, 1913. Other, later, models were used on the transcontinental circuit opened for service in 1915. From Fagen, 1975.

receiver. In about 1910, the Bell System theoretician George Campbell, who also introduced loading coils, invented the so-called "wave filter." The value of this invention was not immediately appreciated, however, and the concept was not patented until 1917. It is of interest to note that today's technology for increasing transmission capacity on a glass fiber is wavelength-division multiplexing (WDM), which, although digital, is nevertheless almost an exact optical analog of the electronic multiplexing developed in the 1920s.

The fundamental physics underlying telephone transmission was fairly well established by 1920. Thus communications-related physics research turned to an exploration of the underlying materials. The primary areas were thermionic emission, noise in vacuum tubes, and the magnetic materials used in transformers and loading coils. During the 1920s physicists at the Western Electric laboratory, which became Bell Telephone Laboratories in 1925, made a number of important physics contributions in these areas. In 1925, J. B. Johnson made the first observation of the thermal noise predicted in 1918 by W. Schottky. Thermionic emission was first described theoretically by O. W. Richardson in Britain in 1901, and the benefit of alkaline-earth oxides in enhancing thermionic emission was observed by A. Wehnelt in Germany in 1904. About 10 years later, C. D. Child at Colgate and Langmuir at GE independently described the physics of space-charge-limited current, which is essential to the operation of a vacuum tube. Nevertheless, by 1920 there was still a long-standing controversy as to whether the enhanced performance of oxide-coated cathodes was due to chemical or physical effects. This was settled by Arnold, who showed in 1920 that the enhanced emission was indeed due to thermionic emission. In 1917, Clinton J. Davisson, a student of Richardson's at Princeton, came to Western Electric in order to understand the fundamental physics of why alkaline-earth oxides had a lower work function than tungsten. As a result of the invention of the ionization vacuum gauge by Oliver E. Buckley in 1916, it was possible to characterize the vacuum conditions necessary for this early surface-physics research, thus allowing Davisson and Lester Germer to do some of the key physics on thermionic emission from oxides and tungsten in the early 1920s. Physics research on magnetic materials was also important at Bell Labs during the 1920s. Permalloy had been developed at Western Electric for use in the loading coils discussed earlier. The first scientific paper on permalloy was given by Arnold and G. W. Elmen at the 1923 Spring Meeting of the American Physical Society. Another highlight of this period was Richard Bozorth's internationally acclaimed work on the Barkhausen effect.

Perhaps the most famous physics experiment at Bell Labs during the 1920s was the observation of the wave nature of the electron, for which Davisson received the 1937 Nobel Prize in Physics. This is indicative of the systematic business practice of recruiting the very best physicists to do applied research in the communications industry. In order to attract and retain top physics talent, it was necessary to allow such researchers the freedom to pursue the fundamental questions uncovered in their applied research. The history of Davisson's Nobel Prize is instructive. At the same time that Davisson was working on thermionic emission, he was also pursuing experiments to study the atomic structure of magnetic metals using inelastic electron scattering. This was relevant to both magnetic materials and the problem of secondary electron emission in vacuum tubes. During these experiments, Davisson noticed that some of the electrons were elastically scattered. While pursuing this unexpected result, he noticed some angular structure in the scattering pattern. In 1925, W. Elsasser suggested that Davisson's data could be evidence of "de Broglie waves," which had been proposed in 1923. During a 1926 trip to Europe, Davisson obtained a copy of Schrödinger's paper on wave mechanics, which had been published earlier that year. Upon returning to Bell Labs, he and Germer repeated the experiments with a single crystal of nickel, looking at the specific angles where electrons would be diffracted according to Schrödinger's equation. The results led to their famous 1927 paper, which also introduced low-energy electron diffraction (LEED) as an important tool for surface physics. Similar stories hold for Karl Jansky, who discovered radio astronomy in 1933 while studying noise in transatlantic radio telephony, and Arno Penzias and Robert Wilson, who discovered the $3K$ microwave background radiation of the big bang in 1964 while studying noise in telecommunications satellites.

The Era of Quantum Mechanics

The fundamental-physics roots of this era began with the explosive growth of quantum mechanics in Europe in the 1920s. The first application of this new theory to solids was Felix Bloch's 1928 quantum theory of metals. The foundations of semiconductor physics quickly followed with Rudolf Peierls's 1929 theory of the positive Hall effect due to holes, Brillouin's 1930 concept that band gaps are related to the Bragg scattering conditions, and Alan Wilson's 1931 band theory of semiconductors, including the effects of doping. A major development in the physics of real materials was E. Wigner and F. Seitz's 1933 approximate method for calculating band structure. This marked the beginning of a shift from the fundamental studies of the 1920s to the practical solid-state physics which would dominate the second half of the 20th century.

The experimental roots of semiconductor physics date from the 19th century. In the 1870s, at almost exactly the same time that Bell was inventing the telephone, physicists working on selenium, copper oxide, and various metallic sulfides (all materials we know today to be semiconductors) were discovering diode rectification behavior, the Hall effect, photoconductivity, and the photovoltaic effect. In fact, even the idea of inventing a solid-state analog of the vacuum tube had occurred to a number of people during the 1920s and 1930s—J. E. Lilienfeld patented the field-effect concept in 1926, and Walter Brattain and Joseph Becker at Bell Labs contemplated putting a grid into copper oxide rectifiers during the 1930s.

By the late 1930s, solid-state physics was well established and had the potential for major applications. In a move reminiscent of earlier eras, Mervin Kelly, Bell Labs' Director of Research, sought out the best of the new breed of solid-state physicists to explore the potential of semiconductors for communications; in 1936 he hired William Shockley from John Slater's group at MIT. However, the effort to make devices of possible use in communications, e.g., solid-state switches or amplifiers, did not start seriously until 1946 when nonmilitary research resumed at Bell Labs after W.W. II. Shockley was put in charge of a new solid-state research group specifically chartered to obtain a fundamental understanding of the device potential of silicon and germanium, which had been developed into excellent microwave detectors during the war. One of his first moves was to hire John Bardeen. The subsequent path to success was as rapid as Arnold's development of the vacuum-tube amplifier in 1912. The point-contact transistor was demonstrated within two years, by the end of 1947. The birth of the transistor is covered in a number of 50th Anniversary reviews (Brinkman et al., 1997; Riordan and Hoddeson, 1997; Ross, 1998), including one in this volume (Riordan et al., 1999, which includes a photograph of the first transistor). Therefore our focus will be to review the relationship of the transistor to the technology changes that have revolutionized communications over the past 50 years.[3]

The application of the transistor to communications occurred in two phases. The first, during the 1950s, was simply the replacement of vacuum tubes in various circuits. The first commercial use of the transistor in the Bell System was in 1954; the first fully "transistorized" product (the $E6$ repeater) was in 1959. There were some benefits of size and power reduction, but the functionality and design of the telephone system was not changed. In the second phase, the transistor made possible digital transmission and switching—an entirely new communications technology that revolutionized the industry. The concept of digital voice communications, known as pulse code modulation (PCM), was first demonstrated in 1947 at Bell Labs. This early demonstration was based on voice coding ideas developed in the 1930s and telephone encryption devices used by the military during the war. Commercial use of PCM, however, was not possible without transistors to make the complex circuits practical. The first digital transmission system, the so-called $T1$ carrier, was introduced in 1962 and carried 24 digital voice channels with an overall bit rate of 1.5 Mbit/sec. Even though a combined digital switching and transmission system was demonstrated in 1959 at Bell Labs, the first commercial use of fully digital switching and transmission was not until the introduction of the $4ESS$ switch for long-distance traffic in 1976.

Digital technology has profoundly affected the communications industry. It could be argued that the 1984 breakup of the Bell System was due to the ease with which various competitors could develop digital telephone systems, as opposed to the complex electromechanical switching systems that required the substantial resources of the Bell System to develop and maintain. An additional factor was the close relationship between the digital technology for communications and for computers. Thus the 1958 invention of the integrated circuit by J. S. Kilby at TI, along with major improvements by R. N. Noyce at Fairchild and the 1960 invention of the MOSFET (metal-oxide-semiconductor field-effect transistor) by D. Kahng and M. M. Attala at Bell Labs, affected both industries in fundamental ways. As a result of the exponential growth in the number of transistors per chip (Moore's Law), the cost and size of electronic devices have changed by orders of magnitude since the 1960s. In communications this made possible the wireless revolution that we are seeing today. Figure 4 shows the reductions in size of cellular telephone equipment since the invention of the concept by Bell Labs in 1960; dramatic reductions have also occurred in the cost. In a similar way, digital electronics and integrated circuits have made the "old" technology of facsimile transmission, introduced by AT&T in 1925, into the practical communication medium of today's ubiquitous Fax machine.

At this moment we stand on the threshold of yet another revolution brought on by the transistor and integrated circuit—the Internet. The widespread use of personal computers and the Internet have made data networking one of the hottest growth industries. Because of digital technology, telecommunications is being redefined to include data and video, as well as voice. It is ironic that after the telephone made the telegraph obsolete, the latest technology is essentially reverting back to an ultrafast version of the telegraph, in which

---

[3]We should note in passing, however, that Bell Labs' broader focus on physics research in areas relevant to communications—magnetism, semiconductors, and surfaces—grew substantially during the postwar period and is still a major effort today. Even though much of this work is directly relevant to communications, it is well beyond the scope of this review to cover it extensively. One of the highlights is P. W. Anderson's theoretical work on magnetism and disordered solids during the 1950s, for which he received the 1977 Nobel Prize in Physics.

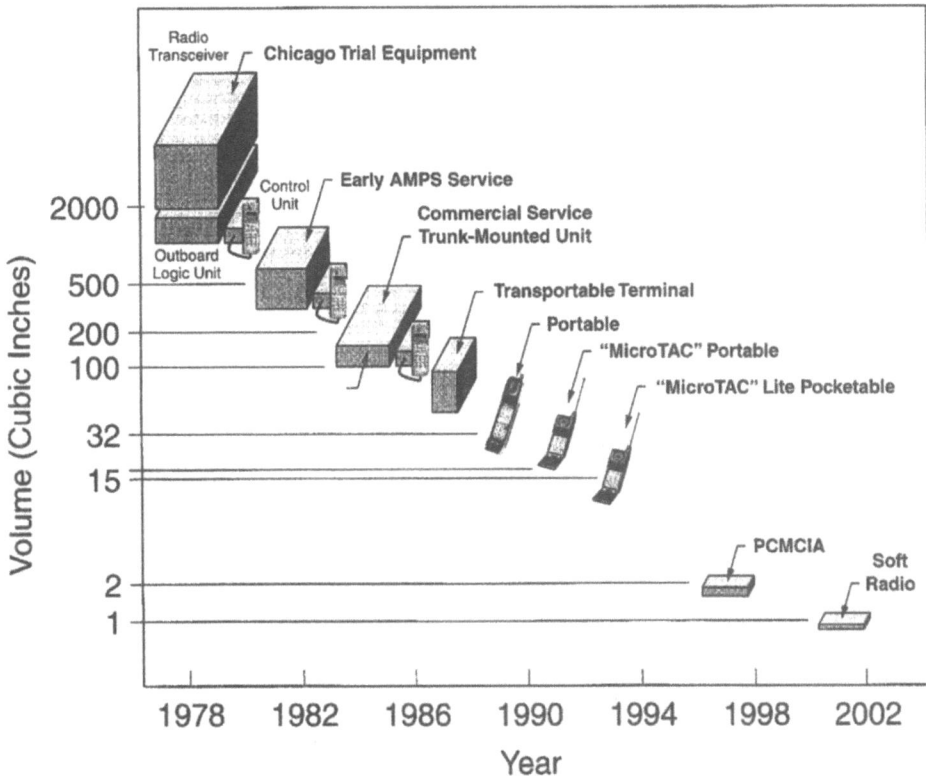

Figure 4. Size reduction of cellular telephones as a result of progress in increasing the number of transistors in an integrated circuit chip, as expressed by Moore's Law.

coded digital messages, not analog voice, dominate the system. However, this revolution is not based solely on the transistor and integrated circuit. We require the ultrahigh transmission bandwidth of fiber optics to complete our story. This brings us to the fourth era in the relationship between physics and communications—learning how to communicate with light.

The Era of Quantum Optics

Alexander Graham Bell invented the photophone in 1880, just a few years after the telephone. This device used a mirrored diaphragm to modulate a beam of sunlight with speech vibrations, analogous to the modulation of an electric current in Bell's telephone. The receiver was a selenium photocell, which had been discovered in 1876. The fiber-optic communications systems of today differ in only three major respects: (1) a glass waveguide, invented by the British physicist John Tyndall in 1870, replaces free-space propagation; (2) high-speed, multichannel digital modulation replaces the single analog voice channel; and (3) a coherent light source replaces the sun. The coherence of the source is critical for two reasons. First, coherent light has much less noise than incoherent, and second, coherent light can be focused into a high-power beam without much loss in intensity. In fact, the Bell Labs communications engineer W. A. Tyrrell pointed out in

1951 the advantages of using optical frequencies instead of microwaves for communications, but noted the lack of the required coherent source. The major impact of physics has been to provide the coherent light source—the laser.

The date commonly cited for the invention of the laser is the theory of A. Schawlow and C. H. Townes in 1958 (see the articles in this issue by Lamb et al. and by Slusher). However, the fundamental physics of stimulated emission was first recognized in 1917 by Einstein. Stimulated emission was first observed at microwave frequencies (24 GHz) in the ammonia beam maser by Townes in 1955. N. Bloembergen demonstrated in 1956 the importance of a three-level system in obtaining the population inversion for maser amplification. In 1957, Bell Labs developed the first solid-state maser, which was used as a low-noise microwave amplifier in the Telstar communications satellite in 1962. With such a background, Schawlow and Townes's 1958 theory for extending stimulated emission to optical frequencies was not terribly surprising. However, it was two years before the first laser (the pulsed ruby laser) was demonstrated by Theodore Maiman at Hughes Research Labs in 1960. Subsequent lasers of various types came at a rapid pace in the early 1960s.

The laser most directly relevant to communications is the semiconductor laser, demonstrated independently in 1962 by GE, IBM, and Lincoln Laboratory. This type of laser can be fabricated from a variety of direct-band-gap III-V semiconductors, which were introduced by H. Welker in 1952. However, it was not until 1970 that laser diodes could be made to operate continuously at room temperature, a prerequisite for optical communications. In order to explain this and other improvements in laser diodes for communications, we must return to the previous era of semiconductor physics and the concept of the heterojunction (see Riordan et al., 1999). A heterojunction is formed when two semiconductors of different band gap are joined together. William Shockley pointed out in 1951 that the performance of a bipolar transistor would be enhanced if the emitter had a wider band gap than the base, i.e., the emitter-base junction was also a heterojunction. In 1963, it was independently suggested by Herbert Kroemer and by Zh. I. Alferov and R. F. Kazarinov that a laser having such heterojunctions would be superior to the homojunction diodes first demonstrated. Materials of different band gaps, however, generally also have different lattice constants, making good crystal growth difficult, if not impossible. Therefore heterojunctions remained a theoretical curiosity until 1967 when J. M. Woodall, H. Rupprecht, and G. D. Pettit at IBM produced good-quality $GaAs/Al_xGa_{1-x}As$ heterojunctions by liquid-phase epitaxy. In 1970, I. Hayashi and M. B. Panish at Bell Labs and Alferov in Russia obtained continuous operation at room temperature using double-heterojunction lasers consisting of a thin layer of GaAs sandwiched between two layers of $Al_xGa_{1-x}As$. This design achieved better performance by confining both the injected carriers (by the band-gap discontinuity) and emitted photons (by the refractive-index discontinuity). The double-heterojunction concept has been modified and improved over the years, but the central idea of confining both the carriers and the photons by heterojunctions is the fundamental approach used in all semiconductor lasers.

The second essential ingredient for optical communications is low-loss silica fiber. Fiber-optic illuminators were developed in the mid 1950s, but this type of glass has an attenuation of about 100 dB/km and would only work for communications systems a few hundred meters in length. A major breakthrough occurred in 1970, when F. P. Kapron and co-workers at Corning produced the first fiber with a loss less than 20 dB/km at 850 nm, the $GaAs/Al_xGa_{1-x}As$ laser wavelength. This marked the beginning of optical fiber communications technology. By 1976, fiber loss was reduced to 1.0 dB/km at 1300 nm, increasing the distance between repeaters to tens of miles. The first commercial trial of optical communications was carried out in Chicago in 1978; by 1982, fifty systems were installed in the Bell System, corresponding to 25 000 total miles of fiber. In 1985, the loss in silica fiber reached a low of 0.15 dB/km at a wavelength of 1550 nm. By this time, the systems had migrated to the lower-loss long-wavelength region of 1300 to 1550 nm, and the laser diodes were being fabricated in the InP/InGaAsP materials system. The first transatlantic optical fiber system (TAT-8), operating at 1300 nm, was installed in 1988.

The capacity of commercial optical fiber communications systems has increased about as fast as Moore's Law for integrated circuits, doubling every two years, as shown in Fig. 5. The recent rate of increase for experimental systems has been even faster. In the latest of a sequence of major advances, Bell Labs reported in 1998 the transmission of 1.0 Terabit/sec over a distance of 400 km on a single fiber using 100 different wavelengths, each modulated at 10 Gb/sec—the so-called dense wavelength-division-multiplexing (DWDM) technology. Three additional elements of physics are necessary, however, to complete our story and bring us to the technology of the 1990s: quantum wells, optical amplifiers, and nonlinear optics.

A quantum well is a double-heterojunction sandwich of semiconductors of different band gap, discussed above, with the central, lower-band-gap layer so thin (typically less than 100 Å) that the quantum states of the confined carriers dominate the properties of the material. This was first demonstrated by R. Dingle, W. Wiegmann, and C. H. Henry at

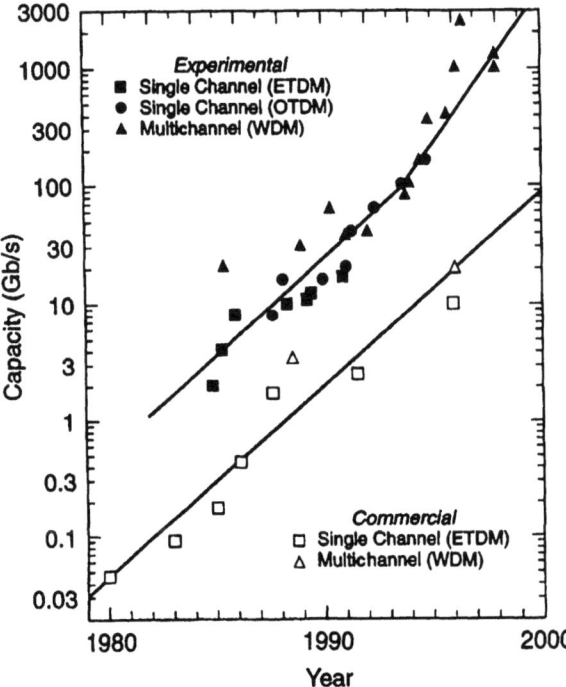

Figure 5. Progress in optical fiber transmission capacity. The growth in commercial capacity is due to the increasing bit rate of electronic-time-division multiplexing (ETDM) and the introduction of multichannel wavelength-division multiplexing (WDM). Experimental systems used ETDM, WDM, and optical-time-division multiplexing (OTDM) to achieve record-setting results.

Bell Labs in 1974. The first quantum-well laser was made at Bell Labs in 1975, and a patent on the quantum-well laser was granted to Dingle and Henry in 1976. Such structures were made possible by the crystal-growth technique of molecular-beam epitaxy, developed into a high-quality epitaxy method by A. Y. Cho in the early 1970s. By using multiple quantum wells in the active layer of a laser diode, the properties needed for communications systems are dramatically improved. In fact, most multiple-quantum-well lasers today also use intentionally strained quantum-well layers, a method first proposed by G. C. Osbourn at Sandia in 1982 to modify the band structure and further enhance device performance. The most recent device, which is essential for wavelength-division multiplexing systems, is a laser and electroabsorption modulator integrated on the same chip. In this case both the laser and modulator are based on quantum wells, with the modulator using the quantum-confined Stark effect discovered at Bell Labs by D. A. B. Miller and co-workers in 1985.

The physics of quantum wells is another example in which communications research has impacted the broad physics community. The 1978 invention of modulation doping by Stormer and co-workers at Bell Labs made possible ultrahigh-mobility two-dimensional electron systems which have dominated the research on mesoscopic and quantum-confined systems over the past twenty years. In 1982, D. C. Tsui, H. L. Stormer, and A. C. Gossard at Bell Labs discovered the fractional quantum Hall effect using high-mobility quantum-well structures. Research on the fractional quantum Hall effect is still a hot topic in fundamental physics today (see Stormer and Tsui, 1999). Another recent example of fundamental quantum-well research is the quantum cascade laser, invented by J. Faist, F. Capasso, and Cho at Bell Labs in 1994. This is the first semiconductor laser to use the quantum-well states, rather than the band gap, to define the wavelength of laser emission. Such lasers currently operate at wavelengths longer than those needed for optical communications; however, the insights gained will most likely impact communications systems in the future.

The optical amplifier is a major mid 1980s advance in communication systems. Such amplifiers are, of course, based on the same physics of stimulated emission as the laser and maser, but it was not at all clear whether such an amplifier would have the required low noise and low cross-talk properties essential for communications systems. Several amplifier designs were explored, including laser diodes with antireflective coatings and Raman amplifiers, but the best amplifier for communications systems proved to be optically pumped erbium-doped silica fiber. The considerable body of physics knowledge developed in connection with rare-earth ion lasers (such as Nd:YAG) invented in the mid 1960s greatly accelerated the development of the erbium-doped fiber amplifier (EDFA). Basic research at Bell Labs and elsewhere on the spectroscopy of rare-earth ions in various matrices, including silica glass, was a necessary precursor for the rapid development of EDFAs. By the late 1980s, EDFAs were widely used in experimental systems. Optical amplifiers make WDM systems possible, since the parallel wavelength channels can be simultaneously boosted without each having to be separated and routed through an expensive optoelectronic repeater typical of older fiber-optic systems.

Nonlinear optics was discovered in 1961 with the observation of two-photon absorption, almost immediately after the first laser was constructed in 1960. Even though the optical nonlinearities of silica are very small, such effects become important in communications systems because of the long distances involved. In 1975, Roger Stolen observed four-photon mixing in silica fibers. This third-order nonlinear effect, analogous to intermodulation distortion in electrical systems, emerged as a serious problem for WDM systems in the 1990s. The problem arose from the typical practice of designing optically amplified transmission systems with the chromatic dispersion of the fiber carefully adjusted to zero, to prevent pulse spreading as a result of the finite spectral width of the modulated source. Such a system creates good phase matching between the WDM channels and, hence, generates four-wave mixing products that seriously degrade the performance of very long systems. The solution was to introduce a new fiber design, the so-called True Wave fiber invented by Bell Labs in 1993, which introduces a small, controlled amount of dispersion to break up the phase matching and prevent four-wave mixing. This makes WDM technology practical.

The extremely high transmission capacity possible with today's optical fiber systems (a Terabit/sec corresponds to over 15 million simultaneous phone calls) along with the extremely large number of transistors on a single chip (over one billion expected in 2001) will undoubtedly lead us into future eras of communications technology which will be poised to benefit from future discoveries of physics.

Conclusions

We have shown the critical impact of the four major eras of physics on the communications industry over the past 125 years. The industry rapidly applied the major physics discoveries during this period and thus made dramatic improvements in communications technology, with demonstrable benefits to society. We note that in all four major eras of physics—electromagnetism, the electron, quantum mechanics, and quantum optics—the fundamental discoveries were applied by the communications industry within 15–20 years. Further, it is evident that the communications industry's practice of employing the best physicists to do both basic and applied research resulted in the successes noted in this review. Indeed, the development of marketable communications technology would most certainly not have occurred so rapidly had this not been the case. This reality was well understood by such visionary communications research leaders as Hayes, Jewett, Buckley, and Kelly. They recognized the potential of applying the latest physics discoveries to enhance communications technology. The current leaders of the communications industry continue in this tradition.

References

Agrawal, G. P., and N. K. Dutta, 1993, *Semiconductor Lasers* (Van Nostrand Reinhold, New York).
Brinkman, W. F., D. E. Haggan, and W. W. Troutman, 1997, IEEE J. Solid-State Circuits **32**, 1858.
Fagen, M. D., 1975, *A History of Engineering and Science in the Bell System: The Early Years (1875–1925)* (Bell Telephone Laboratories, New York).
Hoddeson, L., E. Braun, J. Teichmann, and S. Weart, 1992, *Out of the Crystal Maze* (Oxford, New York).
Kaminow, I. P., and T. L. Koch, 1997, *Optical Fiber Telecommunications IIIA* (Academic, San Diego).
Millman, S., 1983, *A History of Engineering and Science in the Bell System: Physical Sciences (1925–1980)* (Bell Telephone Laboratories, Murray Hill, NJ).
Millman, S., 1984, *A History of Engineering and Science in the Bell System: Communications Sciences (1925–1980)* (AT&T Bell Laboratories, Murray Hill, NJ).
Riordan, M., and L. Hoddeson, 1997, *Crystal Fire: the Birth of the Information Age* (Norton, New York).

Riordan, M., L. Hoddeson, and C. Herring, 1999, Rev. Mod. Phys. **71**, 336; pp. 563–578 in this book.
Ross, I. M., 1998, Proc. IEEE **86**, 7.
Stormer, H. L., and D. C. Tsui, and A. C. Gossard, 1999, Rev. Mod. Phys. **71**, 298; pp. 501–514 in this book.
Whinnery, J. R., 1987, *Lasers: Invention to Application* (National Academy, Washington, D.C.).

Index

ABM (anti-ballistic missile) systems, 785–789
Accelerator-based atomic collisions, 380–392
Accelerators, 19
    future possibilities for, 211–212
    growth patterns of, 203–206
    limitations of, 210–211
    particle physics and, 203–221
    principles, categorization, and evolution of, 206–209
Acceptor defects, 113–114
Accretion theory, 306
Action
    quantum of, 98–99
    unit of, 148
Action potential timing, 736
Action potentials, 728, 730
Aharonov-Bohm effect, 122–123
Algorithms, 743
α particles, 12–13
    emission of, 17
Annihilation catastrophe, 264
Anomalous perihelion advance, 79–80
Anti-ballistic missile (ABM) systems, 785–789
Antiferromagnetism, 109
Antimatter, 264
    neutral, 401
Antiparticles, 34, 145
Antiprotonic atoms, 400–401
Antiquarks, 351
Antisymmetry, 6
Arrhenius law, 710
Artificial atoms, 123
Artificial neural networks (ANN), 732–735
Astronomical unit, 77
Astronomy, cosmic-ray, 286–287
Astrophysical neutrinos, 239–242
Astrophysical sources of gravitational radiation, 324–326
Astrophysics, 56–67
    lasers in, 809
    nuclear, 360–363, 370
Asymptotic expansion methods, 394–396
Asymptotic freedom, 152, 257
Atom cooling, 427
Atom interferometers, 428
    inertial measurements with, 437
Atom lasers, 457
Atom optics, 138, 427–428

Atom trapping, 426, 429
Atomic bomb, 768
Atomic clocks, 133, 436
Atomic collisions, 382
    accelerator-based, 380–392
    cold, 378–380
    complete experiments in, 384
    fundamental measurements in, 382–392
Atomic force microscope (AFM), 546–548
Atomic hydrogen targets, 383
Atomic nucleus, 131
Atomic physics, 6, 377–403
    quantum electrodynamics and, 396–397
    in the twentieth century, 129–139
Atomic properties, measurement of, 438–439
Atomic scattering experiments, 383–390
Atomic theory, high-precision, 392–399
Atoms, 129
    manipulating position and velocity of, 427–429
    scanning tunneling microscope and, 543–553
    single, quantum measurements on, 434
Axion field, 149
Axions, 259, 267, 271, 335–336
Axons, 734

Ballistic electron-emission microscopy, 546
Band index, 101
Band-structure paradigm, 101–105
Bardeen-Cooper-Schrieffer (BCS)
    microscopic theory, 527–529
    superconductivity, 100
Bare couplings, 151
Baryon density, 259
Baryon-lepton asymmetry, 264–265
Baryonic dark matter
    need for, 330
    searches for, 331–335
Baryons, 37, 348
Bell Telephone System, 820
Bell's inequality, 489, 490–491
Bell's theorem, 489
Beta decay, 32–33, 364–365
β rays, 12
Betatrons, 207
Bifurcations, 667–668
Big bang, 62
    heat of, 265
    hot, 246

Big bang (cont.)
   initial conditions and, 587–588
   origin of, 265
Big-bang nucleosynthesis, 251–252
Binding energy, 16
Bioenergetics, 717
   physics and, 717–719
Biological electron transfer, 718
Biological physics, 706–723
Biological systems
   complexity of, 709–712
   dynamics of, 712–715
   fluctuations and relaxations in, 712
   pattern formation in, 675
   reaction theory for, 715–717
   structures of, 707–708
Biology
   lasers in, 806–808
   light transduction in, 718–719
   physics and, 706
Biomolecules
   energy landscape of, 711–712
   forces between, 719–721
Biophysics, 706
Black-hole thermodynamics, 310
Black holes, 58, 59, 84–85, 188, 201, 302–311
   conclusive evidence for, 306–307
   cosmic censorship and, 308–309
   dark matter, 335
   gravitational radiation and, 325
   observational evidence for, 303–307
   observational signatures of, 304
   quantum, 309–311
   supermassive, see Supermassive black holes
   uniqueness of, 307–308
   in x-ray binaries, 305–306
Blackbody radiation, 74
   spectral distribution of, 3
Bloch insulator, 111
Bloch paradigm, 124
Bloch waves, 101
Block copolymers, 560
Blood-oxygen-level dependent images, 762
Bogolmonyi-Prasad-Sommerfield (BPS) states, 310
Bohr radius, 103
Boltzmann distribution, 743
Born-Oppenheimer approximation, 99
Bose-Einstein condensation, 118, 138, 414
   in dilute gases, 431–433
Bose-Einstein statistics, 6
Boson-fermion cancellation, 190
Bosons, 145
Bottom quark, 41, 169
Bow shocks, 685
   cometary, 687
Bragg-Williams model, 114
Brain
   as computer, 726–727
   mind and, 726
Breeder reactors, 773
Breit operator, 396
Broken symmetry, 35
Bubble chambers, 213
Buckminsterfullerene, 698
Bulk photoemission, 105

Cabibbo angle, 166
Cabibbo-Kobayashi-Maskawa (CKM)matrix, 181
Calibi-Yau manifold, 197
Calorimetry, 217–218
Cancer, treatment of, 765
Cantilever probes, 551
Carbon, balls and tubes of, 698–699
Carbon dating, 278
Carbon nanotubes, 561, 698–699
Cavity magnetrons, 89
Cavity quantum electrodynamics, 138, 417–419
"Center manifold theorem," 667–668
Center-of-mass energy, 203, 205
Ceramics, 557
Cerenkov radiation detectors, 215–216
Cesium atomic clock, 133
Cesium fountain, 436
Chain reaction, 18–19
Chandrasekhar limit, 84
Chaos
   deterministic, 665–666
   spatiotemporal, in fluids, 668–671
Charge carriers, flow of, 568–570
Charge conjugation, 38
Charge distribution, 17
Charge independence, 35
Charge transport in molecular systems, 717–718
Charged massive particles (CHAMPs), 335
Charm quark, 41, 166
Chemical lasers, 800
Chemical physics, 693–704
Chemical reaction pathways, laser control of, 703–704
Chemical reactions
   dynamic effects in, 715–716
   in interstellar molecular clouds, 694–697
Chiral perturbation theory, 349
Chiral symmetry, 165, 348
Chirality, 165
Chirped pulse amplification, 479
Circuit theory, 516
Circular electron-positron colliders, 209
Classical lattice vibrations, 97
Classical liquids, 116
Classical simulations, 743–744
   mixed quantum and, 748
Closed shells, 23–24
Closure models, 648
Cloud chambers, 44, 213
Cluster masses, 259–261

Clusters of galaxies, 247
Coherence factors, 530
Coherence length, 527
Coherent structures, large-scale, 652–653
Cold atomic collisions, 378–380
Cold dark matter (CDM) inflation and, 265–267
    model, 297–298
Cold-target recoil-ion momentum spectroscopy, 380–381
Collective excitations, 110–111
Collider detectors, 216–221
Collisionless nature, 679
    of high-temperature plasmas, 679–688
Collisionless shocks
    dissipation mechanism, 686
    in space plasmas, 685–688
Collisionless skin depth, 685
Collisions, analysis of, 8
Colloidal aggregates, 626
Color, 40, 162
Color centers, 113
Cometary bow shocks, 687
Communication systems, fiber optic, 803
Communications
    lasers in, 802–804
    optical, 823
    transistor in, 819–821
Communications industry, 812
    physics and, 812–825
Compaction, 630–631
Complete experiments, 384
Composite fermion pairs, 511–512
Composite fermions, 505–508
Compositeness scale, 183
Comprehensive Test Ban Treaty (CTBT), 789–793
Compton effect, 4
Computational neurobiology, 736
Computational physics, 741
Computed tomography, 759–760
Computers, 727
    brains as, 726–727
    as dynamical systems, 727
Condensed matter, 96–127
    electronic, 626
    soft, see Soft condensed matter
Conductance fluctuations, 122
    spatial variation and, 519–521
Conductance viewed as transmission, 515–524
Confinement, 347
Conformal therapy, 765
Connectivity threshold, 609
Continuous complementarity, 488
Continuum systems, 595–597
Cooper pairs, 120, 541
Coordinate waves, 314
Copolymers, 621–623
Copper-oxide rectifier, 564
Corona satellite, 783–785

Correlation lengths, 638
Cosmic censorship, black holes and, 308–309
Cosmic microwave background radiation (CMBR), 62, 249–250, 291–300
    anisotropy in, 294–297
    future experiments in, 298–300
    inflation and, 273–274
    interpretation of measurements of, 297–298
    measurements of, 292–296
    polarization of, 298
    significant events for, in standard model, 291–292
    temperature of, 292–294
Cosmic-ray acceleration by supernova shocks, 688
Cosmic-ray astronomy, 286–287
Cosmic-ray spectrum, 280–281
Cosmic rays, 43–45, 278–288
    high-energy, 282–284
    history of research in, 279–280
    new experiments on, 287–288
    techniques of measurement of, 281–282
Cosmic scale factor, 252–254
Cosmological constant, 83, 254–256
Cosmological parameters, 272
Cosmology
    current, 245–275
    foundations of, 246–252
    in next hundred years, 274–275
    open problems in, 246
    precision, 267–274
    standard, 245, 252–256
Coulomb blockade, 123, 522
Counterion entropy, 623
Coupling unification, 185–186
CP-violating effects, 39
CP violation, observed, 184
CPT theorem, 39, 145
Critical density, 259
Critical-density universe, 254
Critical phenomena, 116, 601–602, 613–614
Crystal systems, 557
CTBT (Comprehensive Test Ban Treaty), 789–793
Cuban missile crisis, 783
Current algebra, 39
Current drive, 684
Cutoff, coupling, 151
Cyclotron period, 103
Cyclotrons, 19, 207

D-branes, 197, 201
Dark clouds, 694, 695
Dark energy, dark matter and, 259–262
Dark matter, 64, 191, 245
    baryonic, see Baryonic dark matter
    cold, see Cold dark matter
    dark energy and, 259–262
    deciphering nature of, 329–341
    dynamic effect of, 330

Dark Matter (contd.)
  evidence for, 329–331
  neutrinos and, 239
  nonbaryonic, see Nonbaryonic dark matter
  structure formation and, 258–264
Dark matter black holes, 335
Dark-matter parameters, 272
Dark stars, 261
Data collapse, 602
de Haas-van Alphen-Shubnikoff oscillation, 104
Debye-Waller factor, 712
Deceleration parameter, 83, 84, 261
Deep-inelastic scattering, 169–171
Defect model, 263–264
Dendrites in brain, 727–728
Dendritic pattern formation, 673–675
Dense coding, 424
Density-functional theory, 103, 109–110, 748
Detector components, 213–216
Detector systems, 216–221
Deterministic chaos, 665–666
Deuteron, 15
Diamagnetic (screening) term, 152
Diatomic potential-energy curves, 696
Diblock copolymer, 621
Diboson production reactions, 181
Differential phase shifts, 123
Diffuse interstellar clouds, 694
Diffusion chamber, 53
Digital video disks, 801
Digital x-ray image receptors, 757–758
Dilatancy, 637
Dilation symmetry, 619–620
Dilaton, 195
Dimensional transmutation, 152
Dimensionality, 107
Dimensions, quantum, 189
Dimerization, 106
Dirac hole theory, 145
Dirac neutrinos, 237
Dirac wave function, 5
Dirichlet p-branes, 201
Disentanglement rheology, 625
Dislocations, 114–115
"Disoriented chiral condensate," 364
Displacement field, 631
Displacive excitations, 477
Display technologies, flat-panel, 757
DNA molecules, 708
Donor defects, 113–114
Doppler cooling, 137
Double beta decay, 365
Double slit experiments
  one particle, 483–484
  two particles, 484–487
Down quark, 162
Drip lines, 358
Dye lasers, 135

Dynamical model of neural activity, 727–732
Dynamical sypersymmetry breaking, 191
Dynamical systems, 667
  computers as, 727
  pattern formation and, 667–668
Dynamics
  of biological systems, 712–715
  of chemical reactions, 715–716
  of synapses, 731–732

Eddington limit, 304
Edge dislocations, 114–115
Effective magnetic field, 506
Eightfold Way, 164, 168
Einstein-Podolsky-Rosen (EPR), 488–490
Einstein ring, 86
Elastomers, 559
Electric degrees of freedom, 194–195
Electric dipole barrier, 107
Electric-magnetic duality, 194–195
Electromagnetism, era of, 813–816
Electromagnets, 813
Electron, era of the, 816–819
Electron coherence, 522–523
Electron degeneracy, 57
Electron density, 110
Electron dynamics, 100
Electron-electron interaction
  moderate effects of, 107–112
  radical effects of, 111–112
Electrons, 14
  field of, 144
  g values, 223, 224–229
  $g-2$, 224–229
  helicity, 165
  impact excitation, 385
  interacting, Hubbard model of, 111
  interactions, 521–522
  –ion recombination, 381–382
  movement of, in hydrogen atoms, 3
  neutrino, 165
  –phonon interaction, 100
  –positron pair creation, 146
  scattering, 353
  scattering by atoms in laser fields, 386–387
  self-energy, 8
  spin, 5
  spin magnetic moment of, 223
  transfer, biological, 718
  transport theory, 515
  wave nature of, 818–819
Electronic condensed matter, 626
Electronic detectors, 214–216
Electronic energies, total, of metals, 102
Electronic materials, 560–562
Electronic states, 7
Electrostatic accelerators, 206
Electrostatic storage ring for polar molecules, 702

Electroweak interaction, establishing, 175–182
Electroweak model, Z properties and, 178–180
Electroweak theory, path to, 165–166
Electroweak unification, 40–41
Elementary particle physics, 43–53
Elementary particles of standard model, 163
Elements, origin of, 61–62
Emission tomography, 760–761
Energy bands, 101
Energy barriers, 711
Energy gap, 529–530
Energy landscape, 711
  of biomolecules, 711–712
Energy scale of unification, 156
Entropic interactions, 622
Enzyme catalysis, 719
Equation of state, 606
Equivalence, principle of, 74–75
Erbium-doped fiber amplifier, 824
Ethical dilemma of scientists, 795–796
Event horizon, 85, 302, 307, 310
Excimer lasers, 804–805
Excitons, 110–111
Exclusion principle, Pauli, 4
Exotic atoms, 399–403
Exotic lasers, 456–457

F-center, 113
"Fabric tensor," 637
Families of vacua, 193
Fast shocks, 686–687
Feed-forward networks, 732–734
Feedback networks, 734–735
Femtosecond laser sources, 409
Femtosecond light pulse generation, 474–480
Fermi-Dirac statistics, 6
Fermi-edge singularity, 522
Fermi surfaces, 108
  of metals, 103–105
Fermi wave vector, 508
Fermion field, families of, 162
Fermion pairs, composite, 511–512
Fermionic dimensions of space-time, 189–190
Fermions, 145
  composite, 505–508
Ferroelectrics, 561
Ferromagnetic state, 109
Ferry relation, 710
Few-body problems, 392–399
Few-electron ions, theory of, 398
Fiber lasers, 804
Fiber optic communication systems, 803
Fibers, 559
Field concept, 144
Field-effect amplifier, 566
Field-effect transistor, 565–566
Filling, even-denominator, 508–511
Filling factor, Landau-level, 502

Fine-structure constant, 9, 161
  fine-structure splittings and, 397–398
Fission, 18, 767–774
Fixed-node method, 747
Fixed points of renormalization transformation, 610–612
Flat-panel display technologies, 757
Flatness problem, 265
Flavor, 40
Flow cytometers, 805–806
Fluctuating barriers, theory of, 717
Fluctuation phenomena, 712
Fluctuations of local loads, 638–640
Fluid turbulence, 644–661
Fluidized bed, 630
Fluids, spatiotemporal chaos in, 668–671
Fluoroscopy, x-ray, 756–757
Force channels, 638
Forces between biomolecules, 719–721
Form factors, 170
Förster transfer, 718
Fourth-order interference, 463–464
"Fractal objects," 623
Fractals, 125
Fractional quantum Hall effect, 122, 501–512
Frame dragging, 73, 81
Free-dendrite problem, 673
Free-electron lasers, 456, 810
Frenkel exciton, 110
Frequency chains, 412
Frequency-resolved optical gating, 476
Frozen conformation, 631
Fullerene molecules, 698
Fusion, 775–780

g values of electrons and muons, 223–232
Galactic nuclei, supermassive black holes in, 304–305
Galaxies, 60–63, 245
  clusters of, 247
Galaxy, age of, 57, 248
Gamma-ray bursts, 60
Gas molecules, 700–701
Gated reactions, 716–717
Gauge bosons, 167
Gauge couplings, 149
  trilinear, 181
Gauge hierarchy problem, 190–191
Gauge invariance, 166
General relativity
  applications of, 83–87
  review of, 72–88
Geonium atom, 224
Germanium, 564
Gibbs entropy, 585–586
Gibbs measure, 593
Glass transition, 117
  in proteins, 713

Glasses, 117
  soft, 625–626
Gluinos, 189
Gluon jets, 171–172
Gluons, 40, 162, 168, 171, 189, 349
Goldstone bosons, 167
Grain boundaries, 115
Grand unification, 257–258, *see also* Unification
Grand unified theory, 41, 186, 192
Granular matter, 629–641
  microscopic features of, 638–640
  pattern formation in, 671–673
Gravitational collapse, 302
Gravitational constant, possible variation of, 80–81
Gravitational lensing, 85–87, 269, 332–334
Gravitational potential, 313
  time-delay by, 76–78
Gravitational radiation, 81–83, 313–327
  astrophysical sources of, 324–326
  black holes and, 325
  detection criteria for, 327
  history of research in, 313–314
  stochastic background of, 326
  strong indirect evidence for, 314–315
  techniques for detection of, 317–324
  wave kinematics and, 315–317
Gravitational-wave interferometer, 318–319
Gravitinos, 189
Gravitons, 189
Gravity, 161
  quantum theory of, 157, 164, 265
  solar, deflection of light by, 76
Greenberger-Horne-Zeilinger states, 496
Griffiths equality, 605
Ground-state energy, 147
Group theory, 6–7

Hadron colliders, 205
Hadron physics, 348–352
Hadrons, 347
  in nuclear medium, 358–360
Hall effect, 122
  quantum, *see* Quantum Hall effect
Hall resistance, 502
Halo stars, 61
Hartree-Fock approximation, 393
Hartree-Fock system of differential equations, 7
Heavy-fermion systems, 112
Hebbian synapse, 731
Heisenberg uncertainty principle, 5, 10, 133
Helicity, 165
Helium clusters, reactions and spectra in, 699–700
Helium spectrum, 6
Hess-Fairbank effect, 540
Heterojunctions, 822
Hidden variables, 10
Hierarchy problem, 258
Higgs boson mass, 180–181

Higgs field, 162, 164
  origin of, 164
Higgs mass parameter, 155
Higgs phenomenon, 41
Higgs scalars, 167
High-energy cosmic rays, 282–284
High energy densities, matter at, 363–364
High-energy neutrinos, 278
High-energy physics, theoretical, 29–42
High flux densities, phenomena at, 478–480
High-impedance microwave devices, 209
High-precision atomic theory, 392–399
High-temperature plasmas, collisionless nature of, 679–688
Hole theory, Dirac, 145
Homogeneous and isotropic turbulence, 649
Hot big bang, 246
Hoyle paradox, 715
Hubbard model of interacting electrons, 111
Hubble constant, 61, 83–84, 87, 247, 269–270
Hubble radius, 265
Hubble time, 247, 256
Hydrodynamics, 116
Hydrogen atoms
  cosmic abundance of, 694
  hyperfine structure of, 132
  movement of electrons in, 3
  precision spectroscopy of, 410–412
Hydrogen bomb, 781–782
Hyperfine structure of hydrogen, 132
Hyperons, 37, 52

Identical-band phenomenon, 356
Image intensifier, 756
Image management, archiving and communication system, 763
Image networking, 763–764
Imaginary-time path integrals, 745
Impulsive excitations, 477
Inertial-confinement fusion, 775, 779
Inertial measurements with atom interferometers, 437
Inflation, 266, 296–297
  cold dark matter and, 265–267
  cosmic microwave background radiation and, 273–274
Initial conditions
  big bang and, 587–588
  in statistical mechanics, 586–588
Instability, 666
Instrumentation, 30
Integral quantum Hall effect, 122, 503–504
Interacting electrons, Hubbard model of, 111
Interactions, association of, with particle exchange, 146
Interface physics, 106
Interference, 460–461
  fourth-order, 463–464

measurement of time interval between two photons by, 470–473
quantum, 420, 484
second-order, 462–463
Interference experiments
with one parametric downconverter source, 464–465
one-photon, 468–469
with two parametric downconverter sources, 465–470
Interference phenomena, quantum, 420, 484
Interferometers, 452
Intermittency, 654
Internet, 802, 820
Interstellar medium, star formation and, 63–65
Interstellar molecular clouds
chemistry in, 694–697
diffuse, 694
Interstellar molecules, synthesis of, 695
Interstitials, 113
Inverse photoemission, 105
Ion asymmetry, 389
Ising model, 114, 591, 744
Isomerization, photoinduced, 718
Isotopes, 12
Isotopic spin, 15
Isotropic turbulence, 649

Janssen model for silos, 632–634
JASON conclusions, 793
Jets, 63
Josephson effect, 120, 531
Junction transistor, 565, 570–574

Kadanoff cells, 609–610
Kaluza-Klein excitations, 197
Kaons, 37
Kerr-cell shutters, 474
Kerr metric, 307
Kolmogorov turbulence, 649–651
Kondo resonance, 522
Kramers equation, 716

Lagrangian densities, 148–149
Lamb dip, 449–450
Lamb shift, 9, 36, 136
Lambda temperature, 534
Landau damping, 680–681
Landau Fermi-liquid theory, 108, 521
Landau-level filling factor, 502
Langmuir oscillations, 679
Langmuir waves, 680
Large numbers, law of, 80
Laser
applications, medical, 801
beam, 798
control of chemical reaction pathways, 703–704
cooling and trapping, 137, 413–414
fields, electron scattering by atoms in, 386–387
frequency chains, 412
light, 798
manipulating matter with, 413
linewidth, 450–452
phase transition analogy, 455–456
physics, 442–457
ranging, 802
sources, femtosecond, 409
spectroscopy, 408–415
tweezers, 806–807
Laser Interferometer Space Antenna (LISA), 321–324
Laser technology, 798–810
current, 800–802
future, 810
history of, 799–800
in materials processing and lithography, 804–805
Lasers, 134, 135, 822
in astrophysics, 809
in biology, 806–808
in communications, 802–804
exotic, 456–457
history of, 448–456
in medicine, 801, 805–806
in physics, 809
quantum effects in, 450–456
without inversion, 456–457
Lattice gauge theory of quantum chromodynamics, 747
Lattice models, 742
Lattice periodicity, 101
breakdowns of, 112–117
Lattice systems, 590–595
Lattice vibrations, classical, 97
Lee-Yang theorem, 593
Left-handed neutrinos, 237
Lense-Thirring effect, 306, 325
Lenses, gravitational, 85–87, 269, 332–334
Lepton colliders, 205
Leptons, 36–37
Light, 798
deflection of, by solar gravity, 76
Light massive neutrinos, 336–337
Light quanta, 4, 32, 130
Light transduction in biology, 718–719
Light-water reactors, 769–770
Linear induction accelerators, 207
Liquid crystals, 126
Liquid drop model of nucleus, 16–17
Local-probe concept, 543–544
Locality, concept of, 144
Localization, 520–521
Localized voltage drop, 519
London penetration depth, 119, 526
Lorentz transformation, 69
Luminosity, 203–204

Luttinger liquid, 121–122
Luttinger theorem, 108

M theory, 199, 258
MACHOs (massive halo compact objects), 87, 331, 332–334
Macroions, 623
Macroscopic behavior, irreversible, micro-scopic origins of, 582–584
Macroscopic stress fields, 632–638
Macrostate of universe, 587
Magnetic bottles, 429
Magnetic degrees of freedom, 194–195
Magnetic dynamo, 778–779
Magnetic fusion research, 776–779
Magnetic materials, 562
Magnetic moment anomaly, 136
Magnetic resonance, 761–763
Magnetic-resonance imaging, 94, 761–762, 763
Magnetic-resonance spectroscopy, 763
Magnetism, strong, 108–109
Magnetization density, 97
Magneto-optical trap (MOT), 137, 378–379, 429
Magnetrons, 89, 443–444
Main-sequence stars, 57
Majorana neutrinos, 237
Many-body systems, simulations of, 749
Maser operation, 134
Masers, 799
    history of, 442–448
    one-atom, 454–455
    quantum effects in, 450–456
Mass density of universe, 254
Mass excess, 17
Mass function, 305
Mass spectrometer, 135
Mass-to-light ratio, 269
Massive halo compact objects, *see* MACHOs
Materials physics, 555–562
Matrix mechanics, 4–5, 131
Matter
    granular, *see* Granular matter
    at high energy densities, 363–364
    manipulating, with laser light, 413
Matter-antimatter symmetry, 264
Matter/energy in universe, 263
Mean-field model, 601
Measurement
    of atomic and molecular properties, 438–439
    inertial, with atom interferometers, 437
    precision, 435–439
    of time interval between two photons by interference, 470–473
Medical images, 764
    physics and, 755–766
    in radiation treatment planning, 764–765
Medical laser applications, 801, 805–806
Membranes, 125–126

Memory, 726
Meson production, 21
Mesons, 35, 46–53, 348
Mesoscopic systems, 121
Metals
    Fermi surfaces of, 103–105
    quantum-mechanical free-electron model of, 100
    total electronic energies of, 102
Metastable superflow, 540
Microelectromechanical systems (MEMS), 561–562
Microemulsion, 618
Microlasers, 455, 800
Micromasers, 138, 418–419, 454–455
Microscopic origins of irreversible macroscopic behavior, 582–584
Microscopic simulations in physics, 741–750
Microwave background, cosmic, *see* Cosmic microwave background radiation
Microwave spectroscopy, 446
Milky Way, 60
Millipede storage device, 551
Mind, brain and, 726
Minkowski space-time, 11-dimensional, 199
MIRVs (multiple independently targetable reentry vehicles), 786
Mode locking, 475
Moduli space of vacua, 193
Molecular-beam epitaxy, 121
Molecular-beam magnetic resonance, 132–133
Molecular clouds, interstellar, chemistry in, 694–697
Molecular clusters, 697
Molecular dynamics, 743
Molecular properties, measurement of, 438–439
Molecular structure, 7
Molecular systems, charge transport in, 717–718
Molecular trapping, 702–703
Monomers, 124
Monte Carlo, 743
Moral dilemma of scientists, 795–796
Mössbauer effect, 75
Mott insulator, 111
Multiphoton scanning confocal microscopy, 808
Multiphoton transitions, 136
Multiple independently targetable reentry vehicles, *see* MIRVs
Muon $g-2$, theory of, 229
Muon $g-2$ experiments, 225–227
Muon neutrino, 165
Muonic helium, 400
Muonium, 133, 399, 402–403
Muons, 37, 345
Myoglobin, 708, 709

Naked singularities, 308–309
Nanomechanics, 543, 552

Nanoscience, 121–123
Nanosecond light pulse generation, 474–480
Nanotubes, carbon, 561, 698–699
National Ignition Facility (NIF), 792
National security, 782
  U.S., physics and, 781–796
Natural-uranium-based reactors, 770
Navier-Stokes equations, 647
"Negative absorption," 444
Neural activity, dynamical model of, 727–732
Neurobiology, 726
  computational, 736
Neurons, 727–730
Neutral antimatter, 401
Neutral atom traps, 438
Neutral-current reactions, 236
Neutral currents in neutrino scattering, 175–176
Neutralinos, 258, 259, 267, 270
Neutrino mass, 41, 168, 237–239
  nonzero, 184
Neutrino oscillations, 185, 237–239, 271
Neutrino physics, 234–242
Neutrino scattering, neutral currents in, 175–176
Neutrinos, 14, 234–236
  astrophysical, 239–242
  dark matter and, 239
  high-energy, 278
  light, 271
  light massive, 336–337
  solar, 58–59, 184, 239–241, 360–361
  supernova, 240
Neutron binding, limit of, 358
Neutron stars, 59, 302, 361–363
  maximum mass of, 303–304
  surfaces of, nuclear reactions in, 372–373
Neutrons, 14–15
Non-Proliferation Treaty (NPT), 789
Nonbaryonic dark matter, 259
  need for, 330–331
  searches for, 335–337
Nonequilibrium physics, pattern formation in, 665–675
Nonleptonic weak interaction, 367
Nonlinear optics, 824
Nonlinear waves, 668–671
Nonperiodic systems, 115–117
Nonperturbative string theory, 198–201
Nonrelativistic Schrödinger equation, 392–393
Nonrenormalizable interactions, 149
Nonzero neutrino mass, 184
Nuclear astrophysics, 360–363, 370
Nuclear binding, limits of, 357–358
Nuclear capacity, 771
Nuclear conflict, 788
Nuclear dynamics, 100
Nuclear fisson reactors, 767–774
Nuclear forces, 351–352
Nuclear fusion, 775–780

Nuclear magnetic resonance, 90–95, 133, 707, 761
Nuclear matter, 22–23
Nuclear medium, hadrons in, 358–360
Nuclear physics, 12–28, 35–36, 345
  current, 345–368
  experimental facilities for, 367–368
  present perspective on, 347–348
Nuclear reactions in white dwarf and neutron star surfaces, 372–373
Nuclear structure, 353–360
  at high angular momentum, 354–357
Nuclear testing, 789–793
Nuclear waste problem, 774
Nucleic acids, 708
Nucleon structure, 349–351
Nucleons, 22, 162
Nucleosynthesis, 26–27, 363
  big-bang, 251–252
  in red giant stars, 370–371
  stellar, 370–373
Nucleus
  liquid drop model of, 16–17
  radius of, 13

Omega-minus particle, 164
One-component plasmas, 430–431
Onsager-Feynman quantization condition, 538
Open shells, 24
Optical amplifier, 824
Optical coherence tomography, 806
Optical communications, 823
Optical double resonance, 134
Optical fibers, 558
Optical frequency metrology, 412–413
Optical molasses, 137, 414, 427
Optical pumping, 139
Optical tweezers, 137
Optoelectronic materials, 561
Order parameter, 116–117
Organic adhesives, 559
Oscillator strength, 4
Oscillators, perfect, 443
Oscillons, 671

p-branes, 200–201
Paramagnetic spin susceptibility, 153
Paramagnetic state, 109
Parametric down conversion, 485, 491–492, 493
Parametric downconverter source(s)
  interference experiments with, 464–470
Parity, 38
Parity-nonconservation studies, 365–367
Particle detection, physical processesin, 212–213
Particle exchange, association of interactions with, 146
Particle mixtures, 38
Particle physics
  accelerators and, 203–221

Partical physics (contd.)
   standard model of, see Standard model
Partons, 40, 165, 170, see also Quarks
   strong-interaction scattering of, 172–173
Path integrals, imaginary-time, 745
Pattern formation, 665
   in biological systems, 675
   dendritic, 673–675
   dynamical systems and, 667–668
   in granular materials, 671–673
   in nonequilibrium physics, 665–675
Paul trap, 429
Pauli exclusion principle, 4
Pendular orientation and alignment, 701–702
Penning ion trap, 429
Penrose tilings, 115, 116
Percolation problem, 609
Perihelion advance, anomalous, 79–80
Periodicity paradigm, 112–113
Persistent currents, 523, 540
Perturbation theory, 5, 35
Perturbative string theory, 195–198
Petawatt-class lasers, 810
"Pfaffian state," 512
Phase diagrams, 589
Phase shift analysis, 20
Phase shifts, 15
   differential, 123
Phase stability, 207–208
Photoassociative spectroscopy, 379, 438
Photodynamic therapy, 805
Photoelectric effect, 3, 98, 130
Photoemission, 105
Photoinduced isomerization, 718
Photoionization, 382, 390
Photon bunching, antibunching, 416–417, 434
Photon interference experiments, 420
Photon statistics, 416, 453–454
Photon-Z interference, 176–177
Photonic band-gap crystals, 561
Photons, 32, 162
   time interval between two, measurement of, by interference, 470–473
Photophone, 821
Photoreconnaissance from space, 782–785
Photosynthesis, 718
Physical chemistry, 693
Physics, see also specific branches of physics
   bioenergetics and, 717–719
   biology and, 706
   communications industry and, 812–825
   lasers in, 809
   medical imaging and, 755–766
   microscopic simulations in, 741–750
   U.S. national security and, 781–796
Pictorial detectors, 213–214
Picture archiving and communications system, 763–764

Pion production, 21
Pionium, 403
Pions, 37, 348–349
Pirogov-Sinai theory, 595
Planck energy, 188
Planck length, 41, 189
Planck mass, 157, 190
Planck scale, 157
Planck time, times earlier than, 258
Planck's hypothesis, 130
Plane-rotator model, 608
Plasma wave accelerator, 682
Plasma wave echoes, 682
Plasmas, 679
   high-temperature, collisionless nature of, 679–688
   one-component, 430–431
   space, collisionless shocks in, 685–688
Plastic deformation, 115
Plastics, 559
Point-contact transistor, 565, 567, 568
Point crystal defects, 113–114
Polar molecules, electrostatic storage ring for, 702
Polarized-electron/polarized-photon interactions with atoms, 388–390
Polyethylene, 124
Polymers, 124–125, 558–560, 617
Pomeron, 174
Positron tomography, 760
Positronium, 133, 399, 401–402
Positrons, 5, 34, 44, 46
Potential-energy curves, diatomic, 696
Potential-energy surfaces, triatomic, 696
Potts model, Q-state, 606
Powders, 630
Precision cosmology, 267–274
Precision measurements, 435–439
Precision spectroscopy of atomic hydrogen, 410–412
Predictability, 741
President's Science Advisory Committee (PSAC), 794
Primeval inhomogeneity, structure formation and, 262–264
Probability, 5, 9–10
Probability theory, 131, 585
Projection radiography, 755–758
Proportional wire systems, 214
Protective films, 559
Protein folding, 713–715
Proteins, glass transition in, 713
Proton linear accelerators, 208
Proton radioactivity, 358
Proton spin puzzle, 350
Proton spin system, 92

Protons, 13
Pulse code modulation, 820

Q-state Potts model, 606
Q switching, 475
Quanta, 3–11
Quantization, 130
Quantization condition, Onsager-Feynman, 538
Quantum black holes, 309–311
Quantum cascade laser, 824
Quantum chromodynamics, 40, 143, 347
   establishing, 169–175
   lattice gauge theory of, 747
   nonperturbative, 173–175
   path to, 164–165
Quantum complementarity, 487–488
Quantum computers, 422–424
Quantum corral, 548
Quantum cryptography, 424
Quantum dimensions, 189
Quantum dots, 123
Quantum effects in masers and lasers, 450–456
Quantum electrodynamics, 8–9, 34–35, 132, 143, 161
   atomic physics and, 396–397
Quantum engineering, 415, 434–435
Quantum entanglement, 493–496
Quantum field theory, 33–35, 143–159
   formulation of, 148–149
   intellectual history of, 158–159
   limitations of, 156–159
Quantum Hall effect, integral, 503–504
Quantum information, 422–424
Quantum interference phenomena, 420, 484
Quantum liquids, 116
Quantum logic gates, 423
Quantum measurements on single atoms, 434
Quantum-mechanical free-electron model of metals, 100
Quantum-mechanical transport equation, 102
Quantum mechanics, 10, 130
   era of, 819–821
Quantum nondemolition measurements, 420–421
Quantum of action, 98–99
Quantum optics, 138, 415–424
   era of, 821–825
Quantum physics, 482
   experiment and foundations of, 482–496
Quantum simulations, 745–748
   mixed classical and, 748
Quantum states, measurement of, 421–422
Quantum teleportation, 424, 494–495
Quantum theory, 3–11
   of gravity, 157, 164, 265
Quantum-well lasers, 824
Quantum wells, 561, 823–824
Quark helicity, 165
Quark jets, 171–172

Quark mixing matrix, 181–182
Quark model, 164
Quarks, 38, 159, 168, 189, 349–351, *see also* Partons
   absence of further substructure of, 173
   confined, 40
Quasars, 63, 304
Quasielastic description, 632
Quasielastic medium, 634
Quasielastic model, 634–635
Quasilinear theory, 684
Quasiparallel shocks, 687
Quasiperiodicity, 115
Quasiprobability distribution, 422
Quiver energy, 479

Radar, 89
Radiation
   blackbody, *see* Blackbody radiation
   cosmic microwave background, *see* Cosmic microwave background radiation
   gravitational, *see* Gravitational radiation
   squeezed, 421
Radiation Laboratory, 89–90
Radiation pressure cooling, 427
Radiation treatment planning, medical images in, 764–765
Radio astronomy, 56
Radio-carbon dating, 278
Radio-frequency coils, 762
Radio sources, 62
Radioactivity, 12, 30–32
Radium, 29
Random loose-stacked limit, 630
Rayleigh-Bénard convection, 670–671
Realizability condition, 652
Recoil limit, 137
Red giant stars, nucleosynthesis in, 370–371
Redshift of spectral lines, 75–76
Relativistic wave equation, 34
Relativity, 31
   general, *see* General relativity
   review of, 69–88
   special, *see* Special relativity
Relaxation processes, 712
Renormalizable interactions, 149
Renormalizable theories, search for, 166–168
Renormalization, 36, 603–604, 608–613
   breakdown of, 188
Renormalization-group theory, 117
Renormalization transformation, 609–610
   fixed points of, 610–612
Reptation hypothesis, 625
Resonance fluorescence, 415, 416
Reynolds' equations, 648
Reynolds number, 645, 647
Rheological properties of soft condensed matter, 624–625

Rheology, disentanglement, 625
Rhoades-Ruffini calculation, 303
Right-handed neutrinos, 237
RNA molecules, 708
Rotons, 119, 535
Ruderman-Kittel-Kasuya-Yosida (RKKY) interaction, 112
Running couplings, 150–156
Rushbrooke inequality, 605
Rydberg constant, 3, 131, 412
Rydberg states, 417

S duality, 199
S-matrix methods, 39
Safeguard deployment, 786–787
Saturation spectroscopy, 135
Scale factor, cosmic, 252–254
Scaled magnetization, 606
Scaled temperature, 606
Scaling function, 606
Scaling hypothesis, 602, 604
Scaling laws, 125, 602, 604–606
Scanning near-field optical microscope (SNOM), 546
Scanning tunneling microscope (STM), atoms and, 543–553
"Schrödinger-cat" state, 435
Schrödinger equation, 4–5, 99
    nonrelativistic, 392–393
Schwarzschild radius, 59, 304
Schwarzschild singularity, 302
Science advising, 793–795
Scientists, ethical dilemma of, 795–796
Scintillation counting, 214
Second-order interference, 462–463
Second sound, 119
Secrecy, need for, 782
See-saw mechanism, 237, 271
Seebeck effect, 102
Self-energy, electron, 8
Semiconductor detectors, 215
Semiconductor physics, 819
Semiconductors, 90, 561, 564
Semileptonic parity-nonconservation studies, 365–367
Shell model, 23–24, 346
Short-pulse generation and detection, 475–476
Shot noise, 319
Shot noise limit, 421
Silicon, 564
Similarity arguments, 648–649
Simulations, 742
    classical, 743–744
    of many-body systems, 749
    microscopic, in physics, 741–750
    mixed quantum and classical, 748
    quantum, 745–748
Single-atom masers, 454–455

Single-electron transistor, 522
Single-electron tunneling transistor, superconducting, 532
Single-molecule experiments, 721–722
Single-photon emission computed tomography, 760
Single-photon interference experiments, 468–469
Slip bands, 637–638
Small-scale turbulence, 653–656
Smoothness problem, 265
"Snowflake problem," 673–675
Soft breaking terms, 191
Soft condensed matter, 617–627
    in biology, 627
    in chemistry, 626
    dilational order of, 619–620
    flow in, 623–624
    principles of, 618–619
    rheological properties of, 624–625
    self-organization of, 619–624
        from entropic interactions, 621–623
        kinetically driven, 623
        within molecules, 620–621
Soft glasses, 625–626
Soft matter, 123–126
Solar gravity, deflection of light by, 76
Solar neutrino detectors, 241
Solar neutrinos, 58–59, 184, 239–241, 360–361
Solar wind, 685
Solid state physics, 96
Solitons, 804
Solvability theory, 674
Space, photoreconnaissance from, 782–785
Space dimensions, compactification of, 197
Space plasmas, collisionless shocks in, 685–688
Space-time, fermionic dimensions of, 189–190
Spark chambers, 213
Spatial variation, conductance fluctuations and, 519–521
Spatiotemporal chaos in fluids, 668–671
Special relativity, 27–28
    review of, 69–72
Spectral distribution of blackbodyradiation, 3
Spectral lines, redshift of, 75–76
Speed of light, 70
Spin glasses, 117
    theory of, 714
Spin magnetic moment, 5
    of electron, 223
Spin-statistics theorem, 145
Spin waves, 110
Spontaneous symmetry breaking, 154, 257, 594
Squarks, 189
Squeezed radiation, 421
SQUID magnetometers, 531
Standard cosmology, 245, 252–256
Standard model, 43, 161–187, 257
    elementary particles of, 163

events for cosmic microwave background radiation in, 291–292
experimental establishment of, 169–182
extending, 265–267
limits of, 188
more fundamental theory underlying, 169
supersymmetric extension of, 155, 156
tests of, 364–367
unresolved issues with, 182–187
Standard-model elements, brief summary of, 168–169
Standard model gauge group, 168
Stanford Linear Collider, 219–220
Star formation, interstellar medium and, 63–65
Star Wars, 787
Statistical mechanics, 581–597
Stellar energy and evolution, 56–59
Stellar nucleosynthesis, 370–373
Stellarator, 777
Stimulated emission, 444–445
Strange attractors, 668
Strange particles, 51
Strangeness, 37
Streamer chambers, 213
Stress distribution in heaps, 635–637
Stress fields, macroscopic, 632–638
String coupling constant, 195
String theory, 189, 310
Stringy geometry, 197
Strobelike probes, 555
Strong-coupling constant, 172
Strong deformations, 637–638
Strong focusing, 208–209
Strong interactions, 33, 35
    scattering of partons, 172–173
Strong nuclear force, 13, 161
Structural materials, 556–558
Structure formation
    dark matter and, 258–264
    primeval inhomogeneity and, 262–264
Structure functions, 170
Substitutional alloys, 114
Sunyaev-Zel'dovich effect, 300
Supercharges, 199
Superclusters, 247
Supercoils, 620
Superconducting accelerators, 209
Superconducting materials, 562
Superconducting single-electron tunneling transistor, 532
Superconductivity, 7, 98, 117–121, 526–532
    Bardeen-Cooper-Schrieffer (BCS) theory, 100
Superconductors, type-II, 530–531
Supercritical shocks, 687
"Superdeformed" bands, 356
Superfluidity, 117–121, 534–542, 745
Supergiant stars, 57
Supergravity, 258

"Superheavy" nuclei, 358
Superluminal motion, 71–72
Supermassive black holes, 303
    in galactic nuclei, 304–305
    merging, 307
Supernova explosions, 27, 58, 302, 361
Supernova neutrinos, 240
Supernova shocks, cosmic ray acceleration by, 688
Supersonic nozzle, 697
Superspace, 189
Superstring theory, 41, 186, 188, 195–201
    open problems in, 201–202
Supersymmetric extension of standard model, 155, 156
Supersymmetric grand unification, 192
Supersymmetric quantum field theories, 192–195
Supersymmetric standard model, 191
Supersymmetry, 155, 184, 188–195, 258
    in TeV range, 190–192
Surface acoustic waves, 508
Surface energy, 107
Surface gravity, 309
Surface science, 105–106
Surfactants, 617–618
Symmetry breaking, spontaneous, 154
Synapse plasticity, 735–736
Synapses, dynamics of, 731–732
Synchrotrons, 208

T duality, 197
Tau-theta puzzle, 53
Technicolor, 183–184
Telephone amplifier, 816–817
Telephone attenuation problem, 814–815
Telephones, 813–814
Telescopes, building new, 67
Thermal agitation, 629
Thermal noise, 320, 818
Thermal relaxation time, 93
"Thermal" scaling power, 612–613
Thermionic emission, 98, 818
Thermodynamic limit, 589
Thomson effect, 102
Three-body interaction, 21
Three-dimensional structure of universe, 269
Time-delay by gravitational potential, 76–78
Time-delayed perturbed angular correlations, 95
Time dilation effect, 70–71
Time interval between two photons, measurement of, by interference, 470–473
Time-projection chambers, 215
Time-reversal invariance, 38, 367
Tokamaks, 684, 776–777, 778
Tomography
    analog, 759
    emission, 760–761

Tomography (contd.)
  optical coherence, 806
  x-ray, 758–760
Top quark, 41, 169, 180–181
Transistor, 563
  in communications, 819–821
  field-effect, 565–566
Transistor (contd.)
  invention of, 563–576
  junction, 565, 570–574
  point-contact, 565, 567, 568
  preliminary investigations on, 563–565
  single-electron, 522
  superconducting single-electron tunneling, 532
Transition radiation detectors, 216
Transition-state theory, 715
Transmission, conductance viewed as, 515–524
Transport equation, quantum-mechanical, 102
Trapped particle oscillations, 681–682
Triatomic potential-energy surfaces, 696
Triaxial tests, 632
Trilinear gauge couplings, 181
True Wave fiber, 825
Tunnel effects, 716
Tunneling spectroscopy, 546
Tunneling transistor, superconducting single-electron, 532
Turbulence, 644–645
  small-scale, 653–656
Turbulence measurements, 651–652
Turbulent drag reduction, 625
Turbulent flow, 645–647
Two-nucleon interaction, 19
Two-photon optical microscopy, 807–808
Type-II superconductors, 530–531
Typicality, notion of, 585

U2 aircraft, 783
Ultrafast spectroscopy, 476–478
Ultrarelativistic ion energies, 381
Ultrasensitive spectroscopy, 409–410
Uncertainty principle, 5, 10, 133
Undulation forces, 720
Unification, 73, 189
  coupling, 185–186
  electroweak, 40–41
  energy scale of, 156
Unification scale, 192
Unified Method, 398
Universality, 602–603, 606–608
Universality classes, 602
Universe, 245–246
  critical-density, 254
  evolution of, 255–256
  macrostate of, 587
  mass density of, 254
  matter/energy in, 263
  three-dimensional structure of, 269

Up quark, 162
Uranium, 771
U.S. national security, physics and, 781–796

V particles, 51
Vacancies, 113
Vacua
  families of, 193
  moduli space of, 193
Vacuum energy, 255, 261
Vacuum tubes, 817–818
Very large scale integrated circuits, 737
Very massive objects (VMOs), 335
Viscosity dependence, 716
Vlasov equation, 778
Voltage drop, localized, 519
Vortices, 504

W particles, 25–26, 162
  discovery of, 177–178
Wake fields, 210
Warfare, 781
Warheads, 790–791
Wave equation, relativistic, 34
Wave function, Dirac, 5
Wave kinematics, gravitational radiation and, 315–317
Wave mechanics, 5
Wave nature of electrons, 818–819
Wave packet, 10
Wave-particle duality, 4
  interactions, 679–685
Weak interactions, 25–26, 32–33
  nonleptonic, 367
Weak nuclear force, 161
Weakly interactive massive particles, see WIMPs
Weiss field, 98
Whistler wave trains, 686
White dwarf star surfaces, nuclear reactions in, 372–373
White dwarfs, 57, 59
Widom equality, 605
Wiedemann-Franz constant, 102
Wigner function, 422
Wigner lattice, 111
Wigner-Seitz radius, 103
Wilson cloud chamber, 44
WIMPs (weakly interactive massive particles), 337–341
Winding-mode excitations, 197
Wire drift chambers, 214
Work function, 99
World sheet, 196
Worm holes, 73

X-ray binaries, 59
  black holes in, 305–306
X-ray fluoroscopy, 756–757

X-ray image receptors
    analog, 755–756
    digital, 757–758
X-ray tomography, 758–760
X rays, 96–97, 707, 755
Xogen, 695

Yrast line, 355
Yukawa function, 21

Z exchange, 175
Z particles, 26, 162
    discovery of, 177–178
Z-photon interference, 176–177
Z properties, electroweak model and, 178–180
0-branes, 200
Zero-point motion, 147
Zeroth sound mode, 108
Zone refining, 113

MIX
Papier aus verantwortungsvollen Quellen
Paper from responsible sources
FSC® C105338

If you have any concerns about our products,
you can contact us on
ProductSafety@springernature.com

In case Publisher is established outside the EU,
the EU authorized representative is:
Springer Nature Customer Service Center GmbH
Europaplatz 3, 69115 Heidelberg, Germany

Printed by Libri Plureos GmbH
in Hamburg, Germany